W9-CAR-703

Encyclopedia of Nanotechnology

Bharat Bhushan
Editor

Encyclopedia of Nanotechnology

Volume IV

Q–Y

With 1976 Figures and 124 Tables

Editor
Professor Bharat Bhushan
Ohio Eminent Scholar and
The Howard D. Winbigler Professor,
Director, Nanoprobe Laboratory for Bio- & Nanotechnology
and Biomimetics (NLB^2)
Ohio State University
201 W. 19th Avenue
Columbus, Ohio, 43210-1142
USA

ISBN 978-90-481-9750-7 ISBN 978-90-481-9751-4 (eBook)
DOI 10.1007/978-90-481-9751-4
ISBN 978-90-481-9752-1 (print and electronic bundle)
Springer Dordrecht Heidelberg New York London

Library of Congress Control Number: 2012940716

Printed on acid-free paper

Springer is part of Springer Science+Business Media (www.springer.com)

Preface

On December 29, 1959, at the California Institute of Technology, Nobel Laureate Richard P. Feynman gave a speech at the Annual meeting of the American Physical Society that has become one of the twentieth-century classic science lectures, titled "There's Plenty of Room at the Bottom." He presented a technological vision of extreme miniaturization in 1959, several years before the word "chip" became part of the lexicon. He spoke about the problem of manipulating and controlling things on a small scale. Extrapolating from known physical laws, Feynman envisioned a technology using the ultimate toolbox of nature, building nanoobjects atom by atom or molecule by molecule. Since the 1980s, many inventions and discoveries in the fabrication of nanoobjects have been testament to his vision. In recognition of this reality, National Science and Technology Council (NSTC) of the White House created the Interagency Working Group on Nanoscience, Engineering and Technology (IWGN) in 1998. In a January 2000 speech at the same institute, former President W. J. Clinton spoke about the exciting promise of "nanotechnology" and the importance of expanding research in nanoscale science and technology, more broadly. Later that month, he announced in his State of the Union Address an ambitious $497 million federal, multiagency National Nanotechnology Initiative (NNI) in the fiscal year 2001 budget, and made the NNI a top science and technology priority. The objective of this initiative was to form a broad-based coalition in which the academe, the private sector, and local, state, and federal governments work together to push the envelope of nanoscience and nanoengineering to reap nanotechnology's potential social and economic benefits.

The funding in the USA has continued to increase. In January 2003, the US senate introduced a bill to establish a National Nanotechnology Program. On December 3, 2003, President George W. Bush signed into law the 21st Century Nanotechnology Research and Development Act. The legislation put into law programs and activities supported by the (NNI). The bill gave nanotechnology a permanent home in the federal government and authorized $3.7 billion to be spent in the 4-year period beginning in October 2005, for nanotechnology initiatives at five federal agencies. The funds have provided grants to researchers, coordinated R&D across five federal agencies (National Science Foundation (NSF), Department of Energy (DOE), NASA, National Institute of Standards and Technology (NIST), and Environmental Protection Agency (EPA)), established interdisciplinary research centers, and accelerated technology transfer into the private sector. In addition, the Department of Defense (DOD), Homeland Security, Agriculture and Justice, as well as the National Institutes of Health (NIH) also fund large R&D activities. They currently account for more than one-third of the federal budget for nanotechnology.

The European Union (EU) made nanosciences and nanotechnologies a priority in the Sixth Framework Program (FP6) in 2002 for the period 2003–2006. They had dedicated small funds in FP4 and FP5 before. FP6 was tailored to help better structure European research and to cope with the strategic objectives set out in Lisbon in 2000. Japan identified nanotechnology as one of its main research priorities in 2001. The funding levels increased sharply from $400 million in 2001 to around $950 million in 2004. In 2003, South Korea embarked upon a 10-year program with around $2 billion of public funding, and Taiwan has committed around $600 million of public funding over 6 years. Singapore and China are also investing on a large scale. Russia is well funded as well.

Nanotechnology literally means any technology done on a nanoscale that has applications in the real world. Nanotechnology encompasses production and application of physical, chemical, and biological systems at scales, ranging from individual atoms or molecules to submicron dimensions, as well as the integration of the resulting nanostructures into larger systems. Nanotechnology is likely to have a profound impact on our economy and society in the early twenty-first century, comparable to that of semiconductor technology, information technology, or cellular and molecular biology. Science and technology research in nanotechnology is leading to breakthroughs in areas such as materials and manufacturing, nanoelectronics, medicine and healthcare, energy, biotechnology, information technology, and national security. It is widely felt that nanotechnology will be the next industrial revolution.

There is an increasing need for a multidisciplinary, system-oriented approach toward designing and manufacturing micro/nanodevices which function reliably. This can only be achieved through the cross-fertilization of ideas from different disciplines and the systematic flow of information and people among research groups. Reliability is a critical technology for many micro- and nanosystems and nanostructured materials. The first edition of a broad-based *Handbook of Nanotechnology* from Springer was published in April 2004, the second edition in 2007, and the third edition in 2010. It presents an overview of nanomaterial synthesis, micro/nanofabrication, micro- and nanocomponents and systems, scanning probe microscopy, reliability issues (including nanotribology and nanomechanics) for nanotechnology, and various industrial including biomedical applications.

The field of nanotechnology is getting a strong foothold. It attracts people from various disciplines including science and engineering. Given the explosive growth in nanoscience and nanotechnology, this *Encyclopedia of Nanotechnology* is being launched with essays written by experts in the field from academia and industry.

The objective of this encyclopedia is to introduce a large number of terms, devices, and processes. For each entry, a brief description is provided by experts in the field. The entries have been written by a large number of internationally recognized experts in the field, from academia, national research labs, and industry.

The *Encyclopedia of Nanotechnology* is expected to provide a comprehensive and multidisciplinary reference to the many fields relevant to the general field of nanotechnology. The encyclopedia focuses on engineering and applications with some coverage of the science of nanotechnology. It aims to be a comprehensive and genuinely international reference work and is aimed at graduate students, researchers, and practitioners.

The development of the encyclopedia was undertaken by an Editorial Board, who were responsible for the focus and quality of contributions, and an Advisory Board, who were responsible for advising about the selection of topics.

The print version of the encyclopedia contains four volumes with a total of **325** entries and about **3068** pages. The encyclopedia is also available online. Editors expect to update it periodically. The editor-in-chief and all the editors thank a large number of authors for making contributions to this major reference work. We also thank the referees who meticulously read the entries and made their recommendations.

Powell, Ohio
USA
May 2012

Bharat Bhushan

Biography

Bharat Bhushan
Ohio Eminent Scholar
The Howard D. Winbigler Professor, Director
Nanoprobe Laboratory for Bio- & Nanotechnology and Biomimetics (NLB^2)
Ohio State University
201 W. 19th Avenue
Columbus, Ohio, 43210-1142
USA
bhushan.2@osu.edu

Dr. Bharat Bhushan received an M.S. in mechanical engineering from the Massachusetts Institute of Technology in 1971; an M.S. in mechanics and a Ph.D. in mechanical engineering from the University of Colorado at Boulder in 1973 and 1976, respectively; an MBA from Rensselaer Polytechnic Institute at Troy, NY, in 1980; a Doctor Technicae from the University of Trondheim at Trondheim, Norway, in 1990; a Doctor of Technical Sciences from the Warsaw University of Technology at Warsaw, Poland, in 1996; and Doctor Honouris Causa from the National Academy of Sciences at Gomel, Belarus, in 2000 and University of Kragujevac, Serbia, in 2011. He is a registered professional engineer. He is presently an Ohio Eminent Scholar and The Howard D. Winbigler Professor in the College of Engineering and the Director of the Nanoprobe Laboratory for Bio- & Nanotechnology and Biomimetics (NLB^2) at the Ohio State University, Columbus, Ohio. His research interests include fundamental studies with a focus on scanning probe techniques in the interdisciplinary areas of bio/nanotribology, bio/nanomechanics, and bio/nanomaterials characterization and applications to bio/nanotechnology and biomimetics. He is an internationally recognized expert of bio/nanotribology and bio/nanomechanics using scanning probe microscopy and is one of the most prolific authors. He is considered by some a pioneer of the tribology and mechanics of magnetic storage devices. He has authored 8 scientific books, approximately 90 handbook chapters, 700 scientific papers (h-index – 57; ISI Highly Cited in Materials Science, since 2007; ISI Top 5% Cited Authors for Journals in Chemistry since 2011), and 60 technical reports. He has also edited more than 50 books and holds 17 US and foreign patents. He is coeditor of Springer NanoScience and Technology Series and coeditor of Microsystem Technologies. He has given more than 400 invited presentations on six continents and more than 200 keynote/plenary addresses at major international conferences.

Dr. Bhushan is an accomplished organizer. He organized the 1st Symposium on Tribology and Mechanics of Magnetic Storage Systems in 1984 and the 1st International Symposium on Advances in Information Storage Systems in 1990, both of which are now held annually. He is the founder of an ASME Information Storage and Processing Systems Division founded in 1993 and served as the founding chair during the period 1993–1998. His biography has been listed in over two dozen *Who's Who* books including *Who's Who in the World* and has received more than two dozen awards for his contributions to science and technology from professional societies, industry, and US government agencies. He is also the recipient of various international fellowships including the Alexander von Humboldt Research Prize for Senior Scientists, Max Planck Foundation Research Award for Outstanding Foreign Scientists, and the Fulbright Senior Scholar Award. He is a foreign member of the International Academy of Engineering (Russia), Byelorussian Academy of Engineering and Technology, and the Academy of Triboengineering of Ukraine; an honorary member of the Society of Tribologists of Belarus; a fellow of ASME, IEEE, STLE, and the New York Academy of Sciences; and a member of ASEE, Sigma Xi, and Tau Beta Pi.

Dr. Bhushan has previously worked for Mechanical Technology Inc., Latham, NY; SKF Industries Inc., King of Prussia, PA; IBM, Tucson, AZ; and IBM Almaden Research Center, San Jose, CA. He has held visiting professorship at University of California at Berkeley; University of Cambridge, UK; Technical University Vienna, Austria; University of Paris, Orsay; ETH, Zurich; and EPFL, Lausanne. He is currently a visiting professor at KFUPM, Saudi Arabia; Harbin Inst., China; University of Kragujevac, Serbia, and University of Southampton, UK.

Section Editors

Bharat Bhushan Nanoprobe Laboratory for Bio- & Nanotechnology and Biomimetics, The Ohio State University, Columbus, OH, USA

Donald Brenner Department of Materials Science and Engineering, North Carolina State University, Raleigh, NC, USA

Jason V. Clark Department of Mechanical Engineering, Purdue University, W. Lafayette, IN, USA

Robert J. Davis The Ohio State University, Columbus, OH, USA

Maarten deBoer Mechanical Engineering, Carnegie-Mellon University, Pittsburgh, PA, USA

Paolo Decuzzi Department of Translational Imaging and Department of Nanomedicine, The Methodist Hospital Research Institute, Houston, TX, USA

Lixin Dong Electrical & Computer Engineering, Michigan State University, East Lansing, MI, USA

Walter Federle Department of Zoology, University of Cambridge, Cambridge CB2 3EJ, UK

Emmanuel Flahaut Cirimat/Lcmie, Umr Cnrs 5085 Bâtiment, Universite Paul Sabatier, France

Enrico Gnecco Campus Universitario de Cantoblanco, IMDEA Nanociencia, Madrid, Spain

Jian He Department of Physics and Astronomy, Clemson University, Clemson, SC, USA

Michael Heller Department of Bioengineering, University of California at San Diego, La Jolla, CA, USA

David Holmes London Centre for Nanotechnology, University College London, London, UK

Ali Khademhosseini Harvard Medical School, Cambridge, MA, USA

Philippe Lutz The Université de Franche-Comté, Besançon, France

Marc Madou Department of Mechanical and Aerospace Engineering, University of California at Irvine, Irvine, CA, USA

Ernest Mendoza Nanomaterials Aplicats Centre de Recerca en Nanoenginyeria, Universitat Politècnica de Catalunya, Barcelona, Spain

Mohammad Mofrad Director, Molecular Cell Biomechanics Lab, University of California Berkeley, Berkeley, CA, USA

Michael Nosonovsky Department of Mechanical Engineering, University of Wisconsin at Milwaukee, Milwaukee, WI, USA

Gianluca ("Gian") Piazza Department of Electrical and Systems Engineering, University of Pennsylvania, Philadelphia, PA, USA

Fabrizio Pirri Dipartimento di Scienza dei Materiali e Ing. Chimica Laboratorio Materiali e Microsistemi / LATEMAR, Torino, Italy

Shaurya Prakash Department of Mechanical Engineering, The Ohio State University, Columbus, OH, USA

Themistoklis Prodromakis Institute of Biomedical Engineering, Imperial College, London, UK

Yabing Qi Energy Materials and Surface Sciences (EMSS) Unit, Okinawa Institute of Science and Technology Graduate University (OIST), Okinawa, Japan

Apparao M. Rao Department of Physics and Astronomy, Clemson University, Clemson, SC, USA

Maxim Sukharev Department of Applied Sciences and Mathematics, Arizona State University, Polytechnic Campus, Mesa, AZ, USA

Srinivas Tadigadapa Department of Electrical Engineering, The Pennsylvania State University, University Park, PA, USA

Lorenzo Valdevit Department of Mechanical and Aerospace Engineering, University of California, Irvine, Irvine, CA, USA

Chunlei (Peggy) Wang Mechanical and Materials Engineering, Florida International University, Miami, FL, USA

Steve Wereley Department of Mechanical Engineering, Purdue University, West Lafayette, IN, USA

Huikai Xie Biophotonics & Microsystems Lab Department of Electrical & Computer Engineering, University of Florida, Gainesville, FL, USA

Yi Zhao Department of Biomedical Engineering, The Ohio State University, Columbus, OH, USA

Advisory Board

Contributors

Ahmad Nabil Abbas Department of Electrical and Computer Engineering, Florida International University, Miami, FL, USA

Patrick Abgrall Formulaction, L'Union, France

Angelo Accardo Lab. BIONEM, Dipartimento di Medicina Sperimentale e Clinica, Università "Magna Grecia" di Catanzaro, Catanzaro, Italy

Soft Matter Structures Group ID13 – MICROFOCUS Beamline, European Synchrotron Radiation Facility, Grenoble Cedex, France

Wafa Achouak iCEINT, International Consortium for the Environmental Implications of Nanotechnology, Center for the Environmental Implications of NanoTechnology, Aix–En–Provence, Cedex 4, France

Laboratoire d'Ecologie Microbienne de la Rhizosphère et d'Environnements Extrême, UMR 6191 CNRS–CEA–Aix–Marseille Université de la Méditerranée, St Paul lez Durance, France

Ranjeet Agarwala Department of Mechanical & Aerospace Engineering, North Carolina State University, Raleigh, NC, USA

Alessandro Alabastri Nanobiotech Facility, Istituto Italiano di Tecnologia, Genoa, Italy

Muhammad A. Alam School of Electrical and Computer Engineering, Purdue University, West Lafayette, IN, USA

Tuncay Alan Mechanical and Aerospace Engineering Department, Monash University, Victoria, Australia

Antonio Aliano Department of Physics, Politecnico di Torino, Torino, Italy

Wafa' T. Al-Jamal Nanomedicine Laboratory, Centre for Drug Delivery Research, The School of Pharmacy, University of London, London, UK

Paolo Allia Materials Science and Chemical Engineering Department, Politecnico di Torino, Torino, Italy

N. R. Aluru Department of Mechanical Science and Engineering, Beckman Institute for Advanced Science and Technology, University of Illinois at Urbana – Champaign, Urbana, IL, USA

Giampiero Amato Quantum Research Laboratory, Electromagnetism Division, Istituto Nazionale di Ricerca Metrologica, Torino, Italy

Benoy Anand Department of Physics, Sri Sathya Sai Institute of Higher Learning, Prashanthinilayam, Andhra Pradesh, India

Francesco De Angelis Nanobiotech Facility, Istituto Italiano di Tecnologia, Genoa, Italy

José V. Anguita Nano Electronics Center, Advanced Technology Institute, University of Surrey, Guildford, Surrey, UK

Wadih Arap David H. Koch Center, The University of Texas M. D. Anderson Cancer Center, Houston, TX, USA

Departments of Genitourinary Medical Oncology and Cancer Biology, The University of Texas M. D. Anderson Cancer Center, Houston, TX, USA

Walter Arnold Department of Material Science and Technology, Saarland University, Saarbrücken, Germany

Physikalisches Institut, Göttingen University, Göttingen, Germany

Christopher Arntsen Department of Chemistry and Biochemistry, UCLA, Los Angeles, CA, USA

Eduard Arzt INM – Leibniz Institute for New Materials and Saarland University, Saarbrücken, Germany

Burcu Aslan Department of Experimental Therapeutics, M.D. Anderson Cancer Center, The University of Texas, Houston, TX, USA

Elena Astanina University of Torino, Department of Oncological Sciences, Candiolo, Torino, Italy

Orlando Auciello Materials Science Division, Argonne National Laboratory, Argonne, IL, USA

Mélanie Auffan CEREGE, UMR 6635 CNRS/Aix–Marseille Université, Aix–en–Provence, France

iCEINT, International Consortium for the Environmental Implications of Nanotechnology, Center for the Environmental Implications of NanoTechnology, Aix–En–Provence, Cedex 4, France

Thomas Bachmann Institute for Fluid Mechanics and Aerodynamics, Technische Universität Darmstadt, Darmstadt, Germany

Armelle Baeza-Squiban Laboratory of Molecular and Cellular Responses to Xenobiotics, University Paris Diderot - Paris 7, Unit of Functional and Adaptive Biology (BFA) CNRS EAC 4413, Paris cedex 13, France

Darren M. Bagnall Nano Group, Electronics and Computer Science, University of Southampton, Highfield, Southampton, UK

Xuedong Bai Beijing National Laboratory for Condensed Matter Physics, Institute of Physics, Chinese Academy of Sciences, Beijing, P. R. China

Anisullah Baig Department of Electrical and Computer Engineering, University of California, Davis, USA

David J. Bakewell Department of Electrical Engineering and Electronics, University of Liverpool, Liverpool, UK

Mirko Ballarini DIFIS – Dipartimento di Fisica, Politecnico di Torino, Torino, Italy

Belén Ballesteros CIN2 (ICN-CSIC), Catalan Institute of Nanotechnology, Bellaterra, Barcelona, Spain

Antoine Barbier CEA-Saclay, DSM/IRAMIS/SPCSI, Gif-sur-Yvette, France

Robert Barchfeld Department of Electrical and Computer Engineering, University of California, Davis, USA

Yoseph Bar-Cohen Jet Propulsion Laboratory (JPL), California Institute of Technology, Pasadena, CA, USA

W. Jon. P. Barnes Centre for Cell Engineering, University of Glasgow, Glasgow, Scotland, UK

Larry R. Barnett Department of Electrical and Computer Engineering, University of California, Davis, USA

Friedrich G. Barth Life Sciences, Department of Neurobiology, University of Vienna, Vienna, Austria

Michael H. Bartl Department of Chemistry, University of Utah, Salt Lake City, UT, USA

R. Baskaran Department of Electrical Engineering, University of Washington, Seattle, WA, USA

Components Research, Intel Corporation, USA

Werner Baumgartner Department of Cellular Neurobionics, Institut für Biologie II RWTH-Aachen University, Aachen, Germany

Pierre Becker Laboratoire de Biologie des Organismes Marins et Biomimétisme, Université de Mons – UMONS, Mons, Belgium

Novid Beheshti Swedish Biomimetics 3000® Ltd, Birmingham, UK

Rachid Belkhou Synchrotron SOLEIL, L'Orme des Merisiers, Gif-sur-Yvette, France

Laila Benameur Laboratoire de Biogénotoxicologie et Mutagenèse Environnementale (EA 1784/FR CNRS 3098 ECCOREV), Aix-Marseille Université, Marseille Cedex 5, France

Rüdiger Berger Max Planck Institute for Polymer Research, Mainz, Germany

Magnus Berggren Department of Science and Technology, Linköpings University, Norrköping, Sweden

Pierre Berini School of Information Technology and Engineering (SITE), University of Ottawa, Ottawa, ON, Canada

Shekhar Bhansali Department of Electrical and Computer Engineering, Florida International University, Miami, FL, USA

Vikram Bhatia Science and Technology, Corning Incorporated SP-PR-02-1, Corning, NY, USA

Rustom B. Bhiladvala Department of Mechanical Engineering, University of Victoria, Victoria, BC, Canada

Bharat Bhushan Nanoprobe Laboratory for Bio- & Nanotechnology and Biomimetics, The Ohio State University, Columbus, OH, USA

Stefano Bianco Center for Space Human Robotics, Fondazione Istituto Italiano di Tecnologia, Torino, Italy

Ion Bita Qualcomm MEMS Technologies, Inc., San Jose, CA, USA

Martin G. Blaber Department of Chemistry and International Institute for Nanotechnology, Northwestern University, Evanston, IL, USA

Luca Boarino Quantum Research Laboratory, Electromagnetism Division, Istituto Nazionale di Ricerca Metrologica, Torino, Italy

Jorge Boczkowski INSERM U955 Eq04, Créteil, France

University Paris Est Val de Marne (UPEC), Créteil, France

Stuart A. Boden Nano Group, Electronics and Computer Science, University of Southampton, Highfield, Southampton, UK

Peter Bøggild DTU Nanotech – Department of Micro- and Nanotechnology, Technical University of Denmark, Lyngby, Denmark

K. F. Böhringer Department of Electrical Engineering, University of Washington, Seattle, WA, USA

Maarten P. de Boer Department of Mechanical Engineering, Carnegie Mellon University, Pittsburgh, USA

Ardemis A. Boghossian Department of Chemical Engineering, Massachusetts Institute of Technology, Cambridge, MA, USA

Paul W. Bohn Department of Chemical and Biomolecular Engineering, University of Notre Dame, Notre Dame, IN, USA

Department of Chemistry and Biochemistry, University of Notre Dame, Notre Dame, IN, USA

Sonja Boland Laboratory of Molecular and Cellular Responses to Xenobiotics, University Paris Diderot - Paris 7, Unit of Functional and Adaptive Biology (BFA) CNRS EAC 4413, Paris cedex 13, France

Robert D. Bolskar TDA Research, Inc., Wheat Ridge, CO, USA

Alexander Booth Energy and Resources Research Institute, University of Leeds, Leeds, West Yorkshire, UK

Garry J. Bordonaro Cornell NanoScale Science and Technology Facility, Cornell University, Ithaca, NY, USA

Edward Bormashenko Laboratory of Polymers, Applied Physics Department, Ariel University Center of Samaria, Ariel, Israel

Céline Botta CEREGE, UMR 6635, CNRS, Aix-Marseille Université, Aix en Provence, Cedex 04, France

Alain Botta Laboratoire de Biogénotoxicologie et Mutagenèse Environnementale (EA 1784/FR CNRS 3098 ECCOREV), Aix-Marseille Université, Marseille Cedex 5, France

Jean-Yves Bottero CEREGE, UMR 6635 CNRS/Aix–Marseille Université, Aix–en–Provence, France

iCEINT, International Consortium for the Environmental Implications of Nanotechnology, Center for the Environmental Implications of NanoTechnology, Aix–En–Provence, Cedex 4, France

Alessia Bottos University of Torino, Department of Oncological Sciences, Candiolo, Torino, Italy

Floriane Bourdiol EcoLab – Laboratoire d'écologie fonctionnelle et environnement, Université de Toulouse, INP, UPS, Castanet Tolosan, France

CNRS UMR 5245, EcoLab, Castanet Tolosan, France

Institut Carnot Cirimat, Université de Toulouse, UPS, INP, Toulouse cedex 9, France

Olivier Bourgeois Institut Néel CNRS-UJF, Grenoble, France

Herbert Bousack Peter Grünberg Institut, Forschungszentrum Jülich GmbH, Jülich, Germany

Thomas Braschler Laboratory of stem cell dynamics, SV-EPFL, Lausanne, Switzerland

Graham Bratzel Laboratory for Atomistic and Molecular Mechanics, Department of Civil and Environmental Engineering, Massachusetts Institute of Technology, Cambridge, MA, USA

Department of Mechanical Engineering, Massachusetts Institute of Technology, Cambridge, MA, USA

Donald W. Brenner Department of Materials Science and Engineering, North Carolina State University, Raleigh, NC, USA

Victor M. Bright Department of Mechanical Engineering, University of Colorado, Boulder, CO, USA

Lawrence F. Bronk David H. Koch Center, The University of Texas M. D. Anderson Cancer Center, Houston, TX, USA

Joseph J. Brown Department of Mechanical Engineering, University of Colorado, Boulder, CO, USA

Dorothea Brüggemann The Naughton Institute, School of Physics, Trinity College Dublin, CRANN, Dublin, Ireland

Markus J. Buehler Laboratory for Atomistic and Molecular Mechanics, Department of Civil and Environmental Engineering, Massachusetts Institute of Technology, Cambridge, MA, USA

Sven Burger Zuse Institute Berlin, Berlin, Germany

Federico Bussolino University of Torino, Department of Oncological Sciences, Candiolo, Torino, Italy

Donald P. Butler The Department of Electrical Engineering, The University of Texas at Arlington, Arlington, TX, USA

Hans-Jürgen Butt Max Planck Institute for Polymer Research, Mainz, Germany

Javier Calvo Fuentes NANOGAP SUB-NM-POWDER S.A., Milladoiro – Ames (A Coruña), Spain

Mary Cano-Sarabia CIN2 (ICN-CSIC), Catalan Institute of Nanotechnology, Bellaterra, Barcelona, Spain

Andrés Cantarero Materials Science Institute, University of Valencia, Valencia, Spain

Francesca Carpino Department of Chemical and Biomolecular Engineering, University of Notre Dame, Notre Dame, IN, USA

Marie Carrière Laboratoire Lésions des Acides Nucléiques, Commissariat à l'Energie Atomique, SCIB, UMR-E 3 CEA/UJF-Grenoble 1, INAC, Grenoble, Cedex, France

Jérôme Casas Institut de Recherche en Biologie de l'Insecte, IRBI UMR CNRS 6035, Université of Tours, Tours, France

Zeynep Celik-Butler The Department of Electrical Engineering, The University of Texas at Arlington, Arlington, TX, USA

Frederik Ceyssens Department ESAT-MICAS, KULeuven, Leuven, Belgium

Nicolas Chaillet FEMTO-ST/UFC-ENSMM-UTBM-CNRS, Besançon, France

Audrey M. Chamoire Department of Mechanical and Aerospace Engineering, The Ohio State University, E443 Scott Laboratory, Columbus, OH, USA

Peggy Chan Micro/Nanophysics Research Laboratory, RMIT University, Melbourne, VIC, Australia

Munish Chanana Departamento de Química Física, Universidade de Vigo, Vigo, Spain

Pen-Shan Chao Chung-Shan Institute of Science and Technology, Taoyuan, Taiwan ROC

Satish C. Chaparala Science and Technology, Corning Incorporated SP-PR-02-1, Corning, NY, USA

Wei Chen Department of Mechanical and Materials Engineering, Florida International University, Miami, FL, USA

Jian Chen Department of Mechanical and Industrial Engineering, University of Toronto, Toronto, ON, Canada

Department of Chemistry and Biochemistry, University of Wisconsin-Milwaukee, Milwaukee, WI, USA

Angelica Chiodoni Center for Space Human Robotics, Fondazione Istituto Italiano di Tecnologia, Torino, Italy

Alessandro Chiolerio Physics Department, Politecnico di Torino, Torino, Italy

Yoon-Kyoung Cho School of Nano-Bioscience and Chemical Engineering, Ulsan National Institute of Science and Technology (UNIST), Ulsan, Republic of Korea

Jong Hyun Choi School of Mechanical Engineering, Birck Nanotechnology Center, Bindley Bioscience Center, Purdue University, West Lafayette, IN, USA

Michael Chu Advanced Pharmaceutics and Drug Delivery Laboratory, University of Toronto, Toronto, ON, Canada

Han-Sheng Chuang Department of Biomedical Engineering, National Cheng Kung University, Tainan, Taiwan

Medical Device Innovation Center, National Cheng Kung University, Taiwan

Giancarlo Cicero Materials Science and Chemical Engineering Department, Politecnico di Torino, Torino, Italy

C. Cierpka Institute of Fluid Mechanics and Aerodynamics, Universität der Bundeswehr München, Neubiberg, Germany

Dominique Collard LIMMS/CNRS-IIS (UMI 2820), Institute of Industrial Science, The University of Tokyo, Tokyo, Japan

Lucio Colombi Ciacchi Hybrid Materials Interfaces Group, Faculty of Production Engineering and Bremen Center for Computational Materials Science, University of Bremen, Bremen, Germany

Maria Laura Coluccio Nanobiotech Facility, Istituto Italiano di Tecnologia, Genoa, Italy

Lab. BIONEM, Dipartimento di Medicina Sperimentale e Clinica, Università "Magna Grecia" di Catanzaro, Catanzaro, Italy

A. T. Conlisk Department of Mechanical Engineering, The Ohio State University, Columbus, OH, USA

Andrew Copestake Swedish Biomimetics 3000® Ltd, Southampton, UK

Miguel A. Correa-Duarte Departamento de Química Física, Universidade de Vigo, Vigo, Spain

Giovanni Costantini Department of Chemistry, The University of Warwick, Coventry, UK

Claire Coutris Department of Plant and Environmental Sciences, Norwegian University of Life Sciences, Ås, Norway

Eduardo Cruz-Silva Department of Polymer Science and Engineering, University of Massachusetts Amherst, Amherst, MA, USA

Steven Curley Department of Surgical Oncology, The University of Texas M. D. Anderson Cancer Center, Houston, TX, USA

Department of Mechanical Engineering and Materials Science, Rice University, Houston, TX, USA

Christian Dahmen Division Microrobotics and Control Engineering (AMiR), Department of Computing Science, University of Oldenburg, Oldenburg, Germany

Lu Dai The Key Laboratory of Remodeling-related Cardiovascular Diseases, Capital Medical University, Ministry of Education, Beijing, China

Gobind Das Nanobiotech Facility, Istituto Italiano di Tecnologia, Genoa, Italy

Bakul C. Dave Department of Chemistry and Biochemistry, Southern Illinois University Carbondale, Carbondale, IL, USA

Enrica De Rosa Department of Nanomedicine, The Methodist Hospital Research Institute, Houston, TX, USA

Paolo Decuzzi Dept of Translational Imaging, and Nanomedicine, The Methodist Hospital Research Institute, Houston, TX, USA

Christian L. Degen Department of Physics, ETH Zurich, Zurich, Switzerland

Ada Della Pia Department of Chemistry, The University of Warwick, Coventry, UK

Gregory Denbeaux College of Nanoscale Science and Engineering, University at Albany, Albany, NY, USA

Parag B. Deotare Electrical Engineering, Harvard School of Engineering and Applied Sciences, Cambridge, MA, USA

Emiliano Descrovi DISMIC – Dipartimento di Scienza dei Materiali e Ingegneria Chimica, Politecnico di Torino, Torino, Italy

Joseph M. DeSimone Department of Chemistry, University of North Carolina, Chapel Hill, NC, USA

Department of Pharmacology, Eshelman School of Pharmacy, University of North Carolina, Chapel Hill, NC, USA

Carolina Center of Cancer Nanotechnology Excellence, University of North Carolina, Chapel Hill, NC, USA

Institute for Advanced Materials, University of North Carolina, Chapel Hill, NC, USA

Institute for Nanomedicine, University of North Carolina, Chapel Hill, NC, USA

Lineberger Comprehensive Cancer Center, University of North Carolina, Chapel Hill, NC, USA

Department of Chemical and Biomolecular Engineering, North Carolina State University, Raleigh, NC, USA

Sloan–Kettering Institute for Cancer Research, Memorial Sloan–Kettering Cancer Center, New York, NY, USA

Hans Deyhle Biomaterials Science Center (BMC), University of Basel, Basel, Switzerland

Charles L. Dezelah IV Picosun USA, LLC, Detroit, MI, USA

Nathan Doble The New England College of Optometry, Boston, MA, USA

Mitchel J. Doktycz Biosciences Division, Oak Ridge National Laboratory, Oak Ridge, TN, USA

Center for Nanophase Materials Sciences, Oak Ridge National Laboratory, Oak Ridge, TN, USA

Calvin Domier Department of Electrical and Computer Engineering, University of California, Davis, USA

Lixin Dong Electrical and Computer Engineering, Michigan State University, East Lansing, MI, USA

Avinash M. Dongare Department of Materials Science and Engineering, North Carolina State University, Raleigh, NC, USA

Emmanuel M. Drakakis Department of Bioengineering, The Sir Leon Bagrit Centre, Imperial College London, London, UK

Wouter H. P. Driessen David H. Koch Center, The University of Texas M. D. Anderson Cancer Center, Houston, TX, USA

Carlos Drummond Centre de Recherche Paul Pascal, CNRS–Université Bordeaux 1, Pessac, France

Jie Du Beijing Institute of Heart Lung and Blood Vessel Diseases, Beijing Anzhen Hospital, Beijing, China

Jean-Marie Dupret Laboratory of Molecular and Cellular Responses to Xenobiotics, University Paris Diderot - Paris 7, Unit of Functional and Adaptive Biology (BFA) CNRS EAC 4413, Paris cedex 13, France

Julianna K. Edwards David H. Koch Center, The University of Texas M. D. Anderson Cancer Center, Houston, TX, USA

Volkmar Eichhorn Division Microrobotics and Control Engineering (AMiR), Department of Computing Science, University of Oldenburg, Oldenburg, Germany

Masayoshi Esashi The World Premier International Research Center Initiative for Atom Molecule Materials, Tohoku University, Aramaki, Aoba-ku Sendai, Japan

Mikael Evander Department of Measurement Technology and Industrial Electrical Engineering, Division of Nanobiotechnology, Lund University, Lund, Sweden

Enzo Di Fabrizio Nanobiotech Facility, Istituto Italiano di Tecnologia, Genoa, Italy

Lab. BIONEM, Dipartimento di Medicina Sperimentale e Clinica, Università "Magna Grecia" di Catanzaro, Catanzaro, Italy

Yubo Fan Center for Bioengineering and Informatics, Department of Systems Medicine and Bioengineering, The Methodist Hospital Research Institute, Weill Cornell Medical College, Houston, TX, USA

Zheng Fan Electrical and Computer Engineering, Michigan State University, East Lansing, MI, USA

Sergej Fatikow Division Microrobotics and Control Engineering (AMiR), Department of Computing Science, University of Oldenburg, Oldenburg, Germany

Henry O. Fatoyinbo Centre for Biomedical Engineering, University of Surrey, Guildford, Surrey, UK

Joseph Fernandez-Moure The Methodist Hospital Research Institute, Houston, TX, USA

Mauro Ferrari Department of NanoMedicine, The Methodist Hospital Research Institute, Houston, TX, USA

Benjamin M. Finio School of Engineering and Applied Sciences, Harvard University, Cambridge, MA, USA

Emmanuel Flahaut Institut Carnot Cirimat, Université de Toulouse, UPS, INP, Toulouse cedex 9, France

CNRS, Institut Carnot Cirimat, Toulouse, France

Patrick Flammang Laboratoire de Biologie des Organismes Marins et Biomimétisme, Université de Mons – UMONS, Mons, Belgium

Richard G. Forbes Advanced Technology Institute, Faculty of Engineering and Physical Sciences, University of Surrey, Guildford, UK

Isabelle Fourquaux CMEAB, Centre de Microscopie Electronique Appliquée à la Biologie, Université Paul Sabatier, Faculté de Médecine Rangueil, Toulouse cedex 4, France

Marco Francardi Lab. BIONEM, Dipartimento di Medicina Sperimentale e Clinica, Università "Magna Grecia" di Catanzaro, Catanzaro, Italy

International School for Advanced Studies (SISSA), Edificio Q1 Trieste, Italy

Francesca Frascella DISMIC – Dipartimento di Scienza dei Materiali e Ingegneria Chimica, Politecnico di Torino, Torino, Italy

Roger H. French Department of Materials Science and Engineering, Case Western Reserve University, Cleveland, OH, USA

James Friend Micro/Nanophysics Research Laboratory, RMIT University, Melbourne, VIC, Australia

Hiroyuki Fujita Center for International Research on MicroMechatronics (CIRMM), Institute of Industrial Science, The University of Tokyo, Meguro-ku, Tokyo, Japan

Kenji Fukuzawa Department of Micro System Engineering, Nagoya University, Nagoya, Chikusa-ku, Japan

Diana Gamzina Department of Electrical and Computer Engineering, University of California, Davis, USA

Xuefeng Gao Suzhou Institute of Nano-Tech and Nano-Bionics, Chinese Academy of Sciences, Suzhou, PR China

Pablo García-Sánchez Departamento de Electrónica y Electromagnetismo, Universidad de Sevilla, Sevilla, Spain

Jean-Luc Garden Institut Néel CNRS-UJF, Grenoble, France

Paolo Gasco Nanovector srl, Torino, Italy

Laury Gauthier EcoLab – Laboratoire d'écologie fonctionnelle et environnement, Université de Toulouse, INP, UPS, Castanet Tolosan, France

CNRS UMR 5245, EcoLab, Castanet Tolosan, France

Shady Gawad MEAS Switzerland, Bevaix, Switzerland

Denis Gebauer Department of Chemistry, Physical Chemistry, University of Konstanz, Konstanz, Germany

Ille C. Gebeshuber Institute of Microengineering and Nanoelectronics (IMEN), Universiti Kebangsaan Malaysia, Bangi, Selangor, Malaysia

Institute of Applied Physics, Vienna University of Technology, Vienna, Austria

Francesco Gentile Nanobiotech Facility, Istituto Italiano di Tecnologia, Genoa, Italy

Lab. BIONEM, Dipartimento di Medicina Sperimentale e Clinica, Università "Magna Grecia" di Catanzaro, Catanzaro, Italy

Claudio Gerbaldi Center for Space Human Robotics, Fondazione Istituto Italiano di Tecnologia, Torino, Italy

Amitabha Ghosh Bengal Engineering & Science University, Howrah, India

Ranajay Ghosh Department of Mechanical, Aerospace and Nuclear Engineering, Rensselaer Polytechnic Institute, Troy, NY, USA

Larry R. Gibson II Department of Chemical and Biomolecular Engineering, University of Notre Dame, Notre Dame, IN, USA

Jason P. Gleghorn Department of Chemical and Biological Engineering, Princeton University, Princeton, NJ, USA

Biana Godin Department of Nanomedicine, The Methodist Hospital Research Institute, Houston, TX, USA

Irene González-Valls Laboratory of Nanostructured Materials for Photovoltaic Energy, Escola Tecnica Superior d Enginyeria (ETSE), Centre d'Investigació en Nanociència i Nanotecnología (CIN2, CSIC), Bellaterra (Barcelona), Spain

Ashwini Gopal Department of Biomedical Engineering, The University of Texas at Austin, Austin, TX, USA

Claudia R. Gordijo Advanced Pharmaceutics and Drug Delivery Laboratory, University of Toronto, Toronto, ON, Canada

Yann Le Gorrec FEMTO-ST/UFC-ENSMM-UTBM-CNRS, Besançon, France

Alok Govil Qualcomm MEMS Technologies, Inc., San Jose, CA, USA

Paul Graham Centre for Computational Neuroscience and Robotics, University of Sussex, Brighton, UK

Dmitri K. Gramotnev Nanophotonics Pty Ltd, Brisbane, QLD, Australia

Nicolas G. Green School of Electronics and Computer Science, University of Southampton, Highfield, Southampton, UK

Robert J. Greenberg Second Sight Medical Products (SSMP), Sylmar, CA, USA

Julia R. Greer Division of Engineering and Applied Sciences, California Institute of Technology, Pasadena, CA, USA

Dane A. Grismer Department of Chemical and Biomolecular Engineering, University of Notre Dame, Notre Dame, IN, USA

Petra Gruber Transarch - Biomimetics and Transdisciplinary Architecture, Vienna, Austria

Rina Guadagnini Laboratory of Molecular and Cellular Responses to Xenobiotics, University Paris Diderot - Paris 7, Unit of Functional and Adaptive Biology (BFA) CNRS EAC 4413, Paris cedex 13, France

Vladimir Gubala Biomedical Diagnostics Institute, Dublin City University, Glasnevin, Dublin, Ireland

Pablo Gurman Materials Science Division, Argonne National Laboratory, Argonne, IL, USA

Evgeni Gusev Qualcomm MEMS Technologies, Inc., San Jose, CA, USA

MEMS Research and Innovation Center, Qualcomm MEMS Technologies, Inc., San Jose, CA, USA

Maria Laura Habegger Department of Integrative Biology, University of South Florida, Tampa, FL, USA

Yassine Haddab FEMTO-ST/UFC-ENSMM-UTBM-CNRS, Besançon, France

Neal A. Hall Electrical and Computer Engineering, University of Texas at Austin, Austin, TX, USA

Moon-Ho Ham Department of Chemical Engineering, Massachusetts Institute of Technology, Cambridge, MA, USA

Hee Dong Han Gynecologic Oncology, M.D. Anderson Cancer Center, The University of Texas, Houston, TX, USA

Center for RNA Interference and Non–coding RNA, M.D. Anderson Cancer Center, The University of Texas, Houston, TX, USA

Xiaodong Han Institute of Microstructure and Property of Advanced Materials, Beijing University of Technology, Beijing, People's Republic of China

Aeraj ul Haque Biodetection Technologies Section, Energy Systems Division, Argonne National Laboratory, Lemont, IL, USA

Nadine Harris Department of Chemistry and International Institute for Nanotechnology, Northwestern University, Evanston, IL, USA

Judith A. Harrison Department of Chemistry, United States Naval Academy, Annapolis, MD, USA

Achim Hartschuh Department Chemie and CeNS, Ludwig-Maximilians-Universität München, Munich, Germany

Jian He Department of Physics and Astronomy, Clemson University, Clemson, SC, USA

Martin Hegner The Naughton Institute, School of Physics, Trinity College Dublin, CRANN, Dublin, Ireland

Michael G. Helander Department of Materials Science and Engineering, University of Toronto, Toronto, Ontario, Canada

Michael J. Heller Department of Nanoengineering, University of California San Diego, La Jolla, CA, USA

Department of Bioengineering, University of California San Diego, La Jolla, CA, USA

Simon J. Henley Nano Electronics Center, Advanced Technology Institute, University of Surrey, Guildford, Surrey, UK

Elise Hennebert Laboratoire de Biologie des Organismes Marins et Biomimétisme, Université de Mons – UMONS, Mons, Belgium

Joseph P. Heremans Department of Mechanical and Aerospace Engineering, The Ohio State University, E443 Scott Laboratory, Columbus, OH, USA

Simone Hieber Biomaterials Science Center (BMC), University of Basel, Basel, Switzerland

Dale Hitchcock Department of Physics and Astronomy, Clemson University, Clemson, SC, USA

David Holmes London Centre for Nanotechnology, University College London, London, UK

Hendrik Hölscher Karlsruher Institut für Technologie (KIT), Institut für Mikrostrukturtechnik, Karlsruhe, Germany

J. H. Hoo Department of Electrical Engineering, University of Washington, Seattle, WA, USA

Bart W. Hoogenboom London Centre for Nanotechnology and Department of Physics and Astronomy, University College London, London, UK

Kazunori Hoshino Department of Biomedical Engineering, The University of Texas at Austin, Austin, TX, USA

Larry L. Howell Department of Mechanical Engineering, Brigham Young University, Provo, UT, USA

Hou-Jun Hsu C.C.P. Contact Probes CO., LTD, New Taipei City, Taiwan ROC

Jung-Tang Huang National Taipei University of Technology, Taipei, Taiwan ROC

Michael P. Hughes Centre for Biomedical Engineering, University of Surrey, Guildford, Surrey, UK

Shelby B. Hutchens California Institute of Technology MC 309-81, Pasadena, CA, USA

John W. Hutchinson School of Engineering and Applied Sciences, Harvard University, Cambridge, MA, USA

Gilgueng Hwang Laboratoire de Photonique et de Nanostructures (LPN-CNRS), Site Alcatel de Marcoussis, Route de Nozay, Marcoussis, France

Hyundoo Hwang School of Nano-Bioscience and Chemical Engineering, Ulsan National Institute of Science and Technology (UNIST), Ulsan, Republic of Korea

Barbara Imhof LIQUIFER Systems Group, Austria

Hiromi Inada Hitachi High-Technologies America, Pleasanton, CA, USA

M. Saif Islam Electrical and Computer Engineering, University of California - Davis Integrated Nanodevices & Nanosystems Lab, Davis, CA, USA

Mitsumasa Iwamoto Department of Physical Electronics, Tokyo Institute of Technology, Meguro-ku, Tokyo, Japan

Esmaiel Jabbari Biomimetic Materials and Tissue Engineering Laboratory, Department of Chemical Engineering, Swearingen Engineering Center, Rm 2C11, University of South Carolina, Columbia, SC, USA

Laurent Jalabert LIMMS/CNRS-IIS (UMI 2820), Institute of Industrial Science, The University of Tokyo, Tokyo, Japan

Dongchan Jang Department of Applied Physics and Materials Science, California Institute of Technology, Pasadena, CA, USA

Daniel Jasper Division Microrobotics and Control Engineering (AMiR), Department of Computing Science, University of Oldenburg, Oldenburg, Germany

Debdeep Jena Department of Electrical Engineering, University of Notre Dame, Notre Dame, IN, USA

Taeksoo Ji School of Electronics and Computer Engineering, Chonnam National University, Gwangju, Korea

Lixin Jia The Key Laboratory of Remodeling-related Cardiovascular Diseases, Capital Medical University, Ministry of Education, Beijing, China

Lei Jiang Center of Molecular Sciences, Institute of Chemistry Chinese Academy of Sciences, Beijing, People's Republic of China

Dilip S. Joag Centre for Advanced Studies in Materials Science and Condensed Matter Physics, Department of Physics, University of Pune, Pune, Maharashtra, India

Erik J. Joner Bioforsk Soil and Environment, Ås, Norway

Suhas S. Joshi Department of Mechanical Engineering, Indian Institute of Technology Bombay, Mumbai, Maharastra, India

Gabriela Juarez-Martinez Centeo Biosciences Limited, Dumbarton, UK

Soyoun Jung Samsung Mobile Display Co., LTD, Young-in, Korea

C. J. Kähler Institute of Fluid Mechanics and Aerodynamics, Universität der Bundeswehr München, Neubiberg, Germany

Sergei V. Kalinin Center for Nanophase Materials Sciences, Oak Ridge National Laboratory, Oak Ridge, TN, USA

Ping Kao The Pennsylvania State University, University Park, PA, USA

Swastik Kar Department of Physics, Northeastern University, Boston, MA, USA

Mustafa Karabiyik Department of Electrical and Computer Engineering, Florida International University, Miami, FL, USA

Sinan Karaveli School of Engineering, Brown University, Providence, RI, USA

David Karig Center for Nanophase Materials Sciences, Oak Ridge National Laboratory, Oak Ridge, TN, USA

Michael Karpelson School of Engineering and Applied Sciences, Harvard University, Cambridge, MA, USA

Andreas G. Katsiamis Toumaz Technology Limited, Abingdon, UK

Christine D. Keating Department of Chemistry, Penn State University, University Park, PA, USA

Pamela L. Keating Department of Chemistry, United States Naval Academy, Annapolis, MD, USA

John B. Ketterson Department of Physics and Astronomy, Northwestern University, Evanston, IL, USA

S. M. Khaled Department of Nanomedicine, The Methodist Hospital Research Institute, Houston, TX, USA

Arash Kheyraddini Mousavi Department of Mechanical Engineering, University of New Mexico, Albuquerque, NM, USA

Andrei L. Kholkin Center for Research in Ceramic and Composite Materias (CICECO) & DECV, University of Aveiro, Aveiro, Portugal

Seonghwan Kim Department of Chemical and Materials Engineering, University of Alberta, Edmonton, AB, Canada

Seong H. Kim Department of Chemical Engineering, Pennsylvania State University, University Park, PA, USA

Moon Suk Kim Department of Molecular Science and Technology, Ajou University, Suwon, South Korea

CJ Kim Mechanical and Aerospace Engineering Department, University of California, Los Angeles (UCLA), Los Angeles, CA, USA

Bongsang Kim Advanced MEMS, Sandia National Laboratories, Albuquerque, NM, USA

Pilhan Kim Graduate School of Nanoscience and Technology, Korea Advanced Institute of Science and Technology (KAIST), Daejeon, South Korea

M. Todd Knippenberg Department of Chemistry, High Point University, High Point, NC, USA

Yee Kan Koh Department of Mechanical Engineering, National University of Singapore room: E2 #02-29, Singapore, Singapore

Helmut Kohl Physikalisches Institut, Westfälische Wilhelms-Universität Münster, Wilhelm-Klemm-Straße 10, Münster, Germany

Mathias Kolle Harvard School of Engineering and Applied Sciences, Cambridge, MA, USA

Susan Köppen Hybrid Materials Interfaces Group, Faculty of Production Engineering and Bremen Center for Computational Materials Science, University of Bremen, Bremen, Germany

Kostas Kostarelos Nanomedicine Laboratory, Centre for Drug Delivery Research, The School of Pharmacy, University of London, London, UK

Roman Krahne Nanobiotech Facility, Istituto Italiano di Tecnologia, Genoa, Italy

Gijs Krijnen Transducers Science & Technology group, MESA + Research Institute for Nanotechnology, University of Twente, Enschede, The Netherlands

Rajaram Krishnan Biological Dynamics, Inc., University of California San Diego, San Diego, CA, USA

Florian Krohs Division Microrobotics and Control Engineering (AMiR), Department of Computing Science, University of Oldenburg, Oldenburg, Germany

Elmar Kroner INM – Leibniz Institute for New Materials, Saarbrücken, Germany

Tom N. Krupenkin Department of Mechanical Engineering, The University of Wisconsin-Madison, Madison, WI, USA

Satish Kumar G. W. Woodruff School of Mechanical Engineering, Georgia Institute of Technology, Atlanta, GA, USA

Aloke Kumar Biosciences Division, Oak Ridge National Laboratory, Oak Ridge, TN, USA

Momoko Kumemura LIMMS/CNRS-IIS (UMI 2820), Institute of Industrial Science, The University of Tokyo, Tokyo, Japan

Harry Kwok Department of Electrical and Computer Engineering, University of Victoria, Victoria, Canada

Jae-Sung Kwon School of Mechanical Engineering and Birck Nanotechnology Center, Purdue University, West Lafayette, IN, USA

Jérôme Labille CEREGE, UMR 6635, CNRS, Aix-Marseille Université, Aix en Provence, Cedex 04, France

Jean Christophe Lacroix Interfaces, Traitements, Organisation et Dynamique des Systemes Universite Paris 7-Denis Diderot, Paris Cedex 13, France

Nicolas Lafitte LIMMS/CNRS-IIS (UMI 2820), Institute of Industrial Science, The University of Tokyo, Tokyo, Japan

Jao van de Lagemaat National Renewable Energy Laboratory, Golden, CO, USA

Renewable and Sustainable Energy Institute, Boulder, CO, USA

Akhlesh Lakhtakia Department of Engineering Science and Mechanics, Pennsylvania State University, University Park, PA, USA

Périne Landois Institut Carnot Cirimat, Université de Toulouse, UPS, INP, Toulouse cedex 9, France

Amy Lang Department of Aerospace Engineering & Mechanics, University of Alabama, Tuscaloosa, AL, USA

Sophie Lanone Inserm U955, Équipe 4, Université Paris Est Val de Marne (UPEC), Créteil, France

Hôpital Intercommunal de Créteil, Service de pneumologie et pathologie professionnelle, Créteil, France

Gregory M. Lanza C-TRAIN Labs, Washington University, St. Louis, MO, USA

Lars Uno Larsson Swedish Biomimetics 3000® AB, Stockholm, Sweden

Camille Larue Laboratoire Lésions des Acides Nucléiques, Commissariat à l'Energie Atomique, SCIB, UMR-E 3 CEA/UJF-Grenoble 1, INAC, Grenoble, Cedex, France

Michael J. Laudenslager Department of Materials Science and Engineering, University of Florida, Gainesville, FL, USA

Thomas Laurell Department of Measurement Technology and Industrial Electrical Engineering, Division of Nanobiotechnology, Lund University, Lund, Sweden

M. Laver Laboratory for Neutron Scattering, Paul Scherrer Institut, Villigen, Switzerland

Materials Research Division, Risø DTU, Technical University of Denmark, Roskilde, Denmark

Nano–Science Center, Niels Bohr Institute, University of Copenhagen, Copenhagen, Denmark

Department of Materials Science and Engineering, University of Maryland, Maryland, USA

Falk Lederer Institute of Condensed Matter Theory and Solid State Optics, Abbe Center of Photonics, Friedrich-Schiller-Universität Jena, Jena, Germany

Kuo-Yu Lee National Taipei University of Technology, Taipei, Taiwan ROC

David W. Lee Department of Biological Sciences, Florida International University Modesto Maidique Campus, Miami, FL, USA

Andreas Lenshof Department of Measurement Technology and Industrial Electrical Engineering, Division of Nanobiotechnology, Lund University, Lund, Sweden

Donald J. Leo Mechanical Engineering, Center for Intelligent Material Systems and Structures, Virginia Tech, Virginia Polytechnic Institute and State University, Arlington, VA, USA

Zayd Chad Leseman Department of Mechanical Engineering, University of New Mexico, Albuquerque, NM, USA

Nastassja A. Lewinski Department of Bioengineering, Rice University, Houston, TX, USA

Mo Li Department of Electrical and Computer Engineering, University of Minnesota, Minneapolis, MN, USA

Jason Li Department of Mechanical and Industrial Engineering and Institute of Biomaterials and Biomedical Engineering, University of Toronto, Toronto, ON, Canada

Wenzhi Li Department of Physics, Florida International University, Miami, FL, USA

Chen Li Department of Chemistry and Biochemistry, University of Bern, Bern, Switzerland

Chun Li Department of Experimental Diagnostic Imaging-Unit 59, The University of Texas MD Anderson Cancer Center, Houston, TX, USA

King C. Li Department of Bioengineering, Rice university, Houston, TX, USA

Weicong Li Department of Electrical and Computer Engineering, University of Victoria, Victoria, Canada

Meng Lian Biosciences Division, Oak Ridge National Laboratory, Oak Ridge, TN, USA

Carlo Liberale Nanobiotech Facility, Istituto Italiano di Tecnologia, Genoa, Italy

Ling Lin Beijing National Laboratory for Molecular Sciences (BNLMS), Key Laboratory of Organic Solids, Institute of Chemistry Chinese Academy of Sciences, Beijing, People's Republic of China

Chung-Yi Lin FormFactor, Inc., Taiwan, Hsinchu, Taiwan ROC

Lih Y. Lin Department of Electrical Engineering, University of Washington, Seattle, WA, USA

Mónica Lira-Cantú Laboratory of Nanostructured Materials for Photovoltaic Energy, Escola Tecnica Superior d Enginyeria (ETSE), Centre d'Investigació en Nanociència i Nanotecnología (CIN2, CSIC), Bellaterra (Barcelona), Spain

Shawn Litster Department of Mechanical Engineering, Carnegie Mellon University, Pittsburgh, PA, USA

Ying Liu School of Materials Science and Engineering, Georgia Institute of Technology, Atlanta, GA, USA

Chang Liu Tech Institute, Northwestern University, ME/EECS, Room L288, Evanston, IL, USA

Gang Logan Liu Micro and Nanotechnology Laboratory, Department of Electrical and Computer Engineering, University of Illinois at Urbana-Champaign, Urbana, IL, USA

Matthew T. Lloyd National Renewable Energy Laboratory MS 3211/SERF W100-43, Golden, CO, USA

Sarah B. Lockwood Department of Chemistry and Biochemistry, Southern Illinois University Carbondale, Carbondale, IL, USA

V. J. Logeeswaran Electrical and Computer Engineering, University of California - Davis Integrated Nanodevices & Nanosystems Lab, Davis, CA, USA

Mariangela Lombardi IIT - Italian Institute of Technology @ POLITO - Centre for Space Human Robotics, Torino, Italy

Marko Loncar Electrical Engineering, Harvard School of Engineering and Applied Sciences, Cambridge, MA, USA

Kenneth A. Lopata William R. Wiley Environmental Molecular Sciences Laboratory, Pacific Northwest National Laboratory, Richland, WA, USA

Gabriel Lopez-Berestein Department of Experimental Therapeutics, M.D. Anderson Cancer Center, The University of Texas, Houston, TX, USA

Cancer Biology, M.D. Anderson Cancer Center, The University of Texas, Houston, TX, USA

Center for RNA Interference and Non–coding RNA, M.D. Anderson Cancer Center, The University of Texas, Houston, TX, USA

The Department of Nanomedicine and Bioengineering, UTHealth, Houston, TX, USA

Alejandro Lopez-Bezanilla National Center for Computational Sciences, Oak Ridge National Laboratory, Oak Ridge, TN, USA

Jun Lou Department of Mechanical Engineering and Materials Science, Rice University 223 MEB, Houston, TX, USA

M. Arturo López-Quintela Laboratory of Magnetism and Nanotechnology, Institute for Technological Research, University of Santiago de Compostela, Santiago de Compostela, Spain

Yang Lu Department of Materials Science and Engineering, Massachusetts Institute of Technology, Cambridge, MA, USA

Zheng-Hong Lu Department of Physics, Yunnan University, Yunnan, Kunming, PR China

Department of Materials Science and Engineering, University of Toronto, Toronto, Ontario, Canada

Michael S.-C. Lu Department of Electrical Engineering, Institute of Electronics Engineering, and Institute of NanoEngineering and MicroSystems, National Tsing Hua University, Hsinchu, Taiwan, Republic of China

Wei Lu Department of Mechanical Engineering, University of Michigan, Ann Arbor, MI, USA

Jia Grace Lu Departments of Physics and Electrophysics, University of Southern California Office: SSC 215B, Los Angeles, CA, USA

Vanni Lughi DI3 – Department of Industrial Engineering and Information Technology, University of Trieste, Trieste, Italy

Neville C. Luhmann Jr. Department of Electrical and Computer Engineering, University of California, Davis, USA

Lorenzo Lunelli Biofunctional Surfaces and Interfaces, FBK-CMM Bruno Kessler Foundation and CNR-IBF, Povo, TN, Italy

Richard F. Lyon Google Inc, Santa Clara, CA, USA

Kuo-Sheng Ma IsaCal Technology, Inc., Riverside, CA, USA

Marc Madou Department of Mechanical and Aerospace Engineering & Biomedical Engineering, University of California at Irvine, Irvine, CA, USA

Daniele Malleo Fluxion Biosciences, South San Francisco, CA, USA

Supone Manakasettharn Department of Mechanical Engineering, The University of Wisconsin-Madison, Madison, WI, USA

Richard P. Mann Centre for Interdisciplinary Mathematics, Uppsala University, Uppsala, Sweden

Liberato Manna Nanobiotech Facility, Istituto Italiano di Tecnologia, Genoa, Italy

Shengcheng Mao Institute of Microstructure and Property of Advanced Materials, Beijing University of Technology, Beijing, People's Republic of China

Francelyne Marano Laboratory of Molecular and Cellular Responses to Xenobiotics, University Paris Diderot - Paris 7, Unit of Functional and Adaptive Biology (BFA) CNRS EAC 4413, Paris cedex 13, France

Sylvain Martel NanoRobotics Laboratory, Department of Computer and Software Engineering, and Institute of Biomedical Engineering, École Polytechnique de Montréal (EPM), Montréal, QC, Canada

Pascal Martin University Paris 7-Denis Diderot, ITODYS, Nanoelectrochemistry Group, UMR CNRS 7086, Batiment Lavoisier, Paris cedex 13, France

Paola Martino Politronica inkjet printing technologies S.r.l., Torino, Italy

Armand Masion CEREGE, UMR 6635 CNRS/Aix–Marseille Université, Aix–en–Provence, France

iCEINT, International Consortium for the Environmental Implications of Nanotechnology, Center for the Environmental Implications of NanoTechnology, Aix–En–Provence, Cedex 4, France

Daniel Maspoch CIN2 (ICN-CSIC), Catalan Institute of Nanotechnology, Bellaterra, Barcelona, Spain

Cintia Mateo Departamento de Química Física, Universidade de Vigo, Vigo, Spain

Shinji Matsui Laboratory of Advanced Science and Technology for Industry, University of Hyogo, Hyogo, Japan

Theresa S. Mayer Department of Electrical Engineering and Materials Science and Engineering, The Pennsylvania State University, University Park, PA, USA

Jeffrey S. Mayer Department of Electrical Engineering, Penn State University, University Park, PA, USA

Chimaobi Mbanaso College of Nanoscale Science and Engineering, University at Albany, Albany, NY, USA

Eva McGuire Department of Materials, Imperial College London, London, UK

Andy C. McIntosh Energy and Resources Research Institute, University of Leeds, Leeds, West Yorkshire, UK

Federico Mecarini Nanobiotech Facility, Istituto Italiano di Tecnologia, Genoa, Italy

Ernest Mendoza Centre de Recerca en Nanoenginyeria, Universitat Politècnica de Catalunya, Barcelona, Spain

Christoph Menzel Institute of Condensed Matter Theory and Solid State Optics, Abbe Center of Photonics, Friedrich-Schiller-Universität Jena, Jena, Germany

Timothy J. Merkel Department of Chemistry, University of North Carolina, Chapel Hill, NC, USA

Vincent Meunier Department of Physics, Applied Physics, and Astronomy, Rensselaer Polytechnic Institute, Troy, NY, USA

Paul T. Mikulski Department of Physics, United States Naval Academy, Annapolis, MD, USA

Hwall Min The Pennsylvania State University, University Park, PA, USA

Rodolfo Miranda Dep. Física de la Materia Condensada, Universidad Autónoma de Madrid and Instituto Madrileño de Estudios Avanzados en Nanociencia (IMDEA-Nanociencia), Madrid, Spain

Sushanta K. Mitra Micro and Nano-scale Transport Laboratory, Department of Mechanical Engineering, University of Alberta, Edmonton, AB, Canada

Cristian Mocuta Synchrotron SOLEIL, L'Orme des Merisiers, Gif-sur-Yvette, France

Mohammad R. K. Mofrad Molecular Cell Biomechanics Lab, Department of Bioengineering, University of California, Berkeley, CA, USA

Seyed Moein Moghimi Centre for Pharmaceutical Nanotechnology and Nanotoxicology, Department of Pharmaceutics and Analytical Chemistry, University of Copenhagen, Copenhagen, Denmark

Farghalli A. Mohamed Department of Chemical Engineering and Materials Science, University of California, Irvine The Henry Samueli School of Engineering, Irvine, CA, USA

Nancy A. Monteiro-Riviere Center for Chemical Toxicological Research and Pharmacokinetics, North Carolina State University, Raleigh, NC, USA

Philip Motta Department of Integrative Biology, University of South Florida, Tampa, FL, USA

Florence Mouchet EcoLab – Laboratoire d'écologie fonctionnelle et environnement, Université de Toulouse, INP, UPS, Castanet Tolosan, France

CNRS UMR 5245, EcoLab, Castanet Tolosan, France

Prachya Mruetusatorn Department of Electrical Engineering and Computer Science, University of Tennessee Knoxville, Knoxville, TN, USA

Oakridge National Laboratory, Oak Ridge, TN, USA

Weiqiang Mu Department of Physics and Astronomy, Northwestern University, Evanston, IL, USA

Stefan Mühlig Institute of Condensed Matter Theory and Solid State Optics, Abbe Center of Photonics, Friedrich-Schiller-Universität Jena, Jena, Germany

Partha P. Mukherjee Computer Science and Mathematics Division, Oak Ridge National Laboratory, Oak Ridge, TN, USA

Bert Müller Biomaterials Science Center (BMC), University of Basel, Basel, Switzerland

Claudia Musicanti Nanovector srl, Torino, Italy

Jit Muthuswamy School of Biological and Health Systems Engineering, Arizona State University, Tempe, AZ, USA

Shrikant C. Nagpure Nanoprobe Laboratory for Bio- & Nanotechnology and Biomimetics, The Ohio State University, Columbus, OH, USA

Vishal V. R. Nandigana Department of Mechanical Science and Engineering, Beckman Institute for Advanced Science and Technology, University of Illinois at Urbana – Champaign, Urbana, IL, USA

Avinash P. Nayak Electrical and Computer Engineering, University of California - Davis Integrated Nanodevices & Nanosystems Lab, Davis, CA, USA

Suresh Neethirajan School of Engineering, University of Guelph, Guelph, ON, Canada

Celeste M. Nelson Department of Chemical and Biological Engineering, Princeton University, Princeton, NJ, USA

Department of Molecular Biology, Princeton University, Princeton, NJ, USA

Bradley J. Nelson Institute of Robotics and Intelligent Systems, ETH Zurich, Zurich, Switzerland

Gilbert Daniel Nessim Chemistry department, Bar-Ilan Institute of Nanotechnology and Advanced Materials (BINA), Bar-Ilan University, Ramat Gan, Israel

Daniel Neuhauser Department of Chemistry and Biochemistry, UCLA, Los Angeles, CA, USA

J. Tanner Nevill Fluxion Biosciences, South San Francisco, CA, USA

Nam-Trung Nguyen School of Mechanical and Aerospace Engineering, Nanyang Technological Univeristy, Singapore, Singapore

Hossein Nili School of Electronics and Computer Science, University of Southampton, Highfield, Southampton, UK

Nano Research Group, University of Southampton, Highfield, Southampton, UK

Vincent Niviere GDRI ICEINT: International Center for the Environmental Implications of Nanotechnology, CNRS–CEA, Europôle de l'Arbois BP 80, Aix–en–Provence Cedex 4, France

Laboratoire de Chimie et Biologie des Métaux, UMR 5249, iRTSV–CEA Bat. K', 17 avenue des Martyrs, Grenoble Cedex 9, France

Michael Nosonovsky Department of Mechanical Engineering, University of Wisconsin-Milwaukee, Milwaukee, WI, USA

Thomas Nowotny School of Informatics, University of Sussex, Falmer, Brighton, UK

Seajin Oh Department of Mechanical and Aerospace Engineering & Biomedical Engineering, University of California at Irvine, Irvine, CA, USA

Murat Okandan Advanced MEMS and Novel Silicon Technologies, Sandia National Laboratories, Albuquerque, NM, USA

Brian E. O'Neill Department of Radiology Research, The Methodist Hospital Research Institute, Houston, TX, USA

Takahito Ono Department of Mechanical Systems and Design, Graduate School of Engineering, Tohoku University, Aramaki, Aoba-ku Sendai, Japan

Clifford W. Padgett Chemistry & Physics, Armstrong Atlantic State University, Savannah, GA, USA

Christine Paillès CEREGE, UMR 6635 CNRS/Aix–Marseille Université, Aix–en–Provence, France

iCEINT, International Consortium for the Environmental Implications of Nanotechnology, Center for the Environmental Implications of NanoTechnology, Aix–En–Provence, Cedex 4, France

Nezih Pala Department of Electrical and Computer Engineering, Florida International University, Miami, FL, USA

Manuel L. B. Palacio Nanoprobe Laboratory for Bio- & Nanotechnology and Biomimetics, The Ohio State University, Columbus, OH, USA

Jeong Young Park Graduate School of EEWS (WCU), Korea Advanced Institute of Science and Technology (KAIST), Daejeon, Republic of Korea

K. S. Park Department of Electrical Engineering, University of Washington, Seattle, WA, USA

Kinam Park Departments of Biomedical Engineering and Pharmaceutics, Purdue University, West Lafayette, IN, USA

Woo-Tae Park Seoul National University of Science and Technology, Seoul, Korea

Andrew R. Parker Department of Zoology, The Natural History Museum, London, UK

Green Templeton College, University of Oxford, Oxford, UK

Alessandro Parodi Department of Nanomedicine, The Methodist Hospital Research Institute, Houston, TX, USA

Renata Pasqualini David H. Koch Center, The University of Texas M. D. Anderson Cancer Center, Houston, TX, USA

Laura Pasquardini Biofunctional Surfaces and Interfaces, FBK-CMM Bruno Kessler Foundation, Povo, TN, Italy

Melissa A. Pasquinelli Fiber and Polymer Science, Textile Engineering, Chemistry and Science, North Carolina State University, Raleigh, NC, USA

Siddhartha Pathak California Institute of Technology MC 309-81, Pasadena, CA, USA

Cecilia Pederzolli Biofunctional Surfaces and Interfaces, FBK-CMM Bruno Kessler Foundation, Povo, TN, Italy

Natalia Pelinovskaya CEREGE, UMR 6635, CNRS, Aix-Marseille Université, Aix en Provence, Cedex 04, France

Néstor O. Pérez-Arancibia School of Engineering and Applied Sciences, Harvard University, Cambridge, MA, USA

Dimitrios Peroulis School of Electrical and Computer Engineering, Birck Nanotechnology Center, Purdue University, West Lafayette, IN, USA

Vinh-Nguyen Phan School of Mechanical and Aerospace Engineering, Nanyang Technological University, Singapore, Singapore

Reji Philip Light and Matter Physics Group, Raman Research Institute, Sadashivanagar, Bangalore, India

Andrew Philippides Centre for Computational Neuroscience and Robotics, University of Sussex, Brighton, UK

Gianluca Piazza Department of Electrical and Systems Engineering, University of Pennsylvania, Philadelphia, PA, USA

Remigio Picone Dana-Farber Cancer Institute – Harvard Medical School, David Pellman Lab – Department of Pediatric Oncology, Boston, MA, USA

Ilya V. Pobelov Department of Chemistry and Biochemistry, University of Bern, Bern, Switzerland

Ryan M. Pocratsky Department of Mechanical Engineering, Carnegie Mellon University, Pittsburgh, USA

Ramakrishna Podila Department of Physics and Astronomy, Clemson University, Clemson, SC, USA

Martino Poggio Department of Physics, University of Basel, Basel, Switzerland

R. G. Polcawich US Army Research Laboratory RDRL-SER-L, Adelphi, MD, USA

Jan Pomplun Zuse Institute Berlin, Berlin, Germany

Alexandra Porter Department of Materials, Imperial College London, London, UK

Cristina Potrich Biofunctional Surfaces and Interfaces, FBK-CMM Bruno Kessler Foundation and CNR-IBF, Povo, TN, Italy

Siavash Pourkamali Department of Electrical and Computer Engineering, University of Denver, Denver, CO, USA

Shaurya Prakash Department of Mechanical and Aerospace Engineering, The Ohio State University, Columbus, OH, USA

Luigi Preziosi Dipartimento di Matematica, Politecnico di Torino, Torino, Italy

Luca Primo University of Torino, Department of Clinical and Biological Sciences, Candiolo, Italy

R. M. Proie US Army Research Laboratory RDRL-SER-E, Adelphi, MD, USA

Olivier Proux GDRI ICEINT: International Center for the Environmental Implications of Nanotechnology, CNRS–CEA, Europôle de l'Arbois BP 80, Aix–en–Provence Cedex 4, France

OSUG, Université Joseph Fourier BP 53, Grenoble, France

Pascal Puech CEMES, Toulouse Cedex 4, France

Robert Puers Department ESAT-MICAS, KULeuven, Leuven, Belgium

J. S. Pulskamp US Army Research Laboratory RDRL-SER-L, Adelphi, MD, USA

Yongfen Qi The Key Laboratory of Remodeling-related Cardiovascular Diseases, Capital Medical University, Ministry of Education, Beijing, China

Qiquan Qiao Department of Electrical Engineering and Computer Science, South Dakota State University, Brookings, USA

Aisha Qi Micro/Nanophysics Research Laboratory, RMIT University, Melbourne, VIC, Australia

Yabing Qi Energy Materials and Surface Sciences (EMSS) Unit, Okinawa Institute of Science and Technology, Kunigami-gun, Okinawa, Japan

Hongwei Qu Department of Electrical and Computer Engineering, Oakland University, Rochester, MI, USA

Marzia Quaglio Center for Space Human Robotics, Fondazione Istituto Italiano di Tecnologia, Torino, Italy

Regina Ragan The Henry Samueli School of Engineering, Chemical Engineering and Materials Science University of California, Irvine, Irvine, CA, USA

Melur K. Ramasubramanian Department of Mechanical & Aerospace Engineering, North Carolina State University, Raleigh, NC, USA

Antonio Ramos Departamento de Electrónica y Electromagnetismo, Universidad de Sevilla, Sevilla, Spain

Hyacinthe Randriamahazaka University Paris 7-Denis Diderot, ITODYS, Nanoelectrochemistry Group, UMR CNRS 7086, Batiment Lavoisier, Paris cedex 13, France

Apparao M. Rao Department of Physics and Astronomy, Clemson University, Clemson, SC, USA

Center for Optical Materials Science & Engineering Technologies, Clemson University, Clemson, SC, USA

E. Reina-Romo School of Engineering, University of Seville, Seville, Spain

Stéphane Régnier Institut des Systèmes Intelligents et de Robotique, Université Pierre et Marie Curie, CNRS UMR7222, Paris, France

Philippe Renaud Microsystems Laboratory, Ecole Polytechnique Federale de Lausanne (EPFL), Lausanne, Switzerland

Tian-Ling Ren Institute of Microelectronics, Tsinghua University, Beijing, China

Scott T. Retterer Biosciences Division, Oak Ridge National Laboratory, Oak Ridge, TN, USA

Center for Nanophase Materials Sciences, Oak Ridge National Laboratory, Oak Ridge, TN, USA

Department of Electrical Engineering and Computer Science, University of Tennessee Knoxville, Knoxville, TN, USA

Roberto de la Rica MESA + Institute for Nanotechnology, University of Twente, Enschede, The Netherlands

Agneta Richter-Dahlfors Swedish Medical Nanoscience Center, Department of Neuroscience, Karolinska Institutet, Stockholm, Sweden

Michèle Riesen Department of Genetics Evolution and Environment, Institute of Healthy Ageing, University College London, London, UK

Matteo Rinaldi Department of Electrical and Computer Engineering, Northeastern University, Boston, MA, USA

Aditi Risbud Molecular Foundry, Lawrence Berkeley National Laboratory, Berkeley, CA, USA

José Rivas Laboratory of Magnetism and Nanotechnology, Institute for Technological Research, University of Santiago de Compostela, Santiago de Compostela, Spain

INL – International Iberian Nanotechnology Laboratory, Braga, Portugal

Paola Rivolo Dipartimento di Scienza dei Materiali e Ingegneria Chimica – Politecnico di Torino, Torino, Italy

Stephan Roche CIN2 (ICN–CSIC), Catalan Institute of Nanotechnology, Universidad Autónoma de Barcelona, Bellaterra (Barcelona), Spain

Institució Catalana de Recerca i Estudis Avançats (ICREA), Barcelona, Spain

Carsten Rockstuhl Institute of Condensed Matter Theory and Solid State Optics, Abbe Center of Photonics, Friedrich-Schiller-Universität Jena, Jena, Germany

Fernando Rodrigues-Lima Laboratory of Molecular and Cellular Responses to Xenobiotics, University Paris Diderot - Paris 7, Unit of Functional and Adaptive Biology (BFA) CNRS EAC 4413, Paris cedex 13, France

Brian J. Rodriguez Conway Institute of Biomolecular and Biomedical Research, University College Dublin, Belfield, Dublin 4, Ireland

Jérôme Rose CEREGE UMR 6635– CNRS–Université Paul Cézanne Aix–Marseille III, Aix–Marseille Université, Europôle de l'Arbois BP 80, Aix–en–Provence Cedex 4, France

GDRI ICEINT: International Center for the Environmental Implications of Nanotechnology, CNRS–CEA, Europôle de l'Arbois BP 80, Aix–en–Provence Cedex 4, France

Yitzhak Rosen Superior NanoBiosystems LLC, Washington, DC, USA

M. Rossi Institute of Fluid Mechanics and Aerodynamics, Universität der Bundeswehr München, Neubiberg, Germany

Marina Ruths Department of Chemistry, University of Massachusetts Lowell, Lowell, MA, USA

Kathleen E. Ryan Department of Chemistry, United States Naval Academy, Annapolis, MD, USA

Malgorzata J. Rybak-Smith Department of Pharmacology, University of Oxford, Oxford, UK

V. Sai Muthukumar Department of Physics, Sri Sathya Sai Institute of Higher Learning, Prashanthinilayam, Andhra Pradesh, India

Verónica Salgueiriño Departamento de Física Aplicada, Universidade de Vigo, Vigo, Spain

Meghan E. Samberg Center for Chemical Toxicological Research and Pharmacokinetics, North Carolina State University, Raleigh, NC, USA

Florence Sanchez Department of Civil and Environmental Engineering, Vanderbilt University, Nashville, USA

Catherine Santaella iCEINT, International Consortium for the Environmental Implications of Nanotechnology, Center for the Environmental Implications of NanoTechnology, Aix–En–Provence, Cedex 4, France

Laboratoire d'Ecologie Microbienne de la Rhizosphère et d'Environnements Extrême, UMR 6191 CNRS–CEA–Aix–Marseille Université de la Méditerranée, St Paul lez Durance, France

J. A. Sanz-Herrera School of Engineering, University of Seville, Seville, Spain

Stephen Andrew Sarles Mechanical Aerospace and Biomedical Engineering, University of Tennessee, Knoxville, TN, USA

J. David Schall Department of Mechanical Engineering, Oakland University, Rochester, MI, USA

George C. Schatz Department of Chemistry and International Institute for Nanotechnology, Northwestern University, Evanston, IL, USA

André Schirmeisen Institute of Applied Physics, Justus-Liebig-University Giessen, Giessen, Germany

Frank Schmidt Zuse Institute Berlin, Berlin, Germany

Helmut Schmitz Institute of Zoology, University of Bonn Poppelsdorfer Schloss, Bonn, Germany

Scott R. Schricker College of Dentistry, The Ohio State University, Columbus, OH, USA

Georg Schulz Biomaterials Science Center (BMC), University of Basel, Basel, Switzerland

Udo D. Schwarz Department of Mechanical Engineering, Yale University, New Haven, USA

Praveen Kumar Sekhar Electrical Engineering, School of Engineering and Computer Science, Washington State University Vancouver, Vancouver, WA, USA

Rita E. Serda Department of NanoMedicine, The Methodist Hospital Research Institute, Houston, TX, USA

Nika Shakiba Department of Mechanical and Industrial Engineering, University of Toronto, Toronto, ON, Canada

Karthik Shankar Department of Electrical and Computer Engineering, W2-083 ECERF, University of Alberta, Edmonton, AB, Canada

Yunfeng Shi Department of Materials Science and Engineering, MRC RM114, Rensselaer Polytechnic Institute, Troy, NY, USA

Li Shi Department of Mechanical Engineering, The University of Texas at Austin, Austin, TX, USA

Youngmin Shin Department of Electrical and Computer Engineering, University of California, Davis, USA

Wolfgang M. Sigmund Department of Materials Science and Engineering, University of Florida, Gainesville, FL, USA

Department of Energy Engineering, Hanyang University, Seoul, Republic of Korea

S. Ravi P. Silva Nano Electronics Center, Advanced Technology Institute, University of Surrey, Guildford, Surrey, UK

Nipun Sinha Department of Mechanical Engineering and Applied Mechanics, Penn Micro and Nano Systems (PMaNS) Lab, University of Pennsylvania, Philadelphia, PA, USA

S. Siva Sankara Sai Department of Physics, Sri Sathya Sai Institute of Higher Learning, Prashanthinilayam, Andhra Pradesh, India

Dunja Skoko The Naughton Institute, School of Physics, Trinity College Dublin, CRANN, Dublin, Ireland

Craig Snoeyink Birck Nanotechnology Center, Mechanical Engineering, Purdue University, West Lafayette, IN, USA

Konstantin Sobolev Department of Civil Engineering and Mechanics, University of Wisconsin-Milwaukee, Milwaukee, WI, USA

Helmut Soltner Zentralabteilung Technologie, Forschungszentrum Jülich GmbH, Jülich, Germany

Thomas Søndergaard Department of Physics and Nanotechnology, Aalborg University, Aalborg Øst, Denmark

Youngjun Song Department of Electrical and Computer Engineering, University of California San Diego, San Diego, CA, USA

Anil K. Sood Gynecologic Oncology, M.D. Anderson Cancer Center, The University of Texas, Houston, TX, USA

Cancer Biology, M.D. Anderson Cancer Center, The University of Texas, Houston, TX, USA

Center for RNA Interference and Non–coding RNA, M.D. Anderson Cancer Center, The University of Texas, Houston, TX, USA

The Department of Nanomedicine and Bioengineering, UTHealth, Houston, TX, USA

Pratheev S. Sreetharan School of Engineering and Applied Sciences, Harvard University, Cambridge, MA, USA

Bernadeta Srijanto Center for Nanophase Materials Sciences, Oak Ridge National Laboratory, Oak Ridge, TN, USA

Tomasz Stapinski Department of Electronics, AGH University of Science and Technology, Krakow, Poland

Ullrich Steiner Department of Physics, Cavendish Laboratories, University of Cambridge, Cambridge, UK

Michael S. Strano Department of Chemical Engineering, Massachusetts Institute of Technology, Cambridge, MA, USA

Arunkumar Subramanian Department of Mechanical and Nuclear Engineering, Virginia Commonwealth University, Richmond, VA, USA

Maxim Sukharev Department of Applied Sciences and Mathematics, Arizona State University, Mesa, AZ, USA

Bobby G. Sumpter Computer Science and Mathematics Division and Center for Nanophase Materials Sciences, Oak Ridge National Laboratory, Oak Ridge, TN, USA

Yu Sun Department of Mechanical and Industrial Engineering and Institute of Biomaterials and Biomedical Engineering and Department of Electrical and Computer Engineering, University of Toronto, Toronto, ON, Canada

Tao Sun Research Laboratory of Electronics, Department of Electrical Engineering and Computer Science, Massachusetts Institute of Technology, Cambridge, MA, USA

Vishnu-Baba Sundaresan Mechanical and Nuclear Engineering, Virginia Commonwealth University, Richmond, VA, USA

Srinivas Tadigadapa Department of Electrical Engineering, The Pennsylvania State University, University Park, PA, USA

Saikat Talapatra Department of Physics, Southern Illinois University Carbondale, Carbondale, IL, USA

Qingyuan Tan Department of Mechanical and Industrial Engineering, University of Toronto, Toronto, ON, Canada

Hiroto Tanaka School of Engineering and Applied Sciences, Harvard University, Cambridge, MA, USA

Xinyong Tao College of Chemical Engineering and Materials Science, Zhejiang University of Technology, Hangzhou, China

Ennio Tasciotti Department of Nanomedicine, The Methodist Hospital Research Institute, Houston, TX, USA

J. Ashley Taylor Department of Mechanical Engineering, The University of Wisconsin-Madison, Madison, WI, USA

Raviraj Thakur School of Mechanical Engineering and Birck Nanotechnology Center, Purdue University, West Lafayette, IN, USA

Antoine Thill iCEINT, International Consortium for the Environmental Implications of Nanotechnology, Center for the Environmental Implications of NanoTechnology, Aix–En–Provence, Cedex 4, France

Laboratoire Interdisciplinaire sur l'Organisation Nanométrique et Supramoléculaire, UMR 3299 CEA/CNRS SIS2M, Gif–surYvette, France

Alain Thiéry iCEINT, International Consortium for the Environmental Implications of Nanotechnology, Center for the Environmental Implications of NanoTechnology, Aix–En–Provence, Cedex 4, France

IMEP, UMR 6116 CNRS/IRD, Aix–Marseille Université, Marseille, Cedex 03, France

Thomas Thundat Department of Chemical and Materials Engineering, University of Alberta, Edmonton, AB, Canada

Katarzyna Tkacz–Smiech Faculty of Materials Science and Ceramics, AGH University of Science and Technology, Krakow, Poland

Steve To Department of Electrical and Computer Engineering, University of Toronto, Toronto, ON, Canada

Gerard Tobias Institut de Ciència de Materials de Barcelona (ICMAB-CSIC), Bellaterra, Barcelona, Spain

Andrea Toma Nanobiotech Facility, Istituto Italiano di Tecnologia, Genoa, Italy

Katja Tonisch Institut für Mikro- und Nanotechnologien, Technische Universität Ilmenau Fachgebiet Nanotechnologie, Ilmenau, Germany

Elka Touitou Institute of Drug Research, School of Pharmacy, The Hebrew University of Jerusalem, Jerusalem, Israel

Lesa A. Tran Department of Chemistry and the Richard E. Smalley Institute for Nanoscale Science and Technology, Rice University, Houston, TX, USA

Alexander A. Trusov Department of Mechanical and Aerospace Engineering, The University of California, Irvine, CA, USA

Soichiro Tsuda Exploratory Research for Advanced Technology, Japan Science and Technology Agency, Osaka, Japan

Lorenzo Valdevit Department of Mechanical and Aerospace Engineering, University of California, Irvine, CA, USA

Ana Valero Microsystems Laboratory, Ecole Polytechnique Federale de Lausanne (EPFL), Lausanne, Switzerland

Pablo Varona Dpto. de Ingenieria Informatica, Universidad Autónoma de Madrid, Madrid, Spain

Amadeo L. Vázquez de Parga Dep. Física de la Materia Condensada, Universidad Autónoma de Madrid and Instituto Madrileño de Estudios Avanzados en Nanociencia (IMDEA-Nanociencia), Madrid, Spain

K. Venkataramaniah Department of Physics, Sri Sathya Sai Institute of Higher Learning, Prashanthinilayam, Andhra Pradesh, India

Nuria Vergara-Irigaray Department of Genetics Evolution and Environment, Institute of Healthy Ageing, University College London, London, UK

Georgios Veronis Department of Electrical and Computer Engineering and Center for Computation and Technology, Louisiana State University, Baton Rouge, LA, USA

Alexey N. Volkov Department of Materials Science and Engineering, University of Virginia, Charlottesville, VA, USA

Frank Vollmer Laboratory of Biophotonics and Biosensing, Max Planck Institute for the Science of Light, Erlangen, Germany

Fritz Vollrath Department of Zoology, University of Oxford, Oxford, UK

Prashant R. Waghmare Micro and Nano-scale Transport Laboratory, Department of Mechanical Engineering, University of Alberta, Edmonton, AB, Canada

Hermann Wagner Institute for Biology II, RWTH Aachen University, Aachen, Germany

Richard Walker ICON plc, Marlow, Buckinghamshire, UK

Thomas Wandlowski Department of Chemistry and Biochemistry, University of Bern, Bern, Switzerland

Wenlong Wang Beijing National Laboratory for Condensed Matter Physics, Institute of Physics, Chinese Academy of Sciences, Beijing, P. R. China

Feng-Chao Wang State Key Laboratory of Nonlinear Mechanics (LNM), Institute of Mechanics, Chinese Academy of Sciences, Beijing, China

Yu-Feng Wang Institute of Microelectronics, Tsinghua University, Beijing, China

Szu-Wen Wang Chemical Engineering and Materials Science, The Henry Samueli School of Engineering, University of California, Irvine, CA, USA

Chunlei Wang Department of Mechanical and Materials Engineering, Florida International University, Miami, FL, USA

Zhibin Wang Department of Materials Science and Engineering, University of Toronto, Toronto, Ontario, Canada

Enge Wang International Center for Quantum Materials, School of Physics, Peking University, Beijing, China

Guoxing Wang School of Microelectronics, Shanghai Jiao Tong University (SJTU), Minhang, Shanghai P R, China

Zhong Lin Wang School of Materials Science and Engineering, Georgia Institute of Technology, Atlanta, GA, USA

Reinhold Wannemacher Madrid Institute for Advanced Studies, IMDEA Nanociencia, Madrid, Spain

Benjamin L. J. Webb Division of Infection & Immunity, University College London, London, UK

Liu Wei CEREGE UMR 6635-CNRS, Aix-Marseille Université, Europôle de l'Arbois, Aix-en-Provence, France

Michael Weigel-Jech Division Microrobotics and Control Engineering (AMiR), Department of Computing Science, University of Oldenburg, Oldenburg, Germany

Steven T. Wereley School of Mechanical Engineering, Birck Nanotechnology Center, Purdue University, West Lafayette, IN, USA

Tad S. Whiteside Savannah River National Laboratory, Aiken, SC, USA

John P. Whitney School of Engineering and Applied Sciences, Harvard University, Cambridge, MA, USA

Samuel A. Wickline C-TRAIN Labs, Washington University, St. Louis, MO, USA

Mark Wiesner iCEINT, International Consortium for the Environmental Implications of Nanotechnology, Center for the Environmental Implications of NanoTechnology, Aix–En–Provence, Cedex 4, France

CEINT, Center for the Environmental Implications of NanoTechnology, Duke University, Durham, NC, USA

Stuart Williams Department of Mechanical Engineering, University of Louisville, Louisville, KY, USA

Kerry Allan Wilson London Centre For Nanotechnology, University College London, London, UK

Lon J. Wilson Department of Chemistry and the Richard E. Smalley Institute for Nanoscale Science and Technology, Rice University, Houston, TX, USA

Patrick M. Winter Department of Radiology, Imaging Research Center, Cincinnati Children's Hospital Medical Center, Cincinnati, OH, USA

Stephen TC Wong Center for Bioengineering and Informatics, Department of Systems Medicine and Bioengineering, The Methodist Hospital Research Institute, Weill Cornell Medical College, Houston, TX, USA

Robert J. Wood School of Engineering and Applied Sciences, Wyss Institute for Biologically Inspired Engineering, Harvard University, Cambridge, MA, USA

Matthew Wright NanoSight Limited, Amesbury, Wiltshire, UK

Wei Wu Department of Materials Science, Fudan University, Shanghai, China

Xiao Yu Wu Advanced Pharmaceutics and Drug Delivery Laboratory, University of Toronto, Toronto, ON, Canada

H. Xie The State Key Laboratory of Robotics and Systems, Harbin Institute of Technology, Harbin, China

Guoqiang Xie Institute for Materials Research, Tohoku University, Sendai, Japan

Huikai Xie Department of Electrical and Computer Engineering, University of Florida, Gainesville, FL, USA

Tingting Xu Department of Electrical Engineering and Computer Science, South Dakota State University, Brookings, USA

Didi Xu Institute of Robotics and Intelligent Systems, ETH Zurich, Zurich, Switzerland

Christophe Yamahata Microsystem Lab., Ecole Polytechnique Fédérale de Lausanne, Lausanne, Switzerland

Keqin Yang Department of Physics and Astronomy, Clemson University, Clemson, SC, USA

Yuehai Yang Department of Physics, Florida International University, Miami, FL, USA

Yi Yang Institute of Microelectronics, Tsinghua University, Beijing, China

Chun Yang Division of Thermal Fluids Engineering, School of Mechanical and Aerospace Engineering, School of Chemical and Biomedical Engineering (joint appointment), Nanyang Technological University, Singapore

Yoke Khin Yap Department of Physics, Michigan Technological University, Houghton, MI, USA

Leslie Yeo Micro/Nanophysics Research Laboratory, RMIT University, Melbourne, VIC, Australia

Zheng Yin Center for Bioengineering and Informatics, Department of Systems Medicine and Bioengineering, The Methodist Hospital Research Institute, Weill Cornell Medical College, Houston, TX, USA

Yaroslava G. Yingling Materials Science and Engineering, North Carolina State University, Raleigh, NC, USA

Minami Yoda G. W. Woodruff School of Mechanical Engineering, Georgia Institute of Technology, Atlanta, GA, USA

Sang-Hee Yoon Department of Mechanical Engineering, University of California, Berkeley, CA, USA

Molecular Cell Biomechanics Lab, Department of Bioengineering, University of California, Berkeley, CA, USA

Lidan You Department of Mechanical and Industrial Engineering and Institute of Biomaterials and Biomedical Engineering, University of Toronto, Toronto, ON, Canada

Yanlei Yu Department of Materials Science, Fudan University, Shanghai, China

Wei Yu The Key Laboratory of Remodeling-related Cardiovascular Diseases, Capital Medical University, Ministry of Education, Beijing, China

Seok H. Yun Graduate School of Nanoscience and Technology, Korea Advanced Institute of Science and Technology (KAIST), Daejeon, South Korea

Wellman Center for Photomedicine, Department of Dermatology, Harvard Medical School and Massachusetts General Hospital, Boston, MA, USA

The Harvard–Massachusetts Institute of Technology Division of Health Science and Technology, Cambridge, MA, USA

Remo Proietti Zaccaria Nanobiotech Facility, Istituto Italiano di Tecnologia, Genoa, Italy

Li Zhang Institute of Robotics and Intelligent Systems, ETH Zurich, Zurich, Switzerland

Ze Zhang Institute of Microstructure and Property of Advanced Materials, Beijing University of Technology, Beijing, People's Republic of China

State Key Laboratory of Silicon Materials and Department of Materials Science and Engineering, Zhejiang University, Hangzhou, China

Yi Zhang Shanghai Institute of Applied Physics, Chinese Academy of Sciences, Shanghai, China

Jin Z. Zhang Department of Chemistry and Biochemistry, University of California, Santa Cruz, CA, USA

Xiaobin Zhang Department of Materials Science and Engineering, Zhejiang University, Hangzhou, China

John Xiaojing Zhang Department of Biomedical Engineering, The University of Texas at Austin, Austin, TX, USA

Jianqiang Zhao Department of Materials Science, Fudan University, Shanghai, China

Ya-Pu Zhao State Key Laboratory of Nonlinear Mechanics (LNM), Institute of Mechanics, Chinese Academy of Sciences, Beijing, China

Jinfeng Zhao Department of Electrical and Computer Engineering, University of California, Davis, USA

Leonid V. Zhigilei Department of Materials Science and Engineering, University of Virginia, Charlottesville, VA, USA

Ying Zhou Department of Electrical & Computer Engineering, University of Florida, Gainesville, FL, USA

Menghan Zhou Department of Physics and Astronomy, Clemson University, Clemson, SC, USA

Yimei Zhu Center for Functional Nanomaterials, Brookhaven National Lab, Upton, NY, USA

Wenguang Zhu Department of Physics and Astronomy, The University of Tennessee, Knoxville, TN, USA

Rashid Zia School of Engineering, Brown University, Providence, RI, USA

Q

Quantum Cluster

► Synthesis of Subnanometric Metal Nanoparticles

Quantum Dot

► Synthesis of Subnanometric Metal Nanoparticles

Quantum Dot Nanophotonic Integrated Circuits

Lih Y. Lin
Department of Electrical Engineering,
University of Washington, Seattle, WA, USA

Definition

Quantum dot nanophotonic integrated circuits are photonic integrated circuits composed of nanoscale optoelectronic components made with semiconductor quantum dots.

The modern electronics industry was revolutionized by the invention of the transistor. Since then, the transistor has been the fundamental building block of all modern electronic devices widely deployed in various places including communication and computer systems. The number of transistors on an integrated circuit has roughly doubled every 2 years, following Moore's Law. Moore himself, however, has stated that this law cannot be sustained indefinitely due to the fundamental limits to miniaturization.

Although the demand for higher processor chip speeds in a computer system continues to drive the bandwidth and density requirements and performance for on-chip and chip-to-chip interconnects, using electrons as the information carrier is now facing the challenges of speed, power consumption, and crosstalk due to electromagnetic interference between channels. The limits of Moore's law call for fundamentally new ways to transmit, process, and compute information, not electronically, but perhaps optically. In general, smaller and faster photonic integrated circuit systems with versatile functionalities are desirable for applications not only in computation, but also in other areas such as optical communication and sensing. The need has fueled the development of new photonic devices, new fabrication methods like self-assembly, and the synthesis of new optoelectronic materials like nanocrystals and nanowires. With these advances, it is now possible to realize lasing, waveguiding, optical modulation, and detection in submicron scale devices. Furthermore, it is possible to pick and place nanoparticles on silicon substrates by merely exposing the substrate surface to a liquid droplet of suspended particles. With these new capabilities and tools, building nanoscale photonic integrated circuits has begun to be widely investigated, in an effort to achieve an optical analogy of VLSI with higher speed, lower power consumption, and higher integration density.

The development of photonic integrated circuits has nonetheless been hindered by the diffraction limit, which restricts the extent of optical wave confinement to $\sim\lambda/2$ for guiding and transmitting optical energy in

B. Bhushan (ed.), *Encyclopedia of Nanotechnology*, DOI 10.1007/978-90-481-9751-4,

conventional optical waveguides, and constrains the integration density one can achieve using photons as information carriers. While electronic integrated circuit technologies are advancing fast in the sub-100 nm lithographic resolution regime, the diffraction limit must be overcome for photonics to successfully complement and perhaps replace some of the electrical components. The advances in nanofabrication technologies have facilitated the miniaturization and integration density of device technologies, with semiconductor quantum dots (QD), also referred to as nanocrystals (NC), as materials with high promise.

Since the early 1980s, the quantum dot has gained increasing notice and visibility. Combined with modern material science techniques, the focus has shifted from modeling behavior to implementing and taking advantage of the QD's unique properties for various applications. Originally, the quantum dot was conceived as the next logical step in lasing devices after quantum well and quantum film structures. Progressing from one- to two- and then to three-dimensional confinement of electron and hole wave functions narrows the gain spectrum and increases maximum gain, which reduces the threshold current to begin laser operation. Furthermore, the quantization of the electron and hole energy states and the material and size-specific absorption characteristics are responsible for enabling the high extinction ratios of QD optical labels in biomedical research, and provide a basis for several photonic devices, including transistors, switches, and, naturally, lasers.

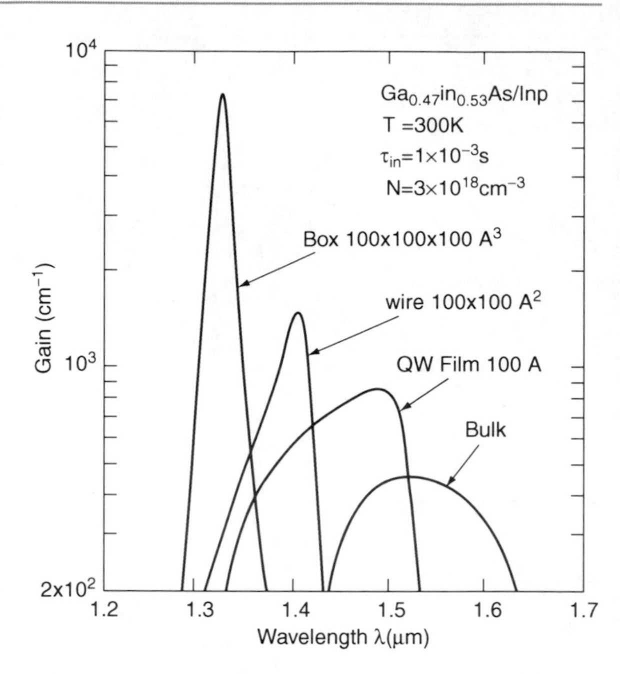

Quantum Dot Nanophotonic Integrated Circuits, Fig. 1 Gain spectra for $Ga_{0.47}In_{0.53}As/InP$ 100 Å × 100 Å × 100 Å cubic QD, 100 Å × 100 Å quantum wire, 100 Å-thick quantum well film, and bulk crystal. The calculation was done assuming $T = 300$ K, intraband relaxation time $\tau_{in} = 1 \times 10^{-13}$ s, and electron–hole density $N = 3 \times 10^{18}$ cm^{-3} [1]

Characteristics of QDs

In general terms, the quantum dot is a form of nanocrystals, or a monodisperse crystalline metal or semiconductor particle whose diameter is adjustable on the nanoscale. QDs are of particular interest from a device standpoint as the electron and photon behavior in the materials differ greatly from that of bulk materials. When deposited in an ordered manner and formed into a solid, collective traits may arise that are also distinctive. As an individual particle, the performance of a QD may be tuned through variations in size and shape, leading to changes in electron–hole generation and recombination dynamics, tunneling behavior, and energy transfer. As in bulk semiconductors, population inversion can be achieved with a sufficient excitation rate and energy, which results in an emission coefficient greater than the absorption coefficient leading to net optical gain. Furthermore, the three-dimensional confinement of electron–hole pairs (exciton wave function) leads to discretized energy levels and delta-function-like distribution of the density of states. Under excitation, the emission of QDs is spectrally much sharper than their bulk material counterparts, which results in a sharp gain spectrum and therefore high gain, low loss, and high quantum efficiency (Fig. 1) [1]. The spectral response of QDs can be easily tuned by particle size and composition. These unique properties make QDs ideal materials for optoelectronic applications.

Conventional QD photonic devices utilize self-organization at the interface of strained epitaxially grown III–V or II–VI semiconductor layers to create pyramidal QD nanostructures due to lattice mismatch. The fabrication is often done through metal-organic chemical vapor deposition (MOCVD). The Stranski–Krastanov process is the basis for manufacture of the first QD lasers, which have been theoretically and experimentally demonstrated with improved gain characteristics and reduced lasing current thresholds compared to quantum well and quantum film structures. Although high-quality optoelectronic devices have been obtained using the self-organization method, the process itself is often not compatible with silicon CMOS fabrication. Another process to fabricate QDs is through chemical synthesis, or solution-based processes. The resulting QDs are in colloidal form and have been widely used as fluorescent tags in biomedicine. The colloidal QDs can then be subsequently deposited on a variety of substrates including silicon and mixed with different materials using various methods to be described later, resulting in rapid expansion in applications such as whispering gallery lasers, sensors, organic LEDs, and solar cells.

In the colloidal case, where individual nanoparticles are fabricated in a solution suspension, the composition materials often fall in the semiconductor category such as II–VI and III–V. The process often involves controlled injection of reagents into the solvent to nucleate and grow the compound of interest. Recently, synthesis of group IV semiconductor QDs has also been investigated. In the case of CdSe nanocrystals, dimethylcadmium, and organophosphine selenide combine to form a core QD structure. A shell layer, such as ZnS, may be subsequently added by solution growth to suppress surface oxidation and improve the fluorescence behavior. The diameters of the core and shell thicknesses determine the energy levels in the QD, which affect the absorption spectrum and correspond to the emission wavelength of the photon generated after the excitation and recombination of an electron–hole pair [2]. The shell layer also allows further surface chemical treatment to enable specific reactions with surrounding media. By design, a shell material with a larger bandgap is commonly selected to enable better electron–hole pair confinement, whereas the surface is treated with stabilizers and can be conjugated with molecules for specialized binding activity, which is dependent on the application.

Colloidal QDs offer high fabrication and integration flexibility. This article focuses on the fabrication, deposition, and photonic devices made of colloidal QDs.

QD Fabrication

Solution-based fabrication of QDs involves techniques to achieve colloids in a controlled manner [3]. To begin, a solvent containing stabilizing and reacting chemicals is poured into a vessel and typically heated to a temperature ideal for stimulating NC formation. Then, a rapid injection of metal-organic material supersaturates the solution and breaks above the nucleation threshold. Subsequently, small particles form, or nucleate, to reduce the energy within the system. Over time, if the rate of injection is below the consumption rate of reagents, then no new nuclei are created and the existing NCs incorporate more metal-organic precursors to become larger.

With regards to specific compounds, long alkane chains attached to phosphorous or phosphorous oxide provide the active ingredients in the solvent mixture. For II–VI semiconductor NC fabrication, the precursors include metal alkyls consisting of the group II metal element and organophosphine chalcogenides or bistrimethylsilylchalcogenides incorporating the group VI chalcogenide element. Taking the CdSe QD as an example, dimethylcadmium and organophosphine selenide are commonly used.

As for ensuring monodisperse nanocrystals, the size distribution will tend toward homogeneity given that the initial nucleation event produced particles whose eventual dimensions eclipse the original volume. Moreover, a secondary growth phase known as Ostwald ripening may aid the process by dissolving smaller NCs with high surface energy in favor of material deposition on the larger ones. Overall, the procedure may be calibrated and timed to synthesize particles of a desired size. Additionally, reactants and NCs are removed from the solution as appropriate.

For creating core/shell QDs, the materials must have compatible deposition conditions and surface energies at the interface. Another requirement is that the nucleation threshold of the shell material be lower than the core, and interdiffusion between the core and shell elements must not occur. The production method of bilayer QDs generally appends to the core synthesis procedure by selecting out NCs of the desired size and adding them to a new reaction solution. Upon reaching the optimal temperature, precursors with the shell material are added at a rate which allows for gradual deposition onto the original particles but avoids a secondary nucleation [3]. To add the final capping layer, an exchange process is used whereby the stabilizing molecules attached to the NCs are interchanged through competition with an excess of another molecule, molecule group, or ion. Furthermore, clumping of particles is avoided if the repulsion strength of outermost layer is greater than the van der Waals attraction force.

On the whole, production of core and core/shell structures results in a size distribution of 5–10% standard deviation from the mean. Further filtration and separation steps may be carried out to identify and collect nanocrystals of better uniformity. Some standard structural characterization methods for NCs include high resolution transmission electron microscopy (TEM) and both small- and wide-angle X-ray scattering (SAXS, WAXS) while analysis of the chemical constitution may be accomplished through nuclear magnetic resonance (NMR), X-ray fluorescence, and photoelectron spectroscopy.

Due to its nontoxicity and lower material cost, silicon (Si) QD is an actively pursued research topic and much effort in developing-related technologies that lead to future commercialization is being invested. Development in colloidal Si QD synthesis thus far lags behind the progress of II–VI QDs; nevertheless, processes that lead to Si QDs with good material characteristics have been demonstrated. Hydrogen-terminated colloidal Si QDs can be synthesized using a low temperature solution-based process [4]. The Si QDs can be prepared by a solution reaction using sodium silicide and ammonium bromide in dimethoxyethane (DME, 80°C) and dioctyl ether (DOE, 260°C). The chemical reaction of the process follows $NaSi + NH_4Br \rightarrow NaBr + NH_3 + Si/H + H_2$, where the desired product, the hydrogen-terminated Si NCs, is noted as Si/H. Once the surface of the Si QDs is hydrogen passivated, further functionalization via photoinitiated hydrosilation can be performed.

Si QDs can also be obtained using electrochemical etching [5]. The process starts with placing p-type boron-doped Si wafers with (100) orientation and 5–20 Ω-cm resistivity in a mixture of hydrofluidic acid (HF), methanol, hydrogen peroxide (H_2O_2), and polyoxometalates (POMs), with the latter two functions as catalysts. At high current density (>10 mA/cm^2) and an etching time of several hours, typical etched surface structures are micropores of diameter around 2 μm, while at low current density (<10 mA/cm^2) and short etching time, nanoscale porous surface is formed. Ultrasonication is used to disintegrate the etched wafer into colloidal Si QDs. Prior to this step, hydrosilylation reaction with chloroplatinic acid as catalyst can result in an octane-modified Si surface, which prevents aggregation of the QDs.

QD Deposition Process

Once the QDs are synthesized, they need to be positioned onto a substrate for further processing into nanophotonic devices. This can be achieved by various methods. In this section, a few examples are listed: Deoxyribonucleic acid (DNA)-directed self-assembly, APTES-mediated molecular self-assembly, solution-based drop-casting, and electrostatic layer-by-layer deposition.

DNA-Directed Self-assembly

The process flow to assemble QDs through the DNA-mediated method is illustrated in Fig. 2 [6]. The area for QD deposition is defined in a layer of polymethylmethacrylate (PMMA) on an oxidized Si substrate using electron-beam lithography (EBL). The sample is then treated with oxygen plasma to generate hydroxyl (–OH) groups, which covalently bind to the dangling ends of silicon oxide and act as hooks for the next chemical reaction. Next, the first self-assembled monolayer (SAM) is formed by gas phase deposition of an organo-silane molecule, 3′mercaptotri-methoxysilane (MPTMS). The silane terminal anchors to the oxygen element of the hydroxyl groups, thus freeing the hydrogen as an energy minimization step while the mercapto, or thiol (SH), group becomes oriented to the outer surface.

The next step is to deposit the second monolayer, composed of a 12 base pair (5′acrydite-ATCCTGAATGCG-3′) chain of DNA terminated by acrylamide on the sample in solution form. This often involves diluting the base DNA with phosphate buffer solution (PBS) and $MgCl_2$. Immersing the sample in buffered acrylic acid followed by PBS rinse leads to passivation of inactivated MPTMS molecules and readies the sample for addition of the complementary DNA (cDNA). Biotin-linked cDNA is used to establish the third SAM. Now, the sample is ready for QD deposition. Streptavidin-bound QDs are combined in PBS and dropped on the surface. The QD assembly step utilizes the well-known biological recognition avidin-biotin binding pair where streptavidin, which is a protein, has four sites that preferentially attract and bind to biotin molecules. The last step is to remove the PMMA layer using toluene submersion.

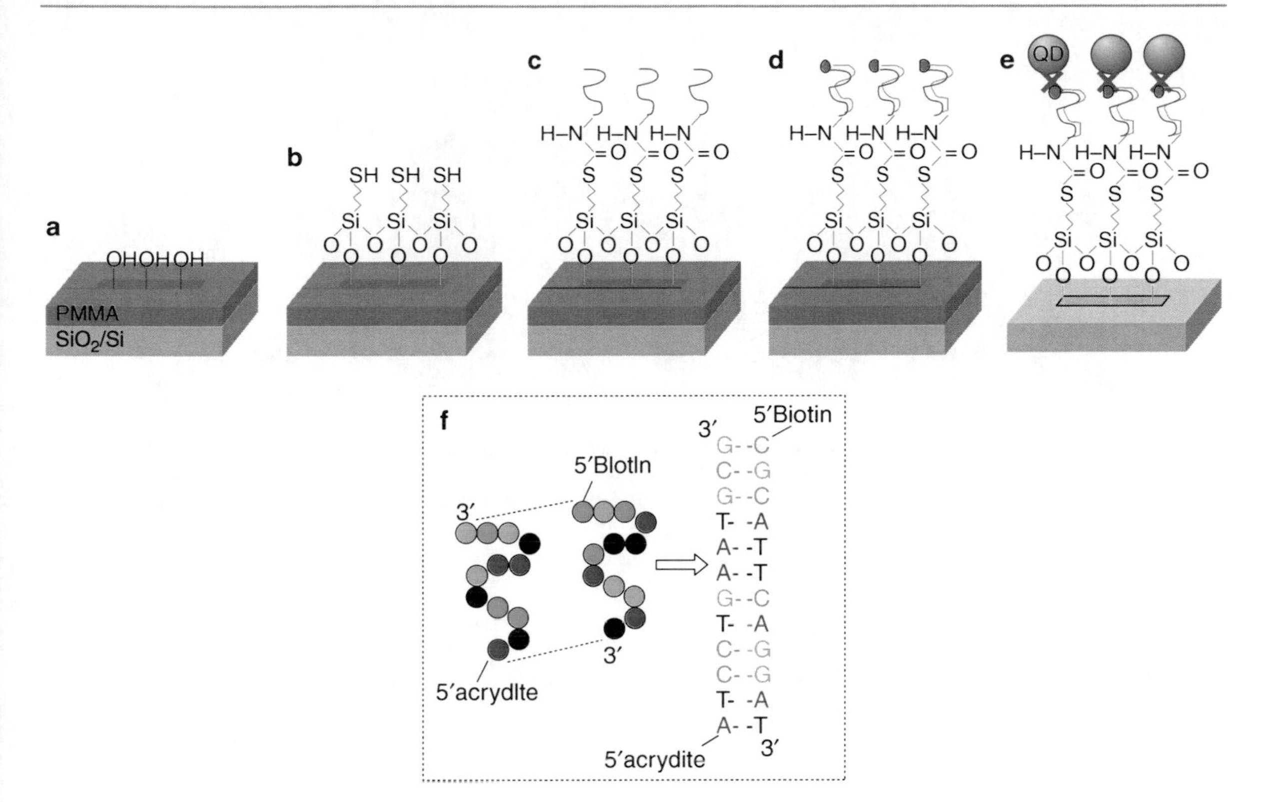

Quantum Dot Nanophotonic Integrated Circuits, Fig. 2 DNA-directed QD deposition process: (**a**) EBL pattern PMMA-coated substrate; (**b**) deposit MPTMS monolayer; (**c**) covalently bind with 5′acrydite-DNA; (**d**) hybridize with biotin-modified cDNA; (**e**) bind streptavidin-QDs to biotin-cDNA sites, remove PMMA; (**f**) detail of DNA chains as hybridized

One chief advantage of DNA-directed self-assembly approach is simultaneous deposition of different types of QDs to form a diverse range of QD structures through specific DNA-cDNA binding recognition. As streptavidin-conjugated structures will preferentially bind to biotin, the critical step is to prepare the QDs with the corresponding biotin-cDNA molecules before reacting them en masse to the base DNA on the substrate. The details of the procedure are much like before, but a few additional steps are required. In particular, use of EBL is needed to open up each of the sections upon which different QDs will be bound. Consequently, after the initial EBL round, the substrate would be treated with MPTMS followed by the passivation buffer and the first acrydite-terminated DNA. With adequate concentration and exposure time, the packing of the base layer with quench molecules will be dense and prevent assembly of subsequent DNA strands. The next EBL writing creates new patterns that are developed and the second round of MPTMS and acrydite-DNA sequence SAMs are added. The process is repeated for as many QD types, and hence DNA-cDNA pairings, that are desired. After the final set of base DNA is added and prior to assembly with the complementary strands, the biotin-cDNA chains are mixed with the matching QD solutions specific to the device design. Subsequently, the QD-cDNA solutions are deposited on the sample as a mass self-assembly step.

APTES-Mediated Molecular Self-assembly

Similar to the DNA-directed technique, the APTES-mediated molecular self-assembly method (Fig. 3) starts with oxygen plasma treatment of an oxidized Si substrate following EBL. Then, the process diverges with sample immersion into 3′aminopropyltriethoxysilane (APTES) thoroughly mixed in an isopropyl alcohol (IPA) and deionized water solvent. In contrast to MPTMS, the chemisorption of APTES leads to an amine ($-NH_3$) rather than a mercapto ($-SH$) group presented at the terminal. Following this, a droplet of carboxylated QDs mixed in deionized water with 1-ethyl-3-(3′dimethylaminopropyl)-carbodiimide (EDC), a

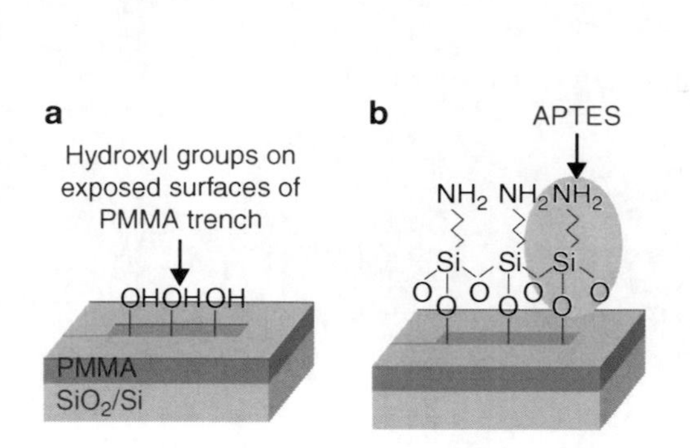

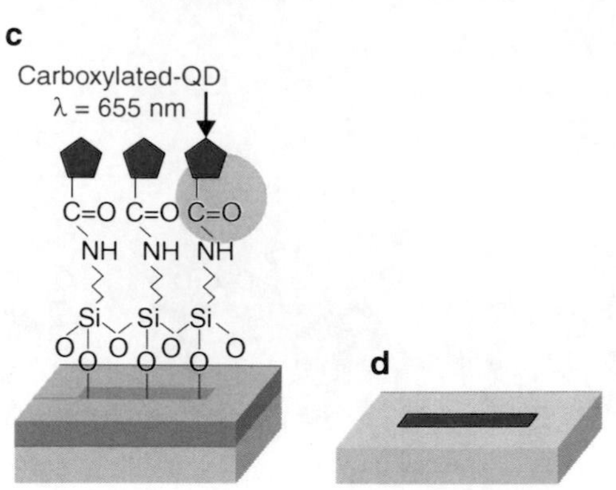

Quantum Dot Nanophotonic Integrated Circuits, Fig. 3 APTES-mediated molecular self-assembly of QDs: (**a**) EBL pattern a SiO_2/Si sample coated with PMMA; develop PMMA; treat with oxygen plasma to create hydroxyl groups on surface; (**b**) solution phase deposition of APTES; (**c**) carboxylated QDs covalently bind to the amine groups of APTES; (**d**) strip PMMA to leave the patterned QDs region

coupling reagent that aids in binding amine and carboxyl groups, is deposited directly on the patterned region.

Solution-Based Drop-Casting

Drop-casting is a method previously used in large-area thin-film devices as a convenient and cost-effective means of making a QD thin film [7]. The method involves dispersing a concentration of QDs in a solvent and then micropipetting a liquid droplet onto a substrate. As the solvent evaporates, the QDs pack together to form a thin membrane on the surface of the substrate. The resultant thin film forms a nanocrystal solid while retaining some of the unique optoelectronic properties of the individual QDs. The quality and thickness of the QD thin film is determined by several factors, including the type of solvent used, the concentration of QDs, the liquid volume of the droplet, the hydrophobicity of the substrate, the type of atmosphere, and the drying temperature. To achieve high-quality nanoscale thickness, using a mixture of hexane/octane as the solvent often achieves a better result. For larger scale thin-film photodetectors, the most common solvent is toluene. Typically, commercial QDs are also suspended in toluene. The hexane/octane mixture prevents ring formations on the resultant film, in part due to the higher rate of evaporation. A high concentration of surface ligands, which prevent the QDs from clumping together, can also result in an opaque or "dirty" thin film. To prevent this, the QDs can be washed by precipitating out of the host solvent with a reagent-like ethanol or methanol. The ethanol acts to strip the organic capping ligands from the shell of the QDs. Once the solution is centrifuged, the supernatant liquid is poured off and the QDs are allowed to dry in a desiccator. The dried QDs can then be redispersed in a solvent of choice. For example, commercially available QDs originally suspended in toluene can be washed and then resuspended in the hexane/octane mixture.

Electrostatic Layer-by-Layer Deposition

The deposition methods described previously achieves either a single layer of QDs or a QD film with unspecified thickness. To achieve controllable film thickness, electrostatic layer-by-layer deposition of QDs has been developed [8]. The patterned areas are pre-covered with a monolayer of charged molecules which serve as traps to capture the QDs by electrostatic bonding. Utilizing a similar electrostatic process, the QDs carrying opposite charges are deposited alternately, forming a multilayer film. By virtue of the layer-by-layer self-assembly, the film thickness can be controlled by the number of layers. Such fabrication process can be performed on a variety of substrates once their surfaces are modified with dangling hydroxyl groups, including silicon oxide and flexible substrates.

The length of ligands on the QD surface is critical to achieving short tunneling barrier and therefore high carrier mobility in the QD thin film. Extremely short ligands can be achieved using 2-mercaptoethylamine (MA) for positive capping and thioglycolic acid (TGA) for negative capping. The lengths of MA (~0.4 nm) and TGA (~0.3 nm) are short enough such that adjacent QDs' surfaces almost contact each other. TGA and MA also work as negative and positive thiol-stabilizers during CdTe QD synthesis. The precursors are formed in the ratio of Cd^{2+}: Te^{2-}: thiol stabilizer = 1: 0.5: 2.4 under appropriate pH conditions (11.5 for TGA and 5.75 for MA). The nanocrystals are grown by refluxing the reaction mixture at 100°C and the particle size is controlled by refluxing time.

To prepare the substrate for electrostatic layer-by-layer deposition, an oxide-covered substrate is treated with oxygen plasma and dipped into NaOH to increase the density of hydroxide on the surface. Subsequently, the sample is immersed in (3-Aminopropyl) triethoxysilane (APTES) solution, followed by protonating the amines in diluted HCl solution for a few seconds. Upon this step, a high area density of positively charged amines on the surface of the substrate can be achieved. The layer-by-layer deposition of QDs is performed by immersing the chip first into TGA (negative)-CdTe QDs solution, then MA (positive)-CdTe QDs solutions. After washing away excess QDs with deionized water, each time of self-assembly will result in deposition of a monolayer of CdTe QDs. Repeating this process leads to a thin film of QDs with a controlled number of layers.

QD Nanophotonic Waveguide

The quantum dot waveguide consists of an array of densely packed QDs anchored to a substrate (Fig. 4a) [6]. A pump light illuminates the sample from above to create excited electron–hole pairs within the QD, which are then triggered by a signal light placed at the waveguide edge to recombine and emit a photon. The stimulated emission of photons occurs in a cascade fashion downstream to generate a response at the output edge. The stimulated emitted photons are identical to the input signal photons in terms of wavelength, polarization, phase, and propagation direction. The overall transmission efficiency at the end of the waveguide depends on the inter-QD coupling efficiency and the gain available in each QD. To avoid wavelength mixing, the pump light energy is set at a higher level than the signal light energy. The finite-difference time-domain (FDTD) simulation of the Poynting vector distribution, representing the power flow and indicating photon transmission behavior in a QD waveguide, is shown in Fig. 4b. Upon absorption of the pump light and triggering of the signal light, the accepting QD produces energy as photons via stimulated emission, with the majority of energy aimed in the propagation direction to interact with the neighboring QD, causing a cascade of energy transmission.

Fabrication of the QD nanophotonic waveguides has been accomplished through either the DNA-directed or the APTES-mediated molecular self-assembly fabrication described earlier. Fluorescence micrographs and atomic force microscope (AFM) images, in Fig. 5a, b, show attachment of 655 nm emission QDs aligned in the 100 and 500 nm width lines. An additional AFM image is shown as an inset in Fig. 5a to provide more details on the QD shape and distribution within the waveguide. Measurement results on straight waveguides and waveguides with 90° bend demonstrated low-loss light transmission in sub-diffraction limit scale [6].

By repeating the fabrication process of the APTES-mediated self-assembly process, it is possible to create waveguides using QDs of different emission wavelengths to operate at targeted transmission wavelengths. Figure 6 shows the fluorescence microscope and AFM images of multiple QD waveguides with red (655 nm) and green (565 nm) emission wavelengths. The AFM images represent the topological profile of the sample, showing well-packed monolayer coverage of QDs. The height is taller for the red patterns, corresponding to larger nanoparticles, as demonstrated by the brighter coloring consistent within the confines of the line.

Nanoscale QD Photodetector

As the integration density of photonic integrated circuits progresses toward the submicron regime, it becomes imperative to detect the optical signal with nanometer resolution as well in order to preserve the on-chip integration density. The requisite photodetection devices are critical for integration of nanophotonics with electronics. The devices should also have high fabrication and integration flexibility with other nanophotonic

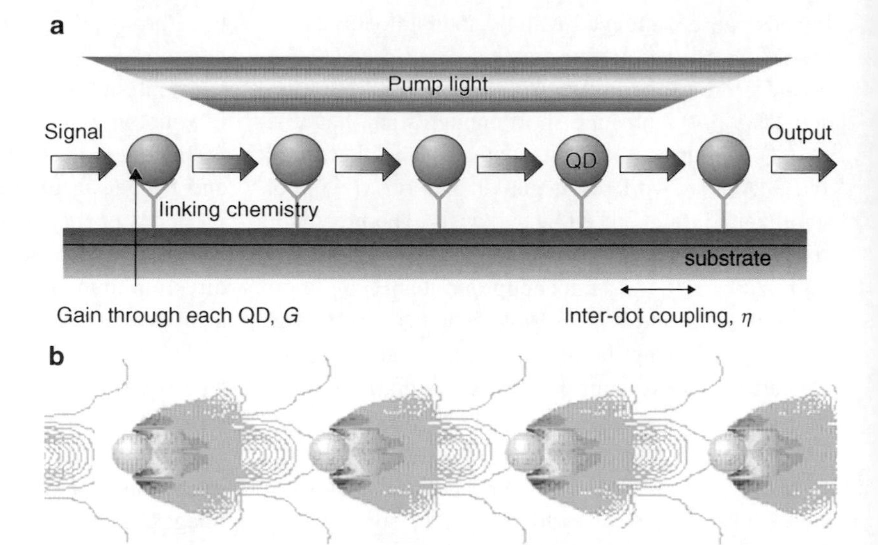

Quantum Dot Nanophotonic Integrated Circuits, Fig. 4 (**a**) Schematic drawing of the QD nanophotonic waveguide. (**b**) Poynting vector distribution within the QD waveguide

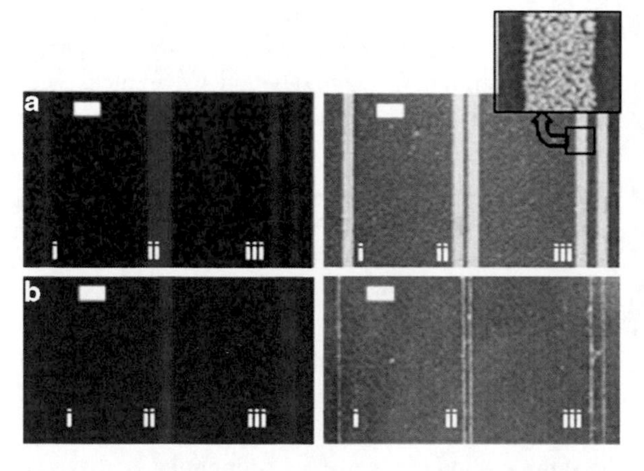

Quantum Dot Nanophotonic Integrated Circuits, Fig. 5 Fluorescence and corresponding AFM images of (**a**) QD waveguide with 500 nm width; the inset shows granular QD shape; and (**b**) 100 nm-wide QD waveguides in (i) single and pair formations spaced, (ii) 200 nm, and (iii) 500 nm apart. Scale bar is 1 μm in length

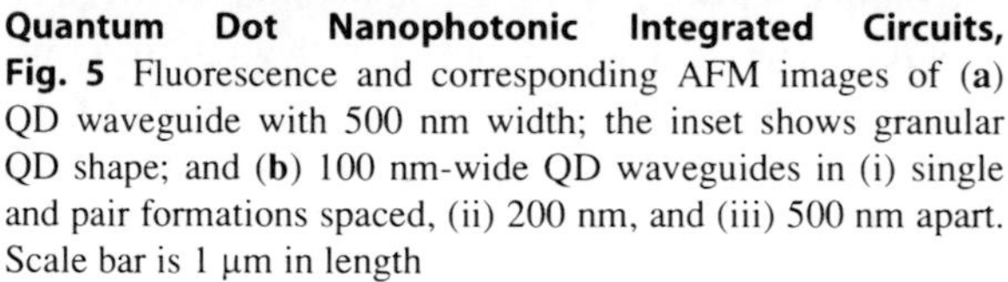

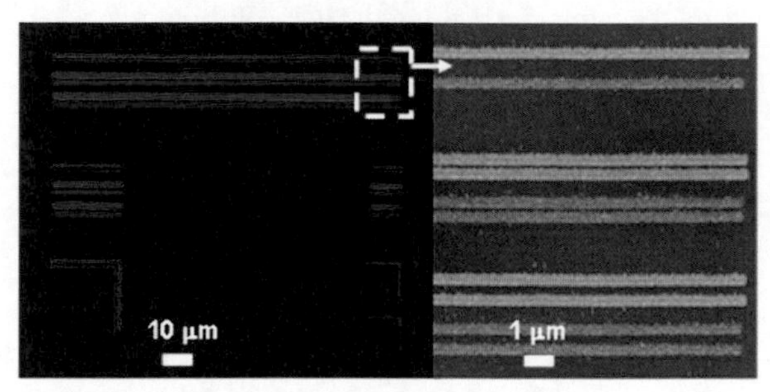

Quantum Dot Nanophotonic Integrated Circuits, Fig. 6 Fluorescence micrograph (*left*) and AFM image of zoomed-in region (*right*) of multiple QD-type waveguides. The emission wavelengths are 655 and 565 nm, corresponding to larger and smaller QDs. The different colors in the AFM image demonstrate the height difference between the two types of QDs

devices. This can be achieved by a nanoscale QD photodetector with a simple configuration schematically depicted in Fig. 7a [9]. Colloidal QDs are placed between two electrodes spaced with a nanogap, so that electrons can tunnel between the electrodes through the QDs. Electron transport in colloidal QDs embedded in a metal-QD-metal structure typically involves significant potential barriers at the QD–QD interface as well as the QD–electrode interface. Transport between QDs can, therefore, be considered as a problem of hopping between localized states. Carriers with higher energies have a higher probability of tunneling through the QDs and contributing to the tunneling current. The effect of optical excitation on the electrical transport of such structures is to reduce the effective potential barrier from the QD to the electrode by exciting photo-generated carriers into higher energy states. The tunneling current increases when the QDs are photo-excited, thus achieving photodetection.

The metal electrodes, spaced at nanometers apart, can be fabricated using electron-beam lithography (EBL). The QDs can be attached to the substrate through the various deposition methods described earlier. The fabrication process of the QD photodetectors consists of two simple steps: patterning of the metal electrodes and attachment of QDs. Both steps, either in

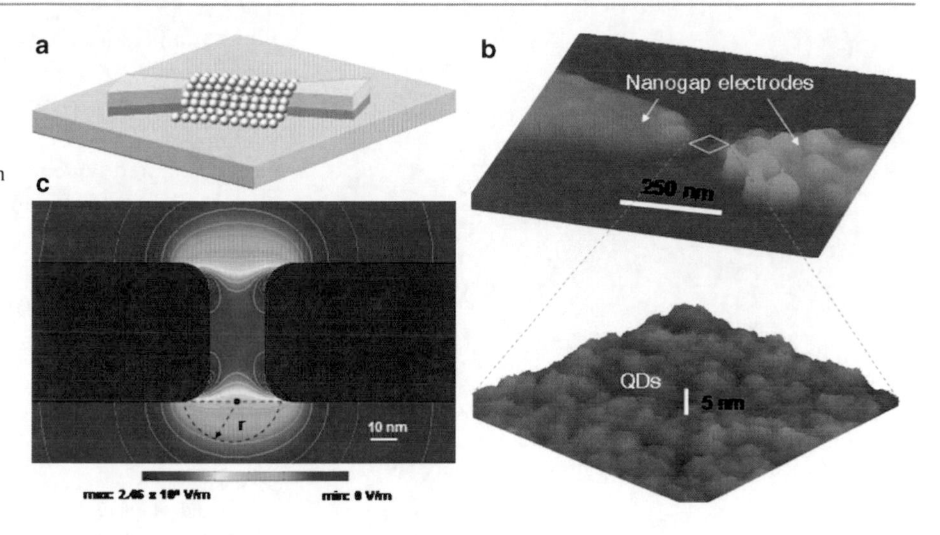

Quantum Dot Nanophotonic Integrated Circuits, Fig. 7 (**a**) Illustration of the nanoscale QD photodetector device structure. (**b**) AFM image with three-dimensional rendering of the device. (**c**) Finite element simulation of the electric field intensity in and around a 20 nm nanogap

sequence or separated, can be readily integrated with the fabrication of a wide range of nanophotonic components and devices. Figure 7b shows an AFM image in three-dimensional view of a device, and zoom-in of the QD film in the nanogap region.

Electron transport through the device is limited by the field-induced ionization rate of the photogenerated excitons, which in this case is proportional to the electron-tunneling rate between neighboring nanocrystals. Thus, for high responsivity, it is necessary to have high field intensities across the junction, which can be obtained most easily in a nanoscale gap. Figure 7c shows finite element simulation of the DC electric field distribution in the gap that is highly concentrated in and around the nanogap. The short distance between electrodes limits the number of tunneling steps required for electrons to traverse the gap, which also increases responsivity and enhances the speed of the device.

QD photodetectors are expected to achieve high sensitivity due to the high quantum efficiency of the nanocrystals. To achieve high-speed operation of the devices remains challenging, due to the tunneling barriers between QDs. Less than 500 fW light detection sensitivity and >125 kHz modulation bandwidth have been reported for the nanoscale QD photodetectors [9].

Integrated fabrication of the nanoscale QD photodetector with the QD waveguides is straightforward due to the nature of self-assembly deposition of QDs. The fabrication result in Fig. 8a shows the AFM topology of a 500 nm-wide QD waveguide aligned to the

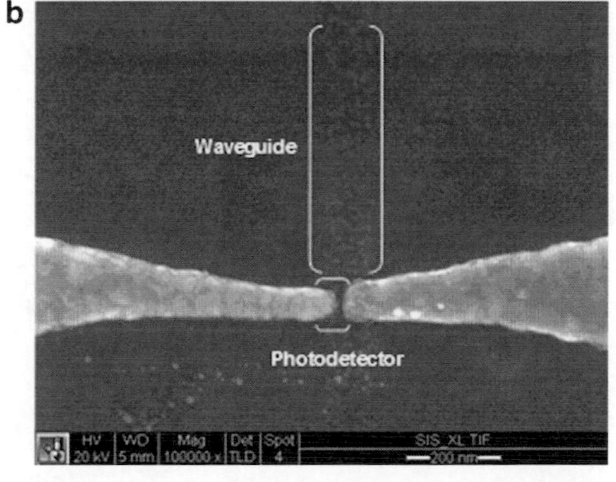

Quantum Dot Nanophotonic Integrated Circuits, Fig. 8 (**a**) AFM image of a 500 nm-wide QD waveguide integrated with a 22 nm break-junction nanogap electrode. (**b**) SEM image of a 50 nm-wide QD waveguide integrated with a nanogap electrode spaced 12 nm apart

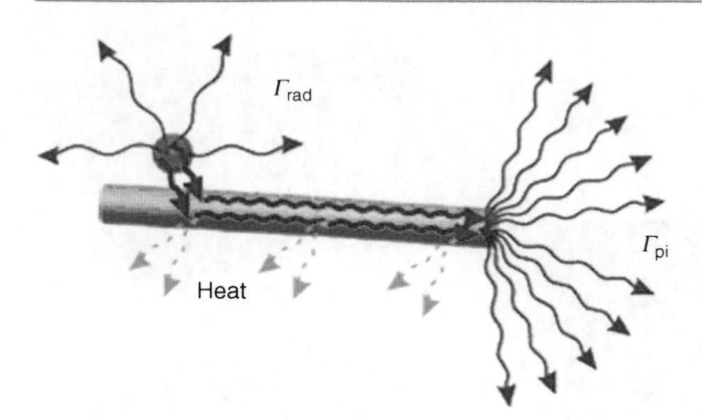

Quantum Dot Nanophotonic Integrated Circuits, Fig. 9 The emission from a single QD can be coupled to a plasmonic nanowire, creating a single plasmon source [10]

center of metal electrodes created by a break-junction process spaced approximately 22 nm apart. Figure 8b shows a SEM image of a 50 nm-wide QD waveguide aligned to the center of metal electrodes created using EBL spaced about 12 nm apart.

Quantum Dot – Plasmonic Waveguide Integration

Plasmonic structures are nanoscale metallic structures that can concentrate light into deep-subwavelength volumes. Therefore, they provide a promising approach to interface diffraction-limited optical components with nanophotonic structures. Integration of a single QD with a plasmonic waveguide has been explored [10]. A CdSe QD is placed in close proximity to a silver nanowire (Fig. 9). When optically excited, the emission from the QD couples directly to guided surface plasmons in the nanowire, causing the end of the wire to light up and acting as a source of single, quantized plasmons. This approach can realize strong nonlinear interactions at the single-photon level, making a single-photon transistor possible [11].

Cross-References

► Active Plasmonic Devices
► AFM
► Atomic Force Microscopy
► Electron Beam Lithography (EBL)
► Finite Element Methods for Computational Nano-optics
► Finite-Difference Time-Domain Technique
► Nanoparticles
► Nanostructured Functionalized Surfaces
► Nanostructures for Photonics
► Nanotechnology
► Scanning Electron Microscopy
► Self-assembled Monolayers
► Self-assembly

References

1. Asada, M., Miyamoto, Y., Suematsu, Y.: Gain and the threshold of three-dimensional quantum-box lasers. IEEE J. Quantum Electron. **22**, 1915–1921 (1986)
2. Bruchez, M. Jr., Moronne, M., Gin, P., Weiss, S., Alivisatos, A.P.: Semiconductor nanocrystals as fluorescent biological labels. Science **281**, 2013–2016 (1998)
3. Murray, C.B., Kagan, C.R., Bawendi, M.G.: Synthesis and characterization of monodisperse nanocrystals and close-packed nanocrystal assemblies. Ann. Rev. Mater. Sci. **30**, 545–610 (2000)
4. Neiner, D., Chiu, H.W., Kauzlarich, S.M.: Low-temperature solution route to macroscopic amounts of hydrogen terminated silicon nanoparticles. J. Am. Chem. Soc. **128**, 11016–11017 (2006)
5. Tu, C.-C., Tang, L., Huang, J., Voutsas, A., Lin, L.Y.: Solution-processed photodetectors from colloidal silicon nano/micro particle composite. Opt. Express **18**, 21622–21627 (2010)
6. Lin, L.Y., Wang, C.-J.: Quantum dot nanophotonic waveguides. In: Ohtsu, M. (ed.) Nanophotonics and nanofabrication, pp. 215–240. Wiley, Weinheim (2009)
7. Konstantatos, G., Sargent, E.H.: Solution-processed quantum dot photodetectors. Proc. IEEE **97**, 1666–1683 (2009)
8. Tu, C.-C., Lin, L.Y.: High efficiency photodetectors fabricated by electrostatic layer-by-layer self-assembly of CdTe quantum dots. Appl. Phys. Lett. **93**, 163107 (2008)
9. Hegg, M.C., Horning, M.P., Baehr-Jones, T., Hochberg, M., Lin, L.Y.: Nanogap quantum dot photodetectors with high sensitivity and bandwidth. Appl. Phys. Lett. **96**, 101118 (2010)
10. Akimov, A.V., Mukherjee, A., Yu, C.L., Chang, D.E., Zibrov, A.S., Hemmer, P.R., Park, H., Lukin, M.D.: Generation of single optical plasmons in metallic nanowires coupled to quantum dots. Nature **450**, 402–406 (2007)
11. Chang, D.E., Sørensen, A.S., Demler, E.A., Lukin, M.D.: A single-photon transistor using nanoscale surface plasmons. Nat. Phys. **3**, 807–812 (2007)

Quantum Dot Solar Cells

► Nanomaterials for Excitonic Solar Cells

Quantum-Dot Toxicity

Wafa' T. Al-Jamal and Kostas Kostarelos
Nanomedicine Laboratory, Centre for Drug Delivery Research, The School of Pharmacy, University of London, London, UK

Synonyms

Semiconductor nanocrystals

Definition

Semiconductor quantum dots (QD) are 1–10 nm nanocrystals consisting mainly of a semiconductor core with or without an inorganic passivation shell usually made of zinc sulfide (ZnS) [1]. The shell is generally used to protect the QD core from oxidation and photolysis and to minimize the associated toxicity related to the release of Cd^{2+} ions.

Currently QD are not only composed of CdS or CdSe but of many different semiconducting materials. These are derived from the II–VI elemental groups (e.g., zinc sulfide [ZnS], zinc selenide [ZnSe], and cadmium telluride [CdTe]), or the III–V elemental groups (e.g., indium phosphide [InP], indium arsenate [InAs]), or the IV–VI elemental groups (e.g., lead selenide [PbSe], lead sulfide [PbS]). In addition to the huge advances in semiconductor synthesis, novel QD such as CdTe/CdSe (core/shell), CdSe/ZnTe (core/shell), cadmium-free QD, and Mn-doped ZnSe have also been developed [2].

In biological sciences QD exhibit great potential as fluorescent probes for optical imaging and diagnostic applications in vitro and in vivo. Due to their bright fluorescence, pronounced photostability and broad excitation spectra, simultaneous detection of QD in different colors has enabled multiplexed imaging for tracking cancer cell metastasis and identification of tumor nodules in vivo [3–6].

QD Solubility and Ligation

QD can be prepared in either water-based or organic solvent environments. The latter approach produces uniformly dispersed QD with a range of emission colors ranging from ultraviolet to infrared. However, QD synthesized in apolar solvents contain organic shells that compromise their water solubility and consequently their compatibility with the biological milieu. Many strategies have been developed to overcome this limitation based on the surface ligation of QD with hydrophilic groups. Several hydrophilic ligands have been utilized to exchange the hydrophobic trioctylphosphine oxide (TOPO) coat on the QD surface with hydrophilic moieties, including: (1) thiol-containing molecules, such as mercaptoacetic acid (MAA), dihydrolipoic acid (DHLA) and mercaptopropyltris (methoxy) silane (MPS); (2) peptides; (3) dendrons; and (4) oligomeric phosphine. In spite of maintaining small mean diameters, surface modification methods generally tend to cause QD aggregation and decrease the efficiency of their optical properties (e.g., fluorescence intensity). Moreover, when labile ligands detach from the QD surface, QD-induced toxicity will increase correspondingly due to exposure to the QD core.

An alternative approach to make QD water dispersible and therefore more biocompatible is based on the hydrophobic interactions between the organic QD surface and the self-assembly of various amphiphilic molecules. Embedding or encapsulating organic QD into phospholipid micelles, lipid bilayers, and amphiphilic diblock or triblock copolymers was found to significantly increase QD diameter, but more importantly to preserve QD photostability, colloidal stability and enable QD application to physiological conditions without the release of the toxic Cd^{2+} ions.

QD Toxicity In Vitro

The toxicity of QD is mainly derived from their intrinsic core composition such as CdSe and CdTe [7, 8]. Cd^{2+} ions have been shown to be toxic upon their release from the QD core due to photolysis and/or oxidation (Fig. 1A). Other mechanisms contributing to QD-induced cytotoxicity have also been identified including: the formation of reactive oxygen species (ROS) that induce cell damage [9] (Fig. 1B) and the interaction of QD nanoparticles with the individual cell components (e.g., DNA or proteins) (Fig. 1C) or with the cell membrane (Fig. 1D).

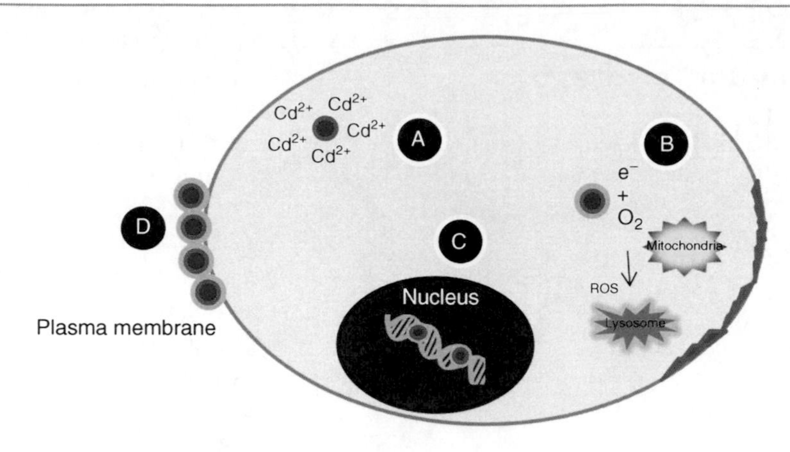

Quantum-Dot Toxicity, Fig. 1 Mechanism of QD toxicity

A Degradation of QD in acidic pH of the endosomes leading to surface coating desorption and Cd^{2+} leakage

B Production of reactive oxygen species (ROS) which affect organelles and plasma membrane integrity

C Interaction of QD with DNA leading to DNA damage

D Precipitation of QD on the plasma membrane

QD-induced cell dysfunction is accompanied by apoptotic and necrotic biochemical changes including: morphological alteration in the plasma membrane; mitochondria and nucleus damage; lysosome enlargement; reduction in cytochrome C concentration; loss of mitochondrial membrane potential; and upregulation of peroxidized lipids [10].

The correlation between cytotoxicity and free Cd^{2+} ions has been established with the occurrence of significant cell death in the range of 100–400 μM Cd^{2+} ions. The blue shift in QD fluorescence spectrum has been proposed as an indication of QD size reduction and Cd^{2+} release. In addition, QD-induced cytotoxicity dramatically increased in the case of QD exposure to oxygen or ultraviolet (UV) light. In fact the principle of QD phototoxicity has been exploited for the development of photodynamic therapeutics [6].

Several attempts to reduce the inherent QD toxicity have been described. For instance, ZnS coating protects the QD core from oxidation, which minimizes Cd^{2+} leakage and reduces the QD-induced cytotoxicity. In addition, the use of antioxidants, such as N-acetylcysteine (NAC) has been shown to be effective in reducing QD cytotoxicity. To date, most cytotoxicity studies have used QD solubilized by direct exchange of the organic coat (TOPO) with hydrophilic ligands, such as mercaptopropionic acid (MPA-QD), mercaptoacetic acid (MAA-QD), mercaptoundecanoic acid (MUA-QD), cysteamine (QD-NH_2), and thioglycerol (QD-OH). In such cases, ligand detachment from the QD surface may also occur due to the weak interaction between the QD surface and the ligands, especially under unfavorable conditions like those in the endosomal compartment. Several studies have indicated that cell incubation with QD solubilized by ligand surface exchange has mostly been associated with severe cell death with increasing QD concentrations, similar to that induced by Cd^{2+} ions, which indicates the need for more stable QD. In contrast, lower cytotoxicity has been observed with polymer-coated QD.

Overall, the in vitro toxicity studies carried out so far have shown that the key determinants of QD toxicity are the nanocrystal composition and the type of surface functionalization. However, other factors including: cell type, QD size, concentration, and QD exposure to oxygen and UV light have also been found to influence QD cytotoxicity.

QD Toxicity In Vivo

The concern about the potential toxicity of QD in vivo is mainly due to the well-established toxicological profile of cadmium, the physicochemical characteristics of certain QD surface coatings, and the size of QD. In general, the small diameter of nanoparticles provides a large available surface for interaction with different biological molecules. Moreover, nanocrystals of a few nanometers in size, such as QD, can enter vital organs such as the heart, lung, liver, and brain after intravenous administration. Therefore, studies that will help determine the toxicological risk and tolerated exposure levels of QD in vivo will be needed for any type of QD developed for biomedical use.

It has been previously described that QD nanocrystals and other nanoparticles could be inhaled and deposited in the respiratory tract, mainly depending on their size [11, 12]. It has also been shown that QD are capable of penetrating intact skin [13]. These preliminary studies suggest that extra care should be taken during QD handling especially during their manufacturing, since systemic toxicity can occur via QD inhalation and direct skin contact.

Preclinical research has shown that QD can be used to visualize lymph nodes and solid tumors in vivo following local or systemic administration. Many studies have also shown that intravenously administered QD mainly reside in the reticuloendothelial system tissues (liver, spleen, and lymph nodes) with no degradation or excretion for up to 2 years; however, no clinical signs of toxicity have been reported. More recent studies have shown that only QD of 5–6 nm in diameter, which is below the renal filtration threshold, were found to be excreted via the urine 4 h post-injection, while QD of larger diameters undesirably remained in the liver up to a few months, which may increase the potential toxicological risks from these nanoparticles in the long term. Interestingly, a recent study has reported that the QD dose in combination with the surface coating and charge play an important role in QD toxicity in vivo. For instance, carboxylated QD injected at a high dose in mice can activate blood coagulation and induce thrombus formation in the lung that can be lethal. Such studies further emphasize the need to investigate the effect of surface coatings on QD toxicity.

In conclusion, QD offer tremendous potential based on their optical properties. Even though a lot of interest and improvements have been achieved in terms of aqueous dispersion and compatibility in physiological environments, there is an inherent risk associated with exposure to QD nanocrystals, in particular those consisting of cadmium. Since the most meaningful clinical translation of QD technology is as probes for biomedical imaging, the acceptable toxicological risks should be minimal or nonexistent. This seems the most formidable challenge that QD technology has to overcome, because as can be seen from the available pharmacological studies, QD can reside in different organs when injected in living animals depending on various QD characteristics (size, dose, surface charge, and coating) and the route of administration. Furthermore, such studies have indicated that QD can accumulate in some tissues for extensive periods of time which requires further investigation to identify the long-term toxicological risks of QD before embarking on any clinical use of QD. In addition, even if QD are excreted from the body the toxicity profile of these nanocrystals on the environment (water and soil) should also be investigated.

Cross-References

▶ Cellular Mechanisms of Nanoparticle's Toxicity
▶ Effect of Surface Modification on Toxicity of Nanoparticles
▶ Genotoxicity of Nanoparticles
▶ Nanoparticle Cytotoxicity

References

1. Rogach, A.L.: Semiconductor Nanocrystal Quantum Dots: Synthesis, Assembly, Spectroscopy, 1st edn. Springer, New York (2008)
2. Nagarajan, R., Hatton, T.A.: Nanoparticles synthesis, stabilization, passivation and functionalization. Oxford University Press, New York (2011)
3. Hotz, C.Z., Bruchez, M.: Quantum Dots: Applications in Biology. Humana Press, New York (2007)
4. Osinski, M., Yamamoto, K., Jovin, T.M.: Colloidal Quantum Dots for Biomedical Applications II. SPIE, San Jose (2007)
5. Pinaud, F.: Peptide-Coated Quantum Dots: Applications to Biological Imaging of Single Molecules in Live Cells and Organisms. VDM Verlag Dr. Müller, Saarbrücken (2009)
6. Kuma, C.S.S.R.: Semiconductor Nanomaterials (Nanomaterials for Life Sciences (VCH). Wiley, Weinheim (2010)
7. Jin, Y., Zhao, X.: Cytotoxicity of photoactive nanoparticles. In: Webster, T.J. (ed.) Safety of Nanoparticles – From Manufacturing to Medical Applications, pp. 19–29. Springer, New York (2009)
8. Sahu, S.C., Casciano, D.A.: Nanotoxicity: From In Vivo and In Vitro Models to Health Risks. Wiley, Chichester (2009)

9. Nadeau, J.: Quantum dot reactive oxygen species generation and toxicity in bacteria:mechanisms and experimental pitfalls. In Callan, J., Raymo, F. (eds.) Quantum Dot Sensors: Technology and Commercial Applications. Pan Stanford Publishing, Singapore (2011)
10. Dave, S. R., White, C. C., Kavanagh, T. J., Gao, X: Luminescent quantum dots for molecular toxicology. In: Balls, M., Combes, R., Bhogal, N. (eds.) New Technologies for Toxicity Testing. Springer, New York (2011)
11. Gehr, P., Mühlfeld, C., Rothen-Rutishauser, B., Blank, F.: Particle-Lung Interactions, (Lung Biology in Health and Disease). Informa Healthcare, London (2009)
12. Marijnissen, J.C., Gradon, L.: Nanoparticles in Medicine and Environment: Inhalation and Health Effects. Springer, Berlin (2009)
13. Ryman-Rasmussen, J.P., Riviere, J.E., Monteiro-Riviere, N.A.: Penetration of intact skin by quantum dots with diverse physicochemical properties. Toxicol. Sci. **91**(1), 159–165 (2006)

R

Radiofrequency

► Thermal Cancer Ablation Therapies Using Nanoparticles

Rapid Electrokinetic Patterning

Jae-Sung Kwon, Raviraj Thakur and
Steven T. Wereley
School of Mechanical Engineering and Birck Nanotechnology Center, Purdue University, West Lafayette, IN, USA

Synonyms

Hybrid opto-electric technique (in viewpoint of driving sources used); Opto-electrokinetic technique (in viewpoint of mechanisms involved); Optofluidics

Definition

Rapid electrokinetic patterning (REP) is a noncontact manipulation technique that concentrates and manipulates colloidal particles on an electrode surface through the simultaneous application of a uniform alternating current (AC) electric field (<200 kHz) and a nonuniform laser illumination. These two phenomena enable REP-based manipulation by creating electric double layer (EDL) polarization of the particles and long-range electrohydrodynamic and toroidal electrothermal motion of the suspending medium induced.

Overview

On-chip manipulation of colloidal micro- and nanoparticles is a key issue in the areas of biologics, diagnostics, therapeutics, and biochemistry [1]. Rapid and precise particle manipulation not only increases the efficiency and accuracy of biological/clinical analysis but also further enables the development of high performance lab-on-a-chip (LOC) systems. These potential benefits have led to the creation of various manipulation techniques and the application of the techniques to a wide range of research fields over the past 10 years: dielectrophoresis [2], electrophoresis [3], hydrophoresis [4], magnetophoresis [5], and optical tweezers [6]. But such efforts also exposed the limitations of each of the manipulation techniques. Dielectrophoresis is the most commonly used electrokinetic method for particle manipulation. This method has been paid enormous attention in many research fields because it can perform high-throughput analysis and works well for both dielectric and conductive particles [1, 7]. However dielectrophoresis basically needs the complicated configuration and design of an electrode structure and its integration on a small chip by microfabrication process. While electrophoresis, another electrokinetic technique, can easily control particles with high separation efficiency, it not only cannot be applied to electrically neutral dielectric particles (or species) but also has several restrictions such as time-consuming and poor reproducibility [3]. Optical tweezers have been often used to study single molecules due to their high precision in handling the molecule [6]. However, they cannot satisfy high throughput, which is a prerequisite for the development of massively parallel analytical systems.

B. Bhushan (ed.), *Encyclopedia of Nanotechnology*, DOI 10.1007/978-90-481-9751-4,

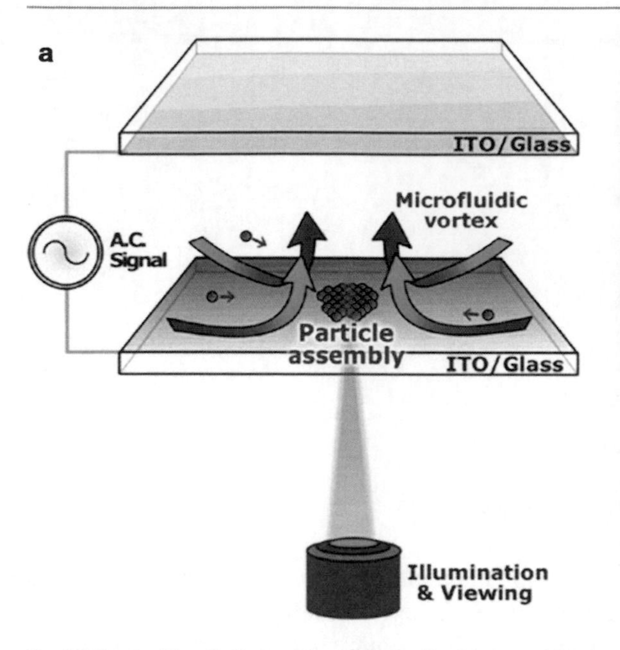

Rapid Electrokinetic Patterning, Fig. 1 Rapid electrokinetic patterning (REP) technique. (**a**) Illustration of particle accumulation by REP technique. In the REP process, particles suspended in a fluid are attracted to an electrode through an interfacial polarization process and a long-range electrohydrodynamic flow and at the same time are transported toward the illuminated region on the electrode surface by an electrothermal vortex (Reprinted with permission from Kumar et al. Copyright 2010 America Chemical Society [11]). (**b**) Assembly of 0.69 μm polystyrene particles using REP technique. Applied electrical frequency, electrical potential, and laser power are 1.6 kHz, $2.0V_{pp}$, and 20 mW, respectively

Magnetophoresis is regarded as a safest manipulation method for biochemical analysis since the involved thermal effects are relatively small compared with the other techniques [1, 5]. But it requires the pretreatment of target particles with magnetic beads or the use of magnetic fluids as suspending media for the technique operation. Hence the recent tendency in the development of particle manipulation technique is going toward combining two different techniques so that their respective advantages complement each other [1, 8].

Rapid electrokinetic patterning (REP) is a novel particle manipulation method suggested to meet the requirements of high speed, high throughput, and high accuracy of biochemical analyses [9–11]. REP can rapidly and dynamically concentrate and manipulate a large quantity of colloidal particles on or near an electrode by applying a laser illumination to an electrode surface biased with a uniform AC electric field (<200 kHz) (Fig. 1) [9, 11]. In the REP technique, an AC electric field plays a role of inducing polarization of particles suspended in a fluid and an electrohydrodynamic flow around the particles and an electrode [10–12]. A nonuniform optical illumination generates nonuniform temperature distribution which interacts with the applied electrical field to generate a toroidal electrothermal microvortex with the center at the middle of the illuminated region on the electrode [13]. The combination of these three physical phenomena enables the rapid concentration and manipulation of colloidal particles on an electrode surface. And the particle clusters can be translated to a desired location simply by changing the spatial location of a laser spot. The details of REP mechanism will be provided in the following section.

In a microfluidic chip where the two driving sources are present together, REP can carry out particle manipulation in a variety of forms (Fig. 2) [9]. It can not only assemble and pattern colloidal particles at single, multiple, or arbitrarily shaped illuminated sites but also translate the particle assemblies on the electrode surface through dynamic control of the laser illumination. In addition, trapping the particles in the continuous flow of a suspending medium is also achievable by REP technique. These abilities of REP offer significant benefits to the performance of biological or chemical analysis on a LOC system. The rapid aggregation and the precise patterning allow analyses to be conducted

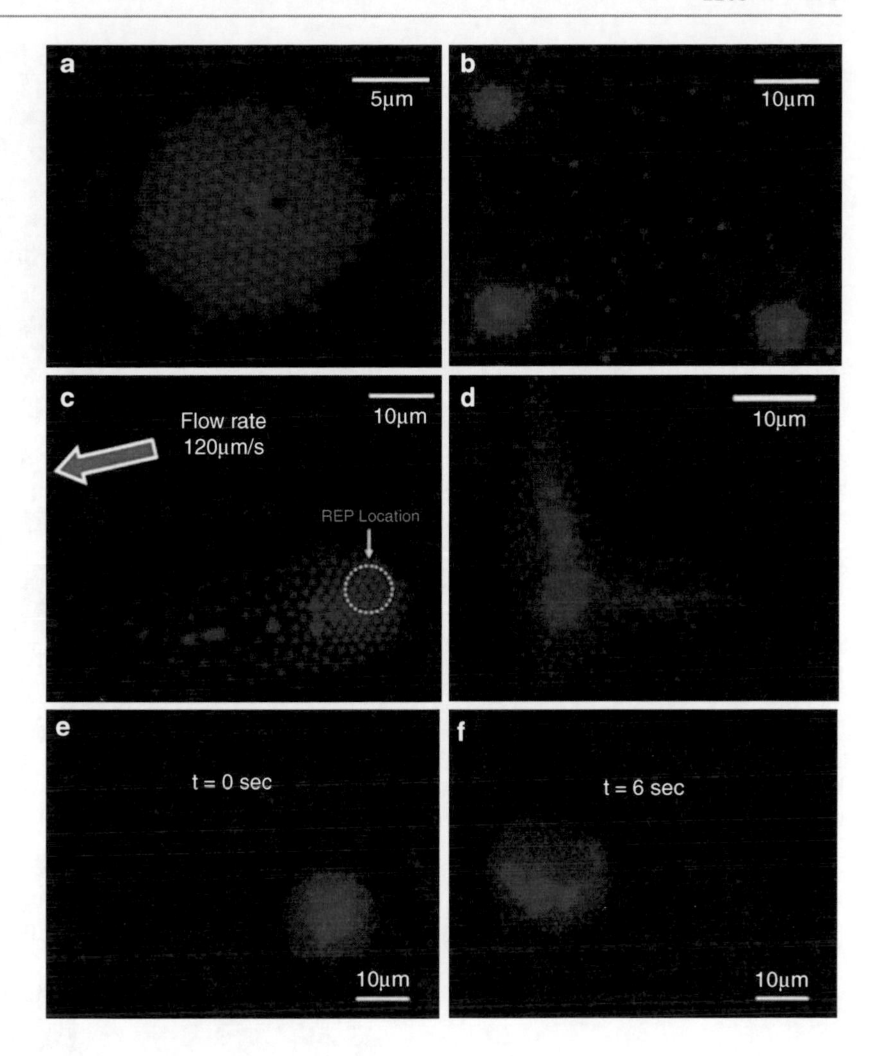

Rapid Electrokinetic Patterning, Fig. 2 Particle manipulations by REP technique. (**a**) REP-based aggregation. The applied electrical signal and laser power are 7.5 kHz, 8.3V_{pp}, and 20 mW, respectively. (**b**) REP-based patterning using multiple laser spots. The electric field of 4.6 kHz and 3.8V_{pp} is applied and the laser power of 40 mW is supplied to the chip. (**c**) REP-based trapping in the continuous flow of a suspending medium. The applied electrical signal and laser power are 5.0 kHz, 10.5V_{pp}, and 40 mW, respectively. (**d**) REP-based patterning with "L" shape. The applied electric field is 1.6 kHz and 2.0V_{pp}, and the laser power is 20 mW. (**e**, **f**) REP-based translation. The translation of the particle assembly from *right* to *left* is achieved by moving a laser illumination under a uniform electric field. The electric signal of 1.6 kHz and 2.0V_{pp} and the laser of 20 mW were applied to the chip. (**c-f**) Williams et al. Reproduced with permission of the Royal Society of Chemistry [9])

with high throughput and high accuracy. The high resolution and dynamic controllability of the technique enhances the sensitivity and monitoring ability of biochemical sensors used in the analytical process. Therefore REP is a promising manipulation technique which can make great contribution to a variety of research fields including biology, chemistry, medicine, and pharmaceuticals.

Physics of REP Technique

REP is a hybrid opto-electrokinetic technique, in the sense that the simultaneous application of a uniform AC electric field and an optical laser should be involved for particle manipulation [9–11]. This characteristic of REP technique is experimentally demonstrated in Fig. 3. When a uniform AC electric field and an optical laser are simultaneously supplied to the chip, particles aggregation is initiated on the electrode surface by REP technique (Fig. 3a). The assembly can happen on either the electrode surfaces depending on where the laser is focused. However, when only the laser is deactivated, the particles consisting of the assembly are immediately scattered on the electrode surface and no particle clusters are observed (Fig. 3b). When only the electric field is switched off, the particles are carried away from the electrode by natural convective flow due to laser heating (Fig. 3c). This shows that REP technique cannot be carried out with the sole application of either the uniform electric field or the optical laser illumination.

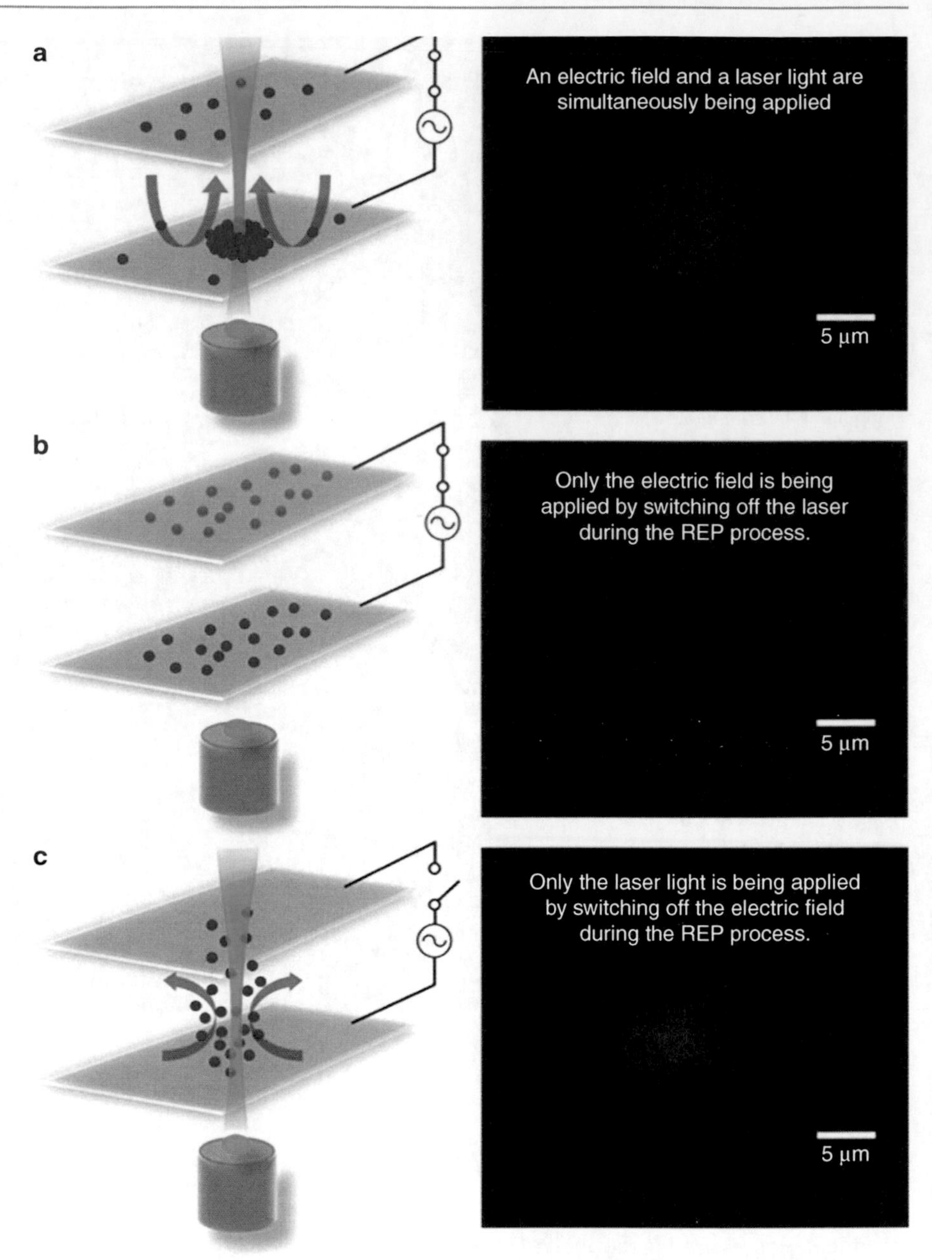

Rapid Electrokinetic Patterning, Fig. 3 Characteristic of REP technique. REP-based manipulation is performed through the simultaneous application of a uniform AC electric field and a laser illumination. Each figure consists of a schematic to describe and an image to experimentally demonstrate the REP nature. The electrical frequency and voltage applied to the chip are 37.3 kHz and 3.3V_{pp} and the power of the laser focused on the bottom electrode is 20 mW. And the used particles are 1 μm polystyrene beads. (**a**) Particle aggregation resulted from the simultaneous application of the uniform AC electric field and the laser illumination. (**b**) Random distribution of the particles appearing when only the electric field is applied by switching off the laser during the REP process. (**c**) Convecting away of the particles caused when only the laser is applied by deactivating the electric field during the REP process

Hence particle manipulation in REP process is performed directly by several physical phenomena that the two driving sources give rise to.

A uniform electric field induces interfacial polarization in the electric double layer (EDL) of a particle, and at the same time causes a electrohydrodynamic (EHD) flow around the particle and electrode surface through the interaction with the polarized EDL [10, 11, 14]. Interfacial polarization typically is understood by Maxwell-Wagner relaxation frequency (MW frequency) [14]. In the frequency region below the MW frequency, positive and negative charges of particles are completely separated, forming dipole moments and which in turn leads to constant lateral repulsive force between the particles once they are on the electrode surface in the presence of an AC electric field. In addition to the interfacial polarization, a recent study suggested the existence of another relaxation mechanism in the frequency band [11]. Kumar et al. found out using Delaunay triangulation that the repulsive force decreases with frequencies even in the region below MW frequency. This anomalous phenomenon is attributed to nonequilibrium polarization process in the EDL of particles [15] and it is closely associated with critical frequency of REP which will be introduced later. In contrast an EHD flow exerts attractive forces to the particles. It scales with the square of electric field strength ($|E|^2$) and the inverse of electrical frequency (ω^{-1}), and shows good agreement in frequencies above 500 Hz [16]. Generally since REP is mostly operated in the frequency band of 2–200 kHz, the attractive EHD force and the repulsive polarization–based force mentioned above is involved in the particle manipulation process by REP.

An optical illumination produces nonuniform temperature distribution inside a suspending medium by laser heating (Fig. 4) and the temperature gradient in turn induces a gradient in density, permittivity, and conductivity of the fluid [14]. While the density gradient causes fluid flow based on natural convection, gradient of electrical permittivity and conductivity generates electrothermal flow in applied electric field. Since the natural convective flow is generally negligible in a system where electric forces exist together, a primary flow responsible for particle transport in REP process is a toroidal electrothermal microvortex [10–12]. The vortex is characterized by time-averaged electrothermal body force equation, as follows [14]:

$$\langle f_e \rangle = \frac{1}{2}\mathrm{Re}\left[\frac{\sigma\varepsilon(\alpha-\beta)}{\sigma+i\omega\varepsilon}(\nabla T\cdot E)E^* - \frac{1}{2}\varepsilon\alpha|E|^2\nabla T\right]$$

where Re[] represents the real part of the bracket [] and not the Reynolds number, E is an electric field, E^* its complex conjugate, T the temperature, ω the applied angular frequency, σ and ε are the conductivity and permittivity of the fluid, respectively, α is $(1/\varepsilon)(\partial\varepsilon/\partial T)$ and β is $(1/\sigma)(\partial\sigma/\partial T)$. The first and the second term on the right-hand side of the equation is Coulomb force and dielectric force, respectively, and the latter term dominates as the applied AC frequency increases. Meanwhile Joule heating also might be considered as a factor contributing to the temperature distribution in the fluid. But because the electric field is uniform, it cannot induce the rapid temperature gradient such as Fig. 4 in a chip for REP consisting of two parallel plate electrodes. Therefore the thermal field for REP operation is generated by an optical laser and it drives electrothermal vortex to carry suspended particles toward the illuminated site on an electrode surface through the coupling with applied electric field.

The relationships among the physical phenomena involved in REP technique can be illustrated as shown in Fig. 5. There are three primary electrokinetic mechanisms: electrokinetic forces existing between induced dipole-dipole moments of particles and between the polarized particles and electrode surface ($F_{\text{particle-particle}}$ and $F_{\text{particle-electrode}}$), electrohydrodynamic drag force locally gathering the particles each other on the electrode ($F_{\text{EHD}}^{\text{lateral}}$ and $F_{\text{EHD}}^{\text{vertical}}$), and electrothermal body force sweeping the particles toward laser-illuminated site across the electrode surface ($F_{\text{ET}}^{\text{lateral}}$ and $F_{\text{ET}}^{\text{vertical}}$) (Fig. 5a). When a uniform AC electric field is supplied to a chip where a suspension containing particles is introduced, the particles on the electrode surface show irregular configurations in the horizontal competition of the repulsive particle-particle force ($F_{\text{particle-particle}}$) and the lateral attractive EHD force ($F_{\text{EHD}}^{\text{lateral}}$) and the vertical balance of the attractive particle-electrode force ($F_{\text{particle-electrode}}$) and the vertical lift-up EHD force ($F_{\text{EHD}}^{\text{vertical}}$). Then the additional application of a laser illumination under the applied electric field causes strong electrothermal microvortex, and while the lateral component of the resulted electrothermal force ($F_{\text{ET}}^{\text{lateral}}$) carries the particles toward the center of the illumination, the vertical component ($F_{\text{ET}}^{\text{vertical}}$) competes with both the attractive particle-electrode force and the vertical EHD force. When the vertical forces balance each other at a certain frequency, the particles transported by the lateral electrothermal force are aggregated at the illumination region on the electrode. However as the electrical signal approaches higher AC frequencies, the REP cluster becomes unstable and the number of particles in the cluster is gradually decreased. And when the frequency is further increased, the particles are not collected in the illumination region anymore and move along the strong electrothermal flow. Such transient process of REP cluster is attributed to the unbalance of the vertical forces resulted from the polarization change in EDLs of the particles, and it shows that there exists a critical point for REP cluster maintenance in AC frequency range. The point termed with a critical frequency varies with particle surface charge density and hence, sets an operating frequency band of REP technique along with the material properties and

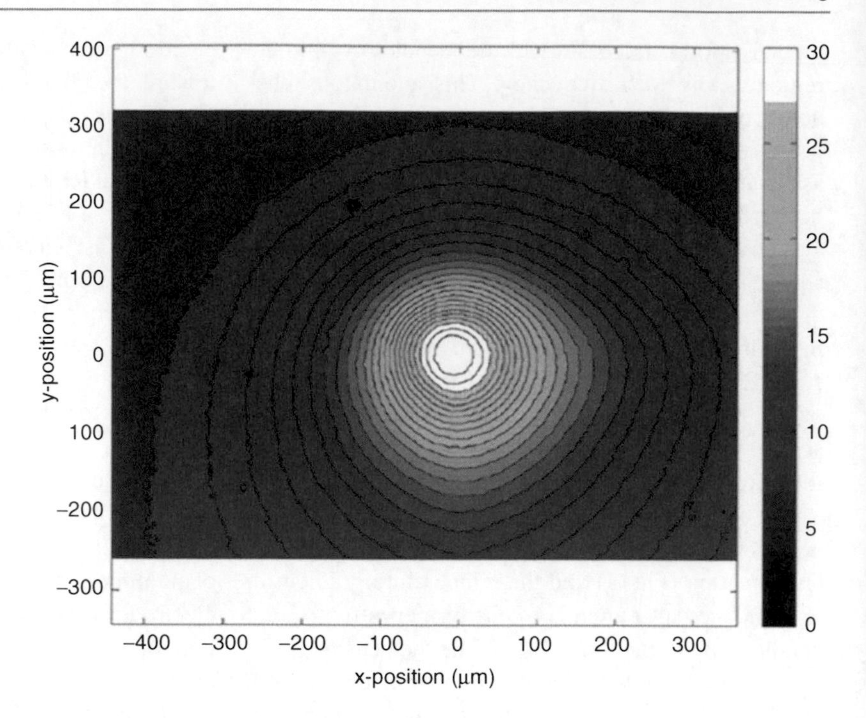

Rapid Electrokinetic Patterning, Fig. 4 Temperature gradient on an ITO electrode surface generated by a 1,064 nm near-infrared laser. Applied laser power is 150 mW and the focus of the laser is positioned at the origin (0, 0) (Reprinted with permission from Kumar et al. Copyright 2010 America Chemical Society [11])

sizes of particles (Fig. 5b). This characteristic of a critical frequency is useful especially for separation (or sorting) of two colloids with the same shape/material but different size. Detailed information of REP mechanisms including the critical frequency is available in references [9–12].

Methodology for REP Operation

The REP technique can be realized in a simple microfluidic chip structure [9–11]. The chip basically consists of two parallel plate electrodes and several chambers including an inlet, outlet, and a microchannel. A pair of the electrodes is constructed with conductive-coated substrates and mostly a transparent indium tin oxide (ITO) glass slide or cover slip is used as an electrode for observation, imaging, and laser irradiation. However, considering relatively high electrical resistance of the ITO coating, a gold or chrome-coated electrode also may be used in the microfluidic chip [13]. The chambers are established by the direct combination of the electrode structure and the insulating spacer patterned for the inlet, outlet, and microchannel. The channel base and lid serve as electrodes. The spacer may be made conveniently using a commercial adhesive double-sided tape or be fabricated precisely with polymers such as PDMS by microfabrication technique.

As an illumination source for REP technique, a near-infrared (NIR) laser (1,064 nm) has often been used because it provides strong energy to the chip and can create a rapid temperature gradient in a fluid [13]. However the illumination source may be replaced with optical lasers or diodes of different wavelength, depending on how much thermal energy from the source can be delivered into the fluid through conductive-coated electrodes used in the chip [11]. This feature is significant as it allows REP to be replicated at different optical wavelengths, provided that the proper combination of electrode-laser wavelength is chosen to generate the necessary optically induced temperature gradients. The laser in REP operation is provided in the form of a single spot or multiple spots through a hologram generated from a spatial light modulator (SLM), being focused by high numerical aperture (N.A.) objective lens.

As an electrical signal for REP technique, AC electrical frequencies of below $<$200 kHz with a certain electrical potential are supplied to the chip [9–12]. Higher frequencies relax the polarization of particles, reducing the effectiveness of REP. Meanwhile particles suspended in a fluid during REP process may be adhered on the electrode biased with AC electric field

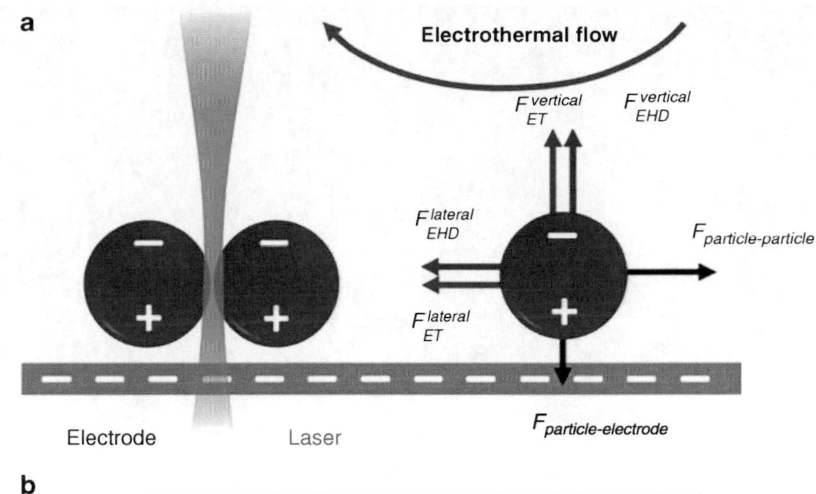

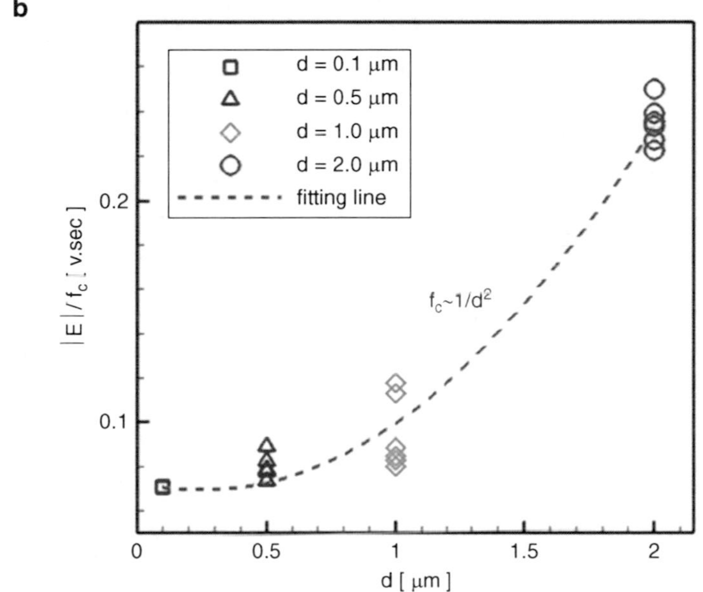

Rapid Electrokinetic Patterning, Fig. 5 Physics of REP technique. (**a**) Various forces exerted on particles in REP-based aggregation. $F_{\text{particle-particle}}$ and $F_{\text{particle-electrode}}$ denote the electrokinetic forces existing between particles and between the particles and an electrode, respectively. $F^{\text{lateral}}_{\text{EHD}}$ and $F^{\text{vertical}}_{\text{EHD}}$ are the attractive electrohydrodynamic forces locally gathering the particles each other onto the electrode. $F^{\text{lateral}}_{\text{ET}}$ and $F^{\text{vertical}}_{\text{ET}}$ are the electrothermal drag force sweeping toward the illumination region across the electrode surface. (**b**) Dependence of a critical AC frequency on particle size. The critical frequency is defined as the minimum AC frequency above which REP-based particle cluster cannot be maintained anymore and is scaled as the inverse of particle diameter squared ($f_c \sim 1/d^2$) (Reproduced with permission from Kumar et al. Copyright 2010 American Chemical Society [11])

because of their nonuniform surface charging [17]. This adhesion eventually disturbs particle manipulation by REP. The adhesion problem can be avoided by pretreatment of the electrode surfaces with a nonionic surfactant like TWEEN [18].

Suspending solutions used for REP are mostly deionized (DI) water ($\sigma \sim$ 1mS/m) or low electrical conductivity electrolytes (σ<50 mS/m) [9–12]. High conductivity solutions provide more surface charges on the particle (i.e., in electrical double layer) than low ones and which in turn enhances the nonuniformity of particle surface charging once the electric field is applied. This increases the possibility of the particle adhesion mentioned previously. Therefore DI water or low conductivity electrolytes should be used for REP operation.

Potential Applications of REP Technique

An enormous advantage of REP technique is the rapid aggregation and patterning of colloidal particles that it enables, and dynamic manipulation of the particle assemblies by movement of laser location. The promising ability of REP has led many preliminary researches for application. Williams et al. have separated size different particles using REP technique [10]. Figure 6 shows the sorting of 0.5, 1, and 2 μm particles that they performed, and the particles are separated

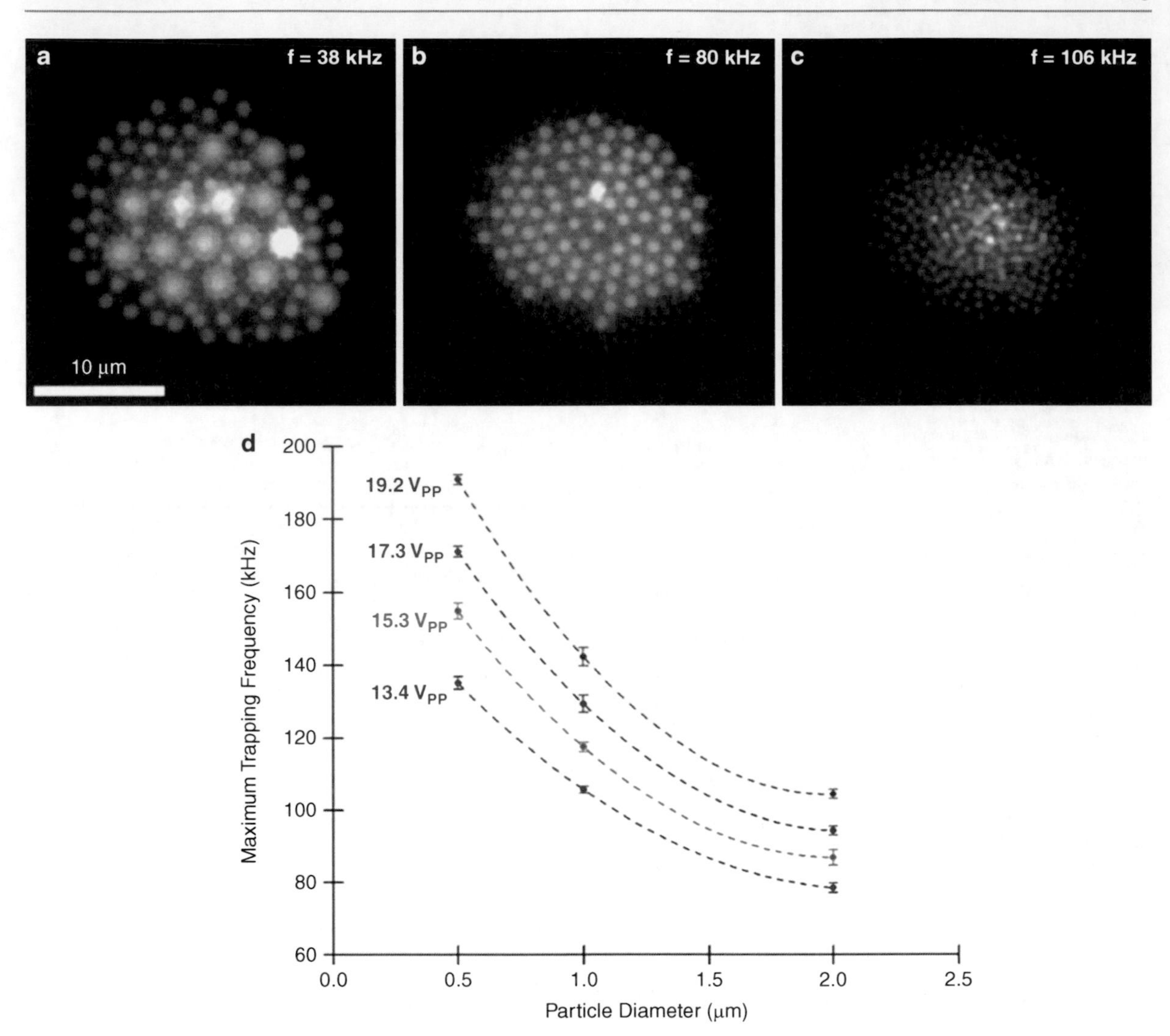

Rapid Electrokinetic Patterning, Fig. 6 Separation of like particles based on their size. (**a**) Aggregation of 2.0 μm, 1 μm, and 0.5 μm polystyrene particles at 38 kHz. (**b**) Aggregation of 1 μm and 0.5 μm polystyrene particles at 80 kHz. As the frequency is increased from 30 kHz, 2.0 μm particles are carried away with the vortex out of the aggregation. (**c**) Aggregation of 0.5 μm polystyrene particles at 106 kHz. Increasing frequency up to 106 kHz results in aggregation of only 0.5 μm particles. The 1 μm particles are no longer in the cluster. (**d**) Maximum trapping frequency for different sized polystyrene particles. It shows second-order polynomial fit in agreement with theory (Williams et al. Reproduced with permission of IOP publishing [10])

successively as AC frequency increases from 38 to 106 kHz. The mechanism involved in the sorting is the dependence of a critical frequency on particle size ($f_c \sim 1/d^2$) mentioned previously. Hence the bigger particles in the particle assembly are swept away by electrothermal flow earlier than the smaller particles due to their lower critical frequency. Additionally Williams et al. also have performed EDL polarization-based separation of unlike particle populations using REP (Fig. 7) [10]. They separated 1 μm polystyrene and silica particles in the frequency range of 90–150 kHz. These results demonstrate that REP is a useful tool for size- and polarization-based separation. In another application, recently Kumar et al. proposed new force spectroscopy using REP technique [11]. They showed through repetitive aggregation and disintegration of particle assemblies by REP that the particles experience not only interfacial polarization but also nonequilibrium polarization of EDL under an electric field below 200 kHz. Such attempts for REP

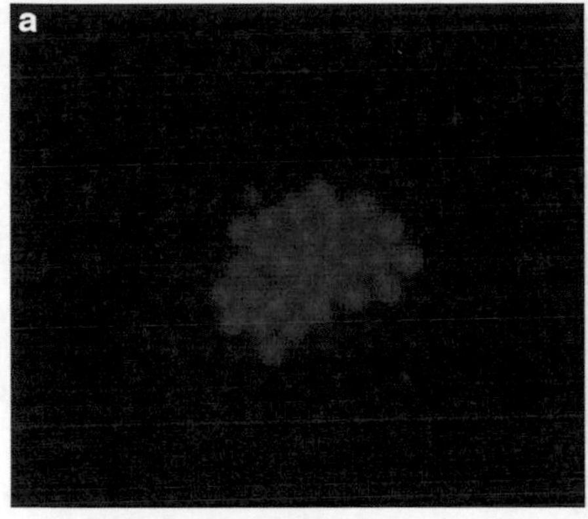

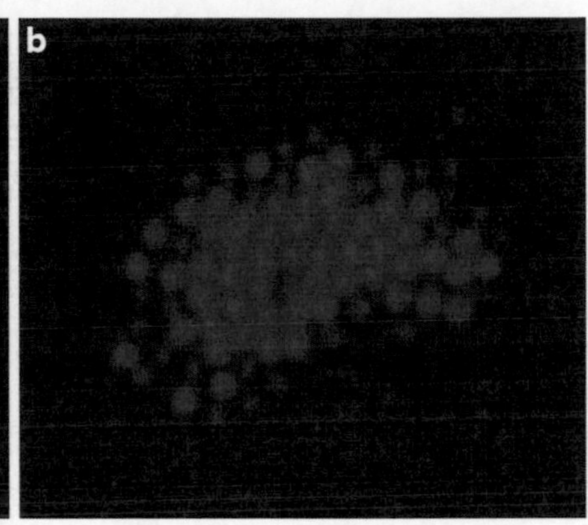

Rapid Electrokinetic Patterning, Fig. 7 Separation of unlike particles based on their polarization behavior. The unlike particles indicate fluorescent polystyrene and nonfluorescent silica particles each 1.0 μm in diameter. (**a**) At 23 mW laser illumination power and 19.8V_{pp}, only polystyrene beads are trapped at 150 kHz frequency. (**b**) When the frequency is reduced to 90 kHz, both polystyrene and silica beads are trapped (Williams et al. Reproduced with permission of IOP publishing [10])

application are currently being extended to manipulation of bacteria and nonspherical beads, transport and merge of emulsion droplets, etc. [19, 20]. It will motivate the discovery of more various REP applications in extensive research fields including colloidal science and microbiology.

Cross-References

- ► AC Electrokinetics
- ► Dielectrophoresis
- ► Lab-on-a-Chip

References

1. Franke, T.A., Wixforth, A.: Microfluidics for miniaturized laboratories on a chip. Chem. Phys. Chem. **9**(15), 2140–2156 (2008)
2. Morgan, H., Hughes, M.P., Green, N.G.: Separation of submicron bioparticles by dielectrophoresis. Biophys. J. **77**, 516–525 (1999)
3. Minden, J.: Comparative proteomics and difference gel electrophoresis. Biotechniques **43**(6), 739–745 (2007)
4. Yamada, M., Seki, M.: Hydrodynamic filtration for on-chip particle concentration and classification utilizing microfluidics. Lab Chip **5**, 1233–1239 (2005)
5. Pamme, N., Manz, A.: On-chip free-flow magnetophoresis: continuous flow separation of magnetic particles and agglomerates. Anal. Chem. **76**, 7250–7256 (2004)
6. Arai, F., Ng, C., Maruyama, H., Ichikawa, A., El-Shimy, H., Fukuda, T.: On chip single-cell separation and immobilization using optical tweezers and thermosensitive hydrogel. Lab Chip **5**, 1399–1403 (2005)
7. Jones, T.B.: Electromechanics of Particles. Cambridge University Press, Cambridge (1995)
8. Chiou, P.Y., Ohta, A.T., Wu, M.C.: Massively parallel manipulation of single cells and microparticles using optical images. Nature **436**, 370–372 (2005)
9. Williams, S.J., Kumar, A., Wereley, S.T.: Electrokinetic patterning of colloidal particles with optical landscapes. Lab Chip **8**, 1879–1882 (2008)
10. Williams, S.J., Kumar, A., Green, N.G., Wereley, S.T.: Optically induced electrokinetic concentration and sorting of colloids. J. Micromech. Microeng. **20**(015022), 1–11 (2010)
11. Kumar, A., Kwon, J.-S., Williams, S.J., Green, N.G., Yip, N.K., Wereley, S.T.: Optically modulated electrokinetic manipulation and concentration of colloidal particles near an electrode surface. Langmuir **26**(7), 5262–5272 (2010)
12. Williams, S.J., Kumar, A., Green, N.G., Wereley, S.T.: A simple, optically induced electrokinetic method to concentrate and pattern nanoparticles. Nanoscale **1**, 133–137 (2009)
13. Kumar, A., Cierpka, C., Williams, S.J., Kähler, C.J., Wereley, S.T.: 3D3C velocimetry measurements of an electrothermal microvortex using wavefront deformation PTV and a single camera. Microfluid. Nanofluidics **10**(2), 355–365 (2011)
14. Morgan, H., Green, N.G.: AC Electrokinetics: Colloids and Nanoparticles. Research Studies Press, Baldock (2002)
15. Dukhin, S.S.: Non-equilibrium electric surface phenomena. Adv. Colloid Interface Sci. **44**(24), 1–134 (1993)
16. Ristenpart, W.D., Aksay, I.A., Saville, D.A.: Assembly of colloidal aggregates by electrohydrodynamic flow: kinetic experiments and scaling analysis. Phys. Rev. E **69**, 0214051–0214058 (2004)
17. Zhou, H., Götzinger, M., Peukert, W.: The influence of particle charge and roughness on particle – substrate adhesion. Powder Technol. **135–136**(2), 82–91 (2003)
18. Ha, J.-W., Yang, S.-M.: Effect of nonionic surfactant on the deformation and breakup of a drop in an electric field. J. Colloid Interface Sci. **206**, 195–204 (1998)

19. Kwon, J.-S., Ravindranath, S., Aloke, K., Irudayaraj, J., Wereley, S.T.: Application of opto-electrokinetic manipulation technique to bacteria. In: Proceedings of the American Society of Mechanical Engineers – International Mechanical Engineering Congress and Exposition (ASME-IMECE), Vancouver (2010)
20. Thakur, R.V., Wereley, S.T.: Optically induced rapid electrokinetic patterning of non-spherical particles- study of colloidal phase transition. In: Proceedings of the American Society of Mechanical Engineers – International Mechanical Engineering Congress and Exposition (ASME-IMECE), Vancouver (2010)

Rate Sensors

► Gyroscopes

Reactive Current Clamp

► Dynamic Clamp

Reactive Empirical Bond-Order Potentials

J. David Schall[1], Paul T. Mikulski[2], Kathleen E. Ryan[3], Pamela L. Keating[3], M. Todd Knippenberg[4] and Judith A. Harrison[3]
[1]Department of Mechanical Engineering, Oakland University, Rochester, MI, USA
[2]Department of Physics, United States Naval Academy, Annapolis, MD, USA
[3]Department of Chemistry, United States Naval Academy, Annapolis, MD, USA
[4]Department of Chemistry, High Point University, High Point, NC, USA

Synonyms

Bond-order potential; REBO

Definition

Reactive empirical bond-order potentials are interatomic energy functions used in molecular dynamics simulation and modeling of nanosystems where an accurate but efficient description of chemical reactivity is required.

Genesis of Reactive Bond-Order Potentials

Atomistic simulation of a large number of atoms using molecular dynamics (MD) is a powerful tool for understanding the fundamental mechanisms present in nanomaterials systems. The ability to accurately model chemistry including reactivity, bonding, charge transfer, polarizability, mixing, etc., is central to this understanding. Underpinning these calculations is the atomic interaction potential. Ideally, the atomic interactions would be obtained directly from first-principles, i.e., through solution of Schrodinger's equation. However, such calculations are orders of magnitude too slow for the large number of energy evaluations required to study nanosystems of reasonable size and practical interest. To obtain useful information in a reasonable amount of time, empirical, and semiempirical approximations to the atomic potentials have been developed. In this entry, a survey of reactive empirical bond-order and related interatomic potentials is presented.

An effective analytic potential should have a relatively simple functional form, which captures the essential essence of the underlying quantum mechanical bonding, and will be able to mimic experimental quantities by utilizing empirically derived functions and parameters. Empirical potentials are likely to include a wide range of fitting parameters, which are selected to reproduce cohesive energies, elastic constants, lattice constants, and surface energies. Ideally, the potential will also have some degree of transferability, having the ability to describe structures not included in the fitting database, at least in a qualitative sense. Finally, the resulting function should be relatively efficient computationally. The process of developing a potential, sometimes referred to as an art as well as a science, requires a combination of chemical insight, trial and error, and tenacity on the part of its developer.

Covalent materials, such as silicon and carbon, form strong directional bonds. This poses a challenge for potential development for this important class of materials. Many standard potential functions, such as the Lennard-Jones (LJ) potential or the embedded atom method (EAM), do not include any bond directionality. The Stillinger–Weber (SW) potential for solid and liquid phases of silicon was one of the first attempts to use a classic potential to overcome this challenge [1]. Stillinger and Weber based their potential model on a many-body approach. In this approach, the total energy is given as a linear superposition of terms representing different types of interatomic interactions, e.g., stretching, bending, rotation, and torsion. For the SW potential, the total energy includes only two of these terms, a pair- and triplet-term based on the geometric factors of bond length and bond angle, respectively. The total potential energy, E_{tot}, is given by

$$E_{tot} = \frac{1}{2}\sum_{\substack{ij\\(i\neq j)}} \phi_{ij}(R_{ij}) + \sum_{ijk} g(R_{ij})g(R_{ik})\left(\cos\theta_{jik} + \frac{1}{3}\right)^2. \quad (1)$$

Here ϕ_{ij} is the pair-term (bond stretching) representing electrostatic interactions between atoms i and j. The second summation is a three-body term, which represents bond bending between atoms i, j, and k. The term $g(r_{ij})$ is a decaying function with a cutoff between the first- and second-neighbor shell, and θ_{jik} is the bond angle described by two neighbors j and k of atom i. The inclusion of the three-body term allows the potential to achieve an acceptable description of short-range order and of atom-exchanging diffusive motion in the liquid phase. While this potential is reasonably accurate when used for modeling solid silicon in the diamond cubic phase, it is biased toward the tetrahedral bond angle through the explicit inclusion of the factor $(\cos\theta_{jik} + 1/3)$ in the three-body term. The inclusion of this trigonometric factor discriminates in favor of pairs of bonds with the tetrahedral geometry, i.e., $(\cos\theta_{jik} = -1/3)$, and limits the transferability of the potential. For instance, it cannot accurately predict the energies for various non-tetrahedrally bonded, high-pressure phases of silicon, it does not correctly predict relative energies of surface structures, and the coordination of liquid-phase silicon is too low.

The family of potentials, CHARMM, AMBER, OPLS, etc., derived from the many-body approach described above for the Stillinger–Weber potential is often used in simulation of organic systems. These so-called force-field models are able to model structural and dynamical properties of very large molecules with a high degree of accuracy. However, it is important to note that the force-field methods generally do not allow for bond-forming or breaking to occur during simulation. Connectivity must be determined a priori. For a review of these potentials see for example Mackerell [2].

Abell presented a very general description of bonding based on the observation of a universal relationship between binding energy and bond length in 1985 [3]. This relationship placed the bonding in crystalline solids and molecules on the same footing. The only criterion for bonding preference was the optimization of the binding energy with respect to the local coordination. The local coordination is, in turn, the dominant topological variable in the determination of binding energy. Soon after the publication of Abell, Tersoff developed a potential constructed to guarantee that this universal behavior was obtained [4]. The Tersoff potential was the first to attempt to incorporate the structural chemistry of covalently bonded systems into empirical potential energy function. The general form of the potential is

$$E = \sum_i E_i = \frac{1}{2}\sum_{i\neq j} V_{ij}, \qquad V_{ij} = f_c[V_R(R_{ij}) - B_{ij}V_A(R_{ij})], \quad (2)$$

where E is the total energy of the system, E_i is the site energy for site i, V_{ij} is the interaction energy between atoms i and j, and R_{ij} is the distance between these atoms. The sum is over the j nearest neighbors of i; $V_R(R)$ and $V_A(R)$ are pair-additive repulsive and attractive interactions, respectively; f_c is a cutoff function to limit the range of the potential. For the Tersoff potential, the repulsive and attractive terms are represented by the Morse-type functions

$$V_R = A\exp(-\lambda_1 R_{ij}), \qquad V_A = B\exp(-\lambda_2 R_{ij}), \quad (3)$$

where A, B, λ_1, and λ_2 are all positive constants with $\lambda_1 > \lambda_2$. All deviations from the simple-pair potential are ascribed to the dependence of the bond-order B_{ij} on the local bonding environment or coordination. More specifically, the bonding strength for a bonded pair should be a monotonically decreasing function of the number of competing bonds, the strength of the competing bonds, and the cosines of the angles of the competing bonds. The innovative feature of the Tersoff potential is that it does not assume different forms for the angular functions for different hybridizations. In his work, B_{ij} has the form

$$\begin{aligned} B_{ij} &= (1+\beta^n \xi_{ij}^n)^{-1/2n}, \\ \xi_{ij} &= \sum_{k\neq i,j} f_c(r_{ik}) g(\theta_{jik}) \exp[\lambda_3^3 (r_{ij}-r_{ik})^3], \\ g(\theta) &= 1 + c^2/d^2 - c^2/[d^2 + (h-\cos\theta)^2], \end{aligned} \tag{4}$$

where θ_{jik} is the bond angle between bonds ij and ik. The angular function $g(\theta)$ is determined globally by a fit to solid structures with different coordination. This gives the function a high degree of transferability. Tersoff extended his original silicon potential to include germanium, carbon, and combinations thereof. This potential form has also been extended to a variety of III-V semiconductors.

An empirical bond-order expression that described both hydrocarbon molecules and solid-state carbon was developed by Brenner [5] to model chemical vapor deposition of diamond films. Taking inspiration from the Tersoff potential, the form of the reactive empirical bond-order potential (REBO) potential allows for bonds to form and break with changes in hybridization. This first-generation Brenner potential is very similar to the Tersoff potential. Pair repulsive and attractive terms are again represented by Morse-type potentials. The primary difference is the way in which the bond order is handled for hydrocarbon molecules. The expression for the total bond order is

$$\bar{B}_{ij} = \frac{B_{ij}+B_{ji}}{2} + F_{ij}(N_i^{\mathrm{t}}, N_j^{\mathrm{t}}, N_{ij}^{\mathrm{conj}}) \tag{5}$$

where

$$B_{ij} = \left[1 + \sum G_i(\theta_{jik}) f_c(r_{ik}) e^{\alpha[(r_{ij}-R_{ij}^E)-(r_{ik}-R_{ik}^E)]} + H_{ij}(N_i^{\mathrm{H}}, N_i^{\mathrm{C}}) \right]^{-\delta}, \tag{6}$$

The quantities N_i^C and N_j^H are the number of carbon and hydrogen atoms bonded to atom i. The total number of neighbors, N_i^{t} of atom i is $(N_i^{\mathrm{C}} + N_j^{\mathrm{H}})$, N_{ij}^{conj} depends on whether a bond between carbon atoms i and j is part of a conjugated system, $G(\theta)$ is a function of the angle between bonds i-j and i-k and has the same form as the Tersoff potential. The two- and three-dimensional cubic splines, H_{ij} and F_{ij}, have discrete nodal values that are modified so that the potential reproduces the energies of various hydrocarbon molecules. The resulting potential successfully describes the different bonding characteristics of hydrogen and carbon radicals and nonconjugated double and triple carbon-carbon bonds.

Extensions of the First-Generation REBO Potential

Because the Tersoff-Brenner formalism has been used successfully to simulate a wide range of processes such as chemical vapor deposition, diamond growth, carbon nanotubes, graphene, diamond-like carbon, etc., it was extended to other systems by various groups. In what follows, some examples of ways in which the first-generation REBO potential has been extended are provided. This analysis is not meant to include all such examples, however.

Parameterization of Additional Atom Types

Because bond-order potentials were designed to model chemical reactivity in covalent systems, it was logical to adopt the first-generation REBO formalism to model Si-H systems [6]. Parameterizations also exist for Si-F and Si-Cl. The first attempts to construct bond-order potentials capable of modeling three different atom types were two independently developed C-Si-H potentials. The C-Si-H potential of Beardmore and Smith is a hybrid potential which makes use of the C-H, Si-H, and Si-C parameters developed by Brenner [5], Murty and Atwater [6], and Tersoff, respectively. Because it is a hybrid potential, it inherits the strengths and weaknesses of the potentials on which it is based. The C-Si-H potential of Dyson and Smith, known as the extended Brenner (XB) potential, adopts the C-H parameters of Brenner and develops three new sets of parameters for Si-Si, Si-C, and Si-H [7] while maintaining the same functional form.

When chemical reactions occur during a simulation, the bond order of atoms must change in a continuous manner from reactant to product values. This is accomplished in the first-generation REBO hydrocarbon potential through the use of a bicubic spline $H_{ij}(N_i^{\mathrm{H}}, N_i^{\mathrm{C}})$ (6), which is a function of the number of neighbors surrounding atom i. To extend a bond-order potential to three atom types, such as Si-C-H, the spline function can be extended into ways. First, the cubic-spline function can be extended to three dimensions, such that it has the form $H_{ij}(N_i^{\mathrm{H}}, N_i^{\mathrm{C}}, N_i^{\mathrm{Si}})$. This method provides the most flexibility in adjusting the bond order but can be difficult to code. Alternatively, because C and Si have similar electronegativities and possess similar bonding characteristics, a two-dimensional cubic spline that considers the number of neighbors of C and Si together, i.e., $H_{ij}(N_i^{\mathrm{H}}, (N_i^{\mathrm{C}} + N_i^{\mathrm{Si}}))$, could also be used. This approach was adopted in both the Si-C-H potentials discussed above. This simplification has the advantage of simplifying the MD code. However, it is at the expense of a considerable loss in flexibility in fitting and, therefore, accuracy in reproducing a wide range of properties. For example, the errors in the energies of organosilane molecules calculated with the XB potential systematically increase with increasing molecule size (Fig. 1).

The performance of the XB potential was evaluated by Sbraccia et al. in 2002 [8]. Several problems, such as spurious minima in some potential energy curves, the incorrect energetic ordering for chemisorption of hydrocarbon molecules on Si surfaces, and over-coordinated carbon and silicon atoms were identified. These problems were corrected by altering a minimal set of parameters and the C–C interaction cutoff values. This parameterization is known as the modified extended Brenner (mXB) potential. It should be noted, however, that these changes do not improve the elastic properties of the solid phases of carbon or silicon [9] or the energies of organosilane molecules shown in Fig. 1.

Extension to Metals

Recently, a classical many-body potential for transition metal (Fe, Co, Ni) carbide clusters based on the bond-order type potential was developed [10]. The covalent interactions are based on the first-generation REBO potential. Deviations from the REBO potential formalism occur in the bond-order terms for carbon–metal interactions. The bond order for carbon–metal interactions has the form $B^* = (1 + b(N^C - 1))^{\delta}$, where b and δ are fitting parameters. For metal–metal

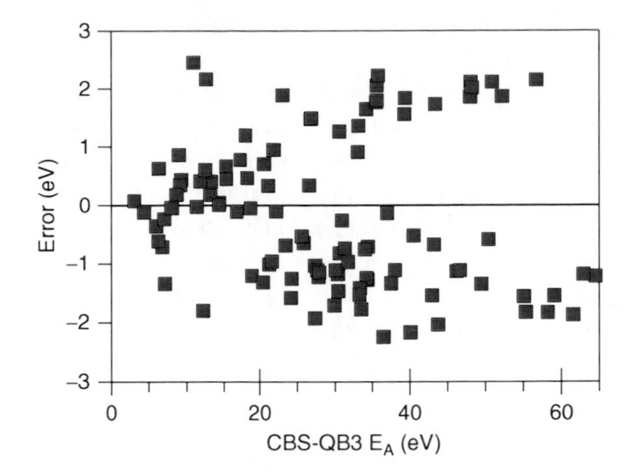

Reactive Empirical Bond-Order Potentials, Fig. 1 Error in atomization energies of $Si_xC_yH_z$ molecules predicted using the mXB potential relative to the atomization energy calculated using the CBS-QB3 method in the ab initio code Gaussian03 illustrating increasing error with atomization energy E_A (increasing molecule size)

interactions, instead of multiplying V^A by B^*, the binding energy D_e and equilibrium bond length R_e are expressed as functions of the metal coordination number N^M and are given by

$$D_e = D_{e1} + D_{e2} exp\{-C_D(N^M - 1)\} \tag{7}$$

$$R_e = R_{e1} + R_{e2} exp\{-C_R(N^M - 1)\}. \tag{8}$$

The reader is referred to Shibuta and Maruyama [10] for all parameter values appearing in (7 and 8).

Limitations

Before undertaking MD simulations with empirical potentials, careful thought should be given to the selected potential's inherent strengths and weaknesses and how these may impact the phenomena to be simulated. For example, multiple potential energy functions exist for modeling Si, including three versions of Tersoff's bond-order potential [4, 11], SW [1], and the environment-dependent interatomic potential (EDIP) [12]. The majority of empirical potentials for Si do not accurately describe the three elastic constants for diamond cubic silicon [13]. In contrast, the EDIP and the 2B-Si [9] potentials do a good job describing elastic constants. However, the bond lengths for many of the crystalline silicon structures are overestimated when using the EDIP potential. For a more complete comparison of the strengths and weaknesses of Si potentials the reader is referred to Balamane et al. [13] and Schall et al. [9].

In general, four limitations have been identified with the first-generation REBO potential. First, the functional form of the pair potentials was not flexible enough to allow a parameter set that could simultaneously reproduce structural energies and force constants. In his original work, Brenner provided two parameter sets, one that provided an excellent fit to bond energies and a second that provided a to fit force constants [5]. Second, because the zero-Kelvin elastic constants C_{11}, C_{12}, and C_{44} for diamond were not included in the fitting database, they are not accurately reproduced by the potential [14]. Third, both the attractive and repulsive pair terms have finite values at zero separation. Thus, in highly energetic atomic collisions, it may be possible for an atom pass through another atom without being repulsed. Finally, the derivatives of the cutoff function $f_{ij}(r)$ are nonrealistic and may lead to spurious minima in energies for certain structures, particularly amorphous carbon [5].

Second-Generation REBO Potentials

To correct the shortcomings of the original potential, Brenner and coworkers developed a second generation of the REBO potential [15]. The repulsive and attractive pair terms have the forms

$$V^R(r) = f^c(r)(1 + Q/r)Ae^{-\alpha r} \tag{9}$$

and

$$V^A(r) = f^c(r) \sum_{n=1,3} B_n e^{-\beta_n r} \tag{10}$$

where A, Q, α, B_n and β_n are fitting parameters. The screened Coulomb function used for the repulsive pair interaction goes to infinity as interatomic distances approach zero, and the attractive term has sufficient flexibility to simultaneously fit the bond properties that could not be fit with Morse-type terms. The bond-order B_{ij} term is also significantly different from either the Tersoff or first-generation REBO potential. Separate terms are included that depend on local coordination and bond angles, radical character and conjugation, and dihedral angle for C–C double bonds.

By altering the functional form and expanding the fitting database, the second-generation REBO potential is able to provide significantly better descriptions of bond energies, lengths, and force constants for hydrocarbon molecules relative to the first-generation REBO potential. In addition, the zero-Kelvin elastic properties, interstitial defect energies, and surface energies for diamond are well reproduced. This potential also qualitatively reproduces the correct behavior of the elastic constants C_{11} and C_{44} of diamond with temperature. Coincidentally, the relatively poor agreement of the calculated C_{12} with the experimentally determined data combined with a fast softening of C_{11} with temperature results in bulk moduli that agree fairly well with the experimental values (Fig. 2) [14].

As a first step toward developing a Si–C–H bond-order potential, Schall et al. recently published parameters for silicon potential (2B-Si) based on the second-generation REBO formalism [9]. This potential does a reasonable job producing a broad range of properties of diamond cubic and amorphous silicon. In principle, the second-generation REBO formalism could be extended to a wide range of interatomic interactions to model chemical reactivity in nanoscale systems that contain multiple atom types. For example, Sinnott and coworkers used density functional theory to develop parameters within the second-generation REBO formalism for C–O, O–O, and O–H [16]. Because this potential is based on the second-energy REBO potential, it possesses all of its inherent strengths and weaknesses, such as a lack of intermolecular forces. The addition of a third atom type complicates the functional form of the bond order as discussed above. This potential suffers from the limitation discussed above by making use of a two-dimensional cubic spline instead of a three-dimensional cubic spline in the bond-order function. In addition, O is treated in a similar manner as C, despite the inherent differences in electronegativities of the two atoms. Additional parameter sets have also been developed recently by this group for Mo–S and C–F.

Covalent+Intermolecular Forces

To model chemical reactions in condensed phases, such as liquids, graphite, and self-assembled monolayers (SAMs), intermolecular forces must be added to the REBO potential. Goddard and coworkers added intermolecular forces to the first-generation REBO potential for hydrocarbons. Distance-based switching

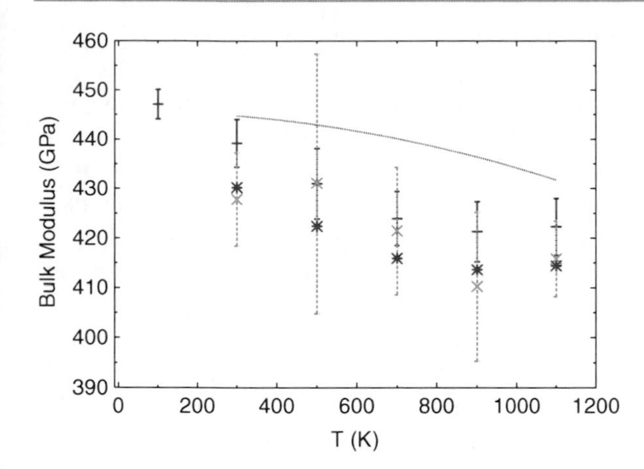

Reactive Empirical Bond-Order Potentials, Fig. 2 Bulk modulus of diamond as a function of temperature. Bulk values were calculated from the elastic constants using the relationship $B = \frac{1}{3}(C_{11} + 2C_{12})$. (+ signs), (crosses), and (asterisks) symbols represent values calculated with the direct, strain fluctuation, and stress fluctuation methods, respectively

functions were used to turn the intermolecular forces "on" and "off." Intermolecular forces and torsional interactions have also been added to the second-generation REBO potential [17]. This potential, known as the adaptive intermolecular reactive empirical bond-order (AIREBO) potential, uses a unique algorithm to determine the conditions under which the intermolecular forces should be "on" and "off."

The AIREBO potential includes intermolecular interactions between nonbonded atoms and torsional interactions associated with a connected sequence of three bonds. This extension makes the AIREBO potential a particularly attractive tool for studying interfacial/tribological systems and liquids. The discussion below focuses on how intermolecular interactions are introduced without compromising the reactivity of the potential. For a discussion of how torsional interactions are modeled, the reader is referred to the original publication [17].

Intermolecular interactions are modeled through a Lennard-Jones (LJ) potential,

$$V^{LJ} = 4\varepsilon\left(\left(\frac{\sigma}{r}\right)^{12} - \left(\frac{\sigma}{r}\right)^{6}\right). \tag{11}$$

Only four parameters are taken as independent: σ_{CC}, σ_{HH}, ε_{CC} and ε_{HH}. The heterogeneous parameters are fixed by Lorentz-Berthelot combining rules,

$$\sigma_{CH} = \frac{1}{2}(\sigma_{CC} + \sigma_{HH}) \tag{12}$$

and

$$\varepsilon_{CH} = \sqrt{\varepsilon_{CC}\varepsilon_{HH}}. \tag{13}$$

To merge this with a reactive potential, the LJ potential may be either completely or partially "turned off" in response to the chemical environment of the interacting pair. The key feature of the AIREBO potential is that it provides a means of smoothly interpolating between purely bonded and nonbonded interactions through a set of three switching functions S_{distance}, S_{bond}, and $S_{\text{connectivity}}$. Each one of these switching functions may turn off the LJ interaction partially or entirely,

$$E^{\text{LJ}} = (1 - S_{\text{distance}}S_{\text{bond}})(1 - S_{\text{connectivity}})V^{LJ}. \tag{14}$$

For each switching function, a value of 1 is associated with turning off the LJ interaction completely, 0 is associated with a full LJ interaction, and values in between are associated with partial LJ interactions.

The distance-based switch, S_{distance}, is 1 for distances below r^{LJmin} and 0 for distances above r^{LJmax}. A cubic-spline function is used to interpolate at distances falling between the setpoints r^{LJmin} and r^{LJmax} to give a value for the switch between 0 and 1. The setpoints are fixed by requiring that the LJ potential minimum remain unchanged and that no artificial repulsive barrier is present as the LJ potential is switched off. These constraints require that $r^{\text{LJmin}} = \sigma_{ij}$ and $r^{\text{LJmax}} = 2^{1/6}\sigma_{ij}$.

The hypothetical bond-order-based switch S_{bond} interpolates to give values between 0 and 1 in the region $b^{\text{min}} < b^* < b^{\text{max}}$ using a cubic spline, 0 for a bond-order below this range and 1 for a bond-order above this range. The bond-order b^* is evaluated at r^{LJmin} for pairs separated by intermolecular distances (distances larger than r^{LJmin}), and thus at these distances b^* is a *hypothetical* bond-order that assesses the potential for these atoms to bond they approach one another. In the calculation of b^*, the distances of each atom in the pair to its neighbors remain unchanged.

The distance-based switch $S_{\text{connectivity}}$ takes into account the set of all one-, two-, and three-bond sequences that connect the pair of atoms under consideration. For each atom pair in each sequence of bonds, a weight is calculated that is 1 for distances below r^{LJmin} and 0 for distances above r^{LJmax}. These setpoints are the same as in S_{distance}; however, for compatibility with the original REBO potential, a shifted half-period cosine function is used to interpolate between r^{LJmin} and r^{LJmax}. A weight for a sequence of bonds is calculated as the product of the sequence's individual bond weights. The $S_{\mathrm{connectivity}}$ switch is taken to be the largest sequence-weight found in the total set of sequences. In the adopted form for E^{LJ}, the switches S_{distance} and S_{bond} appear together as a product while $S_{\mathrm{connectivity}}$ appears on its own. Consequently, a full LJ interaction will be included for atom pairs that are not (1,2), (1,3), or (1,4) neighbors and are *either* beyond the cutoff distance r^{LJmax} *or* have a bond-order below b^{min}.

The treatment of the van der Waals interactions can be improved in the AIREBO potential by treating the long-range forces so that the chemical environment is taken into account [18]. In the spirit of the REBO potential, the LJ parameters for carbon are functions of the number of N^C and N^H, or $\sigma_C\left(N^C, N^H\right)$ and $\varepsilon_C\left(N^C, N^H\right)$. Two-dimensional cubic-spline functions are used to interpolate between values with integer numbers of neighbors.

Modeling Reactivity in Systems Containing Charge

The electronic properties and nature of bonding in molecules and solids stems from the distribution of charges within the system of interest. The simplest way to add charges to MD simulations is to assign a fixed charge to each atom in the simulation. Fixed-charge schemes have several limitations; however, the most important of which is their inability to redistribute charges in a physically realistic manner. Charge clouds can be distorted when molecules are brought into close proximity, without reacting. When chemical reactions occur, charge needs to be transferred. For instance, the charge states of hydrogen and chlorine change dramatically when they approach each other to form an HCl molecule.

The first atomic-level, charge transfer model was developed by Rappé and Goddard [19] and is known as the electronegativity equalization (QEq) method. In this method, the charges in the system distribute themselves so that the electronegativity at each atomic site is equalized for a given nuclear configuration. This is equivalent to saying that the driving force to shift charge from one atom to another is zero, or the derivatives of the potential with respect to charge are zero. The self-energy $E_i^s(q_i)$ on particle i is a function of the charge of the ion and is expressed in terms of the charge, electronegativity χ, and atomic hardness (a term related to the change in energy as a function of charge). Obtaining the charge on each atom for each nuclear configuration of N atoms requires taking the inverse of an N by N matrix. This method was utilized by Streitz and Mintmire in their examination of the aluminum/alumina interface using their ES+EAM potential [20]. Inversion of an N by N matrix at each MD time step usually scales as $O(N^2)$ and becomes computationally intractable as the size of the system increases. Several alternative numerical methods have been developed to solve the N linear equations in charge associated with the N by N matrix. To date, these methods have not found widespread application.

To avoid the matrix inversion, Rick et al. developed an approximate method to equalize the electronegativity based on an extended Lagrangian approach. In this method, charges are treated as dynamic variables that evolve explicitly with time [21]. Equations of motion for the positions and charges of the atoms can be shown to be:

$$m_i \ddot{r}_i = \frac{\partial}{\partial r_i} E_T(\{r_i\}\{q_i\}) \tag{15}$$

and

$$s_i \ddot{q}_i = \frac{\partial}{\partial q_{r_i}} E_T(\{r_i\}\{q_i\}) \tag{16}$$

where m_i, $\mathrm{r_i}$, and q_{r_i} are the mass, position, and charge of the ith atom. The term s_i is a fictitious mass of the charge, which is chosen to provide numerical stability. The individual charges respond to deviations of the electronegativity equilibration by moving to a new charge which more closely satisfies the equilibrium condition. One advantage of this approach is that both equations of motion (15 and 16) can be integrated using standard numerical integration algorithms. In addition, because the matrix-based solution for determining charge is not used, the code remains $O(N)$, a crucial requirement for large-scale simulations.

The Lagrangian-based dynamic-charge method, which amounts to an approximate implementation of the electronegativity equalization condition, suffers from one significant drawback, however. The dynamic-charge method is closely related to a single iteration, steepest-descent search for the electrostatic energy minimum. The magnitude of the charge transfer is controlled by a constant factor for all atoms and simulation steps. Thus, the charge "lags" behind the atomic positions used in minimizing the energy and, as a result, the electronegativity equalization condition is not satisfied. To remedy this situation, Ma and Garofalini have introduced the iterative fluctuation charge (IFC) model [22]. The IFC model is able to satisfy the electronegativity equalization condition by performing multiple iterations for electrons (charge) while keeping the atomic positions fixed. The exact numerical criterion, or convergence factor Δ, for electronegativity equalization is $\Delta\left(\frac{eV}{atom}\right) = \frac{\sqrt{(\chi_i-\chi_{eq})^2)}}{N} = 0$, where χ_i and χ_{eq} are the electronegativities on the ith atom and the equilibrium electronegativity. In practice, a small convergence factor, such as $\Delta = 1.0 \times 10^{-8}$ is used in MD simulations. In addition, charges will oscillate about their fully converged, equilibrium values unless a damping factor is used. The selection of the damping factor and the fictitious charge-mass values can be optimized to achieve optimal damping while not preventing thermal coupling between atomic and electronic degrees of freedom.

Recently, the extended Lagrangian approach with the damping factor discussed by Ma and Garofalini [22] was used to examine the interface between Si and SiO_2 surfaces using a charge optimized many-body (COMB) potential [23]. The potential has the form:

$$E_T = \sum_i \left[E_i^S + \frac{1}{2} \sum_{j \neq i} V_{ij}(r_{ij}, q_i, q_j) + E_i^{BB} \right], \quad (17)$$

where E_T is the total potential energy of the system, E_i^S is the self-energy term of atom i, V_{ij} is the interatomic potential between the ith and jth atoms, r_{ij} is the distance of the atoms i and j, and q_i and q_j are charges of the atoms, and E_i^{BB} is the bond-bending term of atom i. The interatomic potential energy V_{ij} consists of four components: short-range repulsion, U_{ij}^R, short-range attraction, U_{ij}^A, long-range Coulombic interaction, U_{ij}^I, and long-range van der Waals energy, U_{ij}^V, which are defined as

$$V_{ij}(r_{ij}, q_i, q_j) = U_{ij}^R(r_{ij}) + U_{ij}^A(r_{ij}, q_i, q_j) + U_{ij}^I(r_{ij}, q_i, q_j) + U_{ij}^V(r_{ij}), \quad (18)$$

The repulsive and attractive terms in (18) are of the Tersoff form:

$$U_{ij}^R = f(r)Ae^{-\lambda r} \text{ and } U_{ij}^A = f(r)b_{ij}Be^{-\alpha r} \quad (19)$$

In Tersoff's original work, A, B, λ, and α are constants and $f(r)$ is a distance-based cutoff function, which limits the range of the potential. In shorthand notation, the energy contribution due the repulsive and attractive terms (i.e., the short-range pair potential) can be written as

$$E_{pair} = \sum_{j \neq i} f(r)Ae^{-\lambda r} + \sum_{j \neq i} f(r)b_{ij}Be^{-\alpha r} = V^R + bV^A \quad (20)$$

where b_{ij} is the bond-order function.

Modeling Charge Transfer During Chemical Reactions

Model fluctuating charges using the extended Lagrangian approach seems ideally suited to examine chemical reactions using MD. However, the use of this approach requires the definition of a charge-neutral entity, such as a molecule or the simulation cell. For example, Yu et al. recognized the problem with defining a charge-neutral simulation cell when simulating Si/SiO_2 interfaces. To circumvent this problem, the self-energy terms of the COMB potential were modified [23]. Rick et al. defined each water molecule as a charge-neutral entity in their examination of the structural properties of water [21]. Charge equilibration (QE) methods have been applied successfully to relatively homogeneous small-molecule systems near equilibrium. However, application of these methods to large molecules yields atomic charges and molecular dipole moments that are too large. This problem has been traditionally addressed by constraining subentities of large molecules to be charge neutral. This is problematic in a reactive environment where it would not be possible to meaningfully adhere to a fixed set of molecule-based charge-neutral entities.

Recently, Mikulski et al. developed a method to integrate charge equilibration with bond-order potentials that does not require the assignment of charge neutrality to molecules [24]. The bond-order potential/charge equilibration (BOP/SQE) method is a bond-centered approach, where the charge of atom i is the sum of all charges transferred to it across each of its bonds. This has been referred to as a split-charge equilibration (SQE). The problem of the overestimation of dipole moments of large molecules is not solved by merely adopting the bond-centered (SQE) approach. A novel method is needed whereby each bond partly decouples from its bond network and settles into its own equilibrium. The BOP/SQE method connects the bond order of each bond to an amount of shared charge in each bond and interprets the split charge as an imbalance in where the shared charge is concentrated. Split charges cannot grow in size beyond this shared-charge limit, and if the nature of the bond is covalent, the size of the split charges should not get close to this limit. Thus, a fractional split charge is defined as the ratio of the split charge $\overline{f_{ij}}$ on atom i transferred from atom j to its shared charge $\overline{q_{ij}^{max}}$, charge on atom i is expressed as

$$Q_i = \sum\nolimits_j \overline{f_{ij}}\,\overline{q_{ij}^{max}}. \tag{21}$$

The shared-charge limit $\overline{q_{ij}^{max}}$ is calculated directly from the bond order b_{ij} assigned to the bond from the BOP, $\overline{q_{ij}^{max}} = \overline{q_{ij}^{max}}(b_{ij})$. Focus is now shifted from equilibrating split charges to equilibrating fractional split charges.

The equilibration process for bond ij is controlled by three factors: (1) The constant pull to increase charge separation associated with the traditional constant electronegativity difference $|\chi_i - \chi_j|$; (2) Coulomb interactions within the bond and with the surrounding environment; (3) A restoring force that approaches and infinite wall as the shared-charge limit is approached. The first two factors are analogous to the SQE condition but are applied to $\overline{f_{ij}}$ rather than $\overline{q_{ij}}$. The third, unique, element is implemented through inverse hyperbolic tangent function added to the charge-dependent electronegativity of each fractional split charge,

$$\overline{\chi_{ij}} = \left[\chi_i + J_{ii}^o Q_i + \sum\nolimits_{k \neq i} J_{ik} Q_k\right] + \overline{c_{ij}} \tan h^{-1}\left(\overline{f_{ij}}\right). \tag{22}$$

The scale parameter $\overline{c_{ij}}$ can be tailored to each pair of atom types. Large values of $\overline{c_{ij}}$ cause the wall to gradually rise as soon as the split charges grow. The last term in this equation, or the charge-asymmetry penalty, serves to physically constrain the growth of the split charges and partially decouples each bond from its bond network. Therefore, the BOP/SQE formalism becomes a locally driven process that does not map back to the global equilibration of an entire molecule. With the selection of the scale parameter $\overline{c_{ij}}$ to reproduce the dipole moment of water, this method was shown recently to reproduce the correct trends in molecular dipole moments of long-chain alcohols (Fig. 3).

It should be noted, that other methods for handling chemical reactions where charge is included are being developed [25]. Because these methods are not bond-order potentials, they are not included in this overview.

Improved Description of Alloys

In a recent study of the thermodynamics an empirical potential description of Fe–Cu alloys, Caro et al. concluded that to study alloys in an empirical framework, heteronuclear interactions require additional modifications to the empirical formalism to incorporate concentration dependent interactions that reproduce the magnitude of excess enthalpy of mixing and asymmetry around the equi-atomic composition [26]. To illustrate this point, data for the heat of mixing in the Fe-Cr system as determined in an ab initio study conducted by Olsson has been replotted in Fig. 4. This figure clearly illustrates the deviation from ideality with both positive and negative heat of mixing in evidence.

Traditionally, empirical models for alloys use the Lorentz-Berthelot mixing rules (12 and 13). These rules reproduce the ideal heat of mixing, which is, strictly speaking, only accurate for dilute solutions of solute atoms in the host matrix. Caro et al. has proposed a method for reproducing the thermodynamic properties of concentrated alloys through a modification of the embedded atom method (EAM) formalism. The basic EAM description gives the total energy as a the sum of a pair potential, V, and an embedding function, F, which in turn depends on the electron density, ρ

$$E_{EAM} = \frac{1}{2}\sum V(r) + \sum F\left(\sum \rho(r)\right) \tag{23}$$

Consideration of first-principles calculations gives important information about the behavior of these functions. First, the embedding energy relative to the

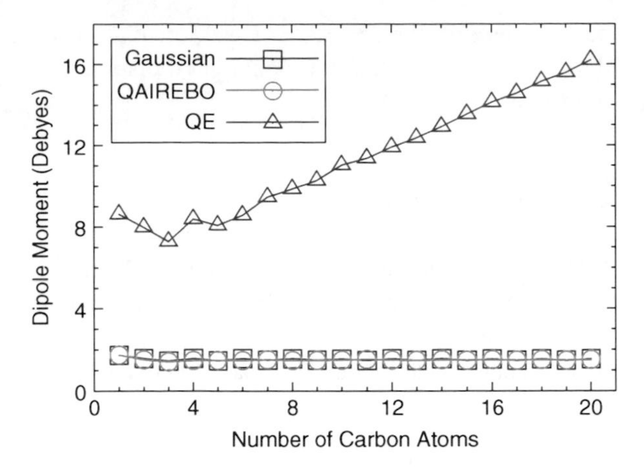

Reactive Empirical Bond-Order Potentials, Fig. 3 Molecular dipole moments of straight chain alcohols calculated using Gaussian, charge equilibration (QE), and a fluctuating-charge MD code that uses the BOP/QE method of charge equilibration and qAIREBO potential

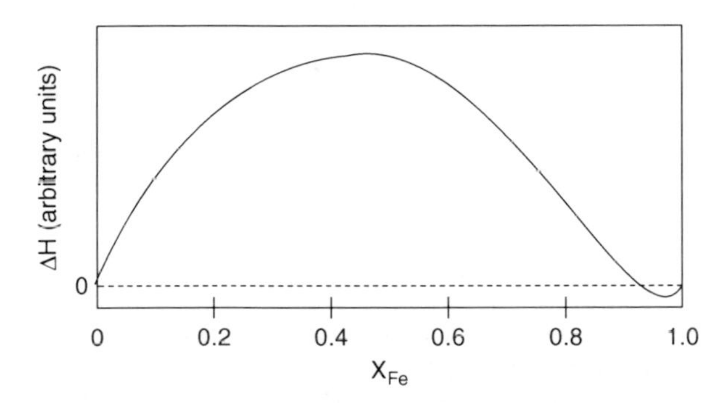

Reactive Empirical Bond-Order Potentials, Fig. 4 The enthalpy of mixing ΔH for the Fe–Cr system as a function of Fe fraction. (replotted from the work of Olsson) The solid line was calculated via first-principles. The dashed line indicates a linear interpolation corresponding to an ideal solution

free atom energy goes to zero for zero electron density and should have a negative slope and positive curvature for the background electron density in metals. Second, the pair interaction term $V(r)$ is purely repulsive.

Caro et al. eschewed the commonly used Lorentz-Berthelot mixing rules and developed a new functional form for the cross potential for random alloy mixtures:

$$E_{RAND} = x_A^2 V_{AA} + x_B^2 V_{BB} + 2x_A x_B V_{AB} + x_A F_A(\tilde{\rho}) + x_B F_B(\tilde{\rho}). \tag{24}$$

Here x_A and x_B are the atomic fractions of alloy element A and B, respectively. The pair potential for each element denoted as V_{AA} or V_{BB}. The model adopts the assumption that atoms are embedded in the same average environment, through $F(\tilde{\rho})$ where $\tilde{\rho}$ is the average electron density seen by either atom type A or B in the host material. Caro et al. found that the contributions to the formation energy from the embedding term were negligible, leaving only the pair potential as the sole contributor to the formation energy for alloys. By assuming V_{AB} is a function of both composition x and distance r and can be separated into a product $h(x)u_{AB}(r)$, an appropriate choice of the cross term is

$$V_{AB}(x,r) = h(x)\frac{1}{2}[V_A(r) + V_B(r)] \tag{25}$$

Where $h(x)$ is a fourth-order polynomial. This selection of the cross term allows for the description of any type of formation energy curves, with an ideal solution of $h(x) = 1$, for a regular solution. Through appropriate fitting of the polynomial coefficients, a positive or negative heat of mixing can be achieved. Stukowski et al. recently published a version of this potential function.

The methodology described above can also be extended to Tersoff-type potentials. The foundation of this approach was laid out by Brenner who recognized that the EAM potential and Tersoff many-body potential are formally identical in the appropriate limits. The essential result is that it is reasonable to assume that embedding function $F(\rho(r))$in the EAM formalism is equal to the attractive part of the Tersoff pair potential, bV^A. Similarly, it is a logical to assume that the repulsive terms in both EAM and Tersoff are equivalent. With these points in mind, a combination of the Tersoff potential with Caro's compositional dependence method can be expressed in shorthand notation as

$$E_{RAND} = x_A^2 V_{AA}^R + x_B^2 V_{BB}^R + 2x_A x_B V_{AB}^R + x_A \tilde{b} V_{AA}^A + x_B \tilde{b} V_{BB}^A \tag{26}$$

where V_{AA} and V_{BB} are the repulsive and attractive pair terms, denoted by superscripts ^{R} or ^{A}, respectively, for the pure elements A or B as given in (3). The repulsive cross term V_{AB} is given in (26). Schall et al. have fit a multicomponent bond-order potential for Si–C–H systems, and it was found that the average bond-order

$\tilde{b}$ must be normalized using a scaling factor to ensure compatibility between bond orders for derived for different pure element systems. The average bond order is then given as

$$\tilde{b} = \chi_{ij}\frac{b_{ij} + b_{ji}}{2}. \quad (27)$$

The normalizing factor χ_{ij} is equal to one for homonuclear atom pairs. For heteronuclear atom pairs, χ_{ij} is determined by fixing the average bond order to a value of one for a specified crystal structure. For example, in the Si–C system, χ_{ij} was determined by setting the average bond order for the β-Zn phase of SiC solving (27) for χ_{ij}. A similar scheme was utilized by Tersoff for multicomponent systems.

Summary and Future Directions

In the foregoing sections, a discussion of reactive empirical bond-order potentials was given. Such potentials were developed to describe the universal relationship between binding energy and bond length and have been extended from simple descriptions of single elements to multicomponent systems and then to a combination of solids and molecules. The motivation for the development of these potentials was to provide a computationally efficient means to model chemical reactivity, including an accurate description of bond-breaking and forming processes, while still using sound quantum mechanical principles. The original formalism has since been expanded to include descriptions of long-range interactions, torsion, charge transfer, and descriptions of alloys. This family of potentials has since been used to describe a wide variety of processes in nanomaterials systems. One serious limitation of the family of reactive empirical bond-order potentials is that only a relative few materials systems have been parameterized. As mentioned previously, parameterization of a potential involves quite a bit of art as well as science. In the future, semi-automated methods such as force matching that combine ab initio force calculations systematically with parameter fitting could help overcome this hurtle. In addition, the incorporation of more than three elements presents a significant challenge due to the complexity of the spline functions used to interpolate the bond order correction terms. To overcome this barrier, either compromises in the accuracy of the potential or new developments in multidimensional interpolation functions must be made.

Cross-References

▶ Computational Study of Nanomaterials: From Large-Scale Atomistic Simulations to Mesoscopic Modeling
▶ Molecular Modeling on Artificial Molecular Motors

References

1. Stillinger, F.H., Weber, T.A.: Computer simulation of local order in condensed phases of silicon. Phys. Rev. B **31**, 5262 (1985)
2. Mackerell, A.D.: Empirical force fields for biological macromolecules: Overview and issues. J. Comput. Chem. **25**, 1584 (2004)
3. Abell, G.C.: Empirical chemical pseudopotential theory of molecular and metallic bonding. Phys. Rev. B **31**, 6184 (1985)
4. Tersoff, J.: New empirical model for the structural properties of silicon. Phys. Rev. Lett. **56**, 632 (1986)
5. Brenner, D.W.: Empirical potential for hydrocarbons for use in simulating chemical vapor deposition of diamond films. Phys. Rev. B **42**, 9458 (1990)
6. Murty, M.V.R., Atwater, H.A.: Empirical interatomic potential for Si-H interactions. Phys. Rev. B **51**, 4889 (1995)
7. Dyson, A.J., Smith, P.V.: Extension of the brenner empirical interatomic potential to C-Si-H systems. Surf. Sci. **355**, 140 (1996)
8. Sbraccia, C., Silvestrelli, P.L., Ancilotto, F.: Modified XB potential for simulating interactions of organic molecules with Si surfaces. Surf. Sci. **516**, 147 (2002)
9. Schall, J.D., Gao, G., Harrison, J.A.: Elastic constants of silicon materials calculated as a function of temperature using a parametrization of the second-generation reactive empirical bond-order potential. Phys. Rev. B **77**, 115209 (2008)
10. Shibuta, Y., Maruyama, S.: Bond-order potential for transition metal carbide cluster for the growth simulation of a single-walled carbon nanotube. Comp. Mater. Sci. **39**, 842 (2007)
11. Tersoff, J.: Empirical interatomic potential for silicon with improved elastic properties. Phys. Rev. B **38**, 9902 (1988)
12. Justo, J.F., Bazant, M.Z., Kaxiras, E., Bulatov, V.V., Yip, S.: Interatomic potential for silicon defects and disordered phases. Phys. Rev. B **58**, 2539 (1998)
13. Balamane, H., Halicioglu, T., Tiller, W.A.: Comparative study of silicon empirical potentials. Phys. Rev. B **46**, 2250 (1992)
14. Gao, G.T., Van Workum, K., Schall, J.D., Harrison, J.A.: Elastic constants of diamond from molecular dynamics simulations. J. Phys. Condens. Matter **18**, S1737 (2006)

15. Brenner, D.W., Shenderova, O.A., Harrison, J.A., Stuart, S.J., Ni, B., Sinnott, S.B.: Second generation reactive empirical bond order (REBO) potential energy expression for hydrocarbons. J. Phys. C. **14**, 783 (2002)
16. Ni, B., Lee, K.H., Sinnott, S.B.: A reactive empirical bond order (REBO) potential for hydrocarbon-oxygen interactions. J. Phys. Condens. Matter **16**, 7261 (2004)
17. Stuart, S.J., Tutein, A.B., Harrison, J.A.: A reactive potential for hydrocarbons with intermolecular interactions. J. Chem. Phys. **112**, 6472 (2000)
18. Liu, A., Stuart, S.J.: Empirical bond-order potential for hydrocarbons: Adaptive treatment of van der Waals interactions. J. Comput. Chem. **29**, 601 (2008)
19. Rappe, A.K., Goddard, W.A.: Charge equilibration for molecular-dynamics simulations. J. Phys. Chem. **95**, 3358 (1991)
20. Streitz, F.H., Mintmire, J.W.: Electrostatic potentials for metal-oxide surfaces and interfaces. Phys. Rev. B **50**, 11996 (1994)
21. Rick, S.W., Stuart, S.J., Berne, B.J.: Dynamical fluctuating charge force-fields: Applicatoin to liquid water. J. Chem. Phys. **1001**, 6141 (1994)
22. Ma, Y., Garofalini, S.H.: Iterative fluctuation charge model: A new variable charge molecular dynamics method. J. Chem. Phys. **124**, 234102 (2007)
23. Yu, J.G., Sinnott, S.B., Phillpot, S.R.: Charge optimized many-body potential for the Si/SiO2 system. Phys. Rev. B **75**, 085311 (2007)
24. Mikulski, P.T., Knippenberg, M.T., Harrison, J.A.: Merging bond-order potentials with charge equilibration. J. Chem. Phys. **131**, 241105 (2009)
25. van Duin, A.C.T., Dasgupta, S., Lorant, F., Goddard, W.A.: ReaxFF: A reactive force field for hydrocarbons. J. Phys. Chem. A **105**, 9396 (2001)
26. Caro, A., Turchi, P.E.A., Caro, M., Lopasso, E.M.: Thermodynamics of an empirical potential description of Fe-Cu alloys. J. Nucl. Mater. **336**, 233 (2005)

REBO

▸ Reactive Empirical Bond-Order Potentials

Relaxation

▸ Nanomechanical Properties of Nanostructures

Relaxation Calorimetry

▸ Nanocalorimetry

Relay

▸ Piezoelectric MEMS Switches

Relay Logic

▸ Piezoelectric MEMS Switches

Reliability of Nanostructures

Tuncay Alan
Mechanical and Aerospace Engineering Department, Monash University, Victoria, Australia

Synonyms

Mechanical characterization of nanostructures; Nanomechanics

Definition

The continuous reduction in the size of structural device components promises improved efficiency and, often, new functionalities, which would otherwise be impossible with conventional macroscale devices. As a result, in the past decades numerous microsystem devices with nanoscale mechanical parts have been demonstrated and used in a variety of engineering applications, ranging from electrical circuit components to sensors, actuators, and biomedical applications [1]. This ongoing drive to create smaller new devices to perform critical functions requires significant attention to the mechanical reliability of materials used at nanoscales. To make use of the full mechanical potential of materials at decreasing sizes, fundamental properties, such as the elastic modulus, yield, fracture and fatigue properties, adhesion and intrinsic stresses, should be determined by testing relevant sized samples that have been fabricated or processed using the same procedures as the structure of interest and subjected to the same environmental conditions.

Overview

Even though bulk elastic properties are generally expected to remain constant down to a length scale of several nanometers, the mechanical reliability of materials depends significantly on the specimen size [2]. Despite their high theoretical strengths, most materials used in practical applications cannot sustain very high stresses due to intrinsic defects. An estimate for the minimum defect size that will cause failure can be obtained by an order of magnitude calculation considering Griffith fracture criterion (which suggests that an infinitesimal crack will propagate when the available strain energy is larger than that needed to create new surfaces) [3]. Accordingly, the plane-stress fracture strength of the material is given by

$$\sigma_f = \frac{1}{1.12}\sqrt{\frac{2E\gamma}{\pi l_c}}, \tag{1}$$

where l_c is the characteristic size of an ideal defect, γ is the surface energy per unit area, and E is the elastic modulus. Considering the theoretical strength of Si (calculated by atomistic simulations), $\sigma_f = 23$ GPa [4], $\gamma = 1{:}25$ J/m^2 [5] and E = 130 GPa, Eq. (1) suggests that *a single 1-nm-long edge crack will reduce the strength of an otherwise defect-free Si component by 60% to 9 GPa!*

Key Research Findings

As the dimensions of a material decrease down to nanometers, approaching the typical size of critical defects, the strength is expected to increase significantly with a potential to reach the theoretical limit. Recent experiments demonstrated that near-theoretical strength values that are more than 25 times higher than typical macroscale strengths could indeed be achieved for 190-nm-thick single crystal silicon structures [6, 7] and 15-nm-thick low-pressure chemical vapor deposited low-stress SiN [8]. These ultrahigh strengths are very promising as they enable novel MEMS devices, which can reliably operate under very high stresses [8]. Yet, at decreasing size scales, the number of intrinsic volume defects decreases and surface-to-volume ratio increases. Hence, surfaces and process parameters gain importance. The above experiments on silicon [6] as well as stochastic simulations [7] show that atomic scale surface features, which would otherwise be negligible in larger structures start acting as defects (Eq. 1) and that they may significantly compromise the reliability of the nano-material. For instance, a slight increase in process induced root-mean-square (rms) surface roughness (from 0.4 nm to 1.5 nm) decreases the strength of Si nanobeams by approximately 20% (Fig. 1). A similar effect is observed when the samples are exposed to air for an extended period. When surfaces are not effectively protected against oxidation, 1-nm-thick native oxide islands that form randomly on the surface have been shown to reduce the strength of structures by 30% [9].

The above examples emphasize that at decreasing size scales, mechanical performance of structures is often dominated by a competition between size, grain structure, surface defects, and environmental conditions. How would failure statistics change as the average defect size approaches the size of the device? At what size scales would the environmental factors start dominating failure? Could the mechanical service life of structures be improved by modifying process parameters? Answering these three questions is crucial to effectively characterize mechanical reliability of any material used at the nanoscale.

Methodology

So far, various nanomechanical test procedures have been proposed, and a few commercial devices have appeared in the market as ultimate nanoscale test platforms. Initial studies have focused on commonly used MEMS materials, and more recently, nanomechanical experiments were also adapted to characterize biological nanostructures such as cells and proteins. Due to differences in experimental techniques, geometric properties of the samples used and fabrication procedures, there is a significant variation in the failure statistics reported by different groups.

Because of the number of parameters affecting mechanical response, there are still no standard techniques and it is difficult to come up with a universal method. The available experiments generally involve variations to well-established macroscale tensile, bulge, and bending tests. The choice of procedure depends on several factors among which, the most important are: size (a micron and nano-sized sample cannot always be tested by the same instrument),

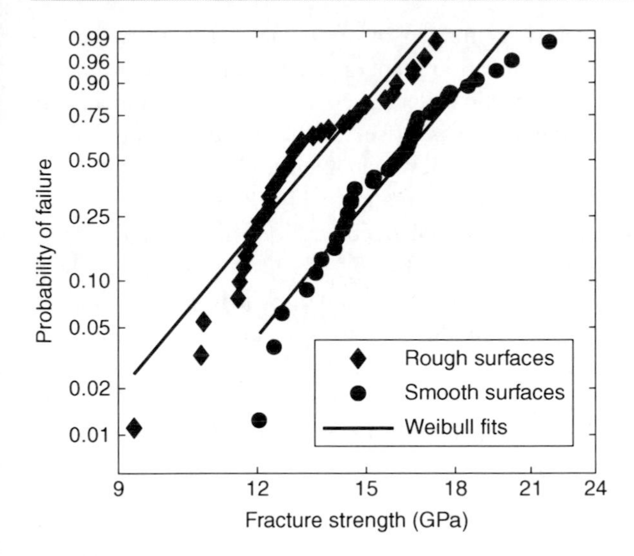

Reliability of Nanostructures, Fig. 1 Fracture strength statistics of nanobeams with different surface roughnesses. The estimated strength of the beams increases from 14.2 GPa to 16.8 GPa as the surface roughness decreases from 1.5 nm to 0.4 nm

required statistical accuracy (due to their complexity some methods do not allow easy repetition), tested material (not all materials are compatible with the fabrication procedures necessary for the preparation of the samples).

Tensile Tests

The main advantage of a tensile test is its analytical simplicity. Since the gauge length of the specimen is under uniform stress and strain fields, load–displacement curves can easily be used to determine elastic modulus and failure properties directly. Required displacement and load resolutions can generally be accommodated by commercially available piezoelectric or capacitive actuators and high-precision load cells [2]. Yet, handling tensile test specimens remains a major challenge: It is very difficult to grip an individual micron-sized sample from both ends to apply a uniform load. Sharpe and collaborators got around this problem by electrostatic gripping. They used an external probe that was temporarily attached to hundreds of micrometer long test samples by electrostatic attraction forces. The probe, which was connected to a load cell was then manipulated to load the samples to failure, while displacements were monitored via image processing. There have been many variations of this procedure as reviewed in detail elsewhere [2]. Using a different approach, Gaspar and coworkers [10] developed a homebuilt setup which uses a commercial load sensor and an automated stage to test micron-scale tensile samples that were supported by MEMS springs. The biggest advantage of the method, despite the relatively large length scales, is that it enables a very large number of tests to be performed at a short time.

The experiments that focused on smaller devices have generally employed specifically designed MEMS-based test setups, where the sample is integrated with the test device. For instance, Haque and coworkers [11] developed a tensile tester that allows in situ analysis of deformation inside an electron microscope while the sample is being loaded by MEMS beams. Similarly, some recent commercial devices by Hysitron [12] and others have promised similar functionalities. A comprehensive review of the test methods can be found in [2, 13, 14].

Bulge Tests

As another approach, bulge tests can effectively be used to test membranes with thicknesses approaching several nanometers while keeping the lateral dimensions in the order of hundreds of micrometers. This well-established method consists of measuring the out-of-plane deflection of a thin membrane under the influence of increasing pressure, applied from its backside [14]. The central deflection of the membrane is monitored continuously and the maximum stress at failure can be inferred from the corresponding pressure–deflection curves. Shrinking only the thickness (while keeping the other dimensions at μm/mm scales relaxes the experimental constraints on detection considerably. Recently, a homebuilt bulge test setup was successfully used to statistically characterize membranes that are as thin as 10 nm [8], while displacements were measured optically. Bulge tests also provide a very good alternative to study 2D materials such as graphene. However, the major disadvantage of the technique is the complicated analysis that is required to interpret the experimental data accurately.

Nanoscale Bending Tests with AFM

The third approach, which is reviewed in more detail here, is the nanoscale bending test. Bending tests offer many benefits over the above methods. Most importantly, gripping of the test samples is no longer a problem and the experimental procedures are generally simpler than in tensile tests. Since the test samples and test setup are independent from each other, there are fewer process restrictions and the effects of different parameters on reliability can be more easily studied. Finally, in a bending test, the area of fracture initiation can be effectively controlled without needing external notches. For an un-notched tensile test sample, fracture load can be reached anywhere along the beam: Sidewall surfaces and beam top or undersides have equal stresses and hence the defect distribution on all sides should be carefully considered. On the other hand, bending samples can be designed so that maximum stresses occur on the critical sites, hence, simplifying the corresponding statistical analysis. This point proves to be very useful when studying the effects of specific defects on mechanical performance [7]. Yet, the major drawback of the method is that the stress–strain distributions within the sample and its failure characteristics must be acquired indirectly, by processing the experimental load–displacement measurements.

Overview of the Method

Among all of the available methods, atomic force microscopy (AFM) is arguably the most popular one for nanoscale bending tests, being extensively employed to characterize both mechanical device components and biological samples. AFM is a powerful analytical tool used mainly to determine the surface topography of structures with the help of a (usually) compliant cantilever, which is attached to a piezo scanner. For instance, in contact mode operation, the cantilever scans the surface of the material while it is in contact with the sample. A feedback control loop moves the piezo scanner vertically at each scan point to maintain a fixed force between the tip and the sample. This vertical piezo displacement can then be transformed into a topographic image of the surface. A nanomechanical AFM-based test consists of using an AFM cantilever, with known stiffness, to deflect a test sample until it fails. The applied force, F, and the sample deflection, δ_{sample}, are simultaneously measured during loading and the fracture strength of the tested structure is obtained from either a simple theoretical calculation or finite element analysis (FEA).

The first step in the experiment is to determine the position of the AFM cantilever tip relative to the test sample accurately. For this purpose, the sample of interest is imaged by a contact mode scan and the load application point is picked from the AFM image. Using the piezoelectric controls of the instruments, the cantilever tip is offset to the loading point with an accuracy of a few nanometers. Next, the sample is loaded. The schematic in Fig. 2 shows a cantilever loading a micron-sized test beam. When the piezo scanner is extended by δ_{piezo}, the AFM cantilever pushes against the sample, which is deflected by δ_{sample}, while the cantilever undergoes an upward displacement of $\delta_{cantilever}$. Since the total piezo extension, δ_{piezo} equals the sum of the cantilever and sample displacements, deflection of the test structure can be expressed as

$$\delta_{sample} = \delta_{piezo} - \delta_{cantilever} \quad (2)$$

Deflection Measurement To obtain load–deflection curves for individual samples, $\delta_{cantilever}$ is continuously recorded throughout the experiment by monitoring the position of the laser light, which is reflected off the back of the AFM cantilever and collected by a position-sensitive photodetector. The upward cantilever deflection results in a change in the photodetector output signal (S_p). Hence, every time a cantilever is attached to the piezo scanner, the relation between $\delta_{cantilever}$ and the S_p (in Volts) must be carefully calibrated. To relate S_p (V) to $\delta_{cantilever}$ (m), the cantilever is brought into contact with a noncompliant surface, such as a rigid Si piece. The piezo is extended by a known distance, δ_{piezo}, pushing the cantilever against the surface and the corresponding change in S_p is monitored. Throughout the calibration, since the surface remains rigid (neglecting elastic contact deformations), $\delta_{cantilever} = \delta_{piezo}$. Considering the nonlinearity often observed at higher deflections, this relation is best characterized by a third-order polynomial ($\delta_{cantilever} = c_1\, S_p^{\,3} + c_2\, S_p^{\,2} + c_3\, S_p$), the constants of which are determined by a least-squares fit.

Force Measurement The force, F, applied by the cantilever on the test structure is equal in magnitude and opposite in direction to the load that causes the cantilever to deflect by $\delta_{cantilever}$. As seen in Fig. 2, the AFM cantilever makes a 10° angle with the horizontal. Hence, any load, F, applied on the test structure will have a vertical component, $F_v = F\cos(10°)$ and a much

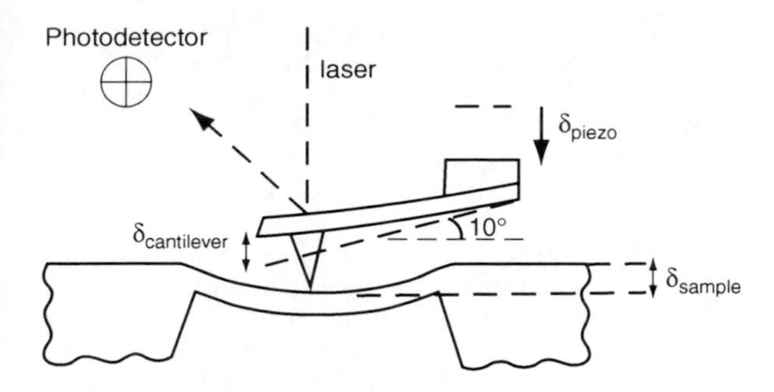

Reliability of Nanostructures, Fig. 2 Schematic of a nanoscale bending test with an AFM cantilever loading a double clamped test beam

smaller, negligible, horizontal component, $F_h = F\sin(10°)$. The vertical force applied on the beam is given by

$$F_v = k_{\text{cantilever}}\delta_{\text{cantilever}}, \tag{3}$$

where $k_{\text{cantilever}}$ is the stiffness of the AFM cantilever.

Accuracy of the force measurements strongly depends on $k_{\text{cantilever}}$. Using commercially available cantilevers may save lengthy fabrication steps; however, since the specifications are not precisely known, a careful calibration process is needed to accurately determine $k_{\text{cantilever}}$.

AFM cantilever stiffness calibration. When uncoated single-crystal Si cantilevers are used, the stiffness can be obtained from the cantilever dimensions and known elastic properties of Si. An SEM can be used to measure the lateral dimensions and the offset of the tip, which defines the effective length of the cantilever. Similarly, the tip height can be obtained using an optical surface profilometer. Measuring the cantilever thickness accurately is often the most important, yet, most difficult step, due to the inclined sidewalls resulting from the fabrication process.

An iterative method based on resonant measurements was recently proposed to accurately determine h and calculate $k_{\text{cantilever}}$ [6]. The resonant frequency of the cantilever in air, f_{air}, is measured with the AFM and the vacuum frequency, f_{vac}, is estimated by correcting for the effects of the air damping.

$$f_{\text{vac}} = [1 + \frac{\pi b \rho_{\text{air}}}{4h\rho_{\text{cantilever}}}]^{0.5} f_{\text{air}}, \tag{4}$$

where ρ_{air} and $\rho_{\text{cantilever}}$ are the densities of air and the cantilever material respectively, b is the average width, and h is the thickness of the cantilever. (For typical cantilevers used in the experiments in [6], $f_{\text{vac}} \sim 1.002 f_{\text{air}}$.) The vibration of the cantilever is then simulated by FEA while the thickness, h, is iterated until simulated and measured frequencies match. The stiffness, $k_{\text{cantilever}}$, is then calculated from a static FEA, which incorporates known elastic properties and the exact cantilever geometry.

Determining the Fracture Strength Finally, the experimental load–deflection curves are processed to obtain the Young's modulus, E, of the samples and the deformation of each sample is modeled with a nonlinear FEA simulation. The FEA model takes into account the maximum load at fracture, load application point, geometric properties, and calculated E values to determine the fracture stress, *maximum tensile stress along the length of the sample*, for each sample. Results from repeat tests can then be analyzed statistically.

Accuracy of the Measurements

The precision of the strength measurements strongly depends on the accuracy of the force measurement, since the fracture load is used to extract fracture strength from FEA. An approximate expression for the error in strength measurements is given by:

$$\frac{\Delta\sigma}{\sigma};\frac{\Delta F_v}{F_v};\frac{\Delta k_{\text{cantilever}}}{k_{\text{cantilever}}} + \frac{\Delta\delta_{\text{cantilever}}}{\delta_{\text{cantilever}}} \tag{5}$$

For this analysis, the cantilever stiffness can be expressed as $k \approx 3EI/L^3$, where $I = bh^3/12$, with a cantilever width, b, thickness, h, and length, L. Similarly, the thickness $h \approx cf_{\text{vac}}L^2$, where c is a constant and f_{vac} is the resonant frequency in vacuum. Hence, Eq. (5) becomes:

$$\frac{\Delta\sigma}{\sigma} = \frac{\Delta b}{b} + \frac{3\Delta f_{\text{vac}}}{f_{\text{vac}}} + \frac{6\Delta L}{L} - \frac{3\Delta L}{L} + \frac{\Delta\delta_{\text{cantilever}}}{\delta_{\text{cantilever}}} \tag{6}$$

The relative errors in the width and length measurements are related to the resolution of the SEM images, while $\frac{\Delta f_{\text{vac}}}{f_{\text{vac}}}$ depends on the sensitivity of the frequency measurement. All of these are much smaller than 1%. On the other hand, the precision of the AFM cantilever displacement is directly related to the precision of the AFM piezo scanner and the photodetector readout and will dominate the measurement error, depending on the setup being used.

Future Directions in Research

AFM has great potential to become the standard testing method for a large class of materials. No doubt, the biggest advantage of the method is the use of commonly available standard instruments, hence allowing similar structures to be tested at different locations without needing specialized test apparatus. Being highly flexible, it can effectively characterize the effects of process defects and environmental degradation on mechanical reliability, hence addressing the questions postulated in the "Key Research Findings" section. Moreover, there are no material restrictions, and, upon correct choice of cantilevers, samples ranging from hundreds of nanometers to a few micrometers can repetitively be tested with very good-to-excellent accuracy. However, further studies are still needed to better understand the interaction between the tip and the sample during loading, determine the optimum tip and cantilever geometry to test different samples, and decide upon standard calibration procedures. One of the current challenges is also to develop new generation of AFM cantilevers in order to adapt the existing methodology to perform tensile tests at the same resolutions.

Cross-References

- ▶ Ab Initio DFT Simulations of Nanostructures
- ▶ AFM
- ▶ AFM Probes
- ▶ Atomic Force Microscopy
- ▶ Finite Element Methods for Computational Nano-optics
- ▶ Mechanical Properties of Hierarchical Protein Materials
- ▶ Nanomechanical Properties of Nanostructures

References

1. Senturia, S.D.: Microsystem Design. Springer, New York (2004)
2. Hemker, K.J., Sharpe Jr., W.N.: Microscale characterization of mechanical properties. Annu. Rev. Mat. Res. **37**, 93–126 (2007)
3. Anderson, T.J.: Fracture Mechanics: Fundamentals and Applications, 3rd edn. CRC Press, Boca Raton (2004)
4. Roundy, M., Cohen, M.L.: Ideal strength of diamond, Si, and Ge. Phys. Rev. B **64**, 212103 (2001)
5. McCarty, A., Chasiotis, I.: Description of brittle failure of non-uniform MEMS geometries. Thin Solid Films **515**(6), 3267–3276 (2006)
6. Alan, T., Hines, M.A., Zehnder, A.T.: Effect of surface morphology on the fracture strength of silicon nanobeams. Appl. Phys. Lett. **89**, 091901 (2006)
7. Alan, T., Zehnder, A.T.: A Monte-Carlo simulation of the effect of surface morphology on the fracture of nanobeams. Int. J. Fract. **148**, 2 (2007)
8. Alan, T., Yokosawa, T., Gaspar, J., Pandraud, G., Paul, O., Creemer, F., Sarro, P.M., Zandbergen, H.W., Microfabricated channel with ultra-thin yet ultra-strong windows enables electron microscopy under 4-bar pressure. Appl. Phys. Lett. **100**, 081903 (2012)
9. Alan, T., Zehnder, A.T., Sengupta, D., Hines, M.A.: Methyl Monolayers Improve the Fracture Strength and Durability of Silicon nanobeams. Appl. Phys. Lett. **89**, 231905 (2006)
10. Gaspar, J., Schmidt, M.E., Held, J., Paul, O.: Wafer-scale microtensile testing of thin films. J. Microelectromech. Syst. **18**(5), 1062–1076 (2009)
11. Haque, M.A., Espinosa, H.D., Lee, H.J.: MEMS for in situ testing-handling, actuation, loading, and displacement measurements. MRS Bull. **35**(5), 375–381 (2010)
12. Kiener, D.A., Minor, A.M.: Source truncation and exhaustion: insights from quantitative insitu TEM tensile testing. Nano Lett. **11** (9), 3816–3820 (2011)
13. Haque, M.A., Saif, T.: Mechanical testing at the micro/nanoscale. In: Sharpe, W.N. (ed.) Springer Handbook of Experimental Solid Mechanics. Springer, New York (2008)
14. Tabata, O., Tsuchiya, T.: Reliability of MEMS: Testing of Materials and Devices. Wiley-VCH, Weinheim (2007)
15. Vlassak, J.J., Nix, W.D.: A new bulge test technique for the determination of Young's Modulus and Poisson's ratio of thin films. J. Mater. Res. **7**(12), 3242–3249 (1992)

Remotely Powered Propulsion of Helical Nanobelts

Gilgueng Hwang[1] and Stéphane Régnier[2]
[1]Laboratoire de Photonique et de Nanostructures (LPN-CNRS), Site Alcatel de Marcoussis, Route de Nozay, Marcoussis, France
[2]Institut des Systèmes Intelligents et de Robotique, Université Pierre et Marie Curie, CNRS UMR7222, Paris, France

Synonyms

Wireless propulsion of helical nanobelt swimmers

Definition

Remotely powered propulsion of helical nanobelts is the swimming propulsion of nanoswimmers created by helical nanobelts.

Overview

Since swimming mobile micro-agents at low Reynolds number experience severe viscous drag, their swimming performances are much limited. Considering the most of in vivo environments are dynamic system, highly dynamic microswimmers would have a big technological impact in many fields. Engineering artificial flagella is inspired from nature's bacteria having their excellent motility. They are created from self-scrolled helical nanobelts. These swimmers are designed with a head and a tail, similar to nature's micro-organisms such as bacteria and their flagella. In this entry, the highly dynamic propulsion of biologically inspired artificial flagella which are remotely powered by electroosmotic force is introduced. They can swim as fast as 24 times of their body lengths per second and apply pressure higher than 300 Pa from their high energy conversion efficiency. Their swimming performance outran the other inorganic microscale swimmers and even nature's bacteria like *Escherichia coli*. These electroosmotic propelled helical nanobelt swimmers could be the potential biomedical carriers, wireless manipulators, and as local probes for rheological measurements.

Introduction

Rapid advancement in semiconductor electronics and microelectromechanical systems (MEMS) has accelerated the size reduction of functional systems like sensors or actuators. Considering much increased needs in less invasive biomedical or clinical tools to prolong human life, micro- or nanoscale systems have many promising features. One of the most important features is to reduce their size and disconnect them from wires. In this context, recent interest toward developing untethered biomedical micro- and nanorobots could be an alternative way to achieve less invasive and targeted medical therapy. Although they seem to increase the complexity of conventional medications, they are still promising due to their controllability, local therapy, and repeatability. Recent developments of in vivo biomedical nanorobots draw much interest in discovering new and efficient wireless power transfer methods and in controlled locomotion mechanisms. This kind of novel tools would have many potential applications, such as targeting, diagnosing, and treating blood clots or cancer cells, for neural cell probing and so on.

Although they are promising to such applications, they are still remained in laboratory experiments due to several reasons. The micro- or nanoscale mobile agents experience much more viscous drag due to the decreased Reynolds number which results in severely limited swimming performance. Considering the capillary of human's circulatory system has the average flow speed which can go up to 1 mm/s, mobile microagents should swim faster than this flow speed for their survival and task continuity. It should be noted that nature's bacteria have well adapted in such a harsh environment.

Many approaches aim to mimic natural bacteria for the locomotion of artificial nanostructures because of their excellent motility, around 5–10 times of body sizes per second [1]. The Nature has adopted several different propulsion techniques. The swimming of real bacteria is mainly divided into corkscrew-type rotating propulsion and the oscillation of flagella tail [2]. Several studies exploring similar propulsion techniques exist in the literature, with different efficiencies. Purcell uses a definition of swimming efficiency which compares the power used to propel a body at a given velocity to the power required to simply pull the body through the fluid at the same velocity (ε = power required to simply pull body/power consumed during propulsion) [2]. Under this definition, direct pulling with field gradient is 100% efficient (i.e. $\varepsilon = 1$), and other methods would always perform less. Recently, direct pulling of nanowires, nanotubes and microtubes with excellent swimming performances (2 mm/s^{-1} which is approximately 50–100 times per body size) was demonstrated by catalytic decomposition of hydrogen peroxide (H_2O_2). However, this technique only works in a solution like H_2O_2, allowing a catalytic reaction with Pt or Ag [3].

Among different propulsion methods, the use of magnetic effect is the most widely encountered. A rotating magnetic field was used to drive the macro-scale flagella-like structure [4]. Microbeads were attached to natural bacteria and moved by flagella propulsion [5], 30-μm-scale bacterial flagella–like propulsion was demonstrated by an attached ferromagnetic metal pad under an external rotating magnetic field [6]. A linear chain of colloidal magnetic particles linked by DNA and attached to a red blood cell can act as a flexible artificial flagellum [7]. Further size reduction and applying same principle to drive artificial bacteria was achieved [8]. Lately, helical propellers are shown as an efficient solution at low Reynolds numbers [9].

Concerning the driving power source, there are several limitations of using an external field. The magnetic field gradient rapidly decreases with the distance from the source. Therefore, although the helical morphology is advantageous at low Reynolds numbers by reducing the viscous drag, the swimming performance of artificial structures is still much lower than the natural bacteria. To improve the swimming performance, increasing field intensity would be necessary.

However, the major challenge toward potential in vivo biomedical applications is the lack of closed loop motion control either by teleoperation or autonomous navigation. This lack is related to position-tracking issues. Widely used noninvasive biomedical sensors or imaging devices include radiography, computed tomography (CT), ultrasound echography, and magnetic resonance imaging (MRI). Considering the real-time navigation and high spatial resolution requirements, MRI is one of the most promising solutions. Ferromagnetic objects, or even magnetotactic bacteria, were tracked and navigated under real-time MRI [10]. In case of smaller objects at low Reynolds numbers, an additional propulsion-dedicated field gradient generator would probably be required. However, magnetic field gradient–based propulsion would be largely limited when it has to be used with an observation device such as MRI because of the conflict between the imaging and propelling magnetic fields. This problem can be avoided by using nonmagnetic propulsion.

As nonmagnetic approaches, the flapping or undulating motion of flagella was achieved in mesoscale using a commercial piezoelectric bimorph [11]. Microrobots that harness natural bacteria have also been demonstrated [5]. There are very few works on electric field–based wireless micro/nanorobots. For example, diodes were actuated using electroosmotic pumping [12]. It showed that electroosmotic field can be a good candidate for actuating micro/nano robots. However, the work was only demonstrated in macro scale and not scaled down to micro or nano.

In this entry, remote propulsion of artificial micro swimmers with high dynamic mobility (5–30 body lengths per second) at low Reynolds numbers by electroosmotic effect is described. These swimmers are based on helical nanobelt (HNB) structures. As different types of HNBs showed different swimming performances, choices of surface material and geometry is discussed along with the fabrication processes. The electroosmotic propulsion principle and experimental setups are presented. Finally, the experimental results on electroosmotic propulsion of HNBs in comparison with different locomotion techniques are described.

Self-Scrolled Helical Nanobelts as Artificial Flagella

Advantages of HNBs as Artificial Flagella

There are several different types of bacterial locomotion, such as rotating and undulating motions. Their major features, especially in terms of surface condition, geometry, and motility have evolved to swim efficiently at low Reynolds numbers. For example, *E. coli* bacteria consist of a highly elastic body formed of protein. Therefore, they can self-transform their morphology to increase their swimming velocity in a viscous liquid environment. It is considered that such ability of natural bacteria is mainly attributed to the nonlinear mechanics of their flagella. To mimic similar performances of natural bacteria, the ultraflexible nanostructures with large nonconstant mechanics are considered to be necessary. The mechanical durability is also an important issue especially in harsh environments.

HNBs are one of the most flexible nanostructures that can be fabricated using standard microfabrication techniques with controlled geometry. Helical morphology appears as an adapted solution for electroosmotic propulsion because of the large surface-to-volume ratio to maximize the external energy reception. Another advantage is their elasticity to passively self-adapt their morphology in thick liquid environments by reducing viscous drag. Different types of nanobelts were synthesized and mechanically characterized [13]. Recently, HNBs revealed very large nonlinear mechanical properties along with giant piezoelectricity

[14]. In case of electroosmotic propulsion, different surface coatings with thin films can also be achieved by standard microfabrication techniques to improve electrokinetic energy. The high surface-to-volume ratio of HNBs can disadvantageously increase friction at the liquid/solid interface. HNBs with hydrophobic surface coatings can reduce this friction to improve the swimming performance.

Additionally, harvesting energy from the environment would be an important advantage for mobile micro/nanoagents. Piezoelectric nanogenerators based on zinc oxide nanowire arrays were firstly proposed [15]. Nanowire-based, self-powered nanosensor has also been demonstrated [16]. Passive motions of HNBs during their swimming at low Reynolds number can self-harvest electrokinetic energy due to their giant piezoelectricity [14].

Therefore, considering the limitations of conventional propulsion mechanisms, remotely powered gradient pulling of HNBs should be efficient to swim in low Reynolds number liquid [9].

Fabrication Process

Four different types of HNB-based nanoswimmers, varying the materials and geometries, summarized in Table 1 are produced for testing purposes. Their propulsion efficiency is then compared in electroosmotic propulsion experiments.

Figure 1 illustrates the fabrication process. The initial layers were grown on semiinsulating GaAs (AXT wafers) using a molecular beam epitaxy system (VEECO, Gen II MBE) equipped with a valved cracker for As and solid sources for Ga and In (Fig. 1a). For n-type doping, a Si source was used. Substrate temperature was measured by a pyrometer. After a thin GaAs buffer layer, the 500-nm thick sacrificial AlGaAs layer is deposited. The layer contains 20% Ga in order to prevent oxidation. The sacrificial layer in previous designs was made of AlAs. The oxidation of these layers after exposure damaged the InGaAs/GaAs bilayer. On top of the sacrificial layer the InGaAs/GaAs layer is deposited which later self-forms into a nanospring. An In content of 15% in the InGaAs layer was determined by X-ray diffraction (XRD) measurements. The thickness of this layer must be smaller than the critical thickness to maintain elastic strain. The layer properties along with other specifications of the structures are summarized in Table 1.

During the deposition of the InGaAs/GaAs bilayer on this wafer for HNB 1, it was attempted to get a slightly lower doping concentration than on the wafer that was used for HNB 2, 3, and 4. From the results, it seems that the doping concentration was too high which resulted in a partial self-compensation and, therefore, in a decrease in the effective doping concentration. The doping concentrations of the structures are 4.4×10^{18} cm^{-3} and 8.5×10^{18} cm^{-3}. The initial pattern can be created through photolithography. Reversible photoresist AZ5214 was used as a resist. After the development of the resist, reactive ion etching (RIE) with a $SiCl_4$ gas was used to transfer the pattern to the InGaAs/GaAs bilayers. For HNB4, the thickness of photoresist was reduced around 50 nm with O_2 plasmas. For metallic HNB3, the Cr/Ni layers on the surface and on the heads to which the structures are fixed at the end were created before RIE through a lift-off process with negative photoresist AZ5214.

The Cr layer is 10 nm thick and serves as an adhesion layer. The 10-nm thick Ni layer is used for metal HNB3. Finally, a 2% HF aqueous solution at 4°C was used to selectively etch the AlGaAs sacrificial layer under the InGaAs/GaAs heterostructures for the self-forming of the nanostructures (Fig. 1a). During this wet etch, the patterned bilayer curled up along a <100> direction, releasing the internal strain and forming 3-D structures. The direction of the scrolling is determined by the anisotropy in stiffness of the InGaAs/GaAs bilayer. After the wet etch release, the chips were rinsed in deionized water and subsequently in isopropyl alcohol [14]. Samples are also conserved in isopropyl alcohol in order to prevent oxidation by water of heads to which the structures are fixed.

Electromechanical Property

To understand the electrokinetic function of the applied strain by liquid environment, the piezoresistivity of HNBs is characterized. For the intrinsic property characterizations, the HNBs with metal pads attached on both sides were fabricated using microfabrication techniques [14]. With the metal connectors, good electrical contact can be achieved. Besides the electromechanical characterization, such connectors also allow for the integration of these structures into more complex assemblies. Nanomanipulation inside an SEM was used for their electromechanical property characterization. The experimental results showed that the structures showed unusually high piezoresistive response.

Remotely Powered Propulsion of Helical Nanobelts, Table 1 Specifications of different types of artificial bacteria (HNB)

	HNB 1 low	HNB 2 high	HNB 3 Cr/Ni	HNB 4 photoresist
Thickness $In_{0.15}Ga_{0.85}As$/GaAs (nm)	11.6/15.6	8 ~ 10/15.6	11.6/15.6	
Thickness Cr/Ni or photoresist (nm)	NA		Cr/Ni 10/10	Photoresist ~50
Diameter (tail = head), pitch and width (tail) (μm)	2.1, 14 and 2.5			
Number of turns (tail)	4.5 turns			
Length (head/total) (μm)	12/74			
InGaAs/GaAs doping (N_D)	4.4×10^{18} cm^{-3}	8.5×10^{18} cm^{-3}		

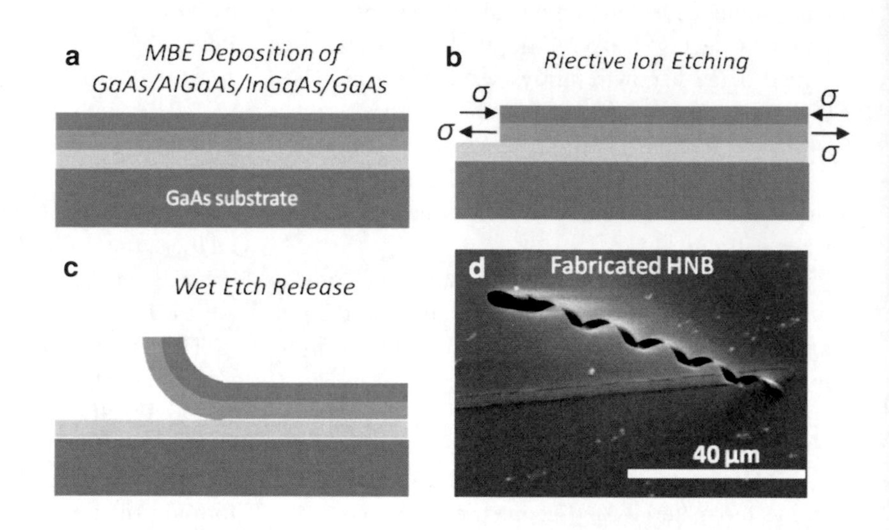

Remotely Powered Propulsion of Helical Nanobelts, Fig. 1 Basic fabrication process sequence of an HNB. (**a–c**) Initial planar bilayer, patterned through conventional microfabrication techniques, folds itself into a 3-D nanostructure during wet etch release. (**d**) Produced HNB with a tail and a head (SEM photo)

Moreover, electrostatic actuation was used to excite the structures at their resonance frequency and investigate their resistance to fatigue. With their low stiffness, high strain capability, and good fatigue resistance, the HNBs can be used as high-resolution and large-range force sensors. By variation of design parameters, such as the number of turns, thickness, diameter, or pitch, an HNB with the required stiffness can be designed through simulation.

For the quantitative comparison with other piezoresistors, the piezoresistance coefficient can be obtained as,

$$\pi_l^\sigma = \frac{1}{X}\frac{\Delta\sigma}{\sigma_0} \tag{1}$$

where σ_0 is the conductivity under zero stress and X is the stress. The piezoresistance coefficient was defined with resistivity as $\pi_l^\rho = \Delta\rho/\rho_0$. The conversion is $\pi_l^\sigma = -\pi_l^\rho$ for small σ. Uniaxial stresses were applied on HNB along their lengths. Since the linearity of each cycle from the measured data is almost constant, the widest measured region of the cycles was chosen to calculate the piezoresistance coefficient. Geometry information from Table 1 was used as the parameters.

These obtained piezoresistance coefficients of HNBs based on Eq. 1 are compared with other elements of the piezoresistors. Each reported piezoresistance coefficients of Bulk Si, boron-doped Si, and Si-NW, and CNT were summarized in Fig. 2 in comparison with HNB [14]. Especially, Si-NW was recently reported as showing giant piezoresistive effect. The revealed piezoresistance coefficients of the HNBs are even much higher than the one of Si-NW. These results are straightforward from the benefit of HNB's flexibility considering the Eq. 1 since the tiny input stress (X) can cause the resistance change ($\Delta\sigma/\sigma_0$). At least the torsional effect seems to affect much to the high piezoresistive response of the HNBs. As was discussed on the difficulty in exact modeling of the piezoresistive HNBs, mechanism behind such a high piezoresistive response was not understood yet. However, there is the fact that the HNBs are the promising nanostructures to be used as the elements of ultraflexible electromechanical devices such as force sensors and bioinspired systems.

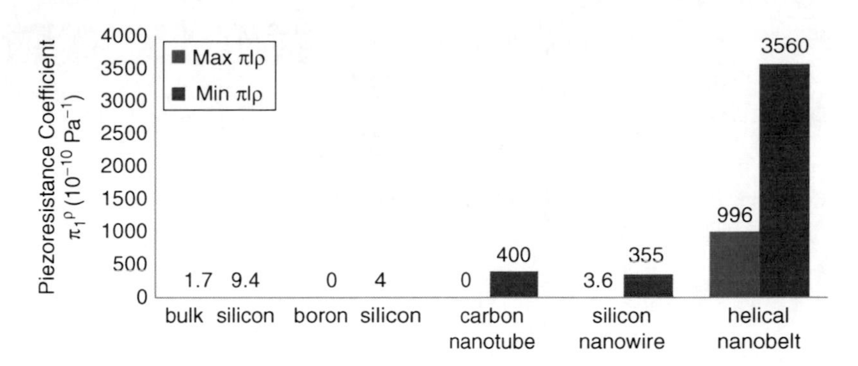

Remotely Powered Propulsion of Helical Nanobelts, Fig. 2 Comparison of longitudinal piezoresistance coefficients in different nanostructures (bulk silicon, boron silicon, carbon nanotube, silicon nanowire, and helical nanobelt) [14]

Mechanical Property

In addition to piezoresistivity of HNBs, they can perform like a spring to save energy from large-range mechanical force based on their nonconstant mechanics. Tuning fork force calibration tool was utilized to characterize the mechanics of HNBs in full-range tensile elongations. The tuning fork as force interaction tools have been proposed mainly to replace the AFM optical cantilevers for imaging [17]. The mechanical property (spring constant) of HNBs was measured by attaching them to the tuning fork tip. The force is obtained from the stiffness of HNBs measured by tuning fork gradient force sensor and by the displacement measurement of HNBs with SEM imaging. Figure 3 summarizes the measured mechanical property while HNBs are under tensile elongation. The nonconstant stiffness behavior of HNBs during their controlled tensile elongation was clearly revealed in their full-range elongation. The obtained stiffness ranges from 0.009 to 0.297 N/m with a resolution of 0.0031 N/m during full elongation and 0.011 N/m to 0.378 N/m with a resolution of 0.0006 N/m for the nondestructive method. It was transformed into full elongation tensile forces as high as 2.95 μN for the first experiment and 1.56 μN for the second.

Wireless Propulsion of Helical Nanobelts

Principle of HNB Electroosmotic Propulsion

The propulsion of HNBs involves two conversions: transforming energy from an external source to a force and transforming this force into a motion. The proposed method to supply power remotely to HNBs is an electrokinetic process. Its main advantages for in vivo applications are the good tolerance of living organisms to electric fields and its compatibility with medical MRI imaging.

In general, the surface charge that develops at the solid-liquid interfaces plays an important role in a number of electrokinetic processes. Due to the presence of a thin interfacial layer named the Stern layer, it results in a nonzero electric potential at the liquid interface, also known as the zeta potential ζ. This potential is screened by mobile counterions diffusing in the liquid in a layer generally named Electrical Diffuse Layer (EDL), which has a typical thickness of a few tens of nanometers depending on the solution ionic strength. Under an electric potential difference, the resulting electric field puts in motion the excess counterions in the EDL. This moving layer drags the whole liquid inside the fluidic medium, resulting in "pluglike" flows also referred to as electroosmotic flows (EOF). In addition, one should also consider the surface charge developing at the HNB surface. In the case of a HNB floating in liquid, its own Stern layer generates a flow which applies a hydrodynamic pressure on the surface of the robot, propelling it in the opposite direction.

Figure 4 depicts the electroosmotic propulsion mechanism through the interface between HNB's thin membrane surface and liquid medium solution. Since the propulsion force occurs through the whole surface, the large surface-to-volume ratio of HNBs is very advantageous.

HNBs' motility results from a local electro-osmotic flux powered by an external field. The specific direction of HNB's propulsion along the cathode and anode probes indicates that a DC field gradient along the HNB is responsible for this propulsion. The equivalent circuit can predict that the electroosmotic propulsion force is affected by the difference of electrical resistivity between medium solution and the HNB (Fig. 4). The electric voltage applied to the HNB by the external field can be estimated from a serially connected

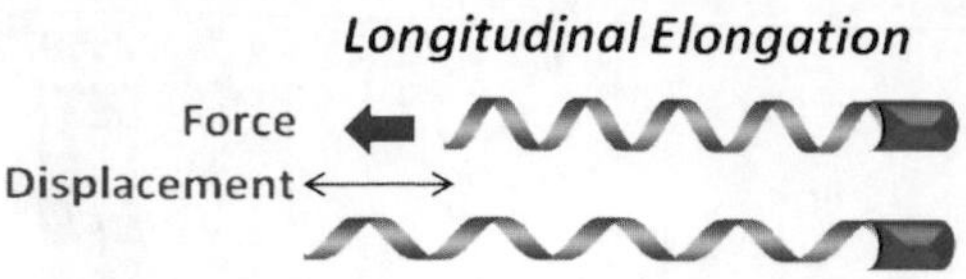

Stiffness K (N/m)	Displacement (μm)	Force (μN)
~ 0.009	0	0
0.297 ~ 0.378	4.13 ~ 9.95	1.56 ~ 2.95

Remotely Powered Propulsion of Helical Nanobelts, Fig. 3 Mechanical property of helical nanobelt with nonconstant stiffness for large-range force sensing

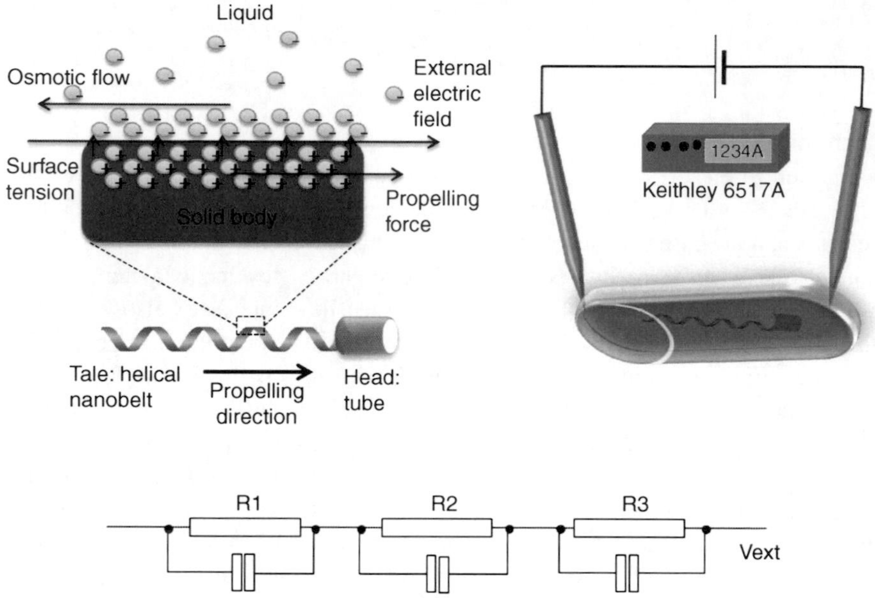

Remotely Powered Propulsion of Helical Nanobelts, Fig. 4 Schematic diagrams of the experiments for HNB swimming. A DC bias was applied between two electrodes located through axial direction of HNB body. Thus, the propulsion direction can be controlled and predicted by the configuration of electrodes. For controlling the direction of electric field, micromanipulators with tungsten probes were utilized to form electrodes

resistors model describing the ionic conductance through the liquid medium and the capacitors for the ionic layers. In a DC electric field, the resistance is the leading contribution. The resulting DC voltage of magnitude V_d induced in the HNB is modeled as

$$V_d = \frac{R_2}{R_1 + R_2 + R_3} V_{ext} \tag{2}$$

where V_{ext} is the DC input voltage applied to the probes in the petri dish, and R_1 (resistance between left electrode and HNB), R_2 (resistance of HNB), and R_3 (resistance between HNB and right electrode) are depicted in Fig. 4. Assuming that the resistance of the liquid is linearly proportional to the distance between the probes, Eq. 2 can be simplified as $V_d = E_{ext}\, l_d$, where l_d is the length of the HNB body and E_{ext} is the external DC field. The external DC field between the electrodes makes electroosmotic fluid flow along the HNB body.

Considering a liquid electrolyte, consisting of positive and negative particles in liquid solution, the interface between the electrolyte and the container wall generally forms a double layer in equilibrium, where a nonzero surface charge is screened by a very thin diffuse layer of excess ionic charge of width λ the Debye screening length (typically 1–100 nm). The double layer is effectively a capacitor skin at the interface, which has a small voltage ζ across called the zeta potential. The effect on the HNB and also the environment under electric field is further detailed in the following model which decomposes the part of intrinsic electrophoretic force directly generated to HNB and the electroosmotic flow from the surface interaction. Considering a tangential electric field E_{ii} applied in parallel to a flat surface, the electric field acts on ions in the diffuse part of the double layer, which drags the fluid to produce an effective slip velocity outside the double layer by the Helmholtz-Smoluchowski formula (Eq. 3) [18],

$$u_{11} = \left(\frac{2\varepsilon_0 \varepsilon_r \zeta_{HNB} H}{3\eta} - \frac{\varepsilon_0 \varepsilon_r \zeta}{\eta} \right) \vec{E} \tag{3}$$

where $\varepsilon_0\varepsilon_r$ and η are the electrolyte permittivity and fluid viscosity, which are constants (environment parameters). Increasing the speed of the HNB requires increasing ζ_{HNB} and H which are the zeta potential around the surface of the HNB and the Henri function, which depends on the geometry and hydrodynamic properties and the external field (E). On the negative term of the equation, the Zeta potential ζ on the borders of the surface is inversely proportional to the resulting speed, thus it should be minimized and controlled by applying experiments under closed microfluidic channel.

An important application of electroosmotic flow is capillary electroosmosis, where an electric field is applied down a capillary tube to generate a uniform plug flow (Eq. 3), driven by the slip at the surface. Typical flow speeds of 100 μm/s are produced by a field of 10 V/mm. This value will be used to compare the swimming performance of HNBs. It should be noted here that the electroosmotic propulsion does not depend on the size of the microagents. Thus, microscopic HNB could move theoretically about as fast as a macrosized one.

Electroosmotic Propulsion Experimental Setup

HNBs are detached from the substrate by micromanipulation. At the start of the experiments, the fixed end of produced HNBs are cut and released from chips using a setup with two probes with high positioning accuracy under an optical microscope. Same probes are also used to generate the electric field through the liquid medium and HNBs. Total experimental setup is shown in Fig. 5. Two nanorobotic manipulators (Kleindiek, MM3A) are installed under optical microscope; each has three degrees of freedom and, respectively, 5, 3.5, and 0.25 nm resolutions in X, Y, and Z directions. For the application of the electrical field and high resistance measurement, an electrometer with a DC power supply (Keithley 6517A) is used. Two probes attached to each side of manipulators are positioned in less than 1 mm distance. To avoid the optical reflection from the meniscus between probes and liquid medium interface, longer probes (~5 mm) with sharp tips (diameter around 100 nm) are used.

Controlled Swimming Experiments

Figure 6 demonstrates the forward and rotating propulsion of HNBs. Experiments are depicted in Fig. 6b and e. The swimming propulsions of HNBs are achieved by modifying the electrodes' configuration. A series of snap shots during forward (Fig. 6a) and rotating (Fig. 6d) propulsions were taken during the propulsion experiments. The trajectories of each swimming experiment are displayed in Fig. 6c and f. For trajectories, video analysis software and a custom-made C++ program are utilized. Rotation of HNBs was achieved by their aligning to the applied field direction. The probes attached to manipulators can change the configuration to generate the desired electric field gradient. Note that in this case the rotation was achieved using an HNB fixed in one extremity, while the other is free. The electrode configuration was fixed to make the HNB rotate. Furthermore, the forward and backward swimming direction can be converted without the complicated tumbling motion of *E. coli* bacteria.

It is confirmed that the swimming directions of HNBs could be easily controlled by differentiating the field gradient. For the steering control, the field gradient needs to be aligned to the rotational axes of the HNBs. In case of an untethered HNB, the rotation is achieved by controlling the field gradient. The high surface-to-volume ratio of HNBs increases the swimming efficiency by increasing the charged area in a given volume and mass.

The maneuverability of the currently demonstrated electroosmotic HNBs is still limited to the basic motions compared to the magnetic field based approaches. However, this issue can be addressed using an array of embedded electrodes.

Comparison of Swimming Performance

Electrokinetic Effect

To further understand the repeatable swimming behaviors due to an electroosmotic force and to find the optimized design parameters, especially the surface profile, the swimming performances of four different types of HNB were characterized. High- and low-doping HNB, a metallic (Cr/Ni) one and a dielectric counterpart were analyzed. Figure 7 shows the average and maximum propulsion velocity and acceleration normalized by the applied electric field. HNB performance was investigated in several steps (6–11 steps) ranging from 24 to 401 V/m. It was difficult to compare four HNBs in identical single field intensity because the experimental configuration of probes and HNB is

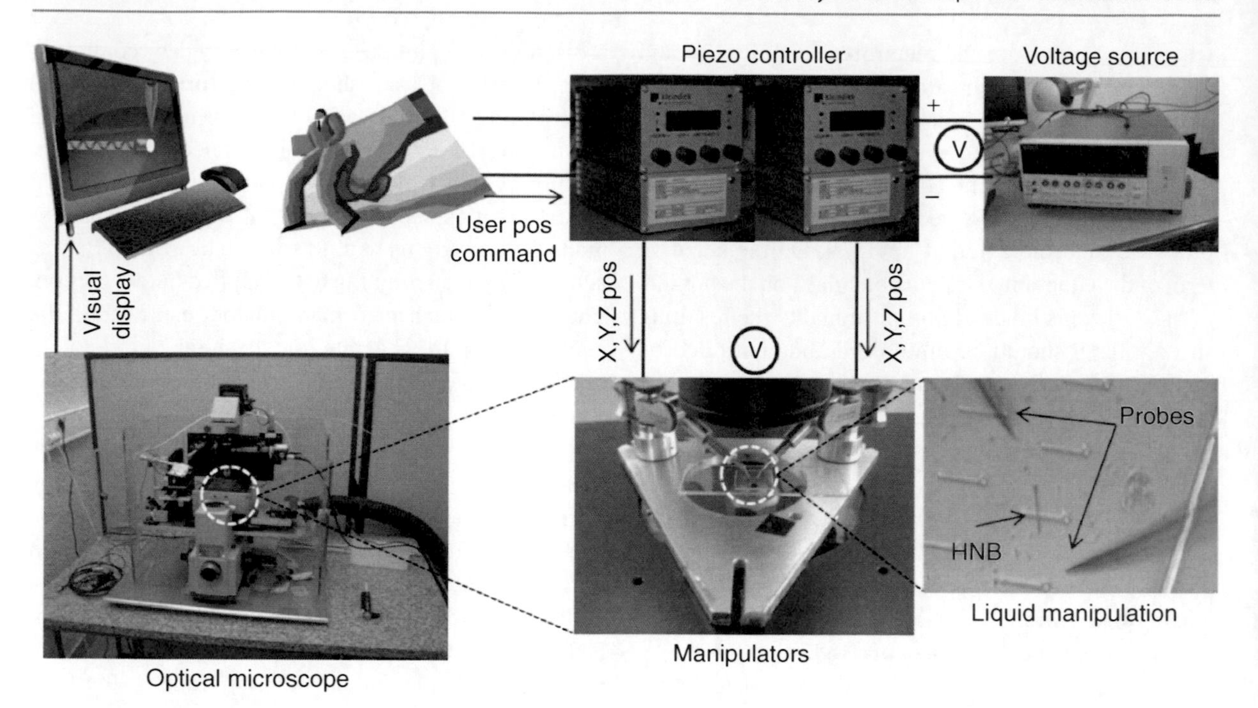

Remotely Powered Propulsion of Helical Nanobelts, Fig. 5 System setup for electroosmotic propulsion experiments: Optical microscope with a camera is used to give user visual feedback and thus proper manipulation in liquid. User operates the piezo controller to manipulate tungsten probes, which are used to manipulate HNBs and to apply the electric field generated by an external voltage source

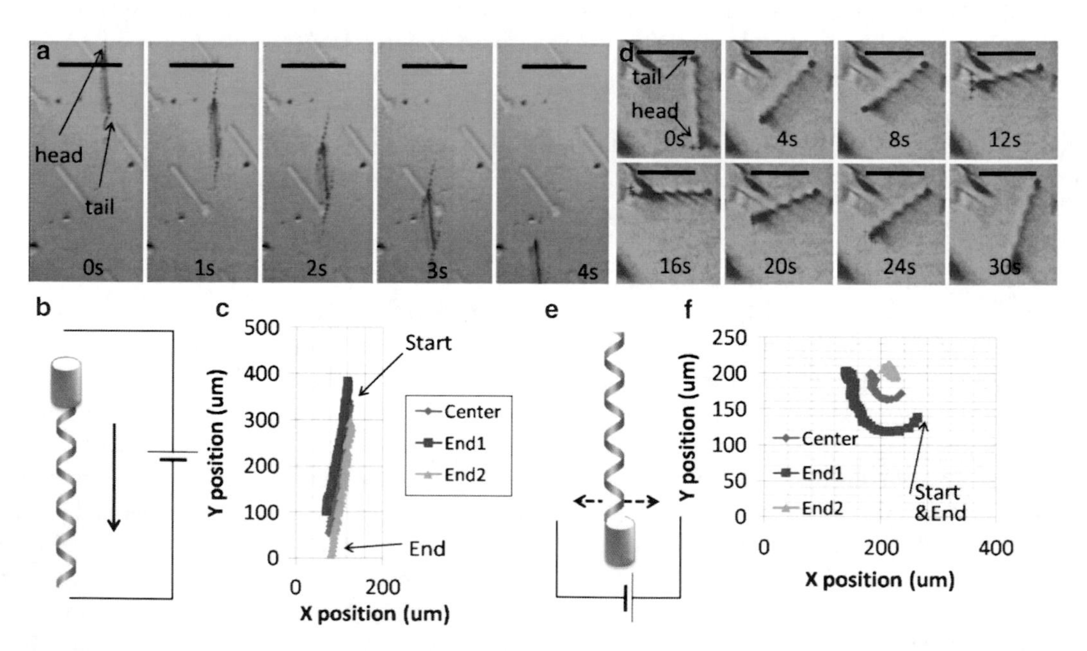

Remotely Powered Propulsion of Helical Nanobelts, Fig. 6 Different swimming propulsions of HNBs are achieved by modifying electrode configuration. Series of snap shots during backward (**a**) and rotating, (**d**) propulsions are taken during the experiments (**b** and **e**). The trajectories of each swimming experiment are displayed in (**c**) and (**f**). The scale bars in (**a**) and (**d**) are, respectively, 80 and 60 μm

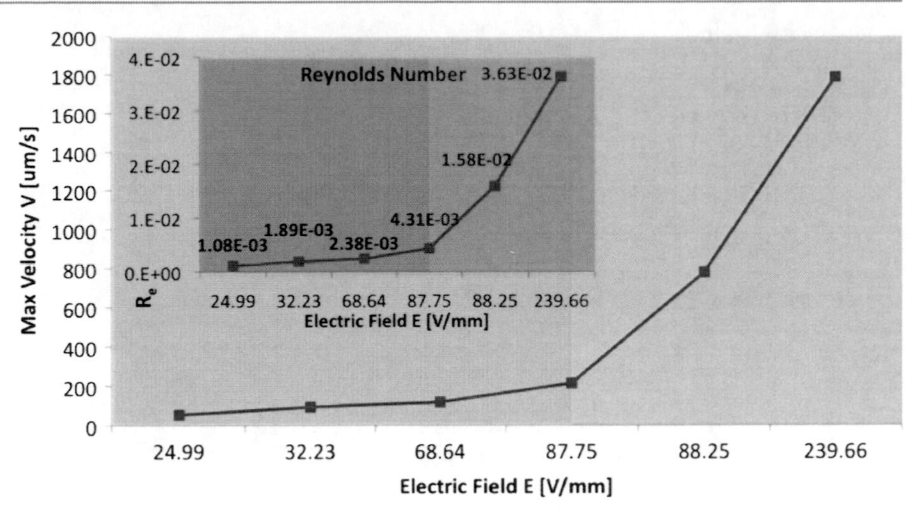

Remotely Powered Propulsion of Helical Nanobelts, Fig. 7 Dependence of HNB velocity on applied electric field intensity. The plot was obtained during the swimming experiments of a 74-μm-long HNB in isopropyl alcohol medium. The nonlinearity of the swimming performance is demonstrated at the region with higher electric field intensity than 87.75 V/mm. The *inset* figure shows the Reynolds numbers in different velocities versus applied electric field

difficult to maintain between each experiment. Therefore, a statistical average velocity was chosen to describe the general behavior of different HNB. It should also be noted that experiments on different types of HNBs were conducted in different days which might slightly change electrical resistivity of the liquid medium. Even though the exact resistance values of the utilized HNBs were not verified because of the difficulty in collecting them after experiments, the intrinsic resistance of HNB can be predicted as $R_{HNB4} > R_{HNB1} > R_{HNB2} > R_{HNB3}$. Hence, the average and maximum velocities and accelerations depend on the resistance of HNB as was expected from the Eq. 3. Highly doped HNB2 showed better swimming performance than low-doped HNB1. This can be explained by the fact that the InGaAs layer thickness (estimated as 8–10 nm) of HNB2 is slightly thinner than the one of HNB1 (11 nm). It results in better performance in HNB2 than HNB1 from the reduced gravity effect and hydrodynamic friction, regardless of the higher conductivity. It reveals that the doping concentration does not play much role in the swimming performance compared to the geometry or volume effect. It should also be noted that the HNB3 and HNB4 are slightly thicker than HNB1 and HNB2 because of additional metallic or dielectric layers (Cr/Ni 10/10 nm, AZ5214 photoresist around 50 nm Table 1).

However, the HNB3 had lower performance, while HNB4 was better than the others. In the case of HNB3, this can be explained by the increase in hydrodynamic friction during swimming which slow down HNB. Furthermore, in case of metallic HNB3, the gravitational force can also increase the surface interaction force by dragging it down to the substrate. The detailed calculation and comparison of buoyancy force and gravitational force are not described here; however, the resulting nonfluidic forces (as the difference between the gravitational force and the buoyancy force) on HNB3 is 0.94 pN, which is two times higher than on HNB1 (0.43 pN). These negative effects on HNB3 reduce its swimming performance compared to the others (Fig. 8). The result of HNB4 is contrary to the HNB3 while additional layer was deposited onto the surface. It can be explained by the fact that the nonfluidic force of HNB4 was decreased by the increased buoyancy force from the photoresist, which has much lower density than both gold and nickel. The result clearly shows the high dependency on the surface charge and thus zeta potential in dielectric surface coating to HNB compared to conductive or semiconductor materials (Fig. 8).

Swimming Performance Comparison with Other Swimmers

The force generated by the HNB's swimming is estimated from the drag it overcomes, considering its shape approximately as an ellipsoid [8].

$$F = \frac{4\eta\pi a v}{\left[\ln\left(\frac{2a}{b}\right) - \frac{1}{2}\right]} \tag{4}$$

where a and b are the dimensions along the longitudinal and lateral axes respectively, η is the viscosity of the liquid medium; and v is speed of the HNB. For

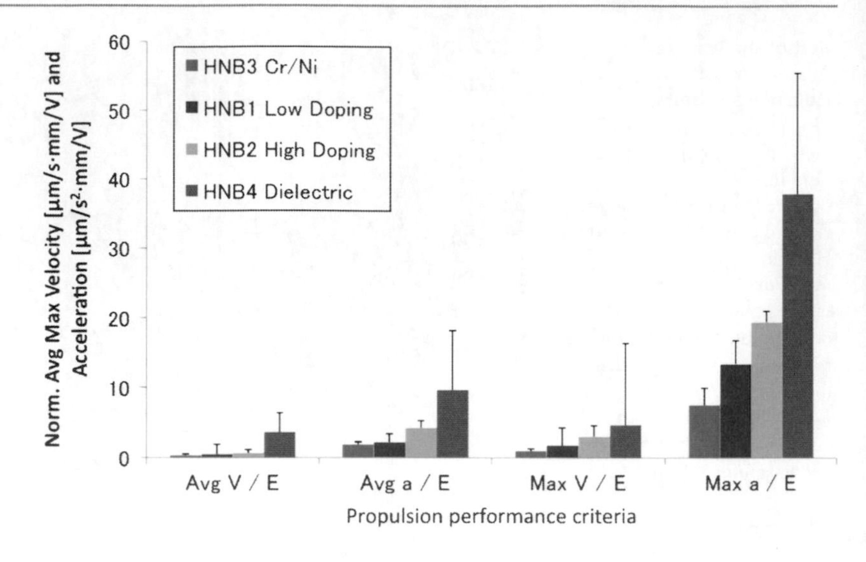

Remotely Powered Propulsion of Helical Nanobelts, Fig. 8 Dependence of normalized velocity and acceleration of propulsion on four types of HNBs. High- and low-doped semiconductor HNB showed higher swimming performance than metal (Cr/Ni)-coated HNB, and best performances are achieved with type 4 (dielectric)

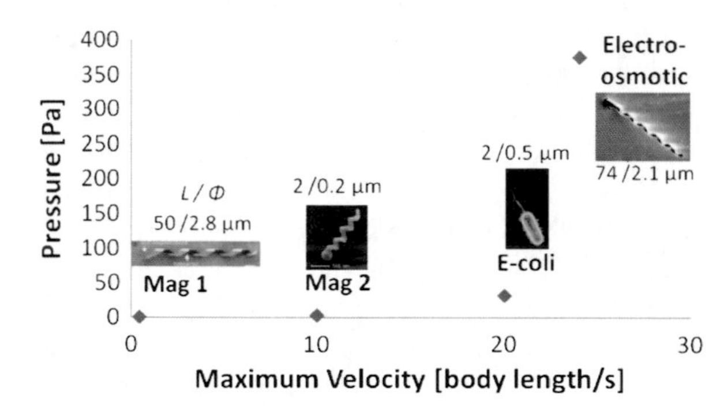

Remotely Powered Propulsion of Helical Nanobelts, Fig. 9 Comparison of swimming performances of swimmers: Magnetic1 [6], Magnetic 2 [8], *E. coli* [1, 19] and Electroosmotic HNB

a 74-μm-long HNB with a diameter of 2.1 μm moving at 1.8 mm/s in isopropyl alcohol medium, the generated force of 1.3 nN (Eq. 4) is an order of magnitude higher than the conventional optical trap force which is less than 200 pN. Furthermore, HNBs can apply a large pressure up to 375.5 Pa, which can be advantageous for local manipulation, such as on biological membranes.

Compared to two rotating magnetic propulsions of helical nanostructures, the demonstrated electroosmotic propulsion of HNBs showed much higher swimming performance in terms of maximum swimming velocity and manipulation force. The demonstrated velocity (1,785 μm/s) can reach 90 times faster and the manipulation force and pressure reaching at around 1.3 nN and 375.5 Pa which are much higher than others respectively (Fig. 9).

Conclusion

This entry introduced biologically inspired microswimmers based on helical nanobelts. Helical nanobelts microswimmers have excellent electromechanical and mechanical properties which can mimic their original model in nature. The demonstrated direct pulling of HNBs by electro-osmotic pumping was proven to be an efficient energy conversion mechanism for microscopic artificial swimming objects and overcomes easily the viscous drag and gravitational forces in an aqueous environment at low Reynolds numbers. Swimming velocities as fast as 1.8 mm/s (24 times the body length per second) and pressure as high as 375.5 Pa are achieved by HNBs propelled by an electro-osmotic force under an external electric field. These performances are higher than other state-of-the-art microscale artificial swimmers with physical energy conversion. Note that the velocity of 24 body lengths per second is even faster than natural bacterial propulsion which is their original model. Although helical morphology with a thin film is beneficial to such a fast swimming at low Reynolds numbers, the swimming performance appears to depend highly on the surface electrokinetic effects. Among the tested four different types, dielectric polymer coated HNB showed the best swimming performance compared to others with metallic or semiconductor surfaces. These HNBs can be used as wireless liquid manipulators, assemblers and nanorobots for biomedical or MEMS/

NEMS applications or remote physical or chemical detection by functionalizing with proper readouts.

Cross-References

- ► Biomimetics
- ► Dielectrophoresis
- ► Microfabricated Probe Technology

References

1. Berg, H.C.: *E. coli* in Motion. Springer, New York (2004)
2. Purcell, E.M.: Lift at low Reynolds number. Am. J. Phys. **45**, 3–11 (1977)
3. Sundararajan, S., Lammert, P.E., Zudans, A.W., Crespi, V.H., Sen, A.: Catalytic motors for transport of colloidal cargo. Nano Lett. **8**, 1271–1276 (2008)
4. Honda, T., Arai, K.I., Ishiyama, K.: Micro swimming mechanisms propelled by external magnetic fields. IEEE Trans. Mag. **32**, 5085–5087 (1996)
5. Behkam, B., Sitti, M.: Bacterial flagella-based propulsion and on/off motion control of microscale objects. Appl. Phys. Lett. **90**, 023902(1–3) (2006)
6. Zhang, L., Abbott, J.J., Dong, L.X., Peyer, K.E., Kratochvil, B.E., Zhang, H., Bergeles, C., Nelson, B.J.: Characterizing the swimming properties of artificial bacterial flagella. Nano Lett. **9**, 3663–3667 (2009)
7. Dreyfus, R., Baudry, J., Roper, M.L., Fermigier, M., Stone, H.A., Bibette, J.: Microscopic artificial swimmers. Nature **437**, 862–865 (2005)
8. Ghosh, A., Fischer, P.: Controlled propulsion of artificial magnetic. Nanostructured propellers. Nano Lett. **9**, 2243–2245 (2009)
9. Abbott, J.J., Peyer, K.E., Lagomarsino, M.C., Zhang, L., Dong, L.X., Kaliakatsos, I.K., Nelson, B.J.: How should microrobots swim? Intl. J. Rob. Res. **28**, 1434–1447 (2009)
10. Martel, S., Tremblay, C., Ngakeng, S., Langlois, G.: Flagellated magnetotactic bacteria as controlled MRI-trackable propulsion and steering systems for medical nanorobots operating in the human microvasculature. Intl. J. Rob. Res. **28**, 571–582 (2009)
11. Kosa, G., Jakab, P., Hata, N., Jolesz, F., Neubach, Z., Shoham, M., Zaaroor, M. and Szekely, G.: Flagella swimming for medical micro robots: theory, experiments and application. Proceeding of the 2nd IEEE RAS-EMBS International Conference on Biomedical Robotics and Biomechatronics, Scottsdale (2008), pp. 258–263
12. Chang, S.T., Paunov, V.N., Petsev, D.N., Velev, O.D.: Remotely powered self-propelling particles and micropumps based on miniature diodes. Nat. Mater. **6**, 235–240 (2007)
13. Gao, P.X., Mai, W., Wang, Z.L.: Superelasticity and nanofracture mechanics of ZnO nanohelices. Nano Lett. **6**, 2536–2543 (2006)
14. Hwang, G., Hashimoto, H., Bell, D.J., Dong, L.X., Nelson, B.J., Schon, S.: Piezoresistive InGaAs/GaAs nanosprings with metal connectors. Nano Lett. **2**, 554–561 (2009)
15. Wang, Z.L., Song, J.H.: Piezoelectric nanogenerators based on zinc oxide nanowire arrays. Science **312**, 242–246 (2006)
16. Xu, S., Qin, Y., Xu, C., Wei, Y., Yang, R., Wang, Z.L.: Self-powered nanowire devices. Nat. Nanotechnol. **5**, 366–373 (2010)
17. Karrai, K., Grober, R.D.: Piezoelectric tip-sample distance control for near field optical microscopes. Appl. Phys. Lett. **66**, 1842 (1995)
18. Hunter, R.J.: Foundations of Colloid Science. Oxford University Press, New York (2010)
19. Berg, H.C., Brown, D.A.: Chemotaxis in *Escherichia coli* analysed by three-dimensional tracking. Nature **239**, 500–504 (1972)

Replica Molding

► Microcontact Printing
► Nanoscale Printing

Resonant Evanescent Wave Biosensor

► Whispering Gallery Mode Resonator Biosensors

Retina Implant

► Artificial Retina: Focus on Clinical and Fabrication Considerations

---

Retina Silicon Chip

► Artificial Retina: Focus on Clinical and Fabrication Considerations

---

Retinal Prosthesis

► Epiretinal Prosthesis

Reversible Adhesion

► Gecko Effect

RF MEMS Switches

► Capacitive MEMS Switches

RF-MEMS/NEMS

► Nanotechnology

Riblets

► Shark Skin Drag Reduction

Rigorous Maxwell Solver

► Finite Element Methods for Computational Nano-optics

RNA Interference (RNAi)

► RNAi in Biomedicine and Drug Delivery

RNAi in Biomedicine and Drug Delivery

Michèle Riesen and Nuria Vergara-Irigaray
Department of Genetics Evolution and Environment, Institute of Healthy Ageing, University College London, London, UK

Synonyms

RNA interference (RNAi); Short-interfering RNA (siRNA); siRNA delivery

Definitions

RNA interference (RNAi) is the targeted down regulation of gene expression, most commonly on the translational level by introducing synthetic double-stranded or short interference RNA (dsRNA, siRNA) into the cell. Gene expression is then reduced by employing an existing cellular mechanism for processing the dsRNA into the active, single-stranded form incorporated in a complex called RISC, which targets and leads to cleavage of the specific messenger RNA (mRNA) substrate.

Biomedicine is the field of medicine that develops concepts or mechanisms discovered or used in basic research into clinical applications. This process is called *From Bench to Bedside*. In the context of RNAi or drugs that are delivered to the organism or target tissue by using nanoparticles, biomedicine is sometimes referred to as nanomedicine. The size of the particles ranges between 1 and 300 nm.

Overview

The recent advances in the understanding of the molecular mechanisms of RNA interference (RNAi) have led to the concept that dsRNA can be delivered into cells to modulate pathogenic functions of genes. This mechanism opens new possibilities to cure diseases since the underlying cause of nearly every pathological process is the dysregulation of one or several genes at different steps in a physiological process. This therapeutic application of RNAi as a new type of drug has been quickly embraced by many researchers as a promising strategy for diseases which were hitherto considered to be difficult or nigh impossible to treat. However, the delivery of dsRNA into cells is not a trivial task due to their large size of approximately 14 kDa and negative surface charge. Additionally, RNA has a rather short lifespan under physiological conditions due to the efficient removal by nucleases.

In order to successfully employ RNAi to target pathological processes, the stability of siRNA has to be addressed. This is generally achieved by introducing chemical modifications to the nucleotides without inducing toxicity or through the selected drug vesicle. Since RNAi is highly sequence-specific, an appropriate sequence in the gene of interest has to be found and preferentially tested using different siRNAs against the same target. Further important factors which should be controlled for are off-target effects (OTEs), for example, cleavage of other mRNA than the sequence-defined target, which can arise despite the precision of the mechanism, and activation of the immune response which can be induced both sequence dependently and independently. The major challenge is still the delivery of the siRNA to the target organ for the treatment. This then expands the list of considerations to the choice of delivery vehicles and routes of administrations. For an excellent and comprehensive introduction into the field please see [11, 16].

Some of these issues, in particular the delivery of the siRNA, are solved by combining the expertise of material, chemical, and physical science to devise efficient vectors for delivery of the dsRNA into the target tissue. The development of carrier molecules in the nano-range, such as polymers, liposomes, peptides, and aptamers, goes hand in hand with the design of new therapeutics for various pathologies, such as cancer, genetic, and viral diseases. Notably, in less than 10 years since the discovery of RNAi, several drugs have been successfully developed and are currently tested in Phase I–III clinical trials in the United States (www.clinicaltrials.gov).

It is worth mentioning that RNAi can originate from either synthetic siRNA, which has to reach the cytosol for final processing and activation, or as plasmid-borne short hairpin RNA (shRNA). These plasmids must be delivered to the nucleus where the RNA is transcribed and undergoes a first step of processing before it is transported to the cytosol. This type of RNAi is generally carried by viral vectors and has been subject of extensive investigation. However, the focus of this review is nonviral delivery of siRNA; hence, the reader is referred to elsewhere for a further discussion of this area. Finally, bacteria have recently been employed as delivery vehicles for shRNA or siRNA in a process termed transkingdom RNAi (tkRNAi) or bacteria-mediated RNAi (bmRNAi).

In this entry, the basic cellular mechanism of RNAi will be introduced. Important criteria for the design of siRNA such as sequence design and chemical modifications will be discussed together with observed off-target effects (OTEs). This will be followed by an overview over physiological barriers and approaches chosen to deal with these challenges. Further, different types of delivery strategies will be presented which are currently employed in clinical applications according. Finally, how nanotechnology is integrated with RNAi will be discussed with an overview and discussion of a selection of particles presently utilized for in vitro, in vivo, and clinical trials.

Basic Mechanisms

RNA Interference: A Shared Strategy to Defend the Genome

In 2006, Andrew Fire and Craig Mello were awarded the Nobel Prize in Medicine or Physiology for their discovery of gene silencing through double-stranded RNA (dsRNA) in *Caenorhabditis elegans (C. elegans)*. Notably, their work describing this phenomenon had only been published for 8 years before they received the honors [5]. The observation that gene silencing could be achieved by the introduction of plasmid DNA into plants was first described some 20 years ago, but the importance of this curious finding was not recognized until it was reported by Andrew Fire and colleagues in one of the most widely used model organisms, the nematode worm *C. elegans*.

Although the exact mechanisms of RNA interference (RNAi) were unknown when it was first reported, not much time passed before it became clear that the strategies and functions of RNAi are conserved in plants, invertebrates and vertebrates, as well as many of the components of the pathway. Once long dsRNA has entered the cell, it is trimmed into small duplexes of approximately 20–25 nucleotides in the cytosol by an RNase III enzyme called DICER1. This step is followed by integration into a complex called pre-RNA induced silencing complex (pre-RISC), also thought to be mediated by DICER1, where further processing into single-stranded RNA (ssRNA) takes place. RISC contains DICER1 and several proteins from the Argonaute family, AGO1–4. The passenger (sense) strand is then degraded and the mature RISC is formed with the guide (antisense) strand integrated into RISC. The guide strand is complementary to mRNA and determines the specificity of RISC. When RISC encounters its target mRNA and perfect Watson–Crick base-pairing occurs, the mRNA is cleaved by the RISC-member AGO2. If the base-pairing is imperfect, the protein translation is disabled, but the mRNA is not cleaved by AGO2. This complex can be assembled stably over an extended period of time and catalytically regulate gene expression on the translational level (Fig. 1).

Since the discovery of RNAi, the knowledge of the field has progressed tremendously. Species possess a vast arsenal of their own small RNAs, for example, microRNAs (miRNA) or PIWI-interacting RNAs (piRNAs). The different classes of RNAs fulfill

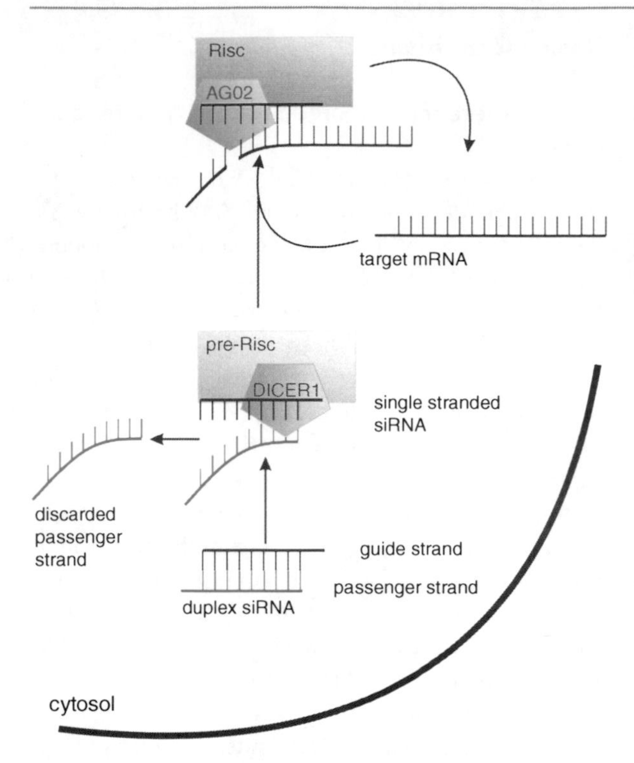

RNAi in Biomedicine and Drug Delivery, Fig. 1 siRNA-induced silencing. Double-stranded siRNA is delivered to the cytosol, where it is recognized by DICER1, and incorporated into a pre-RISC complex upon which the duplex is separated and the passenger strand discarded. This processing step is also mediated by DICER1. The mature RISC complex is then formed and guided to target mRNA. Perfect match interactions between the guide (antisense) strand and mRNA lead to cleavage of the substrate mRNA by AGO2. A fully assembled RISC complex is stable and can catalytically inactivate a large number of mRNA. The model is a simplified depiction of the siRNA pathway only and omits transport of siRNA into the cell as well as other small RNA processing pathways. The nomenclature refers to human proteins

different functions such as regulating gene expression at the translational level, or they are expressed in specific tissues, for example, the germ line or neurons. Furthermore, RNA was found in the nucleus where it was found to regulate gene silencing at the transcriptional level by modulating histone modifications or DNA methylation levels. Further evidence proposes RNAi as some kind of nuclear pest control to defend the genome against insertion of mobile elements such as L1 transposons and viral DNA. Despite the rapid succession of discoveries of RNAi-governed processes, the detailed mechanisms by which these regulations are executed remain yet to be elucidated. Nevertheless, the impact the study of RNAi has had on the understanding of gene regulatory processes over the past decade must not be diminished. For comprehensive and recent reviews over the different pathways, processes, and the components of RNAi please refer to [12, 19].

Design of Synthetic Short Interference RNA for Clinical Applications

Sequence Design

Intensive research into sequence design resulted in a set of criteria that greatly facilitates the development of suitable RNA duplexes for any given gene. There are a number of web tools available from either commercial or academic sources in order to identify appropriate RNAi target sequences [15]. The basic structure of the siRNA generally contains a central domain of 19 basepairs with an overhang of two bases at the 3′-end and a 5′-phosphate group. The reason for this type of synthetic siRNA to be common is because it is identical with the end product of the cellular RNAi pathway and DICER1. However, other siRNA varying in length between 19 and 27 basepairs and with either blunt or asymmetric ends have been shown to be efficient in vitro and to be processed by DICER1. Secondary structures of the siRNA as well as the target sequence should be considered together with the overall thermodynamic properties (T_m), which influence loading into RISC. As a rule, siRNA is more efficient when the 5′-end of the guide strand has a lower melting temperature than the 5′-end of the passenger strand by adjusting the A/U and G/C content, respectively. The positioning of individual bases has also been found to have an effect on efficiency of siRNA [1, 15].

Chemical Modifications to the Sugar or Phosphodiester Backbone

Chemical modifications of the dsRNA in order to increase the stability (T_m) under physiological conditions or reduce the susceptibility to RNase attacks include the introduction or replacement of various reactive groups on the base, sugar or phosphodiester backbone of the RNA. In general, the sugar backbone is modified at the 2′*O*- position, for example, to 2′-Fluoro. This modification does not interfere with loading into RISC or cleavage by DICER1. Another commonly used modification is 2′-*O*-Methyl. This nucleotide is occurring naturally in tRNA or ribosomal components and insertion of these modified nucleotides has been shown to abolish the induction of the innate immune system in mice. Locked nucleic acids (LNA), where the sugar ring is locked in the 3′-*endo* conformation due to the presence of a methylene bridge between the 2′-*O* and the 4′-C of the ribose, have been reported to increase stability successfully.

Modifications to the phosphodiester backbone include, for example, phosphorothioate, which has been shown to increase T_m and resistance to nuclease-mediated degradation. This modification can be easily integrated at defined positions, but it has been related to toxicity, loss of function, and nonspecific protein binding if large numbers of nucleotides in the siRNA are replaced with this modification. The replacement of the phosphodiester bond with a boranophosphate greatly increases stability of the dsRNA, but synthesis of this type of backbone requires in vitro transcription and the position of the insertion of boranophosphate can therefore not be defined. For excellent and detailed reviews of the design of siRNA and modifications, please refer to [1, 2, 8].

Off-Target Effects

A serious concern when developing siRNA as therapeutics are OTEs. They arise most commonly for the following reasons: activation of the innate immune system, sequence homology with other mRNAs additionally to the specific target, and when siRNA enters the cellular miRNA pathway. Initially, it was assumed that the short RNA duplexes could not induce an immune response. However, with the increasing number of work employing siRNA in vitro and in vivo, siRNAs were reported to stimulate the innate immune system and induce a strong interferon response. This activation has been shown to be both sequence dependent and independent, as well as cell-type specific. Intracellular receptors from the Toll-like family, in particular TLR3, 7, and 8 recognize RNA, either dsRNA (TLR3) or ssRNA (TLR7 and 8), and mediate the induction of IFN-α or IFN-β. These receptors are found either on intracellular organelles such as the endosome or lysosomes, as well as cells that constitute the innate immune system such as B cells or monocytes. Under given circumstances, siRNA can therefore induce the innate immune system [22].

In order to avoid OTEs based on sequence homology, it is recommended to carefully screen the transcriptome for similarities with the required siRNA. It is important to note that although single mismatches have been reported to abolish the efficiency of siRNA, short similarities of five to eight bases can direct RISC to unintentional targets. If siRNA engage in the miRNA pathway, the specific effect of siRNA on one gene is replaced by effects on many genes due to the nature of miRNA-mediated gene regulation. Generally, OTEs are dose-dependent and can therefore be circumvented by establishing dose–response curves and choosing the lowest efficient dose. Since it has been shown that downregulations of the targeted gene transcripts were in fact due to OTEs and not to the direct mediation of the administered siRNA, it is strongly recommended to test at least two or three different siRNAs against the same gene. This should result in reproducible effects and confirm that the expected and observed downregulation of the transcripts is due to the direct effect of siRNA [8, 15].

Delivery of Synthetic Short Interference RNA for Clinical Applications

Physiological Barriers

After the successful design, synthesis, and in vitro tests of a suitable siRNA compound, the next and likely more challenging obstacle to be overcome is the delivery of siRNA to its site of action. There is a consensus in the field that the vast potential of RNAi cannot be fully exploited yet because of the lack of efficient delivery vehicles to the target tissue. The body has various highly efficient strategies for clearance of foreign particles from the system, which are briefly described below in order to understand the difficulties in delivery and how they are addressed by the choice of appropriate vesicles or siRNA modification or both combined. Tremendous efforts are directed toward the development of a range of particles with different physicochemical properties which are equipped to reach the target organ and pass the physiological barriers [7].

If siRNAs are not immediately degraded by free nucleases then the next barrier is renal clearance, which is thought to determine the half-life of modified or naked siRNA in vivo. The small size of around 14 kDa, which is far below the cutoff of 40 kDa for glomerular filtration, and the hydrophilic nature of siRNA target these molecules for removal by the kidneys. The next trial is the clearance via the reticuloendothelial system (RES). The RES consists of mobile, circulating phagocytic cells, such as monocytes, or resident tissue macrophages that remove pathogens and cellular debris which remains from apoptosis, wound healing, or tissue remodeling. Large numbers of RES-associated cells reside in the liver, the Kupffer cells, and in the spleen or generally in other tissues with high blood flow and fenestrated vasculature.

Should carriers have successfully dodged clearance by any of the above-described pathways, they will then encounter the endothelial barrier lining the vascular lumen. Endothelial cells are tightly connected with the underlying extracellular matrix through integrins and they are joined together via a number of strictly regulated junctions. Endothelial cells are also perforated with a large number of small pores of about 45 Å and rare large pores of 250 Å. Most of the transport of vesicles or siRNA is thought to be mediated via leaks in junctions rather than the pores themselves; however, there is some controversy regarding this matter. In many tissues, particles larger than 5 nm will not be able to extravasate through the endothelial lining, whereas organs such as the liver or spleen are easily accessible because of the extensive vasculature of these organs.

One specific type of endothelial barrier is the blood brain barrier (BBB) at the boundary to the astrocytes and pericytes where the endothelium forms an exceptionally tight lining to protect the brain. As progress in research into neurodegenerative diseases such as Alzheimer's reveals an ever-increasing number of genes which are potentially "drugable," the brain becomes an attractive target for siRNA. Under normal physiological conditions, this barrier is intact. However, the integrity of the BBB can be compromised during disease or through administration of certain drugs. Also, specific endothelial reporters have been described which could be employed to aid passage of particles across the BBB to the brain. Therefore, development of carriers that can deliver drugs across the BBB will become increasingly important to treat this kind of diseases.

The final step for successful delivery is the cellular uptake and, importantly, endosomal release into the cytosol. In general, particles enter a cell via receptor-mediated endocytosis. Recognition and binding to the receptor induces the formation of a vesicle which leaves the cell membrane on the cytosolic site. From there, the vesicles undergo a series of intracellular trafficking steps and maturation until, finally, late endosomes or lysosomes have been formed. The last challenge the particles face is the release from the endosome into the cell. The *proton sponge effect* has been described as a mechanism by which the trapped drug can exit the endosome. Briefly, acidification of the endosome leads to protonation of amine groups on the vesicle. This leads to an influx of protons and chloride ions, followed by water to equilibrate the changes in osmolarity, eventually causing rupture of the vesicle and release of its contents into the cytosol. For further reading, please see [7, 20].

Types of Delivery

In vivo application of siRNA can be classified, for example, into active or passive systemic and local/cellular delivery. Passive systemic delivery does not require any specific targeting sequences or molecules. It is based on the property of nanoparticles to accumulate in certain tissues under certain conditions. For example, nanoparticles or naked siRNA have been shown to accumulate in liver or spleen because of the specific structure of the endothelium, with large fenestrations of the microvasculature of these organs. Delivery of vesicles to inflamed tissue is thought to be mediated by the RES. Accumulation of particles in tumors has been attributed to the *enhanced permeability and retention effect* (EPR) in tumors. This is thought to be caused by reduced lymphatic drainage, leaky microvasculature, and enhanced angiogenesis of cancerous tissue.

Active or targeted systemic delivery is more complex because it requires the carrier to be equipped with components such as antibodies, ligands against cell-surface receptors, or peptides from viral proteins targeting the nanoparticle to the desired cell type or tissue. Antibodies are used as whole entities, or fragments thereof, and they are generally directed against receptors which are increasingly expressed on tumor cells. Receptors that have been targeted include, for example, the epidermal growth factor receptors (EGFR) or human epidermal growth factor receptor-2 (HER-2), the iron-binding protein transferrin (Tf), which targets the transferrin receptor (TfR) responsible for iron uptake, as well as Toll-like receptors (e.g., TLR3). Successfully employed peptides are the rabies virus glycoprotein (RVG) or arginine-glycine-asparagine (RGD). Sugars, i.e., glucose or mannose, are recognized by a number of cell-type specific receptors and have also been successfully employed in delivery to specific cell types or across the BBB.

Naked siRNA has been injected using the hydrodynamic delivery method into mouse tail veins and was subsequently found to accumulate in the liver and to a lesser extent in other organs such as the lungs. This method requires large volumes to be rapidly injected, and although successfully employed in mice and pigs, it is doubtful whether this way of administering systemic siRNA would be of therapeutic use in humans. The volume for injection needed is around 10% of the body weight, and this would most likely lead to heart failure and death. Direct injection of siRNA is feasible if the target is easily accessible, such as is used, for example, for treatment of wet age-related macular disease or other diseases of the eyes where siRNA is injected directly into the eye. Several Phase I–III clinical trials have been approved by the US FDA. Other means of delivery that have been tested include topical administration which is useful for exposed sites, such as the skin. Inhalation is the means of choice for treatment of respiratory tract diseases. Electroporation has been used for delivery into muscle tissue. For more detailed information, please see [14, 18, 21, 22].

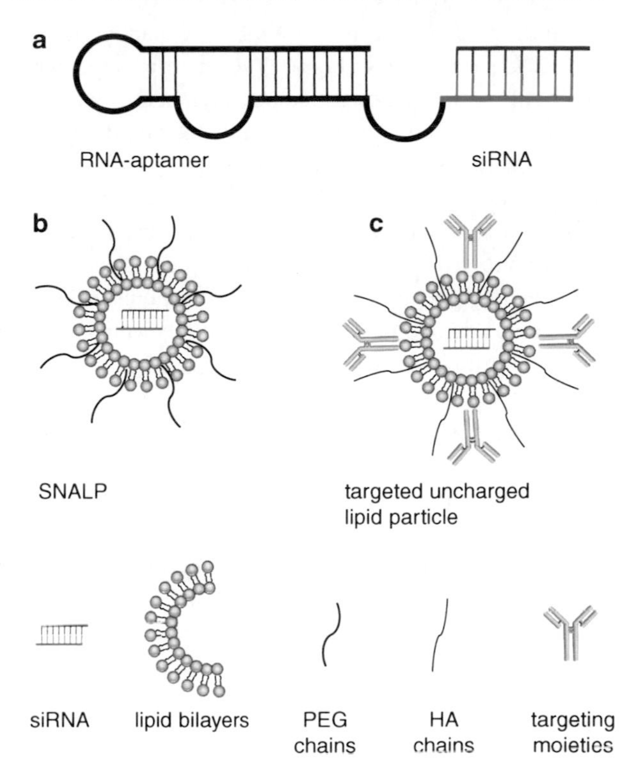

RNAi in Biomedicine and Drug Delivery, Fig. 2 Delivery vehicles for siRNA administration. Three types of vesicles are shown for delivery of siRNA. (**a**) RNA-aptamer with siRNA ligated. (**b**) Stable nucleic acid particle decorated with PEG (SNALP). (**c**) Uncharged lipid particle decorated with targeting moieties, i.e., single chain antibody fragment directed against Tf and HA chains

Delivery Vesicles and Nanoparticles

Considerations for the Design of an siRNA Carrier Investigation into suitable carriers over the past decade has resulted in a vast amount of materials and combinations of several components which all have advantages and disadvantages for delivery of specific cargo, the route of administration, and the target tissues. Criteria to be considered are the size and size distribution of particles, which is, for example, inversely related to cellular uptake and delivery to tumor tissue. Since the smallest microvessels have a diameter of 5–6 μm, particles must be considerably below this diameter in order to avoid accumulation and clogging which could lead to embolism. The surface properties of nanoparticles dictate whether they will be recognized and removed by the host immune system; non-modified hydrophobic particles are generally swiftly removed, whereas vesicles decorated with biodegradable materials which are hydrophilic such as polyethylene glycol (PEG) or polysorbate 80 (Tween 80) are better tolerated.

Drug loading capacities and the release characteristics are further important points. Loading can be achieved by two methods. The drug can be incorporated at the time of nanoparticle assembly or it can be absorbed after the formation of the carrier by incubation in a concentrated drug-solution. Highly efficient loading has been shown for siRNA due to the negative charge of the oligonucleotide and positively charged vesicles, such as cationic lipids or polymers. Drug release depends, for example, on the degradation of the carrier or how fast the drug can diffuse across the polymer membrane, as well as the solubility of the drug. A concise selection of the most frequently and successfully employed carriers is briefly described below. For a detailed description of these design considerations, please see [4, 18]. The reader is also referred to cross-referenced entries which elaborate in more detail the properties and design of nanoparticles specifically for drug delivery.

Aptamers Among biological macromolecules RNA has a unique place because of the ease of manipulation and synthesis similar to DNA, which is combined with a structural versatility found in proteins. These useful properties predestine RNA to be used itself as a building block for the assembly of nanoparticles called RNA-aptamers (Fig. 2a). Generally, aptamers are highly structured oligonucleotides that are obtained by SELEX (systematic evolution of ligands by exponential enrichment) for the required purpose. This is a protocol whereby a library of RNA-aptamers is subjected to iterative rounds of counter-selection and selection to yield aptamers which recognize, for example, specific receptors on target cells, proteins, as well as organic compounds. The prime features which make aptamers attractive carriers are their small size of a few nanometers compared to cationic polymers or liposomal carriers (upto 300 nm), that modified nucleotides can easily be integrated to increase stability in physiological conditions, they have low toxicity and are biodegradable. Instances where aptamers have been employed as carriers for siRNA include murine xenografts of prostate cancer where tumor growth and regression was mediated by aptamer-mediated siRNA delivery, or treatment against viral infection by either targeting HIV gp120, a glycoprotein which recognizes and utilizes the CD4 receptors on T lymphocytes for entry, or the CD4 receptor itself [6, 10, 23].

Cationic Polymers and Dendrimers The cationic polymer cyclodextrin (CDP) has been widely and successfully used as a carrier directed in particular against solid tumors. Advantages of this material are the low toxicity; it is not metabolized in humans and does not induce an immune response. CDPs are commonly of a tri-composite structure, where CDP is employed as the carrier for siRNA due to strong ionic interactions between the cationic polymer and the anionic nucleic acids. The particle is stabilized with the hydrophilic cycloalkane adamantane conjugated to PEG (AD-PEG). Targeting of the vesicle to tumors is achieved by adding an AD-PEG-Transferrin (AD-PEG-Tf) moiety because Tf is expressed abundantly in a number of tumors. This formulation carrying siRNA against the ribonucleotide reductase subunit 2 (RRM2), called CALAA-01 by Calanda Pharmaceuticals, was approved for Phase I clinical trials by the FDA in 2008. The reader is referred to [3, 9] for a comprehensive review on the development of CALAA-01.

Another cationic species that is frequently used for similar properties is polyethylene-imine (PEI), which can be synthesized to be of linear or branched structure and is then classified as a dendrimer. PEI efficiently forms stable nanoparticles with oligonucleotides and promotes endosomal release via the *proton sponge effect*. Because of its branched structure, it can carry a large payload. Despite the supreme quality of PEI as a siRNA delivery vesicle, its identified toxicity toward most cells is not a desirable side effect. Three approaches have been taken to address the toxicity issue of PEI. First, the dendrimer is equipped with targeting moieties such as RGD to reduce the toxicity to nontarget tissue. Second, maintaining the particle size below 10 kDa substantially reduced toxicity compared to larger particles of 25 kDa. Finally, ketalized carriers have been synthesized with the added advantage that these linkages are sensitive to acidic pH, and hence readily dissolve in endosome to release the cargo [13].

Liposomal Carriers Lipids as gene delivery vehicles have been used since the early 1980s. Cationic lipids are often used as vesicle due to the facile handling and preparation since they readily associate with the anionic nucleic acids. Moreover, routinely used reagents for in vitro transfection of cell culture are based on cationic liposomes, such as *Lipofectamine 2000* by Invitrogen or other equivalent products. The most commonly used cationic lipid is 1,2-dioleoyl-3-trimethylammonium-propane (DOTAP) because it efficiently associates with nucleic acids. However, DOTAP was again shown to be cytotoxic; hence, this carrier was equipped with modifications which reduce this side effect, such as lactosylation. Another strategy to inhibit the growth of pancreatic xenografts on a murine model involved DOTAP decorated with histidine-lysine peptides. This facilitated endosomal escape and single-chain antibody fragment directed against Tf, and this vesicle did not induce an IFN-response and could be administered at low doses due to the target specificity.

Another promising type of vesicle based on cationic lipids is the stable nucleic acid-lipid particle (SNALP) (Figure 2b). The siRNA is embedded within the lipid envelope and the composition of SNALPs contains reduced cationic lipid content and is PEG-ylated. SNALPs have been subjected to extensive study as siRNA carrier against in vivo viral models with considerable success of full protection against infection of guinea pigs with the deadly Zaire strain of the Ebola virus. Another trial employed SNALPs with siRNA against apolipoprotein B levels in nonhuman primates, which was well tolerated and showed reduction of Apolipoprotein B (ApoB) and serum cholesterol among other physiological improvements. Following on from these promising results, several therapeutic siRNA-SNALP formulations against tumors or metabolic diseases are now in preclinical or have been admitted to Phase I clinical trials, and targets include VEGF and ApoB among others [8]. Further information on these trials can be found on www.clinicaltrials.gov or the web sites listed in section Further Reading.

Since cationic liposomes have been associated with significant toxicities, development of vesicles composed of uncharged lipids has been resumed with considerable success. One lipid in particular has emerged as a promising carrier: 1,2-dioleoyl-sn-glycero-3-phosphatidylcholine (DOPC). These vesicles successfully reduced tumor xenografts of different origins. Targeted vesicles against the thrombin receptor (TR) expressed on some tumors greatly reduced not only TR but also angiogenic and invasive tumor markers, such as vascular enodothelial growth factor (VEGF) and matrix metalloproteinase 2 (MMP2). Techniques developed in order to maximize the payload of these nanoparticles utilize lyophilization followed by rehydration with siRNA. The half-life in circulation under physiological conditions of these particles was increased by surface coating with hyaluronic acid (HA), a naturally occurring glucosaminoglycan with properties equivalent to PEG (Figure 2c). Further information can be found in [17].

Other Types of Carriers: Atellocollagen, Cationic Cell Penetrating Peptides (CPPs), Polylysine There are a number of other options for building siRNA-nanocarriers. Atellocollagen is a preparation of digested cationic calf dermal collagen successfully employed and well tolerated in animal models of inflammatory disease or tumor xenografts. Despite the fact that the studied atellogen-siRNA nanoparticles were not decorated with a specific targeting moiety, they were found to accumulate in tumors and inflamed areas presumably due to the EPR effect in these tissues. CPPs are short peptides with a high content of positively charged amino acids, in particular arginine, which can be synthesized or are derived from naturally occurring peptides, e.g., TAT from HIV-1 to facilitate cellular uptake. Initially thought to be fully soluble, it appears that their efficiency is due to aggregation into nanoparticles. Successful trials in vivo used polyarginine fused to cholesterol and siRNA against VEGF in a tumor xenograft. Another polyarginine-based CPP coupled with RVG was found to cross the BBB and reduce green fluorescent protein expression in a mouse model. Another type of carrier is based on polylysine (K) stretches. Initial problems with toxicity such as complement activation were overcome by PEG-modifications and the addition of histidine (H), which facilitates endosomal escape. Further, it was found that the efficiency as siRNA carriers was improved by increasing the histidine ratio compared to lysine [9].

Further Reading

Currently, 20 studies are listed on www.clinicaltrials.gov when searched for the keyword siRNA. This clearly shows that the challenges of delivery posed by the physiological barriers or physicochemical properties of the particles can be overcome. The marriage between advances in material sciences and the developing understanding of the physiological and pathophysiological processes culminates in the development of novel RNAi-based therapeutics for the treatment of diseases ranging from viral to inherited and age-related pathologies.

Due to space restraints, a number of additional excellent reviews could not be referenced in this entry; however, they can be found when searching Pubmed or Web of Science with keywords such as "siRNA AND nanotechnology." Further information on research and development of siRNA-based therapeutics can be found on the web sites of the leading companies.

http://www.alnylam.com/
http://www.calandopharma.com/
http://www.opko.com/
http://www.quarkpharma.com/
http://www.silence-therapeutics.com/
http://www.sirna.com/
http://www.tekmirapharm.com/Home.asp

An insightful and curated web site on the recent development of RNAi therapeutics, career opportunities, consulting, and overview over the market and enterprise portfolios can be found here:

http://rnaitherapeutics.blogspot.com/

Cross-References

- ▶ Genotoxicity of Nanoparticles
- ▶ Lab-on-a-Chip for Studies in *C. elegans*
- ▶ Liposomes
- ▶ Microfabricated Probe Technology
- ▶ Nanomedicine
- ▶ Nanoparticle Cytotoxicity
- ▶ Nanoparticles
- ▶ Nanotechnology
- ▶ Nanotechnology in Cardiovascular Diseases

References

1. Behlke, M. A.: Chemical modification of siRNAs for in vivo use. Oligonucleotides **18**, 305–319 (2008)
2. Behlke, M. A.: Progress towards in vivo use of siRNAs. Mol. Ther. **13**, 644–670 (2006)
3. Davis, M. E.: The first targeted delivery of siRNA in humans via a self-assembling, cyclodextrin polym. based nanopart: from concept clinic. Mol. Pharm. **6**, 659–668 (2009)
4. Fattal, E., Barratt, G.: Nanotechnologies and controlled release systems for the delivery of antisense oligonucleotides and small interfering RNA. Br. J. Pharmacol. **157**, 179–194 (2009)
5. Fire, A., Xu, S., Montgomery, M. K., Kostas, S. A., Driver, S. E., Mello, C. C.: Potent and specific genetic interference by double-stranded RNA in Caenorhabditis elegans. Nature **391**, 806–811 (1998)
6. Guo, P.: RNA nanotechnology: engineering, assembly and applications in detection, gene delivery and therapy. J. Nanosci. Nanotechnol. **5**, 1964–1982 (2005)
7. Juliano, R., Bauman, J., Kang, H., Ming, X.: Biological barriers to therapy with antisense and siRNA oligonucleotides. Mol. Pharm. **6**, 686–695 (2009)
8. Lares, M. R., Rossi, J. J., Ouellet, D. L.: RNAi and small interfering RNAs in human disease therapeutic applications. Trends Biotechnol. **28**, 570–579 (2010)
9. Leng, Q., Woodle, M. C., Lu, P. Y., Mixson, A. J.: Advances in systemic siRNA delivery. Drugs Future **34**, 721 (2009)
10. Levy-Nissenbaum, E., Radovic-Moreno, A. F., Wang, A. Z., Langer, R., Farokhzad, O. C.: Nanotechnology and aptamers: applications in drug delivery. Trends Biotechnol. **26**, 442–449 (2008)
11. Martin, S. E., Caplen, N. J.: Applications of RNA interference in mammalian systems. Annu. Rev. Genomics Hum. Genet. **8**, 81–108 (2007)
12. Moazed, D.: Small RNAs in transcriptional gene silencing and genome defence. Nature **457**, 413–420 (2009)
13. Mok, H., Park, T. G.: Functional polymers for targeted delivery of nucleic acid drugs. Macromol. Biosci. **9**, 731–743 (2009)
14. Oliveira, S., Storm, G., Schiffelers, R. M.: Targeted delivery of siRNA. J. Biomed. Biotechnol. **2006**, 63675 (2006)
15. Peek, A. S., Behlke, M. A.: Design of active small interfering RNAs. Curr. Opin. Mol. Ther. **9**, 110–118 (2007)
16. Raemdonck, K., Vandenbroucke, R. E., Demeester, J., Sanders, N. N., De Smedt, S. C.: Maintaining the silence: reflections on long-term RNAi. Drug Discov. Today **13**, 917–931 (2008)
17. Schroeder, A., Levins, C. G., Cortez, C., Langer, R., Anderson, D. G.: Lipid-based nanotherapeutics for siRNA delivery. J. Intern. Med. **267**, 9–21 (2010)
18. Singh, R., Lillard Jr., J. W.: Nanoparticle-based targeted drug delivery. Exp. Mol. Pathol. **86**, 215–223 (2009)
19. Siomi, H., Siomi, M. C.: On the road to reading the RNA-interference code. Nature **457**, 396–404 (2009)
20. Tiemann, K., Rossi, J. J.: RNAi-based therapeutics-current status, challenges and prospects. EMBO Mol. Med. **1**, 142–151 (2009)
21. Tokatlian, T., Segura, T.: siRNA applications in nanomedicine. Wiley Interdiscip. Rev. Nanomed. Nanobiotechnol. **2**, 305–315 (2010)
22. Weinstein, S., Peer, D.: RNAi nanomedicines: challenges and opportunities within the immune system. Nanotechnology **21**, 232001 (2010)
23. Zhou, J., Rossi, J. J.: Aptamer-targeted cell-specific RNA interference. Silence **1**, 4 (2010)

Robot-Based Automation on the Nanoscale

Sergej Fatikow, Daniel Jasper, Christian Dahmen, Florian Krohs, Volkmar Eichhorn and Michael Weigel-Jech
Division Microrobotics and Control Engineering (AMiR), Department of Computing Science, University of Oldenburg, Oldenburg, Germany

Synonyms

AFM; Atomic force microscopy; Carbon-nanotubes; DNA manipulation; Manipulating; Nanorobotics; Nanorobotic assembly; Nanorobotic manipulation of biological cells; Nanorobotics for bioengineering; Nanorobotics for MEMS and NEMS; Scanning electron microscopy; Visual servoing for SEM

Definition

The handling of micro- and nanoscale objects is an important application field of robotic technology. Nanohandling includes the finding, grasping, tracking, cutting, etc., of objects as well as different characterization methods such as indenting or scratching on the nanoscale. Measurement of different features of the object, probe positioning with nanometer accuracy, structuring or shaping of nanostructures, and generally all kinds of changes to matter at the nanolevel could also be defined as nanohandling in the broadest sense. This entry addresses several approaches that can be automated with the help of nanohandling robots. The development of nanohandling robot systems working in constricted work spaces (in a vacuum chamber of an SEM or under an optical microscope) and with a nanometer resolution and accuracy is a big technological challenge for the robotics research community. Advanced actuator and sensor technologies suitable for nanohandling have to be investigated and implemented as well as vision-based feedback methods for tracking and data acquisition. Another crucial issue is the development of real-time robot control methods that meet the demands of automated nanomanipulation. This entry presents several promising solutions and applications of robot-based nanohandling.

Robot-Based Automation on the Nanoscale

Introduction to the Topic

The handling of micro- and nanoscale objects is an important application field of robotic technology. It is often referred to as nanohandling, having in mind the range of aspired positioning accuracy for the manipulation of micro- and nanoscale objects of different nature. The nanohandling of objects may include their finding, grasping, moving, tracking, releasing, positioning, pushing, pulling, cutting, bending, twisting, etc. Additionally, different characterization methods like indenting or scratching on the nanoscale, measurement of different features of the object, probe positioning with nanometer accuracy, structuring or shaping of nanostructures, and generally all kinds of changes to matter at the nanolevel could also be defined as nanohandling in the broadest sense. As in the field of industrial robotics, where humans leave hard, unacceptable work to robots, robots with nanohandling capabilities can help humans to handle extremely small objects with very high accuracy. The size of these robots also plays an important role in many applications [1, 2]. Highly miniaturized robots, often referred to as microrobots, are able to operate in constricted work spaces, for example, under a light microscope or in the vacuum chamber of a scanning electron microscope (SEM). Microsystem technology (MST) and nanotechnology require this kind of robot, since humans lack sensing capabilities, precision, and direct manipulation at those scales. Automated nanohandling by microrobots will have a great impact on both these technologies [3]. This entry will address several aspects of nanohandling automation.

The development of nanohandling robot systems is a big technological challenge for the robotics research community. Advanced actuator and sensor technologies suitable for nanohandling have to be investigated and implemented. Another crucial task is the development of real-time robot control methods that meet the demands of automated nanomanipulation. The latter is challenging, especially due to the difficulty of getting real-time visual feedback and the lack of advanced control strategies able to deal with changing and uncertain physical parameters and disturbances.

Trends in Nanohandling

The following three approaches are being pursued in the nanohandling research community:

- Top-down approach utilizing serial nanohandling by microrobotic systems. The main goal is the miniaturization of robots, manipulators, and their tools as well as the adaptation of robotic technology (sensing, actuation, control, automation) to the demands of MST and nanotechnology. This approach is the major topic of this entry.
- Bottom-up approach or self-assembly, utilizing parallel nanohandling by autonomous organization of micro- and nanoobjects into patterns or structures.
- The use of a scanning probe microscope (SPM) as a nanohandling robot. In this approach, the (functionalized) tip of an atomic force microscope (AFM) probe or of a scanning tunneling microscope (STM) probe acts as a robot end-effector that is applied to a nanoscale part. This approach has been actively investigated in recent years [1, 2] and is discussed in section "AFM as a Nanohandling Robot."

Several other approaches like the use of optical tweezers or electrophoresis might also be adapted for automated nanohandling.

Automated Microrobot-Based Nanohandling

Microrobotics for handling micro- and nanoscale parts has been established as a self-contained research field for nearly 15 years and, in recent years, a trend toward the microrobot-based automation of nanohandling processes has emerged [4, 5]. Process feedback is the most crucial aspect of nanohandling automation. It is difficult to obtain reliable information when handling microscale and especially nanoscale parts. Vision feedback often is the only way to control a nanohandling process. For this reason, the vacuum chamber of a scanning electron microscope is, for many applications, the best place for a nanohandling robot. It provides an ample work space, very high resolution up to 1 nm, and a large depth of field. However, real-time visual feedback from changing work scenes in the SEM containing moving nanorobots is a challenging problem.

Automated Microrobot-Based Nanohandling Station (AMNS)

Figure 1 presents a generic concept of an automated microrobot-based nanohandling station (AMNS).

The microrobots are usually driven by piezoactuators that enable positioning resolutions down to subnanometer ranges. The microrobots of the station have a nanomanipulator integrated in their mobile platform, which makes them capable both of moving over longer distances and of manipulating with nanometer accuracy. This leads to more flexibility as the robots can be deployed anywhere inside the SEM vacuum chamber. Stationary robots can also be used in an AMNS. They are easier to control compared to mobile robots, which makes them more suitable for high-throughput automation. Depending on the application, different combinations of both robot types can be implemented.

The sensor system of the AMNS includes SEM, video cameras, force sensors, as well as position sensors integrated into the robots axes. The sensor data is sent to the station's control system for real-time signal processing. The control systems task is to calculate the positions of the robots and their tools as well as the positions of the parts to be handled or other objects of interest. The calculated positions serve as input data for the closed-loop robot control.

Teleoperation often is the first step on the way to automation. For this reason, a user interface is an important component of the AMNS. The user can influence the handling process by a haptic interface and/or by a graphical user interface (GUI). A good overview of state of the art teleoperation techniques and applications is given in [6].

The positioning accuracy of the microrobots during automated nanohandling is affected by several factors, so that a powerful robot control system is required. On the low level of the control system, driving voltages for the robot actuators are calculated in real time, which keep the robot and its end-effector on the desired path. The high-level control system is responsible among others for path planning, error handling, and the time-saving parallel execution of tasks. GUI and haptic interface are supported by the high-level control system as well.

The AMNS concept has been implemented in different application fields for (semi-) automated nanomanipulation [2, 3]. Relevant aspects of this work will be introduced in the following sections.

Structure of the Essay

The following is a brief summary of the topics covered in the following sections.

Section "Real-Time Vision Feedback" covers the vision feedback in nanohandling robot stations. The focus is on the use of SEM in combination with real-time image processing algorithms, which is proposed as the near-field sensor for the automation of nanohandling tasks.

Section "Position Control Inside SEM" introduces fundamental considerations for automated positioning inside the SEM. The section encompasses actuation types, motion principles, closed-loop control mechanisms for nanorobots, SEM-based visual servoing, as well as the difference between macro- and microscale control engineering.

Section "AFM as a Nanohandling Robot" introduces the atomic force microscope as a robotic system for manipulation at the nanoscale. The AFM tip acts as a nanohandling tool to change the position or orientation of a nanoobject or to modify the surface of a substrate with nanometric precision. Different application fields are introduced and current challenges in this field of research are discussed.

Section "Application Example: Handling of CNT" describes nanorobotic strategies for the microgripper-based pick-and-place handling of individual carbon nanotubes. For a reliable alignment in z-direction, a shadow-based depth detection technique is presented. Furthermore, special handling strategies are introduced for placing and releasing the CNT on the target structure. As an example, the automated

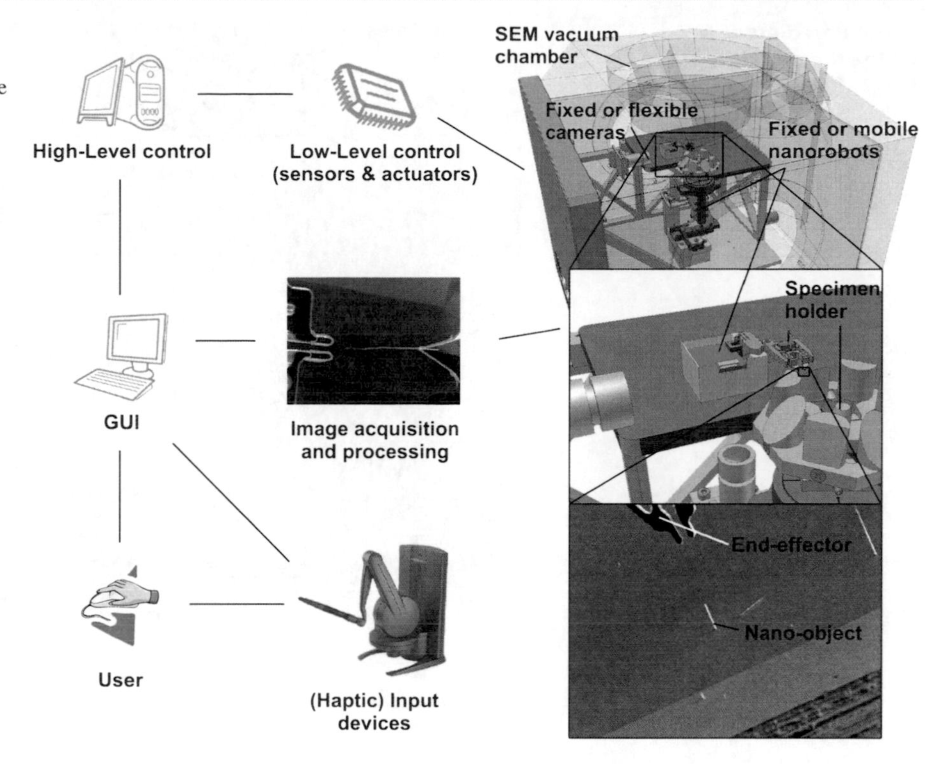

Robot-Based Automation on the Nanoscale, Fig. 1 Generic concept of the AMNS [2]

assembly of so-called CNT-enhanced AFM probes is demonstrated.

Section "Application Example: Handling of Biological Materials" deals with the characterization and manipulation of biological objects by an atomic force microscope. An overview of relevant parameters for AFM measurements of samples in liquids and of soft samples is given. After a short introduction of the biological background, the current state of the art in automated AFM-based handling of biomaterials is given, ranging from the design of DNA-based nanoelectric circuits to the structuring of biosensor chips for medical and forensic use.

Real-Time Vision Feedback

Manipulation and handling of objects on the nanoscale is subject to strong uncertainty on position and status of objects. This is due to the low mass of the objects of interest and the relatively high parasitic and interacting forces. Additionally, the choice of imaging modalities available at this scale is limited. Optical microscopes cannot deliver resolutions sufficient for imaging nanosized objects. Scanning probe microscopes like the AFM, though having the advantageous possibility to be used as a manipulator, do not exhibit the necessary speed for closed-loop feedback. The most flexible solution for imaging in this context is the SEM. It allows for observing processes at the nanometer range and still is fast enough to enable closed-loop control feedback. Drawbacks of the SEM are the sensitivity to noise and movement artifacts during fast scans, the possibility of distortions due to magnetic or electric influences generated by the AMNS, and the lack of three-dimensional information. Various approaches have been developed to deal with these conditions. The important properties of the SEM imaging and the tasks and approaches to solve the problems associated with it will be described in the following sections.

Imaging Source Properties

Certain properties of the SEM and its imaging process are important for the applied computer vision approaches. The most important negative property is the high level of noise. Noise can be reduced by frame averaging or pixel averaging which, however, also reduces the frame rate. Figure 2 shows the SEM images with low speed and high speed and the corresponding noise level.

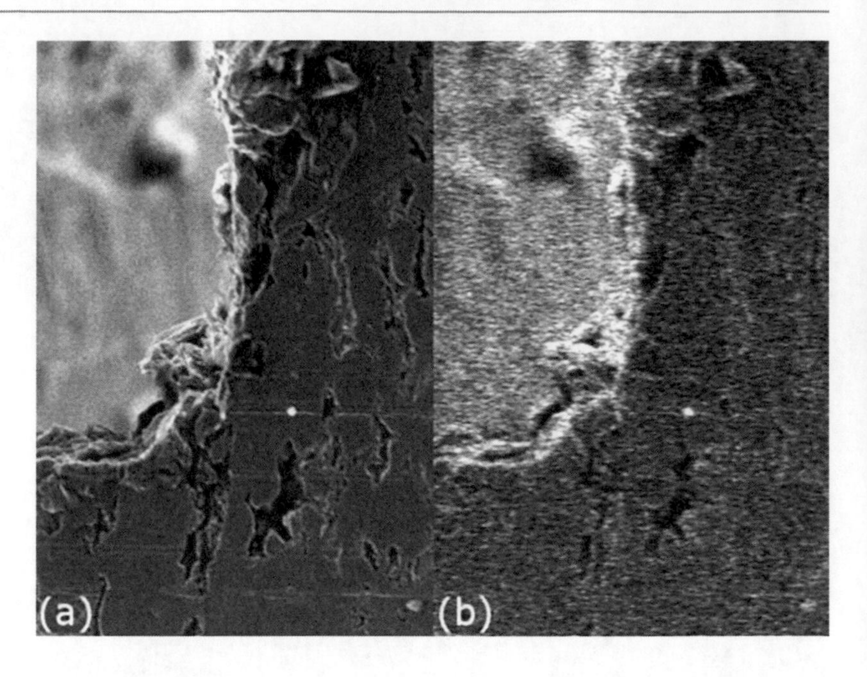

Robot-Based Automation on the Nanoscale, Fig. 2 SEM images at different scan speeds, acquired with a LEO 1,450 tungsten emitter SEM, (**a**) slow scanning, (**b**) fast scanning

The noise in the SEM can be modeled as additive noise:

$$g_{\mathrm{SEM}} = g + g_{\mathrm{noise}}$$

With g the true gray level value, g_{noise} the noise component, and g_{SEM} the acquired gray level value.

Because the generation of secondary electrons is a statistic process and the number of electrons generated and detected can be assumed as Poisson distributed, the gray level fluctuation also has a Poisson characteristic.

Due to the scanning principle of the SEM, fast moving objects appear distorted in the images. This is due to the fact that the object has moved during the time needed to finish each line of the scan. This can significantly reduce the performance of computer vision algorithms. Therefore, the speed of robots has to be restricted if computer vision algorithms are required.

Typical Visual Feedback Tasks

For a successful nanohandling automation, visual feedback is needed for specific tasks. These can be described as object recognition, classification, and position determination. The latter can be executed in two dimensions and in three dimensions, respectively. Important especially for the position determination is speed in order to allow for closed-loop control. Still, even with fast imaging, control is often restricted to a look-then-move strategy, because the actuation is orders of magnitude faster than the imaging and subsequent visual feedback. When using SEM image acquisition with computer vision approaches as an image-based sensor, this is a limitation which cannot be overcome easily.

Object Recognition

Object recognition has the task of determining if an object of interest is in the view field or not. Many different approaches are known in traditional macro- or microscale imaging and computer vision. The approaches used at the nanoscale include template matching, statistical pattern recognition, and recognition based on local features. Object recognition algorithms used for nanohandling automation in many cases do not have special and elevated needs for noise robustness, because the recognition task is not time critical and low frame rates with high averaging can be used. Approaches used in nanohandling applications use support vector machines and various feature extraction algorithms. Applications include CNT-detection and detection of TEM lamellae [7].

Two-Dimensional Position Determination and Tracking

The positions of objects and tools are not known precisely through any sensor except for the SEM. Object position determination and tracking allows image-based visual servoing approaches to work. Example methods which have been developed to work on SEM images rely on template matching [8], active contours [3], and rigid body models [9]. What is common in all these methods is that some measures have been included to diminish the effect of noise on the algorithm performance, either by use of inherent noise robustness like in the case of correlation-based template matching, the use of regions to cancel noise influence, or the usage of multiple or even manifold features and the selection and weighting of these features according to their expected significance and robustness.

Different algorithms may be suitable for different needs, depending on the degrees of freedom and the boundary conditions imposed by the system setup and the object properties. If the objects of interest are rigid, their appearance constant and known, their expected movement restricted to translations, and no magnification changes are required, template matching using cross correlation has turned out to be a robust and powerful yet simple approach. The idea is to calculate the cross correlation of a template image T and an input image I:

$$C(x, y) = T(x, y) * I(x, y)$$

In the resulting matrix, each value corresponds to the similarity of the image part to the template. Assuming that the object of interest is in the image, the position is with very high probability at the position of the maximal value in this matrix:

$$C(x_0, y_0) = \max(C(x, y))$$

The algorithm delivers a single position. It may well determine the position with sub-pixel accuracy, using weighting or interpolating techniques. The speed can be improved by using the frequency domain for calculation. In Fig. 3, the result of the algorithm can be seen. The template is marked at the place the best match is found.

If the objects may rotate or change in size, active contours or rigid body models are more suitable, depending on whether an a priori model of the object is available or not. Also, the occurrence of occlusions is less problematic with these algorithms. The principle behind active contours is that a contour represented by a polygon or spline is combined with an energy function which will determine its shape and position. The energy function consists of two parts:

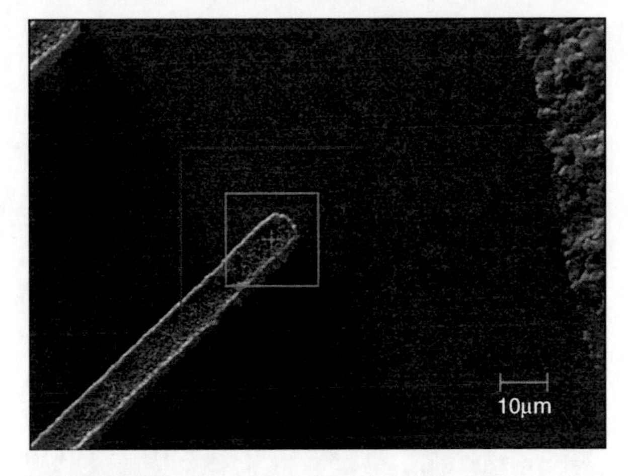

Robot-Based Automation on the Nanoscale, Fig. 3 Tip of a microgripper tracked with correlation tracking

$$J = J_{int} + J_{ext}$$

J_{int} is the internal energy which purely depends on contour properties like length, compactness, or smoothness. For compactness, the internal energy J_{int} can be formulated as:

$$J_{int} = \frac{L^2}{A}$$

The second part is the external energy, which depends on the image content or image features. Possible implementations of this evaluate the distance to edges or the region statistics. The energy function is minimized for each frame by applying transformations to the contour. These transformations can be either single point translations or Euclidean transformations of the whole contour.

Figure 4 shows an active contour tracking an object in SEM images.

Using single point translations, deformable and nonrigid objects can be tracked, but occlusion is problematic. Using Euclidean transformations of the entire contour, the algorithm is robust against occlusion at the expense of trackability of deformable objects. Using these kinds of algorithms, full automation of nanohandling applications has been performed.

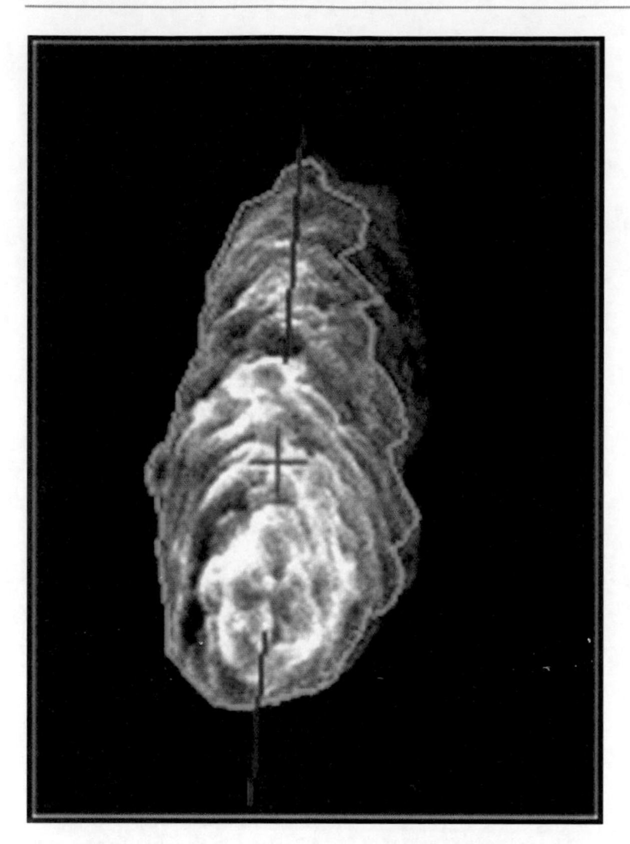

Robot-Based Automation on the Nanoscale, Fig. 4 Object tracked with active contours in SEM images

Three-Dimensional Position Determination

A difficult issue with image-based position determination using SEM images is the extraction of the third dimension. Any setup used for nanohandling and nanomanipulation is essentially a three-dimensional one, which means that in many cases, a two-dimensional position extracted from an image is insufficient for reliable and flexible automation. Several approaches exist to solve this problem:

- Restrict the setup or degrees of freedom to a plane, which is an unpractical solution.
- Use different image sources to enable depth estimation from multiple views. This includes the usage of additional microscope cameras to roughly estimate the information missing in the SEM image, as well as the use of beam deflection mechanisms to acquire SEM images from different angles [3].
- Use models of the SEM image generation and the lens distortions to extract the third dimension. This has been successfully done with smaller magnifications by [9].
- Use additional information from the images, such as defocus [20] or shadow to estimate the missing information or at least determine if the object or tool is at the desired location in space.

All these approaches have different advantages and disadvantages. Restricting the setup or degrees of freedom as mentioned has the advantage that no additional measures are needed to deal with the three-dimensional problem. This restricts the possibilities for handling and manipulation to an unacceptable extent. The use of different sources may deliver everything from rough estimates in case of microscope cameras to very precise (nm resolution) measurements in case of well-designed stereo imaging SEMs or SEM/FIB combinations. Drawbacks here are the reduction in speed when using stereo SEM and the complex correspondence problem in very dissimilar images when using a camera or the SEM/FIB combination. The use of models for the SEM imaging has produced quite good results, but depends strongly on the accuracy of the imaging model and on the accuracy of the rigid body model of the object which is tracked. This restricts the application to some extent to objects for which a model can be assumed, and to a certain parameter range of the SEM imaging. The use of additional information from the image mainly concentrates on defocus and shadow. Using defocus of objects can deliver an accuracy up to 5 μm using time-consuming focus sweeps, and it can also be used to estimate the out-of-focus displacement and therefore the position of any object for a given frame, but with reduced accuracy and reliability. This is partly due to the complex nature of the electron beam optics and the high number of factors influencing the image, especially image and object sharpness.

Position Control Inside SEM

A common challenge for the automation of handling and assembly tasks in the SEM is the precise alignment of two or more objects carried by different robots. Such objects can be wafers with micro- or nanostructured components as well as different tools including grippers, probes, and cantilevers. Dependent on the size of the involved objects, the accuracy of this alignment has to be in the micrometer or even nanometer range. Accuracy in this sense stands for the precision of the relative alignment, that is, a nanoobject needs to be brought precisely into the opening of a gripper.

For automated positioning, a closed-loop controller is necessary.

Actuation Types and Motion Principles

To facilitate closed-loop motion control inside SEMs, actuators capable of generating the required resolution need to be designed. Although there are efforts to scale conventional actuators such as electromotors to enable such a resolution, for example, employing backlash-free gears, novel actuation principles are better suited. In general, an actuator capitalizes on a *physical effect* in order to create motion. Most of the currently employed actuators for nanohandling are based on the inverse piezoelectric effect, that is, the controlled deformation of a piezoelectric element using an electric field. Piezoelectric actuators are popular owing to their large forces, high precision, and fast response time. Such actuators can create forces in the kN-range and respond to an input signal within microseconds. Major downsides are the limited deformation of about 0.1% as well as a strongly nonlinear behavior affected by hysteresis and creep. Other physical effects that are used include electrostatic, thermal, and magnetostrictive actuators as well as shape memory alloys [4].

Piezoelectric actuators can be further categorized into scanning and step-wise actuators. *Scanning actuators* use a backlash-free cinematic structure based on flexure hinges to amplify the relatively small displacements of the piezoelectric elements. Although commercial actuators can reach working ranges of up to 500 μm, the working range is a severe limitation and usually a second actuator is required for coarse positioning. Furthermore, the electric field used for actuation cannot be generated with arbitrary precision and without noise. Thus, there is a trade-off between working range and resolution. *Step-wise actuators* are a solution to overcome these downsides. Instead of amplifying the piezoelectric element's motion using a mechanical structure, the stroke is used to perform small steps. Step-wise actuators have a virtually unlimited working range. Most step-wise actuators are hybrid actuators, so that the piezoelectric element can also be used in a scanning mode to create movements smaller than single steps.

Scanning actuators have the advantage of generating a smooth motion over the entire working range. Furthermore, due to the high force that can be generated by a piezoelectric actuator, very high accelerations and speeds are possible. Another advantage is that a rough knowledge of the actuator's position can be directly derived from the applied voltage. Homing is not required for scanning actuators and even open-loop movements are reliable to a certain extent. A downside is that a scanning actuator always returns to its initial position on power loss. Thus, microrobotic systems have to be designed in a way that this movement cannot result in a severe damage to the employed robots, tools, or objects. Furthermore, the nonlinear characteristics including hysteresis and creep are scaled with respect to the full working range. Thus, these effects have a significant influence and make sensor feedback necessary.

In contrast, step-wise actuators have opposed properties. Their working range is only restricted by the mechanical guides, usually in the centimeter range. They require a sensor because the executed steps are not sufficiently repeatable. Furthermore, dependent on the sensor, homing is required to find the initial location of the actuators. In addition to the large working ranges, major advantages of step-wise actuators are that they do not perform a significant movement if the power is lost and that hysteresis and creep are scaled with the step length.

The most commonly applied step-wise motion principle based on piezoelectric actuators is the stick–slip actuation principle [4]. This principle makes use of the high movement speed and acceleration of a piezoelectric actuator. A mobile part is connected to a piezoelectric actuator mechanically using a contact with a specific friction. If the piezoelectric actuator is deformed slowly, the friction forces the mobile part to perform the same motion. If the actuator is deformed rapidly, however, the inertia of the mobile part keeps it in place and the friction force is overcome generating a net displacement between actuator and mobile part. Alternating slow and rapid deformations into opposite directions generate a step-wise motion over long travel ranges.

Vision-Based Closed-Loop Control

On the macroscale, the location of the manipulated objects in a world coordinate system is either known from the process design or can be measured by a variety of available sensors. To perform closed-loop positioning, a robot can move to this location with sufficient accuracy using its internal sensors. On the nanoscale, the position of objects has to be derived by processing images of the employed SEM (see section Real-Time Vision Feedback). SEMs, however, are designed to create highly magnified images and not to create highly accurate measurements. Thus, the resulting position is often time variant and nonlinear, so that the object's location in a world coordinate system cannot be accurately derived. In addition, the positioning itself has limited accuracy due to the size ratio between the robotic system and the handled objects. Commonly, a robot with a size of a few centimeters is used to manipulate an object with a precision of a few nanometers, that is, the robot is about six orders of magnitude larger than the required accuracy. Effects such as thermally- induced drift and limited mechanical stiffness lead to a nonlinear and time-variant relation between the position measured by a robot's internal sensors and the real position of the robot's tool. Thus, internal sensors are insufficient for reliable nanopositioning.

As known from the macroworld, the limitations of positioning based on internal sensors can be overcome by using the information generated by a visual sensor. Using vision feedback as a sensor for robotic manipulation is promising as it is a noncontact measurement and mimics human behavior. The initial technique, which is still used for most SEM-based positioning, extracts a position deviation from an image and then moves the robot either open loop or based on internal sensors to decrease this deviation. When the robot stops, another image is taken, again extracting and compensating for the deviation. This is called the *look-then-move* approach [10]. Two advantages of the approach are that low throughput rates of image processing are acceptable and that the entire setup can be handled in a quasi-static fashion neglecting dynamic aspects. The effectiveness, however, entirely depends of the accuracy of the visual sensor and the robot's open-loop control.

An alternative is to use the visual feedback as a low-level element in the motion control loop of a robot. This is called *visual servoing* [10]. Two challenges need to be solved to achieve effective visual servoing. First, the image processing needs to be fast. As the available processing power has significantly increased over the recent years, several image processing algorithms can be used for visual servoing. More complex algorithms, however, still require computation times on the order of seconds and do not allow for effective servoing. Second, the dynamic aspects of visual servoing need to be considered. The robot's movement as well as resulting vibrations leads to movement during image acquisition. This movement creates image distortions rendering most of the current image processing algorithms useless. Enabling an effective visual servoing, however, is a necessary step on the way of high-speed micro- and nanohandling viable for industrial application.

Closed-Loop Control

Based either on an internal sensor or on vision feedback, a closed-loop controller can be implemented. Closed-loop control on the micro- and nanoscale is significantly different from macroscale control [1]. Macroscale control mostly solves challenges arising from the dynamic behavior of systems, for example, inertia. On the micro- and nanoscale, these effects become negligible. As an example, a microrobot might carry a mass m of 30 g and can create a force F of approximately 0.02 N. Thus, its maximum acceleration a is:

$$a = F/m = 2/3\left[\mathrm{m/s^2}\right]$$

If it needs to accelerate to a velocity v of 500 μm/s, which is high for a micro- or nano-operation, it requires:

$$t = v/a = 0.375[\mathrm{ms}]$$

This is significantly faster than the update rate of any visual sensor. Thus, acceleration effects can be neglected. Therefore, for many applications and actuator types, a simple proportional controller is sufficient to perform an effective alignment.

AFM as a Nanohandling Robot

Since its development by Binnig, Quate, and Gerber [11], AFM has been increasingly recognized not only as a tool for imaging purposes, but also for manipulating at the micro-, nano-, and atomic scale. The main advantages of the AFM are its high resolution in the sub-nanometer range as well as its flexibility for operating under physiological conditions and in vacuum or liquid environments, depending on the specific application. In addition to its visualization capabilities, the AFM can be used for force spectroscopy and, with functionalized tips, for measuring electric, magnetic, or chemical properties. Moreover, the AFM tip can be

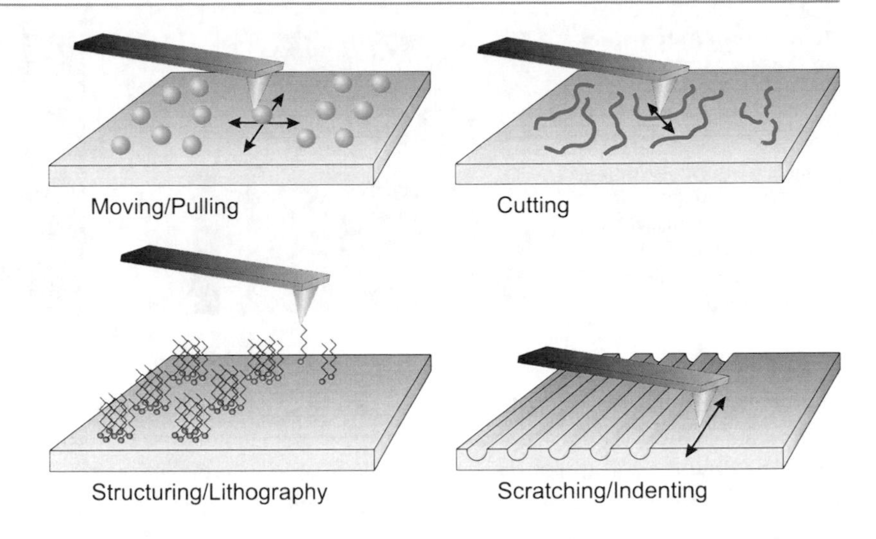

Robot-Based Automation on the Nanoscale, Fig. 5 Schematic view of manipulation tasks that can be performed by AFM-based robotic systems [1, 2]

used to manipulate objects or materials with nanometric precision and thus the AFM can also be seen as a nanohandling robot. AFM-based manipulation benefits from the fact that the AFM directly measures forces that are exerted during the manipulation and thus gives important feedback on the progress of the manipulation. Potential applications of AFM-based manipulations are manifold, ranging from the fabrication of nanoscale devices, reparation or modification of nanostructures to the characterization and handling of biological samples. Different types of manipulation can be conducted with the AFM including pushing, pulling, cutting, indenting, and structuring of a wide range of materials (Fig. 5).

Starting with unintended manipulations of Au-Clusters, reported 1987 by Baró et al. in [12], the AFM as a tool for manipulating single nanoentities became subject of increasing interest in the research community. A controlled manipulation of nanoscale objects by an AFM has been actively investigated over the last decade. Many publications exist describing the utilization of the AFM as an instrument for manipulating spherical nanoobjects such as gold nanoparticles as well as longish nanoobjects such as carbon nanotubes (CNTs), nanorods, or also biological objects such as DNA [2]. Figure 6 illustrates a successful manipulation of multiwalled CNTs that was conducted by the AFM tip. A commercially available dynamic-mode cantilever was used for both – the image acquisition as well as the manipulation. For imaging, conventional intermittent mode was used, whereas for manipulation the tip was brought into proximity with the surface and then moved laterally to mobilize the CNTs.

To perform a controlled movement of a nanoscopic object, it has to be localized and identified on the substrate first. A major drawback of AFM-based manipulation is the lack of an additional, visual sensor. Therefore, before any manipulation is conducted, the AFM usually has to be utilized to obtain a detailed image of the scenery in order to plan the execution of the manipulation. This imaging scan is most often performed in dynamic mode (intermittent or noncontact) in order to avoid any damage or unintended manipulation of the sample. For the manipulation, the AFM tip is brought into proximity with the object and moved – either in dynamic or contact mode, but without any feedback to control the z-direction – toward a predefined position. As a result, the object is then pushed forward by the repulsive forces. After a manipulation step, reimaging of the area of interest often becomes necessary to reveal the result of the manipulation. The disadvantage of this "look and manipulate" scheme is that the image acquisition process usually needs several minutes depending on the required quality. Moreover, the manipulation itself cannot be monitored by means other than the measured forces exerted by the cantilever and thus have to be performed in a "blind" way. Unfortunately, due to a variety of uncertainties, the results of the manipulation are often not satisfying and require frequent trial and error experiments. Real-time imaging would greatly help to make the manipulation process faster, more robust, and accurate. However, optical microscopy cannot be used as an additional real-time capable imaging method because its resolution does not fit the requirements for the manipulation of objects with

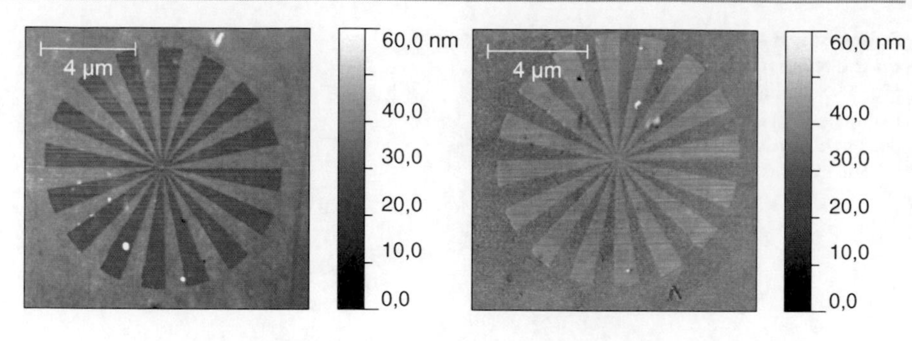

Robot-Based Automation on the Nanoscale, Fig. 6 AFM topography star with diameter of 10 µm, fabricated on a gold layer (*left*) using mechanical AFM-lithography and the replica of this structure fabricated on polypropylene (PP) using the microinjection molding technique (*right*)

nanometric dimensions. An alternative is given by the integration of the AFM into the vacuum chamber of an SEM to allow for real-time observation of the manipulated object and the AFM probe. However, this entails the disadvantage of restriction to vacuum and to some extent conductive sample materials, which is not feasible for many applications.

Besides its ability to manipulate nanoscale objects, the AFM tip can also be used to modify surfaces with nanometric precision or to change an object's shape, for example, by scratching, indenting, cutting, and dissecting. For nanomachining purposes, the AFM tip can be exploited as a nanohandling tool in order to act as, for example, a milling cutter, nanoscalpel, or nanoindenter. However, the type of interaction between AFM tip and substrate is not only mechanical, but may also involve chemical, thermal, or electrical impact. Over the last two decades, a large number of promising nanolithographic technologies have evolved and are still emerging that are based on such AFM-based nanomanipulation of surfaces. These techniques can be used among others for mask-free lithography on the nanoscale and provide a promising alternative to conventional technologies. Using AFM-based surface modification, material can either be removed (e.g., through direct mechanical scratching), chemically changed (e.g., resist exposure techniques or oxidation), or deposited (e.g., dip-pen lithography). A detailed review of such SPM-based lithography can be found [13]. An example for a novel process chain for small series production of nanostructures that is based on AFM-based mechanical lithography, which is the most basic type of AFM lithography, was introduced [14]. Figure 7 depicts how the mechanical AFM-lithography is utilized to fabricate a nanostructure that was subsequently replicated several times by means of the microinjection molding technique.

Up to now, the AFM as a robot for the nanomanipulation can only be found in laboratories for prototypical issues. For industry, the serial AFM-based approach is still not reliable and fast enough to be successfully applied for productive use. To increase throughput and reliability of this sequential process, automation can be considered as a desirable goal. To reach this goal, however, a large number of challenging prerequisites have to be fulfilled.

Besides the lack of real-time imaging of the manipulation process discussed above, spatial uncertainties constitute an import error source. Although positioning inaccuracies caused by creep and hysteresis of piezoactuated scanning stages can nowadays most often be compensated by closed-loop operation, thermal drift of the AFM components lead to spatial positioning errors that are challenging to counteract. Solutions to overcome this problem have been addressed [1, 2, 15]. Moreover, sticking effects also play a crucial role in AFM-based nanohandling and therefore, substrate, cantilever, and tip material have to be carefully chosen in order to minimize these effects. A promising strategy that is pursued by several research groups aims at the precise modeling of the nanohandling task. To compensate for the uncertainty introduced by the inevitable "blind" manipulation, a model of the manipulation process is utilized enabling a manipulation in open-loop mode. Having a valid model of the nanomanipulation, including all relevant interactions between nanoobject, tip, and substrate, it is possible to calculate the expected position of the nanoobject during manipulation in real time. Using such a model for an augmented reality system, where a virtual reality is augmented by the calculated nanoworld situation, enables an additional visual feedback. However, this approach requires exact knowledge of the nanomanipulation phenomena, which is not yet available in the current state of nanosciences.

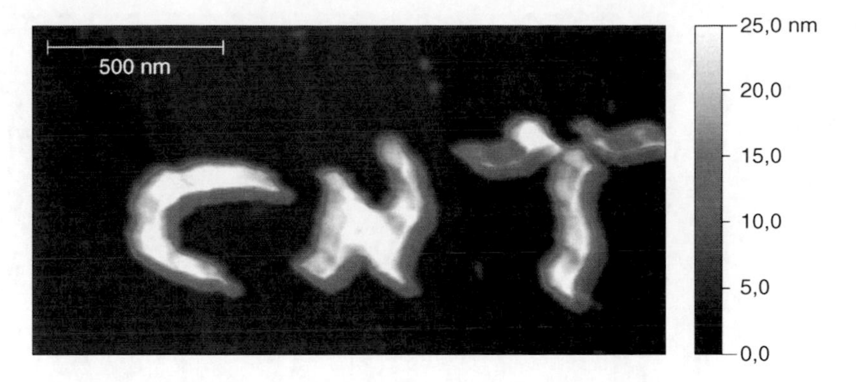

Robot-Based Automation on the Nanoscale, Fig. 7 Multiwalled carbon nanotubes on HOPG substrate manipulated by the AFM. When moving the AFM tip laterally, the height control loop was turned off in order to avoid slipping of the AFM tip over the CNT. It could be observed that, depending on the parameters chosen for manipulation (tip velocity, damping amplitude), cutting as well as pushing and pulling of the CNTs can be accomplished

Application Example: Handling of CNT

Nanomaterials such as different kinds of nanowires and especially carbon nanotubes (CNTs) [16] have unique physical properties and can thus improve and enhance existing microsystems in a multitude of applications. Two main application areas have been identified where in particular the integration of CNTs into known microsystems is of high impact: CNT-enhanced AFM probes [17] for the analysis of high-aspect ratio structures and CNT-interconnects in microchip technology [18]. Even though parallel fabrication techniques will be required for a future bulk production of such devices, enabling technologies for the reliable manipulation of individual nanostructures and their prototypic integration into microsystems have to be developed. Automated robot-based assembly techniques are very promising to implement a fast and systematic prototyping of CNT-based devices. In the following section, a microgripper-based strategy for the automated assembly of CNT-enhanced AFM probes is presented.

Nanorobotic Strategies for Microgripper-Based Handling of CNTs

In prior experiments, electrothermally actuated microgrippers have already been used for teleoperated pick-and-place handling of individual CNTs [19]. The CNTs used for the experiments have a typical diameter of 200 nm and are grown vertically aligned on a substrate with a spacing of 50 μm between nanotubes. In order to automate the microgripper-based handling of CNTs, special nanorobotic strategies have been developed. Most important are methods for the z-approach of gripper and CNT and for the reliable gripping and placing of CNTs that lead to a well-defined system state.

Methods for z-Approach of Microgripper and CNT An AMNS has been developed that can be integrated into the SEM [2], so that visual feedback from the microscope is available during the nanohandling tasks in addition to the integrated sensors of the positioning systems. Image processing algorithms for object recognition and tracking have been developed (compare section Real-Time Vision Feedback) providing the x,y-position of gripper and nanotube. This two-dimensional position information can be used to automate the approach in x,y-direction. However, to receive information on the gripper's and CNT's relative z-position, additional contact sensors or three-dimensional systems are required. Such systems cause additional costs and reduce the available space within the SEM's vacuum chamber [20].

A novel so-called shadow-based depth detection method has been developed that, together with the focus information of the SEM, facilitates the automated z-approach of microgripper and CNT in any available SEM. First of all, the "depth-from-focus" technique [20] is used to coarsely align both objects in z-direction down to 50 μm. Then, for the fine approach of gripper and CNT, a shadow can be detected as soon as the CNT is located exactly between the gripper jaws. The gripper jaws block the secondary electrons from the detector. Figure 8 shows SEM

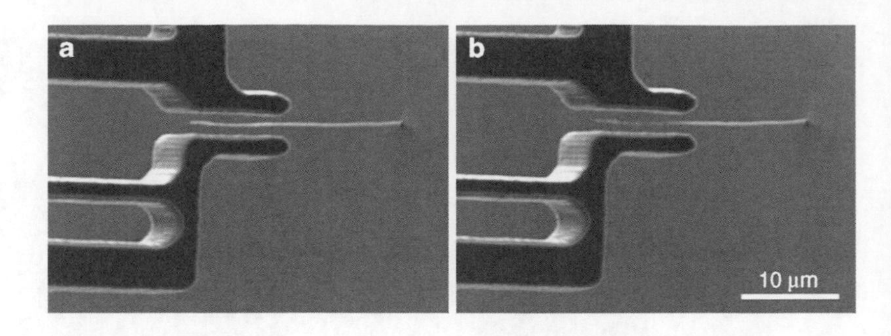

Robot-Based Automation on the Nanoscale, Fig. 8 Shadow-based depth detection for the alignment of microgripper and CNT in z-direction. (**a**) CNT is located below the gripper jaws, no shadow is detectable. (**b**) CNT is between the gripper jaws, a shadow is clearly visible

images illustrating the shadow-based depth detection method. The clearly detectable shadow on the CNT in Fig. 8b is used as a stop criterion for the z-approach.

Gripping and Removing CNTs from the Growth Substrate Different gripping techniques for the reliable picking and removing of individual CNTs that are grown on a silicon substrate by chemical vapor deposition (CVD) have been discussed in [21]. The so-called shear gripping technique has been identified as the most reliable approach, requiring smaller gripping forces compared to tensile gripping. The microgripper is closed around the nanotube and moved sideways to break the CNT off the substrate. After that, the gripped CNT can be transferred to the target structure.

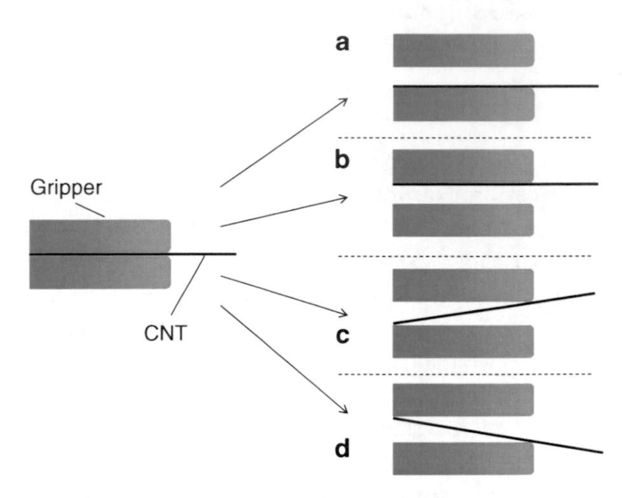

Robot-Based Automation on the Nanoscale, Fig. 9 The four possible configurations after reopening the microgripper. (**a**), (**b**) Line contact between CNT and gripper jaw. (**c**), (**d**) Two-point contact between CNT and gripper jaws

Placing and Releasing CNTs on the Target Structure A novel handling strategy for placing and releasing CNTs onto the target structure has been developed. This strategy relies on the adhesion forces between CNT and microgripper jaws and thus avoids the need of additional joining techniques. After reopening the microgripper, the nanotube can be oriented in four different configurations. Either the CNT is connected by a line contact to one gripper jaw (see Fig. 9a, b) or the CNT is connected by two-point contacts to the gripper jaws (see Fig. 9c, d).

In case of the two-point contacts, the adhesion forces are smaller compared to the line contacts. For this reason, the two-point contact is preferred in order to easily place and release the CNT on the target structure. In addition, obtaining a well-defined system state after reopening the gripper is required to allow full automation of the placing and releasing process. For the automated microgripper-based mounting of an individual CNT onto the tip of an AFM probe, the strategy shown in Fig. 10 has been developed.

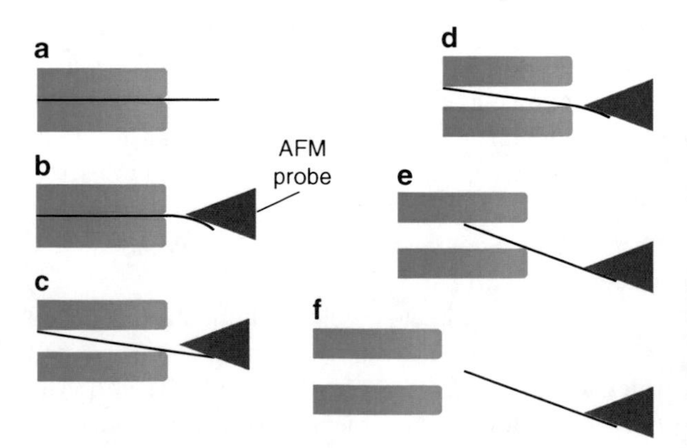

Robot-Based Automation on the Nanoscale, Fig. 10 Nanorobotic handling strategy for automated placing a CNT onto the tip of an AFM probe and releasing it from the microgripper

The closed microgripper holding the CNT (Fig. 10a) is approached to the tip of an AFM probe until a bending of the nanotube can be detected (Fig. 10b). This bending is detected automatically and used as the stop criterion of the z-approach. At the same time, the exerted force assures that CNT and gripper jaws are forming a two-point contact with defined orientation after reopening the microgripper (Fig. 10c). After this, the CNT is moved toward the AFM tip until a line contact is achieved (Fig. 10d). Because the adhesion forces of the line contact between CNT and AFM tip are much higher than the forces of the two-point contacts between CNT and gripper jaws, the nanotube can be easily released from the gripper jaws by retracting the microgripper (Fig. 10e–f).

Automated Assembly of CNT-Enhanced AFM Probes

The position control techniques described in section "Position Control Inside SEM" have been used to fully automate the nanohandling task of mounting a CNT onto the tip of an AFM probe. For this purpose, the whole automation sequence has been divided into three subsequences: The initialization process, the gripping and removing process, and the process of placing and releasing the CNT. In the following, these three subsequences are described in detail with respect to automation purposes.

Initialization Sequence The first step of an automated robotic micro-nano-integration is the initialization. For this purpose, all actuators are moved to a well-known starting position and the coordinate system of the internal position sensors are mapped to the coordinates of the SEM image. Then, the electrothermal microgripper mounted to the mobile microrobots is calibrated. A pattern of the microgripper is recorded that allows for tracking the position of the gripper by applying cross-correlation algorithms to the SEM images (compare section "Real-Time Vision Feedback"). After this, the CNT substrate is scanned automatically to index all CNTs. For each CNT, length and orientation are checked as a quality criterion and only the positions of suitable CNTs are stored to the database. Finally, the AFM probes are indexed by saving the exact position of the probes' tips to the database.

Picking Sequence As described in section "Gripping and Removing CNTs from the Growth Substrate", the shear gripping technique is used to grip and remove an individual CNT from the sample substrate. Figure 11 shows SEM images of the gripping sequence illustrating the process of breaking the CNT off the substrate. First, so-called coarse and fine positioning steps are performed placing the CNT between the gripper jaws (see Fig. 11a). Second, the microgripper is actuated and the CNT is detached by moving the gripper sideways (see Fig. 11b). Third, the gripped CNT can be transferred to the target structure.

Placing Sequence For the automated placing of the CNT onto the tip of an AFM probe, the nanorobotic strategy described in section "Placing and Releasing CNTs on the Target Structure" is used. Figure 12 shows SEM images of the automated placing sequence. After the coarse and fine approach between CNT and AFM probe has been realized in x,y-direction (see Fig. 12a), the final z-approach is performed until a bending of the CNT is detected (see Fig. 12b). Then, the gripper is reopened and a two-point contact between CNT and gripper jaws as well as a line contact between CNT and AFM probe tip is established (see Fig. 12c). Finally, the CNT can be easily released from the gripper jaws and placed onto the tip of the AFM probe by retracting the microgripper (see Fig. 12d).

Application Example: Handling of Biological Materials

Current research areas in molecular and cell biology, medicine, and process sensor technology (sensor systems monitoring various processes in industry and science) often require advances in the nanoengineering technologies to look for molecular phenomena with the highest possible resolution in order to enlighten the "black-box" which still shades most metabolic reactions. The aim of present work, starting from current medical problems, is to propose novel studies especially in molecular and cellular microbiology as well as virology with direct applications in medicine and industry, closing the bottom-up cycle. Thanks to advances in micro- and nanofabrication and micro- and nanorobotics, robotic systems which offer the possibility of characterizing and manipulating biological objects can be developed today. The use of such systems enable new studies of single cell phenomena with nanometer resolution (e.g., studying the local mechanical and electrical properties of single bacteria for characterization and evaluation of the resistance to

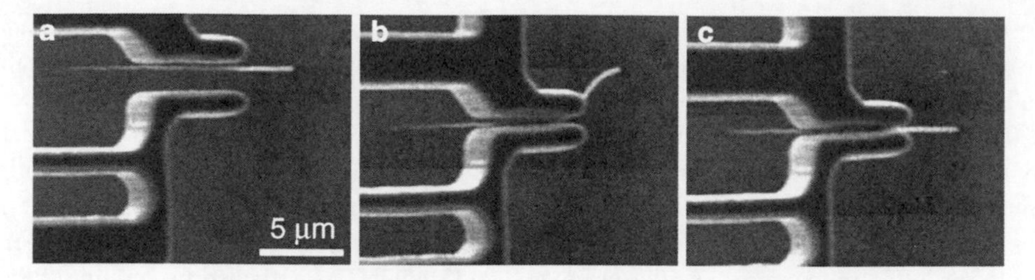

Robot-Based Automation on the Nanoscale, Fig. 11 SEM images of the automated gripping sequence. (**a**) Coarse and fine alignment of gripper of CNT. (**b**) Removing the CNT from the substrate by shear gripping. (**c**) Closed gripper holding the detached CNT

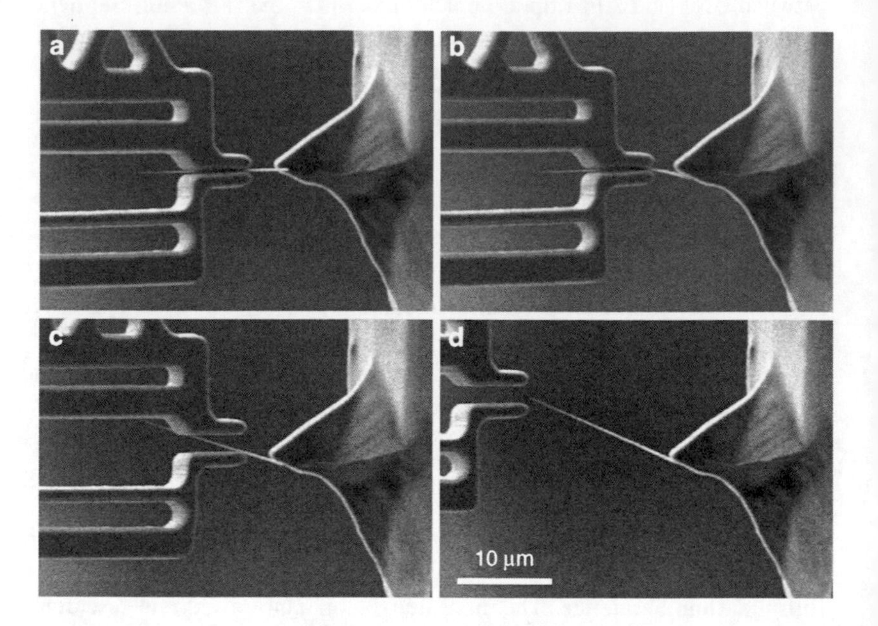

Robot-Based Automation on the Nanoscale, Fig. 12 SEM images of the automated assembly. (**a**) Coarse and fine alignment of gripped CNT and AFM tip. (**b**) Final z-approach until a bending of the CNT is detected. (**c**) Creating a two-point contact between CNT and gripper jaws and a line contact between CNT and AFM probe. (**d**) Releasing the CNT by retracting the gripper

antibiotics or to the spread of infections) up to complete cell compounds (e.g., studying the mechanical properties of bacterial biofilms) and deepen the appreciation of the processes at the nanoscale. The following section will give a brief overview of the handling of biological objects using automated AFM-based approaches.

Manipulation and Handling of Biomaterials Using Atomic Force Microscopy

Section "AFM as a Nanohandling Robot" has already shown the basics of the AFM for the automated handling of nanoscale materials. In this subsection, a brief overview of the differences for biomaterials is given.

For characterization or manipulation, samples have to be immobilized on an adequate substrate. A common method is the immobilization on freshly cleaved mica. Mica has a hydrophilic, charged surface that binds proteins and other biomolecules easily. Occasionally, mica is treated with buffer systems, which change the charging of the surface. Gold substrates can also be used if coated with protein reactive monolayers.

The characteristics of the used cantilevers are determined by material, tip shape, and mechanical parameters such as spring constant, resonant frequency, and Q factor. For biological applications, cantilevers are usually made of silicon or silicon nitride. These materials are chemically inert and can be doped to induce electrical charges. Soft cantilevers ($C < 1$ N/m) are used for the contact mode and hard cantilevers ($C \approx 42$ N/m) are used for the tapping mode. If operated in liquids, the pH value and the presence of electrolytes influence the forces between tip and sample. Although a smaller tip radius results in more accurate images, it may be advantageous for soft biological samples, to use a duller tip reducing the risk of accidentally damaging the sample.

Compared to other methods (e.g., optical tweezers, dielectrophoresis) the AFM has some major advantages for the handling of biomaterials. The most important points are [1, 2, 22]:

- The AFM can be used in liquids, in vacuum, and at ambient conditions meaning that biological samples can be characterized and manipulated at their necessary physiological conditions. Furthermore, living samples and their reaction to mechanical stimulation can be studied.
- The AFM can be used for imaging as well as for manipulation nearly at the same time.
- The possible resolution lies in the sub-nanometer range.
- Three-dimensional topological information can be obtained.
- No special preparation methods are necessary.

However, compared to the other methods AFM-based manipulation entails major drawbacks and limitations. Knowing these limitations is important to choose the right method or tool to handle or characterize a specific sample. The most important limitations are:

- A three-dimensional force measurement is not yet possible.
- The scanning speed is generally too low for real-time mapping.
- The scanning range for high-speed AFMs is still very low (about 3 μm × 3 μm).

Examples of Automated Handling of Biomaterials with an AFM

In the following sections, current work on the field of handling, characterization, and manipulation of biomaterials is described. Current research mainly deals with the development of fully automated solutions for handling and manipulation of DNA, for solving of packaging problems in the future of nanoelectronics, and for structuring of biosensors for medical and forensic use.

Handling and Manipulation of DNA In accordance with the progressing miniaturization of electric circuits, new ways to realize smaller channel widths are needed. To overcome the limitations of silicon-based processing, the use of objects like CNTs or biomolecules like DNA is a promising solution. Nanowires produced via the metallization of DNA are promising candidates for nanoelectronic devices of future generations. DNA can be considered a basic building block for nanostructure fabrication because it provides unique self-recognition properties. DNA constitutes an ideal template for the organization of metallic and semiconductor particles into wire-like assemblies. Nanoscopic (thickness below 10 nm) and regular wire-like metallic structures can be efficiently produced because DNA molecules have a diameter of only two nanometers and a length in the micrometer range. Moreover, both the ends of DNA molecules and the surface of solid substrates can be functionalized through a variety of methods to create specific links between DNA and substrate, allowing for the integration of DNA molecules into specific, predefined sites of microstructured electronic circuits. By exploiting the specific recognition of complementary nucleotide sequences, complex structures made of DNA can be fabricated and metalized at a later stage. Packaging problems of next-generation nanoelectronics technologies, (e.g., single-electron transistors, quantum automata, molecular electronics, etc.) might also be addressed by using DNA strands.

Recently, many researchers have proven the feasibility of DNA manipulation by an AFM tip, primarily in liquid conditions. The next step toward high-throughput fabrication is automation of DNA handling. Figure 13 show two alternative immobilization procedures of DNA, which offer the potential of automated DNA manipulation in dry ambient conditions. Figure 13a shows a topographic scan of a DNA strand. The height of the DNA strand is about 1.5 nm. Figure 13b shows the same DNA strand after several manipulations by the AFM cantilever tip. The DNA was manipulated and successfully moved about 25–100 nm. Figure 13c, d shows an experiment to demonstrate the transport of DNA. Two separate DNA strands are shown. The upper DNA strand (Fig. 13c) was moved by several AFM strokes up to 300 nm. As a result, this DNA strand was removed from its location (Fig. 13d), without affecting the position of the lower DNA.

The presented methods for the handling of biomaterials show the potential for different usage scenarios in industry and scientific research. In case of the handling of DNA, the experiments demonstrate that a DNA manipulation and handling can be performed in dry ambient conditions.

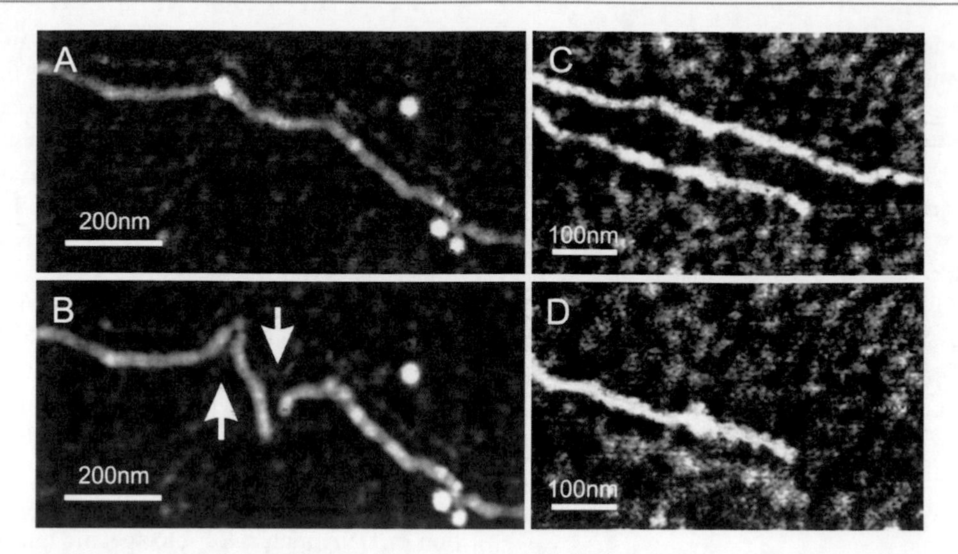

Robot-Based Automation on the Nanoscale, Fig. 13 (**a**, **b**) DNA immobilized on silicon without surface modifications (**a**) and the same DNA after the manipulation process (**b**). On the *left* side of the strand, two short distance movements from bottom to top were implemented. As a result, the corresponding DNA parts have been pushed about 25 nm. On the *right* side, two long distance movements were implemented, resulting in DNA displacement by 100 nm. (**c**, **d**) DNAs immobilized on silicon without surface modifications (**a**) and the same substrate with upper DNA strand pushed away on the surface by the AFM tip, without affecting the lower strand (**b**)

Automated Structuring of Biological Materials for Biosensors Biosensors consist of a sensitive biological component, a transducer, or detector element, and associated electronics or signal processors. An accurate and well-defined preparation of the sensitive components made, for example, of microorganisms, cell receptors, proteins, enzymes, anti-bodies, or nucleic acids is crucial part of the sensor design. Commonly used methods are microstamping and microprinting, both of which are unable to decrease the structures of the biological component down to the nanometer range. However, higher density is necessary to increase the measurement accuracy and the number of different, simultaneously useable biocomponents. In this section, current work on automated design of high-density sensitive components using AFM-based lithography is introduced.

The compensation of spatial uncertainties plays an important role in automation of AFM-based handling. Although the effects of hysteresis and creep of the PZT-based scanner can be reduced by closed-loop control, spatial uncertainties caused by thermal drift are crucial and less straightforward to deal with. Especially in life science applications, where the AFM is often operated in humid environments, the effect becomes strong and often prevents successful nanohandling, let alone automation of the handling task. To achieve reliable results, the recently developed method for real-time drift compensation described in section "AFM as a Nanohandling Robot" and in [1, 15] was applied, both for the presented manipulation of DNA as well as for the AFM-based nanotooling described below.

As a first step toward the automated fabrication of biosensor components, rectangular areas were successfully structured into an APTES monolayer on mica (Fig. 14). To further investigate the correlation between the forces applied during manipulation and the resulting machining depths, a series of experiments was performed to scratch multiple areas.

By using the AFM, an automated structuring of biological materials for biosensors can be done. The automation of the necessary processes will also lead to a feasible industrial use, as well as for clinical and research laboratories. The size of the resulting structures correlates with the resolution of the commonly used read-out techniques such as fluorescence microscopy or surface plasmon resonance spectroscopy.

Future Trends

In the future, closed-loop motion control in SEMs needs to be changed from the quasi-static

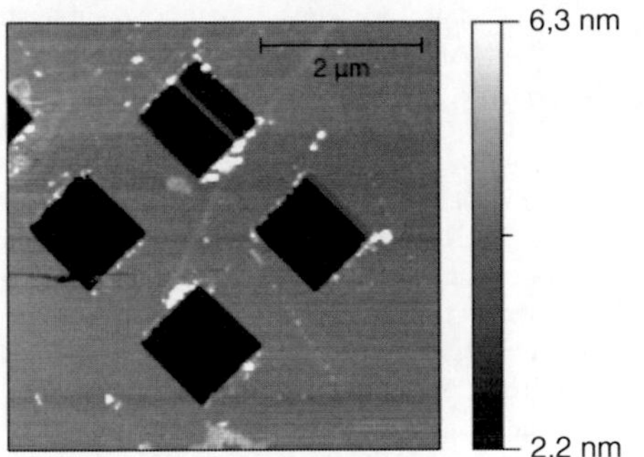

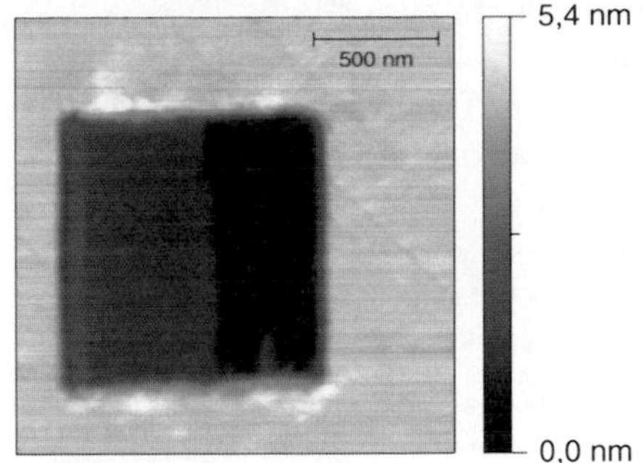

Robot-Based Automation on the Nanoscale, Fig. 14 Height image of 1 × 1 µm area scratched into APTES monolayer by AFM machining. The two visible machining depths arise from different force set points used during the processing (*left*) and AFM height image of a series of 1 × 1 µm sized scratched areas (*right*)

look-then-move-approach to an effective visual servoing in order to implement high-speed micro- and nanohandling viable for industrial application. Furthermore, to allow for a more effective avoidance of both collisions and vibrations, the commonly implemented position control needs to be extended to trajectory control so that the robot moves with a predetermined velocity, acceleration and jerk at all times. Considering visual feedback, an important future task is to enable more reliable and faster feedback especially in case of fast moving objects, because of the artifacts and distortions generated by the object movement. Also speed, accuracy, and resolution of the z-position estimation still is a topic where significant improvements can be made.

The AFM has emerged as an important tool for the fabrication of nanoscale structures, as well as the handling and manipulation of biological objects and for the manipulation of nanoscale objects. To broaden the applicability of those techniques for productive use, it will become necessary to increase throughput significantly. For the lithographic techniques used to structure surfaces, the usage of multiple AFM tips in parallel will greatly enhance throughput. To speed up the assembly of individual nanoentities, achieving a high level of automation is an important research goal for the future. Further miniaturization of gripping tools can facilitate the nanorobotic pick-and-place handling of CNTs with smaller diameters. The optimization of the presented automated assembly sequence of CNT-enhanced AFM probes will lead to a high-throughput prototyping of CNT-based devices, which is an important pre-stage toward a direct integration of CNTs into microstructures by CMOS-compatible CVD-based fabrication techniques.

Cross-References

- ▶ AFM in Liquids
- ▶ AFM Probes
- ▶ AFM
- ▶ Atomic Force Microscopy
- ▶ Basic MEMS Actuators
- ▶ BioPatterning
- ▶ Biosensors
- ▶ Carbon Nanotubes for Chip Interconnections
- ▶ Carbon-Nanotubes
- ▶ Chemical Vapor Deposition (CVD)
- ▶ Dielectrophoresis
- ▶ Dielectrophoretic Nanoassembly of Nanotubes onto Nanoelectrodes
- ▶ Dip-Pen Nanolithography
- ▶ DNA Manipulation Based on Nanotweezers
- ▶ Electric-Field-Assisted Deterministic Nanowire Assembly
- ▶ Electron-Beam-Induced Deposition
- ▶ Manipulating
- ▶ Microfabricated Probe Technology
- ▶ Nanogrippers
- ▶ Nanoparticles
- ▶ Nanorobotic Assembly
- ▶ Nanorobotic Manipulation of Biological Cells
- ▶ Nanorobotic Spot Welding
- ▶ Nanorobotics
- ▶ Nanorobotics for Bioengineering

- ► Nanorobotics for NEMS using Helical Nanostructures
- ► Nanostructured Functionalized Surfaces
- ► Nanostructured Materials for Sensing
- ► Optical Tweezers
- ► Organic Bioelectronics
- ► Piezoelectric Effect at Nanoscale
- ► Piezoelectric MEMS Switch
- ► Scanning Electron Microscopy
- ► Scanning Tunneling Microscopy
- ► Self-assembled Monolayers
- ► Self-assembly
- ► Self-Assembly for Heterogeneous Integration of Microsystems
- ► Self-assembly of Nanostructures
- ► Thermal Actuators
- ► Transmission Electron Microscopy

References

1. Sattler, K.D.: Handbook of Nanophysics – Nanomedicine and Nanorobotics. Taylor & Francis/CRC Press, Boca Raton (2010)
2. Fatikow, S., Wich, T., Dahmen, C., Jasper, D., Stolle, C., Eichhorn, V., Hagemann, S., Weigel-Jech, M.: Nanohandling robot cells. In: Sattler, K.D. (ed.) Handbook of Nanophysics – Nanomedicine and Nanorobotics, pp. 47.1–47.31. Taylor & Francis/CRC Press, Boca Raton (2010)
3. Fatikow, S. (ed.): Automated Nanohandling by Microrobots. Springer, London (2008)
4. Fatikow, S., Rembold, U.: Microsystem Technology and Microrobotics. Springer, Berlin/Heidelberg/New York (1997)
5. Fatikow, S.: Mikroroboter und Mikromontage (in german). B.G. Teubner, Stuttgart/Leipzig (2000)
6. Ferreira, A., Mavroidis, C.: Virtual reality and haptics for nanorobotics. IEEE Robot. Autom. Mag. **13**, 78–92 (2006)
7. Fatikow, S., Dahmen, C., Wortmann, T., Tunnell, R.: Visual feedback methods for nanohandling automation. Int. J. Inf. Acquis. **6**, 159–169 (2009)
8. Sievers, T., Fatikow, S.: Real-time object tracking for the robot-based nanohandling in a scanning electron microscope. J. Micromechatron **3**(3–4), 267–284 (2006). Special Issue on Micro/Nanohandling
9. Kratochvil, B.E., Dong, L.X., Nelson, B.J.: Real-time rigid-body visual tracking in a scanning electron microscope. Int. J. Robot. Res. **28**, 498–511 (2009)
10. Hutchinson, S., Hager, G.D., Corke, P.I.: A tutorial on visual servo control. IEEE Trans. Robot. Autom. **12**, 651–670 (1996)
11. Binnig, G., Quate, C.F., Gerber, C.: Atomic force microscope. Phys. Rev. Lett. **56**, 930–933 (1986)
12. Baró, A.M., Bartolome, A., Vazquez, L., García, N., Reifenberger, R., Choi, E., Andres, R.P.: Direct imaging of 13-Å-diam Au clusters using scanning tunneling microscopy. Appl. Phys. Lett. **51**, 1594–1596 (1987)
13. Tseng, A.A., Notargiacomo, A., Chen, T.P.: Nanofabrication by scanning probe microscope lithography: a review. J. Vac. Sci. Technol. B **23**, 877–894 (2005)
14. Brousseau, E., Krohs, F., Caillaud, E., Dimov, S., Gibaru, O., Fatikow, S.: Development of a novel process chain based on atomic force microscopy scratching for small and medium series production of polymer nanostructured components. ASME J. Manuf. Sci. Eng. **132**, 030901 (2010)
15. Krohs, F., Onal, C., Sitti, M., Fatikow, S.: Towards automated nanoassembly with the atomic force microscope: a versatile drift compensation procedure. ASME J. Dyn. Syst., Meas, Control **131**, 061106 (2009)
16. Iijima, S.: Helical microtubules of graphitic carbon. Nature **354**, 56–58 (1991)
17. Dai, H., Hafner, J.H., Rinzler, A.G., Colbert, D.T., Smalley, R.E.: Nanotubes as nanoprobes in scanning probe microscopy. Nature **384**, 147–150 (1996)
18. Srivastava, N., Li, H., Kreupl, F., Banerjee, K.: On the applicability of single-walled carbon nanotubes as VLSI interconnects. IEEE Trans. Nanotechnol. **8**, 542–559 (2009)
19. Sardan, O., Eichhorn, V., Petersen, D.H., Fatikow, S., Sigmund, O., Bøggild, P.: Rapid prototyping of nanotube-based devices using topology-optimized microgrippers. Nanotechnology **19**, 495503 (2008)
20. Eichhorn, V., Fatikow, S., Wich, T., Dahmen, C., Sievers, T., Andersen, K.N., Carlson, K., Bøggild, P.: Depth-detection methods for microgripper based CNT manipulation in a scanning electron microscope. J. Micro-Nano Mechatron. **4**, 27–36 (2008)
21. Andersen, K.N., Petersen, D.H., Carlson, K., Mølhave, K., Sardan, O., Horsewell, A., Eichhorn, V., Fatikow, S., Bøggild, P.: Multimodal electrothermal silicon microgrippers for nanotube manipulation. IEEE Trans. Nanotechnol. **8**, 76–85 (2009)
22. Castillo, J., Dimaki, M., Svendsen, W.E.: Manipulation of biological samples using micro and nano techniques. Integr. Biol. **1**, 30–42 (2009)

Robotic Insects

► Insect Flight and Micro Air Vehicles (MAVs)

Rolled-Up Nanostructure

► Nanorobotics for NEMS Using Helical Nanostructures

Rolled-Up Nanotechnology

► Electrical Impedance Tomography for Single Cell Imaging

Rose Petal Effect

Bharat Bhushan[1] and Michael Nosonovsky[2]
[1]Nanoprobe Laboratory for Bio- & Nanotechnology and Biomimetics, The Ohio State University, Columbus, OH, USA
[2]Department of Mechanical Engineering, University of Wisconsin-Milwaukee, Milwaukee, WI, USA

Synonyms

Petal effect

Definition

Rose petal effect is the ability of certain rough surfaces to have a high contact angle with water simultaneously with high adhesion (large contact angle hysteresis) with water.

Theoretical

Recent experimental findings and theoretical analyses made it clear that the early Wenzel [1] and Cassie and Baxter [2] models do not explain the complexity of interactions during wetting of a rough surface which can follow several different scenarios. As a result, there are several modes of wetting of a rough surface, and therefore, wetting cannot be characterized by a single number such as the contact angle (CA). Contact angle hysteresis (the difference between the advancing and receding CA) is another important parameter which characterizes wetting (Fig. 1).

Wetting of a solid is dependent upon the adhesion of water molecules to the solid. On the one hand, a high CA is a sign of low liquid–solid adhesion. On the other hand, low CA hysteresis is a sign of low liquid–solid adhesion as well. It is now widely believed that a surface can be superhydrophobic and at the same time strongly adhesive to water [3–9]. The so-called petal effect is exhibited by a surface that has a high CA,

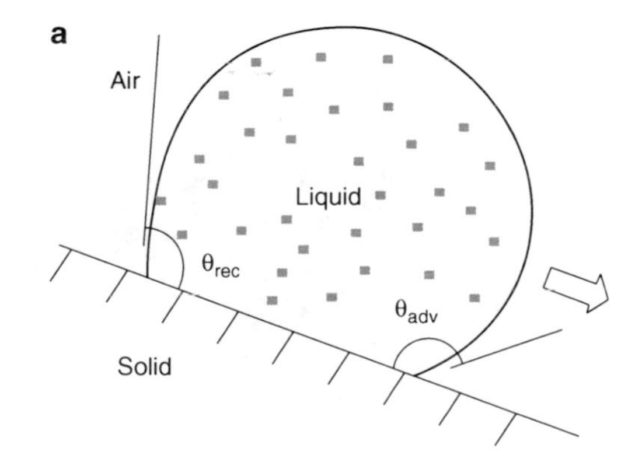

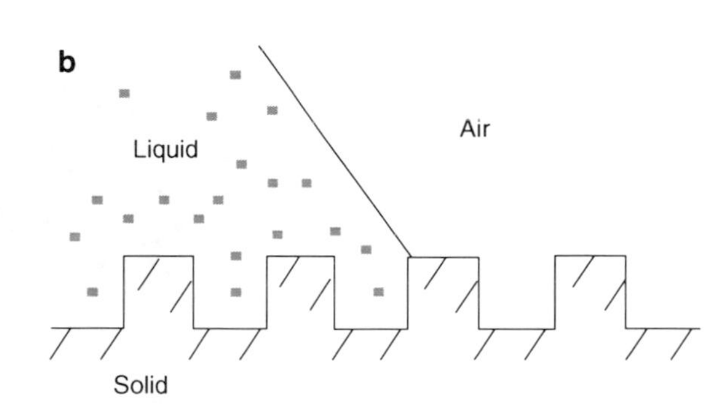

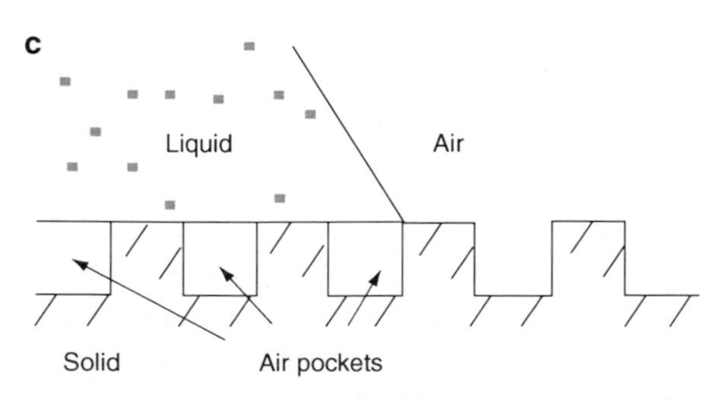

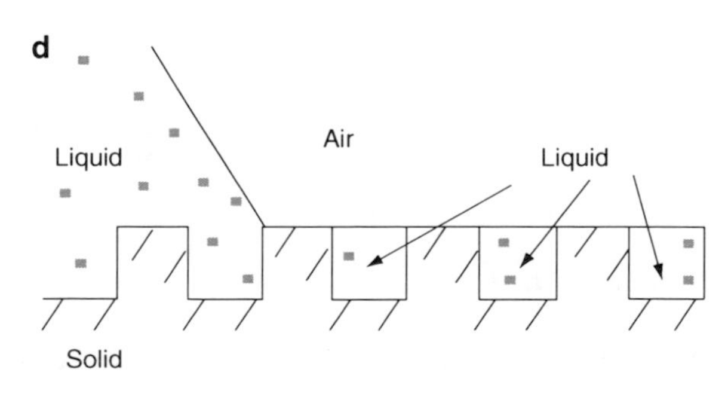

Rose Petal Effect, Fig. 1 (**a**) Schematics of a droplet on a tilted substrate showing advancing (θ_{adv}) and receding (θ_{rec}) contact angles. The difference between these angles constitutes the contact angle hysteresis. Configurations described by (**b**) the Wenzel equation for the homogeneous interface, (**c**) Cassie–Baxter equation for the composite interface with air pockets, and (**d**) the Cassie equation for the homogeneous interface

Rose Petal Effect, Fig. 2 (**a**) Optical images, (**b**) scanning microscope micrographs, and (**c**) atomic force microscope roughness maps of petals of two roses (Rosa Hybrid Tea, cv. Bairage) (Rosa, cv. Bairage), and Rosa Hybrid Tea, cv. Showtime (Rosa, cv. Showtime) (Adapted from Bhushan and Her [3])

but also a large CA hysteresis and strong adhesion to water.

Main Findings

Plant leaves and petals provide an example of surfaces with high CA, and high and low CA hysteresis. Bhushan and Her [3] studied two kinds of superhydrophobic rose petals: (1) Rosa Hybrid Tea, cv. Bairage and (2) Rosa Hybrid Tea, cv. Showtime, referred to as Rosa, cv. Bairage and Rosa, cv. Showtime, respectively. Figure 2 shows optical micrographs, scanning electron microscopy (SEM) images, and atomic force microscope (AFM) surface height maps of two rose petals. Figure 3 shows a sessile and a suspending water droplet on Rosa, cv. Bairage, demonstrating that it can simultaneously have high CA, high adhesion, and high CA hysteresis.

The surface roughness of the two rose petals was measured with the AFM, and the results for the peak-to-base height of bumps, the mid-width, peak radius,

Rose Petal Effect, Fig. 3 Optical micrographs of water droplets on Rosa, cv. Bairage at 0° and 180° tilt angles. Droplet is still suspended when the petal is turned upside down ([3])

Rose Petal Effect, Table 1 Surface roughness statistics for the two rose petals [3]

	Peak-to-base height (μm)	Mid-width (μm)	Peak radius (μm)	Bump density (1/10,000 μm^2)
Rosa, cv. Bairage (high adhesion)	6.8	16.7	5.8	23
Rosa, cv. Showtime (low adhesion)	8.4	15.3	4.8	34

Rose Petal Effect, Table 2 Wetting regimes of a surface with a single level of hierarchy of roughness

State	Cassie–Baxter	Wenzel	Impregnating Cassie
Cavities	Air	Water under droplet	Water everywhere
CA	High	High	High
CA hysteresis	Low	Can be high	Low

Rose Petal Effect, Table 3 Different regimes of wetting of a surface with dual roughness

	Air in microstructure	Water under droplet in microstructure	Water impregnating microstructure
Air in nanostructure	Lotus, high CA, low CA hysteresis	Rose, high CA, high CA hysteresis	Rose-filled microstructure
Water under droplet in nanostructure	Cassie (air-filled microstructure, water in nanostructure), high CA, low CA hysteresis	Wenzel (water in micro- and nanostructure), high CA, high or low CA hysteresis	Wenzel-filled microstructure
Water impregnating nanostructure	Cassie-filled nanostructure	Wenzel-filled nanostructure	Wenzel-filled micro and nanostructure

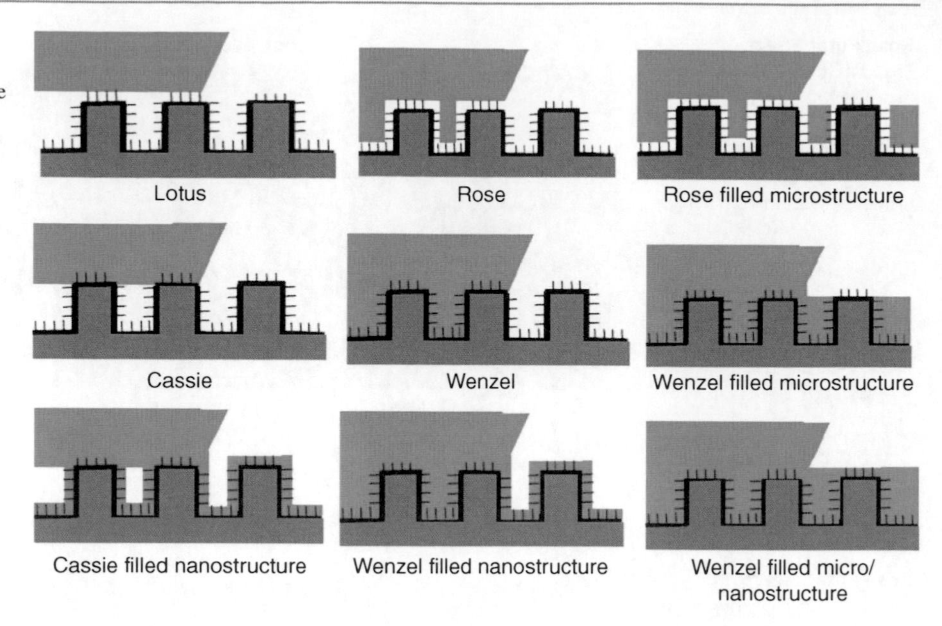

Rose Petal Effect, Fig. 4 Schematics of nine wetting scenarios for a surface with hierarchical roughness

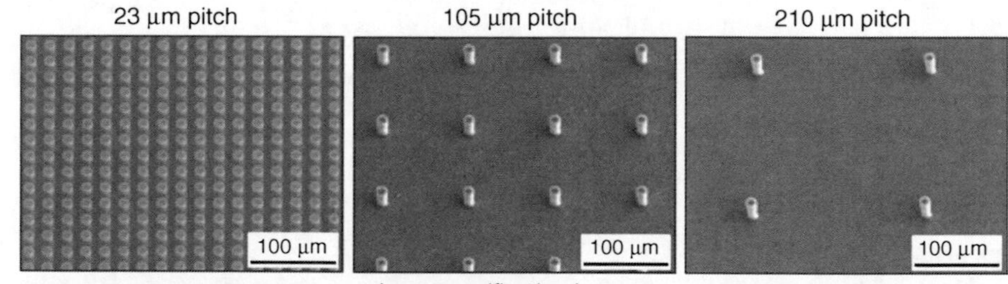

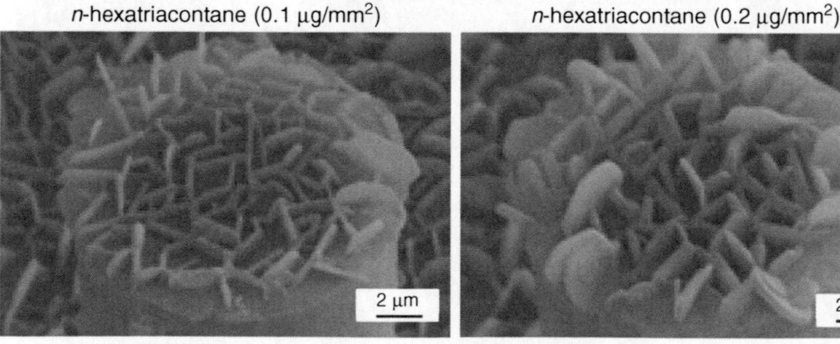

Rose Petal Effect, Fig. 5 SEM micrographs of the microstructures and nanostructures fabricated with two different masses of n-hexatriacontane for hierarchical structure. All images were taken at 45° tilt angle. All samples are positive replicas, obtained from negative replica with dental wax and Si micropatterned master template (14 μm diameter and 30 μm height) fabricated with epoxy resin coated with n-hexatriacontane [3]

and bump density are summarized in Table 1. The data indicates that the low adhesion specimen (Rosa, cv. Showtime) has higher density and height of the bumps, indicating that the penetration of water between the microbumps is less likely. Wetting of a rough surface with a single level of hierarchy of roughness details can follow several scenarios (Table 2).

For a hierarchical structure with small bumps on top of larger bumps, a larger number of scenarios is available, and they are summarized in Table 3 and Fig. 4.

Rose Petal Effect, Table 4 CA and CA hysteresis for surfaces with various micro- and nanoroughness (Based on Bhushan and Her [3])

Mass of n-hexatriacontane ($\mu g/mm^2$)	Pitch					
	23 μm		105 μm		210 μm	
	CA	CA hysteresis	CA	CA hysteresis	CA	CA hysteresis
0.1	164°	3°	152	87	135	45
0.12	165°	3°	153	20	135	42
0.16	166°	3°	160	5	150	12
0.2	167°	3°	168	4	166	3

Water can penetrate either into the micro- or nanostructure, or into both. In addition, the micro- or nanostructure can be impregnated by water or air. The regimes with water penetrating into the microstructure can have high solid–water adhesion and therefore, high CA hysteresis.

Bhushan and Her [3] conducted a series of carefully designed experiments to decouple the effects of the micro- and nanostructures. They synthesized microstructured surfaces with pillars out of epoxy resin. The epoxy surfaces were reproduced from model Si templates and were created by a two-step molding process producing a dual replica (first a negative replica and then a positive replica of the original Si template). Surfaces with a pitch (the periodicity of the structure of the pillars) of 23, 105, and 210 μm and with the same diameter (14 μm) and height (30 μm) of the pillars were produced. After that, nanostructures were created on the microstructured sample by self-assembly of the alkane n-hexatriacontane ($CH_3(CH_2)_{34}CH_3$) deposited by a thermal evaporation method. Alkanes of varying chain lengths are common hydrophobic compounds of plant waxes. On smooth surfaces, alkanes can cause a large contact angle and a small contact angle hysteresis for water droplets. To fabricate the nanostructure, various masses of n-hexatriacontane were coated on a microstructure. The nanostructure is formed by three-dimensional platelets of n-hexatriacontane. Platelets are flat crystals, grown perpendicular to the surface. They are randomly distributed on the surface, and their shapes and sizes show some variation. Figure 5 shows selected images. When different masses of wax are applied, the density of the nanostructure is changed.

For surfaces with a small pitch of 23 μm, while the mass of n-hexatriacontane is changed, there are only small changes in the static contact angle and contact angle hysteresis values, which means that they are always in the "Lotus" wetting regime. On the surface with a 210 μm pitch value, as the mass of n-hexatriacontane is increased, the static contact angle is increased, and the reverse trend was found for the contact angle hysteresis. This was interpreted as evidence that the nanostructure is responsible for the CA hysteresis and low adhesion between water and the solid surface. The results are summarized in Table 4. The wetting regimes are shown schematically in Fig. 6 as a function of the pitch of the microstructure and the mass of n-hexatriacontane. Small mass of the nanostructure material correspond to the Cassie and Wenzel regimes, whereas high mass of the nanostructure corresponds to the Lotus and Rose regimes. The Lotus regime is more likely for larger masses of the nanostructure material. Figure 7 shows a droplet on a horizontal surface of a hierarchical structure with 23 and

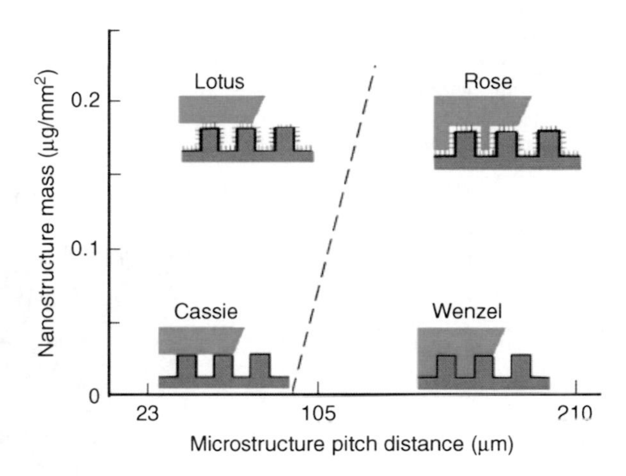

Rose Petal Effect, Fig. 6 Schematic of a wetting regime map as a function of microstructure pitch and the mass of nanostructure material. The mass of nanostructure material equal to zero corresponds to microstructure only (with the Wenzel and Cassie regimes). Higher mass of the nanostructure material corresponds to higher values of pitch, at which the transition occurs

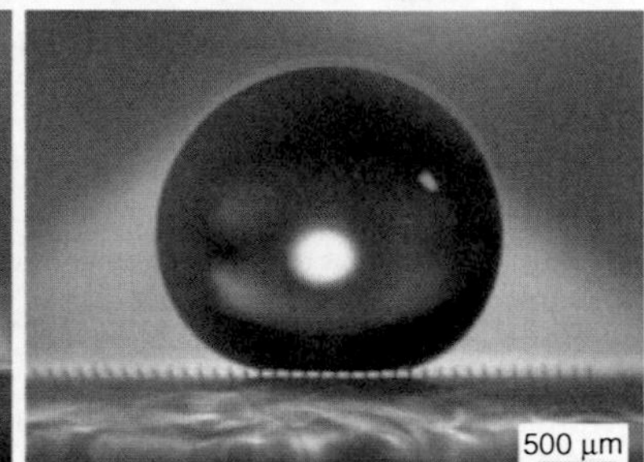

Rose Petal Effect, Fig. 7 (**a**) Droplet on a horizontal surface of hierarchical structure with 23 μm pitch and n-hexatriacontane (0.1 μg/mm^2) showing air pocket formation and (**b**) droplet on a hierarchical structure with 105 μm pitch and n-hexatriacontane (0.1 μg/mm^2) and 0.2 μg/mm^2 showing no air pocket and air pocket formation, respectively. Also shown is the image taken on the inclined surface with hierarchical structure with 0.1 μg/mm^2 showing that droplet is still suspended [3].

105 μm pitches and n-hexatriacontane (0.1 μg/mm^2). Air pockets are observed in the first case and not observed in the second case, indicating the difference between the two regimes [3].

To further verify the effect of wetting states on the surfaces, evaporation experiments with a droplet on a hierarchical structure coated with two different amounts of n-hexatriacontane were performed. Figure 8 shows the optical micrographs of a droplet evaporating on two different hierarchical structured surfaces. On the n-hexatriacontane (0.1 μg/mm^2)-coated surface, an air pocket was not visible at the bottom area of the droplet. However, the droplet on the surface has a high static contact angle (152°) since the droplet still cannot completely impregnate the nanostructure. The footprint size of the droplet on the surface has only small changes from 1820 to 1791 μm. During evaporation, the initial contact area between

Evaporation of a droplet on the surface with 105 μm pitch value and different mass of *n*-hexatriacontane

Hierarchical structure with *n*-hexatriacontane (0.1 μg/mm^2) -- Regime A

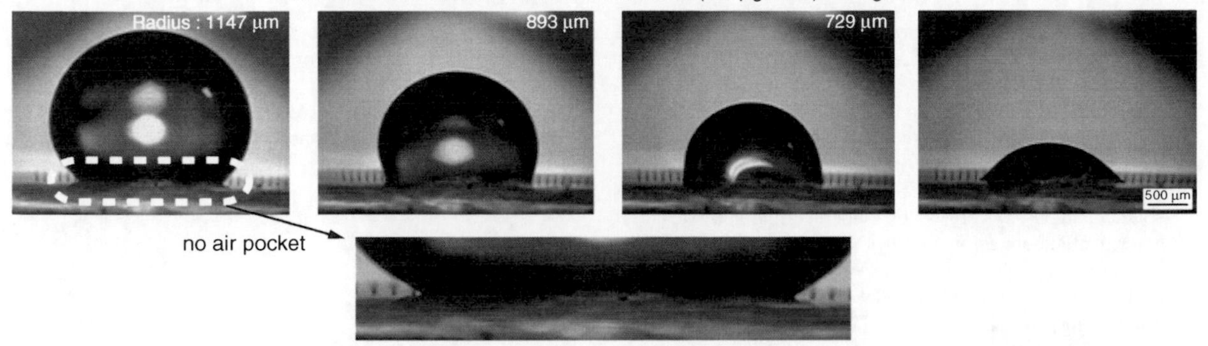

Hierarchical structure with *n*-hexatriacontane (0.2 μg/mm^2) -- Regime B_2

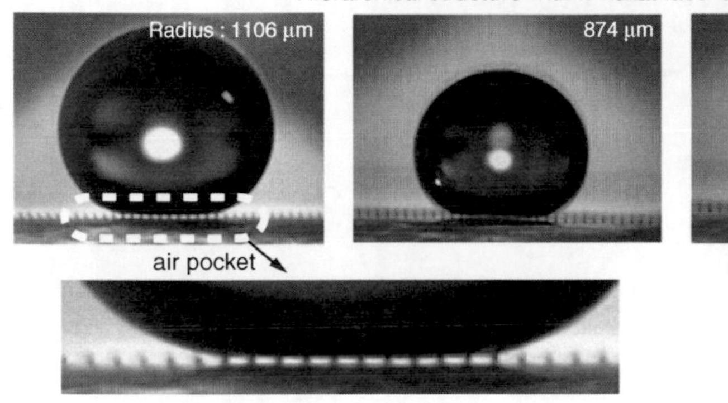

Rose Petal Effect, Fig. 8 Optical micrographs of droplet evaporation on the hierarchical structured surfaces with 105 μm pitch value. n-Hexatriacontane (0.1 μg/mm^2)-coated sample has no air pocket formed between the pillars in the entire contact area until evaporation was completed. Hierarchical structure with n-hexatriacontane (0.2 μg/mm^2) has air pocket, and then the transition from the "Lotus" regime to the "Rose petal" regime occurred [3].

the droplet and hierarchical structured surface does not decrease until the droplet evaporates completely, which means complete wetting between droplet and microstructures. For the n-hexatriacontane (0.2 μg/mm^2)-coated surface, the light passes below the droplet, and air pockets can be seen, so to start with the droplet is in the Cassie-Baxter regime. When the radius of the droplet decreased to 381 μm, the air pockets are not visible anymore. The footprint size of the droplet on the surface is changed from 1177 to 641 μm, since droplet remained on only a few pillars until the end of the evaporation process [9].

The experimental observations of the two types of rose petals show that hierarchically structured plant surfaces can have both adhesive and non-adhesive properties at the same time with high CA. This is due to the existence of various modes of wetting of a hierarchical surface, so that water can penetrate either into macro- or nanoroughness, or into both. Water penetration into the microroughness tends to result in high adhesion with the solid surface, whereas the presence of the nanoroughness still provides high CA. As a result, two distinct modes of wetting are observed. One can be called the "Lotus" mode (with low adhesion) and another is the "rose" mode with high adhesion.

Cross-References

- ▶ Biomimetics
- ▶ Lotus Effect

References

1. Wenzel, R.N.: Resistance of solid surfaces to wetting by water. Indust. Eng. Chem. **28**, 988–994 (1936)

2. Cassie, A., Baxter, S.: Wettability of porous surfaces. Trans. Faraday Soc. **40**, 546–551 (1944)
3. Bhushan, B., Her, E.K.: Fabrication of superhydrophobic surfaces with high and low adhesion inspired from rose petal. Langmuir **26**, 8207–8217 (2010)
4. Bormashenko, E., Stein, T., Pogreb, R., Aurbach, D.: "Petal Effect" on surfaces based on lycopodium: high-stick surfaces demonstrating high apparent contact angles. J. Phys. Chem. C **113**, 5568–5572 (2009)
5. Chang, F.M., Hong, S.J., Sheng, Y.J., Tsao, H.K.: High contact angle hysteresis of superhydrophobic surfaces: hydrophobic defects. Appl. Phys. Lett. **95**, 064102 (2009)
6. Feng, L., Zhang, Y., Xi, J., Zhu, Y., Wang, N., Xia, F., Jiang, L.: Petal effect: a superhydrophobic state with high adhesive force. Langmuir **24**, 4114–4119 (2008)
7. Gao, L., McCarthy, T.J.: Teflon is hydrophilic. Comments on definitions of hydrophobic, shear versus tensile hydrophobicity, and wettability characterization. Langmuir **24**, 9184–9188 (2008)
8. Jin, M.H., Feng, X.L., Feng, L., Sun, T.L., Zhai, J., Li, T.J., Jiang, L.: Superhydrophobic aligned polystyrene nanotube films with high adhesive force. Adv. Mater. **17**, 1977–1981 (2005)
9. Bhushan, B., Nosonovsky, M.: The rose petal effect and the modes of superhydrophobicity. Philos. Trans. Royal. Soc. A. **368**, 4713–4728 (2010)

Rotation Sensors

▶ Gyroscopes

S

Sand Skink

► Friction-Reducing Sandfish Skin

SANS

► Small-Angle Scattering of Nanostructures and Nanomaterials

SAXS

► Small-Angle Scattering of Nanostructures and Nanomaterials

Scanning Electron Microscopy

Yimei Zhu[1] and Hiromi Inada[2]
[1]Center for Functional Nanomaterials, Brookhaven National Lab, Upton, NY, USA
[2]Hitachi High-Technologies America, Pleasanton, CA, USA

Synonyms

Scanning electron microscopy and secondary-electron imaging microscopy

Definition

A scanning electron microscope (SEM) is an instrument that uses high-energy electrons in a raster-scan pattern to form images, or collect other signals, from the three-dimensional surface of a sample.

Introduction

The scanning electron microscope (SEM) is one of the most popular and user-friendly imaging tools that reveal the surface topography of a sample. It is also widely used for structural characterization of materials and devices, especially in the field of nanotechnology. Today there are in excess of 50,000 SEMs worldwide and it is often seen as a "must-have" apparatus for research institutes and industry laboratories. In the SEM, incident electrons interact with the atoms that make up the sample-producing signals that contain information about the sample's surface morphology, composition, and other physical and chemical properties. The most common imaging mode in SEM lies in using secondary electrons. Since secondary electrons have very low energies, they are generated in, and escape from regions near the sample surface. Combined with various detection systems, a SEM also can be used to determine the sample's chemical composition through energy-dispersive x-ray spectroscopy and Auger electron spectroscopy, and identify its phases through analyzing electron-diffraction patterns, mostly via high-energy backscattered electrons. Besides backscattered electrons, Auger electrons, characteristic x-rays, other signals are generated from the interactions of the incident

B. Bhushan (ed.), *Encyclopedia of Nanotechnology*, DOI 10.1007/978-90-481-9751-4,

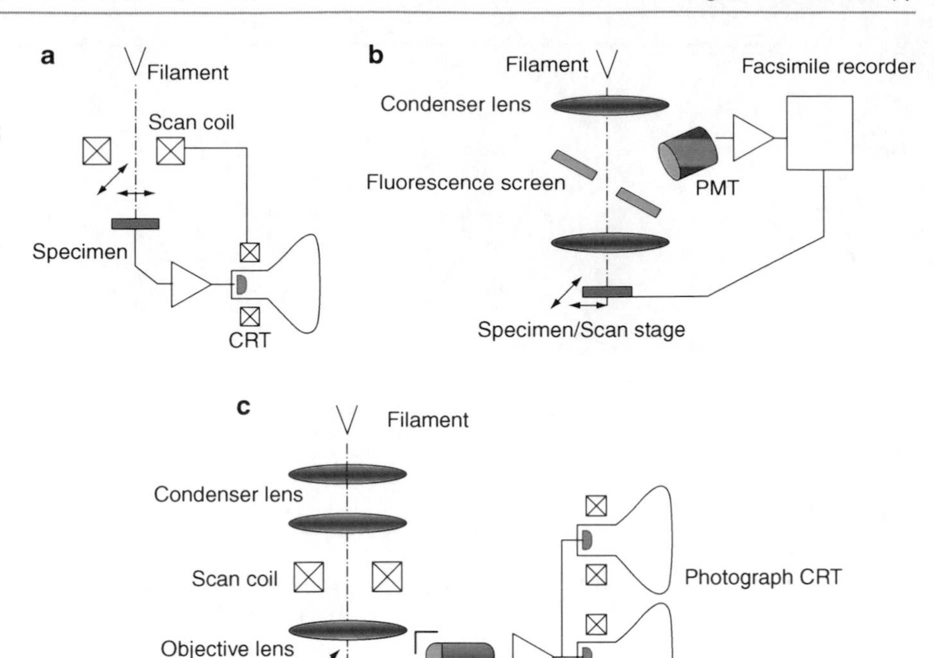

Scanning Electron Microscopy, Fig. 1 Schematics of configurations of the scanning electron microscopes proposed and developed by some notable pioneers in the SEM history. (**a**) Knoll in 1935, (**b**) Zworykin in 1945, and, (**c**) Cambridge Instruments in 1965

beam and the sample under a SEM include plasmons, bremsstrahlung radiation (noncharacteristic x-rays), cathodoluminescence, and electron-beam-induced current. This entry is focused on secondary-electron imaging related instrumentation, signal-generation processes, and state-of-the-art imaging capabilities in SEM.

A Brief History

SEM was invented by Max Knoll in 1935 in Germany [1] to study the targets of television tubes. The instrument consisted of electron-beam deflector coils that scan the beam on a plate as the sample in a cathode ray tube (CRT), and an amplifier that boosts the plate current to display the signal on another CRT (Fig. 1a). Two years later Manfred von Ardenne built an electron microscope with a highly demagnified probe using two condenser lenses for scanning transmission electron microscopy and also tried it as an SEM. Zworykin and his coworkers of the RCA Laboratories in the USA designed and built a dedicated SEM in 1942. Its electron optics includes three electrostatic lenses with scan coils placed between the second and third ones. A photomultiplier was first used to detect secondary electrons (Fig. 1b). The essential components of this apparatus are similar to those used in modern SEMs. The probe size of the incident, or primary, electron beam had a diameter of about 10 nm. However, compared with transmission electron microscopes (TEMs) at that period, it could not image secondary electrons satisfactorily due to the poor signal-to-noise ratios of the images [3]. Sir Charles Oatley and Dennis McMullan built their first SEM at Cambridge University in 1948. The SEM technology was further pioneered by many postgraduate students at Cambridge including Gary Stewart. The first commercial instrument, named as "Stereoscan," was launched in 1965 by the Cambridge Scientific Instrument Company for DuPont [4]. The instrument consists of electron multiplier detector with beryllium–copper dynodes to detect scattered electrons from the specimen surface. Images were displayed on a CRT, while another synchronized CRT recorded them on camera film. An Everhart-Thornley type secondary-electron detector [5] significantly improved the detection efficiency of the low-energy secondary-electron signals (Fig. 1c). JEOL produced first commercial Japanese SEM, JSM-1, in 1966, while Hitachi commercialized its SEM, HSM-2, in 1969.

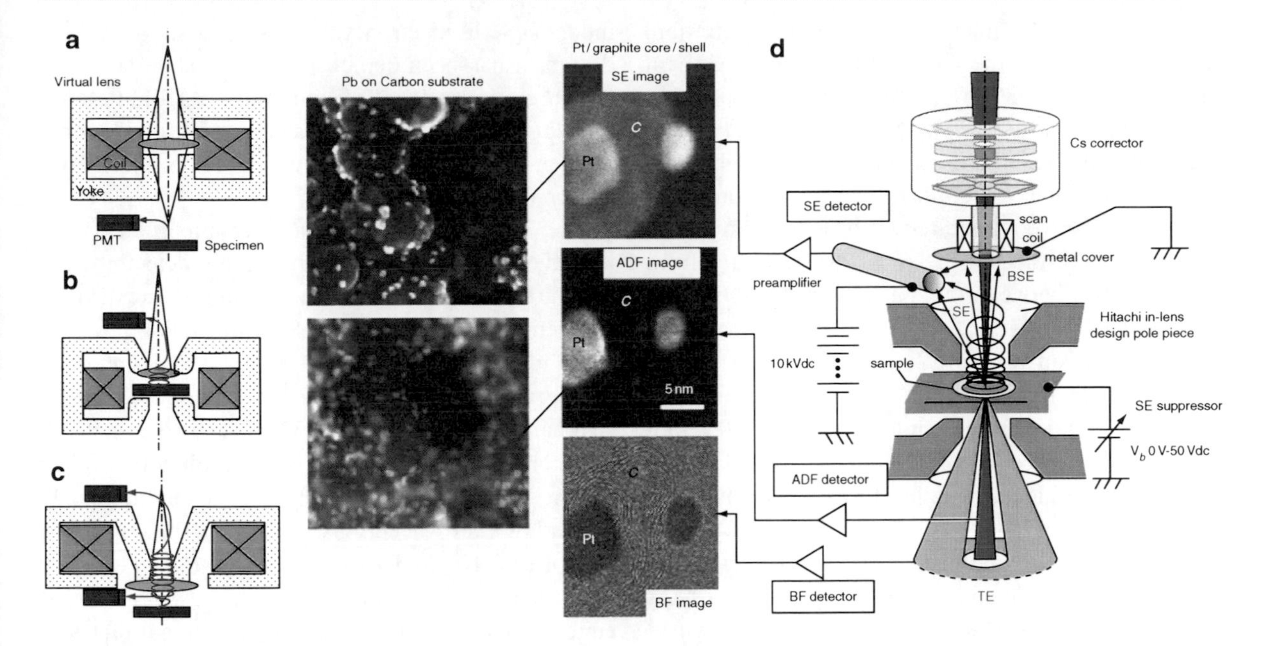

Scanning Electron Microscopy, Fig. 2 (**a**–**c**) Three different objective lens designs. (**a**) Out-lens, (**b**) in-lens, and (**c**) semi-in-lens. (**d**) Schematic of the lens-detector configuration of the aberration-corrected scanning electron microscope, Hitachi HD2700C, that routinely achieves an atomic resolution in imaging using secondary electrons and transmitted electrons. Two examples are shown on the *left*; one is Pb particles on a carbon support, the other is Pt particles with carbon-graphite shells. *BF* bright-field, *ADF* annular dark-field, *SE* secondary electrons, *TE* transmitted electrons, and *BES* backscattered electrons. The SE images clearly give a topographic view of the area, and higher brightness of the light element C, compared with the corresponding ADF images

Instrumentation

A typical SEM consists of an electron gun, an electron lens system, various electron-beam deflection coils, electron detectors, and display and recording devices [6]. In a SEM, an electron beam is emitted from an electron gun sitting on a thermionic filament cathode, or from a field-emission needle tip. Tungsten is often used for thermionic electron guns due to its low cost, high melting point, and low vapor pressure so that it tolerates heating to about 2,800 K for electron emission. Other types of electron emitters include lanthanum hexaboride (LaB6) cathodes, which offer higher brightness but require a better vacuum to avoid oxidizing the gun. Field emission guns (FEGs) often used in modern SEMs can be the thermally assisted Schottky type, using emitters of zirconium oxide (ZrO), or the cold-cathode type using tungsten <310 > single crystal emitters and operated at room temperature. A cold field-emission gun has a much smaller source size ($5 \sim 10$ nm) than a tungsten filament ($1 \sim 10$ μm) with three to four orders of magnitude larger current density and brightness [7].

The typical energy range of the electron beam used in SEM is from 0.5 to 40 keV. The electron-condenser lens system usually demagnifies the electron source more than hundreds of times to form a small probe on the sample. The beam passes through pairs of deflection coils, or scanning coils, in the electron column, typically in the final lens, which deflect the beam in the x and y axes by applying an incremental current into the scan coils, so that it scans in a raster fashion over a rectangular area of the sample surface.

Historically, SEM incorporates an objective lens. Unlike the objective lens in optical microscopes or transmission electron microscopes (TEMs), its purpose in SEM is not to image the sample, but to focus the small probe on the sample. There are three types of objective lenses: out-lens (Fig. 2a); in-lens (Fig. 2b); and, semi in-lens (Fig. 2c). Most early SEMs had the simplest out-lens design, in which the sample sits beneath the lens leaving a large area available in the sample chamber. However, the yoke gap across the optical axis acts as a lens with leakage field, thus yielding significant imaging aberration. The in-lens pole piece, originally designed for TEMs, was adopted for SEM to reduce spherical aberration. Hitachi developed the first commercial SEM with such a design in 1985 [8]. Their microscope operating at 30 kV reaches a probe size of 0.5 nm with spherical aberration coefficient C_s of 1.6 mm. The drawback of the design is that a conventional thin sample, similar to that used in TEM, is required because the sample sits inside the pole-piece gap. A design resulting from a compromise between the out-lens and in-lens is the semi in-lens pole piece (Fig. 2c), offering reasonable spatial resolution but with larger open space for the sample chamber so that a thick sample can be used. Such a design allows the sample to be placed a few mm from the pole piece, and a large magnetic field to be applied to produce a smaller focal length and less spherical aberration. SEMs with the semi in-lens design played a significant role in characterizing devices for the semiconductor industry in early 1990s.

It is noteworthy that the in-lens design shown in Fig. 2b is very similar to that used in conventional TEM and/or STEM (scanning transmission electron microscope). Thus, with an efficient secondary-electron detector, a TEM, or an STEM also can image a sample surface [9, 10]. Figure 2d shows a schematic of the objective lens (in-lens design), the sample, and the arrangement of detectors in the Hitachi HD2700C aberration-corrected SEM/STEM [11]. The instrument operates at 80 ~ 200 kV, allowing simultaneous acquisition of bright-field (BF), annular dark-field (ADF), and secondary-electron (SE) images. In the BF and ADF modes, the transmitted electrons are used to form the images, thus providing structural information from the sample's interior. In contrast, in the SE mode electrons emerging from the surface with low energies, or short escape length, are used, and thus, the signals are surface sensitive. Different imaging modes have their own advantages and limitations. Since the BF signals are close to the phase contrast seen in TEM, they offer high spatial resolution, but are difficult to interpret. ADF images, on the other hand, are based on Rutherford scattering, and thus their image intensity is directly related to the sample's atomic number Z (the so-called Z-contrast imaging). Since BF- and ADF-images are projected images they give little structural information in the direction of the beam's trajectory. In contrast, SE imaging offers depth information on the surface topography, and is more sensitive to the light elements than is ADF imaging (see Fig. 2d). Combining these different imaging modes in an electron microscope with an in-lens design, it has been demonstrated that simultaneously imaging both surface (SE) and bulk (ADF) at atomic resolution is possible in thin samples for a wide range of elements, from uranium and gold to silicon and carbon [11, 12].

The electron detector is another important part of the SEM instrumentation. Although detector itself does not determine the image resolution in SEMs, it is essential to improving the SEM resolving power in terms of the signal-to-noise ratio. The commonest detector used in SEMs today still is that developed by Everhart and Thornley in 1957 [5], the so-called E-T detector. The detector consists of a photomultiplier, a light guide, and a positively biased scintillator. To attract low energy electrons effectively, the scintillator is applied a 10 kV dc bias to accelerate the electrons. The energized electrons cause the scintillator to emit flashes of light (cathodoluminescence) that then are transmitted to the photomultiplier. The amplified output of the electrical signals by the photomultiplier is displayed as a two-dimensional intensity distribution that can be viewed and photographed on an analogue video display. The Everhart-Thornley detector, which normally is positioned to one side above the specimen, exhibits low efficiency in detecting backscattered electrons because few such electrons are emitted in the solid angle subtended by the detector. Furthermore, its positive bias cannot readily attract the high-energy backscattered electrons (close to the energy of the incident electrons). Backscattered electrons are usually collected above the sample in a "doughnut type" arrangement, concentric with the electron beam, to maximize the solid angle of collection.

Secondary-Electron Signal Generation

Secondary-electron (SE) imaging is the most frequently used mode of imaging in SEM. Secondary electrons, defined as the electrons with energy below 50 eV, are generated along the primary electrons' trajectories within the sample, but are subject to elastic and inelastic scattering during their passage through the sample. These electrons can be valence electrons or are ejected from the orbits of the inner shells (most likely the k-shells) of the sample. The consequence of their low kinetic energy is their shallow escape depth, which is about 1 nm for metals, and up to 10 nm for

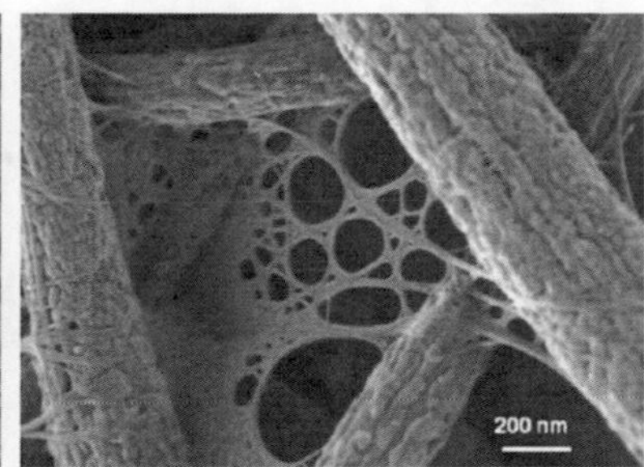

Scanning Electron Microscopy, Fig. 3 SEM micrographs of a polymer membrane that consists of electrospun fibrous scaffold for water filtration. The fibrous scaffold has an average diameter of 200 nm. The high-resolution SEM image on the right shows cellulose nanofibers (5–10 nm in diameter) infused into the scaffold form a cellulose network to enhance the membrane's ability to remove bacteria and viruses. The images were taken with JEOL7600 SEM at operation voltage of 0.5 eV. A thin layer of carbon was coated on the sample to avoid charging. Note the bright contrast at the edge of the fibers generates a topological view of the fibrous structure

insulators. The probability of escape decreases continuously with the increase of the depth below the surface. However, there is a nonzero probability of secondary-electron emission arising from inelastic scattering below the escape depth, as, for example, when a primary electron creates a fast secondary electron (energy > 50 eV) that travels toward the surface and generates a lower-energy secondary electron within the escape depth. The production of the secondary electron signals involves the generation, propagation, escape from the surface, and arrival at the detector. These four processes are detailed in the atomic-imaging section.

The secondary electrons discussed here are often referred to as SE_I, i.e., the secondary electrons generated by the incident beam upon entering the sample. Secondary electrons generated by backscattered electrons when leaving the sample are termed SE_{II}. Secondary electrons generated when the backscattered electrons strike a lens pole piece or the sample chamber's wall, and by primary electrons hitting the aperture are, respectively called SE_{III} and SE_{IV}. Although SE_{III} contain the information on the sample, SE_{IV} do not. Furthermore, there are fast secondary electrons, with energy higher than 50 eV. Bias experiments, wherein a positive dc voltage is applied to the sample to suppress the emission of the secondary electrons, were mainly designed to separate the SE_I from backscattered electrons; they cannot distinguish SE_I from SE_{II}, which for a very thin specimen should be negligible. Measuring other types of secondary electrons, including those with high energy, would require a different bias experiment. Heavy elements (high atomic number) that backscatter electrons more strongly than do light elements (low atomic number), and thus appear brighter in the image, can be used to yield contrast that contains information of a sample's chemical composition.

Figure 3 shows SEM images of a nanofiber-containing polymer membrane developed for water filtration with enhanced ability to remove bacteria and viruses. The image on the right shows the structural network formed by the 5–10 nm diameter cellulose nanofibers. It reveals astonishing topographic details on how the cellulose nanofibers are interwoven with scaffold. In SEM images flat surfaces give even contrast while curved surfaces and sharp edges often appear brighter (high image intensity). This is due to the increased escape of secondary electrons from the top and side surfaces when the interaction volume intercepts them. Since the secondary electron detector is biased with a high acceleration voltage, a surface facing away from the detector still can be imaged. Surfaces tilted away from the normal to the beam allow more secondary electrons to escape [13].

Atomic Imaging Using Secondary Electrons

In the last decade or so, high-resolution SEM has proven an indispensable critical dimension metrology tool for the semiconductor industry. The roadmap for semiconductor nanotechnology identifies the need for ultra-high-resolution SEM in the quest for ever-decreasing device sizes. In a SEM, the size of

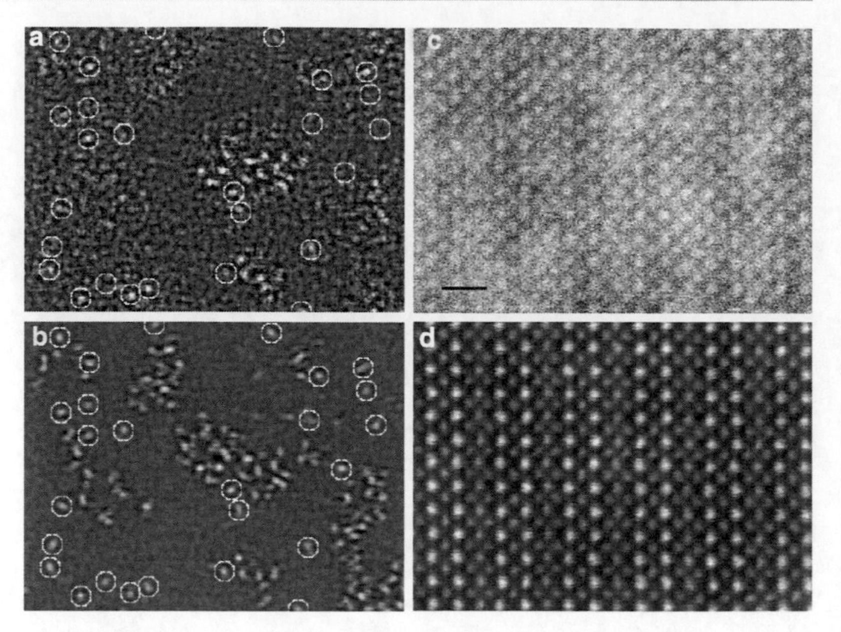

Scanning Electron Microscopy, Fig. 4 (a–b) Simultaneous atomic imaging using secondary electrons (secondary-electron mode, **a**) and transmitted electrons (annual dark-field mode, **b**) of uranium individual atoms on a carbon support (raw data). The circles mark the single uranium atoms. The atoms shown in (**b**) but not in (**a**) are presumably those on the back side of the support. (**c–d**) Simultaneous atomic imaging using secondary electrons (**c**) and transmitted electrons (**d**) of $YBa_2Cu_3O_7$ superconductor viewing along the [010] direction (raw data). The scale bar in (**c**) is 0.7 nm

the imaging probe often determines the instrument's resolution power. The probe size d (measured in full-width and half maximum) is a function of the beam convergence half-angle α is an incoherent sum of contributions from source size, diffraction limit, spherical aberration, and chromatic aberration, and is given by

$$d^2 = \left(\frac{4i_p}{\beta\pi^2\alpha^2}\right)^2 + \left(\frac{0.6\lambda}{\alpha}\right)^2 + (0.5C_s\alpha^3)^2 + \left(C_c\alpha\frac{\Delta E}{E}\right)^2$$

where $i_p$is the probe current, α is the convergence half-angle, β is the source brightness, λ is the electron wavelength at beam energy E, ΔE is the energy spread, and C_s and C_c are, respectively, the spherical- and chromatic-aberration coefficients of the probe-forming lens. The first term on the right side of the equation is the probe size defined by the source size, the second term is due to the diffraction limit, and the third and the fourth terms are due, respectively, to the spherical and chromatic aberrations. Recent advancement on correcting spherical aberration in SEM and STEM diminishes C_s to zero, thus eliminating the third term and produces a small probe. It is important to note that the probe size also depends on the energy spread in the fourth term that includes the energy distribution of the primary electron beam and the fluctuation of the instruments' acceleration voltage.

In SEM, the image resolution depends not only on the instrument, but also on the sample, or, more accurately, on the sampling volume of the sample from which the signal is generated When the incident electrons impinge on a point of the sample's surface, they interact with atoms in the sample via elastic- (change trajectory) and inelastic- (lose energy) interactions. The size of the interaction volume depends on the electron's landing energy, the atomic number of the sample, and its density. The simultaneous energy loss and change in trajectory spreads beam into the bulk of the sample and produces an interaction volume therein, that, for a thick sample, can extend from less than 100 nm to around 5 μm into the surface, i.e., more than an order of magnitude larger than the original probe size. For a very thin sample (a few nm thick), the interaction volume, as the first approximation, might be defined as the probe size.

Figure 4 illustrates the atomic resolution images using secondary electrons on single uranium atoms (a) and the (010) surface of a $YBa_2Cu_3O_7$ crystal (c) recorded on the Hitachi HD2700C SEM/STEM (Fig. 2d). For comparison, the corresponding annular dark-field images using transmitted electrons, (b) and (d), respectively, are included. The equally sharp images in SE and ADF suggest negligible imaging delocalization. Such an attainable resolution was attributed to the combination of several factors: Better design of the electro-optics of the instrument (including ultrahigh electric and mechanical stabilities) and the detector; aberration correction that reduces the probe size and increases the probe current; and, the higher operation voltage that beneficially assures a very small volume of beam interaction for a thin sample.

The physical mechanism of producing the low-energy secondary electrons traditionally is ascribed to inelastic scattering and decay of collective electron excitation with the incident electron giving up, say 20 eV, to produce a secondary electron with energy 20 eV minus the work function of the surface. Since the momentum transfer (scattering angle) of the scattering with 20 eV energy loss is small, the transfer should be delocalized to an area >1 nm; thus atomic imaging using secondary electrons was not considered possible. Recent studies suggest that this is not the primary mechanism for secondary-electron imaging at least on a thin sample [11]. The secondary electrons responsible for atomic scale resolution are generated by inelastic scattering events with large momentum transfer, including those from inner shell orbitals, which give rise to a sharp central peak in the point-spread function for signal generation.

In general, four steps are involved in producing the signal that is used to form a secondary electron (SE_I) image [12]. (1) The generation of secondary electrons through the inelastic scattering of primary electrons in the sample, at a generation rate, G; (2) random motion of these secondary electrons, which are scattered by atoms of the specimen both elastically and inelastically (potentially creating other secondary electrons of lower energies), such that, on average, T electrons reach the sample surface for each secondary electron generated; (3) the escape of secondary electrons over the potential barrier at the sample surface of the specimen, with an average probability P; and, (4) the acceleration of the emitted electrons in vacuum, such that a fraction D reaches the electron detector.

The secondary-electron signal S is a product of these four factors: $S = G \cdot T \cdot P \cdot D$. To generate contrast in a scanned-probe image, one or more of the above steps must depend on the x-coordinate of the electron probe in the scan direction, i.e.,

$$\frac{dS}{dx} = I_0 TPD\frac{dG}{dx} + I_0 GPD\frac{dT}{dx} + I_0 GTD\frac{dP}{dx} + I_0 GTP\frac{dD}{dx}$$

For most non-atomically resolved SE images obtained in an SEM, $\frac{dT}{dx}$ provides the main contrast mechanism: Secondary electrons created at an inclined surface or close to a surface step have an increased probability of escape, resulting in surface-topography contrast [14]. Less commonly, variations in surface work function contribute additional contrast by providing a nonzero $\frac{dP}{dx}$. In voltage-contrast applications, changes in surface voltage provide a nonzero $\frac{dD}{dx}$. Atomic-number contrast is possible if the specimen is chemically inhomogeneous and G varies with atomic number, yielding a nonzero $\frac{dG}{dx}$. However, for atomic imaging in a thin sample (Fig. 4), the dominant mechanism can be quite different. $\frac{dG}{dx}$ is likely to play an important role in the atomic-scale contrast as a consequence of Z-dependence inelastic scattering cross section and channeling effect for crystals (for thickness in the order of extinction distance it should be minor). Because secondary electrons are generated through inelastic scattering of the incident electrons, $\frac{dG}{dx}$ is limited by the delocalization of the scattering process, which may be described by the point-spread function for inelastic scattering. The term $\frac{dT}{dx}$ should be small, for adatoms or surface atoms that lie on the detector side of the sample. $\frac{dP}{dx}$ would not become important unless the effective work function varies on an atomic scale, and $\frac{dD}{dx}$ must be also negligible at atomic scale. These assumptions are reasonable because the scattering process disperses secondary electrons over a range of x that is comparable to the escape depth, typically 1–2 nm. Consequently, T, P, and D are x-averages that vary little with x on an atomic scale. For uranium atoms on a carbon substrate, the argument is even simpler; these atoms lie outside the solid, so the terms T and P are not applicable.

Future Remarks

Secondary-electron imaging is the most popular mode of operation of the scanning electron microscope (SEM) and traditionally is used to reveal surface topography. Nevertheless, this imaging method never was regarded as being on the cutting edge of performance, due to its perceived limited spatial resolution in comparison with its TEM or STEM counterparts using transmitted electrons. Recent work using aberration-corrected electron microscopes demonstrated that secondary electron signals in the SEM can resolve both crystal lattices and individual atoms, showing SEM's unprecedented and previously unrealized imaging capabilities. Furthermore, the work demonstrates the incompleteness of present understanding of the formation of secondary-electron images. Secondary-electron imaging using high acceleration voltage with thin samples can now compete with TEM on spatial resolution, and provide new capabilities, such as depth-resolved profiles, at the atomic level. There seems to be no fundamental reason why atomic resolution in secondary-electron imaging could not be achieved at the accelerating voltages of 0.5–40 keV that are currently used in conventional SEMs. It remains to be seen whether the integrated spherical and chromatic aberration of a probe-forming objective lens can be corrected to a sufficient degree at low operation voltages. One clear outcome thus far is the importance of preparing samples with clean surfaces in order to obtain interpretable and reproducible results.

Acknowledgments The work was supported by the US Department of Energy, Office of Basic Energy Science, Materials Science and Engineering Division, under Contract Number DEAC02–98CH10886.

Cross-References

- ► Robot-Based Automation on the Nanoscale
- ► Scanning Tunneling Microscopy
- ► Transmission Electron Microscopy

References

1. Knoll, M.: Aufladepotentiel und sekundäremission elektronenbestrahlter körper. Z. Tech. Phys. **16**, 467–475 (1935)
2. von Ardenne, M.: Das Elektronen-Rastermikroskop. Praktische Ausführung. Z. Tech. Phys. **19**, 407–416 (1938) (in German)
3. Wells, O.C., Joy, D.C.: The early history and future of the SEM. Surf. Interface Anal. **38**, 1738–1742 (2006)
4. Oatley, C.W.: The early history of the scanning electron microscope. J. Appl. Phys. **53**, R1–R13 (1982)
5. Everhart, T.E., Thornley, R.F.M.: Wide-band detector for micro-microampere low-energy electron currents. J. Sci. Instrum. **37**, 246–248 (1960)
6. Goldstein, G.I., Newbury, D.E., Echlin, P., Joy, D.C., Fiori, C., Lifshin, E.: Scanning Electron Microscopy and x-ray Microanalysis. Plenum, New York (1981)
7. Pawley, J.: The development of field-emission scanning electron microscopy for imaging biological surfaces. Scanning **19**, 324–336 (1997)
8. Tanaka, K., Mitsushima, A., Kashima, Y., Osatake, H.: A new high resolution scanning electron microscope and its application to biological materials. In: Proceedings of the 11th International Congress On Electron Microscopy, Kyoto, pp. 2097–2100 (1986)
9. Liu, J., Cowley, J.M.: High resolution SEM in a STEM instrument. Scanning Microsc. **2**, 65–81 (1988)
10. Howie, A.: Recent developments in secondary electron imaging. J. Microsc. **180**, 192–203 (1995)
11. Zhu, Y., Inada, H., Nakamura, K., Wall, J.: Imaging single atoms using secondary electrons with an aberration-corrected electron microscope. Nat. Mater. **8**, 808–812 (2009)
12. Inada, H., Su, D., Egerton, R.F., Konno, M., Wu, L., Ciston, J., Wall, J., Zhu, Y.: Atomic imaging using secondary electrons in a scanning transmission electron microscope: experimental observations and possible mechanisms. Ultramicroscopy **111**(7), 865–876 (2011). doi:10.1016/j.ultramic.2010.10.002. Invited articles for the special issue in honor of John Spence
13. Joy, D.C.: Beam interactions, contrast, and resolution in the SEM. J. Microsc. **136**, 241–258 (1984)
14. Reimer, L.: Scanning Electron Microscopy, 2nd edn. Springer, New York (1998)

Scanning Electron Microscopy and Secondary-Electron Imaging Microscopy

► Scanning Electron Microscopy

Scanning Force Microscopy in Liquids

► AFM in Liquids

Scanning Kelvin Probe Force Microscopy

► Kelvin Probe Force Microscopy

Scanning Near-Field Optical Microscopy

Achim Hartschuh
Department Chemie and CeNS,
Ludwig-Maximilians-Universität München,
Munich, Germany

Synonyms

Near-field scanning optical microscopy (NSOM)

Definition

Scanning near-field optical microscopy (SNOM) is a microscopic technique for nanostructure investigation that achieves sub-wavelength spatial resolution by exploiting short-ranged interactions between a sharply pointed probe and the sample mediated by evanescent waves. In general, the resolution of SNOM is determined by the lateral probe dimensions and the probe-sample distance. Images are obtained by raster-scanning the probe with respect to the sample surface corresponding to other scanning-probe techniques. As in conventional optical microscopy, the contrast mechanism can be combined with a broad range of spectroscopic techniques to study different sample properties, such as chemical structure and composition, local stress, electromagnetic field distributions, and the dynamics of excited states.

Introduction

Optical microscopy forms the basis of most of the natural sciences. In particular, life sciences have benefited from the fascinating possibility to study smallest structures and processes in living cells and tissue. Optical techniques feature extremely high detection sensitivity reaching single molecule sensitivity in fluorescence, Raman scattering and absorption spectroscopy. Besides the direct visualization, chemically specific information is obtained through Raman spectroscopy.

The resolution of conventional optical microscopes, however, is limited by diffraction, a consequence of the wave nature of light, to about half the wavelength. Concepts extending optical microscopy down to nanometer length scales below the diffraction-limit are distinguished into far-field and near-field techniques. Far-field techniques rely on the detection of propagating waves at distances from the source larger than the wavelength, while near-field techniques exploit short ranged evanescent waves.

Scanning near-field optical microscopy, initiated by pioneering work of Pohl, Lewis and others in the late 1980s and the beginning of 1990s gave access to nanoscale resolution for the first time. The history of near-field optics is reviewed in [1]. In addition numerous review articles and books exist describing fundamentals and applications (see e.g., [2–4]).

This entry introduces first the key physical principles beginning with the role of evanescent and propagating waves, and the loss of spatial information upon light propagation. Concepts of near-field detection are shown using different pointed probes. The next sections describe the experimental realization and present several applications of SNOM. The outlook addresses future prospects of SNOM and remaining challenges.

Key Principles and Concepts

Near-field optics has its origin in the effort of overcoming the diffraction limit of optical imaging. The physical origin of this limit is sketched in the following starting with the distinction between propagating waves that form the optical far-field of a radiation source and their evanescent counterpart that dominate the optical near-field. A powerful tool to describe wave propagation is the so-called angular spectrum representation of fields expressing the electromagnetic field E in the detector plane at z as the superposition of harmonic plane waves of the form $exp\,(i\vec{k}\vec{r} - i\omega t)$ with amplitudes $\bar{E}(k_x, k_y, z = 0)$ emanating from the source plane at $z = 0$ [2].

$$E(x, y, z) = \iint_{-\infty}^{+\infty} \bar{E}(k_x, k_y, z = 0) e^{i(k_x x + k_y y)} e^{\pm i k_z z} dk_x dk_y \qquad (1)$$

The wave vector $\vec{k}$ describing the propagation direction of the wave is represented by its components $\vec{k} = (k_x, k_y, k_z)$ while its length is fixed by the wavelength of light λ and the refractive index of the medium n through $\left|\vec{k}\right| = \sqrt{k_x^2 + k_y^2 + k_z^2} = 2\pi n/\lambda$. In Eq. 1, the time dependence of the fields has been omitted for clarity. For simplicity the following discussion is limited to the x-z-plane and $n = 1$ such that $k_z = \sqrt{4\pi^2/\lambda^2 - k_x^2}$. In Eq. 1 the term $e^{\pm i k_z z}$ controls the propagation of the associated wave: For $k_x \leq 2\pi/\lambda$ the component k_z is real and the corresponding wave with amplitude $\bar{E}(k_x, z = 0)$ propagates along the z-axis oscillating with $e^{-ik_z z}$. If $k_x > 2\pi/\lambda$ the component k_z becomes complex and $e^{-|k_z z|}$ describes an exponential decay of the associated wave that is therefore evanescent. As a result, only waves with $k_x \leq 2\pi/\lambda$ can

S

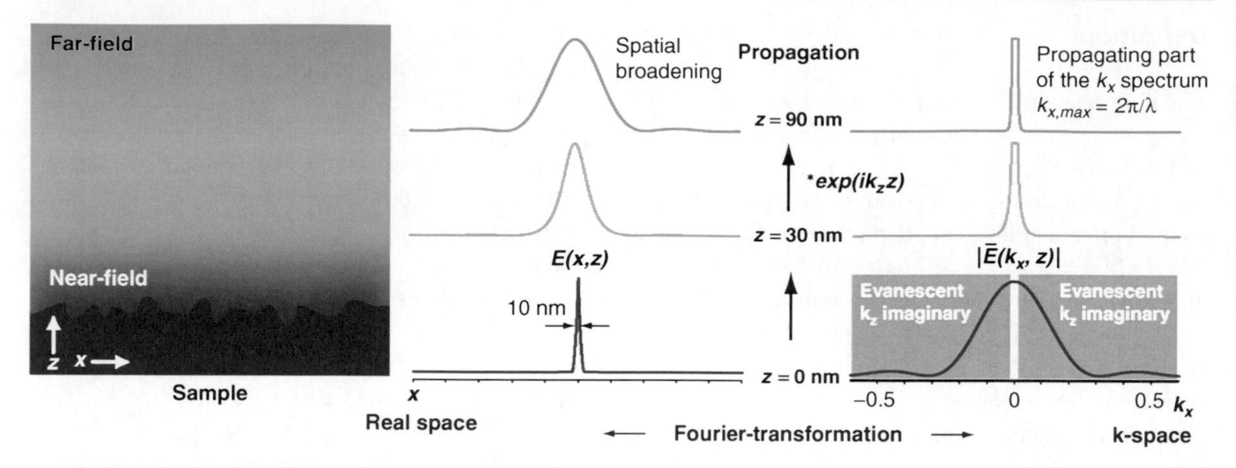

Scanning Near-Field Optical Microscopy, Fig. 1 Scheme illustrating the propagation of waves and the loss of spatial information. Initial field distribution $E(x, z = 0)$ at a 10-nm wide source in the x-z-plane (*center*) and corresponding angular spectrum $|\bar{E}(k_x, z = 0)|$ (*right*). Near the source the spectrum contains both evanescent and propagating waves. Upper panels illustrate the evolution of the fields at $z = 30$ nm and $z = 90$ nm distance for a source wavelength of $\lambda = 500$ nm in vacuum. Only waves with $k_x \leq 2\pi/\lambda$ propagate. Evanescent waves decay exponentially following $e^{-|k_z|z}$. The decay of high spatial frequencies leads to spatial broadening and loss of spatial information in the far-field

propagate and contribute to the field far from the source forming the far-field. Figure 1 schematically illustrates this behavior: In the center, the electric field $E(x, z)$ emanating from a narrow sub-wavelength source at $z = 0$ is shown together with its angular spectrum $\bar{E}(k_x, z = 0)$ calculated by the inverse of Eq. 1. The wave amplitudes $\bar{E}$ result from the Fourier-transformation of E with respect to the spatial coordinate x. As for the correlation between time and frequency domain, where a short optical pulse requires a broad frequency spectrum, a sharp field distribution requires a broad spectrum of spatial frequencies k_x.

Since only waves with limited spatial frequencies can propagate, the spectral width rapidly decreases with increasing distance from the source z leading to fast broadening of the electric field distribution in real space. In other words, propagation corresponds to low-pass filtering with frequency limit $k_{x,\,max} = 2\pi/\lambda$. The far-field thus contains limited spatial frequencies equivalent to limited spatial information. To overcome this limitation different near-field concepts have been developed that are outlined in the following.

The key concept of SNOM is the probing of the sample near-field that contains the evanescent waves using a sharply pointed probe. Since evanescent waves decay rapidly for increasing distance to the source, the probe needs to be in close proximity to the sample. Waves with large k_x components that carry high spatial information decay most rapidly following $e^{-|k_z|z}$ as can be seen from Fig. 1. Hence the spatial resolution obtained in an SNOM experiment drops fast with increasing tip-sample distance z. As for other scanning probe techniques that exploit short-ranged interactions, such as atomic force and scanning tunneling microscopy, AFM and STM, respectively, the lateral resolution is also determined by the lateral dimension of the probe. Two conceptually different types of probes can be distinguished: The first confines and samples electromagnetic fields using an aperture with sub-wavelength diameter (Fig. 2a). The second exploits the antenna concept that couples locally enhanced near-fields to propagating waves and vice versa (Fig. 2b–d). The two types, termed aperture and antenna probe, respectively, are illustrated in the following.

Aperture Probes

Aperture probes confine light by squeezing it through a sub-wavelength hole (Fig. 2a). This approach, termed aperture-SNOM, provides an enormous flexibility regarding signal formation. Different operation modes can be used that are capable of local sample excitation and/or local light collection. Depending on which step of the experiment exploits near-field interactions to obtain sub-wavelength resolution, aperture-SNOM can be implemented in excitation ①, collection ② and excitation-collection ③ mode (Fig. 2a).

The original scheme was proposed by Synge in 1928. He suggested to use a strong light source behind

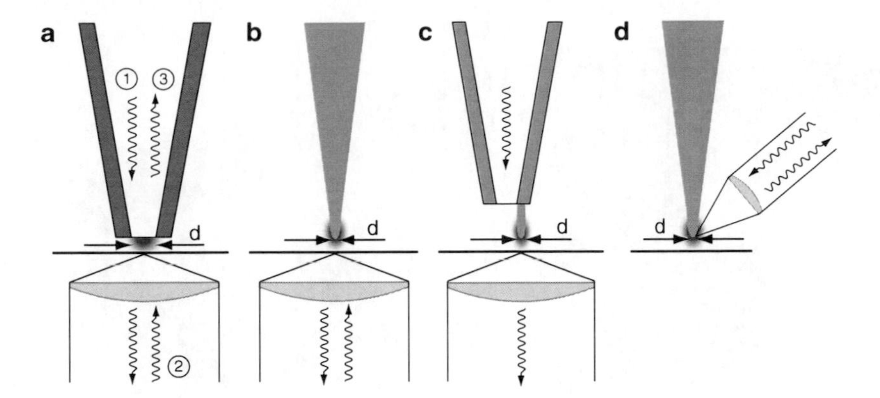

Scanning Near-Field Optical Microscopy, Fig. 2 Schematics of the most common SNOM probes and configurations: (**a**) Aperture probes are utilized in different modes: Near-field excitation – far-field collection ①, far-field excitation – near-field collection ② and near-field excitation – collection mode ③. (**b**) Tip-enhanced near-field optical microscopy (TENOM) using local field enhancement at a sharp metal antenna probe upon far-field excitation. (**c**) Tip-on-aperture (TOA) probe using local field enhancement at an antenna probe that is excited in the near-field of an aperture probe. (**d**) Scattering-SNOM (s-SNOM) based on far-field illumination and detection of local scattering at a sharp antenna probe. The label *d* indicates the structural parameter that determines the achievable spatial resolution

a thin, opaque metal film with a 100-nm diameter hole in it as a very small light source. In 1984, two groups adopted this scheme and presented the first experimental realizations in the optical regime [1]. The aperture was formed at the apex of a sharply pointed transparent probe tip coated with metal. Raster-scanning the probe was made possible by the scanning technology developed in the context of scanning tunneling microscopy (STM). In Fig. 3a, a scanning electron microscopy (SEM) image of an aperture probe consisting of a metal-coated tapered glass fiber is shown. At the front surface a well-defined aperture with diameter of 70 nm is seen.

Analytical expressions for the electric field distribution in a sub-wavelength aperture in a metallic screen were already presented in 1944 and 1950 by Bethe and Bouwkamp. Numerical simulations for metal-coated tapered fiber probes show that strong fields pointing in axial direction occur at the rim due to local field-enhancement by the metal coating (see Fig. 4). The center of the aperture is dominated by a weaker horizontal component [2]. The optical field distribution can be determined experimentally by raster-scanning single fluorescent molecules that act as point-like dipoles across the aperture while recording the fluorescence intensity [5] (see section "Fluorescence Microscopy" and Fig. 3).

Tapered aperture probes suffer from low light transmission due to the cut-off of propagating wave-guide modes. For probe diameters below the cut-off diameter only evanescent waves remain and the intensity decays exponentially toward the aperture (Fig. 2c, d). Probe designs, therefore, aim at maximizing the cone angle that determines the distance between aperture and cut-off diameter. Hollow-cantilever probes feature relatively large cone-angles as compared to fiber-based probes (Fig. 2b). On the other hand, the input power needs to be limited because of the damage threshold of the metal coating in case of fiber-based probes. Due to the limited transmission and the skin-depth of the optical fields on the order of several tens of nanometers, most aperture-SNOM measurements are carried out with apertures of 50–100 nm.

Antenna Probes

Antenna probes act as transmitter and receiver coupling locally enhanced near-fields to propagating waves and vice versa (Fig. 2b–d) [8]. To distinguish this approach from the earlier implementations based on apertures, it is also termed apertureless-SNOM or a-SNOM. Antenna probes can be used in two different techniques: (1) Scattering type microscopy [9, 10], also termed scattering-SNOM or s-SNOM, in which the tip locally perturbs the fields near a sample surface. The response to this perturbation is detected in the far-field at the frequency of the incident light corresponding to elastic scattering (Fig. 2d). (2) Tip-enhanced near-field optical microscopy (TENOM) in which locally enhanced fields at laser-illuminated metal structures are used to increase the spectroscopic response of the system at frequencies different from

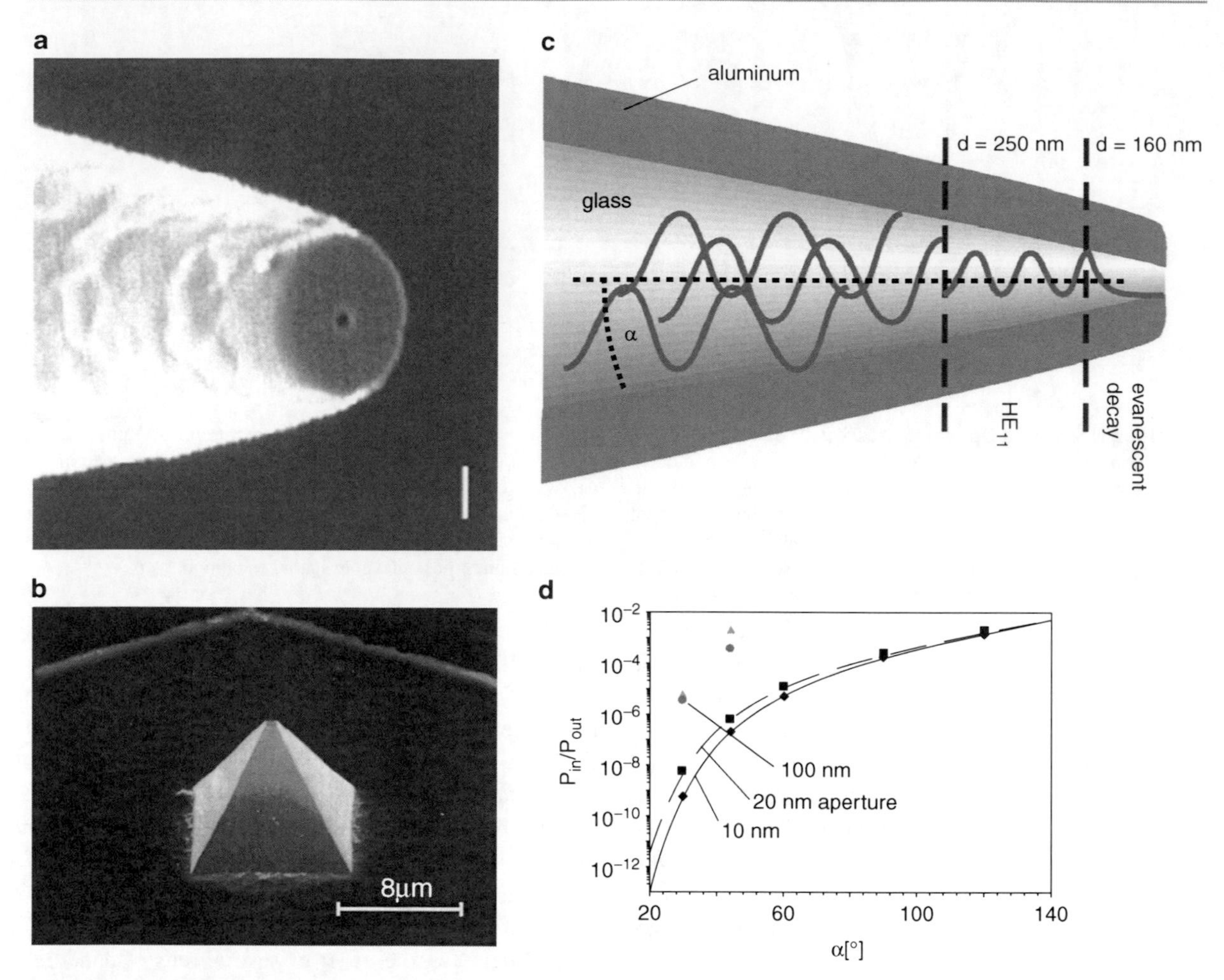

Scanning Near-Field Optical Microscopy, Fig. 3 (**a**) Scanning-electron microscopy (SEM) image of an aperture probe formed by a metal-coated tapered fiber with an aperture diameter of 70 nm (scale bar 200 nm) (Reprinted with permission from Veerman et al. [5]. Copyright 1999, John Wiley and Sons) (**b**) SEM image of a metallic hollow aperture probe microfabricated on a Si cantilever. The aperture diameter is about 130 nm (Reprinted with permission from Mihalcea et al. [6]. Copyright 1996, American Institute of Physics) (**c**) Schematic of the mode propagation in a tapered aperture probe. For probe diameters below the cut-off diameter, here $d = 160$ nm, the intensity decays exponentially toward the aperture. (**d**) Transmission of tapered probes determined as the ratio of input versus output power P_{in}/P_{out} as a function of the cone angle α defined in (**c**). For smaller cone angles, the distance between cut-off and aperture increases leading to extremely low transmission (**c** and **d**) (Reprinted with permission from Hecht et al. [7]. Copyright 2000, American Institute of Physics)

that of the incident light [8, 11] (Fig. 2b, c). The flexibility of this technique allows the study of a variety of spectroscopic signals including Raman scattering (tip-enhanced Raman spectroscopy (TERS)), and fluorescence as well as time-resolved measurements. In the following, the signal formation in the case of optical antennas is sketched.

Elastic scattering signal. The near-field interaction can be treated within a simplified model in which the tip is replaced by a polarizable sphere. Due to the antenna properties of the tip, laser excitation with incident fiel E_i creates a dominating dipole oriented along the tip axis in z-direction normal to the sample surface. This dipole induces a mirror dipole in the sample depending on its dielectric properties. The mirror dipole's field, decreasing with the third power of distance, interacts with the tip dipole. Solving the system of electrostatic equations that describes the multiple interaction between tip and mirror dipoles neglecting retardation yields an effective polarizability of the coupled tip-sample system which fully expresses the influence of the sample.

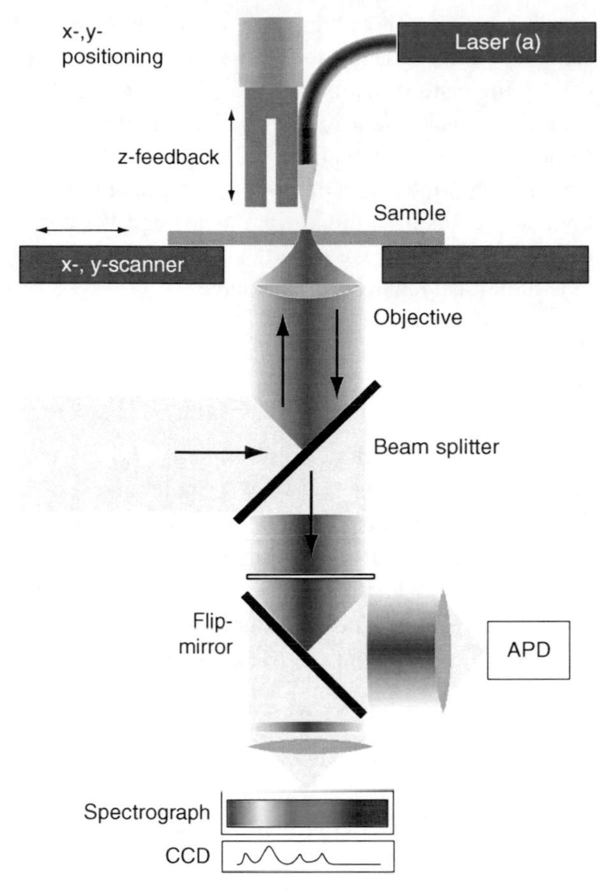

Scanning Near-Field Optical Microscopy, Fig. 4 Schematic of an experimental setup applicable to aperture-probe SNOM (laser (**a**), excitation mode) and TENOM (laser (**b**)) in case of transparent samples. In the case of TENOM the fiber probe would be replaced by an optical antenna, for example, an etched metal wire. The probe tip is positioned in the focus of the microscope objective by piezo-electric actuators. The sample is raster-scanned using a closed-loop *x-y*-scanner while both laser and probe position remain fixed. The tip-sample distance is controlled using a tuning-fork shear-force feedback scheme. The optical signal is collected by the objective and detected either by a highly sensitive avalanche photodiode (APD) or energetically resolved using a spectrograph and a CCD

$$\alpha_{\text{eff}} = \left(\frac{(\alpha (1 + \beta) \;/\; 1 - (\alpha\beta))}{(16\pi (\alpha + z)^3)} \right) \qquad (2)$$

The scattered field can then be calculated from $E_s \propto \alpha_{\text{eff}} E_i$ reflecting the short-ranged interaction required for sub-diffraction resolution. Since the laser illuminates a greater part of the tip and the sample, elimination of background-scattering contributions is crucial. Efficient background suppression can be achieved by demodulating the detected intensity signal at higher harmonics of the tapping mode frequency (see section "Instrumentation"). Besides the amplitude of the scattered field, its phase can be retrieved using interferometric heterodyne detection [9].

Raman scattering signal. In the case of Raman scattering, the total signal depends on the product of the excitation and emission rates $k^{\text{ex}}(\lambda^{\text{ex}})\, k^{\text{rad}}(\lambda^{\text{rad}})$. As a consequence, the total signal enhancement scales with the fourth power of the field enhancement for small differences between the excitation λ^{ex} and emission wavelength λ^{rad} and assuming that the field enhancement at the tip does not depend sensitively on the wavelength.

$$M_{\text{Raman}} = \frac{k_{\text{tip}}^{\text{ex}}}{k_0^{\text{ex}}} \cdot \frac{k_{\text{tip}}^{\text{rad}}}{k_0^{\text{rad}}} \approx f^4 \qquad (3)$$

The factor f measures the ratio between tip-enhanced E_{tip} and non-enhanced electric field E_0 in the absence of the tip. For the general case of surface-enhanced Raman scattering (SERS), Raman enhancement factors are reported reaching up to 12 orders of magnitude for particular multiple particle configurations involving interstitial sites between particles or outside sharp surface protrusions. Since the signal scales with the fourth power, already moderate field enhancement, predicted for a single spherical particle to be in the range of $f = 10$–100 is sufficient for substantial signal enhancement.

Fluorescence signal. The fluorescence intensity depends on the excitation rate k^{ex} and the quantum yield η denoting the fraction of transitions from excited state to ground state that give rise to an emitted photon. The quantum yield is expressed in terms of the radiative rate k^{rad} and the non-radiative rate k^{nonrad} through $\eta = k^{\text{rad}}/(k^{\text{rad}} + k^{\text{nonrad}})$. Accordingly, the fluorescence enhancement due to the presence of the metal tip can be written as

$$M_{\text{Flu}} = \left(\frac{E_{\text{tip}}}{E_0}\right)^2 \left(\frac{\eta_{\text{tip}}}{\eta_0}\right) = f^2 \left(\frac{\eta_{\text{tip}}}{\eta_0}\right). \qquad (4)$$

Here, it is assumed that the system is excited far from saturation. From Eq. 4 it is clear that TENOM works most efficiently for samples with small fluorescence quantum yield η_0 such as semiconducting single-walled carbon nanotubes [11]. For highly fluorescent samples such as dye molecules, the quantum yield η_0 is already close to unity and cannot be enhanced further.

Because of the small separation between emitter and metal tip required for high spatial resolution, non-radiative transfer of energy from the electronically excited state to the metal followed by non-radiative dissipation in the metal has to be taken into account. This process represents an additional competing non-radiative relaxation channel and reduces the number of detected fluorescence photons. Metal-induced fluorescence quenching can also be exploited for image contrast formation (see e.g., [12]). In this case M_{Flu} in Eq. 4 becomes smaller than unity.

While the theory of energy transfer between molecules and flat metal interfaces is well understood in the framework of phenomenological classical theory, nanometer-sized objects are more difficult to quantify [2]. Tip- and particle-induced radiative rate enhancement and quenching has been studied in literature both experimentally and theoretically. Experiments on model systems formed by single dipole emitters such as molecules and semiconductor nanocrystals and spherical metal particles revealed a distance-dependent interplay between competing enhancement and quenching processes. While semiconducting tips cause less efficient quenching, they also provided weaker enhancement because of their lower conductivity at optical frequencies.

Polarization and angular-resolved detection of the fluorescence signal of single emitters demonstrated that the fluorescence rate enhancement provided by the optical antenna also results in a spatial redistribution of the emission [13]. The same redistribution can be expected to occur for tip-enhanced Raman scattering. The spatial distribution of the enhanced electric field follows approximately the outer dimensions of the tip apex. Since the signal enhancement scales with higher orders of field enhancement, the optical resolution can surpass the size of the tip [11]. Stronger fields and field confinement are observed for so-called gap modes formed by metal tips on top of metal substrates.

The scheme depicted in Fig. 2b shows that in addition to the signal resulting from the near-field tip–sample interaction the confocal far-field signal contribution is detected, representing a background. This background originates from a diffraction-limited sample volume that is far larger than the volume probed in the near-field. The near-field signal has to compete with this background, and strong enhancement is required to obtain clear image contrast. This requirement is relaxed in case of low-dimensional sample structures such as spatially isolated molecules or one-dimensional nanostructures [11]. The near-field signal to background ratio can be improved by exploiting the non-linear optical response of sample and tip. Examples include two-photon excitation of fluorescence using a metal tip antenna and the application of the four-wave mixing signal of a metal particle dimer as local excitation source.

Instrumentation

Near-field optical microscopy exploits short-ranged near-field interactions between sample and probe. SNOM instruments thus require a mechanism for tip-sample distance control working on the scale of nanometers. Typical implementations utilize other non-optical short-ranged probe–sample interactions such as force or tunneling current used for topography measurements in AFM and STM, respectively. During image acquisition by raster-scanning the sample with respect to the tip, optical and topographic data are thus obtained simultaneously. Due to the strong tip-sample distance dependence of the near-field signal, cross-talk from topographic variations is possible that can lead to artifacts in the optical contrast. Tip-sample distance curves need to be measured to prove unequivocally the near-field origin of the observed image contrast. Since the optical signal results from a single sample spot only, SNOM instruments are often based on a confocal laser scanning optical microscope equipped with sensitive photodetectors.

Scattering-SNOM is often implemented with intermittent-contact-mode AFM in which the tip-sample distance is modulated sinusoidally at frequencies typically in the range of 10–500 kHz. The elastically scattered laser light intensity is demodulated by the tapping-mode frequency using lock-in detection for background suppression.

Since Raman and fluorescence signals are typically weak requiring longer acquisition times, signal demodulation is more challenging. In the case of single-photon counting time-tagging can be used to retrieve the signal phase with respect to the tapping oscillation. While this can be applied to single-color experiments, spectrum acquisition using CCD cameras in combination with spectrometers is not feasible at typical tapping frequencies.

In most aperture-SNOM and TENOM experiments, the tip-sample distance is kept constant by either using STM, contact/non-contact AFM, or shear-force feedback. Sensitive piezoelectric tuning-fork detection schemes operating at small interaction forces have been developed and are used for fragile fiber and antenna probes [2].

Probe Fabrication

The near-field probe forms the crucial part of an SNOM setup since both optical and topographic signals are determined by its short-ranged interactions with the sample. Instrument development, therefore, focuses mainly on the design of optimized tip concepts and tip geometries as well as on fabrication procedures for sharp and well-defined probes with high reproducibility. These continuing efforts benefit substantially from improving capabilities regarding nanostructuring and nanocharacterization, using, for example, focused ion beam milling (FIB) or electron beam lithography (EBL).

Aperture Probes

Optical fiber probes. Sharply pointed optical fiber probes were the first to provide sub-diffraction spatial resolution and are widely used as nanoscale light sources, light collectors or scatterers (Fig. 2a and 3a). Probe fabrication requires a number of steps starting with the formation of a tapered optical fiber. Typically two different methods are used. Chemical etching of a bare glass fiber dipped into hydro-fluoric (HF) acid yields sharp tips [4, 7]. The surface tension of the liquid forms a meniscus at the interface between air, glass, and acid. A taper is formed due to the variation of the contact angle at the meniscus, while the fiber is etched and its diameter decreased. Chemical etching allows reproducible production of larger quantities of probes in a single step. A specific advantage is that the taper angle can be tuned and optical probes with correspondingly large transmission coefficient can be produced.

The second method combines local heating using a CO_2 laser or a filament and subsequent pulling until the fiber is split apart. The resulting tip shapes depend heavily on the temperature and the timing of the heating and pulling, as well as on the dimensions of the heated area. The pulling method has the advantage of producing tapers with very smooth surfaces, which positively influences the quality of the evaporated metal layer. Etched probes, on the other hand, typically feature rough surfaces. Pulled fibers, however, have small cone angles and thus reduced optical transmission as well as flat end-faces limiting the minimum aperture size.

The aperture is formed during the evaporation of aluminum. Since the evaporation takes place under an angle slightly from behind, the deposition rate of metal at the apex is much smaller than on the sides. This geometrical shadowing effect leads to the self-aligned formation of an aperture at the apex.

The ideal aperture probe should have a perfectly flat end face to position a sample as close as possible into the near-field of the aperture. Conventional probes generally have a roughness determined by the grain size of the aluminum coating, which is around 20 nm at best. Due to the corrugated end face the distance between aperture and sample increases, which lowers the optical resolution and decreases the light intensity on the sample. Furthermore, these grains often obscure the aperture, which makes the probe ill-defined and not suited for quantitative measurements [4, 5, 7]. Subsequent focused ion beam (FIB) milling can be used to form high definition SNOM probes with well-defined end face (Fig. 3).

Microfabricated cantilevered probes. Hollow-pyramid cantilevered probes can be batch-fabricated with large taper angles. Lithographic patterning of an oxidized silicon wafer first defines the position of the aperture and the dimensions of the cantilever beam by structuring the oxide layer [6]. Anisotropic etching of the exposed silicon with buffered HF forms a pyramidal groove for the tip and trenches for the cantilever beam. After removing the oxide layer on the opposite side anisotropic etching is used to open a small aperture in the pyramidal groove. A 120-nm chromium layer is deposited on the back side forming the hollow pyramidal tip that is finally freed by isotropic reactive ion etching.

Si_3N_4 tips can be fabricated by dry etching in a CF_4 plasma and covering with thin aluminum films. Microfabricated probes based on quartz tips attached to silicon cantilevers have also been reported [2, 7]. First sharp quartz tips were produced in hydrofluoric acid followed by coating with thin films of aluminum and silicon nitride. Reactive ion etching was then used to selectively remove the silicon nitride from the tip apex, while the remaining film served as a mask for wet-etching of the protruding aluminum in a standard

S

Al-etching solution, leaving a small aperture on the apex of the tip. For light coupling, windows are etched into the backside of the levers at the position of the tip.

Microfabricated tips provide several advantages over fiber-based probes. The mechanical stability is typically increased and often sufficient to measure also in contact AFM mode without destroying the tip. A reproducible fabrication of the tips leads to a well-defined aperture shape. Large taper angles of around 70° shift the position of the mode cut-off closer to the aperture, resulting in higher transmission.

Antenna Probes

Sharp metal tip are fabricated in a single step by simple electrochemical etching. In the case of gold, pulsed or continuous etching in hydro-chloric acid (HCL) or a mixture of HCL and ethanol routinely yields tips with diameters in the range of 30–50 nm. These tips can either be used in shear-force mode after gluing to the prong of a tuning-fork or in STM-mode. Silicon or silicon nitride cantilevered probes can be coated by a thin metal film through evaporation. Subsequent nanostructuring by FIB-milling can be used to optimize tip parameters.

Tip-on-aperture probes need to be fabricated in a series of sequential steps [14] (Fig. 2c). First, a fiber-based aperture probe is produced on which a well-defined end face is formed by FIB-milling. In the second step, a nanoscale tip is grown by electron beam–induced deposition of carbon. Next, the carbon tip is coated by a thin layer of chromium to improve adhesion of an aluminum layer evaporated during the final preparation step. Since the length of the tip can be controlled during electron beam deposition, TOA probes can be tailored to provide optimum antenna enhancement for a chosen wavelength [8].

Applications in Nanoscience

The energy of light quanta – photons – is in the range of electronic and vibrational excitations of materials. These excitations are directly determined by the chemical and structural composition of matter. Optical spectroscopy, the energy-resolved probing of the material response to light exposure, thus provides a wealth of information on the static and dynamical properties of materials. Combining spectroscopy with near-field microscopy is particularly interesting since spectral information is obtained spatially resolved at the nanoscale. In the following several representative examples covering different material responses including fluorescence, Raman scattering and elastic scattering are briefly illustrated to highlight the capabilities of SNOM techniques.

Fluorescence Microscopy

Fluorescence measurements were among the first applications of SNOM. In these experiments, aperture-based probes were used to probe highly fluorescent dye-molecules on substrates. Aluminum-coated fiber tips were used for excitation while the fluorescence was collected in the far-field by a high-numerical aperture objective. A molecule is excited only if the optical electric field is polarized parallel to its transition dipole moment. The resulting fluorescence patterns rendered by a single molecule with known orientation can thus be used to visualize the local electric field distribution at the probe. Conversely, the molecular orientation can be determined for known field distributions. These experiments showed that the strongest electrical fields do not occur in the center of the aperture, but at the rims of the metal coating. This is the result of local field enhancement at the thin metal rim (see section "Aperture Probes"). Two lobes with strong fields oriented in axial direction occur located on opposite sides of the aperture in the direction of the polarization of the incident linearly polarized light. Molecules with a transition dipole moment oriented parallel to the tip axis are excited efficiently by these field components as can be seen in Fig. 5 [5]. Rotating the polarization of the incident linearly polarized light is seen to rotate the resulting double lobe pattern that indicates the area with strongest fields.

Fluorescence measurements typically do not require high excitation densities and in many cases, the small light transmission of aperture probes represents no major drawback. In fact, fluorescence imaging of single dye molecules with 32 nm spatial resolution has been demonstrated using a microfabricated cantilevered glass tip covered with a 60-nm-thick aluminum film. Due to the thickness of the virtually opaque film, possible contributions from surface plasmons propagating on the outside of the film have been discussed [4]. Examples of fluorescence microscopy measurements based on aperture probes also include, for example, studies of single nuclear pore

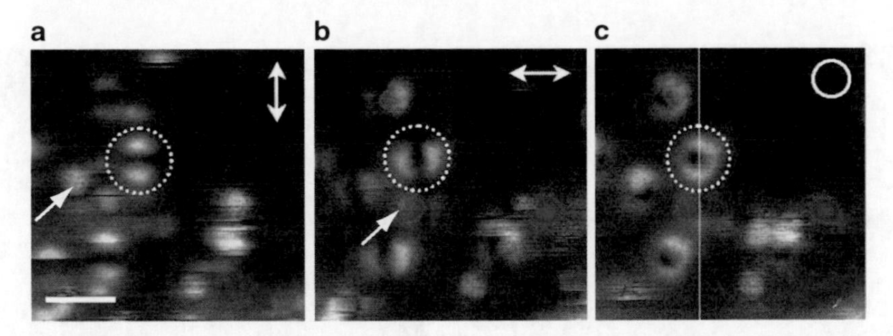

Scanning Near-Field Optical Microscopy, Fig. 5 Series of three successive aperture-SNOM fluorescence images of the same area (1.2 × 1.2 μm) of a sample of dye molecules embedded in a thin transparent polymer film. The excitation polarization, measured in the far-field, was rotated from linear vertical (**a**) to linear horizontal (**b**) and then changed to circular polarization (**c**). Circular features marked by arrows result from molecules with transition dipole moments oriented parallel to the sample plane. The double lobe structure marked by the dashed circle results from a molecule with perpendicularly oriented transition dipole moment. This molecule senses the strong electrical fields that occur at the rim of the metal aperture at positions determined by the far-field polarization. Scale bar 300 nm (Reprinted with permission from Veerman et al. [5]. Copyright 1999, John Wiley and Sons)

complexes and the kinetics of protein transport under physiological conditions.

Optical antennas have been used for a variety of samples and materials including photosynthetic proteins, polymers, semiconductor quantum dots, and carbon nano-tubes [11]. The image contrast was based on local field enhancement provided by the tip. The spatial resolution achieved in these experiments was essentially determined by the diameter of the tip and ranged between 10 and 20 nm. Imaging can be combined with local spectroscopy to visualize emission energies on the nanoscale. Single-molecule experiments revealed the field distribution at the tip-antenna in analogy to the discussion made above for aperture probes [14].

While most of the reported studies were exploiting local signal enhancement, distance-dependent metal-induced quenching of fluorescence can also be used for high-resolution imaging. This approach provides sub 10 nm spatial resolution and has been applied to single fluorescent organic molecules and inorganic semiconductor nanorods (see e.g., [12]). In these experiments, the spectrally integrated fluorescence signal was demodulated by the tapping-mode frequency of the AFM cantilever after recording photon-arrival times.

Raman Microscopy

Raman scattering probes the unique vibrational spectrum of a sample and directly reflects its chemical composition and molecular structure. A main drawback of Raman scattering is the extremely low scattering cross-section which is typically 10–14 orders of magnitude smaller than the cross-section of fluorescence in the case of organic molecules. Raman measurements thus require higher laser intensities and in many cases the low transmission of aperture probes prohibits their application. The signal enhancement provided by the antenna tip in TENOM is substantial for the detection of nanoscale sample volumina. In the following, a review of selected examples is given to illustrate the possibilities of tip-enhanced Raman scattering (TERS) (see e.g., [11, 15]).

In Fig. 6, simultaneous near-field Raman and topographic imaging of individual single-walled carbon nanotubes is shown. The optical image in (a) reflects the intensity of the G' band, a particular Raman-active vibrational mode of carbon nanotubes. The optical resolution obtained in this experiment was about 25 nm as can be seen from the width of the peaks in the cross-section in Fig. 6c.

The strong fields required for sufficient enhancement of the Raman scattering signal can cause laser-induced decomposition and photochemical reactions in the presence of oxygen. TERS of single electronically resonant molecules has been demonstrated for ultra-high vacuum conditions. A review focusing on single-molecule surface- and tip-enhanced Raman scattering can be found in [15].

Elastic Scattering Microscopy

Elastic scattering SNOM probes the dielectric properties of the sample and has been used from the visible to the microwave regime of the electromagnetic

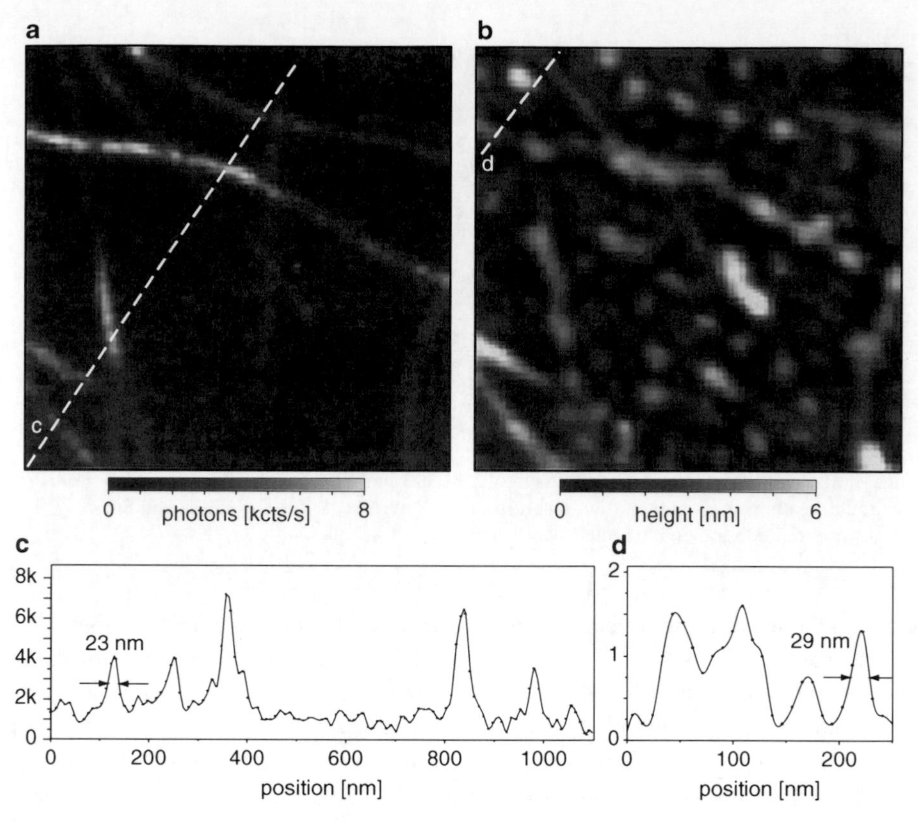

Scanning Near-Field Optical Microscopy, Fig. 6 Tip-enhanced Raman spectroscopy (TERS) of single-walled carbon nanotubes on glass. Simultaneous near-field Raman image (**a**) and topographic image (**b**). Scan area 1 × 1 μm^2. The Raman image is acquired by detecting the intensity of the G' band upon laser excitation at 633 nm. No Raman scattering signal is detected from humidity-related circular features present in the topographic image. (**c**) Cross-section taken along the *dashed line* in the Raman image indicating a spatial resolution around 25 nm. (**d**) Cross-section taken along the indicated dashed line in the topographic image. Vertical units are photon counts per second for c and nanometer for d (Reprinted with permission from Hartschuh et al. [16]. Copyright 2003, American Physical Society)

spectrum. Reviews of the fundamentals of the technique and representative applications can be found in [9, 10]. The majority of s-SNOM experiments have been reported for the IR to THz spectral range. Applications include detection of the Mott-transition in nanodomains, mapping of the doping concentration in semiconductors, surface characterization with a sensitivity of a single monolayer, strain-field mapping, and infrared spectroscopy of a single virus.

As an example nanoscale infrared spectroscopic near-field mapping of single nano-transistors is shown in Fig. 7. A cantilevered metallized Si-tip operating in tapping-mode with an oscillation frequency of 300 kHz and an amplitude of about 60 nm was used [17]. The data clearly demonstrates the potential of s-SNOM for infrared spectroscopic recognition of materials within individual semiconductor nanodevices.

Based on the antenna approach, s-SNOM typically provides 10–20 nm spatial resolution determined by the diameter of the tip-apex. In most of the s-SNOM experiments to date, monochromatic laser sources were used. Since only the optical response at this frequency is determined, the acquisition of scattering spectra or spectrally resolved images can only be obtained sequentially with a series of image scans at different laser frequencies. New developments exploiting broadband NIR laser sources aim at overcoming this limitation, and recently obtained a spectral bandwidth exceeding 400 cm^{-1}.

Plasmonics and Photonic Nanostructures

SNOM plays a vital role in the field of plasmonics which deals with the study of optical phenomena related to the electromagnetic response of metals

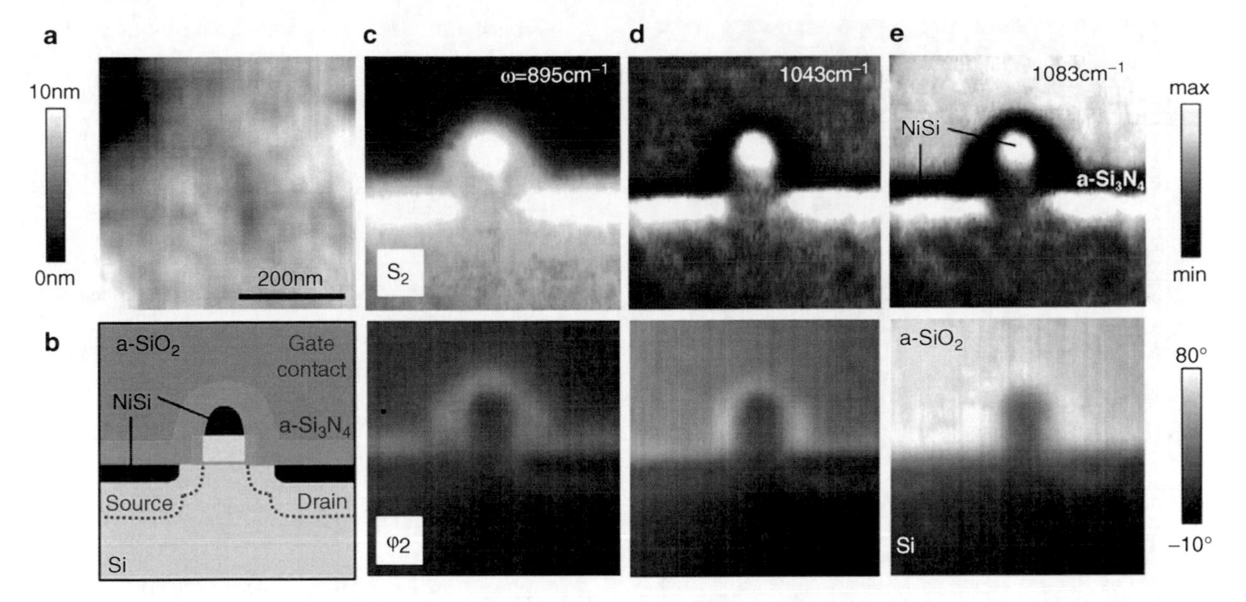

Scanning Near-Field Optical Microscopy, Fig. 7 Material-specific mapping of transistor components using s-SNOM: Cross-sectional images of a single transistor fabricated at the 65 nm technology node. (**a**) Topography. (**b**) Sketch of the transistor with materials indicated. (**c–e**) Near-field amplitude and phase images recorded at three different laser frequencies. Amorphous SiO_2 and Si_3N_4 render reversed optical contrast and are clearly distinguished. A spatial resolution better than 20 nm has been achieved (Reprinted with permission from Huber et al. [17]. Copyright 2010, IOP)

[18]. Near-field optical probes are particularly important for two reasons. First, they provide a means to locally excite propagating surface plasmon polaritons (SPPs) in metal films, a process that is not possible in case of propagating light waves because of momentum (k-vector) mismatch. The broad k-spectrum associated with the near-field of the probe contains sufficient bandwidth for efficient SPP-excitation (see Fig. 1). Second, near-field probes can be used simultaneously to convert SPPs back into propagating waves, thereby probing the local distribution of electromagnetic fields in the vicinity of metallic nanostructures. As an example, the near-field associated with SPPs has been visualized along gold nanowires using an aperture probe in collection mode (see Fig. 2a, [4]). This approach is also termed photon scanning tunneling microscopy (PSTM) to illustrate the analogy between evanescent electromagnetic waves and the corresponding exponentially decaying electron wavefunctions within the tunnel barrier of an STM. PSTM has been widely used to spatially resolve light wave propagation also in dielectric photonic nanostructures [4].

Besides the visualization of static field distributions, optical spectroscopy also allows for the study of their temporal evolution and the propagation of pulses. Figure 8 illustrates ultrafast and phase-sensitive imaging of the plasmon propagation in a metallic waveguide by PSTM [19]. In this case the near-field microscope uses an aperture-probe in collection mode and incorporates a Mach-Zehnder–type interferometer enabling heterodyne time-resolved detection.

Scattering-SNOM with antenna tips has been used extensively to study localized surface plasmon polaritons (LSPP) in different metal nanostructures. By varying the laser excitation frequency near-field optical imaging allowed for distinguishing higher order plasmonic resonances [20].

Perspectives

During the last 25 years SNOM has demonstrated its capabilities for sub-wavelength optical imaging and spectroscopy of surfaces and sub-surface features. The strength of SNOM results from its enormous flexibility with respect to sample types as well as measurement configurations and in particular, from its combination with a broad range of spectroscopic techniques. Ongoing developments aim at increasing antenna efficiencies and new aperture-type schemes [18]. In addition, the combination of nano-optical approaches and ultrafast laser technique is explored

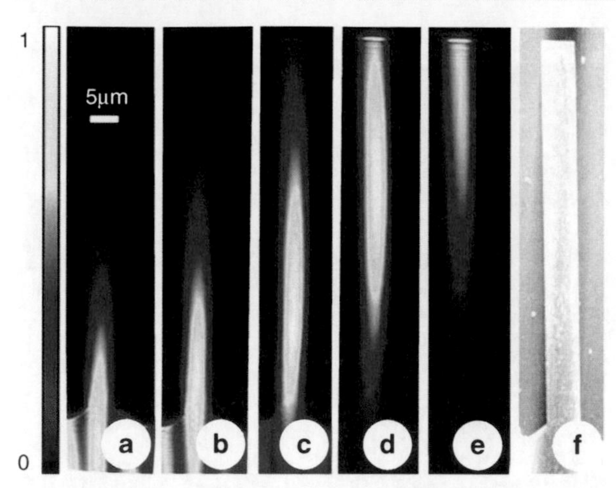

Scanning Near-Field Optical Microscopy, Fig. 8 Phase-sensitive and ultrafast near-field microscopy of a surface plasmon polariton (SPP) waveguide. The local electric field is collected by an aperture-probe and detected interferometrically in a Mach-Zehnder–type configuration. (**a–e**) Normalized amplitude information of the SPP wavepacket E-field. Succeeding frames are new scans of the probe. In between the frames the delay line is lengthened to 14.4 μm. Therefore, the time between two frames is 48 fs. The scan frame is $15 \times 110\ \mu m^2$, scan lines run from *top* to *bottom*. (**f**) Topography of the SPP waveguide obtained by shear-force feedback (Reprinted with permission from Sandtke et al. [19]. Copyright 2008, American Institute of Physics)

to achieve enhanced light localization and the control of optical near-fields on the time scale of few optical cycles.

Cross-References

- ▶ Atomic Force Microscopy
- ▶ Confocal Laser Scanning Microscopy
- ▶ Light Localization for Nano-optical Devices
- ▶ Nanostructures for Photonics
- ▶ Scanning Tunneling Microscopy

References

1. Novotny, L.: The history of near-field optics. In: Wolf, E. (ed.) Progress in Optics, vol. 50, pp. 137–184. Elsevier, Amsterdam (2007)
2. Novotny, L., Hecht, B.: Principles of Nano-optics. Cambridge, Cambridge University Press (2006)
3. Kawata, S., Shalaev, V.M. (eds.): Advances in Nano-Optics and Nano-Photonics Tip Enhancement. Elsevier, Amsterdam (2007)
4. Kawata, S., Shalaev, V.M. (ed.): Handbook of Microscopy for Nanotechnology. Kluwer, Dordrecht (2006)
5. Veerman, J.A., Garcia-Parajo, M.F., Kuipers, L., van Hulst, N.F.: Single molecule mapping of the optical field distribution of probes for near-field microscopy. J. Microsc. **194**, 477–482 (1999)
6. Mihalcea, C., Scholz, W., Werner, S., Münster, S., Oesterschulze, E., Kassing, R.: Multipurpose sensor tips for scanning near-field microscopy. Appl. Phys. Lett. **68**, 3531–3533 (1996)
7. Hecht, B., Sick, B., Wild, U.P., Deckert, V., Zenobi, R., Martin, O.J.F., Pohl, D.E.: Scanning near-field optical microscopy with aperture probes: fundamentals and applications. J. Chem. Phys. **112**, 7761–7774 (2000)
8. Bharadwaj, P., Deutsch, B., Novotny, L.: Optical antennas. Adv. Opt. Photon. **1**, 438–483 (2009)
9. Keilmann, F., Hillenbrand, R.: Near-field microscopy by elastic light scattering from a tip. Phil. Trans. R. Soc. Lond. **A 362**, 787–805 (2004)
10. Bründermann, E., Havenith, M.: SNIM: scanning near-field infrared microscopy. Annu Rep Prog Chem Sect. C Phys. Chem. **104**, 235–255 (2008)
11. Hartschuh, A.: Tip-enhanced near-field optical microscopy. Angew. Chem. Int. Ed. **47**, 8178–8198 (2008)
12. Ma, Z., Gerton, J.M., Wade, L.A., Quake, S.R.: Fluorescence near-field microscopy of DNA at sub-10 nm resolution. Phys. Rev. Lett. **97**, 260801–260804 (2006)
13. Taminniau, T.H., Stefani, F.D., Segerink, F.B., van Hulst, N.F.: Optical antennas direct single-molecule emission. Nat. Photon. **2**, 234–237 (2008)
14. Frey, H.G., Witt, S., Felderer, K., Guckenberger, R.: High resolution imaging of single fluorescent molecules with the optical near field of a metal tip. Phys. Rev. Lett. **93**, 200801–200804 (2004)
15. Pettinger, B.: Single-molecule surface-and tip-enhanced Raman spectroscopy. Mol. Phys. **108**, 2039–2059 (2010)
16. Hartschuh, A., Sánchez, E.J., Xie, X.S., Novotny, L.: High-resolution nearfield Raman microscopy of single-walled carbon nanotubes. Phys. Rev. Lett. **90**, 095503–4 (2003)
17. Huber, A.J., Wittenborn, J., Hillenbrand, R.: Infrared spectroscopic near-field mapping of single nanotransistors. Nanotechnology 21, 235702–6 (2010)
18. Schuller, J.A., Barnard, E.S., Cai, W., Jun, Y.C., White, S. W., Brongersma, M.L.: Plasmonics for extreme light concentration and manipulation. Nat. Mater. **9**, 193–204 (2010)
19. Sandtke, M., Engelen, R.J.P., Schoenmaker, H., Attema, I., Dekker, H., Cerjak, I., Korterik, J.P., Segerink, F.B., Kuipers, L.: Novel instrument for surface plasmon polariton tracking in space and time. Rev. Sci. Instrum. **79**, 013704–10 (2008)
20. Dorfmüller, J., Vogelgesang, R., Khunsin, W., Rockstuhl, C., Etrich, C., Kern, K.: Plasmonic nanowire antennas: experiment, simulation, and theory. Nano. Lett. **10**(9), 3596–3603 (2010)

Scanning Probe Microscopy

▶ Atomic Force Microscopy

Scanning Surface Potential Microscopy

► Kelvin Probe Force Microscopy

Scanning Thermal Microscopy

Li Shi
Department of Mechanical Engineering,
The University of Texas at Austin, Austin, TX, USA

Synonyms

Micro-thermal analysis; Scanning thermal profiler

Definition

Scanning Thermal Microscopy (SThM) is a class of experimental methods for high spatial resolution mapping of the surface temperature distribution of an operating device or the thermal property variation of a structure with the use of a sensor fabricated on a scanning probe.

Thermal Probes

In 1986, Williams and Wickramasinghe [1] pioneered a so-called scanning thermal profiler technique based on a thermocouple sensor fabricated at the end of a probe tip for Scanning Tunneling Microscopy (STM). The thermocouple sensor consisted of two dissimilar conductors that made a junction at the end of the STM tip. An insulator separated the two conductors in all areas remote from the tip. The size of the thermocouple junction could be made as small as 100 nm. When a temperature difference (ΔT) exists between the thermocouple junction and the ends of the two lead wires, thermal diffusion of electrons through the two dissimilar wires results in a thermoelectric voltage between the two lead wires. The magnitude of the thermovoltage measured with the use of a voltmeter is

$$V = (S_1 - S_2)\Delta T \tag{1}$$

where S_1 and S_2 are the Seebeck coefficient of the two thermocouple wires.

The main purpose of this first scanning thermal profiler was not for mapping temperature distribution of a surface but to regulate the tip-sample distance using the relationship between the thermocouple tip temperature and the distance between the tip and the sample. The work of Williams and Wickramasinghe stimulated intense efforts to develop a scanning probe for high spatial resolution mapping of the temperature distribution or thermal properties on a surface. In 1993, Majumdar and co-workers [2] introduced a wire thermocouple probe that could be used in an atomic force microscope (AFM) for simultaneous mapping of topography and temperature. Since then, different designs of scanning thermal probes have been fabricated. A common feature of these thermal probes is a thermal sensor fabricated at the end of an AFM or STM tip. The thermal sensor can be a thermocouple, a resistance thermometer, a Schottky diode, or a fluorescence particle [3–5].

As discussed above, a thermocouple measures the temperature difference between the junction of its two constituent metals and the other ends of the two metal wires. The current-voltage (I-V) characteristics of a Schottky junction depends on the temperature, and can thus serve as a local temperature sensor. A resistance thermometer is usually made of a metal or degenerately doped semiconductor with a relatively large and constant temperature coefficient of resistance. Because the resistance of a nanoscale resistor can be too small for sensitive electronic detection, resistance thermometer cannot be miniaturized as readily as a thermocouple sensor. Another approach is to use the emission band of some fluorescence particles, which shifts with increasing temperature. This feature has been used for making a SThM probe with a fluorescence particle located at the end of a scanning probe tip [5]. Besides these temperature-sensing techniques, the thermally induced bending of a biomaterial AFM cantilever has been used for temperature measurements [4]. Moreover, the thermal expansion of a sample has been measured with a regular AFM tip, and used to infer the sample temperature [4].

Different methods have been reported for fabricating different SThM probes. Some of the methods use sequential (or one at a time) fabrication methods [4], whereas batch fabrication processes have been developed for wafer scale fabrication of the SThM probes

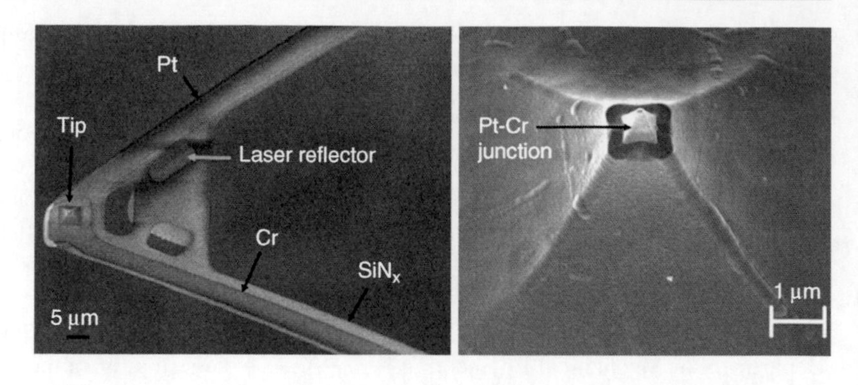

Scanning Thermal Microscopy, Fig. 1 Scanning electron micrographs of a SiN_x AFM cantilever probe (**a**) with a sub-micron Pt-Cr thermocouple junction (**b**) formed at the apex of the SiO_2 tip

[6]. While the probe batch fabrication processes are somewhat similar to those that have been developed for the manufacturing of ultra large scale integrated (USLI) devices, they often involve additional etching steps to make free standing cantilever probes, as well as novel processes to form a thermal sensor such as a thermocouple at the end of the cantilever probe tip. Figure 1 shows a batch-fabricated thermocouple SThM cantilever probe that consists of a 0.5 μm thick low stress silicon nitride (SiN_x) cantilever, a 8 μm tall silicon dioxide (SiO_2) pyramid tip with a ~20 nm tip radius, and a sub-micron Pt-Cr thermocouple junction formed at the apex of the SiO_2 tip [6]. Several other unique fabrication processes have been developed for wafer scale fabrication of a thermocouple junction, Schottky diode, Pt-C or doped silicon resistance thermometer [4].

Special probe holders with necessary electrical contacts or optical access can allow for the use of the SThM sensor probes in a commercial AFM or STM. STM requires the sample surface to be conducting because the tip-sample gap is controlled based on the tunneling current. This issue has limited the use of SThM probes in STM. In comparison, AFM with a cantilever SThM probe can allow simultaneous topography and thermal mapping of both conducting and non-conducting samples. In a typical contact-mode AFM, the tip-sample spacing is regulated by the force acting on the probe tip. The tip-sample interaction force can be obtained from the AFM cantilever bending that is measured with the use of several methods. One of these methods is to measure the laser beam reflected by the cantilever with the use of a position-sensitive detector. Another method is to employ a built-in piezoresistive sensor in the cantilever to measure the cantilever bending. Although the optical detection method provides superior sensitivity, the laser heating of the thermal sensor at the end of the cantilever needs to be minimized.

Theoretical Analysis of Heat Transfer Mechanisms

For many of the various SThM sensor probes, the sensitivity and spatial resolution of the thermal imaging technique depend on the mechanisms of heat transfer between the thermal probe and the sample. For SThM measurements conducted in air, heat is transferred between the probe and the sample via conduction through the solid–solid contact, a liquid meniscus at the tip-sample contact, and the air gap between the probe, and via radiation, as illustrated in Fig. 2 heat transfer through the air gap and via radiation is not localized at the tip-sample contact, and can deteriorate the spatial resolution of the SThM probes.

For a circular contact area with radius b, the spreading thermal resistance of the sample is obtained as [7]

$$R_s = \frac{1}{4\kappa_s b} \tag{2}$$

where κ_s is the thermal conductivity of the sample. For $\kappa_s \approx 5$ W/m-K, $b \approx 30$ nm, $R_s \approx 1.7 \times 10^6$ K/W.

The thermal interface resistance through the solid–solid contact is given as [8]

$$R_{i,s} \approx \frac{4}{\alpha C_s v_s \pi b^2} = \frac{4K}{3\alpha\kappa_s \pi b} \tag{3}$$

where C_s and v_s are the specific heat and phonon group velocity of the sample, α is the phonon transmission coefficient from the sample into the tip, $K \equiv l_s/b$ is the Knudsen number, and l_s is the phonon mean free path

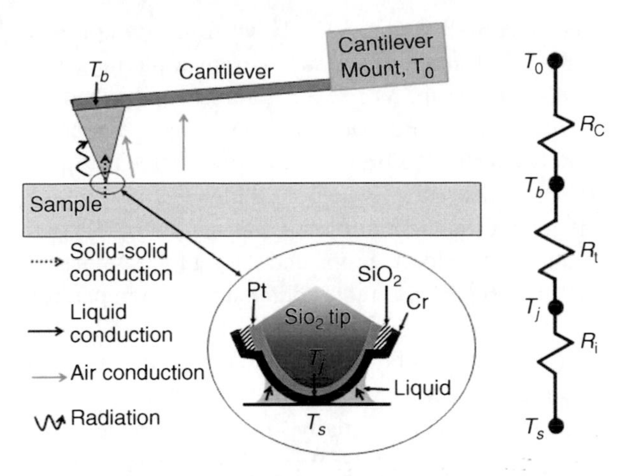

Scanning Thermal Microscopy, Fig. 2 Schematic diagram showing the heat transfer mechanisms between the thermal probe and the sample as well as a thermal resistance circuit where the heat transfer through the air gap is ignored (Reproduced from [6] with permission by ASME)

in the sample. The second equation is obtained with the use of the kinetic theory, $\kappa_s = C_s v_s l/3$ [8]. For $\alpha = 0.3$, $\kappa_s = 5$ W/m-K, $b = 30$ nm, and $l_s = 5$ nm, $R_{i,s} = 1.5 \times 10^6$ K/W.

A liquid meniscus often forms at the tip-sample junction when the tip is scanned on the sample in air. The interface thermal resistance through the liquid meniscus ($R_{i,\,l}$) at the tip-sample junction depends on the humidity and surface properties. The value of $R_{i,\,l}$ has been estimated by Majumdar to be on the order of 10^5 K/W [4], which is lower than the solid-solid thermal interface resistance. The total interface tip-sample thermal resistance is

$$R_i = (R_{i,s}^{-1} + R_{i,l}^{-1})^{-1} \tag{4}$$

The spreading thermal resistance of the conical tip can be estimated as

$$R_t \approx \frac{1}{\pi \kappa_t b \tan\theta} \tag{5}$$

where κ_t is the thermal conductivity of the tip and θ is the half angle of the conical tip. For $\kappa_t \approx 5$ W/m-K, $b = 30$ nm, $\theta = \pi/8$, $R_t = 5 \times 10^6$ K/W. Hence, if the thermocouple junction can be made as small as the contact size, the high spreading thermal resistance of the tip can be utilized to thermally isolate the junction from the cantilever.

An approximation expression for the tip-sample air thermal conductance per unit area is given here as

$$g_a \approx \frac{\kappa_a}{z + l_a} \tag{6}$$

where κ_a and l_a are the thermal conductivity and mean free path of air molecules, and z is the air gap. In the ballistic limit of $z \ll l_a$, this expression is reduced to $g_a = \kappa_a / l_a$. Based on the kinetic theory, $\kappa_a = C_a v_a l_a/3$, where C_a and v_a are the specific heat and velocity of air molecules, so that $g_a = C_a v_a/3$ for the ballistic limit. This result is close to the ballistic thermal conductance per unit area [8], $C_a V_a/4$, for heat transfer of gas molecules between two close parallel plates when the thermal accommodation coefficient is unity, corresponding to the case that the scattered molecules take the temperature of the surface. In the diffusive limit of $z \gg l_a$, $g_a = \kappa_a/z$ is the diffusive thermal conductance per unit area of the air gap. The total tip-sample thermal conductance through the air gap can be obtained as

$$\begin{aligned} G_{g,t-s} &= \int_b^{b+H\tan\theta} \frac{\kappa_a}{l_a + (r-b)/\tan\theta} 2\pi r dr \\ &= 2\pi\kappa_a H \tan^2\theta \left[1 + \frac{b - l_a\tan\theta}{H\tan\theta} \ln\frac{l_a + H}{H}\right] \end{aligned} \tag{7}$$

where H is the tip height. For $H = 8$ μm and $\theta = \pi/8$, $G_{g,\,t\text{-}s} \approx 2.2 \times 10^{-7}$ W/K, corresponding to a resistance $R_{g,\,t\text{-}s} \equiv 1/G_{g,\,t-s} = 4.4 \times 10^6$ K/W [6].

Far-field radiation conductance between the tip and the sample is approximated as that of a two-surface enclosure as

$$G_{\text{rad},t-s} \approx \frac{4A_t\sigma T^3}{(1-\varepsilon_t)/\varepsilon_t + 1/F_{t-s} + A_t(1-\varepsilon_s)/A_s\varepsilon_s} \tag{8}$$

where σ is the Stefan–Boltzmann constant, A_t and A_s are the surface areas of the tip and sample, ε_t and ε_s are the surface emissivity of the tip and sample, and F_{t-s} and T are the view factor and the average temperature between the tip and sample. When the temperature is close to 300 K, $G_{\text{rad},t-s}$ is on the order of 1×10^{-10} W/K. Both the far-field and near-field radiation heat transfer between the tip and the sample can be ignored compared to those via the air gap, solid–solid interface, and liquid meniscus [4], unless for a very high tip or sample temperature.

Because of heating by the sample surface, the temperature of the air molecules surrounding the cantilever, T_g, can be different from the temperature at the cantilever mount, T_0, which is usually close to the room temperature. In this case, heat transfer from the tip into the cantilever can be obtained as the fin heat transfer rate expressed as [7]

$$q_{t-c} = M \frac{\cosh mL - \theta_0/\theta_b}{\sinh mL} \quad (9)$$

where $M = (hP\kappa_c A_c)^{1/2}\theta_b$, $m = (hP/\kappa_c A_c)^{1/2}$, h is the heat transfer coefficient between the cantilever and the surrounding gas, κ_c and A_c are the thermal conductivity and cross-section of the cantilever, P is the perimeter of the cantilever cross-section, $\theta_0 = T_0 - T_g$, $\theta_b = T_b - T_g$, and T_b is the temperature at the joint between the tip and the cantilever. When only a small sample area is at a temperature different from the room temperature so that $T_g \approx T_0$, the thermal resistance of the cantilever can be defined as

$$R_c \equiv \frac{T_b - T_0}{q_{t-c}} = \frac{\tanh mL}{\sqrt{hp\kappa_c A_c}} \quad (10)$$

For $\kappa_c \approx 5$ W/m-K, and 0.5 μm by 10 μm cantilever cross-section, $L = 150$ μm, $h \approx k_a(\pi/A_c)^{1/2}$, $R_c = 3 \times 10^5$ K/W.

On the other hand, when a large area of the sample is at a temperature higher than T_0, the air molecules surrounding the cantilever can be heated by the sample surface to a temperature T_g that is considerably higher than T_0. In this case, the cantilever is heated by the surrounding air, resulting in an additional heat transfer path from the sample to the probe.

Experimental Characterization of Heat Transfer Mechanisms

Experimentally, the heat transfer path through the air gap between the sample and the cantilever was found to be significant when the size of the heated zone on the sample surface was not small [6]. In the experiment [6], a 350-nm wide metal line was joule heated to 5.3 K above room temperature. The cantilever deflection and temperature rise of the thermocouple junction were recorded simultaneously when the sample was approached and then retracted from a thermocouple SThM probe tip. When the sample approached the tip, the cantilever deflection signal remained approximately constant before the sample contacted the tip, as shown in the deflection curve in Fig. 3. In this region, the junction temperature rise was caused by air conduction between the probe and the sample. As the tip-sample distance was reduced, the junction temperature rise due to air conduction increased slowly. Before the sample made solid–solid contact to the tip, the adsorbed liquid layers on the tip and the sample bridged each other. Initially, this liquid bridge pulled the tip down by a van der Waals force, as being seen in the dip labeled as "jump to contact" in the deflection curve. Coincidentally, there was a small jump in the junction temperature due to heat conduction through the liquid bridge. As the sample was raised further, both the solid–solid contact force and the junction temperature increased gradually, until the cantilever was deflected for more than 100 nm. After this point, the junction temperature remained almost constant as the contact force increased.

As the sample was retracted from the tip, the junction temperature remained almost constant until at a cantilever deflection of 100 nm, after which the junction temperature rise decreased roughly linearly but at a smaller slope than that found in the approaching cycle. As the sample was lowered further, the tip was pulled down together with the sample by surface tension of the liquid bridge until after a certain point, the restoring spring force of the cantilever exceeded the surface tension, and the tip "snapped out of contact" with the sample. Associated with the breaking of the liquid bridge, there was a small drop in the junction temperature.

This experiment shows several mechanisms. First, before the tip contacted the sample, air conduction contributed to a junction temperature rise up to 0.03 K per K sample temperature rise, which was about 60% of the maximum junction temperature rise at the maximum contact force of the experiment. Second, conduction through a liquid meniscus was responsible for the sudden jump and drop in junction temperature when the tip "jumped to contact" to and "snapped out of contact" from the sample, respectively. Third, solid–solid conduction resulted in the almost linear relationship between the junction temperature rise and the contact force. This is a well

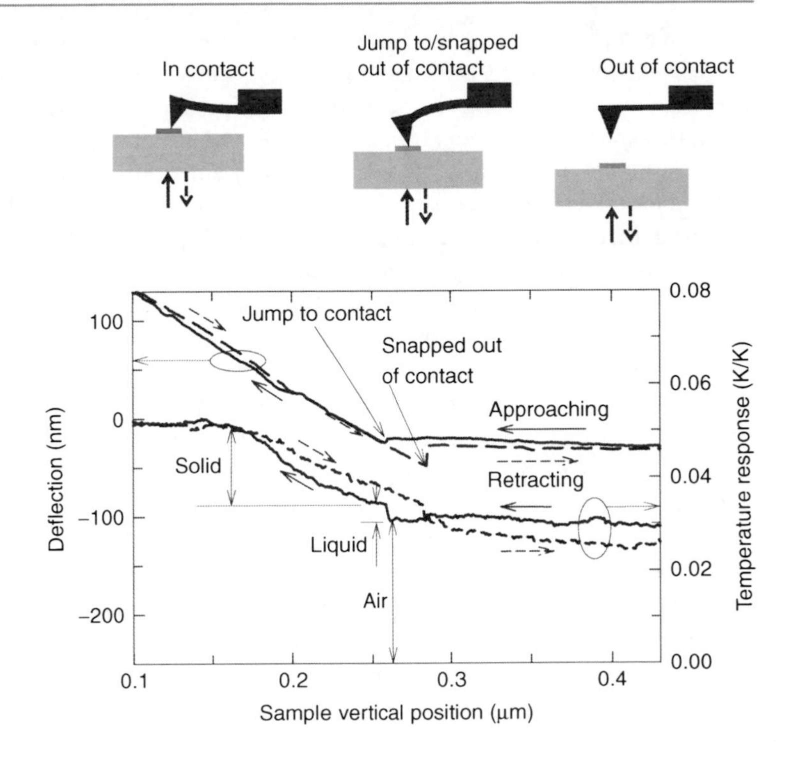

Scanning Thermal Microscopy, Fig. 3 Cantilever deflection and temperature response of a thermocouple probe as a function of the sample vertical position when a 350-nm wide heater line sample was raised toward and then retracted from the tip (Reproduced from [6] with permission by ASME)

understood feature for macroscopic solid–solid contacts. When the contact force was increased further, the junction temperature approached a constant likely because the contact size approached the maximum possible value limited by the diameter of an asperity located at the tip-sample interface.

The point contact experiment was repeated for a 5.8 μm wide and 2,000 μm long heater line on an oxidized silicon wafer. While the increase of the normalized junction temperature due to solid and liquid conduction was similar in magnitude to those observed in the narrower line, the normalized junction temperature rise due to air conduction was one order of magnitude higher than the corresponding one in the narrower line. In addition, the temperature rise of the thermocouple junction was measured when the tip was in contact with three heater lines of different line widths. It was found that the temperature rise at the thermocouple junction of the tip was about 53%, 46%, and 5% of that of a 50, 3, and 0.3 μm wide line, respectively, showing a trend of increasing temperature rise in the SThM probe with increasing heater line width. These results suggest that the cantilever was heated more by air conduction between the tip and the larger hot area of the wider heater lines. This trend indicates that air conduction plays an important role in probe-sample heat transfer, especially when the hot area on the sample surface is large in comparison to the tip size.

The unwanted heat transfer through the air gap can be eliminated by conducting SThM measurements in vacuum. In this case, the tip-sample air conductance $G_{g,\,t-s} = 0$, and the thermal resistance of the cantilever is the conduction thermal resistance given as

$$R_c = \frac{L}{\kappa_c A_c} \tag{11}$$

For a cantilever length $L = 150$ μm, thermal conductivity $\kappa_c = 5$ W/m-K, and cross-section $A_c = 0.5$ μm × 10 μm, $R_c = 6 \times 10^6$ W/K based on the above equation. In addition, the heat transfer from the sample to the tip in vacuum can be described with the use of the thermal resistance circuit shown in Fig. 2.

However, vacuum operation of the SThM probe is considerably much more challenging than operation in open air environment. Moreover, the liquid meniscus at the tip-sample junction is also eliminated by the vacuum environment. Although this can further enhance the spatial resolution of the thermal imaging technique, the tip-sample thermal interface conductance could be reduced considerably. Consequently, the thermocouple junction temperature at the tip end could become rather different from the sample temperature, resulting in low sensitivity and large uncertainty in the measured temperature.

An alternative method to eliminate the unwanted air conductance while keeping the heat transfer via the liquid meniscus is based on a dual scan measurement approach [9]. In this approach, after a line scan of the SThM probe on the sample surface in the contact mode, the probe is scanned on the same line in the lift mode with a small fixed tip-sample distance between the tip and the sample. The difference between the two thermal sensor signals obtained in the contact mode and the lift mode is attributed to the local heat transfer through the solid–solid contact and the liquid meniscus at the tip-sample junction, and can be used to obtain the local sample surface temperature at the contact point, as discussed below.

Surface Temperature Mapping

When the heat transfer through the probe-sample gap is eliminated by the dual scan technique, the difference in the measured thermocouple junction temperature values in the contact mode and the lift mode is still different from that for vacuum measurement given by the thermal resistance circuit of Fig. 2 as

$$T_j = (T_s - T_0)\left[1 + \frac{R_i}{R_c + R_t}\right]^{-1} + T_0 \qquad (12)$$

where T_s is the local sample temperature at the contact point, T_0 is the ambient temperature.

It is important to increase R_t and R_c relative to R_i and R_s, so as to maximize ΔT_j and minimize the sample temperature change due to the contact by the tip. For this reason, it is desirable to use low-thermal conductivity dielectrics such as SiO_2 and SiN_x to fabricate the pyramid tip and the cantilever, and low-thermal conductivity conductors such as Cr and Pt to fabricate the thermocouple sensor. The geometry of the sensor, tip, and cantilever also needs to be optimized to minimize heat loss.

Even for state-of-the-art thermocouple SThM probes, R_t and R_c are still not much larger than R_i. Hence, the ΔT_j obtained from the differential dual scan measurement needs to be calibrated in a separate experiment, where the sample is heated with an external heater and the sample surface temperature is determined using a different method, such as a resistance thermometer fabricated on the sample surface [10].

SThM methods operated in the contact mode or the dual scan mode have been employed to measure the surface temperature distribution of silicon, carbon nanotube, and graphene nanoelectronic devices, with a spatial resolution better than 100 nm [10, 11]. The very high electric field in electrically biased nanoelectronic devices can result in non-equilibrium transport of electrons, optical phonons, and acoustic phonons [12]. Energetic hot electrons accelerated in the high field lose their energy to lattice vibration via emission of high frequency optical phonons. The optical phonons are further scattered with acoustic phonons that make a larger contribution to the lattice thermal conductivity because of a much higher group velocity than the optical phonons. The scattering time between optical phonons and acoustic phonons can be much longer than that between hot electrons and optical phonons in nanoelectronic devices made of silicon and carbon nanotubes. Consequently, energy transfer from optical phonons and acoustic phonons becomes a bottle neck in the heat dissipation process. This bottle neck can result in different temperatures for acoustic phonons and optical phonons. The local temperatures of different energy carriers in operating nanoelectronic devices can be measured with different methods. For example, infrared spectroscopy has been used to measure the electronic temperature of electrically biased graphene devices by fitting the infrared emission from the hot electrons in the device with the Planck distribution [13]. The intensity ratio between the anti-Stokes peak to the Stokes peak in the Raman spectrum can be used to probe the local temperature of the zone center or zone boundary optical phonons that are active in the Raman scattering processes [14]. Complementary to these optical thermometry methods, the thermocouple SThM probe can be used to map the low frequency phonon temperature distribution because the phonon transmission coefficient at the tip-sample interface increases with decreasing phonon frequency. This capability has been employed to observe bias-dependant and asymmetric acoustic phonon temperature distribution in electrically biased graphene with a superior spatial resolution of about 100 nm [10], as illustrated in Fig. 4.

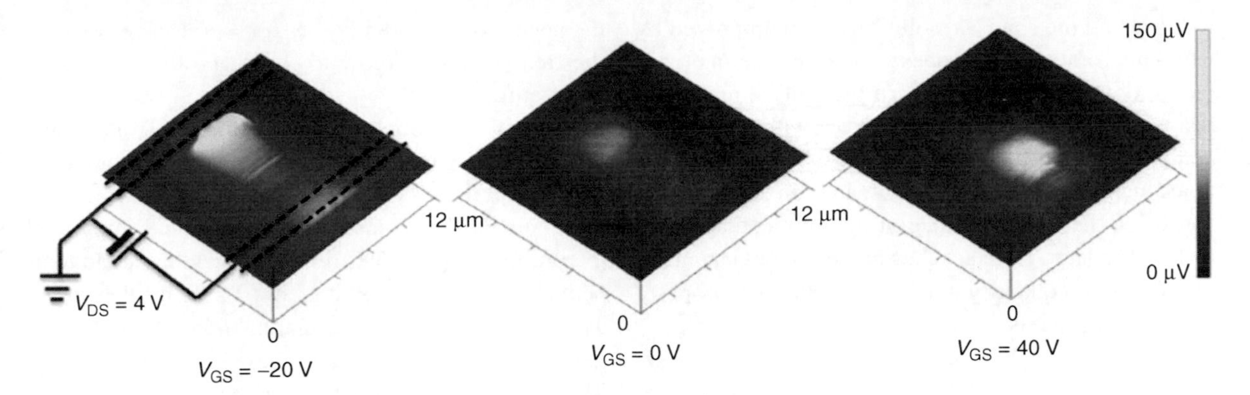

Scanning Thermal Microscopy, Fig. 4 Measured SThM thermovoltage maps of an electrically biased graphene device for a constant drain-source bias ($V_{DS} = 4$ V) and different gate-source voltage of −20 V, 0 V, and 40 V, respectively. The scan size is 12 μm × 12 μm

Besides the dual scan method, an active SThM method has been reported by Nakabeppu and Suzuki [15] to obtain quantitative temperature map of a sample surface in vacuum. Their probe for active SThM consists of two thin film thermocouples (TCs) and a micro-heater fabricated on a cantilever. When the cantilever was scanned on a heated sample, heat flow from the sample into the cantilever resulted in a temperature drop (ΔT) along the cantilever, which was measured using the differential thermocouple. The micro-heater power was adjusted by using a feedback loop until the differential thermocouple reads $\Delta T = 0$. Under this condition, the heat flow from the sample to the cantilever was zero so that the measured tip temperature is the same as the sample surface temperature. Alternatively, the active SThM method can also be implemented using the dual scan operation in air [9]. In this implementation, the SThM probe is heated to different temperatures and is scanned on a hot sample at each heating rate for the probe. When the measured sensor temperature of the heated SThM probe is the same during the contact-mode and lift-mode scanning of the probe, the sensor temperature is taken as the same as the local sample temperature, and the heat flow between the tip and the sample is nullified.

It is worth emphasizing that the key issue in SThM measurements is that the thermal sensor temperature can be quite different from the sample temperature. This issue was addressed in the aforementioned measurements via either a detailed calibration or actively heating the sensor until the sensor temperature was the same as the sample temperature. While four-point measurement methods with minimum current leakage into the voltage measurement devices are commonly used to address problems caused by contact electrical resistance in electrical measurements, a thermal analogue of this approach has not been effective for addressing the problem caused by thermal contact resistance in SThM measurements, because it is difficult to eliminate heat loss into a temperature measurement device, or the thermal probe in SThM.

Thermal Property Mapping

Scanning thermal probes consisting of a resistance heater and thermometer at the tip end have been used to measure the local thermal property near the surface in a solid sample. For this measurement, the temperature of the heated probe tip decreases as the tip touches the sample. A larger drop in the tip temperature is expected when the tip touches a higher thermal conductivity region of the sample because of a smaller spreading thermal resistance in the sample. Alternatively, the probe heating rate can be feedback-controlled to maintain a constant tip temperature during tip scanning. In this case, a higher heating power is needed to achieve a constant tip temperature when the tip touches a higher thermal conductivity region. Although thermal conductivity contrast has been observed under dc operations, accurate thermal conductivity measurement remains a challenge especially for low-thermal conductivity samples [4]. This is in part because a significant fraction of the power may be dissipated through the cantilever instead of into the sample. The heat loss through the cantilever can be reduced and the spatial resolution can be improved by using microfabricated probes with a low-thermal conductivity cantilever and a sharp pyramid Si resistance heater and thermometer tip. However, with a decreased contact radius b, the tip-sample thermal interface resistance increases according to $1/b^2$, and can dominate the spreading thermal resistance of the sample that scales as $1/b$. The large interface thermal resistance can make it difficult to quantify the sample thermal property when the tip size is small.

The resistance thermometer probe can also be operated in the ac mode, as shown by Pollock, Hammiche, and coworkers [3]. The sensor probe used in their measurements was 5-μm-diameter Pt resistance wire bent to form a tip at the bent. An ac electrical heating current was used to modulate the probe temperature by 5°C at 10 kHz around a dc temperature of 40°C, the measured amplitude and the phase shift in the ac voltage drop across their Pt wire thermometer probe depended on the thermal properties of the sample. One advantage of measuring the phase lag is that it is independent of the temperature dependence of the electrical resistivity of Pt wire and the power input to the probe, because the phase lag is given as [4]

$$\tan\phi = -\frac{G_{\mathrm{im}}}{G_{\mathrm{r}}} \tag{13}$$

where G_{r} and G_{im} are the real and imaginary parts of the complex AC thermal conductance, which can be expressed as

$$\begin{aligned} G_{\mathrm{r}} &\approx 2\pi\kappa_s b\left(1 + b\sqrt{\frac{\omega}{2\alpha_s}}\right) + A_c\kappa_c\sqrt{\frac{\omega}{2\alpha_c}} \\ G_{\mathrm{im}} &\approx \omega mC + 2\pi\kappa_s b^2\sqrt{\frac{\omega}{2\alpha_c}} \end{aligned} \tag{14}$$

where b is the contact radius, κ_s and α_s are the thermal conductivity and thermal diffusivity of the sample, κ_c and α_c are the thermal conductivity and thermal diffusivity of the cantilever probe, and the mC product is the thermal mass of the temperature sensor. For simplicity, the thermal interface resistance is ignored in this analysis [4]. For the phase lag to be sensitive to the thermal properties of the sample, the thermal conductance of the cantilever probe cannot be much larger than that of the sample. In addition, it should be noted that both the amplitude and the phase of the signal are influenced by the penetration depth of $\sqrt{2\alpha_s/\omega} = \delta$. Hence, varying the frequency can allow for depth profiling and subsurface imaging.

Pollock, Hammiche, and coworkers [3] have also used their Pt wire resistance heater and thermometer probe to perform localized ac calorimetry [16]. Here they ramped the temperature of both the sample and the probe at about 15°C/min while adding an ac temperature modulation of 1°C at 10 kHz by the probe. They found that any phase transition in the sample gave rise to a change in the phase signal in ac mode. This is observed more clearly in the first derivative of the phase as a function of temperature. By scanning the sample under this mode, calorimetric analysis can be performed locally.

Summary and Future Directions

A number of SThM probes have been designed and fabricated. Different operating methods including the dual scan and the zero heat flux techniques have been employed for quantitative mapping of the surface temperature distribution of operating devices. Both dc and ac operations of a resistance heater and thermometer probe have allowed the mapping of local thermal property variation of nanostructured materials. Further enhancement of the spatial resolution and thermal measurement accuracy of the SThM methods can be made by continuous miniaturization of the sensor size at the probe tip as well as improved thermal isolation of the sensor. Detecting the phase lag in the ac SThM signal can potentially allow for locating sub-surface defects that generate localized heating in operating electronic devices, as suggested in [17].

The SThM probes based on a thermal-to-electrical signal transduction method can be complemented with non-contact optical methods for mapping the local temperature distribution of different phonon populations or different thermal properties. Although the spatial resolution of conventional far-field optical imaging methods is limited by diffraction to be on the order of the wavelength, the diffraction barrier has been broken by near-field scanning optical microscopy (NSOM) methods [18] as well as several far-field optical nanoscopy methods [19]. Near-field infrared evanescence wave emitted from a sample surface has been collected using a solid immersion lens [11] and/or scattered into far-field signal by an apertureless metal tip [20]. These techniques have the potential for profiling the surface temperature distribution of devices with a spatial resolution of a fraction of the IR wavelength. Similarly, the spatial resolution of Raman thermograph or thermal reflectance techniques can potentially be improved with either a near-field or far-field nanoscopy technique.

Cross-References

▶ Atomic Force Microscopy
▶ Carbon-Nanotubes
▶ Graphene
▶ Scanning Tunneling Microscopy
▶ Thermal Conductivity and Phonon Transport

References

1. Williams, C.C., Wickramasinghe, H.K.: Scanning thermal profiler. Appl. Phys. Lett. **49**, 1587–1589 (1986)
2. Majumdar, A., Carrejo, J.P., Lai, J.: Thermal imaging using the atomic force microscope. Appl. Phys. Lett. **62**, 2501 2503 (1993)
3. Pollock, H.M., Hammiche, A.: Micro-thermal analysis: techniques and applications. J. Phys. D Appl. Phys. **34**, R23–R53 (2001)
4. Majumdar, A.: Scanning thermal microscopy. Annu. Rev. Mater. Sci. **29**, 505–585 (1999)
5. Aigouy, L., Tessier, G., Mortier, M., Charlot, B.: Scanning thermal imaging of microelectronic circuits with a fluorescent nanoprobe. Appl. Phys. Lett. **87**, 3 (2005)
6. Shi, L., Majumdar, A.: Thermal transport mechanisms at nanoscale point contacts. J. Heat Trans-T. ASME. **124**, 329–337 (2002)
7. Incropera, F.P., Dewitt, D.P., Bergman, T.L., Lavine, A.S.: Fundamentals of Heat and Mass Transfer. John Wiley, Hoboken (2007)
8. Chen, G.: Nanoscale Energy Transport and Conversion: A Parallel Treatment of Electrons, Molecules, Phonons, and Photons. Oxford University Press, New York (2005)
9. Chung, J., Kim, K., Hwang, G., Kwon, O., Jung, S., Lee, J., Lee, J.W., Kim, G.T.: Quantitative temperature measurement of an electrically heated carbon nanotube using the null-point method. Rev. Sci. Instrum. 81, 5 (2010)
10. Jo, I., Hsu, I.-K., Lee, Y.J., Sadeghi, M.M., Kim, S., Cronin, S., Tutuc, E., Banerjee, S.K., Yao, Z., Shi, L.: Low-frequency acoustic phonon temperature distribution in electrically biased graphene. Nano Lett. (2010). doi:10.1021/nl102858c
11. Cahill, D.G., Goodson, K., Majumdar, A.: Thermometry and thermal transport in micro/nanoscale solid-state devices and structures. J. Heat Trans-T. ASME. **124**, 223–241 (2002)
12. Tien, C.L., Majumdar, A., Gerner, F.M.: Microscale Energy Transport. Taylor & Francis, Washington, DC (1998)
13. Berciaud, S., Han, M.Y., Mak, K.F., Brus, L.E., Kim, P., Heinz, T.F.: Electron and optical phonon temperatures in electrically biased graphene. Phys. Rev. Lett. **104**, 227401 (2010)
14. Chae, D.H., Krauss, B., von Klitzing, K., Smet, J.H.: Hot phonons in an electrically biased graphene constriction. Nano Lett. **10**, 466–471 (2010)
15. Nakabeppu, O., Suzuki, T.: Microscale temperature measurement by scanning thermal microscopy. J. Therm. Anal. Calorim. **69**, 727–737 (2002)
16. Price, D.M., Reading, M., Hammiche, A., Pollock, H.M.: Micro-thermal analysis: scanning thermal microscopy and localised thermal analysis. Int. J. Pharm. **192**, 85–96 (1999)
17. Kwon, O., Shi, L., Majumdar, A.: Scanning thermal wave microscopy (STWM). J. Heat Trans-T. ASME. **125**, 156–163 (2003)
18. Dunn, R.C.: Near field scanning optical microscopy. Chem. Rev. **99**, 2891–2928 (1999)
19. Hell, S.W.: Far-field optical nanoscopy. Science **316**, 1153–1158 (2007)
20. De Wilde, Y., Formanek, F., Carminati, R., Gralak, B., Lemoine, P.A., Joulain, K., Mulet, J.P., Chen, Y., Greffet, J.J.: Thermal radiation scanning tunneling microscopy. Nature **444**, 740–743 (2006)

Scanning Thermal Profiler

▶ Scanning Thermal Microscopy

Scanning Tunneling Microscopy

Ada Della Pia and Giovanni Costantini
Department of Chemistry, The University of Warwick, Coventry, UK

Definition

A scanning tunneling microscope (STM) is a device for imaging surfaces with atomic resolution. In STM, a sharp metallic tip is scanned over a conductive sample at distances of a few Å while applying a voltage between them. The resulting tunneling current depends exponentially on the tip-sample separation and can be used for generating two-dimensional maps of the surface topography. The tunneling current also depends on the sample electronic density of states, thereby allowing to analyze the electronic properties of surfaces with sub-nm lateral resolution.

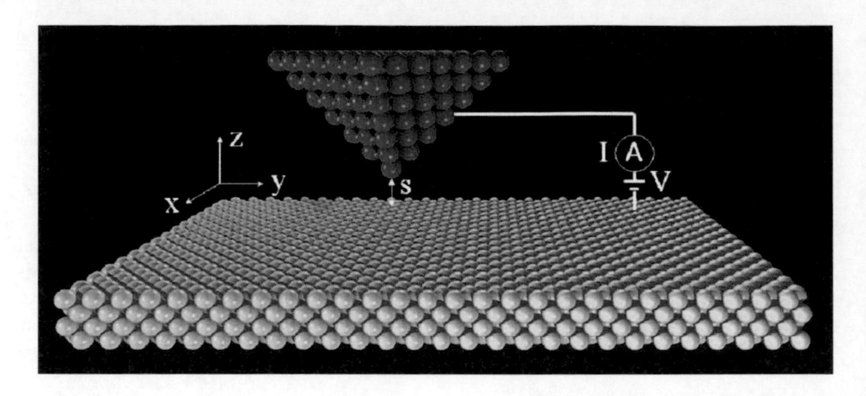

Scanning Tunneling Microscopy, Fig. 1 Schematic representation of a STM. Tip and sample are held at a distance s of a few Å and a bias voltage V is applied between them. The resulting tunneling current I is recorded while the tip is moved across the surface. The coordinate system is also shown

Overview and Definitions

If two electrodes are held a few Å apart and a bias voltage is applied between them, a current flows even though they are not in contact, due to the quantum mechanical process of electron tunneling. This current depends exponentially on the electrode separation, and even minute, subatomic variations produce measurable current changes. In 1981, Gerd Binnig and Heinrich Rohrer at IBM in Zürich realized that this phenomenon can be used to build a microscope with ultrahigh spatial resolution [1], if one of the electrodes is shaped as a sharp tip and is scanned across the surface of the other (Fig. 1). Moreover, since the tunneling current depends also on the electronic properties of the electrodes, this microscope has the ability to probe the electronic density of states of surfaces at the atomic scale. A few years later, Don Eigler at IBM in Almaden, showed that, due to the extremely localized interaction between tip and sample, it is also possible to use this instrument to manipulate individual atoms, to position them at arbitrary locations and therefore to build artificial structures atom-by-atom [2]. This remarkable achievement brought to reality the visionary predictions made by Richard Feynman in his famous 1959 lecture "*There's plenty of room at the bottom*" [3].

The construction of this instrument, dubbed the scanning tunneling microscope (STM), was awarded the 1986 Nobel Prize in Physics and has since then revolutionized contemporary science and technology. The STM has enabled individual atoms and molecules to be imaged, probed, and handled with an unprecedented precision, thereby essentially contributing to our current understanding of the world at the nanoscale. Together with its offspring, the atomic force microscope (AFM) [▶ AFM], the STM is considered as the main innovation behind the birth of nanotechnology.

This entry will start with a discussion of the physical principles and processes at the heart of STM in section Theory of Tunneling. This will be followed by a description of the experimental setup and the technical requirements needed for actually operating such a microscope in section Experimental Setup. Section STM Imaging is dedicated to the most frequent use of STM, namely imaging of surfaces, while section Scanning Tunneling Spectroscopy gives a brief account of the spectroscopic capabilities of this instrument. Finally, section Applications discusses several applications and possible uses of STM.

Theory of Tunneling

Figure 2 is a schematic representation of the energy landscape experienced by an electron when moving along the z axis of a metallic-substrate/insulator/metallic-tip tunneling junction. The following treatment can easily be extended to include also semiconducting tips or samples. Usually, the tip and the sample are not made of the same material and therefore have different work functions, ϕ_T and ϕ_S, respectively. At equilibrium, the two metals have a common Fermi level, resulting in an electric field being established across the gap region and in different local vacuum levels, depending on the difference $\phi_T - \phi_S$ (Fig. 2a). Since the work functions in metals are of the order of several eV, the potential in the gap region is typically much higher than the thermal energy kT and thus acts as a barrier for sample and tip electrons. A classical particle cannot penetrate into any region where the potential energy is greater than its total energy because this requires a negative kinetic energy. However, this is

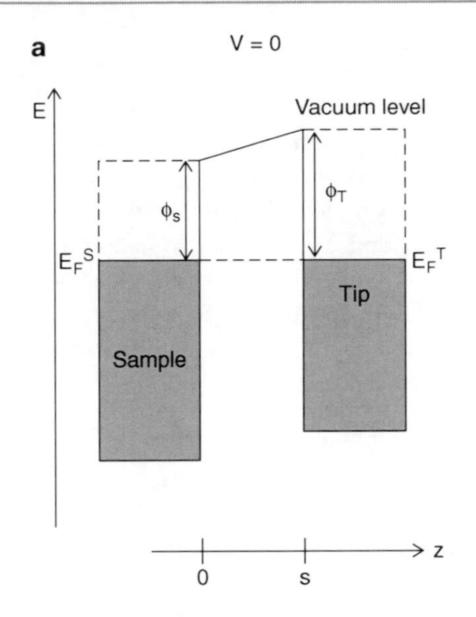

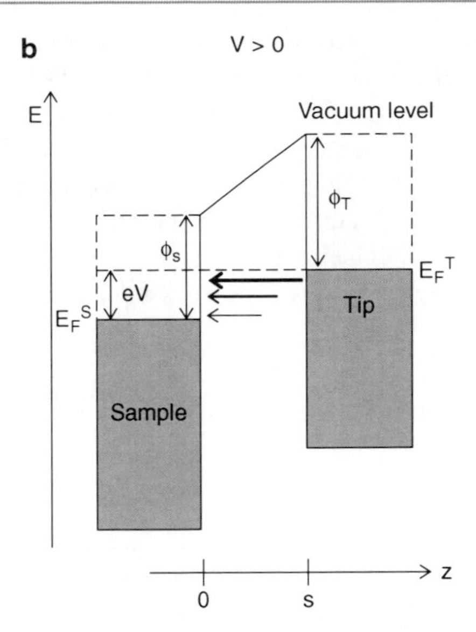

Scanning Tunneling Microscopy, Fig. 2 Energy potential perpendicular to the surface plane for an electron in a tip-vacuum-sample junction. z is the surface normal direction, and s is the tip-sample distance. The gray boxes represent the Fermi–Dirac distribution at 0 K. $\phi_{T,S}$ and $E_F^{T,S}$ are the work functions and the Fermi levels of tip and sample, respectively. (**a**) Tip and sample in electrical equilibrium: a trapezoidal potential barrier is created. (**b**) Positive sample voltage V: the electrons tunnel from occupied states of the tip into unoccupied states of the sample. The thickness and the length of the *arrows* indicate the exponentially decreasing probability that an electron with the corresponding energy tunnels through the barrier

possible for electrons which, being quantum mechanical objects, are described by delocalized wave functions. This phenomenon goes under the name of *quantum tunneling*. In an unpolarized tip-sample junction the electrons can tunnel from the tip to the sample and vice versa, but there is no net tunneling current. On the contrary, if a voltage V is applied between sample and tip, the Fermi level of the former is shifted by $-eV$ and a net tunneling current occurs, whose direction depends on the sign of V (Fig. 2b). Here the convention is adopted to take the tip as a reference since experimentally the voltage is often applied to the sample while the tip is grounded. If V is the bias voltage, the energy for an electron in the sample will change by $-eV$, that is, it will decrease for positive values of V.

The tunneling current can be evaluated by following the time-dependent perturbation approach developed by Bardeen [4, 5]. The basic idea is to consider the isolated sample and tip as the unperturbed system described by the stationary Schrödinger equations:

$$(\mathcal{T} + \mathcal{U}_S)\psi_\mu = E_\mu \psi_\mu \quad (1)$$

and

$$(\mathcal{T} + \mathcal{U}_T)\chi_\nu = E_\nu \chi_\nu, \quad (2)$$

where $\mathcal{T}$ is the electron kinetic energy. The electron potentials $\mathcal{U}_S$ and $\mathcal{U}_T$ and the unperturbed wavefunctions ψ_μ and χ_ν are nonzero only in the sample and in the tip, respectively. Based on this, it can be shown [5] that the transition probability per unit time $w_{\mu\nu}$ of an electron from the sample state ψ_μ to the tip state χ_ν is given by Fermi's golden rule:

$$w_{\mu\nu} = \frac{2\pi}{\hbar}|M_{\mu\nu}|^2 \delta(E_\nu - E_\mu), \quad (3)$$

where the matrix element is:

$$M_{\mu\nu} = \int \chi_\nu^*(\vec{x})\mathcal{U}_T(\vec{x})\psi_\mu(\vec{x})d^3\vec{x}. \quad (4)$$

The δ function in Eq. (3) implies that the electrons can tunnel only between levels with equal energy, that is, (3) accounts only for an *elastic tunneling* process. The case of an inelastic tunneling process will be considered in section Scanning Tunneling Spectroscopy. The total current is obtained by summing $w_{\mu\nu}$ over all the possible tip and sample states and by multiplying this by the electron charge e. The sum over the states can be changed into an energy integral by considering the density of states (DOS) $\rho(E) : \Sigma \rightarrow 2\int f(\varepsilon)\rho(\varepsilon)d\varepsilon$, where the factor 2 accounts for the spin degeneracy while f, the Fermi–Dirac distribution function, takes into consideration Pauli's exclusion principle and the electronic state population at finite temperatures.

As a consequence, the total current can be written as:

$$I = \frac{4\pi e}{\hbar}\int_{-\infty}^{\infty}[f_T(E_F^T - eV + \varepsilon) - f_S(E_F^S + \varepsilon)] \times \rho_T(E_F^T - eV + \varepsilon)\rho_S(E_F^S + \varepsilon)|M|^2 d\varepsilon, \quad (5)$$

where E_F is the Fermi energy and the indexes T and S refer to the tip and the sample, respectively. Equation 5 already accounts for the movement of electrons from the sample to the tip and vice versa.

Several approximations can be made to simplify Eq. 5 and to obtain a manageable analytical expression for I. If the thermal energy $k_BT << eV$, the Fermi–Dirac distributions can be approximated by step functions and the total current reduces to:

$$I = \frac{4\pi e}{\hbar}\int_{0}^{eV}\rho_T(E_F^T - eV + \varepsilon)\rho_S(E_F^S + \varepsilon)|M|^2 d\varepsilon. \quad (6)$$

(Note that Eq. 6 is valid only for $V > 0$. For $V < 0$ the integrand remains identical but the integration limits become $-e|V|$ and 0). In this case, only electrons with an energy differing from E_F by less than eV can participate to the tunneling current. This can be directly seen in Fig. 2b for the case of positive sample bias: tip electrons whose energy is lower than $E_F^T - eV$ cannot move because of Pauli's exclusion principle, while there are no electrons at energies higher than E_F^T. The main problem in determining expression (5) is, however, the calculation of the tunneling matrix elements M since this requires a knowledge of the sample and the tip wave functions, which can be very complicated. On the other hand, for relatively small bias voltages (in the ± 2 V range), Lang [6] showed that a satisfactory approximation of $|M|^2$ is given by a simple one-dimensional WKB tunneling probability. In the WKB approximation [7], the probability $D(\varepsilon)$ that an electron with energy ε tunnels through a potential barrier $U(z)$ of arbitrary shape is expressed as:

$$D(\varepsilon) = \exp\left\{-\frac{2}{\hbar}\int_0^s [2m(U(z) - \varepsilon)]^{\frac{1}{2}} dz\right\}. \quad (7)$$

This semiclassical approximation is applicable if ($\varepsilon << U$) which is generally satisfied in the case of metal samples where the work function is of the order of several eV. In order to obtain a simple analytical expression for D, the trapezoidal potential barrier of a biased tip-sample junction (see Fig. 2b) is further approximated with a square barrier of average height $\phi_{eff}(V) = (\phi_T + \phi_S + eV)/2$. By using this, the integral in Eq. 7 becomes:

$$D(\varepsilon, V, s) = \exp(-2ks), \quad (8)$$

where

$$k = \sqrt{\frac{2m}{\hbar^2}(\phi_{eff} - \varepsilon)}. \quad (9)$$

In order to evaluate k, it must be noted that electrons closest to the Fermi level experience the lowest potential barrier and are therefore characterized by an exponentially larger tunneling probability (see Fig. 2b). Thus, in a first approximation, it can be assumed that only these electrons contribute to the tunneling current which, for positive bias, is equivalent to set $\varepsilon \approx eV$ in (Eq. 9). Moreover, if the bias is much smaller than the work functions, eV can be neglected, resulting in

$$k \cong \frac{\sqrt{m(\phi_T + \phi_S)}}{\hbar} = 5.1\sqrt{\frac{\phi_T + \phi_S}{2}}\ \mathrm{nm}^{-1}, \quad (10)$$

where the work functions are expressed in eV. Using typical numbers for metallic work functions, the numerical value of the inverse decay length $2k$ in Eq. 8 becomes of the order of 20 nm^{-1}. Therefore, variations in s of 1 Å correspond to one order of magnitude changes in the tunneling probability and, as a consequence, in the measured current. This very high sensitivity provides the STM with a vertical resolution in the picometer regime. The lateral resolution of STM depends on how different points of the tip contribute to the total tunneling current. By considering a spherical tip shape with radius R, most of the current originates from the central position since this is closest to the surface. A point laterally displaced by Δx from the tip center is $\Delta z \approx \frac{\Delta x^2}{2R}$ further away from the substrate (higher order Δx terms are neglected in this evaluation). As a consequence, with respect to the tip center, the corresponding tunneling probability is reduced by a factor:

$$\exp\left(-2k\frac{\Delta x^2}{2R}\right). \tag{11}$$

By considering a tip radius $R \approx 1$ nm, the current changes by one order of magnitude for variations $\Delta x = 3$ Å. The actual lateral resolution is typically smaller than this upper limit and can reach down to fractions of an Å. Its specific value however depends on the precise shape of the tip which is unknown a priori. These values, together with the vertical resolution discussed above, lie at the basis of the STM atomic imaging capabilities.

Finally, if the tunneling probability (Eq. 8) is substituted for the tunneling matrix $|M|^2$ in Eq. 6, the total tunneling current can be expressed as:

$$I = \frac{4\pi e}{\hbar}\int_0^{eV} \rho_T(E_F^T - eV + \varepsilon)\rho_S(E_F^S + \varepsilon)e^{-2ks}d\varepsilon. \tag{12}$$

Therefore, for a fixed lateral position of the tip above the sample, the tunneling current I depends on the tip-sample distance s, the applied voltage V and the tip and sample density of states ρ_T and ρ_S, respectively.

Experimental Setup

As seen in the previous section, variations of 1 Å in s induce changes in the tunneling probability of one order of magnitude. The exponential dependence in Eq. 8 is responsible for the ultimate spatial resolution of STM but places stringent constraints on the precision by which s must be controlled, as well as on the suppression of vibrational noise and thermal drift. Moreover, typical tunneling currents are in the 0.01–10 nA range, requiring high gain and low noise electronic components. The following subsections are dedicated to a general overview of technologies and methods used to meet these specifications.

Scanner and Coarse Positioner

The extremely fine movements of the tip relative to the sample required for operating an STM are realized by using piezoelectric (► Piezoresistivity) ceramic actuators (*scanners*) which expand or retract depending on the voltage difference applied to their ends. In a first approximation, the voltage-expansion relation can be considered as linear with a proportionality factor (piezo constant) usually of few nanometer/Volt. The main requirements for a good scanner are: high mechanical resonance frequencies, so as to minimize noise vibrations in the frequency region where the feedback electronics operates (see section Electronics and Control System); high scan speeds; high spatial resolution; decoupling between x, y, and z motions; minimal hysteresis and creep; and low thermal drift. Although several types of STM scanner have been developed, including the bar or tube tripod, the unimorph disk and the bimorph [8], the most frequently used is a single piezoelectric tube whose outer surface is divided into four electrode sections of equal area. By applying opposite voltages between the inner electrode and opposite sections of the outer electrode, the tube bends and a lateral displacement is obtained. The z motion is realized by polarizing with the same voltage the inner electrode in respect to all four outer electrodes. By applying several hundred Volts to the scanner, lateral scan widths up to 10 μm and vertical ones up to 1 μm can be obtained, while retaining typical lateral and vertical resolutions of 0.1 nm and 0.01 nm, respectively.

While scanning is typically done by one individual piezoelectric element, larger displacements up to several millimeters are needed to bring the tip in close proximity to the sample, to move it to different regions of the surface or to exchange samples or tips. These are achieved by mounting the scanner onto a coarse position device. Several designs have been developed to this aim including micrometric screws driven either manually or by a stepper motor, piezoelectric walkers like the louse used in the first STM [9] or the inchworm [10], magnetic walkers where the movement is obtained by applying voltage pulses to a coil with a permanent magnet inside and piezoelectric driven stick-slip motors, as the Besocke-beetle [11] or the Pan motor [12].

Electronics and Control System

The voltages driving the piezoelectric actuators and their temporal succession and duration are generated by an electronic control system. The electronics are also used to bias the tunneling junction, to record the tunneling current and to generate the STM images.

In most of the modern instruments, these tasks are digitally implemented by a computer interfaced with digital to analog (DAC) and analog to digital (ADC) converters. The tunneling current is amplified by a high gain I–V converter (10^8–10^{10} V/A) usually positioned in close proximity of the tip, so as to reduce possible sources of electronic interference. This signal is then acquired by an ADC and processed by the control system. DACs are used to apply the bias voltage (from a few mV to a few V) between tip and sample and, in conjunction with high voltage amplifiers, to polarize the piezo elements. A feedback loop is integrated into the control system and is activated during the frequently used *constant current* imaging mode (see section STM Imaging). By acting on the z motion of the scanner, the feedback varies s to keep the tunneling current constant. This is controlled by a proportional-integral and derivative (PID) filter whose parameters can be set by the operator. Finally, a lock-in amplifier is often used to improve the signal-to-noise ratio in scanning tunneling spectroscopy (STS) measurements (see section Scanning Tunneling Spectroscopy).

Tip

Sharp metal tips with a low aspect ratio are essential to optimize the resolution of the STM images and to minimize flexural vibrations of the tip, respectively. Ideally, in order to obtain atomically resolved topographies and accurate spectroscopic measurements, the tip should be terminated by a single atom. In this case, because of the strong dependence on the tip-sample separation (see section Theory of Tunneling), most of the tunneling current would originate from this last atom, whose position and local DOS would precisely determine the tunneling conditions. In practice, however, it is almost impossible to determine the exact atomic configuration of the tip and the actual current is often due to a number of different atoms. This is still compatible with good tunneling conditions as long as these atoms are sufficiently localized (in order to avoid "multiple tip effects") and their structural and chemical state remains constant during scanning.

The most commonly used methods to produce STM tips are to manually cut or to electrochemically etch thin wires of platinum-iridium and tungsten, respectively. These materials are chosen because of their hardness, in order to prevent tips becoming irreversible damaged after an accidental crash. Other metallic elements and even semiconductor materials have been used as tips for specific STM applications. Due to their chemical inertness, Pt-Ir tips are often used to scan in air on atomically flat surfaces without the need of any further processing. However, they typically have inconsistent radii, while etched W tips are characterized by a more reproducible shape. These latter have the drawback that a surface oxide up to 20 nm thick is formed during etching or exposure to air. For this reason, W tips are mostly used in ultrahigh vacuum (UHV) where the oxide layer can be removed through ion sputtering and annealing cycles. Prior to use, tips are often checked by optical microscopy, scanning electron microscopy (► SEM), and field ion microscopy or transmission electron microscopy (► TEM). The quality of a tip can be further improved during scanning by using "*tip forming*" procedures, including pulsing and controlled crashing into metal surfaces. These processes work because the desorption of adsorbed molecules or the coating with atoms of the metallic substrate can produce a more stable tip apex. If STM is performed in polar liquids (► EC-STM), electrochemical processes might generate Faradaic or non-Faradaic currents which can be of the same order of magnitude or even larger than the tunneling current. In order to minimize these effects, the tip, except for its very apex, must be coated with an insulating material.

Vibration Isolation

A low level of mechanical noise is an essential requirement for any type of scanning probe microscopy. For this reason, the core of a STM, where the tip-sample junction is located, is always equipped with one or several types of vibration damping systems. These can be stacks of metal plates separated by elastic spacers, suspension springs, or eddy current dampers composed of copper elements and permanent magnets. The low-frequency components of mechanical noise (<10 Hz), which are the most difficult to eliminate, are minimized by building a small and rigid STM with a high resonance frequency. Depending on the overall size and weight of the microscope, further noise damping strategies can be adopted. Smaller, typically ambient conditions STMs can be placed on metal or granite slabs suspended by springs or bungee cords or floating on pneumatic isolators. Sometimes, piezo-driven, feedback-controlled active vibration suppressors are also combined with passive systems. Larger versions of pneumatic isolators and active damping are often used to float the frames and the chambers of big UHV STMs. The laboratory where a STM instrument is located also plays an essential role for its performance. Ground floor rooms are always preferred since they minimize low-frequency natural building oscillations, which can be very difficult to counteract. High-resolution instruments are sometimes placed on large concrete blocks which are separated from the rest of the laboratory floor and rest either on a sand bed, an elastomer barrier or on second-stage pneumatic isolators. Moreover, they are also often surrounded by an acoustically insulating box. All these systems essentially act as low-pass mechanical filters whose effectiveness improves with decreasing cutoff frequencies, that is, with increasing mass and decreasing rigidity. For this reason, the body of a STM is typically a relatively heavy block of metal and the frames, slabs, and vacuum chambers supporting or containing the microscope often have a considerable weight.

Setups for Different Environments and Temperatures

Different types of STMs have been developed that can operate in various environments such as air, inert atmosphere (N_2, Ar), vacuum, high pressure, liquid, or in an electrochemical cell. The core of the different instruments is essentially the same, although the experimental chambers and setups in which they are located can vary substantially. Ambient condition STMs are typically quite compact and rigid and do not need elaborated anti-vibrational mechanisms. On the other hand, since sound waves represent a major problem, atmospheric pressure STMs are usually contained in an acoustic enclosure. A STM operating in vacuum must be hosted in a chamber with vibration-free pumps (typically ionic pumps for UHV) and must be equipped with sophisticated sample and tip manipulation mechanisms. Such systems often also have an in situ surface preparation stage allowing the handling of samples without air exposure.

STM can be performed at high pressures (1–30 bar) by installing the microscope head into gas manifolds under conditions similar to those used in industrial catalytic processes. Also in this case, sample and tip manipulation and preparation stages are mandatory parts of the system. Since these types of studies are typically performed at elevated temperatures (up to 600 K) and in the presence of highly reactive gases, the metallic parts of the STM scanner and of the chamber are often gold plated, the volume of the STM chamber is kept as small as possible and the tip material is chosen to be inert toward the gases [13]. Moreover, low voltages are used for polarizing the piezos in order to avoid gas discharges at intermediate pressures (10^{-3}–10 mbar) and shields are added to protect the STM from the deposition of conductive materials which could create electrical shorts.

STM at the liquid/solid interface and electrochemical STM (EC-STM) (► EC-STM) need the tip and sample to be inside a liquid cell which, in turn, may be placed in a humidity-controlled atmosphere. In the case of low vapor pressure liquids, the STM can be simply operated under ambient conditions by dipping the tip into a liquid droplet deposited on the sample. A special coating must be applied to the tip when working with polar liquids (see section Tip).

STM can also be performed at different temperatures (in vacuum or controlled atmosphere chambers): variable temperature STM (VT-STM) able to cover the 5–700 K range, low temperature STM (LT-STM) operating at 77 K or 5 K and even milli-Kelvin STM instruments are currently available. A VT-STM is typically used to study thermally activated processes such as diffusion and growth, phase transitions, etc. These systems have sample heating and cooling stages which can be operated in a combined way so as to achieve a very precise temperature stabilization. Resistive heating is normally employed to increase the temperature, while both flow and bath cryostats with liquid nitrogen or helium as cryogenic fluids are used to reduce it. Continuous flow cryostats offer a high flexibility in temperature but are characterized by lower thermal stability, by inherent mechanical vibrations and do not easily attain temperatures below 20 K. Bath cryostats are more stable, are able to reach lower temperatures but are often also much bulkier (e.g., in order to limit the He consumption rate, a liquid He cryostat is actually a double-stage cryostat with an outer liquid nitrogen mantle). For most of these instruments the variable temperature capabilities refer to the possibility of choosing different (fixed) temperatures at which the microscope is run. However, few systems endowed with specific position tracking and drift compensating capabilities allow a "true" variable temperature operation where the same surface area can be imaged with atomic resolution while its temperature is changed. LT-STMs are operated at a fixed temperature and are typically inserted inside double stage cryostats which significantly complicates the tip and sample access. However, these instruments are extremely stable with a very low thermal drift and are therefore the best choice for STS and manipulation experiments (see sections Scanning Tunneling Spectroscopy and Applications). Milli-Kelvin STMs enable temperatures to be reached where extremely interesting magnetic, quantum Hall physics and superconductivity phenomena occur. Moreover, the thermal broadening of electronic features is strongly reduced, which is required for high-resolution measurements. These systems operate based on the evaporative cooling of liquid ^{3}He to temperatures of about 300 mK or liquid ^{3}He and ^{4}He mixtures below 10 mK. The STM heads can be further placed inside large-bore superconducting magnets (up to 15 T), enabling the low temperature and high magnetic field conditions necessary to access superconductive phase transitions or to detect single spin flip processes.

STM Imaging

STM images are generated by recording the tunneling current as a function of the tip position while the tip is scanned across the sample surface. This can be done in two different ways which define the two main STM imaging modes:

- *Constant height mode*. The z section of the piezo scanner is kept fixed while the tip is moved over the substrate at a constant bias voltage (Fig. 3a). Variations of the tip-sample distance due to the surface topography produce a corresponding variation of the tunneling current which is recorded point-by-point and used to build the STM gray-level image. This mode is employed only in small areas of extremely flat surfaces, where the probability of crashing into protrusions such as steps or defects is relatively small. Very high scanning speeds can be used because of the absence of a feedback control.
- *Constant current mode*. While the x and y sections of the piezo scanner are used to laterally move the tip across the surface, the z section is driven by the electronic feedback so as to maintain a constant tunneling current (Fig. 3b). The corresponding z-voltage applied to the scanner (feedback signal) is recorded point-by-point and used to build the STM gray-level image. This mode can be employed for any type of surface topography and is therefore the most frequently used.

Since the constant height mode is applied to atomically flat surfaces with sub-Å height variations, the exponential $I - s$ relation derived from Eq. 12 can be approximated by a linear dependence. As a consequence, constant height STM images are a good representation of flat surfaces. On the other hand, for less planar substrates, one must use the constant current mode which directly reproduces the surface height due to the linear voltage-extension relation of piezoelectric materials. However, even constant current STM images are a reliable representation of the "true" surface topography only if the sample local DOS does not vary across the scanned area. If this is not the case, a constant current profile corresponds to a complex convolution of topographical and electronic features (see Eq. 12) which can be particularly relevant for surfaces covered with molecular adsorbates.

Scanning Tunneling Spectroscopy

Besides complicating the interpretation of STM images, the dependence of the tunneling current on the sample DOS also offers the unique opportunity of probing the electronic characteristics of surfaces with sub-nm spacial resolution. Having fixed the tip lateral position, the tunneling current I is a function of the applied bias voltage V and the tip-sample separation s only, the precise relation being established by Eq. 12. In a STS experiment, the relation between two of these three parameters is measured while the remaining one is kept constant (STS). $I(V)$ spectroscopy, where the tunneling current is measured as a function of the bias voltage for a constant tip-sample separation, is the most widely used technique because it provides indications about the DOS of the sample.

Due to the spatial localization of the tunneling current (see section Theory of Tunneling), STS enables the characterization of the electronic properties of individual atoms and molecules in relation to their structure, bonding and local environment. Moreover, STS can also be used to create 2D maps of the sample DOS with sub-nm resolution. Such measurements are particularly interesting for quantum confined

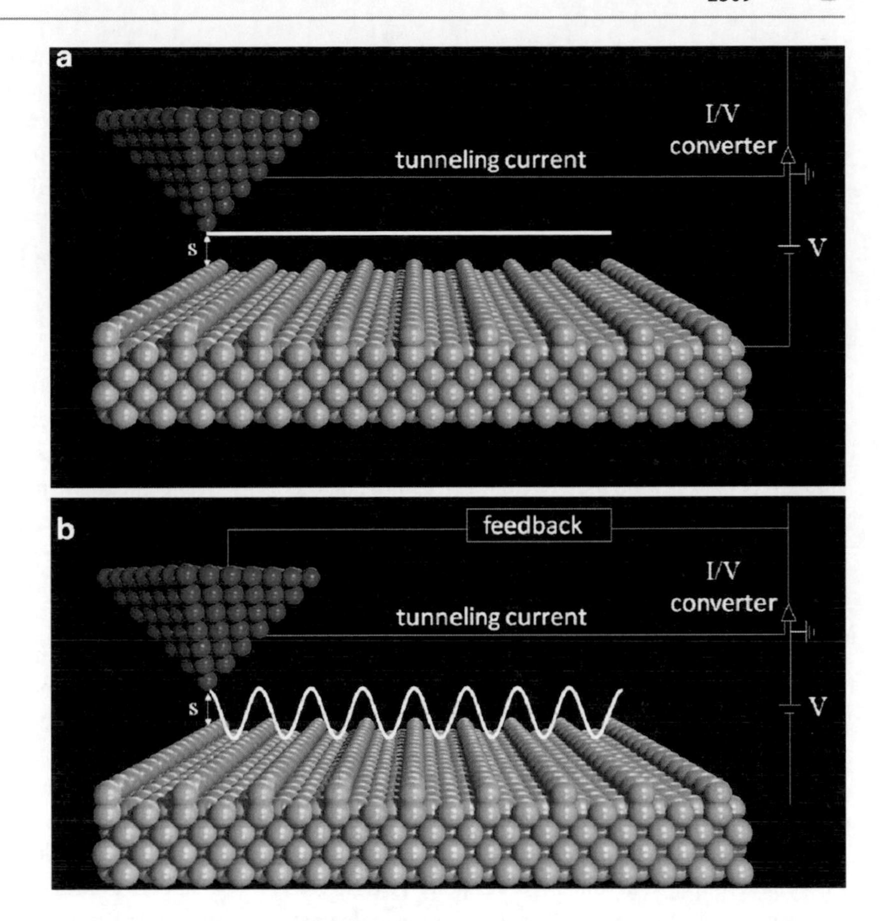

Scanning Tunneling Microscopy, Fig. 3 Schematic representation of (**a**) the constant height and (**b**) the constant current imaging modes, respectively. The *thick lines* represent the trajectory followed by the tip

electronic systems (e.g., quantum dots or quantum corrals) or for determining the shape of molecular orbitals [14] (*wavefunction mapping*). By changing the polarity of the bias voltage, STS gives access to both the occupied and the unoccupied states of the sample. In this sense, it is often considered as complementary to ultraviolet photoemission spectroscopy (UPS), inverse photoemission spectroscopy (IPS) and electron energy loss spectroscopy (EELS), where the signal is averaged over a large area of the surface (between 0.1 and 2 mm in diameter). On the other hand, STS does not provide direct chemical information and tip artifacts can strongly influence the spectroscopic data.

So far we have assumed that electrons conserve their energy during the tunneling process (see Eq. 3). However, electrons can also tunnel inelastically between the tip and the sample by exchanging energy and inducing the excitation of vibrational modes, spin-flips, magnons, plasmons, excitons, etc. These extra tunneling channels become available only above specific voltage thresholds since only beyond these values a part of the electron energy can be converted into the excitation. The additional inelastic pathways increase the overall tunneling probability and therefore show up as discrete step-like features in the tunneling conductivity or as slope changes in $I(V)$ curves. This technique is called inelastic electron tunneling spectroscopy (IETS) and benefits from the same spatial resolution as STM and STS. IETS has been used to measure vibrational modes of individual molecules, spin excitations of single magnetic atoms, collective plasmon excitations in 2D materials and magnons in ferromagnets.

A different way of detecting tunneling-induced molecular vibrations by means of a STM is to rely on their coupling with dynamical processes such as molecular motions. In particular, by measuring the frequency of molecular hopping events as a function of the applied bias voltage, it is possible to create so-called *action spectra* which reflect the vibrational spectrum of an individual molecule in a quantitative manner [15]. Optical excitations can also be revealed in an alternative way by coupling the STM with a photon detection system able to collect and analyze the luminescence stimulated by inelastically tunneling electrons [16]. Such a setup has been used to characterize plasmon emission from metallic surfaces and luminescence from semiconductor quantum structures and adsorbed molecules.

Applications

Since the first STM images of the surfaces of $CaIrSn_4$ and Au [1] were published back in 1982, STM has been used to analyze a wide range of materials: clean and adsorbate covered metal surfaces, semiconductors, superconductors, thin insulating layers, small and large organic molecules, individual atoms, liquid-solid interfaces, magnetic layers and surfaces, quasi-crystals, polymers, biomolecules, nanoclusters, and carbon nanotubes. Imaging is the most frequent application of STM used to determine the structural properties of substrates and their reconstructions, the presence of defects, sites of adsorption for adatoms and molecules and the symmetry and periodicity of adsorbate superstructures. Nevertheless, right from the beginning, it became clear that the ultimate spatial resolution of STM, in combination with its dependence on the electronic properties of tip and sample, could allow a much wider range of applications of this instrument. These include the characterization of surface electronic, vibrational, optical, and magnetic properties, the measurement of single molecule conductivities, and the study of dynamic processes. In the following, we will only touch upon some of the most frequent applications, without any presumption of being exhaustive.

Equation 12 shows that the tunneling current depends on the sample DOS close to the Fermi energy E_F. As a consequence, at a typical bias of a few Volts, it should be possible to image conductors, superconductors and small-gap or doped semiconductors but not molecules and insulating materials due to the vanishing DOS in the probed energy range (for most molecules the highest occupied and the lowest unoccupied orbitals are separated by an energy gap of several eV). However, the great majority of molecules adsorbed on metallic substrates can be easily imaged at moderate bias voltages. This is due to the formation of a metal-organic interface which can modify the molecular electronic properties leading to a broadening of the initial discrete energy levels, to a reduction of the gas-phase energy gap and even to the development of new states if covalent molecule-substrate bonds are established. All these effects contribute to the DOS at E_F and allow the imaging process. Regarding insulating materials, STM can only be done on films deposited onto conductive substrates if they are thin enough to allow the tunneling of electrons. These films are often used to electronically decouple organic adsorbates from metallic substrates.

The mechanism which allows the imaging of biomolecules such as DNA and proteins is currently still under debate [17]. As these molecules have a very large energy gap (5–7 eV) they can be considered as insulating materials and the current measured in STM experiments might be mediated by the thin water layer surrounding the molecules in air. Metalloproteins have also been imaged in their "natural" environment by using EC-STM (► EC-STM). Several reports have shown that when these redox active molecules are imaged under potentiostatic control, the tunneling current can be mediated by their metal redox-center, with enhanced conductivities measured for bias voltages close to the redox potential.

Although STM is a surface sensitive method, it can be also used to analyze buried interfaces and structures in cross-sectional STM (XSTM) [18]. The specific sample preparation in this technique requires brittle materials such as oxide samples or semiconductor wafers. A cross section of the structure to be analyzed is prepared by cleaving the sample and positioning the STM tip onto the exposed edge. In this way, various physical properties can be probed, including the morphology and abruptness of buried nanostructures and interfaces, the alloying in epitaxial layers, the spatial distribution of dopants and their electronic configuration and the band offsets in semiconductor heterojunctions. XSTM has also been used to study, in real time, the changes occurring in semiconductor quantum well laser devices under operating conditions.

STM can further be employed for tracking dynamic surface processes, provided that the corresponding characteristic times are longer than the acquisition time. By choosing optimized designs for the piezo scanners and the electronic feedback, video-rate instruments have been developed able to record several tens of images per second and thereby to follow mobility and assembly processes in real time [19, 20].

When a tip with spin polarized electrons is used in STM, besides the parameters already indicated in Eq. 12, the local sample magnetization also influences the tunneling current. In fact, due to the different density of states at E_F of "spin-up" and "spin-down" electrons in magnetic materials, a spin polarized tip causes a tunnel magnetoresistance effect which results in a further contrast mechanism. This technique, called spin-polarized STM (SP-STM), has been used both in the presence and absence of an external magnetic field for detecting magnetic domain structures and boundaries in ferro- and antiferromagnetic materials, visualizing atomic-scale spin structures and determining spatially resolved spin-dependent DOS. An essential aspect of SP-STM is the ability to control the magnetization direction of the tip which can be achieved by evaporating different types of ferromagnetic or antiferromagnetic thin films on nonmagnetic tips. This technique is preferred to the use of bulk magnetic tips since it reduces the magnetic stray fields which can significantly modify the sample magnetization [21].

The ability of STM to identify and address individual nano-objects has been used to measure the conductivity of single molecules absorbed on metal surfaces. While the tip is approached to the molecule of interest at constant bias voltage, the current flowing in the junction can be measured, thereby generating an $I(s)$ curve. Alternatively, $I(V)$ curves can be recorded at different s values. Since tip and substrate act as electrodes, both methods enable information to be obtained about the conductance of the individual molecule embedded in the junction. These measurements are often complemented by IETS experiments in the same configuration. IETS might in fact help to determine the arrangement and the coupling of the junction, which has a significant influence on the electronic and structural properties of the molecule. Single molecule STM conductance experiments represent an important source of information for understanding mechanisms of electron transport in organic molecules with applications in organic electronics and photovoltaics. They complement narrow gap electrode and break junction techniques, having the significant advantage of a highly localized electrode which allows to address and characterize individual molecules.

A similar type of application, although typically not aimed at individual molecules, is at the basis of the four point probe STM, where four STM tips, in addition to imaging, are used for local four point electric conduction measurements. A scanning electron microscope is installed above the STM enabling the positioning of the tips on the contact. The purpose of such very complex instruments is to measure the charge transport through individual nanoelectronic components (in particular self-assembled ones) and to correlate this information with a local high-resolution structural characterization.

Tip-Induced Modification

Besides being an extraordinary instrument for the characterization of structural, electronic, vibrational, optical, and magnetic properties of surfaces with subnanometer resolution, STM has also developed as a tool to modify and nanoengineer matter at the single molecule and atom scale.

By decreasing the distance between the tip and the sample in a controlled way, indentations can be produced in the substrate with lateral sizes down to a few nm. Nanolithography can also be performed by tunneling electrons into a layer of e-beam photoresist (▶ SU-8 Photoresist), thereby reaching a better resolution compared to standard electron beam lithography (EBL). Many other STM-based nanopatterning and nanofabrication techniques have been developed based on a number of physical and chemical principles including anodic oxidation, field evaporation, selective chemical vapor deposition, selective molecular desorption, electron-beam induced effects, and mechanical contact. All these methods exploit the extreme lateral localization of the tunneling current and can be applied in air, liquids and vacuum.

However, the nanotechnological application that gained most attention is the ability to manipulate individual atoms and molecules on a substrate. This is possible due to a controlled use of tip-particle forces and is typically done in UHV and at low temperatures. The first atomic manipulation experiment was performed by Eigler and Schweizer in 1989 [2]. This phenomenal result fulfilled Richard Feynman's prophecy that "ultimately-in the great future-we can arrange the atoms the way we want; the very atoms, all the way down!" [3].

During a lateral manipulation experiment, the tip is first placed above the particle to be moved (for example an atom) and the tunneling current is increased while keeping a constant voltage. This results in a movement of the tip toward the atom, see Eq. 12. If their separation is reduced below 0.5 nm, Van der Waals forces start to come into play together with attractive and repulsive chemical interactions. When these forces equal the diffusion energy barrier, a lateral displacement of the tip can induce a movement of the atom parallel to the surface. After the desired final position is reached, the tip is retracted by reducing the tunneling current to the initial value, leaving the atom in the selected place. Depending on the tip-particle distance and therefore on the strength and nature of the interaction, different manipulation modes including pulling, pushing, and sliding [22] were identified and used to move different types of atoms and molecules.

Thanks to this technique, it was possible to fabricate artificial nanostructures such as the *quantum corral* [23] and to probe quantum mechanical effects like the quantum confinement of surface state electrons or the *quantum mirage*. Lateral STM manipulation has also been used to switch between different adsorption configurations and conformations of molecules on surfaces and to modify their electronic properties in a controlled way [24].

A further application of STM manipulation is the synthesis of new molecular species based on the ability of STM to form and break chemical bonds with atomic precision. Reactants are brought close together on the surface and the actual reaction is realized by applying a voltage pulse or by exciting vibrational modes through inelastically tunneling electrons. Examples of this technique include the dissociation of diatomic molecules, the Ullmann reaction, the isomerization of dichlorobenzene and the creation of metal-ligand complexes.

The STM tip has also been used to perform vertical manipulations of nanoparticles where an atom (or molecule) is deliberately transfered from the surface to the tip and vice versa by using the electric field generated by the bias voltage. In contrast to the lateral manipulation, here the bonds between the surface and the atom are broken and re-created [25]. By approaching the tip at distances of a few Å from the chosen particle chemical interactions are established that reduce the atom-surface binding energy. If a voltage pulse is applied under these conditions, the resulting electric field (of the order of 10^8 V/cm) can be enough to induce the particle desorption. The vertical manipulation technique has also been used as a means to increase the lateral resolution of STM. In fact, the controlled adsorption of a specific molecule onto the tip often makes it "sharper" and can add a chemical resolution capability if the DOS of the extra molecule acts as an "energy filter."

A related effect is exploited in the recently proposed scanning tunneling hydrogen microscopy (STHM) technique. In STHM, the experimental chamber is flooded with molecular hydrogen while the tip is scanned in constant height mode at very close distances over the surface. H_2 can get trapped in the tip-sample junction and its rearrangement during scanning of the surface generates a new contrast mechanism based on the short-range Pauli repulsion. This is extremely sensitive to the total electron density, thereby endowing the STM with similar imaging capabilities to non-contact AFM (► AFM, Non-contact Mode) and making it able to resolve the inner structure of complex organic molecules [26].

Acknowledgments This work was supported by EPSRC (EP/D000165/1); A. Della Pia was funded through a WPRS scholarship of the University of Warwick. J. V. Macpherson, T. White, and B. Moreton are gratefully thanked for their critical reading of the manuscript.

Cross-References

- ► AFM, Non-contact Mode
- ► Atomic Force Microscopy
- ► Electrochemical Scanning Tunneling Microscopy
- ► Electron Beam Lithography (EBL)
- ► Piezoresistivity
- ► Scanning Electron Microscopy
- ► Scanning Tunneling Microscopy
- ► SU-8 Photoresist
- ► Transmission Electron Microscopy

References

1. Binnig, G., Rohrer, H., Gerber, C., Weibel, E.: Surface studies by scanning tunneling microscopy. Phys. Rev. Lett. **49**, 57 (1982)

2. Eigler, D.M., Schweizer, E.K.: Positioning single atoms with a scanning tunneling microscope. Nature **344**, 524 (1990)
3. Feynman, R.P.: There's plenty of room at the bottom: an invitation to enter a new field of physics. Engineering and Science. **23**, 22 (1960)
4. Bardeen, J.: Tunneling from a many-particle point of view. Phys. Rev. Lett. **6**, 57 (1961)
5. Gottlieb, A.D., Wesoloski, L.: Bardeen's tunnelling theory as applied to scanning tunnelling microscopy: a technical guide to the traditional interpretation. Nanotech. **17**, R57 (2006)
6. Lang, N.D.: Spectroscopy of single atoms in the scanning tunneling microscope. Phys. Rev. B **34**, 5947 (1986)
7. Landau, L.D., Lifshitz, E.M.: Quantum mechanics: non-relativistic theory. Pergamon Press, Oxford, (1977)
8. Chen, C.J.: Introduction to scanning tunneling microscopy. Oxford University Press, Oxford (2008)
9. Binnig, G., Rohrer, H.: Scanning tunneling microscope. Helv. Phys. Acta **55**, 726 (1982)
10. Okumura, A., Miyamura, K., Gohshi, Y.: The STM system constructed for analytical application. J. Microsc. **152**, 631 (1988)
11. Besocke, K.: An easily operable scanning tunneling microscope. Surf. Sci. **181**, 145 (1987)
12. Pan, S.H., Hudson, E.W., Davis, J.C.: ^{3}He refrigerator based very low temperature scanning tunneling microscope. Rev. Sci. Instrum. **70**, 1459 (1999)
13. Laegsgaard, E., et al.: A high-pressure scanning tunneling microscope. Rev. Sci. Instrum. **72**, 3537 (2001)
14. Repp, J., Meyer, G., Stojkovic, S.M., Gourdon, A., Joachim, C.: Molecules on insulating films: scanning-tunneling microscopy imaging of individual molecular orbitals. Phys. Rev. Lett. **94**, 026803 (2005)
15. Sainoo, Y., et al.: Excitation of molecular vibrational modes with inelastic scanning tunneling microscopy processes: examination through action spectra of cis-2-butene on Pd(110). Phys. Rev. Lett. **95**, 246102 (2005)
16. Gimzewski, J.K., Reihl, B., Coombs, J.H., Schlittler, R.R.: Photon emission with the scanning tunneling microscope. Z. Phys. B:Condens. Matter **72**, 497 (1988)
17. Davis, J.J.: Molecular bioelectronics. Philos. Trans. Roy. Soc. A **361**, 2807 (2003)
18. Feenstra, R.M.: Cross-sectional scanning-tunneling-microscopy of III-V semiconductor structures. Semicond. Sci. Technol. **9**, 2157 (1994)
19. Rost, M.J., et al.: Scanning probe microscopes go video rate and beyond. Rev. Sci. Instrum. **76**, 053710 (2005)
20. Petersen, L., et al.: A fast-scanning, low- and variable-temperature scanning tunneling microscope. Rev. Sci. Instrum. **72**, 1438 (2001)
21. Wiesendanger, R.: Spin mapping at the nanoscale and atomic scale. Rev. Mod. Phys. **81**, 1495 (2009)
22. Bartels, L., et al.: Dynamics of electron-induced manipulation of individual CO molecules on Cu(111). Phys. Rev. Lett. **80**, 2004 (1998)
23. Crommie, M.F., Lutz, C.P., Eigler, D.M.: Confinement of electrons to quantum corrals on a metal-surface. Science **262**, 218 (1993)
24. Moresco, F., et al.: Conformational changes of single molecules induced by scanning tunneling microscopy manipulation: a route to molecular switching. Phys. Rev. Lett. **86**, 672 (2001)
25. Avouris, P.: Manipulation of matter at the atomic and molecular-levels. Acc. Chem. Res. **28**, 95 (1995)
26. Weiss, C., et al.: Imaging Pauli repulsion in scanning tunneling microscopy. Phys. Rev. Lett. **105**, 086103 (2010)

Scanning Tunneling Spectroscopy

Amadeo L. Vázquez de Parga and Rodolfo Miranda
Dep. Física de la Materia Condensada, Universidad Autónoma de Madrid and Instituto Madrileño de Estudios Avanzados en Nanociencia (IMDEA-Nanociencia), Madrid, Spain

Definition

Scanning tunneling spectroscopy (STS) is a technique that allows the study of the electronic structure of surfaces with atomic resolution.

Overview

Scanning tunneling microscopy (STM) was historically the second technique that could image individual atoms one by one. It was invented in 1981–1982 by Gerd Binnig and Heindrich Rohrer [1], long after the technique of field ion microscopy (FIM) developed in 1951 by Erwin Müller [2].

In STM a sharp tip probes the surface of interest by allowing electrons to tunnel quantum-mechanically between the tip and the surface. Because such tunneling is extremely sensitive to the distance between tip and surface, one gets high resolution perpendicular to the surface. Assuming a constant density of states on the surface, when the STM tip is scanned over the sample surface while keeping the tunneling current constant, the tip movement depicts the surface topography, because the separation between the tip apex and the sample surface is always constant. It is worth noting that STM not only converts the spatial change in the tunneling current into a highly detailed topographic image of surfaces with constant density of states, but also the tunneling current changes with the available surface electronic states. This dependence of the tunneling current on the surface electronic structure together with the high spatial resolution of STM allows us to study the electronic structure of the

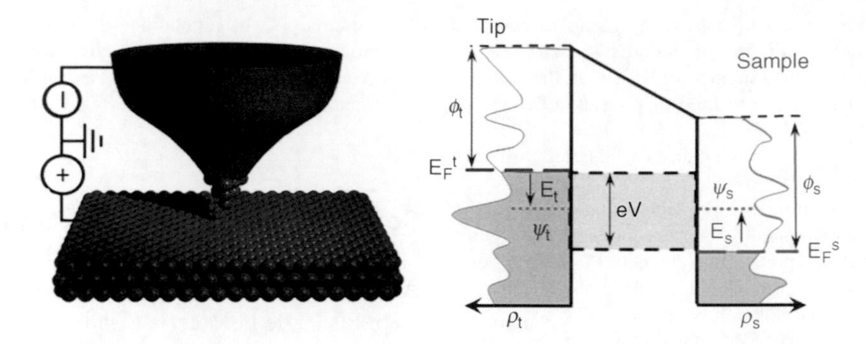

Scanning Tunneling Spectroscopy, Fig. 1 *Left panel*: Configuration of a positively biased tunnel junction in a STM. The bias voltage is applied to the sample and the tunneling current is measured on the tip. *Right panel*: Energy diagram of an STM tunnel junction. In this diagram the vertical axis represents energy and the horizontal axes distance between tip and sample and density of states. E_t, Ψ_t and E_s, Ψ_s are the energy and wave function of the states of the electrode tip and sample respectively. ϕ_t, ϕ_s, E_F^t, E_F^s, ρ_t, ρ_s are the work functions, Fermi energies and densities of states (DOS) of electrode tip and sample, respectively, and V is the voltage applied to electrode sample

surfaces with atomic resolution. The technique is known as scanning tunneling spectroscopy (STS).

Scanning Tunneling Spectroscopy Theory

Most theoretical treatments applied today to describe the tunneling process in an STM start from the formalism of the transfer Hamiltonian developed by Bardeen in 1961 [3] for the study of superconducting tunnel junctions. In this approach, the electronic structure and electron wave functions of both electrodes are calculated assuming no interaction between them and afterward the tunneling current is calculated [3]. Figure 1a shows the scheme of the tunnel junction in a STM where one of the electrodes is a tip. Figure 1b shows the energy diagram of the tunnel junction. In this diagram the vertical axis represents energy. E_t and Ψ_t are the energy and wave function of the states of the electrode "tip" in the absence of electrode "sample." E_s and Ψ_s are the energy and wave function of the states of the electrode "sample" in the absence of electrode "tip." ϕ_t, ϕ_s, $E_F{}^t$, $E_F{}^s$, ρ_t, ρ_s are the work functions, Fermi energies, and densities of states (DOS) of electrode "tip" and "sample," respectively, and V is the voltage applied to electrode "sample." When the distance between the electrodes is small enough, the overlap between their wave functions is significant and the probability of electron transfer between the two electrodes by tunneling starts to be noticeable. In the absence of applied voltage, the Fermi levels of the two electrodes are aligned and no net tunneling current flows. However, by applying a voltage V the Fermi levels move with respect to each other opening an energy window, eV, where electrons from one electrode can tunnel to the empty states of the other and, thus, the tunneling current starts to flow.

In 1983, Tersoff and Hamann applied the Bardeen's formalism to the STM, replacing one of the electrodes by a point [4, 5, 6]. The tip was shaped like an s orbital centered at the tip position and the calculated tunneling matrix elements proved to be proportional to the amplitude of the wave functions of the sample at the position of the tip. If the distance between tip and sample is not very large (few ångstroms), the bias voltage small and the temperature low, the tunneling current can be written as follows:

$$I \propto \int_o^{eV} \rho_s(\vec{r}_s, E)\rho_t(\vec{r}_s, E - eV)T(E, eV, d, \phi)dE \quad (1)$$

where $\vec{r}_s$ is the tip position over the sample surface, d is the distance between tip and sample, and T is the transmission probability that depends on the energy of the states involved, the bias voltage applied between tip and sample, the distance between tip and sample, and the tunneling barrier height, which is related with the tip and surface work functions. This equation indicates that the tunneling process depends, for a given energy, on three interconnected parameters, i.e., the tunneling current I, the bias voltage V, and the tip sample separation d. Almost all attempts to explore these complex dependences of the tunneling current (and simultaneously extend the performance of STM) have been demonstrated by the late 1980s [7]. Scanning tunneling spectroscopy measures the relation between any two of while keeping fixed the third one. This gives three modes of spectroscopy measurements: (1) I-V curves, where the variation of the tunneling current with the bias voltage is measured for a fixed distance between tip and sample, (2) I-z curves, where the variation of the tunneling current with the distance between tip and sample is measured for a fixed bias voltage V, and (3) V-z curves, where the variations in the tip sample distance are measured as function of the bias voltage for a fixed tunneling current.

In these three modes energy conservation for the tunneling electrons is assumed. If electrons change their energy during the tunneling process by an inelastic process, the inelastic electron tunneling spectroscopy (IETS) mode is possible. Finally, if the STM tip and the sample are magnetic, the tunneling current depends on the relative orientation of the magnetization of both tip and sample. This mode is called spin polarized scanning tunneling microscope (SP-STM) and allows the study of magnetic properties with atomic resolution. In the following the five modes will be discuss in detail.

Scanning Tunneling Spectroscopy Modes

I-V Curves

I-V measurements are the most widely used spectroscopic technique in STM experiments. If the tunneling current (Eq. 1) is differentiated with respect to the bias voltage the following expression is obtained:

$$\frac{dI}{dV} \propto \rho_t(0)\rho_s(eV)T(eV,eV,d,\phi) + \int_0^{eV} \rho_t(E-eV)\rho_s(E)\frac{dT(E,eV,d,\phi)}{dV}dE + \int_0^{eV} \rho_s(E)\frac{d\rho_t(E-eV)}{dV}T(E,eV,d,\phi)dE$$

Assuming a constant density of states for the tip, the third term is zero, but, it should be mentioned that, often, the tip electronic states have a strong influence in the STS spectra [8]. Another common simplification is to assume that the transmission coefficients are constant in the voltage range explored in the measurement. Then the second term also vanishes and the expression becomes:

$$\frac{dI}{dV} \propto \rho_t(0)\rho_s(eV)T(eV,eV,d,\phi)$$

In general the experimentally determined differential tunneling conductance is widely accepted as a good approximation to the DOS of the surface (modulated by the specific transmission of the barrier) at an energy value of eV, with V being the bias voltage applied between tip and sample.

Figure 2 shows an STS experiment performed on the Cu(111) surface in which the transmission probability (see Eq. 1) can be seen to depend on the energy parallel to the surface. Figure 2a shows the bulk band projection of Cu(111) along the $\overline{\Gamma M}$ direction of the surface Brillouin zone. Bulk bands are represented in gray and the projectional bandgap in white. The gray line corresponds to the dispersion relation of Shockley surface state of Cu(111). The surface state can be seen as a two-dimensional electron gas with the bottom of the band at −0.44 eV and an effective mass $m^* = 0.42$ m_e. Therefore, disregarding any contribution from the bulk electronic structure, the LDOS expected around the Fermi level of Cu(111) is a step function centered at the bottom of the band. This step function is shown in Fig. 2b. Figure 2c shows an STS spectrum taken on the Cu(111) surface. The experimental data correspond to the dots. The spectrum has a peak at −0.38 eV superimposed on a background which decays with increasing energy. The bottom of the surface state band corresponds to the point halfway up the peak (dashed line in Fig. 2c). The line in the graphic is a fit to the background that reflects the contribution of the bulk states. As the bias voltage approaches the Fermi level from below, the $k_{\|}$ of the accessible bulk states increases, so that they present a smaller effective perpendicular energy which means a smaller transmission probability and, thus, they contribute less to the spectrum. On the other hand, the peak at −0.38 eV corresponds to the sharp increase in the LDOS associated to the bottom of the surface state, superimposed on the background due to the bulk bands. Although the LDOS of a 2D is a step function to a constant value, the peak in the spectrum instead of staying constant for energies higher than the bottom of the band (−0.44 eV), decreases with increasing energy. This reduction in the signal with energy reflects the dispersion of the surface state, i.e., the fact that the $k_{\|}$ of the surface state increases also when the energy increases. Accordingly, the transmission probability (and the signal in the spectrum) gets smaller.

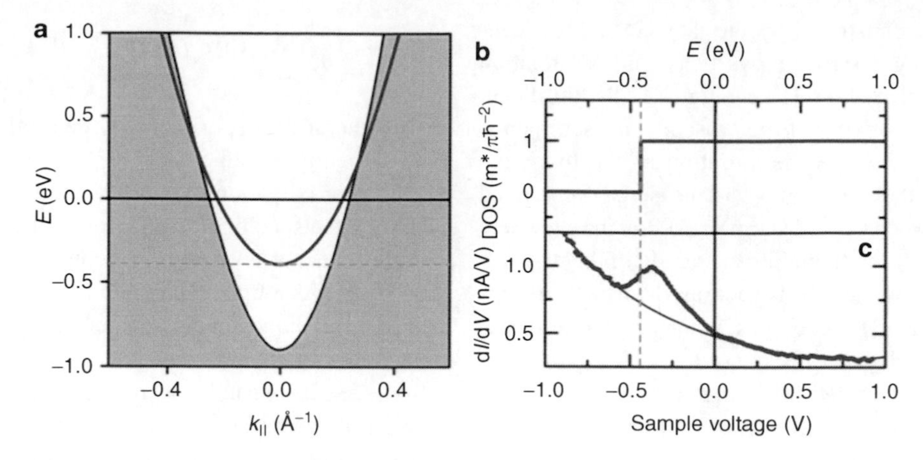

Scanning Tunneling Spectroscopy, Fig. 2 (**a**) Shows the bulk band projection of Cu(111) along the $\overline{\Gamma M}$ direction of the surface Brillouin zone. Bulk bands are represented in *gray*, and in *white* the projectional bandgap. The *gray line* corresponds to the dispersion relation of surface state of Cu(111). (**b**) Expected density of states for a two-dimensional electron gas. (**c**) An STS spectrum taken on the surface Cu(111). The experimental data correspond to the dots

Experimentally in order to record I-V curves, the distance between tip and sample has to be kept constant during the measurement time. This can be done in different ways. In one of them the tip is placed at a desired position on the surface and at the desired distance from the surface. The distance between tip and sample is dictated by the values of the tunneling current and the bias voltage used in the topographic image. The feedback circuit, that keeps the tunneling current constant adjusting the tip sample distance, is disconnected and then the voltage V is ramped and the tunneling current is recorded over the desired bias voltage range. The dI/dV values are obtained by numerical differentiation of the I-V curves. If the measurement is repeated in every pixel of a topographic image, the method is called *current imaging tunneling spectroscopy* (CITS) and provides with a map of the spatial distribution of the LDOS on the surface.

Another method is to detect directly the dI/dV signal using a lock-in amplifier. In order to do that, a small high-frequency sinusoidal signal, $V_{\mathrm{mod}}\ \sin(\omega t)$, is superimposed on the bias voltage between tip and sample. The modulation causes a sinusoidal response in the tunneling current and the amplitude of the modulated current is sensitive to dI/dV. For a small applied sinusoidal signal, the modulated current can be Fourier decomposed on the applied modulation frequency ω:

$$I(V_{\mathrm{bias}},t) = I(V_{\mathrm{bias}}) + \frac{dI(V_{\mathrm{bias}})}{dV} V_{\mathrm{mod}} \sin(\omega t) + \frac{d^2 I(V_{\mathrm{bias}})}{dV^2} \frac{V_{\mathrm{mod}}^2}{4} \sin(2\omega t) + \cdots \quad (2)$$

The first harmonic, which is proportional to the differential conductance (dI/dV), can be extracted by means of lock-in detection and the spatial variation of the dI/dV signal can be mapped in certain area of the surface. During a constant current topographic image, the dI/dV signal is simultaneously recorded in each point of the image at a certain bias voltage. The result is a map that reflects the LDOS of a surface area at a defined energy eV. Since the feedback loop is connected during the topographic image, the frequency of the modulated signal needs to be higher than the cutoff frequency of the feedback loop response in order to keep the tip distance constant during the acquisition of the data.

I-Z Curves

In the simplest theoretical treatment of the tunneling process for a metal-vacuum-metal tunnel junction [7] where the bias voltage is much smaller than the work function (assumed to be the same for both metals), the problem is reduced to a square potential barrier as shown in Fig. 3. The wave function that describes an electron in the tunneling barrier is given by:

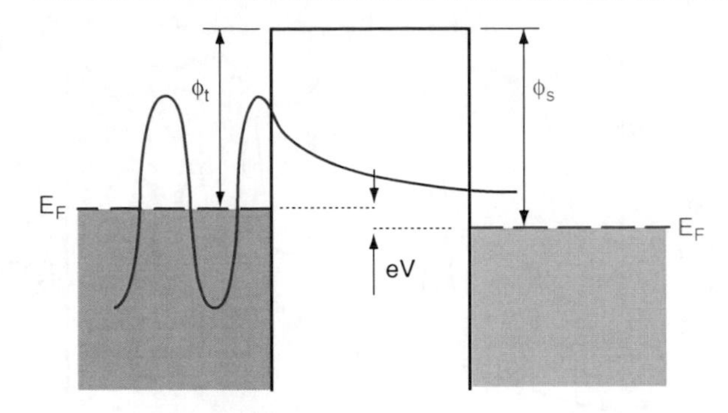

Scanning Tunneling Spectroscopy, Fig. 3 One dimensional metal-vacuum-metal tunnel junction. Both the sample, in *red*, and the tip, in *blue*, are modeled as semi-infinite pieces of free electron metals. The tunneling probability depends on the exponential decay of the electrons wave function into the vacuum barrier. The bias voltage is small enough to consider a square barrier

$$\psi(z) = \psi(0)e^{-\kappa z}$$

where

$$\kappa = \frac{\sqrt{2m(\phi)}}{\hbar}$$

is the decay constant that describes the probability of finding the electron along the $+z$ direction and depends on the surface work function (Φ). The tunneling probability for a given distance, d, between tip and sample is:

$$P \propto |\psi(0)|^2 e^{-2\kappa d}$$

and the tunneling current is proportional to the number of occupied states available in the energy interval defined by the bias voltage applied between tip and sample, and therefore the tunneling current can be written as follows:

$$I \propto \sum_{E_n=E_F-eV}^{E_F} |\psi_n(0)|^2 e^{-2\kappa d}$$

If the bias voltage is small enough to consider the density of states constant, the equation can be written in terms of the Local Density of States (LDOS) at the Fermi level. At a position d and energy E, the LDOS of the sample can be expressed as:

$$\rho_s(d, E) \equiv \frac{1}{\varepsilon} \sum_{E_n=E-\varepsilon}^{E} |\psi_n|^2$$

for small enough ε. The tunneling current in terms of the surface LDOS at the Fermi level is:

$$I \propto V\rho_s(0, E_F)e^{-2\kappa d}$$

Assuming a typical value of 4 eV for the work function, $\kappa = 1.025$ Å^{-1}. From the expression above, the dependence of the logarithm of the tunneling current with respect to distance is a measure of the work function, or more precisely, of the tunneling barrier height. The corresponding expression is:

$$\phi \approx 0.95\left(\frac{dlnI}{dz}\right)^2$$

The measurement of the apparent barrier height can be carried out by approaching or retracting the tip from the sample and recording the tunneling current. In order to measure the spatial change in apparent barrier height, a small modulation in the separation between the tip and the sample is introduced at high frequency and the modulated tunneling current is measured using a lock-in amplifier. This type of measurement gives the apparent barrier height at certain bias voltage and at certain distance from the surface. Measurements performed with different bias voltages or different tip sample distance may give different values for the apparent barrier height [7]. It is important to realize that the apparent barrier height is different from the work function in traditional surface science but is closely related; the apparent barrier height measures the spatial correlation of the overlap between the wave functions of the tip and the sample.

Z-V Curves

In Z-V measurements the bias voltage is ramped at a fixed tunneling current, and the tip-sample separation is constantly adjusted. When the applied voltage exceeds the sample or tip work function (depending on the sign of the bias voltage), there is a transition from the vacuum tunneling regime to the field emission regime. In the field emission regime, a triangular potential well is formed between tip and sample due to the bias voltage applied in the tunneling junction. In this triangular potential well, the existence of quantum well states leads to resonances in the electron transmission at certain energies, as illustrated in Fig. 4.

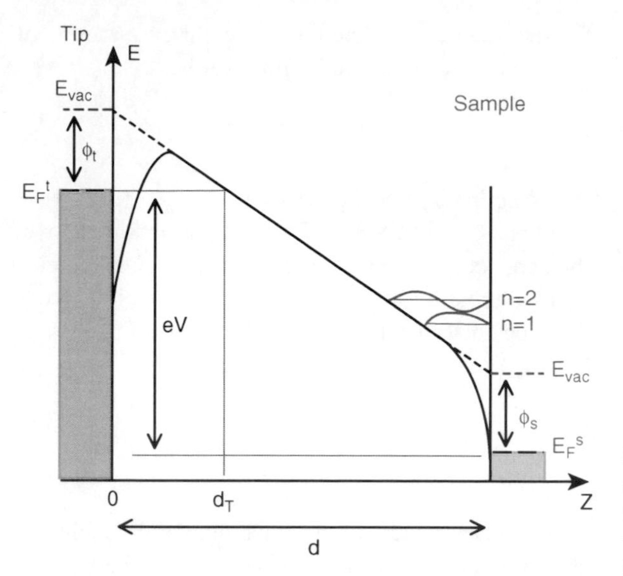

Scanning Tunneling Spectroscopy, Fig. 4 Diagram of the tunnel junction when the positive bias voltage (eV) is larger than the work function of the sample. The electrons from the tip tunnel through a narrow (d_T) triangular potential barrier to be afterward trapped in a triangular potential well. The first two quantum well states are shown

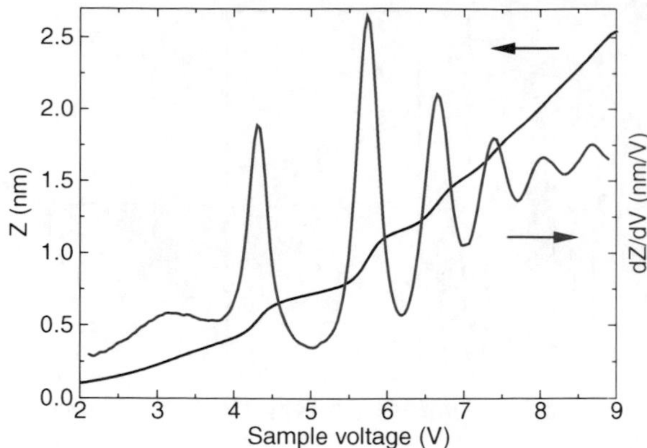

Scanning Tunneling Spectroscopy, Fig. 5 *Black curve*, tip displacement as function of the tunneling bias voltage. The field emission resonances appear as wiggles on the curve. *Gray curve*, dZ/dV, the field emission resonances appear as well-defined peaks that allow a precise determination of the energy

These transmission resonances show up as wiggles in the Z-V data (black curve in Fig. 5) and are closely related to the image states. Image states are unoccupied states bound by the classical image-charge response of metallic surfaces and have a free-electron-like dispersion parallel to the surface. The inverse dependence on distance from the surface of the image potential leads to a Rydberg-like series of states that converges to the continuum at the vacuum level (E_{vac}). Inverse photoemission studies of the image potential states have shown experimentally that the energy position of the Rydberg series is tied to the local surface potential of the material. In STM the electric field across the tunnel junction causes a Stark shift of these states, expanding the image state spectrum into a resonance spectrum associated with the triangular potential well (Fig. 5). Following the analysis performed by Gundlach in the 1960s [9], the resulting energy spectrum can be written as follows:

$$E_n = \phi + \alpha(n - 0.25)^{2/3} F^{2/3}$$

where ϕ is the surface work function, α is a constant, F is the electric field between tip and sample, and n is the quantum number of the states. These field emission resonances (FERs) were experimentally observed in field ion microscopy (FIM) by Jason [10] and with an STM by Binnig et al. [11] and since then have been used to chemically identify different transition metals on surfaces, to obtain atomic resolution on insulating surfaces (e.g., diamond), or to study local changes in the surface work function [12].

The experiments are typically performed with the feedback loop on to keep the current constant. The tip movement is recorded as a function of the bias voltage and afterward the curves are numerically differentiated to obtain the energy position of the field emission resonances, as shown in the Fig. 5 (red curve).

Spin Polarized Tunneling Spectroscopy

In 1975 Julliére [13] discovered spin-dependent tunneling between two planar ferromagnetic electrodes separated by an insulating tunnel barrier, which has become the basis for the development of magnetic random access memories and the spin polarized version of the STM. In fact, the tunneling current between a magnetic sample and an STM tip (covered with a magnetic thin film) shows an asymmetry in the spin population. The magnitude of the tunneling conductance between two magnetic electrodes with directions of the respective magnetization differing by a certain angle depends on the cosine of this angle. Spin polarized tunneling with an STM was observed in 1990s by Wiesendanger [16]. The spectroscopic mode of spin polarized STM is based on using the different intensity of certain features in differential

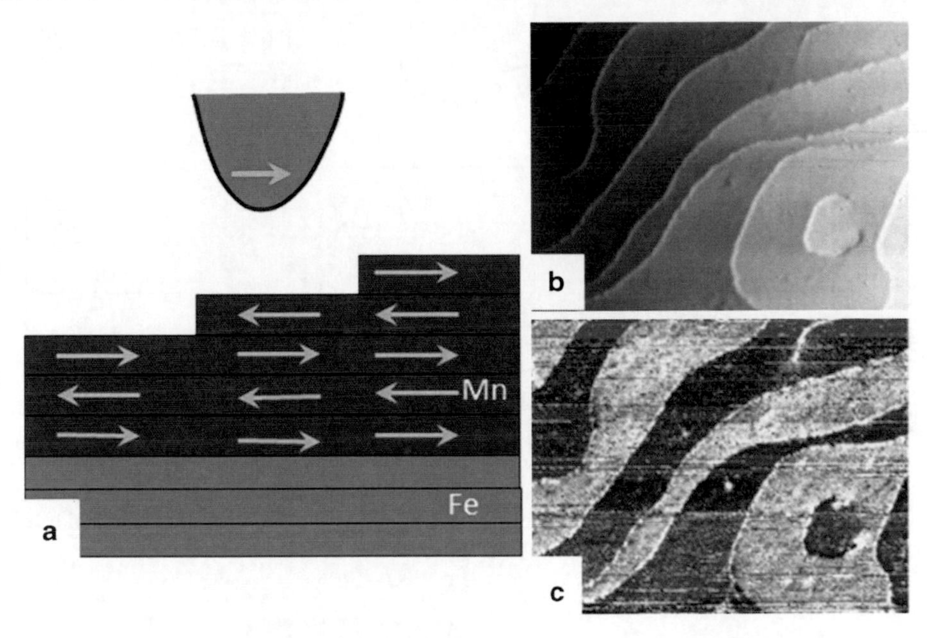

Scanning Tunneling Spectroscopy, Fig. 6 (**a**) Model showing the layered antiferromagnetic structure of Mn films grown on Fe(001). (**b**) STM topographic (100 × 78 nm) image of 6.5 mL of Mn grown on Fe(001). (**c**) Spatially resolved spectroscopic image measured simultaneously with the topography shown in (**b**) where the dI/dV signal shows low and high levels depending on the relative directions of magnetization of tip and surface terrace revealing the topological antiferromagnetic order of the Mn(001) surface

conductance spectra as source of contrast to image magnetic domains and domain walls with atomic resolution [16].

Magnetic domain observations in antiferromagnetic materials have been difficult in the past due to the limited number of experimental techniques that are sensitive to domain states in antiferromagnetic crystals. Another aspect that hampered the studies of antiferromagnetic materials was the limited spatial resolution of the available techniques. Mn and Cr crystals or thin films of these materials exhibit quite complex spin structures. The crystallographic structure of Mn thin films grown on Fe(001) is body-centered tetragonal. In this crystallographic structure, the Mn magnetic structure is layered antiferromagnetic. These means that the magnetization orientation rotates by 180° in every layer, as can be seen in Fig. 6a. The tip magnetization direction is constant in the experiment and the magnetization direction of the sample rotates by 180° every time the tip crosses a step on the surface. The change in the magnitude of the tunneling conductance can be measured on every pixel of the topographic image and an image of the magnetic domains is obtained, as can be seen in Fig. 6b, c. In those domains where the magnetization of tip and sample are aligned, the tunneling conductance is higher (brighter color) and on those areas where are antiparallel the tunneling conductance is lower (darker colors).

Vibrational Spectroscopy (Inelastic Electron Tunneling Spectroscopy)

In the previous sections, elastic tunnel processes, where the tunneling electrons do not change their energy, have been discussed. However, in certain cases, i.e., for molecules adsorbed on surfaces or samples with easy excitation of phonons, there is a small fraction of electrons that lose energy in the tunneling process [14]. For bias voltages larger than corresponding the quantum of vibration, $\hbar\omega$, a new tunneling channel, i.e., inelastic channel, opens up (as illustrated in Fig. 7a). The inelastic channel acts in addition to the elastic channel, and increases slightly the differential conductance (dI/dV) of the junction (Fig. 7b upper panel). Although vibrations of molecules were detected in the 1960s by tunneling in extended tunneling junctions with insulating layers spray coated with molecules, the use of STM facilitates the acquisition of vibrational spectra in single molecules in well-characterized environments and was pioneered by Wilson Ho [14].

In practice the change in conductance is smaller than 10% and can be detected only under very severe conditions of stability of the tunnel junction and energy resolution (i.e., with the STM at low temperatures). The vibrational modes are detected as peaks in the second derivative of the tunneling current (Fig. 7b lower panel) measured by means of lock-in techniques. Equation 2 shows that the magnitude of the second

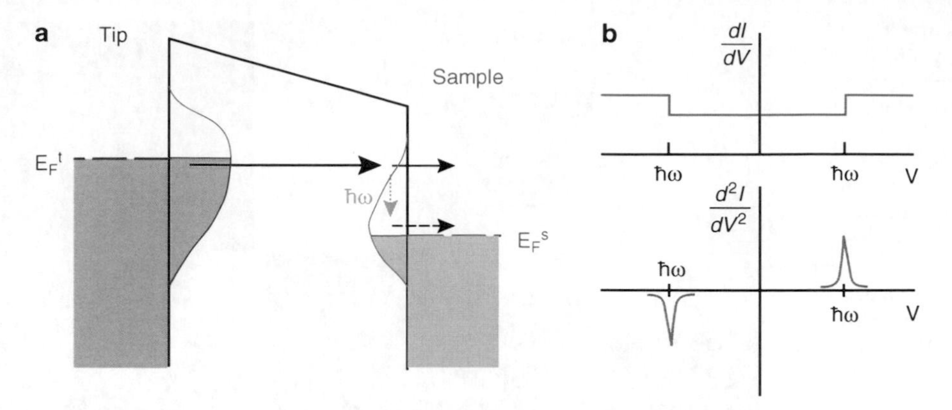

Scanning Tunneling Spectroscopy, Fig. 7 (**a**) Energy distance diagram of the tunneling processes with an applied bias voltage V. When eV is larger than the energy of the molecular vibration (ħω), empty final states at the Fermi level of the sample become accessible and the inelastic channel opens up. (**b**) The opening of the inelastic channel causes a sharp increase in the tunneling conductance (*upper panel*) or peaks in the second derivative (*lower panel*). The activation channel is symmetric with respect to the Fermi level

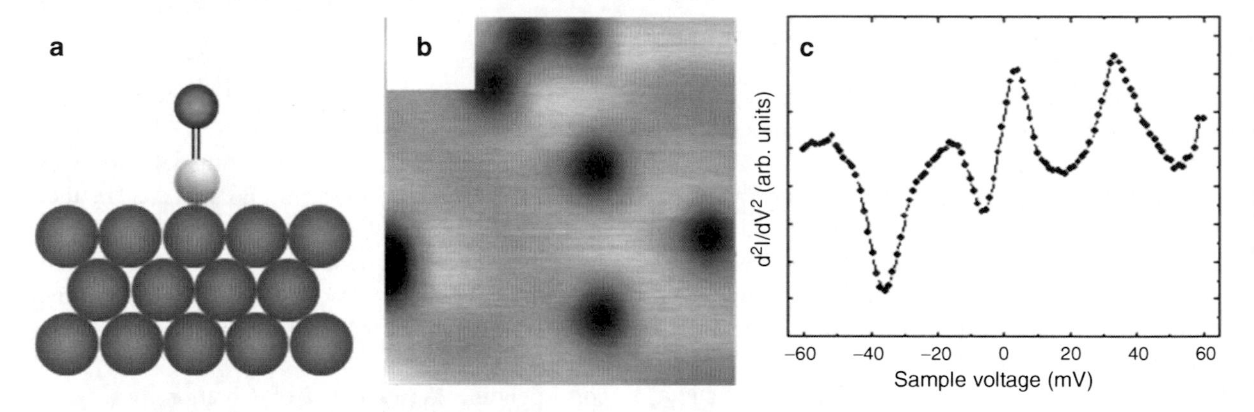

Scanning Tunneling Spectroscopy, Fig. 8 (**a**) Model of the adsorption of CO molecules on Cu(111). The molecule adsorbs perpendicular to the surface with the carbon (in *gray*) chemically bonded to the surface copper atoms. (**b**) STM topographic image measured using a tunneling current of 1 nA and a sample bias voltage of 0.25 V. (**c**) Vibrational spectra of CO on Cu(111): the peaks at ±5 mV and ±35 mV are due to excitation of the CO frustrated rotation and translation with respect to the Cu(111) surface

harmonic of the tunneling current is proportional to d^2I/dV^2. In practice, a small (≈3 mV) ac component is added to the bias voltage, the dc component of the bias voltage is scanned across the selected energy range, and the variations in d^2I/dV^2 are recorded. The width of the peaks is given by the Fermi energy distribution, i.e., the FWHM is $3.5k_BT$ (1.2 mV at 4 K).

A common observation in IETS spectra is that peaks at certain values of positive voltages appear as *dips* at opposite polarity. The symmetry position with respect to the zero bias of the features observed is a fingerprint of their inelastic origin. Differences in the density of states of the electrodes, however, may change the peak intensity. The observed symmetry implies that the inelastic processes are accessible for electrons tunneling on both directions. Selection rules for which modes are detectable, unlike Raman or infrared spectroscopy, seems to depend on the symmetry of the molecular state involved in the tunneling process.

A well-studied system are CO molecules adsorbed on Cu(111). It is known that the CO molecules adsorb on the surface on top of the copper atoms with the oxygen pointing toward the vacuum and the carbon chemically bonded to the copper surface, as shown in Fig. 8a. Figure 8b shows a topographic STM image of several CO molecules adsorbed on Cu(111) at 4.6 K measured with a tunneling current of 1 nA and a bias voltage of −0.25 V. With these values the CO molecules are imaged as a round depression because the chemical bond between the molecule and the Cu(111) surface reduces the electron density around the Fermi level. In order to measure the vibrational spectra, the tip is positioned over the center of the CO molecule. With the feedback off, the sample bias voltage is ramped over the range of the vibrational peaks while a sinusoidal bias modulation is superimposed. The derivative of the conductance exhibits peaks at the molecular vibration energy. For CO on Cu(111) the vibration spectra are characterized by two features at about 5 and 35 meV, as can be seen in Fig. 8c. These peaks are assigned to the two degenerated transverse vibration modes: the frustrated translation and the frustrated rotation, respectively.

Cross-References

- ► AFM
- ► Field Electron Emission from Nanomaterials
- ► Scanning Tunneling Microscopy
- ► Surface Electronic Structure

References

1. Binnig, G., Rohrer, H.: Scanning tunneling microscopy. Helv. Phys. Acta **55**, 726 (1982)
2. Müller, E.W.: Das Feldionenmikroskop. Z. Physik **31**, 136 (1951)
3. Bardeen, J.: Tunneling from many particle point of view. Phys. Rev. Lett. **6**, 57 (1961)
4. Tersoff, J., Hamann, D.R.: Theory and application for the scanning tunneling microscope. Phys. Rev. Lett. **50**, 1998 (1983)
5. Tersoff, J., Hamann, D.R.: Theory of the Scanning tunneling microscope. Phys. Rev. B **31**, 805 (1985)
6. Lang, N.D.: Spectroscopy of single atoms in the Scanning tunneling microscope. Phys. Rev. B **34**, 5947 (1986)
7. Chen, C.J.: Introduction to Scanning Tunneling Microscopy. Oxford University Press, New York (1993)
8. Vázquez de Parga, A.L., Hernán, O.S., Miranda, R., Levi Yeyati, A., Mingo, N., Martín-Rodero, A., Flores, F.: Electron resonances in sharp tips and their role in tunneling spectroscopy. Phys. Rev. Lett. **80**, 357 (1998)
9. Gundlach, K.H.: Zu berechnung des tunnelstroms durch eine trapezförmige potentialstufe. Solid State Electron. **9**, 949 (1966)
10. Jason, A.J.: Field induced resonance states at a surface. Phys. Rev **156**, 266 (1966)
11. Binnig, G., Frank, K.H., Fuchs, H., Garcia, N., Reihl, B., Rohrer, H., Salvan, F., Williams, A.R.: Tunneling spectroscopy and inverse photoemission: image and field states. Phys. Rev. Lett. **55**, 991 (1985)
12. Borca, B., Barja, S., Garnica, M., Sánchez-Portal, D., Silkin, V.M., Chulkov, E.V., Hermanns, C.F., Hinarejos, J.-J., Vázquez de Parga, A.L., Arnau, A., Echenique, P.M., Miranda, R.: Potential energy landscape for hot electrons in periodically nanostructured graphene. Phys. Rev. Lett. **105**, 036804 (2010)
13. Person, B.N.J., Baratoff, A.: Inelastic electron tunneling from a metal tip: the contribution from resonant processes. Phys. Rev. Lett. **59**, 339 (1987)
14. Stipe, B.C., Rezaei, M.A., Ho, W.: Single molecule vibrational spectroscopy and microscopy. Science **280**, 1732 (1998)
15. Julliere, M.: Tunneling between ferromagnetic films. Phys. Lett. **54A**, 225 (1971)
16. Wiesendanger, R.: Soin mapping at the nanoscale and atomic scale. Rev. Modern Phys. **81**, 1495 (2009)

Scanning X-Ray Diffraction Microscopy (SXDM)

► Selected Synchrotron Radiation Techniques

Scanning-Probe Lithography

► Dip-Pen Nanolithography

Scincus officinalis

► Friction-Reducing Sandfish Skin

Scincus scincus

► Friction-Reducing Sandfish Skin

Scrolled Nanostructure

► Nanorobotics for NEMS Using Helical Nanostructures

Seebeck Coefficient

► Thermoelectric Heat Convertors

Seebeck Effect

► Thermoelectric Heat Convertors

Selected Synchrotron Radiation Techniques

Antoine Barbier[1], Cristian Mocuta[2] and Rachid Belkhou[2]
[1]CEA-Saclay, DSM/IRAMIS/SPCSI, Gif-sur-Yvette, France
[2]Synchrotron SOLEIL, L'Orme des Merisiers, Gif-sur-Yvette, France

Synonyms

Scanning X-ray diffraction microscopy (SXDM); Small angle X-ray scattering in grazing incidence geometry; Soft X-ray microscopy; Spectromicroscopy; X-ray diffraction with micron-sized X-ray beams

Definition

Synchrotron radiation is produced when highly energetic charged particles are deviated in a magnetic field. The generated spectrum consists of a large energy range from infra-red to gamma X-rays. The synchrotron ring size determines the accessible photon energy range as well as the energy particles which lie in the several giga electron-volt range for a stable orbit. Harder X-ray ranges require larger rings.

Recent third generation synchrotron radiation rings offer simultaneously energy tunability for photons with high brightness (or brilliance defined as the photon flux per unit source size per unit solid angle and 0.1% energy bandwidth), high-photon fluxes and low divergence at high (ESRF, APS, SPring-8, PETRA-3 etc.) and low (electron/positron) energy (SLS, NSLS, SOLEIL, Diamond, ELETTRA, ALBA, etc.) [1]. This unique combination of characteristics allowed developing genuine experimental methods and techniques particularly suitable to investigate microstructures and nanostructures. For example, high-flux collimated beams impinging a sample surface at glancing angles allow small angle X-ray scattering experiments which provide a reciprocal space picture of an assembly of particles. High brightness enables focalization of the X-rays with reasonable photon fluxes in the focused beam and opens the route to scanning and full field microscopy techniques. For all techniques photon energy tunability and variable polarization provide chemical and/or magnetic sensitivity.

Synchrotron radiation techniques have numerous advantages over conventional laboratory setups; they are intended to complement laboratory experiments in order to obtain in-depth information on specific issues. Photon-in and photon-out techniques are not hampered at all by charge build-up effects. The tunable penetration depth using glancing incident and/or exit scattering angles enables the study of buried interfaces. In a lot of cases, sample preparation requirements are limited to a minimum allowing thus the investigation of a large variety of samples not specifically designed for a dedicated experiment. Importantly, X-ray techniques, especially at high photon energies can be used with various sample environments including (and not limited to) high pressure, high temperatures, ultra-high vacuum, liquids, and low temperatures. However, for low energy photons or when electrons have to be detected ultra-high vacuum may become mandatory. The use of synchrotron radiation has also some drawbacks. The available beam time is limited and access is granted through highly competitive project selection. Organic and biological samples are often unable to bear the highly intense X-ray beams. X-ray diffraction techniques measure scattered intensities and the phase information is lost; unless the coherence of the synchrotron radiation or statistical methods are used the data interpretation requires fit procedures and the input of some models.

Synchrotron radiation proposes a number of mature and/or useful techniques and methods in the field of material science. It is beyond the scope of this entry to discuss all of them or all possible microscopy techniques. The goal here is rather to exemplify some approaches proven to be relevant to investigate nanostructures of current interest. Because of space restrictions, this document is definitely partial and non-exhaustive: the choices made necessarily suffer subjectivity and cover only a part of the possible approaches that might already be described in the literature and some authors/readers might find unfair the choice of topics and cited references. The interested reader may consider more general text books or reviews dedicated to synchrotron radiation techniques [2–5]. Within the field of nanostructure investigations three techniques that are particularly illustrative were chosen: Grazing Incidence Small Angle X-ray Scattering (GISAXS), micro X-Ray Diffraction (μ-XRD), and X-ray Photo-Emission Electron Microscopy (X-PEEM). These techniques are essentially nondestructive, noninvasive and allow investigating surfaces and buried interfaces. GISAXS and μ-XRD allow studying samples in various environments (e.g., temperature, gas, or liquid cells for "real" environments, external magnetic or electric fields, etc.). X-PEEM is more limited within this respect and requires good vacuum conditions in order to detect electrons. To the cost of losing the full field imaging approach, the microscopy approach can however be made more flexible using a well-focused photon beam across which the sample surface is raster scanned [6] (Microspectroscopy: Scanning Transmission X-ray Microscope (STXM), scanning photoemission microscope (SPEM)...). The techniques chosen here can be defined as follows:

1. GISAXS is a powerful technique to study nanostructured surfaces, thin films, and assemblies of nano-objects, combining Small-angle X-ray scattering (SAXS) [7] and the surface sensitivity of Grazing Incidence Diffraction (GID) [8]. The quantity of matter involved in supported particles is small making the use of high brightness synchrotron radiation almost mandatory. The primary photon energy and the geometrical experimental settings (distance between sample and detector, etc.) determine the accessible length scales. The method is fully nondestructive for materials science samples and gives access to morphological (e.g., size/shape, dispersion) statistical information of an assembly of nano-objects averaged over a large area (several square millimeters); it characterizes the "mean" object in reciprocal space. The high collimation and low divergence of the beam allows precise defining and tuning of the X-ray incidence angle, and thus enables the study of buried interfaces and a probed depth range from several 10 nm to several micrometers in the sample. Chemical contrast can be enhanced by performing anomalous scattering (i.e., tuning the photon energy to a given absorption edge). The technique does not require any specific sample preparation other than thin film deposition techniques; ex-situ or in-situ studies are both possible.
2. Microbeam X-ray diffraction experiments use highly intense and very small (focused to micron or sub-micron sizes) X-ray beams in order to address locally the properties of matter by diffraction. The lateral resolution is essentially given by the size of the X-ray spot. It can give access to crystalline structure, strain, defects, composition (by exploiting the variation of the lattice unit cell size with the composition), etc.
3. Spectromicroscopy (X-PEEM) employs parallel imaging techniques making use of special electron optics. The high brightness of third generation synchrotron radiation sources has opened the way to surface and interface imaging with resolution in the 10 nm ranges and with further instrument improvements in the 1 nm range [9–12]. This technique is well suited to investigate modern nanostructures, fabricated using various methods (bottom-up or top-down, lithography, self-organized growth, ... methods), at the nanoscopic level combining good spatial (few nanometers) and time (sub-nanosecond) resolutions. This is particularly the case for magnetic and semiconducting technologies, where the interest for submicrometric range devices has been rapidly increasing during the last decade. Moreover, the increasingly sophistication level of the nanostructures prompted considerable progress in high-resolution microscope probe that are sensitive to the compositional, electronic, chemical, and magnetic microstructure.

Overview

X-ray scattering probes materials properties and, depending on the technique used, can be sensitive to chemical species, magnetic and/or crystallographic ordering and even density fluctuations near surfaces and buried interfaces. The process can be elastic (i.e., photon energy conservative) or inelastic, specular or diffuse [13]. X-ray scattering (inelastic as well as elastic) can be used in spectroscopies that provide chemical information about the sample composition. Photon energy tunability can be used to greatly enhance the chemical contrast between several chemical species (a specific contrast is obtained when the photon energy is tuned to a given electron level edge) i.e., anomalous X-ray scattering conditions [2–4].

For the analysis of materials at the micro- and the nano-scale, well-established microscopy-based techniques can probe different properties at short length scale down to atomic length scale. Techniques like the transmission electron microscopy (► TEM), scanning probe methods (scanning electron microscopy ► SEM, Atomic Force Microscopy ► AFM, micro Photo Luminescence μPL, etc.) have an undeniable contribution in understanding what is happening at the small scale for nanostructured materials. From some of their drawbacks, one could mention here their surface sensitivity, invasive sample preparation methods, presence of vacuum environment, or sensitivity to external fields (magnetic, electric). Alternative X-ray-based techniques can offer complementary insights and overcome some of these drawbacks. Some examples are detailed hereafter.

During a *GISAXS* experiment the incident photon energy is mainly chosen with respect to the reciprocal space resolution that is expected. In case of chemical contrast studies the photon energy may be tuned to the absorption edge of a given element. The incident beam is scattered by an assembly of objects of nanometric sizes and separated by an average distance in the several 10–100 nm range. The pattern is conveniently observed on an area detector (2D camera) because at high-photon energy the Ewald sphere region corresponding to these small scattering angles can reasonably be approximated by the plane tangent to the sphere.

X-ray diffraction (*XRD*) (a detailed description can be found in Ref. [13]) is not only another technique to analyze nanostructures, but in some specific cases it might become the method of choice. Without forgetting the advantages of the microscopy techniques, XRD can unveil aspects difficult to address by other techniques. Besides the advantages mentioned for Synchrotron Radiation X-ray-based techniques, hard X-ray ($\sim$10 keV energy range) based techniques can be the better adapted ones to access properties like the internal strain: for example, the strain is measured in a simple experiment and with an accuracy (10^{-4}) hard to achieve with other techniques. It should also be pointed here some of the drawbacks of the method. The resulting data (images or maps in the so-called "reciprocal space," which is the Fourier transform space) are not straightforward to visualize and understand (contrarily to other imaging/scanning probe techniques, which generally yield to a more intuitive image of the sample surface or inner structure).

In order to access the properties of small sized (nano-)structures, two approaches can be considered:

(a) A first approach in which the sample contains a large number (10^4–10^6) of *identical* objects and is illuminated by a large (several 0.1–1 mm size) X-ray beam [14]. For a long time, the averaging over large surfaces of the sample was presented as one of the major advantages of XRD methods. The measured signal represents the average from the ensemble of objects, yielding thus to averaged properties (and possibly distributions) [14–16]. It is a very good tool to characterize properties over *large* areas of the sample (in contrast to microscopy-based techniques, which remain local). This approach makes sense only if these averaged quantities are meaningful, that is, the sample consists of objects which are quasi-identical or with properties exhibiting simple distributions. The approach is expected to fail for complex samples (multimodal growth modes, lithographed non-identical structures, etc.).

(b) A second approach (μ-XRD) is a scanning microscopy-like approach and consists of investigating a single structure at a time. The signal originating from individual micro/nanostructures is recorded by using local probe techniques and a two-dimensional raster image of the sample is obtained; then the X-ray beam is pointed to illuminate only the object of interest (or part of it) while performing the high-resolution diffraction experiment.

XPEEM (X-ray Photo Emission Electron Microscopy) spectromicroscopy is a derivative of the classical PEEM/LEEM [17]. If a photon energy just above the photothreshold is used, the photoelectron yield is mainly determined by the differences in the work function ϕ of the sample. The local variations of ϕ result in images with high contrast. This Ultra-violet (UV)-PEEM mode of operation is ideally suited to study surface chemical reactions in real time. With the advent of high-brilliance synchrotron radiation from storage rings, a wide and tunable energy range of photons of the illuminating beam has become available, allowing to access well-established techniques like Ultraviolet Photoemission Spectroscopy (UPS), X-ray Photoemission Spectroscopy (XPS), and X-ray Absorption Spectroscopy (XAS) at the nanoscopic level, and thus leading to element-selective imaging. Note that photoemission techniques require an additional analyzer enabling energy filtering of the photoemitted electrons. Moreover, information on the spatial distribution of the electronic structure, chemical composition and nature, or the local magnetization at the surface can be obtained. Dedicated beamlines with high brilliance, variable photon polarization (both circular and linear), and a broad energy range have become recently available at many synchrotron facilities [1]. This opens a wide research area for the XPEEM including surface magnetism, surface and interfaces, surface chemistry, tribology, etc.

Grazing Incidence Small Angle X-Ray Scattering (GISAXS)

Geometry

In a typical GISAXS experiment, the incident beam, of wave vector $\mathbf{k}_i$, makes a small incident angle α_i with respect to the sample surface. The incident beam is supposed to be monochromatic, parallel, and free from background contributions; the beam is cleaned using scattering slits whenever necessary. The direction of the scattered wave vector $\mathbf{k}_f$ spans the (Q_y, Q_z) reciprocal plane (Fig. 1 – top) that includes information about island size, height, and inter-island distance (or correlation). The pattern is observed on a 2D detector as far as possible from the sample in order to increase the reciprocal space resolution. Beam stops hiding the direct and specular beam and rod are necessary because of the limited dynamical range of detectors (Charge Coupled Device, CCD camera in this particular case). The scattered intensity distribution depends on α_i, Q_y, and Q_z.

The experiment can be run in transmission for thinned samples or in grazing incidence conditions. Using focused X-ray beams and transmission geometries can provide a scanning microscopy approach valuable to investigate inhomogeneities in the sample [18–20]. The grazing incidence geometry is better adapted for the study of nano-objects and their correlations without sample preparation.

Grazing Incidence

To understand the usefulness of grazing incidence some elements of photon–matter interaction have to be recalled. Consider a sharp interface between vacuum and a surface of a material of wavelength-dependent index of refraction n.

The incident beam is supposed to be a linearly polarized plane wave that impinges on this surface at a shallow angle α_i with wave vector $\mathbf{k}_i$; the reflected and transmitted beams leave at angles α_f and α_t with wave vectors $\mathbf{k}_f$ and $\mathbf{k}_t$ respectively. Snell's law writes as:

$$\cos(\alpha_t) \cdot n = \cos(\alpha_i) \text{ and } \alpha_f = \alpha_i \tag{1}$$

As long as $n > 1$ total reflection cannot occur when light travels from the vacuum to the material, even if $\alpha_i = 0$. Fortunately, and unlike visible light, hard X-rays have an index of refraction less than unity and can be written as

$$n = 1 - \delta - i\beta \tag{2}$$

with

$$\delta = \frac{\lambda^2}{2\pi} r_e \rho_e$$

and

$$\beta = \frac{\lambda}{4\pi} \mu \tag{3}$$

in which λ is the X-ray wavelength, r_e the Bohr atomic radius, ρ_e the electronic density of the material, and μ the material linear absorption coefficient at the considered wavelength.

Consequently, for hard X-rays, the transmitted beam will be deflected always toward the sample internal surface and total external reflection occurs on the vacuum side with a critical angle for total external reflection $\alpha_c \approx \sqrt{2\delta}$ in the 0.1–0.6° range because δ and β are, respectively, in the 10^{-5} and 10^{-6} ranges respectively.

When $\alpha_i < \alpha_c$, the component of the transmitted wave vector normal to the surface becomes imaginary. The refracted wave is exponentially damped as a function of the distance below the surface and is an evanescent wave traveling parallel to the surface.

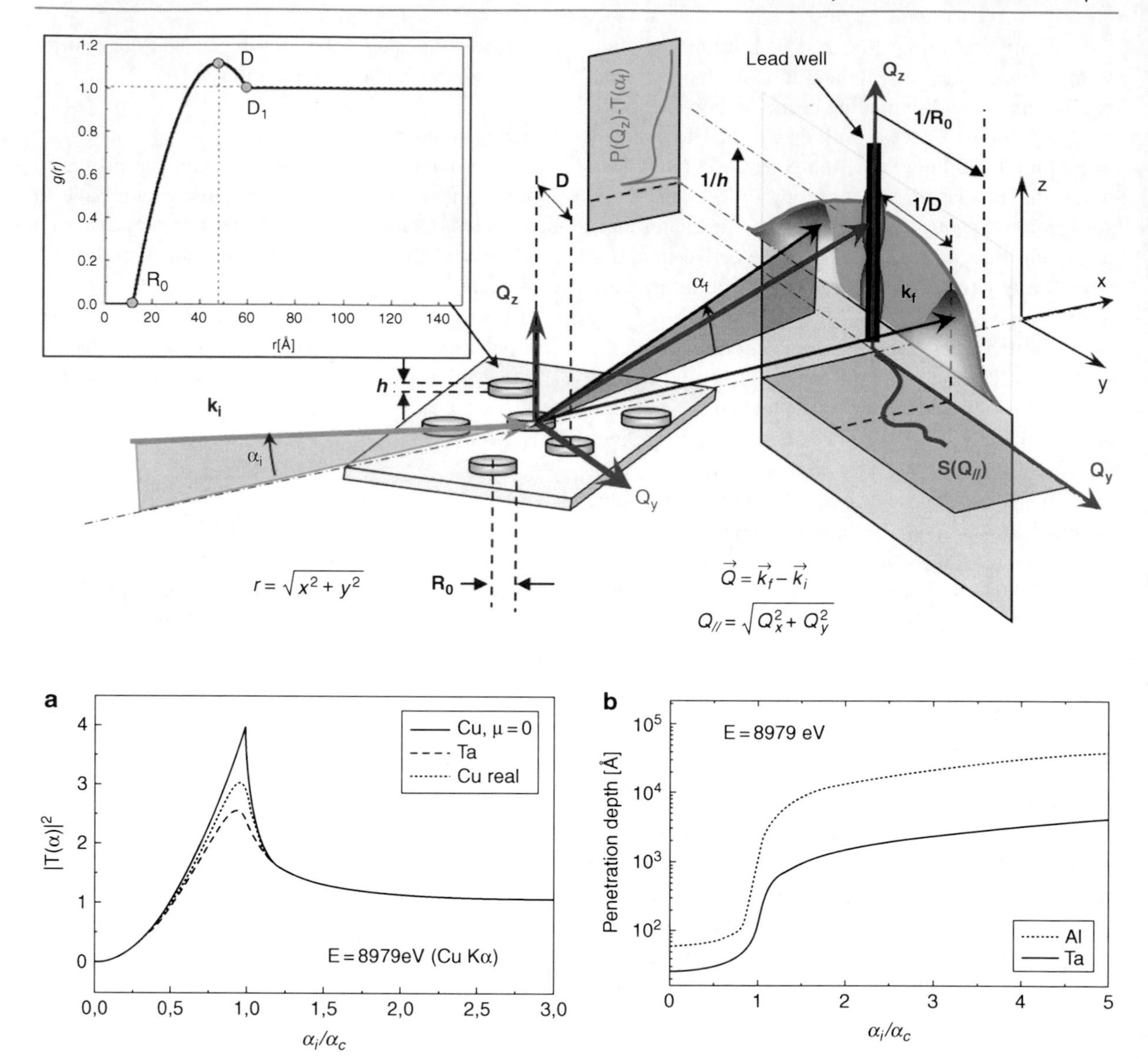

Selected Synchrotron Radiation Techniques, Fig. 1 *(Top)* Typical geometry of a GISAXS experiment. $\mathbf{k}_i$ and $\mathbf{k}_f$ are the incident and scattered wave vectors, respectively, yielding the momentum transfer (i.e., the reciprocal space vector) $\mathbf{Q} = \mathbf{k}_f - \mathbf{k}_i$. The angles α_i, α_f, and $2\theta_f$ are related to the components of the momentum transfer, either parallel (Q_x and Q_y) or perpendicular (Q_z) to the sample surface, by the equations $\mathbf{k}_i = \mathbf{k}_0[\cos(\alpha_i), 0, -\sin(\alpha_i)]$, corresponding to the wavelength $\lambda = 2\pi/k_0$ and $k_f = k_0[\cos(\alpha_f)\cos(2\theta_f), \cos(\alpha_f)\sin(2\theta_f), \sin(\alpha_f)]$ of equal modulus. k_0 is known to be the incident wave vector modulus. *(Bottom)* (**a**) Fresnel transmission coefficients calculated at Cu Kα photon energy for nonabsorbing Cu, Ta, and Cu using real δ and β values. (**b**) Penetration depth of X-rays calculated at Cu Kα photon energy for Al and Ta. The incidence angle scale is normalized by the critical angle of total external reflection

Considering the Fresnel transmission coefficient that writes as:

$$T = |t|^2 = \left|\frac{2\sin(\alpha_i)}{\sin(\alpha_i) - \sqrt{n^2 - \cos(\alpha_i)}}\right|^2$$

and the penetration depth given by:

$$\Lambda(\alpha_i) = \frac{1}{-2k_0\mathrm{Im}\left(\sqrt{\alpha_i^2 - \alpha_c^2 - 2i\beta}\right)} \quad (4)$$

One can see that the surface sensitivity is strongly enhanced when α_i becomes close to α_c because the incident and reflected waves are nearly in phase. The penetration depth of the X-rays becomes nanometric, whatever is the considered material, and the Fresnel transmission coefficient is maximal (see Fig. 1 – bottom). One may note that higher absorption coefficients reduce slightly the Fresnel coefficient maximum and conversely decrease the penetration depth at any angle. Importantly, for $\alpha_i > 3 \times \alpha_c$ surface effects become negligible.

Grazing incidence scattering geometries provide simultaneously enhanced surface sensitivity and limit the unwanted bulk scattering (diffuse or not) that may otherwise overcome the surface or nanostructure feeble signal. It opens additional experimental possibilities as compared to simple specular reflectivity for which incidence and exit angles are kept equal. Grazing incidence is thus a configuration of choice in order to nondestructively investigate nanostructures deposited on a substrate or buried under a substrate.

Pattern Analysis

The interpretation and simulation of the measured scattering patterns is challenging and requires accurate models and, if size/distance distributions are considered, is demanding tedious computing. However, a complete modeling is not always necessary and the data evaluation can go from quite simple for a crude first approximation to tedious when the full pattern reproduction is sought [21]. A practical illustration of all important terms is given in Fig. 1. In an extremely crude and semi-qualitative approach, and neglecting particle shape, correlation and anisotropy, one can deduce from the position and shape (width) of the diffusion lobes directly an approximation of the inter-island distance (D), height (h), and size (R_0) as illustrated in Fig. 1.

The scattering signal contains a coherent contribution and an incoherent scattering term as soon as the size distribution of the nanostructures is not monodisperse. The coherent term is the product of the Fourier transform of the nanostructure shape with the interference function $S(Q_{//})$. For the incoherent scattering the two limit cases are (1) the Decoupling Approximation (DA), assuming no nanostructure correlations, and (2) the Local Monodisperse Approximation (LMA), assuming full correlation between nanostructure sizes at a scale corresponding to the photon coherence length.

Hard X-ray photons interact weakly with matter and, except for perfect crystals or at glancing angles, the kinematical Born Approximation (BA), that neglects multiple scattering effects, is valid. At angles below $3 \times \alpha_c$ dynamical effects of reflection and refraction at interfaces become important and have to be taken into account using the popular Distorted Wave Born Approximation (DWBA) [22].

Within surface physics and nano-science two approaches have to be distinguished depending on the level of interpretation that is sought:

1. The more intuitive Effective Layer Born Approximation (ELBA) which applies for isotropic buried islands with simple shapes. The index of refraction is that of the effective layer in which the islands are supposed to be buried. In ELBA the resulting scattering pattern is the weighted sum of the scattering arising from each sub-element since an assembly of objects can be considered as the sum of the individual scatterings as far as the relative intensities remain linked to the total number of electrons in each object within the X-ray coherence length.
2. The DWBA, which applies in the vicinity of critical scattering geometries for islands deposited on a substrate. In the complete treatment, an anisotropic signal originating from island facets can also be considered. The effective form factor corresponds to the coherent interference of four waves corresponding to the four possible scattering events (weighted by the corresponding reflection coefficients) experienced by the incoming and exiting beams on a given island.

In both cases (ELBA and DWBA) specular reflectivity may be added, eventually evaluated using the Parratt recursive formalism when multilayer structures are involved.

Within the ELBA, to first order and a rapid interpretation strategy assuming weak reflectivity, the scattered intensity I(Q) is proportional to the product of three terms: the form factor P(Q) which is the Fourier transform of the island volume, the interference function S(Q), and the transmission factor $T(\alpha_f)$ that is supposed to simulate the dependence of the scattered intensity as function of α_f.

The interference function S(Q) is related to the correlation length, which characterizes a distribution of islands on the surface. It is also related to the real space correlation function g(r) via the formula:

$$S(Q) = 1 + 2\pi \cdot \rho_s \cdot \int_0^r (g(r) - 1) \cdot J_0 \cdot (Q_c r).r.dr \quad (5)$$

where ρ_s is the particle density per surface unit and J_0 the Bessel function of zero order. Many analytical different pair-correlation functions can be used within models. One may consider, for example, a Gaussian pair-correlation function (illustrated in inset of Fig. 1 – top) given by:

$$g(r) = \begin{cases} 0, \text{for } 0 \leq r \leq R_0 \\ \dfrac{e^{-\frac{(r-D)^2}{\omega^2}} - e^{-\frac{(R_0-D)^2}{\omega^2}}}{e^{-\frac{(D_1-D)^2}{\omega^2}} - e^{-\frac{(R_0-D)^2}{\omega^2}}}, \text{for } R_0 \leq r \leq D_1 \\ 1, \text{for } D_1 \leq r \leq \infty \end{cases} \quad (6)$$

The most reliable input remains the direct evaluation of g(r) from real space measurements made, for example, by complementary microscopy techniques.

Within the DWBA the effective form factor corresponds to the coherent interference of four waves corresponding to the four possible scattering events (weighted by the corresponding reflection coefficients) experienced by the incoming and exiting beams on a given island. The DWBA scattering intensity I_{DW} from islands deposited onto a substrate may be summarized as:

$$I_{DW} = \left\{ \frac{|1 - n_\Delta^2|}{2 \cdot \mathrm{Re}(1 - n_\Delta)} \cdot \left[\tilde{\chi}_\Delta(q_{//}, q_z) + R^f \tilde{\chi}_\Delta(q_{//}, -p_z) + R^i \tilde{\chi}_\Delta(q_{//}, p_z) + R^i R^f \tilde{\chi}_\Delta(q_{//}, -q_z) \right] \right\}^2 \quad (7)$$

where n_Δ is the index of refraction of the islands, $\tilde{\chi}_\Delta(q_{//}, q_\perp)$ is the form factor of the islands, $q_{//} = (q_x, q_y)$ is the component of the momentum transfer q parallel to the surface, q_z corresponds to the net momentum transfer perpendicular to the substrate surface, $p_z = k_z^f + k_z^i$ where k_z^f and k_z^i are the normal to the substrate plane wave vectors of the incoming and exiting waves. The reflectivity coefficients for the incoming and outgoing waves are denoted R^i and R^f, respectively. The Born approximation is restored with $R^i = R^f = 0$.

The specular reflectance of discontinued multilayer systems with interfacial roughness can be calculated by use of a recursive classical exact layer by layer method, based on Fresnel coefficients within the Parratt optical formalism and the dispersion model.

Interestingly, variations of electronic densities are the only requirement mandatory to enable the observation of GISAXS patterns. Objects (of lower electronic densities) in a larger electron density matrix will yield such signals. The extreme case will then be the presence of structured voids or pores. For reactive interfaces like NiO/Cu(111) [23], self-organized holes in the surface or the presence of Ni decorated corrals lead to GISAXS patterns similar to growing nanoparticles.

Size distributions can be included in the pattern evaluation in any framework. The obvious effect of the size distributions is to smooth the scattering pattern. Height distribution and cross correlation between lateral size and height distributions are very difficult to extract from the GISAXS pattern.

Examples of Application

In 1989, GISAXS was introduced to study the dewetting of gold deposited on a glass surface and later, using synchrotron light metal agglomerates on surfaces and in buried interfaces were considered.

Determining the morphology of islands on a substrate, embedded clusters in matrices as well as the fabrication control of nanometer-sized objects can be successfully addressed using GISAXS. It has become a popular technique because it can be applied to a broad range of samples in various sample environments: investigation of samples in standard conditions [21], porous layers, precipitates and quantum dots [24], growing particles in ultra-high vacuum conditions [13, 21], complex nanostructures like carbon nanotubes [25], and shape modifications upon various treatments. The method in itself does not require any particular sample preparation – it is the thin film growth method requiring ultra-high vacuum environment. A major step has been realized in the last decade because of an in-depth understanding of the underlying phenomena and the availability of algorithms enabling the simulation of the experimental patterns. In the most advanced studies coherent GISAXS experiments are considered enabling full particle shape reconstruction [24]. In general, GISAXS can be applied to characterize self-assembly and self-organization at the nanoscale in thin films.

The technique becomes highly interesting in situations where charge build-up is a problem, e.g., for investigating metal/insulating oxide systems. Electron-based techniques are generally hampered by the insulating character of the sample and the charge build-up. Many such systems are characterized by Volmer-Weber or Stranski-Krastanov 3D growth of particles following a nucleation, growth, and coalescence scheme. In GISAXS such a growth sequence leads to diffusion lobes moving toward the reflected beam in reciprocal space with respect to film thickness as illustrated in Fig. 2 for the Co/NiO(111) system. Even without pattern simulation the position and shape changes of the diffusion lobes indicates directly the growth mode.

From sequences of images taken at various film thicknesses and sample conditions (like substrate temperatures or quality) one can extract the overall behavior of a metal/oxide interface and evaluate, for example, the effect of defects in the cluster and/or on the substrate. The study of Co deposits, performed in ultra-high vacuum conditions using substrates of different qualities, that is, with variable amounts of nucleation centers is reported in Fig. 2. The experiment was performed on BM32 beamline (ESRF, Grenoble, France) using a beam energy of 10 keV. GISAXS was also combined with grazing incidence diffraction measurements (not shown here) to access the crystallinity of the sample. When the growth is performed at low temperature ($T = 450$ K) the Co clusters have numerous defects and poor crystalline quality. For Co thicknesses below 0.8 nm the island diameters are similar what so ever is the surface quality or Co island crystalline quality. Thus in the nucleation regime, for cluster sizes below the onset of defects in the clusters, the number of nucleation centers on the surface does not lead to noticeable effects. Above 0.8 nm the diameters fairly diverge indicating that when coalescence starts to play an important role, the island mobility increases with temperature. This effect is consistent with the inter-island distance behavior of the islands. From the island height behavior it appears that the deposition at high temperature (good Co crystalline quality) on a substrate with numerous nucleation centers is equivalent to low-temperature deposition (poor Co crystalline quality) on a high-quality substrate. Such investigations can hardly be made using other techniques.

GISAXS is not only highly efficient to investigate in situ grown particles, it can also tackle very complex ex situ samples like carbon nanotubes (CNT) [25] organized in aligned long tubes perpendicular to the sample surface. Each tube includes on its top a Co nucleation particle used to promote the growth of the CNT. Figure 3 reports measurements made on beamline ID01 at the ESRF (Grenoble, France). The experimental pattern includes GISAXS scattering and reflectivity contribution. The calculated patterns required the use of a core-shell C-Co structure inspired from transmission electron microscopy additional measurements. The height/size distribution was fairly large and the DWBA adequate parameters were retrieved from a first ELBA approximation. Such an analysis enables the evaluation of the effects of preparation condition changes like the ammonia content in the reactive gas mixture [25].

High Resolution Micro-X-ray Diffraction (μ-XRD)

In a local probe diffraction approach (μXRD) is necessary to illuminate with the X-ray beam only a single object. The probe size yields the lateral resolution. As a consequence, the volume of nanostructures probed by the X-rays will be highly reduced (from several 10^4–10^6 objects in a "standard" XRD experiment to a single one), thus the detected scattered signal is reduced. Indeed, this one is proportional to the incident X-ray photon flux *and* the probed volume. In order to compensate for the very small probed volume, the X-ray beam has to be focused to small sizes. Several focusing alternatives are nowadays available.

A Short Introduction to Hard X-Ray Focusing Optics

The following discussion is centered mostly around hard X-ray (several kilo electron volts energy) optics used at synchrotron sources, although laboratory sources with X-ray beams of several 10 μm are available and can be used for certain XRD experiments with local (lateral) resolution. But a stable, highly brilliant and low divergent X-ray source is a "must have" for obtaining very small and intense X-ray spots. The synchrotron X-ray source is generally situated far from the experimental station (several 10 m) which allows for large demagnification factors for the focusing optics. Consequently, major progress was done at synchrotron facilities, where these focusing devices can be used up to their limits: beam sizes of several

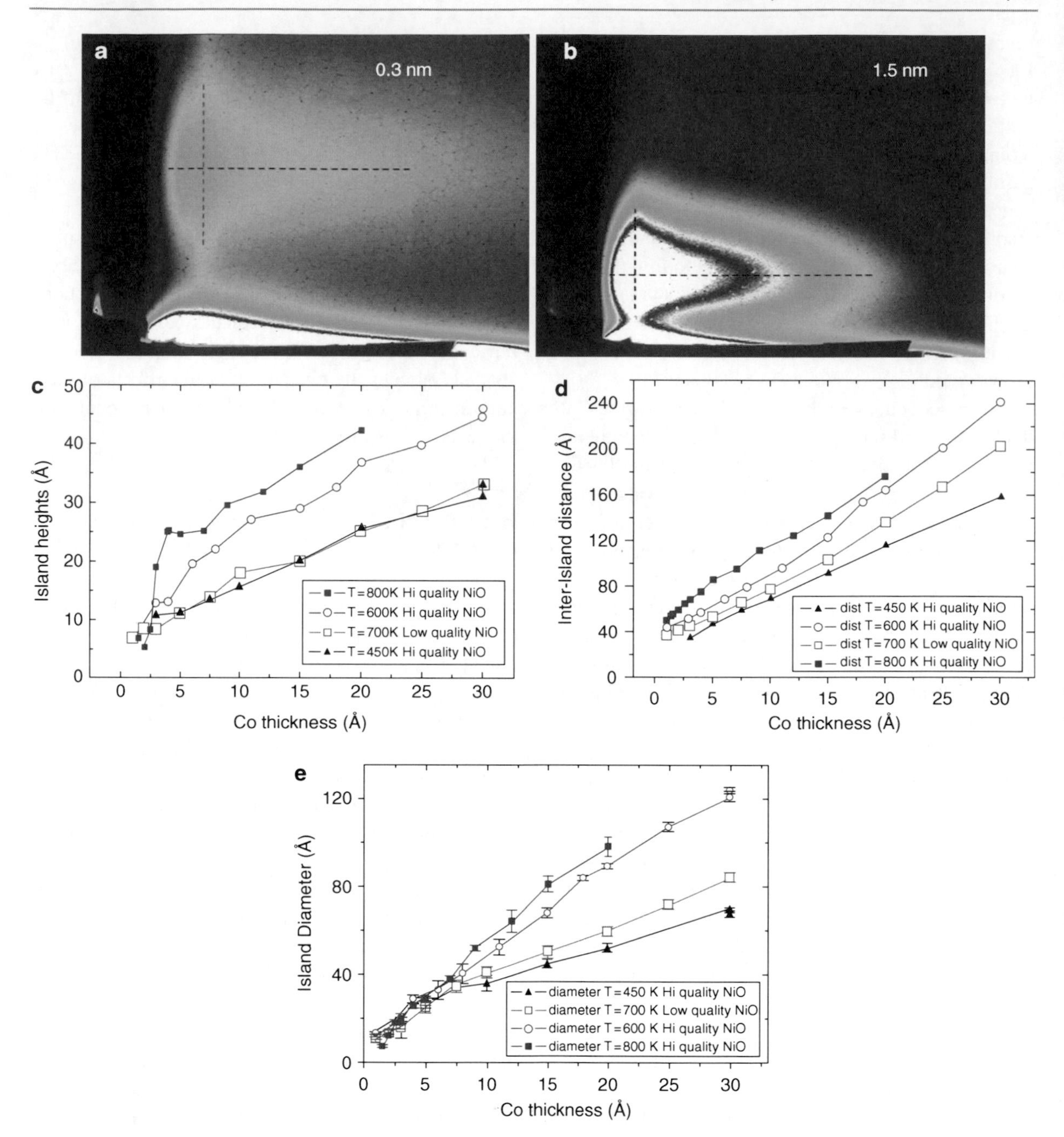

Selected Synchrotron Radiation Techniques, Fig. 2 (**a**, **b**) Part of the GISAXS signal recorded on a CCD camera for (**a**) 0.3 nm and (**b**) 1.5 nm nominal thickness of Co deposited on NiO(111). The beam energy was 10 keV; the vertical of the CCD is parallel to the sample surface and the horizontal direction is perpendicular to it. The lead well at the *bottom* of the pictures hides the direct and reflected beams. The *dashed lines* are guided for the eye to evidence the changes in shape and position of the GISAXS signal. (**c**, **d**, **e**) Morphology of Co islands grown on NiO(111): Island height (**c**), inter-island distance (**d**) and island diameter (**e**) with respect to the deposited Co thickness, as extracted from GISAXS measurements. Hi- and low-quality NiO substrates had mosaic spreads of 0.05° and 0.5° respectively

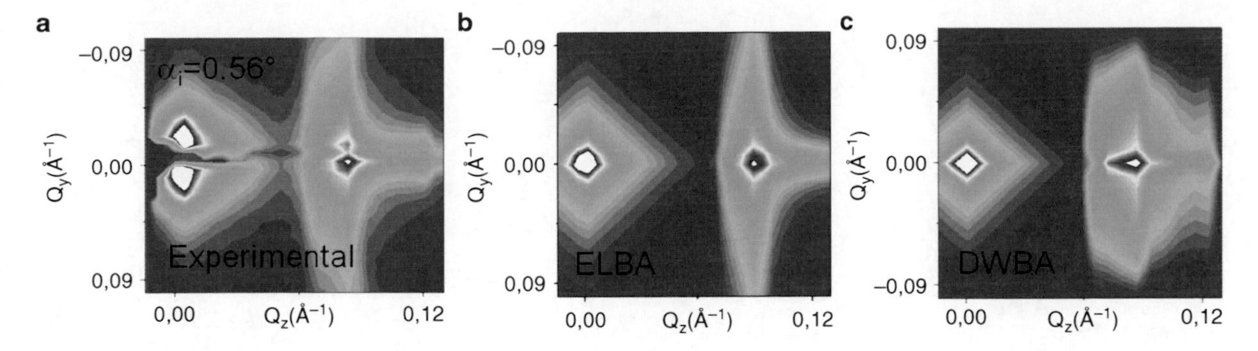

Selected Synchrotron Radiation Techniques, Fig. 3 (**a**) Experimental GISAXS pattern of CNTs prepared with 1% ammonia concentration in the reactive gas mixture, collected at $\alpha_i = 0.56°$. The sample surface is vertical and located on the left side of the pattern. A horizontal beam-stop stops the intense direct and specular beams. (**b**) ELBA pattern including Co cylinders, C tubes, Co–C intermixing (Co–C core-shell 14.8 nm average object and intermixed C–Co intermixed tubes); Co and C of radii 2.5 and 3.3 nm respectively, inter-tube distance 55.6 nm, (**c**) DWBA pattern using the parameters retrieved from the ELBA model. The color scale is logarithmic and identical for all patterns

10 nm have been demonstrated and highly intense beams in the sub-micrometer size range are available at a number of synchrotrons [1].

Various focusing/collimation schemes for X-rays were proposed (for reviews, see Refs. [26, 27] and references therein):

- Refractive focusing optics: compound refractive lenses or planar nano-lenses
- Reflective focusing optics: Kirkpatrick–Baez (KB) crossed bended mirrors; Capillaries (mono and poly-capillaries, single and multi bounce versions)
- Diffractive focusing optics: circular, linear, or crossed linear Fresnel Zone Plates, photon sieves
- Waveguides
- Asymmetrically cut crystals used as beam compressors or collimators

Each of these solutions will have its own advantages and drawbacks [26], which will have to be carefully balanced when designing an experiment. Some key parameters to consider are:

1. Photon flux: the small investigated volumes make the scattered signal very small, so an intense and bright X-ray spot is required
2. Easiness in use: some of these devices are simple to use and align, some other require specific setups. The beam deflection from its initial direction (e.g., reflective optics) can also be a parameter which might complicate the experimental setup
3. Achromaticity: is important only for white beam microdiffraction or micro-spectroscopy experiments. The case and the examples detailed hereinafter require monochromatic beam.
4. Focal (working) distance and beam divergence: the use of bulky environments will need a longer working distance of the optical element, in order to allow for the necessary space. This in return might limit the minimum achievable spot size by reducing the geometrical demagnification factor. If a short-working distance optics is used a smaller spot can be obtained, but there is an increased beam divergence, thus degrading the resolution for strain determination in high-resolution diffraction experiments [14, 15]. A compromise should be found depending on the sample and type of experiment.
5. Stability (mechanical), price, lifetime in the X-ray beam, etc.

μ-XRD Setup and Experimental Approach

Using X-ray focusing optics, it is possible to build various X-ray microscopes working with different contrast mechanisms, or combinations of them: fluorescence, luminescence, transmission, and diffraction. The particular case in which a diffraction contrast is used will be briefly described here. The setup is a combination of an X-ray focusing element and an accurate diffractometer. It consists of a scanner stage used for laterally positioning the sample in the X-ray beam, coupled to precise rotation stages both for the sample angles (incidence, azimuth, tilts) and the X-ray detector (at or close to Bragg diffraction angle) for probing, in the reciprocal space, any desired lattice parameter. Optionally a high-resolution optical microscope pointing at the precise position of the diffractometer's center (and X-ray beam) is used for sample

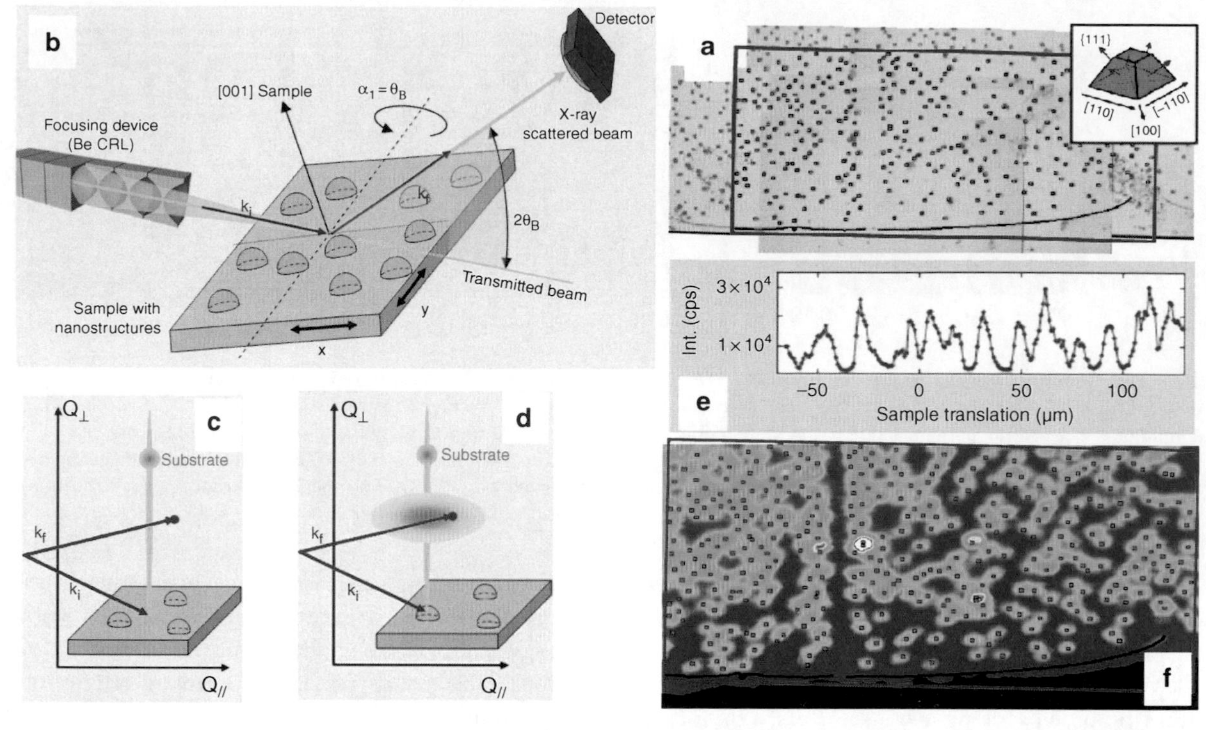

Selected Synchrotron Radiation Techniques, Fig. 4 Principle of the SXDM approach (see also Ref. [28]). (**a**) Optical image of the SiGe/Si pyramids sample. The *inset* shows a sketch of the shape of the SiGe square-based truncated pyramids with the (111) factes highlighted. The large rectangular zone shows the area imaged using the μXRD approach (*panel* **f**). The optical image, obtained in back-reflected light, makes that the side facets are deflecting the light out of the optical axis of the objective, thus appearing dark. Each small square represents a single SiGe island. (**b**) Illustration of the setup, $\mathbf{k}_i$ and $\mathbf{k}_f$ being the scattering vectors. (**c**, **d**) The X-ray spot (represented by the tip of the $\mathbf{k}_i$ *arrow*) illuminates the substrate only or a single island (respectively). Only in the latter case a broad X-ray diffraction signal (different lattice spacing distribution inside the island) is observed, at a different position with respect to the substrate signal. Tuning the diffraction angle to be sensitive in the reciprocal space to a position characteristic to the islands (dot pointed by the tip of the *arrow* $\mathbf{k}_f$) and scanning the sample laterally, higher intensity is observed only when the X-ray spot illuminates an island (*panel* **e**). If this is done in both x and y directions, a map of the surface of the sample (SXDM) is obtained (*panel* **f**), with enhanced intensity (various shades of grey) when an island is probed. Overlaid to this image, the position of the SiGe islands extracted from the optical image (shown in *panel* **a**) is superposed as small squares

pre-aligning and finding of the region of interest. The instrument can function in two ways: one in which the sample surface is *imaged point by point like in a scanning probe technique*, and a second one in which, for particular regions (microstructures) of the sample, detailed *high resolution μ-XRD* data are recorded. Both approaches will be illustrated hereafter. Figure 4 shows the principle of the method. The sample considered here consists of SiGe islands (square-based truncated pyramids, inset Fig. 4, panel a) [16]. Panel (a) shows an optical image of the sample surface, image on which the individual SiGe islands can easily be identified (dark squares). The X-ray beam is focused onto a small spot on the sample. The elastically scattered X-ray photons are collected by the detector. By rotating the sample and detector angles, the intensity distribution around different reciprocal lattice point positions can be probed [13, 15]. This is done by appropriately setting the Bragg angles (panel (b), beam impinging incidence angle given by the $\mathbf{k}_i$ vector, and the position of the detector given by $\mathbf{k}_f$). If the sample is laterally translated in the X-ray beam without changing any of the angles, one expects to see differences in the scattered intensity depending on whether an island or the bare substrate is illuminated. In the case of the substrate, the intensity distribution shows a sharp Bragg peak (panel (c)), and a streak perpendicular to the sample surface, the so-called crystal truncation rod (CTR) [13]. If the focused X-ray beam hits an island, additional and broader features will appear in reciprocal space at different positions (panel (d)), since a different lattice parameter, the one of the SiGe island, is probed. The origin of this different lattice parameter can be multiple, some of the possible causes can be: a various composition, strain, shape, relaxation, etc.

Examples of Applications

Scanning X-Ray Diffraction Microscopy (SXDM): Imaging with Diffraction Contrast

It is now straightforward to perform a "microscopy"-like experiment, the result being an image of the surface of the sample. The diffraction angles are fixed at the expected values to record diffraction from the crystalline planes inside the nanostructures and the position of the sample is scanned laterally while recording, at each point, the scattered intensity (Fig. 4 panel (e)). The resulting contrast is diffraction-based and is related to the presence of islands. If the mapping is performed in both directions (x and y), the resulting map is the result of the Scanning X-ray Diffraction Microscopy (SXDM) approach. Panel (f) shows an SXDM image (color scale) superimposed to the optical image of the same region of the sample (from panel (a)). The agreement is very good and validates the point-by-point microscopy approach, showing that it is possible to find precisely a particular object: in this case, the region in the vicinity of the "defect" exhibited as the vertical zone without any SiGe islands. Markers could also be envisaged. Indeed, it is of utmost importance not only to be able to measure an isolated object out of an ensemble, but to precisely know which particular object was measured: the properties of these SiGe islands can change dramatically from object to object and can be correlated with their shape obtained by ► SEM.

The second example below is showing the power of the method and its sensitivity not only to (local) crystalline changes, but to thickness as well. The sample (metal oxide Magnetic Tunnel Junction, MTJ) consists of a stacking of several metallic and oxide layers on which structures of lateral size in the 10–100 μm range were made by optical lithography [29]. This particular sample suffered a mask misalign during one of the lithography steps. Well that this could be easily detected by simple optical microscopy examination of the surface, μXRD can also bring some information about how this happened and which layers suffered; this sample is a good example to show how the different structures can be imaged and differentiated. The corresponding XRD spectra recorded at these different locations are shown in Fig. 5 panel (b). Signal originating from the various crystalline structures of the layers (Pt, Fe_3O_4, and Co) can easily be differentiated and used as probe in imaging the surface of the sample, after tuning the diffraction angles to these positions (as highlighted by the colored arrows in panel b). The contrast on the surface images changes accordingly. Moreover, sensitivity to thickness is achieved: the arrow pointing to the maximum of the Pt signal also corresponds to a minimum contrast (interference) for the Fe_3O_4 layer. This explains the reversed contrast obtained in the left bottom panel (a) of Fig. 5.

High-Resolution X-Ray Micro-diffraction (HR – MXRD) and Combination with SEM

With the approach depicted above it is possible to measure not only an *individual* object but a very *particular* one. The case of the SiGe islands will be used as example. On the SXDM image of the surface of the sample (Fig. 4), individual objects are chosen and measured in HR-μXRD. Once the lateral positions of the sample are fixed such that the X-ray spot illuminates the object of interest, the angles (sample and detector) are scanned in order to describe the reciprocal space in the vicinity of the chosen Bragg position. The recorded data show characteristics allowing to be grouped in three categories corresponding to three types of SiGe objects: the scattered intensity might vary slightly for the same type of object (depends on the particular imaged object), but, for one family the same characteristics are found. The differences are from one type to another. Figure 6a–h shows such measurements. The differences from the three mentioned types are visible both close to the (004) and (115) Bragg peaks. Since these μXRD measurements were performed at known positions of the sample (cf. image of the surface, as described in Fig. 4), it is possible, at the end of the diffraction experiment, to image in detail precisely the objects in question. The corresponding SEM images of the objects are reported as insets in Fig. 6a–h. Comparing the scattered intensities measured for SiGe pyramids and flat islands, some major differences can be highlighted: the intensity distribution close to the SiGe peak is changing. If the truncated pyramid exhibits scattered intensity concentrated around two positions (two maxima, labeled (1) and (2) in the figure), the flat islands show distributions mostly around the position (2). In the case of the (004) map, the position (1), closer to the Si Bragg

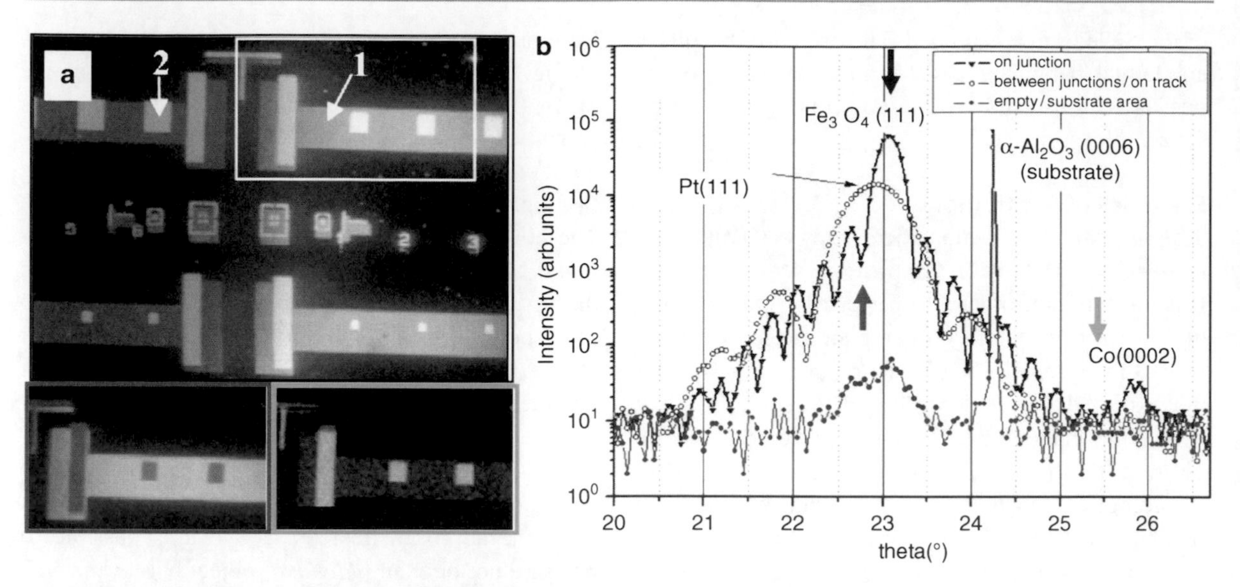

Selected Synchrotron Radiation Techniques, Fig. 5 (a) Scanning μXRD image (SXDM) of the MTJ surface showing defects due to mask align in the optical lithography process. Areas where only the Pt layer (1) remained after the lithography process or the full MJT (2) is present are shown. (**b**) μXRD data (θ-2θ geometry) at different locations on the sample: ▼ on a MTJ; ○ on contact track (Pt buffer); ● on the empty area (substrate). The arrows point to reciprocal space positions characteristic of the different crystalline structures (from left to right, Pt, Fe3O4, and Co), positions for which SXDM images of the sample surface (shown in *panel* **a**) were performed (top Fe_3O_4, left Pt and right Co respectively)

location, shows the presence of a lattice parameter (out of the surface plane) closer to the Si one, that is, indication of a lower Ge concentration. The signal at position (2) indicates a higher Ge concentration and a larger lattice parameter. The availability of two types of maps, around symmetric (004) and asymmetric (115) Bragg peaks, brings complementary information: the peak's position on the (115) maps has also an in-the surface plane component of the SiGe lattice parameter [28].

This example shows the need and the power of combining several methods on the very same micro/nano-object, for detailed characterization and modeling: the shape and the sizes are deduced from SEM images and then injected into a model used to simulate the diffraction data (finite element methods for strain distribution, then semi-kinematical scattering theory). Details can be found in references [14, 15, 26, 28].

In Situ Combination of μXRD with Atomic Force Microscopy (AFM)

The previous example showed the combination of μXRD with (ex situ) SEM (► SEM) images, yielding to an understanding of the structure of microscopic semiconductor samples. The next example shows the in situ combination of a μXRD experiment with an Atomic Force Microscope (AFM) [30, 31]. This approach allows measuring in the same time and for the same individual small object, the mechanical behavior, the internal strain, and the response to external stress. Such measurements, especially when approaching the nanoscale, are of the highest importance for material characterization, both in the elastic and the plastic deformation regimes.

The SiGe island sample described before was used as well as a model sample to prove the principle of this experiment. The experimental setup is schematically shown in Fig. 7a: an ► AFM is mounted on the diffractometer and the focused X-ray beam is aligned with the apex of the tip. During point by point mapping of the sample surface with the X-ray spot (SXDM), the electron current induced in the AFM tip by the X-ray beam is also recorded. The resulting two contrast images (acquired simultaneously) are then combined (green and red colors respectively) to yield the image shown in panel (b). Panel (c) shows precisely the same sample area imaged in SEM (ex situ) previous to the experiment – some of the missing islands in panel (b) were "destroyed" by plastic deformation attempts during the first part of the AFM experiment.

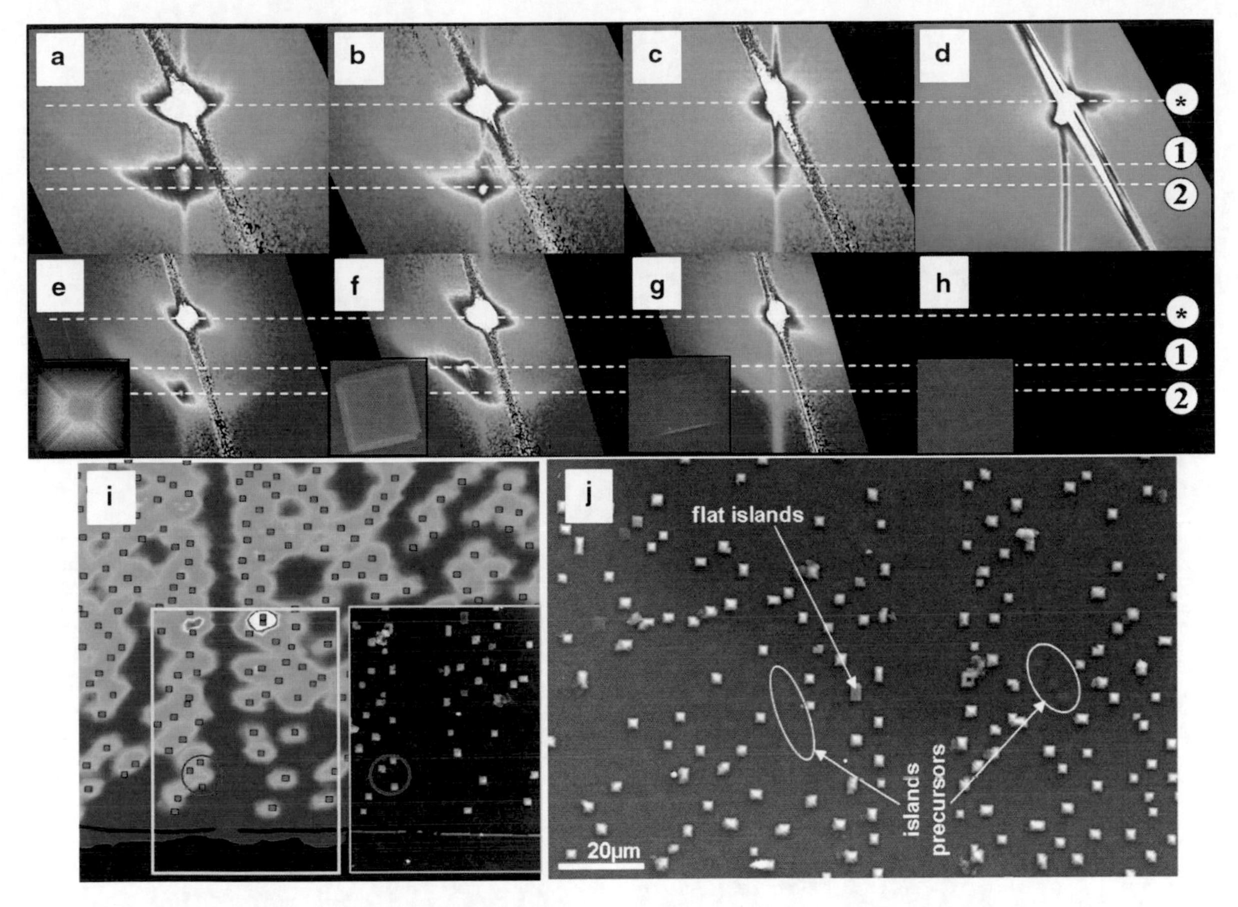

Selected Synchrotron Radiation Techniques, Fig. 6 (a–h) Reciprocal space maps (logarithmic intensity scale) in the vicinity of the (004) and (115) Bragg peaks, recorded using a X-ray focused beam. The representative SEM image for each type of object is shown in the inset. The Si substrate Bragg peak is indicated by (*). (i) Image of the sample surface, performed with diffraction contrast (SXDM), compared to a SEM image of the same area, with particular objects identified (circle). (j) SEM image (details) of the sample, around the region with the "vertical path" free of SiGe islands, in the center of the image

The elastic properties of individual SiGe islands while pushing on them with the AFM tip are well documented in Refs. [30, 31]: the applied force (and thus the pressure) can be extracted from the changes in the resonance frequency of the AFM tip; the lattice parameter (average value over the whole island) is extracted from the Bragg position in μXRD experiments, at each applied external pressure. It was also possible to access the plastic regime. Figure 7 (panels f and g) shows μXRD measurements (reciprocal space maps) close to the (004) Bragg position for an SiGe pyramid before and after pushing. The insets show the corresponding SEM images of the island and AFM tip before and after the experiment. Well that the AFM tip bends, the applied pressure can be large enough to completely destroy the SiGe island, which results in a major change of the μXRD signal.

This example validates the use of the coupled μXRD and AFM not only for studying the elastic properties of small structures, but also for in situ indentation of materials (► Nanoindentation) (a stiffer tip, and with an adapted shape for indentation, should then be used). In such an indentation experiment, by applying laterally resolved μXRD, the resulting strain field around the indented area could be mapped and modeled.

X-Ray Photo Emission Electron Microscopy

XPEEM Instrumentation and Operating Principle

XPEEM microscopy is a parallel imaging method that combines X-ray electron spectroscopy and electron microscopy. It is based on soft X-ray-in/Electrons-out

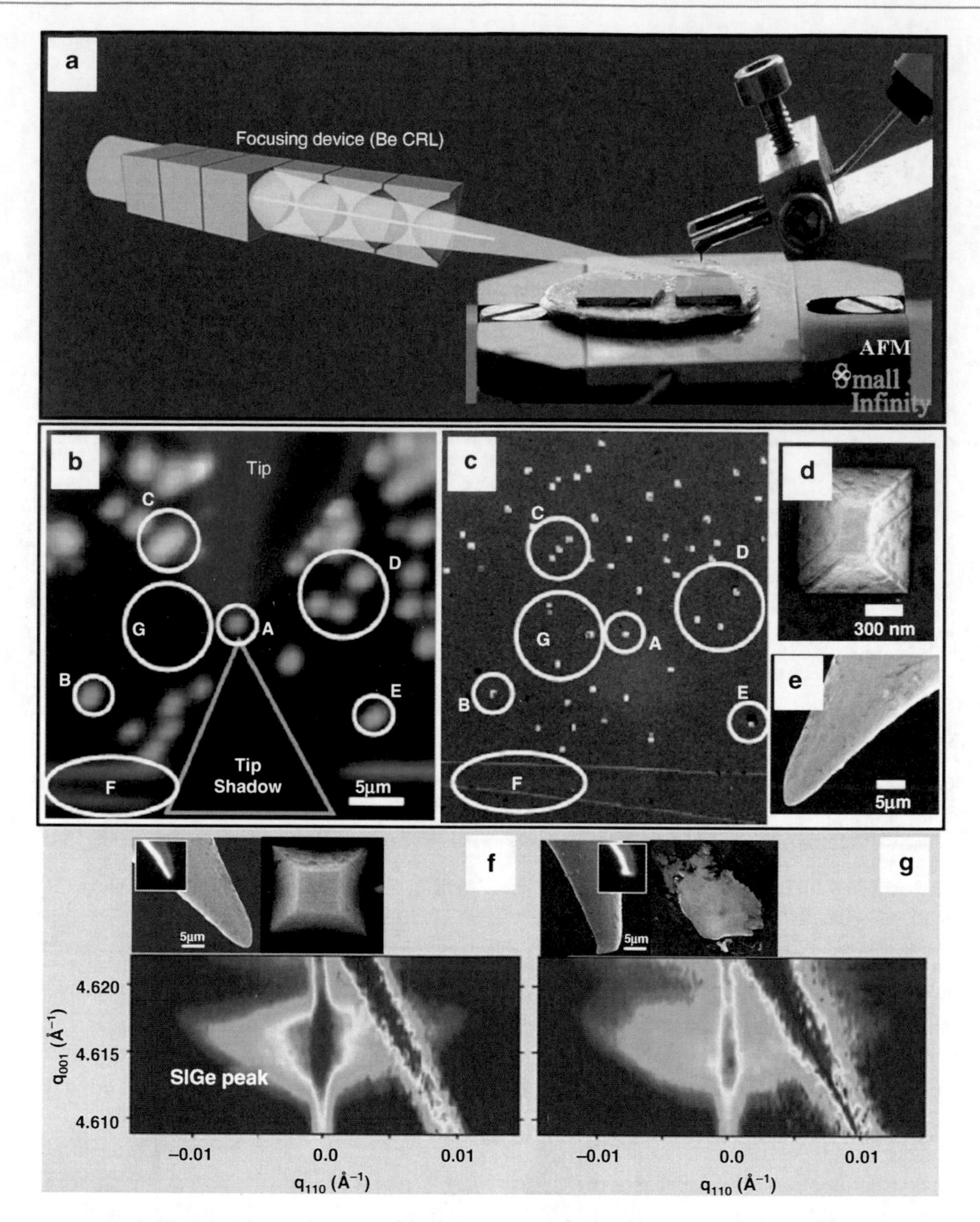

Selected Synchrotron Radiation Techniques, Fig. 7 (**a**) Cartoon of the combination of μXRD with AFM (simplified setup). Once the X-ray focused beam is aligned with the tip of the AFM, it is used to "image" (high-resolution XRD) the object placed beneath the tip. (**b**) superposed maps (images) of the sample, with the X-ray beam aligned on the AFM tip and a single SiGe island placed just beneath the tip: *in green* the detector signal (sensitive to SiGe) is recorded, while *in red* is the electron current induced in the tip when this one is hit by the X-ray beam. (**c**) SEM image of the sample showing precisely the region measured in panel (**b**). The different areas on the sample are labeled. Note the absence of the islands labeled "G" – during the experiment and previously taking this image, these islands were "smashed" (destroyed) or laterally displaced using the AFM tip. (**e**) SEM image of an SiGe single pyramid of ~1 μm lateral size. (**e**) Image of the blunt tip used for pushing experiments. Note the relatively large apex radius of curvature (with respect to standard AFM tips) – this feature makes easier pushing on the small objects. (**f**) Reciprocal space map (close to the (004) Bragg peak) for a single SiGe island. The Si substrate Bragg position is not shown on the map ($q_{Si(004)} = 0.4628\ nm^{-1}$). (g) Same map after pushing with the AFM and reaching the plastic regime (destroying the SiGe island). The intensity distribution changed completely. The *insets* show the SiGe island and the AFM tip (SEM images) before and after plastic deformation. The electron yield images of the tip are also shown for each case

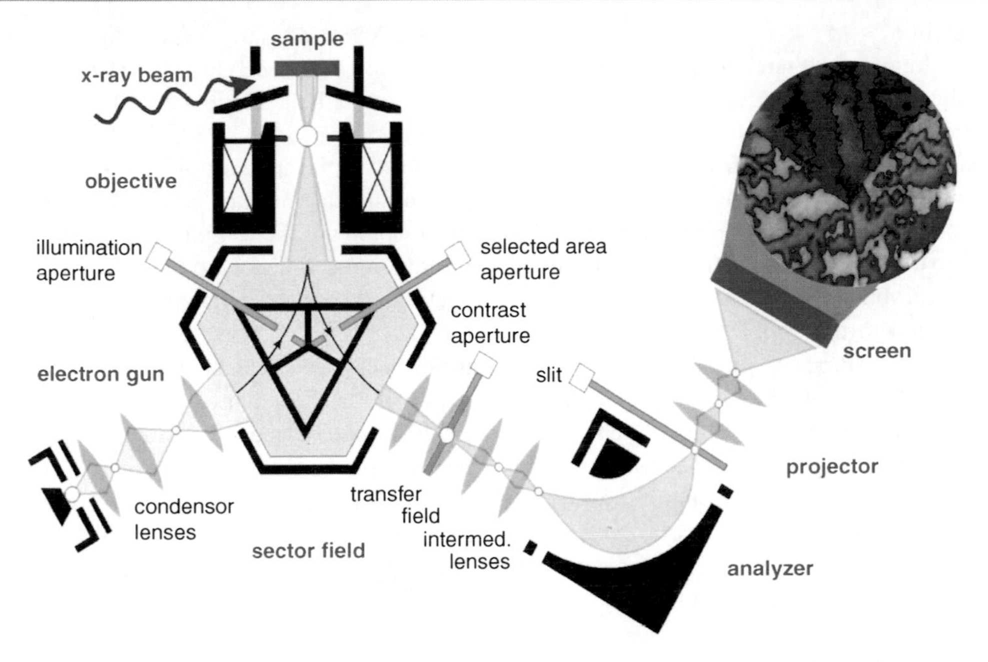

Selected Synchrotron Radiation Techniques, Fig. 8 Schematic sketch of the LEEM-PEEM microscope

principle and was originally pioneered by B. Tonner. The sample is illuminated by a monochromatic soft X-ray beam (<1,500 eV). The method does not require extreme focusing of the X-rays. Moderate focusing (few square micrometers) using dedicated X-ray optics (Kirkpatrick – Baez, Volter mirrors, etc.) is used so that the beam spot matches the useful microscope field of view. After the absorption of the X-ray photons, the emitted photoelectrons are collected, magnified, and projected onto a detector using a dedicated (electrostatic or magnetic) low electron energy microscopy column.

There is a certain number of XPEEM microscopes available commercially. A version that combines PEEM and LEEM microscopy is nowadays the most popular one used with synchrotron radiation sources. This apparatus is very powerful because it allows to combine in a single instrument structural (LEEM) and spectroscopic (PEEM) methods, enabling a true multi-technique approach to study surfaces, buried interfaces, and nanostructures.

The electron column used in the XPEEM microscopy is similar to those used in Electron microscopy, the main difference arising from the use of low kinetic energy photoelectrons. The XPEEM instrument in its simpler version consists of an objective lens, projective lenses, an image detector, and a contrast aperture (see Fig. 8). A first image, collected and magnified by the objective lens, is formed in the back-focal plane of the objective lens. An aperture placed within this plane (contrast aperture) allows to reduce the energy spread of the photoelectrons and limits their angular dispersion. The image is further magnified by a set of projective lenses. The final image is collected via a detector consisting of a Micro-Channel plate, a florescence screen, and a CCD camera. In the most sophisticated instruments, a certain number of lenses are additionally implemented: corrector lenses for astigmatism, retarding grid as high-pass energy filter, field lenses, etc.

Although the incorporation of an energy filter in XPEEM microscope is not mandatory for imaging, its installation in the most recent and advanced microscopes opens up the way for core level and valance band photoemission microscopy. The energy filter is not only helpful for imaging with primary photoelectrons but also useful for secondary electron imaging. It allows to select a narrow energy window around the maximum of the secondary electron energy distribution and, thus, to improve the spatial resolution without unacceptable loss of transmission.

The main requirements for X-ray microscopy, in particular for PEEM/LEEM, are the spatial resolution and the contrast mechanism.

Spatial Resolution

A spatial resolution of 22 nm using synchrotron radiation has already been achieved and a resolution of 6 nm using LEEM. A key limitation to the performance of electron-optical X-ray microscopes is the severe chromatic and spherical aberration of the immersion lens used. This problem constrains the X-PEEM microscope to a fairly small transmission in an attempt to offset these problems. Rempfer's group has demonstrated a solution to this problem, by designing and testing an electron mirror, which has spherical and chromatic aberrations of similar magnitude to the objective lens, but opposite in sign. When used in combination with a properly designed objective lens and a highly symmetric beam separator, the aberration can be highly reduced. Therefore, both transmission and resolution can be increased. Several projects using correcting mirror are under development [32, 33].

Contrast Mechanism

The contrast mechanism used in XPEEM microscopy arises from the spectroscopic capabilities and the characteristic binding energies of the atomic core electrons. The illumination of the specimen with X-rays excites a broad electron spectrum consisting of primary photoelectrons and inelastically scattered secondary photoelectrons. Two operation modes can therefore be performed:

- XPS (X-ray Photoelectron Spectroscopy). The XPEEM detects and selects, using an appropriate energy filter, the primary photoemitted electrons from atomic core levels. The photon energy is fixed and the kinetic electron energy is given by: $E_K = h\nu - E_b - \phi$, where $h\nu$ is the photon energy, E_b is the core level binding energy, and ϕ the work function, that is, the energy barrier that the photoelectrons have to overcome to escape from the sample surface.
- XAS (X-ray Absorption Spectroscopy). In this case the XPEEM detects the secondary photoelectron as function of the impinging photon energy. When the photon energy matches the absorption threshold of the element, the photoelectron spectrum shows strong resonances due to electron transition arising from the core level to the unfilled valence band states.

The choice of XAS or XPS modes is mainly made on the basis of sample depth probe requirements (1/e) which are directly correlated to the electron inelastic mean free path (e) and the photoelectron kinetic energy. XPS allows to achieve a high-surface sensitivity with a probing depth bellow 1 nm, meanwhile the XAS is more bulk sensitive (<10 nm).

In both cases, the XPS or XAS intensities are proportional to the number of emitter atoms within the probing depth, and thus provides direct and quantitative mapping of the chemical composition. Additional information can also be obtained from the line-shape analysis as they are a fingerprint of the emitter chemical state (valence state, site location, etc.). Finally, local spectroscopy can be performed selecting a small region of the specimen: local spectroscopy (μ-XAS or μ-XPS), photoelectron diffraction (μ-XPD), angle resolved photoelectron spectroscopy (μ-ARPES), etc.

The high-spatial resolution of XPEEM microscopy, its chemical sensitivity, and spectroscopic capability open up a broad field of application which extends from thin film studies, nanostructures, and magnetism to more exotic applications for surface science such as in tribology and geology.

Examples of Application: Magnetic Imaging

The interest in magnetic domain imaging in the nanometer range has been rapidly increasing during the last decade. A considerable impetus is coming from the development of high-density magnetic storage devices and from the forthcoming achievement of spin electronics. In order to tailor the magnetic behavior of these systems to specific needs (for instance, a certain response to magnetization reversal), a detailed understanding of the structure and of the dynamics of magnetic domains is mandatory. In addition, the thin film nature of such devices emphasizes the surface aspect of magnetism. This situation requires magnetic domain-imaging techniques which combine surface sensitivity and high-spatial resolution. Moreover, for many applications element specificity is even more important than high-lateral resolution. Magnetic storage media or building elements of spin-electronic devices are often composed of several chemical elements or intermetallic compounds, each of which distinctly contributes to the magnetic behavior. All these requirements pose a considerable challenge to conventional magnetic domain imaging techniques such as magneto-optical Kerr microscopy, Lorentz microscopy, scanning electron microscopy (SEMPA), Magnetic Force Microscopy (MFM), etc.

XPEEM magnetic microscopy is today a good candidate for an ideal surface magnetic imaging technique, as it combines the magnetic and element selectivity with a spatial resolution below the size of the magnetic domains. One may identify three important length scales for magnetic imaging which span over two orders of magnitude. The first one is about 1 μm, set by the size of lithographically manufactured magnetic cells such as in spin valve read heads or magnetic memory cells. The second one is about 10 nm, corresponding to the crystallographic grain size of typical magnetic materials and to domain walls. The last one is 0.1 nm, that is, the atomic size. The actual spatial resolution of the XPEEM microscope allows to image straightforwardly the magnetic domain, the further improvements of the resolution may allow access to the second characteristic length scale of 10 nm.

The elemental specificity in XPEEM magnetic microscopy arises from the characteristic binding energies of the atomic core electron, as mentioned above. Both X-ray photoelectron spectroscopy (XPS) and X-ray absorption (XAS) can be used for magnetic imaging, although the latter is the most used because it is less demanding in terms of instrumentation and photon flux.

The use of polarized synchrotron radiation enables studies of the electronic and magnetic anisotropies, and thus allows magnetic contrast for the XPEEM [34]. A simple description of the photon polarization by a biaxial vector for linear polarization and a vector for left-/right-handed circular polarization is the physical basis for probing various anisotropies of the sample. In general, linearly polarized light can only detect anisotropy of electronic charge. In contrast, helicity resolved circularly polarized light can measure a dipolar or vector quantity, in the present case the modulus and direction of the electron angular moment and spin.

For magnetic XPEEM spectromicroscopy, X-ray Magnetic Circular Dichroism (XMCD) is exploited. This effect, pioneered by Schütz et al. [35], is widely used to determine the size, the direction, and the anisotropy of the atomic magnetic moments in magnetic material at a macroscopic scale. In the simple figure of a 3d transition metal, the magnetization or more precisely the spin moment is given by the imbalance between the spin-up and spin-down electrons (or holes) within the d shell and bellow the Fermi level. The use of circularly polarized light will make the absorption process spin-dependent and therefore, will enable access to the spin moment. In other words, the spin-split valence band will act as a detector for the spin of the excited photoelectron: the XAS intensity is simply proportional to the number of empty d states of a given spin. The measurements of the XAS signal anisotropy with respect to the direction of the light for a given circular polarization enable directly the access to the projection of the magnetization as illustrated in the Fig. 9. In XMCD-PEEM, a set of two images are acquired at respectively the L_3 and L_2 edges for a given light helicity and direction. These two images will contain a mixture of different contrast: magnetic, structural, chemical, etc. In order to cancel all the contrast mechanism except the magnetic one, an asymmetric image is calculated by subtracting the above two mentioned data. This image reflects directly the local variation of the $L_{2,3}$ asymmetries and therefore gives access to the projection of the magnetization moment with respect to the light direction: the black and white regions in the image reflect the domains where the magnetic axis is aligned parallel or antiparallel to the direction of the light at fixed polarization. The gray area corresponds to domains where the magnetic axis is perpendicular to the direction of the light or that have no magnetic moment. It is worth noting that the same XMCD-PEEM image can be obtained by varying the circular light helicity and keeping the photon energy fixed at the same absorption edge (L_3 or L_2).

The study of antiferromagnetic (AF) surfaces and interfaces has posed an even larger challenge because conventional techniques are mainly bulk-sensitive, and antiferromagnet does not carry any net external magnetic dipole moment. These limitations were overcome by the use of X-ray Magnetic Linear Dichroism (XMLD) spectroscopy. In contrast to XMCD which directly measures the magnetic moment, XMLD measures the projected value of the square of the magnetic moment. XMLD can therefore be applied for all uniaxial magnetic system, that is, antiferromagnets as well. Recently it was shown that XMLD spectroscopy in conjunction with X-PEEM microscopy is capable of imaging the detailed antiferromagnetic domain structure of a surface and interface [36]. This has been shown on a NiO(001) thin film and cleaved sample. The XMLD-PEEM images reveal antiferromagnetic contrast corresponding to the different in-plane projection of the antiferromagnetic axis.

The ability to combine in the same instrument XMCD and XMLD XPEEM microscopy with high-spatial resolution and chemical selectivity has opened

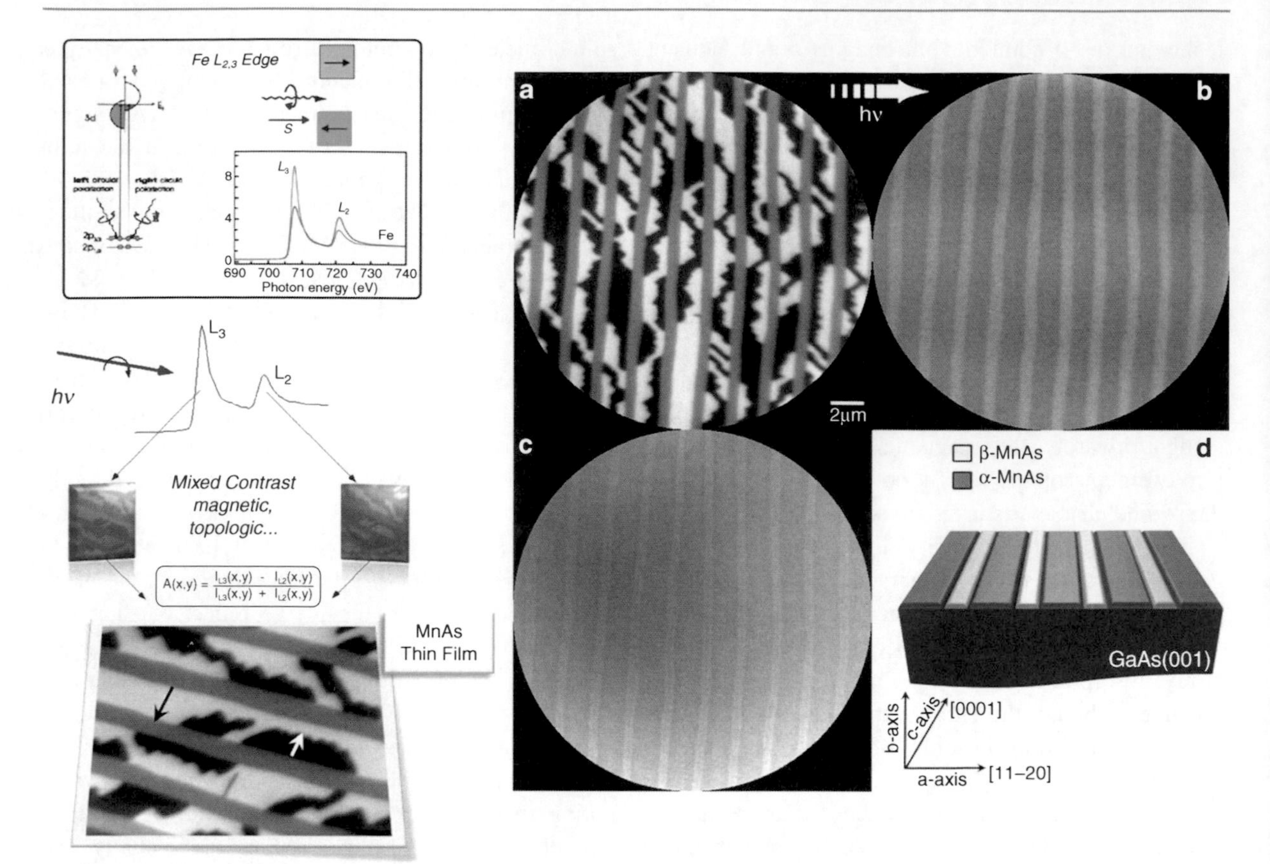

Selected Synchrotron Radiation Techniques, Fig. 9 (*Left panel*) *Top*: Principles of X-ray magnetic circular dichroism (XMCD) in the case of L Fe edge. The XMCD spectrum reflects the anisotropy of L absorption edge with respect to the relative direction between the incident light and the magnetization direction at a given circular polarized light. *Bottom*: Principles of XMCD-PEEM in the case of MnAs thin film. The measurements are performed at the Mn L edge (see text for explanation). (*Right*) XMCD and XMLD-XPEEM images. MnAs sample thickness 330 nm: (**a**) XMCD image at the L_3 Mn edge highlighting the FM Mn domains in the α-phase. (**b**) XMLD image taken using the L_3 multiplets asymmetry at vertical linear polarization evidencing the AF domains. The *gray* level corresponds to the isotropic α-phase and the light region to the orthorhombic β-phase. (**c**) LEEM image taken at the same sample area showing the α-β phase coexistence. (**d**) Schematic sketch of the FM/AF configuration in the MnAs thin films

a new route to investigate magnetic materials, as illustrated in the following examples.

Surface Magnetism in MnAs Thin Films

Ferromagnetic (FM) MnAs is a promising candidate for electrical spin injection into GaAs and Si-based semiconductors, since it exhibits a large carrier spin polarization, small coercive field, and relatively high-saturation magnetization and Curie temperature. Bulk MnAs is FM at Room Temperature (RT) (α phase) and shows close to 40°C a first order transition to the paramagnetic β phase. On the contrary, epitaxial MnAs films on GaAs substrate, which are more appropriate for the injection applications, show at RT the coexistence of both phases. The phase coexistence results in the formation of self-organized stripes of alternating α and β phases. The XMCD-PEEM image of Fig. 9 (right) illustrates this phase coexistence. The black/white regions correspond to ferromagnetic domains in the α-phase, the vertical gray stripes correspond to the non-magnetic α-phase. The easy magnetization direction is perpendicular to the stripes direction, pointing in opposite directions in the black and white regions. The period of the stripes is closely correlated to the film thickness (4.8 × thickness). The phase coexistence is due to anisotropic strain applied by the GaAs substrate. As a consequence the magnetization is pointing not along the stripe direction as expected from shape anisotropy considerations but is perpendicular to it, predominantly in-plane. This leads to a complex, thickness-dependent magnetic domain structure in the interior of the α-MnAs which is reflected in the complexity of magnetic images of the surface of the film.

The nature of the non-magnetic β-phase is still controversial. In the coexistence range exchange bias and giant magnetoresistance effects of possible practical importance have been reported. Until recently β-MnAs has generally been considered to be paramagnetic but the effects just mentioned and first principles calculations suggest antiferromagnetism. A more thorough first principles calculation comes to the conclusion that β-MnAs is paramagnetic. The XMLD-PEEM microscopy enables to elucidate this controversy. Figure 9 evidences a weak antiferromagnetism in the β-phase which explains the reported exchange bias and giant magnetoresistance [37].

Micromagnetism of Patterned Nanostructures and Magnetic Tunnel Junction

Magnetic tunnel junctions (MTJs) are probably one of the most studied devices in the so-called spintronics research field. They are usually composed by two active ferromagnetic electrodes separated by a thin insulating layer (<3 nm) that act as a tunnel barrier for the spin polarized electrons, that is, a spin dependent tunnel current exists and can be induced between the two ferromagnetic electrodes. The most attractive property of MTJs concerns their strong electrical resistance variation with the magnetic configuration of the electrodes, the so-called magnetoresistance. The resistance can vary from more than 100%. The Magnetic Random Access Memories (MRAM) and the read heads of hard drive disks take benefits from this property. In order to increase storage capacity, a strong effort is made toward a size reduction of the MTJs. However, in the course of miniaturizing a unit cell for magnetic device a certain number of questions related to the finite size effect arise. Size effect starts then to play an important role and can affect drastically the electrodes magnetic properties. The results obtained on thin films cannot be straightforwardly extrapolated to patterned objects and nanostructures. Combining the high-spatial resolution of XPEEM magnetic imaging, its chemical sensitivity and micromagnetic simulations allowed to demonstrate the strong influence of a dipolar magnetic coupling on the magnetization reversal of MTJs [38]. Such studies are mandatory because understanding the domain wall formation process is crucial since it affects the magnetic–electrical properties of MTJs.

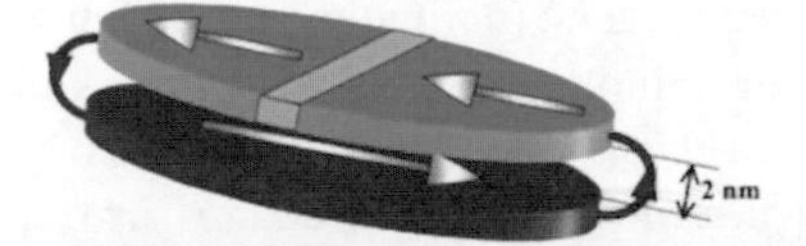

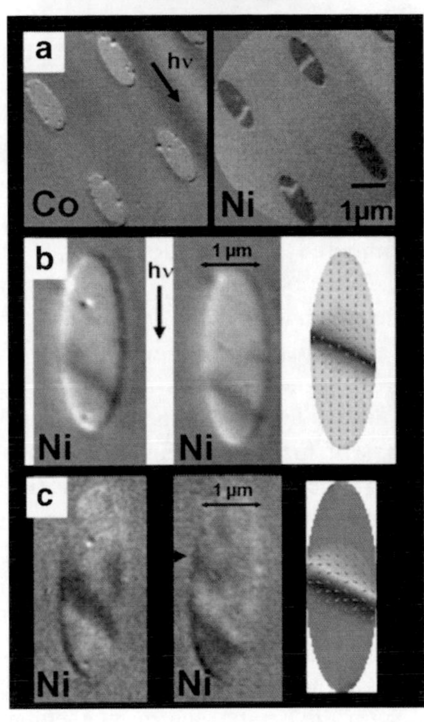

Selected Synchrotron Radiation Techniques, Fig. 10 *Top*: Schematic representation of MTJs. The two magnetic electrodes, composed of a nickel-iron alloy and of cobalt, are respectively represented in *green* and *purple*. They are separated by 2 nm thick layer of oxidized aluminum (not represented for clarity reason). The magnetic stray field sketched in *red* tends to align the magnetization each electrode in an antiparallel configuration. *Bottom*: XMCD-PEEM images of four ellipses recorded at the Co and at the Ni edges. The photons incidence direction is aligned along the ellipses long axis from the *top left corner*. The *white* and *black* contrasts correspond to magnetization components aligned along the long ellipses axis. XMCD-PEEM images of two ellipses (1 × 3 μm) measured at the Ni edge. The *gray* level distribution corresponds to the scalar projection of the local magnetization with respect to the light incidence: (**b**) parallel and (**c**) perpendicular to the ellipses long axis

When the MTJs are structured in micrometer-sized elements, an antiferromagnetic coupling tends to align the magnetization of each electrode in an antiparallel configuration contrary to what is usually observed in the thin film case. This coupling originates from the magnetic stray field at the nanostructure edges (see red arrows in Fig. 10). The MTJs is formed by two ferromagnetic layers: (permalloy and cobalt) separated by a thin aluminum oxide layer: Co(4 nm)/Al_2O_3(2 nm)/$Fe_{20}Ni_{80}$(4 nm). The MTJs have been patterned in ellipse-shaped structures in order to reenforce the stray field effect. The high-spatial resolution of the X-ray Photoemission Electron Microscopy (X-PEEM) combined to X-ray Magnetic Circular Dichroism (XMCD) enables to directly image the magnetic configuration of each electrode of the MTJs. This powerful technique allows to image independently the magnetic configuration in each electrodes if composed of different elements, thanks to its elemental selectivity. Figure 10a shows that, when no magnetic field is applied, both layer magnetizations are mainly in an antiparallel configuration. Moreover, if the Co layer magnetization is uniform, surprisingly this is not always the case in the NiFe layer. The (b) and (c) parts of Fig. 10 present images obtained in geometries where the technique is sensitive to the magnetization component either along the ellipses long axis (b) or along the ellipses short axis (c). The combination of this imaging technique with in silico modeling (right part of the Fig. 2b and c) allows to understand the nature and the formation process of the nonuniform magnetization distribution presents in the NiFe layer. In this region, the magnetization rotates continuously by 360°, forming an object separating two regions of uniform magnetization. This magnetic object is called a 360° domain wall. By simulations, only one chirality is obtained for the wall, although three different cases can be observed experimentally: no wall and two chiralities (Fig. 10). This difference has been attributed to local magnetic anisotropy fluctuations at the ellipses extremities which drive the magnetization curling direction during the reversal process.

Future Directions for Research (Outlook)

Future challenges of GISAXS may lie in the use of the X-ray coherence [24] in order to reconstruct single objects without any model assumptions. As far as the corresponding nano-objects can bear the X-ray beam, soft matter systems like polymers, block copolymers, or even biological applications, such as proteins, peptides, and viruses attached to surfaces or in lipid layers, will be very challenging for GISAXS.

For μXRD experiments aiming at a characterization of individual microstructures and nanostructures [26], the actual demand is that of availability of brighter and smaller X-ray spots, with increased stability. Last generation synchrotron sources can fulfill in a certain measure these demands, and dedicated experimental stations are built [1]. Coupling μXRD with other techniques is achieved ([26] and references therein) and allows obtaining complementary results from various techniques, which completes the understanding about these systems and their properties. Another experimental path using focused X-ray beams is the one using a white (continuous energy spectrum) X-ray microbeam – the Laue microdiffraction experiments. It can give access to grain distributions (orientation, strain) in polycrystalline samples [39]. This particular topic needs achromatic focusing elements (e.g., Kirkpatrick-Baez mirrors, capillaries) and was not detailed here. Alternative techniques like the lensless X-ray microscopy (Coherent Diffraction Imaging CDI, X-ray holography, . . .) using coherent X-ray beams are also a path opened for accessing nondestructively the properties of small structures: 3D imaging and strain determination with spatial resolutions of 10 nm and better were proved. Tremendous gain in photon flux and coherence can be achieved by the use of free electron laser sources. These approaches will not be detailed here; for more details, one can refer to the reviews [26, 40] and the references therein. The noninvasive character of the XRD method and the relatively large penetration depth of the hard X-rays make possible experiments in which the diffraction approach is combined (insitu or ex situ) with other analysis methods. The particular cases of SEM and AFM have been examined. Ultra High Vacuum (UHV) environment with in situ preparation facilities of thin metallic films was also addressed, but other environments (high temperature, clean environment, high pressure, reaction cells, electric or magnetic fields, . . .) can be easily imagined.

One of the main challenges for XPEEM microscope in the future is to combine the high-spatial resolution with time-resolved experiments. In the case of magnetic materials for example, the dynamics of the magnetization reversal in thin magnetic films has become a matter of high interest for the future of magnetic recording and nonvolatile magnetic memories. Parallel to the evolution toward smaller magnetic bits and memory cells, writing and reading times approaching the ns range will be required in a few years from now. A complete understanding of the magnetization dynamics in these structures requires the ability to probe the magnetization of the individual layers as well as their mutual interaction. Time-resolved X-PEEM measurements are very challenging, since the secondary electrons that are used for the image are strongly perturbed by the magnetic field necessary to switch the magnetization direction. Therefore, the time-resolved XPEEM experiment can be only performed in stroboscopic mode. The magnetic pulses can be synchronized with the X-ray pulses coming from the storage ring to perform dynamic measurements in a pump-probe scheme. X-PEEM images cannot be acquired during the field pulses, but time resolution of 50 ps has been already demonstrated [41]. The advent of fourth generation synchrotron and more specifically X-ray Free Electron Laser [40] will certainly open up the way for high-resolution imaging at a time scale of femtosecond.

Finally, many efforts are done to improve the spatial resolution. Although the ultimate resolution will be always limited by the diffraction limit (1 nm), the development of new aberration corrected setups will certainly open up the way to image nanostructure with resolution below 5 nm. Other aberrations correction methods using numerical algorithms or phase retrieval have also been proposed.

Cross-References

▶ AFM
▶ Nanoindentation
▶ SEM

References

1. See for example http://www.lightsources.org/ for links to existing facilities
2. Wiedemann, H.: Synchrotron Radiation. Springer, Berlin (2002). ISBN 13: 978-3540433927
3. Hippert, F., Geissler, E., Hodeau, J.-L., Lelièvre-Berna, E., Regnard, J.-R.: Neutron and X-Ray Spectroscopy. Springer, Dordrecht (2005). ISBN 978-1-4020-3336-0
4. Baruchel, J., Hodeau, J.L., Lehmann, M.S., Regnard, J.R., Schlenker, C. (eds.): Neutron and Synchrotron Radiation for Condensed Matter Studies: HERCULES – Higher European Research Course for Users of Large Experimental Systems, vol. 1–3. Les Editions de Physisque/Springer, Les Ulis/Berlin/Heidelberg (1993). Springer, ISBN 0-387-56561-2, ISBN 2-86883-185-0, ISBN 3-540-56561-2
5. Reimers, W., Pyzalla, A.R., Schreyer, A., Clements, H.: Neutrons and Synchrotron Radiation in Engineering Materials Science: From Fundamentals to Material and Component Characterization. Wiley, Waintheim (2008). ISBN 978-3-527-31533-8
6. Warwick, T., Franck, K., Kortright, J.B., Meigs, G., Moronne, M., Myneni, S., Rotenberg, E., Seal, S., Steele, W.F., Ade, H., Garcia, A., Cerasari, S., Denlinger, J., Hayakawa, S., Hitchcock, A.P., Tyliszczak, T., Kikuma, J., Rightor, E.G., Shin, H.-J., Tonner, B.P.: A scanning transmission x-ray microscope for materials science spectromicroscopy at the advanced light source. Rev. Sci. Instrum. **69**, 2964 (1998)
7. Porod, G.: Small Angle X-Ray Scattering, p. 37. Academic, San Diego (1982)
8. Feidenhans'l, R.: Surface structure determination by X-ray diffraction. Surf. Sci. Rep. **10**, 105 (1989)
9. Warwick, T., Ade, H., Hitchcock, A.P., Padmore, H.A., Tonner, B.P.: Soft x-ray spectromicroscopy development for materials science at the Advanced Light Source, J. Electron Spectros. Relat. Phenom. **84**, 85–98 (1997)
10. Stöhr, J., et al.: Principles of X-Ray Magnetic Dichroism Spectromicroscopy, Surface Review and Letters **5**, 1297–1308 (1998)
11. Locatelli, A., Bauer, E.: Recent advances in chemical and magnetic imaging of surfaces and interfaces by XPEEM, J. Phys.: Condens. Matter **20**, 093002 (2008)
12. Kuch, W.: Imaging magnetic microspectroscopy. Spinger, Berlin (2003)
13. Als-Nielsen, J., McMorrow, D.: Elements of Modern X-ray Physics. Wiley, New York (2001)
14. Stangl, J., Holy, V., Bauer, G.: Structural properties of self-organized semiconductor nanostructures. Rev. Mod. Phys. **76**, 725 (2004)
15. Holy, V., Pietsch, U., Baumbach, T.: High-Resolution X-Ray Scattering from Thin Films and Multilayers. Springer Tracts in Modern Physics, vol. 149. Springer, Berlin Heidelberg (1999). ISBN 3-54o-62o29-X
16. Schmidbauer, M.: X-ray Diffuse Scattering from Self Organized Mesoscopic Semiconductor Structures. Springer Tracts in Modern Physics, vol. 199. Springer, Berlin (2004)
17. Bauer, E.: Low energy electron microscopy. Rep. Prog. Phys. **57**, 895–938 (1994). doi:10.1088/0034-4885/57/9/002
18. Riekel, C.: New avenues in x-ray microbeam experiments. Rep. Prog. Phys. **63**, 233 (2000)
19. Davies, R.J., Burghammer, M., Riekel, C.: Simultaneous microRaman and synchrotron radiation microdiffraction: tools for materials characterization. Appl. Phys. Lett. **87**, 264105 (2005)
20. Davies, R.J., Burghammer, M., Riekel, C.: A combined microRaman and microdiffraction set-up at the European synchrotron radiation facility ID13 beamline. J. Synchr. Rad. **16**, 22 (2009)
21. Renaud, G., Lazzari, R., Leroy, F.: Probing surface and interface morphology with grazing incidence small angle X-ray scattering. Surf. Sci. Rep. **64**, 255–380 (2009)
22. Rauscher, M., et al.: Grazing incidence small angle x-ray scattering from free-standing nanostructures. J. Appl. Phys. **86**, 6763 (1999). doi:10.1063/1.371724
23. Barbier, A., Stanescu, S., Boeglin, C., Deville, J.-P.: Local morphology and correlation lengths of reactive NiO/Cu (111) interfaces. Phys. Rev. B **68**, 245418-1-7 (2003)
24. Zozulya, A.V., Yefanov, O.M., Vartanyants, I.A., Mundboth, K., Mocuta, C., Metzger, T.H., Stangl, J.,

S

Bauer, G., Boeck, T., Schmidbauer, M.: Imaging of nanoislands in coherent grazing-incidence small-angle x-ray scattering experiments. Phys. Rev. B **78**, 121304 (2008)
25. Mane Mane, J., Cojocaru, C.S., Barbier, A., Deville, J.P., Jean, B., Metzger, T.H., Thiodjio Sendja, B., Le Normand, F.: GISAXS study of carbon nanotubes grown on SiO2/Si (100) by CVD. Phys. Stat. Sol. (RRL) **1**, 122–124 (2007)
26. Stangl, J., Mocuta, C., Diaz, A., et al.: X-ray diffraction as a local probe tool. Chem. Phys. Chem. **10**, 2923 (2009)
27. Snigirev, A., Snigireva, I.: High energy X-ray micro-optics. C. R. Phys. **9**, 507 (2008)
28. Mocuta, C., Stangl, J., Mundboth, K., et al.: Beyond the ensemble average: x-ray microdiffraction analysis of single SiGe island. Phys. Rev. B **77**, 245425 (2008)
29. Mocuta, C., Barbier, A., Ramos, A.V., et al.: Effect of optical lithography patterning on the crystalline structure of tunnel junctions. Appl. Phys. Lett. **91**(24), 241917 (2007)
30. Rodrigues, M.S., Cornelius, T.W., Scheler, T., et al.: In situ observation of the elastic deformation of a single epitaxial SiGe crystal by combining atomic force microscopy and micro x-ray diffraction. J. Appl. Phys. **106**, 103525 (2009)
31. Scheler, T., Rodrigues, M., Cornelius, T., Mocuta, C., Malachias, A., Magalhaes-Paniago, R., Comin, F., Chevrier, J., Metzger, T.H.: Probing the elastic properties of individual nanostructures by combining in situ atomic force microscopy and micro-x-ray diffraction. Appl. Phys. Lett. **94**, 023109 (2009)
32. Fink, R., Weiss, M.R., Umbach, E., Preikszas, D., Rose, H., Spehr, R., Hartel, P., Engel, W., Degenhardt, R., Wichtendahl, R., Kuhlenbeck, H., Erlebach, W., Ihmann, K., Schlögl, R., Freund, H.-J., Bradshaw, A. M., Lilienkamp, G., Schmidt, Th., Bauer, E., Benner, G., et al.: SMART: a planned ultrahigh-resolution spectromicroscope for BESSY II, J. Electron Spectros. Relat. Phenom. **84**, 231–250 (1997)
33. Feng, J., Forest, E., MacDowell, A.A., Marcus, M., Padmore, H., Raoux, S., Robin, D., Scholl, A., Schlueter, R., Schmid, P., Stöhr, J.,Wan, W., Wei, D.H., Wu, Y.: An X-ray photoemission electron microscope using an electron mirror aberration corrector for the study of complex materials, J. Phys.: Condens. Matter **17**, S1339 (2005) doi:10.1088/0953-8984/17/16/005
34. Stöhr, J., Wu, Y., Hermsmeier, B.D., Samant, M.G., Harp, G.R., Koranda, S., Dunham, D., Tonner, B.P.: Element-specific magnetic microscopy with circularly polarized X-rays, Science **259**, 658 (1993)
35. Stöhr, J.: Exploring the microscopic origin of magnetic anisotropies with X-ray magnetic circular dichroism (XMCD) spectroscopy, Journal of Magnetism and Magnetic Materials **200**, 470–497 (1999)
36. Ohldag, H., Scholl, A., Nolting, F., Anders, S., Hillebrecht, F.U., Stöhr, J.: Spin Reorientation at the Antiferromagnetic NiO(001) Surface in Response to an Adjacent Ferromagnet, Phys. Rev. Lett. **86**, 2878–2881 (2001)
37. Bauer, E., Belkhou, R., Cherifi, S., Locatelli, A., Pavlovska, A., Rougemaille, N.: Magnetostructure of MnAs on GaAs revisited, J. Vac. Sci. Technol. B **25**, 1470 (2007)
38. Hehn, M., Lacour, D., Montaigne, F., Briones, J., Belkhou, R., El Moussaoui, S., Maccherozzi, F., Rougemaille, N.: 360° domain wall generation in the soft layer of magnetic tunnel junctions, Appl. Phys. Lett. **92**, 072501 (2008)
39. Budai, J.D., Yang, W., Tamura, N., et al.: X-ray microdiffraction study of growth modes and crystallographic tilts in oxide films on metal substrates. Nature Mater. **2**, 487 (2003)
40. Vartanyants, I.A., Robinson, I.K., McNulty, I., et al.: Coherent scattering and lensless imaging at the European XFEL facility. J. Synchr. Rad. **14**, 453 (2007)
41. Choe, S.-B., Acremann, Y., Scholl, A., Bauer, A., Doran, A., Stöhr, J., Padmore, H.A.: Vortex Core-Driven Magnetization Dynamics, Science **304**, 420 (2004)

Self-Assembled Monolayers

► BioPatterning
► Nanostructures for Surface Functionalization and Surface Properties

Self-Assembled Monolayers for Nanotribology

Bharat Bhushan
Nanoprobe Laboratory for Bio- & Nanotechnology and Biomimetics, The Ohio State University, Columbus, OH, USA

Synonyms

Molecularly thick layers; Monolayer lubrication; Organic films; Self-organized layers

Definition

Organized and dense molecular-scale layers of long-chain, organic molecules are referred to as self-assembled monolayers (SAMs). SAMs are molecularly thick, well-organized, and chemically bonded to the substrate. Ordered molecular assemblies with various properties can be engineered using chemical grafting of various polymer molecules with suitable functional head groups, spacer chains, and surface terminal groups.

Overview

Reliability of various micro- and nanodevices, also commonly referred to as micro/nanoelectromechanical systems (MEMS/NEMS) and BioMEMS/BioNEMS, requiring relative motion, as well as magnetic storage devices (which include magnetic rigid disk and tape drives) requires the use of hydrophobic and lubricating films to minimize adhesion, stiction, friction, and wear [2–4, 7, 9–12, 15, 36, 39]. In various applications, surfaces need to be protected from exposure to the operating environment. For example, in various biomedical applications, such as biosensors and implantable biomedical devices, undesirable protein adsorption, biofouling, and biocompatibility are some of the major issues [9–11]. In micro- and nanofluidic based sensors, the fluid drag in micro- and nanochannels can be reduced by using hydrophobic coatings [42]. Selected hydrophobic films are needed for these applications.

For lubrication, an effective approach involves the deposition of organized and dense molecular layers of long-chain molecules. Two common methods to produce monolayers and thin films are the Langmuir-Blodgett (L-B) deposition and self-assembled monolayers (SAMs) by chemical grafting of molecules. LB films are physically bonded to the substrate by weak van der Waals attraction, while SAMs are chemically bonded via covalent bonds to the substrate. Because of the choice of chain length and terminal linking group that SAMs offer, they hold great promise for boundary lubrication of MEMS/NEMS. A number of studies have been conducted to study tribological properties of various SAMs deposited on Si, Al, and Cu substrates [14, 16–18, 21, 24–31, 33–35, 38, 40].

Bhushan and Liu [14], Liu et al. [35], and Liu and Bhushan [8] studied adhesion, friction and wear properties of alkylthiol and biphenylthiol SAMs on Au (111) films. They explained the friction mechanisms using a molecular spring model in which local stiffness and intermolecular forces govern the friction properties. They studied the influence of relative humidity, temperature, and velocity on adhesion and friction. They also investigated the wear mechanisms of SAMs by a continuous microscratch AFM technique.

Fluorinated carbon (fluorocarbon) molecules are known to have low surface energy and are commonly used for lubrication [5, 7, 8, 12]. Bhushan and Cichomski [13] deposited fluorosilane SAMs on polydimethylsiloxane (PDMS). To make a hydrophobic PDMS surface chemically active, PDMS surface was oxygenated using an oxygen plasma, which introduces silanol groups (SiOH). They reported that SAM coated PDMS was more hydrophobic with lower adhesion, friction, and wear. Bhushan et al. [17, 19], Kasai et al. [30], Lee et al. [31], Tambe and Bhushan [38], and Tao and Bhushan [41] studied the adhesion, friction, and wear of methyl- and/ or perfluoro- terminated alkylsilanes on silicon. They reported that perfluoroalkylsilane SAMs exhibited lower surface energy, higher contact angle, lower adhesive force, and lower wear as compared to that of alkylsilanes. Kasai et al. [30] also reported the influence of relative humidity, temperature, and velocity on adhesion and friction. Tao and Bhushan [40] studied degradation mechanisms of alkylsilanes and perfluoroalkylsilane SAMs on Si. They reported that oxygen in the air causes thermal oxidation of SAMs.

Tambe and Bhushan [38], Bhushan et al. [18], Hoque et al. [24, 25] and DeRose et al. [21] studied the nanotribological properties of methyl- and perfluoro-terminated alkylphosphonate, perfluorodecyldimethylchlorosilane, and perfluorodecanoic acid on aluminum, of industrial interest. Hoque et al. [26] and DeRose et al. [21] studied the nanotribological properties of alkylsilanes and perfluroalkylsilanes on aluminum. Hoque et al. [27–29] studied the nanotribological properties of alkylphosphonate and perfluoroalkylsilane SAMs on copper. The authors found that these SAMs on aluminum and copper perform well irrespective of the substrate used. They confirmed the presence of respective films using X-ray photoelectron spectroscopy (XPS).

Hoque et al. [27–29] studied the chemical stability of various SAMs deposited on Cu substrates via exposure to various corrosive conditions. DeRose et al. [21] studied the chemical stability of various SAMs deposited on Al substrates via exposure to corrosive conditions (aqueous nitric acid solutions of a low pH of 1.8 at temperatures ranging from 60°C to 80°C for times ranging from 30 to 70 min). The exposed samples were characterized by XPS and contact angle measurements. They reported that perfluorodecanoic acid/Al is less stable than perfluorodecylphosphonate/Al and octadecylphosphonate/Al, but more stable than perfluorodecyldimethylchlorosilane/Al, which has implications in digital micromirror devices (DMD) applications. In general, chemical stability data of various SAMs deposited on Cu and Al surfaces to corrosive environments has been reported by these authors. Based on these studies, it was concluded that chemisorption occurs at the interface and is responsible for strong interfacial bonds.

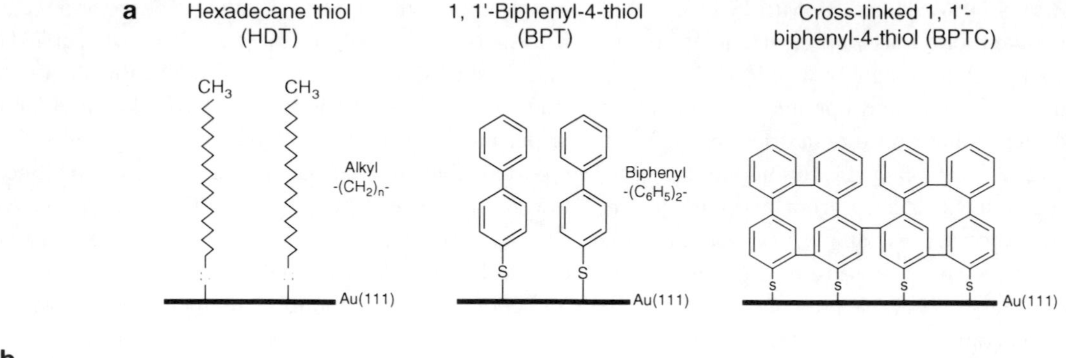

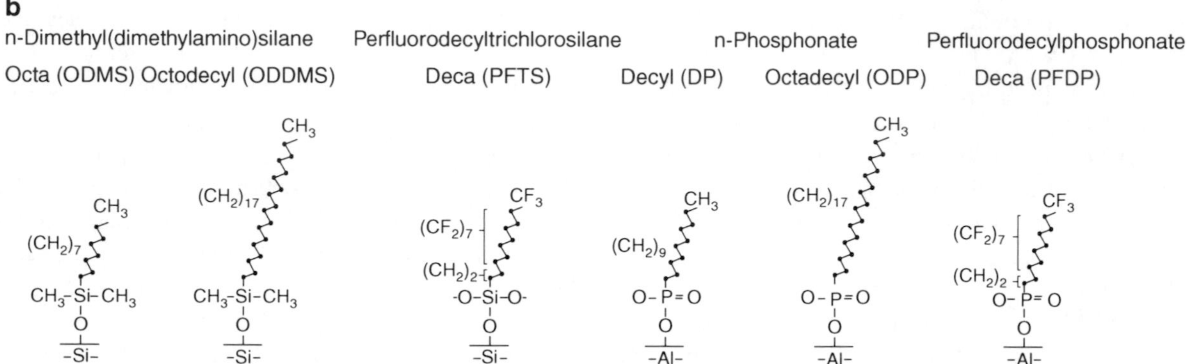

Self-Assembled Monolayers for Nanotribology, Fig. 1 Schematics of the structures of (**a**) hexadecane and biphenyl thiol SAMs on Au(111) substrates, and (**b**) perfluoroalkylsilane and alkylsilane SAMs on Si with native oxide substrates, and perfluoroalkylphosphonate and alkylphosphonate SAMs on Al with native oxide

To date, the contact angle and nanotribological properties of alkanethiol, biphenylthiol, alkylsilane, perfluoroalkylsilane, alkylphosphonate and perfluoroalkylphosphane SAMs have been widely studied. In the following, the nanotribological properties of various SAMs are reviewed having alkyl and biphenyl spacer chains with different surface terminal groups ($-CH_3$, $-CF_3$) and head groups (–S–H, –Si–O–, –OH, and P–O–) which have been investigated by AFM at various operating conditions [14, 17, 18, 30, 33–35, 38, 40]. Hexadecane thiol (HDT), 1, 1′-biphenyl-4-thiol (BPT), and crosslinked BPT (BPTC) were deposited on Au(111) films on Si(111) substrates by immersing the substrate in a solution containing the precursor (ligand) that is reactive to the substrate surface (Fig. 1a). Crosslinked BPTC was produced by irradiation of BPT monolayers with low energy electrons. Perfluoroalkylsilane and alkylsilane SAMs were deposited on Si(100) by exposing the substrate to the vapor of the reactive chemical precursors (Fig. 1b). Perfluoroalkylphosphonate and alkylphosphonate SAMs were deposited on sputtered Al film on Si substrate as well as bulk Al substrates (Fig. 1b).

Hexadecane Thiol and Biphenyl Thiol SAMs on Au (111)

Bhushan and Liu [14] studied the effect of film compliance on adhesion and friction. They used hexadecane thiol (HDT), 1,1,biphenyl-4-thiol (BPT), and crosslinked BPT (BPTC) solvent deposited on Au (111) substrate, Fig. 1a. The average values and standard deviation of the adhesive force and coefficient of friction are presented in Fig. 2. Based on the data, the adhesive force and coefficient of friction of SAMs are less than the corresponding substrates. Among various films, HDT exhibits the lowest values. Based on stiffness measurements of various SAMs, HDT was most compliant, followed by BPT and BPTC. Based on friction and stiffness measurements, SAMs with high-compliance long carbon chains exhibit low friction; chain compliance is desirable for low friction. Friction mechanism of SAMs is explained by a so-called

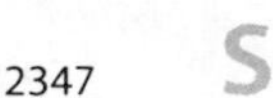

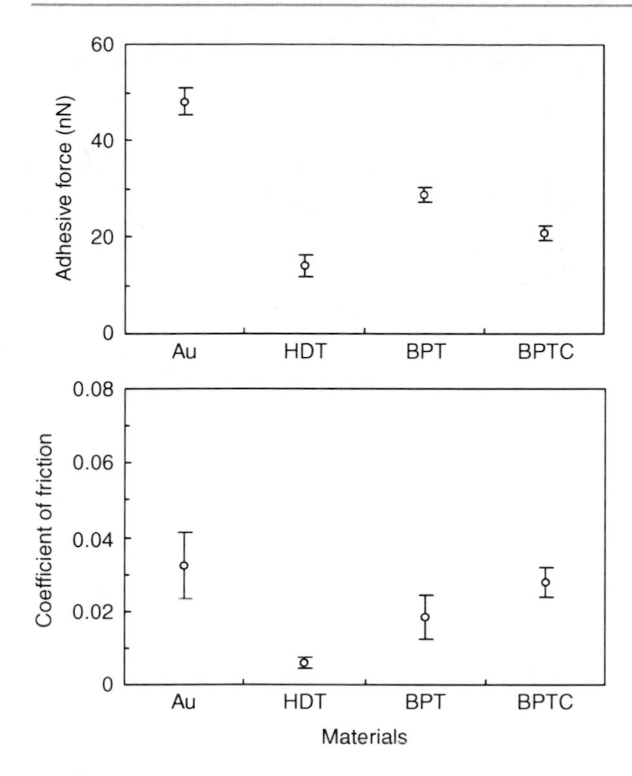

Self-Assembled Monolayers for Nanotribology, Fig. 2 Adhesive forces and coefficients of friction of Au(111) and various SAMs

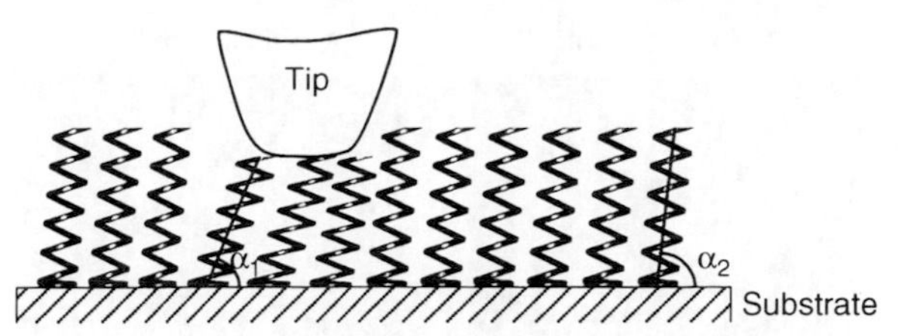

Self-Assembled Monolayers for Nanotribology, Fig. 3 Molecular spring model of SAMs. In this figure, $\alpha_1 < \alpha_2$, which is caused by the orientation under the normal load applied by AFM tip. The orientation of the molecular springs reduces the shearing force at the interface, which in turn reduces the friction force. The molecular spring constant, as well as the inter-molecular forces can determine the magnitude of the coefficients of friction of SAMs. In this figure, the size of the tip and molecular springs do not in the exactly scale [14]

"molecular spring" model (Fig. 3). According to this model, the chemically adsorbed self-assembled molecules on a substrate are just like assembled molecular springs anchored to the substrate. An asperity sliding on the surface of SAMs is like a tip sliding on the top of "molecular springs or brush." The molecular spring assembly has compliant features and can experience orientation and compression under load. The orientation of the molecular springs or brush under normal load reduces the shearing force at the interface, which in turn reduces the friction force. The orientation is determined by the spring constant of a single molecule as well as the interaction between the neighboring molecules, which can be reflected by packing density or packing energy. It should be noted that the orientation can lead to conformational defects along the molecular chains, which lead to energy dissipation.

An elegant way to demonstrate the influence of molecular stiffness on friction is to investigate SAMs with different structures on the same wafer. For this purpose, a micropatterned SAM was prepared. First the biphenyldimethylchlorosilane (BDCS) was deposited on silicon by a typical self-assembly method [33]. Then the film was partially crosslinked using mask technique by low energy electron irradiation. Finally the micropatterned BDCS films were realized, which had the as-deposited and crosslinked coating regions on the same wafer. The local stiffness properties of this micropatterned sample were investigated by force modulation AFM technique [22]. The variation in the deflection amplitude provides a measure of the relative local stiffness of the surface. Surface height, stiffness, and friction images of the micropatterned biphenyldimethylchlorosilane (BDCS) specimen are obtained and presented in Fig. 4 [33]. The circular areas correspond to the as-deposited film, and the remaining area to the crosslinked film. Figure 4a indicates that crosslinking caused by the low energy electron irradiation leads to about 0.5 nm decrease of the surface height of BDCS films. The corresponding stiffness images indicate that the crosslinked area has higher stiffness than the as-deposited area. Figure 4b indicates that the as-deposited area (higher surface height area) has a lower friction force. Obviously, these data of the micropatterned sample prove that the local stiffness of SAMs has an influence to their friction performance. Higher stiffness leads to larger friction force. These results provide a strong proof of the suggested molecular spring model.

The SAMs with high-compliance long carbon chains also exhibit the best wear resistance [14, 33]. In wear experiments, the wear depth as a function of normal load curves show a critical normal load, at which film wears rapidly. A representative curve is shown in Fig. 5. Below the critical normal load, SAMs undergo orientation; at the critical load SAMs wear away from the substrate due to relatively weak

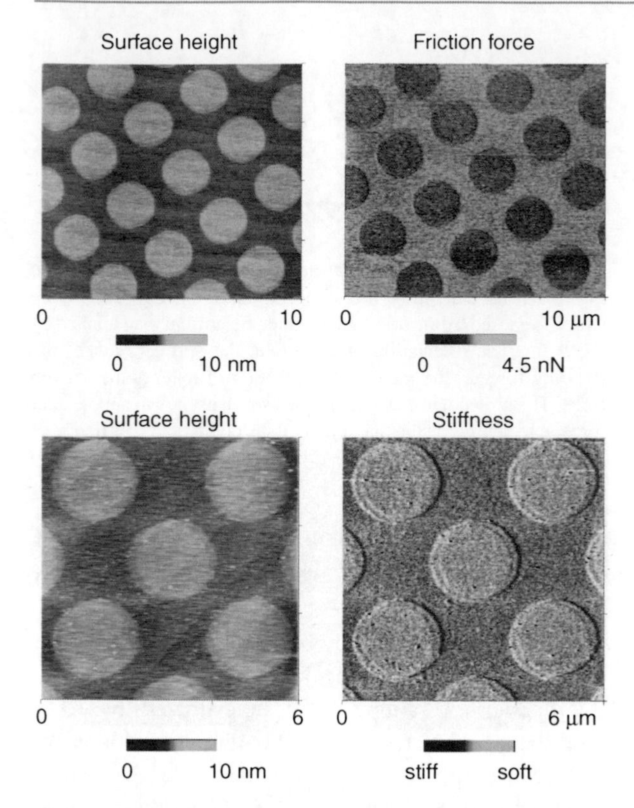

Self-Assembled Monolayers for Nanotribology, Fig. 4 (**a**) AFM Grayscale surface height and stiffness images, and (**b**) AFM grayscale surface height and friction force images of micropatterned BDCS [33]

interface bond strengths, while above the critical normal load severe wear takes place on the substrate.

Perfluoroalkylsilane and Alkylsilane SAMs on Si (100), and Perfluoroalkylphosphonate and Alkylphosphonate SAMS on Al

Perfluorodecyltricholorosilane (PFTS), CF_3–$(CF_2)_7$–$(CH_2)_2$–$SiCl_3$, n-octyldimethyl (dimethylamino)silane (ODMS), CH_3–$(CH_2)_n$–$Si(CH_3)_2$–$N(CH_3)_2$ (n = 7), and n-octadecyldimethyl(dimethylamino)silane (n = 17) (ODDMS) vapor deposited on Si(100) substrate and perfluorodecylphosphonate (PFDP)

$$CF_3-(CF_2)_7-(CH_2)_2-\overset{O}{\underset{O}{\overset{|}{\underset{||}{P}}}}-OH,$$ decylphosphonate

(DP), $$CH_3-(CH_2)_n-\overset{O}{\underset{O}{\overset{|}{\underset{||}{P}}}}-OH \quad (n=9)$$ and

octadecylphosphonate (ODP) (n = 17) by liquid deposition on sputtered Al film on Si substrate were selected

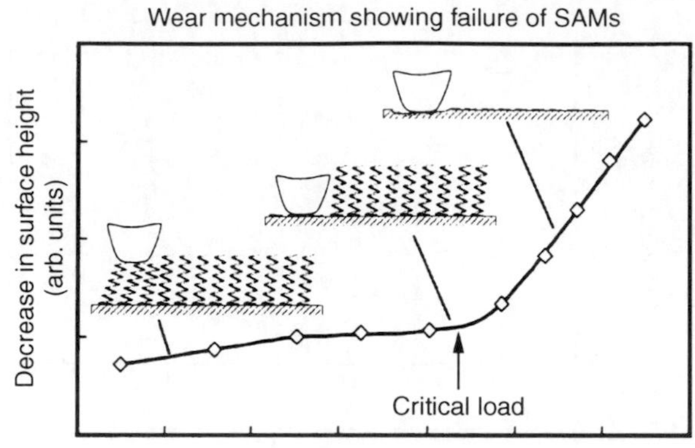

Self-Assembled Monolayers for Nanotribology, Fig. 5 Illustration of the wear mechanisms of SAMs with increasing normal load [33]

(Fig. 1b). Perfluoro SAMs were selected because fluorinated films are known to have low surface energy. Two chain lengths of alkylsilanes (with 8 and 18 carbon atoms) were selected to compare their nanotribological performance with that of the former as well as to study the effect of chain length. Al substrate was selected because of the application of Al micromirrors in digital projection displays. Perfluoroalkylphosphane (with ten carbon atoms) and alkylphosphonate SAMs (with 10 and 18 carbon atoms) on Al were selected.

The measured values are compared among the samples in Fig. 6a [17, 18]. Significant improvement in the water repellent property was observed for perfluorinated SAMs as compared to bare Si and Al substrates. Static contact angles of alkylsilanes and alkylphosphonates were also higher than corresponding substrates, but lower than corresponding perfluorinated films. The contact angle generally increases with a decrease in surface energy [23], which is consistent with the data obtained. The contact angles can be influenced by the packing density as well as the sample roughness [37]. The higher contact angles for the SAMs deposited on Al substrates than those on Si substrate are probably due to this effect. The –CH_3 groups in ODMS, ODDMS, DP, and ODP are non-polar and are known to contribute to the water repellent property. Perfluorinated SAMs exhibited the highest contact angle among the SAMs tested in this study.

Figure 6a shows the adhesive force, friction force, and the coefficient of friction measured under ambient conditions using an AFM, and Fig. 6b shows the

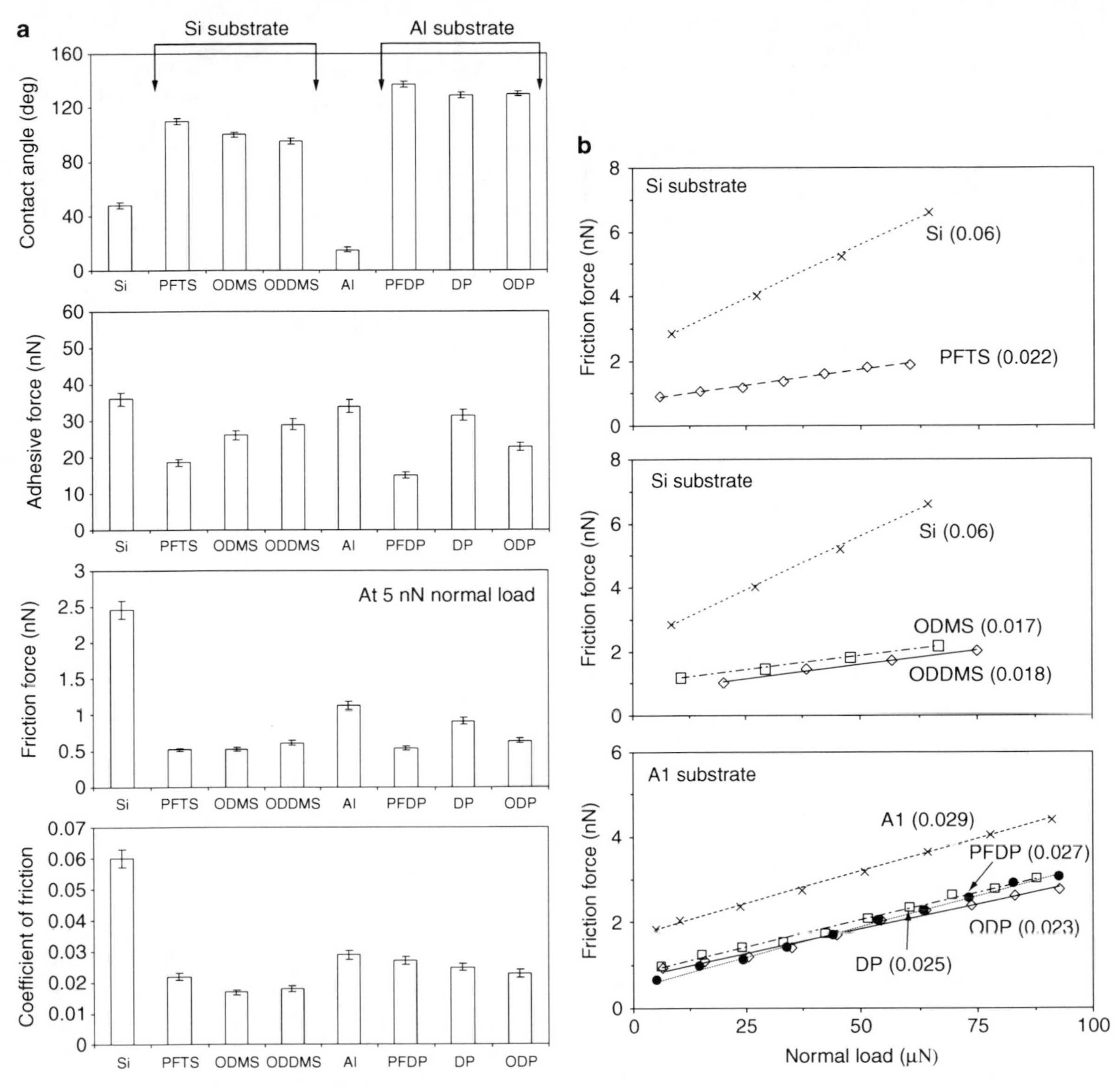

Self-Assembled Monolayers for Nanotribology, Fig. 6 (**a**) The static contact angle, adhesive force, friction force and coefficient of friction measured using an AFM for various SAMs on Si and Al substrates, and (**b**) friction force vs. normal load plots for various SAMs on Si and Al substrates [17, 18]

friction force vs. normal load plots for various SAMs deposited onto the Si and Al substrates [17, 18]. The bare substrates showed higher adhesive force than the SAMs coatings. ODMS and ODDMS shows an adhesive force comparable to DP and ODP despite their lower water contact angles. These SAMs have the same tail groups, and during AFM measurements the AFM tip interacts only with the tail groups, whereas the contact angles can also be influenced by the head groups in these SAMs. This is probably the reason as to why the adhesive forces for these SAMs are comparable. PFTS and PFDP, which have the highest contact angles, showed the lowest adhesion.

The effect of relative humidity for various SAMs on adhesion and friction was studied. Adhesive force, friction force at 5 nN of normal load, coefficient of friction, and microwear data are presented in Fig. 7 [18, 30]. The result of adhesive force for silicon showed an increase with relative humidity, Fig. 7. This is expected since the surface of silicon is

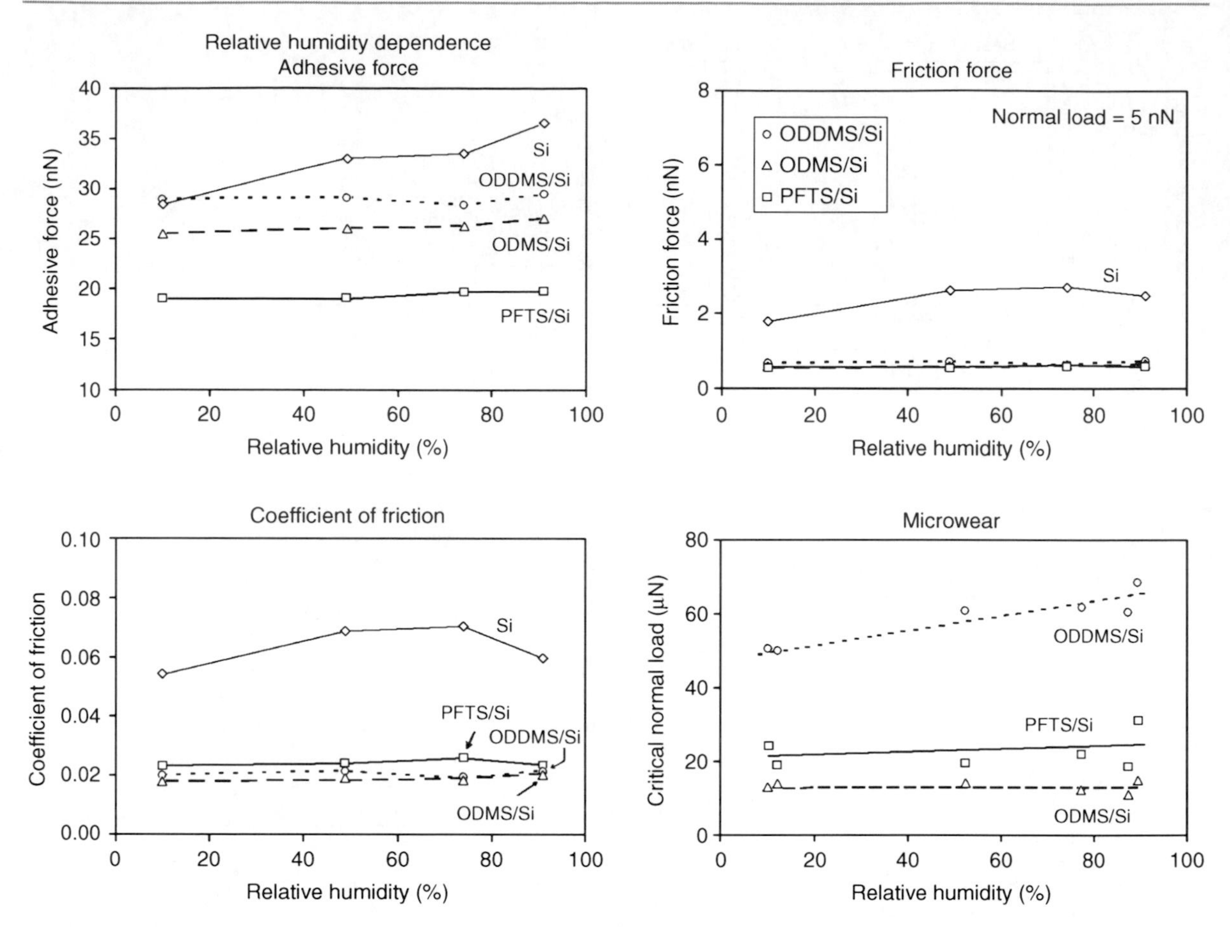

Self-Assembled Monolayers for Nanotribology, Fig. 7 Relative humidity effect on adhesive force, friction force, coefficient of friction and microwear for various SAMs on Si substrates [30]

hydrophilic, as shown in Fig. 6a. More condensation of water at the tip-sample interface at higher humidity increases the adhesive force due to capillary effect. On the other hand, the adhesive force for the SAMs showed a very weak dependency on the change in humidity. This occurs since the surface of the SAMs is hydrophobic. The adhesive force of ODMS/Si and ODDMS/Si showed a slight increase from 75% to 90% RH. Such an increase was absent for PFTS/Si, possibly because of the hydrophobicity of PFTS/Si.

The friction force of silicon showed an increase with relative humidity up to about 75% RH and a slight decrease beyond this point, see Fig. 7. The initial increase can result from the increase in adhesive force. The decrease in friction force at higher humidity could be attributed to the lubricating effect of the water layer. This effect is more pronounced in the coefficient of friction. Since the adhesive force increased and coefficient of friction decreased in this range, those effects cancel each other out and the resulting friction force showed slight changes. On the other hand, the friction force and coefficient of friction of SAMs showed very small changes with relative humidity like that found for adhesive force. This suggests that the adsorbed water layer on the surface maintained a similar thickness throughout the relative humidity range tested. The differences among the SAM types were small within the measurement error, however a closer look at the coefficient of friction for ODMS/Si showed a slight increase from 75% to 90% RH as compared to PFTS/Si, possibly due to the same reason for the adhesive force increment. The inherent hydrophobicity of SAMs means that they did not show much relative humidity dependence.

Figure 8 shows the effect of temperature on adhesive force, friction force at 5 nN of normal load, and coefficient of friction for various SAMs on Si substrate [30]. The adhesive force showed an increase with the

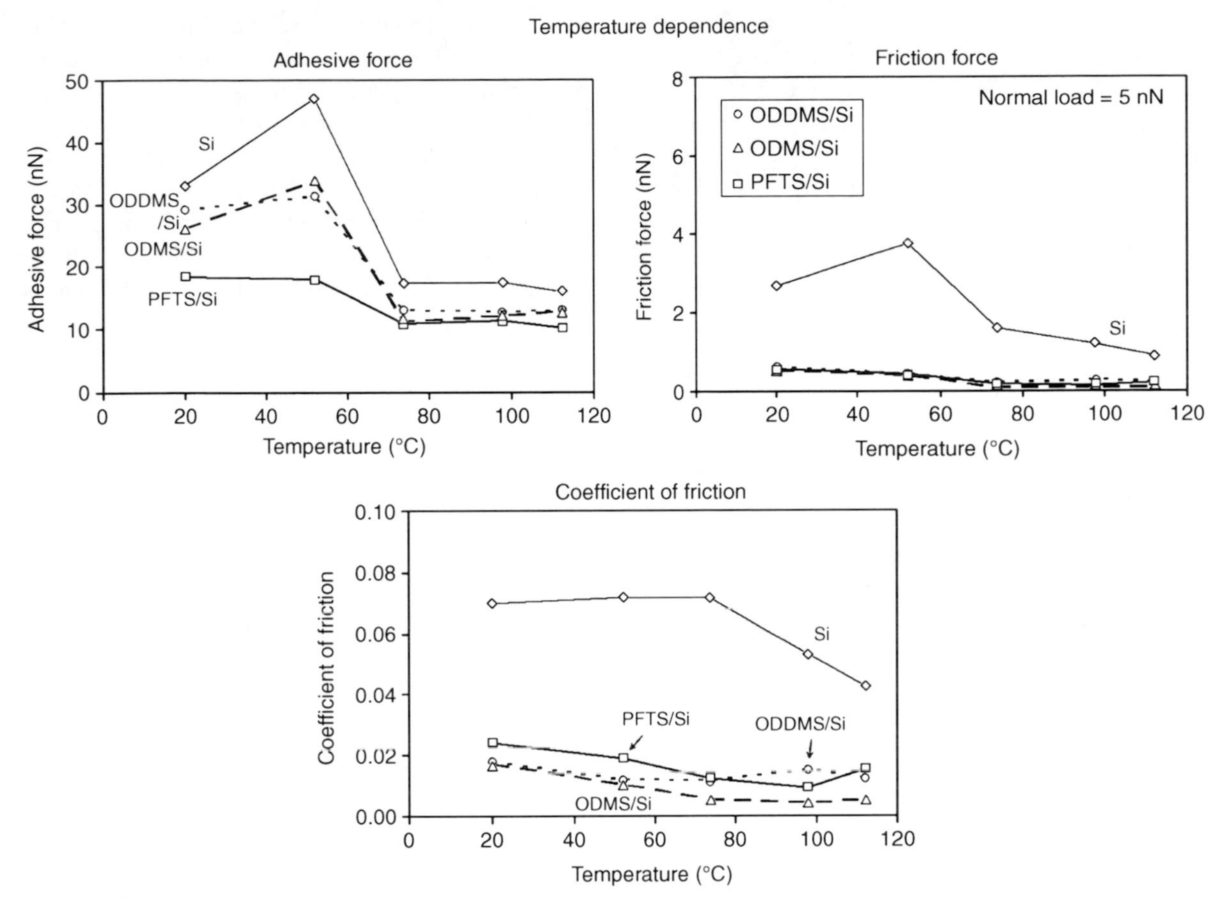

Self-Assembled Monolayers for Nanotribology, Fig. 8 Temperature effect on adhesive force, friction force, and coefficient of friction for various SAMs on Si substrates [30]

temperature, from room temperature (RT) to about 55°C, followed by a decrease from 55°C to 75°C, and eventually leveled off from 75°C to 100°C. The initial increase of adhesive force at lower temperatures is not well understood. The observed decrease could be attributed to the desorption of water molecules on the surface. After almost full depletion of the water layer, the adhesive force remains constant. The SAMs with hydrocarbon backbones showed similar behavior as that of the Si substrates, but the initial increase in the adhesive force with temperature was smaller. The SAMs with fluorocarbon backbone chains showed almost no temperature dependence. For the SAMs with hydrocarbon backbone chains, the initial increase in adhesive force is believed to be caused by the melting of the SAM film. The melting point for a linear carbon chain molecule such as $CH_3(CH_2)_{14}CH_2OH$ is 50°C [32]. With an increase in temperature, the SAM film softens, thereby increasing the real area of contact and consequently the adhesive force. Once the temperature is higher than the melting point, the lubrication regime is changed from boundary lubrication in a solid SAM to liquid lubrication in the melted SAM [33].

The friction force for silicon showed an increase with temperature followed by a steady decrease. The friction force is highly affected by the change in adhesion. The decrease in friction can result from the depletion of the water layer. The coefficient of friction for silicon remained constant followed by a decrease starting at about 80°C. For SAMs, the coefficient of friction exhibited a monotonic decrease with temperature. The decrease in friction and coefficient of friction for SAMs possibly results from the decrease in stiffness. As introduced before, the spring model suggests a smaller friction for more compliant SAMs [33]. The difference among the SAM types was not significant.

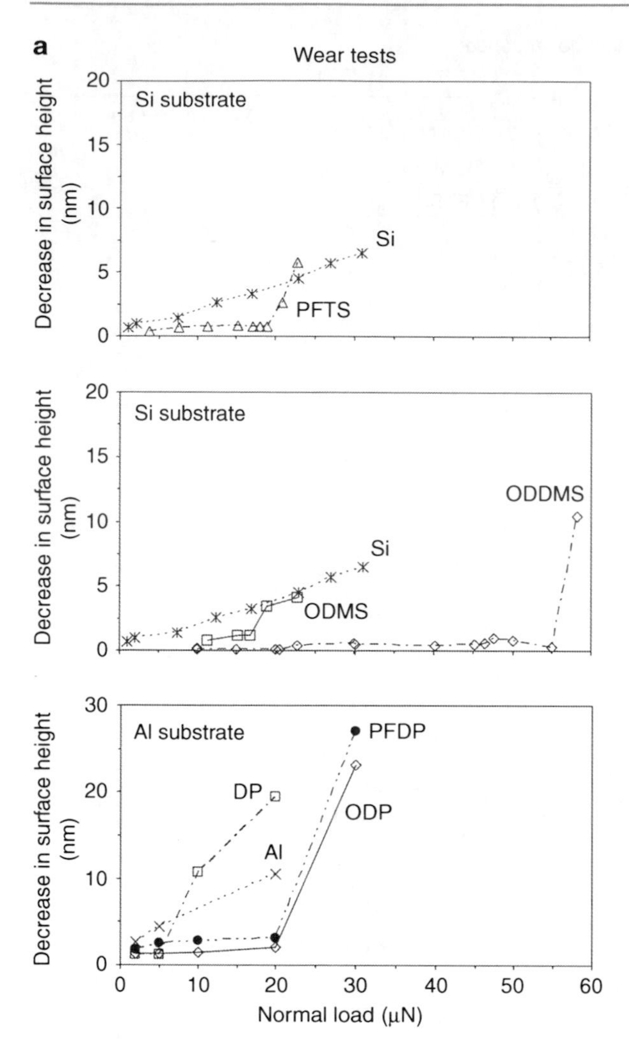

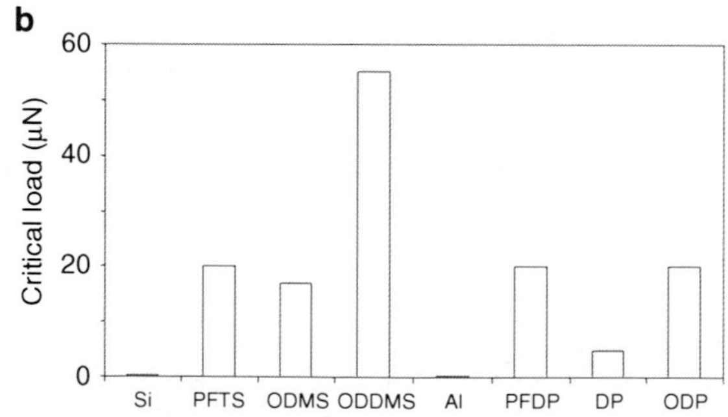

Self-Assembled Monolayers for Nanotribology, Fig. 9 (**a**) Decrease in surface height as a function of normal load after one scan cycle for various SAMs on Si and Al substrates, and (**b**) comparison of critical loads for failure during wear tests for various SAMs [18, 30]

PFTS could maintain its stiffness more than ODMS and ODDMS when temperature is increased [20], however, it was not pronounced in the results.

Figure 9a shows the relationship between the decrease in surface height as a function of the normal load during wear tests [18, 30]. As shown in the figure, the SAMs exhibit a critical normal load, beyond the point of which the surface height drastically decreases. Figure 9a also shows the wear behavior of the Al and Si substrates. Unlike the SAMs, the substrates show a monotonic decrease in surface height with the increasing normal load with wear initiating from the very beginning, i.e., even for low normal loads. Si (Young's modulus of elasticity, E = 130 GPa [1], hardness, H = 11 GPa [6]) is relatively hard in comparison to Al (E = 77 GPa, H = 0.41 GPa), and hence the decrease in surface height for Al is much larger than that for Si for similar normal loads.

The critical loads corresponding to the sudden failure of SAMs are shown in Fig. 9b. Amongst all the SAMs, ODDMS shows the best performance in the wear tests, and this is believed to be because of the longer chain length effect. Fluorinated SAMs – PFTS and PFDP show a higher critical load as compared to ODMS and DP with similar chain lengths. ODP shows a higher critical load as compared to DP because of its longer chain length. The mechanism of failure of compliant SAMs during wear tests has been presented earlier in Fig. 5. It is believed that the SAMs fail mostly due to shearing of the molecule at the head group, that is, by means of shearing of the molecules off the substrate.

Closure

The contact angle, adhesion, friction, and wear properties of SAMs having alkyl, biphenyl, and perfluoroalkyl spacer chains with different surface terminal groups (–CH_3 and –CF_3) and head groups (–SH, –Si–O–, –OH, and P–O–) studied are presented in this article. It is found that the adhesive force varies linearly with the work of adhesion value of SAMs, which indicates that capillary condensation of water plays an important role to the adhesion of SAMs on the nanoscale at ambient conditions. SAMs with high-compliance long carbon spacer chains exhibit the lowest adhesive force and friction force. The friction data are explained using a molecular spring model, in which the local stiffness and intermolecular force govern frictional performance. The results of the stiffness and friction characterization of the micropatterned sample with different structures support this model. Perfluoroalkylsilane and perfluoroalkylphosphonate SAMs exhibit lower surface energy, higher contact angle, and lower adhesive force as compared to that of alkylsilane and alkylphosphonate SAMs, respectively. The substrate had little effect. The coefficient of friction of various SAMs were comparable.

The influence of relative humidity on adhesion and friction of SAMs is dominated by the thickness of the adsorbed water layer. At higher humidity, water increases friction through increased adhesion by meniscus effect in the contact zone. With an increase in temperature, the desorption of the adsorbed water layer and reduction of the surface tension of water reduces the adhesive force and friction force. A decrease in adhesion and friction with temperature was found for all films.

PFTS/Si showed a better wear resistance than ODMS/Si. ODDMS/Si showed a better wear resistance than ODMS/Si due to the chain length effect. Wear behavior of the SAMs is mostly determined by the molecule-substrate bond strengths. Similar trends were observed for films on Al substrates.

Cross-References

▶ Nanotechnology
▶ Nanotribology
▶ Reliability of Nanostructures

References

1. Anonymous: Properties of silicon, EMIS data reviews series No. 4, INSPEC, Institution of Electrical Engineers, London (1988) (see also Anonymous, MEMS materials database, http://www.memsnet.org/material/ (2002))
2. Bhushan, B: Tribology and mechanics of magnetic storage devices, 2nd edn. Springer, New York (1996)
3. Bhushan, B (ed.): Tribology issues and opportunities in MEMS. Kluwer, Dordrecht (1998)
4. Bhushan, B (ed.): Handbook of micro/nanotribology, 2nd edn. CRC Press, Boca Raton (1999)
5. Bhushan, B: Principles and applications of tribology. Wiley, New York (1999)
6. Bhushan, B: Chemical, mechanical and tribological characterization of ultra-thin and hard amorphous carbon coatings as thin as 3.5 nm: recent developments. Diam. Relat. Mater. **8**, 1985–2015 (1999)
7. Bhushan, B (ed.): Modern tribology handbook. Principles of tribology, vol. 1. CRC Press, Boca Raton (2001). Vol. 2 – Materials, coatings, and industrial applications
8. Bhushan, B: Introduction to tribology. Wiley, New York (2002)
9. Bhushan, B: Nanotribology and nanomechanics of MEMS/NEMS and BioMEMS/BioNEMS materials and devices. Microelectron. Eng. **84**, 387–412 (2007)
10. Bhushan, B: Nanotribology and nanomechanics in nano/biotechnology. Philos. Tr. R. Soc. A **366**, 1499–1537 (2008)
11. Bhushan, B: Springer handbook of nanotechnology, 3rd edn. Springer, Heidelberg (2010)
12. Bhushan, B: Nanotribology and nanomechanics I – measurement techniques and nanomechanics, II – Nanotribology, biomimetics, and industrial applications, 3rd edn. Springer, Heidelberg (2011)
13. Bhushan, B, Cichomski, M: Nanotribological characterization of vapor phase deposited fluorosilane self-assembled monolayers deposited on polydimethylsiloxane surfaces for biomedical micro-/nanodevices. J. Vac. Sci. Technol. A **25**, 1285–1293 (2007)
14. Bhushan, B, Liu, H: Nanotribological properties and mechanisms of alkylthiol and biphenyl thiol self-assembled monolayers studied by atomic force microscopy. Phys. Rev. B **63**, 245412-1–245412-11 (2001)
15. Bhushan, B, Israelachvili, J N, Landman, U: Nanotribology: friction, wear and lubrication at the atomic scale. Nature **374**, 607–616 (1995)
16. Bhushan, B, Kulkarni, A V, Koinkar, V N, Boehm, M, Odoni, L, Martelet, C, Belin, M: Microtribological characterization of self-assembled and Langmuir-Blodgett monolayers by atomic and friction force microscopy. Langmuir **11**, 3189–3198 (1995)
17. Bhushan, B, Kasai, T, Kulik, G, Barbieri, L, Hoffmann, P: AFM study of perfluorosilane and alkylsilane self-assembled monolayers for anti-stiction in MEMS/NEMS. Ultramicroscopy **105**, 176–188 (2005)
18. Bhushan, B, Cichomski, M, Hoque, E, DeRose, J A, Hoffmann, P, Mathieu, H J: Nanotribological characterization of perfluoroalkylphosphonate self-assembled monolayers deposited on aluminum-coated silicon substrates. Microsyst. Technol. **12**, 588–596 (2006)
19. Bhushan, B, Hansford, D, Lee, K K: Surface modification of silicon surfaces with vapor phase deposited ultrathin fluorosilane films for biomedical devices. J. Vac. Sci. Technol. A **24**, 1197–1202 (2006)
20. Callister, W D: Materials science and engineering, 4th edn. Wiley, New York (1997)
21. DeRose, J A, Hoque, E, Bhushan, B, Mathieu, H J: Characterization of perfluorodecanote self-assembled monolayers on aluminum and comparison of stability with phosphonate and siloxy self-assembled monolayers. Surf. Sci. **602**, 1360–1367 (2008)
22. DeVecchio, D, Bhushan, B: Localized surface elasticity measurements using an atomic force microscope. Rev. Sci. Instrum. **68**, 4498–4505 (1997)
23. Eustathopoulos, N, Nicholas, M, Drevet, B: Wettability at high temperature. Pergamon, Amsterdam (1999)
24. Hoque, E, DeRose, J A, Hoffmann, P, Mathieu, H J, Bhushan, B, Cichomski, M: Phosphonate self-assembled monolayers on aluminum surfaces. J. Chem. Phys. **124**, 174710 (2006)
25. Hoque, E, DeRose, J A, Kulik, G, Hoffmann, P, Mathieu, H J, Bhushan, B: Alkylphosphonate modified aluminum oxide surfaces. J. Phys. Chem. B **110**, 10855–10861 (2006)
26. Hoque, E, DeRose, J A, Hoffmann, P, Bhushan, B, Mathieu, H J: Alkylperfluorosilane self-assembled monolayers on aluminum: a comparison with alkylphosphonate

self-assembled monolayers. J. Phys. Chem. C **111**, 3956–3962 (2007)
27. Hoque, E, DeRose, J A, Hoffmann, P, Bhushan, B, Mathieu, H J: Chemical stability of nonwetting, low adhesion self-assembled monolayer films formed by perfluoroalkylsilazation of copper. J. Chem. Phys. **126**, 114706 (2007)
28. Hoque, E, DeRose, J A, Bhushan, B, Mathieu, H J: Self-assembled monolayers on aluminum and copper oxide surfaces: surface and interface characteristics, nanotribological properties, and chemical stability. In: Bhushan, B, Fuchs, H, Tomitori, M (eds.) Applied scanning probe methods vol. IX – characterization, pp. 235–281. Springer, Heidelberg (2008)
29. Hoque, E, DeRose, J A, Bhushan, B, Hipps, K W: Low adhesion, non-wetting phosphonate self-assembled monolayer films formed on copper oxide surfaces. Ultramicroscopy **109**, 1015–1022 (2009)
30. Kasai, T, Bhushan, B, Kulik, G, Barbieri, L, Hoffmann, P: Nanotribological study of perfluorosilane SAMs for antistiction and low wear. J. Vac. Sci. Technol. B **23**, 995–1003 (2005)
31. Lee, K K, Bhushan, B, Hansford, D: Nanotribological characterization of perfluoropolymer thin films for BioMEMS applications. J. Vac. Sci. Technol A **23**, 804–810 (2005)
32. Lide, D R: CRC handbook of chemistry and physics, 85th edn. CRC Press, Boca Raton (2004)
33. Liu, H, Bhushan, B: Investigation of nanotribological properties of alkylthiol and biphenyl thiol self-assembled monolayers. Ultramicroscopy **91**, 185–202 (2002)
34. Liu, H, Bhushan, B: Orientation and relocation of biphenyl thiol self-assembled monolayers. Ultramicroscopy **91**, 177–183 (2002)
35. Liu, H, Bhushan, B, Eck, W, Stadler, V: Investigation of the adhesion, friction, and wear properties of biphenyl thiol self-assembled monolayers by atomic force microscopy. J. Vac. Sci. Technol. A **19**, 1234–1240 (2001)
36. Man, K F, Stark, B H, Ramesham, R: A resource handbook for MEMS reliability, rev. A. JPL Press, Jet Propulsion Laboratory, California Institute of Technology, Pasadena (1998). See also, Man, K. F.: MEMS reliability for space applications by elimination of potential failure modes through testing and analysis. http://www-rel.jpl.nasa.gov/Org/5053/atpo/products/Prod-map.html (2002)
37. Ren, S, Yang, S, Zhao, Y, Yu, T, Xiao, X: Preparation and characterization of ultrahydrophobic surface based on a stearic acid self-assembled monolayer over polyethyleneimine thin films. Surf. Sci. **546**, 64–74 (2003)
38. Tambe, N S, Bhushan, B: Nanotribological characterization of self assembled monolayers deposited on silicon and aluminum substrates. Nanotechnology **16**, 1549–1558 (2005)
39. Tanner, D M, Smith, N F, Irwin, L W, et al.: MEMS reliability: infrastructure, test structure, experiments, and failure modes. Sandia National Laboratories, Albuquerque (2000). SAND2000–0091
40. Tao, Z, Bhushan, B: Degradation mechanisms and environmental effects on perfluoropolyether self assembled monolayers and diamond like carbon films. Langmuir **21**, 2391–2399 (2005)
41. Tao, Z, Bhushan, B: Surface modification of AFM silicon probes for adhesion and wear reduction. Tribol. Lett. **21**, 1–16 (2006)
42. Wang, Y, Bhushan, B: Boundary slip and nanobubble study in micro/nanofluidics with atomic force microscope. Soft Matter **6**, 29–66 (2010)

Self-Assembled Nanostructures

▶ Self-assembly of Nanostructures

Self-Assembled Protein Layers

▶ Interfacial Investigation of Protein Films Using Acoustic Waves

Self-Assembly

▶ Self-Assembly for Heterogeneous Integration of Microsystems

Self-Assembly for Heterogeneous Integration of Microsystems

J. H. Hoo[1], K. S. Park[1], R. Baskaran[1,2] and K. F. Böhringer[1]
[1]Department of Electrical Engineering, University of Washington, Seattle, WA, USA
[2]Components Research, Intel Corporation, USA

Synonyms

Heterogeneous integration; Self-assembly; Stochastic assembly

Definition

Self-assembly is defined as the autonomous organization of components into ordered patterns or structures without human intervention. In this entry, the scope of self-assembly will be restricted within the microelectronics field – typically, the process of transporting, aligning, and permanently adhering discrete components onto various substrates, achieving heterogeneous integration.

Introduction

The packaging and integration of microscale modules, such as sensors, actuators and transceivers, and data processing, microfluidic, power management, electromechanical and optoelectronic devices into heterogeneous multifunctional systems is one of the big challenges in the field of microelectromechanical systems (MEMS). The fabrication processes and the material prerequisites for the components are predominantly incompatible, thereby eliminating the possibility of fabricating them all on a common substrate, as it would have been done in a monolithic approach. Consequently, in a heterogeneous approach, the components of a system are fabricated separately under respective optimal conditions before being assembled to construct a functional system (Fig. 1).

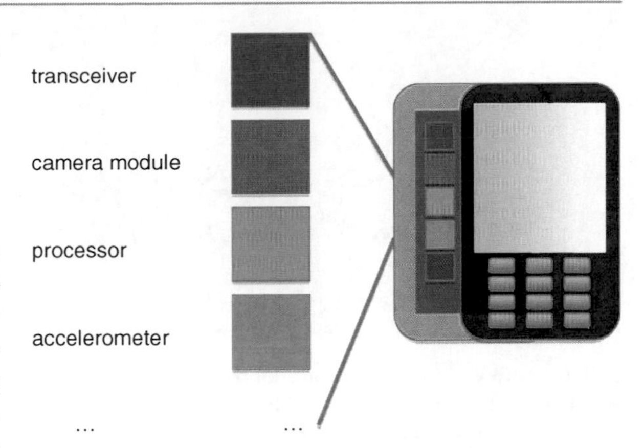

Self-Assembly for Heterogeneous Integration of Microsystems, Fig. 1 A modern cell phone: An example of a multifunctional device put together by discrete components from incompatible fabrication processes

The prevailing method used for assembly and packaging in microelectronic manufacturing is robotic pick-and-place. This technique has proven accurate and reliable for large component scales, and its throughput is satisfying for consumer electronics applications [1]. However, pick-and-place is confronted with a trade-off between throughput and component placement accuracy. Furthermore, the method is serial, it requires closed-loop control, and it becomes more expensive as device dimensions shrink, and registration constraints become more stringent. Furthermore, stiction problems [2] set in for device sizes smaller than 300 μm [1].

Self-assembly, defined as *the autonomous organization of components into ordered patterns or structures without human intervention* [3], provides a promising alternative to pick-and-place machinery. Self-assembly is parallel in nature and can be applied to a wide range of sizes, from the millimeter to nanometer scales. Self-assembly processes can be engineered to avoid direct contact manipulation of components, thereby eliminating the issues with stiction.

Self-assembly is certainly not a suite of techniques exclusive to the microelectronic field. Self-assembling processes (also satisfying the stated definition) from the natural environment and reported in other scientific disciplines range from the noncovalent association of organic molecules to the formation of crystals from precipitates to DNA origami [4]. The scope of this entry will be restricted to the application of self-assembly within the microelectronics field – typically, the process of transporting, aligning, and permanently adhering discrete components onto various substrates (Fig. 2). It is important to note the stochastic nature of the transportation phase; it is impossible, and unnecessary, for a user of self-assembly to determine the specific component that gets delivered to a specific binding location. Table 1 tabulates some notable microscale self-assembly techniques in the field.

Self-assembly techniques can be classified by the underlying mechanisms that drive the transportation and alignment assembly steps (Table 1). Having introduced self-assembly as a possible alternative for prevailing methods of microcomponent assembly and packaging, metrics necessary to the evaluation and comparison of self-assembly techniques are also listed, including the dimensions of the discrete components to be assembled (lateral size, thickness, and aspect ratio), how closely can components be assembled on the substrate (packing density), if there is control over the orientation of the placement of the components (orientation specificity), and the average rate of completeness of the assembly processes (yield).

The successful application of self-assembly requires not only a match between the dimensions of the discrete components reported in Table 1 and that of the components one wishes to assemble. It is also important to consider the assembly mechanisms, which are closely related to the environment in which the self-assembly takes place. For example, one should

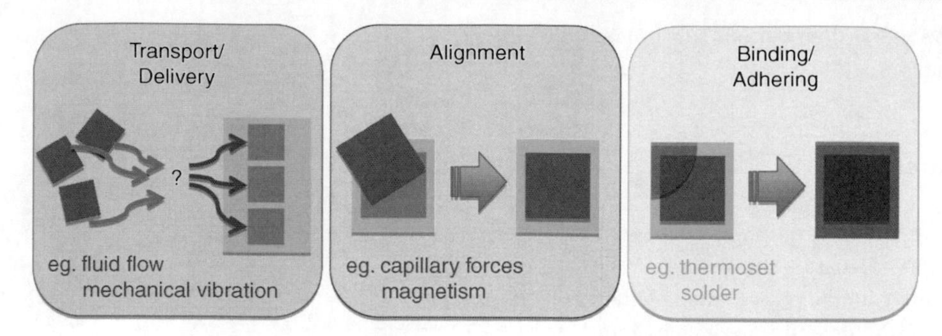

Self-Assembly for Heterogeneous Integration of Microsystems, Fig. 2 Self-assembly process can be separated into the transport, alignment, and permanent binding/adhering phases. During the transport phase, components are brought to the immediate vicinity of binding/assembly sites; the alignment phase corrects the orientation of the component; the binding phase adheres the component onto the binding site permanently

avoid any self-assembly techniques requiring magnetic force should the discrete components be sensitive to magnetic force. In the following sections of the entry, most of the assembly methodologies will be further examined.

Fluidic Transport–Shape-Matching Alignment

Yeh et al. pioneered the work on self-assembly based on fluidic flow and shape matching [5]. Trapezoidal-shaped light-emitting diode (LED) devices were fabricated on GaAs substrates (Fig. 3) and released into ethanol or methanol, the carrier fluids of choice, which inhibits the oxidation and, therefore, degradation of the GaAs structures.

The LEDs were flowed over silicon host wafers fabricated with corresponding trapezoidal holes to capture them (Fig. 3). After the holes were (mostly) filled, the carrier fluid was evaporated, leaving the GaAs LEDs sitting in the holes, attached to the silicon surface only by van der Waals forces. The flowing of LEDs over the silicon substrate is identified as the transportation component of a self-assembly process. The alignment step was realized by shape matching between the trapezoidal LEDs and the corresponding silicon holes. The final procedure of binding was performed by the deposition of metallic connections between the devices and the host substrate, establishing electrical and mechanical connection, permanently adhering the chips and the substrate (Fig. 4). Yeh's self-assembly technique was subsequently commercialized, and has been adapted to be a high volume manufacturing compatible process, specializing in the production of radio frequency identification (RFID) tags [6].

By combining Yeh's concept of shape matching under fluidic flow with capillary forces as drivers of self-assembly, Parviz et al. demonstrates the possibility to assemble multiple device types simultaneously onto a variety of substrates [1], including flexible plastics [7]. Figure 5 summarizes the features of Parviz's multi-device assembly process.

Components of distinct shapes were flowed down the inclined surface of a template. Binding sites patterned on the template correspond to the shapes of components to be assembled. Component–binding site pairs are designed to only be compatible with each other, disallowing unintended assembly. The system is also designed to allow binding sites to only accept components when it happens to approach with the correct surface facing the binding site (Fig. 6).

The assembly process (Figs. 5, 6) was maintained above the melting temperature of the low-melting-point alloy solder deposited at the contact pads of the template, within the binding sites. When the metal contact pads on a single microcomponent comes into contact with the molten solder, the component will be pulled to the template as the molten solder wets the contact pad on the chip. Conversely, if a component approaches the binding site with the incorrect side, capillary forces will not be exerted on the microcomponent, as molten solder will not wet on silicon surfaces. Excess components collected at the bottom of the container will be reused repeatedly until assembly is complete (Fig. 5).

Self-Assembly for Heterogeneous Integration of Microsystems, Table 1. Survey of current self-assembly process. Components are square in shape except where marked by ^{C}, and ^{R}, which are circular and rectangular respectively. Part sizes of components are taken to be the length for square and rectangular parts, and the diameter for circular components

	Assembly mechanism										
	Transport		Aligment								
Researcher, institution	Fluidic	Mechanical agitation	Shape matching	Capillary forces	Magnetic	Part size (lateral dim) (μm)	Thickness (μm)	Aspect ratio (lateral/thickness)	Packing density (in plane) (%)	Orientation specificity	Reported yield (%)
Yeh, Berkeley	✓		✓			18	9.9	1.8	–	No	>90
Parviz, UW	✓		✓	✓		100^{C}	10–20	5–10	20	Yes	97
Srinivasan, Berkeley	✓			✓		150–400	15–50	8–10	16	No	100
Jacobs, UMN & Harvard	✓			✓		20	10	2	25	No	>98
Böhringer, UW	✓			✓		1,000–2,000	100	10–20	25	No	100
Böhringer, UW	✓			✓	✓	2,000	100	20	40	Yes	100
Böhringer, UW		✓	✓			370	150	2.47	80	No	100
Böhringer, UW		✓	✓			400^{R}	200	2	35	No	100
Ramadan, IME		✓	✓		✓	$1,000^{C}$	350	2.86	6	No	97
Fischer, KTH		✓	✓		✓	35^{C}	350^{L}	–	–	No	>95

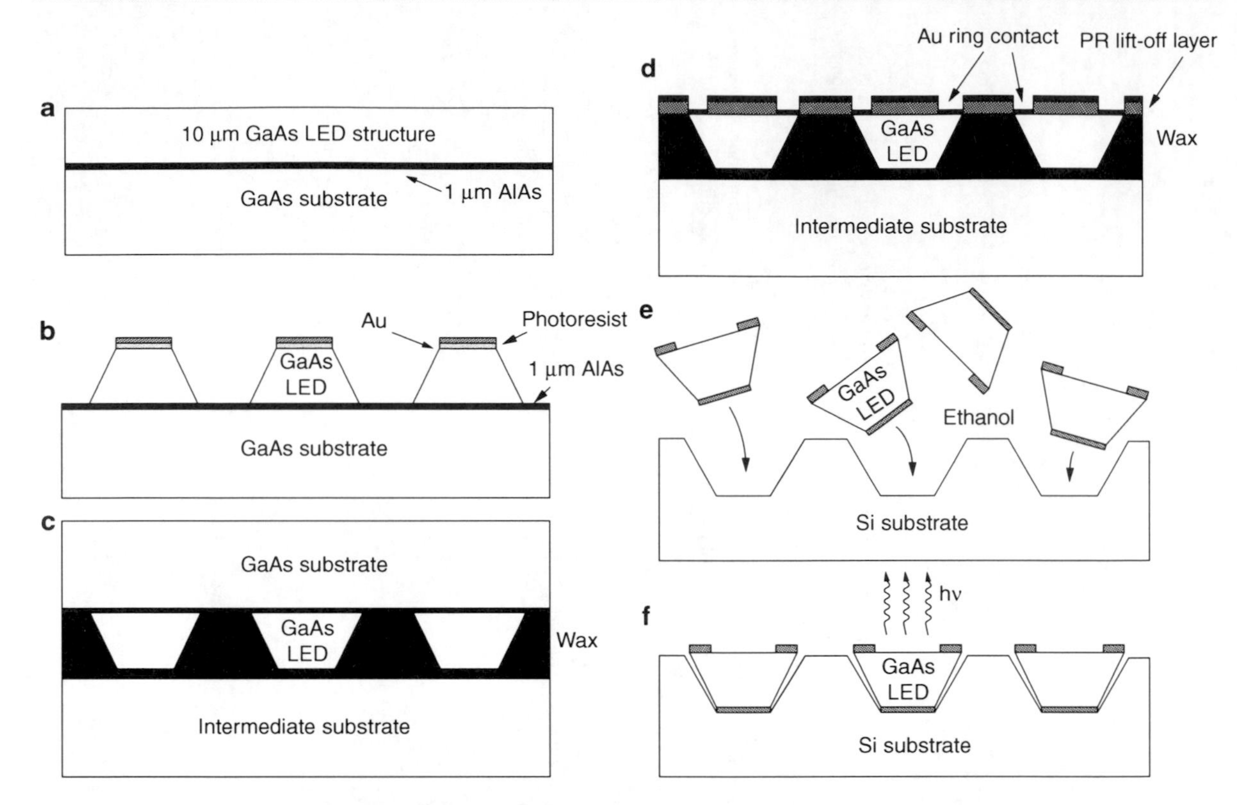

Self-Assembly for Heterogeneous Integration of Microsystems, Fig. 3 Schematic diagram of the fluid self-assembly process: (**a**) Molecular beam epitaxy (MBE) grown structure with 1 μm AlAs etch-stop layer, (**b**) trapezoidal GaAs mesa definition, (**c**) bonding to intermediate substrate with wax, (**d**) top-side ring contact metallization, (**e**) solution containing the GaAs blocks dispensed over patterned Si substrate and (**f**) Si substrate with GaAs, light-emitting diodes integrated by fluidic self-assembly. (Reprinted with permission from [5], © 1994 IEEE)

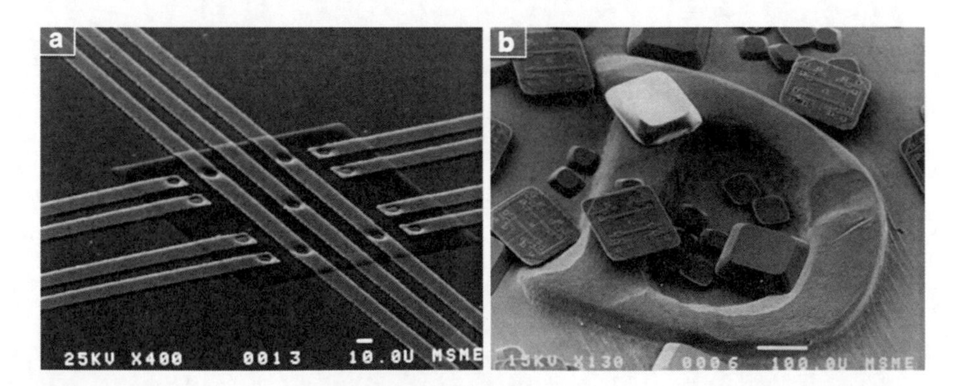

Self-Assembly for Heterogeneous Integration of Microsystems, Fig. 4 Scanning electron microscope (SEM) picture of Yeh's devices: (**a**) a device assembled using fluidic self-assembly techniques – the metallic connections are deposited over the device such that mechanical and electrical connection is established. (**b**) Devices placed on a US dime for scale. (From [6], reprinted with permission from Alien Technology)

When assembly was observed to be complete, the temperature of the self-assembly setup was lowered to room temperature, thereby solidifying the molten solder, establishing electrical and mechanical connection between the template and the discrete components. While the fluidic flow transportation mechanism is similar to Yeh et al., the alignment of microcomponents is achieved through shape-matching and solder capillary action. Finally, binding of the discrete components was achieved through the solidification of the molten solder bumps. Final products are shown in Figs. 7 and 8.

Parviz et al. added programmability to their fluidic shape-matching self-assembly process with a technique that involves the use of photosensitive materials to control the access of binding sites to microcomponents being flowed across the template [8]. Programmable binding locations were created by depositing photoresist within receptor sites that uses the principle of shape matching to capture microcomponents that are flowed over them. To activate a group of receptor sites, the region of interest is selectively exposed to ultraviolet radiation to remove the blocking photoresist material, thereby availing said sites to capture incoming components. By successive steps of site activation and component flowing, different components can be assembled onto a single type of receptor sites (Fig. 9).

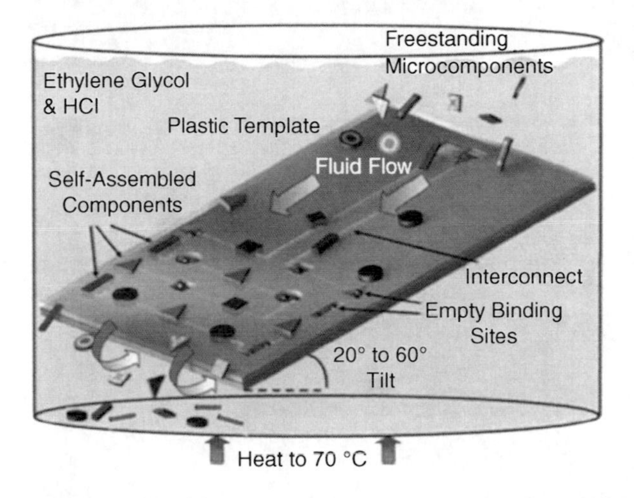

Self-Assembly for Heterogeneous Integration of Microsystems, Fig. 5 The heterogeneous self-assembly process. Microcomponents are introduced over a template submerged in a liquid medium and moved with the fluid flow. Self-assembly occurs as microcomponents first fall into complementarily shaped wells and then become bound by the capillary forces resultant from a molten alloy. (Reprinted from [7], with permission from IOP)

It is instructive, at this point, to definite "programmability" in the context of micro self-assembly as *the ability to direct different reliably reproducible outcomes with the same set of components in a self-assembly process*. Programmability, which will be revisited in other works, will add to the appeal of self-assembly in industry.

Fluidic Transport–Capillary Forces Alignment

Capillary forces dominate inertial forces and gravity as one approaches submillimeter scales, given that capillarity scales with length, while gravity scales with length cubed. There are many studies that concentrate on the alignment properties of capillary forces instead of an entire self-assembly process as defined in Fig. 2.

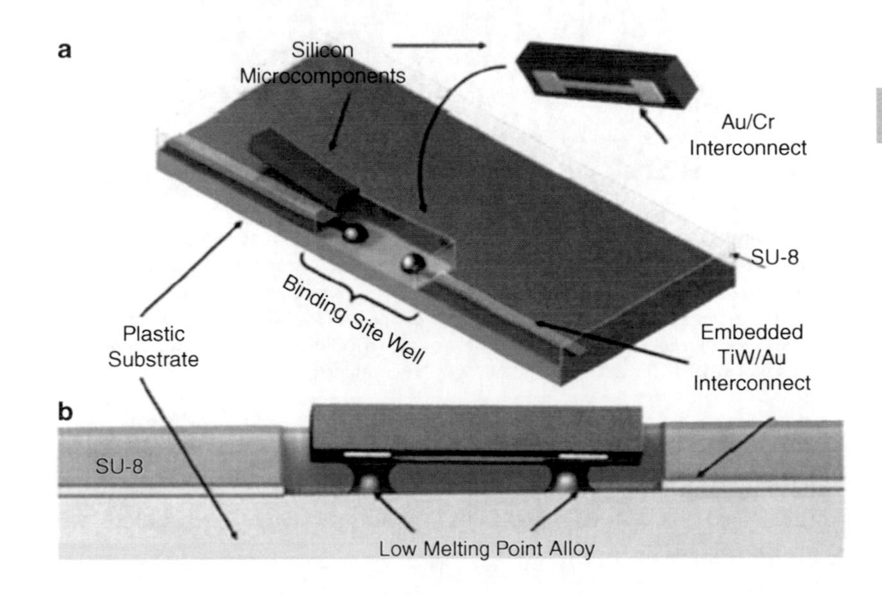

Self-Assembly for Heterogeneous Integration of Microsystems, Fig. 6 Details of the self-assembly process for a single microcomponent. (**a**) The microcomponent approaches a binding site with complementary shape. (**b**) The microcomponent is held by capillary force resultant from molten-alloy-bridging the metal pads positioned on the microcomponent and on the template. (Reprinted from [7], with permission from IOP)

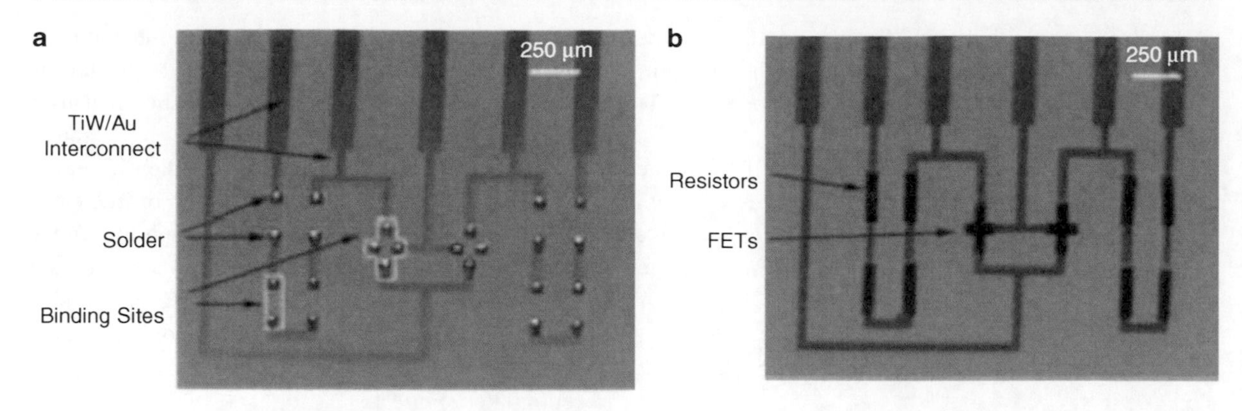

Self-Assembly for Heterogeneous Integration of Microsystems, Fig. 7 Self-assembly of field effect transistors (FETs). (**a**) Optical microscopic image of a plastic substrate with empty binding sites (two outlined with white lines for clarity); (**b**) The template after completion of the self-assembly process showing the position of FETs and diffusion resistors. (Reprinted from [7], with permission from IOP)

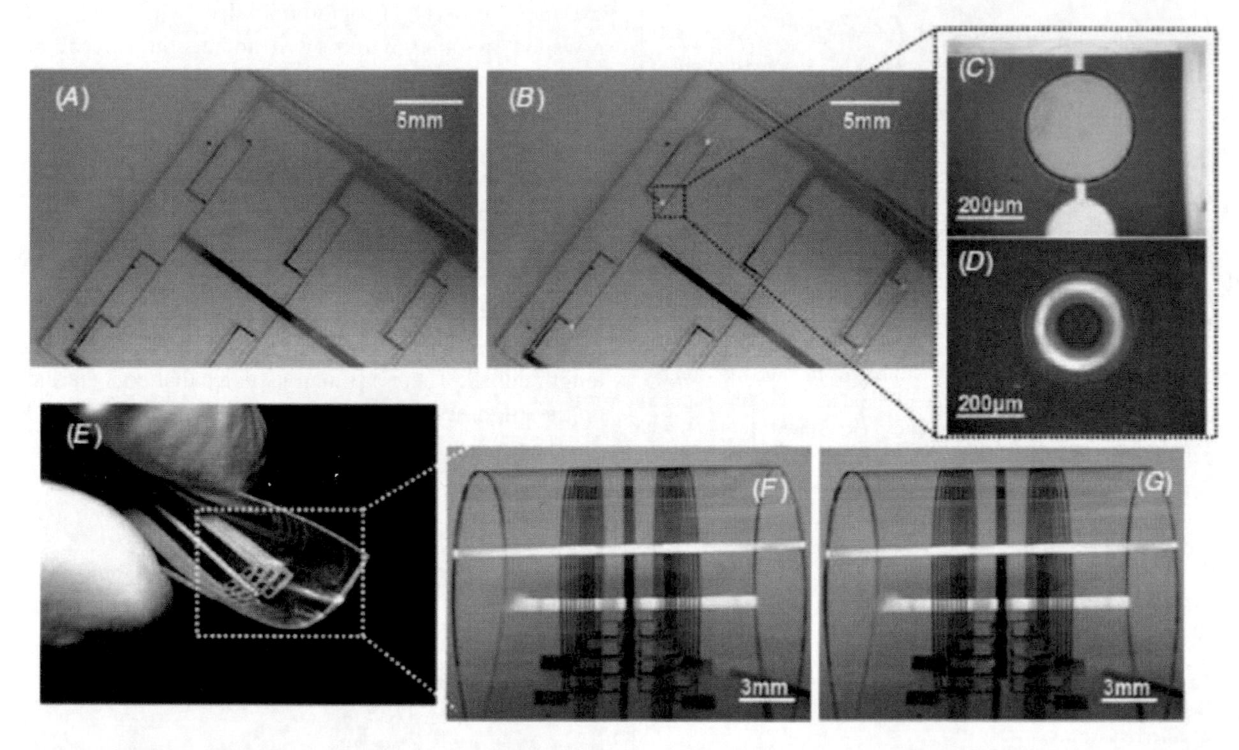

Self-Assembly for Heterogeneous Integration of Microsystems, Fig. 8 (**a**) A glass template after the conclusion of the self-assembly process. The micro (μ)-LEDs are assembled on the template. (**b**) 6 μ-LEDs are turned ON on the same glass template via the application of a 5 V bias. (**c**), (**d**) Close-up images of 1 μ-LED turned OFF and ON, on the template under an optical microscope; (**e**) A flexible plastic template after the completion of the self-assembly process; (**f**, **g**) The same plastic template, bent over itself, with μ-LEDs OFF and then ON upon the application of a 4 V bias. (Reprinted from [7], with permission from IOP)

Notably, work from Koyanagi et al. focuses on the aligning and bonding of silicon chips on microfabricated silicon pedestals with corresponding dimensions [9] (Fig. 10).

Droplets of an aqueous solution of hydrofluoric acid (HF) were placed on the pedestals before introducing silicon chips with a multi-chip vacuum pick system. The faces of the chips and the pedestals were treated to

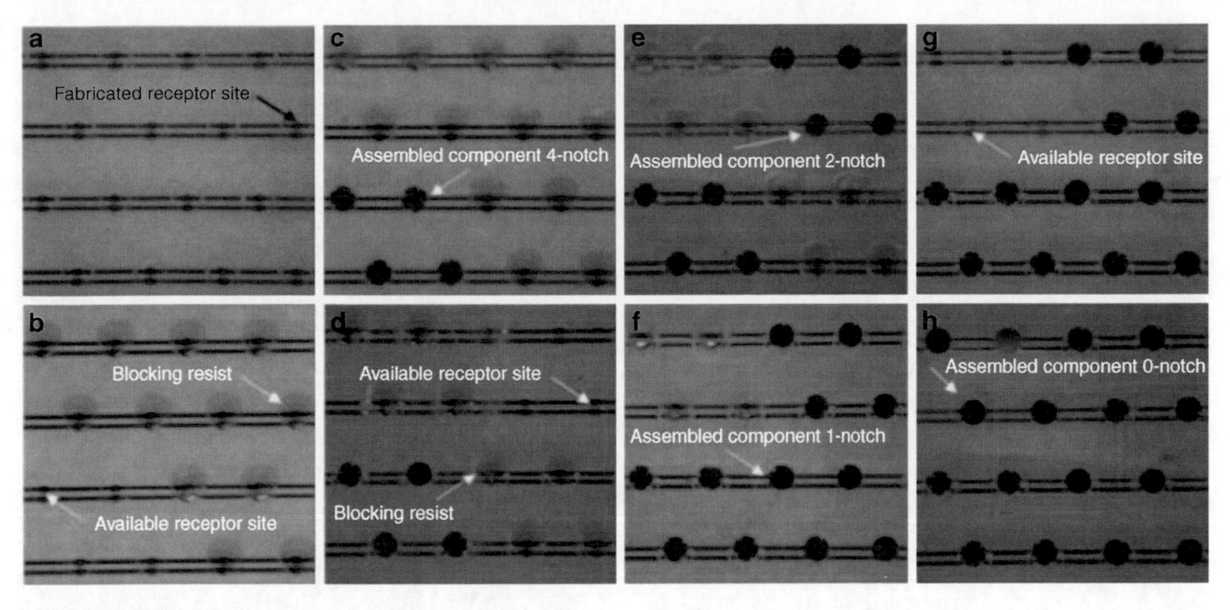

Self-Assembly for Heterogeneous Integration of Microsystems, Fig. 9 Optical microscope images of a sequence of events during the self-assembly process of four different microcomponent types. (**a**) Fabricated receptor site wells and C-shaped traps on a plastic template prior to self-assembly; (**b**) Positive photoresist was patterned on the template to block all the receptor sites, except the four receptor sites located on the *bottom left*. (**c**) The first type of components (circular 4-notch) was assembled within the 2 × 2 parallelogram pattern of available receptor sites. (**d**) A thin layer of photoresist was coated on the whole template to fix the assembled microcomponents. The blocking resist was then removed from the four top right receptor sites designated for the second type of microcomponents. (**e**) The second type of components (circular 2-notch) was assembled in available receptor sites. (**f**) The third type of components (circular 1-notch) was assembled in the *bottom left corner* of the image in a similar fashion. (**g**) The resist was removed from the remaining four receptor sites on the *top left*. (**h**) Final type of components (circular 0-notch) was self-assembled. (Reprinted from [8], with permission from MRS)

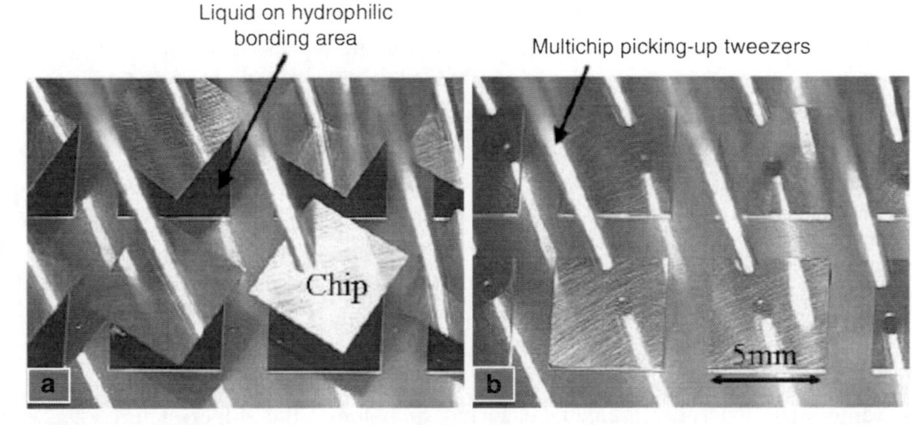

Self-Assembly for Heterogeneous Integration of Microsystems, Fig. 10 Self-alignment process from Koyanagi et al. (**a**) Introduction of chips using a multi-chip vacuum pick system. (**b**) Chips self-aligned after placement on bonding areas. (Reprinted with permission from [9], © 2008 IEEE)

be hydrophilic so that the droplet underneath the chips will wet both surfaces and self-align the chips to the pedestals underneath to minimize the interfacial energies. Alignment will improve as the HF solution evaporates, and eventually dissipates entirely, while bonding chips to pedestals permanently with strong SiO_2–SiO_2 bonds.

There are many other works titled "self-assembly" that deal exclusively with alignment rather than the entire self-assembly process which were presented in Fig. 2. "Self-alignment" is probably a more appropriate title for such studies, drawing the distinction between investigations focused on alignment and works that also include a stochastic transportation

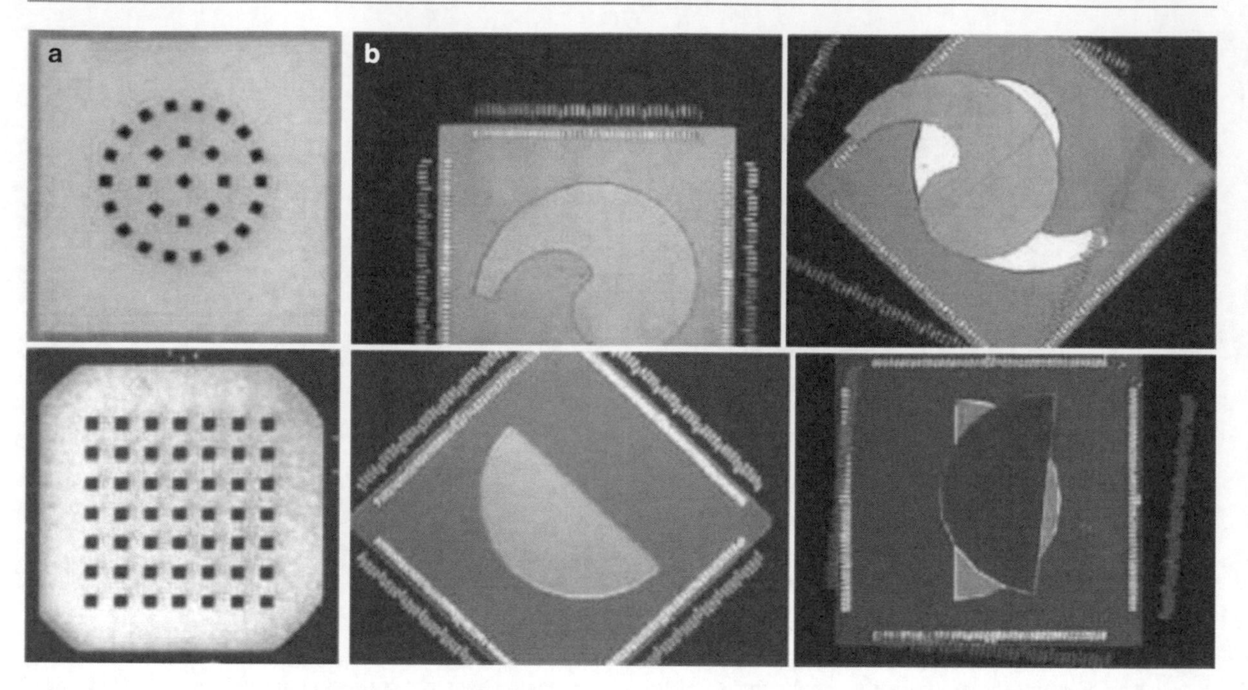

Self-Assembly for Heterogeneous Integration of Microsystems, Fig. 11 Optical micrographs of capillary-driven self-assembly of flat microparts: (**a**) *Square* parts on quartz substrates in *ring* and grid configurations and (**b**) correct and wrong alignments for *semicircle*- and *comma-shaped* binding sites. (Reprinted with permission from [10], © 2001 IEEE)

mechanism that brings discrete components to the vicinity of the intended assemblies.

Srinivasan et al. demonstrated a complete self-assembly technique with fluidic component transportation and capillary force capture and alignment [10]. They patterned a substrate with an array of hydrophobic self-assembled monolayer (SAM)–coated gold-binding sites. By inserting the substrate into water through a film of hydrophobic adhesive floating on the water surface, the hydrophobic binding sites were coated with the adhesive (Fig. 11).

Microparts fabricated from silicon-on-insulator (SOI) wafers were then introduced through a pipette toward the substrate in water. When the hydrophobic pattern on the microparts came into contact with the adhesive, the parts were pulled to the hydrophobic binding sites and self-alignments occurred spontaneously. Finally, when all sites were occupied, the adhesive was polymerized by heat (thermoset) or UV radiation, depending on the type of adhesive used, bonding the chips to the substrate permanently. Binding sites of shapes with in-plane rotational symmetries such as squares gave alignment yields of up to 100%, with translation and rotational misalignments of less than 0.2 μm and 0.3°, respectively. Component–binding site pairs without in-plane rotational symmetries (aimed at specific in-plane alignment) such as semicircles and commas gave alignments yields of approximately 30–40%.

Jacobs et al. developed a similar self-assembly process that uses low-temperature melting solder to mount LED arrays on flexible cylindrical templates [11]. The assembly template was patterned with copper squares, and a simple dip-coating process applied molten solder on these copper squares, given that copper has good wetting characteristics for solders. The dip-coating process was performed in an acidic aqueous environment to prevent oxidation of the solder surface.

Hundreds of LEDs and a flexible assembly template were placed inside a vial; the vial was filled with water and heated to a temperature above the melting point of the solder. The LEDs were tumbled inside the vial, allowing for the LEDs to come into contact with the molten solder, be pulled to a copper square by the solder, and self-aligned (Fig. 12c). The cathode and anode of the LEDs were distinguished by a small round contact and a larger square contact (the size of the copper squares on the template), respectively, on two faces of the discrete LEDs (Fig. 12a). The agitation intensity of the tumbling process was tuned to

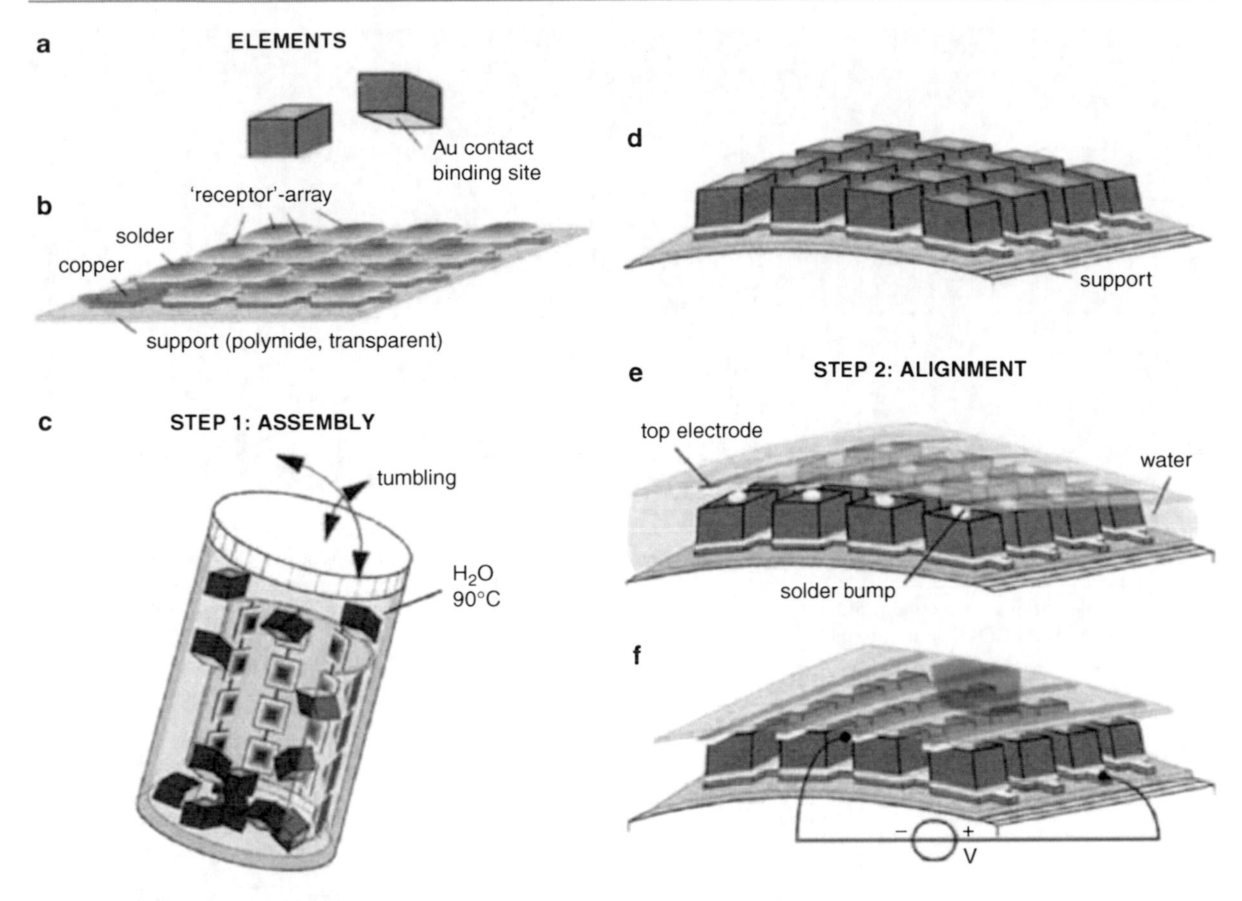

Self-Assembly for Heterogeneous Integration of Microsystems, Fig. 12 Procedure used to assemble a functional cylindrical display. (**a**) *Top* and *bottom* views of an LED segment that has two contacts: a small circular gold contact (cathode) on the *front*, and a large square gold contact (anode) on the *back*. (**b**) Array of solder-coated copper squares supported on a flexible substrate; these squares are of the same size as the anodes of the LEDs and act as receptors for the LEDs during the self-assembly. (**c**) The components are tumbled in a vial at a temperature above the melting point of the solder. (**d**) Two-dimensional array of the assembled LEDs. (**e**) Alignment of a top electrode: The copper wires of the top electrode and the cathodic contacts on the front of the LEDs were first dip coated with solder. The array of wires is prealigned with the array of cathodic solder bumps. At a temperature above the melting point, electrical connections form and the structure self-aligns. (**f**) Test of the self-assembled display-prototype. (From [11], reprinted with permission from AAAS)

dissociate assemblies between the solder and the smaller round contact while not severe enough to affect LEDs that were caught with their square contact face. This gives the assembly process side-selectively.

A permanent bond between each LED and the template was established when the template was allowed to cool, solidifying the solder. To make the LED array functional, a transparent film patterned with electrical circuitry was aligned and bonded to the assembled array, establishing the electrical connections to the cathodes of the LEDs. By activating the LEDs (Fig. 13), the defect rate was visually determined to be approximately 2%.

Jacobs et al. also worked on self-assembly based on the delivery of components on a fluid-fluid interface [12]. Components were fabricated out of silicon and SU-8, with one face coated with gold. The gold surfaces were treated with a mercaptoundecanoic acid (MUA) SAM to render them hydrophilic. The silicon faces were treated to be hydrophobic using 3-glycidoxypropyltrimethoxysilane (GPTMS), creating an orientation preference for components when they are introduced at a silicone oil–water fluid-fluid interface (SU-8 surface are hydrophobic and thus required no further surface modification). By mildly agitating the system, components at the silicone

Self-Assembly for Heterogeneous Integration of Microsystems, Fig. 13 (**a**) Photograph of the display after self-alignment of the top electrode; (**b–d**) Photographs of the operating display after the alignment of the top electrode. The display contains 113 LEDs that are assembled in an interleaved fully addressable array with eight columns of eight receptors interleaved with seven columns of seven on the *bottom* electrode that connect to 15 rows of crossing copper wires on the *top* electrode. (From [11], reprinted with permission from AAAS)

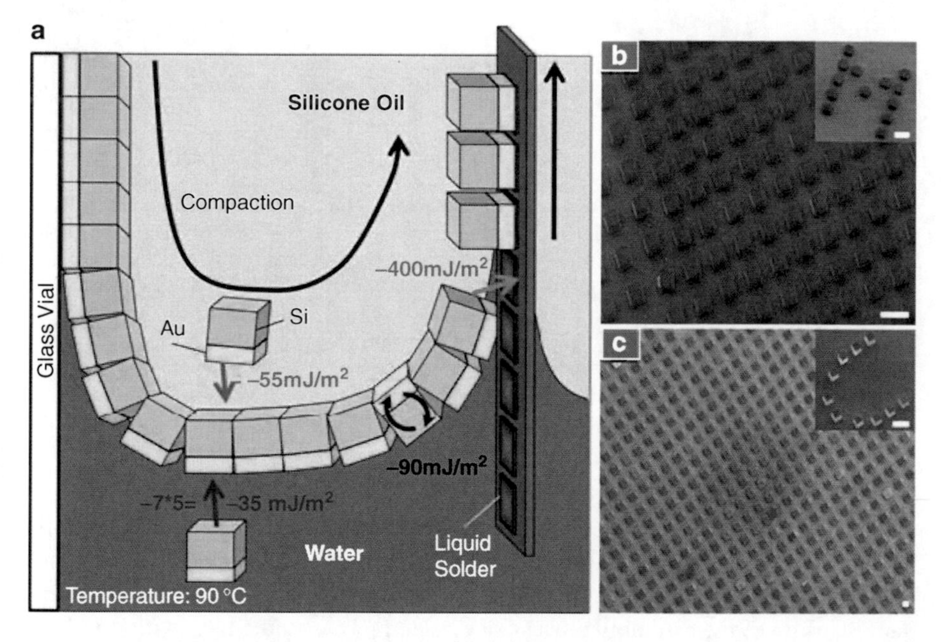

Self-Assembly for Heterogeneous Integration of Microsystems, Fig. 14 (**a**) Procedure of surface tension–directed self-assembly at a liquid-liquid–solid interface. SEM of (**b**) SU-8 (20 μm side length) and (**c**) Si chiplets (20 μm and 60 μm side length) assembling in regular arrays and arbitrary text patterns (insets). (Reprinted from [12], © 2010 National Academy of Sciences, USA)

oil–water interface will predominantly orient themselves such that the hydrophilic gold surface will face toward water, and the hydrophobic silicon/SU-8 will face the direction of the silicone oil.

Two types of substrates have been used: silicon and flexible propylene terephthalate (PET). Copper contact pads, corresponding to the sizes of the components, were patterned on the surface. A dip-coating process was used to deposit low-melting-temperature solder onto the copper pads.

The transfer of the components is shown in Fig. 14. A container containing silicone oil, water, and components on the interface between the two fluids was first heated to a temperature beyond the melting point of the solder. Substrates were pulled upward through the liquid-liquid interface, allowing the components to

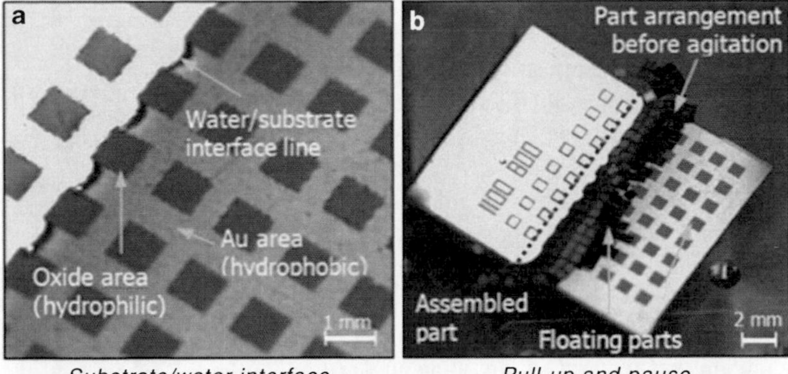

Substrate/water interface
Pull-up and pause

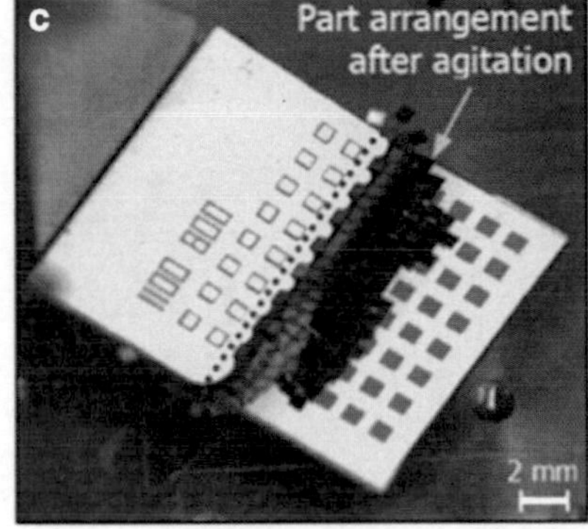

After agitation

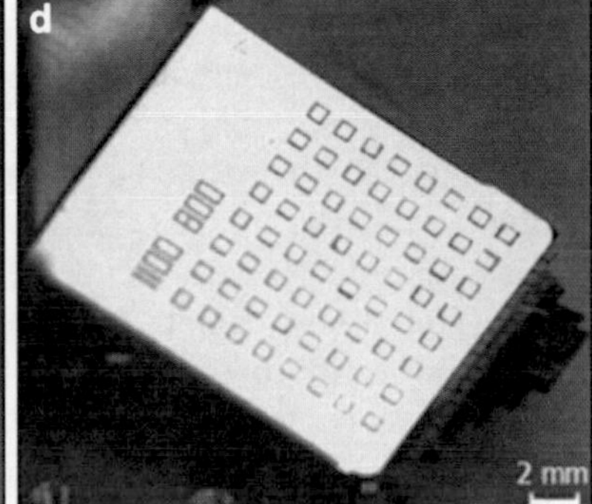

Example of 100% assembly

Self-Assembly for Heterogeneous Integration of Microsystems, Fig. 15 Parts rearrangement by agitation. *Darker squares* are hydrophilic silicon surfaces

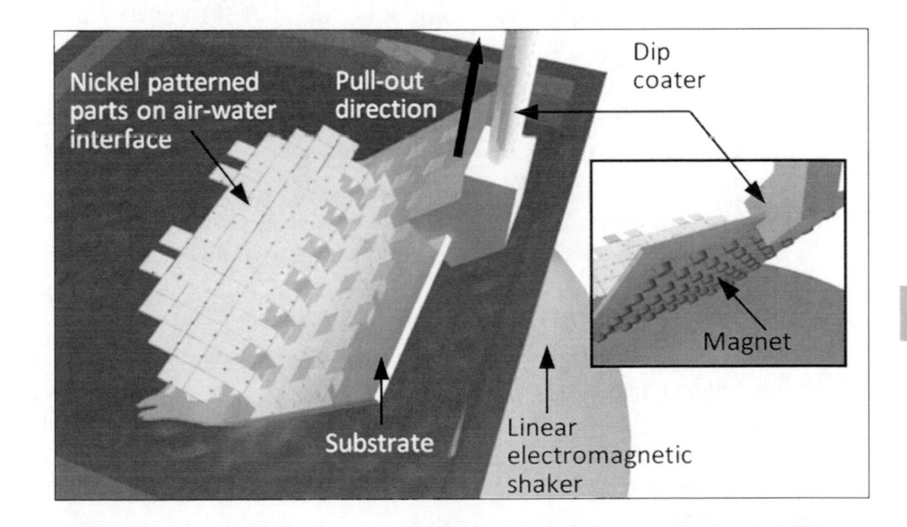

Self-Assembly for Heterogeneous Integration of Microsystems, Fig. 16 The experimental setup consists of a water container, linear electromagnetic actuator, dip coater, assembly substrate, magnets, and test parts floating at the air-water interface

come into contact with the molten solder. Upon contact, the molten solder will pull components in, transferring the components onto the substrates. While it is possible to deliver components to over 90% of the copper pads on a substrate in a single pass, multiple passes were required to achieve full coverage. Using this method, chips $20 \times 20 \times 10$ and $60 \times 60 \times 20\ \mu m^3$ in size were assembled onto rigid and flexible substrates at reported speeds of up to 62,500 chips in 45 s.

Böhringer et al. studied the assembly of thin microcomponents in a fluidic environment, delivering components at the air-water interface [13].

Components ($1{,}000 \times 1{,}000 \times 100\ \mu m^3$) with one face coated in gold are fabricated in silicon. Gold-coated silicon assembly templates were prepared with square openings, exposing silicon, that were designed to be slightly larger than the lateral dimensions of the components. All gold surfaces were then treated to be hydrophobic using dodecanethiol SAM.

Components were introduced onto a water surface. By applying controlled agitation, the components will preferentially orient themselves with their gold faces upward due to the surface treatment. The assembly template was then pulled up, at an angle, through the air-water interface, in the immediate proximity of the floating components.

When the level of the assembly template reaches the top edges of a row of assembly sites (Fig. 15a), controlled agitation was applied to cause the components to arrange themselves such that the edge of a single part was pinned to the edge of a single site (Fig. 15b, c). At this point, the assembly template was pulled up, allowing an entire row of parts to be transferred and self-aligned to the template. By repeating this pull-up, pause and agitate cycle, entire assembly templates can be perfectly filled. The most significant aspect of this work lies in the aspect ratio of the components being self-assembled; no other self-assembly technique has been demonstrated to work well with very thin components.

The programmability of this process was demonstrated by the ability to assemble only on alternate rows of binding sites. After a single (first) row of parts have been transferred onto the assembly substrate, one can avoid assembling on the subsequent row by agitating the water surface while the substrate is extracted through the air-water interface, bypassing the second row and stopping at the assembly position of the third. Using similar methods, arbitrary row assemblies can be reproduced reliably.

Fluidic Transport–Capillary/Magnetic Forces Alignment

By depositing nickel squares near the edges of chips, and positioning a magnet underneath each binding site underneath the assembly substrate, Böhringer et al. extended the previous fluidic transport–capillary forces alignment self-assembly method to include orientation specificity (Fig. 16).

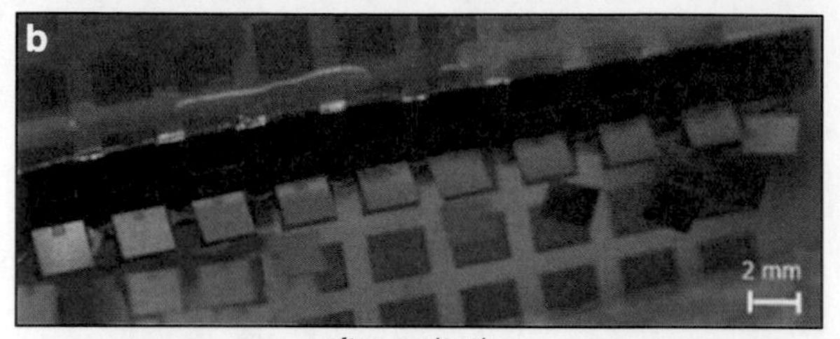

Self-Assembly for Heterogeneous Integration of Microsystems, Fig. 17 Assembly process of $2{,}000 \times 2{,}000 \times 100\ \mu m^3$ components on the assembly substrate: (**a**) A binding site does not yet have a corresponding component, and another has two test parts in its vicinity. (**b**) After applying specific Faraday waves using agitation, every binding site gets a single component. (**c**) Two rows of components assembled onto the assembly substrate

Similar to the previous method, controlled agitation of the water surface ensures one-to-one addressing of parts to binding sites. During this agitation, the interaction of the magnetic forces between the magnets underneath each binding site and the ferromagnetic nickel squares ensures also that chips approach the binding sites with the nickel-patterned edge, allowing for orientation specificity previously unachievable (Fig. 17).

Mechanical Agitation Transport–Capillary Forces Alignment

For discrete components and/or substrates that are adverse to fluidic environments, there are also self-assembly processes that are conducted in dry or "semidry" [14–16] conditions. Dry environment

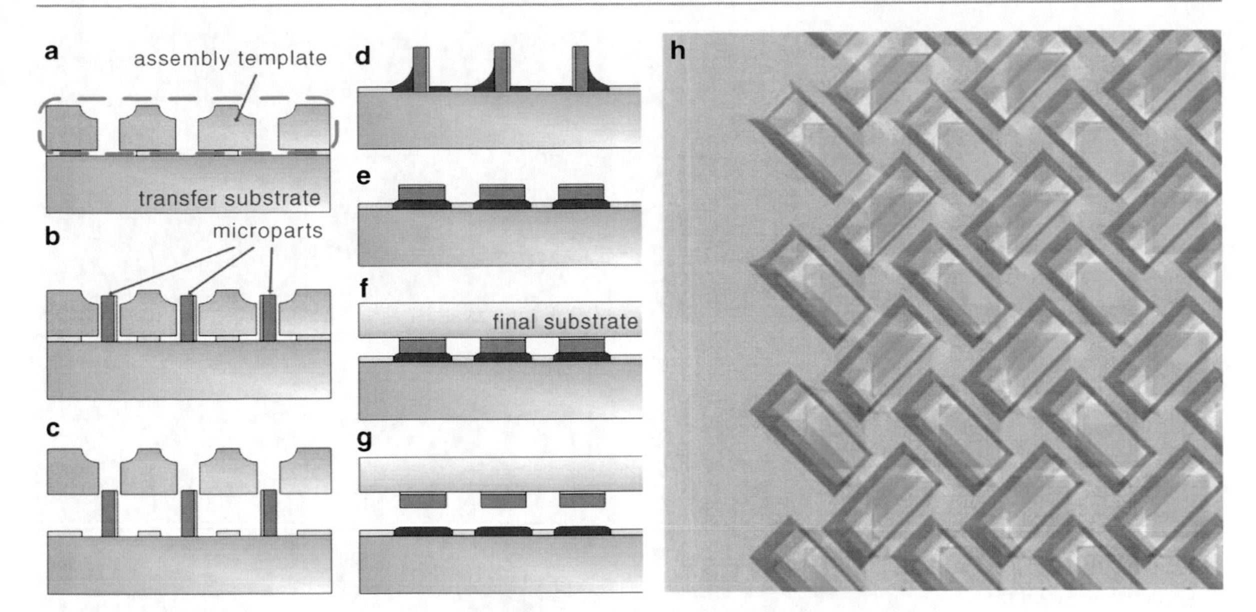

Self-Assembly for Heterogeneous Integration of Microsystems, Fig. 18 Full assembly process flow (**a–g**): (**a**) Mount assembly template on a gold-patterned transfer substrate. (**b**) Assemble parts. (**c**) Remove template. (**d**) Apply moisture. (**e**) Vibrate the setup gently to have the parts fall on hydrophilic side and self-align. (**f**) Attach parts to destination substrate/device. (**g**) Complete the process by removing the assembled substrate/device from transfer template. (**h**) *Top-down* view of assembly template placed over transfer substrate. (Reprinted with permission from [15], © 2009 IEEE)

self-assembly processes predominantly require the use of mechanical vibration/agitation to distribute the microcomponents.

Böhringer et al. devised a suite of self-assembly methods to be performed in an air environment that uses a two-step assembly process consisting of a shape-matching step and a capillary force–driven alignment step [14–16] for square silicon parts as well as surface mount technology (SMT) passive devices.

The latest iteration in the technique for the assembly of silicon chips is shown in Fig. 18. A silicon fabricated assembly template was first aligned on top of a transfer substrate. The transfer substrate was fabricated by first depositing gold onto a silicon wafer. Square openings (silicon) corresponding to the size of the microparts were subsequently developed, and the gold surfaces were treated to be hydrophobic with a SAM coating. Each aperture in the assembly template addresses a single silicon opening underneath (Fig. 18h), and can only contain a single discrete component standing on its side.

Square-shaped silicon microcomponents with one gold-coated (hydrophobic SAM-treated) side were delivered to every aperture on the assembly template. When the delivery was completed, the assembly template was removed, and steam was introduced to the system to form water droplets on the hydrophilic silicon surfaces. When the droplets become large enough, they will exert enough force to pull the silicon face of the microparts flat onto the transfer substrate, and self-align the components to the squares. The aligned microcomponents can then be transferred onto a destination substrate to be permanently mounted. Similar techniques have been tested with parts ranging from $370 \times 370 \times 150\ \mu m^3$ to $790 \times 790 \times 330\ \mu m^3$ in size.

Böhringer et al. also developed a system of control for the precise manipulation of microcomponents on a vibrating surface [15]. Parts can be made to jump or walk (where parts seem to crawl across the assembly template) in predictable directions and speeds. Using these controllable part-motion modes, the transportation phase of the self-assembly process can be programmed to deliver components to arbitrary regions of apertures or even empty out filled apertures. In the case of a programmed "feedback-driven" delivery process [15] (Fig. 19), visual feedback is used to identify empty apertures and direct excess components toward them, guaranteeing the complete delivery of components.

Finally, Böhringer et al. also adapted the template-based methodology (complete with the feedback-driven assembly process) to the assembly of surface mount technology (SMT) thin-film resistors and

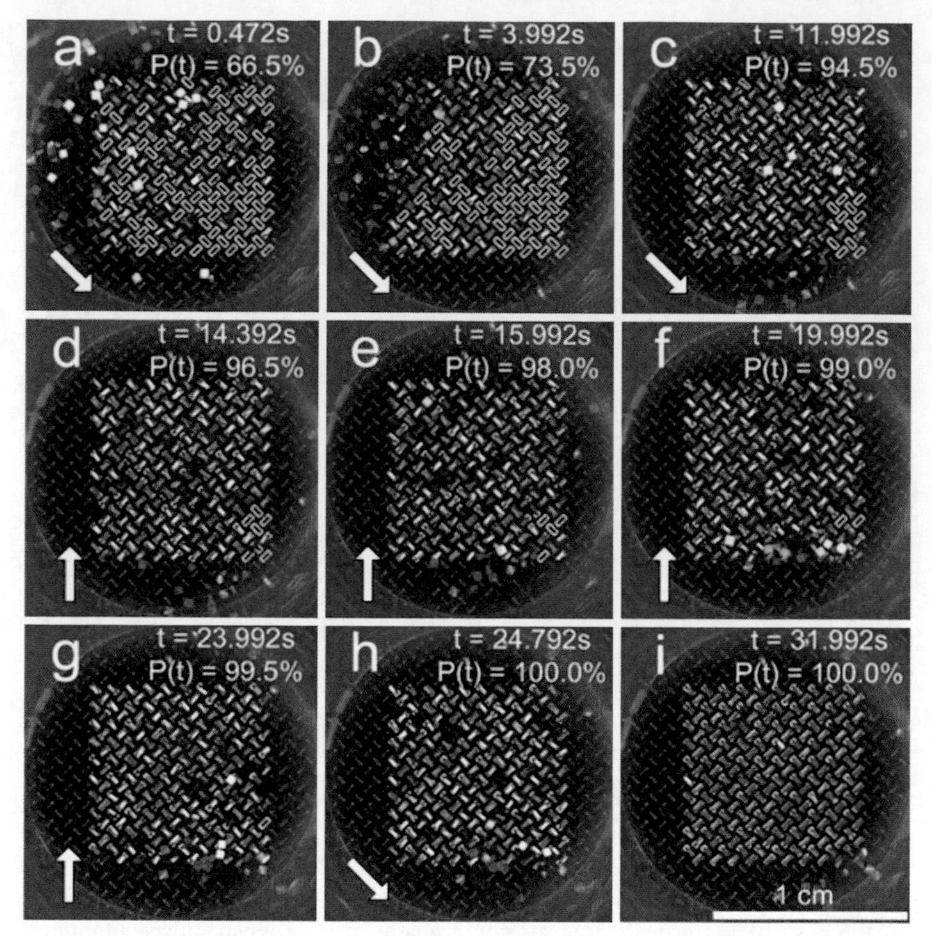

Self-Assembly for Heterogeneous Integration of Microsystems, Fig. 19 Stages of a feedback-driven assembly: (**a**) shows the instance when jumping mode (with a *top left corner* bias) is switched to a walking mode with part-motion direction as indicated by the *white arrow* in the *box*. (**b**) and (**c**) show the progressive filling of the entire assembly area. *Right* after (**c**), when assembly-percentage, P(t), has plateaued, a walking mode in the upward direction as seen in (**d–g**) is activated, moving excess parts below the assembly area to the empty sites. After achieving 100% assembly in (**h**), a walking mode directed toward the lower-right corner is activated, moving excess parts away from the assembly area, finishing the process in (**i**); note: size-scale provided in image (**i**). (Reprinted with permission from [15], © 2009 IEEE)

monolithic ceramic capacitors of the 01005 standard (0.016″ × 0.008″, 0.4 × 0.2 mm) [16]. In this instance, two templates were used, one stacked on top of another. Template 1 (Fig. 20) ensures that only one component will be allowed into template 2, and template 2 guides the in-plane orientation of the component. During the solder reflow process, the molten solder wets the metallized contact-ends of the SMT components and aligns them to the contact pads on the target substrate. At the end of the solder reflow, mechanical and electrical bonds between components and substrate were achieved.

Mechanical Agitation/Magnetic Forces Transport–Shape-Matching Alignment

A dry environment self-assembly technique combining transportation via mechanical vibration and magnetic attractive force with shape-matching alignment was devised by Ramadan et al. [17] (Fig. 21). A host substrate, featuring physical recesses, was superimposed to a master array comprising embedded NdFeB magnets, whose position exactly matched that of the recesses. Target chips, bearing a CoNiP soft magnetic film on their nonfunctional side, were distributed over the substrate, which was mechanically vibrated and tilted. This motion causes chips to be distributed across the entire substrate, until being collected in an enclosure at the lower end of the tilt.

The amplitude of the substrate vibration and the thickness of chips were tailored so that the only stable configuration possible for the chips to be assembled is to have the functional side facing upward. The in-plane alignment of the chips was controlled by shape matching between the chips and the recesses of the host substrate. After assembly, the master array could be removed and reutilized while the assembled substrate could be further processed. An assembly of

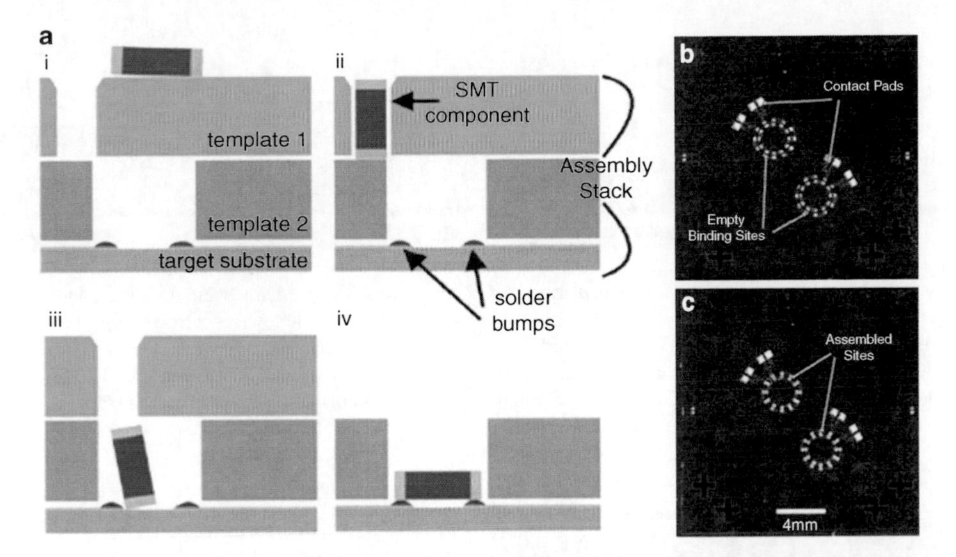

Self-Assembly for Heterogeneous Integration of Microsystems, Fig. 20 (**a**) Assembly process: (*i*) Walking mode [3] component delivery performed on assembly stack in Configuration 1; (*ii*) a single component is captured by template 1 near the binding location; (*iii*) Configuration 2 – component is allowed to drop into template 2; (*iv*) slightly agitating the system aligns the component to the orientation of the aperture on top of the binding location on the target substrate. After step (*iv*), solder reflow is performed to bond the component to the target substrate mechanically and electrically. Assembly of SMT passive components onto a circular test device: (**b**) before assembly; (**c**) after assembly. (Reprinted with permission from [16], © 2009 IEEE)

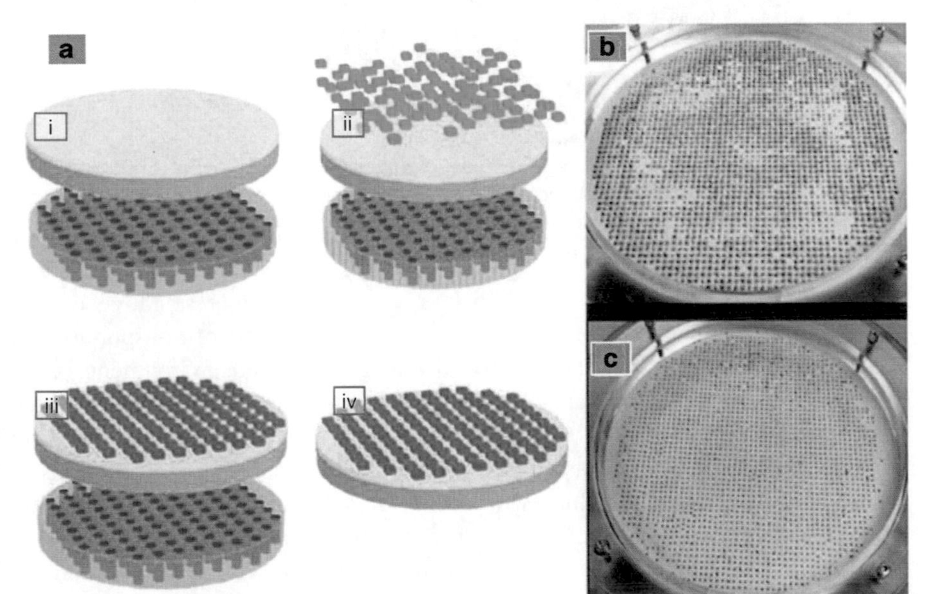

Self-Assembly for Heterogeneous Integration of Microsystems, Fig. 21 Assembly process (**a**): (*i*) Host substrate aligned to the master magnetic array, (*ii*) chip seeding, (*iii*) applying vibration, and (*iv*) chip final alignment and master magnetic array removal. Photo of 1 mm^2 assembly (**b**) 2 mins and (**c**) 5 mins into the vibration process. (Reprinted with permission from [17], © 2007 American Institute of Physics)

2,500 chips in 5 min with 97% yield was reported for this technique.

Fischer et al. proposed a scheme to implement through silicon vias (TSVs) by using magnetic force [18]. High-aspect-ratio holes are first etched on a silicon substrate using deep reactive ion etching (DRIE). Commercially available nickel wires (length: 350 μm, diameter: 35 μm) were deposited on the surface of the silicon substrate. A magnet was introduced at the bottom of the substrate, causing

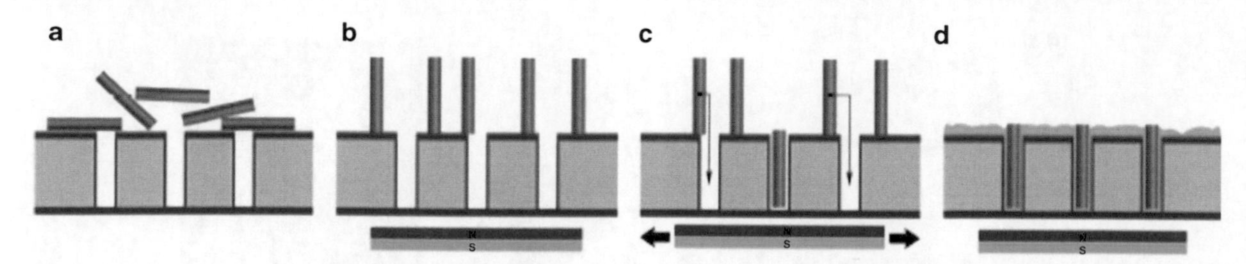

Self-Assembly for Heterogeneous Integration of Microsystems, Fig. 22 Nickel wire assembly process: (**a**) nickel wires are introduced to a substrate etched with corresponding holes, (**b**) a magnet is placed under the substrate, causing the nickel wires to stand upright due to magnetic forces, (**c**) the magnet is moved laterally under the substrate, moving wires to unoccupied holes and trapping them, (**d**) BCB is applied over the assembled wires. Processing steps after (**d**) include curing the BCB, polishing the substrate to expose the ends of the nickel wires, and depositing metal lines for electrical connection. (Reprinted with permission from [18], © 2011 IEEE)

Self-Assembly for Heterogeneous Integration of Microsystems, Fig. 23 SEM image of a 30 × 30 array with a pitch of 120 μm of nickel wires placed in via holes prior to the filling of BCB. (Reprinted with permission from [18], © 2011 IEEE)

the wires to stand upright. By moving the magnet laterally, wires were dragged along the surface (standing), and pulled into the holes (Fig. 22).

After the wires have been assembled into the holes (Fig. 23), the excess space within a wire-occupied hole was filled by a manual application of the thermosetting polymer bisbenzocyclobutene (BCB). The BCB was then cured, permanently adhering the nickel wires within the silicon substrate. Subsequent steps involve steps to expose the ends of the wires on two sides of the silicon, and depositing metal contact lines per specific application. The filling rate of a 30 by 30 array of via holes with a pitch of 120 μm was reported at about 80% in 20 s, and near complete (>95%) in about 2 min.

TSVs are vertical electrical connections passing through die layers. They are important for heterogeneous integration solutions involving the stacking of dies. While seemingly out of place with previously mentioned assembly methodologies for chips/dies and passive devices, it is significant to show that there is a viable self-assembly solution for TSV placements, demonstrating that there are self-assembly solutions for every aspect of the device assembly process.

Conclusion

Microscale self-assembly has been presented as a suite of techniques that can enable the advancement of the electronics industry through heterogeneous integration. A selection of self-assembly methodologies that are representative of techniques employing various modes of transportation and alignment, and the associated environments in which they are performed have been introduced. By identifying self-assembly techniques with suitable discrete component dimensions as well as carefully considering the transport and alignment mechanism and assembly environment, one can harness the parallel and scalable nature of self-assembly, thereby streamlining the existing assembly processes or enabling revolutionary pathways.

References

1. Morris, C.J., Stauth, S.A., Parviz, B.A.: Self-assembly for microscale and nanoscale packaging: steps toward self-packaging. IEEE Trans. Adv. Packag. **28**, 600–611 (2005)

2. Fearing, R.S.: Survey of sticking effects for micro parts handling. Presented at the proceedings of the international conference on intelligent robots and systems, vol. 2, 1995
3. Whitesides, G.M., Grzybowski, B.: Self-assembly at all scales. Science **295**, 2418–2421 (2002)
4. Rothemund, P.W.K.: Folding DNA to create nanoscale shapes and patterns. Nature **440**, 297–302 (2006)
5. Yeh, H.J.J., Smith, J.S.: Fluidic self-assembly for the integration of GaAs light-emitting diodes on Si substrates. IEEE Photonics Technol. Lett. **6**, 706–708 (1994)
6. Alien Technology. Available: http://www.alientechnology.com (2011). Accessed 1 Mar 2011
7. Saeedi, E., Kim, S., Parviz, B.A.: Self-assembled crystalline semiconductor optoelectronics on glass and plastic. J. Micromech. Microeng. **18**, 7 (2008)
8. Saeedi, E., Etzkorn, J.R., Parviz, B.A.: Sequential self-assembly of micron-scale components with light. J. Mater. Res. **26**, 1 (2011)
9. Fukushima, T., Konno, T., Tanaka, T., Koyanagi, M.: Multichip self-assembly technique on flexible polymeric substrate. In: 58th Electronic Components and Technology Conference 2008 (ECTC 2008), pp. 1532–1537 (2008)
10. Srinivasan, U., Liepmann, D., Howe, R.T.: Microstructure to substrate self-assembly using capillary forces. J. Microelectromech. Syst. **10**, 17–24 (2001)
11. Jacobs, H.O., Tao, A.R., Schwartz, A., Gracias, D.H., Whitesides, G.M.: Fabrication of a cylindrical display by patterned assembly. Science **296**, 323–325 (2002)
12. Knuesel, R.J., Jacobs, H.O.: Self-assembly of microscopic chiplets at a liquid-liquid–solid interface forming a flexible segmented monocrystalline solar cell. Proc. Natl. Acad. Sci. U.S.A. **107**, 993–998 (2010)
13. Park, K.S., Xiong, X., Baskaran, R., Böhringer, K.F.: Fluidic self-assembly of millimeter scale thin parts on preprogrammed substrate at air-water interface. In: IEEE International Conference on Micro Electro Mechanical Systems (MEMS), pp. 504–507 (2010)
14. Fang, J., Böhringer, K.F.: Parallel micro component-to-substrate assembly with controlled poses and high surface coverage. J. Micromech. Microeng. **16**, 721–730 (2006)
15. Hoo, J., Baskaran, R., Böhringer, KF.: Programmable batch assembly of microparts with 100% yield. In: Transducers, Denver, pp. 829–832 (2009)
16. Hoo, J., Lingley, A., Baskaran, R., Xiong, X., Böhringer, K.F.: Parallel assembly of 01005 surface mount technology components with 100% yield. In: IEEE International Conference on Micro Electro Mechanical Systems (MEMS), pp. 532–535 (2010)
17. Ramadan, Q., Uk, Y.S., Vaidyanathan, K.: Large scale microcomponents assembly using an external magnetic array. Appl. Phys. Lett. **90**, 172502/1–172502/3 (2007)
18. Fischer, A.C., Roxhed, N., Haraldsson, T., Heinig, N., Stemme, G., Niklaus, F.: Fabrication of high aspect ratio through silicon vias (TSVs) by magnetic assembly of nickel wires. In: IEEE 24st International Conference on Micro Electro Mechanical Systems 2011(MEMS 2011), Cancun (2011)

Self-assembly of Nanostructures

Wei Lu
Department of Mechanical Engineering, University of Michigan, Ann Arbor, MI, USA

Synonyms

Self-assembled nanostructures; Self-organized nanostructures

Definition

Self-assembly of nanostructures is a process where atoms, molecules or nanoscale building blocks spontaneously organize into ordered structures or patterns with nanometer features without any human intervention. It is the most promising practical low-cost and high-throughput approach for nanofabrication.

The Concept of Self-Assembly

The term "self-assembly" can be understood from its two components. The first component "self" implies "spontaneous and on its own," which suggests that it is a process that happens without human intervention or operation from outside of the system. The second component "assembly" indicates "forming or putting together," which suggests that the result is a structure built up by lower level building blocks or parts. Self-assembly of nanostructures refers to structures or patterns with nanometer features that form spontaneously from the basic building blocks such as atoms, molecules, or nanoparticles. Cells and living organisms are perfect examples of fairly sophisticated structures self-assembled in nature. These functional assemblies have motivated the study and design of nonliving systems.

From a scientific point of view, the spontaneous formation of ordered nanostructures from a random disordered state is inherently an intriguing process in any system. The fundamental mechanisms behind these behaviors raise significant research interests. In addition, studying self-assembly in nonliving systems may provide the critical understanding toward decoding the living systems, which is far more complicated. From an engineering point of view, self-assembly as the fundamental principle of a bottom-up approach is possibly the most promising technique for nanofabrication. Thus self-assembly plays a central role in the broad nanotechnology field. The self-assembly of nanostructures may enable a wide range of applications [1], such as nanoelectronic devices, ultrasensitive biosensors, carriers for drug delivery, high capacity lithium batteries, efficient photovoltaic devices, and advanced materials with unique mechanical, electrical, magnetic, or photonic properties.

Discrete System Versus Continuum System

Depending on the type of building blocks, self-assembled systems can be classified into two categories: discrete and continuum. A discrete system uses prefabricated building blocks with fixed sizes and shapes. Figure 1 shows an example, where the building blocks are spherical nanoparticles with engineered bonding sites through surface treatment. When put in a liquid, these nanoparticles collide randomly due to the Brownian motion. When two particles are close to each other with matching orientations so that their bonding sites meet, a bond between them forms. A chain structure as shown in Fig. 1a will appear when the system has a single type of nanoparticles. Figure 1b shows a system composed of two types of nanoparticles. A yellow particle has two red bonding sites, indicating that it will bond to two red particles. A red particle has two yellow bonding sites, indicating that it will bond to two yellow particles. This selective bonding leads to a self-assembled chain of alternating particles. Figure 1c shows a case where each particle has four bonding sites. They self-assemble into a network of alternating particle chains. Similar approach can be used to construct three-dimensional structures as well. Various structures can self-assemble by engineering the bonding properties.

In addition to local bonding properties, electric and magnetic fields as well as shear forces and spatial constraints have been used to direct the assembly of nanoparticles and nanorods into different configurations [2]. The electric and magnetic field method induce a dipole-type long-range interaction to act as the driving force to bring nanoparticles together, while the shear force method uses hydrodynamic interactions. The assembly of nanowire arrays is more challenging than that of nanoparticles and nanorods due to their highly anisotropic shape. Nanowire self-assembly usually results in partially ordered, small superlattices. This issue has been addressed by several new methods to direct the process, including the use of microfluidic channels and electric fields [3, 4].

In contrast, a continuum system exploits the spontaneous formation of nanoscale domains. Examples include self-assembled domain patterns in binary monolayers [5], block copolymers [6], and organic molecular adsorbates on metal surfaces [7, 8]. A binary monolayer on an elastic substrate may separate into two phases, and self-assemble into ordered patterns, such as triangular lattice of dots, parallel stripes, or serpentine stripes. The feature size is on the order of $1 \sim 100$ nm, and stable against annealing. Block copolymers are polymers consist of at least two chemically distinct, immiscible polymer fragments that are joined by a covalent bond. These systems have been shown to develop a variety of regular domain patterns via phase separation [9, 10]. The size and the period of the structures are typically on the order of 10–100 nm, depending on the conditions of preparation and the relative chain lengths of the participating polymers.

Figure 2 shows an example of domain patterns formed by two phases. A notable feature is the dependence of the pattern on the average concentration. Take Cu and Pb monolayer on Cu (111) substrate as an example. The Pb and Cu mixture monolayer on a Cu substrate forms two phases: a Pb overlayer and a disordered Pb–Cu surface solution. The two phases in the monolayer self-assemble into ordered, nanoscale patterns. When the average concentration of Pb atoms increases, the system can experience a series of patterns from (a) to (e), where the bright phase is Pb while the dark phase is the Pb–Cu surface alloy. When the average concentration is very low, isolated dots are obtained, as shown in Fig. 2a. The bright phase is embedded in the continuous matrix of the dark phase. No long-range ordering is observed. When the average concentration increases, the area covered by the bright dots increases. The dots order and form patterns close to a triangular lattice, as shown in Fig. 2b. When the bright phase occupies roughly half of the surface, stripe structure as shown in Fig. 2c forms. "Inverted" dots are observed when the bright phase dominates, as shown in Fig. 2e.

A continuum system offers several unique features. For instance, domains and their patterns self-assemble simultaneously, so that there is no need to pre-synthesize

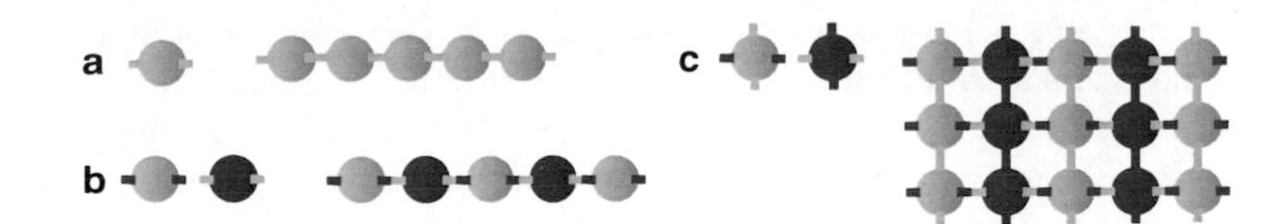

Self-assembly of Nanostructures, Fig. 1 An example shows the self-assembly of a discrete system

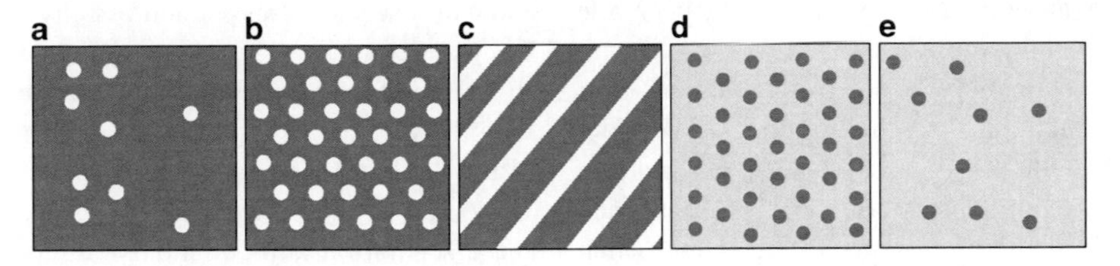

Self-assembly of Nanostructures, Fig. 2 An example shows the self-assembly of a continuum system. The domain patterns evolve from *dots* to *parallel stripes*, and to *inverted dots* with the increasing coverage of the bright phase

the building blocks; a significant degree of process flexibility and control can be achieved; and the approach may be applied to diverse systems.

Static Self-assembly Versus Dynamic Self-assembly

Depending on the nature of interactions, self-assembly systems can be classified into two categories: static and dynamic. Static self-assembly is a process driven by energy minimization to form static structures which are at global or local equilibrium. Many examples fall into this category, such as colloidal crystals, microphase separation, or spontaneous domain patterns. Dynamic self-assembly involves interactions that only occur when the system is dissipating energy. The structure is determined by an energetic minimum relying on an influx of energy to the system. When the energy flux stops, the minimum configuration does not exist anymore, leading to disintegration of the structure. For instance, rotating nanodisks in a liquid can generate hydrodynamic interactions due to the flow of liquid. When the disks are close to each other, the hydrodynamic interaction causes them to form a nicely ordered hexagonal pattern with a regular inter-disk spacing. When the rotation stops, the interaction disappears and the pattern vanishes. Another example is the oscillating reaction-diffusion reaction, where an evolving pattern develops from a quiescent medium upon the influence of stimuli. A living organism is possibly the most typical example of dynamic self-assembly.

Compared to static self-assembly, dynamic self-assembly is even less understood [11]. This process involves interacting building blocks that can adapt or react to a chemical or physical stimulus from the environment. Unlike the forces that drive static self-assembly through a reduction of the free energy toward equilibrium, the interactions in dynamic self-assembly may drive the formation of structures and patterns away from the thermodynamic equilibrium sustained by a continuous energy input. Such patterns can thus be sensitive to external stimuli and adaptive in response to the surrounding conditions [12].

Forces That Drive Self-assembly

Forces of different physical origins may contribute to the interactions between the building blocks. The short-range interaction is generally controlled by forces such as ionic, covalent, or metallic bonding; hydrogen bonding; and van der Waals forces. The long-range interaction has an effective range much longer than the atomic distance, and can originate from elasticity, electrostatic or magnetic field, colloidal and capillary forces, or hydrodynamic interaction. The short-range interaction alone typically leads to either a homogenous structure as shown in Fig. 1a, or a structure with only local modulation whose wavelength is on the order of the size of the building blocks, such as those in Fig. 1b and c. The long-range interaction is essential to form characteristic feature sizes larger than the dimension of the building blocks. Thus long-range interaction is especially important for the self-assembly in a continuum system to form nanoscale feature size, which is much larger than the building block of a single atom or molecule.

In a discrete system of nanoscale building blocks, the feature size is determined by the competition between the attractive and repulsive forces. Thus force balance analysis is a useful approach. While such a concept may still be applied to a continuum system, the connection between the domain size and their interaction is much less direct. Therefore the energy method is usually used to analyze a continuum system. The feature size is determined by two competing actions: a coarsening action which tends to increase the domain size and a refining action which tends to reduce the domain size. The following discusses a few representative forces and their roles in self-assembly.

Surface Stress

Surface stress plays crucial roles in a variety of surface phenomena in solids. To explain the idea, consider a uniform and infinitely large solid shown in Fig. 3a. The body is unstrained and is taken as the reference state. In Fig. 3b, the body is cut into two parts, and the atoms on the surfaces are allowed to find their equilibrium configuration. The energy in state (b) is higher than that in state (a). The energy difference between the two states, G_0, can be written as $G_0 = 2\gamma_0 A_0$, where γ_0 is the excess free energy per unit area owing to the existence of the surface, and A_0 is the surface area. G_0 is called surface energy. γ_0 is usually called surface energy density, which is the reversible work per unit area to create a surface. Figure 3c is the same solid in (a) but under strain ε. In Fig. 3d, the strained body is cut into two parts. The surface energy can be written as $G = 2\gamma A_0$. The two energies, G and G_0, are different. The former depends on the strain, i.e., $G = G(\varepsilon)$. Accordingly, the surface energy density also depends on the strain state, $\gamma = \gamma(\varepsilon)$. Hence the surface energy depends on both the created surface area and the strain state, namely, $d(\gamma A_0) = A_0 f d\varepsilon$, where $f = \partial\gamma/\partial\varepsilon$ is called surface stress. Surface stress becomes a tensor when generalized to biaxial strains.

The physical origin of surface stress is related to the difference of bonding between atoms at the surface and the interior atoms. A typical approach to measure surface stress is the cantilever bending method. Upon deposition of atoms on the top surface, surface stress causes the cantilever to bend. By measuring the curvature and some geometric parameters, such as the thickness of the cantilever, one can calculate surface stress. The surface stress of a domain depends on its composition. For a binary epilayer, a composition modulation will cause surface stress nonuniformity.

Consider a thin binary layer that grows epitaxially on a solid substrate with fixed thickness, say a monolayer. The two species can relocate by diffusion within the layer. Several physical ingredients contribute to the spontaneous domain formation behavior. Phase separation is a commonly observed phenomenon in various material systems, which can be explained in terms of the free energy of mixing. Consider a mixture of two atomic species A and B, and define concentration by the fraction of one species. The free energy density is given by a double well function of the concentration. A tangent line contacts the function at two concentrations, corresponding to the equilibrium A-rich and B-rich phases. A mixture with an average concentration between these two concentrations will separate into the two phases to reduce the energy. The atoms at the phase boundaries have excess free energy. To reduce the free energy, the total area of the phase boundaries must reduce. Consequently, atoms leave small domains, diffuse in the matrix, and join large domains. Over time the small domains disappear, and the large ones become larger. Thus phase boundary energy causes phase coarsening, which cannot form stable patterns. The observed stable periodic patterns suggest that there must be a refining action in the regular nanoscale structures. This action is provided by the surface stress.

Figure 4 shows an important concept: substrate deformation allows the nonuniform surface stress to reduce the total energy. The surface stress is f_0 everywhere except for the portion of the surface between points P and Q, where the surface stress steps up by F. The nonuniformity causes P and Q to move toward each other, giving rise to a strain field in the substrate. Imagine that a pair of forces are applied to move the points P and Q away from each other. When the applied forces reach the magnitude F, the substrate becomes unstrained. Because the forces do work to the body, the unstrained state has a higher energy than the strained state. The energy change depends only on F, but not on f_0. Whether the overall surface stress is tensile, compressive, or vanishing does not

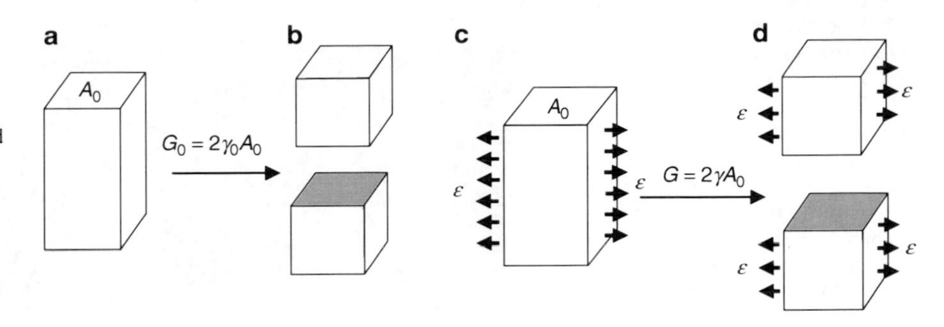

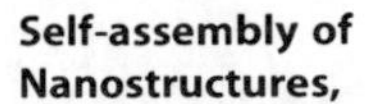
Self-assembly of Nanostructures, Fig. 3 Schematic representation illustrates the concept of surface energy and surface stress

make any difference to the energy reduction. When the domain size is refined, more pairs of F will contribute so that the energy is reduced further. Consequently, the elastic energy in the film-substrate composite tends to refine the domains. The refinement, however, adds more domain boundary, which increases the phase boundary energy. The phase boundary energy tends to coarsen the domains. The two competing actions – refining and coarsening – can select an equilibrium domain size.

These physical ingredients have been incorporated into an energy framework to simulate various nanostructures [13]. Figure 5 clearly shows the refining effect of the surface stress. The average concentration is 0.5. Each phase takes half of the surface area. In sequence (a), shortly after phase separation, the two phases form serpentine stripes. The width of the stripes stabilizes very fast. From $t = 1{,}000$ to $t = 10^6$, the widths are almost invariant. In contrast, in sequence (b), the two phases try to increase their sizes as much as possible. The system finally evolves into a state that one phase takes half of the calculation cell and the other phase takes the rest half. This reproduces the classical spinodal decomposition. The pattern depends on the average concentration. Changing the average concentration away from 0.5 can generate dots instead of stripes. Anisotropic surface stress induces interesting patterns such as parallel stripes and herringbone structures [14].

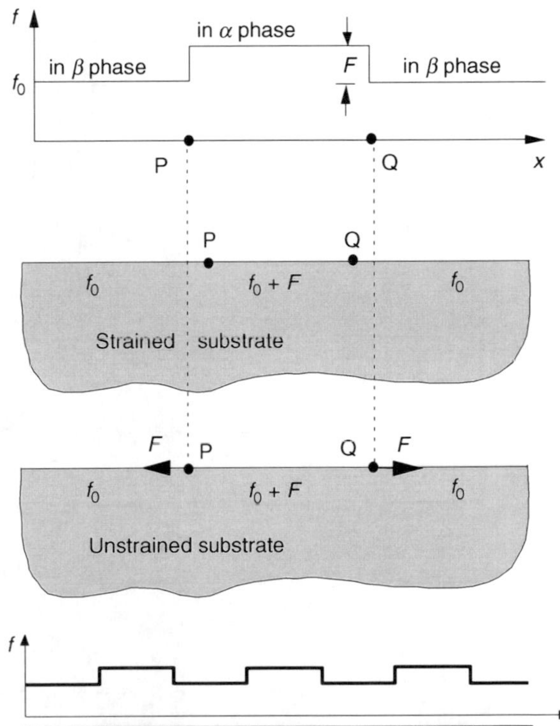

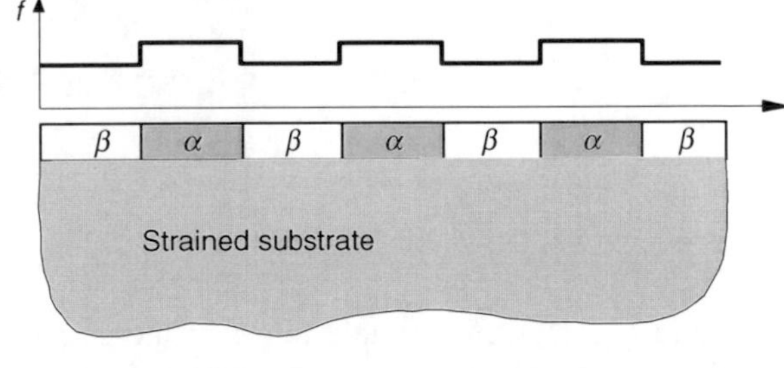

Self-assembly of Nanostructures, Fig. 4 Substrate deformation allows the nonuniform surface stress to reduce the energy, and surface stress provides a refining action

Guided or Templated Self-assembly

Guided or templated self-assembly uses a coarse scale external field to influence the fine scale self-assembly behavior. An essence of the approach is that the wavelength of the guiding field or template can be much larger than the nanoscale structures to be formed. Consequently, preparing the coarse scale field involves low cost and can be easily applied to a very large area. The guiding field can be electrostatic, geometric pre-patterns, surface chemistry, or elastic field [15].

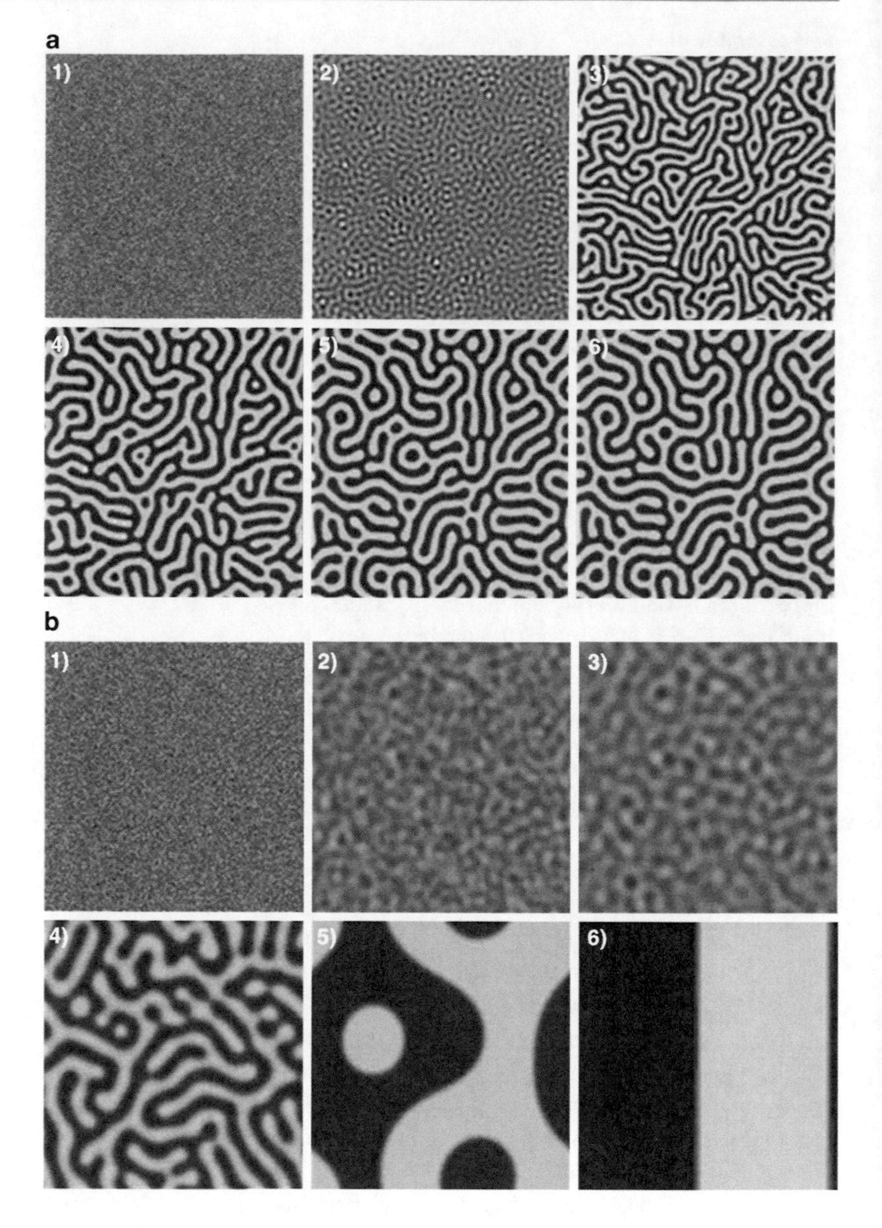

Self-assembly of Nanostructures, Fig. 5 Simulation starts with a random initial condition with an average concentration of 0.5. (**a**) The phases form nanoscale serpentine stripes with surface stress. (**b**) Spinodal decomposition without surface stress. The phases always coarsen. The times are (1) $t = 0$, (2) $t = 10$, (3) $t = 100$, (4) $t = 1{,}000$, (5) $t = 10^5$, (6) $t = 10^6$ (Adapted from [14])

Figure 6 shows an example to guide the surface patterns with strain field on the substrate surface. Surface strain is applied in the black region in Fig. 6a. Three cases are shown: (1) no guidance, (2) guiding strain in wide stripes, and (3) guiding strain in wavy stripes. In practice, there are many ways to induce an elastic field on the substrate surface. In addition to direct mechanical loading, pre-patterning a substrate with different materials by photolithography or applying an electric field to a substrate embedded with piezoelectric particles produce diverse well-defined strain fields. Figure 6b shows patterns for an average concentration of 0.3. The monolayer separates into two phases and evolves into a multi-domained triangular lattice of dots when there is no guidance. The middle column shows that in case [2], the monolayer evolves into pattern colonies. The dots form an almost perfect triangular lattice within each colony, and line up with the close packed direction parallel to the edge of the stripes. The lattices in the white and black stripe areas are self-contained, maintaining their own lattice spacings. The right column shows that in case [3], the dots

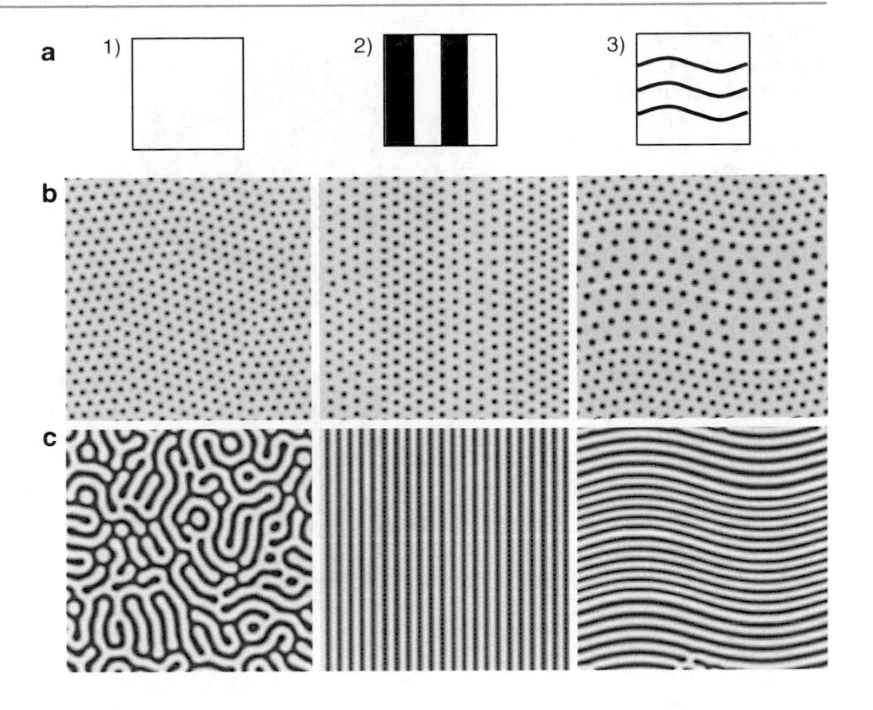

Self-assembly of Nanostructures, Fig. 6 Guided self-assembly with strain field on the substrate surface. (**a**) A constant strain is applied in the *black region* to guide the self-assembly on the substrate surface. (**b**) Patterns for an average concentration of 0.3. (**c**) Patterns for an average concentration of 0.5 (Adapted from [15])

orientate themselves along the wavy edges of the three curves. The narrow curves preclude the formation of any dots inside. Similar guiding effect is also shown for an average concentration of 0.5 in Fig. 6c. These results suggest that external loading can effectively change size, shape, and orientation of self-assembled surface structures. Similar guided self-assembly has also been achieved by surface chemistry [16].

Figure 7 shows another example where electric field is used to guide self-assembly. A three-dimensional model has been developed to allow the simulation of the entire self-assembly process and electric field design [17]. It is shown that a melted thin polymer film subjected to an electrostatic field may lose stability at the polymer-air interface, leading to uniform self-organized pillars emerging out of the film surface. The film thickness can affect the patterns formed. With patterned electrodes, parallel stripes replicating an electrode pattern emerge.

Electric Dipoles

An adsorbate molecule usually carries an electric dipole. Even if the molecules are nonpolar, the act of binding onto a substrate breaks the symmetry and causes the formation of dipoles. A molecule can be engineered to carry a large electric dipole moment by incorporating a polar group. Electric dipole interaction can give rise to domain patterns. Examples include Langmuir films at the air–water interface, ferrofluids in magnetic fields, and organic molecules on metal surface. Despite the difference of these systems, similar phenomenology and mechanism can be identified. The adsorbed molecules are mobile. Domains coarsen to reduce the domain boundary and refine to reduce the dipole interaction energy. The competition leads to equilibrium patterns. This mechanism may be used to make two-dimensional nanostructures. Specifically, molecule monolayers composed of electric dipoles can be manipulated with an electric field induced by an AFM tip, a ceiling above the layer, or an electrode array in the substrate.

Multilayers of dipole molecules have the capability for making complex structures, especially the formation of nanointerfaces and three-dimensional nanocomposites. Studies have revealed self-alignment between layers, reduction of feature sizes, and guided self-assembly by layer–layer interaction and embedded electrodes [18]. Figure 8 gives an example. Each layer is composed of two kinds of molecules with different dipole moments, which form two domains. Figure 8a shows that the first layer self-assembles into a triangular lattice of dots at an average concentration

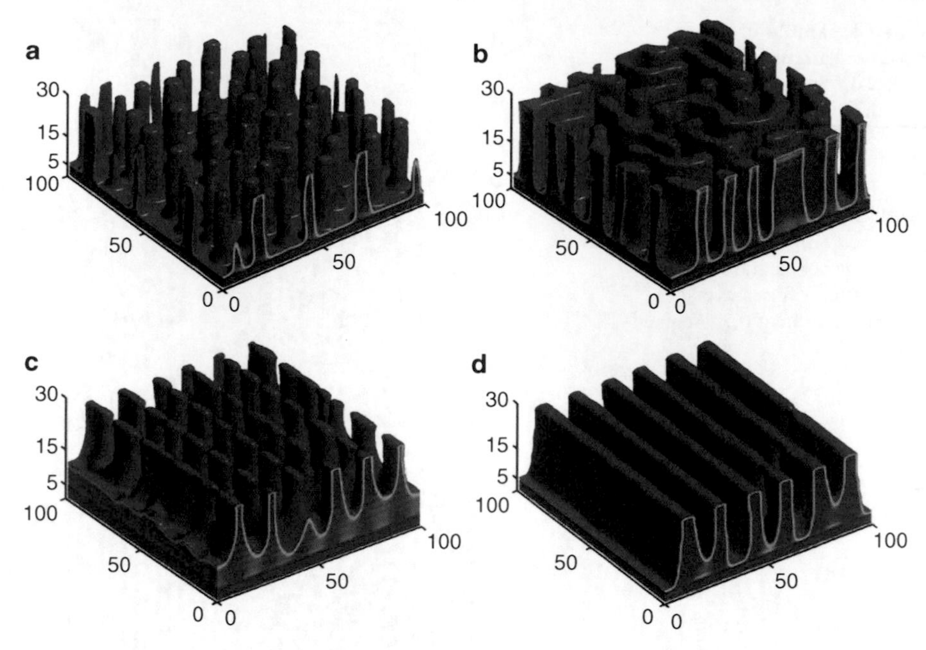

Self-assembly of Nanostructures, Fig. 7 Self-assembled morphological patterns of a melted thin polymer film induced by two electrodes parallel to the initial flat film surface, one above the film and the other beneath the film. (**a**) Induced nanopillars in a thin film. (**b**) A relatively thick film. The film continues to evolve after touching the ceiling electrode and then spread. (**c**) Patterned top electrode causes the nanopillars to line up. (**d**) Replication of the top electrode pattern (Adapted from [17])

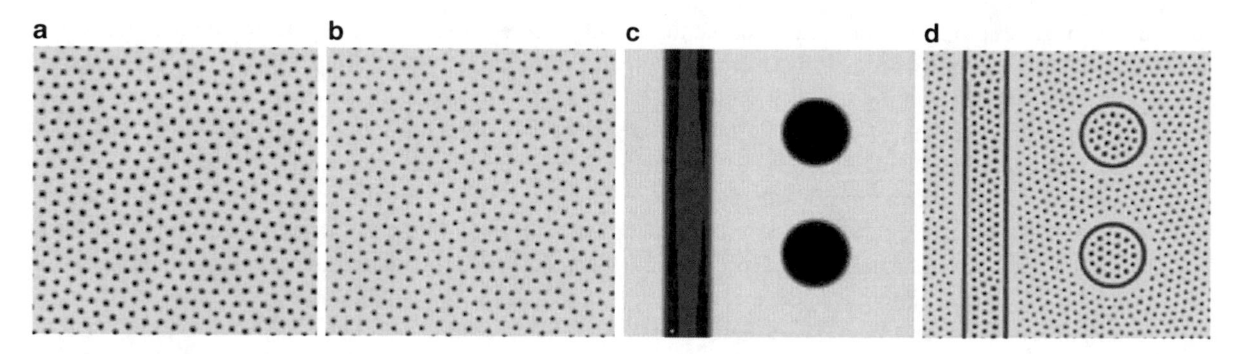

Self-assembly of Nanostructures, Fig. 8 Self-assembled patterns by electric dipole interaction. (**a**) First layer, average concentration 0.3. (**b**) Second layer on top of (**a**). (**c**) First layer pattern replicating the electrode pattern. (**d**) Second layer on top of (**c**) (Adapted from [18])

of 0.3. The dots have uniform size and form multiple grains. Figure 8b shows the second layer pattern at an average concentration of 0.2 and grows on top of the pattern in Fig. 8a. The second layer evolves from a completely different random initial condition. The dots of the second layer stay at exactly the same positions as those in Fig. 8b, suggesting the anchoring effect of the first layer. The dot size of the second layer is smaller due to lower average concentration. The observations suggest that the first layer determines the ordering and lattice spacing, while the second layer determines the feature size. A scaling down of size can be achieved via multilayers. The interesting behavior suggests a potential fabrication method. In additional to self-assembly, the first layer pattern can be defined by embedded electrodes, proximal probe technique, or nanoimprinting. Figure 8c shows the first layer pattern with the application of a high voltage, which sweeps off any self-assembled features so that the monolayer replicates the voltage pattern. The second layer evolves into a pattern shown in Fig. 8d. The dots align themselves along the edges of the first layer pattern, and form triangular lattice. It is interesting to note the formation of pairing

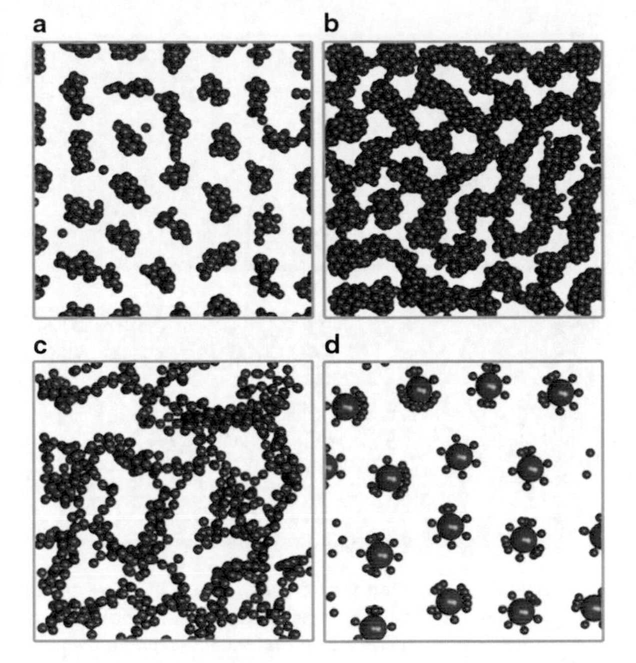

Self-assembly of Nanostructures, Fig. 9 Top view of structures self-assembled in a system of binary nanoparticles. (**a**) Gradient core-shell structure. (**b**) A continuous structure with isolated holes in between. (**c**) Network structure. (**d**) A single layer (Adapted from [19])

dark and white lines following the contour of the first layer pattern. Two nearby regions separated by the lines have different preference for the two dipole species. This causes local accumulation and depletion of the two dipoles, resulting in the observed phenomena.

Another example of electric dipole interaction is the self-assembly of nanoparticles [19]. Under an applied electric field, each particle will acquire an effective induced dipole. The dipole interaction causes particles to move in a liquid medium and line up into chains following the electric field direction. The formation of an ordered three-dimensional structure is further affected by the interaction between the chains. The self-assembly of binary nanoparticles can lead to a wide class of nanocomposite materials with properties not attainable by a single particle component. Figure 9a shows a top view of the structure. Behind each visible particle shown in the figure is a particle chain lining up. The two kinds of particles are denoted by red and blue colors. They have the same diameter and are both more polarizable than the medium, while the red particle is more polarizable than the blue one. The volume fraction of the particles is 10.2%. Under these situations the particles assemble into isolated columns with a core-shell configuration. The core is composed of the more polarizable red particles, while the shell is composed of the blue ones. From an energetic point of view, the attraction between two red particle chains is stronger than the interaction between a red particle chain and a blue particle chain. The latter is strong than the attraction between two blue particle chains. As a result, the red chains aggregate to form columns. The blue chains tend to get close to the red columns as much as possible, leading to the formation of shells. It is exciting to note that this self-organized functionally gradient structure offers a gradual transition of the permittivity from that of the core to that of the medium. This approach could be potentially very useful to construct functional gradient nanocomposites from bottom up.

Figure 9b shows a higher volume fraction of particles, 30.6%. The system evolves into a continuous structure with isolated holes in between. The blue particles enclose the red particles, and form the peripheral regions around the holes. The particle columns demonstrate local BCT structures. Another common feature is that same kind particles tend to aggregate. Figure 9c demonstrates the structure when one kind of particle is more polarizable than the medium while the other is less polarizable. The particles also form pure chains along the field direction with each chain composed of single kind particles. A distinct feature in Fig. 9c is that the red and blue chains are highly dispersed and form a network. This morphology is in contrast to that in Fig. 9a, where same color chains aggregate.

Assembling a single layer of nanoparticles on a substrate has many potential applications. A relevant simulation is shown in Fig. 9d for a layer of two kinds of particles. The repulsion between the red particles causes them to form a nicely ordered triangular lattice. The attraction between the blue and red particles causes the formation of blue rings surrounding the red cores.

Sequential Activation of Self-assembly

The lack of long-range order has been a major challenge for self-assembled nanostructures since high regularity is crucial to many applications. A general

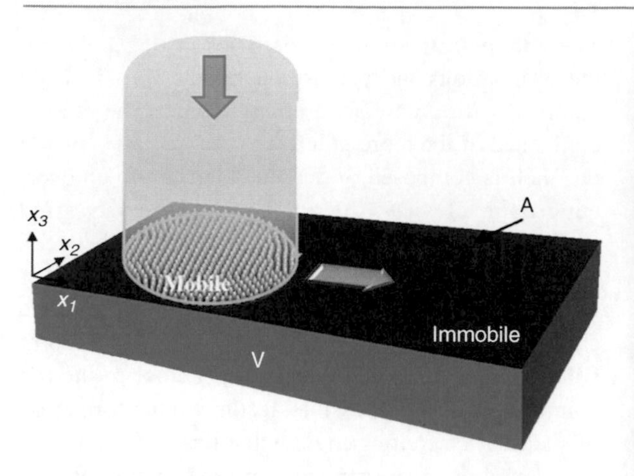

Self-assembly of Nanostructures, Fig. 10 A schematic of sequential activation of self-assembly. Self-assembly is first activated in a finite mobile region, where atoms are allowed to diffuse and form domain patterns. This initial mobile region serves as a "seed." The mobile region is then shifted like scanning. Self-assembly in the newly activated region will be under influence of patterns already formed in the seed (Adapted from [20])

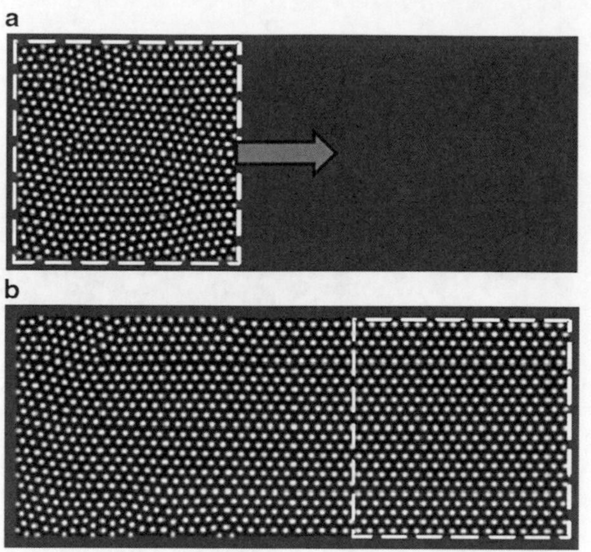

Self-assembly of Nanostructures, Fig. 11 Growth of superlattice from seeds. (**a**) A square seed. (**b**) A band of nicely ordered hexagonal superlattice formed after scanning over the width. The lattice improved to perfect along with the scanning, demonstrating tolerance of defects in the seeds (Adapted from [20])

kinetics-based template-free approach has been developed recently to grow nanostructured superlattices over a very large area [20]. The essence is sequential activation of self-assembly. The idea is illustrated in Fig. 10. A binary monolayer is used as an example. Typically multiple grains of dots will form due to simultaneous self-assembly at different locations, producing a pattern lack of long-range order. Here self-assembly is first activated in a finite mobile region, where atoms are allowed to diffuse and form domain patterns. This initial mobile region will serve as a "seed." The seed does not need to have a perfect lattice. Then the mobile region is shifted like scanning. The self-assembly in the newly activated region will be influenced by the pattern already formed in the seed. In experiments this process can be achieved by laser or ion beam scanning to control the local temperature, so that diffusion is activated sequentially at each spot along the scanning path. This sequential activation will lead to a large long-range ordered domain pattern even if started with an imperfect seed. The pattern quickly improves and converges to a perfect superlattice along with the sequential activation.

Figure 11 demonstrates the growth of long-range ordered superlattices from a seed. Rather than using a perfect lattice directly, the seed is grown on site. Take Fig. 11a as an example. The constraint of kinetics leads to a square seed composed of dots. The seed is larger than the size of a typical single grain. Therefore defects such as misalignment and multiple grains appeared in this seed. Next the mobile region is shifted to the right. This process is called scanning. The shift distance in each step is much smaller compared to the seed size, which creates a continuous scanning effect. The scanning velocity is chosen to ensure ample time for new domains to develop. Figure 11b shows the structure after scanning over the width of the calculation cell, which forms a band of nicely ordered hexagonal superlattice. Noticeably, the lattice improves to perfection along with the scanning, demonstrating the tolerance of defects in the seeds. High output is essential to nanostructure applications. Two schemes have been proposed to facilitate large-scale fabrication. (1) As shown in Fig. 12a, the scanning directions are alternated between left-right scanning and up-down scanning, using the superlattice created in each previous step as a large seed. This scheme allows the growth rate (area of lattice created per unit time) to increase exponentially with time, greatly

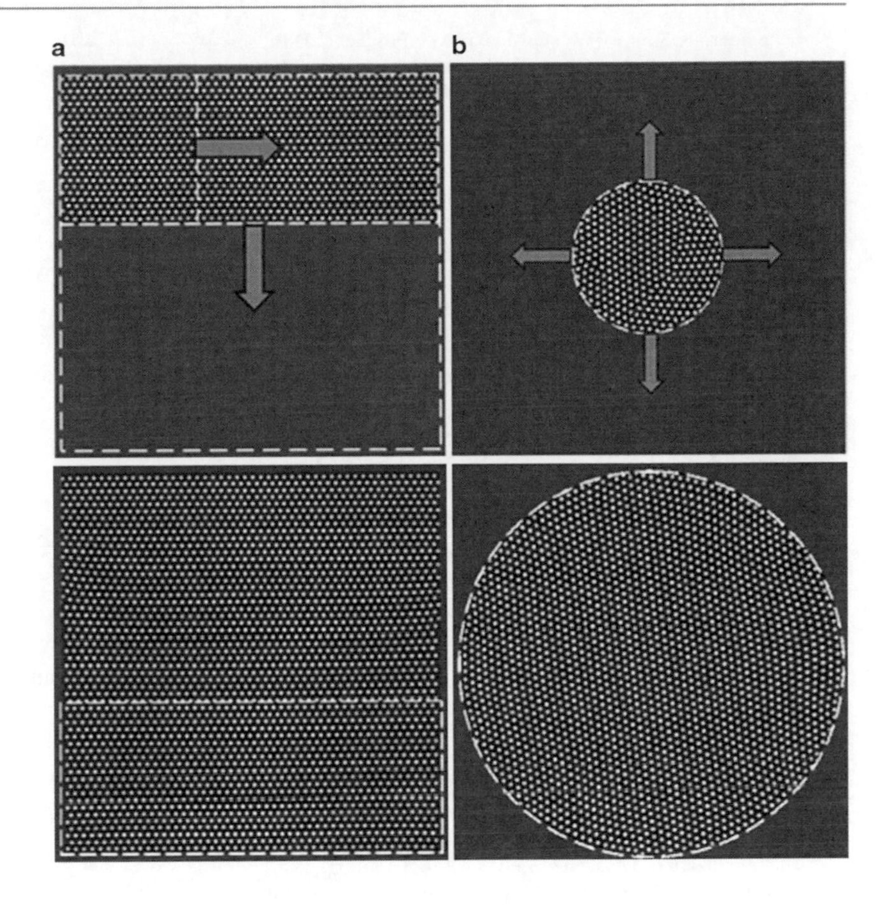

Self-assembly of Nanostructures, Fig. 12 Two schemes for scaling-up growth. (**a**) Alternate the scanning directions, using the superlattice created in previous step as a large seed. (**b**) Increase the size of the mobile in two dimension (Adapted from [20])

accelerating large area fabrication. (2) As shown in Fig. 12b, the size of the mobile area is increased in two dimensions, rather than scanning along one direction. This scheme allows the growth rate to be quadratic of the scanning velocity.

Cross-References

- ▶ Charge Transport in Self-assembled Monolayers
- ▶ Nanomaterials for Electrical Energy Storage Devices
- ▶ Nanomaterials for Excitonic Solar Cells
- ▶ Nanoparticles
- ▶ Nanostructured Functionalized Surfaces
- ▶ Nanotechnology
- ▶ Self-Assembled Monolayers
- ▶ Self-Assembly
- ▶ Self-Assembly for Heterogeneous Integration of Microsystems

References

1. Lu, W., Sastry, A.M.: Self-assembly for semiconductor industry. IEEE Trans. Semicond. Manuf. **20**, 421–431 (2007)
2. Pileni, M.P.: Nanocrystal self-assemblies: fabrication and collective properties. J. Phys. Chem. B **105**, 3358–3371 (2001)
3. Huang, Y., Duan, X., Wei, Q., Lieber, C.M.: Directed assembly of one-dimensional nanostructures into functional networks. Science **26**, 630–633 (2001)
4. Smith, P.A., Nordquist, C.D., Jackson, T.N., Mayer, T.S., Martin, B.R., Mbindyo, J., Mallouk, T.E.: Electric-field assisted assembly and alignment of metallic nanowires. Appl. Phys. Lett. **77**, 1399–1401 (2000)
5. Plass, R., Last, J.A., Bartelt, N.C., Kellogg, G.L.: Nanostructures – self-assembled domain patterns. Nature **412**, 875 (2001)
6. Fasolka, M.J., Mayes, A.M.: Block copolymer thin films: physics and applications. Annu. Rev. Mater. Res. **31**, 323–355 (2001)
7. Love, J.C., Estroff, L.A., Kriebel, J.K., Nuzzo, R.G., Whitesides, G.M.: Self-assembled monolayers of thiolates on metals as a form of nanotechnology. Chem. Rev. **105**, 1103–1169 (2005)

8. Schneider, K.S., Lu, W., Owens, T.M., Fosnacht, D.R., Banaszak Holl, M.M., Orr, B.G.: Monolayer pattern evolution via substrate strain-mediated spinodal decomposition. Phys. Rev. Lett. **93**:166104 (2004)
9. Krausch, G., Magerle, R.: Nanostructured thin films via self-assembly of block copolymers. Adv. Mater. **14**, 1579–1583 (2002)
10. Lopes, W.A., Jaeger, H.M.: Hierarchical self-assembly of metal nanostructures on diblock copolymer scaffolds. Nature **414**, 735–738 (2001)
11. Whitesides, G.M., Boncheva, M.: Beyond molecules: self-assembly of mesoscopic and macroscopic components. Proc. Natl. Acad. Sci. USA **99**, 4769–4774 (2002)
12. Ozin, G.A., Hou, K., Lotsch, B.V., Cademartiri, L., Puzzo, D.P., Scotognella, F., Ghadimi, A., Thomson, J.: Nanofabrication by self-assembly. Mater Today **12**, 12–23 (2009)
13. Lu, W., Suo, Z.: Dynamics of nanoscale pattern formation of an epitaxial monolayer. J. Mech. Phys. Solids **49**, 1937–1950 (2001)
14. Lu, W., Suo, Z.: Symmetry breaking in self-assembled monolayers on solid surfaces: anisotropic surface stress. Phys. Rev. B. **65**, 85401 (2002)
15. Lu, W., Kim, D.: Engineering nanophase self-assembly with elastic field. Acta Mater. **53**, 3689–3694 (2005)
16. Lu, W., Kim, D.: Patterning nanoscale structures by surface chemistry. Nano. Lett. **4**, 313–316 (2004)
17. Kim, D., Lu, W.: Three-dimensional model of electrostatically induced pattern formation in thin polymer films. Phys. Rev. B **73**, 035206 (2006)
18. Lu, W., Salac, D.: Patterning multilayers of molecules via self-organization. Phys. Rev. Lett. **94**, 146103 (2005)
19. Park, J., Lu, W.: Self-assembly of functionally gradient nanoparticle structures. Appl. Phys. Lett. **93**, 243109 (2008)
20. Zhao, Z., Lu, W.: Growing large nanostructured superlattices from a continuum medium by sequential activation of self-assembly. Phys. Rev. E. **83**, 041610 (2011)

Self-Cleaning

▶ Lotus Effect

Self-Healing Materials

▶ Self-Repairing Materials

Self-Organized Layers

▶ Self-Assembled Monolayers for Nanotribology

Self-Organized Nanostructures

▶ Self-assembly of Nanostructures

Self-Regeneration

▶ Self-repairing Photoelectrochemical Complexes Based on Nanoscale Synthetic and Biological Components

Self-Repairing Materials

Michael Nosonovsky
Department of Mechanical Engineering, University of Wisconsin-Milwaukee, Milwaukee, WI, USA

Synonyms

Self-healing materials

Definition

Self-repair or self-healing is the ability of a material (usually, a composite or nanocomposite) to partially or completely repair damage, such as voids and cracks, occurring during its service lifetime [1, 2]. Self-repair is achieved by embedding a nano-, micro-, or multi-scale repair mechanism into the structure of the material (Fig. 1). Autonomous self-healing refers to the ability to repair damage without any external activation. Nonautonomous self-repair refers to healing mechanisms, which require external activation, such as heating the material.

Occurrence and Key Findings

Self-healing is related to the general ability of many systems for self-organization. To facilitate self-organization, usually a special nano-, micro-, or multi-scale structure of a material is needed. Self-healing in biological objects is a source of inspiration for this new

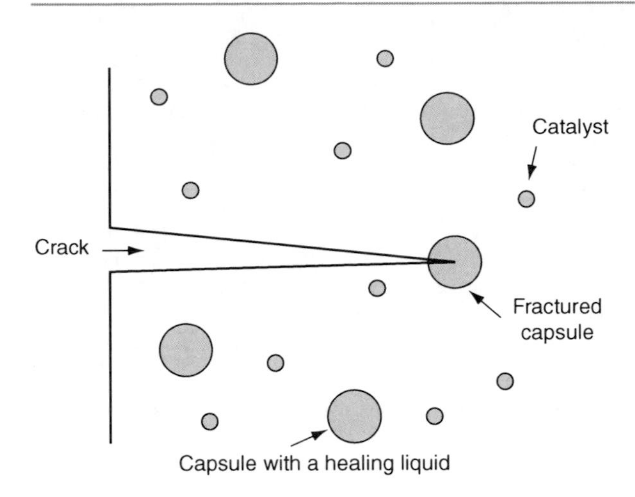

Self-Repairing Materials, Fig. 1 Self-repair by embedding microcapsules with healing agent

research area. Most living tissues or organisms can regenerate and heal themselves, provided the incurred damage is small or moderate. Most engineered materials, however, deteriorate with time irreversibly due to wear, brittle fracture, fatigue, creep, and other modes of degradation, which limits the life of various components and sometimes causes catastrophic damage. Biological mechanisms of healing are very complex and cannot be directly borrowed for artificial materials. Therefore, it is preferable to speak about bio-inspired, rather than biomimetic, self-healing materials.

Polymers

Self-healing has been applied most successfully in polymers since they have relatively high rates of diffusion and plasticity due the presence of the cross-molecular bonds. One way to create self-healing polymers is to use thermosetting polymers that have the ability to cure (toughening or hardening by cross-linking of polymer chains), such as the thermosetting epoxy. Epoxy is a polymer formed by a reaction of an epoxide resin with polyamine hardener. Epoxy can serve as a healing agent that is stored within thin-walled inert brittle macrocapsules embedded into the matrix along with a catalyst or hardener. The catalyst or hardener is also embedded in the matrix, but separately from the healing agent. When a crack propagates, the capsules fracture, the healing agent is released and propagates into the crack due to capillarity. Then the healing agent mixes with the catalyst embedded in the matrix, which triggers the cross-linking reaction and hardening of the epoxy that seals the crack [3].

A different approach involves thermoplastic polymers with various ways of incorporating the healing agent into the material. In this approach, heating is often required to initiate healing [3]. A self-repair method involving two concentric cylinders of conducting material filled with a liquid solution containing electromagnetic particles of polystyrene or silica was suggested. When damage occurs, voltage is applied between the inner pipe and outer pipe, the current density naturally increases at the location of the damage, this increase in current density causes particle coagulation at the damage site, which in a way to heal the damage.

Ceramics

Besides the polymers, ceramic self-healing materials are being developed. For example, a concrete composite was produced with glass fibers containing an air-curing sealant embedded in the concrete matrix. This composite exhibited the self-healing behavior, but it suffered from a significant (10–40%) loss of stiffness compared with standard concrete due to fibers. This is a typical situation, when a compromise between self-healing and mechanical properties should be sought. In another project involving self-healing ceramics, researchers have studied the crack-healing behavior and mechanical properties of a mullite composite toughened by the inclusion of 15% (by volume) SiC whiskers. Self-healing ceramic materials often use oxidative reactions because the volume of oxide exceeds the volume of the original material, and therefore, products of these reactions can be used to fill small cracks [3]. Self-healing nanocomposites constitute another area of research.

Metallic Materials

It is much more difficult to heal metallic materials, than polymers, because metallic atoms are strongly bonded and have small volumes and low diffusion rates. Currently, there are three main directions which have been taken in the development of self-healing metallic systems. First is the formation of precipitates at the defect sites that immobilize further growth until failure. This mechanism is called sometimes "damage prevention" because it prevents the formation of voids by the diffusion of the precipitate. The atomic transport of matter to voids and defects is provided by a supersaturated solid solution in alloys having decreasing solid solubility of solute elements with

Self-Repairing Materials, Table 1 Self-healing mechanisms in metals

Mechanism	Precipitation	SMA reinforcement	Healing agent encapsulation
Type	Damage prevention (solid)	Damage management (solid, possibly liquid-assisted)	Damage management (liquid-assisted)
Matrix material	Al–Cu, Fe–B–Ce, Fe–B–N, etc.	Sn–Bi, Mg–Zn	Al
Reinforcement materials	–	NiTi	Sn–Pb
Microstructure parameter (ψ)	Solute fraction	Concentration of microwires	Concentration of microcapsules or low-melting-point alloy
Degradation measure (ξ)	Volume of voids	Volume of voids	Volume of voids
Healing measure (ζ)	Amount of precipitated solute	SMA strain	Amount of released healing agent
Characteristic length of degradation	Void size (microscale)	Void/crack size (macroscale)	Void/crack size (macroscale)
Characteristic length of the healing mechanism	Atomic scale (atomic diffusion)	Microwires diameter (macro or microscale)	Microcapsule size (microscale)
Phase transition involved	Solute precipitation	Martensite/austenite	Solidification of the solder
Healing temperature	Ambient	Martensite/austenite transition	Melting of the low-melting-point alloy
Property improved	Creep resistance	Restored strength	Restored strength and fracture toughness

decreasing temperature (e.g., Al–Cu). Such an "under-aged" alloy, when quenched from high temperature, becomes supersaturated or metastable. However, in order to facilitate precipitation of the solute, heterogeneous nucleation sites are needed. Sites with high surface energy, such as voids, defects, grain boundaries, and free surfaces, become nucleation sites. The driving mechanism for the diffusion is the excess surface energy of microscopic voids and cracks that serve as nucleation centers of the precipitate that plays the role of the healing agent. As a result, the newly formed void is sealed before it grows and thus minimizing the creep and fatigue [1, 2].

Second is reinforcement of an alloy matrix with microfibers or wires made of a shape-memory alloy (SMA), such as Nitinol (NiTi). SMA wires have the ability to recover their original shape after some deformation has occurred if they are heated above certain critical temperature. If the composite undergoes crack formation, heating the material will activate the shape recovery feature of the SMA wires which then shrink and close the cracks [1, 2].

The third approach is to use a healing agent (such as an alloy with a low-melting temperature) embedded into a metallic solder matrix, similarly to the way it is done with the polymers. However, encapsulation of a healing agent into a metallic material is much more difficult task than in the case of polymers. The healing agent should be encapsulated in microcapsules which serve as diffusion barriers and which fracture when crack propagates [1, 2].

Self-healing mechanisms in metals are summarized in Table 1, showing typical materials for the matrix and reinforcement, parameters that characterize microstructure, degradation, and healing, characteristic length scales for the degradation and healing mechanisms, the type of phase transition involved into the healing and what property is improved in the self-healing alloy, the nature of the healing force and details of healing mechanisms will be discussed in the consequent section.

Surface Healing

Surface is often the most vulnerable area of a material. Several approaches have been suggested to for fixing surface damage. One approach is to mimic the healing of skin that has a network of vessels and veins so that cutting the skin triggers the blood flow, its coagulation, and sealing of the cut. The vascular network is a tree-like hierarchical structure that provides a uniform and continuous distribution of fluids throughout the material volume. The vascularization has been successfully used for polymeric composites [4–6].

A biomimetic approach combining self-healing with the Lotus-effect has been suggested, mimicking healing of a leaf surface. In that approach, micro-reservoirs with coating liquid (e.g., wax paraffin) were attached to the surface with the supply of coating to the damage area.

Various types of self-lubrication (e.g., self-replenishing solid and liquid lubrication, reservoirs for liquid lubricant, lubricant supply facilitated by oxidation or a tribochemical reaction, etc.) can also be viewed as surface healing mechanisms.

Theoretical Considerations

From the thermodynamic point of view, self-repair can be viewed as a nonequilibrium self-organization process that leads to increasing orderliness of the material and thus to decreasing entropy. In most schemes of self-repair, the self-healing material is driven away from the thermodynamic equilibrium either by the deterioration process itself or by an external intervention, such as heating. After that, the composite slowly restores thermodynamic equilibrium, and this process of equilibrium restoration drives the healing [5].

In most situations, damage repair at a certain length scale is achieved at the expense of the deterioration at the lower-length scale, making self-repairing materials multi-scale systems. For example, macroscale cracks are healed at the expense of fracturing macrocapsules of the healing agent embedded in the matrix of the material. In the precipitation-induced damage prevention mechanisms, micro-voids are healed by atomic-scale material transport. Multi-scale organization is typical for biological materials and tissues, it is therefore not surprising that it plays a central role in biomimetic self-repairing materials.

In order to characterize degradation, it is convenient to introduce a so-called "degradation parameter" ξ, to represent, for example, the wear volume, and a corresponding generalized thermodynamic force, $Y^{\deg}$. When a self-healing mechanism is embedded in the system, another generalized coordinate, the healing parameter, ζ, can be introduced, for example, the volume of released healing agent together with the corresponding generalized force, Y^{heal}. The generalized degradation and healing forces are external forces that are applied to the system, and flows are related to the forces by the governing linear Onsager's equations

$$J^{\deg} = LY^{\deg} + MY^{heal}$$

$$J^{heal} = NY^{\deg} + HY^{heal} \tag{1}$$

where L, M, N, H are corresponding Onsager coefficients [5].

The degradation force $Y^{\deg}$ is an externally applied thermodynamic force that results in the degradation. The healing force Y^{heal} is an external thermodynamic force that is applied to the system. In most self-healing mechanisms, the system is placed out of equilibrium and the restoring force emerges, so this restoring force could be identified with Y^{heal}. Since the restoring force is coupled with the degradation parameter ξ by the negative coefficients $N = M$, it also causes degradation decrease or healing.

Measure of Healing

A quantitative measure is required in order to characterize the efficiency of a healing mechanism and to compare different mechanisms. Several parameters have been suggested, such as (1) restored strength divided by the original strength, (2) fracture toughness divided by the original toughness, and (3) creep life of the material (in the case of damage-prevention mechanisms). Most self-repair mechanisms are not capable to provide 100% restoration of the original properties of the material.

Cross-References

▸ Lotus effect

References

1. Ghosh, S.K. (ed.): Self-Healing Materials: Fundamentals, Design Strategies, and Applications. Wiley, Weinheim (2009)
2. van der Zwaag, S. (ed.): Self Healing Materials – An Alternative Approach to 20 Centuries of Materials Science. Springer, Dordrecht (2007)
3. van der Zwaag, S.: Self-healing behaviour in man-made engineering materials: bioinspired but taking into account their intrinsic character. Philos. Trans. R. Soc. A **367**, 1689–1704 (2009)
4. Nosonovsky, M., Amano, R., Lucci, J.M., Rohatgi, P.K.: Physical chemistry of self-organization and self-healing in metals. Phys. Chem. Chem. Phys. **11**, 9530–9536 (2009)
5. Nosonovsky, M.: Self-organization at the frictional interface for green tribology. Philos. Trans R. Soc. A. **368**, 4755–4774 (2010)
6. Nosonovsky, M., Bhushan, B.: Thermodynamics of surface degradation, self-organization, and self-healing for biomimetic surfaces. Philos. Trans. R. Soc. **A367**, 1607–1627 (2009)

S

Self-repairing Photoelectrochemical Complexes Based on Nanoscale Synthetic and Biological Components

Moon-Ho Ham[1], Ardemis A. Boghossian[1], Jong Hyun Choi[2] and Michael S. Strano[1]
[1]Department of Chemical Engineering, Massachusetts Institute of Technology, Cambridge, MA, USA
[2]School of Mechanical Engineering, Birck Nanotechnology Center, Bindley Bioscience Center, Purdue University, West Lafayette, IN, USA

Synonyms

Biomimetic energy conversion devices; Photovoltaic devices; Self-regeneration; Solar cells

Definition

Self-repairing photoelectrochemical cells are synthetic solar energy conversion devices that illustrate principles of self-regeneration and repair through a reversible assembly/disassembly cycle. Motivated by the repair mechanism in living chloroplasts, this cycle can potentially extend the lifetime of a photoelectrochemical cell indefinitely.

Introduction

Advancements in nanotechnology toward the synthesis and manipulation of materials have enabled the direct interfacing of synthetic components with biological complexes to create new functions. Direct coupling of synthetic, nanoscale components with biological complexes has been used to study biological systems [1, 2] as well as develop new biomimetic electronics [3–5]. In most recent developments, protein complexes have been interfaced with nanoparticles and single-walled carbon nanotubes (SWCNTs) in the synthesis of biomimetic solar cells [6–10]. This new generation of nano-bio solar cells often relies on a combination of biologically derived components and biologically inspired mechanisms for device fabrication to minimize costs and enhance device efficiencies. As technology progresses in this field, the mechanisms demonstrated by the new devices increasingly mimic processes found in nature. In this work, the hydrophobicity of SWCNTs for the self-assembly of nanoscale components was employed to mimick the self-repair process used by plants [6]. In this sense, a thorough understanding of the natural processes is essential for developing such biomimetic devices.

Photosynthesis: Chloroplast Structure

Natural light-harvesting mechanisms in plants are primarily carried out in the chloroplast, an organelle responsible for the conversion of sunlight into energy for the plant (Fig. 1) [11, 12]. Like most organelles, the chloroplast is surrounded by an outer as well as an inner membrane. A thick, aqueous fluid, the stroma, fills the organelle. The stroma contains genetic materials, including deoxyribonucleic acid (DNA), ribonucleic acid (RNA), and several ribosomes. The most pronounced containment within the chloroplast is the stacks of thylakoid membranes, or grana stacks. Each of the thylakoid membranes contains four key protein complexes used in photosynthesis: photosystem II (PSII), cytochrome b_6f (cyt b_6f), photosystem I (PSI), and adenosine-5′-triphosphate (ATP) synthase. These complexes are responsible for carrying out the first stage of photosynthesis, the light-dependent reactions [11].

The light-dependent reactions are initiated by the absorption of light in PSII. In PSII, light is primarily absorbed by the P680 site at approximately 680 nm. The light-harvesting antennas surrounding by the protein complexes contain an array of pigments and chlorophyll molecules that absorb light at various wavelengths throughout the solar spectrum, enhancing the overall absorbance of light by the plant. The photons absorbed by these antennas are then transferred to the P680 site via resonance energy transfer. Absorption of light at the P680 site excites an electron, transferring it to the pheophytin (Phe) site of the PSII reaction center (RC). The hole remaining in the P680 site is used in the oxidation of water for the production of oxygen and hydrogen.

$$2H_2O \xrightarrow{hv} O_2 + 4e^- + 4H^+$$

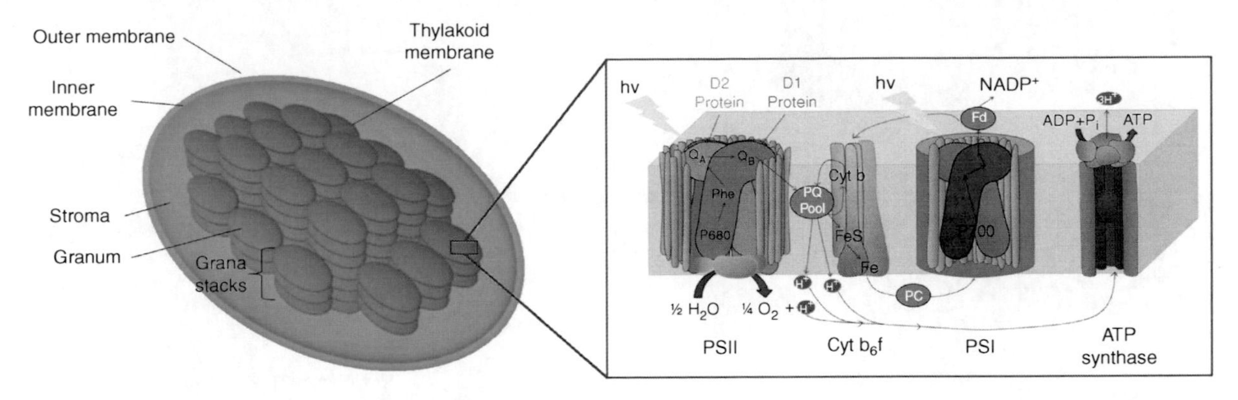

Self-repairing Photoelectrochemical Complexes Based on Nanoscale Synthetic and Biological Components, Fig. 1 Chloroplast structure and function. The chloroplast (left) consists of an outer membrane surrounding stacks of thylakoid membranes, or grana. Each thylakoid membrane is embedded with several protein complexes (right) including PSII, cyt b6f, PSI and ATP synthase. Electron-hole separation occurs in PSI upon light absorption. The hole is used in the oxidation of water into hydrogen and oxygen, while the electron is transmitted through cyt b6f and PSI. Synthesis of ATP for chemical storage of energy ultimately occurs in ATP synthase

Meanwhile, the electron is transferred to the Q_A and Q_B sites of PSII, where it is ultimately shuttled via a redox carrier to the next major transmembrane protein complex, cyt b_6f.

Cyt b_6f is responsible for electron transfer between the two major RCs of the membrane, PSI and PSII. The electron transfer between the two centers is coupled to the establishment of a transmembrane proton gradient between the outer stroma and inner thylakoid space. The electron undergoes a series of redox reactions demoting electron energy in exchange for proton transfer from the outer stroma to the inner thylakoid lumen. At the conclusion of this series of redox reactions, the electron is emitted at a lower energy state.

The electron emitted from cyt b_6f is transferred via a redox carrier to the next major protein complex, PSI. Like PSII, PSI is surrounded by a variety of pigment- and chlorophyll-containing antennas. The P700 site of PSI is responsible for light absorption at approximately 700 nm. As before, the surrounding antennas broaden the absorbance of the chloroplast and transfer this absorbed energy to the P700 site via resonance energy transfer. Upon energy absorption, the photoelectric effect is used to excite the electron at the P700, where it is emitted toward the ferrodoxin site for pickup by the next major complex. Excited electrons emitted by PSI are then subject to a variety of chemical pathways, one of which is nicotinamide adenine dinucleotide phosphate (NADPH) production. In this reduced state, the NADPH's reducing power is used to power the Calvin cycle, or the second stage of photosynthesis in the light-independent series of reactions, which takes place in the stroma.

The fourth, and final, major complex embedded within the thylakoid membrane, ATP synthase, is not directly involved in the preceding series of light-dependent reactions. In ATP synthase, a hydrogen ion is extracted from other sites in the membrane, including the water-oxidizing site of PSII, to produce ATP. ATP, which is also used in the Calvin cycle mentioned above, is the storage of light-absorbed energy in chemical form. Energy storage and release is carried out by phosphate bond formation and scission, respectively.

Natural Self-repair Cycles in Plants

The aforementioned system of light-dependent reactions relies on a series of embedded protein complexes under nearly continuous illumination that become highly susceptible to photodamage. In particular, the D1 subunit, which contains the P680, Phe, and the Q_B sites of PSII, is vulnerable toward protein damage. To address this, plants have evolved sophisticated mechanisms of self-repair wherein the damaged D1 protein is replaced with a newly synthesized protein (Fig. 2a) [13–15]. The initiation of the self-repair cycle is the damage of the D1 protein, which undergoes a change

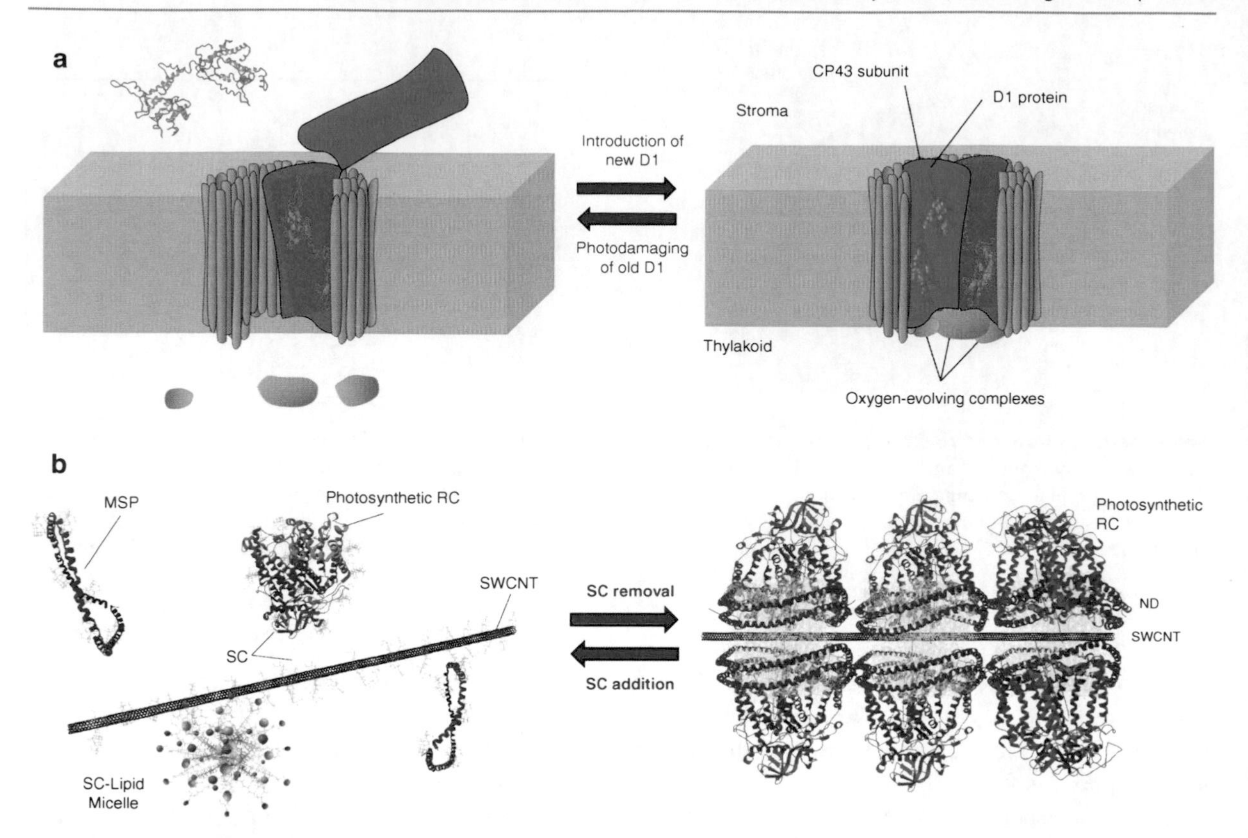

Self-repairing Photoelectrochemical Complexes Based on Nanoscale Synthetic and Biological Components, Fig. 2 Natural and synthetic mechanisms of self-assembly. (**a**) The self-repair in plants relies on the molecular recognition of components and trapping of meta-stable thermodynamic states. In the disassembled state (left), the damage D1 protein is released from the dysfunctional PSII and replaced with a new D1 protein. Introduction of the new protein triggers the self-assembly of PSII into a functional state (right). (**b**) The self-assembly of the RC-ND-SWCNT relies on the removal of the surfactant from the system. In the presence of SC, the systems exists in the disassembled state with surfactant dispersion of individual components (left). Upon SC removal, the proteins, lipids and nanotubes self-assemble into the complex shown (right)

in conformation and phosphorylation levels when denatured. The dysfunctional protein signals the spontaneous, partial disassembly of the PSII complex. With complex disassembly, the damaged protein is released from the system and diffuses toward the unappressed, stromal region of the membrane. Meanwhile, new D1 protein biosynthesized elsewhere in the cell diffuses toward the appressed region of the membrane. The selective diffusion of the damaged protein toward unappressed regions and new protein toward appressed regions of the membrane is attributed to the varying phosphorylation levels of the two proteins, which vary the degree of protein hydrophilicity. The more hydrophilic protein is driven toward the appressed region of the membrane whereas the more hydrophobic, damaged protein is driven toward the unappressed region. Introduction of the new protein to the disassembled complex triggers the spontaneous reassembly of a functional PSII incorporating the new D1 protein.

The described self-repair cycle consists of transitions between a series of metastable states in an energy-minimized fashion. Aside from the synthesis of a new protein, this cycle requires the input of no additional energy; the reversible disassembly and reassembly of PSII is completely driven by the formation of thermodynamically and kinetically trapped states. The self-assembly in particular is driven by hydrophobic/hydrophilic interactions and the assortment of diffusivities demonstrated by the proteins on the molecular scale. Synthetic replication of this reversible self-assembly process therefore hinges on the ability to control molecular interactions on the nanoscale.

Use of Biological Light-Harvesting Components in Solar Conversion Devices

Biological components such as PSI and photosynthetic RCs have external quantum efficiencies of almost 100% and energy yields of approximately 58%. Recent works have demonstrated that the biological light-harvesting components can be incorporated into optoelectronic devices including photovoltaic devices [6–10].

Jennings and coworkers [10] reported that biomimetically inspired energy conversion devices were fabricated with spinach-derived PSIs which were bound to an Au electrode surface through covalent imine bonds. To increase the surface area of the electrode, the nanoporous leaf-like structure of an Au electrode was employed, which led to the threefold enhancement of the photocurrent due to the increased density of PSI complexes compared to planar electrodes.

Carmeli and coworkers [8] directly bound PSI as a monolayer onto an Au surface. The PSIs were isolated from the cyanobacteria *Synechocystis* sp. PCC6803 and modified at cysteine sites near the P700 for covalent attachment. Kelvin probe force microscopy (KPFM) images show distinct photocurrent generation domains corresponding to PSIs covalently bound to the Au surface with unidirectional orientation. They extended their work to improve the photo-efficiency and apply it to other substrates. Orientated multilayers of PSI complexes were fabricated on the Au surface. In addition, they covalently attached PSI directly to carbon nanotubes and indirectly to GaAs surfaces via a small linker molecule.

A self-assembled monolayer of RCs, isolated from the purple bacterium *Rhodobacter (Rb.) sphaeroides*, on an Au electrode with a His tag was incorporated into solid-state photovoltaic cells, as reported by Baldo and coworkers [7]. The photocurrent spectrum of the device matched the absorption spectrum of the RCs, producing internal quantum efficiencies of $\sim$12%. Lebedev and coworkers [9] used cytochrome *c* as a conductive wire between photosynthetic RCs and the electrode, where the RCs were the same as those in Baldo's group, leading to a remarkable enhancement of the photocurrent. They also used arrayed carbon nanotube electrodes to improve the photo-conversion efficiency, where the RCs were encapsulated inside carbon nanotube arrays. The efficiency was considerably improved by increasing the number of RCs attached to the electrode surface by about fivefold compared to that obtained with the same proteins when immobilized on a planar graphite (HOPG) electrode.

Until now, most of these efforts have been centered on binding photoactive components onto an electrode surface in a unidirectional orientation to enhance photo-efficiencies. However, even with such an enhancement, these arrangements, like those often found in nature, undergo photo-degradation and protein damage. To address issues of photo-degradation, plants have evolved self-repair mechanisms whereby damaged proteins are replaced with photoactive, newly synthesized proteins. In fact, without this self-assembling mechanism, plants would produce less than 5% of their typical photosynthetic yields with lifetimes on the order of minutes under intense illumination.

Synthetic Regeneration Cycles in Photoelectrochemical Cells

One approach to mimicking this self-assembly process is to use nanoscaled materials with controllable properties to interface with biologically derived photoactive components to create a photoelectrochemical cell capable of plant-like regeneration. In a recent study [6], this approach was used to develop the first photoelectrochemical cell capable of mimicking key steps in the self-repair cycle. An aqueous solution consisting of SWCNTs, RCs isolated from *R. Sphaeroides*, the phospholipid dimyristoylphosphatidylcholine (DMPC), membrane scaffold proteins (MSPs), and the surfactant sodium cholate (SC) (Fig. 2b, left) is placed within a dialysis bag with a 10,000 molecular weight cutoff and dialyzed against buffer to selectively remove SC from the system. The removal of SC spontaneously triggers the self-assembly of the photoactive complex shown in Fig. 2b (right). This complex contains a series of agglomerates sequentially aligned along the length of a SWCNT. These agglomerates consist of a lipid bilayer disk, or nanodisk (ND) (Fig. 3). In the ND, the DMPC molecules arrange themselves such that their hydrophilic heads face outward toward the aqueous solution and their hydrophobic tails are sandwiched within the bilayer. Wrapped around the circumference of this disk are two strands of MSP,

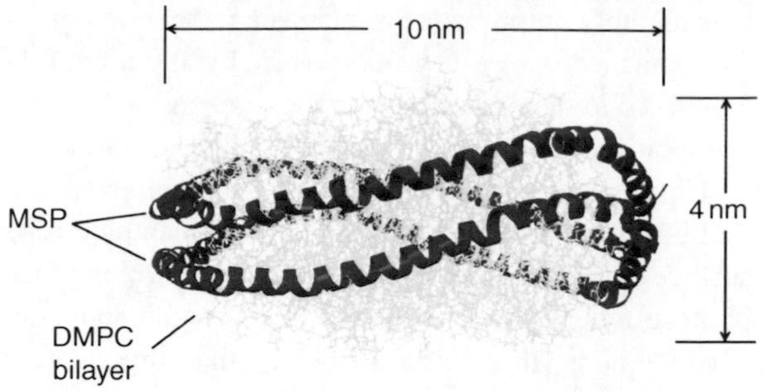

Self-repairing Photoelectrochemical Complexes Based on Nanoscale Synthetic and Biological Components, Fig. 3 ND structure. NDs consist of DMPC lipids arranged into a bilayer surrounded by two strands of MSPs. Atomic force microscopy (AFM) measurements reveal that these lipid bilayer disks are approximately 10 nm wide and 4 nm high

the length of which determines the overall disk diameter. The particular strands of MSP used in this study result in NDs that are approximately 10 nm wide and 4 nm high. These NDs, which align along the length of the nanotube in the assembled state, house photoactive RCs. Although the self-assembly process occurs spontaneously, the RCs are specifically orientated such that their hydrophobic area (the region near the P680 site) faces the SWCNT and their hydrophilic area (the area near the Q_B site) faces outward toward its aqueous surroundings. This self-assembly process is completely reversible, such that the re-addition of surfactant to the solution decomposes the complex back into its initial micellar state, which is monitored using the photoluminescence shift of the nanotubes [6].

Understanding and Quantifying the Self-assembly Process

This spontaneous assembly of nano- and bio-based materials into a precise configuration via chemical signaling alone can be attributed to the comparative scalabilities of the synthetic and biological components and the controlled, molecular interactions that occur on the nanoscale. The removal of surfactant from the system and, most importantly, from the hydrophobic SWCNT surface triggers the formation of (meta)stable hydrophobic/hydrophilic agglomerates that minimizes SWCNT exposure to its aqueous surroundings. The precise locations and sizes of the hydrophobic/hydrophilic regions in the RC, the thickness of the ND bilayer, and the unidimensionality and length of the SWCNT are all factors that contribute to the precise molecular arrangement of the RCs along the nanotube length. Even the slightest alteration in sizes, diffusivities, and hydrophobicities of the nanocomponents are enough to perturb the dynamics of the system and even altogether inhibit the formation of such a photoactive complex. To develop a more thorough, quantitative understanding behind the formation of these complexes, the formation of various agglomerates upon surfactant removal was modeled as a series of reactions [16].

$$A[SC]_{free} \underset{k_{1r}}{\overset{k_{1f}}{\rightleftharpoons}} [SC]_{micelle} \tag{1}$$

$$B[DMPC] + C[SC]_{free} \underset{k_{2r}}{\overset{k_{2f}}{\rightleftharpoons}} [SC - DMPC]_{micelle} \tag{2}$$

$$[MSP] + D[SC]_{free} \underset{k_{3r}}{\overset{k_{3f}}{\rightleftharpoons}} [SC - MSP] \tag{3}$$

$$[SWCNT] + E[SC]_{free} \underset{k_{4r}}{\overset{k_{4f}}{\rightleftharpoons}} [SC - SWCNT] \tag{4}$$

$$F[DMPC] + [SWCNT] \underset{k_{5r}}{\overset{k_{5f}}{\rightleftharpoons}} [DMPC - SWCNT] \tag{5}$$

$$G[MSP] + [SWCNT] \underset{k_{6r}}{\overset{k_{6f}}{\rightleftharpoons}} [MSP - SWCNT] \tag{6}$$

$$2[SWCNT] \underset{k_{7r}}{\overset{k_{7f}}{\rightleftharpoons}} [SWCNT_2] \tag{7}$$

$$[SC] \xrightarrow{k_{8f}} [SC]_{removed} \quad (8)$$

$$H[DMPC] + I[MSP] \underset{k_{9r}}{\overset{k_{9f}}{\rightleftharpoons}} [ND] \quad (9)$$

$$J[ND] + [SWCNT] \underset{k_{10r}}{\overset{k_{10f}}{\rightleftharpoons}} [ND - SWCNT] \quad (10)$$

where $[SC]_{free}$ is free sodium cholate concentration in monomeric form, $[SC]_{micelle}$ is sodium cholate concentration in micellar form, [DMPC] is free lipid concentration in monomeric form, $[SC - DMPC]_{micelle}$ is SC-lipid mixed micelle concentration, [MSP] is MSP concentration, [SC – MSP] is SC-suspended MSP concentration, [SWCNT] is free SWCNT concentration, [SC – SWCNT] is SC-suspended SWCNT concentration, [DMPC – SWCNT] is lipid–SWCNT aggregate concentration, [MSP – SWCNT] is protein–lipid aggregate concentration, $[SWCNT_2]$ designates SWCNT bundle concentration, $[SC]_{removed}$ is the concentration of SC that has been removed from the system, [ND] is a ND concentration, and [ND – SWCNT] is ND–SWCNT concentration. In these reactions, k_f is the forward rate constant, k_r is the reverse rate constant, and A-J are stoichiometric coefficients. Fitting this model to experimental data on ND–SWCNT concentration over the course of dialysis, the best-fit rate constants for ND and ND-SWCNT formation were 79 $mM^{-1}s^{-1}$ and 5.4×10^2 mM^{-1} s^{-1}, respectively. In a diffusion-controlled process, one would expect the ND–SWCNT rate constant to be smaller than that of the ND, since the bulkier NDs and SWCNTs are expected to have smaller diffusivities than the individual lipid and MSP molecules. However, the reverse is the case, and the calculated ND–SWCNT rate constant is larger than that of the ND, indicating that the system is not under diffusion-controlled conditions. Because of the strongly hydrophobic nature of the SWCNT, the adsorption of the ND is expected to be nearly instantaneous, and in this case, the interaction is expected to be closer to diffusion-controlled conditions than ND formation [16].

The hydrophobic nature of SWCNTs not only induces the formation of ND–SWCNT complexes, but also the formation of other agglomerates, including SWCNT bundles. As discussed above, removal of surfactant from the SWCNTs forces the nanotubes to seek relatively hydrophobic surfaces to shield them from the aqueous surroundings. These surfaces may be lipid-formed agglomerates, such as NDs, or other nanotubes. In a diffusion-controlled dialysis, the structure with the largest diffusion coefficient is expected to diffuse toward the exposed nanotube surface more quickly. Large, bulky nanotubes with lengths on the order of microns demonstrate smaller diffusion coefficients than the more agile NDs; hence, they are expected to diffuse more slowly toward one another than the ND, favoring the faster ND-SWCNT formation. However, although diffusion occurs more slowly, the nanotube bundling is nearly irreversible: once nanotube bundles are formed, their dissociation would require the use of high-energy perturbations via sonication. Over extended periods of time, SWCNTs are expected to occupy their thermodynamically favored state in bundles, where strong, continuous, hydrophobic interactions along the entire length of the nanotube maintain irreversible bundle formation. Slower dialysis conditions would thus favor the formation of kinetically slower, but thermodynamically favored nanotube bundles. Faster dialysis conditions, on the other hand, favor the formation of kinetically faster, but thermodynamically metastable, ND–SWCNT complexes over smaller time scales [6].

Photoelectrochemical Measurements of Self-assembled Complexes

To exemplify the advantages of using the kinetically and thermodynamically favored formation of nanotube-based complexes in dynamic solar cells, the self-assembly was used as a means of regeneration in photoelectrochemical measurements. Photoelectrochemical measurements were carried out using a three-electrode system: a casted SWCNT working electrode, a Pt counter-electrode, and a Ag/AgCl standard reference electrode. A ferricyanide/ubiquinone dual mediator was used to optimize device efficiency.

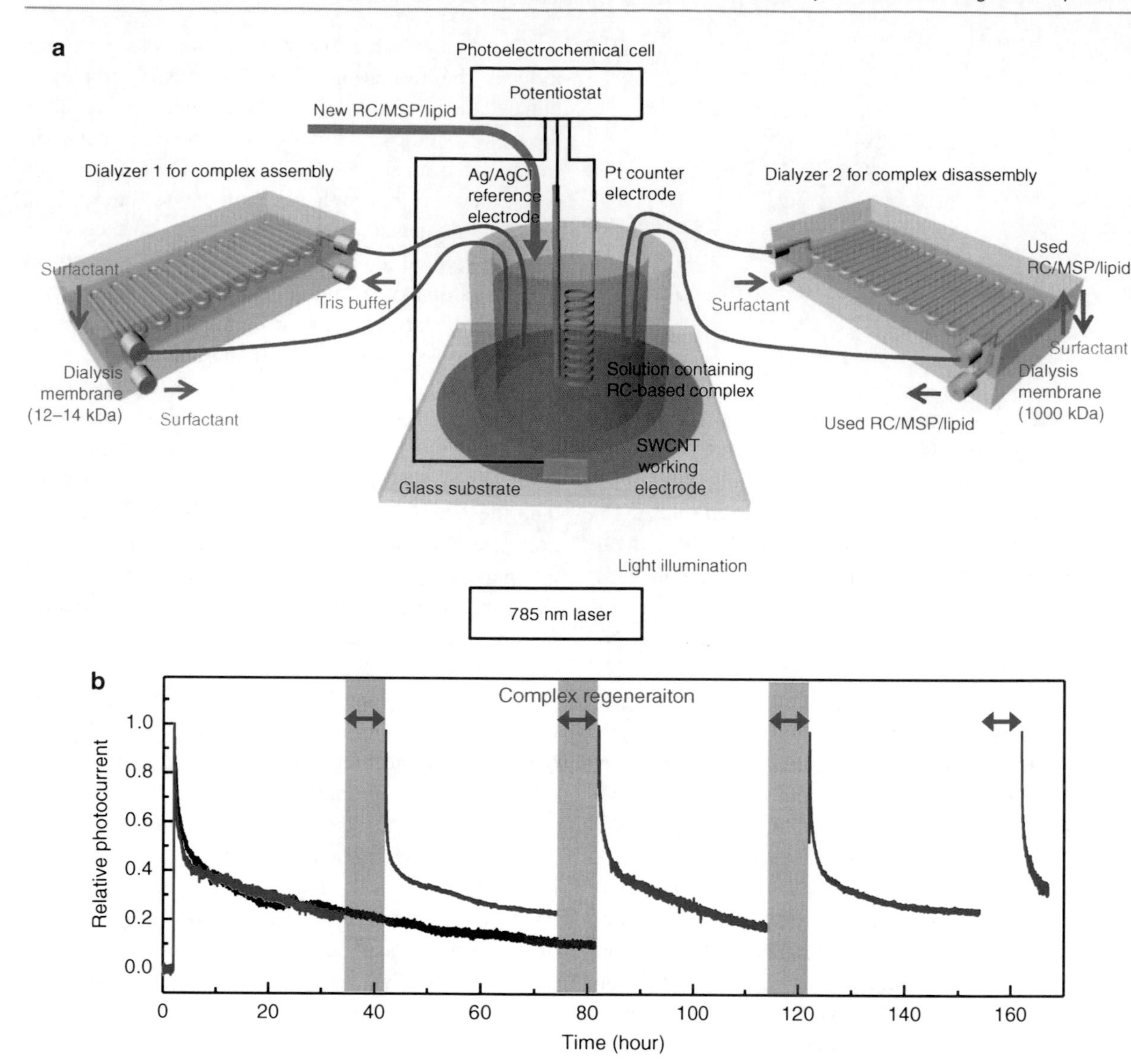

Self-repairing Photoelectrochemical Complexes Based on Nanoscale Synthetic and Biological Components, Fig. 4 Regenerating photoelectrochemical cell with self-assembled RC-ND-SWCNT complexes. (**a**) Schematic of the setup of photoelectrochemical systems which consists of a photoelectrochemical cell incorporated to two re-circulating membrane dialyzers. Dialyzers 1 and 2 are used for assembly and disassembly of the complex, respectively. The photodamaged RCs are removed during dialysis for disassembly of the complex and replaced with new ones while SWCNT are retained. (**b**) Temporal photoresponse of the RC-ND-SWCNT complexes with (red line) and without (black line) regeneration. (Reproduced from6 © 2010 Nature Publishing group)

Illumination of a 700 nM solution of RC-ND-SWCNT yielded a per nanotube external quantum efficiency (EQE) of approximately 40%. Upon introduction of SC to the illuminated solution, the photoactive complex disassembles, and no photocurrent is recovered [6].

Although the exact mechanism behind the demonstrated photoactivity of the assembled complex remains unclear, it is hypothesized that electron-hole generation occurs in the RC in a manner similar to that found in nature, where the hole ultimately occupies the P680 site and the electron is shuttled to the Q_B. In nature, the Q_B site contains ubiquione. Ubiquinone molecularly consists of an electron-withdrawing head and a conjugated tail. In vivo, ubiquinone occupies a thermodynamically favored configuration within the RC wherein the head is docked in the Q_B site and the tail is extended outward toward to the surface of the protein complex. Since photoactivity is only observed in the presence of ubiquinone, it is hypothesized that the ubiquinone docks to the RC in an analogous manner, serving as a means of electron extraction from the Q_B site buried deep within the RC. On the other hand, the P680 site, which is much more easily accessible, faces the SWCNT, which may serves as a hole-conducting wire in this configuration.

Although photoactivity has been demonstrated in the assembled state, it remains to be seen whether reversible self-assembly could be used as a means of solar cell regeneration, as is the case with the plants. To address this, photoelectrochemical measurements were taken over an extended period of continuous illumination, provoking photodamage and diminishing photocurrents (Fig. 4). When the photocurrent approaches approximately 30% of its initial value, SC is introduced to the system to disassemble the complex, the damaged proteins are replaced with new ones, and surfactant is once more removed from the system to reassemble the complex and recover photocurrent to its initial value. The synthetic regeneration cycle is hence conducted in analogy to the regeneration cycle used by plants, whereby chemical signaling is used to transition between the metastable assembled and disassembled states to replace photodamaged proteins with new ones. Using this regeneration cycle in the synthetic device, photoeletrochemical cell lifetime can be extended indefinitely while increasing efficiency by over 300% over 168 h.

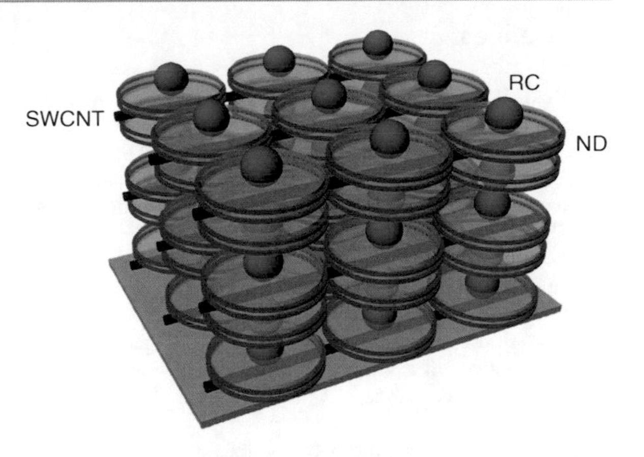

Self-repairing Photoelectrochemical Complexes Based on Nanoscale Synthetic and Biological Components, Fig. 5 High-density stacks for optimal efficiencies. Arrays of RC-ND-SWCNT can be aligned to maximize concentration of these complexes. Alignment of these complexes relies on recent advancements on SWCNT alignment

Nanotechnology as a Means of Modulating Bio-inspired Devices

The development of the regenerable, self-assembling solar device hinges on the ability to interface SWCNTs with biologically derived components. The precise arrangement of hydrophobic and hydrophilic regimes in biological complexes, such as RCs, is confined to regions that are at most a few nanometers apart. To utilize these variations in non-covalent interactions, one must subject these complexes to materials with properties that can be controlled on the nanoscale. For instance, nanotubes offer unidimensional confinement of hydrophobic surfaces that allows for the one-dimensional alignment of light-harvesting complexes. Confinement of these complexes into high-density, unidirectional arrays allows for device optimization and design. In fact, preliminary findings [6] indicate a linear increase in photo-efficiency with increasing RC–ND–SWCNT complexes, suggesting that maximum efficiency (e.g., 40%) could be achieved by maximizing concentration. Maximum concentration can be realized through high-density RC–ND–SWCNT stacks (Fig. 5). The key in synthesizing these stacks is controlling the alignment of not only RC-NDs along the nanotube, but also the nanotubes relative to one another. Recent advancements in nanotube alignment [17–20] have expanded researchers' ability to precisely control device fabrication on the nanoscale, making the fabrication of such high-density stacks feasible.

Cross-References

► Carbon Nanotubes
► Hybrid Solar Cells
► Nanomaterials for Excitonic Solar Cells
► Nanostructures for Energy
► Self-assembly of Nanostructures

References

1. Rosi, N.L., Mirkin, C.A.: Nanostructures in biodiagnostics. Chem. Rev. **105**, 1547–1562 (2005)
2. Wong, S.S., Joselevich, E., Woolley, A.T., Cheung, C.L., Lieber, C.M.: Covalently functionalized nanotubes as nanometre-sized probes in chemistry and biology. Nature **394**, 52–55 (1998)
3. Oregan, B., Gratzel, M.: Low-cost, high-efficiency solar-cell based on dye-sensitized colloidal TiO2 films. Nature **353**, 737–740 (1991)
4. Colfen, H., Mann, S.: Higher-order organization by mesoscale self-assembly and transformation of hybrid nanostructures. Angew. Chem. Int. Ed. **42**, 2350–2365 (2003)
5. Lowe, C.R.: Nanobiotechnology: the fabrication and applications of chemical and biological nanostructures. Curr. Opin. Struct. Biol. **10**, 428–434 (2000)
6. Ham, M.H., Choi, J.H., Boghossian, A.A., Jeng, E.S., Graff, R.A., Heller, D.A., Chang, A.C., Mattis, A., Bayburt, T.H., Grinkova, Y.V., Zeiger, A.S., Van Vliet, K.J., Hobbie, E.K., Sligar, S.G., Wraight, C.A., Strano, M.S.: Photoelectrochemical complexes for solar energy conversion that chemically and autonomously regenerate. Nat. Chem. **2**, 929–936 (2010)
7. Das, R., Kiley, P.J., Segal, M., Norville, J., Yu, A.A., Wang, L.Y., Trammell, S.A., Reddick, L.E., Kumar, R., Stellacci, F., Lebedev, N., Schnur, J., Bruce, B.D., Zhang, S.G., Baldo, M.: Integration of photosynthetic protein molecular complexes in solid-state electronic devices. Nano Lett. **4**, 1079–1083 (2004)
8. Frolov, L., Wilner, O., Carmeli, C., Carmeli, I.: Fabrication of oriented multilayers of photosystem I proteins on solid surfaces by auto-metallization. Adv. Mater. **20**, 263 (2008)
9. Lebedev, N., Trammell, S.A., Spano, A., Lukashev, E., Griva, I., Schnur, J.: Conductive wiring of immobilized photosynthetic reaction center to electrode by cytochrome c. J. Am. Chem. Soc. **128**, 12044–12045 (2006)
10. Ciesielski, P.N., Scott, A.M., Faulkner, C.J., Berron, B.J., Cliffel, D.E., Jennings, G.K.: Functionalized nanoporous gold leaf electrode films for the immobilization of photosystem I. ACS Nano **2**, 2465–2472 (2008)
11. Asimov, I.: Photosynthesis. Basic Books, New York (1968)
12. Stern, K.R., Bidlack, J.E., Jansky, S.: Introductory Plant Biology, 11th edn. McGraw-Hill Higher Education, Boston (2008)
13. Aro, E.M., Virgin, I., Andersson, B.: Photoinhibition of photosystem-2 - inactivation, protein damage and turnover. Biochim. Biophys. Acta **1143**, 113–134 (1993)
14. Melis, A.: Photosystem-II damage and repair cycle in chloroplasts: what modulates the rate of photodamage in vivo? Trends Plant Sci. **4**, 130–135 (1999)
15. Melis, A.: Dynamics of photosynthetic membrane-composition and function. Biochim. Biophys. Acta **1058**, 87–106 (1991)
16. Boghossian, A.A., Choi, J.H., Ham, M.H. Strano, M.S.: Dynamic and reversible self-assembly of photoelectrochemical complexes based on lipid bilayer disks, photosynthetic reaction centers, and single-walled carbon nanotubes. Langmuir **27**, 1599–1609 (2011)
17. Xie, X.L., Mai, Y.W., Zhou, X.P.: Dispersion and alignment of carbon nanotubes in polymer matrix: A review. Mater. Sci. Eng. R Rep. **49**, 89–112 (2005)
18. Vigolo, B., Penicaud, A., Coulon, C., Sauder, C., Pailler, R., Journet, C., Bernier, P., Poulin, P.: Macroscopic fibers and ribbons of oriented carbon nanotubes. Science **290**, 1331–1334 (2000)
19. Haggenmueller, R., Gommans, H.H., Rinzler, A.G., Fischer, J.E., Winey, K.I.: Aligned single-wall carbon nanotubes in composites by melt processing methods. Chem. Phys. Lett. **330**, 219–225 (2000)
20. Murakami, Y., Chiashi, S., Miyauchi, Y., Hu, M.H., Ogura, M., Okubo, T., Maruyama, S.: Growth of vertically aligned single-walled carbon nanotube films on quartz substrates and their optical anisotropy. Chem. Phys. Lett. **385**, 298–303 (2004)

SEM

► Electron Microscopy of Interactions Between Engineered Nanomaterials and Cells

Semiconductor Nanocrystals

► Quantum-Dot Toxicity

Semiconductor Piezoresistance

► Piezoresistivity

Semicrystalline Morphology

► Spider Silk

Sense Organ

► Arthropod Strain Sensors

Sensors

► Arthropod Strain Sensors

Shark Denticles

► Shark Skin Drag Reduction

Shark Skin Drag Reduction

Amy Lang[1], Maria Laura Habegger[2] and Philip Motta[2]
[1]Department of Aerospace Engineering & Mechanics, University of Alabama, Tuscaloosa, AL, USA
[2]Department of Integrative Biology, University of South Florida, Tampa, FL, USA

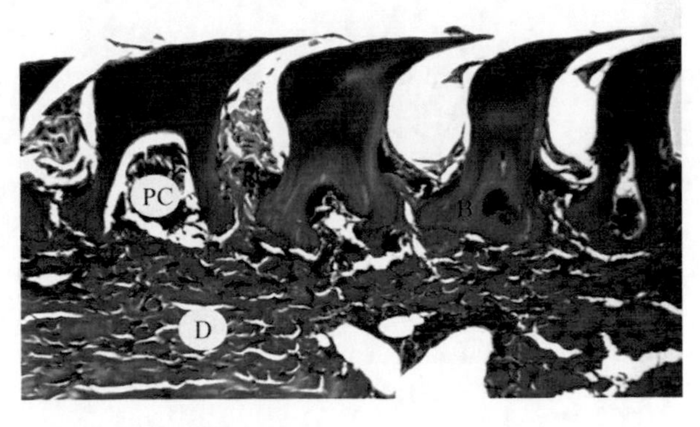

Shark Skin Drag Reduction, Fig. 1 Lateral view of sectioned placoid scales of a shortfin mako shark *Isurus oxyrinchus* in the region midway between the leading and trailing edge of the pectoral fin. Note the relatively long scale base relative to the crown length. The pulp cavity (PC) of the scale on the *left* is visible, and the base of the scales is anchored by collagen fibers to the dermis (D) of the skin

Synonyms

Riblets; Shark denticles; Shark skin separation control

Definition

The scales, or denticles, on fast-swimming sharks have evolved two mechanisms for controlling the boundary layer flow over the skin surface leading to a reduction in drag. The first, and most widely known and studied, consists of the small streamwise keels covering the surface of the scales also known as riblets which reduce turbulent skin friction drag. The second mechanism is attributed to loosely embedded scales that are located on key regions of the body. When actuated to bristle by the flow, these scales potentially act as a means of controlling flow separation, thereby minimizing pressure drag during swimming maneuvers. Shark scales display a wide variation in geometry both across species while also varying with body location, but on faster swimming sharks, they typically range in size from 180 to 500 μm in crown length.

Overview

The Shark Skin

Sharks are covered with minute scales, also known as denticles or placoid scales because of their tooth-like nature. The scales have a pulp cavity and a hard enameloid covering and are anchored at the base of the scale to the collagenous layer of the skin known as the stratum laxum (Fig. 1). The interlocking crowns of each scale make up the surface of the shark exposed to the water, and it is on the crown where many species have developed small riblets, or keels, orientated in the streamwise direction of the flow (Fig. 2). A reduction in the length of the base relative to the length of the crown and a change of shape of the base for some species over certain regions of the body appear to be the means by which certain scales have developed the capability to bristle or erect upon flow reversal (Fig. 3; flow would normally pass over the surface from left to right; flow reversal proceeding right to left can induce scale bristling as shown in Fig. 4 with a schematic shown in Fig. 5). The length of the scales is typically fixed for specific regions of the body within a species but differs among regions and species. Similarly, the number of keels per scale is also consistent per location for a species. For instance, on the fast-swimming shortfin mako (*Isurus oxyrinchus*), the flank scales have a crown length of approximately 0.18 mm; each crown typically has three keels, each having a height of 0.012 mm and a spacing of 0.041 mm. The slower swimming blacktip shark (*Carcharhinus limbatus*) has flank scales typically 0.32 mm in length; each crown typically has five keels with a height of 0.029 mm and a spacing of 0.065 mm (Fig. 2).

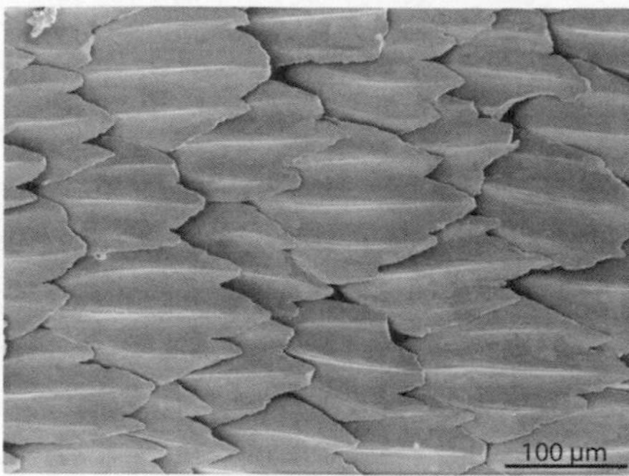

Shark Skin Drag Reduction, Fig. 2 Scanning electron micrograph (200×) of the placoid scales of a shortfin mako shark *Isurus oxyrinchus* (*left*) and blacktip shark *Carcharhinus limbatus* (*right*) from the dorsal body wall anterior to the dorsal fin. The mako shark scale has three keels or riblets, whereas the blacktip shark has five. Anterior is to the *left*

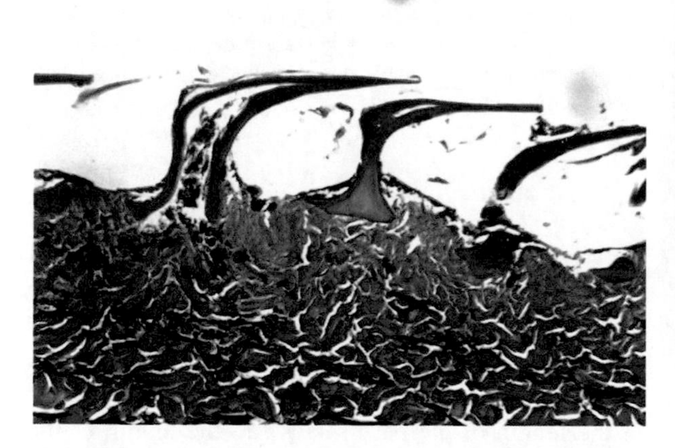

Shark Skin Drag Reduction, Fig. 3 Lateral view of sectioned placoid scales in the flank region of a shortfin mako shark *Isurus oxyrinchus* in the region midway between the dorsal fin and the pectoral fin. Note the relatively long scale crown relative to the shorter base length

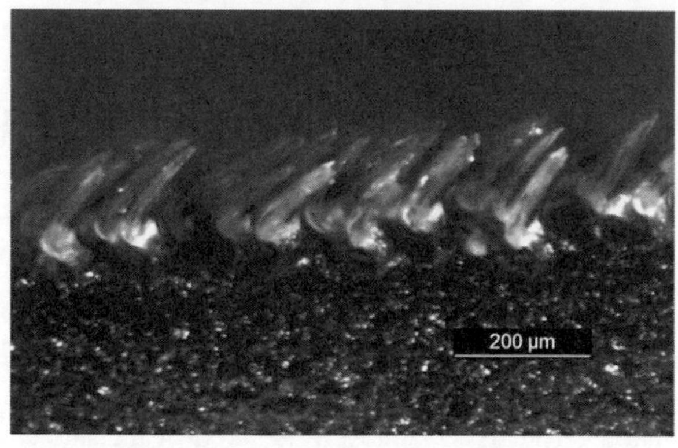

Shark Skin Drag Reduction, Fig. 4 Side view through the shortfin mako skin (from the flexible flank area) showing scales that have been manually erected. Because of the individual manual erection, not all scales are erected to the same degree. Flow would normally pass over the skin from *left* to *right*, and reversed flow, as occurs during separation, is hypothesized to cause bristling as shown

Specifications and Fluid Dynamics

A shark swimming through water experiences two major sources of drag due to the viscous resistance of the fluid flow in the direction of motion. It is generally accepted that faster swimming sharks, such as the shortfin mako (*Isurus oxyrinchus*), have the capability to reduce both types of drag through evolutionary adaptations to their denticles. The first source of drag is skin friction drag; though not necessarily the largest contributor to overall drag, it is the one most associated with friction of the flow moving over the body of the shark. Skin friction drag is a result of the no-slip boundary condition between the water and the surface of the shark which results in the formation of a boundary layer. Because of the high speeds a shark can achieve (often greater than $U = 10$ m/s), this leads to a very high Reynolds number ($Re = Ux/\nu$, where x is length and ν is kinematic viscosity) in the boundary layer forming over most portions of the body and indicates the development of turbulent flow over most of the shark's body. For instance, at this speed, the typical transition location when the local $Re = 5 \times 10^5$ occurs at a location just $x = 5$ cm from the nose. While a turbulent boundary layer is less prone to separate from the body, it will have a skin friction magnitude 5–10 times larger than if the flow were to remain laminar. This riblet drag reduction mechanism has demonstrated the potential to reduce the turbulent skin friction drag by approximately 8–10% in man-made applications [1, 2]. The size and spacing of the

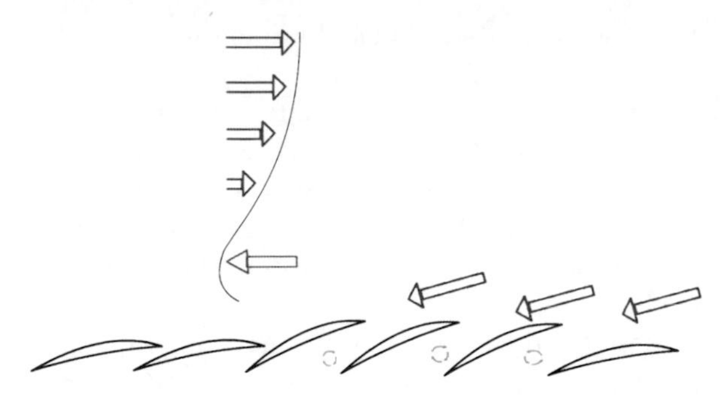

Shark Skin Drag Reduction, Fig. 5 Schematic showing forward flow in the boundary layer (*blue arrows*) subject to an adverse pressure gradient causing flow reversal (*red arrows*). The reversed flow close to the scales causes localized bristling to occur

riblets are indicative of the speed at which a shark may swim with faster species having closer spaced keels.

The second type of drag, and the most important to be controlled, is due to a difference in pressure around the body and is often referred to as form drag. It is highly dependent on whether the flow remains attached or separates during swimming. Flow separation causes regions of low pressure on the downstream portions of the body leading to an imbalance of the pressure fore and aft, and thus a dramatic increase in drag. The first means of decreasing pressure drag is streamlining, which consists of smoothing sharp corners and elongating the body with a tapered downstream portion. As a result, sharks and other marine animals have evolved a very streamlined body shape. Body proportions for fast-swimming sharks such as the shortfin mako are in part determined by its thunniform swimming ability. However, when it undergoes turning maneuvers, the fusiform body will undergo relatively greater body curvature than when swimming in a straight line. Turning may thus induce flow separation, but recent experimental evidence suggests that loosely embedded scales on the flank and behind the gills, in the region of maximum girth and further downstream, may act as a means of flow control.

Flow separation first involves a reversal of the fluid particles in a thin section adjacent to the surface; this is induced by a region of adverse pressure gradient occurring aft of the point of maximum girth in the streamwise direction. In other words, a suction pressure upstream induces the flow with the least momentum, that closest to the surface, to reverse. Scale bristling likely inhibits the process leading to reversed flow which thereby can control flow separation. Preventing separation will also favorably affect the flow further downstream over the caudal fin which can lead to higher thrust production. While for a swimming shark, the production of thrust and drag occur simultaneously over the body and are inexorably linked; evidence signifies that sharks have evolved mechanisms within the structure of their skin to reduce drag leading to increased thrust production and allow greater maneuverability at high speeds.

Reif, a German biologist, working in the late 1970s is generally considered to be the first to report in literature the hypothesis as to the drag-reducing properties of shark skin [3]. Engineering research began at about the same time in both America [4] and Germany [5] as to the functional aspect of riblet surfaces for reducing skin friction drag. Seminal work into riblet design was completed by Bechert et al. in 1997, which demonstrated a maximum turbulent skin friction drag reduction for man-made, blade-like riblets of 9.9% [6]. The bristling of shark scales leading to a mechanism for separation control was always hypothesized to function in a manner similar to a man-made technique known as vortex generators utilized since the 1940s [7, 8]. In more recent years, Lang et al. [9] have posited a new mechanism, whereby flow reversal leads to a region of localized scale bristling, leading to a flow-actuated separation control method which can be derived from the shark skin and for which research is ongoing.

Key Research Findings

Since the 1970s when studies began, riblets have become a well-accepted means of reducing turbulent skin friction drag with an upper limit of reduction just below 10%. As previously stated, the most exhaustive testing of various riblet geometries, with comparisons made to other researchers working in the field, was completed in 1997 by Bechert et al. [6]. Previous work by Walsh [4] had focused on a sawtooth geometry, which resulted in a maximum drag reduction of about 5%. While flow field measurement and visualization were not carried out to fully understand the mechanism behind the drag reduction, an exhaustive series of drag measurement experiments was carried out in a specially designed oil channel facility [6]; these measurements allowed for a determination of the most effectual geometry for man-made riblet applications.

Bernard and Wallace provide a summarization of the basic research that has been performed to map out the key characteristics found in turbulent flows [10]. To understand the mechanism whereby drag is reduced by surface modifications, some basic aspects of a turbulent boundary layer flow field need to be understood. A wall-bounded turbulent flow consists of a layer of vorticity within which fluid located further away from the surface is moving faster than that nearer to the surface due to the no-slip condition, and this results in an overall rotational characteristic of the flow in the clockwise direction for flow moving from left to right where the surface itself is stationary (this is the common reference frame to be used for studying a boundary layer flow). Within this layer, a complex assortment of horseshoe- or hairpin-shaped, vortices of various sizes and stages of growth/decay are formed and interact. Scaling of the flow within the boundary layer can be achieved by considering a viscous length scale defined as $\delta_v = \nu/u^*$, where u^* is the friction velocity. The friction velocity is a function of the shear stress, or skin friction, at the surface (τ) and the fluid density (ρ) such that $u^* = (\tau/\rho)^{1/2}$. The viscous length scale determines the characteristic sizing of the fluid scales found in the boundary layer. For instance, in a given flow with fixed viscosity, ν, and at a particular downstream location, x, within a boundary layer consider a variation in free-stream velocity, U. As U is increased, the thickness of the boundary layer at that location will decrease, and the local average shear stress at that same location will increase. This results in an increase in the friction velocity and thus a decrease in the viscous length scale. The resulting decrease in viscous length scale reduces the characteristic sizing of the vortices forming within the boundary layer.

In the region close to the surface, longitudinal vortices with an axis of rotation in the streamwise direction are found to form and persist. These vortices have a characteristic diameter of approximately $30\delta_v$ and a streamwise length ranging from a few hundred up to 1,000 viscous length scales. When these vortices pair up, a region of low-speed fluid is lifted up from the wall, resulting in the formation of a low-speed streak. It is the instability of these streaks located in a region of high shear that leads to the process known as a turbulent burst and subsequent turbulent sweep. A burst is a sudden ejection of low-speed fluid up into the boundary layer, and a sweep is a sudden injection of high-speed fluid down toward the wall. It is the sweep of this high-momentum fluid onto the surface that results in localized, time-varying patches of high skin friction that are the main causation of increased drag in a turbulent boundary layer. It is the interaction of the shark skin with these components of the boundary layer flow that is essential to controlling the flow.

Results from the work of Bechert et al. [6] plotted the relative change in skin friction ($\Delta\tau/\tau$) (compared to a flat surface) versus spacing (s^+), where riblet spacing (s) is nondimensionalized such that $s^+ = s/\delta_v$. They found that the reduction in skin friction increases and reaches a maximum in the vicinity of $s^+ \sim 16$ and thereupon begins to decrease such that larger spacing can actually result in an increase in drag. It is noteworthy that this value corresponds to half the characteristic diameter of the longitudinal vortices forming close to the wall. Thus the riblets sized correctly restrain these near-wall vortices, which for reduced skin friction and reduced viscous length scale will now form at a slightly larger size, and these in turn induce the formation of low-speed streaks [1]. It was also found that a height of the riblets corresponding to half the spacing ($h = 0.5$ s) gave optimal results. The other key result to garner from these experiments is the variation in maximum decrease with geometry, where a blade-like shape provided the upper limit for skin friction reduction (9.9%). However, for durability of the surface in real applications, a trapezoidal configuration was tested and found to provide improved performance over the sawtooth geometry. Finally, it is remarkable that the geometry found on shark skin scales, with keels of a scalloped shape closely resembling the trapezoidal shape but with smoothed corners, was also tested by Bechert et al. [6] and found to perform comparably to that of the trapezoidal shape. Many shark species also appear to have the approximate $h = 0.5$ s ratio found to be optimal in experiments. Both of these are indicators that sharks have indeed evolved a scale and riblet geometry for turbulent skin friction reduction; however, the keels may play a role in separation control as well.

Experiments to discern the benefits of shark scale bristling began with Bechert et al. [8]; they used a shark skin replica whereby an overlapping array of individual shark scales was built and tested in their oil tunnel facility. In this man-made model of the shark skin, they meticulously built 800 small replicas of a hammerhead (*Sphyrna zygaena*) shark denticle scaled up 100 times in size (resulting in a crown length of 19 mm and riblet spacing of 4 mm). Each scale was anchored with a compliant spring with variation in stiffness to discern if denticle bristling could result in any additional mechanism to further decrease turbulent skin friction drag. The model was used for cases where the scales laid flat or were wholly bristled at a collective angle of attack. Bristled cases only resulted in increased drag, with higher spring stiffness resulting in higher drag. Flat, aligned scales gave comparable results to riblet surfaces with the only difference being that the greatest drag reduction achieved around $s^+ = 15$ had a value of about 3%. This decrease in performance was attributed to the construction of the model with small gaps and other imperfections preventing a smoother surface. Thus if bristling of the scales is to be advantageous to the shark, it must be something that is only activated upon demand and thus concurs with the postulation that bristling is utilized as a means to control flow separation.

Initial speculation began with the hypothesis that bristled shark scales act as vortex generators [8]. These devices produce streamwise vortices which energize the flow close to the surface. Vortex generators need to be placed at a specific downstream location within a boundary layer for maximum performance and typically upstream of the point of separation [9]. Another method of controlling flow separation, used to date at a more global scale, consists of movable flaps; for airfoil applications, these are placed close to the trailing edge (~10% of chord length or larger) and have been shown to delay the onset of stall resulting in greater lift [8]. When the flap itself was given a 3D jagged trailing edge, they were also found to act more effectively.

More recent work by the authors [11–13] has investigated the mechanism by which shark skin bristling may lead to the development of a passive, flow-actuated mechanism for separation control. The current working hypothesis is that the passive, flexible scales of the shark work as microflaps to locally control flow separation as needed on crucial regions of the body where it most often occurs during swimming maneuvers. Recent observations on shark skin bristling angles on the shortfin mako (*Isurus oxyrinchus*) indicate that, for this particular species known for its swimming capability, only certain portions of the body have very flexible scales. Bristling capability was measured on dead specimens, and the effect of body pressurization was also considered as a potential bristling mechanism. However, results showed that subcutaneous skin pressurization did not cause scale bristling and had no effect on bristling angles; furthermore, such scales have no muscles attached to them as they lie in the deeper layer of the skin, the dermis. Because the flank scales are very loosely attached to the skin and highly mobile, these observations led us to infer that the scales are most likely bristled by reverse flow actuation. Sixteen body locations [12, 13] were considered, six on fins and ten on the body. Scales were manipulated by a fine acupuncture needle and remained bristled once manipulated as shown in Fig. 4.

Scale angles vary with body location, but the most flexible scales are found along the flank of the body extending behind the gills to the tail; here scales are found to be easily flexible with slight manipulation on dead specimens to angles of 50° or greater. Highly flexible scales are also found at the trailing edge of the pectoral fins as compared to the leading edge where there was zero scale flexibility; this indicates the scales may be used to control dynamic stall (unsteady separation) to maintain lift forces on these surfaces during swimming and thereby maintain control. Contragility during swimming, or the ability to change direction quickly and easily, requires low pressure drag as well as high musculature control. These recent findings, herein reported as to variation in scale flexibility, corroborate the hypothesis that the flank region, extending from the location of maximum girth to the tail, is where a shark with a side-to-side swimming motion requires separation control to increase contragility. The scale flexibility appears to be a result of a reduction in size of the scale base anchored in the skin and a change in the shape of the base (as can be seen by the histological data shown in Figs. 1 and 3). The reduction appears to occur in the length of the base relative to its width, where the portion of the base that would pivot up and out of the skin at high bristling angles shows a decrease in length. Thus the scales with greater bristling angles are less firmly anchored in the anterior to posterior direction and can pivot more freely within the skin.

Cassel et al. [14] describe the process leading to unsteady flow separation, as would occur in a turbulent boundary layer. In the presence of an adverse pressure gradient, the fluid closest to the wall, which also has the lowest momentum, is where flow reversal is first initiated (Fig. 5). This region of fluid moves back upstream and thickens and then subsequently erupts from the surface leading to a patch where the flow is separated from the surface. In a turbulent boundary layer, the fluid with the lowest momentum is that contained in the low-speed streaks, and separation of a streak results in the formation of a horseshoe-shaped structure as observed in computational studies of a separating turbulent boundary layer [15]. Thus for the shark skin, made up of a staggered array of scales, flow reversal in long, thin patches of fluid may locally bristle the scales, thereby interrupting the flow separation process. This feature of shark skin resulting in a surface with a favored flow direction is likely key to its capability to inhibit flow separation.

Furthermore, previous experiments over a bristled shark skin model established the existence of embedded cavity vortices, axis of rotation in spanwise direction, forming between replicas of the scales [11]. Thus if flow is induced to form between the scales when bristled, there are two supplementary mechanisms that may aid to control the flow. The formation of embedded vortices, similar as occurs for golf ball dimples, would allow the flow to pass over the surface with an ensuing partial slip condition, thereby leading to higher momentum adjacent to the skin. Secondly, with a turbulent boundary layer flow forming above the cavities, there may be added momentum exchange whereby high-momentum fluid is induced at a greater rate to move toward the skin and into the cavities. This latter mechanism, resulting in turbulence augmentation [1], is another possible means to enhance the momentum overall in the flow closest to the wall. These three mechanisms may be working in combination to control flow separation over the skin of the shark.

Past and Future Applications

The application of drag-reducing techniques to vehicles in both air and water has obvious advantages and at the same time limitations. Typical riblet spacing for both water and air applications, at speeds normally encountered, falls in the range of ~0.035 mm. Thin plastic films with adhesive on one side and riblets on the other have been made commercially available; thus a wide array of testing and use on aircraft has already taken place where drag reduction was documented. However, practical limitations from cost, added weight, and maintenance of the riblet film (particularly with respect to particulates clogging the surface) with only an associated total decrease in drag of 1–3% have prevented widespread use in most aircraft applications [1].

Separation control cannot only reduce drag but can also lead to increased maneuverability for vehicles. Decreasing drag overall can lead to increased fuel efficiency, payload and range in both military and commercial applications. Flow separation is also an important issue for maintaining use of control surfaces on vehicles (i.e., prevention of stall), which can also have relevance to helicopter rotors and turbine/compressor blades. Passive control mechanisms, including those found on shark skin, have been and will continue to be applied in all these applications.

Cross-References

► Biomimetics
► BioPatterning
► Shark Skin Effect

References

1. Gad-el Hak, M.: Flow Control: Passive, Active and Reactive Flow Management. Cambridge University Press, Cambridge, UK (2000)
2. Bhushan, B.: Shark skin effect. In: Encyclopedia of Nanotechnology. Springer, Berlin (2012)
3. Reif, W.-E.: Protective and hydrodynamic function of the dermal skeleton of elasmobranchs. Neues Jahrb. Geol. Palaontol. Abh. **157**, 133–141 (1978)
4. Walsh, M.: Drag characteristics of V-groove and transverse curvature riblets. In: Hough, G.R. (ed.) Viscous Flow Drag Reduction. Progress in Astronautics and Aeronautics, vol. 72, pp. 168–184. AIAA, New York (1980)
5. Bechert, D., Hoppe, G., Reif, W.: On the drag reduction of the shark skin. American Institute of Aeronautics and Astronautics (AIAA) Paper No. 85-0546 (1985)
6. Bechert, D., Bruse, M., Hage, W., Van der Hoeven, J., Hoppe, G.: Experiments on drag-reducing surfaces and their optimization with an adjustable geometry. J. Fluid Mech. **338**, 59–87 (1997)
7. Bushnell, D., Moore, K.: Drag reduction in nature. Annu. Rev. Fluid Mech. **23**, 65–79 (1991)
8. Bechert, D., Bruse, M., Hage, W., Meyer, R.: Fluid mechanics of biological surfaces and their technological application. Naturwissenschaften **80**, 157–171 (2000)
9. Lin, J.: Review of research on low-profile vortex generators to control boundary-layer separation. Prog. Aerosp. Sci. **38**, 389–420 (2002)

10. Bernard, P., Wallace, J.: Turbulent Flow: Analysis, Measurement, and Prediction. Wiley, Hoboken (2002)
11. Lang, A., Motta, P., Hidalgo, P., Westcott, M.: Bristled shark skin: a microgeometry for boundary layer control? Bioinspir. Biomim. **3**, 046005 (2008)
12. Lang, A., Habegger, M., Motta, P.: Shark skin boundary layer control. In: Proceedings of the IMA Workshop "Natural Locomotion in Fluids and on Surfaces: Swimming, Flying, and Sliding," 1–5 June 2010. IMA Volumes in Mathematics and its Applications (2012)
13. Lang, A., Motta, P., Habegger, M., Hueter, R., Afroz, F.: Shark skin separation control mechanisms. Mar. Technol. Soc. J. **45**(4), 208–215 (2011)
14. Cassel, K., Smith, F., Walker, J.: The onset of instability in unsteady boundary-layer separation. J. Fluid Mech. **315**, 223–256 (1996)
15. Na, Y., Moin, P.: Direct numerical simulation of a separated turbulent boundary layer. J. Fluid Mech. **374**, 379–405 (1998)

Shark Skin Effect

Bharat Bhushan
Nanoprobe Laboratory for Bio- & Nanotechnology and Biomimetics, The Ohio State University, Columbus, OH, USA

Synonyms

Low fluid drag surface

Definition

The shark skin effect is the reduction of the fluid drag during swimming at fast speeds and protection of its surface against biofouling. The presence of surface microstructure on the skin surface is responsible for this effect.

Overview

Many structures, materials, and surfaces found in nature can be exploited for commercial applications. As an example, nature has created ways of reducing drag in fluid flow, evident in the efficient movement of fish, dolphins, and sharks [10, 14]. The mucus secreted by fish causes a reduction in drag as they move through water, and also protects the fish from abrasion by making the fish slide across objects rather than scrape, and it makes the surface of the fish difficult for microscopic organisms to adhere to [28]. It has been known for many years that by adding as little as a few hundred parts per million guar, a naturally occurring polymer, friction in pipe flow can be reduced by up to two thirds. Other synthetic polymers provide an even larger benefit [18]. The compliant skin of the dolphin has also been studied for drag-reducing properties. By responding to the pressure fluctuations across the surface, a compliant material on the surface of an object in a fluid flow has been shown to be beneficial. Studies have reported 7% drag reduction [12].

Another set of aquatic animals which possess multipurpose skin is fast swimming sharks [14]. The skin of fast swimming sharks reduces the drag experienced by sharks as they swim through water and protects against biofouling. The tiny scales covering the skin of fast swimming sharks, known as dermal denticles (skin teeth), are shaped like small riblets and aligned in the direction of fluid flow (Fig. 1). Shark skin–inspired riblets have been shown to provide a drag reduction benefit up to 9.9% [5]. The spacing between these dermal denticles is such that the riblets may not be very effective against very small (micro-) organisms – they probably work best against larger organisms such as mussels, algae, and barnacles. Prevention of undesirable accumulation of microorganisms protects the surface against biofouling. Slower sharks are covered in dermal denticles as well, but these are not shaped like riblets and do not provide much drag reduction benefits.

The effect of riblet structures on the behavior of fluid drag, as well as the optimization of their morphology, is the focus of this entry.

Mechanisms of Fluid Drag and Role of Riblets in Drag Reduction

Fluid drag comes in several forms, the most basic of which are pressure drag and friction drag [14]. Pressure drag is the drag associated with the energy required to move fluid out from in front of an object in the flow, and then back in place behind the object. Much of the drag associated with walking through water is pressure drag, as the water directly in front of a body must be moved out and around the body before the body can move forward. The magnitude of pressure drag can be

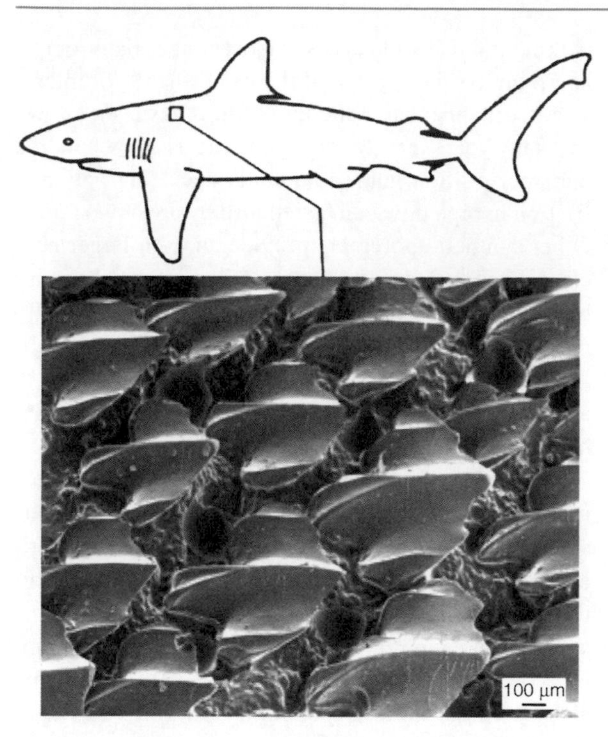

Shark Skin Effect, Fig. 1 Scale patterns on fast-swimming sharks (*Squalus acanthias*) [19]

reduced by creating streamlined shapes. Friction or viscous drag is caused by the interactions between the fluid and a surface parallel to the flow, as well as the attraction between molecules of the fluid. Friction drag is similar to the motion of a deck of cards sliding across a table. Fluids of higher viscosity – the attraction between molecules – have higher apparent friction between fluid layers, which increases the thickness of the fluid layer distorted by an object in a fluid flow. An increase in drag occurs as fluid velocity increases. The drag on an object is in fact a measure of the energy required to transfer momentum between the fluid and the object to create a velocity gradient in the fluid layer between the object and undisturbed fluid away from the object's surface.

The above discussion of friction drag assumes all neighboring fluid molecules move in the same relative direction and momentum transfer occurs between layers of fluid flowing at different velocities. Fully developed turbulent flow is commonly said to exhibit complete randomness in its velocity distribution, but distinct regions exist within fully developed turbulent flow that exhibit different patterns and flow characteristics [20]. As these vortices rotate and flow along the surface, they naturally translate across the surface in the cross-flow direction. The interaction between the vortices and the surface, as well as between neighboring vortices that collide during translation, initiates bursting motions where vortices are rapidly ejected from the surface and into the outer boundary layers. As vortices are ejected, they tangle with other vortices and twist such that transient velocity vectors in the cross-stream direction can become as large as those in the average flow direction [20]. The translation, bursting of vortices out of the viscous sublayer, and chaotic flow in the outer layers of the turbulent boundary layer flow are all forms of momentum transfer and are large factors in fluid drag. Reducing the bursting behavior of the stream-wise vortices is a critical goal of drag reduction, as the drag reduction possibilities presented by this are sizable.

The vortices have been visualized using flow visualization techniques to capture cross-sectional images, shown in Fig. 2, of the stream-wise vortex formations above both flat-plate and riblet surfaces [26]. The average cross-stream wavelength of these high- and low-speed streaks, the added widths of one high-speed streak and one low-speed streak, is equal to the added diameters of two neighboring vortices and has been measured at 70–100 wall units [6, 20, 34]. This corresponds to a vortex diameter of 35–50 wall units. The flow visualizations in Fig. 2 show vortex cross sections and relative length scales, demonstrating vortex diameters smaller than 40 wall units [26]. (As flow properties change, the dimensions of the turbulent flow structures change as well. As such, it is useful to use non-dimensional length values to better compare studies performed in different flow conditions. Dimensionless wall units, marked $^+$, are used for all length scales, which are calculated by multiplying the dimensional length by V_τ/ν. For example, $s^+ = sV_\tau/\nu$, where s^+ is the non-dimensional riblet spacing, s is the dimensional riblet spacing, ν is the kinematic viscosity, and $V_\tau = (\tau_0/\rho)^{0.5}$ is the wall stress velocity, for which ρ is the fluid density and τ_0 is the wall shear stress. Wall shear stress can be estimated for round pipe flow using the equation $\tau_0 = 0.03955\nu^{1/4}\rho V^{7/4}d^{-1/4}$, where V is the average flow velocity and d is the hydraulic diameter. For flow in rectangular pipes, the equation for hydraulic diameter $d = 4A/c$ can be applied, where A is the cross-sectional area and c is the wetted perimeter.)

The small riblets that cover the skin of fast swimming sharks work by impeding the cross-stream translation of the stream-wise vortices in the viscous

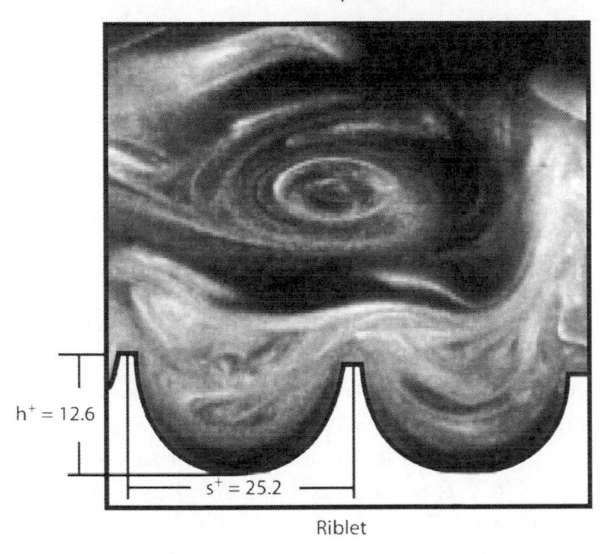

Shark Skin Effect, Fig. 2 Turbulent flow visualization of stream-wise vortices in a vertical cross section over flat-plate and riblet surfaces (Adapted from [26])

sublayer. As vortices form above a riblet surface, they remain above the riblets, interacting with the tips only and rarely causing any high-velocity flow in the valleys of the riblets. Since the higher-velocity vortices interact only with a small surface area at the riblet tips, only this localized area experiences high-shear stresses. The low velocity fluid flow in the valleys of the riblets produces very low shear stresses across the majority of the surface of the riblet. By keeping the vortices above the riblet tips, the cross-stream velocity fluctuations inside the riblet valleys are much lower than the cross-stream velocity fluctuations above a flat plate [26]. This difference in cross-stream velocity fluctuations is the evidence of a reduction in shear stress and momentum transfer near the surface, which compensates the potentially drag-increasing effect of a larger surface area. Though the vortices remain above the riblet tips, some secondary vortex formations do occur that enter the riblet valleys transiently. The flow velocities of these transient secondary vortices are such that the increase in shear stress caused by their interaction with the surface of the riblet valleys is small.

Protruding into the flow without greatly increasing fluid drag allows the riblets to interact with the vortices to reduce the cross-stream translation and related effects. As the riblets protrude into the flow field, they raise the effective flow origin by some distance. The amount by which the height of the riblets is greater than the apparent vertical shift of the flow origin is referred to as the effective protrusion height. By calculating the average stream-wise velocity in laminar flow at heights over riblet surfaces and comparing them to the average stream-wise velocities in laminar flow at heights over a flat plate, the effective stream-wise protrusion height, h_{ps}, is found for laminar flow. The effective cross-stream protrusion height, h_{pc}, is similarly found for laminar flow by comparing the cross-stream velocities over a riblet surface to those over a flat plate. A schematic of stream-wise and cross-stream flow velocity profiles and effective protrusion heights is shown in Fig. 3. The difference between the vertical shifts in the stream-wise and cross-stream origin, $\Delta h = h_{ps} - h_{pc}$, for any riblet geometry has been proposed to be the degree to which that riblet geometry will reduce vortex translation for low Reynolds number (Re) flows [5]. As Re increases, the degree to which increased surface area affects the overall fluid drag increases, and the drag reduction correlation to the laminar flow theories deteriorates.

Optimization of Riblet Geometry

The cross-sectional shape of riblets on fast swimming sharks varies greatly, even at different locations on the same shark. Many types of riblets have been studied, the shapes of which have been chosen for several reasons [14]. Riblet shapes have been chosen for their similarity to natural riblets, for their ease of fabrication, and for purposes of drag reduction optimization.

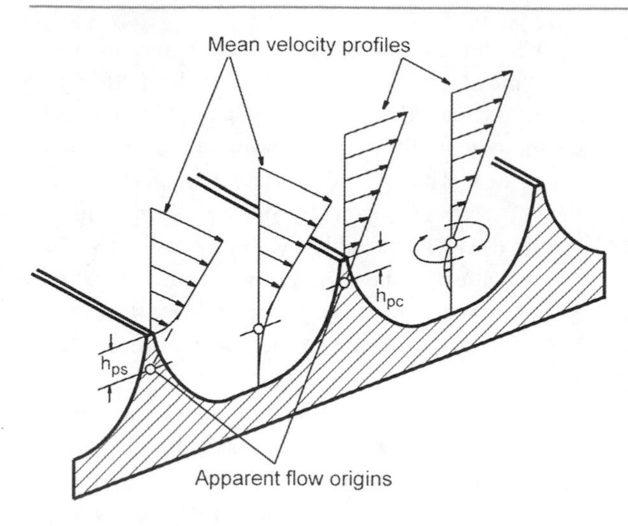

Shark Skin Effect, Fig. 3 Schematic representation of the mean velocity profiles and effective protrusion heights for flow in both the stream-wise direction, h_{ps}, and in the cross-flow direction, h_{pc} (Adapted from [5])

Two-dimensional (2D) riblets, which have a continuous extrusion of a simple cross section in the stream-wise direction, have been most extensively characterized. The most thorough characterization has been completed for symmetrical 2D riblets with sawtooth, scalloped, and blade cross sections as shown in Fig. 4a [1, 2, 4, 6, 7, 29–32, 35, 36]. Alternative riblet geometries have, in general, shown no increased benefit. These riblets, including asymmetrical riblets, hierarchical riblets, and riblets with rounded or notched peaks have been studied in detail and do not improve upon the benefit of standard riblet geometries [29, 30, 32]. Other 2D riblet shapes which have been studied include alternating brother-sister-type riblets [4] and hierarchical riblets with small riblets on top of larger riblets [35]. Three-dimensional (3D) riblets, which include segmented 2D riblets as well as shark skin moldings and replicas have also been studied. Riblet types characterized include aligned segmented-blade riblets [35], offset segmented-blade riblets [6], offset-3D blade riblets [6], and 3D shark skin replicas [7, 19, 24].

Most studies are done by changing the non-dimensionalized spacing, s^+, by varying only fluid velocity and collecting shear stress data from a shear stress balance in a wind tunnel or fluid flow channel. The use of non-dimensional characteristic dimensions for riblet studies, namely, nondimensional spacing, s^+, is important for comparison between studies performed under different flow conditions. Non-dimensionalization accounts for the change in size of flow structures like vortex diameter, which is the critical value to which riblets must be matched. When comparing the optimal drag reduction geometries for sawtooth, scalloped, and blade riblets shown in Fig. 4b, it is clear that blade riblets provide the highest level of drag reduction, scalloped riblets provide the second most, and sawtooth riblets provide the least benefit [14]. A summary of comparison features for sawtooth, scalloped, and blade riblets is presented in Table 1. In general, it can be seen in Fig. 4b that each type of riblet is most beneficial near $s^+ \sim 15$, which is between 1/3 and 1/2, the width of the stream-wise vortices. Larger s^+ will cause vortices to begin falling into the gap between the riblets, which increases the shear stress at the surface between riblets. As s^+ decreases below optimum, the overall size of the riblets decreases to a point below which they cannot adequately impede vortex translation.

An additional concern to the application of riblets is the sensitivity of drag reduction to yaw angle, the angle between the average flow direction and the riblet orientation. Figure 5 shows the effects of yaw angle on riblet performance for flow over sawtooth riblets. Yaw angle has a deleterious effect on the drag reduction benefits of riblet surfaces. Riblet surfaces become drag inducing above $\beta = 30°$, but small drag reductions can still be seen up to $\beta = 15°$ [32].

Riblets on shark skin exist in short segments and groups, not as continuous structures. Riblets with 3D features have been created to better approximate the performance of actual shark skin and to determine if there are methods of drag reduction not yet understood from 2D riblet studies. Studies have explored the effects of compound riblet structures and 3D riblets comprised of aligned, segmented-blade riblets [36]. No improvement in net drag reduction was realized when compared to corresponding performance of continuous riblet geometries. More recently, experiments with similarly shaped segmented-blade riblets at spacing s with a matching set of segmented-blade riblets staggered between each row of blades at a spacing of s/2 from either side have been performed [5]. A schematic and image of staggered trapezoidal blade riblets is shown in Fig. 6a. Using these and other staggered riblets, Bechert et al. [5] hoped to achieve the same vortex elevation and anti-translation effects of continuous riblets with less effect on the flow

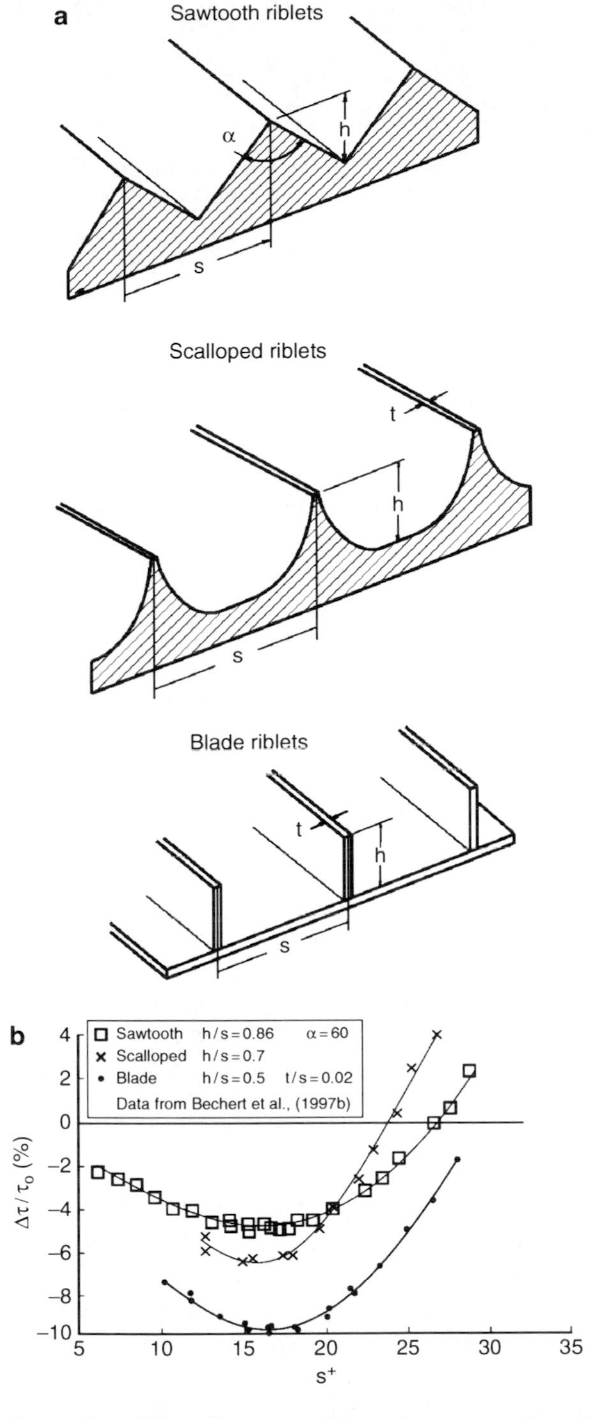

Shark Skin Effect, Fig. 4 (**a**) Schematic representation of riblet dimensions and (**b**) drag reduction comparison for sawtooth, scalloped, and blade riblets (Adapted from [5])

origin. Experimental data comparing the largest drag reduction achieved with staggered segmented-blade riblets to optimum continuous blade riblets can be seen in Fig. 6b. Again, no net benefit in drag reduction was achieved, and after comparison of data, the conclusion was made that it is unlikely that 3D riblets comprised of segmented 2D riblets will greatly outperform continuous 2D riblets.

The scales have been molded in epoxy resin from the skin of the Spiny Dogfish (*Squalus acanthias*) by Jung and Bhushan [19], as shown in Fig. 7a. A decrease in pressure drop corresponding to a decrease in fluid drag was achieved as compared to a smooth surface in a rectangular flow cell experiment, as shown in Fig. 8a. Pressure drop from inlet to outlet of a pipe is a measure of drag with large pressure drop occurring as a result of high drag. Aligned riblets were fabricated on acrylic by using micromachining (Fig. 7b), and a minimal decrease in pressure drop was realized as compared to a smooth acrylic test section (Fig. 8b) [19].

Riblet Fabrication and Applications

Riblets have been fabricated for studies and large-scale applications [14]. Typical microscale manufacturing techniques are ill-fitted for large-scale application due to the associated costs. Even for studies, most researchers have opted for traditional milling or molding methods over the micro-fabrication techniques used in the microtechnology industry. Though nondimensional units allow for comparison between flow fields of different fluids and at different conditions, the accurate microscale manufacture of riblets for experimentation has been a field of study in its own right. The largest difficulty in optimizing riblet geometries has been the fabrication of riblet series with incremental changes in characteristic dimension. Riblets used in airflow require spacings at or below 1 mm due to the low viscosity of air and the high speed at which wind tunnels must operate to create accurately measurable shear stresses on a test surface. Conversely, studies in an oil channel have been carried out in flow that is both highly viscous and slower moving. This allows for riblets to be made with spacings in the 3–10 mm range [3].

Commercial and experimental application of riblets outside wind tunnels and test stands is also limited by the high costs for less-than-optimal riblet performance. Application of riblets on a large scale has been done for several studies as well as for competition and retail purposes. Sawtooth riblets on vinyl films produced by

Shark Skin Effect, Table 1 Summary and comparison of optimum riblet geometry for various riblet shapes [14]

Riblet shape	Relative rank[a]	Maximum drag reduction[b]	Optimum geometry[b]	Comments
Sawtooth	3	5%	h/s ~ 1, α ~ 60°	Most durable
Scalloped	2	6.5%	h/s ~ 0.7	
Blade	1	9.9%	h/s ~ 0.5	Drag reduction increases as riblet thickness, t decreases. Durability is an issue.

[a]1 corresponds to greatest drag reduction
[b]Based on published data in [5]

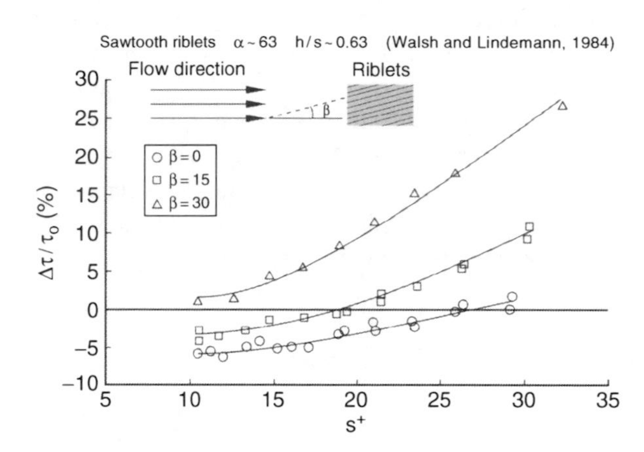

Shark Skin Effect, Fig. 5 Drag reduction dependence on yaw angle, β, of sawtooth riblets in free stream for h/s ≈ 0.62 (Adapted from [32])

3 M riblets have been applied on surfaces ranging from boat hulls to airplanes. Racing swimsuits produced by Speedo and others also employ a riblet pattern on the surface to reduce drag during the streamline portion of each lap of a race [23]. Additionally, a novel surface-scratching technique has been applied to the inside surface of pipelines to create a faux-riblet surface [33].

Beginning in the mid-1980s, vinyl film sawtooth riblets have been applied to boat hulls for racing. Both an Olympic rowing boat and an America's Cup sailing yacht have been covered with riblets during competition. Because skin friction of an airplane accounts for as much as 48% of total drag, vinyl film riblets have also been applied to test planes of both Boeing and Airbus. These films have not seen use on standard commercial flights yet, but the benefits seen in testing should not go unmentioned. Application of riblets to an airplane requires that several concessions are made. Several locations that would be covered by riblets must be left uncovered due to environmental factors; windows are not covered for the sake of visibility; several locations where dust and debris contact the airplane during flight are left bare because the riblets would be eroded during flight; and locations where deicing, fuel, or hydraulic fluid would come in contact with the riblets are left bare. After these concessions, the riblets covering the remaining 70% of the aircraft have provided 3% total drag reduction. This 3% drag reduction correlates to a similar 3% savings in fuel costs [4].

Another large commercial application for riblet technologies is drag reduction in pipe flow. Machining the surface or applying vinyl film riblets proves difficult in the confines of most pipes, and an alternate solution must be used. Experimental application of a scratching technique to the inside surface of pipes has created a riblet-like roughness that has provided more than 5% drag reduction benefit [33]. Stemming from an old sailors' belief that ships sail faster when their hulls are sanded in the longitudinal direction, Weiss fabricated these riblets by using a steel brush moved through the pipeline to create a ridged surface. Studies have shown as much as a 10% reduction in fluid flow with the combined effect of cleaning the pipe and ridging the surface. Tests on a 10-mile gas pipeline section have confirmed this benefit during commercial operation [4].

The dominant and perhaps only commercial market where riblet technology for drag reduction is commercially sold is competitive swimwear [14]. The general population became aware of shark skin's drag reduction benefits with the introduction of the FastSkin® suits by Speedo in 2004. Speedo claimed a drag reduction of several percent in a static test compared to other race suits. However, given the compromises of riblet geometry made during manufacturing, it is hard to believe the full extent of the drag reduction.

It is clear that creating surface structures by weaving threads is difficult. As a result, riblet geometries woven from thread have limited options of feasible riblet shapes. By the pattern woven into the FastSkin® swimsuits, riblets are formed which resemble wide

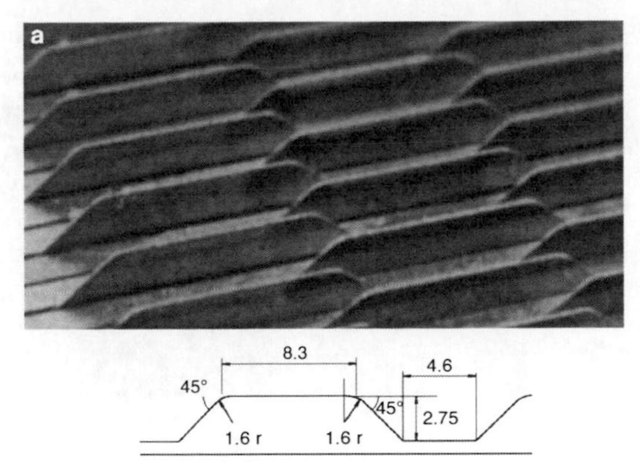

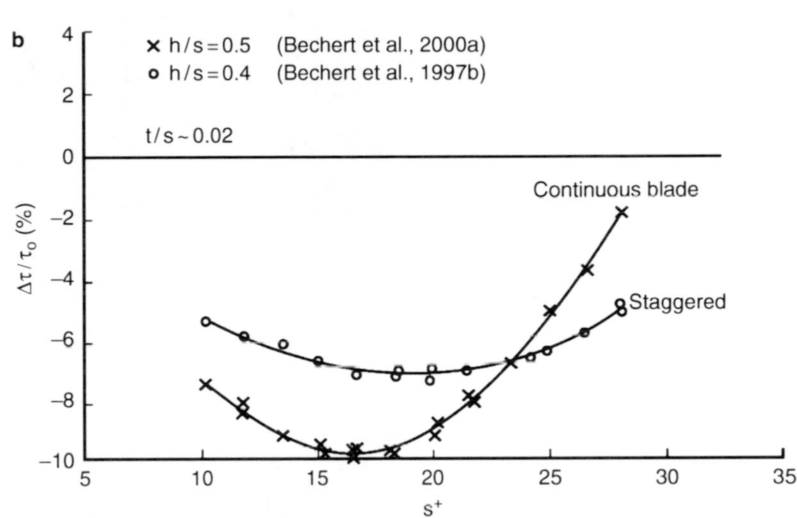

Shark Skin Effect, Fig. 6 Comparison of drag reduction over optimum continuous blade riblets with optimum segmented trapezoidal blade riblets. (**a**) Segmented riblets were staggered as shown. Spacing between offset rows is s/2, while spacing between corresponding rows is s. (**b**) Optimal hour to second ratio for staggered blade riblets is 0.4. Staggered blade riblets provide less drag reduction benefit than continuous blade riblets (Adapted from [5, 6])

blade riblets with small grooves on top. The larger riblets are formed by the macro-weaving pattern, and the smaller riblets are created by the individual weaves of thread aligned with the macro-riblets. Both of these riblet-like shapes are distinguishable in Fig. 9. As shown in Fig. 9a, unstretched riblets are tightly packed. As the fabric stretches, the riblet width and spacing increase (Fig. 9b). The associated decrease in h/s ratio depends on the dimensions of each swimmer's body, which is another compromising factor in the design. Riblet thickness is also a factor considered in the design. Aside from the limitations imposed by the weaving patterns available, flexibility in the riblet tips will hinder the fabric's ability to impede the cross-stream translation of stream-wise vortices. Thicker riblets are probably needed, used for strength, and cause a decrease in the peak drag reduction capability compared to thinner riblets.

3 M vinyl film riblets [27] have been applied to many test surfaces, including the inside of various pipes for pipe flow studies [22], flat plates in flow channels and wind tunnels, boat hulls in towing tanks [11], airplane wings [17], and airplane fuselages. Similar riblet films have been fabricated using bulk micromachining of silicon to create a master for molding of Polydimethylsiloxane (PDMS) to create a thin, flexible riblet film. This film has been used in flow visualization tests [25].

Grinding and rolling methods of riblet fabrication have been studied for application in both research and large-scale application. A profiled grinding wheel has been used to fabricate several riblet geometries based on sawtooth riblets with $h = 20\ \mu m$ and $s = 50\ \mu m$ [15]. Dressing of the grinding wheel was done with diamond-profile roller used in a two-step process in which the profile roller dresses every second tooth on the first pass, shifts axially the distance of one riblet spacing, and dresses the remaining teeth on a second pass. One downside of the grinding process is the lack of hardening on the final riblet surface. Alternatively,

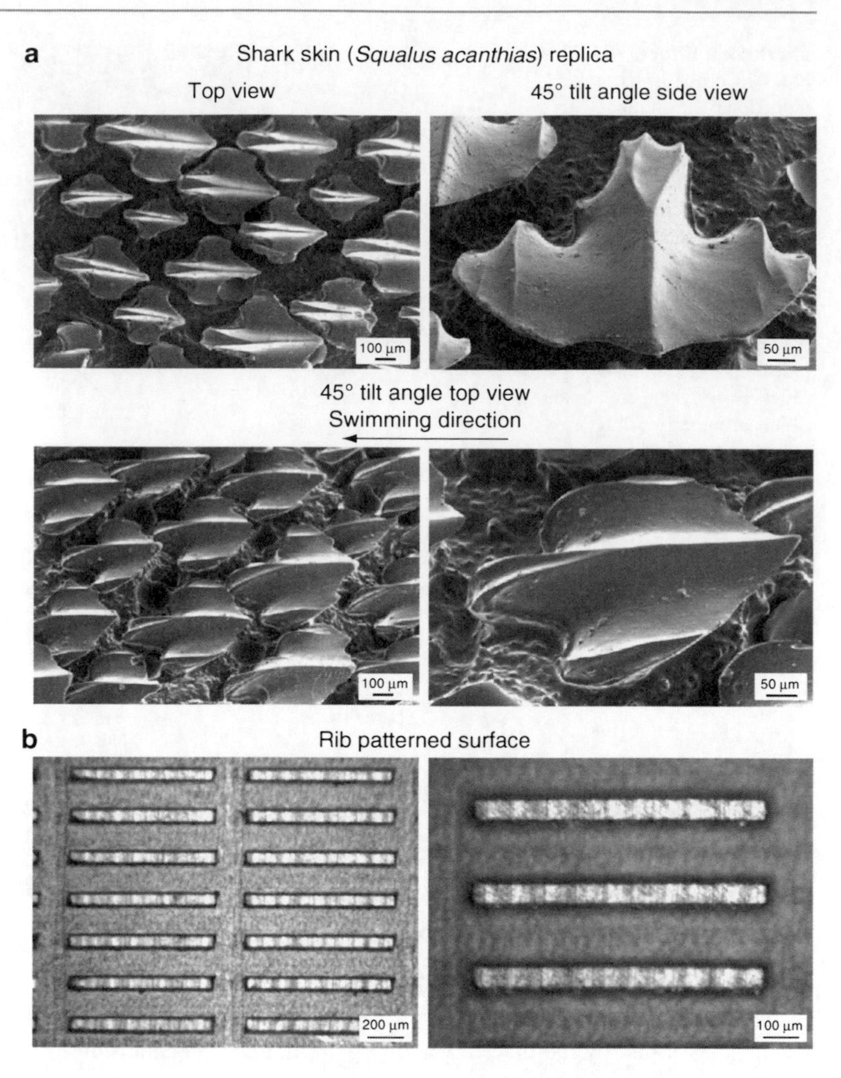

Shark Skin Effect, Fig. 7 (**a**) Scanning electron microscope (SEM) micrographs of shark skin replica patterned in epoxy, and (**b**) segmented bade-style riblets fabricated from acrylic [19]

rolling methods can be used to strain harden the riblets during fabrication. Using a roller with the profile of two riblets on its outer face, a linearly patterned rolling process has been used to fabricate scalloped riblets in a titanium alloy with $h = 162\ \mu m$ and $s = 340\ \mu m$ [21]. The strain hardening, favorable grain patterns, and residual compressive stresses in the riblet surface after fabrication provide advantages in riblet strength for production applications.

Effect of Fish Mucus and Polymers on Fluid Drag

Fish are known to secrete mucus during swimming [14]. Though it is not known whether the mucus is present at all times, it is known that certain environmental factors cause, alter, or enhance the production of mucus. These environmental stressors may present a need for increased swimming speed to catch or avoid becoming prey, for protection against non-predatory threats such as microorganisms, or to resist abrasion while swimming near rocky surfaces. Regardless of which events cause fish to secrete mucus, the drag reduction benefits during mucus-assisted swimming are known. Numerous experiments have demonstrated the drag reduction possible with fish-covered mucus compared to non-mucus-covered shapes [18]. In an experiment comparing the drag on wax models to a mucus-covered fish, a reduction in skin friction drag of 50% was seen [13].

Similar to these fish mucus experiments, polymer additives in pipe flows have been known for many years to reduce the drag in fluid flows by extreme

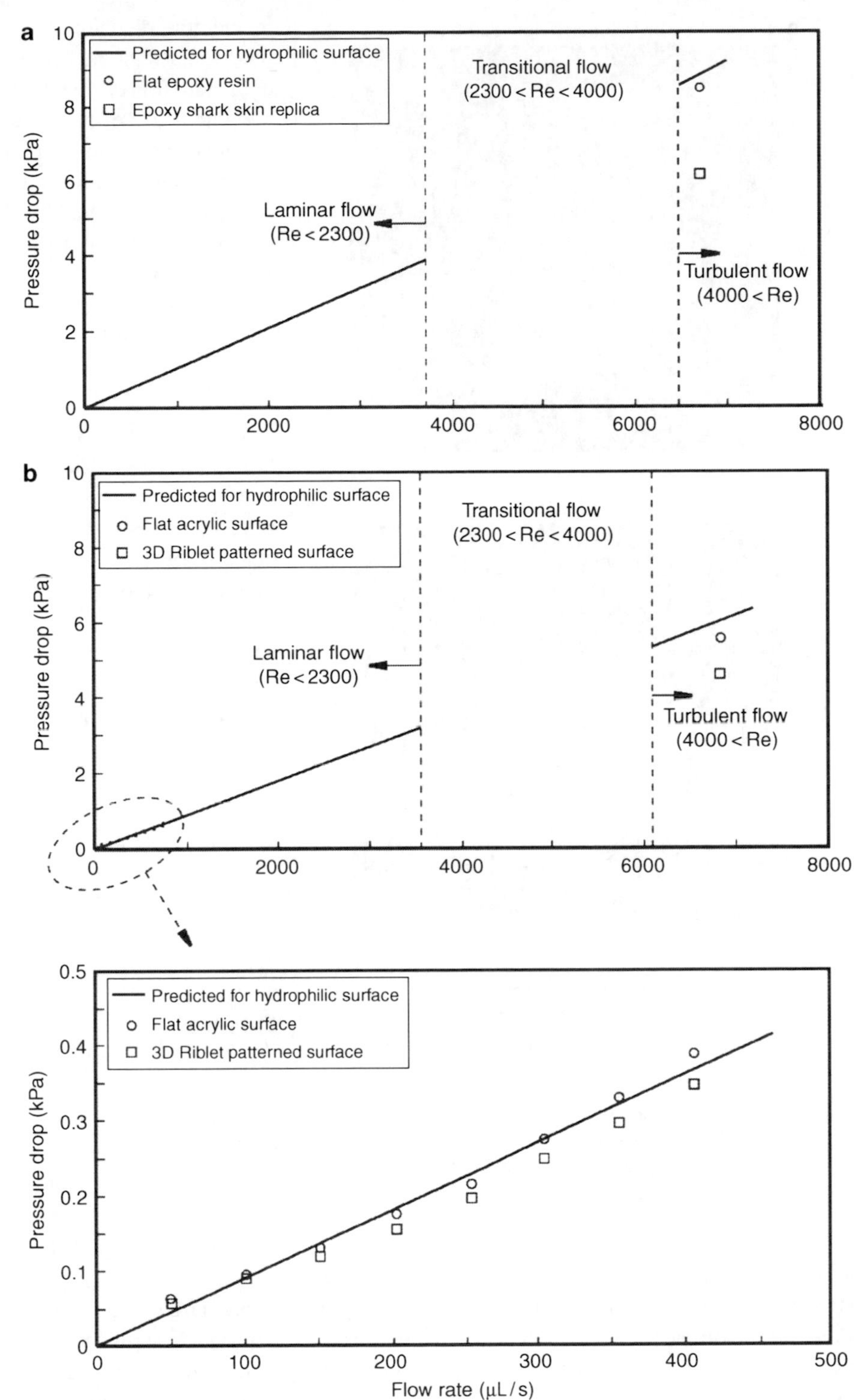

Shark Skin Effect, Fig. 8 (**a**) Comparison of pressure drop in rectangular pipe flow over flat epoxy surface with shark skin replica surface. (**b**) Comparison of pressure drop in rectangular pipe flow over flat acrylic surface and segmented blade riblets. Data are compared with the predicted pressure drop function for a hydrophilic surface (Adapted from [19])

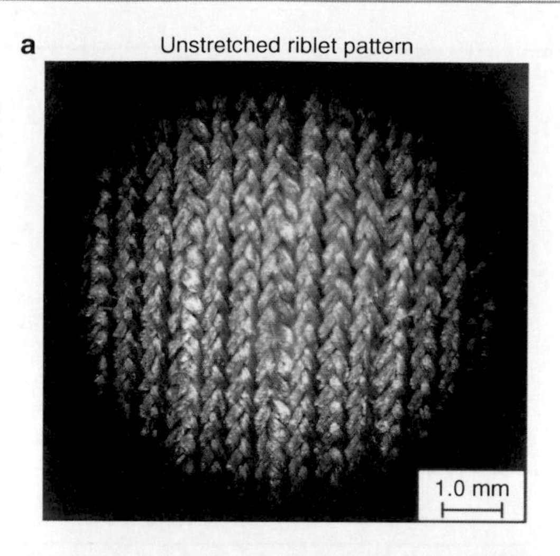

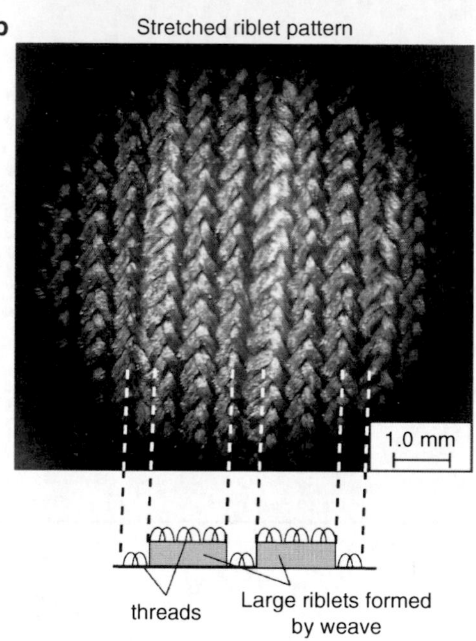

Shark Skin Effect, Fig. 9 Images of riblet geometries on (**a**) unstretched and (**b**) stretched Speedo FastSkin® swimsuit. (**c**) Schematic showing apparent hierarchical riblet structure formed by threads [14]

amounts. Polymers are known to have low shear strength which results in low friction [8, 9]. In a pipe flow study comparing various injection techniques of polymer solutions into water, drag reductions of up to 80% were achieved [16]. Additionally, the drag reduction benefit increases with increased Reynolds number. While this works well for pipe flows, in which the polymer remains mixed and active throughout the length of the pipe, its application to external flows is much more difficult. Mucus on fish does not mix well with water in static contact, but does mix and provide drag reduction during dynamic contact. By this feature, the mucus use of fish is minimized. Unfortunately for any long range application of polymer drag reduction on an external flow, the polymer solution must be continuously injected. This would cause large quantities of the solution to be used and likely render the strategy inefficient in terms of overall energy use.

Though sharks do not secrete enough mucus to use this mechanism for drag reduction, small amounts of mucus are present on the skin of sharks [2]. It is possible that shark skin mucus secretion is similar to fish, where only environmental stressors or swimming causes an increase in output, but the total quantity of mucus on the surface at any given time is likely quite low. One possible mechanism by which this trace quantity of mucus could be useful is in changing the flow characteristics in the riblet valleys or at the riblet peaks, where shear stresses are highest. In the riblet valleys, a trace secretion of mucus could increase flow velocity and decrease the overall momentum transfer from the shark to the surrounding water by condensing the overall structure of the velocity gradient. Alternatively, injection at the riblet peaks may cause a reduction in shear stresses where they are at their maximum and again cause a reduction in drag. These small effects near the surface may propagate into a larger benefit as the lines of constant velocity in the flow shift and condense [14].

Closure

Fluid drag in the turbulent boundary layer is in large part due to the effects of the stream-wise vortices formed in the fluid closest to the surface. Turbulence and associated momentum transfer in the outer boundary layers is in large part due to the translation, ejection, and twisting of these vortices. Additionally, the vortices also cause high velocities at the surface which create large shear stresses on the surface. Riblets impede the translation of the stream-wise vortices, which causes a reduction in vortex ejection and outer layer turbulence. In addition, riblets lift the vortices off the surface and reduce the amount of surface area exposed to the high-velocity flow. By modifying the velocity distribution, riblets facilitate a net reduction in shear stress at the surface.

Various riblet shapes have been studied for their drag-reducing capabilities, but sawtooth, scalloped, and blade riblets are most common. Drag reduction by riblet surfaces has been shown to be as high as nearly 10% given an optimal geometry of h/s ~0.5 for blade riblets with a no-slip condition. The maximum reliable drag reduction provided by scalloped riblets and sawtooth riblets is about 6% at h/s ~0.7 and 5% at ridge angle α ~60°, respectively. Experimentation of other shapes has provided similar benefits of around 5% drag reduction. Though the optimum shape for drag reduction performance is blade riblets, the fragile nature of these blades makes their commercial application of little use. Scalloped and sawtooth riblets, which provide considerably less drag reduction benefit, are much stronger shapes mechanically speaking and should be used for application in environments where contact may occur with non-fluid materials.

Commercial applications of riblets include competition swimsuits, which use a thread-based riblet geometry, as well as experimental applications to airplanes. Drag reductions in riblet application have been accomplished, and flight applications have seen fuel savings of as much as three percent. Manufacturing techniques for riblets must also be chosen specific to their application. Vinyl film riblets are the easiest method, as application of a film to a surface requires less for work small-scale application than other methods. Rolling or grinding methods of riblet application should be investigated for turbine blades or high volume commercially sold pieces.

Cross-References

▶ Biognosis
▶ Biomimetics
▶ Biomimicry

References

1. Becher, D.W., Hoppe, G.: On the drag reduction of the shark skin. Paper # AIAA-85-0564, presented at AIAA Shear Flow Control Conference, Boulder, CO, 12–14 Mar (1985)
2. Bechert, D.W., Bartenwerfe, R.M., Hoppe, G., and Reif, W.-E.: Drag reduction mechanisms derived from shark skin. Paper # ICAS-86-1.8.3, Proceedings of the 15th ICAS congress, vol. 2 (A86-48-97624-01), pp. 1044–1068. AIAA, New York (1986)
3. Bechert, D.W., Hoppe, G., van der Hoven, J.G.T., Makris, R.: The Berlin oil channel for drag reduction research. Exper. Fluids **12**, 251–260 (1992)
4. Bechert, D.W., Bruse, M., Hage, W., and Meyer R.: Biological surfaces and their technological application – laboratory and flight experiments on drag reduction and separation control. Paper # AIAA-1997-1960, presented at AIAA 28th Fluid dynamics conference, Snowmass Village, CO, June 29–Aug 2 (1997)
5. Bechert, D.W., Bruse, M., Hage, W., van der Hoeven, J.G.T., Hoppe, G.: Experiments on drag reducing surfaces and their optimization with an adjustable geometry. J. Fluid Mech. **338**, 59–87 (1997)
6. Bechert, D.W., Bruse, M., Hage, W.: Experiments with three-dimensional riblets as an idealized model of shark skin. Exper. Fluids **28**, 403–412 (2000)
7. Bechert, D.W., Bruse, M., Hage, W., Meyer, R.: Fluid mechanics of biological surfaces and their technological application. Naturwissenschaften **87**, 157–171 (2000)
8. Bhushan, B.: Principles and Applications of Tribology. Wiley, New York (1999)
9. Bhushan, B.: Introduction to Tribology. Wiley, New York (2002)
10. Bhushan, B.: Biomimetics: lessons from nature – an overview. Phil. Trans. R. Soc. A **367**, 1445–1486 (2009)
11. Choi, K.S., Gadd, G.E., Pearcey, H.H., Savill, A.M., Svensson, S.: Tests of drag-reducing polymer coated on a riblet surface. Appl. Sci. Res. **46**, 209–216 (1989)
12. Choi, K.S., Yang, X., Clayton, B.R., Glover, E.J., Altar, M., Semenov, B.N., Kulik, V.M.: Turbulent drag reduction using compliant surfaces. Proc. R. Soc A **453**, 2229–2240 (1997)
13. Daniel, T.L.: Fish mucus: *in situ* measurements of polymer drag reduction. Biol. Bull. **160**, 376–382 (1981)
14. Dean, B., Bhushan, B.: Shark-skin surfaces for fluid-drag reduction in turbulent flow: a review. Phil. Trans. R. Soc. A **368**, 4775–4806;5737 (2010)
15. Denkena, B., de Leon, L., Wang, B.: Grinding of microstructures functional surfaces: a novel strategy for dressing of microprofiles. Prod. Eng. **3**, 41–48 (2009)
16. Frings, B.: Heterogeneous drag reduction in turbulent pipe flows using various injection techniques. Rheologica Acta **27**, 92–110 (1988)
17. Han, M., Huh, J.K., Lee, S.S., and Lee, S.: Micro-riblet film for drag reduction. In: Proceedings of the Pacific Rim Workshop on transducers and micro/nano technologies, Xiamen, China (2002)
18. Hoyt, J.W.: Hydrodynamic drag reduction due to fish slimes. Swim. Fly. Nat. **2**, 653–672 (1975)
19. Jung, Y.C., Bhushan, B.: Biomimetic structures for fluid drag reduction in laminar and turbulent flows. J. Phys.: Condens. Matt. **22**, #035104 (2010)
20. Kline, S.J., Reynolds, W.C., Schraub, F.A., Runstadler, P.W.: The structure of turbulent boundary layers. J. Fluid Mech. **30**, 741–773 (1967)
21. Klocke, F., Feldhaus, B., Mader, S.: Development of an incremental rolling process for the production of defined riblet surface structures. Prod. Eng. **1**, 233–237 (2007)
22. Koury, E., Virk, P.S.: Drag reduction by polymer solutions in a riblet-lined pipe. Appl. Sci. Res. **54**, 323–347 (1995)
23. Krieger, K.: Do pool sharks really swim faster? Science **305**, 636–637 (2004)

24. Lang, A.W., Motta, P., Hidalgo, P., Westcott, M.: Bristled shark skin: a microgeometry for boundary layer control? Bioinspir. Biomim. **3**, 1–9 (2008)
25. Lee, S.J., Choi, Y.S.: Decrement of spanwise vortices by a drag-reducing riblet surface. J. Turbulence **9**, 1–15 (2008)
26. Lee, S.-J., Lee, S.-H.: Flow field analysis of a turbulent boundary layer over a riblet surface. Exp. Fluids **30**, 153–166 (2001)
27. Marentic, F.J., Morris, T.L.: Drag reduction article. United States Patent No 5,133,516, 1992
28. Shephard, K.L.: Functions for fish mucus. Rev. Fish Biol. Fisheries **4**, 401–429 (1994)
29. Walsh, M.J.: Drag characteristics of v-groove and transverse curvature riblets. Viscous Flow Drag. Red. **72**, 169–184 (1980)
30. Walsh, M.J.: Turbulent boundary layer drag reduction using riblets. Paper # AIAA-82-0169, presented at AIAA 20th Aerospace sciences meeting, Orlando FL, 11–14 Jan 1982
31. Walsh, M.J., Anders, J.B.: Riblet/LEBU research at NASA Langley. Appl. Sci. Res. **46**, 255–262 (1989)
32. Walsh, M.J., Lindemann, A.M.: Optimization and application of riblets for turbulent drag reduction. Paper # AIAA-84-0347, presented at AIAA 22nd aerospace sciences meeting, Reno, NV, 9–12 Jan 1984
33. Weiss, M.H.: Implementation of drag reduction techniques in natural gas pipelines. Presented at 10th European drag reduction working meeting, Berlin, Germany, 19–21 March 1997
34. Wilkinson, S.P.: Influence of wall permeability on turbulent boundary-layer properties, Paper # AIAA 83–0294, presented at 21st aerospace sciences meeting of the american institute of aeronautics and astronautics, Reno, NV, 10–13 Jan 1983
35. Wilkinson, S.P., Lazos, B.S.: Direct drag and hot-wire measurements on thin-element riblet arrays. Presented at the IUTAM symposium on turbulence management and relaminarization, Bangalore, India, 19–23 Jan 1987
36. Wilkinson, S.P., Anders, J.B., Lazos, B.S., Bushnell, D.M.: Turbulent drag reduction research at NASA Langley: progress and plans. Inter. J. Heat Fluid Flow **9**, 266–277 (1988)

Shark Skin Separation Control

▶ Shark Skin Drag Reduction

Short (Low Aspect Ratio) Gold Nanowires

▶ Gold Nanorods

Short-Interfering RNA (siRNA)

▶ RNAi in Biomedicine and Drug Delivery

Si Nanotubes

▶ Physical Vapor Deposition

Silent Flight of Owls

▶ Silent Owl Wings

Silent Owl Wings

Thomas Bachmann[1] and Hermann Wagner[2]
[1]Institute for Fluid Mechanics and Aerodynamics, Technische Universität Darmstadt, Darmstadt, Germany
[2]Institute for Biology II, RWTH Aachen University, Aachen, Germany

Synonyms

Silent flight of owls

Definition

Wings of owls are equipped with macro- and microstructured devices that reduce flight noises significantly. These birds of prey hunt in both twilight and darkness. Since visual information is limited at that time of day, owls use acoustic information to detect and track potential prey. The silent flight affords the detection of prey in flight and makes the owl inaudible for its prey. Different wing and feather specializations, such as serrations at the leading edge of the wing, a velvet-like upper surface and fringes along the inner vanes, influence the airflow and thus are important for the reduced flight noise in owls.

Overview

Fundamentals of Bird Flight

In the course of evolution, birds have conquered the air for locomotion. A great variety of wing geometries have evolved for different needs of birds. Wings can be long or short, narrow or broad, thick or thin. Furthermore, wings can be pointed or have a rounded tip, which may be smooth or fingered. Considering this variety, each wing appears to be adapted to the distinct flight conditions of the particular species and the ecological niche it uses. However, all wings have at least one aerodynamic feature in common: They are cambered to the dorsal side. Only curved wings are able to produce enough lift at moderate angles of attack and low flight speeds to allow a bird to become airborne. Bird wings are mostly cambered in proximal regions. Thickness as well as camber decrease towards the wing tip. The resulting reduction of mass at the wing tip causes a decrease of inertia to allow high wing-beat frequencies (2–12 Hz) [1] at low energy consumption [2].

A bird has to defy gravity to become airborne. Hence, flight consumes a lot of energy, especially during takeoff and landing [2]. High lift is achieved either by an intense beating of the wings or by forming wings with high-lift properties [3]. Such wings are characterized by a high camber and a large wing area. Apart from gravity, birds have to deal with an additional force, the drag. Drag is the resistance of the body and wings while moving through the air. It is comprised of three components [3]: first, the friction drag between the airflow and the surface, second, the form drag of the bird's body and the leading edges of the wings, and third, the induced drag. The induced drag is the component of drag force on the bird's wing that is caused by an induced downward velocity. The drag is also influenced by the microstructure of the plumage. Several anatomical specializations of feathers have been reported that influence the airflow around the bird wing, most of them in owls [4–8].

Function of Silent Flight in Owls

Owls have an almost global distribution, which is reflected in the variety of species and subspecies. These birds spend a high proportion of their time searching for prey, either by sitting on a perch, or flying slowly over fields and vegetation [9, 10]. The prey of most owls is mainly active at night. Since visual information is limited in low-light conditions, owls use acoustical cues to detect their prey. The prey emits noise by rustling through the vegetation or through intra- and interspecific communication sounds which can be detected by the owl even in absolute darkness [11]. Several anatomical and behavioral adaptations of the hunting strategy have been discovered in owls. One conspicuous specialization in many owl species is the facial ruff with its specialized feathers that direct sound toward the ears like a dish antenna. The facial ruff is also responsible for the low hearing threshold [12]. Asymmetrically arranged ear flaps in front of the outer ear canals help to localize the elevation of sound sources. Furthermore, nuclei in the brain that process acoustic information are enlarged compared with other similar-sized birds [13]. With this set of adaptations, owls are able to detect and localize their prey precisely. Nevertheless, to be able to utilize this ability during flight, owls should fly silently. If that would not be the case, noise emitted by prey would be masked by the owl's own flight noise and the prey might be warned by the approaching owl. Consequently, owls rely on a silent flight, which is achieved by adaptations of the wing, the plumage, and the flight behavior [4–10].

Specializations of the Owls' Plumage

Feathers are the main aerodynamic components of a bird's wing. Interaction with the airflow during flight and protection against cold, heat, and wetness are essential functions. Furthermore, feathers are used for display or camouflage. A feather consists of a central shaft (rachis) and two laterally attached vanes. The feather vane is made up of parallel barbs that branch off from either side of the rachis. Barbs are interlinked via hook and bow radiates to form a closed surface. While each feather quill is inserted into the integument, all flight feathers (remiges) are additionally associated with the underlying skeleton. Secondary remiges are connected to the posterior edge of the ulna; primary remiges are supported by bones of the hand and fingers. Coverts arise from the anterior integument membrane, forming smooth and closed upper and lower wing surfaces. All wing feathers together form a wing that is light and flexible, but strong enough to meet the bird's requirements to fly.

In owls, several microstructured plumage specializations are known that influence the flow over the wing [4–8]: The dorsal surface of the feathers has an increased roughness caused by elongations of hook radiates, termed pennula. Those feathers that form the leading edge of the distal wing are equipped with serrations at their outer vanes. Every feather of the wing is surrounded by a fringed structure which is due to unconnected barb endings. Furthermore, owl feathers are very flexible and pliant due to a reduced number of radiates and thus a lower number of interlinks between the barbs compared to other birds [8]. Finally, cross-sectional profiles of the rachises are less structured compared to other birds of similar size. These plumage specifications have been implicated in noise suppression during flight by damping noise above 2 kHz [5]. Thus, flight noise is reduced within the typical hearing spectrum of the owls' prey (>3 kHz) [5] and within the owl's own best hearing range (5–9 kHz) [14]. The reduction of noise is mainly achieved by airflow control and friction reduction of single feathers.

Basic Methodology

Morphometrics

Bird wings, together with many other parameters, have been morphometrically characterized in order to classify bird species and to create a systematic order within birds (*Aves*). However, aerodynamic parameters are not included in this approach. Aerodynamic properties such as wing span, wing area, and aspect ratio can be measured in living birds or well-preserved museum specimens. Camber and thickness distribution are more prone to errors due to drying processes or unnatural fixation. Therefore, anesthetized or recently deceased birds yield better results. Best results are gained from experiments with living birds. Hereby, the wing geometry is obtained during free-flight experiments. Computer-aided analyses of three-dimensional reconstructions of the wings provide high-resolution profile data.

Free-flight experiments are conducted to measure the noise emission of flying owls in different flight modes while the birds fly over an extreme sensible microphone array at a defined flight speed and height. The measured noise spectrum can then be correlated with the flight parameters. Finally, the results are compared to other non-silently flying birds [15].

Substructures of the feathers are investigated with different digital imaging techniques having high spatial resolutions. Light microscopy and scanning electron microscopy (SEM) enable to some extent two-dimensional measurements and morphometric characterizations. Confocal laser scanning microscopy (CLSM) and micro-computed-tomography (μ-CT) allow the precise digitization of the three-dimensional shape of microstructures. In all cases, the data are then processed and analyzed with adequate visualization and measuring software to determine relevant morphometric and aerodynamic parameters. After specifying the shape and geometry of the owl-specific structures, these can be artificially manufactured and applied to wing models for further analysis, e.g., in wind tunnel experiments. By this means, their aerodynamic function may be clarified.

Behavioral Studies

Birds use their wings to maneuver through the air. The form and geometry but also the movements of the wings are important for an efficient locomotion. Hence, studying the flight behavior of birds is one fundamental piece of the large puzzle of understanding how bird wings work. The study of wing motion in different flight modes, ideally in combination with flow visualization, helps to understand the functions of each wing component. Flight speed, wing-beat frequency, and wing-beat amplitude are some parameters that influence the flight of birds. Different flight modes like flapping flight, gliding, or soaring exist. Furthermore, some birds adapt their flight behavior during hunting, escaping, courtship, and other modes. Nowadays, high-speed cameras outperform traditional bird watching with binoculars and cameras in many aspects. Images with high temporal and spatial resolution of wing and feather movements during beating of the wings allow a much better investigation of each wing element and its aerodynamic function. Such techniques are used in field observations or wind tunnel experiments with living birds.

Wind Tunnel Experiments

Fixed wings of birds and artificial models of bird wings are investigated in wind tunnel experiments. Performing experiments in wind tunnels guarantee reliable and repeatable results. The use of wing models instead of living animals allows the manipulation of certain parameters of the wing. The impact of several wing geometries can then be compared and general conclusions can be drawn based on these results. Flow visualization techniques such as particle-image velocimetry (PIV), oil-flow pattern or pressure measurements are carried out to understand the influence of the wing profiling and each wing element on the airflow. In case of owl wings, the specific feather adaptations are added separately on a wing model or are removed step by step from fixed owl wings. Their potential function is revealed by comparing wings either with or without the owl's specializations in fluid-mechanical experiments.

Key Research Findings

Owl Wings

Owl wings differ from those of other birds. Figure 1a shows an example wing of a barn owl (*Tyto alba*). The large wing area, in combination with a relatively low body mass, results in a low wing loading allowing the owl to fly slowly and to carry large prey. Specific camber and thickness distributions (Fig. 1b, c) are also characteristic for owl wings [6]. While the proximal wing is highly cambered and anteriorly thickened, the distal wing is thin and less cambered. Anatomical studies revealed the construction plan of the wing (Fig. 1d). Skeletal elements, the muscle mass, and the coverts are responsible for the thick anterior part of the wing. By contrast, the distal wing and the posterior arm part of the wing are formed by remiges. These areas are extremely thin and lightweight. Hand and arm part of the wing have different functions (lift production and thrust) expressed by different geometries and profiling. At both wing parts, however, the airflow tends to separate in wind tunnel experiments when tested with a smooth surface without any surface or edge modifications [5, 16]. Owls, however, evolved microstructures on their feathers to influence the airflow around the wing in such a way that the airflow stays attached especially during critical flight maneuvers [5, 16] such as takeoff, landing, or striking. Hence, the wings produce a huge amount of lift even at low flight speeds.

Owl Feathers

Owl feathers share the anatomy of typical contour feathers [8]. However, they differ in some detail. As feathers are the main aerodynamic components of the bird wing, their geometry and dimensions affect the overall wing geometry. In owls, the feather vanes are large and at the same time extremely light and flexible. Fewer interlinks of the barbs cause these effects [8]. Hereby, owl feathers are very air-transmissive and flexible, whereby they can react rapidly to varying airflow conditions. Their bending behavior is passively driven. It depends on the material properties of feather keratin and the geometry of the rachis and the barbs, respectively [17].

Compared to other birds, owls stand out due to a large count of contour feathers [9]. Specifically, head, feet, and body are densely covered. This upholstery of the body serves the suppression of noise to some extent, but also thermal insulation since owls have only small fat reserves. The coloration of the plumage is mostly adapted to the environment of the relevant owl species with respect to an effective camouflage, but also to sexual display to some extent. Snowy owls (*Bubo scandiacus*), for instance, have a white plumage, while species that live in wooded environment have a brownish patterned plumage (see Fig. 2). Different color intensities between males and females are found, for instance, in barn owls with the females being slightly darker.

The most conspicuous attribute of owl remiges are several anatomical microstructures that are responsible for the silent flight [4, 8]. Interestingly, fishing owls like the Malay fish owl (*Bubo ketupu*) lack such feather specialization [4]. Fish cannot hear the approaching owl. Hence, these owls do not need to fly silently.

Microstructures of Owl Feathers

Owls evolved several surface and edge modifications of their wings that are in turn specializations of the plumage. Three structures are mainly worth mentioning: the leading-edge comb-like serrations, the velvet-like upper surface, and the inner vane fringes [4, 7, 8].

Each feather that functions as a leading edge of the wing is equipped with comb-like serrations (Fig. 3). The number of remiges forming the leading edge varies among owl species, but is in the range of 1–3 remiges. Barb endings separate and bend upward, changing their form and function. Differences in the anatomical phenotype are small within one feather, though small variations occur (Fig. 3) [8]. While long serrations with small spacing are found in basal and central regions of the feathers, shorter serrations with a larger distance are located at the tip of the feather.

Functionally, serrations affect the airflow and noise production only marginally during cruise flight [5]. The stagnation pressure at the leading edge of the wing prevents a functional airflow around the serrations. However, in critical flight maneuvers, such as landing or striking, the angle of attack is much higher and the serrations cause the airflow to stay attached to the wing. This is achieved by dividing the airflow into

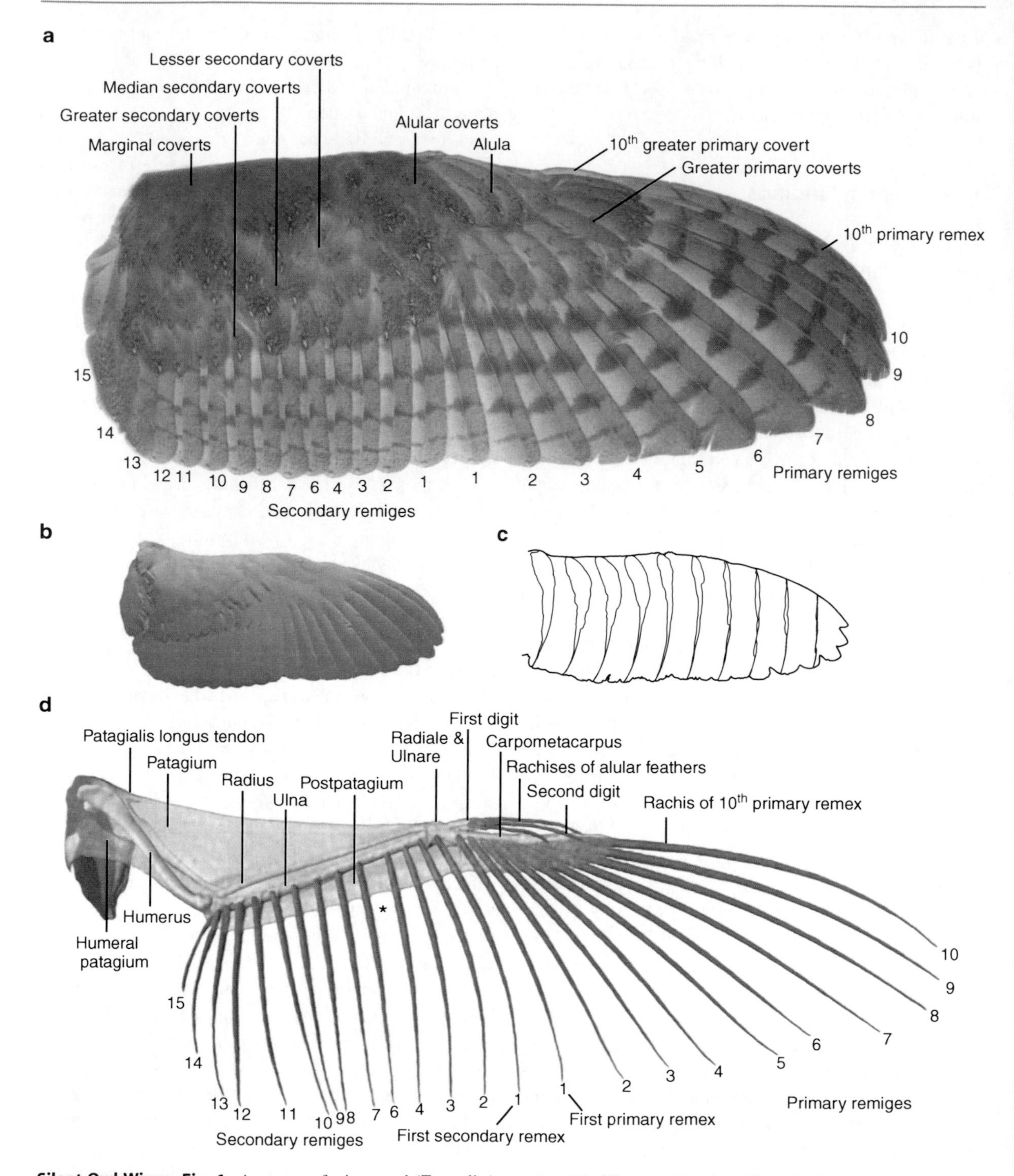

Silent Owl Wings, Fig. 1 Anatomy of a barn owl (*Tyto alba*) wing: (**a**) Topography of the dorsal wing (photograph). (**b**) Surface image of the digitized wing (surface scan). (**c**) Profiles of the owl wing in steps of 10% of the half wing span. (**d**) 3D reconstruction of skeletal elements, skin, and rachises of the remiges (processed CT scan). *The fifth secondary remex is missing in all owl species

Silent Owl Wings, Fig. 2 *F*eathers of different owl species: (**a**) Tenth primary remex of a barn owl (*Tyto alba*) wing. The outer vane is equipped with tiny serrations, the inner vane is lined with fringes. (**b**) Contour feather from the body of a snowy owl (*Bubo scandiacus*). Note the symmetry of the vanes in comparison to the flight feathers (**a**) and (**c**) and the elongated plumulaceous barbs (fringes at the end of the vanes). (**c**) Primary remex of an eagle owl (*Bubo bubo*). As the eagle owl is the largest owl, so are its feathers. Scale bars represent 2.5 cm

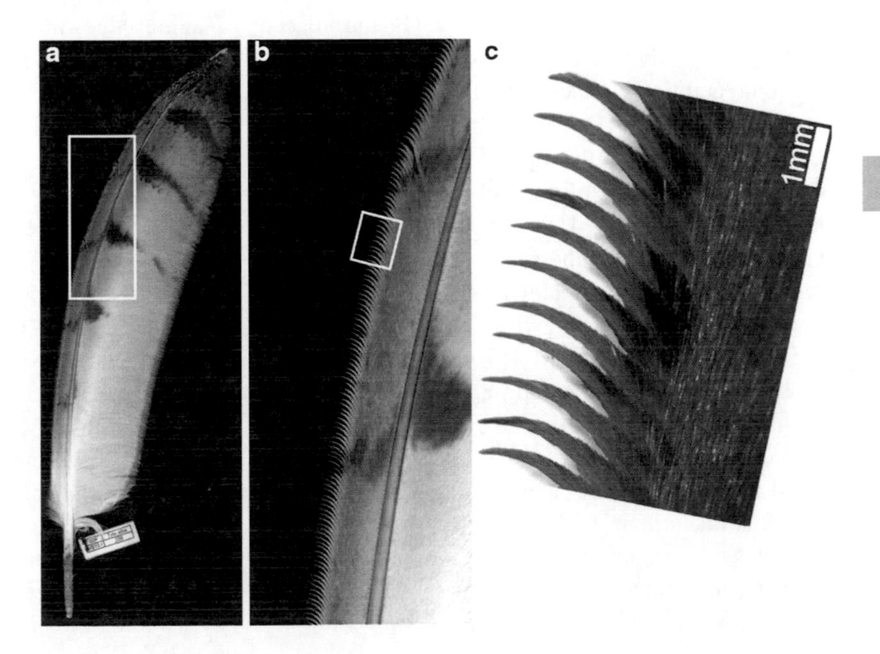

Silent Owl Wings, Fig. 3 Serrations at the leading edge of a tenth primary of a barn owl (*Tyto alba*) in different magnifications: Barb endings separate and bow toward the *upper side* (**a**, **b**, **c**). Note also the fringes at the inner vane (**a**) and the dorsal surface texture

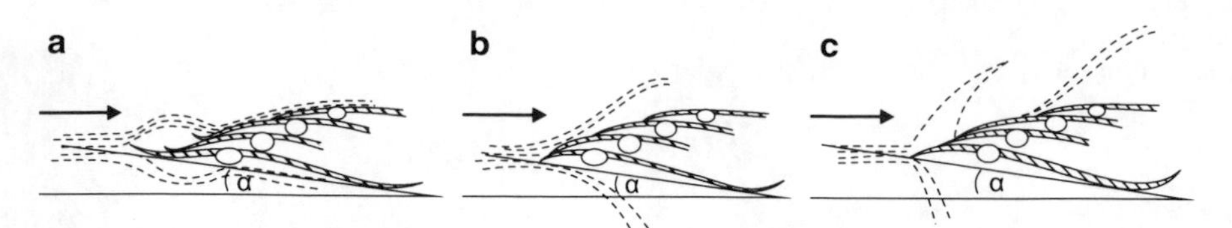

Silent Owl Wings, Fig. 4 Scheme of the airflow at the leading edge of a distal bird wing: (**a**) Intact wing of a tawny owl (*Strix aluco*) in the region of the serrations. Note the laminar flow conditions around the wing. (**b**) After manual removing of the serrations at the wing of the tawny owl, the flow separates early. (**c**) Flow conditions around a wing of a mallard duck (*Anas platyrhynchos*). The *arrow* indicates the flow direction; alpha is the angle of attack (Drawings after Neuhaus et al. [5])

many small micro-vortices that extent chordwise over the upper wing surface. As a consequence, the boundary layer is more energetic and flow separation is prevented (Fig. 4a). By this means, lift production is maintained even at high angles of attack and low flight speed. This effect gets lost after manual removal of the serrations at the leading edge (Fig. 4b). Here, the flow field around the wing resembles that of a mallard duck (Fig. 4c) which produces much more noise [5].

The second specialization is found on the upper wing surface. Each feather of the owl's plumage is covered with a velvet-like dorsal texture (Fig. 5). This structure is formed by elongations of the hook radiates. Each hook radiate terminates in a filament structure, termed pennula. Due to their large number, the surface becomes very fluffy and porous. Functionally, two aspects are at least associated with this structure.

On the one hand, the velvet-like dorsal surface of the inner vane, which is mostly covered by the adjacent feather, is very thick and well developed. The pennula are long and interconnect at large angles with the feather's surface. This in turn causes a friction-reductive device on the upper surface and enables a smooth and silent gliding of the feathers. In all other birds investigated so far, the parallel-oriented barbs of adjacent feathers rub against each other producing a well-noticeable high-frequency sound, especially during flapping flight. In owls, however, such noises are prevented by the velvety surface. Those noises that still occur are dampened by the porous feather texture acting like acoustic absorbers.

On the other hand, the velvet-like structure is also found at the outer vane. Here, the texture affects the aerodynamics of the wing. As mentioned above, the airflow tends to separate at wings with the specific camber and thickness distribution of owls and a smooth wing surface [16]. Figure 6 shows the PIV results of the velocity distribution of the flow around an owl-based airfoil at Reynolds number 40,000, two different angles of attack (0° and 3°), and two different surfaces. The velocity of the flow is color coded. At the wing with a clean and smooth surface (Fig. 6a, c), a separation bubble occurs on the suction side that increases with the higher angle of attack and leads to a complete flow separation beyond 3° angle of attack (not shown) [16].

Increasing the surface roughness as it is found in owls is one means in the direction of flow control. An artificial velvety surface texture applied to the owl-based wing geometry has a dramatic influence on the flow field (Fig. 6b, d). The velvety surface shifts the separation toward higher angles of attack indicated by a smaller separation bubble in comparison to the clean wing model. The boundary layer of the wing is controlled in such a way that occurring separation bubbles are reduced in size and shifted toward the leading edge of the wing. Consequently, the airflow is stabilized even at low flight speeds [16].

A further specialization can be found at the inner vanes of the remiges. Here, barb endings separate by a reduction of the hooklets of the hook radiates. Barbs are no longer interconnected and fringes are formed (Fig. 7). Since the tips of the inner vanes of remiges form the trailing edge of the wing, fringes are also found here. The appearance of fringes is very fluffy and unstable in shape. They are made up of three components: the central barb shaft and the laterally attached hook and bow radiates. The radiates support the formation of fringes by a parallel orientation to the barb shafts [8]. Since the barb shaft does not change its shape, as is the case for the serrations, the structure is pliant and the fringes can float freely resulting in a thin, almost two-dimensional structure. Functionally, fringes are assumed to decrease the sound intensity by a reduction of turbulence at the trailing edge of

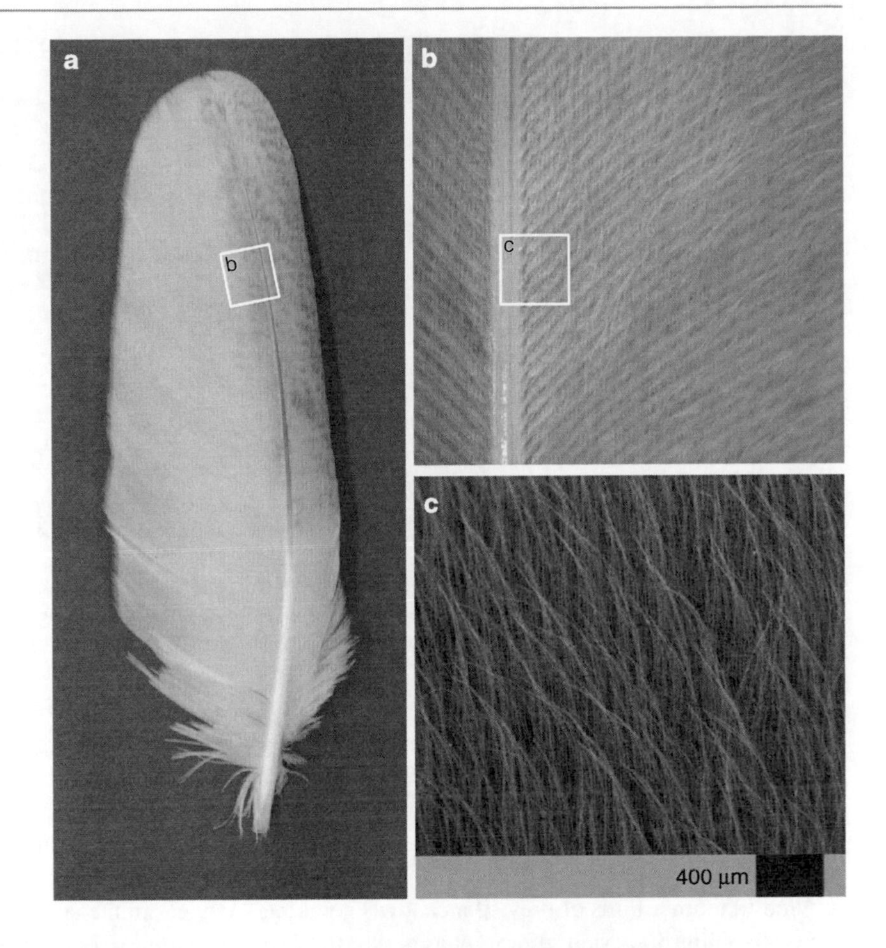

Silent Owl Wings, Fig. 5 Velvet-like dorsal surface of a barn owl's fourth secondary in different magnification: Elongations of the hook radiates create this filigree texture which is supposed to influence the airflow and to reduce friction noise. (**a, b**) Photographs, (**c**) scanning electron microscope image

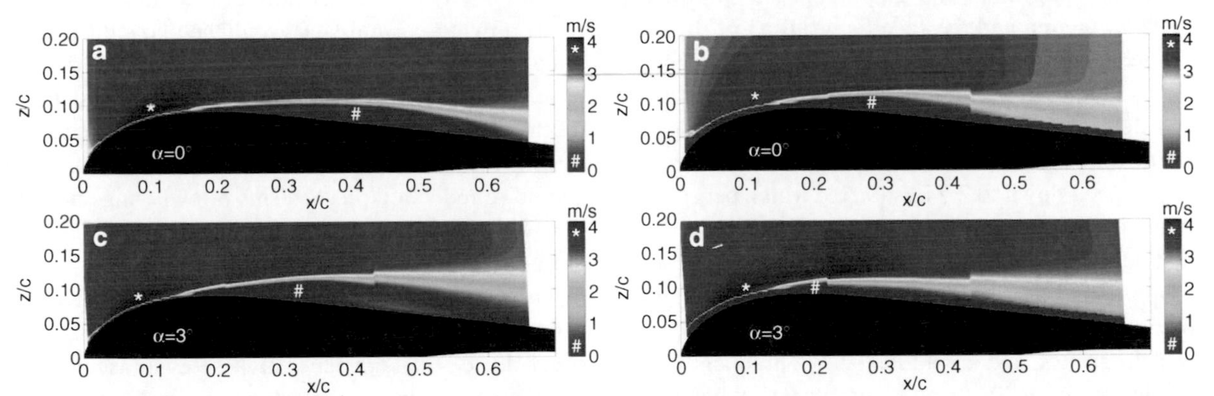

Silent Owl Wings, Fig. 6 Averaged velocity fields at Re = 40,000 and an angle of attack of 0° and 3° for a clean wing (**a, c**) and the same wing geometry with a velvety surface (**b, d**). The velocity of the airflow is *color* coded (online version)

the wing [7, 18]. While sharp edges of wings and flat plates are known to produce a well-noticeable noise even at low Reynolds numbers [19], fringed trailing edges of owl remiges prevent such phenomena. Each wing feather has its own trailing edge that produces noise during flight independently from being part of the wing or separated, for example, during flapping flight. In general, fringes act as a pressure release mechanism that interrupt scattering phenomena which can be found when turbulent eddies pass over the posterior wing region that gradually transitions from the loaded wing to the freestream conditions

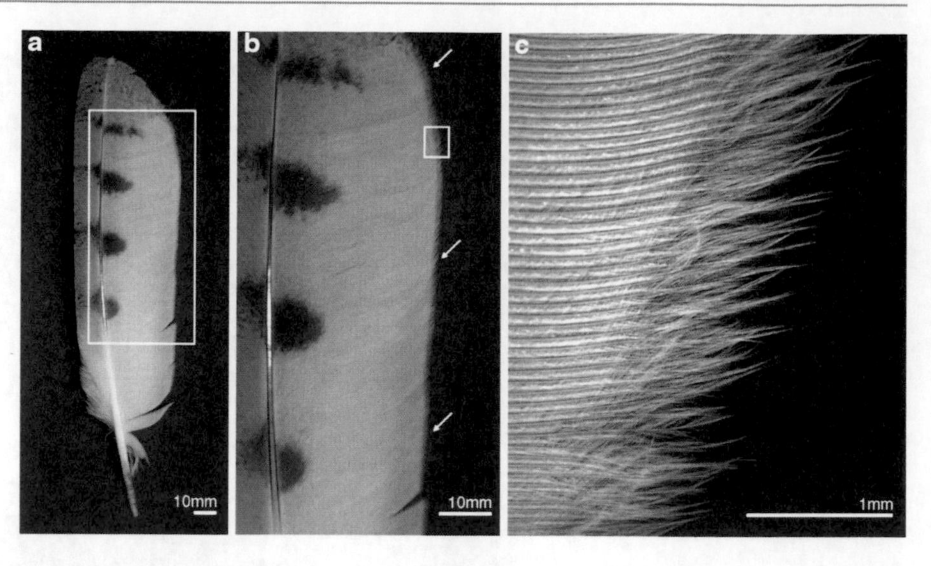

Silent Owl Wings, Fig. 7 Fringes at the inner vane of a barn owl's fifth primary in different magnification:Hooklets of the hook radiates are absent causing the barbs to separate. Hook and bow radiates change their length and thereby support the formation of fringes. (**a**) and (**b**) dorsal, (**c**) ventral

[18, 19]. Without such fringes, turbulent boundary layers being convected past the trailing edge into the wake would generate an intense, broadband scattering noise. In owls, fringes prevent sharp edges at the trailing edge such that the turbulent boundary layer is merged smoothly into the airflow.

Behavioral Adaptations

Nocturnal owls are unable to benefit from thermal upwind like other birds of prey. Hence, owls combine active flapping flight and gliding phases [9, 10]. The large wing area and the specific profiling of the wing enable the owl to glide and maneuver with little movements of the wing. The flight speed is relatively low and depends on the illumination of the environment. Barn owls (*Tyto alba*), for instance, can reduce their flight speed down to 2.5 m/s [9, 10] which per se leads to a silent flight, since only little turbulences occur at that velocity. Apart from the slow gliding flight, some adaptations in the motion of the owl already prevent rising noises. During hunting, owls reduce their wing beat frequency and amplitude [9]. Consequently, less friction between feathers occurs during flight which would cause rubbing noises. Those noises that still rise are efficiently dampened by the porous upper wing surface [4, 7].

Future Directions for Research

Urbanization and the increasing demand for air freight and renewable energy have led to a dramatic increase of noise pollution in the last decades. At ground level, airplanes produce an intense noise by their engines and high-lift configurations of the wings. Furthermore, wind farms are constantly upgraded. Their noise emission does not only annoy humans, but also disturbs migratory animals such as birds, or in case of offshore wind farms, whales, and fishes. How flight noise can efficiently be reduced is demonstrated by owls. These birds evolved noise-reduction and noise-suppression devices in the course of evolution. The implementation of the underlying mechanisms in modern aircraft and rotors of wind-power plants would help to reduce noise pollution. Until then more studies and experiments have to be carried out. In joint biological and fluid-mechanical efforts, the basic aerodynamic characteristics and the fluid-mechanical details that define the drag and noise reduction mechanisms of owl wings have to be investigated in more detail. This will be done using preserved material from zoological collections, experiments with living owls, and wing models in wind tunnel experiments. The knowledge gained will be used to create devices that operate at higher Reynolds numbers, since copying the natural structure alone would not deliver satisfying results. In the long run, designing silent wings of modern aircraft will bring us closer to the goal of a quieter environment.

Cross-References

▶ Biomimetic Flow Sensors
▶ Confocal Laser Scanning Microscopy

► Friction-Reducing Sandfish Skin
► Insect Flight and Micro Air Vehicles (MAVs)
► Scanning Electron Microscopy
► Shark Skin Drag Reduction

References

1. Pennycuik, C.J.: Wing beat frequency of birds in steady cruising flight: new data and improved predictions. J. Exp. Biol. **199**, 1613–1618 (1996)
2. Rayner, J.M.V.: On aerodynamics and the energetics of vertebrate flapping flight. Contemp. Math. **141**, 351–400 (1993)
3. Rüppell, G.: Vogelflug. Rowohlt Taschenbuchverlag GmbH, Hamburg, Germany (1980)
4. Graham, T.: The silent flight of owls. J. Roy. Aero. Soc. **38**, 837–843 (1934)
5. Neuhaus, W., Bretting, H., Schweizer, B.: Morphologische und funktionelle Untersuchungen über den lautlosen Flug der Eule (*Strix aluco*) im Vergleich zum Flug der Ente (*Anas platyrhynchos*). Biol. Zentr. Bl. **92**, 495–512 (1973)
6. Nachtigall, W., Klimbingat, A.: Messungen der Flügelgeometrie mit der Profilkamm-Methode und geometrische Flügelkennzeichnung einheimischer Eulen. In: Nachtigall, W. (ed.) Biona Report 3, pp. 45–86. Gustav Fischer Verlag, Stuttgart und New York (1985)
7. Lilley, G.: A study of the silent flight of the owl. AIAA Paper, pp. 98–2340 (1998)
8. Bachmann, T., Klän, S., Baumgartner, W., Klaas, M., Schröder, W., Wagner, H.: Morphometric characterisation of wing feathers of the barn owl (*Tyto alba*) and the pigeon (*Columba livia*). Front. Zool. **4**, 23 (2007)
9. Mebs, T., Scherzinger, W.: Die Eulen Europas. Franckh-Kosmos, Stuttgart, Germany (2000)
10. Taylor, I.: Barn Owls: Predator – Prey Relationship and Conservation. Cambridge University Press, Cambridge (1994)
11. Knudsen, E.: The hearing of the barn owl. Sci. Am. **245**(69), 113–125 (1981)
12. Hausmann, L., Campenhausen, V.M., Endler, F., Singheiser, M., Wagner, H.: Improvements of sound localization abilities by the facial ruff of the barn owl (*Tyto alba*) as demonstrated by virtual ruff removal. PLoS One **4**(11), e7721 (2009)
13. Konishi, M., Takahashi, T., Wagner, H., Sullivan, W., Carr, C.: Neurophysiological and anatomical substrates of sound localization in the owl. In: Auditory Function – Neurobiological Bases of Hearing, pp. 721–745. Wiley, New York (1988)
14. Konishi, M.: How the owl tracks its prey. Sci. Am. **61**, 414–424 (1973)
15. Sarradj, E., Fritzsche, C., Geyer, T.: Silent owl flight: Bird flyover noise measurements. AIAA Paper 2010–3991 (2010)
16. Klän, S., Bachmann, T., Klaas, M., Wagner, H., Schröder, W.: Experimental analysis of the flow field over a novel owl based airfoil. Exp. Fluids **46**, 975–989 (2008)
17. Bonser, R.H., Purslow, P.: The young's modulus of feather keratin. J. Exp. Biol. **198**, 1029–1033 (1995)
18. Herr, M.: Experimental study on noise reduction through trailing edge brushes. Note. N. Fl. Mech. Mul. D. **92/2006**, 365–372 (2006)
19. Moreau, D.J., Brooks, L.A., Doolan, C.J.: On the aeroacoustic tonal noise generation mechanism of a sharp-edged plate. JASA E. L (2011). doi:10.1121/1.3565472

Silica Gel Processing

► Sol-Gel Method

Silicon Microphone

► Electrostatic MEMS Microphones

Siloxanes

► BioPatterning

Silver (Ag)

► In Vitro and In Vivo Toxicity of Silver Nanoparticles

Simulations of Bulk Nanostructured Solids

Donald W. Brenner
Department of Materials Science and Engineering, North Carolina State University, Raleigh, NC, USA

Synonyms

Plasticity of nanostructured solids

Definition

A bulk nanostructured solid is a material available in macroscopic quantities that has nanometer-scale structural features in at least one dimension.

Computer modeling has emerged to become a third pillar of research alongside experiment and theory. There are many reasons for this emergence, including an exponential increase in processing speed, relatively inexpensive platforms for parallel computing and data storage, new visualization capabilities, and the development of powerful algorithms that take full advantage of these advances in hardware. Advances in simulation methodologies have also made the results of atomic-level computer modeling reliable to the extent that in many cases simulations can replace difficult and expensive experiments. Methods that enable atomistic dynamics using full electronic energies now allow processes involving up to several thousand atoms to be accurately modeled, with significantly larger systems on the horizon. Similarly, advances in materials theory have produced relatively simple analytic potential energy functions that can capture essentials of quantum mechanical bonding for simulations involving well over a billion atoms [1].

Computer modeling has played an especially important role in developing the current understanding of nanometer-scale structures and processes [2]. Indeed, among the many scientific and technological advances coming from nanotechnology is the ability for computer modeling and experiment to characterize phenomena on a common scale. Atomic-level computer modeling is now commonly used to explore and even predict new properties and processes that can be probed experimentally, to suggest new materials and structures with unique and desirable properties, to provide insight into the results of experiments, to generate data for larger-scale analysis, and to test scaling laws and analytic theories. In the case of nanostructured solids, computer modeling is allowing researchers to understand at the atomic level the structure, deformation mechanisms, and thermal-mechanical properties of these new materials with unprecedented detail [3, 4].

The two standard methods for modeling nanometer-scale systems are molecular dynamics and Monte Carlo simulation. In the former, coupled classical equations of motion for the atoms are integrated stepwise in time. Time steps generally range from one-half to tens of femtoseconds depending on the highest frequency vibration of interest, and simulations are typically carried out for picoseconds to tens of nanoseconds depending on system size, the phenomena to be studied, and the method in which interatomic forces are calculated. In a typical equilibrium Monte Carlo simulation, atomic configurations are generated with a probability that is proportional to their Boltzmann factor. Thermodynamic quantities in a Canonical ensemble are then obtained by averaging over the properties of each configuration. Alternatively, in kinetic Monte Carlo simulation, time-ordered configurations are generated, typically using some rate expression.

Equilibrium Monte Carlo modeling requires specifying a potential energy as a function of atomic positions to calculate Boltzmann factors, while molecular dynamics simulations require interatomic forces, which are typically obtained as partial derivatives of the potential energy. In general, two approaches are used to calculate interatomic energies and forces. In the least computationally intensive approach, the interactions between atoms due to the electrons are replaced with effective interactions that are described with an analytic potential energy function. At present, there is no definitive mathematical form for the potential energy function, and forms that range from relatively simple pair-additive expressions to complicated many-body forms are used depending on the system, type of bonding, and phenomena to be studied. In the second approach to calculating potential energies, explicit electronic degrees of freedom are retained, the energy for which is calculated either using first principles methods or through a simplified semiempirical Hamiltonian. The calculation of total energies from first principles is well defined in terms of basis sets, electron correlation for ab initio methods, or choice of density functional expression (and pseudopotential) for density functional theory calculations. This is in contrast to analytic potentials, where a set of parameters (and often entirely new functional forms) must be developed for each system.

Nanostructured materials can be defined as materials that have at least one dimension in the "nanoscale" (typically 1–100 nm). Depending on which dimension this is, they can be classified into nanoparticles, layered or lamellar structures, filamentary structures, and bulk nanostructured materials. This entry is focused on the latter, which can be thought of as a traditional material with grain sizes at the nanometer scale. The nanometer-scale grains introduce unusual properties compared to more traditional micron-scale grain sizes, including unique combinations of strength and ductility as discussed in more detail below. The origin of these differences, and how they are being revealed by computer modeling, makes up the bulk of the material discussed below.

Atomic positions for bulk nanostructured solids have been generated by a number of different methods. Some of the methods are based on geometrical constraints imposed by the simulation conditions, for example, ensuring that active slip systems are properly oriented with respect to periodic boundaries [5, 6]. Crystallization dynamics have also been used to generate nanostructures. These methods can be based on a Johnson-Mehl or a Potts model construction, both of which produces a log normal grain size distribution, or by using a molecular dynamics simulation to model crystallization from a melt [3, 7, 8]. Other researchers have generated nanostructures by simulating compaction and sintering of nanoclusters [9–11]. Other methods use random grain centers, with grain boundaries chosen based on a Voronoi construction. Variations on this method include picking grain orientations to produce a particular range of tilt angle (e.g., low angle grains), or picking grain centers to produce a log normal distribution of grain sizes [4, 12].

Many of the atomic simulations of sintering and grain growth dynamics – and many of the simulation studies in general on nanostructured materials – have focused on understanding how processes and rates at the nanometer scale differ from their counterparts in macroscopic-scale systems. At the macroscopic scale, six distinct mechanisms contribute to the sintering dynamics of crystalline particles: surface diffusion; lattice diffusion from the surface, from grain boundaries and through dislocations; vapor transport; and grain boundary diffusion. There is a much larger degree of surface curvature at the nanometer scale and a much higher ratio of interface to bulk atoms, both of which may lead to very different sintering mechanisms. The details regarding these differences, and some new and unexpected effects at the nanometer scale, have been revealed from computer simulations. Molecular dynamics simulations, for example, have been used to study surface energies, grain boundary mobility, and sintering of metal nanoparticle arrays at different temperatures [13, 14]. The results suggest that of the macroscopic-scale mechanisms associated with sintering, only surface and grain boundary diffusion contribute significantly to nanometer-scale sintering dynamics, and that these two processes are accelerated due to the large interfacial forces. Simulations have also suggested three unconventional mechanisms that contribute to the early stages of nanometer-scale sintering: mechanical rotation, plastic deformation via dislocation generation and transmission, and amorphization of subcritical grains. This grain rotation mechanism has also been observed in structures with columnar grains, where simulations suggest that necessary changes in the grain shape during grain rotation in the nanostructure can be accommodated by diffusion either through the grain boundaries or through the grain interior [15].

In conventional metals with micron-scale grains, plastic deformation occurs by motion of dislocations. This dislocation motion can be inhibited by grain boundaries, which leads to the well-established Hall-Petch relation that the yield strength is proportional to the inverse square root of the grain size. However, there appears to be a threshold grain size below which materials become softer with decreasing grain size. This so-called inverse Hall-Petch behavior has been attributed to a transition from dislocation-mediated plastic deformation to grain boundary sliding for some critical grain size. This transition has been observed in several sets of molecular dynamics simulations that predict a transition grain size of about 10–15 nm in good agreement with experimental measurements. At the same time, the simulations have also revealed a rich and unanticipated set of dynamics near the threshold region that can be related to the fundamental properties of the bulk materials. These unique dynamics include an enhanced role of grain rotation (analogous to sintering dynamics), cooperative intergrain motion, and formation of stacking faults via motion of partial dislocations across grains.

The deformation of strained nanocrystalline copper with grain sizes that average about 5 nm has been studied with molecular dynamics simulations. These simulations showed a material softening for small grain sizes, in agreement with experimental measurements. These simulations suggest that plastic deformation in the inverse Hall-Petch region occurs chiefly by grain boundary sliding with dislocation motion having a minimal influence on deformation mechanisms. In related studies, molecular dynamics simulations have been used to understand the deformation of nanostructured nickel and copper with grain sizes ranging from 3.5 to 12 nm [16]. For grain sizes less than about 10 nm, deformation occurred mainly by grain boundary sliding, while for the larger grain sizes, deformation occurred by a combination of dislocation motion and grain boundary sliding. Detailed mechanisms of strain accommodation have been characterized; these include both single atom motion and correlated motion of several atoms, as well as stress-assisted free volume migration [17].

Molecular simulations of the deformation of columnar structures of aluminum have shown emission of partial dislocations during deformation that form at grain boundaries and triple junctions [18]. Atomic simulations have also suggested that these structures can be reabsorbed upon removal of the applied stress, which may contribute to an inability to observe dislocations experimentally in systems of this type after external stresses are released. In addition, near the grain size where plastic deformation transitions from dislocation-mediated plasticity to grain boundary sliding, the motion of single partial dislocations across nanograins during tensile loading has been observed in simulations. Without emission of a trailing partial dislocation, an intrinsic stacking fault is formed along the width of the nanograin. From the simulations, it appears that the formation of a low-energy ordered grain boundary drives emission of a full dislocation and the resulting absence of a stacking fault.

It has been argued that nucleation of the initial partial dislocation and the atomic rearrangement at the grain boundary associated with its emission sufficiently lowers the grain boundary energy such that emission of the trailing partial dislocation is not always needed to further relax the system [19]. Based on simulations of aluminum with a columnar nanostructure, it has been further suggested that the stacking fault width, and hence the intrinsic stacking fault energy, as determined by the distance between two partial dislocations, is the critical quantity that defines the transition from full to partial dislocation emission as grain sizes approach the critical size for the onset of inverse Hall-Petch behavior [20]. On the other hand, it has been argued that the relation between the emission of partial dislocations does not correlate well with calculated stacking fault energies for nickel, copper, and aluminum. Instead, it has been suggested that full dynamics associated with the nucleation of a partial dislocation from a grain boundary must be considered, and therefore that the full planar fault energy, which includes the stable and unstable stacking fault energy as well as twin fault energies, must be used to understand and ultimately predict the relation between mechanical deformation and grain size.

Extensive simulations of crack propagation in nanostructured metals have also been carried out to better understand how fracture, fatigue, and toughness depend on grain scale [21–23]. These simulations have revealed crack propagation mechanisms that are similar to the plastic response of fully dense samples as well as key differences resulting from the presence of the crack tip. For example, in simulations of nanocrystalline nickel with grain sizes in the inverse Hall-Petch region, mode I crack propagation occurred by inter-grain decohesion via a mechanism involving coalescence of nanovoids that form in front of the crack tip. Plastic deformation leading to both full and partial dislocations was also observed in the neighboring grains.

Compared to their mechanical properties under tensile loading, much less is understood about the influence of grain size on the shock loading properties of nanostructured solids. Atomic simulations that have been carried out, however, suggest a strong coupling between nanostructure and shock loading dynamics, as well as unique and very important properties of shocked-loaded nanostructured metals. It has been noted, for example, that the mechanisms associated with the mechanical deformation of nanostructured metals depend strongly on pressure, temperature, and strain rate, and therefore, these materials may show ultrahigh strength under shock loading depending on the shock loading conditions and system [24]. The fast temperature and pressure rises associated with shock fronts freeze out deformation mechanisms that require diffusion. Similarly, production of dislocations that requires nucleating events is inhibited. In the case of grain boundary accommodation, increasing the pressure results in an increase in the threshold stress for sliding plasticity. Similarly, the threshold for dislocation plasticity increases with increasing pressure because of an increase in the shear modulus with increasing pressure. Taking these effects into account, and assuming that the maximum in hardness as a function of grain size occurs where stress for sliding and dislocation plasticity are equal, it has been shown that ultrahigh hardness can be achieved by shock loading of nanostructured solids. These arguments have been validated by using molecular dynamics simulations to model the shock loading of nanostructured copper with different grain sizes. At relatively low shock velocities (i.e., low stresses), grain boundary sliding is observed, which results in a relatively low hardness value that increases with increasing grain size. At intermediate stresses, the hardness of the copper increases with increasing shock strength for all grain sizes, with a shift in the maximum hardness toward lower grain sizes compared to deformation at lower strain rates. This leads to a net increase in the maximum hardness of the material. At even higher stresses, simulations predict a drop in strength due to an increase in temperature and an associated increase in dislocation nucleation and motion.

It is clear from the discussion in the preceding sections that atomic simulations have provided new and exciting insights into the unique properties of nanosystems in general and nanostructured solids in particular. This remains a very active area of research within which molecular simulation will continue to provide new insights into relations between structural mechanical and thermodynamic properties.

Cross-References

- ▶ Computational Study of Nanomaterials: From Large-scale Atomistic Simulations to Mesoscopic Modeling
- ▶ Molecular Dynamics Simulations of Nano-Bio Materials
- ▶ Nanomechanical Properties of Nanostructures

References

1. Brenner, D.W., Shenderova, O.A., Areshkin, D.A.: Quantum-based analytic interatomic forces and materials simulation. In: Lipkowitz, K.B., Boyd, D.B. (eds.) Reviews in Computational Chemistry, pp. 213–245. VCH, New York (1998)
2. Brenner, D.W., Shenderova, O.A., Schall, J.D., Areshkin, D. A., Adiga, S., Harrison, J.A., Stuart, S.J.: Contributions of molecular modeling to nanometer-scale science and technology, Chapter 24. In: Goddard, W., Brenner, D., Lyshevski, S., Iafrate, G. (eds.) Nanoscience, Engineering and Technology Handbook. CRC Press, Boca Raton (2002)
3. Wolf, D., Yamakov, V., Phillpot, S.R., Mukherjee, A., Gleiter, H.: Deformation of nanocrystalline materials by molecular dynamics simulations: relation to experiment? Acta Mater. **53**, 1 (2005)
4. Kumar, K.S., Van Swygenhoven, H., Suresh, S.: Mechanical behavior of nanocrystalline metals and alloys. Acta Mater. **51**, 5743 (2003)
5. Van Swygenhoven, H., Spaczer, M., Car, A., Farkas, D.: Competing plastic deformation mechanisms in nanophase metals. Phys. Rev. B **60**, 22 (1999)
6. Yanakov, V., Wolf, D., Salazar, M., Phillpot, S.R., Gleiter, H.: Length-scale effects in the nucleation of extended dislocations in nanocrystalline Al by molecular-dynamics simulation. Acta Mater. **49**, 2713 (2001)
7. Phillpot, S.R., Wolf, D., Gleiter, H.: Molecular-dynamics study of the synthesis and characterization of a fully dense, three-dimensional nanocrystalline material. J. Appl. Phys. **78**, 847 (1995)
8. Phillpot, S.R., Wolf, D., Gleiter, H.: A structural model for grain boundaries in nanocrystalline materials. Scripta Metall Mater. **33**, 1245 (1995)
9. Zhang, Y.W., Liu, P., Lu, C.: Molecular dynamics simulations of the preparation and deformation of nanocrystalline copper. Acta Mater. **52**, 5105 (2004)
10. Kodiyalam, S., Kalia, R., Nakano, A., Vashashta, P.: Multiple grains in nanocrystals: effect of initial shape and size on transformed structures under pressure. Phys. Rev. Lett. **93**, 203401 (2004)
11. Chatterjee, A., Kalia, R.K., Nakano, A., Omeltchenko, A., Tsuruta, K., Vashishta, P., Loong, C.-K., Winterer, M., Klein, S.: Sintering, structure, and mechanical properties of nanophase SiC: a molecular-dynamics and neutron scattering study. Appl. Phys. Lett. **77**, 1132 (2000)
12. Schiotz, J., Di Tolla, F.D., Jacobsen, K.W.: Softening of nanocrystalline metals at very small grain sizes. Nature **391**, 1223 (1998)
13. Zeng, P., Zajac, S., Clapp, P.C., Rifkin, J.A.: Nanoparticle sintering simulations. Mater. Sci. Eng. A **252**, 301 (1998)
14. Xiao, S., Hu, W.: Molecular dynamics simulations of grain growth in nanocrystalline Ag. J. Crystal Growth **286**, 512 (2006)
15. Haslam, A.F., Phillpot, S.R., Wolf, D., Moldovan, D., Gleiter, H.: Mechanisms of grain growth in nanocrystalline fcc metals by molecular-dynamics simulation. Mater. Sci. Eng. A **318**, 293 (2001)
16. Van Swygenhoven, H., Spazcer, M., Caro, A., Farkas, D.: Competing plastic deformation mechanisms in nanophase metals. Phys. Rev. B **60**, 22 (1999)
17. Van Swygenhoven, H., Derlet, P.M.: Grain boundary sliding in nanocrystalline fcc metals. Phys. Rev. B **64**, 224105 (2001)
18. Yamakov, V., Wolf, D., Salazar, M., Phillpot, S.R., Gleiter, H.: Length scale effects in the nucleation of extended dislocations in nanocrysalline Al by molecular dynamics simulation. Acta Mater. **49**, 2713 (2001)
19. Van Swygenhoven, H., Derlet, P.M., Froseth, A.G.: Stacking fault energies and slip in nanocrystalline metals. Nat. Mater. **3**, 399 (2004)
20. Yamakov, V., Wolf, D., Phillpot, S.R., Mukherjee, A.K., Gleiter, H.: Deformation mechanism map for nanocrystalline metals by molecular dynamics simulation. Nat. Mater. **3**, 43 (2004)
21. Latapie, A., Farkas, D.: Molecular dynamics investigation of the fracture behavior of nanocrystalline α-Fe. Phys. Rev. B **69**, 134110 (2004)
22. Farkas, D., Van Petegem, S., Derlet, P.M., Van Swygenhoven, H.: Dislocation activity in nano-void formation near crack tips in nanocrystalline Ni. Acta Mater. **53**, 3115 (2005)
23. Farkas, D., Sillemann, M., Hyde, B.: Atomistic mechanisms of fatigue in nanocrystalline metals. Phys. Rev. Lett. **94**, 165502 (2005)

24. Bringa, E.M., Caro, A., Wang, Y., Victoria, M., McNaney, J.M., Remington, B.A., Smith, R.F., Torralva, B.R., Van Swygenhoven, H.: Ultrahigh strength in nanocrystalline materials under shock loading. Science **309**, 1838 (2005)

Single Cell Analysis

▶ Electrical Impedance Tomography for Single Cell Imaging

Single Cell Impedance Spectroscopy

▶ Electrical Impedance Cytometry

Single-Cell Electrical Impedance Spectroscopy

▶ Single-Cell Impedance Spectroscopy

Single-Cell Impedance Spectroscopy

Jian Chen[1], Nika Shakiba[1], Qingyuan Tan[1] and Yu Sun[2]
[1]Department of Mechanical and Industrial Engineering, University of Toronto, Toronto, ON, Canada
[2]Department of Mechanical and Industrial Engineering and Institute of Biomaterials and Biomedical Engineering and Department of Electrical and Computer Engineering, University of Toronto, Toronto, ON, Canada

Synonyms

Impedance measurement of single cells; Impedance spectroscopy for single-cell analysis; Single-cell electrical impedance spectroscopy

Definition

Single-cell impedance spectroscopy is a technique that operates by applying a frequency-dependent excitation signal on a single cell positioned in between two measurement microelectrodes. The current response is measured as a function of frequency. By interpreting the impedance profile, dielectric properties of a single cell such as cell membrane capacitance and cytoplasmic resistance are obtained.

Overview

Historical Development

In conventional cell electrical impedance spectroscopy, the impedance of multiple cells as a whole is measured using an AC excitation signal. The cell suspension is held in a container with two electrodes. Current is measured as a function of frequency to determine the electrical properties of the cells in the suspension.

Recent advances in microfabrication and lab-on-a-chip technologies enable the development of electrical impedance spectroscopic devices to measure impedance profiles of cells at the single-cell level, providing useful biophysical characteristics of single cells and promising potentially noninvasive, label-free approaches for analyzing cells.

Working Principle

Single-cell impedance spectroscopy measures impedance profiles of a cell positioned in between two microelectrodes [1]. A low-magnitude AC voltage, $\tilde{U}(j\omega)$, over a range of frequencies is used as the excitation signal. The current response, $\tilde{I}(j\omega)$, is measured. The impedance $\tilde{Z}(j\omega)$ is

$$\tilde{Z}(j\omega) = \frac{\tilde{U}(j\omega)}{\tilde{I}(j\omega)} = \tilde{Z}_{RE} + j\tilde{Z}_{IM} \tag{1}$$

where $\tilde{Z}_{RE}$ and $\tilde{Z}_{IM}$ are the real and imaginary parts of impedance. The real part is termed resistance while the imaginary part is termed reactance. The magnitude and phase angle are

$$|\tilde{Z}| = \sqrt{\tilde{Z}_{RE}^2 + \tilde{Z}_{IM}^2} \tag{2}$$

and

$$\angle\tilde{Z} = \arctan\left(\frac{\tilde{Z}_{IM}}{\tilde{Z}_{RE}}\right) \tag{3}$$

Advantages and Applications

Compared to impedance measurement on a cell suspension, single-cell impedance spectroscopy interrogates the property of a single cell (vs. a cell population). Theoretical analysis can also become simpler because it is not necessary to take into account electrical interactions among cells [2].

The technique of single-cell impedance spectroscopy has been used as a noninvasive method to quantify the physiological state of single cells. As a biophysical marker, single-cell impedance profiles have also been utilized to distinguish normal cells from abnormal cells (e.g., human cancer cells with different metastatic potential and malaria-infected red blood cells) [3, 4].

Methodology

Equivalent Circuit Model

Figure 1a [1] shows an equivalent circuit for interpreting single-cell impedance measurement data [1]. In this circuit model, the cell is represented by a capacitor in series with a resistor, with elements connected in parallel with the capacitor and resistor (i.e., the cell) representing the suspending medium. At low frequencies, the thin cell membrane gives rise to a large capacitance. As the frequency increases, the reactive component of this element tends to zero out, and the cell internal properties are represented by the resistor. In this circuit model, the membrane is assumed to have a high resistance and the cytoplasmic capacitance is ignored. All the circuit elements are functions of the volume fraction or cell size.

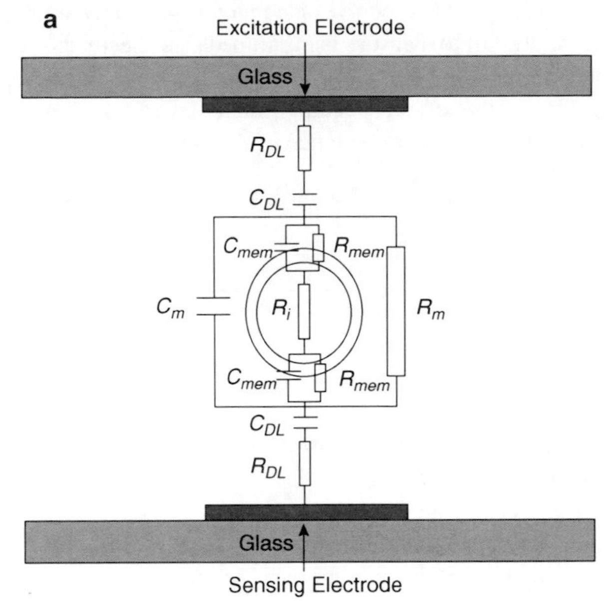

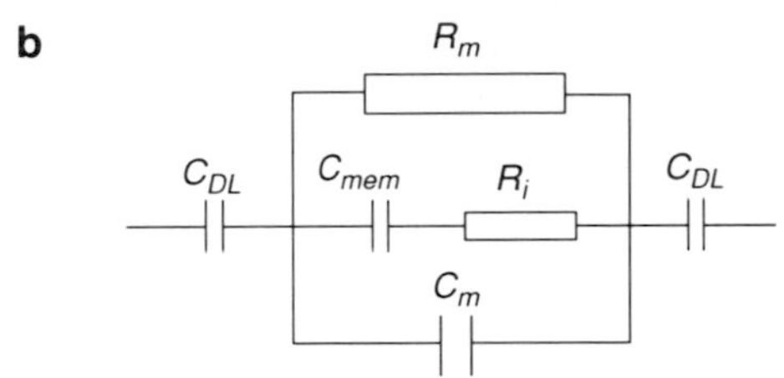

Single-Cell Impedance Spectroscopy, Fig. 1 (a) The equivalent circuit model for single-cell impedance spectroscopy. R_{DL} and C_{DL} represent the resistance and capacitance of the electrical double layer, R_m and C_m the resistance and capacitance of the medium, R_{mem} and C_{mem} the resistance and capacitance of the cell membrane and R_i the resistance of the cytoplasm. (**b**) The simplified equivalent circuit model neglecting the electrical double layer resistance and the membrane resistance (Reproduced with permission from [1])

The electrical double layer has an influence on measurements at low frequencies (below 1 MHz for high-conductivity solutions). This is generally modeled as a constant phase angle, represented by a resistor (R_{DL}) and capacitor (C_{DL}) in series. As shown in Fig. 1a, the double layer is in series with the network model. In the simplified model shown in Fig. 1b, the double layer is assumed to be purely capacitive (C_{DL}), with a value given by the product of the inverse Debye length and the permittivity of the medium.

The simplified circuit in Fig. 1b [1] shows that at very low frequencies current flow is blocked by the double layer capacitor, and only the impedance of the double layer is measured. As the measurement frequency increases, this capacitor is gradually short-circuited and the excitation voltage charges the cell in suspension. It takes a finite amount of time to charge the membrane through the extracellular and intracellular fluid, resulting in two dielectric dispersions in the frequency range of 1–100 MHz. The lower frequency dispersion is governed by the polarization of the cell membrane. Measurement of this parameter provides information about the dielectric properties of the membrane. The higher-frequency dispersion is governed by the polarization of the cytoplasm and the suspending medium as the membrane is short-circuited at these frequencies. This second dispersion is generally small and difficult to measure using impedance spectroscopy techniques.

With the effect of the double layer taken into account, the total impedance of the circuit is

$$\tilde{Z}(j\omega) = \frac{1}{j\omega C_{\mathrm{DL}}} + \frac{R_m(1 + j\omega R_i C_{\mathrm{mem}})}{j\omega R_m C_{\mathrm{mem}} + (1 + j\omega R_i C_{\mathrm{mem}})(1 + j\omega R_m C_m)} \tag{4}$$

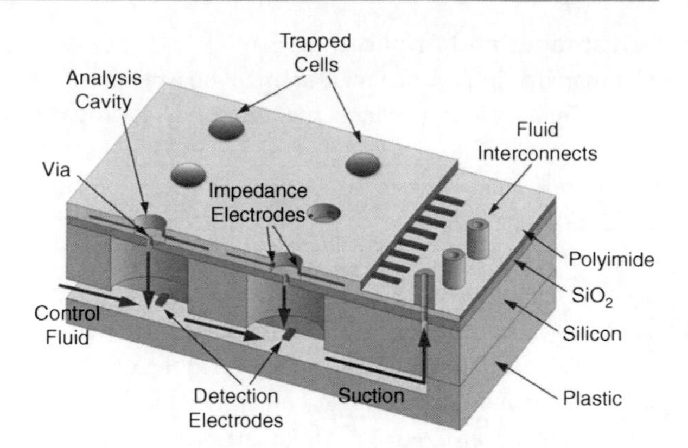

Single-Cell Impedance Spectroscopy, Fig. 2 Schematic view of an aspiration-based single-cell impedance spectroscopy system. The system is composed of an array of analysis cavities, each capable of analyzing a single cell at a time, each containing a mechanical fluid via with flow control to hold the cell in position during the impedance analysis operation, and multiple electrodes surrounding the circumference of the cell to measure the electrical characteristics of the captured cell (Reproduced with permission from [5])

Single-Cell Impedance Spectroscopy on Stationary or Moving Cells

Single-cell impedance spectroscopy in microfluidic devices can be divided into two main categories: measurements on either stationary or moving cells. For measuring a stationary cell, a cell can be positioned/trapped in between microelectrodes using several approaches. Cells can also be controlled to pass the measurement electrodes at the speed of microfluidic flow.

1. Cell aspiration [5]
   (a) Working mechanism: In the aspiration approach, a negative pressure is applied to trap a single cell in a channel opening. Measurement electrodes can be built into such channel openings. A tapped cell is forced by the applied negative pressure to contact the electrodes (Fig. 2) [5]. The impedance profile of the cell is then obtained.
   (b) Advantages: The main advantage of the aspiration approach for capturing a cell is the seal formed between the measurement electrodes and the cell. The suction force is sufficient to hold the cell in place throughout the measurement process, especially when the cell-capturing channel has a more or less circular cross section.
   (c) Potential limitations: It has been suggested that mechanical deformations of a cell due to the suction force may lead to changes in the electrical properties of the cell, thus, affecting the impedance profile measured.
2. Hydrodynamic trapping [6]
   (a) Working mechanism: In hydrodynamic trapping, cells are trapped by microfabricated weirs, used as filters on top of electrodes. By adjusting flow rates, cells can be captured on measurement electrodes for impedance data recording (Fig. 3) [6].
   (b) Advantages: The main advantage of this approach is the feasibility to realize large-array single-cell trapping and potentially high-throughput impedance measurement.
   (c) Potential limitations: Contact between the trapped cell and the measurement microelectrodes underneath can be problematic since there is no force to hold the cell in close contact with the microelectrodes.

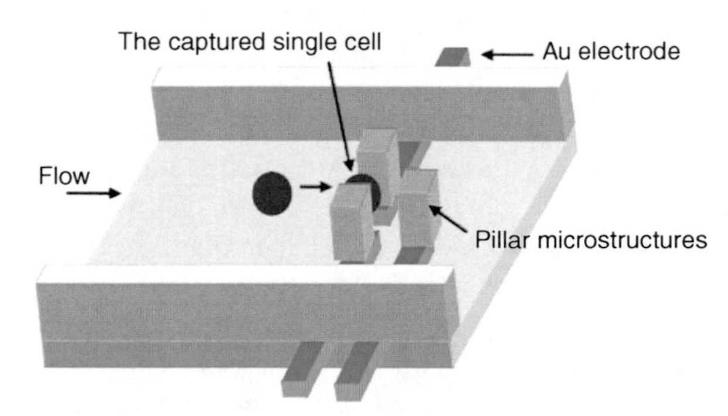

Single-Cell Impedance Spectroscopy, Fig. 3 Schematic view of a microfluidic device for single-cell impedance spectroscopy using the concept of hydrodynamic trapping for single-cell immobilization (Reproduced with permission from [6])

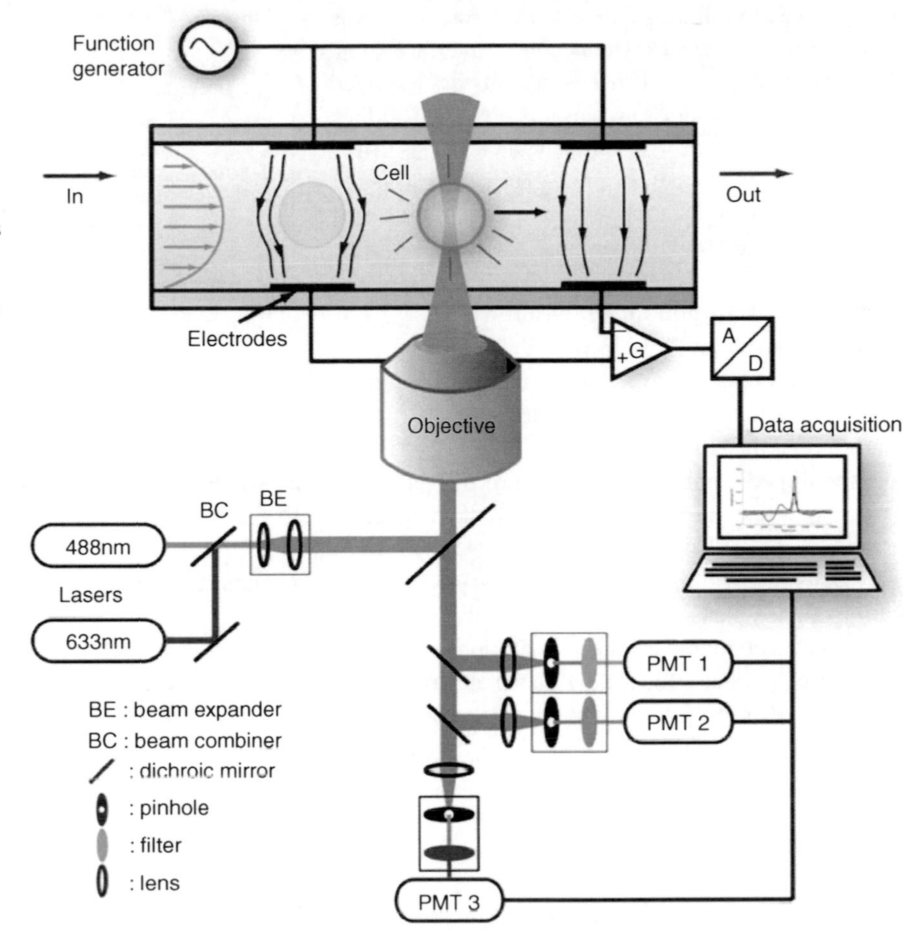

Single-Cell Impedance Spectroscopy, Fig. 4 Schematic view of differential impedance micro flow cytometry. Each cell's impedance signal is recorded by a differential pair of microelectrodes using media without a single cell passing as a reference. In this setup, the effect of the electrical double layer on impedance profiles is canceled (Reproduced with permission from [7])

3. Flow cytometry [7]
   (a) Working mechanism: Flow cytometry is a technique that allows for the analysis of cells in suspension with single-cell precision. A laminar flow carries suspended cells through the measurement location. Each cell's impedance signal is recorded by a differential pair of microelectrodes, using the surrounding media without a cell as reference. Microfabricated devices and electronic circuits allow simultaneous impedance measurements at multiple frequencies (Fig. 4) [7].
   (b) Advantages: Flow cytometry is advantageous in that it allows for rapid analysis of cells in suspension with a high throughput. Furthermore, flow cytometry can also allow for the sorting of cells into subpopulations based on their impedance measurement profiles.
   (c) Potential limitations: There is no contact between the cell and measurement electrodes and thus, the impedimentary contribution of the current leakage in the medium around the cell is significant.

Key Research Findings

Effect of Electrical Double Layer on Impedance Profile

When a polarizable electrode (i.e., an electrode operated at a regime where no charge transfer reaction occurs at the surface) comes into contact with an electrolyte, charges from the electrolyte opposite in sign to the charges present on the surface of the electrode move the electrode/electrolyte interface and provide a localized condition of electroneutrality as well as a layer of charge, termed the electrical double layer.

The electrical double layer has an influence on cell impedance measurements at low frequencies, which is generally represented as a capacitor. The most effective way to minimize the electrical double layer is to maximize the electrolyte-electrode interface area. This interface area enhancement can be achieved either by mechanically roughing the electrode surface to an electrode-electrolyte interface area that is effectively larger than the actual electrode surface, or to use chemical treatments that lead to a high electrode-electrolyte interface area.

Comparison of Impedance Measurement Mechanisms

The essential feature of measurements on a stationary cell trapped/positioned between electrodes is a tight seal between the cell and electrodes. In this configuration, the impedimentary contribution of the cell shunt (i.e., current leakage in the medium around the cell) is insignificant.

The advantage of impedance cytometry is differential impedance measurement that minimizes the effect of the electrical double layer. In an alternating fashion, each measurement electrode-ground electrode pair without a cell serves as a reference to the other pair, over which a cell passes.

Future Work

A single-cell impedance spectroscopy measurement system ideally should have a high testing throughput and the capability of sorting single cells based on impedance measurement results. The potential combination of single-cell impedance spectroscopy with other detection/measurement methods, such as fluorescent detection and the use of biochemical markers, may prove powerful for a range of applications, such as disease diagnostics and rare cell isolation.

References

1. Morgan, H., Sun, T., Holmes, D., Gawad, S., Green, N.G.: Single cell dielectric spectroscopy. J. Phys. D: Appl. Phys. **40**, 61–70 (2007)
2. Valero, A., Braschler, T., Renaud, P.: A unified approach to dielectric single cell analysis: impedance and dielectrophoretic force spectroscopy. Lab. Chip. **10**, 2216–2225 (2010)
3. Sun, T., Morgan, H.: Single-cell microfluidic impedance cytometry: a review. Microfluid. Nanofluid. **8**, 423–443 (2010)
4. Cheung, K.C., Di Berardino, M., Schade-Kampmann, G., Hebeisen, M., Pierzchalski, A., Bocsi, J., Mittag, A., Tarnok, A.: Microfluidic impedance-based flow cytometry. Cytom. A **77A**, 648–666 (2010)
5. Han, K.H., Han, A., Frazer, A.B.: Microsystems for isolation and electrophysiological analysis of breast cancer cells from blood. Biosens. Bioelectron. **21**, 1907–1914 (2006)
6. Jang, L.S., Wang, M.H.: Microfluidic device for cell capture and impedance measurement. Biomed. Microdevices **9**, 737–743 (2007)
7. Holmes, D., Pettigrew, D., Reccius, C.H., Gwyer, J.D., Berkel, C.V., Holloway, J., Daviesb, D.E., Morgan, H.: Leukocyte analysis and differentiation using high speed microfluidic single cell impedance cytometry. Lab Chip **9**, 2881–2889 (2009)

Single-Walled Carbon Nanotubes (SWCNTs)

► Chemical Vapor Deposition (CVD)

siRNA Delivery

► RNAi in Biomedicine and Drug Delivery

Size-Dependent Plasticity of Single Crystalline Metallic Nanostructures

Julia R. Greer
Division of Engineering and Applied Sciences, California Institute of Technology, Pasadena, CA, USA

Definition

"Size-dependent plasticity" here refers to the strength of metallic samples being a strong function of their size when their dimensions are reduced to the micron and below scales. The notion of reduced sample size applies to all three dimensions, i.e., stand-alone, or one-dimensional (1D) nano and microstructures rather than thin films (2D), where only their thicknesses are reduced to nano- and micro-scales, or to the small-scale deformation volumes within a bulk matrix (3D) as would be the case during nanoindentation experiments, for example.

Introduction and Overview of Stress Vs. Strain for Bulk Metals

Pure bulk single crystalline metals like Cu, Ni, Mo, Ti, etc. generally exhibit a convex, continuous stress–strain relationship upon uniaxial deformation – tension or compression – as can be found in any classic "Mechanical Behavior of Materials" book and as is also schematically shown in Fig. 1. For the low-symmetry orientations, i.e., where only a single slip system is activated, the deformation immediately following yield is called the "easy glide" region, which is characterized by a low hardening rate. Such low hardening rate stems from the dislocations traveling large distances unimpeded in their glide planes as only one set of crystallographic slip planes is active, and the dislocations are not forced to overcome closely spaced obstacles such as impurities or other dislocations in the course of their motion. A simple way to think of it is the following: Each dislocation travels a distance L before encountering an obstacle, which pins it. The shear strain associated with this motion is: $d\gamma = bLd\rho$, where b is the Burgers vector, L is the dislocation mean free path, and ρ is the mobile dislocation density [1]. The distance traveled by each dislocation then scales with $1/\sqrt{\rho}$, i.e., $L = \beta/\sqrt{\rho}$, where β may be on the order of 1,000 since the dislocations traveling in parallel crystallographic planes have very limited interactions. The hardening corresponding to this "easy glide" region is a result of dislocation storage through the well-known Taylor relation: $\tau = \alpha\mu b\sqrt{\rho}$, where $\alpha \sim 0.2$, which results in a very low hardening rate $d\tau/d\gamma_{low-symmetry} = 10^{-4}\mu$. In high-symmetry orientations, multiple symmetric slip systems are activated, and dislocations come into close encounters with one another as they are traveling toward one another in the symmetric slip planes, resulting in a pronounced dislocation density increase, which in turn, drives the high rate of hardening: $d\tau/d\gamma_{high-symmetry} = 0.01\mu$. This hardening in bulk single crystals stems from dislocation multiplication processes arising from the well-established dislocation interactions, which are followed by the dislocation networks formation processes. Therefore, in bulk single crystals – regardless of the sample size – the dislocations multiply in the course of deformation, thereby creating dense networks and dislocation substructures, which require the application of higher stresses to move the additional gliding dislocations through these obstacles, thereby carrying out plastic deformation. This holds true for crystals of any size with greater-than-several-microns dimensions. As a result, bulk samples of the same single crystalline material also exhibit identical yield strengths, flow stresses, and hardening rates.

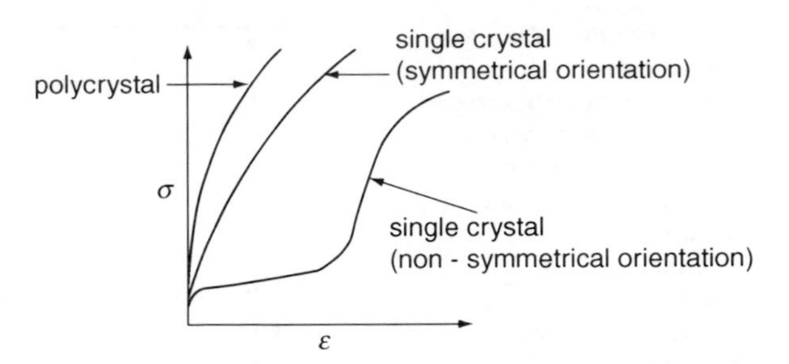

Size-Dependent Plasticity of Single Crystalline Metallic Nanostructures, Fig. 1 Schematic of typical stress–strain curves for bulk metals: comparing polycrystalline and single crystalline (in high- and low-symmetry orientations) microstructures of the same metal

Emergence of Size Effects in Single Crystals at the Nanoscale

In a striking deviation from this classical depiction, in the last 5–6 years, it was ubiquitously demonstrated that at the micron- and sub-micron scales, the sample size dramatically affects crystalline strength – even in the absence of any constraining effects, strain gradients, and grain boundaries – as revealed by room-temperature uniaxial compression experiments on a wide range of single crystalline metallic nano-pillars (for reviews, see [2–4]). In these studies, cylindrical nano-pillars were fabricated mainly by the use of the focused ion beam (FIB), as well as with some FIB-less methodologies, and remarkably, the results of these reports for all metallic single crystals show power law dependence between the flow stress and sample size, regardless of fabrication technique, of the form $\sigma = C \times D^{-k}$ with the exponent k strongly dependent on the initial crystal structure (i.e., face-centered cubic (fcc), body-centered cubic (bcc), hexagonal close-packed (hcp), etc.), experimental aspects (i.e., sample aspect ratio, lateral stiffness of the instrument,etc.),and dislocation density. A particularly auspicious example is the unique scaling of strength with pillar diameter

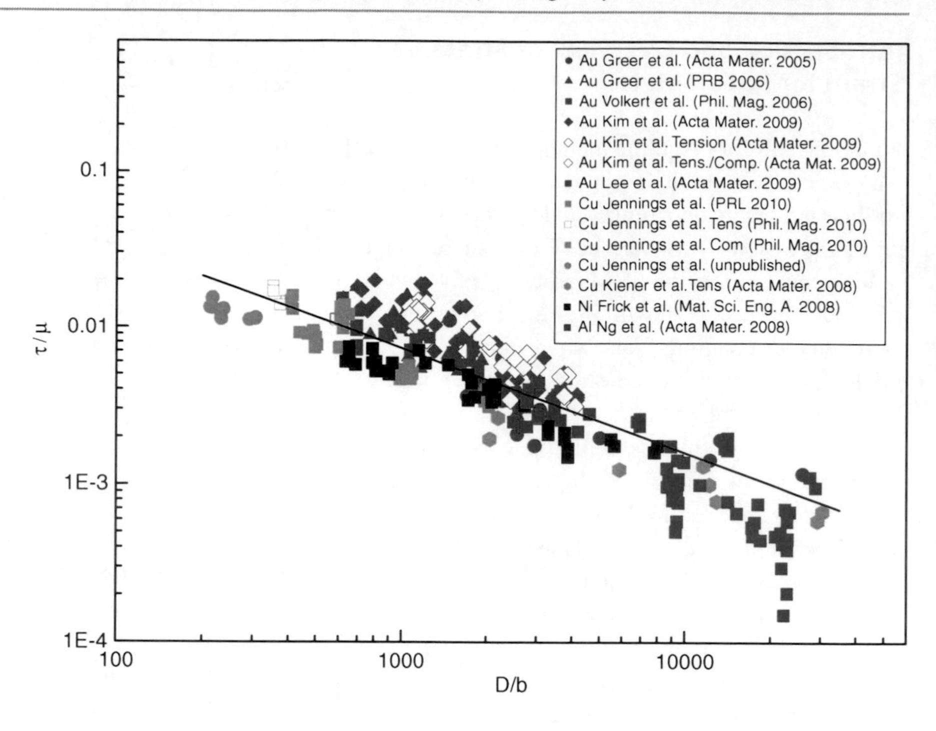

Size-Dependent Plasticity of Single Crystalline Metallic Nanostructures, Fig. 2 Resolved shear strength normalized by the shear modulus as a function of diameter (normalized by the Burgers vector) of experimentally determined compressive strengths of most face-centered cubic metallic nano-pillars published to date (Reprinted with permission from [2])

with the exponent of ~ -0.6 exhibited by nearly all non-pristine (i.e., containing initial dislocations) fcc nano-pillars subjected to uniaxial compression or tension, as illustrated in Fig. 2.

The notion of "size effects" applied specifically to the strength of single crystalline metals dates back to the works of Brenner [5, 6], who demonstrated that dislocation-free Cu and Ag whiskers attained very high strengths upon uniaxial tension. Although significant efforts have since been dedicated to studying single crystalline plasticity in confined dimensions, until ~5 years ago, most of these research thrusts were focused on thin films, whose yield strengths increased with decreasing film thickness (for example, [7]). The renaissance of small-scale mechanical testing on free-standing vertical pillars is largely due to the original work of Uchic et al. who reported higher compressive strengths attained by focused ion beam (FIB)-machined cylindrical Ni micro-pillars [8]. Greer and Nix then extended this robust and elegant methodology into the nanoscale regime, where <001 > −oriented Au nano-pillars with diameters below 1 μm were reported to be 50 times higher than bulk [9], and today numerous groups are pursuing this type of uniaxial nano-mechanical testing of materials ranging from single crystalline metals to lithiated battery anodic materials, ceramics, irradiated materials, shape memory alloys, nano-twinned and nanocrystalline metals, metallic glasses, superalloys, and nanolaminates. The results of many of these studies are overviewed in detail in three existing reviews: [2–4]. Figure 3 shows representative images of before- and after-testing single crystalline Nb nano-pillars, as well as the representative stress–strain curves with clear discrete characteristics and size effect [10]. Further adapting this methodology, the exploration of size effects in plasticity for a large variety of materials ensued, albeit mainly focusing on fcc structures. More recently, investigating plasticity in small volumes has been further advanced through uniaxial tensile experiments, usually conducted inside of in-situ mechanical deformation instruments custom built by some research groups [9–12].

Key Research Findings

Experimental Findings

To date, uniaxial compression and tension tests have been performed on Ni and Ni-based superalloys, Au, Cu, and Al (as-fabricated and intentionally passivated) [2–4]. Beyond single crystalline fcc metals, mechanical behavior of other single crystals has been published to date: bcc metals (W, Nb, Ta, Mo, and V), hcp metals

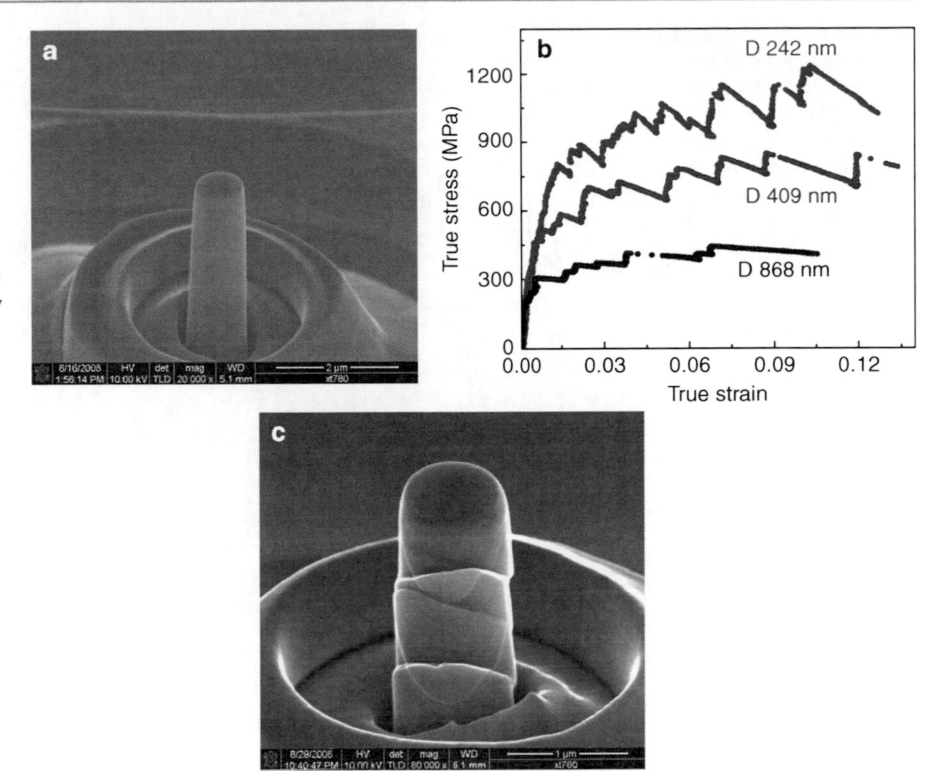

Size-Dependent Plasticity of Single Crystalline Metallic Nanostructures, Fig. 3 SEM images of (**a**) single crystalline Nb nano-pillar as fabricated by the focused ion beam, (**c**) same pillar after compression showing pronounced crystallographic slip lines. (**b**) Stress vs. strain curves clearly revealing the size effect (smaller is stronger) and the stochastic signature of small-scale deformation (Reprinted with permission from [10])

(Mg, Ti-Al alloy, and Ti), tetragonal metals (In) [2–4]. The reports on nearly all single crystalline metals with non-zero initial dislocation densities (i.e., not whiskers or nano-wires) unanimously demonstrate that their strengths significantly rise with reduced size, with ~100 nm-diameter samples sometimes attaining a flow stress ~10× higher than bulk. Intriguingly, unlike in bulk, where the dislocation multiplication processes result in Taylor hardening (introduced in Section Introduction and Overview of Stress Vs. Strain for Bulk Metals), the flow stresses in small structures do not appear to scale with the dislocation density. Rather, the global dislocation density appears to decrease upon mechanical loading, while the applied stress required to deform the structure increases (see, for example [11]). Consistent with this "upside down" behavior in the nano- and micron-sized crystals, whereby samples containing fewer mobile dislocations are stronger than their bulk counterparts, it has been shown that introduction of additional dislocations into the structure (for example, via pre-straining) actually *weakens* these crystals [12, 13]. These findings are antipodal to classical plasticity, where dislocation interactions lead to multiplication, causing higher dislocation densities and requiring higher applied stresses for deformation to continue, as described in the Section Introduction and Overview of Stress Vs. Strain for Bulk Metals. The initial dislocation density indeed plays a critical role in the onset of the size effect, as has now been shown by several research groups [2–4].

Computational Findings

Several models attempting to explain the origins of size-dependent flow stress in the absence of strong strain gradients, as well as the stochastic nature of deformation, have been put forth. There are generally three classes of these models: (1) phenomenological theories, which attempt to describe the physical processes occurring in a nano-single-crystal upon deformation, (2) discrete dislocation dynamics (2- and 3-dimensional) simulations (DD), which rely on the entered initial dislocation density and distribution, as well as on the dislocation mobility laws, and (3) molecular dynamics (MD) simulations, which are generally limited to small (50 nm and below) scales and unrealistically high strain rates ($\sim 10^{-8}\ s^{-1}$). An example of (1) is the "hardening by dislocation starvation" theory which hypothesizes that the mobile dislocations inside a small nano-pillar have a greater probability of annihilating at a free surface than of interacting with one

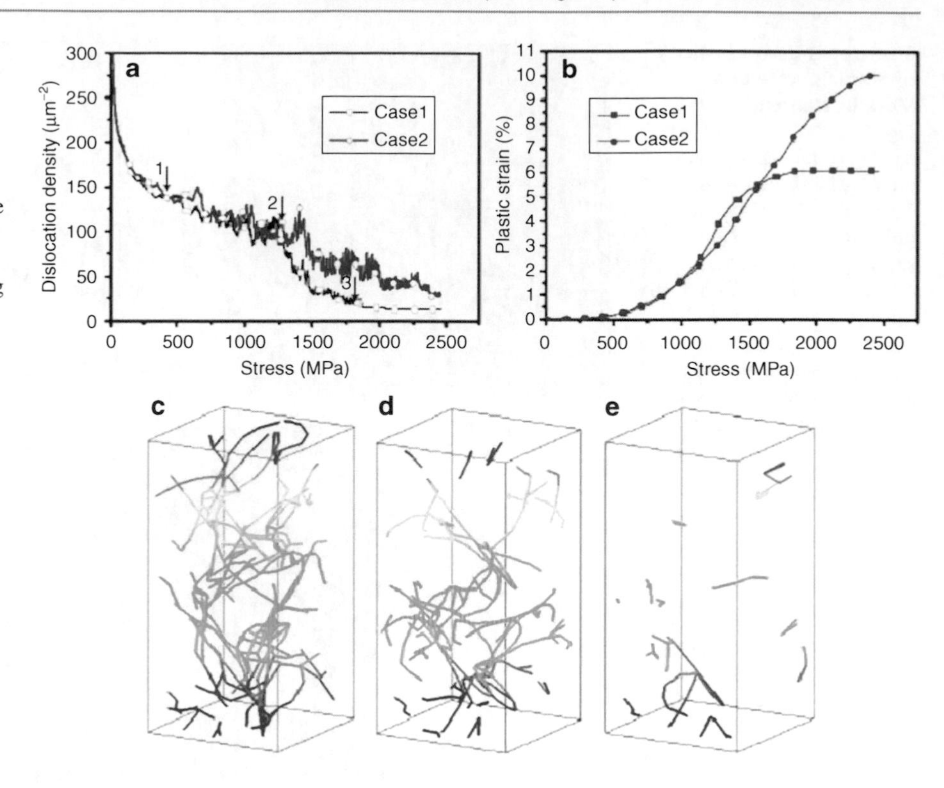

Size-Dependent Plasticity of Single Crystalline Metallic Nanostructures, Fig. 4 Variations in (**a**) dislocation density and (**b**) plastic strain with increasing stress. A dislocation avalanche occurs during the second stage. (**c**), (**d**), and (**e**) are the microstructures corresponding to points 1, 2, and 3 in (**a**), respectively (Reprinted with permission from [20])

another, thereby shifting plasticity into nucleation-controlled regime [14–16]. Other models include source exhaustion hardening, source truncation, and weakest link theory, all of whose general premise involves representing dislocation source operations in a discrete fashion, and then evaluating the effect of sample size on the source lengths, and therefore on the strengths of their operation [17]. These models capture the ubiquitously observed stochastic signature of the experiments results and show either marginal dislocation storage or no storage at all. An example of an evolved microstructure as revealed by dislocation dynamics simulations is shown in Fig. 4 and clearly shows that upon deformation, the dislocation density decreases, rendering multiplication processes unlikely and shifting plasticity into dislocation nucleation-controlled regime.

Discussion: What Causes the Size effect?

Despite the definition of "ultimate tensile strength" according to Wikipedia stating that UTS is an intensive property and "therefore its value does not depend on the size of the test specimen," there is now a large body of work that convincingly demonstrates that this does not hold true in the micron and sub-micron size regime. The ubiquitously reported size-dependent strengths in small-scale single crystalline micro- and nano-pillars show the "smaller is stronger" phenomenon as revealed by the existence of a power law dependence between the attained flow stress (σ_{flow}) and pillar diameter (D): $\sigma_{flow} = C \times D^{-n}$, where n varies with the specific crystal structure, initial dislocation density, and deformation type. Intriguingly, it appears that the power law slope for all fcc metals is unified, on the order of −0.6, as shown in Fig. 2. This is not the case for body-centered cubic metals, whose power law slopes vary within this crystal structure family, possibly due to the unique potential energy landscape of each bcc metal that a gliding screw dislocation has to overcome. The power law slopes of bcc metals appear to correlate with the intrinsic lattice resistance of each metal, a.k.a. the Peierls barrier, and can be separated into groups according to the low, mid- and high Peierls barrier: low-barrier group containing Nb (−0.93, −0.48, −1.07) and V (−0.79); mid-barrier group containing Mo (−0.44, −0.38) and Ta (−0.43, −0.41); and high-barrier group containing W (−0.21, −0.44) [2–4]. Beyond these two cubic crystalline structures, other types of crystals were also found to exhibit size-dependent strengths, with smaller generally being stronger. For example, hexagonal close-packed (hcp) metals, Mg and Ti, oriented for slip (as opposed to twinning) were also characterized by power law size effects: [11–21] Ti oriented for prismatic slip has the critical resolved shear stress (CRSS) scale inversely with the diameter, resulting in the slope of -1, while the [3–964] orientation of Mg (basal slip) has the slope of -0.64 [2].

There is currently no physics-based theory that captures these size effects as a function of metal, size, and initial microstructure. Several semi-empirical theories have been proposed, and nearly all identify the initial dislocation density as a key factor in defining the natural microstructural length scale, which in turn, drives the size effect. In an attempt to explain the emergence of higher strengths in small-scale crystals, multiple deformation mechanisms have been proposed, with two distinct ones: (1) single-arm source (SAS) theory first developed by Parthasarathy and Rao, et al. [17] applicable to micron-sized pillars, and (2) hardening by dislocation starvation proposed by Greer and Nix [9, 14] for the much smaller, nano-sized pillars. In the SAS theory, the creation of dislocations occurs by the operation of truncated Frank-Read spiral sources, a.k.a. single-arm sources (SAS), whose strength, τ_s, is inversely proportional to their average length, λ, through $\tau_s = k_s\mu\frac{\ln(\bar{\lambda}/b)}{\bar{\lambda}/b}$, where, k_s is a source-hardening constant [17]. The smaller micron-sized pillars are not capable of accommodating large single-arm sources, and therefore require the application of higher stresses to activate the stronger, "shorter" sources, giving rise to higher strengths in smaller pillars.

In the dislocation starvation theory, applicable to the much smaller pillars, with diameters deeply in the sub-micron regime, new dislocations are created via surface nucleation rather than through a multiplicative process of SAS operation as is the case in the micron-sized samples. The surface nucleation gets activated only after a significant fraction of the pre-existing gliding dislocations within the pillar has exited through the free pillar surface, i.e., "starving" the crystal of the necessary mobile dislocations to carry plastic strain. A necessary condition for dislocation starvation is that the largest distance that any gliding dislocation has to travel is smaller than the so-called breeding length, or the distance before it replicates itself, as defined by Gilman [18]. This is likely to be the case in the samples with diameters deeply in the submicrometer range since Ag, for example, has the breeding distance on the order of 0.7 μm. The nucleation stress for a surface source may be represented as $\sigma = \frac{Q^*}{\hat{Q}} - \frac{k_B T}{\hat{Q}} \ln\frac{k_B T N \nu_0}{E\dot{\varepsilon}\hat{\Omega}}$, where the first term corresponds to the athermal contribution and the last term represents thermally activated nature of such events [19]. The activation of surface sources has a significant thermal activation component, as is evident from the $T\ln T$ and strain rate ($\dot{\varepsilon}$) dependence of the nucleation stress. Recent experimental studies performed on very small (below 500 nm) Cu nano-pillars are consistent with this surface nucleation model and lay out the single-arm source vs. surface source operation regimes as a function of both the pillar size and strain rate [2]. While these theories may appear to be competing, it is likely that they both take place at different pillar sizes: with SAS strengthening occurring in the micron-sized pillars (perhaps, down to ~500 nm) and with dislocation starvation followed by surface nucleation prevailing at the smaller sizes, deep in the sub-micron regime.

Summary

- Recent experimental and computational results convincingly demonstrate that the strength of nano- and micron-sized single crystalline metals is indeed size dependent and appears to be well represented by a power law of the form $\sigma = C \times D^{-k}$ with the exponent strongly dependent on the initial microstructure (i.e., dislocation density, lattice resistance, and the presence of impurities).
- With the advent of highly capable nano-fabrication and analysis instruments, as well as of the sophisticated computational tools, the society is ever closer to developing a more complete understanding of size-dependent mechanical behavior of small-scale structures. To date, it is generally agreed that non-pristine (i.e., containing dislocations) micron-sized fcc structures attain higher strengths at smaller sizes through dislocation multiplication processes produced by the operation of single-arm dislocation sources, whose lengths scale with the sample size.

The operation of these truncated sources, which become exhausted in the course of deformation, drives the corresponding strength increase in these small-scale structures. On the contrary, when the sample dimensions are further reduced to the nano-sized regime, plasticity likely occurs via dislocation nucleation from surface sources (rather than through a multiplicative process), and the higher strengths arise due to the lower probability of finding weaker dislocation sources in smaller structures.

Acknowledgments JRG gratefully acknowledges the financial support of the National Science Foundation (NSF) CAREER grant (DMR-0748267) and the Office of Naval Research (ONR) Grant No. N00014-09-1-0883. The author is particularly grateful to W.D. Nix, A.T. Jennings, D. Jang, J.-Y. Kim, Q. Sun, A. Ngan, C. Weinberger, J. Li, and D. Gianola for useful discussions.

References

1. Nix, W.D.: MSE 208: Mechanical behavior of materials class notes. (1980)
2. Greer, J.R., de Hosson, J.Th. M.: Critical review: plasticity in small-sized metallic systems: intrinsic versus extrinsic size effect. Progress in Materials Science **56**, 654–724 (2011)
3. Kraft, O., Gruber, P.A., Monig, R., Weygand, D.: Plasticity in confined dimensions. Annu Rev Mater Res **40**, 293 (2010)
4. Uchic, M.D., Shade, P.A., Dimiduk, D.M.: Plasticity of micrometer-scale single crystals in compression. Annu Rev Mater Res **39**, 361–386 (2009)
5. Brenner, S.S.: Tensile strength of whiskers. J. Appl. Phys. **27**, 1484–1490 (1956)
6. Brenner, S.S.: Growth and properties of "whiskers". Science **128**, 568 (1958)
7. Thompson, C.V.: The yield stress of polycrystalline thin films. J. Mater. Res. **8**, 237 (1993)
8. Uchic, M.D., Dimiduk, D.M., Florando, J.N., Nix, W.D.: Sample dimensions influence strength and crystal plasticity. Science **305**, 986 (2004)
9. Greer, J.R., Oliver, W.C., Nix, W.D.: Size dependence of mechanical properties of gold at the micron scale in the absence of strain gradients. Acta Materialia **53**, 1821 (2005)
10. Kim, J.-Y., Jang, D., Greer, J.R.: Insights into deformation behavior and microstructure evolution in Nb single crystalline nano-pillars under uniaxial tension and compression. Scripta Materialia **61**, 300 (2009)
11. Tang, H., Schwartz, K.W., Espinosa, H.D.: Dislocation escape-related size effects in single-crystal micropillars under uniaxial compression. Acta Materialia **55**, 1607 (2007)
12. Bei, H., Shim, S., Pharr, G.M., George, E.P.: Effects of pre-strain on the compressive stress–strain response of Mo-alloy single-crystal micropillars. Acta Materialia **56**, 4762 (2008)
13. Lee, S., Han, S., Nix, W.D.: Uniaxial compression of fcc Au nanopillars on an MgO substrate: the effects of prestraining and annealing. Acta Materialia **57**, 4404 (2009)
14. Greer, J.R., Nix, W.D.: Nanoscale gold pillars strengthened through dislocation starvation. Phys. Rev. B **73**, 245410 (2006)
15. Shan, Z.W., Mishra, R., Syed, S.A., Warren, O.L., Minor, A.M.: Mechanical annealing and source-limited deformation in submicron-diameter Ni crystals. Nat Mater, **7**, 115–119 (2008)
16. Weinberger, C., Cai, W.: Surface controlled dislocation multiplication in metal micro-pillars. Proc. Natl. Acad. Sci. USA **105**, 14304 (2008)
17. Rao, S.I., Dimiduk, D.M., Parthasarathy, T.A., Uchic, M.D., Tang, M., Woodward, C.: Athermal mechanisms of size-dependent crystal flow gleaned from three-dimensional discrete dislocation simulations. Acta Materialia **56**, 3245 (2008)
18. Gilman, J.J.: Micromechanics of flow in solids. McGraw-Hill, New York (1969)
19. Zhu, T., Li, J., Samanta, A., Leach, A., Gall, K.: Temperature and strain rate dependence of surface dislocation nucleation. Phys. Rev. Lett. **100**, 025502 (2008)
20. Liu, Z.L., Liu, X.M., Zhuang, Z., You, X.C.: Atypical three-stage-hardening mechanical behavior of Cu single-crystal micropillars. Scripta Materialia **60**, 594 (2009)

Skin Delivery

► Dermal and Transdermal Delivery

Skin Penetration

► Dermal and Transdermal Delivery

Small angle X-Ray Scattering in Grazing Incidence Geometry

► Selected Synchrotron Radiation Techniques

Small Unilamellar Vesicle (SUV)

► Liposomes

Small-Angle Neutron Scattering

▶ Small-Angle Scattering of Nanostructures and Nanomaterials

Small-Angle Scattering

▶ Small-Angle Scattering of Nanostructures and Nanomaterials

Small-Angle Scattering of Nanostructures and Nanomaterials

M. Laver
Laboratory for Neutron Scattering, Paul Scherrer Institut, Villigen, Switzerland
Materials Research Division, Risø DTU, Technical University of Denmark, Roskilde, Denmark
Nano–Science Center, Niels Bohr Institute, University of Copenhagen, Copenhagen, Denmark
Department of Materials Science and Engineering, University of Maryland, Maryland, USA

Synonyms

SANS; SAXS; Small-angle neutron scattering; Small-angle scattering; Small-angle X-ray scattering

Definition

Small-angle scattering of nanostructures and nanomaterials encompasses the measurements of scattering from structures with length scales ranging between the near-atomic (nanometer) to the near-optical (micrometer), using beams of nanometer wavelengths or less.

Overview

Small-angle scattering reveals structural features on length scales between the near-atomic (nanometer) to the near-optical (micrometer). With such versatility, the technique has made an astounding impact in many fields of research including polymer systems [1, 2], complex fluids [3], biology [4], condensed matter physics, and materials science [1, 5]. The methodology for performing small-angle scattering studies extends back to the 1930s [6], following the first small-angle X-ray scattering (SAXS) studies. Subsequently, synchrotron sources have enabled explorations with ever smaller sample amounts, in complex sample environments. Laboratory SAXS instruments are now routinely used for sample characterization in many research institutions. The creation of large-scale facilities producing neutrons for research has proved to be instrumental for small-angle neutron scattering (SANS) studies. As X-rays and neutrons interact differently with matter, SAXS and SANS are quintessentially complementary, and a particular technique may frequently become an indispensable probe for a particular application.

This entry is organized as follows: in section Key Principles, the general concepts of small-angle scattering are explained. In section X-Rays or Neutrons, the advantages and practical aspects of X-ray and neutron techniques are compared. In section Magnetic Neutron Scattering, the discussion is focused on small-angle neutron scattering from magnetic structures, and neutron polarization analysis (Section Neutron Polarization Analysis) is provided as an example of a developing field for small-angle studies. The concluding section (Section Summary and Outlook) incorporates an outlook toward novel directions for the small-angle scattering technique.

An overview of symbols used repeatedly throughout this entry is provided in Table 1.

Key Principles

Bragg Diffraction

Small-angle scattering patterns are frequently continuous in nature, rather than consisting of crystalline diffraction peaks. Nevertheless, Bragg's law encapsulates a fundamental relationship

$$\lambda = 2d \sin\theta = \frac{4\pi}{q} \sin\theta \tag{1}$$

showing that the angle of diffraction θ varies inversely with the separation d of the diffracting lattice

Small-Angle Scattering of Nanostructures and Nanomaterials, Table 1 List of symbols used throughout this entry

θ	Bragg angle of diffraction
$\vec{k_i}, \vec{k_f}$	Wavevector of incoming, final (scattered) beams
λ	Wavelength of beam
$\vec{q}, q$	Scattering vector, its magnitude
ζ	Angle of q in detector plane
d	Spacing between Bragg planes
D	Characteristic length scale within sample
$\vec{r}, r$	(Real space) position vector in sample or within nano-object, its magnitude
b, b_x, b_n, b_m	Scattering length, that of X-rays, that of neutrons from nuclei, that of neutrons from atomic moments
ρ, ρ_0	Scattering length density, that of solvent or matrix
F	Form factor of single nano-object
I	Scattered intensity
M, N	Magnetic, nuclear parts of (neutron) form factor of single nano-object
P	Form factor $P = \|F\|^2$
S	Structure factor
C	Component of magnetic moment perpendicular to $\vec{k_i}$
ϕ	Angle of magnetic moment in detector plane
Z	Component of magnetic moment parallel to $\vec{k_i}$
$\vec{A}$	Halpern–Johnson vector

planes. It can be seen that the scattering from objects on nano- to micrometer scales falls predominantly in the small-angle regime $<1°$. A beam of $\lambda \simeq 7$ Å wavelength, for example, would scatter from objects with a d-spacing of $\simeq$1000 Å into an angle $2\theta \simeq 0.4°$ separated from the unscattered beam. Practical constraints on the wavelengths used are imposed by the spectrum of the source available and by the absorption of the beam by the sample, which increases with wavelength for both X-rays and neutrons. Typically, "soft" X-rays ($\approx$1 Å, or equivalently, 12 keV) or neutrons ($\approx$5 Å) from a cold ($\lesssim$40 K) moderator are used in small-angle studies. All scattering measurements are made in reciprocal (Fourier transform) space. The scattering vector $\vec{q}$, whose magnitude q satisfies (1), parameterizes the scattering pattern in reciprocal space and is helpful when comparing patterns between different instruments. The vectorial version of (1), $\vec{q} = \vec{k_f} - \vec{k_i}$, is illustrated in Fig. 1, where $\vec{k_i}, \vec{k_f}$ are, respectively, the wavevectors of the incident and final (scattered) beams.

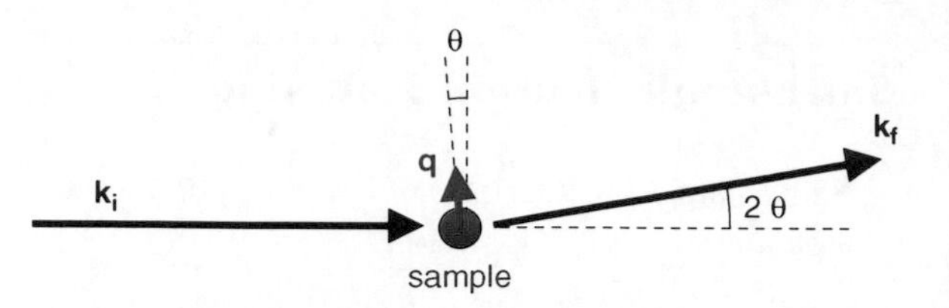

Small-Angle Scattering of Nanostructures and Nanomaterials, Fig. 1 Schematic depicting the relation for the scattering vector $\vec{q} = \vec{k_f} - \vec{k_i}.2\theta$ is the angle between the incident beam $\vec{k_i}$ and the scattered beam $\vec{k_f}$

Experimental Method

Small-Angle Scattering

The fundamental task of a small-angle instrument is to separate the intense, unscattered direct beam from the weaker scattering at small angles 2θ. A common experimental setup, depicted in Fig. 2, resembles the classic pinhole camera. Following the source, the wavelength λ of the incoming beam is selected. Due to the low scattering angles, beam intensity can be maintained using a large wavelength spread $\Delta\lambda/\lambda$ without significant detriment to the q-resolution. On small-angle X-ray scattering (SAXS) instruments, a monochromator crystal such as Si (111) is used if a wavelength spread narrower than the source distribution is required. On small-angle neutron scattering (SANS) instruments at pulsed neutron sources, a neutron's wavelength may be calculated by timing its journey to the detector (known as the *time of flight* technique). At continuous (e.g., reactor) neutron sources, velocity selectors are mainly used. Velocity selectors comprise of channels or slotted disks forming helical pathways about an axle; only neutrons of a particular velocity are able to pass through the rotating helical pathway. The desired neutron wavelength is selected by varying the rotational speed. The opening angle of the helical pathway is typically set to give wavelength spreads of $\Delta\lambda/\lambda \approx 0.06$.

After the wavelength selection, a collimation section that may consist of a pair of apertures or a slit system ensures that the beam has a tight angular spread $<0.1°$ when it is incident on the sample. The precise collimation and resulting angular spread are often tuned to match the desired measurement range in 2θ (or, equivalently, in q) and to optimize beam intensity. A two-dimensional multidetector, comprising an array of pixels in a plane normal to the unscattered beam, is placed at a sufficient distance behind the sample so as to resolve the small-angle scattering in the desired q range. For example, the D11 SANS instrument at the Institut Laue-Langevin in Grenoble has a multidetector of 128 × 128 pixels, each pixel 7.5 × 7.5 mm^2, and features maximum collimation and sample-to-detector distances of 40.5 m and 39 m, respectively, a configuration that permits scattering vectors as low as $q \approx 0.0003$ Å^{-1} to be measured. The 2D area detectors used in SANS or SAXS are designed to have high detection efficiencies (>80%), and may need to be shielded from intense fluxes, namely, the unscattered beam, that may saturate and can even damage the detector. A beamstop may be placed in front of the detector for this purpose.

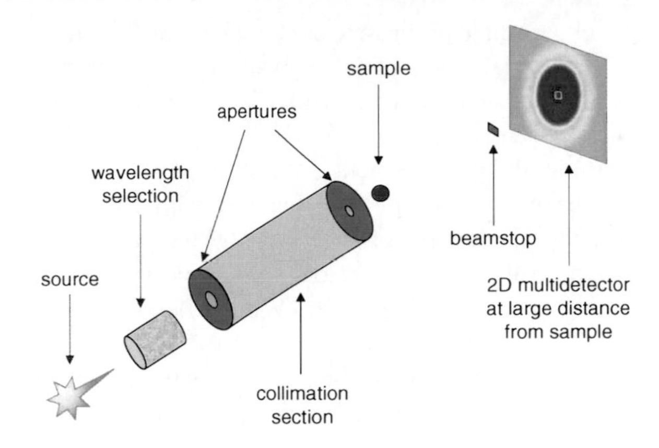

Small-Angle Scattering of Nanostructures and Nanomaterials, Fig. 2 Schematic of a typical small-angle scattering instrument

Ultra-Small-Angle Scattering

For some systems one would like to extend the range of observable structures to length scales larger than those which may be probed using the habitual pinhole setup (Fig. 2) and measure scattering at "ultra" small angles. To achieve this, an experimental geometry developed by Bonse and Hart [7], which is now commonplace, utilizes a pre-sample monochromator comprising of a single crystal out of which a channel has been precisely cut. The beam passes in through one end of the channel at an incident angle θ, and appears at the other end (also at an angle θ) after having bounced multiple times from Bragg reflections off both sides of the channel. The resulting beam has an extremely small angular divergence $\lesssim 10^{-4}$ degrees. A similar channel-cut crystal placed behind the sample analyzes the angles of the scattered beam. With this setup, measurements are possible very close to the unscattered beam, down to scattering vectors $q \lesssim 10^{-5}$ Å^{-1}, in other words, up to length scales in the tens of micrometer range. Unlike the pinhole SANS setup where a single measurement with the 2D detector captures an entire surface of scattering vectors $\vec{q}$, the Bonse-Hart geometry can only probe one point in $\vec{q}$ at a time and the measurement times required with this technique are accordingly much longer.

General Theory

In this and in the following subsection (Section Spherically Symmetric Forms) the concepts of *scattering length density, contrast, structure factor, form factor, polydispersity*, and *Guinier* and *Porod regimes* are introduced. These are also explained in several introductory textbooks [1, 3, 6]. The q-dependence of the measured scattered intensity is described by the "differential cross section," denoted simply as $I(\vec{q})$. For now the scattering is presumed to be elastic ($|\vec{k}_i| = |\vec{k}_f|$ i.e., there is no energy transfer between beam and sample) and coherent (produces interference effects). Detailed treatments of inelastic and incoherent scattering may be found in the textbooks on diffraction methods [8].

Born Approximation and Scattering Length

One presumption of particular pertinence to small-angle scattering experiments is the *Born approximation*, in which the scattering is regarded to be weak (such that the scattering potential may be treated analytically as a perturbation). Within the Born approximation, the photon or neutron scatters no more than once during its passage through the sample. Then the measured scattered intensity is

$$I(\vec{q}) \sim \left| \sum_j b_j e^{i\vec{q}.\vec{R}_j} \right|^2 \tag{2}$$

where $\vec{R}_j$ denotes the position of the jth scatterer in the sample, and b_j is the *scattering length*, epitomizing the capability of the jth object to scatter the beam. Neutrons scatter from nuclei and from magnetic moments [8], while X-rays scatter from electrons (Thomson scattering) [1, 3]: the scattering length of one electron is 2.8×10^{-5} Å. The differences between neutron and X-ray scattering will be discussed in section X-rays or Neutrons. Here and throughout this work, leading prefactors are omitted for simplicity. These prefactors would, for example, ensure that the scattered intensity scales with the illuminated sample volume. In practice $I(\vec{q})$ is scaled onto absolute units through the measurement of either a standard sample or the incident beam flux, together with corrections for the attenuation of the beam by the sample and instrument-dependent factors such as electronic noise and detector efficiency [9]. With $I(\vec{q})$ converted to absolute units, intensities obtained on different small-angle instruments may be directly compared, and sample properties, for example, the stoichiometry of that part of the sample giving rise to scattering, may be estimated quantitatively.

Most model functions for small-angle scattering data presuppose that the Born approximation holds. In practice, for strong scatterers, care should be taken to verify that this applies before fitting models to $I(\vec{q})$. If multiple scattering is a problem, the probability of scattering may be diminished by making the sample thinner or, where possible, diluting it or decreasing the *contrast*. For grazing incidence experiment geometries such as the grazing-incidence SAXS (GISAXS) technique that are used to resolve in-plane structures on films [1], the Born approximation is sufficient for very small or thin scatterers, otherwise quantitative analysis of the GISAXS profile requires what is known as the "distorted-wave" Born approximation, where the scattering potential is divided into two parts such that an exact solution to the scattering problem may be obtained [10]. The grazing incidence geometry is not expounded here; further details may be found in Refs. [1] and [10].

Scattering Length Density and Contrast

Theoretical small-angle scattering profiles can be calculated directly from models comprised of individual atomic scatterers, but for fitting purposes the computational effort required is too great at present. As the objects leading to small-angle scattering are generally on larger scales, a more continuous, "coarse-grained" description of the scattering potential landscape would seem appropriate. A continuous function $\rho(\vec{r})$, known as the *scattering length density*, is accordingly defined [6]:

$$\rho(\vec{r}) = \sum_j b_j \delta(\vec{r} - \vec{R}_j)$$

and the measured scattered intensity becomes

$$I(\vec{q}) \sim \left| \int e^{i\vec{q}.\vec{r}} \rho(\vec{r})\, d\vec{r} \right|^2 \tag{3}$$

$$\sim \left| \int e^{i\vec{q}.\vec{r}} \rho_0\, d\vec{r} + \int e^{i\vec{q}.\vec{r}} (\rho(\vec{r}) - \rho_0)\, d\vec{r} \right|^2 \tag{4}$$

The last equation serves to illustrate the concept of *contrast*. Here it is useful to consider the scattering from nanoparticles in solution, or equally, nanoscale heterogeneities in a solid matrix. ρ_0 would be the scattering length density of the matrix or solvent, while $\rho(\vec{r})$ would be that of the nanoparticles or heterogeneities. The first term in (4) involves just the matrix or solvent and is nonzero only at $q = 0$ [the Fourier transform of a constant is $\delta(0)$], so no small-angle scattering originates from the matrix or solvent. In the second term in (4), the particles or heterogeneities give forth to small-angle intensity proportional to the squared difference $(\rho(\vec{r}) - \rho_0)^2$. This is referred to as the "contrast factor." Systematic alterations of the solvent ρ_0 may thus be used to determine the scattering length density of a solute ρ_1, since the intensity will go to zero at the match point where $\rho_1 = \rho_0$. This is called "contrast matching." In multicomponent systems, separate contrast-variation analyses for different q-regimes or for different features in the scattering profile can also be used with great effect to reveal the sizes and shapes of the underlying components. Contrast methods were recently demonstrated, for example, in a prominent study of the nanoscale phases within cement [11].

Structure Factor and Form Factor

The notions of *structure factor* and *form factor* are familiar concepts in crystallographic work [8]. Considering N identical scattering nano-objects, each centered at a position $\vec{s}_j$, the scattered intensity becomes

$$I(\vec{q}) \sim \left| \sum_j^N e^{i\vec{q}.\vec{s}_j} \right|^2 \times \left| \int_{\text{nano-object}} e^{i\vec{q}.\vec{r}} \rho(\vec{r})\, d\vec{r} \right|^2$$
$$= S(\vec{q}) \times |F(\vec{q})|^2$$

The left side of the multiplication sign defines the *structure factor* $S(\vec{q})$. The integral on the right side ranges over an isolated nano-object, and this term defines the single particle *form factor* $F(\vec{q}) = \int e^{i\vec{q}.\vec{r}} \rho(\vec{r})\, d\vec{r}$. It is also commonplace to refer to the function $P(\vec{q}) = |F(\vec{q})|^2$ as the form factor too. In the limit where the nano-objects arrange into a crystal with positions $\vec{s}_j$ forming a perfect lattice, the structure factor becomes a series of δ-functions $S(\vec{q}) = \Sigma_l\, \delta(\vec{q} - \vec{G}_l)$, where $\vec{G}_l$ are the reciprocal lattice vectors of the crystal [8]. Crystalline Bragg-diffraction peaks appear in the scattering profile, as illustrated in Fig. 3, with intensities modulated by the form factor. To measure each $\vec{G}_l$ peak, the sample must be rotated with respect to the neutron beam in order to satisfy the condition $\vec{q} = \vec{G}_l$. For situations in the opposite limit where nano-objects are well separated, with no interactions between nano-objects, $S(\vec{q}) = 1$ for all $\vec{q}$ and the small-angle intensity depends only upon $P(\vec{q})$, that is to say, upon the form of an isolated particle. Such a situation is called the "dilute" limit because, with small-angle scattering from solutions, it may be attained by sufficient dilution of the solute [4, 9]. The near-dilute situation is also often realized, where the interparticle interactions are finite but small, such that $S(\vec{q})$ is essentially uniform in the q-range of interest. In situations where the structure factor and the form factor are comparable, for example where the volume fraction of nano-objects is high but no long-range order emerges, the $I(\vec{q})$ must be fitted with models for both $S(\vec{q})$ and $P(\vec{q})$. Details of the many model functions available may be found in the literature [1–6, 13–15].

There are also several numerical methods for modeling $P(\vec{q})$ or $S(\vec{q})$. Numerical methods often start from a radial pair distribution function, which can be obtained by an indirect Fourier transform of $I(q)$ [3]. One approach, commonly used to reconstruct the shape of biological macromolecules, imitates the scattering length density profile $\rho(\vec{r})$ of the macromolecule by several hundred close-packed "beads," each of constant ρ. Genetic algorithms and simulated annealing techniques are able to refine the virtual bead assemble to fit the experimental data [4, 16]. Reverse Monte Carlo methods have also been used to recreate the structure factor $S(\vec{q})$ in situations where quasi-long-range order is observed [17].

Spherically Symmetric Forms

Most texts on small-angle scattering begin by considering a spherically centrosymmetric nano-object [3, 6, 13], in other words, a nano-object whose scattering length density $\rho(\vec{r})$ depends only on $r = |\vec{r}|$. Here this scenario is also pursued in order to illustrate concepts such as contrast and polydispersity and furthermore, to highlight the different *regimes* in the scattering profile. To obtain the form factor for a spherical object, it is useful to select a spherical polar coordinate system such that the pole lies along $\vec{q}$; the angular components of the integral within $F(\vec{q})$ may then be evaluated, yielding

$$F(q) = 4\pi \int \rho(r) r^2 \frac{\sin(qr)}{qr}\, dr$$

The form factor is here a function only of the magnitude $q = |\vec{q}|$ of the scattering vector. Even when the nano-object is not spherical, but a large number of randomly oriented nano-objects are contained in the sample, these orientations may be averaged over and the result becomes again a function of $q = |\vec{q}|$. It is usual to average the scattering measured on the 2D detector (cf. Fig. 2) over the detector's azimuthal direction, so reducing the scattering profile to one dimension, i.e., an I versus q dataset. In section Magnetic Neutron Scattering an example is given of where the scattering shows an azimuthal dependence arising for a reason other than crystalline order (cf. Fig. 3). The azimuthal information can impart vital clues as to the nature of the underlying nanostructures.

Core-Shell Form Factor

Core-shell nano-objects are frequently encountered in small-angle scattering studies, e.g., aerosol droplets, polymer micelles [2], inorganic nanoparticles [18]. In addition, core-shell models may provide a useful quantitative starting point even when nano-objects depart from being spherically symmetric. The spherical core-shell model form factor is

$$\begin{aligned} F(q) &= 4\pi \int_0^{r_c} \rho_c r^2 \frac{\sin(qr)}{qr}\, dr + 4\pi \int_{r_c}^{r_s} \rho_s r^2 \frac{\sin(qr)}{qr}\, dr \\ &= 4\pi\,(\rho_c - \rho_s) r_c^3 \frac{j_1(qr_c)}{qr_c} + 4\pi\,\rho_s r_s^3 \frac{j_1(qr_s)}{qr_s} \end{aligned} \tag{5}$$

where the sphere has a uniform core of radius r_c and scattering length density ρ_c, encapsulated by a uniform

S

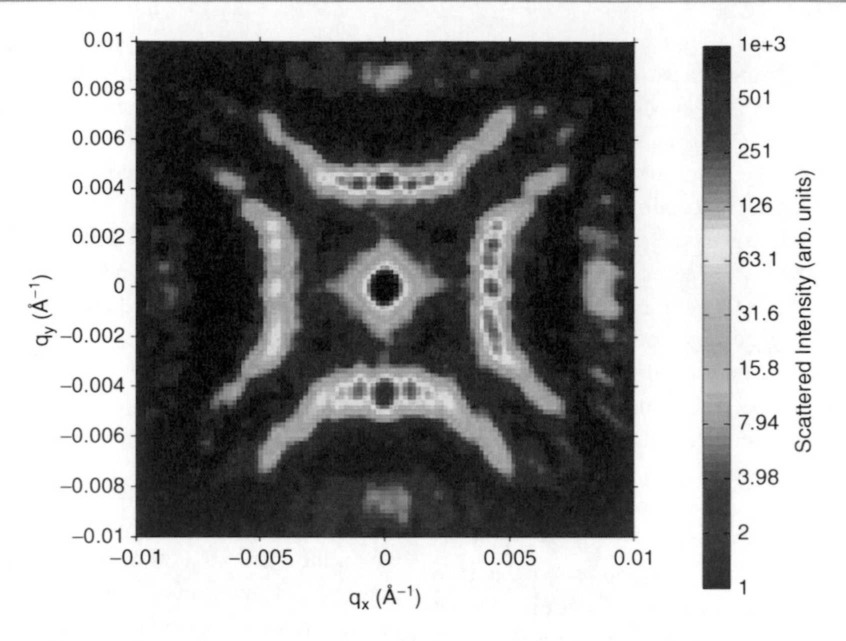

Small-Angle Scattering of Nanostructures and Nanomaterials, Fig. 3 Example of a crystal diffraction pattern on a small-angle neutron detector, here from a flux line lattice [12] in a single crystal of superconducting niobium. To obtain this picture, several detector images were summed as the sample was rotated over a range $\pm 1^\circ$ of rocking angles about the vertical and horizontal axes. Here 28 distinct first order Bragg spots are discernible, due to four scalene triangular and two square flux line lattices coexisting in different regions of the sample. The peaks on the left side of the picture appear weaker due to detector efficiency. The unscattered beam, not covered up by a beamstop in this experiment, appears in the center at $q = 0$

shell of scattering length density ρ_s extending to radius $r_s > r_c$, and $j_1(x) = \frac{\sin x}{x^2} - \frac{\cos x}{x}$ is a spherical Bessel function. For brevity, ρ_c and ρ_s are measured relative to the scattering length density ρ_0 of the surrounding solution or matrix; equivalently, ρ_0 may be considered to be zero. Equation (5) may be simplified by introducing volume terms such as $V_c = \frac{4}{3}\pi r_c^3$. Figure 4b exemplary $P(q) = |F(q)|^2$ curves for this core-shell model are plotted.

Polydispersity

In reality, most scattering systems show a distribution in their sizes and shapes; this is known as *polydispersity*. Form factors can be straightforwardly modified to average over a distribution in size or shape, and several functions defining distributions for the polydispersity are available [19]. The effects of polydispersity on the small-angle scattering profile are now examined briefly. In Fig. 4c the core-shell model is modified to include a distribution in core radii with standard deviation 20% of the mean core radius. By comparing this figure with Fig. 4b, the adverse effects of polydispersity – a smearing in q of the scattering profile – can readily be seen. A smearing of the scattering profile also results from a finite instrument resolution; nonetheless it is clear that sizeable benefits are gained if samples for small-angle scattering studies are as monodisperse as possible. Figure 4 furthermore illustrates how contrast variation may be exploited, where feasible: by matching the solvent ρ_0 to the core ρ_c (dotted green line in Fig. 4), the interference bumps in the profile are more discernible, notably in polydisperse situations.

Regimes in the Scattering Profile

The appearance of distinctive q regimes in the scattering profile may also be seen in Fig. 4. The solid sphere model (continuous blue lines in Fig. 4) features a characteristic length scale D, namely, the sphere radius 80 Å. The scattering features a high-q regime for $qD \gg 1$, where the scattering profile appears to decay algebraically with q, and a regime at low q where the profile falls off more slowly in q. These regimes emerge independently of the absolute intensity and of any model, and invariably provide useful clues as to the size and homogeneity of the nano-object. In the high-q regime $qD \gg 1$, known as the *Porod regime*, the exponent α of the power-law decay $I(q) \sim q^{-\alpha}$ reveals the "fractal dimension" of the

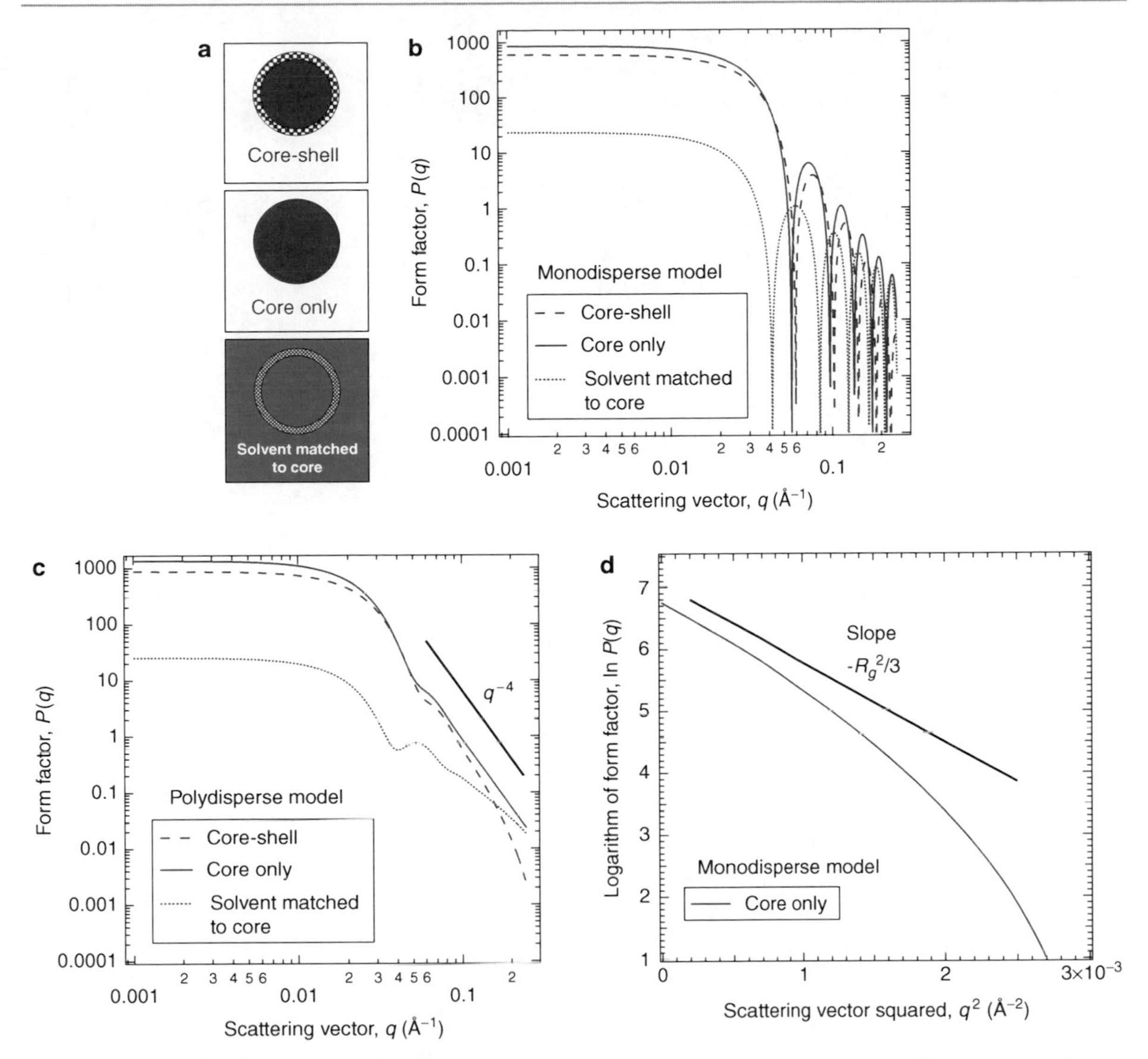

Small-Angle Scattering of Nanostructures and Nanomaterials, Fig. 4 The spherical core-shell form factor $P(q)$. (**a**) Schematic showing the various parameters for the model curves plotted in (b)–(d): in (**b**) the particles are monodisperse; in (**c**) the particles have polydisperse core sizes with the standard deviation of the polydispersity distribution set at 20% of the mean core radius r_c. The *dashed red* and *dotted green lines* indicate $P(q)$ with $r_c = 70$ Å and a shell 10 Å thick; the *dashed red line* shows $P(q)$ when the core scattering length density ρ_c is twice that of the shell ρ_s; the *dotted green line* shows $P(q)$ when the solvent ρ_0 matches ρ_c. The continuous *blue line* shows $P(q)$ for a core only model with $r_c = 80$ Å; in (d), this line is redrawn on ln $P(q)$ versus q^2 axes. *Black lines* are guides to the eye

scattering objects [14]. The $\alpha = 4$ observed in Fig. 4c, for example, indicates smooth surfaces, as expected for the model of uniform spheres with boundaries sharp in $\rho(r)$.

In the *Guinier regime* as $q \to 0$, the scattering follows a Gaussian dependence $\sim \exp(-\frac{1}{3}q^2R_g^2)$ where R_g is the "radius of gyration" of the nano-object [6], and is analogous to the radius of gyration in mechanics. For uniform solid spheres of radius r_c, $R_g^2 = \frac{3}{5}r_c^2$. To reveal this behavior at low q, a "Guinier plot" is useful, where the logarithm of the scattering is plotted versus q^2; the plot will be linear with slope $-\frac{1}{3}R_g^2$ at sufficiently low q. A Guinier plot for the sphere model is shown in Fig. 4d, together with the expected low-q slope [black line in Fig. 4d]. Comparing the two, it may be seen that the

Guinier approximation holds reasonably well when $q^2 \lesssim 0.0008$ Å^{-2}, that is, for $qR_g \lesssim 1.3$. The upper limit for the Guinier regime is found to depend somewhat on the scattering model [13].

Almost every small-angle scattering article incorporates an analysis of the low- or high-q regimes. Extensive examples of Guinier plots and associated methods (e.g., the Zimm and Kratky plots) that quantify scattering regimes without resorting to detailed model fitting may also be found in the literature [3]. There have also been efforts to encapsulate both low-q and high-q regimes in unified models. A common example is the Beaucage model, applicable to situations where a nanostructure exhibits a hierarchy of scales, each discernible scale contributing a pair of Guinier and Porod regions to the scattering profile [15].

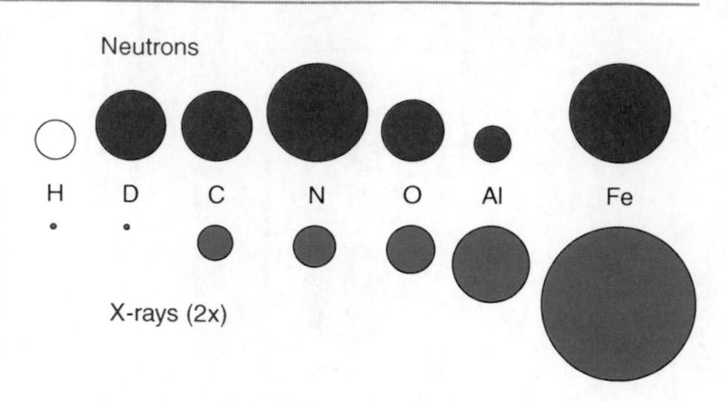

Small-Angle Scattering of Nanostructures and Nanomaterials, Fig. 5 Schematic indicating the propensity for scattering of X-rays or neutrons [20] by selected atoms. The radii of the circles are drawn proportional to the scattering length. For X-rays, the scale has been 2× magnified for clarity. The scattering length of hydrogen for neutrons is negative, in contrast to the other atoms depicted

X-rays or Neutrons?

Scattering Length

The basic principles of small-angle scattering, as overviewed in previous sections, have the same form for both X-rays and neutrons. However, there are fundamental distinctions between the techniques arising from how the X-rays or neutrons interact with matter. One such distinction is patent in the atomic scattering length b [cf. (2)]. For neutrons, b_n depends on the structure of the nucleus and is isotope dependent. For X-rays, b_x depends on the electronic structure and scales with the total number of electrons Z in the atom. At the length scales probed by small-angle scattering, nuclei look like point sources of scattering and b_n is q-independent, whereas for X-rays (atomic absorption edges neglected) $b_x(q)$ falls off slowly with q – this is essentially the Guinier regime of the form factor for the atomic electron clouds, with $b_x(0) = Z \times 2.8 \times 10^{-5}$ Å. Low-resolution modeling of SAXS profiles can be achieved with b_x considered constant. The scattering lengths for selected atoms are pictured in Fig. 5 for both neutrons [20] and X-rays.

Generation of Contrast

The scattering length density ρ of an elementary volume V (e.g., a molecular volume) within the sample is calculated by summing the scattering lengths of atoms within V, that is, $\rho = \Sigma_j b_j/V$. Here, a uniform electron density needs to be assumed for X-rays. The scattering length densities for neutrons of selected

Small-Angle Scattering of Nanostructures and Nanomaterials, Table 2 The scattering properties, for neutrons, of selected substances

Substance	Neutron attenuation length (cm)	Neutron scattering length density ($\times 10^{-6}$ Å^{-2})
H_2O	0.18	−0.56
D_2O	1.55	6.33
C	1.59	7.53
B	0.0036	6.91
^{10}B	0.00067	−0.14
^{11}B	1.35	8.51
Al	7.74	2.08
Fe	0.629	8.02
Cd	0.0031	2.27

For the calculation of these values, room temperature densities are used, and the elements with no isotope specified are deemed to comprise of their isotopes in naturally abundant ratios

substances is listed in Table 2. Comparing with Fig. 5, it is clear that there is an enormous distinction between hydrogen and deuterium. The former has $b_n = -3.74 \times 10^{-5}$ Å, the latter 6.67×10^{-5} Å [20], and this is reflected in ρ as shown for H_2O and for D_2O in Table 2. Great contrast may be achieved by using hydrogenated solutes in deuterated solvents. Polystyrene in toluene, for example, presents a contrast of 0.48×10^{-6} Å^{-2} when both are hydrogenated, whereas the contrast of hydrogenated polystyrene in deuterated toluene is 4.2×10^{-6} Å^{-2}. In contrast-variation studies, tuning the solvent scattering length density ρ_0 may be accomplished by using a binary mixture whose components have widely different scattering length densities; these are mixed in a linear ratio to yield the desired ρ_0. For X-rays, the two components would have different electron densities; for neutrons, they may be deuterated and hydrogenated versions of the same solvent. Isotope substitution may also be used to create SANS contrast *inside* a particle by specific isotope labeling of a particular part of interest, which can be used, for example, to determine how this part fits in with the rest of the particle. In SAXS, a novel way to induce contrast is to measure scattering profiles at different photon energies in the vicinity of an atomic absorption edge of one of the elements within the sample: for this element, the form factor b_x varies across the absorption edge, while the rest of the scattering potential landscape is essentially invariant in the small range of photon energies measured. This contrast-variation technique is known as anomalous SAXS (ASAXS) and may be exploited for the large number of elements having an absorption edge in the range 5–25 keV [1, 9, 21].

Neutron Absorption

Another aspect of neutron scattering that is heavily dependent on the isotope is absorption. A few elements, for example boron and cadmium (c.f. Table 2), are outstanding neutron absorbers in their naturally abundant forms. Not surprisingly, boron-containing substances and cadmium are common shielding materials. For scattering explorations of these and other such materials, samples need to be prepared using elements devoid of the absorbing isotopes; boron-11, for example, is seen from Table 2 to have an acceptable attenuation length. Separating elements into their different isotopes, however, becomes increasingly difficult and expensive for heavier atoms. It is also apparent from Table 2 that some materials, such as aluminum, are rather poor neutron absorbers. This is providential for the construction of bulky or demanding sample environments such as cryostats or pressure cells.

General Considerations

The majority of elements have neutron attenuation lengths on the order of a centimeter, reflecting the weak interaction of neutrons with matter. In general, neutron scattering explores the bulk of samples and, for similar reasons, effectively no radiation damage to the sample occurs with the neutron fluxes currently available (up to $\approx 10^8$ neutrons cm^{-2} s^{-1}) on SANS instruments. For SAXS experiments at synchrotron sources, the high brilliance of the X-ray beam ($\approx 10^{16}$ photons cm^{-2} s^{-1}) can damage soft matter and biological samples within milliseconds [9]. This may affect the validity of the scattering profile measurement. For SAXS the sample thickness may also need to be chosen to obtain the optimum scattered intensity, since at the photon energies ($\approx$10 keV) used in SAXS, the X-ray attenuation length is on the order of millimeters for the lighter elements [9], which make up most soft matter systems. The many orders of magnitude more intense fluxes available on SAXS beamlines are, of course, invaluable for high-resolution, low uncertainty measurements of scattering profiles. The associated fast data acquisition times also facilitate the characterization of time-dependent effects. Likewise, relatively small sample quantities are typically required (<50 μL for SAXS) compared to SANS ($\approx$1 mL).

Neutron Incoherent and Spin Scattering

In previous sections, it has been shown that variations in scattering length density on the nanoscale give rise to coherent small-angle scattering. However, if the scattering length b varies in a disordered fashion for different atoms of the same element, then for neutrons, since the nuclei are effectively point sources, this gives rise to incoherent (q-independent) scattering [8]. This adds to the background of the coherent scattering profile and in a few cases may be significant enough to limit the measurement even when longer count times are used. A variation in b for the same element can arise due to different isotopes, present in sufficient abundance, having widely different scattering lengths. Another reason for a variation in b occurs when the nuclei have nonzero spin and the scattering length depends on the spin state. Neutrons are sensitive to such spin disorder because the neutron itself has spin $\frac{1}{2}$. A notorious example is hydrogen: this has an incoherent scattering length of 25.3×10^{-5} Å, compared to 4.1×10^{-5} Å for deuterium. Deuterated solvents should therefore be used where possible.

The spin of the neutron is rather advantageous for the exploration of magnetic structures. Small-angle scattering can arise from nano- to micrometer-scale variations in the magnetic field, as well as correlations over similar length scales between atomic moments. This is discussed in the following section.

Magnetic Neutron Scattering

The spin-$\frac{1}{2}$ nature of the neutron and its associated magnetic moment make neutron scattering techniques a natural probe of magnetic systems. Indeed, magnetic structures are considerably better understood following the dawn of high-flux neutron sources. The origin of an observed scattering signal – whether it is nonmagnetic (i.e., from nuclei) or magnetic in origin – can be determined by measurements at either side of the appropriate thermodynamic phase transition. In other cases, for example if magnetic and structural order appear simultaneously, the origin of the signal may be determined by analyzing the neutron polarization and/or the azimuthal dependence of the scattering on the 2D detector. Furthermore, neutron polarization analysis can convey additional information as to the orientation of the magnetic moments behind the scattering signal. The small-angle applications of neutron polarization analysis are introduced in the second part of this section (Section Neutron Polarization Analysis.

Unpolarized Neutrons

The interaction potential of the neutron with a magnetic field $\vec{B}$ is $-\vec{\mu}.\vec{B}$, where $\vec{\mu} = \gamma\mu_N\vec{\sigma}$ is the magnetic moment operator for the neutron, γ is the neutron gyromagnetic ratio, and μ_N is the nuclear magneton [8]. $\vec{\sigma}$ is the Pauli spin operator describing the neutron spin state; its average value characterizes the polarization of the neutron beam. All the general principles explained in the previous sections can be used to describe the small-angle scattering from magnetic structures, whether the neutrons are polarized or not. Noting, for example, that the scattering interaction potential $-\vec{\mu}.\vec{B}$ is already continuous, (3) for the unpolarized elastic coherent scattering becomes, with a collinear field B in the sample

$$I(\vec{q}) \sim \int e^{i\vec{q}.\vec{r}} B(\vec{r})\, d\vec{r}$$

This prescription is very convenient for flux line lattices that form in Type-II superconductors [12], where crystalline diffraction peaks are frequently observed, as illustrated in Fig. 3. The long-range order and nanometer d-spacings of magnetic skyrmion lattices also result in SANS diffraction peaks [22].

The magnetic scattered intensity may also be estimated in terms of magnetic scattering lengths of individual atoms. Like the atomic form factors in SAXS, the magnetic scattering length $b_m(q)$ is close to constant at low-q [8], with $b_m = m \times 2.7 \times 10^{-5}$ Å$/\mu_B$ where m is the atomic moment in Bohr magnetons μ_B [5, 23]. Under the proviso that the model chosen to fit the scattered intensity is representative, this relation can be utilized to estimate the underlying moment behind a scattering feature. With SANS, studies may be directed at the magnetism of individual atoms, since in ferri- and ferromagnetic systems in particular, the structure factor at low-q affords a measure of the interactions between individual atomic moments separated by large distances. An example is the collection of "random anisotropy" magnets, where SANS arises from random variations in the magnetic anisotropy fields. SANS studies on these materials are the subject of a recent review [5], in which various structure factor models pertinent to magnetic scattering are also detailed.

Magnetic domain walls in ferri- or ferromagnets also give rise to small-angle scattering. Such walls are normally observed to be spatially uncorrelated so that the scattering is dominated by the form factor. Because the domain sizes D are typically micrometers or above, a pinhole SANS instrument is able only to measure the high-q (Porod) regime of this scattering. The full profile may be accessible with an ultra-small-angle instrument [7, 24]. An example of where structure-factor-dominated and form-factor-dominated magnetic scattering features occur in the same sample is demonstrated in the phase-separated perovskite oxides, which exhibit colossal magnetoresistance [25]. In these systems, micrometer-sized ferromagnetic regions materialize in the nonmagnetic matrix below a critical temperature T_c. The Porod regime of their form factor appears as SANS intensity $I(q) \sim q^{-4}$. At the same time, at temperatures in the vicinity of T_c, quasielastic critical fluctuations give (Ornstein–Zernike) scattering, which has a Lorentzian profile $I(q) \sim 1/[q^2 + (1/\xi)^2]$ where ξ is the correlation length.

The dipole nature of the interaction between the neutron and atomic magnetic moments means that only moments $\vec{m}$ which are perpendicular to the scattering vector $\vec{q}$ are effective in scattering [8, 23]. The relevant part of $\vec{m}$ is expressed by the Halpern–Johnson vector

$$\vec{A} = \hat{\mathbf{q}} \wedge (\vec{m} \wedge \hat{\mathbf{q}}) = \vec{m} - (\vec{m}.\hat{\mathbf{q}})\hat{\mathbf{q}}$$

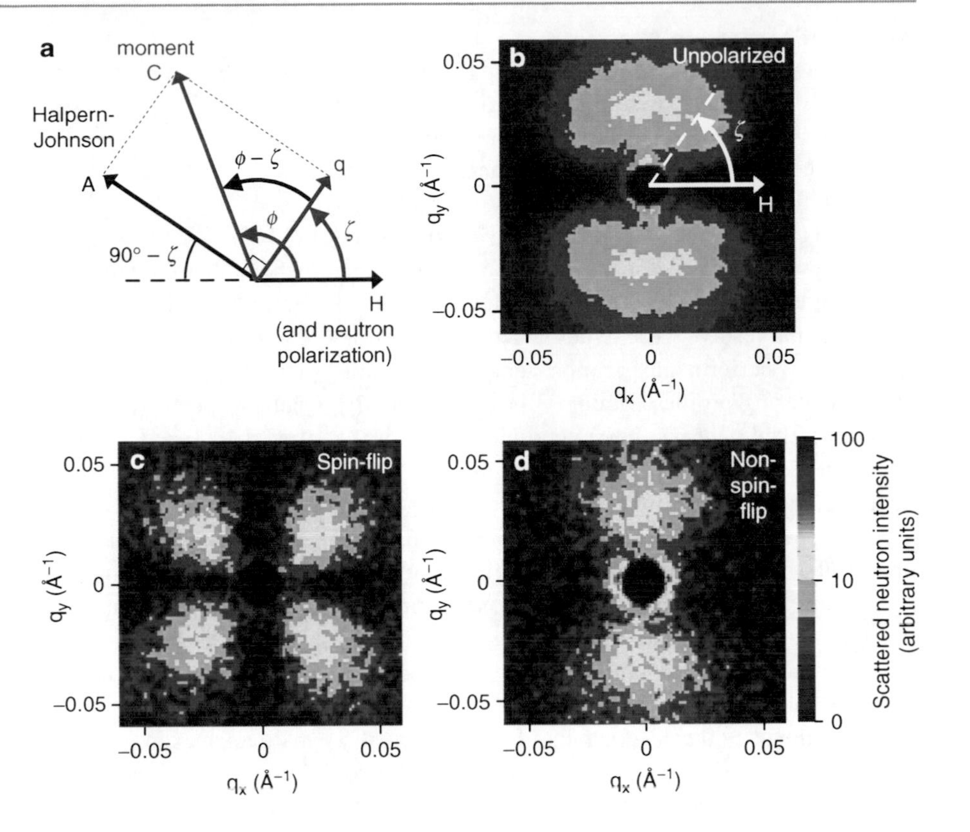

Small-Angle Scattering of Nanostructures and Nanomaterials, Fig. 6 2D SANS detector images from magnetically distinct nanoscale heterogeneities in a ferromagnet. (**a**) Schematic drawn in the detector plane depicting how detector anisotropy arises from the component of the moment C lying at an angle ϕ in the detector plane. (**b**)–(**d**) Measurements from an iron–gallium single crystal under saturating field $\vec{H}$ applied along $\zeta = 0$ in the detector plane such that $\phi = 0$: (**b**) unpolarized neutrons, (**c**) polarized neutrons that scatter with spin flip or (**d**) without spin flip

where $\hat{\mathbf{q}}$ is the unit vector $\vec{q}/|\vec{q}|$. In terms of $\vec{A}$, the unpolarized neutron scattering (2) becomes

$$I(\vec{q}) \sim \sum_{j,l} (b_{nj}b^*_{nl} + b_{mj}b^*_{ml}\vec{A}_j.\vec{A}_l)e^{i\vec{q}.(\vec{R}_j - \vec{R}_l)}$$

Spin-dependent (incoherent) scattering from nuclei has been excluded here; this would enter in the form of terms similar to those in $\vec{A}$ [23]. The Halpern–Johnson relationship can lead to an anisotropy on the SANS detector. To describe this anisotropy, it is convenient to use cylindrical polar coordinates with the longitudinal axis along the beam direction $\vec{k}_i$. The polar plane of this coordinate system is drawn schematically in Fig. 6a. The scattering vector $\vec{q} \simeq (q,\ \zeta,\ 0)$ at small angles. Magnetically distinct heterogeneities within a ferromagnet provide a useful demonstration: the moment of the jth nanoscale heterogeneity, or rather, the contrast of this nanoscale moment with the ferromagnetic matrix, is described by $(C_j,\ \phi_j,\ Z_j)$ in the cylindrical polar coordinates. The unpolarized small-angle scattered intensity is

$$I(\vec{q}) \sim \sum_{j,l} (N_j(\vec{q})N^*_l(\vec{q}) + M_j(\vec{q})M^*_l(\vec{q})(Z_jZ_l + C_jC_l \sin(\zeta - \phi_j)\sin(\zeta - \phi_l))e^{i\vec{q}.(\vec{R}_j - \vec{R}_l)}) \quad (6)$$

where $N(q)$ and $M(q)$ are, respectively, the nuclear and magnetic components of the form factor of a single nanoscale heterogeneity. $(\zeta - \phi)$ is the angle between the moment and $\vec{q}$; the component perpendicular to the scattering vector $\vec{q}$ is $\sin(\zeta - \phi)$, as shown in Fig. 6a. Magnetic heterogeneities with moments co-aligned along a direction ϕ in the detector plane will give rise to a $\sin^2(\zeta - \phi)$ anisotropy on the detector, as illustrated in Fig. 6b. Correspondingly, any observed sine-squared anisotropy on the detector could indicate that the small-angle scattering is magnetic in origin. Nuclear scattering can also be anisotropic, for example, if elongated particles are co-aligned in a direction perpendicular to the beam. For further confirmation that the observed scattering is magnetic, analysis of the neutron polarization can be performed.

Neutron Polarization Analysis

A multilayer or "supermirror" (e.g., of Fe and Si) is commonly used to polarize the incident cold neutron beam. A similar device can in principle be employed to analyze the spin of the scattered neutrons. For SANS, however, the angular acceptance of these devices is not sufficient to cover the 2θ range of scattering angles and these devices may also confer an undesirable small-angle background. The recent development of ^{3}He cells that resolve these issues has made it possible to routinely perform polarization analysis on small-angle scattering profiles. Helium-3 has a particularly spin-dependent neutron absorption: with 100% of the nuclei in the ^{3}He cell polarized, one spin state of the neutrons (that parallel to the ^{3}He spin) is negligibly absorbed, while the other is strongly absorbed. Thus one is able to determine whether, during scattering, a neutron flips its spin ("spin-flip" scattering) or not ("non-spin-flip" scattering). In other words, the polarization is analyzed fully in one direction; this is called "longitudinal" polarization analysis. The scattered intensity breaks down into four components, denoted ++, +−, −+, and −−; the first symbol designates the incident neutron spin, the second the spin of the scattered neutron.

Longitudinal polarization analysis allows an unambiguous distinction to be made between coherent nuclear scattering and spin (incoherent) nuclear scattering, or between magnetic scattering and nuclear scattering [21, 23]. The latter is aptly demonstrated with the example of magnetically distinct heterogeneities in a ferromagnetic matrix. The contribution from nuclear spins is excluded, as before. The non-spin-flip channel is found to be sensitive to nuclear scattering and to magnetic scattering that has a component of the Halpern–Johnson vector $\vec{A}$ parallel to the neutron polarization. Meanwhile, all the scattering observed in the spin-flip channel is magnetic in origin, arising from the components of the Halpern–Johnson vector $\vec{A}$ perpendicular ($\vec{A}_\perp$) to the neutron polarization. The neutron polarization at the sample follows the applied magnetic field $\vec{H}$, that is, along $\zeta = 0$ in this example. Then, the polarized scattered intensity

$$I^{\pm\mp}(\vec{q}) \sim \sum_{j,l} M_j M_l^* (Z_j Z_l + C_j C_l \sin(\zeta - \phi_j) \sin(\zeta - \phi_l)\cos^2\zeta)\, e^{i\vec{q}\cdot(\vec{R}_j - \vec{R}_l)} \tag{7}$$

$$I^{\pm\pm}(\vec{q}) \sim \sum_{j,l} \Big(N_j N_l^* \pm (M_j N_l^* C_j \sin(\zeta - \phi_j) + N_j M_l^* C_l \sin(\zeta - \phi_l))\sin\zeta + M_j M_l^* C_j C_l \sin(\zeta - \phi_j)\sin(\zeta - \phi_l)\sin^2\zeta \Big) e^{i\vec{q}\cdot(\vec{R}_j - \vec{R}_l)} \tag{8}$$

where the form factors are $\vec{q}$-dependent functions $N(\vec{q})$ and $M(\vec{q})$. Comparing (7) and (8) to the unpolarized scattering (6), additional $\sin(90° - \zeta)$ and $\cos(90° - \zeta)$ terms appear in the spin-flip and non-spin-flip channels, respectively. The origin of these terms, as illustrated in Fig. 6a, lies in the sensitivity of polarized neutrons to the components of $\vec{A}$. In the spin-flip channel (7), a $\vec{A}_\perp \wedge \vec{A}_\perp$ term has been omitted that arises from chiral magnetic structures, such as spin helices or skyrmions [26]. The small-angle scattering from magnetic nanoparticles provides an example of where a significant nuclear component $N(\vec{q})$ of the form factor arises in conjunction with a magnetic component $M(\vec{q})$ from the same nano-object [18]. For this particular example of magnetically distinct heterogeneities illustrated in Fig. 6, the nuclear form factor $N(\vec{q})$ is found to be negligible. With moments co-aligned at magnetic saturation along $\phi = 0$, the anisotropy on the detector follows a $\sin^4\zeta$ dependence for the non-spin-flip channel and a $\sin^2\zeta\cos^2\zeta$ dependence for the spin-flip channel (Fig. 6c, d). It is clear that these anisotropies have provided unambiguous confirmation that the origin of the small-angle scattering is magnetic.

Summary and Outlook

Small-angle scattering techniques have made an enormous impact on condensed matter research, spanning across several disciplines. In this short entry, a comprehensive review of the wealth of publications in this realm has been avoided. Instead, the basic principles that form the quoin of topical small-angle scattering articles have been overviewed. The advantages and practical aspects of small-angle neutron and X-ray techniques have also been compared. For further reading, in-depth reviews focusing on selected disciplines are available, in particular for studies of soft matter [1, 2, 4, 9, 10]. These include discussions of the applications of small-angle scattering for biology [4], polymers [1, 2], disordered solids [1], and SAXS on solutions in general [3].

The examples presented in this entry tend toward "hard" condensed matter research, as despite significant topical results, there are presently few reviews in this area. Exceptions may be found in Ref. [5], where the focus is on magnetic SANS, in particular for nanocrystalline materials and magnetic fluids. In particular, magnetic neutron scattering and neutron polarization analysis have been introduced. Neutron polarization analysis has only just become routinely available on small-angle neutron instruments due to the recent development of ^{3}He apparatus to analyze the spin of the scattered neutron. This enables the separation of spin-incoherent from coherent scattering [1, 21]. Furthermore, through the particular character of the interaction between neutron spin and atomic moments [23], small-angle polarized neutron studies can provide tremendous insights into the magnetic nature of nanostructures [5, 18].

In conjunction with advances in beamline apparatus, significant progress is also resulting from the availability of increasingly powerful sources of X-rays and neutrons, which allow the imaging of nanostructures in unprecedented ways. The advent of the free electron laser – a source providing super-intense, femtosecond pulses of X-rays – allows unrivaled scattering studies in real time on ultrashort timescales, from the functional dynamics of biomolecules to magnetic spin-flip processes [27]. Concurrently, advancement steadily continues with the enhancement of existing techniques [28] and the introduction of new scattering models and numerical methods [4, 5, 15–17].

Acknowledgments DanScatt is acknowledged for financial support and Annabel J. Lingham is thanked for a careful reading of the manuscript.

Cross-References

Note: With the exception of 'Synchrotron Radiation Techniques', these cross-references are essentially a list of applications where it is apparent that small-angle scattering techniques have been employed and made a nontrivial contribution to the field. The length of the list reflects that small-angle scattering provides a staple technique across the realm of nanotechnology, and is routinely used for sample characterization.

- ► Carbon Nanotubes
- ► Chitosan Nanoparticles
- ► Dermal and Transdermal Delivery
- ► Electric Field–Directed Assembly of Bioderivatized Nanoparticles
- ► Fullerenes for Drug Delivery
- ► Gas Phase Nanoparticle Formation
- ► Gold Nanorods
- ► Hollow Gold Nanospheres
- ► Light-Element Nanotubes and Related Structures
- ► Liposomes
- ► Macromolecular Crystallization Using Nanovolumes
- ► Magnetic-Field-Based Self-assembly
- ► Nano-Concrete
- ► Nanomaterials for Electrical Energy Storage Devices
- ► Nanomaterials for Excitonic Solar Cells
- ► Nanoparticles
- ► Nanoscale Water Phase Diagram
- ► Nanostructured Thermoelectric Materials
- ► Nanostructures Based on Porous Silicon
- ► Nanostructures for Coloration (Organisms other than Animals)
- ► Nanostructures for Photonics
- ► Optical Properties of Metal Nanoparticles
- ► Prenucleation Clusters
- ► Selected Synchrotron Radiation Techniques
- ► Self-assembly
- ► Self-assembly of Nanostructures
- ► Spider Silk
- ► Structural Color in Animals
- ► Wetting Transitions

References

1. Naudon, A., Schmidt, P.W., Stuhrmann, H.B.: In: Brumberger, H. (ed.) Modern aspects of small-angle scattering, pp. 1–56, 181–220, and 221–254. Kluwer, Dordrecht (1995)
2. Hammouda, B.: SANS from polymers – review of the recent literature. J. Macromol. Sci. Part C: Polymer Reviews **50**, 14–39 (2010); Melnichenko, Y.B., Wignall, G.D.: Small-angle neutron scattering in materials science: recent practical applications. J. Appl. Phys. **102**, 021101 (2007)
3. Glatter, O., Kratky, O. (eds.): Small angle X-ray scattering. Academic, London (1982); Feigin, L.A., Svergun, D.I.:

S

Structure analysis by small-angle X-ray and neutron scattering. Plenum, New York (1987)
4. Putnam, C.D., Hammel, M., Hura, G.L., Tainer, J.A.: X-ray solution scattering (SAXS) combined with crystallography and computation: defining accurate macromolecular structures, conformations and assemblies in solution. Q. Rev. Biophys. **40**, 191–285 (2007); Kasai, N., Kakudo, M.: X-ray diffraction by macromolecules. Springer, Berlin (2005)
5. Michels, A., Weissmüller, J.: Magnetic-field-dependent small-angle neutron scattering on random anisotropy ferromagnets. Rep. Prog. Phys. **71**, 066501 (2008); Avdeev, M.V., Aksenov, V.L.: Small-angle neutron scattering in structure research of magnetic fluids. Phys. – Uspekhi **53**, 971–993 (2010)
6. Guinier, A., Fournet, G.: Small-angle scattering of X-rays. Wiley, New York (1955)
7. Bonse, U., Hart, M.: Tailess X-ray single-crystal reflection curves obtained by multiple reflection. Appl. Phys. Lett **7**, 238–240 (1965)
8. Squires, G.L.: Introduction to the theory of thermal neutron scattering. Cambridge University Press, Cambridge (1978); Lovesey, S.W.: Theory of thermal neutron scattering from condensed matter, vol. 1. Oxford University Press, Oxford (1984); Balcar, E., Lovesey, S.W.: Theory of magnetic neutron and photon scattering. Oxford University Press, Oxford (1989)
9. Narayanan, T., Grillo, I.: In: Borsali, R., Pecora, R. (eds.) Soft matter characterization, pp. 725–782 and 900–952. Springer, Berlin (2008)
10. Tolan, M.: X-ray scattering from soft-matter thin films. Springer, Berlin (1999); Okuda, H., Kato, M., Kuno, K. Ochiai, S., Usami, N., Nakajima, K., Sakata, O.: A grazing incidence small-angle x-ray scattering analysis on capped Ge nanodots in layer structures. J. Phys.: Condens. Matt. **22**, 474003 (2010)
11. Allen, A.J., Thomas, J.J., Jennings, H.M.: Composition and density of nanoscale calcium-silicate-hydrate in cement. Nat. Mater. **6**, 311–316 (2007)
12. Huxley, A.: In: Huebener, P., Schopohl, N., Volovik, G.E. (eds.) Vortices in unconventional superconductors and superfluids, pp. 301–339. Springer, Berlin (2002)
13. Glatter, O., May, R.: Small-angle techniques. In: Prince, E. (ed.) International tables of crystallography, vol. C, pp. 89–112. Kluwer, Dordrecht (2004)
14. Teixeira, J.: Small-angle scattering by fractal systems. J. Appl. Crystallogr. **21**, 781–785 (1988); Bale, H.D., Schmidt, P.W.: Small-angle X-ray-scattering investigation of submicroscopic porosity with fractal properties. Phys. Rev. Lett. **53**, 596–599 (1984); Ruland, W.: Small-angle scattering of two-phase systems: determination and significance of systematic deviations from Porod's law. J. Appl. Crystallogr. **4**, 70–73 (1971)
15. Beaucage, G.: Approximations leading to a unified exponential/power-law approach to small-angle scattering. J. Appl. Crystallogr. **28**, 717–728 (1995); Hammouda, B.: Analysis of the Beaucage model. J. Appl. Crystallogr. **43**, 1474–1478 (2010)
16. Chacón, P., Fernando Díaz, J., Morán, F., Andreu, J.M.: Reconstruction of protein form with X-ray solution scattering and a genetic algorithm. J. Mol. Biol. **299**, 1289–1302 (2000); Svergun, D.I.: Restoring low resolution structure of biological macromolecules from solution scattering using simulated annealing. Biophys. J. **76**, 2879–2886 (1999); Svergun, D.I., Petoukhov, M.V., Koch, M.H.J.: Determination of domain structure of proteins from X-ray solution scattering. Biophys. J. **80**, 2946–2953 (2001)
17. Laver, M., Forgan, E.M., Abrahamsen, A.B., Bowell, C., Geue, Th, Cubitt, R.: Uncovering flux line correlations in superconductors by reverse Monte Carlo refinement of neutron scattering data. Phys. Rev. Lett. **100**, 107001 (2008)
18. Krycka, K.L., Booth, R.A., Hogg, C.R., Ijiri, Y., Borchers, J.A., Chen, W.C., Watson, S.M., Laver, M., Gentile, T.R., Dedon, L.R., Harris, S., Rhyne, J.J., Majetich, S.A.: Core-shell magnetic morphology of structurally uniform magnetite nanoparticles. Phys. Rev. Lett. **104**, 207203 (2010)
19. Walter, G., Kranold, R., Gerber, T., Baldrian, J., Steinhart, M.: Particle size distribution from small-angle X-ray scattering data. J. Appl. Crystallogr. **18**, 205–213 (1985)
20. Sears, V.F.: Neutron scattering lengths and cross-sections. Neutron News **3**, 26–37 (1992)
21. Stuhrmann, H.B.: Contrast variation in X-ray and neutron scattering. J. Appl. Crystallogr. **40**, s23–s27 (2007)
22. Mühlbauer, S., Binz, B., Jonietz, F., Pfleiderer, C., Rosch, A., Neubauer, A., Georgii, R., Böni, P.: Skyrmion lattice in a chiral magnet. Science **323**, 915–919 (2009)
23. Moon, R.M., Riste, T., Koehler, W.C.: Polarization analysis of thermal-neutron scattering. Phys. Rev. B **181**, 920–931 (1969); Halpern, O., Johnson, M.H.: On the magnetic scattering of neutrons. Phys. Rev. **55**, 898–923 (1939)
24. Wagner, W., Bellmann, D.: Bulk domain sizes determined by complementary scattering methods in polycrystalline Fe. Physica B **397**, 27–29 (2007)
25. Wu, J., Lynn, J., Glinka, C.J., Burley, J., Zheng, H., Mithcell, J.F., Leighton, C.: Intergranular giant magnetoresistance in a spontaneously phase separated perovskite oxide. Phys. Rev. Lett. **94**, 037201 (2005)
26. Grigoriev, S.V., Chernyshov, D., Dyadkin, V.A., Dmitriev, V., Maleyev, S.V., Moskvin, E.V., Menzel, D., Schoenes, J., Eckerlebe, H.: Crystal handedness and spin helix chirality in $Fe_{1-x}Co_xSi$. Phys. Rev. Lett. **102**, 037204 (2009); Pappas, C. Lelièvre-Berna, E., Falus, P., Bentley, P.M., Moskvin, E., Grigoriev, S., Fouquet, P., Farago, B.: Chiral paramagnetic skyrmion-like phase in MnSi. Phys. Rev. Lett. **102**, 197202 (2009)
27. Treusch, R., Feldhaus, J.: FLASH: new opportunities for (time-resolved) coherent imaging of nanostructures. New J. Phys. **12**, 035015 (2010)
28. Hura, G., Menon, A.L., Hammel, M., Rambo, R.P., Poole, F.L., Tsutakawa, S.E., Jenney, F.E., Classen, S., Frankel, K.A., Hopkins, R.C., Yang, S., Scott, J.W., Dillard, B.D., Adams, M.W.W., Tainer, J.A.: Robust, high-throughput solution structural analyses by small angle X-ray scattering (SAXS). Nat. Methods **6**, 606–614 (2009)

Small-Angle X-Ray Scattering

► Small-Angle Scattering of Nanostructures and Nanomaterials

Smart Carbon Nanotube-Polymer Composites

► Active Carbon Nanotube-Polymer Composites

Smart Drug Delivery Microchips

► Stimuli-Responsive Drug Delivery Microchips

Smart Hydrogels

S. M. Khaled, Alessandro Parodi and Ennio Tasciotti
Department of Nanomedicine, The Methodist Hospital Research Institute, Houston, TX, USA

Synonyms

Stimulus-responsive polymeric hydrogels

Definition

Smart hydrogel is defined as the polymer network able to respond to external stimuli through abrupt changes in the physical nature of the network.

Summary

Smart hydrogels, a special class of polymer networks, have offered new breakthroughs in the medical field due to their exceptional properties, engineering flexibility, natural abundance and ease of manufacturing. This entry represents a basic overview of classifications and material properties of smart hydrogels for guided drug delivery and regenerative medicine applications. While providing a brief representation on the current trends and advancement of polymer science in medicine, we have also revealed some cutting edge progresses in the field originating from the blending of nanotechnology with polymer science.

Polymer Science in Medicine

The first application of polymers in medical field can date back to the 1940s when polymethyl methacrylate (PMMA) was used for the replacement of damaged corneas. Since then, the mechanical, physical, and chemical properties of polymers have been extensively investigated and utilized for numerous medical applications. In particular, regenerative medicine has benefited greatly from polymer research and development. Polymers can now replace metal devices used in orthopedic settings, and the investigation into the development of new biomaterials to repair and substitute body tissues is proceeding with great momentum. Today, the use of polymers in medicine includes many applications such as replacing damaged bones, increasing the likelihood of wound repair, fabrication of external devices for dialysis, heart valves, vascular grafts, prosthesis, implantable lenses, and dental materials [1]. In 1975, Helmut Ringsdorf introduced the concept of "pharmacologically active polymer," specifically referring to the chemical conjugation between a polymer and a drug, which represented a revolutionary approach in drug delivery [2]. Since then, many efforts have been made within the scientific community to translate Dr. Ringsdorf's theories into materials and products that can be applied to medicine. Biodegradable polymers were the first type of polymers intended for drug delivery because of the wide range of payloads that can be easily loaded and released during the degradation of the polymers [3]. This class of polymers has been extensively studied over the last decades with successful examples such as calcium alginate and PEG-based polymer networks [4]. In addition, the advancement of nanomedicine and of its biomedical applications sparked a new breakthrough in the use of polymers for drug delivery. This application of polymers to drug delivery has been classified into five main categories: polymeric drugs, polymer-drug conjugates, polymer-protein conjugates, polymeric micelles and

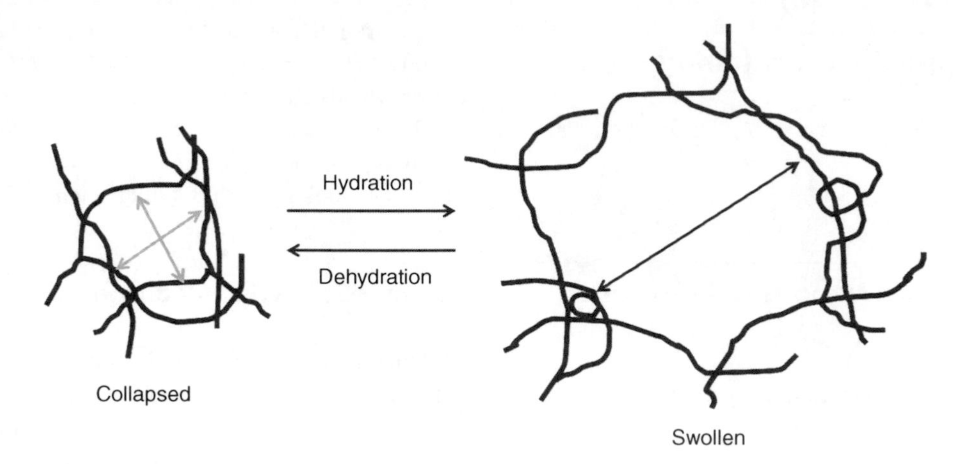

Smart Hydrogels, Fig. 1 Swelling of hydrogel. *Left*: Hydrogel in the un-swollen or dry state. The polymer chains are in close proximity and may interact with each other. However, when fluid enters the hydrogel, the polymer chains undergo hydration that leads to swelling due to this hydrophobic or electrostatic interaction. *Light gray arrows* indicate pressure being exerted on the system that leads the swelling. *Right*: The hydrogel in the swollen or hydrated state where the polymer chains are fully extended and only the cross-links prevent the material from dissolution

poly-nucleic acids [5]. The improvements in synthesis, surface modifications, and characterization techniques have transformed polymers into ideal partners for many of the nanoparticles currently used in drug delivery. The fusion of these two sciences enables nanovectors to exhibit enhanced bioavailability and specificity with respect to the targeted biological sites.

Introduction of Hydrogel in Medical Field

In order to be used in the medical field, polymers must meet strict biocompatibility criteria. They have to exhibit low to no toxic effects, be water soluble, degrade into safe by-products in aqueous environments, and be non-immunogenic to avoid activating the inflammatory response [6]. Hydrogels, also defined as hydrophilic three-dimensional polymeric networks, represent a promising option in this regard. Their ability to form biocompatible and biodegradable three-dimensional structures allowed for the synthesis of scaffolds for tissue engineering. These materials are extensively used in the growth of synthetic tissues (pancreas, cornea, skin, bones) [7], or as controlled drug delivery platforms [8]. Composed of hydrophilic cross-linked polymers, hydrogels do not dissolve in water and hold a large amount of water or biological fluid while maintaining their structure. The ability of hydrogels to absorb water arises from the hydrophilic functional groups attached to their polymeric backbone. Their resistance to dissolution arises from cross-linking between the network chains. Figure 1 represents the structure of the polymeric network of a typical hydrogel interacting with the aqueous phase. While in a dehydrated or de-swollen state, its polymer chains are in close proximity to each other leaving little room for the diffusion of molecules. As the material swells, the polymer chains separate to an extent determined by the properties of the solvent. During the swelling process, the polymer chains of the hydrogel extend, decreasing the interaction that takes place between them. In this state, the swelling of the hydrogel is counteracted by the crosslinkers present within the hydrogel matrix. At its maximum hydration, the diffusion of small molecules (e.g., drugs) approaches the diffusion coefficient in pure fluid.

Smart Hydrogels

Hydrogels are characterized by a dynamic balance with water or other biologic fluids that drives the absorption and release of drugs in the desired environment. The presence of water in the hydrogel matrix determines the physical and chemical properties of these polymeric structures in the drug delivery field and increases their biocompatibility. Hydrogels represent a very interesting tool in regenerative medicine because they reduce the probability of inflammatory responses due to mechanical frictions with the surrounding tissues in implant surgeries. In addition, these matrices can be formulated in injectable solutions, which represent an appreciable advantage over the conventional surgical procedures. During the last decades, hydrogels gained increasing interest in medical field because they can be designed or tailored to undergo discrete or continuous volume transformation in response to infinitesimal changes of environmental stimuli such as pH, temperature, electric field, solvent composition, salt concentrations/ionic strength, light/photon, pressure, coupled magnetic and electric fields. Better known as smart or stimuli-responsive hydrogels, these highly maneuverable and adaptive polymers can sense changes in the environment and respond by inducing structural changes (increasing or decreasing their degree of swelling) without requiring an external driving force. This "volume-changing" phenomenon is particularly useful in drug delivery applications as drug release can be triggered upon these environmental changes. Smart hydrogels are ideal candidates for the development of self-regulated drug delivery systems with enhanced therapeutic efficacy. Temperature and pH are the most commonly used stimuli to trigger the hydrogel's curative action because they have biological and physiologic relevance [9].

Temperature-Sensitive Hydrogels

Temperature-sensitive hydrogels are probably the most universally studied class of environment-sensitive polymer systems in drug delivery research. These hydrogels are able to swell or collapse as a result of changes in the temperature of surrounding fluid and are classified into negatively and positively thermosensitive hydrogels. Thermoresponsive hydrogels are composed of polymer chains that possess either moderately hydrophobic groups or a mixture of hydrophilic and hydrophobic segments. Negative temperature-sensitive hydrogels are characterized by their lower critical solution temperature (LCST). Below the LCST, the hydrogel swells in solution while above this temperature the polymer contracts. On the other hand, positive temperature-sensitive hydrogels are characterized by their upper critical solution temperature (UCST). In this case, below UCST, the polymer shrinks and above it, it swells [10]. Hence, adjusting the critical solution temperature within a physiological range, these hydrogels acquire intriguing biological activities that

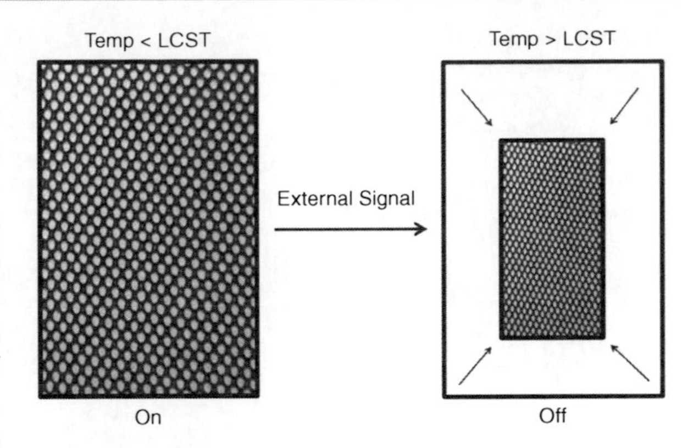

Smart Hydrogels, Fig. 2 Schematic illustration of the "on-off release" from a squeezing hydrogel device for drug delivery

make them reliable candidates for drug delivery in environmental conditions in which the temperature varies. Under the critical point of negative temperature-sensitive hydrogels, hydrogen bonding between hydrophilic segments of the polymer chain and water molecules dominates, leading to enhanced dissolution in water. As the temperature increases, however, interactions among hydrophobic segments are strengthened, while hydrogen bonding becomes weaker. The net result is shrinking of the hydrogels due to inter-polymer chain association, through hydrophobic interactions [9]. The LCST can be changed by adjusting the ratio of hydrophilic and hydrophobic segments in the hydrogel system. It can also be done by developing copolymers of hydrophobic (e.g., *N*-isopropylacrylamide [NIPAAm]) and hydrophilic (e.g., acrylic acid [AA]) monomers. Hydrogels with negative temperature sensitivity have the potential to show an "on-off" drug release with "on" at low temperature and "off" at high temperature allowing pulsatile drug release. Figure 2 shows a schematic illustration of the on-off release.

Moreover, polymers having LCST below human body temperature have the potential for injectable applications and they are often combined with polysaccharides, such as chitosan, alginate, cellulose, and dextran [11], that improve the material's properties in terms of biocompatibility, swelling ratio, pH-sensitivity, swelling dynamics (swelling and de-swelling rate), modulation of LCST, and thermal stability. Another issue of the thermoresponsive polymers is represented by temperature-sensitive peptides often associated with the widely investigated elastin-like polypeptide (ELP) polymers. This polymer is composed of

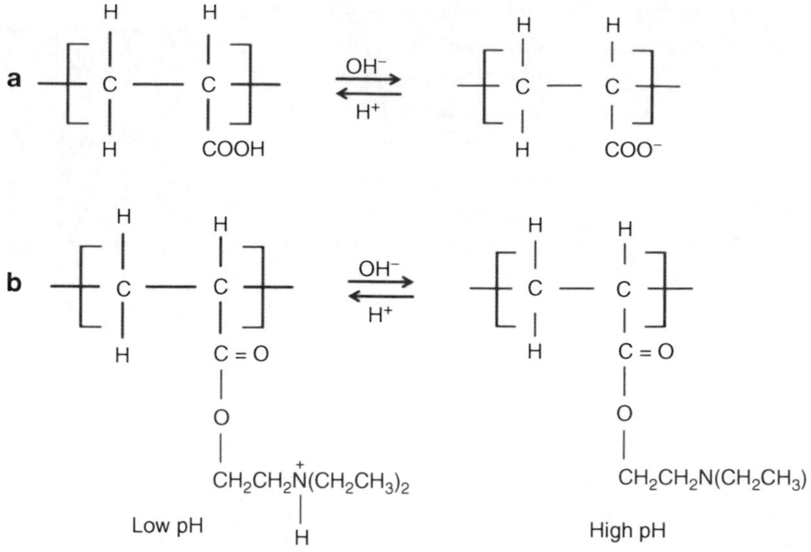

Smart Hydrogels, Fig. 3 pH-dependent ionization of hydrogels. (**a**) Poly (acrylic acid) and (**b**) poly (*N, N*/-diethylamionethyl methacrylate) (adapted from [13])

pentapeptide monomers, derived from human tropoelastin [12], which tend to aggregate in a reversible way under heating, entrapping the payload. This biopolymer showed extraordinary biological properties: it is fully biocompatible, completely biodegradable without any immunogenic reaction, and moreover, it appeared to be ideal for the delivery of protein-based treatments. The degradation of the polymer is constant and slow when compared to the free form of the monomer. The temperature-sensitive peptides represented a breakthrough in polymer science in the biomedical field. The exploitation of standard recombinant DNA techniques via genetic engineering allowed them to be easily synthesized and modified in the primary sequence thereby changing chemical, physical, and biological properties of the polymer while permitting a controlled design of the peptide according to the nature of the cargo. Furthermore, they could easily be attached to a protein-based payload during the synthesis process. The two main features of ELP polymers are directionality and reversibility. Directionality describes the conformational state of the polymer under a specific stimulus. In particular, the ability of the peptide to associate or disassociate under heating is essential if the treatment requires fast or slow drug release. Reversibility regards the capacity of the polymer to return to its initial state after the thermal stimulus [9, 12]. A reversible polymer could be very useful for concentrating the payload in a region of the body and obtaining a targeted release of the drug after heating through an external stimulus. Conversely, an irreversible polymer could allow drug delivery in a controlled fashion through slow degradation. This category of thermosensitive polymers is emerging as the new opportunity in cancer cell therapy [9] and future research aims to increase and optimize the biocompatibility and the tenability of these delivery systems.

pH-Sensitive Hydrogels

Another class of smart hydrogels undergoes a volume transition with the variation of the environmental pH. These hydrogels are characterized by the presence of ionic moieties on the polymeric backbone. These ions either accept or release protons in response to pH changes. Weak acidic groups such as acrylic and cationic acids or weak basic groups such as amines provide pH sensitivity to the polymer chain. Swelling of hydrogels sharply changes in the vicinity of the pKA or pKB values of the acidic or basic functional groups. Carboxylic pendant groups accept hydrogen at low pH but exchange it for other cations above the pKA value, and become ionized at higher pH. The hydrodynamic volume and swelling capability of these polymer chains increases sharply when their carboxylic groups become ionized and reach plateau around pH 7. Amines, on the other hand, accept protons at low pH and become positively ionized at and below their pKB values. Hence, their swelling capability increases sharply in acidic solutions. Figure 3 shows structures of anionic and cationic polyelectrolytes and their pH-dependent ionization. Poly(acrylic acid) (PAA) becomes ionized at high pH, while poly(*N,N*/-diethylaminoethyl methacrylate) (PDEAEM) becomes ionized at low pH [13].

The swelling and pH-responsiveness of polyelectrolyte hydrogels can be adjusted by using neutral comonomers, such as 2-hydroxyethyl methacrylate, methyl methacrylate, and maleic anhydride [9]. Different comonomers provide different hydrophobicity to the polymer chain, leading to different pH-sensitive behavior. pH-responsive polymers were first applied as an oral drug delivery systems due to the variation of the pH that characterizes the different areas of the gastrointestinal tract [13]. Several anionic polymers such as methyl acrylic acid, methyl methacrylate, and hydroxypropyl methylcellulose phthalate have been commercially used as enteric coatings for oral delivery of protein-based drugs. In the acidic environment of the stomach, the carboxylic acid groups of the above-mentioned polymers are unionized and the particle retains its therapeutic cargo. They were shown to release the encapsulated drug molecules into the small intestine in response to the alkaline pH [14] because the carboxylic acid group becomes ionized and the particle swells. PNIPAAm can be classified as a thermal and pH-responsive polymer. It is used for oral drug delivery applications exploiting its degradation characteristics in the alkaline environment of the intestine. It was shown that the polymer synthesized in combination with butyl methacrylate (BMA) and acrylic acid (AAc) is very stable at low pH while resulting in full degradation and release of payload within the high pH environment [15].

Vesicles derived from endocytosis fuse their membranes with vesicles of the early endosomal compartment. Lysosome formation follows after several other membrane fusion steps (with late endosomal and lysosomal vesicles). During these membrane fusions inside the cytoplasm, the vesicles enrich themselves of protons, decreasing their internal pH through an active import of H^+. This proton enrichment is permitted by an ATPase that works using ATP to pump in hydrogen ions and a negative counter ion (Cl^-) to satisfy electroneutrality [16]. Due to the presence of many ionizable groups in the backbone, a pH-responsive polymer is usually able to work as a "proton sponge" with a buffering effect on the surrounding environment. Therefore, the ATPase continues to work with the result of increasing the concentration of the counter ion and affecting the osmotic balance of the vesicles. The final result of this process is the breaking of the vesicle's membrane caused by an increase in osmotic pressure. Moreover, many polymers increase their volume when subjected to an acid environment; this produces a mechanical stress on the lysosomal membranes that affects the integrity of these organelles. Lysosomal escape is very useful for nonviral gene delivery and could potentially be more efficient and less toxic than standard virus-based carriers. These molecules are usually cationic, thus likely efficient in forming macro-complexes with DNA (polyplexes) or new therapies based on RNA (e.g., siRNA, miRNA). Figure 4 represents a schematic of the lysosomal escape performed by pH-responsive cationic hydrogel. The polymer in this case plays several roles: (1) to physically carry the plasmid, (2) to shield the DNA payload from degradation, (3) to perform lysosomal escape and facilitate the nuclear delivery of the cargo. Poly(ethylene imine) (PEI) [17], poly(amidoamine) (PAMAM) [18], Poly (*N,N*-dimethylaminoethyl methacrylate) (PDEAEM) [19], poly (L-Lysine) [20] and modified chitosan represent a few examples of all the polymers studied for this application.

Thermo- and pH-responsive hydrogels allow (with or without the use of copolymers) the loading and the triggered release of many drugs for the treatment of different pathological states. Table 1 reports some examples of these polymer-drug formulations and their potential clinical applications.

Other Responsive Compounds

Electro-Responsive Hydrogels

Recently, electro-responsive hydrogels stimulated the curiosity of the scientific community because they can un-swell or bend, depending on the shape and orientation of the gel along an electric field. The gel bends when it is parallel to the electrodes, whereas unswelling occurs when the hydrogel lies perpendicular to them. This kind of polymer is usually based on a polyelectrolyte matrix (polymers that contain a relatively high concentration of ionizable groups along the backbone chain), so these polymers are also pH-responsive [21]. The electric field generates an interaction between the mobile counter ions and the immobile charged groups of the gel's polymeric network in order to induce the phase transition. For example, in partially hydrolyzed polyacrylamide gels, the

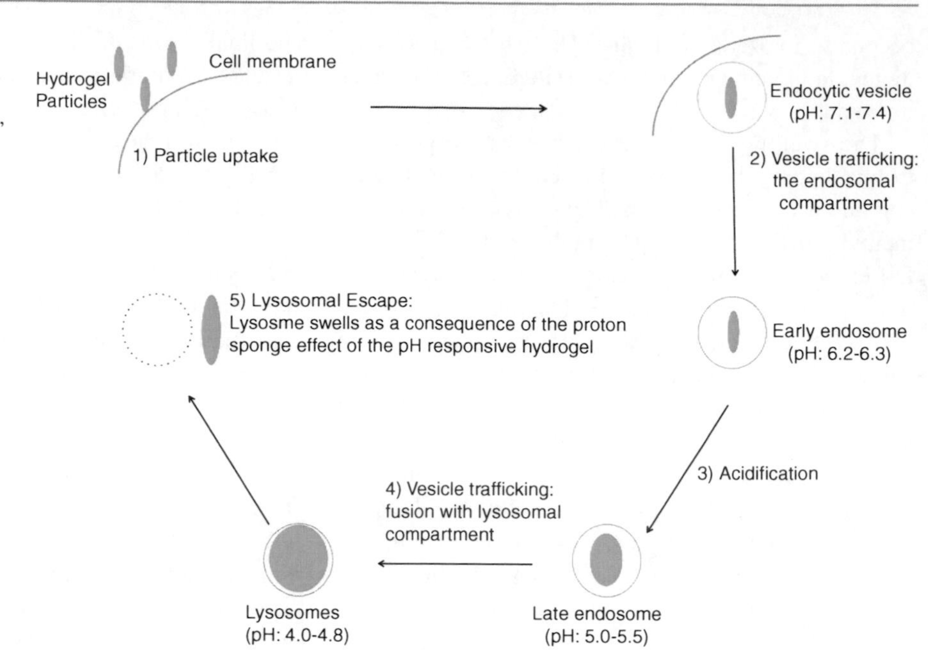

Smart Hydrogels, Fig. 4 Schematic of the lysosomal escape. Upon cell internalization by endocytosis, hydrogel particles are trapped in the endolysosomal compartment. During their transition from early to late endosomes and finally to lysosomes, particles undergo increased swelling due to protonation. The proton sponge effect they exert eventually leads to the breakage of the lysosomal vescicle

Smart Hydrogels, Table 1 Some applications of temperature and pH-responsive polymer

Polymer	Environmental stimulus	Copolymer	Loaded drug	Application
PNIPAAm	Temperature	Chitosan	Diclofenac	Anti-inflammatory
Glycidyl methacrylate	Temperature	Chitosan	Doxorubicin	Chemotherapeutic
β-glycerophosphate	Temperature	Chitosan	Chitosan	Antibiotic
Poly(propylene)-g-AA	Temperature	NIPAAm/Chitosan	Chitosan	Antibiotic
Triblock polymer	Temperature	Poly(lactic-co-glycolic acid)/poly (lactic acid)/poly(ethylene glycol)	Adriamycin	Chemotherapeutic
PNIPAAm	Temperature	Alginate	Indomethacin	Anti-inflammatory
PNIPAAm	pH	Acrylic acid	Growth factors	Wound repair
Poly(ethylene imine)	pH		Nucleic acid	Gene delivery
poly(amidoamine)	pH		Nucleic acid	Gene delivery
Poly (*N,N*-dimethylaminoethyl methacrylate)	pH		Nucleic acid	Gene delivery
poly (L-Lysine)	pH		Nucleic acid	Gene delivery

mobile H^+ ions migrate toward the cathode while the negatively charged immobile acrylate groups in the polymer networks are attracted toward the anode (Fig. 5). Thus, the anionic polymeric gel network is pulled toward the anode. This pull creates a uniaxial stress along the gel axis, being maximal at the anode and minimal at the cathode. This stress gradient contributes to the gel deformation and when triggered by electric filed, these hydrogels undergo the swelling and contraction necessary to release the payload. Electro-responsive polymers represent one of the most complex challenges in biomedical polymer synthesis. Even if the electric stimulus can be finely tuned in terms of strength, duration, and impulse frequency, the use of these techniques is limited by objective safety restrictions. This kind of hydrogel was conceived to obtain a pulsatile delivery of hormones (e.g., insulin, hydrocortisone), proteins, and peptides, and glucose and a few clinical applications for electro-responsive polymers are currently being investigated for dermal and transdermal drug delivery [21, 22].

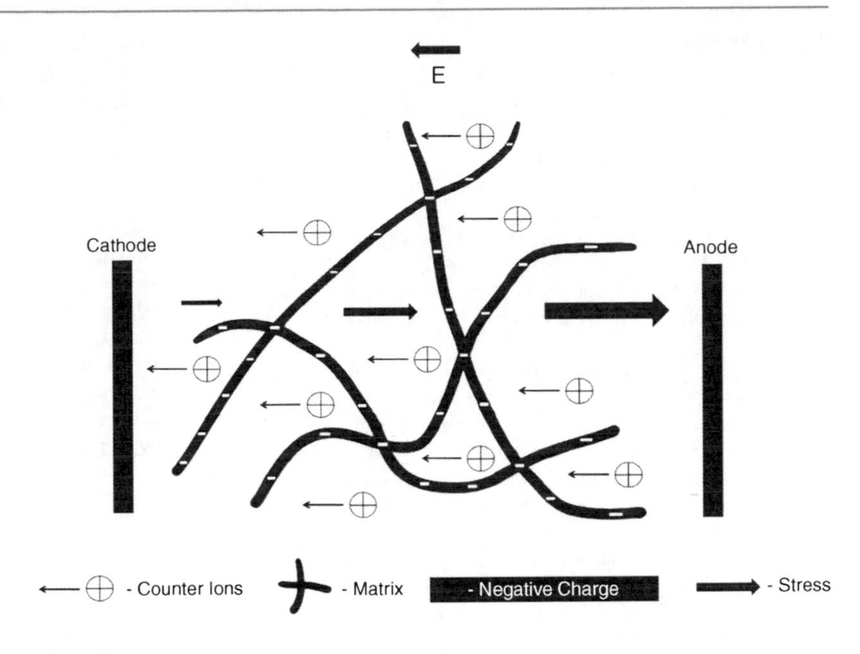

Smart Hydrogels, Fig. 5 The effect of an electric field on a polyelectrolyte gel. Positively charged counterions migrate toward the cathode, while the immobile polymeric anionic groups are attracted toward the anodes (Adapted from Ref. [21])

Stimulus-Sensitive Connection

This approach allows for the development of drug delivery tools responsive to the composition of the biological microenvironment. In this case, the polymer's function is limited to being a carrier for the cargo and targeting the area with the correct environmental conditions for drug release. Polymers like *N*-(2-hydroxypropyl) methacrylate (HPMA), already exploited for steric stabilization of bioactive proteins, were used to coat biological or synthetic particles to provide them protection against antibodies and complement factors [23]. HPMA was also used to bind doxorubicin via a pH-sensitive linker. This polymer-bound drug showed significant improvement in the ability to kill MCF-7 breast cancer cells when compared with the drug alone [24]. Another example of drug delivery system based on using a peptide linker (Pro-Val-Gly-Leu-Ile-Gly) was composed of a chemotherapeutic agent and a dextran-based polymer. This linker is sensitive to the action of the overexpressed cancer metalloproteinases 2 and 9 while remaining stable in blood circulation. Therefore, the drug release is more concentrated at the site of the tumor where these enzymes are usually overexpressed [25].

Future Perspectives

Smart hydrogels, with proper chemical and biological modifications, represent promising alternatives to current controlled release and functional biomaterials applications. Their capability of addressing specific biological and medical challenges will determine the success of the next generation of smart hydrogel materials. Hydrogels that are being used for load-bearing applications in tissue engineering will need to provide enhanced mechanical properties as well as the molecular signals that are present in the native or regenerating tissues. Similarly, the stimuli-guided controlled release of drugs from smart hydrogels is, in most practical cases, too slow due to the thickness of the matrix. Therefore, a fast-acting hydrogel comprised of thinner and smaller matrix is necessary for this purpose. However, this type of hydrogel would be characterized by low mechanical stability and would have limited applications. Synthesis of new polymers, crosslinkers, and reinforcing agents with higher mechanical properties, more biocompatibility, and better biodegradability is essential for success in all aspects of translational medicine. With the development of new materials and novel methods of engineering chemical, mechanical, and biological functionality into the hydrogel matrix, it can be anticipated that future smart hydrogels will provide even greater contributions to the field of drug delivery and regenerative medicine.

Cross-References

► Nanomedicine
► Nanoparticles

References

1. Navarro, M., Michiardi, A., Castaño, O., Planell, J.A.: Biomaterials in orthopaedics. J. R. Soc. Interface **5**, 1137–1158 (2008)
2. Ringsdorf, H.: Structure and properties of pharmacologically active polymers. J. Polym. Sci. Pol. Sym. **51**, 135–153 (1975)
3. Liu, S., Maheshwari, R., Kiick, K.L.: Polymer-based therapeutics. Macromolecules **42**, 3–13 (2009)
4. Keys, K.B., Andreopoulos, F.M., Peppas, N.A.: Poly(ethylene glycol) star polymer hydrogels. Macromolecules **31**, 8149–8156 (1998)
5. Dirk, S.: Thermo- and Ph-Responsive polymers in drug delivery. Adv. Drug Deliv. Rev. **58**, 1655–1670 (2006)
6. Anderson, J.M.: The future of biomedical materials. J. Mater. Sci. Mater. Med. **17**, 1025–1028 (2006)
7. Alijotas-Reig, J., Garcia-Gimenez, V.: Delayed immune-mediated adverse effects related to hyaluronic acid and acrylic hydrogel dermal fillers: clinical findings, long-term follow-up and review of the literature. J. Eur. Acad. Dermatol. Venereol. **22**, 150–161 (2008)
8. Nikolaos, A.P.: Hydrogels and drug delivery. Curr. Opin. Colloid Interface Sci. **2**, 531–537 (1997)
9. Chilkoti, A., Dreher, M.R., Meyer, D.E., Raucher, D.: Targeted drug delivery by thermally responsive polymers. Adv. Drug Deliv. Rev. **54**, 613–630 (2002)
10. Prabaharan, M., Mano, J.F.: Stimuli-responsive hydrogels based on polysaccharides incorporated with thermo-responsive polymers as novel biomaterials. Macromol. Biosci. **6**, 991–1008 (2006)
11. Webber, R.E., Shull, K.R.: Strain dependence of the viscoelastic properties of alginate hydrogels. Macromolecules **37**, 6153–6160 (2004)
12. Urry, D.W.: Physical chemistry of biological free energy transduction as demonstrated by elastic protein-based polymers. J. Phys. Chem. B **101**, 11007–11028 (1997)
13. Qiu, Y., and Park, K.: Environment-sensitive hydrogels for drug delivery. Adv. Drug Deliv. Rev. **53**, 321–339 (2001)
14. Davis, B.K., Noske, I., Chang, M.C.: Reproductive performance of hamsters with polyacrylamide implants containing ethinyloestradiol. Acta Endocrinol. **70**, 385–395 (1972)
15. Serres, A., Baudyš, M., Kim, S.W.: Temperature and Ph-sensitive polymers for human calcitonin delivery. Pharm. Res. **13**, 196–201 (1996)
16. Varkouhi, A.K., Scholte, M., Storm, G., Haisma, H.J.: Endosomal escape pathways for delivery of biologicals. J. Control. Release **151**, 220–228 (2011)
17. Godbey, W.T., Wu, K.K., Mikos, A.G.: Poly (Ethylenimine)-mediated gene delivery affects endothelial cell function and viability. Biomaterials **22**, 471–480 (2001)
18. Dufès, C., Uchegbu, I.F., Schätzlein, A.G.: Dendrimers in gene delivery. Adv. Drug Deliv. Rev. **57**, 2177–2202 (2005)
19. Vicent, M.J., Dieudonné, L., Carbajo, R.J., Pineda-Lucena, A.: Polymer conjugates as therapeutics: future trends, challenges and opportunities. Expert Opin. Drug Deliv. **5**, 593–614 (2008)
20. Brown, M.D., Gray, A.I., Tetley, L., Santovena, A., Rene, J., Schätzlein, A.G., Uchegbu, I.F.: In vitro and in vivo gene transfer with Poly(Amino Acid) vesicles. J. Control. Release **93**, 193–211 (2003)
21. Sudaxshina, M.: Electro-responsive drug delivery from hydrogels. J. Control. Release **92**, 1–17 (2003)
22. Peppas, N.A., Bures, P., Leobandung, W., Ichikawa, H.: Hydrogels in pharmaceutical formulations. Eur. J. Pharm. Biopharm. **50**, 27–46 (2000)
23. Urry, D.W., Luan, C.H., Parker, T.M., Gowda, D.C., Prasad, K.U., Reid, M.C., Safavy, A.: Temperature of polypeptide inverse temperature transition depends on mean residue hydrophobicity. J. Am. Chem. Soc. **113**, 4346–4348 (1991)
24. Fisher, K.D., Seymour, L.W.: Hpma copolymers for masking and retargeting of therapeutic viruses. Adv. Drug Deliv. Rev. **62**, 240–245 (2010)
25. Liu, W., MacKay, J.A., Dreher, M.R., Chen, M., McDaniel, J.R., Simnick, A.J., Callahan, D.J., Zalutsky, M.R., Chilkoti, A.: Injectable intratumoral depot of thermally responsive polypeptide–radionuclide conjugates delays tumor progression in a mouse model. J. Control. Release **144**, 2–9 (2010)

Soft Actuators

► Organic Actuators

Soft Lithography

► BioPatterning
► Microcontact Printing
► Nanoscale Printing

Soft X-Ray Lithography

► EUV Lithography

Soft X-Ray Microscopy

► Selected Synchrotron Radiation Techniques

Soil/Terrestrial Ecosystem/Terrestrial Compartment

► Exposure and Toxicity of Metal and Oxide Nanoparticles to Earthworms

Solar Cells

► Self-repairing Photoelectrochemical Complexes Based on Nanoscale Synthetic and Biological Components

Sol-Gel Method

Bakul C. Dave and Sarah B. Lockwood
Department of Chemistry and Biochemistry, Southern Illinois University Carbondale, Carbondale, IL, USA

Synonyms

Chemical solution deposition; Gel chemical synthesis; Silica gel processing; Wet chemical processing

Definition

The sol-gel method is a wet chemical process of making oxide-based materials starting from hydrolyzable precursors via hydrolysis and condensation. The precursors usually contain weaker ligands as compared to water such as halides, nitrates, sulfates, alkoxides, or carboxylates. The hydrolyzed precursors then condense together to form small colloidal nanoparticles suspended in a liquid called a sol. Further polycondensation of the sol particles leads to an extended oxo-bridged network of polymeric oxide-based materials. The as-formed gels obtained by the sol-gel method are biphasic materials that contain gel network along with a significant amount of liquid phase. Drying of these gels, either under ambient conditions or at elevated temperature, can lead to expulsion of solvent phase to form dense materials. Because the method utilizes molecular precursors, it offers appealing prospects for the design of nanomaterials with a degree of control over their structure, composition, dimension, morphology, organization, geometry, and bulk architecture. While the method can be used to make bulk monoliths, it is particularly suited for fabricating nanomaterials of varied compositions in different shapes, geometries, and dimensionalities including nanoparticles, fibers, thin films, and coatings.

Overview

The sol-gel method is a solution-based method of making ceramic particles, powders, coatings, films, and monolithic objects [1]. The process is commonly used for making oxides, however, silica-based materials constitute a primary archetype system. With highly electropositive metals, the precursor molecules can react rapidly with water to undergo facile hydrolysis under ambient conditions. However, precursors based on alkoxysilanes usually require the use of catalysts, refluxing, or use of ultrasound for the reaction to proceed at an appreciable rate. The liquid sol containing nanoscale particles can be used to form powders, fibers, coatings, or monolithic forms by allowing the reaction to proceed under specific processing conditions. In addition to the formation of pure oxides, the method can also be used for making organic-inorganic hybrids and organically modified silicas as well as bioceramic hybrids [2].

In the sol-gel method, a suitable molecular precursor is hydrolyzed to generate a solid-state polymeric oxide network (Fig. 1). Initial hydrolysis of the precursor generates a liquid sol, which ultimately turns to a solid, porous gel. The gels formed this way are porous and contain a substantial amount of solvent phase. Slow drying of the gels under ambient conditions (or at elevated temperature) leads to evaporation of the solvent phase. The dried gels – termed xerogels – are substantially less porous than wet gels. Since the method begins from a molecular species, it offers a degree of control over structure and properties that can be tailored at the molecular level. The method can be used to prepare a wide range of oxides. However, the most well-known and extensively studied examples are those based on silica chemistry where there is a good understanding of how the chemical parameters can be used to control the properties of the final product.

In general, the applications of the sol-gel method can be loosely divided into three interdependent aspects of synthesis, processing, and properties, which are a direct function of the structural composition, morphology, and microstructure. The sol-gel method for synthesis of materials is particularly appealing because with a chemical modification of the precursor at molecular level, the functional properties of the final material can be changed. Therefore, by selectively integrating specific organic functional

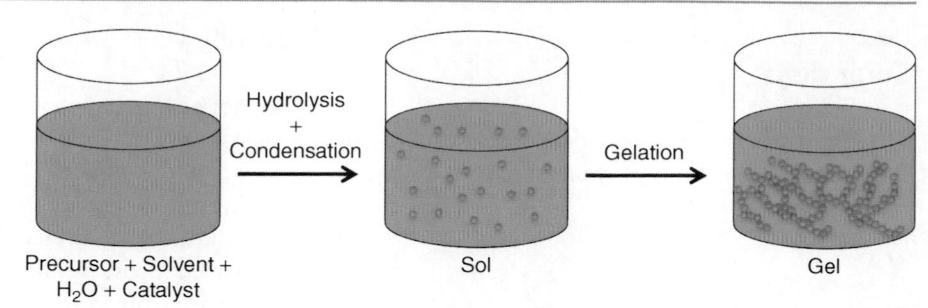

Sol-Gel Method, Fig. 1 Schematic depiction of the sol-gel method

groups into the precursor at the molecular level, it is possible to introduce desired properties into the product sol-gel inorganic-organic hybrid material.

The sol-gel chemistry is feasible with any metal precursor that is hydrolytically unstable. Typical precursors used are alkoxides, halides, or carboxylates. With silica chemistry, alkoxides [$Si(OR_4)$] are predominantly used. In addition, it is possible to use organically modified alkoxides [$(OR)_3SiR'$] to incorporate organic functionalities in glasses. The R′ group can be any organic group, including polymers or oligomers. When a precursor containing an organic group is hydrolyzed, it leads to formation of an ORganically MOdified SILicate (ORMOSIL) glass. The organic R′ groups can be simple hydrocarbons or can include other functional groups. In addition, it is possible to make glasses with precursors of the type [$(OR)_3Si\text{-}R\text{-}Si[(OR)_3]$], where the organic functionality is present as a spacer group. The precursors can be used alone, or alternatively, a combination of precursors can be used to make hybrid or composite materials. Finally, prior to gelation, other species such as organic-, inorganic-, and biomolecules and/or polymers can be added to the liquid sol to make composite materials. When large biomolecules are added to the glass, they usually become physically entrapped in the glass but nonetheless retain their structure and properties and the resultant materials exhibit biological function and activities [3].

The sol-gel method is primarily a method at the interface of synthesis and processing. Synthesis refers to the making of new materials by means of specialized chemical precursors through arrangement of precise functionalities at the molecular level. On the other hand, processing refers to the specific organization of matter at the nano-, micro-, and macroscopic level. Processing of gels is greatly facilitated by the stability of the liquid sol, which enables construction of materials with good control over their nanostructure, shape, and overall geometry. The final gels can be fabricated in a variety of shapes, geometries, and configurations, including monoliths, coatings, thin films, fibers, and powders. The liquid sol can be used to make monolithic geometries whose final shape is determined by the container used. From the perspective of nanotechnology, the method provides a rather facile pathway to construction of different nanomaterials. Formation of oxide-based nanoparticles is a key area in use of the sol-gel method due to the unique opportunities provided by the solution-based method that proceeds via the sol state characterized by dispersed nanoscale particulate matter. The liquid sol can be used for dip-coating, spin-coating, and spray-coating to prepare films and coatings of varied thicknesses on diverse surfaces including glass, metals, composites, or plastics. The versatility of the method has resulted in a burgeoning utility of the sol-gel method in diverse fields.

Applications are the key drivers in the utilization of the sol-gel method in technology. Because of its versatility, the method has found widespread applications in all walks of science and engineering including, optics, photonics, microelectronics, and biotechnology to name a few. The intrinsic properties of sol-gel materials such as optical transparency, porosity, surface area along with chemical and physical stability confer a unique vantage point for their applications in different areas. Different functions and properties can be imparted to the parent glass by incorporation of additional organic, inorganic, or biological entities. Particularly important in this direction are the organosilica sol-gels and hybrid materials, which integrate the properties of both organic and inorganic materials and constitute a novel platform for development of new materials and devices from a molecular perspective. Similarly, the access to formation of bio-hybrid via integration of biological species makes the method particularly suited for the fabrication of biological-inorganic hybrids and composites. For the design of advanced materials, the advantage of using sol-gel derived glasses is that the parent silica material is structurally inert, functionally inactive, and operationally nonresponsive. Therefore, by selectively integrating specific (bio)chemical entities into the glass (either through precursor modification or through encapsulation), it is possible to introduce desired structural, functional, and operational properties in a modular fashion. The introduction of properties can be singular, sequential, or even multiple. The strategy offers a powerful approach for designing a diverse range of sol-gel-derived materials whose properties can be tailored with a degree of control over the compositional, morphological, and functional characteristics of the product materials. This flexibility enables exploring new opportunities, and materials made using the sol-gel method have found a variety of applications in optics, photonics, magnetics, biocatalysis, detection, sensing, controlled release, and separation in addition to providing access to a range of structures and morphologies in the form of particles, fibers, coatings, and other nanoscale, nanostructured, and nanoporous materials.

The sol-gel method has brought about a fundamental change in synthesis and processing of inorganic materials. The fact that final materials in different geometries and morphologies can be prepared from a solution circumvents the need to go via conventional solid-state high temperature processing pathways. The solution-based method provides considerable flexibility in making pure, homogenous materials as well as composites and hybrids. The method can be used to make materials by using a given precursor, or alternatively, by mixing a plurality of different precursors, it is possible to form multicomponent mixed-oxides. Similarly, the method can be used to make hybrids and composites via integration of polymers or other organic/biological components to obtain materials with varied properties and applications (Fig. 2). The low temperature at which the materials can be prepared makes the method economical and environmentally friendly for practical applications.

Basic Methodology

The sol-gel method of making glasses begins with the reaction of hydrolyzable precursors with water. The most commonly used precursors are alkoxides. The process is illustrated herein with the alkoxysilanes as an example; however, the reaction sequences are analogous for other metals. These precursors are dissolved in a compatible solvent such as an alcohol followed by addition of water to initiate hydrolysis. Often a catalyst is used to accelerate the reaction. The nonpolar nature of alkoxysilanes makes them immiscible with water, and therefore, the reactions proceed at very slow rates even in the presence of a catalyst. The mixing can be enhanced by use of ultrasound, or alternatively, the reaction mixture is refluxed at elevated temperature to facilitate the reaction. Under these conditions, first the alkoxides react with water and form hydroxylated species which then condense to form colloidal nanoparticles containing oxo-bridges. As the reactions proceed, the sol particles combine to form an extended network and the reaction mixture becomes viscous, ultimately leading to the formation of a solid gel. The as-prepared gels are biphasic porous materials with interstitial solvent phase. Drying of the gels under ambient conditions results in evaporation of a majority of the solvent phase (water and alcohol) and the gels shrink by up to 80% in volume. These dried glasses – termed xerogels – are mechanically and dimensionally stable materials that contain some solvent phase, which can be further dried at elevated temperatures to form dense glasses. The overall method of making oxides from alkoxides is represented as

$$\text{Precursor} \xrightarrow[\text{Condensation}]{\text{Hydrolysis}} \text{Sol} \xrightarrow{\text{Gelation}} \text{Gel} \xrightarrow{\text{Drying}} \text{Xerogel} \xrightarrow{\text{Densification}} \text{Glass}$$

The sequence is characterized by a series of reactions along with physical and chemical changes that take place along the precursor to sol to gel to xerogel to glass transformation steps. The properties of materials depend upon the conditions employed and by controlling the parameters under which these steps take place one can modulate, regulate, and fine-tune the properties of the materials at each stage.

The Precursor: The nature, composition, and structure of precursor molecules is central to the properties of final product. Controlling the number of hydrolyzable versus nonhydrolyzable groups on the precursor provides an easy pathway to influence the chemical makeup of the materials obtained by the

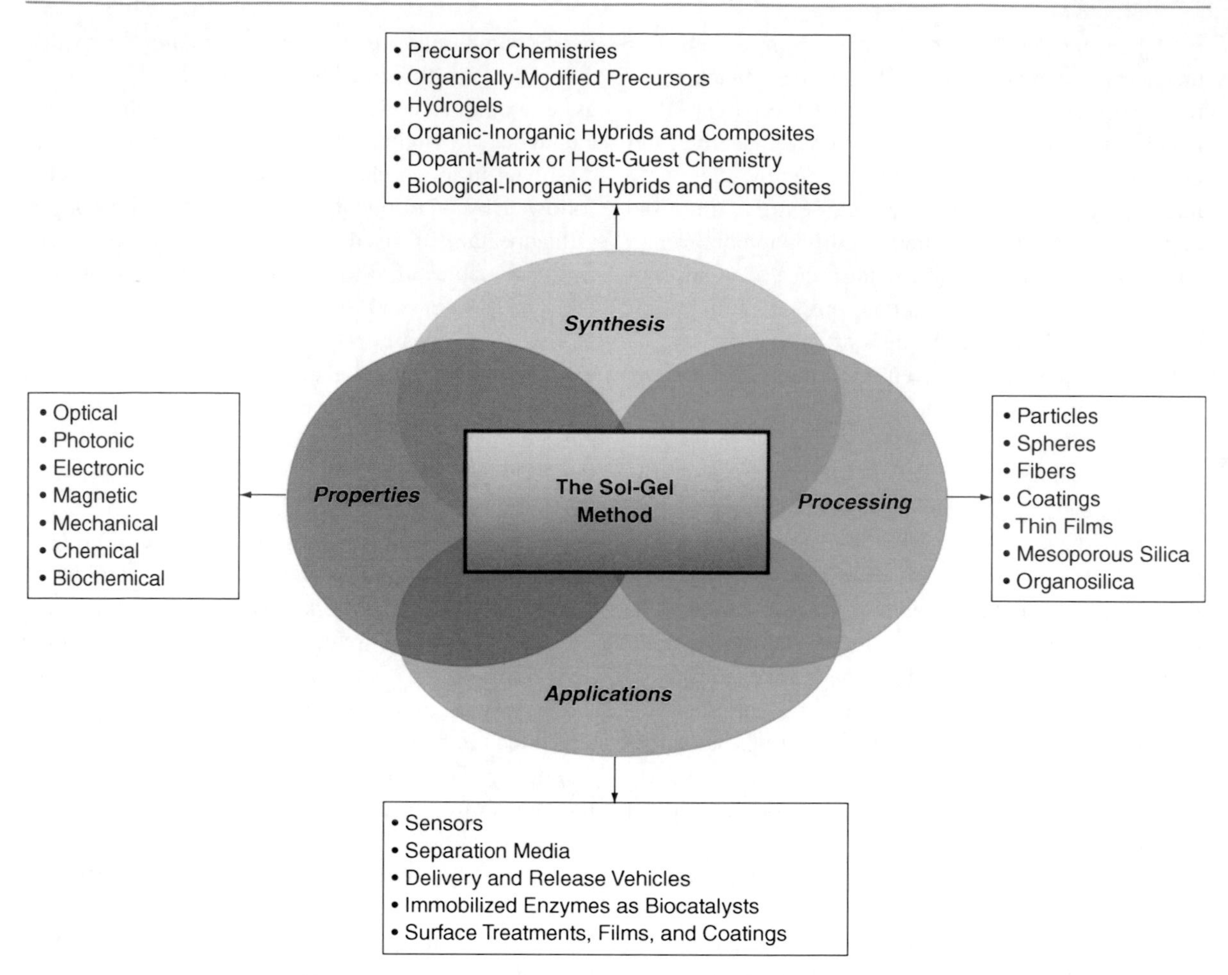

Sol-Gel Method, Fig. 2 Representative summary of significant aspects of the sol-gel method

sol-gel method. Therefore, in order to obtain pure oxide-based materials, one can use precursors of the type $Si(OR)_4$ wherein all the alkoxides are replaced with oxygen atoms in the product due to hydrolysis and condensation. Use of nonhydrolyzable ligands on the precursors ensures that these groups will remain in the final material thereby altering the structural, morphological, and functional characteristics of the product. Precursors of the type $(OR)_3SiR'$ where R' is organic group provide an easy access to introduce organic functional groups that can impart specific properties. Additionally, use of multiple precursors with different functional groups provides further access to tailoring the final properties.

Alkoxides are the preferred precursors of choice due to the fact that the by-product of their hydrolysis is the corresponding alcohol, which can easily be evaporated to obtain relatively pure materials [4]. The rates of hydrolysis and condensation depend upon the nature of the alkoxide groups, which can lead to dramatic variations in the rates of the reactions as well as product morphologies. The rates of hydrolysis of alkoxides depend upon the polarity of M-OR bond as well as the steric environment around the central metal atom which can influence the approach of water molecule for hydrolysis of the bond. In general, more polar alkoxides are more susceptible to hydrolysis in a polar solvent environment. The inductive and steric effects of the alkoxide ligands can also alter the hydrolysis rates by altering the polarity of the M-OR bonds, and by modulating the access of water molecules to effect the hydrolysis step. Usually, the hydrolysis step is much slower with longer chain and/or branched alkoxides which inhibit the access of water molecules to the metal site. A catalyst is typically used, especially with silicon alkoxides, to accelerate the

Sol-Gel Method, Fig. 3 Hydrolysis mechanisms in acid or base catalyzed reactions

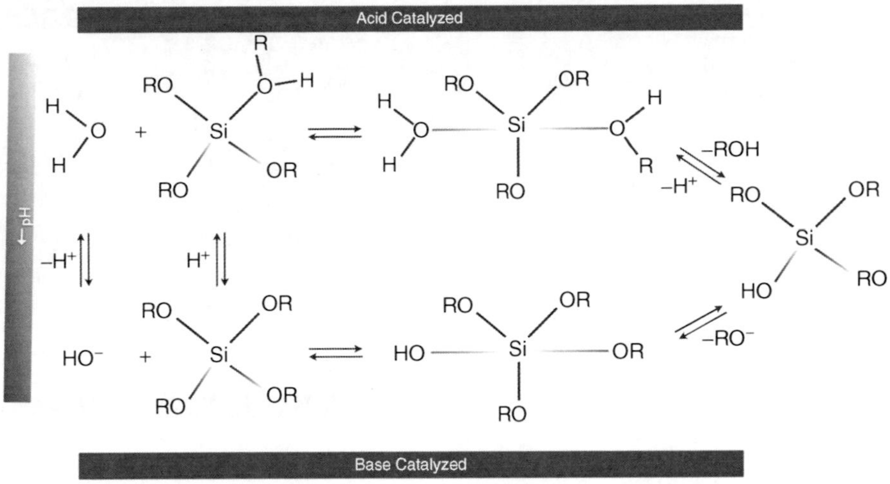

hydrolysis step. On the other hand, alkoxides of transition metals and other highly electropositive elements usually exhibit much faster hydrolysis due to the intrinsic polarity of the M-OR bond along with the ability of these elements to accommodate the incoming water molecules via coordination to the metal site.

Hydrolysis: The first step of the sol-gel reaction is the replacement of alkoxide groups in the precursor molecule with water. The alkoxide groups leave as the corresponding alcohols along with formation of hydroxylated silanol species. Since alcohol is a by-product of the hydrolysis reaction, carrying out the reaction in the same alcohol as the corresponding alkoxide can alter the reaction equilibrium. If a different alcohol is used as a solvent then, due to its predominance, there is usually a replacement of the alkoxide ligand via transesterification. The ease with which the reaction occurs depends upon the miscibility of the precursors with water, its hydration characteristics, and its ability to accommodate water molecules in the primary coordination sphere. The products of the hydrolysis reaction are the hydroxylated precursor and the corresponding alcohol. The general reactions is given as

$$(OR)_4Si + H_2O \rightleftarrows (OR)_3Si - OH + ROH$$

The hydrolysis reaction proceeds via associative (or S_N2) pathways such that the transition state is characterized by an increase in coordination number. The reaction is usually acid or base catalyzed to make it occur at an appreciable rate. The general reaction under both conditions proceeds via associative pathways through the aid of the predominant catalytic species, namely, protons in acidic conditions and the hydroxide ions under basic conditions.

Under acidic conditions, the reaction occurs in three steps (Fig. 3). First, the alkoxide group gets protonated, which makes the metal-alkoxide bond weaker and the silicon atom electron-deficient. Second, the water molecule binds to the silicon atom to form a positively charged pentacoordinate transition state. Finally, the release of alcohol molecule completes the reaction. Under basic conditions, the reaction also occurs in three steps (Fig. 3). First, the water molecule loses its proton to form the hydroxide ion. Second, the hydroxide ion binds to the silicon atom to form a negatively charged transition state. Finally, the alkoxide ion dissociates to form the hydroxylated species.

The replacement of alkoxides to form hydroxylated species proceeds at different rates under acidic and basic conditions. Under acidic conditions, subsequent hydrolysis steps are retarded after the initial substitution of alkoxide, while in basic medium, these follow-on steps are accelerated. In other words, the hydrolysis of $[(OR)_{4-n}Si(OH)_n]$ (where n varies from 0 to 4) species in acidic medium is faster for smaller values of n while in a basic medium the hydrolysis is faster for larger values of n. As such, the formation of multihydroxylated species predominates in the basic medium. The rate and extent of hydrolysis can be altered by use of suitable solvents that can change the polarity of the medium, and consequently, the stability of polar versus nonpolar species to shift the equilibrium. Solvents that interact with water by hydrogen bonding interactions also alter the activity of water in

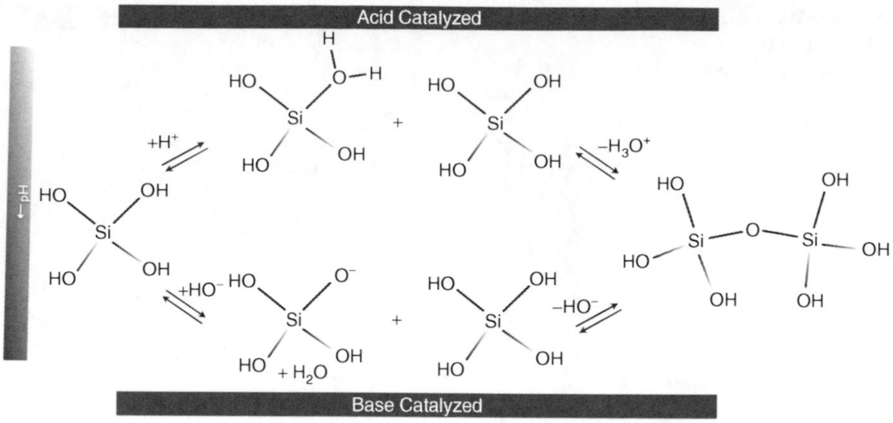

Sol-Gel Method, Fig. 4 Condensation mechanisms in acid or base catalyzed reactions

the solution and affect the rate and extent of hydrolysis. Solvents that can homogenize the reactants to make them more miscible augment the rates and shift the equilibrium toward products. Finally, use of ultrasound can also facilitate mixing of the precursor with water and can accelerate the reaction.

Condensation: Once the reactive hydroxylated species are formed, they begin to undergo condensation reaction to form Si-O-Si linkages to generate oligomers and polymers. The reaction can be represented as

$$(OR)_3Si - OH + OH - Si(OR)_3 \rightleftarrows (OR)_3Si - O - Si(OR)_3 + H_2O$$

As with hydrolysis, the condensation reactions are dependent upon pH of the medium and can be acid- or base-catalyzed (Fig. 4). Under conditions of acid catalysis, the reaction proceeds in three steps. First, there is a rapid protonation of the silanol group to form a positively charged species. Second, the protonated positively charged species is attacked by a neutral hydroxylated species. Finally, an elimination of the hydronium ion completes the reaction. Base catalysis follows a slightly different sequence. First, there is rapid deprotonation of the silanol group to form an anionic silicate species. Second, the negatively charged species coordinates with the neutral hydroxylated species via formation of a pentacoordinate transition state. Finally, the expulsion of hydroxide ions completes the reaction sequence.

Analogous to hydrolysis reactions, the rates of the condensation reaction are dependent upon pH of the medium. Condensation reaction for $[(OR)_{4-n}Si(OH)_n]$ (where n varies from 1 to 4) species in acidic medium is faster for smaller values of n while in a basic medium it is faster for larger values of n. As such, base-catalyzed systems, with a predominance of multihydroxylated species, are characterized by faster gelation times as compared to acid-catalyzed reactions. Silanol (Si-OH) groups become more acidic as the silicon atom gets increasingly surrounded by the electron withdrawing oxo groups. This high acidity of silanols with the increasing extent of condensation results in maximum condensation rates in the intermediate pH range of 3–5. In very high pH region ($pH \geq 10$), condensation reactions occur but gelation is thwarted by repulsion due to surface charges leading to formation of discrete particles. As water is a by-product of condensation reactions, the amount of water in the reaction mixture influences the equilibrium. Reaction media high in water content favor hydrolysis but hinder condensation. Similarly, more polar and hydrogen bonding solvents can stabilize the hydroxylated species to retard the condensation process. Finally, the steric effects of the remaining alkoxo groups can also influence the rate with bulkier groups hindering the condensation reactions.

Gelation: As the hydrolysis and condensation reactions continue, the reaction medium becomes more viscous due to increased degree of polymerization. The viscous sol continues to evolve until the viscosity reaches a critical point where the liquid can no longer maintain its fluidity. Gelation is characterized as the point along the hydrolysis-condensation reaction coordinate when the liquid sol turns to a solid gel. Gelation is typically defined by gelation time (t_g) as the time taken from the start of the reaction till the formation of a solid gel. Gelation times are dependent upon a variety of factors, as discussed above, that control the rate and extent hydrolysis and condensation reactions. At gelation time, the growing polymer structure essentially gets "frozen" into an immobile state. The initial gels formed have high viscosity but they lack the mechanical stability due to lack of an extensive interconnected network. Even though the physical state of the reaction mixture changes from liquid to solid, the hydrolysis and condensation reactions still continue to evolve leading to formation of additional Si-O-Si linkages. The increasing degree of condensation results in enhanced mechanical strength over time even after gelation has occurred.

Aging: The as-formed gel is a porous structure with pendent silanols and even some remaining unhydrolyzed alkoxides on the polymeric network. Aging of the gels under closed conditions, usually extending over a period of hours to days, is employed to enable completion of these reactions. During aging, continuation of hydrolysis and condensation reactions leads to an increase in cross-linking and overall stiffening of the gel network. During aging, the gels shrink and expel the solvent phase due to a decrease in internal pore volume caused by extensive condensation of silanols to siloxane linkages. Aging is a necessary step in processing of bulk gels to ensure mechanical stability of these materials during subsequent stages.

Drying: Drying of the gels either under ambient conditions or elevated temperature is used to expel solvent from the gels. The elimination of solvent from the pores of the gels compacts the network and facilitates further condensation reactions between silanols on the surface of pores. The ambiently dried gels can lose up to 70–80% of the solvent and shrink considerably due to compaction of the pore network. In order to completely dry the gels, use of elevated temperatures is necessary. Drying enhances the mechanical stability of the gels, and the dried gels are considerably stiffer and stronger materials as compared to aged gels. The dried gels have a greater density and are dimensionally invariant under ambient conditions.

Sintering/Densification: Heat treatment of the gels further eliminates the porous structure to obtain ceramics or dense glasses. During heat treatment, adsorbed and hydrogen-bonded solvent is removed and the collapsing pore structure causes further condensation of silanols to siloxane linkages. Heat treatment makes the gels nonporous and harder with densities and structure approaching close to fused silica.

Nonhydrolytic sol-gel method: The sol-gel method depends upon water to initiate the hydrolysis of precursors. This presents a challenge for system that are either unstable in water or too reactive. In order to circumvent these issues, the nonhydrolytic sol-gel method [5] utilizes nonaqueous solvent medium and does not require the hydrolysis step. While there exist many different variants, the general nonhydrolytic sol-gel method can be represented as

$$M(OR)_n + X_nM \rightleftarrows [(OR)_{n-1}M - O - M(OR)_{n-1}] + RX$$

where X = halide, -OR′, or $-O_2CR'$. The reaction bypasses the hydrolysis process altogether and leads to direct formation of oxo-bridged linkages. The reaction is essentially a condensation pathway that is usually carried out at elevated temperatures. The nonhydrolytic process is advantageous in a system wherein a greater degree of control over the relative rates of hydrolysis and condensation to fine tune product microstructure, porosity, and morphology may not be possible with the hydrolytic pathway.

Formation of nanoparticles and powders: The molecule-nanomaterial interface along the precursor to sol reaction coordinate provides a convenient access to a variety of nanoscale systems. Powders such as precipitated or fumed silica have been long used in commercial applications. A key challenge has been to obtain uniform monodisperse nanoscale particles with a control over their nuclearity, shape, and dimensions. The issue of uniformity and homogeneity, to a certain extent, is addressed by utilizing specific reaction conditions to prevent interparticle interactions. Two methods commonly used to stabilize isolated discrete nanoparticles are based on manipulating electrostatic and steric factors. As discussed above, highly basic conditions favor rapid hydrolysis and condensation while producing negative surface charges which favor the formation of particulate structures. The Stober method of forming particles specifically takes advantage of this approach. Uniform size spherical particles can be obtained under basic conditions by using dilute solutions to prevent interparticle aggregation and to allow dissolution-growth phenomena to occur without precipitation. Usually the concentration

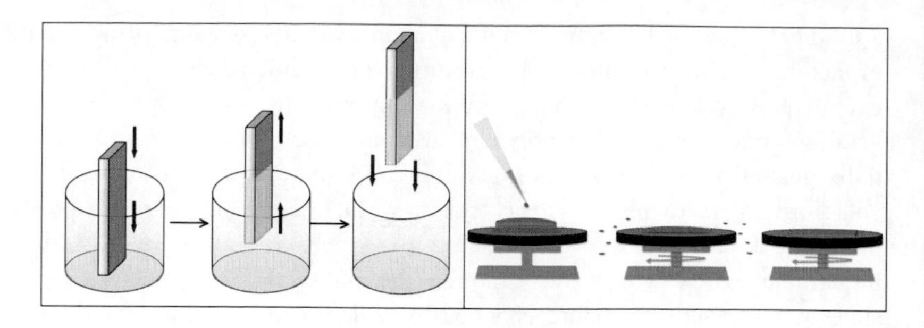

Sol-Gel Method, Fig. 5 Schematic depiction of dip-coating (*left*) and spin-coating (*right*) processes

of the precursor in the reaction mixture is optimized to prevent interparticle polycondensation and prevent formation of larger particles due to particle growth and coalescence. Sometimes, a synthesis strategy of combined acid and base catalysis is used for the initial formation of a stable sol under acidic conditions followed by addition of base to facilitate particle growth. Optimized reaction conditions that favor dynamic surface reorganization and reconstruction to provide uniform spherical particles are commonly utilized. Use of microemulsions has been another method that is employed to form spherical nanoparticles wherein the emulsion provides a sterically enclosing environment for isolated particles to form without aggregation.

Formation of thin films and coatings: The sol-gel method has found wide applicability in the construction and formation of two-dimensional nanostructures such as thin films and coatings. The films made via the sol-gel method provide a better dimensional control with the ability to coat complex shapes, and can even be used to form multilayer coatings. In order to form a coating, a stable sol with controlled viscosity is desired. Usually stable sols can be obtained under acidic conditions and by adjusting the amount of the precursors, solvent, and pH it is possible to control the viscosity of the sol. The films can be formed by dip-coating, spin-coating, or spray-coating methods. The dip-coating method of thin film deposition is perhaps more important technologically since a uniform coating can be deposited onto substrates of large dimensions and complex geometries. In the dip-coating process, a suitable substrate immersed in a low viscosity sol is withdrawn slowly at a constant speed (Fig. 5). The adhesion of the sol layer on the substrate aided by gravitational draining leads to formation of a thin film. In order to deposit uniform homogeneous films, a steady vibration-free withdrawal of the substrate from the sol is necessary. Film thickness is controlled by adjusting the viscosity of the sol and the rate of pulling the substrate. Another method of forming thin films is via spin-coating wherein an excess of sol is placed onto a substrate that is rotated at high speeds to disperse the liquid uniformly (Fig. 5). The film thickness in this case depends upon viscosity of the sol and the speed of rotation along with the rate of evaporation of the solvent. The conversion of deposited sol to a porous gel is caused mainly by solvent evaporation. Unlike the bulk gel system where gelation, aging, and drying occur sequentially over a period of several days, all of these processes typically occur within 30 s to 1 min in the thin-film such that the drying stage overlaps the gelation and aging stages. Loss of solvent molecules accelerates hydrolysis and condensation steps such that the two steps occur concurrently leading to the eventual formation of a porous xerogel structure. The important microstructural characteristics of coatings and films such as porosity, surface area, and pore size are determined by pH, concentration, and nature of precursors, viscosity of the initial sol, the solvent and its rate of evaporation, the rate of sol dispersion on the substrate, the ambient temperature, the rate of drying, and the relative rates of evaporation and condensation reactions during drying.

Organically modified materials: Perhaps the most useful aspect of the sol-gel method is the ability to make materials with covalently integrated organic functionalities via hydrolysis of precursors containing hydrolytically stable organic groups. This enables a wide range of functional groups to be incorporated within the oxide-based network to from a diverse array of new materials with unique, novel, and interesting properties. More than one type of precursors with different functional groups can also be used to include a multiplicity of functional groups within a given material. An important consideration in use of multiple precursors is their differential reactivities toward hydrolysis and condensation which can lead to phase separation and/or spatially isolated nanodomains. Strategies to overcome these are based on prehydrolysis of the more reactive precursor followed by addition of the other precursors as well as use of different alkoxides groups on the precursors to modulate their reactivities. The organically modified materials contain less than four siloxane linkages per silicon atom and are more elastic as compared to pure oxides. The ability to tune functionality in the organically modified materials provides appealing prospects for applications based on the nature of organic group incorporated.

Hybrid Materials: Different types of hybrids can be made via the sol-gel method. These include organic-inorganic hybrids and nanocomposites. The organic-inorganic hybrids combine the properties of organic materials with the mechanical strength of the oxides. The main issue in formation of hybrids relates to their miscibility which can be enhanced by means of suitable solvents or by use of functional groups that can noncovalently interact with silica pore structure via hydrogen bonding or electrostatic interactions. The simplest way to make a hybrid material is by encapsulation of organic molecules in the pores of the gels. The organic molecules in the pores of gels can be used to alter the properties and structure of the inorganic network. Typical examples of these include glasses containing encapsulated catalysts, dyes, and other optically active organic molecules. Another approach is based on forming interpenetrating networks of organic polymer with the silica network. These nanocomposites combine the properties of both inorganic and organic polymers, and usually influence the structure, morphology, and properties of each other. These synergistic interactions make the hybrid materials structurally and functionally superior as compared to their component parts.

An important development in formation of hybrid materials by the sol-gel method centers on formation of bio-hybrids. The solution based sol-gel pathway can be used for direct integration of biological species such as proteins, enzymes, DNA, RNA, and even whole cells. Optimized pH regions are necessary during preparation to stabilize the biological component. Typical procedure of enzyme immobilization involves initial acid hydrolysis followed by raising the pH to neutral region. The biological macromolecules are usually added to a buffered sol and get encapsulated in the growing gel network as the sol turns to a gel. The water-filled porous structure of the gel along with electrostatically charged and hydrogen bonding pore walls provide a noncovalently interacting environment for the biological molecules. The encapsulated biological entity retains its structural and functional characteristics and is able to interact with exogenous molecules that can diffuse in and out of the porous structure. The last two decades have seen tremendous advances in the design and development of bio-silica hybrids made by the sol-gel method [3].

Key Research Findings

Structure and Synthesis: The sol-gel method has been used for several decades to make silica-based materials [6]. Oxides of different metals including mixed oxides have been prepared [7]. Initial research in this area focused on mainly pure and mixed oxides but in recent times the focus has shifted to organically modified, hybrid, and composite system. A classification of organic-inorganic materials synthesized using the sol-gel method has been developed to categorize them [8]. Nanotechnology has also spurred an active interest fabrication of sol-gel-derived particles [9]. These particles are usually synthesized in a microemulsion and can be solid, hollow, or contain core-shell structures.

Synthesis of organically modified glasses has been studied in much detail and a wide range of precursor chemistries have been identified. The organically modified precursors provide access to a diverse range of materials with tunable properties including chemical reactivity, optical properties, electronic properties, enhanced mechanical stability, and elasticity. A major area of research using these precursors has been in the construction of mesoporous materials by means of using suitable templates wherein the assembly is aided by electrostatic and hydrogen bonding interactions between the template and the growing silica network [10]. Gels made from organosilicate precursors made with spacered linkers containing both hydrophobic and hydrophilic groups exhibit hydrogel-like properties and have been shown to undergo structural and volume change with respect to different environmental stimuli. These materials are characterized by enlarged pores due the presence of spacer units along with elasticity which confers the necessary flexibility to undergo shape and volume changes [11]. These materials also provide a suitable matrix for immobilization of biological molecules.

An important aspect of protein and enzyme immobilization has been elucidation of local environment of the immobilized biomolecule and its interactions with the sol-gel matrix [12]. The immobilizing environment of the pore restricts translational diffusion, rotational mobility, and conformational flexibility. Proteins and enzymes have been shown to exhibit enhanced stability against denaturation when encapsulated in the sol-gel-derived glasses. The enzymes in the gel matrix retain their catalytic activities and the materials have been used as solid-state catalysts for various transformations. Studies have shown that when large proteins get encapsulated in the growing gel network, they act as self-specific templates for the formation of their own pore environment. Recent research also shows that cells can be used for directing assembly of sol-gel materials. Furthermore, proteolytic enzymes have been shown to catalyze the sol-gel reactions leading to formation of gels starting from alkoxide precursors.

Function and Properties: The transparency of the sol-gel-derived glasses has been widely utilized for the development of functions related to modulation of photon flux through the materials [13]. Incorporation of dyes and luminescent fluorophores has been used as a means to develop materials for lasers, luminescent materials, and solar concentrators. Incorporation of photochromic molecules in the pores of the gels has been employed as a means to impart photochromic properties to the resulting glasses. Electronic properties have been developed in the sol-gel glasses via incorporation of conducting polymers, metal particles, or formation of mixed oxides. The nanopores of the gels have been used to form nanoscale superparamagnetic particles of transition metal oxides. The use of organically modified precursors also provides access to tailoring of mechanical properties such as strength, elasticity, and hardness depending upon the functional groups used. In addition to development of physical properties, use of precursors with specific functional groups has also been used to tune their chemical properties.

Applications: The applications of the sol-gel-derived systems have been continuously increasing [14]. The transparency of silica-based systems makes them especially suitable for development of optical applications and sol-gel-derived materials have been used as optical sensors based on changes in color or luminescence upon reaction with an analyte [15]. The analytical applications of sol-gel materials as electrochemical sensors have also been developed [16]. Additionally, the presence of specific functional groups on organosilica materials confers selectivity in their interactions and these materials have been used as separation media in chromatography columns [17]. The use of porous particles for controlled release applications has opened up new opportunities in market and commercial products have been designed for acne treatment and sunscreen release based on particles made using the sol-gel method. A new area in this direction has been applications in fabric-care systems and environmentally sensitive organosilica hydrogels have been demonstrated as controlled release vehicles for water-triggered release of enzymes.

Coatings made from sol-gel pathways have been utilized for an array of applications including corrosion protection and surface passivation, tailoring water affinity of surfaces, enhancing optical gloss, thermal insulation, scratch resistance, and fingerprint resistance among a host of other applications [18]. The ability to modify hydrophobicity and surface microstructure of these coatings has enabled construction of surface features that mimic the lotus leaf effects. An offshoot of this research has also resulted in textile treatments to make them stain- and water-repellent.

In recent times, the use of bio-sol-gel hybrids has been an area of much activity. Initial applications related to immobilization of protein and enzymes inside the pores of silica sol-gels to make solid-state biocatalysts for various chemical transformations. The enzyme sol-gel hybrids have also been used as optical and electrochemical sensors, and as media for bio-affinity chromatography. The use of sol-gel-derived materials has also been established for immobilization of whole cells and microorganisms [19]. Organosilica gels as matrices for bio-immobilization have provided access to a wide range of biocatalysts whose activities can be exploited in practical applications. The applications domain of materials incorporating biological species continues to grow and the biological-inorganic composite area remains a topic of continued development.

Future Directions for Research

The sol-gel method of making materials provides a versatile pathway to not only making new materials, but also making known materials more useful by obtaining them in different morphologies and length scales. The last decade has witnessed significant advances in construction of nanoscale particles, assemblies, and organized structures made using the sol-gel method. Fabrication of different oxides, nanohybrids, and nanocomposites characterized by distinct nanoscale structures, surfaces, and interfaces that can be tailored to obtain the desired functional responses would continue to remain an area of vital investigation in the near future. Use of nanoscale building blocks obtained by the sol-gel method and their organization into higher-order structures along with control, manipulation, and regulation of the emergent properties at different length scales to generate specifically tailored responses would provide an appealing paradigm for future studies. An additional key issue to address in the future development of nanoscale systems made via the sol-gel method would be resolving, manipulating, and controlling the functional dichotomy of properties originating from the nanoscale as well as collective properties due to the ensemble nature of the materials.

Being able to tailor properties of materials made from molecular precursors with specific chemistries offers enticing prospects for controlling and manipulating their organization as they undergo their structural evolution from molecules to clusters to particles and other higher order structures. While there have been some advances in supramolecular assemblies of sol-gel-derived materials, the complexity and sophistication to achieve self-assembling structures with distinct interactions, interfaces, and organization over varied length scales remain as yet elusive. An area of ambitious exploration, given the formidable challenges, would be the development of new precursors and chemistries to fine-tune covalent and noncovalent interactions to generate supramolecular systems with programmed assembly and organization. Tuning this dynamic interplay of structure and organization would require new methods and techniques of sol-gel chemistry that would provide access to far from equilibrium structures. In the near term, the structural-organizational paradigms from nature would continue to provide the necessary inspiration toward achieving these objectives.

The ability to fabricate bio-inorganic hybrids and composites has opened another exciting avenue in the design of novel materials with useful properties and applications. The integration of biological entities into the nanopores of materials, to a certain extent, restores the spatial and supramolecular organization that exists in the native cellular environment which is lacking in freely mobile biomolecules dispersed in an isotropic solution phase. The next extension of this method would generate organized assemblies that would take advantage of the specialized interactions at the biological-inorganic interfaces. Consolidated nanostructures integrating biological materials would become increasingly significant in creating a new generation of bio-hybrids that would combine the structural integrity of the sol-gel-derived materials with the functional and operational diversity of biomaterials.

During the last decade there have been significant advances in mimicking nature with sol-gel-derived materials including biocatalysis, immunodetection, surfaces that mimic lotus leaf, and biomineralized supramolecular architectures. The opportunities provided by the facile access to nanomaterials, internal porous architecture, surfaces, and interfaces would continue to evolve as chemists, materials scientists, biochemists, and physicists attempt to develop correlations between structure, morphology, form, and function. The next evolution of these materials lies in the design of nanomaterials with adaptive capabilities that can adapt to their environment and exhibit active functional responses to generate dynamically active systems capable of self-diagnosis and self-correction. The interplay of structure, form, function, and operation that exists at different length scale would provide the guidelines for the development of next generation of materials capable of intelligent responses. While recent developments have seen the emergence of materials with smart functions, many of the systems, for the most part, remain monofunctional and passive. In the long range, design and development of new materials with hierarchical functions that span a range of functional length scales and tunable active responses would generate the next wave in this line of research.

The last decade has also seen utilization of sol-gel methods to fabricate devices for detection, diagnosis, separation, sensing, controlled release, and delivery to name a few. The real potential of these materials as nanoscale devices remains to be explored. The ability to tailor size, porosity, and the surfaces of sol-gel nanomaterials would find new avenues of research activity in the development of nanodevice technology based on a combination of inorganic, organic, and biological components. Multifunctional systems that can detect, sense, separate, sort, release, and deliver molecules to targeted sites and in a predetermined fashion would constitute the long range evolution of nanodevices made using sol-gel-derived systems. In the near term, effective utilization of strategies to adjust and modulate physical and chemical interactions of exogenous molecules with sol-gel systems would provide the necessary means of control and regulation to design nanoscale sensors, filters, separators, sorters, and delivery agents. Development of nanodevice systems for biomedical applications would also constitute an important area of research and development.

Finally, another area that would become increasingly important is the development of new "green" technologies utilizing the sol-gel method. The opportunities offered by the sol-gel method would provide the ideal framework for development of novel (bio) catalytic systems, molecular filters, sorters, or separators to reduce, remove, and/or remediate toxic and harmful substances. The sol-gel method is a greener alternative to conventional methods, however, use of volatile solvents and formation of by-products remain as issues to be addressed. Development of new processing techniques that minimize waste and improve atom economy by eliminating by-products would be essential to the development of eco-friendly manufacturing processes based on sol-gel methods. In the long range, sustained development and refinement of the sol-gel method would be critical to displace the current manufacturing processes in large-scale production technologies.

Cross-References

▶ Bioinspired Synthesis of Nanomaterials
▶ Biosensors
▶ Self-Assembly of Nanostructures
▶ Smart Hydrogels

References

1. Wright, J.D., Sommerdijk, A.J.M.: Sol-Gel Materials—Chemistry and Applications. CRC Press, Boca Raton (2001)
2. Sanchez, C., Belleville, P., Popall, M., Nicole, L.: Applications of advanced hybrid organic–inorganic nanomaterials: from laboratory to market. Chem. Soc. Rev. **40**, 696–753 (2011)
3. Avnir, A., Coradin, T., Lev, O., Livage, J.: Recent bio-applications of sol–gel materials. J. Mater. Chem. **16**, 1013–1030 (2006)
4. Hubert-Pfalzgraf, L.G.: Alkoxides as molecular precursors for oxide-based inorganic materials – opportunities for new materials. New J. Chem. **11**, 663–675 (1987)
5. Mutin, P.H., Vioux, A.: Nonhydrolytic processing of oxide-based materials – simple routes to control homogeneity, morphology, and nanostructure. Chem. Mater. **21**, 582–596 (2009)
6. Hench, L.L., West, J.K.: The sol-gel process. Chem. Rev. **90**, 33–72 (1990)
7. Livage, J., Henry, M., Sanchez, C.: Sol-gel chemistry of transition metal oxides. Prog. Solid State Chem. **18**, 259–341 (1988)
8. Sanchez, C., Ribot, F.: Design of hybrid organic-inorganic materials synthesized via sol-gel chemistry. New. J. Chem. **18**, 1007–1047 (1994)
9. Ciriminna, R., Sciortino, M., Alonzo, G., de Schrijver, A., Pagliaro, M.: From molecules to systems – sol-gel microencapsulation in silica based materials. Chem. Rev. **111**, 765–789 (2011)
10. RIvera-Munoz, E.M., Huirache-Acuna, R.: Sol-gel-dervied SBA-16 mesoporous material. Int. J. Mol. Sci. **11**, 3069–3086 (2010)
11. Rao, M.S., Grey, J., Dave, B.C.: Smart glasses – molecular programming of dynamic responses in organosilica sol-gels. J. Sol-Gel. Sci. Technol. **26**, 553–560 (2003)
12. Menaa, B., Menaa, F., Aiolfi-Guimaraes, C., Sharts, O.: Silica-based nanoporous glasses – from bioencapsulation to protein folding studies. Int. J. Nanotechnol. **7**, 1–45 (2010)
13. Penard, A.-P., Thierry, G., Boilot, J.-P.: Functionalized sol-gel coatings for optical applications. Acc. Chem. Res. **40**, 895–902 (2007)
14. Aegerter, M.A., Menning, M.: Sol-gel Technologies for Glass Producers and Users. Kluwer, Boston (2010)
15. Tran-Thi, T.-H., Dagnelie, R., Crunaire, S., Nicole, L.: Optical chemical sensors based on hybrid organic–inorganic sol–gel. Chem. Soc. Rev. **40**, 62–639 (2011)
16. Walcarius, A., Collinson, M.M.: Analytical chemistry with silica sol-gels – traditional routes to new materials for chemical analysis. Annu. Rev. Anal. Chem. **2**, 121–143 (2009)
17. Kloskowski, A., Pilarczyk, M., Chrzanowski, W., Namiesnik, J.: Sol-gel technique – a versatile tool for adsorbent preparation. Crit. Rev. Anal. Chem. **40**, 172–186 (2010)
18. Aegerter, M.A., Almeida, R., Soutar, A., Tadanaga, K., Yang, H., Watanabe, T.: Coatings made by sol–gel and chemical nanotechnology. J. Sol-Gel. Sci. Technol. **47**, 203–236 (2008)
19. Livage, J., Coradin, T.: Living cells in oxide glasses. Med. Mineral. Geochem. **64**, 315–332 (2006)

Solid Lipid Nanocarriers

▶ Solid Lipid Nanoparticles - SLN

Solid Lipid Nanoparticles - SLN

Claudia Musicanti and Paolo Gasco
Nanovector srl, Torino, Italy

Synonyms

Lipospheres; Solid lipid nanocarriers; Solid lipid nanospheres; Solid lipid-based nanoparticles

Definition

Solid lipid nanoparticles (SLN) are drug carriers in submicron size range (50–500 nm) made of biocompatible and biodegradable lipids solid at room and body temperature.

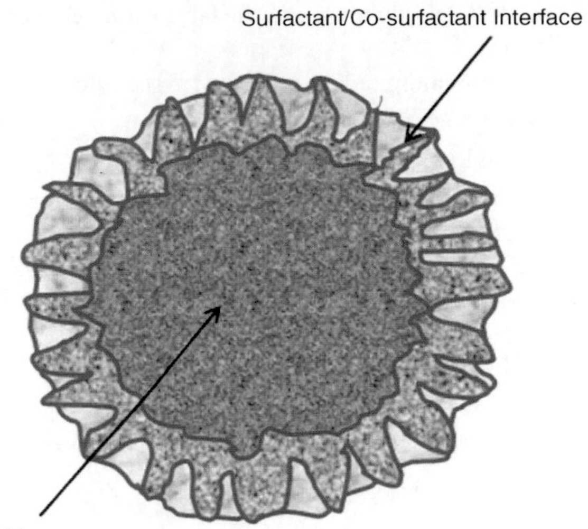

Solid Lipid Nanoparticles - SLN, Fig. 1 Model representation of general structure of SLN

Introduction

Solid lipid nanoparticles (SLN) were developed at the beginning of the 1990s as alternative colloidal carriers to emulsions, liposomes, and polymeric nanoparticles. SLN have attracted increasing attention as delivery system for hydrophobic drugs: prepared with lipids, they can be administered by different routes of administration such as oral, parenteral, dermal, transdermal, ocular and pulmonary.

A typical model of the structure of SLN consists of a solid lipid core surrounded by an emulsifier interface which stabilizes the particle (Fig. 1).

SLN can offer several advantages for drug delivery

1. Possibility of controlled release of drug from lipid matrix
2. Possibility of drug targeting (active and passive)
3. Protection of incorporated drug from degradation
4. Increasing drug bioavailability
5. Feasible incorporation of hydrophobic and also hydrophilic drugs (lipid drug conjugates or other techniques)
6. Very low when absent toxicity
7. Affinity for biological barriers
8. Modification of pharmacokinetic parameters and drug distribution in organs
9. Possibility of large-scale production

On the other hand SLN show certain possible limitations:

1. Poor drug loading capacity
2. Possible drug release during storage
3. Possible stability problems during long-term storage (aggregation, component degradation, tendency to form gel)

The principal components used for the preparation of SLN are:

1. Lipids, such as triglycerides (trilaurin, trimyristin, tripalmitin, tristearin), mono/di/triglycerides mixtures (glyceryl stearate, glyceryl palmitostearate, glyceryl behenate), fatty acids (stearic acid, palmitic acid, behenic acid), waxes (cetyl palmitate), cholesterol esters
2. Emulsifiers (surfactants), such as phospholipids (e.g., lecithins), polysorbates (Tween) and sorbitan esters (Span), polymers (e.g., poloxamers), generally used to stabilize interface between water and lipid
3. Co-emulsifiers (cosurfactants), such as bile salts, short-chain fatty acids and alcohols

Emulsifiers are amphiphilic molecules which possess both hydrophobic and hydrophilic portions; they are able to stabilize emulsion system by placing at the oil/water (o/w) interface (orienting the hydrophobic portion to the oil phase and the hydrophilic portion to the aqueous phase) and lowering the interfacial tension between the two phases. Co-emulsifiers intercalate between the emulsifier molecules and contribute to reduce the interfacial tension between oil and water.

Additional ingredients are used to specifically modify surface of SLN:

1. Stealth agents, such as polyethylene glycol (PEG) for improving circulation time
2. Charge modifiers to modify surface charge
3. Targeting molecules, such as peptide or antibody fragment, for linking to specific site receptor (active targeting)

Antioxidants, antimicrobial agents, and thickening agents can be usually added to SLN (dried or in dispersion) as for other pharmaceutical products.

Preparation Methods of SLN

General issue in the preparation of SLN is to obtain liquid phase where lipid and water can be emulsified before SLN formation: lipids, which are solid at room temperature, are melted or they are dissolved into solvents. Main methods to prepare SLN are reported below.

High-Pressure Homogenization (HPH)

High-pressure homogenization is a well-established technique on the large scale and it is already available in the pharmaceutical industry. High-pressure homogenizers have been used extensively in the production of nanoemulsions for parenteral nutrition.

High-pressure homogenizers push a liquid with high pressure (100–2,000 bar) through a micron size gap: by applying high pressure, the liquid accelerates to high velocity (over 1,000 km/h), and the resulting shear stress and cavitation forces break down the accelerated particles to submicron size.

HPH technique can be applied by two main different approaches: hot high-pressure homogenization (HHPH) technique and cold high-pressure homogenization (CHPH) [1].

For both techniques, the drug is dissolved or dispersed in the melted lipid at approximately 5–10°C above its melting point.

For the HHPH, the drug-loaded lipid is dispersed under high-speed stirring in a hot aqueous surfactant solution maintained at same temperature, to form a pre-emulsion. The obtained pre-emulsion is then subjected to high-pressure homogenization, still at temperature above melting point of lipid, to produce a hot o/w nanoemulsion. The homogenization process can be repeated several times: in most cases three to five homogenization cycles at 500–1,500 bar are performed. Increasing the homogenization pressure or the number of cycles often results in an increase of the particle size due to particle coalescence occurring for the high kinetic energy of the particles.

Obtained o/w nanoemulsion is then cooled down to room temperature where SLN solidify.

In CHPH, the drug-loaded lipid is rapidly cooled using liquid nitrogen or dry ice to obtain a solid solution which is then ground (milled) to obtain microparticles, approximately in the range 50–100 μm, then the solid microparticles are dispersed in a cold surfactant solution to obtain a pre-suspension. The pre-suspension is subjected to high-pressure homogenization at room temperature or below to obtain solid lipid nanoparticles: usually larger particle size and broader size distributions are obtained compared to hot homogenization.

Cold homogenization minimizes the thermal exposure of the drug, although it does not avoid it, due to dissolution of drug in melted lipid in the initial step.

Main reported disadvantages of HPH are: high energy-intensive operating conditions and potential thermal degradation of drug and excipients during production process.

Warm Microemulsion Technique

Microemulsions are transparent (clear), optically isotropic and thermodynamically stable systems. They are dispersions of two immiscible liquids (oil and water) stabilized by surfactants and optionally by cosurfactants.

Depending on the composition of the system different structures can be formed:

1. Oil dispersion in water medium – o/w microemulsion
2. Water dispersion in oil medium – w/o microemulsion
3. Bicontinuous structure

In o/w and w/o microemulsions, the dispersed phases are in shape of very small drops (10–100 nm) stabilized by an interfacial film of surfactant and cosurfactant molecules.

In bicontinuous structures, regions of water and oil are interdispersed with no spherical geometry: it occurs when the amount of water and oil are comparable.

The thermodynamic stability of microemulsion is due to the very low interfacial tension which allows the spontaneous formation of microemulsions without high energy input.

Warm microemulsions obtained with melted lipids have been used as precursor for the preparation of SLN: after their preparation at temperature ranging from 55°C to 85°C depending on melted lipid used, they are then dispersed in cold aqueous medium where melted lipid drops solidify into SLN [2].

The hydrophobic drug is dissolved in mixture composed of melted lipid and surfactant maintained at warm temperature, the cosurfactant is dissolved in water and the obtained solution is heated at the same temperature of the lipid mixture; cosurfactant aqueous solution is then added to the mixture composed of lipid, drug and surfactant and by mild mixing an o/w microemulsion is obtained and successively dispersed in cold water at 2–3°C under mechanical stirring.

Also hydrophilic drugs have been loaded into SLN by water in oil in water (w/o/w) double microemulsion technique: water, where hydrophilic drug is dissolved, constitutes the internal phase of w/o microemulsion, and the melted lipid (55–85°C) the external phase; the w/o warm microemulsion is then added to an external aqueous solution of surfactant and cosurfactant maintained at the same temperature and after mixing w/o/w microemulsion is obtained and dispersed in cold water to obtain SLN.

The SLN dispersion, obtained by warm microemulsion technique, is usually purified by tangential ultrafiltration in order to remove surfactant and cosurfactants not incorporated into SLN, and to concentrate dispersion when needed.

Advantage of the production of SLN by warm microemulsion include: low mechanical energy input, flexibility of interphase composition that allow surface functionalization, regular spherical shape. Potential drawbacks are thermal exposure, which can cause degradation of drug and of excipients, use of higher amount of surfactant and cosurfactant, which sometimes need to be removed.

High Shear Homogenization and Sonication

Following this process, drug dispersed in melted lipid phase is homogenized by rotor-stator homogenizer with a hot surfactant water solution to obtain nanoemulsion that is then cooled to form SLN.

Process parameters that can affect particle size are: emulsification time, stirring rate, and cooling conditions.

Homogenization can be used in association with sonication: melted lipid phase loaded with drug is homogenized with a hot surfactant solution to obtain a coarse emulsion that is then ultrasonicated to obtain final nanoemulsion before SLN are formed upon its cooling to room temperature [3].

Main possible disadvantages of this method include high energy input, broad particle size distribution, and potential damage of sensitive biomolecules.

When sonication is used, potential metal contamination and temperature increase have to be considered.

Solvent Emulsification-Evaporation Method

In this method, drug, lipid matrix, and emulsifier are dissolved in a water-immiscible organic solvent which is then emulsified in an aqueous phase containing a water-soluble cosurfactant and homogenized. Upon evaporation of the solvent under reduced pressure, SLN are formed by precipitation of the lipid in the aqueous medium [4].

Parameters that affect mean particle size of nanoparticles obtained by this method are: surfactant/cosurfactant blend and lipid concentration in the organic phase.

Advantages of this technique are: avoidance of heat application, production of small size nanoparticles that can be sterilized by filtration through 0.2 μm pores.

The major possible disadvantage of this method is the presence of organic solvent residues which can cause toxicity.

Solvent Emulsification-Diffusion Method

The process consists in dissolving the lipid in a partially water-soluble solvent (previously saturated with water) at room or controlled temperature, depending on solubility of lipid in the solvent. This organic phase is emulsified with an aqueous solution (saturated with solvent) containing the stabilizing agent and maintained at the same temperature, by a rotor-stator homogenizer. This o/w emulsion is then diluted with water maintaining a constant stirring and controlled temperature for promoting the diffusion of solvent of the internal phase toward the external phase, causing lipid aggregation and precipitation in form of SLN.

Depending on its boiling point, solvent can be eliminated from SLN dispersion under reduced pressure or by washing (ultrafiltration).

The selection of the water-miscible solvent and the stabilizers are critical parameters to obtain lipid particles in the nanometric range: usually solvents with high water miscibility and stabilizers able to form stable emulsions are preferred.

It is possible to reduce the particle size by increasing the process temperature, the stirring rate, the amount of stabilizer, and by lowering the amount of lipids [5].

This technique is efficient and easy to be implemented, so feasible for industrial upscaling: no high energy is required and low physical (thermal and mechanical) stress is applied.

Possible drawbacks of this method are solvent residues which need to be cleaned up and large dilution produced to obtain diffusion of solvent which needs further concentration steps.

Solvent Injection Method

In this procedure, solid lipid is dissolved in a water-miscible solvent or solvents mixture, and then rapidly injected through an injection needle into a stirred aqueous phase with or without surfactants.

The SLN production by this technique is based on the rapid diffusion of the solvent across the solvent-lipid and solvent-water interface, causing precipitation of nanoparticles.

Two simultaneous effects contribute to the effective formation of SLN:

1. Gradual solvent diffusion out of lipid-solvent droplets into water causes reduction of droplet size and simultaneously increases lipid concentration
2. Diffusion of pure solvent from the lipid-solvent droplet causes local variations in the interfacial tension at droplet surface, inducing reduction of size of droplets

In this process particle size of SLN can be influenced and controlled by variation of process parameters such as injected solvent, lipid concentration, injected volume of solvent, lipid concentration in the solvent phase and viscosity of the aqueous phase [6].

This technique is simple and fast, without the need of sophisticated equipment; possible disadvantage of this method is the use of organic solvent which has to be considered for pharmaceutical applications and for this reason a complete removal of the solvents by ultrafiltration, evaporation, or freeze-drying is required.

Coacervation Method

This process allows to obtain SLN made of fatty acids by acidification of micellar solution of their alkaline salts: fatty acids sodium salts are dispersed in a polymer water solution (use of steric stabilizer polymer is essential to avoid particle aggregation) and heated under stirring just above the Krafft point of the fatty acid sodium salt to obtain a clear solution (Krafft point is the temperature at which the solubility of sodium soap increases dramatically and the solution becomes isotropic and transparent): an acidifying solution is then added drop-wise until pH about 4 is reached and the obtained suspension is then cooled down to 15°C in a water bath under stirring.

Parameters that can influence size of nanoparticles and polydispersity index are: type and molecular weight of polymer and lipid concentration in the dispersion [7].

Although average dimensions are reported in a bigger range of 260–500 nm, this process presents several advantages like absence of solvents and need for common apparatus: the method is feasible and suitable for laboratory production and it is easy to be scaled up; studies are ongoing for loading of different drugs.

Membrane Contactor Method

This method employs a cylindrical membrane module: aqueous phase containing a surfactant is circulated in internal channel of membrane and melted lipid is pressed through pores of membrane into internal water flow allowing the formation of small droplets which are swept away by the aqueous phase; water is maintained at lipid melting temperature. SLN are then formed by cooling the preparation to room temperature.

The membrane contactor allows the preparation of SLN with a lipid phase flux between 0.15 and 0.35 $m^3/h/m^2$ [8].

The advantages of this new process are its simple use, the control of the SLN size by an appropriate choice of process and membrane parameters, and its scaling-up abilities. Drug loading and surface modification, for targeting needs, have to be fully assessed and developed.

Characterization of SLN

Particle Size, Size Distribution and Shape

Particle size, shape and stability over time may influence several important pharmaceutical features of SLN

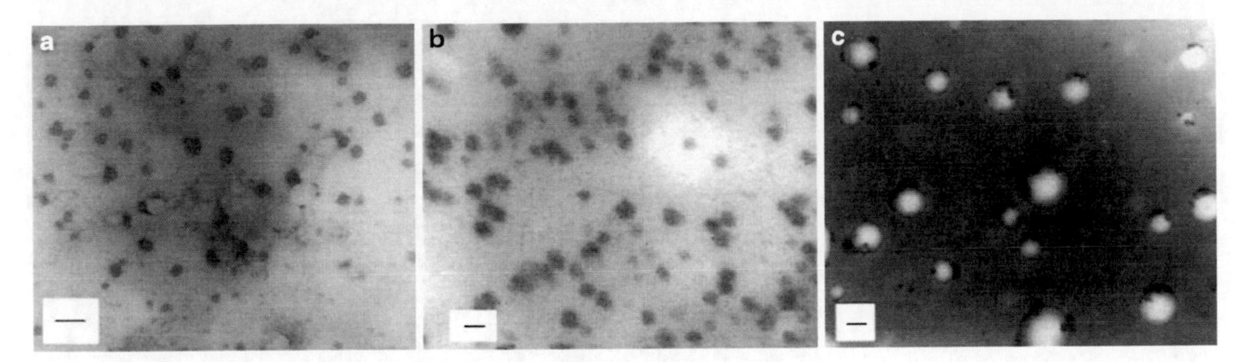

Solid Lipid Nanoparticles - SLN, Fig. 2 TEM micrographs of SLN obtained by warm microemulsion method and containing 2.5% of tobramycin in different physiological media: (**a**) aqueous dispersion before administration to rats, (**b**) lymph, and (**c**) plasma after duodenal administration to rats (bar = 100 nm) [9]

such as drug release, suitable administration route (e.g., particle size is a critical issue for intravenous administration, where particle size must not aggregate or enlarge by protein binding to exceed diameter of blood vessels) and in vivo biodistribution.

Particle size and morphology of SLN are mainly affected by method of production and composition (lipid matrix, type and amount of emulsifiers, viscosity of aqueous phase, etc.).

Main techniques commonly used for characterization of SLN are summarized below.

Photon correlation spectroscopy (PCS) also known as dynamic light scattering (DLS), is widely used method for the measurement of particle size in range from nanometer to few micrometers.

This technique determines hydrodynamic diameter of particles by measuring fluctuation of the intensity of scattered light which is due to Brownian motion of particles.

Instruments usually report hydrodynamic diameter (Zaverage) and polydispersity index (PI), which give estimation of width of particle size distribution.

Electron microscopy methods are very useful techniques for determining shape, morphology and size of lipid nanoparticles.

The most commonly reported method for morphological characterization is *transmission electron microscopy* (TEM): different techniques can be applied for sample preparation (negative staining, freeze-fracture, sample vitrification (cryo-TEM)) and can provide different information about colloidal lipid structures; sample preparation is a crucial point as it can lead to structural alterations of the sample that need to be taken into consideration. As an example, TEM micrographs of SLN (obtained by warm microemulsion method) are reported in Fig. 2 [9]: images show SLN in water dispersion before administration and same SLN in biological fluids coming from in vivo sampling after administration.

Other methods to determine SLN particle size and morphology are based on *scanning electron microscopy* (SEM).

Nonconductive samples, such as lipid-based nanoparticles, may be visualized without metal coating using specialized SEM instrumentation such as *environmental SEM* (*ESEM*) in which the sample is placed in a internal chamber at higher pressure than vacuum: positively charged ions generated by beam interactions with the gas, present in the chamber, help to neutralize the negative charge on the surface of sample.

In *field emission SEM* (*FESEM*), the electron beam is produced with a cold cathode field emitter instead of a thermoionic emitter (tungsten filament heated with an electric current) as in conventional SEM: this allows for much higher resolution showing particle shape and size. As an example, images obtained from two different formulations of SLN, produced by warm microemulsion technique, are reported in Fig. 3.

Atomic force microscopy (AFM) is another technique used to determine particle size and shape. AFM can be operated in different modalities depending on the specific application requirements: sample does not require any special treatments (such as coating) and technique also allows to work in ambient conditions, even in liquid environment. As an example, AFM characterization of SLN produced by solvent-diffusion method is reported in Fig. 4 [10]: SLN

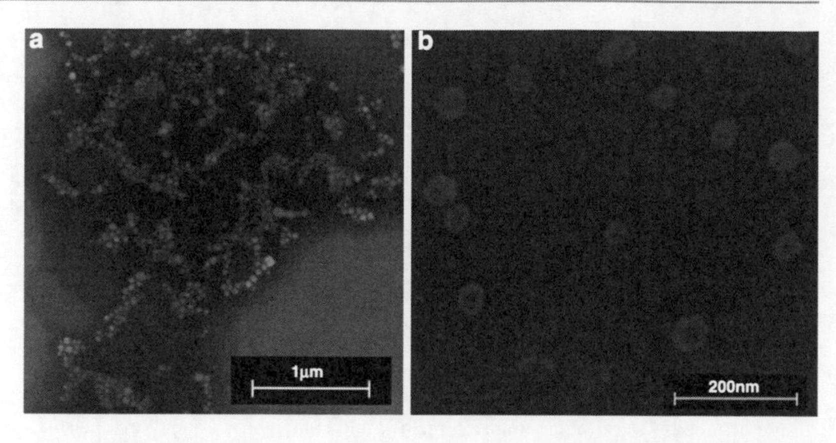

Solid Lipid Nanoparticles - SLN, Fig. 3 FESEM image of SLN obtained by warm microemulsion method: (**a**) SLN with lipid matrix of Stearic acid, (**b**) SLN with lipid matrix of cholesteryl butyrate

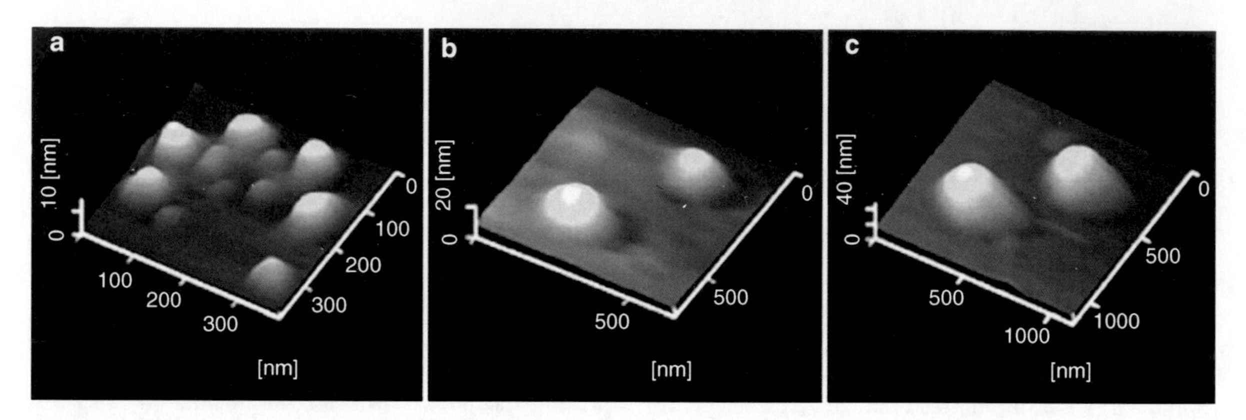

Solid Lipid Nanoparticles - SLN, Fig. 4 Atomic force microscopy images of SLN prepared by solvent diffusion method in a nanoreactor system. (**a**) Blank SLN, (**b**) SLN with 10% of clobetasol propionate charged (**c**) SLN prepared by conventional method with 10% of clobetasol propionate charged [10]

produced in miniemulsion used as nanoreactor system is compared with SLN obtained by conventional method.

Surface Charge

Nanoparticles dispersed in a liquid medium are surrounded by an electrical double layer composed by an inner layer (stern layer) where the ions are strongly bound to particle surface and an outer layer (diffuse layer) where ions are less firmly associated with the particle.

Within the diffuse layer there is a boundary (shear plane) inside of the ions and the particle form a stable entity: when the particle moves, only ions within the boundary move with the particle, while ions beyond the boundary remain in the bulk dispersant.

Zeta potential (Z potential) is the electric potential at this boundary and it gives measureof surface charge. It is used to assess stability of colloidal systems as it can express tendency of particles to repulse or to attract: higher values of zeta potential (absolute value > 30 mV) usually imply more stable dispersions due to electrical repulsion, while low values can indicate possible colloid instability which could lead to aggregation.

Z potential affects not only dispersion stability but also interaction with the biological surroundings (interaction with proteins and cells) and the in vivo fate of nanoparticles (biodistribution); it depends on the composition of the particle and on its surrounding medium: pH, electrolyte concentration, and concentration of components of the formulation.

Z potential is determined by measuring the velocity of the particles dispersed in a liquid medium under the influence of an applied electric field (electrophoresis measurements): charged particles move toward the electrode under applied electric field with velocity dependent on the strength of electric field or voltage gradient, the dielectric constant and the viscosity of the medium, and the Z potential of the particle.

Electrophoretic mobility, defined as the velocity of the particle in a unit electric field, is related to Z potential by Henry equation and is usually measured by light scattering technique.

Crystallinity, Polymorphism, Structure, and Stability

Crystallinity and polymorphic transitions of lipids in the dispersed state may differ from bulk material due to the small particle size of colloidal system, the presence of emulsifiers used for the stabilization of SLN and the preparation method. In thermal analysis SLN dispersions usually show lipid melting point slightly shifted to lower temperatures and broader melting peak than the bulk lipid, polymorphic transitions are generally faster in SLN than in the bulk material [1, 11].

Crystallization and polymorphic behavior of SLN are correlated with drug incorporation and drug release: highly ordered crystal lattices (monoacid triglycerides) in SLN matrix can lead to drug expulsion, while less ordered crystal lattice (mixture of mono-, di-, and triglycerides), because imperfections which provide space to accommodate the drug, can allow better drug incorporation.

Polymorphic transitions may cause alterations of lipid packing and thus of the internal structure of the nanoparticles, which may have negative consequences for drug loading. The course of polymorphic transitions depends on the type of lipid matrix and can be modified by other components of the dispersions [11].

As an example, triglycerides crystallize in three main polymorphic forms: α, β', and β; the β form is highly ordered with high thermodynamic stability, the α form is a less ordered form with low thermodynamic stability, the β' form has intermediate characteristics between α and β forms. SLN based on triglyceride matrix tend to crystallize in the metastable α form: generally thermodynamically unstable configurations allow lipid molecules to have higher mobility causing lower density and a higher capability to incorporate drug molecules, but during storage the α form transform via β' form in the more thermodynamically stable configuration β, characterized by higher packing density which can cause expulsion of the drug form structure.

SLN may change their shape during polymorphic transitions: as an example, tripalmitin SLN have a spherical shape in α form, but they have a platelet shape in the β form. During storage the polymorphic transition from the α to the β configuration is connected with an increase of the particle surface due to preferred formation of platelets; surfactant molecules could not provide any more complete coverage of the lipid surface and particle aggregation can occur, leading to gelation (irreversible transformation of SLN dispersion into a viscous gel) [1].

SLN can display a lower crystallization tendency than the bulk material and may not crystallize properly after preparation and storage, at a temperature below melting point of lipid, forming super-cooled melts, which are liquid lipid nanoemulsions and not solid lipid dispersion. This phenomenon is particularly pronounced in SLN made from short-chain monoacid triglycerides such as trimyristin and trilaurin, and it is mainly due to size dependency of crystallization process that requires a critical number of nuclei to start, which cannot be reached in small droplets: tendency to formation of super-cooled melts increases with decreasing of droplet size [1, 11].

Hence, the physical state is a very important parameter which affects performance of SLN as drug delivery system: the techniques usually used to investigate physical state of SLN are *differential scanning calorimetry* (DSC) and *X-ray diffraction* (XRD) techniques.

DSC is a thermal analysis technique which measures the difference in heat flow between the sample and a reference when they are both subjected to the same controlled temperature program: DSC quantifies the enthalpy changes during endothermic and exothermic process and it is a useful technique to determine melting point, melting enthalpy, degree of crystallinity, glass transitions of amorphous material, and to identify different polymorphic forms.

In XRD techniques, an X-ray beam is focused on the sample, the scattered X-rays interfere with each other giving particular diffraction patterns: crystalline materials display many diffractions bands which depend on the disposition of the atoms within the unit cell of the crystalline lattice, while amorphous compounds present more or less a regular baseline.

XRD techniques allow to differentiate between crystalline and amorphous material and to identify different polymorphic forms in SLN dispersions: the most commonly used are *powder X-ray diffraction* (PXRD), *small angle X-ray scattering* (SAXS), and *wide angle X-ray scattering* (WAXS).

Nuclear magnetic resonance (NMR) is another useful technique to investigate structure of SLN for surfactant distribution on particle surface, drug distribution in the particle and potential formation of super-cooled melts.

Due to different chemical shifts it is possible to attribute the NMR signals to particular molecules or their segments (functional groups); NMR active nuclei of interest are ^{1}H, ^{13}C, ^{19}F, and ^{31}P. Width and amplitude of NMR signals can be related to liquid and semi-solid/solid state of nuclei: nuclei in the liquid state give sharp signals with high amplitude, while nuclei in the semisolid/solid state give weak and broad signals [1].

Drug Incorporation

A broad range of drugs, mainly with hydrophobic properties, has been already incorporated into SLN, although the crystalline nature of the lipid matrix can offer limited space for drug incorporation inside the particle core [11].

Drug loading capacity of SLN is defined as the percentage of drug incorporated into nanoparticles referred to the total weight of the lipid phase or to the content of dispersed material; value of this parameter is usually in range 1–20%.

Drug entrapment efficiency is defined as the percentage of drug incorporated into nanoparticles referred to the total drug added for the SLN preparation; values of this parameter is usually relatively high (80–99%).

The distribution (localization) of the drug within SLN structure (Fig. 5) can vary considerably and three *drug incorporation* models have been proposed [12]:

1. *Solid solution model.* The drug is molecularly dispersed in the solid lipid matrix (SLN matrix as a solid solution).
2. *Drug-enriched shell model.* The drug is concentrated in the outer shell of the SLN. This situation can be explained by a higher solubility of the drug in the aqueous-surfactant outer phase at increased temperature in the production process; during cooling, lipid in the core starts to solidify and becomes less/not accessible for the repartition of the drug into the lipid inner phase leading to enrichment in the particle shell.

   High temperatures employed in the preparation process can increase solubility of the drug in the aqueous-surfactant phase promoting drug localization at the surface region.
3. *Drug-enriched core model.* The drug is concentrated in the core of the lipid particle and it is surrounded by a lipid shell. This situation can be explained by a precipitation of the drug before the lipid crystallizes during cooling: this takes place preferentially when drug concentration in the lipid at process temperature is at its saturation solubility, so during cooling a super saturation and subsequent drug precipitation are achieved

Drug-loading capacity of SLN is affected by different factors:

1. Solubility of the drug in the melted lipid. High solubility is a prerequisite to obtain sufficient drug loading; the amount of drug that can be dissolved in the lipid formulation may exceed the pure value of its solubility in lipid alone and a greater solubility in the whole formulation suggests that the drug can localize both in inner lipid phase and in external part on SLN (interfacial region) where surfactants are present. Solubility of drugs in the lipid can be improved by preparing lipid prodrugs such as stearic or palmitic acid derivatives (lipid drug conjugates).
2. Physical structure of solid lipid matrix. Complex lipids which form less ordered crystals with many imperfections can favor drug incorporation.
3. Polymorphic state of lipid matrix. Lipid transformation in more stable form reduces the number of imperfections in crystal lattice and can determine drug expulsion.

Determination of drug-loading capacity can be usually performed by separation of free drug from the dispersion medium by employing different techniques such as ultrafiltration (tangential or centrifugal), ultracentrifugation, gel filtration chromatography, and dialysis.

Drug Release

Drugs incorporated into SLN are released by degradation and surface erosion of the lipid matrix and by diffusion of drug molecules through the lipid matrix. SLN are composed of physiological lipids for which metabolic pathways exist in the body: most important enzymes involved in in vivo SLN degradation are lipases, which are present in various organs and tissues [1].

Drug release from SLN is mainly affected by localization of the drug [12]:

1. Localization of the drug within the core of solid lipid matrix offers the possibility to obtain a prolonged drug release.

Solid Lipid Nanoparticles - SLN, Fig. 5 Schematic representation of drug incorporation models for SLN

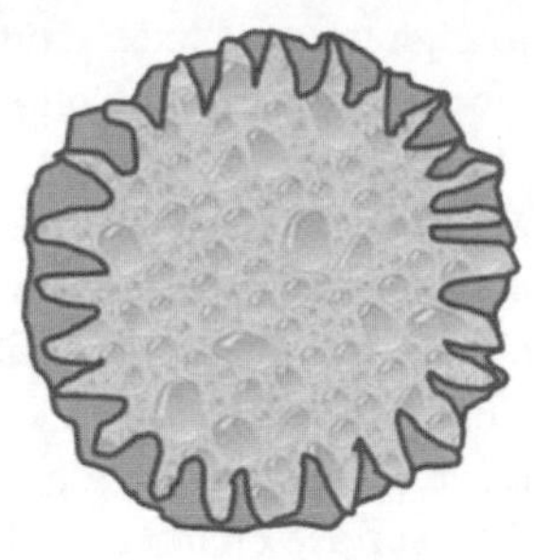

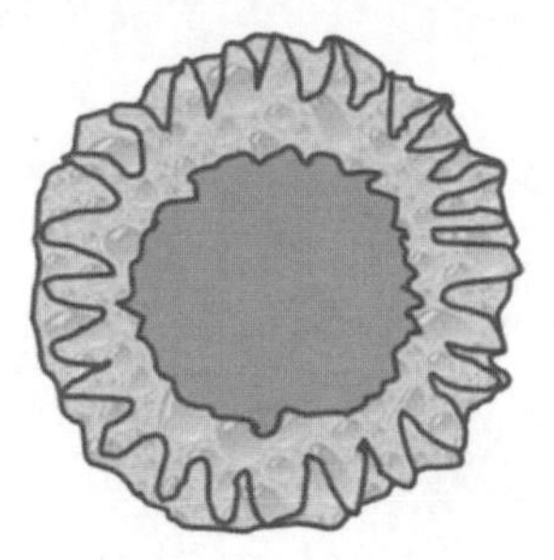

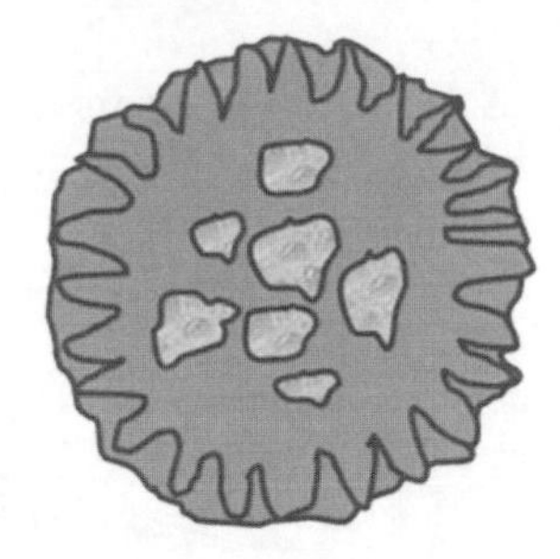

2. Localization of drug molecules on particle surface often leads to burst effect (fast initial drug release). SLN can show a biphasic drug release profile: an initial burst release, due to the drug localized at the surface, is followed by a more gradual release due to the drug localized in the lipid matrix.

Possible presence of alternative colloidal species always has to be taken into account for drug release characterization: stabilizing agent cannot be localized exclusively on the lipid surface as expected, but also in the aqueous phase forming micelles, mixed micelles or liposomes that can solubilize the drugs and constitute alternative drug incorporation sites [1].

Composition of SLN is other important parameter that can affect drug release: in vitro experiments indicate that SLN show different degradation rates by lipases as a function of their composition (kind of lipid matrix and emulsifier). Degradation of triglycerides SLN show dependence to the length of the fatty acid chain: longer fatty acid chain shows slower degradation, while degradation of SLN made of waxes (cetylpalmitate) is slower compared to glyceride matrices [1].

Emulsifiers containing polyethylene glycol (PEG) chains can reduce SLN degradation by a hindering effect which reduces the anchoring of the enzyme on the SLN surface [1].

Since SLN are degraded by surface erosion, in SLN with small size and higher surface area, drug release is expected to be more rapid; nonspherical particles, such as thin platelet shape, are characterized by larger surface area and require shorter time for degradation and drug diffusion to particle surface [11].

Drug release can be studied by different techniques such as dialysis membranes (flat or bag) and Franz diffusion cells. Release kinetics is affected by in vitro release conditions: sink or non-sink conditions, release medium, dilution of nanoparticles dispersion that has to be considered in particular if the formulation is intended for oral and intravenous administration; release experiments have to be performed mimicking in vivo conditions in order to avoid distorted release profile [11].

As an example of in vitro study [9], drug release was investigated, working in sink conditions, for three types of SLN incorporating different percentage of tobramycin (1.25%, 2.50%, and 5.00%) by using a multicells rotating tool, where donor and acceptor compartments were separated by double hydrophilic/hydrophobic membranes: results showed that release kinetic was pseudo-zero order for all three types of SLN and the amount of drug released was higher for Tobra-SLN 5.00% than for Tobra-SLN 2.50% which was higher than for Tobra-SLN 1.25%, showing that in vitro release can be modified by changing amount of drug loaded into SLN and consequently their physical characteristics (Fig. 6) [9].

Routes of Administration/Applications

SLN have been tested for drug delivery applications by different routes of administration including intravenous, oral, ocular, topical dermal, transdermal, and pulmonary.

Parenteral Administration: Intravenous Route (IV)

Overview

The evaluation of SLN interaction with blood and other tissues is very important for IV administration: many studies have been performed to assess toxicity of SLN both in in vitro and in vivo tests.

Cellular toxicity of SLN has been investigated using several cell lines: in vitro experiments demonstrate that toxicity is dependent on composition of SLN (nature of lipid matrix and type of surfactant used) and on concentration of SLN in the culture medium [13].

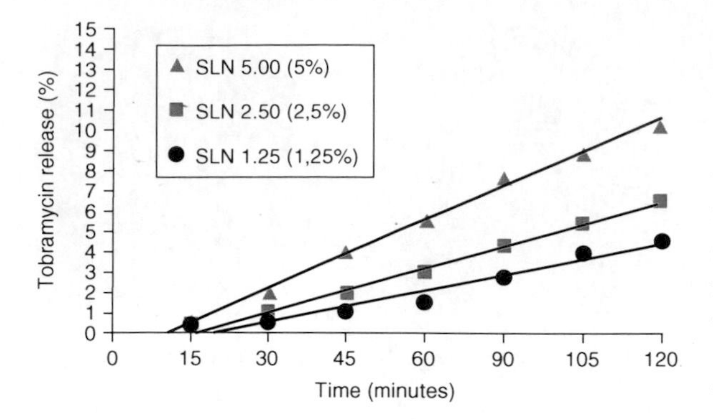

Solid Lipid Nanoparticles - SLN, Fig. 6 In vitro release – percentage of tobramycin diffused through a double membrane versus time for SLN loaded at three different percentages of tobramycin [9]

In vivo toxicity studies confirmed that SLN are well tolerated in living system after their iv administration [13].

In order to obtain SLN suitable for parenteral administration, pharmaceutically acceptable excipients must be employed, and sterility must be assessed. Studies showed some SLN formulations can be sterilized by sterilizing filtration (such filtration is applicable for particles with size <0.2 μm), some other can be sterilized by autoclaving generally at 121°C for at least 15 min (SLN melt during autoclaving and recrystallize during cooling); physicochemical properties of SLN formulation should not change during the sterilization process. Autoclaving temperature can promote chemical degradation and can affect physical stability of SLN because their structure can be lost when particles melt and recrystallize in noncontrolled way; performed studies show that critical parameters for SLN sterilization by autoclaving are temperature and timing of sterilizing process and composition of formulation [1].

Generally, all in vivo studies performed showed drug-loaded SLN change pharmacokinetic parameter of carried drug when compared to reference drug solution: much higher mean residence time (MRT) and area under the curve (AUC) induced by SLN formulation mean higher drug availability and consequent possible higher efficacy.

Performed studies confirmed as well that important limiting factor for IV administration of SLN is the uptake of particles by macrophages of reticuloendothelial system (RES). SLN are recognized as foreign substances and quickly removed from blood circulation. Colloidal particles interact with blood plasma proteins called opsonins which are adsorbed on particle surface (opsonization): opsonins mediate RES recognition by interacting with specific membrane receptors of macrophages. The capacity of nanoparticles to avoid opsonization and consequent macrophages uptake depends on size of particles, on their surface charge, and hydrophobic characteristics.

In order to avoid RES recognition, SLN surface can be functionalized with different molecules having hydrophilic and flexible chains which form a hydrophilic steric barrier that cover and protects nanoparticles from interaction with plasma proteins and prolongs their blood circulation time. Such sterically stabilized nanoparticles (now reported as *stealth*) are mainly obtained by using lipid derivatives of polyethylene glycol (PEG): amount of stealth agents and length of their chain can affect degree of surface coverage and rate of uptake by macrophages.

Macromolecular drugs and colloidal drug carriers such as SLN, liposomes, and polymeric nanoparticles, can accumulate preferentially in tumor tissue, a phenomenon called enhanced permeation and retention effect (EPR effect), due to the presence of a discontinuous endothelium and to the lack of efficient lymphatic drainage in the tumor tissue: by reducing RES uptake, nanoparticles can further passively accumulate inside such tumor tissues (*passive targeting*).

In order to further increase drug accumulation in target tissue or to produce higher and more selective therapeutic activity, specific functionalization of colloidal drug carriers has been developed (*active targeting*) by decorating SLN surface with ligand molecules able to recognize specific site on target cells inducing internalization of carrier after binding. Targeting molecules include: monoclonal antibodies or fragments, peptides, glycoproteins, and receptor ligands.

Main Applications of SLN by Parenteral/IV Administration

Cancer Therapy

SLN have been investigated as drug delivery system by IV administration mainly for cancer treatment: several antitumor drugs have been incorporated into SLN such as doxorubicin, camptothecin, paclitaxel, etoposide, idarubicin, vinorelbine, mitoxantrone, 5-fluorouracil, and their in vitro and in vivo distribution and efficacy have been evaluated [13].

SLN are considered suitable potential carriers in oncology for ability to incorporate antitumor drugs with different chemical properties for improvement of drug stability (drug protection), enhancement of drug efficacy, reduction of drug toxicity (lowering of side effects), improvement of pharmacokinetics parameters, passive targeting, and the possibility to reach difficult districts like the brain [2, 13].

One of the first in vivo studies on these carriers in oncology has been performed on camptothecin (CA) loaded into SLN prepared by high pressure homogenization method and administered intravenously into mice: the concentration of camptothecin was determined in various organs and compared to a camptothecin control solution [13]. The results showed that in tested organs, the area under the curve (AUC) versus dose and the mean residence times (MRT), pharmacokinetic parameters relevant to drug bioavailability, of CA-SLN were much higher than those of CA-Solution (Fig. 7) [13].

As another example, in vivo tests have investigated pharmacokinetics and tissue distribution of doxorubicin incorporated in both stealth and not stealth SLN prepared by warm microemulsion technique and injected intravenously into rabbits [2]. AUC increased as a function of the amount of stealth agent present in SLN and doxorubicin was still present in the blood 6 h after the injection of both stealth and non-stealth SLN, while none was any more detectable after IV injection of doxorubicin solution.

The pharmacokinetics and tissue distribution of doxorubicin SLN were also studied in rat animal model after intravenous administration and compared with commercial solution of doxorubicin [2]: doxorubicin was still present in the blood 24 h after the injection of stealth and non-stealth SLN, while it was not detectable after the injection of the commercial solution. The results confirmed the prolonged circulation time of SLN, in particular SLN decreased the concentration of doxorubicin in the heart indicating lower cardiotoxicity compared to doxorubicin solution.

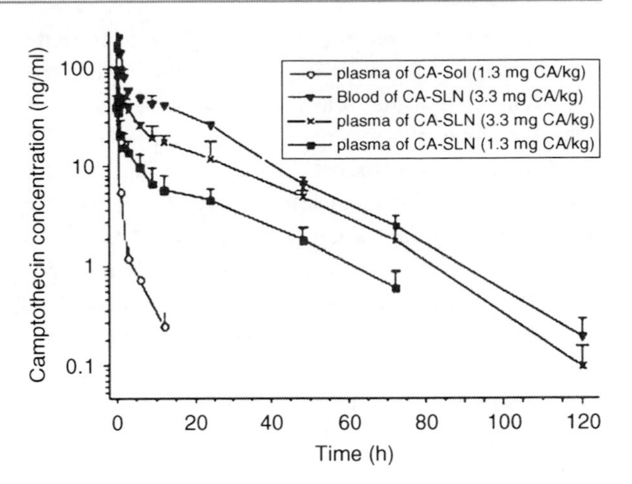

Solid Lipid Nanoparticles - SLN, Fig. 7 Concentration-time curves of camptothecin after IV administration of CA-SLN with doses of 1.3 (■) and 3.3 (x) mg CA/kg in plasma and 3.3 (▼) mg CA/kg in blood, and CA-SOL with a dose of 1.3 (○) mg CA/kg in plasma. Results represent means ± SD of four animals [13]

Diseases of the Central Nervous System (CNS)

Pharmacological treatment of diseases of the CNS, such as brain tumor, neurological, and neurodegenerative diseases, is limited by the presence of the *blood-brain barrier* (BBB) that restricts enormously the transport of many important drugs from the blood into the brain. The BBB is formed by the endothelial cells of the cerebral capillaries which differ from endothelial cells in the rest of the body for the absence of fenestrations, for the presence of particular and more extensive tight junctions, and for minimal pinocytic vescicular transport.

In order to evaluate transport of SLN across the BBB, both stealth and non-stealth radiolabeled SLN, prepared by warm microemulsion method, were injected intravenously into rats [2], and tissue distribution was monitored for 60 min: radioactivity in liver and in lung was lower for stealth SLN than for non-stealth SLN confirming difference in their uptake. Both types of SLN were detected in brain and cerebrospinal fluid, although at low percentage.

SLN incorporating baclofen have been investigated as new pharmaceutical preparation of this drug which is used in treatment of spasticity [2]: SLN were prepared by warm microemulsion method and were injected intraperitoneally to rats at increasing dosage. As an important finding, drug effects were detectable with lower doses of baclofen when loaded into SLN, in comparison with the needed amount of baclofen when in solution.

In vivo results showed a good correlation with plasma and tissue concentration of baclofen: after 2 and 4 h, only baclofen-SLN produced detectable baclofen plasma concentrations, while 2 h after the administration of baclofen solution, the amount of baclofen in plasma was undetectable. Moreover, baclofen concentration in the brain 2 h after SLN administration was almost double than after baclofen solution, suggesting that baclofen may pass the BBB in much higher amount when formulated in SLN.

In order to improve brain uptake, SLN modified with specific ligands on surface have also been tested (active targeting): the presence in the BBB of receptor-mediated transport systems for endogenous molecules can be exploited to gain access to the brain.

Transferrin-conjugated SLN have been investigated for their ability to target quinine hydrochloride to brain [13]: biodistribution studies showed that quinine hydrochloride concentration in brain was significantly higher in case of transferrin-conjugated SLN as compared to that of quinine-SLN not functionalized and of quinine hydrochloride solution.

Imaging

Imaging technologies are important tools for diagnosis of diseases and for monitoring of therapies.

SLN have been proposed as carriers for hydrophobic imaging agents in order to increase concentration of the agent in the target tissue.

SLN have been investigated as carrier of contrast agents for magnetic resonance imaging (MRI) such as iron oxides nanoparticles (NP) [2]. SLN loaded with iron oxides NP were prepared by warm microemulsion method and were studied in in vitro and in vivo tests and compared to Endorem®, commercial preparation of iron oxides NP. Iron oxide-SLN showed in vitro relaxometric properties similar to those of Endorem® and after intravenous administration to rats showed to have slower blood clearance than Endorem®.

Gene Therapy

A potential approach for the treatment of human genetic disorders is gene therapy. This is a technique whereby the absent or faulty gene is replaced by a working gene, so that the body can make the correct enzyme or protein and consequently eliminate the root cause of the disease. Process of introducing nucleic acids into cells is generally called transfection.

Polynucleotides molecules (DNA or RNA) are large, hydrophilic, and negatively charged molecules. They are very labile in biological environment and their spontaneous entry inside cells is a very inefficient process.

Nucleic acids can be delivered by two main classes of vectors: viral vectors and nonviral vectors [14]. Although viral vectors are very efficient in nucleic acid delivery, their immunological risk is high (insertional mutagenesis, adverse immunogenic responses, inflammatory reactions) and there is limitation on dimension of nucleic acid to be incorporated: for these reasons nonviral carriers can offer preferable alternative. Most important tested nonviral vectors are liposomes, complexes made by DNA or RNA and cationic polymers or cationic lipids, and SLN [14].

The first obstacle for systemic delivery of DNA or RNA molecules is the extracellular environment, extreme pH, proteases and endonucleases, and immune defense.

The second barrier is cellular membrane; nucleic acids alone are unable to cross it because of their high negative charge.

One of the main routes of internalization of vectors carrying RNA or DNA is endocytosis: the internalized particle exists in endosomes, compartments of the endocytic membrane transport pathway, that either fuse with lysosomes, the main hydrolytic compartment of the cell where RNA or DNA are degraded losing their activity. Therefore, escape from endosomes is crucial for efficient transfection. Certain lipids have the ability to destabilize endosomal membranes favoring the escape of nucleic acids from this compartment: those lipids are called fusogenic lipids (e.g., dioleoyl phosphatidylethanolamine DOPE) and they are usually included in SLN formulation for this reason.

For their interference with lysosomes, lysosomotropic agents, such as cloroquine, are reported as well to enhance gene expression [14].

Another barrier against DNA transfection is nuclear envelope, containing pore regulating passive transport for molecules with maximum mass of 70 kDa or mean diameter size of 10 nm.

Cationic SLN, with positive surface charge, carry the nucleic acids by means of electrostatic interaction: different in vitro studies showed their efficacious transport of nucleic acid through cellular barriers into the cytoplasm or nucleus [12, 14]

Cationic SLN can be obtained by using different cationic lipids such as: dioleoyl trimethylammonium-propane (DOTAP), cetylpyridinium chloride, dimethyldioctadecylammonium bromide (DDAB), DC-cholesterol.

The cationic SLN are incubated with negatively charged nucleic acids; electrostatic interactions occur resulting in the formation of SLN-DNA or SLN-RNA complexes.

Positive charge of SLN-DNA/SLA-RNA surface is usually preferable for in vitro transfection, as it helps to promote the cellular uptake of the complex because of negative charge of cell membranes.

On the other side, use of some cationic lipids for in vivo delivery is limited by their toxicity [15].

The capacity of SLN-DNA vectors to in vivo transfect after intravenous administration in mice has been evaluated: as an example, cationic SLN were prepared by solvent emulsification evaporation method and by using DOTAP as cationic lipid, SLN-DNA vectors were obtained by mixing SLN with the plasmid pCMS-EGFP, which encodes the enhanced green fluorescent protein (EGFP). The intravenous administration in mice led to transfection in hepatic tissue and spleen, protein expression was detected from the third day after administration and it was maintained for at least 7 days; the results showed the capacity of SLN-DNA vectors to induce expression of foreign proteins in the spleen and in the liver assessing the potential of SLN for gene therapy [15].

SLN have been formulated also to incorporate nucleic acids inside their solid lipid matrix: this kind of nanoparticles are mainly prepared by (w/o/w) double warm microemulsion technique.

Efficacy of SLN incorporating vascular endothelial growth factor antisense oligonucleotide (VEGF-AS-ODN) to downregulate VEGF expression has been evaluated in rat glioma cells (in vitro) and in experimental murine model of glioma (in vivo) [2]: both experiments demonstrated that cellular VEGF expression was significantly reduced in tumor cells with SLN carrying VEGF-AS-ODN.

Oral Administration

SLN have been studied for oral administration of several drugs such as cyclosporin A, clozapine, tobramycin, idarubicin, and apomorphine.

SLN can improve oral bioavailability of drugs by different mechanisms [12]:

1. Protection of incorporated drugs from gastrointestinal fluids and enzymatic degradation
2. Transportation into lymphatic system avoiding first-pass metabolism
3. Absorption enhancing effect of lipids: lipids can promote the absorption of poorly water soluble active compounds
4. Bioadhesion of nanoparticles (due to their small size) to the intestinal wall increasing the residence time in intestinal tract
5. Effect of surfactants which may contribute to increase permeability of the intestinal membrane or to improve affinity between lipid particles and intestinal membrane

Gastrointestinal uptake of radiolabeled SLN administered to duodenum of rats by canula was studied [2]. SLN were observed in the lymph by using electron microscopy; the radioactivity data confirmed targeting of particles to lymph and blood.

In other in vivo study, in order to evaluate gastrointestinal absorption of drugs incorporated into SLN, tobramycin was selected as model drug because it is not absorbed by the gastrointestinal tract. SLN containing different percentages of tobramycin were administered in rats by canula to duodenum: the time-concentration curves showed different profiles (Fig. 8) [9] and pharmacokinetics parameters varied considerably among the three types of tobramycin-SLN. Possible reasons for the different behavior are number of SLN administered, particle size, total surface area, and drug concentration in each nanoparticle [9].

The results of this study confirmed transmucosal transport of drug when loaded in SLN to lymphatic system and the lymphatic uptake of tobramycin-SLN was detected also by TEM micrographs (Fig. 2) [9].

A critical parameter that has to be considered in oral administration is physicochemical stability of SLN into gastrointestinal fluids. SLN aggregation can occur in the stomach due to the acidity and high ionic strength of the gastric environment [1].

Drying of SLN dispersion into a dry powder (by freeze-drying and spray-drying process) can be necessary for the administration of solid dosage forms such as capsules and tablets.

Skin Application

SLN show many advantageous features for skin application. They can provide controlled release profiles, they are composed of physiological and biodegradable lipids with good tolerability, their small size allows a close contact with stratum corneum (horny layer) and causes formation of an adhesive film on the skin; film formation leads to an occlusive effect

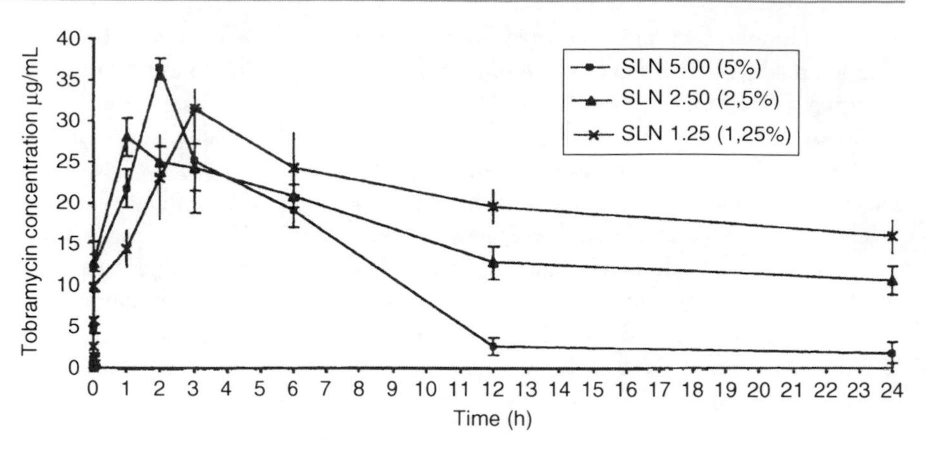

Solid Lipid Nanoparticles - SLN, Fig. 8 Tobramycin plasma concentrations versus time ± SD after duodenal administration of Tobra-SLN, loaded at different percentage of tobramycin [9]

which increases skin hydration, improving drug penetration into the skin. Furthermore, SLN can enhance chemical stability of compounds sensitive to light, oxidation, and hydrolysis [16].

The horny layer is an efficient barrier that protects humans from excessive water loss, toxic agents, and microorganisms: it is formed by corneocytes embedded into epidermal lipids forming highly structured layers. Below the horny layer there is the epidermis followed by the dermis.

In general lipid nanoparticles do not penetrate the horny layer, but a follicular uptake has been reported for particulate systems.

Incorporation of drugs into SLN can be exploited for both topical administration and for transdermal administration [2, 16].

SLN have been proposed for improvement of treatment of skin diseases (such as atopic eczema, psoriasis, acne, skin mycosis, and inflammations): drugs incorporated into SLN and investigated for dermal application have been glucocorticoids, retinoids, and antimycotics.

Experimental data showed that SLN can enhance drug penetration into the skin increasing treatment efficiency: improved bioavailability at the site of action reduces the required dose and reduces dose-dependent side effects of the drug. SLN can act as a drug reservoir and this can be an important tool when it is necessary to supply the drug over a prolonged period of time and when drug produces irritation in high concentration [16].

SLN have been investigated also as carriers in cosmetics, especially for UV blockers. SLN by themselves have a sun-protective effect, due to their solid particulate character, and are capable of reflecting UV radiation leading to photoprotection: molecular sunscreens showed to be much more effective after incorporation into SLN for synergistic effects [1, 16].

In order to allow skin application, SLN dispersion need to be incorporated into a cream or gel base which must not induce dissolution nor aggregation of SLN.

Ocular Topical Administration

Ocular bioavailability of drugs is limited by the complex structure of the eye: for ophthalmic application, drugs are usually formulated as eyedrops and administered topically but limited permeability of the cornea and other different mechanisms (such as lachrymal secretions, nasolacrimal drainage, drug adsorption into systemic circulation, drug spillage due to the limited capacity of human cul-de-sac, and drug metabolism caused by enzymes) strongly contribute to limit efficacy of the applied drug. Due also to this removal, several drug applications in a day are required to achieve therapeutic effect [17].

SLN can be useful system to enhance ocular bioavailability of both hydrophilic and hydrophobic drugs.

Possible advantages of SLN for improvement in ocular drug delivery include:

1. Bioadhesive properties, which can prolong ocular surface residence time, with better drug penetration and reduced drainage, increasing drug ocular bioavailability and decreasing dosage and relevant side effects
2. Controlled release of drugs avoiding frequent administrations
3. Protection of incorporated drug from action of metabolic enzymes present on the ocular surface

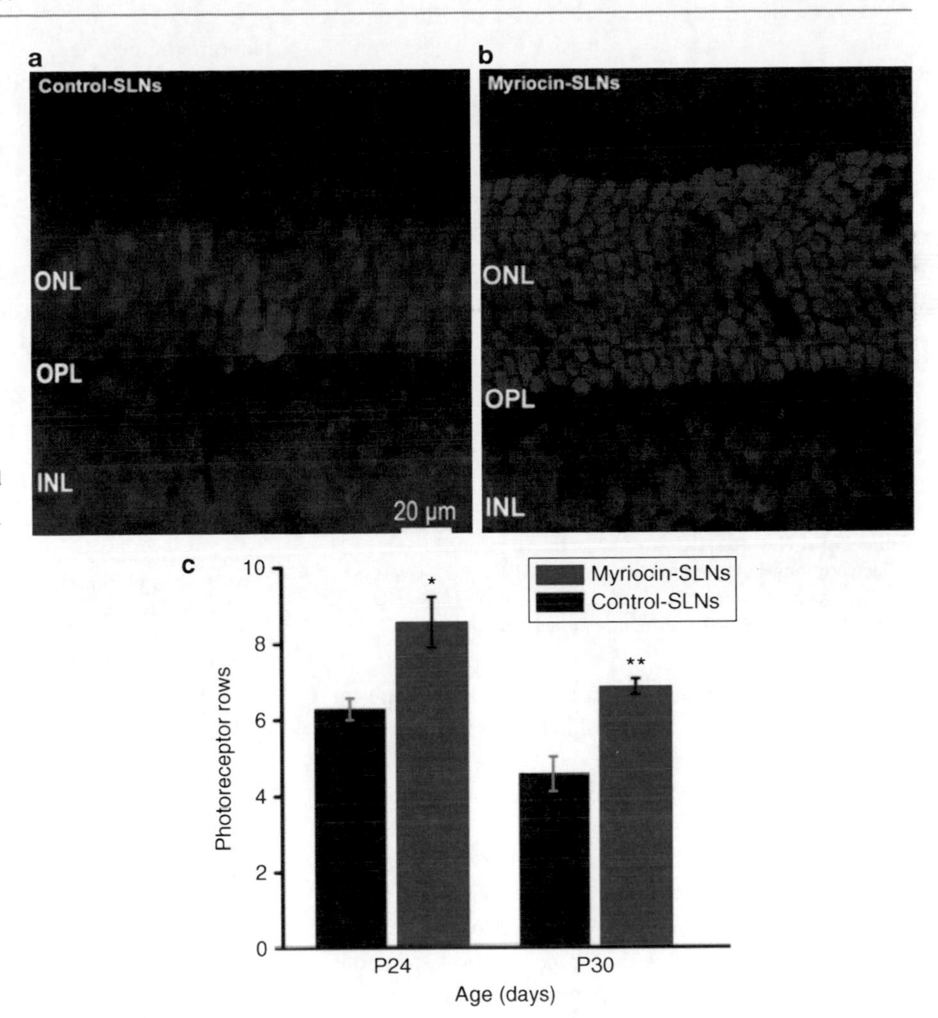

Solid Lipid Nanoparticles - SLN, Fig. 9 Effects of myriocin-SLN on retinal morphology: Vertical retinal sections from rd10 mice treated with control SLN (**a**) and myriocin-SLN (**b**) for 10 days (from P14 to P24). The outer nuclear layer (*ONL*) of the myriocin-treated retina is thicker because it contains more photoreceptor rows than the control retina. *INL* inner nuclear layer, *OPL* outer plexiform layer. (**c**) Quantification of photoreceptor rows at P24 and P30 in rd10 mice treated with control SLN or myriocin-SLN [18]

4. Good tolerability as SLN can provide low irritation and good compatibility with ocular tissues
5. Eyedrops formulation, which is self-administered by patients, improving compliance

SLN have been investigated in animal model for ocular delivery of different drugs such as tobramycin, diclofenac, timolol, and cyclosporin A: specific SLN formulation allowed to assess efficacy of myriocin (MYR) in retinitis pigmentosa animal model [18].

MYR is a hydrophobic drug, it can inhibit intracellular synthesis of ceramide, a physiological lipid involved in apoptotic cascade. MYR has been proposed as drug for treatment of retinitis pigmentosa (RP), a disease where degeneration of photoreceptor causes blindness: there is not efficacious treatment approved for RP at this time. Efficacy of MYR loaded SLN has been showed in RP animal model (RD10 mouse model): three times a day administration by eyedrops of MYR-SLN, lasting a period of 35 days, did not show toxicity in mice and physiological registration of activity of photoreceptors by electroretinography (ERG) showed reduced degeneration occurred in animals treated with MYR-SLN (Figs. 9, 10) [18].

Pulmonary Administration

The large inner surface of the lung and the thin alveolar epithelium allow rapid drug absorption avoiding first-pass metabolism; pulmonary administration using solid lipid nanoparticles represents an alternative for both local and systemic drug delivery.

SLN show different advantages for pulmonary drug delivery, such as the possibility to have controlled release profile, high tolerability, and possibility to reach lymphatic system through the action of alveolar macrophages. For pulmonary administration, SLN dispersions can be nebulized; inhalation device (design of

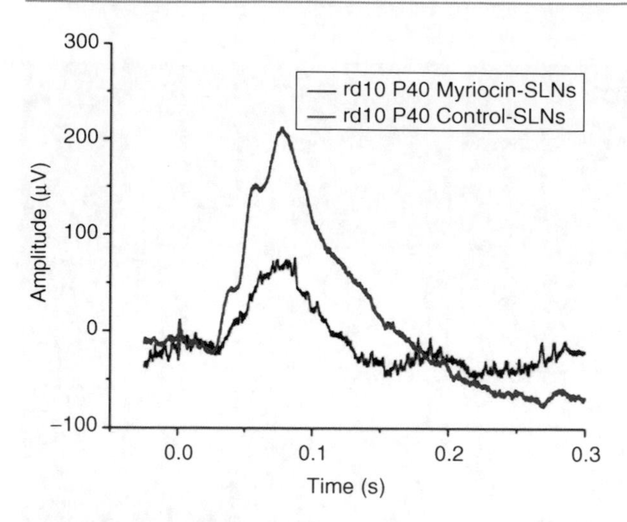

Solid Lipid Nanoparticles - SLN, Fig. 10 Cone-driven ERG (Electroretinography) responses from rd10 mice (*red* trace) treated with myriocin-SLN or control SLN (*black* trace). Animals were age P40 and treatment had initiated at P14 [18]

nebulizer, flow rate) and physicochemical properties of formulations can affect aerosol nebulization efficiency and site of aerosol deposition.

Toxicological studies have been performed by using in vitro and in vivo tests [19]: A549 cells and murine precision-cut lung slices (PCLS) were exposed to increasing concentrations of SLN in order to estimate the toxic dose of SLN. The in vitro experiments showed toxic effects begin at concentrations of about 500 μg/mL, while for in vivo experiments toxicological potential of SLN was determined in a 16-days repeated dose inhalation study using mice which were daily exposed to different concentration of SLN (1–200 μg deposit dose): results showed that repeated inhalation exposure to SLN is safe in murine inhalation model.

Biodistribution of inhaled radiolabeled SLN in rats was studied [20]. Results show an important and significant SLN uptake into the lymph, few minutes after inhalation SLN translocate and accumulate into regional lymph nodes suggesting translocation mechanism of SLN may involve phagocytosis by bronchoalveolar macrophages followed by migration to lymphatic system.

Cross-References

- ▶ Liposomes
- ▶ Nanoencapsulation
- ▶ Nanomedicine
- ▶ Nanoparticle Cytotoxicity
- ▶ Nanoparticles
- ▶ RNAi in Biomedicine and Drug Delivery

References

1. Mehnert, W., Mäder, K.: Solid lipid nanoparticles: production, characterization and applications. Adv. Drug Deliv. Rev. **47**, 165–196 (2001)
2. Gasco, M.R., Mauro, A., Zara, G.P.: In vivo evaluations of solid lipid nanoparticles and microemulsions. In: Drug Delivery Nanoparticles Formulation and Characterization, pp. 219–238. Informa Healthcare, New York (2009)
3. Manjunath, K., Venkateswarlu, V.: Pharmacokinetics, tissue distribution and bioavailability of clozapine solid lipid nanoparticles after intravenous and intraduodenal administration. J. Control Release **107**, 215–228 (2005)
4. Siekmann, B., Westesen, K.: Investigations on solid lipid nanoparticles prepared by precipitation in o/w emulsions. Eur. J. Pharm. Biopharm. **43**, 104–109 (1996)
5. Quintanar-Guerrero, D., Tamayo-Esquivel, D., Ganem-Quintanar, A., Allemann, E., Doelker, E.: Adaptation and optimization of the emulsification-diffusion technique to prepare lipidic nanospheres. Eur. J. Pharm. Sci. **26**, 211–218 (2005)
6. Schubert, M.A., Müller-Goymann, C.C.: Solvent injection as a new approach for manufacturing lipid nanoparticles – evaluation of the method and process parameters. Eur. J. Pharm. Biopharm. **55**, 125–131 (2003)
7. Battaglia, L., Gallarate, M., Cavalli, R., Trotta, M.: Solid lipid nanoparticles produced through a coacervation method. J. Microencapsul. **27**, 78–85 (2010)
8. Charcosset, C., El-Harati, A., Fessi, H.: Preparation of solid lipid nanoparticles using a membrane contactor. J. Control Release **108**, 112–120 (2005)
9. Cavalli, R., Bargoni, A., Podio, V., Muntoni, E., Zara, G.P., Gasco, M.R.: Duodenal administration of solid lipid nanoparticles with different percentages of tobramycin. J. Pharm. Sci. **92**, 1085–1094 (2003)
10. Yuan, H., Huang, L.-F., Du, Y.Z., Ying, X.Y., You, J., Hu, F.Q., Zeng, S.: Solid lipid nanoparticles prepared by solvent diffusion method in a nanoreactor system. Colloids Surf. B Biointerfaces **61**, 132–137 (2008)
11. Bunjes, H.: Lipid nanoparticles for the delivery of poorly water-soluble drugs. J. Pharm. Pharmacol. **62**, 1637–1645 (2010)
12. Muchow, M., Maincent, P., Muller, R.H.: Lipid nanoparticles with a solid matrix (SLN, NLC, LDC) for oral drug delivery. Drug Dev. Ind. Pharm. **34**, 1394–1405 (2008)
13. Joshi, M.D., Muller, R.H.: Lipid nanoparticles for parenteral delivery of actives. Eur. J. Pharm. Biopharm. **71**, 161–172 (2009)
14. Bondì, M.L., Craparo, E.F.: Solid lipid nanoparticles for applications in gene therapy: a review of the state of the art. Expert Opin. Drug Deliv. **7**, 7–18 (2010)
15. Del Pozo-Rodriguez, A., Delgado, D., Solinis, M.A., Pedraz, J.L., Echevarria, E., Rodriguez, J.M., Gascon, A.R.:

Solid lipid nanoparticles as potential tools for gene therapy: in vivo protein expression after intravenous administration. Int. J. Pharm. **385**, 157–162 (2010)
16. Pardeike, J., Hommoss, A., Müller, R.H.: Lipid nanoparticles (SLN, NLC) in cosmetic and pharmaceutical dermal products. Int. J. Pharm. **366**, 170–184 (2009)
17. Seyfoddin, A., Shaw, J., Al-Kassas, R.: Solid lipid nanoparticles for ocular drug delivery. Drug Deliv. **17**, 467–489 (2010)
18. Strettoi, E., Gargini, C., Novelli, E., Sala, G., Ilaria, P., Gasco, P., Ghidoni, R.: Inhibition of ceramide biosynthesis preserves photoreceptor structure and function in a mouse model of retinitis pigmentosa. Proc. Natl. Acad. Sci. U. S. A. **107**, 18706–18711 (2010)
19. Nassimi, M., Schleh, C., Lauenstein, H.D., Hussein, R., Hoymann, H.G., Koch, W., Pohlmann, G., Krug, N., Sewald, K., Rittinghausen, S., Braun, A., Muller-Goymann, C.: A toxicological evaluation of inhaled solid lipid nanoparticles used as potential drug delivery system for the lung. Eur. J. Pharm. Biopharm. **75**, 107–116 (2010)
20. Videira, M.A., Botelho, M.F., Santos, A.C., Gouveia, L.F., De Lima, J.J., Almeida, A.J.: Lymphatic uptake of pulmonary delivered radiolabelled solid lipid nanoparticles. J. Drug Target. **10**, 607–613 (2002)

Solid Lipid Nanospheres

► Solid Lipid Nanoparticles - SLN

Solid Lipid-Based Nanoparticles

► Solid Lipid Nanoparticles - SLN

Solid–Liquid Interfaces

► Nanoscale Properties of Solid–Liquid Interfaces

Sound Propagation in Fluids

► Acoustic Nanoparticle Synthesis for Applications in Nanomedicine

Spectromicroscopy

► Selected Synchrotron Radiation Techniques

Spectroscopic Techniques

► Optical Techniques for Nanostructure Characterization

Spider Silk

Fritz Vollrath
Department of Zoology, University of Oxford, Oxford, UK

Synonyms

Biopolymer; Cob web; Extrusion spinning; Semicrystalline morphology

Definitions

Silks are animal fibers (or more rarely ribbons or sheets) of proteinaceous biomaterials that are, by definition, extrusion spun [1]. While the evolutionary origins and taxonomic placement of silk feedstocks can differ widely across the arthropods, filaments can be surprisingly similar [2]. *Capture silks* are sticky materials deploying either nanoscale filaments or aqueous glycoprotein glues that have evolved from dry silks [3].

Outline

Silks are fascinating biological products and have evolved several times independently in the arthropods. Spiders and moths are the best known and best studied silk spinners but there are others ranging from mites to bees [2]. In each taxon the diversity of silks has evolved in only one ancestor but then radiated quickly (given time frames of millions of years) into many different types fit for the purpose required by the animal – be it integration into a cocoon composite or use as a single safety line. Spiders are unique in that an individual has a real armory of silk-glands with each silk being tailored to a specific use. Accordingly the silks of the spiders show a surprising diversity ranging from tough dry threads to soft wet fibers and from sticky aqueous glue to adhesive nanoscale filaments [4]. This very diversity provides a window into the interaction of a silk's molecular structure with the animal's ecology and thus delivers deep insights into the material's function–structure relationship on the macroscopic, organismic scale. In addition, and of relevance to the subject of nanotechnology, such diversity also offers examples for function–structure relationships on the micro- and nanoscale, and these are the dimensional domains explored here.

However, in order to better understand the scope of the problem, and to better appreciate the range of possible solutions, it may be helpful to first briefly outline the biology of the system, and in this way to examine the constraints as well as demands imposed on the material "silk." After this more generic introduction, I will explore specific lessons learned so far from the study of spider silks and from comparisons to the commercially much more important mulberry and wild insect silks. I shall conclude my chapter with the briefest of outlooks into the future of silk as a model material for academic investigations into nanoscale processes.

Key Principles

Importantly, silks have evolved to function in the dead state, i.e., detached from the animal and typically dehydrated/denatured [1]. This means that silks provide a material in which one can study under realistic functional conditions and in great detail the thermomechanical properties of the material. Indeed, there are very few (if any) other biological materials providing comparable access to in-depth investigations under fully natural conditions. This feature of "natural function even in the fully detached state" provides detailed as well as biologically realistic data about a given silk's material properties. By applying general biological knowledge one can speculate about properties–function relationship (▶ Plasticity Theory at Small Scales). But in order to understand a silk's structure, and its potential relationships with the material's properties, one requires more than biological insight, i.e., structural data are needed. Such data would come from fine-grained structural analysis such as NMR, X-ray, and neutron scattering as well as Raman, IR, and CD/LD spectroscopy as well as data on gene and molecular sequence and insights into biological constraints such as the overbearing role of water in biological systems and the evolutionary history of the organism and the silks under investigation. Not surprisingly, while a good understanding of a specific silk's material properties is emerging, the links to function and structure are still rather weak [5].

Accordingly it might be more realistic, at this stage, to view the study of silks as research aiming to link properties to function and on to structure – rather than taking the more common approach of exploring a "straight" structure–function relationship. This view notwithstanding, with modern molecular techniques (ranging from sequencing to scattering and modeling) the protein and bond interactions that underlie the functional properties are rapidly being uncovered and a composite picture is beginning to emerge with clear-cut hypothesis that are ready for rigorous testing [6].

Why Study Spider Silks?

The threads of the commercial mulberry silkworm *Bombyx mori* are the fibers spun by the larva of a lepidopteran moth as part of its cocoon composite constructed to shelter the larva during its metamorphosis. These fibers have been the mainstay of a key textile industry for over 6,000 years and as such have seen thousands of years of ever more sophisticated R&D. However, as the worm spins from converted salivary glands through the mouth, it can cut the thread, which (as we will see) poses certain problems to studying the extrusion process itself. Moreover, lepidopteran silks have evolved for integration into multilayered cocoon complexes and thus have never been optimized for single fiber strength or toughness but for the specific qualities required of threads in a composite.

Spider silks, on the other hand, have evolved for a much wider range of applications (Fig. 1). The best studied of these silks, the dragline fibers of the Major Ampulate Glands, are optimized for effective and efficient deployment as single-thread safety lines and for integration as key structural members in a web's open-mesh network. In both applications, these threads' primary function is always the mitigation of kinetic energy be it for body support or prey impact and restraint. Many spider silks, therefore, are excellent examples for naturally evolved protein (bio)materials optimized for toughness. Moreover, spiders spin from

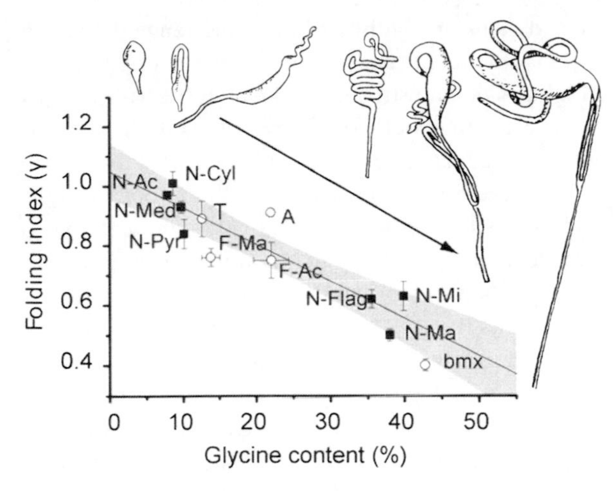

Spider Silk, Fig. 1 Relationship between the glycine content of a silk and the "folding index" γ of its proteins in the prespun liquid stage. γ is taken to be indicative of the protein intrinsic disorder. The correlation and model are taken to quantitatively explain the structure–function relationship by describing the molecular conformation, i.e., β-sheet propensity. The data suggest that, in order to achieve specialization and performance, silks require higher structural flexibility at the expense of reduced stability and increased conversion energy. γ is calculated from the ratio of the circular dichroism spectrum bands at 208 nm and 220 nm (at 20°C); a γ value near 1 would denote helix-type folding while γ values <0.5 would signify mostly unfolded chains. The arrow shows the direction of gland evolution and the insets depict schematically the overall gland shape (not to scale). *Legend*: *Nephila edulis* (golden orb spider) major ampullate (N-Ma), minor ampullate (N-Mi), flagelliform (NFlag), cylindriform (N-Cyl), aciniform (N-Ac), pyriform (N-Pyr), median (N-Med), *Kukulkania hibernalis* (Filistatidae) major ampullate (F-Ma) and acinous (F-Ac). *Antrodiaetus unicolor* (Antrodiaetidae) single-type glands (A) and *Aphonopelma chalcoldes* (Theraphosidae) acinous (T) as well as the commercial mulberry silk from *Bombyx mori* (Insecta: Bombicidae) (bmx – silkworm silk) (for details see ref. [11])

modified, abdominal leg glands through specialist spinnerets, far away from any cutting mouthpart. Hence spiders can be "silked" easily and under controlled experimental conditions [7].

Such experimental access allows to explore the contribution of environmental key parameters (such as temperature, pH, and flow characteristics during extrusion spinning) on the conformation and interactions of supramolecular structures using (even on-line as the animal spins) high-resolution analytical tools such as X-ray scattering or Raman spectroscopy [7]. Hence much of what is known to date about the molecular structure–function relations in the biological elastomer "silk" originated from studies of spider silks rather than insect silks. Not surprisingly, all relevant studies showed that the key determinant for the silk's outstanding mechanical properties seems to be its notable structure, which, in turn, relays on the scaling of its semicrystalline morphology [8].

The Importance of Extrusion Spinning

Importantly, the morphology of a silk is the outcome of extrusion processing not growth processes. Extrusion "spinning" relies on the refolding of molecules under physical forces such as flow elongation and shear in a tubular duct typically combined with shifts in the chemical environment [9]. Chaperoning molecules might provide additional guidance. Whatever the details of the individual extrusion processing of specific silks, as the fiber is drawn, the solvent water is spontaneously ejected in a self-denaturing sol–gel conversion with the molecules refolding and rearranging themselves from one conformation into another. While the molecular conformation in the feedstock is adapted to safe storage (sometimes for weeks) in an aqueous solution, the final conformation in the thread is adapted to stability (sometimes for years) against chemical, physical, and biological agents of the environment. The transition between the two conformations typically happens in milliseconds and appears to require very little energy [9]. At present it is known *where* it happens (in the duct), it is known roughly *what* happens (molecular alignment in the flow accompanied by unfolding and hydrogen bond "cross-linking" within and between molecules) but it is not known exactly *how* it happens [10].

One may assume that a silk, be it in the liquid or solid state, adopts shapes and conformations that are dictated by the interactions between polar and nonpolar moieties [11]. A fiber's macroscale properties are determined by bulk orientations with optimal axial stiffness on the nanoscale. To achieve this, the molecules must be allowed (during the transition from solution to solid) to self-organize into their extended configurations and be accompanied by an efficient intermolecular lock-in. However, in order to maintain flow viscosity, the number of cross-links must be limited. α-helical structures are stable as isolated secondary structures (while β-structures are not) and tend to engage and stabilize interactions with neighboring strands thus forming an intermolecular gelation network. Control of the silk molecules seems to be achieved by a number of reactions. For example, calcium ions stabilize silk proteins at high concentrations but they destabilize (and apparently promote β-sheet structures) at low concentrations. Sodium ions also stabilize the silk proteins (important for storage) and are exchanged during the extrusion process by β-sheet promoting potassium ions. Interestingly, similar conditions will produce the formation of amyloids in globular and other fibrous proteins [11].

These (and other) processes during extrusion "spinning" emphasize the importance of a molecular structure and conformation that, while still in the prespun phase, has its future conformations as solid fiber embedded in the amino acid sequence and their accompanying hydrogen bonds [12]. Importantly for the commercialization of bio-inspired ideas, extrusion spinning is a process that is well established in industrial applications and commercial polymer production while highly controlled biological growth processes are still outside industrial exploitation (other than via microbial intermediaries). Polymer theory provides a possible tool to explore the various nanoscale structures that self-assemble in the silks during spinning [1]. Ideally, the lessons thus learned can then be used by practitioners to design and manufacture new polymers using novel (and hopefully more sustainable) production methods (▶ Bioinspired Synthesis of Nanomaterials, ▶ Self-assembly, Sol Gel Based Nanostructures). After all, silks (and spider silks especially) have many properties that are not only highly eco-friendly but also economically interesting, ranging from full biocompatibility to exceptional mechanical properties.

Mechanical Properties

Spider silks show an enormous diversity of morphologies and mechanical properties, which is not surprising given the great diversity and age of the material [13]. Moreover spider silks have adapted to a wide range of purposes ranging from "wall papering" burrows to aerial nets and range in fiber diameters (which is a key parameter of actual strength) from about 10 μm down to a few tens of nanometers [4]. As a rough guide one could take spider silk fibers to have engineering properties (i.e., mechanical properties compensated for fiber dimension) with moduli in the range 1 kPa for a highly hydrated "gel" to about 20 GPa for the stiffest dragline silk, and strengths from almost zero values of yield stress to about 1.6 GPa respectively. In comparison, a decent commercial silk fiber (as used in textiles) would have a modulus and strength of about 10 GPa and 400 MPa, respectively [1]. The combination of excellent strength and high strain tolerance gives spider silks their exceptional toughness.

Importantly for the polymer scientist, silks offer a wide range of combinations of stiffness and toughness that might provide, if the underlying principles are understood, attractive opportunities for the bio-inspired design of novel materials. For example, a dragline of *Nephila clavipes* with a fiber diameter of about 6 μm has a strength to break of about 1.6 GPa at strains of around 30%. The estimated yield stress (ca. 0.2–0.3 GPa for a 10 GPa modulus) suggests that the spider deploys post-yield plastic flow and strain hardening to dampen its movement in a sophisticated combination of viscoelastic properties including the interesting feature of shape memory [1]. Models and simulations allow the exploration of the molecular processes underlying the stress–strain curves that, for spider silks, can easily be obtained under a wide range of experimental conditions [1, 7, 9], as outlined earlier.

Modeling Approaches

Surprising fits between observation and interpretation can be achieved by suitable models and simulations that set out to predict ab initio the energy interactions between properties and structure in silk feedstocks and fibers (▶ Ab Initio DFT Simulations of Nanostructures). For example, quantum mechanics can be used to predict elastic instability energy conditions for hydrogen-bonded interactions between water and the silk proteins. Such energy conditions can then be transformed into parameters such as temperature and stress by using clearly defined relations for zero point and thermal energies that contribute to the denaturation and glass transition temperatures. A realistic model (e.g., 14) for an energy instability criterion for protein denaturation would have four key stages: (1) calculate the potential energy of water–amide hydrogen bonding as a function of bond length, (2) calculate the energy at which water become free to move away from the amide segments, (3) calculate the kinetic/entropic energy countering the bond potential energy, and (4) calculate the probability of water–amide bond dissociation for denaturation as a function of time, temperature, and applied stress.

A specific simulation of stress–strain processes in a spider's dragline silk might serve to demonstrate here how the general mechanical properties of a dragline silk can be related to internal structure [14]. Molecular dynamics can be used to develop a model that is able to match observed stress–strain profiles with hypothetical "snapshots" of the deforming structures [1]. One insightful model leads to the conclusion that the extraordinary toughness of spider draglines derives from an energy equilibration, as does the post-yield strain hardening, which in turn is critical for the strength–toughness balance [14]. This model takes a poly(glycine) "string-of-beads" structure with periodic boundary conditions in an energy minimized structure to have a density of 1.3 g/cc at 300 K. The purely elastic reference modulus B_e would be given by the cohesive binding energy density with the molecular interactions derived from the Lennard-Jones potential function providing a proportionality constant for the molecular interactions of 18. Consequently, a cohesive energy of about 40 kJ/mol and a volume of 50 cc/mol for a generic silk peptide segment would give $B_e \approx$ 14 GPa. The energy dissipated through the broad secondary relaxation due to hydrocarbon segments would result in an isotropic tensile modulus of about 9 GPa at a Poisson's ratio of about 0.4. The evaluation of such molecular dynamics simulations suggest that the gradual post-yield regeneration of hydrogen bonding under strain can absorb large amounts of energy of deformation, which would increase the elastic modulus to its low-strain initial value [14]. There are other approaches to molecular modeling and the field is moving forward rapidly [8].

Of paramount importance for any modeling approach aiming to relate predictions of silk properties to structural features is the reduction of complex protein/polypeptide structures into "ordered" and "disordered" fractions [15]. The ordered domains are more rigid and deform significantly less than the disordered domains, which (like in any amorphous, viscoelastic polymer) would be responsible mainly for energy dissipation, i.e., toughness. Experiments have shown that silk properties are fine-tuned by water acting (as plasticizer) on the disordered fraction where it binds specifically to polar groups and thus reduces the glass transition temperature. As polar amide–amide bonding determines the glass transition temperature, T_g (of ca. 200°C in dry silk), increasing hydration of a silk will reduce not only T_g but also modulus and yield stress [8].

Clearly, as in other organic and inorganic materials, the nanoscale order–disorder organization coupled with the degree of hydration are the two key features that define property–structure relationships. An experimental window into these relationships is provided by the comparison of natural silks and silk-like materials prepared from silks that have been dissolved in chaotropic agents to be re-spun using dehydrating agents. These materials (often called reconstituted silks) share molecular sequence but not, apparently, molecular structure with native silks. Nor do they share mechanical properties, which opens the window for studying the details of properties–structure relationships experimentally and with the aim of testing specific hypotheses [15] (Fig. 2).

Functional Properties

Energy is the currency of living organisms. And water is the primary commodity for life and thus the key to understanding biological interactions. Silk molecules are formed in a fully hydrated state and they are stored in an aqueous solution. Extraction of water obliges them to go through a sol–gel transition, become immobile and form the fiber. The rate and ratio of dehydration tunes the mechanical properties of the material via the nature of the hydrogen bonds and their intra-, as well as intermolecular interactions.

There is no question that spider silks and their properties can be defined by their sensitivity to hydration and, indeed, even more powerful chaotropic agents. Some silks swell and shrink when exposed to high humidity or submersion in water while others show no response at all. It seems that the proportion of the amino acid proline is a major contributor to such high water sensitivity – also called supercontraction – reflecting the silk's observable behavior. Proline, given its shape, would be a side group that brings "disorder" to the folding of the molecular chain. Its presence would thus affect a silk's material properties by skewing the key ratio of order and disorder. Interestingly, high-proline content silks not only take up water readily but also become rather "stiff" when re-dried. Moreover, if such a wetted and re-dried silk is

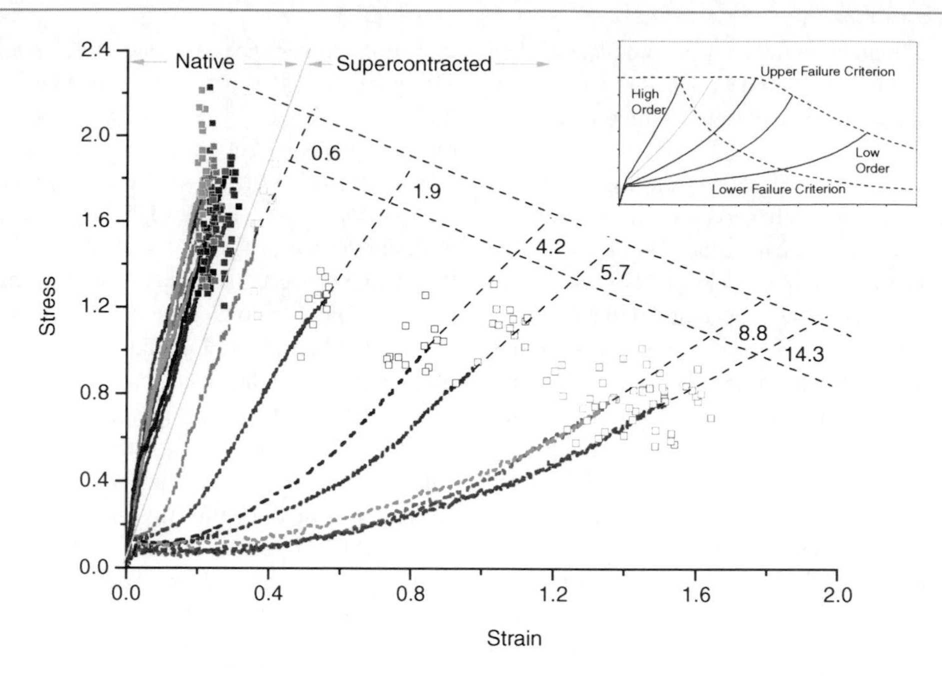

Spider Silk, Fig. 2 Representative stress–strain curves of silks with different percentages of proline, relating to different nanoscale structural arrangements. The silks were all major ampullate gland MAA dragline silks forcibly reeled from a range of species. The reeling condition had been tuned to produce silk samples with a native breaking strain of 25% and the silks were tested either in the native (*solid lines/squares*) or the supercontracted state (*dashed lines/open squares*). The proline content and ordered fraction of supercontracted MAA silks are marked. The straight black line defines the upper limit of native MAA silk, where the ordered fraction is 1.0; the straight gray line roughly separates the stress–strain curves of native and supercontracted silks (for details see ref. [1]). The inset shows the calculated mechanical properties of threads with different relationships of order/disorder (for details see ref. [18])

stretched and re-stretched repeatedly (i.e., cyclically loaded) it reverts to its original stress–strain behavior. This suggests that there is a reversible change in the order–disorder ratio even in the finished fiber, which reflects the level of hydration. It could be argued that the hydrogen bonds of the molecular chain segments around the proline segment are being forced apart under hydration only to be fixed into new positions during dehydration. Rehydration would allow water molecules to re-access the protein opened up at the proline site and assist in the formation of new hydrogen bonds. This would relax the mechanical strain "frozen" into the material and open the fiber for ingress of water and the exhibition of supercontraction with its accompanying changes in mechanical properties.

Clearly water in its interaction with the different motifs in the molecular chain of the silk protein can control a silk's material properties – presumably via the hydrogen bonds at occupied and potential bond sites.

The Importance of Spinning

Generically speaking, the "typical" spider silk consists of two proteins, one rather large (in the hundreds of kD) and one significantly smaller. There are notable exceptions, such as the silks of bees, which have three or four smallish proteins (in the tens of kD) or the silks of some midges, which seem to consist of hundreds of even smaller proteins [2]. Together, these and some of the unusual spider and lepidopteran silks should provide important test grounds for hypotheses (derived from the "standard" spider silks outlined earlier in this chapter) aiming to probe the relationships between structure, function, and properties of silks in general.

The principal hypothesis of nanoscale order–disorder as key to silk properties centers around the importance of a number of processes during the formation of the fiber. Indeed, the observable nanoscale structures and interactions appear highly evolved and, by conjecture and experiment, seem to confer on silks the impressive properties that, from the human standpoint, are highly desirable. Importantly, the molecular interactions have a fixed component (i.e., genetically transferred) as well as a flexible component (i.e., environmentally responsive). Thus, to use a human analogy, a spun silk fiber embodies both nature (the genetic blueprint) and nurture (the milieu during maturation).

Structure (having both fixed and a flexible component) provides the opportunity to experimentally probe the interactions between structure, properties, and function on the molecular and supramolecular, i.e., the nanoscale level. Firstly, sequencing the silk genes exposes the genetic blueprint and gives the exact positioning of the amino acids that make up the side chains of all silk-polymer macromolecules. Secondly, extracted liquid silk precursor (feedstock or dope) from the gland of the animal allows examination of molecular conformations in that prespinning state. Thirdly, molecular conformation in the spun fiber state can be studied in the fiber. Importantly, it is also possible to examine the molecular transformations from dope to fiber, although these kinds of study are much more difficult than the others given the small dimensions and quick timescales involved.

But it is during extrusion where the protein molecules self-assemble into the complex containing the correct intra- and intermolecular assemblies [9]. Clearly, parameters associated with flow through very fine ducts are important and, not surprisingly given this constraint, the block-copolymer molecules of the silk proteins display characteristics of liquid crystals. However, it is still debated whether they really are true liquid crystals or semicrystalline complexes behaving as such [1]. However this may be, already during the synthesis of the silk protein the molecular complex takes on the distinct morphology of a tightly folded nanoscale rod. Minute changes in pH imposed during the travel down the duct, in combination with mechanical forces, are exerted by the accelerating flow and elongate the silk fiber as it moves through the hyperbolic section of the extruder [9]. Water is pressed out as the proteins align and interact. Further down in the duct, which is now tubular, additional mechanisms (about which little is known) impose further interactions that give the thread its final shape, inside and out, as well as determining the mechanical properties, which are of course tightly correlated with internal and external structure.

Molecular Sequences

Silks are made up mostly of proteins and contain a range of amino acid combinations. The peptide segments/motifs have a core configuration of –NH–CO–CHR– where R represents one of 20 different side groups, which may exhibit highly divergent chemical functionalities. Spider silk proteins typically have very high fractions of the amino acid glycine, which has the simplest (smallest) of all sidegroups (G, R=H) and which, because the lack of steric hindrance, allows a wide range of folding conformations. The next important, and still rather simple, peptide groups are alanine (A, $R=CH_3$) and serin (S, $R=CH_2OH$) introducing hydrophobicity and polarity, respectively. Finally, there is proline (P, $R=CH_2\ CH_2\ CH_2$), which with its large cyclic and rather rigid side group intrinsically disorders the chains by twisting them out of their usual highly regular conformer torsional angles.

Importantly, it appears that in spiders the more "advanced" silks seem to have higher ratios of glycine coupled with rather complex spinning glands (Fig. 1). This suggests that molecular flexibility in a silk feedstock requires more control during extrusion; however, the inverse is also possible, i.e., that increased complexity during processing has led to increased molecular flexibility [11].

Proline is another key component of silk nanostructure by forcing the molecular chain that contains it in regular spacings into the string-of-beads morphology mentioned earlier. This conformation (as yet hypothetical but based on strong evidence) consists of "beads" of beta folds typically 10 units long and with a chain axis length of about 4 nm each [8]. Such a "string of beads" would facilitate processing in the flow field of the narrow extrusion duct [1] by allowing reorientation but not entanglement thus conferring liquid-crystalline non-Newtonian flow behavior [9].

Clearly, analysis of the primary structure already allows for the formulation of testable hypotheses and related interpretations of the secondary structures. And these are the structures that form most silks' semicrystalline, nanometer scale morphology of "hard," highly ordered domains in a "soft" matrix of more disordered polymer chains (Fig. 3). Degree of hydration is the key as the secondary conformations are held in place by hydrogen bonds connecting the amide groups of the main-chain backbone either directly (hard) or via a water molecule (soft).

	Soft/Disordered		Hard/Ordered
Spidroin I	GQG GYG GLG SQG A GRG GLG GQG A GAAAAAAAGG A		
	- (G X G)$_7$-	α-helix	β-sheet
Spidroin II	GPGGY GPGQQ GPGGY GPGQQ GPGGY GPGQQ GPSGPGS AAAAAAAAAA		
	- (GPGGX)$_7$-	random coil	β-sheet

Spider Silk, Fig. 3 Schematic of the two principal proteins identified in a benchmark spider dragline silk outlining the link between sequence and properties. Amino acid codes are A Alanine, G Glycine, P Proline, Y Thyrosine, L Leucine, and Q Glutamine

It must be assumed, although the final verdict is still out, that strong spider dragline fibers have a hybrid structure of about 50/50 hard/ordered to soft/disordered domains whereby the hard beta sheet-folded domains are embedded in a matrix of chains in various degrees of helical conformations thus acting like an amorphous matrix. Indeed, perhaps it is wise to hedge about the minute details of structural function–properties relationships until better data has come in from fine-grained analytical spectroscopic techniques such X-ray and neutron scattering, nuclear magnetic resonance Fourier transform infrared and ultraviolet spectroscopy, as well as dynamic mechanical thermal analysis, differential scanning calorimetry, and thermogravimetric analysis.

Conclusions

More than 400 million years of evolution will have seen to it that the hierarchical morphology in the patterning and mesophase assembly of spider silk molecular motifs is spatially optimized for each of the many specific functions required for survival of the spider. It has been argued that the evolutionary optimization of energy resources provides a perfect framework for multiscale models probing the intrinsic property–structure relationships of silks. In order to investigate the distribution and exchange of energy at the nanoscale, such a model would deploy the control of energy storage (strength) and dissipation (toughness) at the molecular level. If correct, then the model should predict the full stress–strain profile of silks to failure – covering the full range right from the strongest dragline threads to the most compliant capture threads in the web. Importantly, it appears that this can be done [8].

The model explaining key aspects of the behavior of finished fibers can be extended to integrate also key aspects of the formation of the fiber from the dope [7]. Experimental data demonstrate that "live" (i.e., native) silk dope differs significantly from "dead" (i.e., spun denatured then reconstituted) silk [16]. Studies that theoretically examine this conversion from unspun to spun *native* silk seem to indicate that a silk, once spun, cannot be "un-spun" and spun again [12]. This insight, if correct, will have important implications for the goal of producing spinnable silk dopes either by reconstitution from fibers [6] or by protein extraction from microbial expression systems [17]. According to present insights [12], both ways of producing seminative dope would fatally deconfigure the natural molecular structures; yet these (if the models are correct) would be key requirements for the natural spinning process to function. And that, after all, is one of the most amazing traits of silks and their energetically so efficient and clean production : it uses only water as solvent at ambient pressures and temperatures yet produces a biopolymer fiber that can hold its own proudly in comparison with the mechanical properties of all man-made polymer fibers.

As I have outlined in this chapter, spiders provide a generic class of natural materials, silks, which lend themselves as a models for a wide range of other biological elastomers. Unlike these, which naturally perform in the hydrated state, silks have evolved to operate in a wide range of states of hydration; from dry webs and cocoons to air sacs for underwater spiders, and with the subtle control of tightening sagging webs by supercontraction with dew condensation. Understanding the interaction of biological elastomers with water is a key requirement if we are to produce synthetic biomimetic analogues. Biological functionality, after all, relies on wet engineering coupled with nanoscale dimensions often integrated into structural "hierarchies" of differing degrees of order.

Cross-References

▶ Ab Initio DFT Simulations of Nanostructures
▶ Bioinspired Synthesis of Nanomaterials
▶ Plasticity Theory at Small Scales

References

1. Vollrath, F., Porter, D.: Silks as ancient models for modern polymers. Polymer **50**, 5623–5632 (2009)
2. Sutherland, T.D., Young, J., Weisman, S., Hayashi, C.Y., Merrit, D.: Insect silk: one name, many materials. Annu. Rev. Entomol. **55**, 171–188 (2010)
3. Brunetta, L., Craig, C.: Spider silk: evolution and 400 million years of spinning, waiting, snagging, and mating, pp. 1–229. Yale University Press, New Haven (2010)
4. Vollrath, F.: Spider webs and silks. Sci. Am. **266**(3), 46–52 (1992)
5. Fu, C., Shao, Z., Vollrath, F.: Animal silks: their structures, properties and artificial production. Chem. Commun. **43**, 6515–6529 (2009)
6. Omenetto, F., Kaplan, D.L.: New opportunities for an ancient material. Science **329**, 528–531 (2010)
7. Vollrath, F., Porter, D., Dicko, C.: The structure of silk. In: Eichhorn, S.J., Hearlem, J.W.S., Jaffe, M., Kikutan, T. (eds.) Handbook of textile fibre structure, vol. 2, pp. 146–198. Woodhead Publishing, Oxford (2009)
8. Porter, D., Vollrath, F.: Silk as a biomimetic ideal for structural polymers. Adv. Mater. **21**, 487–492 (2009)
9. Vollrath, F., Knight, D.P.: Liquid crystal silk spinning in Nature. Nature **410**, 541–548 (2001)
10. Aldo Leal-Egaña, A., Scheibel, T.: Silk-based materials for biomedical applications. Biotechnol. Appl. Biochem. **55**, 155–167 (2010)
11. Dicko, C., Porter, D., Vollrath, F.: Silk: relevance to amyloids. In: Riggaci, S., Bucciantini, M. (eds.) Functional amyloid aggregation. Research Sign Post, pp. 51–70 (2010)
12. Porter, D., Vollrath, F.: The role of kinetics of water and amide bonding in protein stability. Soft. Matter. **4**, 328–336 (2008)
13. Harmer, A.M., Blackledge, T.A., Madin, J.S., Herberstein, M.E.: High-performance spider webs: integrating biomechanics, ecology and behaviour. J R Soc Interface. **8**(57), 457–471 (2011)
14. Porter, D.: Group interaction modelling of polymers. Marcel Dekker, New York (1995)
15. Holland, C., Vollrath, F.: Biomimetic principles of spider silk for high-performance fibres. Chapter 7. In: Ellison, M.S., Abbott, A.G. (eds.) Biologically inspired textiles. Woodhead Publishing, Cambridge (2008)
16. Holland, C., Terry, E.A., Porter, D., Vollrath, F.: Rheological characterisation of native spider and silkworm dope. Nat. Mater. **5**, 870–874 (2006)
17. Spiess, K., Lammel, A., Scheibel, T.: Recombinant spider silk proteins for applications in biomaterials. Macromol. Biosci. **10**, 998–1007 (2010)
18. Liu, Y., Sponner, A., Porter, D., Vollrath, F.: Proline and processing of spider silks. Biomacromolecules **9**, 116–121 (2008)

Spiders

▶ Arthropod Strain Sensors

Spintronic Devices

▶ Magnetic Nanostructures and Spintronics

Spontaneous Polarization

▶ Polarization-Induced Effects in Heterostructures

Spray Technologies Inspired by Bombardier Beetle

Alexander Booth[1], Andy C. McIntosh[1], Novid Beheshti[2], Richard Walker[3], Lars Uno Larsson[4] and Andrew Copestake[5]
[1]Energy and Resources Research Institute, University of Leeds, Leeds, West Yorkshire, UK
[2]Swedish Biomimetics 3000® Ltd, Birmingham, UK
[3]ICON plc, Marlow, Buckinghamshire, UK
[4]Swedish Biomimetics 3000® AB, Stockholm, Sweden
[5]Swedish Biomimetics 3000® Ltd, Southampton, UK

Synonyms

Flash evaporation liquid atomization; Liquid atomization through vapor explosion; μMist®

Definition

μMist® is a liquid spray and atomization system inspired by a method that is used as a defense mechanism by several types of bombardier beetle. The liquid atomization, spray formation, and its propulsion/emanation are achieved by causing a liquid in a sealed chamber to flash evaporate on the release of an exhaust valve. This flash evaporation in the confined space of the chamber causes a vapor explosion, the force of which overcomes the surface tension of the liquid, causing it to break down into small droplets and exit the valve with a considerable momentum as a spray.

The μMist® System Development

Initial investigations were inspired by the entomological research of Professor Thomas Eisner in to the bombardier beetle's defensive spray system. A typical example of a bombardier beetle from Africa can be seen in Fig. 1. Computer simulations of the spray system, using computational fluid dynamics (CFD), then led to the understanding of the physical principles governing the beetle's unique spray facility, and the μMist® system is based upon these principles. When threatened, the bombardier beetle produces a hot liquid spray from its abdomen, which is ejected at its attacker. Eisner's work focused on how this spray was generated by the beetle. He found that at the rear of the beetle's abdomen was a complex series of glands and chambers in which an exothermic chemical reaction occurred [1]. This chemical reaction heated its constituent liquid to a very high temperature in a reaction chamber, linked directly to an exhaust orifice sealed by a biological pressure-triggered exhaust valve. When the liquid in the reaction chamber reached a certain pressure, this exhaust valve opened, causing the liquid in the reaction chamber to be ejected as a spray.

Eisner's work inspired McIntosh and Beheshti [2–5], at Leeds University, to investigate the physical process in the bombardier beetles which generated this spray [2, 3] and found that it is a major advance on established liquid atomization methods. This initial work was funded by the Engineering and Physical Sciences Research Council (EPSRC) and also by Swedish Biomimetics 3000® Ltd. All further work mentioned in this article and all continuing work on the μMist® system has been and is currently being supported and advanced by Swedish Biomimetics 3000® Ltd under its V^{2}IO innovation acceleration model. Additional funding has been given by partners for particular projects, including Carbon Connections and the NIHR i4i program.

The initial investigation conducted by McIntosh and Beheshti consisted of a series of Computational Fluid Dynamics (CFD) simulations to better determine the thermodynamic and fluid dynamic processes which

Spray Technologies Inspired by Bombardier Beetle, Fig. 1 African bombardier beetle

were occurring in the chamber of the bombardier beetle. These simulations did not include the associated chemical reaction seen in the bombardier beetle, as in this context it simply functions as a heat source to the chamber. The results of this study showed very good correlation between the simulated behavior of the bombardier beetle's spray mechanism and the behavior observed in the actual beetle [2, 3]. It was these results which inspired the ideas that led to the development of the initial physical μMist® system.

The Bombardier Beetle System and Simulation Work

The series of glands and chambers which make up the spray mechanism of the bombardier beetle are shown in Fig. 2, and the most important section is shown in diagram form in Fig. 3. The core of the beetle's spray mechanism is the chamber and the hard cuticle of the outlet tube, which acts as a pressure triggered exhaust valve. Initially the chamber is mostly empty, the cuticle is closed, blocking the outlet tube, and the inlet tube is open. When the beetle decides to spray, reactants are fed through the inlet tube into the chamber in the form of a dilute aqueous solution. These reactants are a toxic mixture of hydroquinone and hydrogen peroxide which react exothermically due to the addition of a catalyst believed to be in the inner surface collagen of the beetle's combustion chamber. This causes an

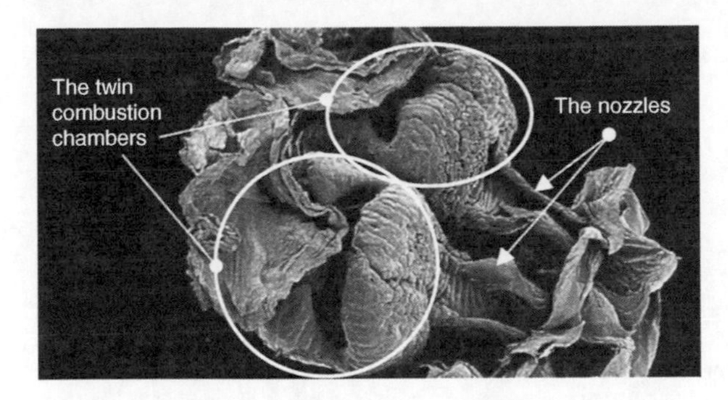

Spray Technologies Inspired by Bombardier Beetle, Fig. 2 An electron micrograph of the bombardier beetle's spray mechanism (Courtesy of Prof. T. Eisner, Cornell University)

increase of both temperature and pressure in the chamber. As the pressure in the chamber increases, it causes the inlet tube to be pinched shut by the chamber's extremities, forming a sealed volume of liquid (mainly water, plus some quinones from the products of the reaction). The ongoing exothermic reaction further heats the liquid in the chamber above its usual boiling point. Only a small fraction of the liquid is able to evaporate due to both the inlet and outlet tubes now being blocked. Eventually, the chamber pressure becomes high enough that it can force open a cuticle flap which holds the outlet tube shut. This opening occurs very suddenly and is entirely due to the buildup of pressure in the chamber. Thus, the cuticle flap is essentially a pressure relief exhaust valve which is passively triggered in the outlet tube. As the liquid in the chamber has been heated to a point high above its normal boiling point, and the opening of the exhaust valve is so sudden, a very rapid evaporation of the liquid occurs, known as flash evaporation. This flash evaporation does not affect the whole volume of liquid in the chamber, so liquid is caused to be forced out of the outlet tube by a rapidly expanding mass of vapor. The evaporation is rapid due to a vapor explosion, as sudden expansion takes place because of the change in bulk volume of water to vapor. This then shatters the remaining water into small droplets and the mixture is ejected with great force, with shear forces also contributing to the break up into a cascade of ever smaller droplets. The size of the droplets produced are inversely proportional to the magnitude of the forces exerted on and in the liquid volume, so larger forces produce smaller droplets as the surface tension and viscosity of the liquid are more easily overcome. As the liquid is ejected from the bombardier beetle's chamber, the chamber pressure drops, which causes chamber physiology of the outlet tube to now be closed again by the collapse of the cuticle valve. Since the chamber is now partially empty the inlet port is open once again and the process repeats itself. This is exactly the same principle used in modern pulse combustors.

This process was modeled by using Computational Fluid Dynamics (CFD) starting the simulation from the moment at which the cuticle valve on the outlet tube is forced open by the chamber pressure. The chamber is full of liquid (mainly water) already at a suitably high temperature. The inlet tube was not included in the CFD model. The outlet tube however was included in the model and, at the beginning of the simulation, the liquid in the chamber is allowed to flash evaporate. A diagram showing the chamber model used in the simulation can be seen in (Fig. 4) – a slice of the cylindrical chamber is simulated (assuming an axial symmetry within the chamber). The chamber dimensions were such that the volume was similar to that of the bombardier beetle, around 0.1 mm^3. Good correlation was found between the results of the simulation work and the results of tests on the spray produced by the beetle [3] in terms of characteristics such as spray velocity, throw ratio (i.e. ejection distance divided by chamber length), ejection duration, and mass discharged per pulse of spray. This showed that the method of spray atomization used by the beetle was correctly modeled. With this natural process now understood, the research was taken further by building a scaled-up artificial chamber to develop and thereby assess the suitability of this mechanism as a practical approach to liquid atomization and to conduct further research into the process itself.

Overview of the Physical μMist® System

Based upon the simulation work conducted at Leeds University, an experimental demonstration facility (Fig. 5) was built, which implemented the principles of this liquid atomization method as seen in the bombardier beetle. This system is called μMist®. In a similar way to the bombardier beetle and simulation work, the core of this system is the chamber and valves, as well as a heat source which replaces the heat generated through the bombardier beetle's

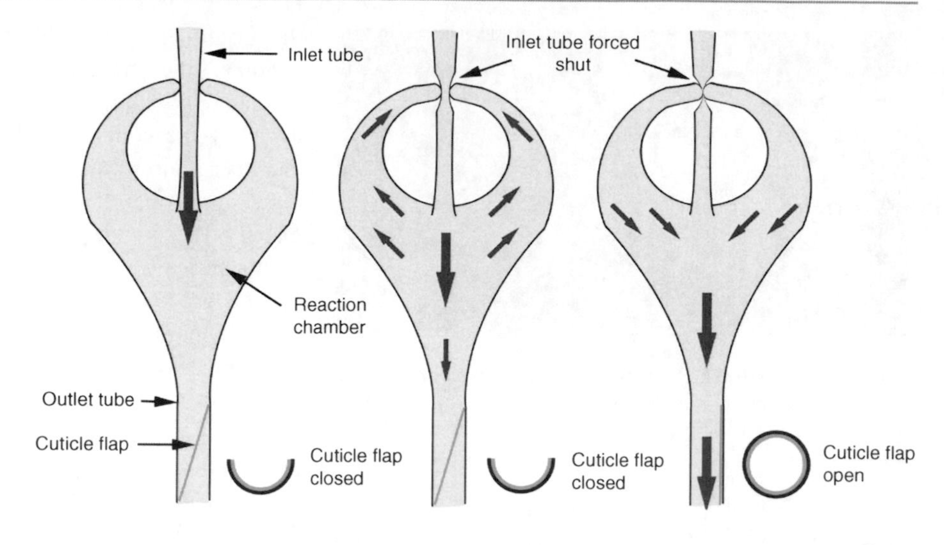

Spray Technologies Inspired by Bombardier Beetle, Fig. 3 A diagram of the spray mechanism of the bombardier beetle at three different stages in the ejection cycle

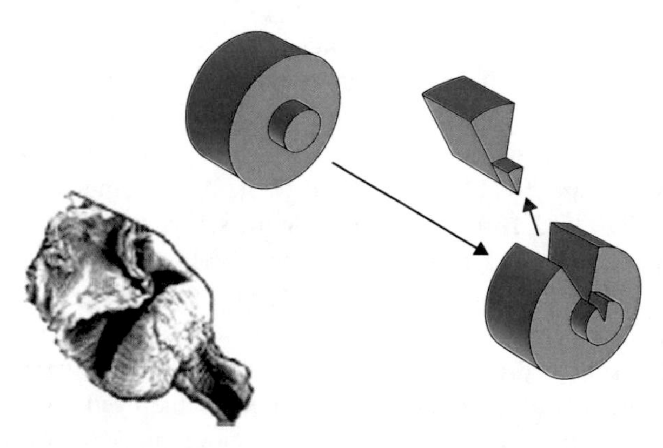

Spray Technologies Inspired by Bombardier Beetle, Fig. 4 An electron micrograph of a bombardier beetle's reaction chamber alongside diagram of simulated version used in CFD modeling (Electron micrograph courtesy of Prof. T. Eisner, Cornell University)

Spray Technologies Inspired by Bombardier Beetle, Fig. 5 The pre-prototype μMist® system in operation

chemical reaction. Unlike the bombardier beetle and the early simulation work, the exhaust valve on the physical μMist® system is not opened due to a buildup of pressure in the chamber (that is, passively), but instead, all valves are activated electronically, resulting in an actively controlled pulsed spray. There are also other valves in the system, which control the refill flow into the chamber between ejections, similar to the chamber extremities which control flow into the beetle's chamber through an inlet tube. Opening or closing these valves and the exhaust valve then controls the flow of liquid as seen in the bombardier beetle and results in atomized spray generation. As with the beetle and CFD simulation work, this is due to the formation of a sealed volume of liquid at high temperature in the chamber just before the exhaust valve is opened. It is important for this high-temperature volume of liquid in the chamber to be sealed to allow the required temperature rise with minimal associated vaporization. This ensures that the chamber liquid temperature is allowed to rise above its boiling point and that a portion of liquid will flash evaporate once the exhaust valve is opened. This produces liquid atomization at much lower pressures than utilized in most liquid atomization devices (which use pressure atomization – forcing the liquid through small holes at high pressure) and is energetically efficient. The fact that flash evaporation is used to achieve fast ejection of the water and steam, combined with appropriate control systems, means the system lends itself to a fine control of spray characteristics, particularly droplet size. This control is further improved by the decoupling of the exhaust valve from the chamber pressure. In an ideal system, there would be no vaporization of the chamber liquid during the heating phase; however, in practice, some vaporization will always occur, as governed by the equation for vapor pressure (Eq. 1). This is the major cause for the rise in chamber pressure seen in the bombardier beetle, which in that case triggers the opening of the passive pressure relief exhaust valve. Therefore in the bombardier beetle system, this limits the atomization level achieved, since the extent of flash evaporation will be roughly similar at each ejection. The liquid is heated a similar amount and exhausted at a similar time due to the correlation between chamber pressure and chamber temperature. This follows empirical equations [6] such as Eq. 1.

$$\ln P = C_1 + \frac{C_2}{T} + C_3 \ln T + C_4 T^{C5} \qquad (1)$$

Where:
P = Vapor pressure of the liquid
T = Temperature of the liquid
C_1, C_2, C_3, C_4, C_5 = A series of constants depending on the liquid; values for these constants for a range of liquids can be found in [6].

As the exhaust valve on the physical μMist® system is not pressure triggered, but actively controlled electronically, the extent of flash evaporation can be controlled to a greater degree than that of the bombardier beetle.

The physical scale of the first pre-prototype of the μMist® system is much larger than that of the system found in the bombardier beetle. Whereas the bombardier beetle system has a chamber of approximately 1×10^{-3}m in length, the first μMist® pre-prototype system has a much larger chamber, 0.02 m in length. This level of scaling is not limited to the chamber of the system and extends to other parameters, such as the diameter of the feed and outlet tubes. The process in the μMist® system however significantly increases the atomization of the liquid as it exits the chamber. The μMist® system can not only be used to atomize very small liquid volumes like those found in the bombardier beetle, but also much larger volumes with a wider range of practical uses. Some key performance parameters are also scaled with the increase in chamber size, such as the throw ratio. A bombardier beetle can throw liquid 0.2 m, using a chamber 1×10^{-3}m in length. This gives a throw ratio of 200 (Eq. 2). Comparatively, it was found that the chamber of 0.02 m in length on the μMist® system could also achieve a throw ratio of 200, throwing liquid a distance of up to 4 m [4, 5].

$$R_T = \frac{D_T}{L_C} \qquad (2)$$

Where:
R_T = Throw Ratio (Dimensionless)
D_T = Distance of throw (m)
L_C = Characteristic length of Chamber (m)

In summary, despite the differences between the physical μMist® system and the liquid atomization seen in the bombardier beetle and the simulation work, the core atomization method is the same. A liquid, heated past its boiling point, and at a constant volume, is suddenly allowed to vaporize though the rapid opening of an exhaust valve. This causes a flash evaporation of a portion of the liquid, which generates a very large force, which then ejects vapor and liquid out through the valve. The flash evaporation is such that as the liquid is rapidly pushed out, large shear forces are generated on and within the volume of the liquid which repeatedly shatter it by overcoming its surface tension and viscosity in a cascade, producing a range of droplet sizes in the emanating spray. Many advantages of the liquid atomization seen in the bombardier beetle, such as high throw ratio, low chamber pressure, and overall energetic efficiency of the system are also seen in the physical μMist® system, despite it being on a much larger scale. The μMist® pre-prototype system also expands upon the liquid atomization method by decoupling the exhaust valve from the chamber pressure, allowing more control over an already versatile system.

Performance of the μMist® Spray System

One of the key features of the μMist® system as a liquid atomization system is that it can be controlled to produce large variation in the characteristics of the

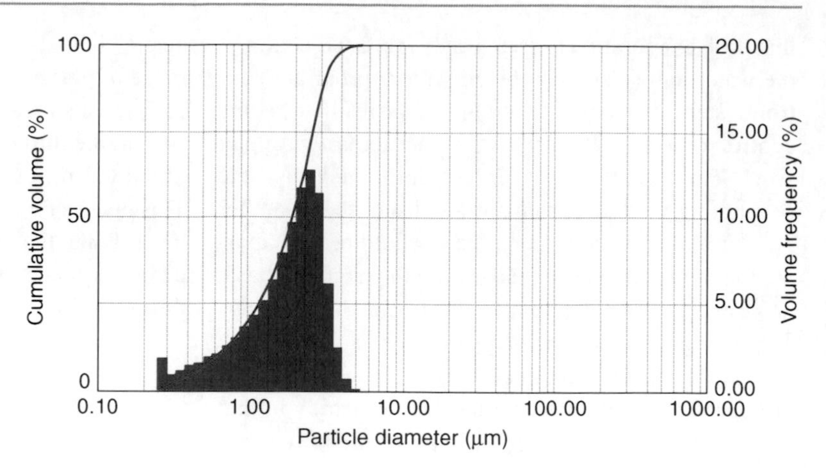

Spray Technologies Inspired by Bombardier Beetle, Fig. 6 Typical droplet size distribution when pre-prototype μMist® system is producing its smallest droplets

spray produced. The type of spray produced is controlled by a number of factors including pressure and temperature in the chamber. Consequently, a range of spray characteristics can be achieved which is much wider than possible with other liquid atomization systems. Experimental work on the μMist® system has shown that it can generate a very wide range of spray characteristics such as droplet size distribution, the ejection velocity of the spray, the mass ejection rate, and the temperature of the ejected spray.

One of the most significant characteristic for many spray applications is the droplet size distribution of the spray produced. The volume of droplets in a particular droplet size band is also important. This is usually categorized graphically, such as is shown in Figs. 6–8. A usual indicator of size of droplets is given by calculating the droplet size below which the total volume of droplets account for 90%, 50%, and 10% of the overall spray volume. These indicators are termed Dv (90), Dv (50), and Dv (10), respectively. The graphs in Figs. 6–8 show typical droplet size distributions recorded during experimental work on the first μMist® pre-prototype system under different input conditions [4]. These results were recorded using a Malvern Spraytec system, which measures droplet sizes through laser diffraction. The graphs give an idea of the range of droplet sizes that can be achieved by altering basic input parameters of the system. Figure 7 is a good example of bimodality in the droplet size distribution, where the majority of the liquid volume is contained within two peaks, rather than one. This result is significant as some spray applications require two or more ranges of droplet sizes to make up the majority of the spray volume, each of which are designed to perform different tasks and when combined produce a more effective delivery than a unimodal distribution. A μMist® system would be particularly suited to these sorts of applications. The ability to generate such a wide range of droplet size distribution shown in Figs. 6–8 shows that the μMist® system could potentially provide a single technology which can be used to satisfy a wide range of spray applications.

Applications of the μMist® Spray System

There are many potential applications for the μMist® system in the global market of sprays. This is primarily due to the wide range of spray characteristics which can be achieved using this technology and also due to it being much more environmentally friendly than current spray technologies. There are also related energy efficiencies inherent in the μMist® system, while using water can replace environmentally unfriendly substances currently in use for aerosols and some other spray applications. Some of the potential applications are currently being actively pursued in industrial development programs led by Swedish Biomimetics 3000® Ltd using its V^2 IO innovation acceleration business model. In this section, a few examples of potential applications for the type of droplet size distributions seen in Figs. 6–8 will be examined, showing how the μMist® system is suited meet to a wide range of requirements.

Fuel Injectors

A very important application of the μMist® technology is to fuel injectors, and this is one of the potential

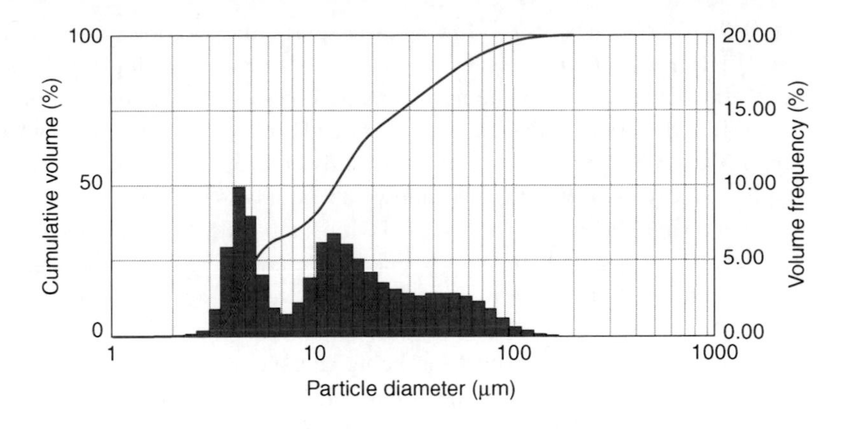

Spray Technologies Inspired by Bombardier Beetle, Fig. 7 Typical droplet size distribution when pre-prototype μMist® system is producing medium sized droplets

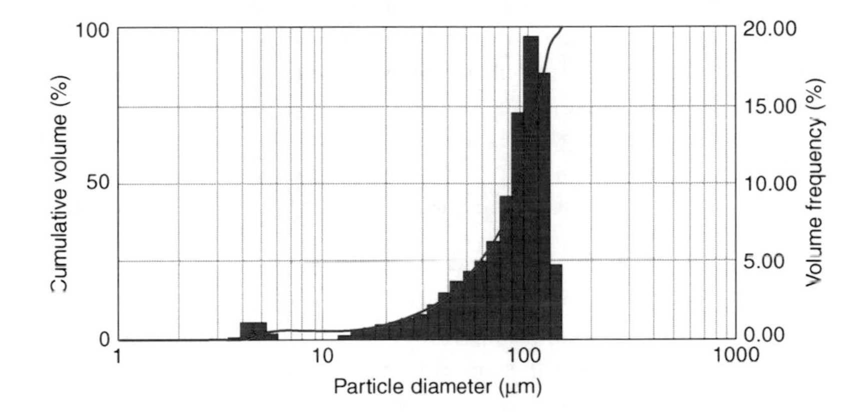

Spray Technologies Inspired by Bombardier Beetle, Fig. 8 Typical droplet size distribution when pre-prototype μMist® system is producing its largest droplets

applications currently being pursued by Swedish Biomimetics 3000® Ltd, with initial support from Carbon Connections. The droplet size distribution in Fig. 6 at the lower end of the range of the μMist® system shows why μMist® is a potential fuel injector technology. Here it can be seen that very small droplet sizes can be produced, resulting in droplets with a high surface area to volume ratio. This droplet size distribution is particularly suited to combustion applications, such as fuel injection, etc., as the larger overall surface area means that the fuel burns much more efficiently. Current fuel injectors primarily work on the principle of pressure atomization, where liquid is atomized by using high pressure to force it through a small opening to create a fine spray. It is this that generates the stresses on the surface of the liquid which cause it to break down into small droplets [7], rather than the flash evaporation seen in the μMist® system. Generating these high pressures requires a significant amount of energy due to the large pressures used (over 1000 Bar) to ensure proper atomization of the liquid. The energy requirements are large even to produce relatively large droplets. The μMist® system would require substantially less energy to properly atomize the liquid, as all that is required is to heat the system to a relatively modest temperature. Additionally, typical fuel injectors produce droplet sizes in the region of 30–80 μm. It can be seen from Fig. 6 that a μMist® system is capable of producing droplet sizes far smaller than this, in the region of 1–3 μm with some fluids. This would be a significant improvement over current fuel injection technologies, as not only do the smaller droplets burn much more efficiently, increasing engine efficiency and decreasing fuel consumption, but the more complete burning of the fuel in the engine would also lead to fewer harmful emissions being released into the atmosphere. This is in addition to the much lower energy consumed to atomize the fuel compared to current technologies.

Drug Delivery Systems

Another potential application of the droplet size distribution shown in Fig. 6 – again being pursued by

Swedish Biomimetics 3000® Ltd, with support from the NIHR i4i program – is in the design of next-generation drug delivery systems. For many illnesses, the preferred method of drug delivery is directly to the lungs. This is best achieved using small droplet sizes (typically less than 10 μm) that allow the liquid to travel deep into the lungs, where it can be most rapidly absorbed into the blood stream. The inherent efficacy of inhaled therapeutic drugs makes this a key development area. μMist® could potentially provide a generic drug delivery system, with one unit being capable of delivering a wide range of drugs, or a personalized medicine device, which is built to satisfy a particular patient's unmet medical needs. Current investigations are also being made to address the pharmaceutical industry's need for innovative drug delivery systems for the administration of novel compounds, including, but not limited to, peptides and oligonucleotides. This would be a significant step forward in drug delivery technology. Other potential drug delivery applications of the μMist® system include needleless injection and nasal drug delivery.

Consumer Aerosols

The μMist® system could also have significant impact as a consumer aerosol generator, primarily due to its technically advanced performance, but most importantly due to environmental benefits. Standard spray/aerosol cans, as well as most medical inhalers, generally use Volatile Organic Compounds (VOCs) such as propane and butane to generate the high pressures required to atomize the delivered liquid as it exits the can through the nozzle. It is thought that these VOCs cause environmentally damaging atmospheric compounds such as carbon dioxide and other atmosphere borne reactive species. They are also highly flammable and therefore are a considerable safety risk at the point of use. The μMist® system has the capability of delivering consumer spray aerosols as well or even better than the spray systems currently used without the need for possibly harmful dangerous VOCs. The only by-product of this system, besides the active ingredient itself, is water.

Fire Extinguishers and Fire Suppressants

The droplet size distribution seen in Fig. 8 primarily features large droplets (around 100 μm) with a wide overall droplet size distribution (down to around 4 or 5 μm). This sort of droplet size distribution would be well suited to a fire-fighting system as the different droplet sizes achieved both have a significant role in fighting fires. Large droplets, maybe ~100 μm, are effective at cooling the fire to a level below its reaction temperature, whereas small droplets evaporate very quickly and move oxygen away from the source of the fire – fire suppression. This range of droplets being produced by one system would allow fires to be extinguished and suppressed efficiently as the fire progresses. Another significant feature of the μMist® system when producing droplets in this range is that it has a very high throw ratio, even for small droplets, and thus targeting is possible. As mentioned previously, the current μMist® system has achieved a throw distance of up to 4 m under these conditions, which is a throw ratio with respect to the characteristic chamber length of 200 [4, 5]. With larger chamber volumes, it is envisaged that an even greater throw distance can be achieved, which would be particularly relevant to fire-fighting applications.

In summary, the wide range of spray characteristics and the significantly reduced environmental impact of the μMist® spray system allow it to be well suited to many applications, in particular fuel injection, drug delivery, consumer aerosol systems, fire-fighting systems, and fire-suppressant systems. In general, the μMist® spray system offers improvements in overall performance, efficiency, and environmental impact. Additionally, the system is economic, being very efficient in energy use – especially in comparison to pressure atomization systems.

Conclusions

The μMist® system is a versatile new liquid atomization technology inspired by the unique self-defense mechanism found in the bombardier beetle. The liquid atomization occurs when a sufficiently heated liquid in a sealed chamber is caused to flash evaporate due to the sudden release of an exhaust valve. The flash evaporation mechanism is a new method of providing the large forces necessary to rapidly eject the liquid from the chamber, with shear forces shattering it into a cascade of small droplets. Initial work on the understanding of the bombardier beetle's defense system as a general liquid atomization system was undertaken by McIntosh and Beheshti [2–5] at Leeds University, taking the form of a series of CFD simulations to better understand the key thermodynamic and fluid dynamic principles of the atomization process. This understanding of the thermodynamic and fluid dynamic principles of the atomization process was then used to inspire a design and then construct the first pre-prototype μMist® system as a base for experimental work and further research.

The experimental work on the μMist® system showed it to be highly versatile, producing a wide range of variation in the resulting spray characteristics. For example, it was shown experimentally that the system could be designed to produce droplets with diameters of anywhere between ~1 and 100 μm. Other spray characteristics which showed similar levels of variation due to the design of the μMist® system included the ejection velocity of the spray, modality cone angle or divergence, the mass ejection rate, and the temperature of the ejected spray.

The large variation in the spray characteristics seen in the experimental work on the μMist® system shows it to be adaptable to a wide range of applications, including fuel injection, drug delivery, consumer aerosols, fire fighting, and fire suppression. In some cases, the performance benefits of using the μMist® system would create significant efficiency savings and related environmental and economic benefits. In addition, further environmental benefits are possible as the μMist® system removes the need for harmful VOCs in those spray systems that currently use them, outputting only water, as well as requiring much less energy than standard pressure atomization systems.

Swedish Biomimetics has a worldwide exclusive licensing agreement with the University of Leeds to research, develop, and to commercialize the μMist® platform technology and its various potential applications. Patent pending publication Nos. US11/528,297 and WO2007/0342307 are in place for the μMist® platform technology and its application into fuel injection, respectively.

Cross-References

- ▶ Bio-inspired CMOS Cochlea
- ▶ Biomimetic Flow Sensors
- ▶ Biomimetic Mosquito-like Microneedles
- ▶ Biomimetic Muscles and Actuators
- ▶ Biomimetics
- ▶ Biomimetics of Marine Adhesives
- ▶ Toward Bioreplicated Texturing of Solar-Cell Surfaces

References

1. Eisner, T.: For Love of Insects. Harvard University Press, Cambridge (2005)
2. Beheshti, N., McIntosh, A.C.: A biomimetic study of the explosive discharge of the bombardier beetle. J Des. Nat **1**(1), 61–69 (2007)
3. Beheshti, N., McIntosh, A.C.: The bombardier beetle and its use of a pressure relief valve system to deliver a periodic pulsed spray. Bioinspir. Biomim. **2**, 57–64 (2007)
4. Beheshti, N., McIntosh, A.C.: A novel spray system inspired by the bombardier beetle. Presented Weds 25th June 2008 as Invited Presentation, Design and Nature IV, Proceedings of 4th Int. Conference on Design and Nature, Tivoli Almansor, Algarve, Portugal, 24–26 June 2008. In: Brebbia, C.A. (ed.), Design and Nature IV. WIT Transactions on Ecology and the Environment, vol. 114, pp. 13–21. WIT Press, Southampton/Boston (2008)
5. McIntosh, A.C., Beheshti, N.: Insect inspiration. Phys. World Inst. Phys. **21**(4), 29–31 (2008)
6. Perry, R.H., Green, D.W.: Perry's Chemical Engineers' Handbook. McGraw-Hill, New York (2008)
7. Lefebvre, A.: Atomization and Sprays. Taylor & Francis, London (1989)

Sputtering

▶ Physical Vapor Deposition

Stable Ion Clusters

▶ Prenucleation Clusters

Stereolithography

Hyundoo Hwang and Yoon-Kyoung Cho
School of Nano-Bioscience and Chemical Engineering, Ulsan National Institute of Science and Technology (UNIST), Ulsan, Republic of Korea

Synonyms

Microstereolithography

Definition

Stereolithography (SL) is an additive manufacturing process based on photopolymerization of a photo-curable polymer. The solid polymer part can be built via localized polymerization by its selective exposure to light in layer-by-layer format. SL is a kind of solid freeform fabrication (SFF) process for manufacturing solid objects by the sequential delivery of energy or material to specified points in space. SFF is sometimes referred to as rapid prototyping, rapid manufacturing, additive manufacturing, additive fabrication, or layered manufacturing. Microstereolithography or nanostereolithography simply refers to SL which possesses feature resolutions significantly higher than several tens of micrometers or nanometers precision, respectively.

Overview

Stereolithography (SL), which was first introduced in 1981 [1], is a three-dimensional (3D) manufacturing technology based on a layer-by-layer photopolymerization of a photo-curable polymer, so-called resin. A light irradiates or traces in a specific cross-sectional pattern on each layer of the photo-curable polymer, which is initially a liquid state, resulting in the solidification of the resin in the exposed area by the photopolymerization process. After every exposure step for a single layer, the other sliced two-dimensional (2D) pattern is exposed and polymerized on the below layer by moving the vertical position of the resin chamber or the light. This layer-by-layer 2D fabrication process is repetitively continued until the formation of 3D structure is completed. The performance of SL depends on various factors including the nature of resin, apparatus for fabrication process, strength of light energy, drawing speed of light beam, focusing angle of light beam, depth of focus, and constrained method of liquid-state resin.

Due to the ardent wish of many researchers for reducing the minimum feature of structures that one can manufacture as small as possible, several kinds of stereolithographic techniques, which use much more tightly focused light to initiate polymerization and cure the part in micro- or nanoscale, have arisen. Microstereolithography (MSL), which allows much higher spatial resolution than conventional SL, has been one of the most promising technologies for manufacturing 3D structures in microscale due to its capability for direct fabrication of complicated 3D structures with high accuracy up to 1 or 2 μm. Most of the MSL techniques have been based on the conventional 2D photolithography using ultraviolet (UV) light source and UV curable polymer. Therein, UV exposure for hardening the liquid-state resin is repetitively conducted on each layer for finally forming 3D microstructures by stacking one layer at a time. Depending on the methods of light exposure, there are two big categories in SL for micro-/nanofabrication: (1) beam scanning-based [2–4]; and (2) image projection-based approaches [5–7]. The former uses a focused light beam tracing a specific 2D pattern over or within the liquid-state resin for each layer, while, in the latter, an image pattern generated by a dynamic pattern generator irradiates only one time for each layer.

Methodology

UV Beam Scanning Stereolithography

In the scanning-based SL process, a very tightly focused laser beam is scanned over or within the liquid-state polymer to produce a 3D solid object. There are several ways to construct a 3D solid part using this scanning-based SL scheme in the perspectives of (1) scanning component – a mirror or an *X-Y* stage (see Fig. 1); (2) light beam direction – from upper or bottom; (3) surface conditions of resin – free or constraint (see Fig. 2); and (4) number of light beam – single or multiple.

With the mirror scanning method, it is very important to maintain the focal point of the light beam over the planar surface of the liquid resin. For this purpose, a dynamic focusing lens controlled by a computer should be required in a classical SL system (see Fig. 1a). However, it is very difficult to achieve high resolution using this dynamic light focusing. To deal with this problem, not only a Z-stage, but also an *X-Y* stage for scanning the resin chamber has been applied instead of the dynamic focusing lens (see Fig. 1b). While the chamber containing the liquid resin is translated in the *X-Y* plane, the focal point of the light beam keeps its position in the plane to achieve the tightest focus as possible. However, the maximum scanning speed of this *X-Y* stage scanning method is lower than that of the method using a rotating mirror.

Four types of SL system which can be distinguished on the basis of two criteria – the direction of light beam and the surface conditions of resin – are shown in Fig. 2. In the free surface configurations shown in Fig. 2a, b, the surface of the liquid-state resin is free, while the

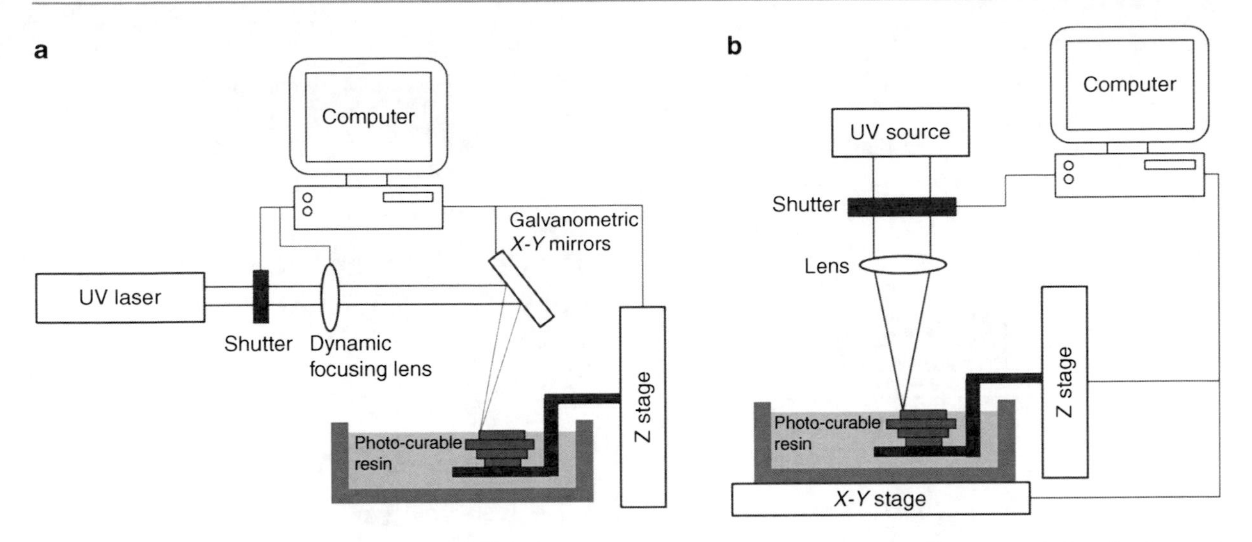

Stereolithography, Fig. 1 Schematic illustrations of (**a**) a classical stereolithography system using a dynamic focusing lens and (**b**) a microstereolithography system using a *X-Y* stage

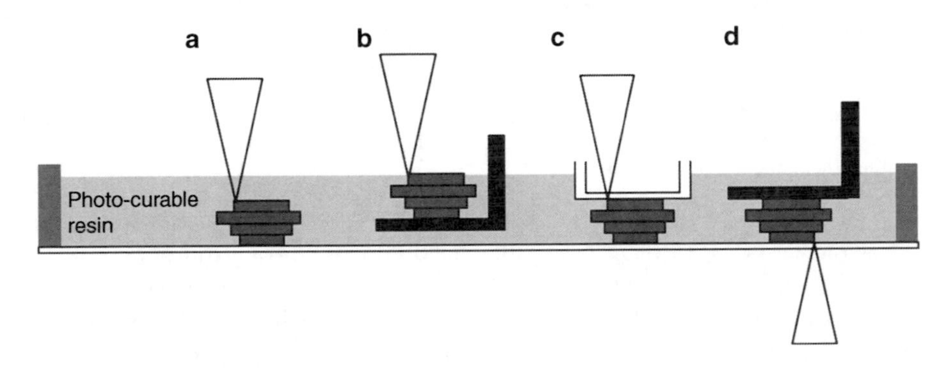

Stereolithography, Fig. 2 Schematic illustrations of four basic configurations for a stereolithography process. Free surface configurations with a light beam projected from upper direction (**a**) within and (**b**) over the surface of the liquid photo-curable resin. Constraint surface configuration with a light beam projected from (**c**) upper and (**d**) bottom directions

liquid surface is covered by the transparent plate, which allows UV beam passage with little energy loss, in the constraint surface configurations shown in Fig. 2c, d. The free surface configurations are relatively easy to construct, but it is difficult to control the thickness of each layer due to the viscosity of the liquid resin causing a relatively long time to be stabilized. On the other hand, the constraint surface configurations allow more accurate control of the layer thickness by controlling the height of the plate. However, the photopolymerized resin can be often adhered to the plate and broken in this constraint surface method.

In 1993, Ikuta and his colleagues have first demonstrated the MSL system using this constraint surface configuration as shown in Fig. 3a [2]. They scanned the resin chamber in the *X-Y* plane and the focused light beam in the *Z*-axis. They could achieve the minimum feature size of 5 μm in the *X-Y* plane and 3 μm in the *Z*-axis. However, it took about 30 min to produce a microstructure, the dimension of which is only 10 μm × 10 μm × 1,000 μm. To obtain a large number of 3D microstructures at a time using MSL, Ikuta and coworkers reported an advanced MSL system using multiple optical fibers in 1996 [3]. In this system, the light source was tightly focused onto a fiber-optic bundle, and each fiber end was placed in an ordered array for generating a series of exposure point sources in the liquid resin chamber, which is scanned in *X-Y-Z* directions (see Fig. 3b).

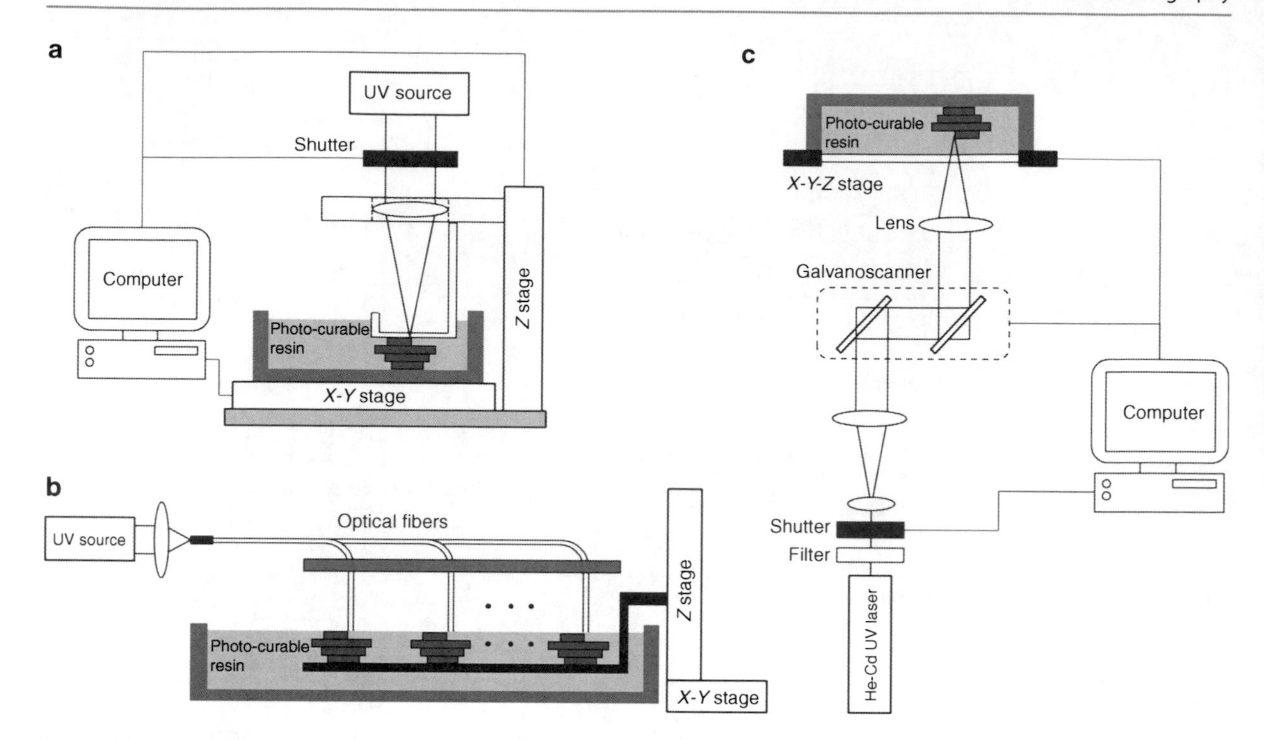

Stereolithography, Fig. 3 Schematic illustrations of several types of microstereolithography (MSL) systems. (**a**) The first-generation MSL system based on constraint surface condition, a light from upper direction, and an *X-Y* stage. (**b**) Fiber-optic MSL system for massive manufacturing in large area. (**c**) High-resolution MSL system using a tightly focused UV laser source irradiating inside the photo-curable resin

Despite these advances in MSL, there have still been some challenging issues caused by the basic scheme of original SL process, which is a layer-by-layer process. In the layer-by-layer process, the vertical resolution of the polymerized objects is limited by the thickness of the layer piled up. In addition, surface tension of viscous liquid resin can deform and destruct the solidified parts in microscale during the fabrication process. To overcome these limitations, Ikuta and coworkers have developed a more advanced MSL system, in which a UV laser source is tightly focused into the liquid resin, not over its surface, to induce the photopolymerization only at the focal spot of the beam, in 1998 (see Fig. 3c) [4]. Since the photopolymerization occurs only around the focal spot of the beam inside the liquid resin, not the surface, the problems related to the layer-by-layer process, which are mentioned above, can be reduced. They could produce freely moved 3D microstructures with 500 nm resolution using this technique. However, extensive polymerization in the defocused regions, where the beam passes through, is an unavoidable issue associated with this technique.

Two-Photon Stereolithography

Two-photon stereolithography (TPS), which is based on two-photon induced photopolymerization, is one of typical SL techniques for the 3D micro-/nanofabrication [8–13]. The TPS technology is based on the polymerization of liquid-state polymer via two-photon absorption (TPA) within very local regions inside the focused high-intensity laser beam tracing in a specific cross-sectional pattern (see Fig. 4). TPA is the simultaneous absorption of two photons by a molecule resulting in its excitation from a low-energy electronic state to a higher energy electronic state. The TPA-initiated polymerization of liquid-state polymer occurs only at a very narrow focal spot of the beam where the light intensity is highest. Thus, the spatial resolution achievable by the TPS is almost 100 nm, which is smaller than the diffraction limit of the beam, while that of other SL systems based on single-photon polymerization is limited by the optical diffraction limit. Based on the fascinated advantages of TPS technology, SL could meet 3D nanofabrication since the 2000s. Not only photonic devices [9] and micromachines [10], but also microsculptures of bull

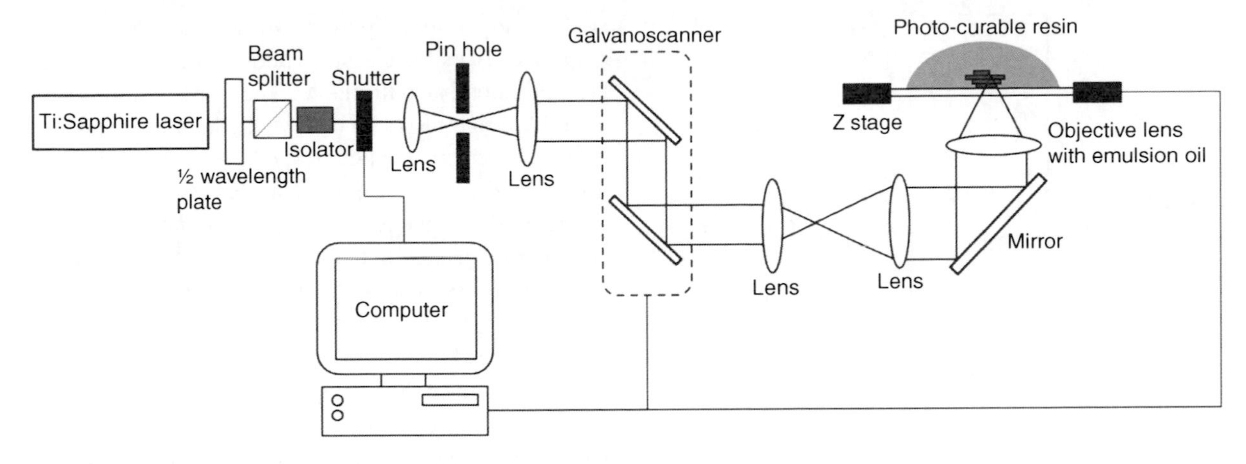

Stereolithography, Fig. 4 Schematic illustration of two-photon stereolithography system

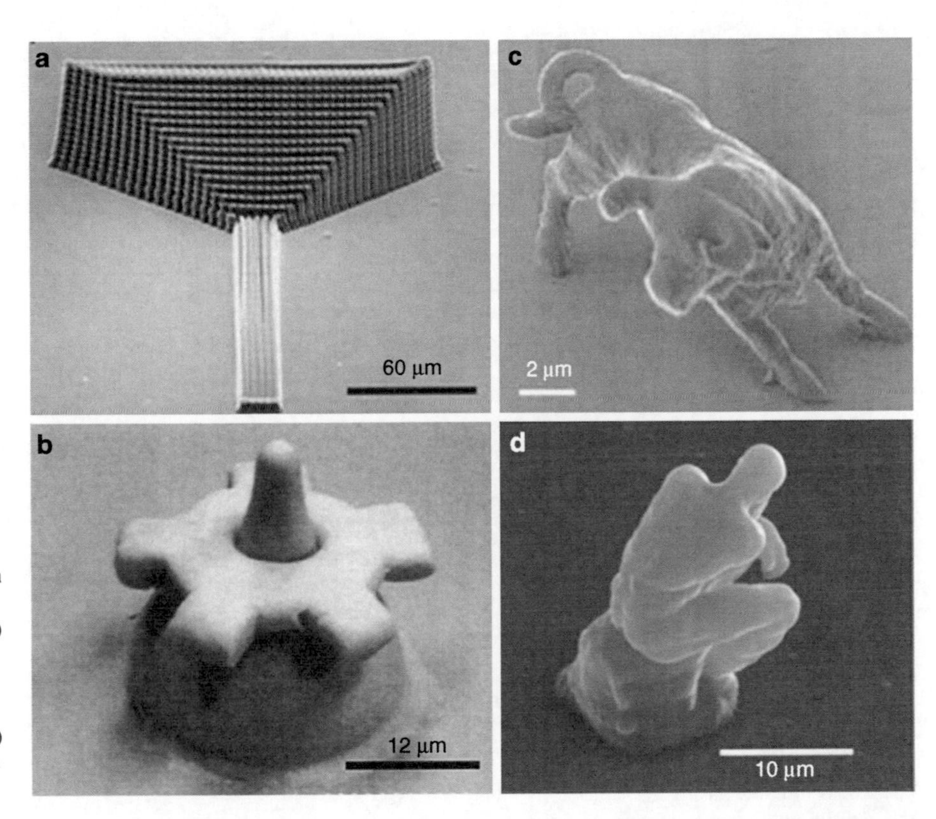

Stereolithography, Fig. 5 Microstructures fabricated by two-photon stereolithography. Various structures including **(a)** a photonic device (Reprinted by permission from Macmillan Publishers Ltd: Nature [9], copyright (1999)) **(b)** a micromachine (Reprinted with permission from [10]. Copyright 2000, The Optical Society), and microsculptures of **(c)** bull (Reprinted by permission from Macmillan Publishers Ltd: Nature [11], copyright (2001)) and **(d)** The Thinker (Reprinted with permission from [12]. Copyright 2007, American Institute of Physics) with nanoscale precision have been fabricated

[11] and The Thinker [12] with nanoscale precision have been fabricated using this technology as shown in Fig. 5. To date, the TPS systems have been able to produce the smallest objects based on MSL.

Image Projection Stereolithography

Another category of the approaches for MSL is the image projection-based one. In this method, the photopolymerization and solidification of an object is carried out by irradiating the photo-curable resin with an image pattern in one time for one layer. The image pattern for each layer can be generated by a conventional photo-mask or a spatial light modulator (SLM), which is capable of generating and modulating dynamic image patterns in a continuous way. Typical types of the SLM include a liquid crystal display

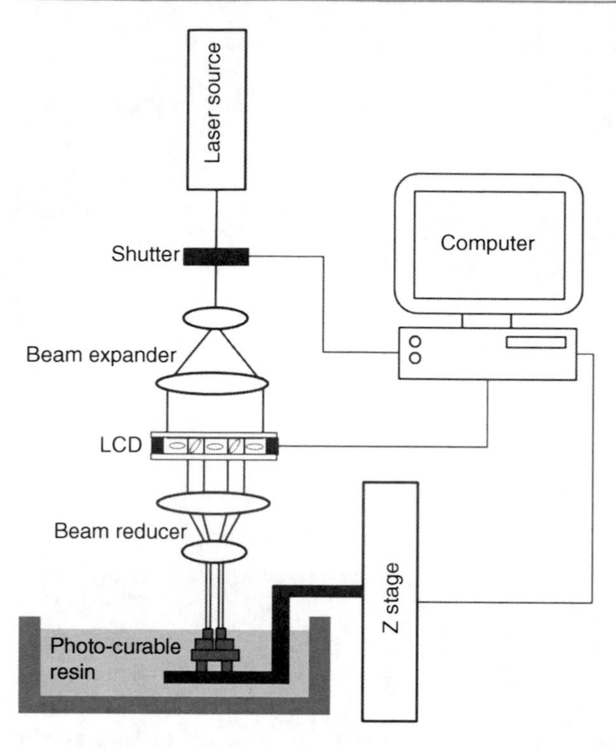

Stereolithography, Fig. 6 Schematic illustration of an image projection microstereolithography system using a dynamic image generated by a liquid crystal display (LCD)

(LCD) and a digital micromirror device (DMD). Both display devices have a large number of pixels, which are individually controllable. The image pattern generated through the LCD or DMD, which are transmissive and reflective types, respectively, is focused by a lens onto the liquid resin surface. By continuously changing the image pattern using a computer, the resin chamber is scanned in the *Z*-axis to construct a 3D part (see Fig. 6). Bertsch and colleagues have first demonstrated the image projection MSL system using an LCD as the dynamic pattern generator in 1997 [5]. Since the mid-2000s, several systems using a DMD have been developed and commercialized because the DMD provides better performance – fill factor of each pixel and reflectivity – compared to the LCD [6, 7]. These image projection-based approaches take much shorter time to fabricate each layer because a large area of liquid resin can be simultaneously exposed in a predesigned 2D pattern differently from the beam scanning methods, which scan all the area with a tightly focused light. However, these techniques still have the problems due to the layer-by-layer process mentioned above – vertical resolution dependent on the thickness of the layer stacked together.

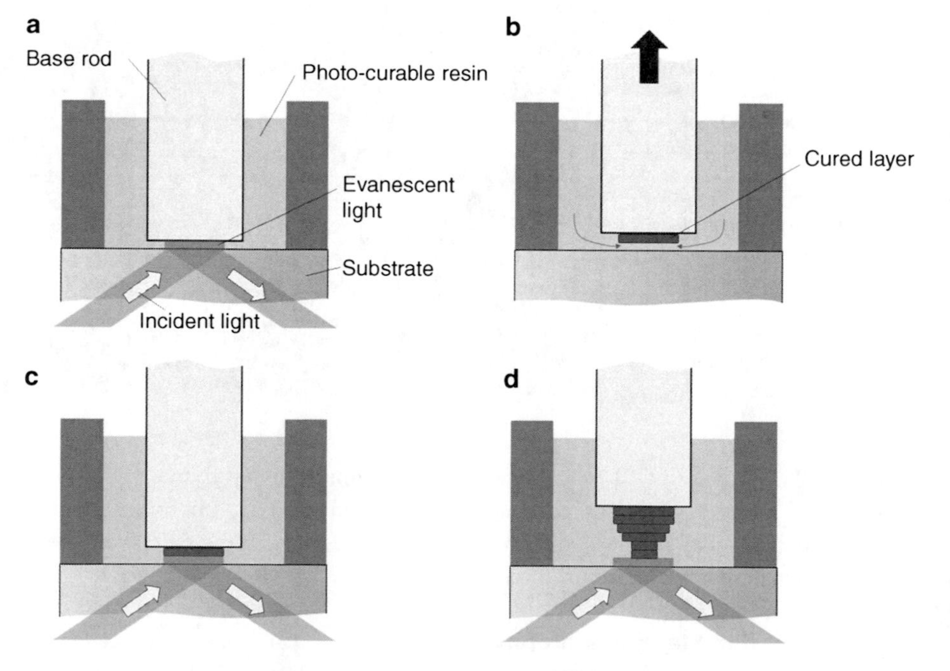

Stereolithography, Fig. 7 Schematic illustration of evanescent light stereolithography processes. (**a**) First of all, an incident light exposes and cures the photo-curable polymer. (**b**) Next, the cured layer adhering on the base rod is lifted by raising the rod and filling the space with a new layer of the liquid-state polymer. (**c**) Then a light beam with different specific 2D patterns exposes and cures the next layer. (**d**) Finally, the desired object can be fabricated by repeating this loop continuously

Evanescent Light Stereolithography

Evanescent light stereolithography, which uses an evanescent light for the photopolymerization of the liquid-state polymer differently from the conventional stereolithography technologies based on a propagating light, has also been developed [14] (see Fig. 7). When light travels across materials with different refractive indices at an incident angle greater than the so-called critical angle, above which total internal reflection of the light occurs at the interface, the evanescent light is generated within a localized area from the boundary of the substrate and the photo-curable polymer. As a result, the liquid-state polymer is hardened by the evanescent light in the sub-micrometer-thickness layer. By repeating the processes, which includes refilling the boundary with the liquid-state polymer and its photopolymerization by the evanescent light, 3D structures with sub-micrometer scale in vertical axis are formed. However, the lateral resolution of the objects formed by this technique is not very fine yet.

Conclusions

SL has been a promising technology for the direct fabrication of complicated 3D structures in micro-/nanoscale. Although there has been a great deal of achievement in the stereolithographic technologies for several decades, there are still challenging issues, which should be explored for their practical uses: massive production, large-area fabrication, reliability of processes, and improvement of photopolymerizable materials. Nevertheless, the SL has been one of the most trustworthy technologies for the fabrication of 3D nano-/microstructures and would take a place as a powerful 3D fabrication technology, which is applicable to diverse research fields.

References

1. Kodama, H.: Automatic method for fabricating a three-dimensional plastic model with photo-hardening polymer. Rev. Sci. Instrum. **52**, 1770–1773 (1981)
2. Ikuta, K., Hirowatari, K.: Real three dimensional micro fabrication using stereo lithography and metal molding. In: Proceedings of IEEE Micro Electro Mechanical Systems, pp. 42–47. Fort Lauderdale, FL, USA (1993)
3. Ikuta, K., Ogata, T., Tsubio M., Kojima, S.: Development of mass productive micro stereo lithography (Mass-IH Process). In: Proceedings of IEEE Micro Electro Mechanical Systems, pp. 42–47. Fort Lauderdale, FL, USA (1993)
4. Ikuta, K., Maruo, S., Kojima, S.: New microstereolithography for freely movable 3D micro structure-super IH process with submicron resolution. In: Proceedings of IEEE Micro Electro Mechanical Systems, pp. 290–295. Heidelberg, Germany (1998)
5. Bertsch, A., Zissi, S., Jezequel, J.-Y., Corbel, S., Andre, J. C.: Microstereolithography using a liquid crystal display as dynamic mask-generator. Microsys. Technol. **3**, 42–47 (1997)
6. Hadipoespito, G., Yang, Y., Choi, H., Ning, G., Li, X.: Digital micromirror device based microstereolithography for micro structures of transparent photopolymer and nanocomposites. In: Proceedings of the Solid Freeform Fabrication Symposium, pp. 13–24. Austin, TX, USA (2003)
7. EnvisionTEC: http://www.envisiontec.de. Accessed 16 March 2011
8. Maruo, S., Nakamura, O., Kawata, S.: Three-dimensional microfabrication with two-photon-absorbed photopolymerization. Opt. Lett. **22**, 132–134 (1997)
9. Cumpston, B.H., Ananthavel, S.P., Barlow, S., Dyer, D.L., Ehrlich, J.E., Erskine, L.L., Heikal, A.A., Kuebler, S.M., Lee, I.-Y.S., McCord-Maughon, D., Qin, J., Rockel, H., Rumi, M., Wu, X.-L., Marder, S.R., Perry, J.W.: Two-photon polymerization initiators for three-dimensional optical data storage and microfabrication. Nature **398**, 51–54 (1999)
10. Sun, H.-B., Kawakami, T., Xu, Y., Ye, J.-Y., Matuso, S., Misawa, H., Miwa, M., Kaneko, R.: Real three-dimensional microstructures fabricated by photopolymerization of resins through two-photon absorption. Opt. Lett. **25**, 1110–1112 (2000)
11. Kawata, S., Sun, H.-B., Tanaka, T., Takada, K.: Finer features for functional microdevices. Nature **412**, 697–698 (2001)
12. Yang, D.-Y., Park, S.H., Lim, T.W., Kong, H.-J., Yi, S.W., Yang, H.K., Lee, K.-S.: Ultraprecise microreproduction of a three-dimensional artistic sculpture by multipath scanning method in two-photon photopolymerization. Appl. Phys. Lett. **90**, 013113 (2007)
13. Park, S.-H., Yang, D.-Y., Lee, K.-S.: Two-photon stereolithography for realizing ultraprecise three-dimensional nano/microdevices. Laser Photonics. Rev. **3**, 1–11 (2009)
14. Kajihara, Y., Inazuki, Y., Takahashi, S., Takamasu, K.: Study of nano-stereolithography using evanescent light. In: Proceedings of the American Society for Precision Engineering (ASPE) Annual Meeting, 149–152. Orlando, FL, USA (2004)

Stimuli-Responsive Drug Delivery Microchips

Jian Chen[1], Michael Chu[2], Claudia R. Gordijo[2], Xiao Yu Wu[2] and Yu Sun[3]
[1]Department of Mechanical and Industrial Engineering, University of Toronto, Toronto, ON, Canada
[2]Advanced Pharmaceutics and Drug Delivery Laboratory, University of Toronto, Toronto, ON, Canada
[3]Department of Mechanical and Industrial Engineering and Institute of Biomaterials and Biomedical Engineering and Department of Electrical and Computer Engineering, University of Toronto, Toronto, ON, Canada

Synonyms

Intelligent drug delivery microchips; MEMS-based drug delivery devices; Smart drug delivery microchips

Definition

Stimuli-responsive drug delivery microchips are MEMS-based "smart" drug delivery devices composed of individually sealed drug reservoirs that could be opened selectively for complex drug release by various stimuli, targeting long-term implantation applications.

Overview

Overview of Working Mechanism

Advances in MEMS technology have enabled the precise fabrication of miniature biomedical devices with micrometer-sized features for implantable drug delivery. Drug delivery microchips contain small reservoirs that are loaded with drugs and separated from the outside environment by a drug release barrier. Examples of reported MEMS devices use an electrochemical dissolution approach, electrothermal activation approach, chemical degradation approach, or self-regulated approach to modulate the permeability of the drug release barrier for controlled drug delivery.

Advantage

These microreservoir devices are well suited for applications in chronotherapy due to their ability to achieve complex release with well-defined temporal profiles of drug depots. In addition, microchip-based implantable drug delivery devices allow localized delivery by direct placement of the device at the treatment site. Another benefit of these microchip devices is that they contain no moving parts and are capable of delivering reservoir drugs in the solid, liquid, or gel state. The microchip design also protects the drug depot from the outside environment before release so that stability of the drug inside drug reservoirs can be maintained more effectively in its controlled environment.

Potential Concerns

Implantable microchips require minor surgery to implant and remove and need to be as unobtrusive as possible. There is a fundamental limitation of device size imposed by the need to have sufficient storage capacity for a chronic dosing regimen. Suitable candidate drugs will be potent and will be formulated as high-concentration preparations stable for extended periods at body temperature. Long-term stability of the devices, both chemically and physically, need to be monitored for effective treatment.

Methodology

Currently available drug delivery microchips can be divided into three main categories, based on drug release methods: active drug release (e.g., electrochemical dissolution approach and electrothermal activation approach), passive drug release (e.g., chemical degradation approach), and self-regulated drug release (e.g., pH-responsive delivery).

Electrochemical Dissolution Approach

As shown in Fig. 1, the device was fabricated by the sequential processing of a silicon wafer using microelectronic processing techniques including UV photolithography, chemical vapor deposition, electron beam evaporation, and reactive ion etching. Samples of reservoirs were sealed at one end by a thin membrane of gold to serve as an anode in an electrochemical reaction. Electrodes were placed on the device to serve as a cathode.

When the microfabricated reservoir system is submerged in an electrolyte solution, ions form

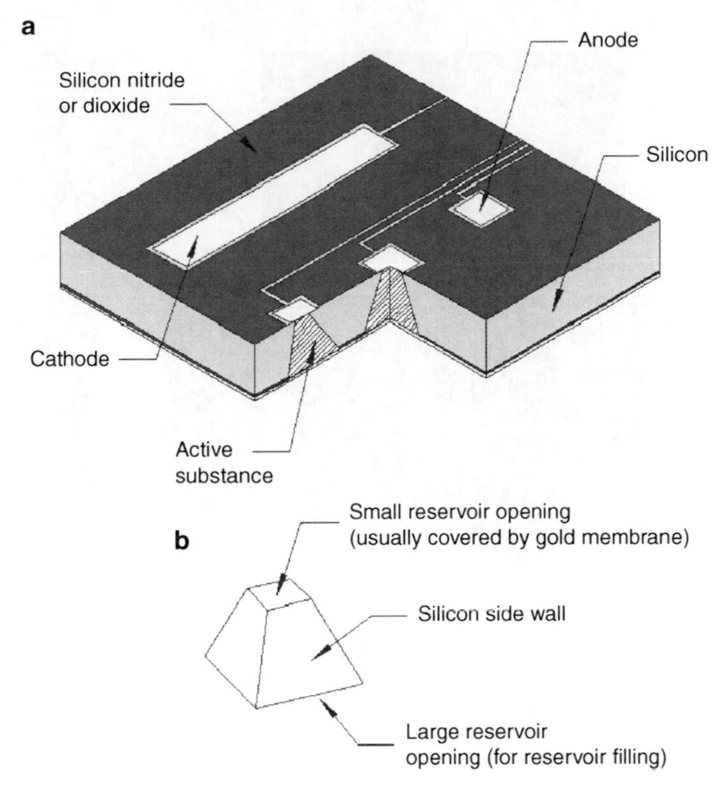

Stimuli-Responsive Drug Delivery Microchips, Fig. 1 Schematic illustration of the first implantable drug delivery microchip via electrochemical dissolution approach. (**a**) Fabrication of these microchips began by depositing low stress, silicon-rich nitride on both sides of prime grade silicon wafers using a vertical tube reactor. The silicon nitride layer on one side of the wafer was patterned by photolithography and electron cyclotron resonance–enhanced reactive ion etching to give a square device containing square reservoirs. The silicon nitride served as an etch mask for potassium hydroxide solution, which anisotropically etched square pyramidal reservoirs (**b**) into the silicon along the (111) crystal planes until the silicon nitride film on the opposite side of the wafer was reached. The newly fabricated silicon nitride membranes completely covered the *square* openings of the reservoir. Gold electrodes were deposited and patterned over the silicon nitride membranes by electron beam evaporation and lift-off. A layer of plasma-enhanced chemical vapor deposition silicon dioxide was deposited over the entire electrode-containing surface. The silicon dioxide located over portions of the anode, cathode, and bonding pads were etched with reactive ion etching to expose the underlying gold film; this technique was then used to remove the thin silicon nitride and chromium membranes located in the reservoir underneath the gold anode (Reproduced with permission from [1])

a soluble complex with the anode material in its ionic form. An applied electric potential oxidizes the anode membrane, forming a soluble complex with the electrolyte ions. The complex dissolves in the electrolyte, the membrane disappears, and the solution within the reservoir is released. The release time from each individual reservoir is determined by the time at which the reservoir's anode membrane is removed.

The release studies demonstrate that the activation of each reservoir can be controlled individually, creating a possibility for achieving many complex release patterns. Varying amounts of chemical substances in solid, liquid, or gel form can be released into solution in either a pulsatile manner, a continuous manner, or a combination of both, either sequentially or simultaneously from a single device. Such a device has additional potential advantages including small size, quick response times, and low power consumption. In addition, all chemical substances to be released are stored in the reservoirs of the microchip itself, creating a possibility for the future development of autonomous devices.

Electrothermal Activation Approach

The devices for this study were produced using standard microfabrication processes which contain an array of individually sealed and actuated reservoirs, each capped by a thin metal membrane comprised of either gold or multiple layers of titanium and platinum (see Fig. 2). The passage of a threshold level of electric current through the membrane causes it to disintegrate,

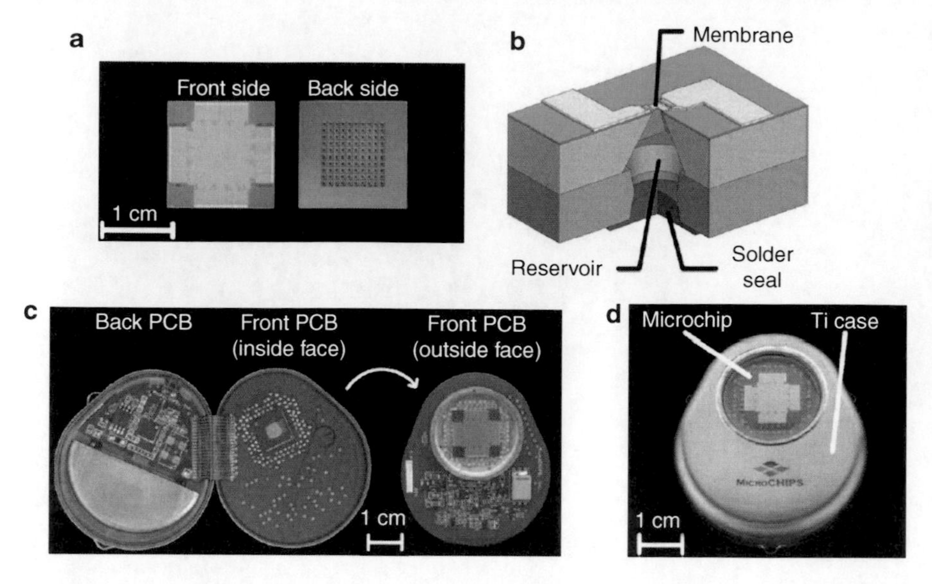

Stimuli-Responsive Drug Delivery Microchips, Fig. 2 Schematic illustration of the implantable drug delivery microdevice via electrothermal dissolution approach under remove control. (**a**) Front and back of the 100-reservoir microchip. (**b**) Representation of a single reservoir. (**c**) Electronic components on the printed circuit board (PCB) in the device package. (**d**) The assembled implantable device (Reproduced with permission from [2])

thereby exposing the protected drugs of the reservoir to the surrounding environment. Compared to the electrochemical dissolution approach, electrothermal reservoir opening is more reliable and repeatable than the gradual opening achieved by the corrosion-based method.

In addition, reservoirs were aseptically sealed with spheres of indium-tin eutectic solder by thermocompression bonding. Filled and sealed microchips were electrically connected to the wireless communication hardware, power supply, and circuit boards of the in vivo implant, which were hermetically sealed inside a laser welded titanium case. Pulsatile release of a therapeutic polypeptide on demand, in response to telemetry between an external device controller and the implant, was demonstrated. Importantly, although a tissue capsule formed around the device (as expected), the capsule did not significantly affect the drug release profile during the 6-month implantation. This result addresses a major perceived risk of implanted drug delivery devices, that a tissue barrier will compromise the effectiveness and control of release.

Chemical Degradation Approach

Biodegradable polymer version multireservoir drug delivery microchips have also been designed to investigate the feasibility of achieving multipulse drug release from a polymeric system over periods of several months without requiring the application of a stimulus to trigger the drug release. A prototype device is shown in Fig. 3, in which chemicals can be released at different times on the basis of the characteristics of the reservoir membranes that affect their degradation rate, such as the material used, the molecular mass, the composition, or the thickness.

Separation of the release formulation (the reservoir membranes) from the drug formulation (which is loaded into the reservoirs behind the membranes) might enable greater flexibility in adapting this system for a desired application than is currently achievable with existing methods. An advantage of biodegradable polymeric microchips is the elimination of a requirement for a second surgery to remove the device. In addition, the lack of electronics reduces any size restrictions in terms of device manufacture.

Self-regulated Approach

Compared to these microchips with release mechanisms of electrochemical dissolution, electrothermal activation, and chemical degradation, which are not capable of regulating drug delivery rates in response to in vivo environmental parameters, a self-regulated microdevice enables drug release in response to local pH or glucose variations by integrating pH or glucose-responsive nanohydrogel composite membranes functioning as intelligent nanovalves.

As shown in Fig. 4, the patterned PDMS structure forms a drug reservoir and provides physical support for the thin nanohydrogel-embedded composite membrane. The embedded hydrogel nanoparticles in the composite membrane detect environmental pH changes as intelligent nanovalves. Corresponding volumetric swelling and shrinking response of the nanoparticles controls

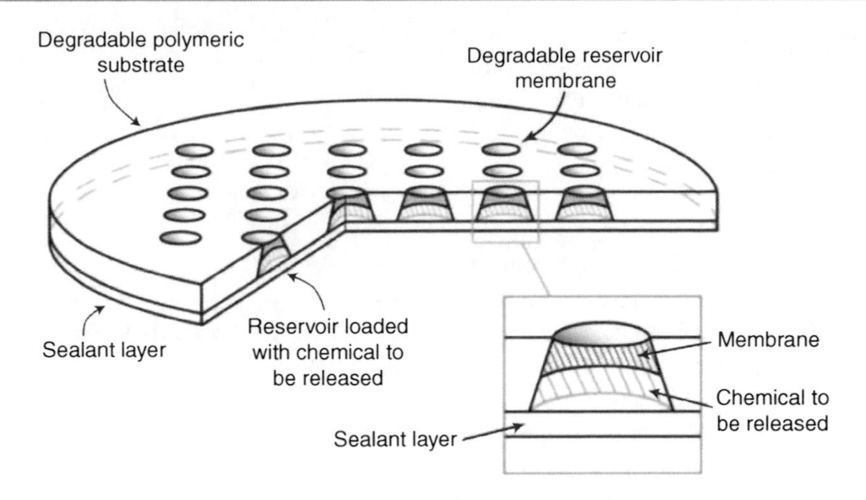

Stimuli-Responsive Drug Delivery Microchips, Fig. 3 Schematic illustration of the first implantable drug delivery microdevice with biodegradable membranes. The main body of the device is composed of a reservoir-containing substrate that is fabricated from a degradable polymer. Truncated conical reservoirs in the substrate are loaded with the chemical to be released, and sealed with polymeric degradable reservoir membranes on one end and a sealant layer (polyester tape) on the opposite end. Inset, close-up of a reservoir, reservoir membrane, sealant layer, and chemical to be released (Reproduced with permission from [3])

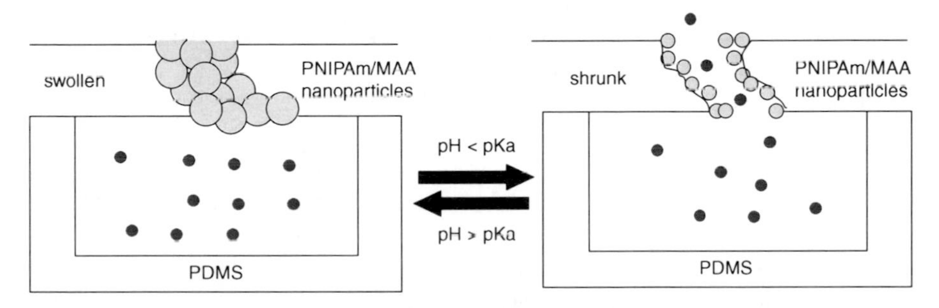

Stimuli-Responsive Drug Delivery Microchips, Fig. 4 Illustration of the mechanism for pH-responsive drug release out of the microdevice. *Left*: Nanoparticles are in the swollen state when the surrounding pH value is higher than pKa (acid dissociation constant) of the nanoparticles. *Right*: Nanoparticles are in the shrunk state when the surrounding pH value is lower than pKa. Resulting volumetric swelling and shrinking of the nanoparticles control drug-release rates [4]

drug-release rates. The polymeric microchips are monolithic without requiring peripheral control hardware or additional components for controlling drug-release rates. By adjusting nanoparticle percentages, drug reservoir shape and size, and drug loading concentrations, complex drug release profiles in response to local pH changes can possibly be achieved, functioning as a platform technology for intelligent drug delivery.

Key Research Findings

Device Dimension

For versatility and acceptability, overall device size is a major consideration. To be commercially attractive, the device should be small enough to implant subcutaneously in a doctor's office, with local anesthesia and a profile that is not obtrusive or apparent. The shape and materials of the device must allow easy removal when the device is not biodegradable. The implanted units must be designed with an exterior constructed of biocompatible materials, and any materials exposed after releasing drug must also be biocompatible.

MEMS-based implants must accommodate drug doses on the device, creating a design hurdle for incorporating a sufficient quantity of drug on the device. The minimum size will constrain the potential drug payload, which may be in one or multiple reservoirs. This constraint and the importance of being able to deposit an adequate number of doses in the device require the nondrug components of the device to occupy minimal volume. Suitable candidate drugs should be potent, should be formulatable as high-concentration preparations, and should be stable for extended periods at body temperature.

Materials

The materials of construction of an implantable device need to meet a wide range of requirements. Originally, the substrate was silicon, but the category is no longer limited to silicon-based devices. Polymers have superior properties to silicon for BioMEMS applications with regard to cost and versatility of physical properties. To provide more design flexibility, biocompatible polymers like polymethylmethacrylate (PMMA) and polydimethylsiloxane (PDMS) are being investigated as alternative materials to silicon.

The exterior of the device must be biocompatible. The primary materials and associated leachables must be nontoxic over long device use periods. The materials must be stable in contact with physiological fluids and tissues. In particular, the metal and/or polymer in contact with the body must have minimal leachables and be hypoallergenic.

Future Work

MEMS and miniaturization technologies have been used for producing novel stimuli-responsive drug delivery microchips that require minimal invasiveness of the medical procedures. Future development of ideal implantable drug delivery microchips requires several subsystems to provide all necessary functions: electrical components for selectively addressing individual reservoirs, biosensors for information acquisition, wireless communication hardware for remote control and a power source. These drug delivery microchips will be tailored for personalized medicine and other unmet medical need.

References

1. Santini, J.T., Cima, M.J., Langer, R.: A controlled-release microchip. Nature **397**, 335–338 (1999)
2. Prescott, J.H., Lipka, S., Baldwin, S., Sheppard, N.F., Maloney, J.M., Coppeta, J., Yomtov, B., Staples, M.A., Santini, J.T.: Chronic, programmed polypeptide delivery from an implanted, multireservoir microchip device. Nat. Biotechnol. **24**, 437–438 (2006)
3. Grayson, A.C.R., Choi, I.S., Tyler, B.M., Wang, P.P., Brem, H., Cima, M.J., Langer, R.: Multi-pulse drug delivery from a resorbable polymeric microchip device. Nat. Mater. **2**, 767–772 (2003)
4. Chen, J., Chu, M., Koulajian, K., Wu, X.Y., Giacca, A., Sun, Y.: A monolithic polymeric microdevice for pH-responsive drug delivery. Biomed. Microdevices **11**, 1251–1257 (2009)

Stimulus-Responsive Polymeric Hydrogels

► Smart Hydrogels

Stochastic Assembly

► Self-Assembly for Heterogeneous Integration of Microsystems

Strain Gradient Plasticity Theory

► Plasticity Theory at Small Scales

Structural Color in Animals

Mathias Kolle[1] and Ullrich Steiner[2]
[1]Harvard School of Engineering and Applied Sciences, Cambridge, MA, USA
[2]Department of Physics, Cavendish Laboratories, University of Cambridge, Cambridge, UK

Synonyms

Animal coloration; Biological photonic structures; Biological structural color; Bio-optics; Bio-photonics

Definition

Intense and bright colors result from the interaction of light with periodic micro- and nanostructures that cause color by interference, coherent scattering, or diffraction. These colors are termed structural colors, and structures that cause color by modulation of light are called photonic structures. Photonic structures are usually composed of regular lattices with periodicities on the order of the wavelength of light. Various organisms in nature are known to use intriguingly diverse photonic structures.

Introduction

Structural Colors and Photonic Structures

Structural colors in the animal kingdom have attracted increasing research interest in recent years. Biological organisms offer an enormous variety of periodic micro- and nanostructures that by specific interaction with light provide distinct coloration. This sometimes dynamic reflectivity is tailor-made for the organism's purpose within its natural illumination environment. Intriguing photonic structures have been identified on the wing cases and armors of beetles, the scales of butterflies, the feathers of birds, in the shells of marine animals, or even within the skin of mammals. Nature offers a huge choice of blueprints for novel artificial optical materials and photonic structures. The strongest color contrasts are achieved by a combination of different physical effects, including multilayer interference, diffraction, coherent scattering, and spatially confined absorption [1–3]. A common design concept in natural photonic systems is the complex interplay of structural regularity on the length scale of the wavelength of visible light combined with structural disorder on a larger scale [4]. In many cases, the complex interaction of these hierarchical structures with incident light lead to an outstanding, dynamic coloration, bright reflectivity that is perceivable in a wide angular range, brilliant whiteness, or enhanced transmission [5, 6]. While photonic structures can be made entirely from transparent materials, the incorporation of absorbing pigments deposited under or incorporated into a photonic structure is frequently used in nature to prevent spurious reflections. This improves the contrast leading to an enhancement of the color perceived from the photonic system [7]. Biological micro- and nanostructures that cause structural color are very diverse and often show periodicities on several length scales. This makes the optical characterization of a biological photonic system and the determination of the optical properties of its constituent materials (e.g., refractive index) very challenging. A set of useful techniques for the determination of the complex refractive indices of materials in biological photonic structures has recently been reported and successful attempts have been made to determine the complex refractive index of the organic cuticle material in the scales of butterflies and the armor of beetles [8].

Occurrence and Purpose of Structural Colors in Nature

Structural colors can be found in the feathers and skin of various birds [9–11], in the shells, spines, and scales of marine animals [12, 13], and in the skin of some mammalian species [14]. They are probably most abundant in species of the insect orders *Lepidoptera* and *Choleoptera* that comprise butterflies, moths, and beetles [15–19]. The different purposes of structural colors are as diverse as the organisms that use them [20]. Intense colors with stark contrast to their environment can serve in interspecies interaction including agonistic displays to confuse or scare away potential predators, while structural colors are also applied to induce a cryptic coloration for camouflage. For other animals, they are playing an import role for intraspecies communication such as competition between males of the same species for territory and/or females. Structural colors often provide sexual dichroism between males and females and are believed to have a function in sexual selection. They might also play a role in temperature regulation. While the physics and functioning of photonic structures found in many different organisms are mostly understood, it is frequently very challenging to clearly identify the specific benefit for the animal, which often remains mysterious.

The Physical Effects Underlying Structural Colors

Bright, pure, and intense colors arise from strong reflectivity in narrow spectral bands caused by highly ordered structures including multilayers, surface diffraction gratings, and photonic crystals. While many biological photonic structures are based on multilayer assemblies and two- or three-dimensional photonic crystals [1], diffraction gratings are rare in nature, possibly because they do not display a specific color but give rise to a range of colors depending on light incidence and observation direction [4]. The contrast of structural colors in their environment is often increased by the placement of absorbing elements spatially under or around the photonic structure. Intense blacks are achieved in nature by structurally assisted pigmentation as in the case of the butterfly *Papilio ulysses* [7]. In some avian species the pigments are incorporated directly into the photonic structure [9]. Brilliant white is achieved by multiple, incoherent scattering caused by highly disordered structures of random size, aperiodically arranged on the length scale of the wavelength of visible light [6]. Intense, spatially homogenous, angle-independent colors arise from coherent scattering caused by structures of well-defined size comparable to the wavelength of light, such as air pores of a narrow size distribution dispersed randomly in the volume of a material. This, for example, gives rise to the strong blue of the feathers of several different bird species [11]. High transparency is achieved by graded refractive index surfaces, involving arrays of conical protrusions found in moth eyes [5].

In the following, the structural designs, which give rise to strong coloration, ultrahigh blackness, or brilliant whites are discussed with emphasis on the interplay of different structural, hierarchically assembled elements. Specific natural organisms that apply these particular structural combinations are presented in an exemplary manner, without recounting all the animals known to apply the same or very similar concepts. First, static systems that induce structural colors based on "simple" multilayer elements are discussed before progressing to more and more sophisticated two- and three-dimensional structural arrangements. Attention is paid to combinations of different photonic elements and to the interplay of order and disorder in hierarchical structures on the nano- and microscale, a powerful concept occurring in natural structural colors.

Static Structural Colors in Animals

Multilayer Structures of Varying Complexity in Natural Photonic Systems

Thin film interference and multilayer interference are among the most common phenomena inducing structural color in animals. Light reflected from periodic stacks of multiple planar, optically distinct, transparent layers (often called Bragg mirrors) is colored, provided the thickness of the individual layers is comparable to the wavelength of light in the visible spectrum. The color, its intensity, and purity depend on the refractive index contrast of the multilayer materials, the individual layer thicknesses, and the number of layers in the stack [2]. Furthermore, the perceived color strongly varies with the angle of light incidence and the observation direction. The peak wavelength λ_{max} of light reflected from a multilayer in air composed of two materials with refractive index n_1, n_2 and film thicknesses d_1, d_2 upon incidence at an angle θ is given by the relation

$$m \cdot \lambda = 2 \cdot (n_1 d_1 \cos\theta_1 + n_2 d_2 \cos\theta_2), \qquad (1)$$

where m is a positive integer and the angles θ_1, θ_2 of the light path in the materials are given by Snell's law $n_{air} \sin\theta = n_1 \sin\theta_1 = n_2 \sin\theta_2$. In this case the rays reflected from the family of $n_1 - n_2$-interfaces interfere constructively with each other, if the light has the wavelength λ_{max}. The same holds for light rays reflected from the $n_2 - n_1$-interfaces in the stack. If a second relation given by

$$\left(m + \frac{1}{2}\right) \cdot \lambda = 2n_1 d_1 \cos\theta_1 = 2n_2 d_2 \cos\theta_2 \qquad (2)$$

is also satisfied, the multilayer stack is referred to as an ideal multilayer. This relation signifies the fact that in an ideal multilayer the light rays reflected from the $n_1 - n_2$-interfaces interfere constructively with the rays reflected from the $n_2 - n_1$-interfaces, which is not the case for a "nonideal" multilayer. The reflectance of a nonideal multilayer is therefore lower compared to an ideal multilayer made of the same materials, with an equal overall layer number and equal periodicity $d_1 + d_2$. Most multilayer arrangements found in nature are nonideal in the sense of this classification. In some cases this may even be beneficial for the organism since nonideal multilayers have a narrower bandwidth compared to ideal multilayers, which increases the purity of the reflected color.

The wood-boring Japanese jewel beetle *Chrysochroa fulgidissima* found in the forests of Japan during summer displays a vivid ventral and dorsal coloration (Fig. 1a). This color originates from interference of light in a multilayer stack of about 20 alternating layers (Fig. 1d) with refractive indices of 1.5 and 1.7 incorporated in the epicuticle of the beetle's wing cases [4]. The perceived coloration strongly depends on the orientation of the beetle with respect to the incident light and the observation direction, as expected for a multilayer structure (Fig. 1a). Nevertheless, the beetle's surface does not act like a colored mirror, as expected from a perfectly flat multilayer

Structural Color in Animals, Fig. 1 Natural photonic systems based on multilayer arrangements. (**a**) The wing cases and the ventral side of the Japanese Jewel beetle *Chrysochroa fulgidissima* display bright and iridescent colors that show a strong angular color variation (**b**, **c**). The surface of the wing cases shows irregular corrugations on the 100 μm scale (**b**, scale bar ~100 μm) and more regular indentations on the 10 μm scale (**c**, scale bar 10 μm). (**d**) Cross-sectional transmission electron micrographs reveal the stack of alternating high and low refractive index layers that causes multilayer interference, resulting in the beetle's iridescent color, scale bar 400 nm (Pictures (**a–d**) reproduced with permission of S. Kinoshita and IOP Publishing © 2008 Kinoshita et al. [1]). (**e**) The Madagascan moth, *Chrysiridia rhipheus*, displays a range of bright shimmering colors on its wings. (**f**) Different wing regions are covered by patterns of colorful scales reflecting green, yellow, red, or violet light. The scales are highly curved so that the observer only perceives light reflected from part of the scales, which creates the impression of texture caused by the juxtaposition of bright and dark regions, scale bar 200 μm. (**g**) The scales are spanned by parallel micro-ribbed ridges, scale bar 2 μm. (**h**) Cuticle layers of well-defined thickness that are spaced by small struts extend between and under the ridges, forming a regular multilayer arrangement with air as the low refractive index material, scale bar 500 nm. (**i**) The butterfly *Papilio blumei* displays lucid green stripes on its upper and lower wing pairs. (**j**) At higher magnification, the green scales show regions of distinct blue or yellow color, scale bar 100 μm. (**k**) The yellow color from the centers of concavely shaped surface corrugations disappears when the scales are imaged between crossed polarizers while the blue from the edges of the concave shapes persists due to polarization rotation upon reflection, scale bar 5 μm. (**l**) Concave surface corrugations on the scales, scale bar 2 μm. (**m**) A cross-section through one of the concavities reveals cuticle layers of well-defined thickness and regular spacing forming a multilayer reflector, scale bar 1 μm (Figure (**m**) reproduced with permission of P. Vukusic and Springer Science+Business Media © 2009 Vukusic [19]). (**n**) The South American butterfly *Morpho rhetenor*. (**o**) Bright blue scales cover the wing membrane like tiles on a roof top, scale bar 100 μm. (**p**) Top and cross-sectional view of the ridges running along each scale, scale bars 500 and 100 nm. (**q**) A cross-section through a scale exposes the intricate design of the ridges, scale bar 1 μm. The horizontal extensions on each ridge act as a multilayer reflector tuned for the blue spectral range with air as the low refractive index medium (Figures (**o** and **q**) reproduced with permission of P. Vukusic and Springer Science+Business Media © 2009 Vukusic [19])

arrangement. Minute hexagonally arranged holes and indentations are distributed in the multilayer on the 10 μm length scale (Fig. 1c) and the areas bordered by the micro-holes vary in their inclination. On the 100 μm length scale, the surface shows a strongly irregular corrugation (Fig. 1b). This structural irregularity leads to the scattering of light reflected from the multilayer, giving the beetle its characteristic color in a larger angular range around the specular reflection direction thus increasing its visibility.

The scales of the Madagascan moth *Chrysiridia rhipheus* (also called sunset moth because of the two bright yellow-orange-violet spots on the lower wing pair) display intense, beautiful colors ranging from green to red, similar to the jewel beetle's hues (Fig. 1e, f). The different, similarly bright colors result from a multilayer incorporated into the body of the moth's wing scales (Fig. 1h). The multilayer is made from layers of one single material (cuticle) spaced by perpendicular struts, effectively creating a controlled air spacing between the cuticle layers. Including air as the second component in the multilayer stack leads to the maximization of the refractive index contrast, providing intense reflection with a much smaller number of layers compared to the Japanese jewel beetle, where the refractive index contrast is much lower. The air spacing varies in the differently colored regions of the moth's wing with the smallest spacing found in the green areas and the largest spacing seen in the red wing spots. The multilayered scales show a strong curvature along their long axis. This curvature superimposed onto the photonic multilayer structure leads to a high visibility of the colors from various directions and also introduces multiple reflection effects between adjacent scales, leading to an increased purity of the reflected color and to interesting polarization effects [21]. Overall, caused by the interplay of the highly reflective air-cuticle multilayer and the pronounced scale curvature, the moth wing displays bright colors with a velvety, textile-like shimmer.

Air-cuticle multilayers of even more complex shapes can be found on the wing scales of butterflies of the genus *Papilio*. The South-East Asian Emerald Swallowtail *Papilio palinurus* and the Green Swallowtail *Papilio blumei* (Fig. 1i) display bright green spots on their wings resulting from a concavely shaped multilayer architecture on the individual scales (Fig. 1l, m). The vivid green originates from a superposition of blue and yellow reflected from the edges and the centers of the multilayer concavities, respectively (Fig. 1j, k) [22]. Different colors result from distinct regions of the concavely shaped multilayer due to the spatially varying angle of light incidence on the edges and in the center of the concavities. Light reflected from the edges can undergo multiple reflections within a single concavity, inducing a change in polarization. The butterfly *Papilio ulysses* displays concavely shaped multilayer structures on its scales that have less curvature and smaller cuticle and air gap thicknesses, resulting in a lucid blue color. While in the case of *Papilio palinurus* and *Papilio blumei* the higher local concavity wall curvature results in the juxtaposition of yellow and blue reflected from different regions forming green (a color that might have a purpose in camouflage), the shallower concavities of *Papilio ulysses* do not produce the same effect but rather diffuse the reflected light into a wide angular range for higher visibility [19].

The possibly most sophisticated and intensely studied multilayer structure giving rise to bright animal coloration is found on the wing scales of butterflies belonging to the genus *Morpho*. These butterflies that are native to South and Central America are known for their intense, widely visible coloration. Among them, *Morpho rhetenor* is the most striking example of vivid bright blue structural color (Fig. 1n, o). The intriguing hue of its wings results from the interference of light caused by ridge structures (Fig. 1p) on the wing scales that have a Christmas-tree like cross-section (Fig. 1p, q). The horizontal periodic protrusions of the ridges cause "quasi"-multilayer interference [4, 5]. The term "quasi" is used in this context, because light not only interferes when reflected from protrusions of the same ridge but also when reflected from protrusions on different adjacent ridges. As for the moth *Chrysiridia rhipheus* and the butterflies of the genus *Papilio*, the incorporation of air layers into the multilayer structure leads to a large refractive index contrast and consequently to a bright reflection in a wide spectral band. The ridges are spaced by $\lesssim 1$ μm on average, which in the past lead to the assumption that they also act as a diffraction grating. However, local variation between the ridges in height and distance and in the orientation of entire scales confines the diffraction to individual ridges and inhibits the manifestation of a pronounced diffraction grating effect in reflection [4]. Nevertheless, this randomization seems to have a beneficial side effect: despite its origin from multilayer reflection, the

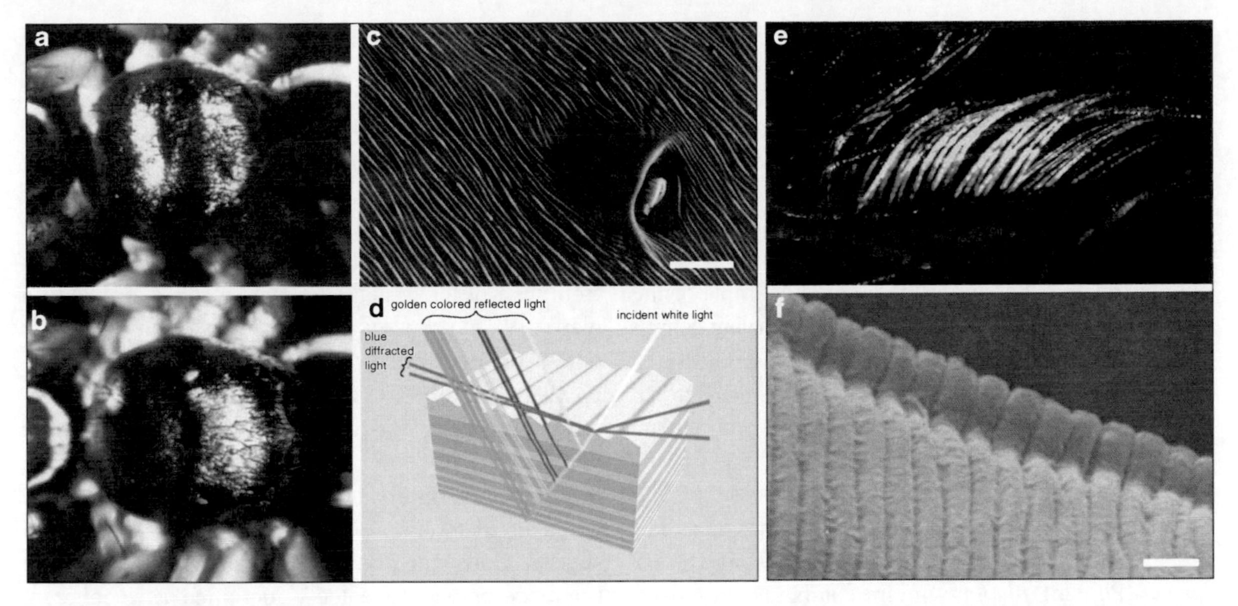

Structural Color in Animals, Fig. 2 Natural photonic systems based on diffraction. (**a**) The torso of the spider *Cosmophasis thalassina* shows two broad metallic golden-greenish stripes in air. (**b**) The color changes to purple-silvery when immersed in water. (**c**) The structure on the spider torso that causes these colors consists of a combination of surface striations superimposed on a chirped multilayer reflector (not shown here), scale bar 5 μm. (**d**) A schematic of the interaction of light with the spider's photonic structure. Blue light is very efficiently scattered by the striations while light of higher wavelength interferes with the multilayer stack. Red is specularly reflected in the top section of the chirped mirror where the layers are thicker while green is reflected by the thinner layers further down the stack. (**e**) The hairs of the first antenna of the seed shrimp *Azygocypridina lowryi* show strong iridescence caused by grating diffraction. (**f**) The diffraction grating on a single *setule* (hair) of the antenna, scale bar 1 μm (Images (**d** modified) reproduced with permission of A. R. Parker, IOP Publishing © 2003 Parker and Hegedus [23] and The Royal Society © 2005 Parker [13])

blue color of the *Morpho rhetenor* is very illumination and observation angle insensitive and perceivable from all points in space above the wing plane, only changing to violet at very high angles of light incidence or observation.

Diffractive Elements

Diffraction from surface gratings is believed to be rare in nature. This might be due to the fact that diffraction gratings with periodicities of 400 nm-2 μm need long-range order on the 60 μm scale (the spatial coherence of sun light) in order to be efficient, a criterium that is hard to meet in the natural context where irregularity is usually predominant at this length scale. During the past decade some animals, mostly invertebrates, have been shown to employ diffraction gratings, however [12]. As opposed to multilayer reflectors, which reflect light in a relatively narrow spectral band, diffraction gratings split incident light into its spectral components and redirect each color into a different direction. Consequently, diffraction gratings are less suitable for providing spectrally well-defined colors in a wide angular range (which is better achieved with a multilayer reflector combined with some irregularity on the micron-scale). While diffraction gratings made from transparent materials efficiently create vivid colors in transmission, they have to be made from (or backed by) reflective materials to produce strong colors in reflection, which seems to be less efficient for the organism.

Periodic grating-like surface structures on an underlying chirped broadband multilayer reflector have been found on the torso of spiders [23]. In a chirped multilayer the thicknesses of the individual layers decrease or increase gradually from the top to the bottom of the stack. Constructive interference occurs for the reflection of light of a particular color/wavelength range at a specific depth in the stack (Fig. 2d) where the layer thicknesses fulfill the condition discussed above (Eq. 1) thereby leading to a broadband reflection usually resulting in a silver or golden color. In the case of the spider *Cosmophasis thalassina* (Fig. 2a, b), the striations (Fig. 2c) mainly serve the purpose to disperse blue light before it interacts with the underlying reflector (Fig. 2d), thereby biasing the color of the spider's body toward a golden appearance, instead of the more silvery shine which it would have without the striations. Fourier analysis shows that there is no pronounced long-range order in these striations and consequently the striations do not create strong grating diffraction. The striations are likely to disperse light reflected from the multilayer reflector into a wider angular range. This represents a good example of the interplay of different structures with varying length scales to achieve a specific optical response.

The use of periodic diffraction gratings as surface structures seems particularly beneficial on narrow cylindrical geometries where the implementation of multilayers might be impractical or impossible. The thin hairs (*setae*) of some Crustacean species including the antenna of the male seed shrimp, *Azygocypridina lowryi* (Fig. 2e), display vibrant colors resulting from periodic surface structures [12, 13]. In the antenna of *Azygocypridina lowryi* the grating is formed by regular undulations of the hair thickness with a periodicity of 600–700 nm (Fig. 2f).

A curious structure causing diffraction with a reverse angular color sequence was recently found on the wings of the male butterfly *Pierella luna* [24]. The ends of the scales in the central portions of this butterfly's forewings are curled upward, enabling regularly arranged, on average 440 nm spaced cross-ribs to act as an upward-directed diffraction grating in transmission, resulting in this intriguing phenomenon of "inverse" diffraction. As opposed to a conventional diffraction grating, these structures make the central forewing regions appear red for small angles of observation (measured from the surface normal), changing to yellow, green, and finally blue as the observation angle is increased.

Two-Dimensional Photonic Crystals

Two-dimensional photonic crystals that cause structural colors have been found in marine animals [13, 25, 26], birds [9, 10], and mammals [14]. This part of the review focuses on two-dimensional photonic systems found in nature that induce the striking color in some animals. The *setae* of the polychaete worm *Pherusa* sp. (Fig. 3a, b) show a vivid play of colors originating from the periodic two-dimensional hexagonal arrangement of cylindrical channels (Fig. 3c, d) with a well-defined lattice constant within a single *seta* [26]. This system is strikingly similar to artificial photonic crystal fibers. A very similar arrangement was found earlier in the spines of the sea mouse *Aphrodita* sp. [25] (Fig. 3e–h).

The feathers of kingfishers, peacocks, ducks, pigeons, and trogons among various other birds display strong blue and green colors that reportedly have attracted the interest of scientists for more than 300 years [1, 4, 18]. The bright blues and greens result from well-ordered melanin granules in a dense keratin matrix [9] or a spongy keratin network with regular air pores [11] of varying dimensions and structural complexity within the birds' feather barbules. These structures have been classified and investigated in great detail in the last three decades [9]. The order of the layers of the solid or hollow melanin granules or the pores in the keratin matrix varies from species to species. Consequently, the physical effect causing the bright color was repeatedly identified as incoherent scattering for the less ordered randomly distributed structural elements with well-defined sizes or as coherent scattering (for instance, multilayer interference) in structures of quasi-ordered arrangements with higher spatial order. In many bird species not only the feather barbules but also the skin on different body parts show strong color. These colors are based on coherent scattering caused by two-dimensional photonic crystal structures, consisting of regular arrays of collagen fibers in the dermis of the birds (Fig. 3i–l). The dimensions of the collagen fibers are very well defined with a narrow distribution in fiber diameter for each species, while the extent of spatial order varies from species to species (compare Fig. 3j, l).

Three-Dimensional Photonic Crystals

The first discovered biological three-dimensional photonic crystal is a solid, hexagonally close-packed array of transparent spheres of 250 nm diameter in a matrix of hydrated chitinous material found inside the scales of the Australian weevil *Pachyrhynchus argus* [28] (Fig. 4a–c). Similar to the ordered arrangement of silica spheres in precious opal, this structure gives the beetle a metallic coloration visible in any direction. Recently, a similar structure was also observed in the Asian longhorn beetle *Pseudomyagrus waterhousei* (Fig. 4d–f).

Inverse opal photonic structures have been identified in various butterfly and beetle species. The beetle *Pachyrrhynchus congestus pavonius* [30] displays vivid patches of orange color on its body (Fig. 4g, h)

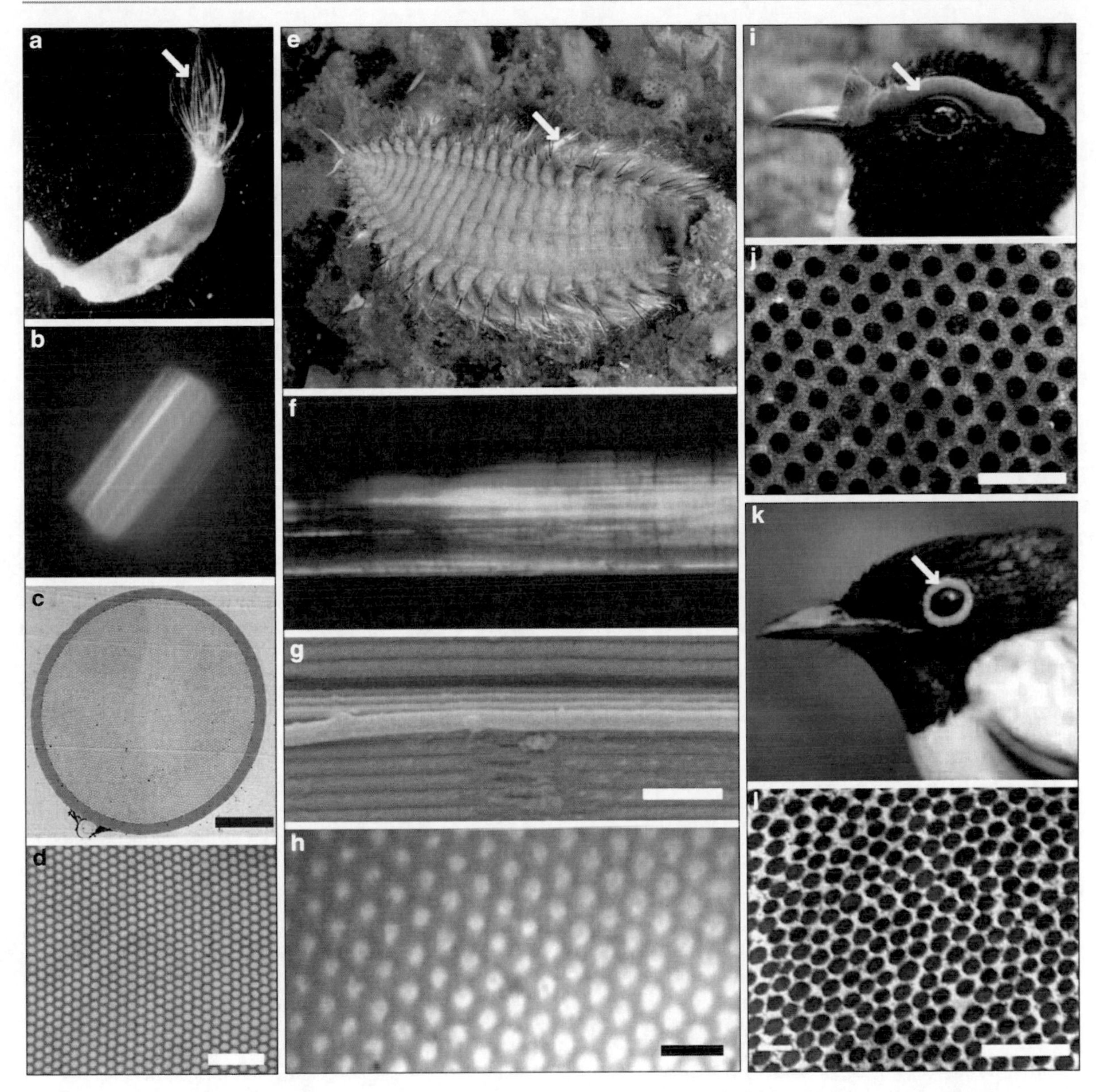

Structural Color in Animals, Fig. 3 Structural colors based on two-dimensional photonic crystals. (**a**) The *setae* of the polychaete worm *Pherusa* sp. show a remarkable structural color (*white arrow*). (**b**) Micrograph of a *seta* of the worm. (**c**) A cross-sectional transmission electron micrograph of a *seta*, scale bar 5 μm. (**d**) High-magnification image of the regular structure in the *seta*, scale bar 2 μm (Image (**a**) reproduced with permission of A. R. Parker, 1995. Images (**b–d**) reproduced with permission of P. Vukusic and The American Physical Society © 2009 Trzeciak and Vukusic [26]). (**e**) The colorful spines (*white arrow*) of the sea mouse *Aphrodita* sp. (Image courtesy of D. Harasti). (**f**) Optical micrograph of one of the colorful spines. (**g**, **h**) Side view and cross-section of the tubular structures in the spine, scale bars 2 μm and 1 μm (Images (**f–h**) reproduced with permission of A. R. Parker and The Royal Society © 2004 Parker [27]). (**i**) The bright green wattle (*white arrow*) of the Madagascan bird *Philepitta castanea* (velvet asity). (**j**) A transmission electron micrograph of a cross-section of the nanostructured arrays of dermal collagen fibers responsible for the wattle's green color, scale bar 500 nm. (**k**) The blue eye spot (*white arrow*) of the Madagascan bird *Terpsiphone mutata* (Madagascar paradise flycatcher). (**l**) A cross-section of the collagen fiber array that causes the blue color around the bird's eye, scale bar 500 nm (Images (**i–l**) adopted from Prum and Torres [10] with permission of R. O. Prum and T. Schulenberg (**k**))

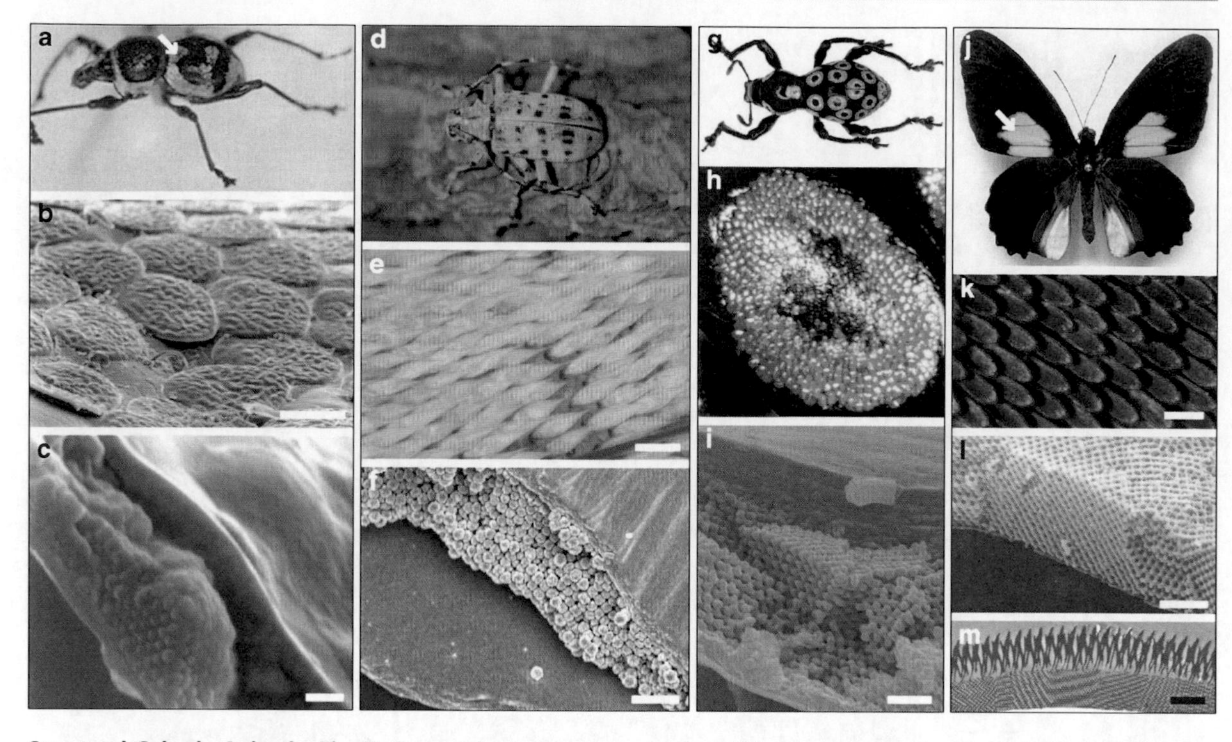

Structural Color in Animals, Fig. 4 Structural colors based on three-dimensional photonic crystals. (**a**) Dorsal view of the Australian weevil *Pachyrhynchus argus* showing the green metallic spots (*white arrow*) on its back side. (**b**) Scanning electron micrograph of the beetles scales, scale bar 50 μm. (**c**) Cross-section of a scale revealing the opaline structure responsible for the beetle's metallic coloration, scale bar 1 μm (Images (**a–c**) reproduced with permission of A.R. Parker and The Royal Society © 2004 Parker [27]). (**d**) The South-East Asian bright blue longhorn beetle *Pseudomyagrus waterhousei*. (**e**) Micrograph of the iridescent drop-shaped scales on the beetle's body, scale bar 50 μm. (**f**) Scanning electron micrograph of a cross-section through a single scale revealing the origin of the iridescent blue color, regularly sized, closely packed spherules forming a direct opal, scale bar 1 μm (Images (**d–f**) reproduced with permission of J. P. Vigneron and The American Physical Society © 2011 Simonis and Vigneron [29]). (**g**) The beetle *Pachyrrhynchus congestus pavonius*. (**h**) Micrograph of the highly conspicuous annular spots on the beetle's thorax. The scales of different colors are clearly visible. (**i**) Scanning electron micrograph of the cross-section of a scale, revealing the photonic structure consisting of a combination of a multilayer reflector on top of a three-dimensional face-centered cubic photonic crystal, scale bar 1 μm (Images (**g–i**) reproduced with permission of J. P. Vigneron and The American Physical Society © 2007 Welch et al. [30]). (**j**) The South American butterfly *Parides sesostris*, commonly called Emerald-patched Cattleheart due to the bright green patches on its upper wing pairs (*white arrow*). (**k**) Scales in the green areas of the butterfly's wings, scale bar 100 μm. (**l**) SEM image of a cross-section of the three-dimensional photonic structure found within a single scale, scale bar 1 μm. (**m**) Transmission electron micrograph of the cross-section of a scale showing the superficial ridging and the domains of the underlying of photonic crystal, scale bar 2 μm (Images (**j, l, m**) reproduced with permission of P. Vukusic and Springer Science+Business Media B.V © 2009 Vukusic [19]. Image (**k**) courtesy of M. Doolittle)

that originate from the interference of light within the beetle scales, which consist of a multilayered cortex surrounding an inverse opal photonic crystal (Fig. 4i). This is an example of the superposition of different structural geometries, in this case a multilayer mirror and a three-dimensional photonic crystal. The light reflected from a regular multilayer mirror results in an iridescent color while the interference of light with a polycrystalline three-dimensional photonic crystal usually leads to a uniform color over wide angular ranges. However, the precise influence of each structural component on the overall appearance of this particular beetle, *Pachyrrhynchus congestus pavonius*, remains to be investigated in detail. The wings of the butterfly *Parides sesostris* display patches of bright green angle-independent color (Fig. 4j, k), resulting from the interaction of light with a polycrystalline, three-dimensional photonic crystal structure that is buried under the superficial ridging in the body of the butterfly's wing scales (Fig. 4l, m). The individual crystallites were shown to have cubic symmetry and recent research suggests that they consist in fact of

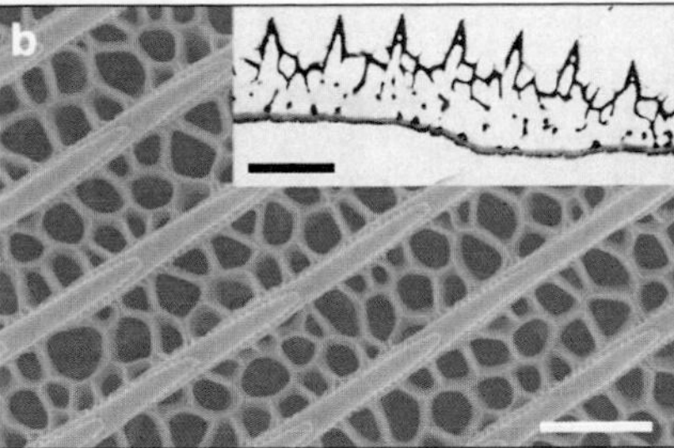

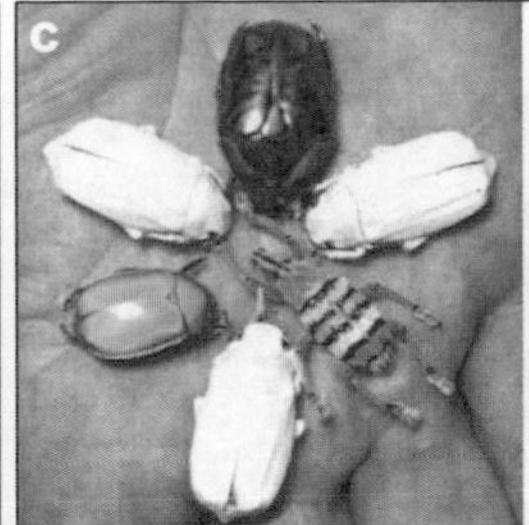

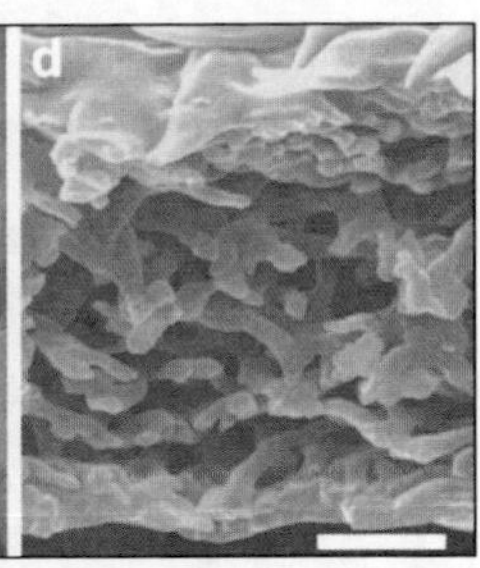

Structural Color in Animals, Fig. 5 Deep blackness and brilliant whiteness. (**a**) The butterfly *Papilio ulysses* shows regions of deep black (*white arrow*) surrounding the bright blue spots on its wings. (**b**) The scales in the black regions carry cuticle microstructures that trap light. The inset shows a cross-section of a scale revealing the dense network of pigment-loaded cuticle, scale bars 2 μm. (**c**) The brilliant white beetle *Cyphochilus* spp. compared to other beetles. (**d**) The intense white results from incoherent diffuse scattering from an interconnected network of filaments within the body of each beetle scale, scale bar 1 μm (Inset in (**b**), images (**c**) and (**d**) reproduced with permission of P. Vukusic, The Royal Society © 2004 Vukusic et al. [7] and The Optical Society of America © 2009 Hallam et al. [32])

a bi-continuous regular gyroid network [31]. The different orientations of the photonic crystal domains and the light scattering induced by the superposed ridge structure ensure that the reflected color is independent of observation angle.

Structurally Assisted Blackness and Brilliant Whiteness

The natural photonic structures presented so far serve the organisms to display a distinct coloration. The contrast of a color is often enhanced in nature by placing it against a black background and color purity is ensured by absorption of undesired spectral light components in regions beneath or around the color creating photonic elements. Consequently, optimized absorption plays an important role for natural structural colors. The scales of various butterflies that display structural colors are supported on a surface containing melanin pigments. The pigments absorb the light that passes through photonic structures of the scales, suppressing spurious reflection and consequently preventing the desaturation of the reflected color.

Very strong blackness can be achieved by an optimized interplay of pigment absorption and structure. A particularly good example is the butterfly *Papilio ulysses* [7]. The bright blue patches on the butterfly's wings are surrounded by very dark scales that show strong light absorption induced by melanin-containing microstructures (Fig. 5a). These pigment-filled structures act as light traps that are tailored to prevent light from escaping once it enters the scales (Fig. 5b). This maximizes the interaction with the pigment, resulting in the extremely low reflectivity of the butterfly's black scales.

Materials with a high structural irregularity, a random distribution, orientation, and size of individual transparent or highly reflective scattering elements on the scale of the wavelength of light are known to equally scatter light of any color into a broad angular range, thereby creating an intense white appearance. An extra-ordinary example of brilliant structural whiteness has been discovered in the case of the *Cyphochilus* spp. beetle [6]. The insect's intense white (Fig. 5c) is based on the incoherent scattering of light from a random network of about 250 nm thick interconnected filaments that are randomly distributed within the ultrathin beetle scales (Fig. 5d). Despite the fact that the scales are only about 5 μm thick, an optimized void fraction, high aperiodicity, and the large refractive index contrast of about 0.56 ensure that light of all colors is very efficiently scattered in all directions.

Dynamically Variable Structural Colors in Animals

The most interesting structural color systems in nature are the ones that show aspects of intentional color tuning, like the photonic structure employed by the beetle *Charidotella egregia* (Fig. 6a). In situations of distress, for instance, caused by predator attacks [33], the beetle can change its appearance from a bright golden shine to a striking red color (Fig. 6c) within 2 min. The structure inducing this phenomenon is

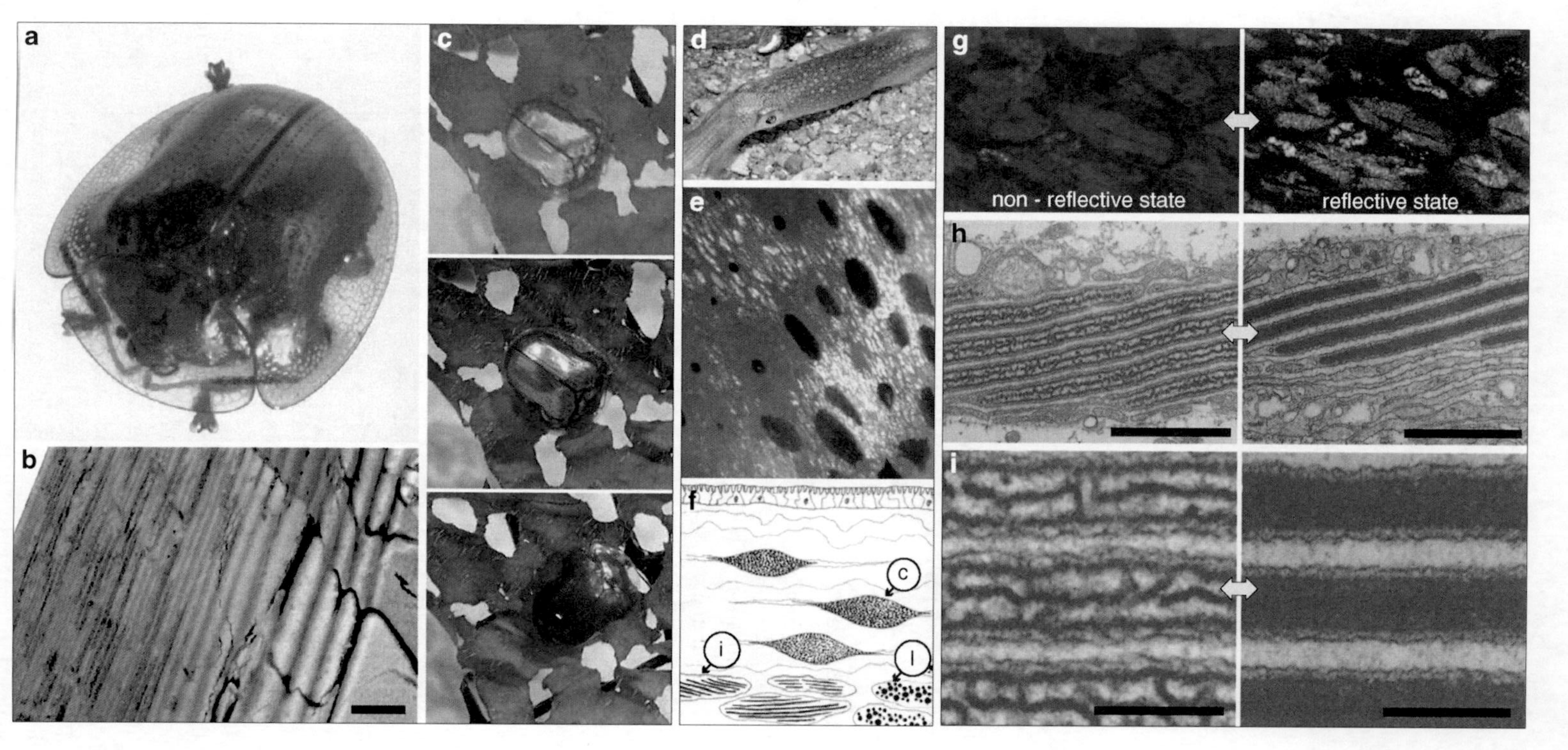

Structural Color in Animals, Fig. 6 Dynamically tunable structural color. (**a**) The golden beetle *Charidotella egregia*. (**b**) Transmission electron micrograph of the chirped multilayer reflector in the outer layer of the beetle's armor, showing the density variation of irregularly distributed porous and solid areas within the stack, scale bar 1 μm. (**c**) When threatened the beetle can change its golden color to a striking red within less than 2 min (Images (**a–c**) reproduced with permission of J.P. Vigneron and The American Physical Society © 2007 Vigneron et al. [33]). (**d**) The squid *Logilo pealeii.* (**e**) Microscope images of the squid's skin show the brown and reddish pigmented chromatophores and the underlying iridescent iridophores shimmering in colors from green to orange. (**f**) A sketch of the arrangement of chromatophores (**c**), iridophores (**i**), and leucophores (**l**) in the skin of cephalopods (Images (**d–f**) reproduced with permission of L. Mäthger, Springer Science+Business Media © 2007 Mäthger and Hanlon [34] and The Royal Society © 2009 Mäthger et al. [35]). (**g**) Optical microscope images, and (**h**, **i**) transmission electron micrographs of active iridophores of the squid *Lolliguncula brevis* reveal the variation in reflection and the ultrastructural changes in the iridophore platelets when switching between non-reflective (*left*) and reflective (*right*) states, scale bars 1 μm (**h**), 250 nm (**i**) (Images (**g–i**) reproduced with permission of R. T. Hanlon and Springer Science+Business Media © 1990 Cooper et al. [36])

a multilayer buried in the exocuticle of the beetle's transparent armor (Fig. 6b). The multilayer is chirped which results in reflection of light in a broad wavelength range, leading to the golden appearance of the beetle. Internally, the multilayer structure shows a random distribution of porous regions and channels within the layer planes and perpendicular to the layers. In the default state, these irregularly distributed voids are filled with liquid leading to a homogenous refractive index within each layer thereby suppressing scattering and allowing interference of light in the chirped multilayer, which results in the golden metallic color. Beneath the multilayer resides a layer of bright red pigment. When under a potential threat the beetle resorts to an aposematic protection mechanism. It can withdraw the liquid from the porous regions in the multilayer which strongly increases the scattering thereby rendering the layer stack translucent and revealing the bright red color of the underlying pigments, which serves as a warning signal.

The impressive camouflage and signaling capacities of cephalopods (squid, cuttlefish, and octopus) result from an intricate interplay of mainly three different functional elements within the skin of the animals [35]. Chromatophores are small pigment-filled organs that can be stretched and compressed by radially attached muscle strands. The squid *Logilo pealeii* (Fig. 6d) can vary the size of its chromatophores from 1.5 mm in the expanded state to about 0.1 mm when retracted [34]. A distinct layer of iridophores buried in the skin beneath the chromatophores provide spectrally selective reflection (Fig. 6e, f). The iridophores are colorless cells, which contain stacks of thin platelets (iridosomes) that reflect light by multilayer interference. Among the iridophores, cuttlefish and octopus employ additional structural elements, so-called leucophores (Fig. 6f). These cells are made up of disordered spherical assemblies with particles ranging from 250 to 1,250 nm in diameter (leucosomes). They induce diffuse broadband scattering and are responsible for the white patterns on cuttlefish and octopus. While the leucophores are passive elements, cephalopods have a high physiological control over chromatophores and iridophores [35]. The chromatophores can be expanded or retracted within a fraction of a second. In the expanded state they display the pigment color (red, yellow/orange, or brown/black depending on the species) and hide the underlying iridophores and leucophores from interfering with the incident light. When the chromatophores are retracted iridophores and leucophores are revealed and determine the reflected color. Squids are able to change the iridescence of the iridophores with shifts of over 100 nm in the reflected wavelengths observed for some species. This reflectance change progresses much slower than the actuation of the chromatophores and can take several seconds to minutes. The change in iridophore reflection can result from two different processes. The platelets in the iridophores, which are made of a protein called reflectin, can change their refractive index by a change in state of the protein conformation [36] (Fig. 6g–i). Furthermore, the thickness of the plates can change to tune the reflection. In addition to the described passive optical elements, some cephalopod species make use of structurally very complex light-emitting photophores that can be highly directional in their emission. The emission of photophores is often enhanced by the incorporation of multilayer back-reflectors. More complex photophores contain filters and light guides that help to channel and direct the emitted light. In summary, cephalopods use an extensive repertoire of actively tunable optical elements that rely on a range of physical effects in order to manipulate incident light, including absorption, light interference, bioluminescence, and scattering, which makes them the uncontested masters of color, light manipulation, and camouflage in nature.

Conclusions

In the course of evolution, various organisms in nature have developed a huge variety of photonic systems that by interference, diffraction, coherent, and incoherent scattering cause distinct color. While the bearers of such photonic structures come from very different animal orders, common design principles can be identified across the distinct taxonomic groups. Regular periodic multilayer arrangements and two- or three-dimensional photonic crystals build the base for strong color, while disordered arrays of particles and filaments cause bright whiteness. Pigment-loaded structures with corrugations on different length scales act as efficient light traps that render surfaces deep black. Strong blackness and brilliant whiteness are frequently employed in nature to provide contrast for intriguing patterns of color on wings and bodies of insects, scales and shells of marine animals, and feathers of birds. The coordinated interplay of regularity and irregularity on different length scales plays an important role in the function of many natural photonic systems [4, 9]. Well-defined, structural regularity and periodicity on the submicron scale ensures the reflection of strong bright colors, while irregularity on the scale of several microns inducing random scattering often mediates color stability and conspicuousness in a wide angular range.

By looking at organisms in nature new insight and knowledge can be gained for the design of materials that show specific and efficient interaction with light. This fact is widely acknowledged in the scientific community [15, 27]. Researchers show increased interest in the development of bio-inspired photonic systems. The current techniques and tools used in the industry, research, and everyday life for light harvesting, optical signaling, data transfer, and processing might soon benefit from a better understanding of the composition and functioning of biological photonic structures.

Cross-References

► Biomimetics of Optical Nanostructures
► Moth-Eye Antireflective Structures
► Nanostructures for Coloration (Organisms other than Animals)
► Nanostructures for Photonics

References

1. Kinoshita, S., Yoshioka, S., Miyazaki, J.: Physics of structural colors. Rep. Prog. Phys. **71**, 076401 (2008)
2. Land, M.F.: The physics and biology of animal reflectors. Prog. Biophys. Mol. Bio. **24**, 75–106 (1972)
3. Parker, A.R., Martini, N.: Structural colour in animals – simple to complex optics. Opt. Laser Technol. **38**, 315–322 (2006)
4. Kinoshita, S., Yoshioka, S.: Structural colors in nature: the role of regularity and irregularity in the structure. Chemphyschem **6**, 1442–1459 (2005)
5. Vukusic, P., Sambles, J.R.: Photonic structures in biology. Nature **424**, 852–855 (2003)
6. Vukusic, P., Hallam, B., Noyes, J.: Brilliant whiteness in ultrathin beetle scales. Science **315**, 348 (2007)
7. Vukusic, P., Sambles, J.R., Lawrence, C.R.: Structurally assisted blackness in butterfly scales. Proc. R. Soc. B **271**, S237–S239 (2004)
8. Vukusic, P., Stavenga, D.G.: Physical methods for investigating structural colours in biological systems. J. R. Soc. Interface **6**, S133–S148 (2009)
9. Kinoshita, S.: Structural Colors in the Realm of Nature. World Scientific, Singapore (2008)
10. Prum, R.O., Torres, R.: Structural colouration of avian skin: convergent evolution of coherently scattering dermal collagen arrays. J. Exp. Biol. **206**, 2409–2429 (2003)
11. Prum, R.O., Torres, R.H., Williamson, S., Dyck, J.: Coherent light scattering by blue feather barbs. Nature **396**, 28–29 (1998)
12. Parker, A.R.: 515 million years of structural colour. J. Opt. A Pure Appl. Opt. **2**, R15–R28 (2000)
13. Parker, A.R.: A geological history of reflecting optics. J. R. Soc. Interface **2**, 1–17 (2005)
14. Prum, R.O., Torres, R.H.: Structural colouration of mammalian skin: convergent evolution of coherently scattering dermal collagen arrays. J. Exp. Biol. **207**, 2157–2172 (2004)
15. Biró, L.P., Vigneron, J.P.: Photonic nanoarchitectures in butterflies and beetles: valuable sources for bioinspiration. Laser Photonics Rev. **5**, 27–51 (2011)
16. Ghiradella, H.: Shining armor: structural colors in insects. Opt. Photonics News **10**, 46–48 (1999)
17. Seago, A.E., Brady, P., Vigneron, J.P., Schultz, T.D.: Gold bugs and beyond: a review of iridescence and structural colour mechanisms in beetles (*Coleoptera*). J. R. Soc. Interface **6**, S165–S184 (2009)
18. Srinivasarao, M.: Nano-optics in the biological world: beetles, butterflies, birds, and moths. Chem. Rev. **99**, 1935–1962 (1999)
19. Vukusic, P.: Advanced photonic systems on the wing-scales of *Lepidoptera*. In: Gorb, S.N. (ed.) Functional Surfaces in Biology: Little Structures with Big Effects. Springer Science+Business Media, Dordrecht (2009)
20. Doucet, S., Meadows, M.: Iridescence: a functional perspective. J. R. Soc. Interface **6**, S115–S132 (2009)
21. Yoshioka, S., Kinoshita, S.: Polarization-sensitive color mixing in the wing of the Madagascan sunset moth. Opt. Express **15**, 2691–2701 (2007)
22. Vukusic, P., Sambles, J.R., Lawrence, C.R.: Structural colour: colour mixing in wing scales of a butterfly. Nature **404**, 457 (2000)
23. Parker, A.R., Hegedus, Z.: Diffractive optics in spiders. J. Opt. A Pure Appl. Opt. **5**, S111–S116 (2003)
24. Vigneron, J.P., et al.: Reverse color sequence in the diffraction of white light by the wing of the male butterfly *Pierella luna* (Nymphalidae: Satyrinae). Phys. Rev. E **82**, 021903 (2010)
25. Parker, A.R., McPhedran, R.C., McKenzie, D.R., Botten, L.C., Nicorovici, N.P.: Aphrodites iridescence. Nature **409**, 36–37 (2001)
26. Trzeciak, T.M., Vukusic, P.: Photonic crystal fiber in the polychaete worm *Pherusa* sp. Phys. Rev. E **80**, 061908 (2009)
27. Parker, A.R.: A vision for natural photonics. Phil. Trans. R. Soc. A **362**, 2709–2720 (2004)
28. Parker, A.R., Welch, V.L., Driver, D., Martini, N.: Structural colour: opal analogue discovered in a weevil. Nature **426**, 786–787 (2003)
29. Simonis, P., Vigneron, J.P.: Structural color produced by a three-dimensional photonic poly-crystal in the scales of a longhorn beetle: *Pseudomyagrus waterhousei* (Coleoptera: Cerambicidae). Phys. Rev. E **83**, 011908 (2011)
30. Welch, V., Lousse, V., Deparis, O., Parker, A.R., Vigneron, J.P.: Orange reflection from a three-dimensional photonic crystal in the scales of the weevil *Pachyrrhynchus congestus pavonius* (Curculionidae). Phys. Rev. E **75**, 041919 (2007)

31. Saranathana, V., et al.: Structure, function, and self-assembly of single network gyroid ($I4_132$) photonic crystals in butterfly wing scales. PNAS **107**, 11676–11681 (2010).
32. Hallam, B.T., Hiorns, A.G., Vukusic, P.: Developing optical efficiency through optimized coating structure: biomimetic inspiration from white beetles. Appl. Opt. **48**, 3243–3249 (2009)
33. Vigneron, J.P., et al.: Switchable reflector in the Panamanian tortoise beetle *Charidotella egregia* (Chrysomelidae: Cassidinae). Phys. Rev E **76**, 031907 (2007)
34. Mäthger, L.M., Hanlon, R.T.: Malleable skin coloration in cephalopods: selective reflectance, transmission and absorbance of light by chromatophores and iridophores. Cell Tissue Res. **329**, 179–186 (2007)
35. Mäthger, L.M., Denton, E.J., Marshall, N.J., Hanlon, R.T.: Mechanisms and behavioural functions of structural coloration in cephalopods. J. R. Soc. Interface **6**, S149–S163 (2009)
36. Cooper, K.M., Hanlon, R.T., Budelmann, B.U.: Physiological color change in squid iridophores. II. Ultrastructural mechanisms in *Lolliguncula brevis*. Cell Tissue Res. **259**, 15–24 (1990)

Structural Color in Nature

▶ Biomimetics of Optical Nanostructures

Structural Colors

▶ Nanostructures for Coloration (Organisms other than Animals)

Structure and Stability of Protein Materials

Szu-Wen Wang
Chemical Engineering and Materials Science,
The Henry Samueli School of Engineering,
University of California, Irvine, CA, USA

Synonyms

Nanostructure; Thermostability

Definition

Protein-based materials are polymeric biomaterials comprising amino acid subunits that are connected together by peptide bonds. These materials are usually biomimetic, self-assemble into higher-order nanometer-scale architectures, and can interact with biological entities. Determination of their structure and stability is an important component of assessing their utility and function.

Protein-Based Nanomaterials

The control of architecture at the nanoscale is a challenge in which nature has been highly successful. Since genetic manipulation enables the definition of every monomer in a polymeric protein structure, giving far greater control than conventional chemical synthesis, one approach in materials synthesis is the use of protein engineering to create biologically inspired materials [1]. By combining natural scaffolds, structural elements, and biologically reactive sites, materials with novel architectures and properties can be obtained. These materials are fabricated in microorganisms as recombinant proteins comprising amino acid units. Furthermore, proteins often self-assemble into complex, higher-order nanostructures, which include fibrous and spherical architectures [2, 3]. Protein-based materials have been evaluated for a wide variety of applications, such as tissue engineering scaffolds, templates for nanomaterials synthesis, and drug delivery carriers. The evaluation of structure and stability in these protein-based nanomaterials is primarily established using techniques developed for general proteins.

Structure

The structures of proteins span several length scales, and different methods are used depending on the type of structure being evaluated [4]. The *primary structure* of proteins describes the sequence of individual amino acids, or monomer residues, covalently coupled by peptide bonds to form a polypeptide chain. There are 20 known natural amino acids, each differing in their side chain identity, thereby imparting different physicochemical characteristics at each location along the polymer. One can also incorporate "unnatural" amino acids, which expands the functionalization of the protein-based material with useful chemical components that are not native [5, 6]. When the protein materials have been created in a recombinant system (such as in microorganisms), the DNA encodes for the protein sequence. Primary

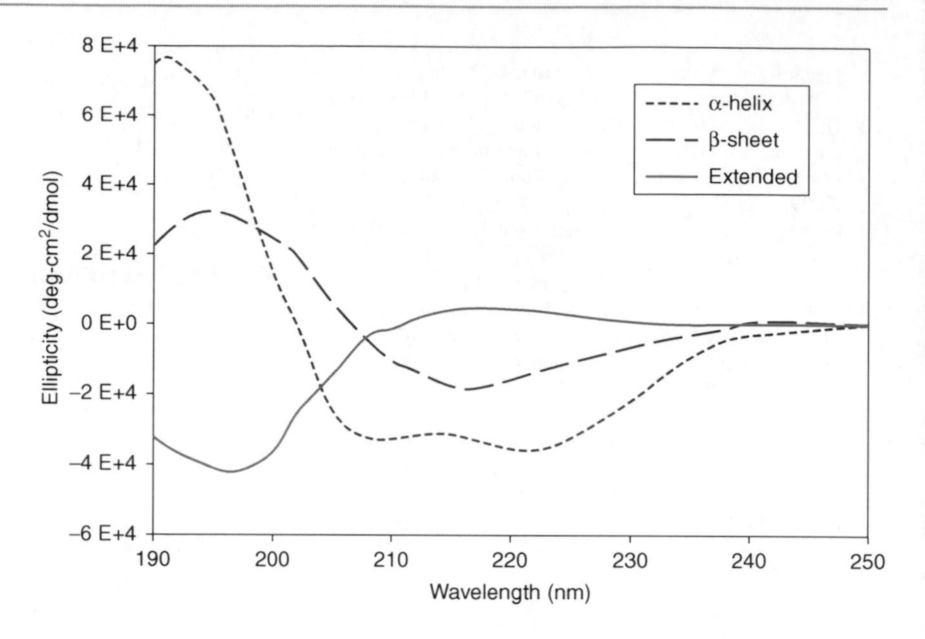

Structure and Stability of Protein Materials, Fig. 1 Circular dichroism spectra of polypeptides with representative α-helix (*dotted line*), β-sheet (*dashed line*), and extended helical (*solid line*) conformations (Data replotted from [8]. Reprinted with permission, copyright 1969, American Chemical Society)

structure can also be determined by N-terminal peptide sequencing or proteolytic mass spectrometry.

The local primary structure affects the *secondary structure*, which are regions of local folding. Secondary structure of protein materials is commonly determined by circular dichroism (CD) [7]. Asymmetric molecules, such as polypeptides, will absorb left and right circularly polarized light to different degrees. Circular dichroism is this absorbance difference, and different types of secondary folding (such as α-helices, β-sheets, or extended helical conformations) have characteristic CD spectra (Fig. 1, [8]). The degree of folding can be used to monitor conformational changes of protein materials.

Tertiary structure is the three-dimensional conformation of the folded protein. The *quaternary structure* is the assembly of the individual polypeptide or protein subunits. Global size or assembly properties, such as hydrodynamic diameters or molecular weights of the protein complexes, can be evaluated by dynamic light scattering or ultracentrifugation. While these techniques can give useful information, they do not elucidate structural details. The most precise structural information can be obtained by techniques such as X-ray diffraction or nuclear magnetic resonance spectroscopy. However, these methods require highly pure protein, a crystalline form of the protein, and/or extensive data analysis, and these conditions are not usually available or feasible for protein-based materials used in nanotechnology applications. Therefore, alternative methods providing intermediate amounts of information are used to probe tertiary and quaternary structure.

Microscopy strategies enable visualization of protein nanostructures ranging from nanometer-to-micron length scales. In transmission and scanning transmission electron microscopy (TEM and STEM, respectively), an electron beam is transmitted through a sample and information regarding the scattered electrons is obtained [9]. Proteins are first deposited in a thin layer onto carbon-coated electron microscope grids. These grids are then frozen at liquid nitrogen temperatures in cryogenic preparation, or proteins can also be negatively stained with a heavy ion salt (e.g., uranyl acetate) to provide contrast. Due to the insulating nature of proteins, the resolution using TEM or STEM is low relative to conducting samples and are typically on the order of 5–10 nm (Fig. 2, [10]).

Alternatively, atomic force microscopy (AFM) can yield lateral resolutions of less than 1 nm for protein structures (Fig. 3, [9]). Proteins are adsorbed onto an atomically flat substrate (typically mica or graphite) and are imaged dried or hydrated in buffer. As the AFM probe is scanned over the surface of a sample, deflections (due to interactions between the probe tip and sample) are measured. Wet samples reflect native protein conditions, and therefore AFM potentially can yield more accurate structural information relative to electron microscopy. One challenge of AFM relative to electron microscopy, however, is the time-intensive nature of imaging.

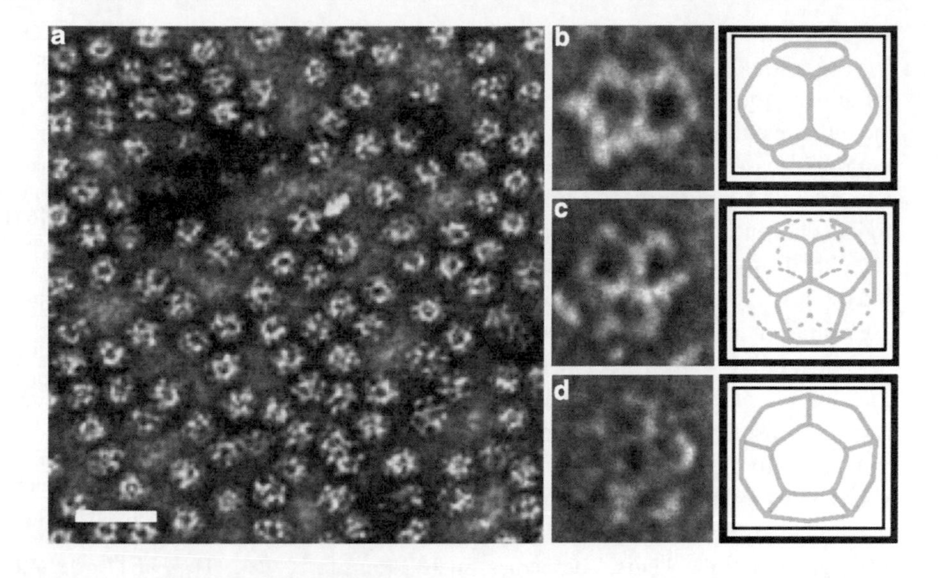

Structure and Stability of Protein Materials, Fig. 2 Transmission electron microscopy image of dodecahedral protein complex. (**a**) Negatively stained protein assembly of the E2 subunit from pyruvate dehydrogenase. Scale bar is 50 nm. Projection views of (**b**) twofold, (**c**) threefold, and (**d**) fivefold axes of symmetry verify icosahedral assembly (Figure from [10]. Reprinted with permission, copyright 2008, Wiley)

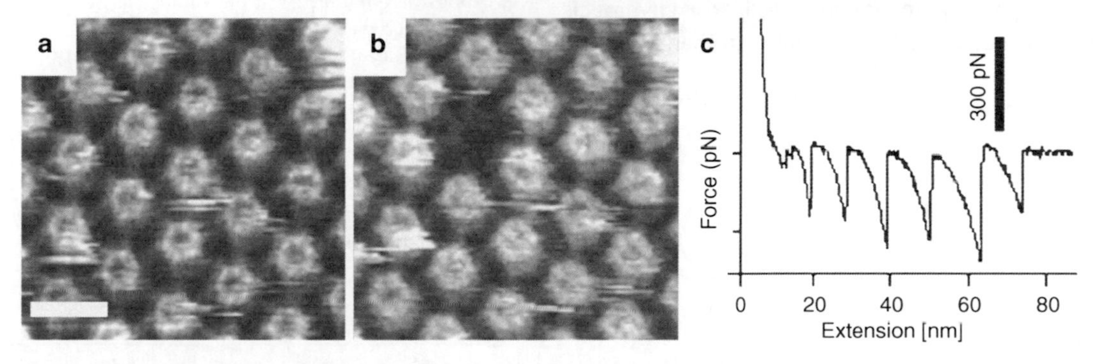

Structure and Stability of Protein Materials, Fig. 3 Atomic force microscopy analysis of a hexagonally packed intermediate (HPI) monolayer of protein. (**a**) Extracellular HPI layer. Scale bar is 20 nm. (**b**) A single HPI core can be attached to an AFM tip and extracted by tip retraction, as shown by the vacant hole in the image. (**c**) Force-distance curve exhibits six peaks, corresponding to successively unfolding of the linked HPI proteins (From [9]. Reprinted with permission, copyright 2011, Elsevier)

Stability and Mechanical Properties

The stability of protein materials can be evaluated as the structure and assembly are probed while stressing conditions (such as temperature, pH, denaturants, or ionic strength) are changed. For example, the thermostability of protein materials can be evaluated by CD measurement at a characteristic wavelength as the temperature of the sample is increased [11]. Using this data, thermodynamic and thermostability values for protein unfolding (e.g., enthalpy, entropy, free energy, midpoint of unfolding temperature, and onset of unfolding temperatures) can be calculated. Differential scanning calorimetry (DSC), which measures the heat capacity of a protein solution as a function of temperature, can also determine unfolding and thermodynamic parameters of protein-based materials [12]. DSC can also be used to gain insight into kinetic stability and conditions in which protein materials exist at a local energetic minimum with a nonideal structure [13]. Although more direct thermodynamic information can be obtained from DSC, advantages of CD over DSC include the significant lower amounts of protein required for analysis and greater accessibility to spectropolarimeters.

Applying force to single protein molecules gives insight into how the materials may be assembled and the stability of these interactions. Using single-molecule force spectroscopy, AFM can be used to pull out proteins within structured assemblies, and force-distance curves yield information about the forces required to disrupt folded proteins and protein complexes [9, 14]. Subsequent imaging by AFM can show the resulting effects on the complex (Fig. 3, [9]).

Protein-based materials often demonstrate both viscous and elastic properties, particularly if they exist at high concentrations or as hydrogels. The storage modulus (G') describes the elastic, solid-like properties and the loss modulus (G") measures its viscous, liquid-like character. These dynamic moduli can be determined experimentally with a rheometer or with particle-tracking microrheology [15]. The latter technique measures the displacement of microparticles mixed with protein solution over time and is particularly useful when only small amounts of protein are available, which is often the case in early development of protein materials. Static mechanical properties, such as tensile stress and tensile strain, of bulk protein-based hydrogels can be measured using a conventional mechanical tester.

Cross-References

- ► AFM in Liquids
- ► AFM, Tapping Mode
- ► Atomic Force Microscopy
- ► Bioinspired Synthesis of Nanomaterials
- ► Biomimetic Synthesis of Nanomaterials
- ► Biomimetics
- ► Biomimetics of Marine Adhesives
- ► Mechanical Properties of Hierarchical Protein Materials
- ► Spider Silk
- ► Transmission Electron Microscopy

References

1. van Hest, J.C.M., Tirrell, D.A.: Protein-based materials, toward a new level of structural control. Chem. Commun. (19), 1897–1904 (2001)
2. Kluge, J.A., Rabotyagova, U., Leisk, G.G., Kaplan, D.L.: Spider silks and their applications. Trends Biotechnol. **26**, 244–251 (2008)
3. Uchida, M., Klem, M.T., Allen, M., Suci, P., Flenniken, M., Gillitzer, E., Varpness, Z., Liepold, L.O., Young, M., Douglas, T.: Biological containers: protein cages as multifunctional nanoplatforms. Adv. Mater. **19**, 1025–1042 (2007)
4. Creighton, T.E.: Proteins: structures and molecular properties, 2nd edn. Freeman, New York (1993)
5. Xie, J.M., Schultz, P.G.: Innovation: a chemical toolkit for proteins – an expanded genetic code. Nat. Rev. Mol. Cell Biol. **7**, 775–782 (2006)
6. Connor, R.E., Tirrell, D.A.: Non-canonical amino acids in protein polymer design. Polym. Rev. **47**, 9–28 (2007)
7. Greenfield, N.J.: Using circular dichroism spectra to estimate protein secondary structure. Nat. Protoc. **1**, 2876–2890 (2006)
8. Greenfield, N.J., Fasman, G.D.: Computed circular dichroism spectra for evaluation of protein conformation. Biochemistry **8**, 4108–4116 (1969)
9. Muller, S.A., Muller, D.J., Engel, A.: Assessing the structure and function of single biomolecules with scanning transmission electron and atomic force microscopes. Micron **42**, 186–195 (2011)
10. Dalmau, M., Lim, S., Chen, H.C., Ruiz, C., Wang, S.W.: Thermostability and molecular encapsulation within an engineered caged protein scaffold. Biotechnol. Bioeng. **101**, 654–664 (2008)
11. Greenfield, N.J.: Using circular dichroism collected as a function of temperature to determine the thermodynamics of protein unfolding and binding interactions. Nat. Protoc. **1**, 2527–2535 (2006)
12. Ladbury, J.E., Doyle, M.L. (eds.): Biocalorimetry 2: Applications of Calorimetry in the Biological Sciences. Wiley, Hoboken (2004)
13. Sanchez-Ruiz, J.M.: Protein kinetic stability. Biophys. Chem. **148**, 1–15 (2010)
14. Muller, D.J., Dufrene, Y.F.: Atomic force microscopy as a multifunctional molecular toolbox in nanobiotechnology. Nat. Nanotechnol. **3**, 261–269 (2008)
15. Wirtz, D.: Particle-tracking microrheology of living cells: principles and applications. Annu. Rev. Biophys. **38**, 301–326 (2009)

SU-8 Photoresist

Frederik Ceyssens and Robert Puers
Department ESAT-MICAS, KULeuven,
Leuven, Belgium

Synonyms

Gamma-butyrolactone (GBL); Hardbake (HB); Lithographie, galvanoformung, abformung (LIGA); N-Methyl-2-pyrrolidone (NMP); Polyethylene carbonate (PEC); Polymethyl methacrylate (PMMA); Polypropylene carbonate (PPC); Post-exposure Bake (PEB); Propylene glycol methyl ether acetate (PGMEA)

Definition

SU-8 is a high aspect ratio epoxy-based negative photoresist commonly used as structural material in lithographic fabrication.

Introduction

SU-8 was developed by IBM as a thick negative photoresist targeted to the fabrication of molds for electroplating. The epoxy-based negative photoresist has some remarkable properties. First of all it has a wide range of coating thicknesses: layers from several hundreds of nanometers up to several hundreds of microns can be deposited by a single standard spin coating step, using the appropriate dilution of the SU-8 resin. Even thicker layers can be deposited and photopatterned successfully as will be discussed later. Second, structures featuring almost straight sidewalls can be created in the SU-8 layer by simple UV exposure through a contact mask. Aspect ratios of about 15 are routinely obtained. Using optimized processes in layers over a few hundreds of microns thick, aspect ratios of over 100 are obtainable [20].

The fabrication of such polymer structures was up to that time only possible by the expensive LIGA technique which requires synchrotron radiation. Though LIGA is still superior in terms of achievable aspect ratio and wall straightness, SU-8 can be humoristically referred to as a "poor man's LIGA." This must not be taken pejoratively per se as "cheap" can be read as "opening new possibilities" just as well.

SU-8 has high chemical resistance and because of the high chemical functionality of the monomers its mechanical properties are good enough to render it useful as structural material in a large number of cases, as further discussed in the material properties section.

SU-8 is an insulator, but covered later it can be made conductive, for example, by metal deposition. Also, it can of course be used as an electroplating mold to create thick, high aspect ratio metal structures.

It is transparent, enabling easy inspection of the underlying structures and certain optical readout principles, for example, based on fluorescence. Furthermore, the processing of SU-8 does not require a large investment in equipment or consumables.

This all leads to the fact that SU-8 has undoubtedly become one of the most widely used structural polymer materials in microsystems. It is not only frequently used in work of the traditional micromechanics community, but has also spread into groups working on labs-on-chip, integrated optics, packaging, and others. This is illustrated by the large number of research papers mentioning SU-8 in the title that have appeared over the last decade (Fig. 1).

Commercially available SU-8 resist has a solid fraction consisting of SU-8 monomer (Fig. 2), which contains eight epoxy functional groups, and a mixture of triaryl sulfonium hexafluoroantimonate salts as photoinitiator (4.8 wt%).

The solvent added hereto is either Gamma-butyrolactone (GBL) or cyclopentanone. Both types are available from Microchem in a wide range of dilutions under the names "SU-8" or "SU-8 2000," respectively. Another supplier is Gersteltec.

Material Properties

The mechanical and electrical properties of SU-8 are not unlike those of a typical epoxy. Some material data are summarized in Table 1. It can be seen that creep is rather low, which is a good indication of the suitability of the material for mechanical applications.

For microfluidic applications the contact angle of the material is often important. Untreated SU-8 is rather hydrophobic, displaying a contact angle of 75° with water. The material can be made more hydrophilic by several means such as a short exposure to oxygen plasma. A treatment at 600 W for 8 s at a pressure of 27 Pa reduces the contact angle to 25.6°. After 4 min at 400 W the contact angle is down to 3.2° but goes up again to 25° over a period of 40 days [7]. Another way is the addition of surfactants to the SU-8 itself. Bohl et al. [3] used 10 wt% trisiloxane alkoxylate to obtain a contact angle of 28°. Combined with plasma activation this becomes a mere 10°. Mostly, adhesion is sufficient (Table 2), except when using glass substrates. An adhesion promoter based on the deposition of a titanium oxide monolayer (such as AP-300) can then be used.

Processing of Thick High Aspect Ratio Structures

The procedure to deposit a single layer of SU-8 comprises seven major steps. In each of these steps several parameters can be selected, each having an effect on one or more properties of the final structure. Especially to attain thick, high aspect ratio features such as shown in Fig. 3, careful processing is mandatory.

The main process sequence and its most significant parameters will be discussed below. Most important in

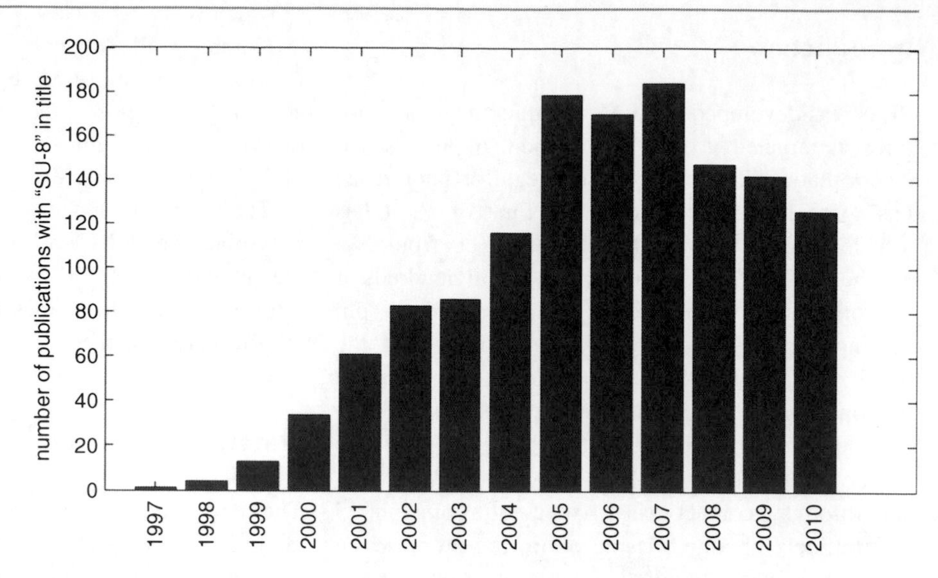

SU-8 Photoresist, Fig. 1 Number of research papers with "SU-8" in their title per calendar year according to Google Scholar on February 15, 2011. As SU-8 is becoming a mature and commonly used technology, it being mentioned in the title of research papers is expected to become less frequent again

SU-8 Photoresist, Fig. 2 SU-8 monomer

process design is attaining a high ratio between the mechanical strength and the built-in mechanical stress in order to avoid deformation of the fabricated structures. This is done by controlling process parameters such as the solvent concentration during the softbake step and a few additions to the standard process such as the incorporation of relaxation steps and ramped heating [5].

Substrate Preparation

Standard cleaning procedures and a dehydration bake are recommended. On silicon, a HF dip improves adhesion and uniformity [16]. The substrate type has a large influence on uniformity and adhesion as well (Table 2).

Layer Deposition

The deposition of a uniform film of non-cross-linked SU-8 on the surface is of course of great importance for the successful outcome of the entire process. Not only is the uniformity of the height of the structures produced determined by this step, the lateral uniformity can be affected as well: an uneven distance between resist surface and mask changes the diffraction

SU-8 Photoresist, Table 1 Material properties of SU-8 [5]. Creep and fracture strength data from [15] who are using the following modified Voight-Kelvin model to model creep in SU-8: $\frac{\varepsilon(t)}{\sigma_{applied}} = D_0 + (D_e - D_0)\left(1 - e^{-\left(\frac{t}{\tau}\right)^m}\right)$. With ε as the mechanical strain and t time. The mechanical stress $\sigma_{applied}$ used in these experiments was 13.2MPa

Property	Value	Remarks
Young's modulus (GPa)	2.1 ± 0.2	HB @ 150°C for 1 h
Maximum strain to break (%)	1.2 ± 0.5	HB @ 150°C for 1 h
Fracture strength (MPa)	73.3 ± 9.2	HB @ 180°C for 30′
Transient creep compliance De (GPa^{-1})	0.310 ± 2.5%	HB @ 180°C for 30′
Initial creep compliance D_0 (GPa^{-1})	0.251 ± 13.8%	HB @ 180°C for 30′
Creep time constant τ (s)	1.84×10^{10} ± 3.5%	HB @ 180°C for 30′
Creep coupling parameter m	0.208 ± 5.3%	HB @ 180°C for 30′
CTE (ppm/K)	52 std. 5.1	After PEB @ 95°C
Thermal conductivity (W/m K)	0.3	
Breakdown strength (V/m)	200	Tested on a 3 μm thick layer
Refraction index	1.658	λ = 365 nm
	1.638	λ = 405 nm
Contact angle with water	75°	Untreated

SU-8 Photoresist, Table 2 Shear adhesion test for SU-8 2000 [13]. Test conditions: RCA clean for glass, plasma clean for other surfaces. Test tool: Dage 4,000 working on 100 × 100 × 50 μm SU-8 posts. SU-8 was hardbaked at 150°C

Substrate	Adhesion (MPa)
Si	53
SiN	43
Ni	45
Au	29
Al/Cu (99–1)	23
Cu	38
Cu with AP-300	56
Glass	Poor
Glass with HDMS	Poor
Glass with AP-300	92
Quartz	61

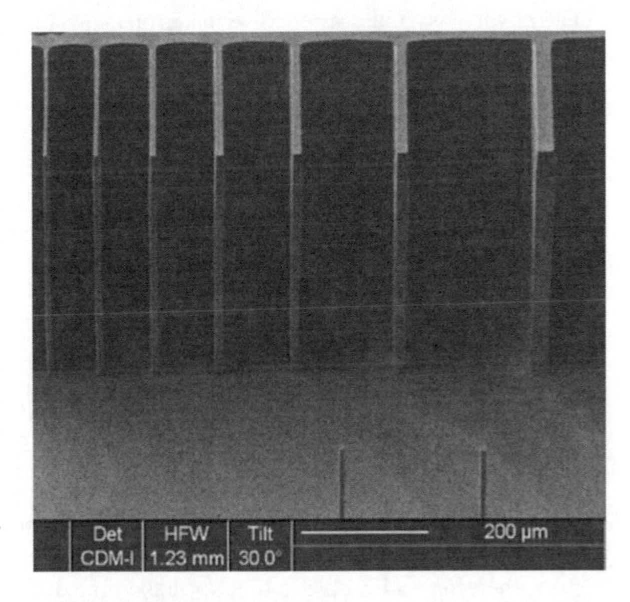

SU-8 Photoresist, Fig. 3 SU-8 test structures, 700 um high, down to 10 um wide

conditions during contact exposure, the most common method of UV exposing SU-8.

Spin coating is the most straightforward deposition technique for SU-8. During spinning, irregularities are introduced. The largest and always appearing irregularity is the so-called edge bead. This is a ring near the edge of the wafer where the resist is thicker. The edge bead can be negligible for thinner (<10 μm) SU-8 layers but can be over a centimeter for thick (>100 μm) layers. In the latter case, the edge bead is typically over 50 μm higher than the inner resist layer. Smaller irregularities disappear when the SU-8 layer is given enough time to reflow, preferably in a solvent-saturated atmosphere. The edge bead can be removed mechanically or by a directed spray of a solvent, a feature common in modern coating equipment.

Casting is another method commonly used to deposit SU-8 layers. During casting, a known weight or volume of low viscosity SU-8 is poured on a wafer resting on a leveled hot plate. Reflow and resist spreading in a solvent-rich environment (e.g., by putting a Petri dish over the wafer) is typically necessary. This way, for layers over a few hundreds of microns thick, superior thickness control and uniformity can be achieved. Also, no edge bead formation occurs.

In order to determine the final layer thickness the shrinkage of the material due to solvent evaporation in the softbake step following the casting step has to be accounted for. A cast layer of SU-8 2007 was determined experimentally to shrink 44% after a softbake of 6 h at 65°C followed by 7 h at 95°C. For SU-8 2010, this is about 32% and for SU-8 2002 61% [5].

Exceptionally thick layers (>1 mm) can be deposited by casting too. However, in such layers a significant solvent concentration gradient was observed to remain after the subsequent softbake step. This tends to weaken fine structures. It was observed that depositing and softbaking several layers on top of each other does not alleviate this significantly, as solvent from an upper layer tends to diffuse into the dry baked lower layers. A better solution is to put a known mass of pre-dried SU-8 flakes on the wafer surface, and melt it into a smooth layer [2].

Softbake

The purpose of the softbake is to lower the solvent concentration, preventing adhesion to a contact mask and improving the lithographic performance, that is, the ability to produce high aspect ratio structures with straight sidewalls, by reducing the diffusion of the photoinitiator. The softbake temperature must be below 120°C, which is the onset of thermally induced cross-linking.

A too high solvent content causes the photoacid generated during exposure to diffuse easily to unexposed regions, creating bulges or protrusions in the structure. A too low solvent concentration can cause cracks and peeling off of structures.

It is, therefore, advisable to tune softbake time and temperature to reach a certain solvent concentration. In thick layers, the solvent concentration can be easily monitored by weighing the wafer on a balance with milligram precision. An advised solvent concentration is 5–7%. An even lower solvent content decreases cross-linking [18] and increases stress deformations in high aspect ratio features [5]. After softbaking, a relaxation period of several hours is advised.

Exposure

Typically, SU-8 is exposed using contact exposure on a mask aligner equipped with a standard mercury arc UV lamp. During exposure, the photoinitiator's triarylsulfonium hexafluoroantimonate salts are split, releasing a Lewis acid [16]. The acid generated serves as a catalyst, allowing a two-step reaction to occur that links the epoxy groups of different SU-8 monomers together. In practice, the latter two-step reaction occurs only very slowly at room temperature. To speed up the reaction rate, the SU-8 is heated up after exposure, during the post-exposure bake (PEB) step.

SU-8 Photoresist, Table 3 Absorbance (inverse of penetration depth) of unexposed SU-8 for typical wavelengths [24]

Wavelength (nm)	Absorbance (μm^{-1})
313	1.19×10^{-1}
334	5.85×10^{-2}
365	2.71×10^{-3}
405	1.41×10^{-4}
436	6.90×10^{-5}

For any practical application, both the exposure dose and the wavelength of the light used are important. From the critical dose up, parts that are unsolvable in developer will be formed. The critical dose for i-line (365 nm) exposure was determined to be 30 ± 0.5 mJ/cm^2 [20], for a PEB of 6′ at 65°C and 3′ at 95°C. This dose was observed to be by far insufficient for a good lithographic result: The *adhesion* of the structures formed to the substrate becomes only sufficiently strong to survive development at a much higher exposure dose. A second problem of a low dose is the permanent *deformation* of vulnerable and weakly cross-linked structures caused by internal stress during the development step. On the other hand, it is not possible to increase the exposure dose indefinitely as two problems arise with overexposure: the broadening of the features compared to their designed sizes and the appearance of effects caused by stress in larger parts. Careful choice of process parameters is therefore required. A typical i-line exposure dose is 200 mJ/cm^2.

Furthermore, the choice of exposure wavelength is important. Some UV wavelengths produced by a standard pressurized mercury arc lamp are absorbed too quickly and are not suitable to exposure of relatively thick SU-8 evenly (Table 3: absorbance). The transmission of light down to a certain depth is illustrated in Fig. 2. This shows that for even exposure of a layer thicker than a few microns, the 313-nm line should always be filtered out. For layers thicker than a few hundred microns, filtering out all light below 400 nm is recommended. As the photoinitiator is less

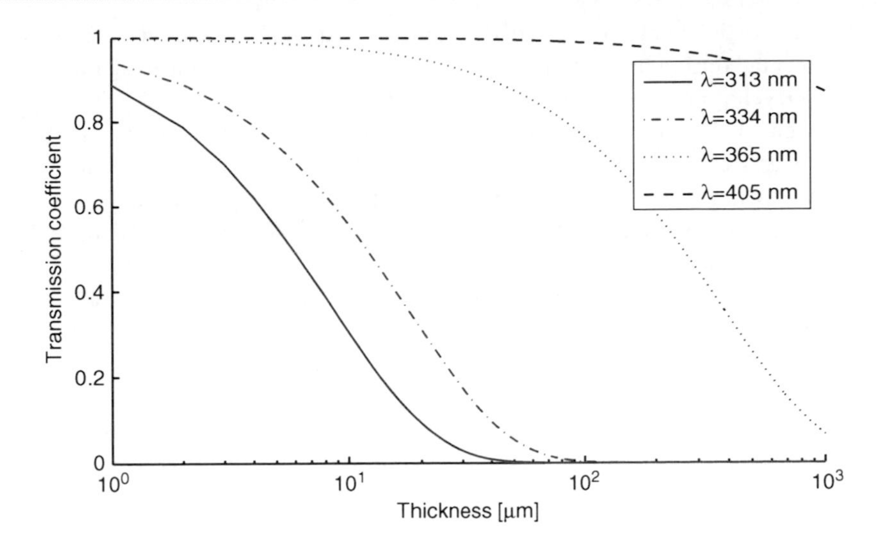

SU-8 Photoresist, Fig. 4 Transmission of typical UV lines to the bottom of layer versus thickness of that layer, assuming no reflection of light and using the absorbances quoted in Table 3. transmission coefficient = $e^{\text{absorbance} \times \text{thickness}}$

sensitive for 405-nm light, the recommended exposure dose is higher. A dose of 10 J/cm^2 suffices for most applications.

When exposing layers of irregular thickness, the use of an index matching fluid such as glycerin to fill up the air gap between mask and wafer can be beneficial to reduce diffraction effects and increase resolution (Fig. 4).

Post-exposure Bake (PEB)

The baking step following the exposure of the resist causes a cross-linking reaction in the exposed parts, rendering those parts insoluble during the subsequent development step. Meanwhile, a few side effects occur. These effects are shrinking of the resist by the cross-linking reaction, the thermal expansion of the resist during baking, and the (dissimilar) thermal expansion of the substrate during baking. The situation is clearly different for the in-plane and the out-of-plane direction.

In the out-of-plane direction, the SU-8 layer is unconstrained and shrinking can occur freely. A shrinkage of approximately 4.5% (std 0.25%) was measured over a PEB of 1 h at 65°C [5]. In-plane, although some relaxation occurs, an internal stress is built up (~4 MPa for a PEB temperature of 95°C in a standard process [5]).

The combination of exposure energy and PEB time and temperature must cater for enough cross-linking for the structures to resist delamination and deformation during the subsequent development step. Typically, post-exposure baking raises the glass transition temperature, that is, the temperature above which plastic deformation is possible, to a value slightly larger than the PEB temperature. After a PEB of 20 min, the glass transition temperature lies exactly at the PEB temperature. After that, saturation sets in quickly [10]. Therefore, it makes sense to use a minimum PEB time of 20 min. Typical PEB temperatures are between 80 and 100°C.

After PEB, there is some stress relaxation over time. Hammacher et al. [11] notice a 7.5% drop in stress during the first 8 h after the PEB. After 4 days, there is an additional drop of 2%. This indicates that it is beneficial to build a waiting period between PEB and development, diminishing adhesion loss and deformations caused by stress. Also, a ramp up and slow ramp down of the PEB temperature are beneficial. For example, a cool down step of four hours in an oven can be employed [21].

Development

During development, the non-cross-linked parts of the SU-8 layer are removed in a solvent. Typically, propylene glycol methyl ether acetate (PGMEA) is employed, though ethyl lactate or diacetone alcohol can be used as healthier alternatives with lower vapor pressure [13]. Exposed structures will, even after post-exposure baking, adsorb solvents and thus swell during development. Structures in the 10 μm range are typically saturated within a minute [18]. As the solvent evaporates again after development is over, this should not be a problem. However, swollen structures that are not cross-linked hard enough can become permanently deformed during development.

Development is dependent on agitation and on the design of the structures fabricated, and development rates as well as development times are therefore to be taken with caution. A development time of 8 min for a 50-mm-thick layer without agitation is typically sufficient. Ultrasonic agitation was found to increase the development rate by approximately eightfold. However, it can damage fragile structures. Megasonic actuation does not cause this problem. Another option is developing the wafer in an upside down position, which was found to decrease the development time by a factor 3.5 without inducing extra damage [5].

Hardbake

The hardbake is an optional extra step at the end of the SU-8 process. During the hardbake, the material is heated well above the threshold for thermally induced cross-linking. This causes, also in regions where no photoacid is present, additional cross-linking reactions to occur until saturation. Hardbaking is done for increasing the long-term stability of the material and to close small cracks that arise in some process sequences.

The additional cross-linking causes a higher chemical resistance as well: it was found impossible to remove hardbaked SU-8 structures from their substrates with solvents such as acetone or NMP (N-methyl-2-pyrrolidone). For this reason, a hardbake is often not advised when the SU-8 layer will have to be removed later in the fabrication process.

For the hardbake, typically a temperature between 150°C and 250°C is used. Following the observations of Hammacher et al. on the PEB, it can be assumed that most of the extra cross-linking is complete after 30 min. Hardbaking at 150°C for 60 min increases internal stress to 12 MPa (tensile) in 45% humidity air [18]. In the latter work, the built-in stress was shown to be strongly dependent on the humidity in the environment.

Removal of SU-8

The downside of the high chemical resistance of SU-8 is the difficulty to remove the material if needed. Non-hardbaked SU-8 can still be cracked up in strong solvents such as warm NMP. When this step is followed by rinsing or ultrasonification to remove the remaining flakes, a surface having a not too high topography can be clean again.

SU-8 can be cleanly removed in piranha acid, though this is aggressive toward many metals and polymers that may already be present on the wafer. Silicon and glass are not attacked at all. Also, platinum, tantalum and gold are inert to piranha. Chromium is etched only very slowly, allowing the cleaning of contact masks in piranha after contamination with SU-8. Another application of piranha is for cleaning SU-8 off wafers for reuse.

Reactive ion etching (RIE) in oxygen plasma with a small fluorine content is used as well, though traces of a nonvolatile antimony containing residue may be formed.

Typically, 4–5% of fluorine-containing gas such as SF_6 or CF_4 is added to the reactor atmosphere to boost the etch rate significantly.

Other less widely used options are ashing in an air or oxygen atmosphere and molten salt removal. These need a rather high temperature, 450–500°C and 350°C, respectively.

Finally, one can consider downstream chemical etching (DCE). DCE is an organic removal technique in which reactants produced in a plasma are blown on the wafer, which is not exposed to the plasma itself. With a wafer temperature of 225°C and a gas composition of 98% oxygen and 2% CF_4,, an etch rate of 6.8 μm/min can be obtained [9]. Given the reasonable working temperature, the etch speed, and the low metal etch rate, DCE is likely the most promising method for industrial use.

Alternative Exposure Methods

Next to classic UV lithography, a number of other methods to selectively cross-link SU-8 exist. Though much less commonly used, they open some extra possibilities and therefore, deserve consideration. A distinction can be made between direct write methods in which structures are written pixel by pixel or voxel by voxel, and wafer-scale methods using masks which are generally much faster as writing is done in parallel over a large surface.

The photoinitiator in SU-8 can be activated by the simultaneous absorption of two 800-nm photons, instead of by a single 400-nm photon [17]. The probability of such activation rises with the square of the

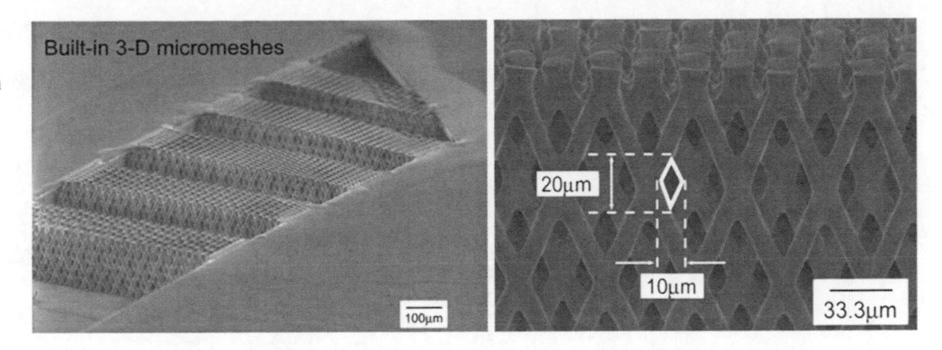

SU-8 Photoresist, Fig. 5 Microsieve structure in microchannel, fabricated in SU-8 using inclined UV exposure ([14], used with permission) *Left*: overview. *Right*: detail of sieve

intensity of the exposure. This is the basis for so-called *two-photon lithography*. The nonlinear dependence makes it possible, by focusing a laser to create a high-intensity spot that is small with respect to the layer thickness, to write true 3D structures such as photonic crystals directly in a single SU-8 layer.

Proton beam writing is another direct write methods found in literature. The penetration depth of the exposing proton beam is dependent on its energy. Thus, bridge-like structures can be written in a single SU-8 layer, but not the more complicated structures that are possible with two-proton lithography.

X-ray exposure has been tested as well. Becnel et al. [22] used SU-8 as a more sensitive alternative for the PMMA normally used as photoresist in LIGA processing. The exposure time needed was in the order of 10 min, two orders of magnitude smaller than for PMMA. The very limited diffraction and low absorption of the X-rays allows for a higher resolution in thick layers than that can be offered by UV-based processing, and aspect ratios greater than 100.

Finally, *inclined light UV exposure* is worth mentioning. Some researchers have experimented with exposing SU-8 with inclined UV light through a contact mask. Thus, cylinders and beams with inclined orientation can be realized. Unless countermeasures are taken, the UV light will reflect from the substrate and a mirror image of the structure will be created by the light reflecting back from the substrate surface. At intersection of two cylinders a ball-shaped structure forms that can be used as an in-plane microlens. Others use the technique to construct microsieves (Fig. 5). Due to the high refractive index of SU-8 it is hard to attain large inclinations as explained by Snell's law: Light entering at an oblique angle from the air will be bent toward the perpendicular axis of the higher refractive index material it enters. A solution for this is to immerse wafer and mask in a higher refractive index fluid such as glycerin. When X-rays are used instead of UV, this is not important as refractive indices do not vary much.

Multilayered, Freestanding Structures in SU-8

A few methods that can be used to fabricated freestanding structures such as cantilevers and microchannels out of SU-8 will be discussed here.

Modulation of Exposure Light

The most straightforward method is to modulate the exposure energy, that is, to use a high UV dose (i-line) to expose a single SU-8 layer from top to bottom where the anchors of the freestanding structures must come and a lower dose to define the freestanding structures (Fig. 6). However, as SU-8 is quite transparent the resulting layer thickness is very sensitive to small variations in the exposure energy and other process parameters. Also, the absorbance of SU-8 tends to increase during exposure.

Tight control of all the different parameters is required which makes it hard to achieve good tolerances. In practice, the method is only useful for fabricating thick freestanding structures with reasonable accuracy in relatively thick layers (>250 μm). Using an antireflective layer or a low-reflectivity substrate reduces the sensitivity of the process for exposure energy variations [6]. The minimum layer thickness is then reduced to 100 μm.

The above process can be improved by using different wavelengths in the two UV exposure steps. By using a wavelength that is quickly absorbed for the second exposure step, thinner freestanding structures can be made with increased repeatability. It is straightforward to use the 313-nm wavelength for this as it is

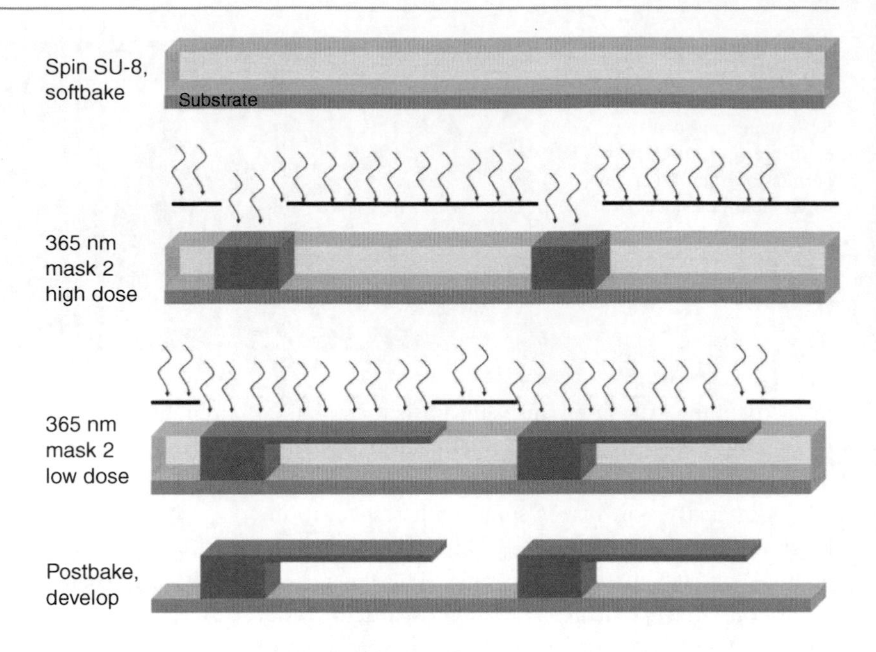

SU-8 Photoresist, Fig. 6 Principle of the creation of freestanding structures by exposure time control

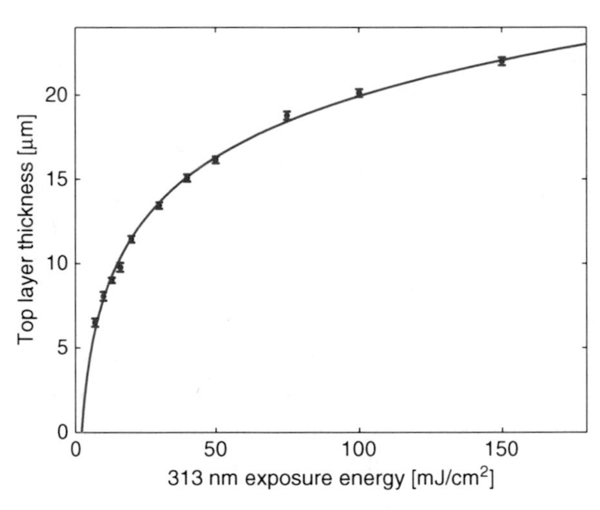

SU-8 Photoresist, Fig. 7 UV exposure energy at 313-nm wavelength versus resulting freestanding layer thickness showing LSE fit. Data points from measurement with (tiny) error *bars* showing standard deviation are plotted as well [5]

already present in the spectrum of standard exposure tools. The practical thickness range that can be reached this way is between 6 and 25 μm (Fig. 7). Due to uneven exposure between the top and bottom parts of the freestanding structure, single-clamped beams were observed to bend down [5].

Buried Mask Process

In this process two SU-8 layers are used, as illustrated in Fig. 8. Thus, a larger thickness range and a better control of layer thicknesses are obtained at the cost of a higher process complexity. First, the lower layer is processed as usual up to the PEB step. Then, a UV blocking layer is deposited. The purpose of this layer is to shield unexposed parts of the lower layer from the UV light used to expose the upper layer, later in the process. Then, the upper layer is spun on and processed. The two layers are developed together at the end of the process sequence. Halfway in the development step, the UV blocking layer must be removed when it is not dissolving in the normal SU-8 developer, for example, when a metal UV blocking layer is used. It is important to deposit the UV blocking layer such that no excess UV or heat is produced, which would cross-link SU-8 unwantedly. Also, temperature must be kept at a minimum when processing the second SU-8 layer to avoid wrinkling [5]. Figure 9 shows some example structures fabricated with the process.

Evaporation by resistive heating of a low melting temperature metal such as aluminum, zinc, or magnesium is a suitable method. Another author uses a metal layer transfer process based on a gold-covered silicone stamp [8].

Sacrificial Layer–Based Processes

A third major category of processes used to fabricate freestanding SU-8 structures is based on the deposition of a sacrificial layer in another material. After the SU-8 layer is processed on top of the sacrificial layer the sacrificial layer is dissolved, creating freestanding

SU-8 Photoresist, Fig. 8 Buried mask process

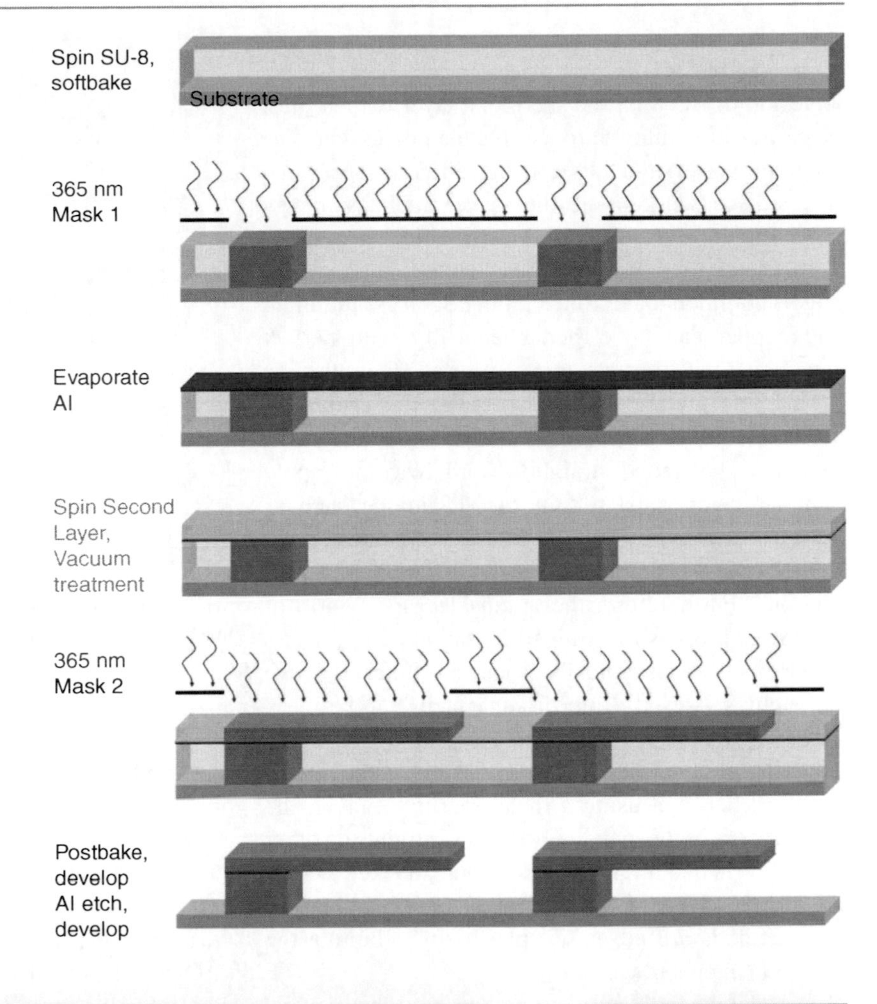

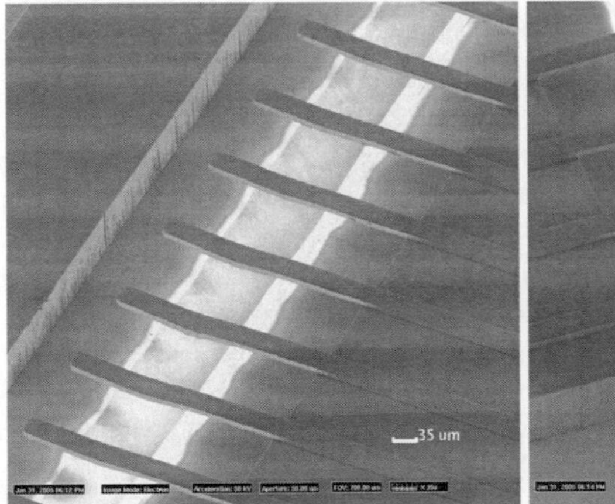

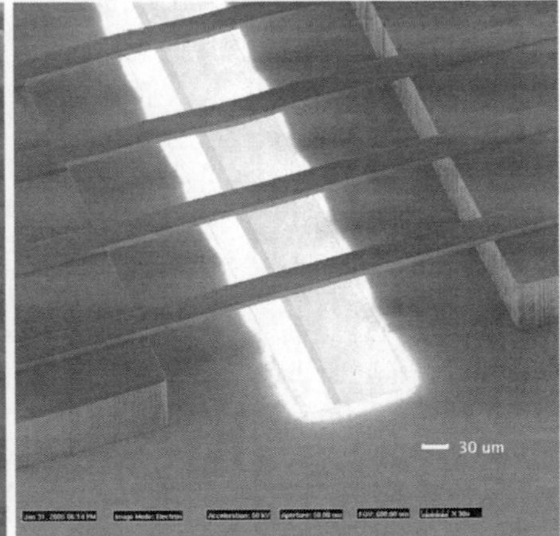

SU-8 Photoresist, Fig. 9 Single- and double-clamped freestanding beams, fabricated with the buried mask method, lying over a KOH-etched U-groove

SU-8 structures. The sacrificial layer can be patterned before SU-8 deposition to create anchors defining where the SU-8 structures that will be formed in the next step will be attached to the substrate. When no anchor sites are defined either a well-timed etch stop can define anchors or loose SU-8 components are created.

The main property a sacrificial layer must have is that there must be a way to remove it without damaging SU-8. Reversely, the solvent in SU-8 must not dissolve the sacrificial layer to prevent layer intermixing and unpredictable results. Another point of interest is the adhesion of SU-8 on the sacrificial layer, which must be sufficient to allow it to survive the processing. The other properties that discern the different sacrificial layers discussed here are their layer thickness range, processing cost and time, and achievable aspect ratio.

A metal sacrificial layer is a first option. Many metals common to micromachining, such as aluminum and copper can be etched chemically with perfect selectivity with respect to SU-8. Also, the adhesion of SU-8 to metals is generally satisfying (Table 2). There are some disadvantages to the use of metals, though. The most straightforward way to apply a metal layer is by a thin film technique such as sputtering. Due to stress and relatively slow speed, these processes are limited to a thickness of a few microns. Furthermore, the attainable aspect ratio of the lower layer is determined by the etching process of the metal, and is typically around one unless reactive ion etching is used. Both disadvantages can be overcome at the cost of an extra process step by the use of a sacrificial layer that is electroplated in a photoresist mold. Still, when using a metal sacrificial layer, the etch process of the metal may pose constraints on other metal layers that might be present and attacked as well by the release etch. A final point is that metal deposition methods are never self-planarizing, limiting the use on uneven surfaces.

A second candidate is polyimide sacrificial layers. Polyimides resist the solvent in SU-8 very well once properly cross-linked. However, their removal is a problem. Polydimethylglutarimide (PMGI), on the other hand, is etchable in alkaline solutions such as positive photoresist developer while still being resistant to solvents. Thus, it can be removed selectively with respect to cross-linked SU-8 and is easily patternable. Disadvantages are the limited lower layer thickness, and the fact that the weak alkaline solution will still etch aluminum and, after an incubation time in which native oxide is removed, even (slowly) silicon. In Ref. [5], comb actuator structures were fabricated based on this process (Fig. 10).

Positive photoresists can also be considered as sacrificial layer. They are inexpensive, can be quickly applied, and are available in a thickness range from below 1 μm to about 100 μm. Therefore, they would be ideally suited as sacrificial layer for a wide range of applications. The main problem hindering their applicability as sacrificial layer for SU-8 is that they dissolve in the solvent (cyclopentanone or GBL) present in SU-8. There are two ways to prevent this. The first is to deposit a thin metal layer on top of the positive resist. The second one is to strongly hardbake the positive photoresist. However, in the latter case resist reflow typically occurs, which severely limits the attainable aspect ratio of the sacrificial layer.

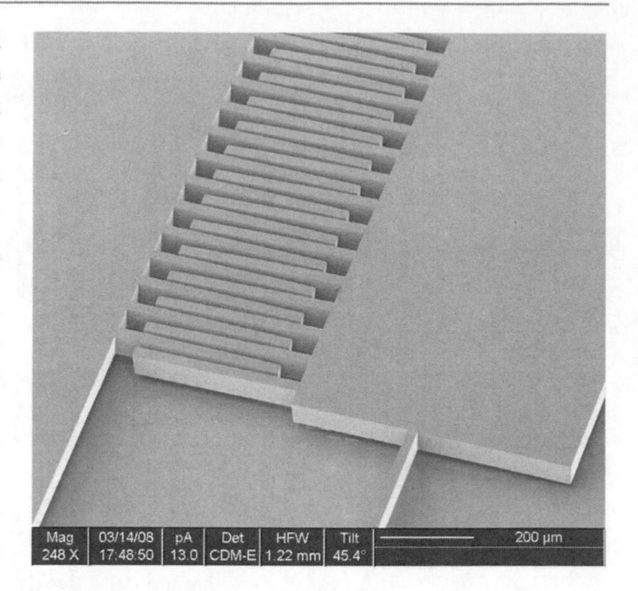

SU-8 Photoresist, Fig. 10 Comb actuator structure fabricated in SU-8 using a PMGI sacrificial layer

Other sacrificial polymers used by workers in the field are polystyrene (PS) and PMMA. Toluene can be used as a solvent for spincoating and release, as cured SU-8 is resistant to several hours of immersion in toluene.

It is possible to use RIE to pattern the polymers or to include photosensitizers in the PMMA, such as is done for deep-UV photoresists.

A final interesting type of sacrificial polymers is polymers that decompose into volatile components cleanly at a temperature SU-8 can withstand. Examples are polypropylene carbonate (PPC) and polyethylene carbonate (PEC) [12]. Dissolved in NMP (N-Methyl-2-pyrrolidone), they can be deposited by spin coating. Structuring is possible using oxygen plasma. PEC starts to decompose thermally around 220°C in nitrogen at atmospheric pressure. For PPC this is around 240°C. The most remarkable property of this process is that the volatile decomposition products can diffuse throughout SU-8. Thus, the sacrificial layer can even be removed from closed cavities, and long channels can be cleared in a time independent of channel length. For a 30-mm-thick SU-8 layer, the sacrificial layer removal process time is about 10 min. For very thick upper layers, this time might be considerably longer.

Layer Transfer Processes

A final major approach to fabricating freestanding structures in SU-8 is to fabricate a single SU-8 layer on a carrier substrate and then transfer that layer to a second substrate on which another SU-8 layer is already present.

This has the advantage that closed cavities or long channels can be fabricated. As the sealed cavities may contain released micromechanical structures, these processes can be used for capping purposes as well.

The different variations of the processes can be further categorized according to the process step after which the layer transfer and optionally the removal of the carrier substrate is done (Fig. 11).

When the transfer is done after softbaking, typically by applying limited heat and pressure, the material is still thermoplastic. Therefore, in order to achieve good control over the shape of the fabricated structure care must be taken to select process parameters such that little flowing of the non-cross-linked layer occurs. A typical sign of too much reflow is the clogging of channels that were sealed by the process and rounded corners.

Another way to limit channel clogging is to use a micron-thin softbaked SU-8 layer to bond two substrates together. Bonding is done at 2 bars in a vacuum bonder [4]. However, the advantage that the upper layer can be readily patterned, for example, integrating vertical microfluidic connectors, is thereby lost.

As the transferred SU-8 layer still must be cross-linked, either the carrier wafer must be UV transparent or it must be removed before exposure. After exposure and PEB, the carrier wafer must be removed in any case, unless the SU-8 layer is used as a simple adhesive. The removal is not straightforward with common hard substrates covered with a sacrificial material as this involves under etching of that sacrificial material for several centimeters. A better way is to use a foil as carrier, which can be laminated on with relatively low pressure, avoiding air entrapment and channel clogging. Furthermore, a foil can be removed by simply peeling it off.

A polyimide foil should be peeled off before exposure as it is not UV transparent. Teflon foil, on the other hand, is transparent to UV.

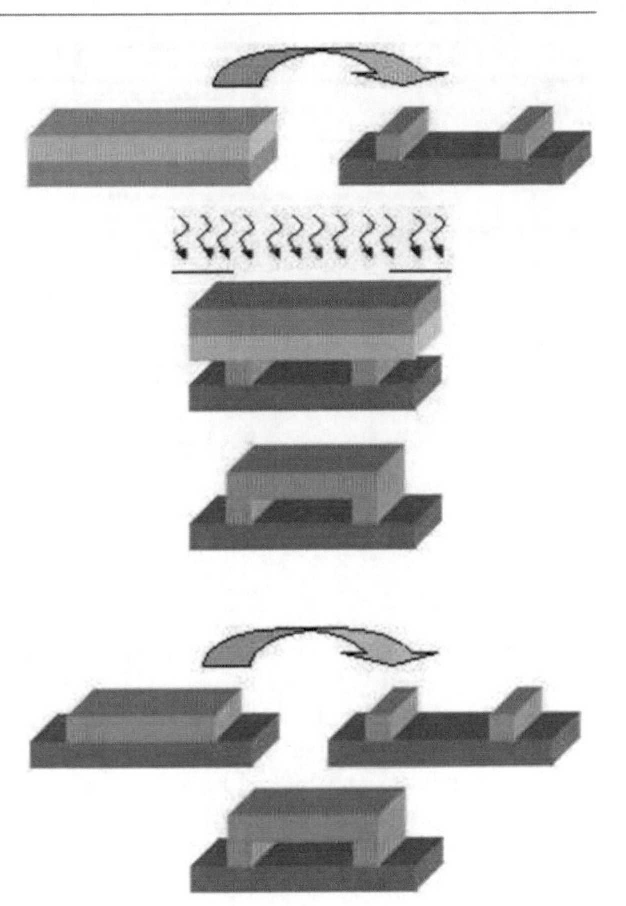

SU-8 Photoresist, Fig. 11 Layer transfer processes. *Top*: cross-linking after transfer. Sometimes the carrier wafer is not removed, and SU-8 serves as an adhesive between the two substrates. *Bottom*: transfer of partially cross-linked layers

Due to the rather rough nature of the carrier wafer removal process it is best used when the top layer contains relatively large features. Examples are device capping and the sealing of microchannels.

Deformation of the transferred layer can be avoided altogether by cross-linking it first before transferring it. The transferred layer can then also be developed before transfer.

To make sure sufficient reactivity remains available to enable thermocompression bonding, the exposure energy and PEB must be scaled down. Arroyo et al. [1] determined the optimal parameters as 140 mJ/cm^2 for i-line exposure followed by a PEB of 4′ at 85°C. Due to the decreased deformability of the SU-8, a much higher bonding force is needed during thermocompression bonding: good results (95% yield) were achieved with a pressure 3.25 bar applied at a temperature of 88°C for 12 min.

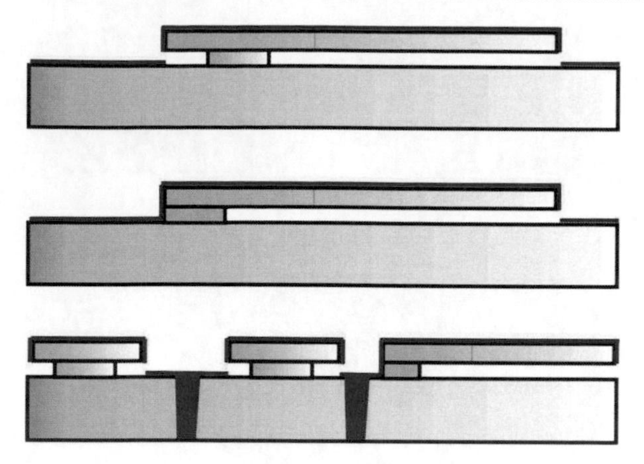

SU-8 Photoresist, Fig. 12 Metallization of SU-8 by sputter deposition of a thin metal layer (*red*). *Top*: isolated freestanding structure. *Middle*: connection to substrate. *Bottom*: isolation of bonding pad (*left*)

Furthermore, it is of course essential to have good thickness uniformity. The edge bead must certainly be removed, be it mechanically, chemically, or by designing the mask such that the rim of the wafer is not exposed. In this case too, it can be advantageous to use a Teflon or polyimide foil as carrier for the second layer as it can be easily pulled off.

Electrically Active Structures in SU-8

In this paragraph, a few strategies will be discussed to render SU-8 conductive, which is mandatory for many actuation and sensing applications such as comb drives and capacitive accelerometers.

Conductive polymers such as polyaniline (PANI) can be added to the SU-8 resin. However, it was shown that high PANI quantities (up to 50%) are required in order to achieve useful conductance, a point at which many SU-8 properties such as adhesion degrade.

A different route to create conductive SU-8 is to use a silver nanoparticle-based filler. A silver content of 6 vol% is enough to reach to the percolation threshold, and to create a conductivity of over 10 S cm^{-1}, which can go up a factor 100 with a higher silver content [23]. Such blends are commercially available. However, because of light reflected by the filler, the available aspect ratio and achievable thickness are limited to about 1 μm and 10 μm, respectively. Carbon nanotubes have been explored as filler material as well. However, conductivities are typically quite limited.

A straightforward method to achieve conductive SU-8 structures is to sputter deposit a thin metal layer. As during sputtering material gets deposited on top of as well as on sidewalls of structures, the mask layout must be designed such as not to cause short circuits, for example, by incorporating overhanging structures such as illustrated in Fig. 12 [5].

Conclusion

Concluding this oversight of SU-8 photoepoxy processing, there is little doubt that this material will find – and indeed already has found – its way into a wide range of applications that benefit from the unique combination of low cost wafer-scale fabrication and the high aspect ratios and very high layer thicknesses attainable. Examples of those are the fabrication of microchannels, labs-on-a-chip, integrated optics, ink jet heads, and electroplating molds for fabricating metal structures. For the latter, electroplating in SU-8 is a low cost alternative for the LIGA process, provided minimum features are several microns wide and more relaxed tolerances can be acceptable. As SU-8 is a low-temperature process, it even shows potential for the fabrication of sensors directly on CMOS wafers.

Cross-References

► Lab-on-a-Chip for Studies in *C. elegans*
► Microfludic Whole-cell Biosensor
► Plating
► Stereolithography

References

1. Arroyo, M.T., Fernández, L.J., Agirregabiria, M., Ibañez, N., Aurrekoetxea, J., Blanco, F.J.: Novel all-polymer microfluidic devices monolithically integrated within metallic electrodes for SDS-CGE of proteins. J. Micromech. Microeng. **17**, 1289–1298 (2007)
2. Becnel, C., Desta, Y., Kelly, K.: Ultra-deep x-ray lithography of densely packed SU-8 features: I. An SU-8 casting procedure to obtain uniform solvent content with accompanying experimental results. J. Micromech. Microeng. **15**, 1242–1248 (2005)
3. Bohl, B., Steger, R., Zengerle, R., Koltav, P.: Multilayer SU-8 lift-off technology for microfluidic devices. J. Micromech. Microeng. **15**, 1125–1130 (2005)

4. Carlier, J., Arscott, S., Thomy, V., Fourrier, J.C., Caron, F., Camart, J.C., Druon, C., Tabourier, P.: Integrated microfluidics based on multi-layered SU-8 for mass spectrometry analysis. J. Micromech. Microeng. **14**, 619–624 (2004)
5. Ceyssens, F.: Micromachining in polymers and glass: process development and applications. PhD thesis, KULeuven, Belgium (2009)
6. Chuang, Y., Tseng, F., Cheng, J., Lin, W.: A novel fabrication method of embedded micro-channels by using SU-8 thick-film photoresists. Sensor Actuat A-Phys **103**, 64–69 (2003)
7. Chung, C.K., Hong, Y.Z.: Surface modification of SU8 photoresist for shrinkage improvement in a monolithic MEMS microstructure. J. Micromech. Microeng. **17**, 207–212 (2007)
8. del Campo, A., Greiner, C.: SU-8: a photoresist for high-aspect-ratio and 3D submicron lithography. J. Micromech. Microeng. **17**, R81–R95 (2007)
9. Dentinger, P.M., Miles, C., Goods, S.H.: Removal of SU-8 photoresist for thick film applications. Microelectron.Eng. **61–62**, 993–1000 (2002)
10. Feng, R., Farris, R.: Influence of processing conditions on the thermal and mechanical properties of SU8 negative photoresist coatings. J. Micromech. Microeng. **13**, 80–88 (2003)
11. Hammacher, J., Fuelle, A., Flaemig, J., Saupe, J., Loechel, B., Grimm, J.: Stress engineering and mechanical properties of SU-8-layers for mechanical applications. J. Microsyst. Technol. **14**, 1515–1523 (2007)
12. Metz, S., Jiguet, S., Bertsch, A., Renaud, Ph: Polyimide and SU-8 microfluidic devices manufactured by heat-depolymerizable sacrificial material technique. Lab Chip **4**, 114–120 (2004)
13. MicroChem: SU-8 datasheet and adhesion results-shear analysis. www.microchem.com. Accessed 18 Nov 2011. (2007)
14. Sato, H., Matsumura, H., Keino, S., Shoji, S.: An all SU-8 microfluidic chip with built-in 3D fine microstructures. J. Micromech. Microeng. **16**, 2318–2322 (2006)
15. Schoeberle, B., Wendlandt, M., Hierold, C.: Long-term creep behavior of SU-8 membranes: application of the time-stress superposition principle to determine the master creep compliance curve. Sensor Actuat A-Phys **142**, 242–249 (2008)
16. Teh, W.H., Dürig, U., Drechsler, U., Smith, C.G., Gntherodt, H.-J.: Effect of low numerical-aperture femtosecond two-photon absorption on SU-8 resist for ultrahigh-aspect-ratio microstereolithography. J. Appl. Phys. **97**, 4095 (2005)
17. Witzgall, G., Vrijen, R., Yablonovitch, E., Doan, V., Schwartz, B.J.: Single-shot two-photon exposure of commercial photoresist for the production of threedimensional structures. Opt. Lett. **23**, 1745 (1998)
18. Wouters, K., Robert, P.R.: Diffusing and swelling in SU-8: insight in material properties and processing. J. Micromech. Microeng. **20**, 095013 (2010)
19. Yang, R., Wang, W.: A numerical and experimental study on gap compensation and wavelength selection in UV-lithography of ultra-high aspect ratio SU-8 microstructures. Sensor Actuat B **110**, 279–288 (2005)
20. Zhang, J., Tan, K.L., Hong, G.D., Yang, L.J., Gong, H.Q.: Polymerization optimization of SU-8 photoresist and its applications in microfluidic systems and MEMS. J. Micromech. Microeng. **11**, 20–26 (2002)
21. Williams, J.D., Wang, W.: Using megasonic development of SU-8 to yield ultrahigh aspect ratio microstructures with UV lithography. J. Microsyst. Technol. **10**, 694–698 (2004)
22. Becnel, C., Desta, Y., Kelly, K.: Ultra-deep x-ray lithography of densely packed SU-8 features: II. Process performance as a function of dose, feature height and post exposure bake temperature. J. Micromech. Microeng. **15**, 1249–1259 (2005)
23. Jiguet, S., Bertsch, A., Hofmann, H., Renaud, P.: Conductive SU8-Silver composite photopolymer. Proceedings of Micro Electro Mechanical Systems, 125–128 (2004)
24. Reznikova, E.F., Mohr, J., Hein, H.: Deep photo-lithography characterization of SU-8 resist layers. J. Microsyst. Technol. **11**, 282–291 (2005)

Submicron Structures in Nature

► Biomimetics of Optical Nanostructures

Sub-retinal Implant

► Artificial Retina: Focus on Clinical and Fabrication Considerations

Subwavelength Antireflective Surfaces Arrays

► Moth-Eye Antireflective Structures

Subwavelength Structures

► Moth-Eye Antireflective Structures

Sub-wavelength Waveguiding

► Light Localization for Nano-optical Devices

Superelasticity and the Shape Memory Effect

Xiaodong Han[1], Shengcheng Mao[1] and Ze Zhang[1,2]
[1]Institute of Microstructure and Property of Advanced Materials, Beijing University of Technology, Beijing, People's Republic of China
[2]State Key Laboratory of Silicon Materials and Department of Materials Science and Engineering, Zhejiang University, Hangzhou, China

Synonyms

Hyperelasticity; Pseudoelasticity; Theoretical elasticity; Ultralarge strain elasticity

Definition

Superelasticity, or pseudoelasticity, is a unique property of shape memory alloys (SMAs), wherein up to 13% deformation strain can be sustained and the material can recover its original shape after removing the stress. The shape memory effect occurs in SMA and is defined as when a material can remember its original shape upon heating or cooling.

Overview

Superelasticity (SE), also called pseudoelasticity, normally refers to a phenomenon observed in shape memory alloys (SMAs), a kind of metal that can remember its original shape after heating or cooling [1–3]. SE is different from true elasticity in bulk metals, which is a reflection of the interatomic spacing variation within a material (Hooke's law). Regular elasticity in bulk metals is normally less than 0.5%, whereas the elasticity of shape memory alloys can be up to 13% in Fe-based SMA [4]. This SE occurs through reversible martensitic transformations from austenite to martensite and vice versa. Upon external mechanical loading or temperature change (or a mixture of the two), the SMA deforms by reversible martensitic phase transformations rather than by irreversible plastic dislocation glide. The martensitic transformation occurs through shear on an invariable plane (habit plane) along the shear direction. SE occurs when the metals are deformed in the austenite state, and the deformation shape is recoverable when the applied load is removed.

Large strain elasticity (LSE) can be achieved in nanomaterials due to their defect-free structure, particularly in nanowires. The true elasticity of metallic nanowires can even approach the theoretical elastic limit (8%) [5], and this is typically associated to extremely high strength [6]. In theory and in computer simulations, some metallic nanowires, such as Cu nanowires, can reversibly deform up to 50% by a combination of large strain elasticity and pseudoelasticity [7]. Some of these exceptional mechanical properties have already been experimentally demonstrated in metallic nanowires.

The *shape memory effect* (SME) also relates to SMAs, and it refers to the phenomenon where a metallic specimen changes shape in response to temperature changes [1–3]. Figure 1 shows a sketch comparing SME and SE. A deformed SMA can return to its original high-temperature shape upon heating or go back to the low-temperature shape upon cooling (see Fig. 1a). If a SMA is deformed by mechanical loading at a constant temperature higher than a critical temperature (the austenite finish temperature, defined below), the initial shape can be recovered by unloading, through a reverse martensitic transformation (Fig. 1b). The SME was first discovered by A. Olander in Au-Cd alloy in 1932. Investigations into SMA and their applications have been further promoted since the discovery of the SME in the near-equiatomic NiTi alloys, known as "Nitinol." The SME has also been found in other metallic alloys such as Cu-Al-Ni, Cu-Zn, Fe-Mn-Si, Ni-Mn-Ga, and Ti-Ni-Hf. Because of these unique properties, SMAs have been used for a variety of applications, including orthodontic wire, stents, microactuators, pipe coupling, eyeglass frames, microelectromechanical systems, and a variety of biomedical devices [8, 9]. *Superelasticity and the shape memory effect are* correlated with thermoelastic and stress/strain-induced martensitic transformations. In the following, we introduce some basic concepts of *martensitic transformations* in NiTi SMAs.

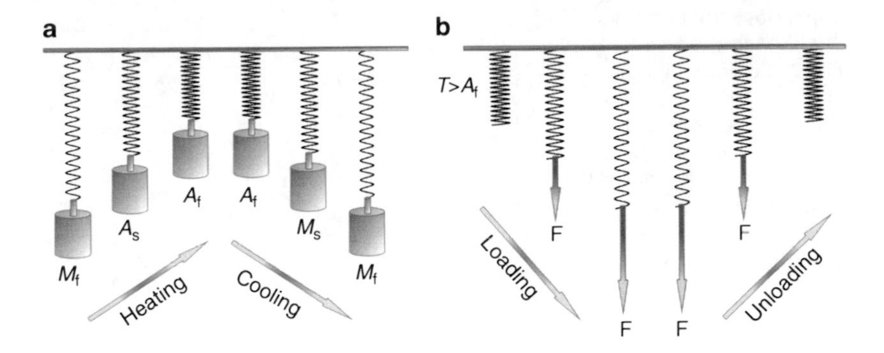

Superelasticity and the Shape Memory Effect, Fig. 1 A schematic illustration of (**a**) the shape memory effect and (**b**) superelasticity

Martensitic Transformation

Thermodynamic Consideration of the Thermoelastic Martensitic Transformation

The martensitic transformation is a transformation between austenite (stable at high temperature) and martensite (stable at low temperature), and it is usually treated as a diffusionless first-order phase transformation. The Gibbs free energy, defined as $\Delta G = \Delta H - T\Delta S$ (where ΔH and $T\Delta S$ are the changes in chemical enthalpy and entropy), is the pertinent chemical potential of a system, which is at the minimum when reaches equilibrium at constant pressure and temperature. The martensite and austenite Gibbs free energies (ΔG^{M} and ΔG^{A}) gradually decrease, following different slopes, with increasing temperature (T), as illustrated in Fig. 2a. At temperature T_o, the Gibbs free energies of the two phases reach a thermodynamic equilibrium ($\Delta G^{M} = \Delta G^{A}$). To drive the forward or reverse martensitic transformation, the energy gap between the two phases ($\Delta G^{A \to M}$ or $\Delta G^{M \to A}$) needs to be overcome by either temperature or mechanical loading, as expressed in Eqs. 1 and 2: [10].

$$\begin{aligned}\Delta G^{A\to M} &= \Delta H^{A\to M} - T\Delta S^{A\to M} - \Delta E_{mech} \\ &= \Delta H^{A\to M} - T\Delta S^{A\to M} - \frac{\sigma\varepsilon}{\rho}\end{aligned} \quad (1)$$

$$\Delta G^{M\to A} = \Delta H^{M\to A} - T\Delta S^{M\to A} + \frac{\sigma\varepsilon}{\rho} \quad (2)$$

where ρ, σ, and ε are the density of materials, the external stress, and strain, respectively.

Figure 2b shows schematically the evolution of the martensite volume fraction with temperature during forward and reverse transformations. There are four characteristic temperatures defining a thermoelastic martensitic transformation: (1) The martensite start temperature, M_s, at which martensite starts to nucleate; (2) the martensite finish temperature, M_f, with $M_f < M_s$, at which the transformation has completed; (3) the austenite start temperature, A_s, at which the system starts moving from martensite to austenite upon heating; and (4) the austenite finish temperature, A_f, with $A_f > A_s$, at which the martensitic phase has fully transformed into the austenite phase.

Aside from the thermally induced martensitic transformation, martensite can also be obtained by applying external stress. Figure 3a shows a typical stress–strain curve of NiTi SMA showing superelasticity. There are two stress plateaus, defined as the upper stress plateau and the lower stress plateau, corresponding to the forward and reverse martensitic transformation, as indicated in Fig. 3a. The martensite starts to nucleate after reaching a critical transformation stress, σ_c. The martensitic transformation then proceeds at nearly constant stress upon further straining (the physical mechanisms are described below). As shown in Fig. 3a, the upper plateau stress is lower than the critical transformation stress. As a result, the growth energy is smaller than that of the nucleation energy of martensite. The austenite nearly fully transforms to martensite at the end of the upper stress plateau. Upon unloading to a critical reverse stress, σ_r, martensite starts transforming back to austenite. The martensite phase is nearly fully reverted to the austenite phase at the end of the lower stress plateau. According to Eq. 1, higher stress is required to trigger

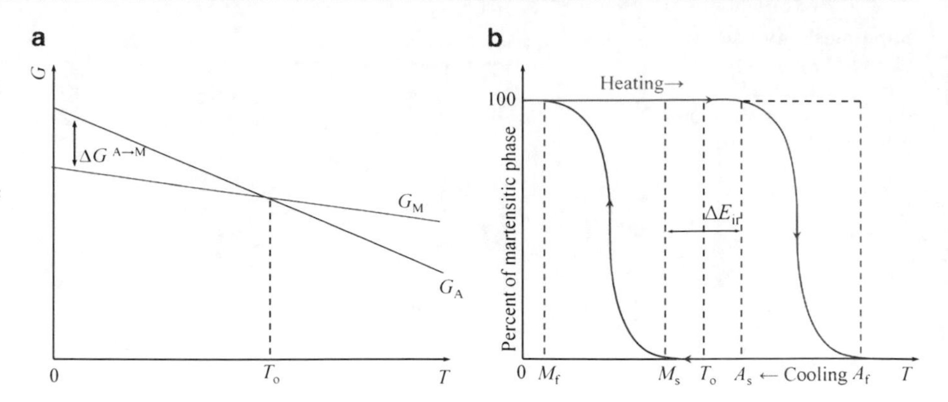

Superelasticity and the Shape Memory Effect, Fig. 2 Schematic illustration of the evolutions of (**a**) the Gibbs free energies of the martensite and austenite phases and (**b**) the martensitic fraction with temperature

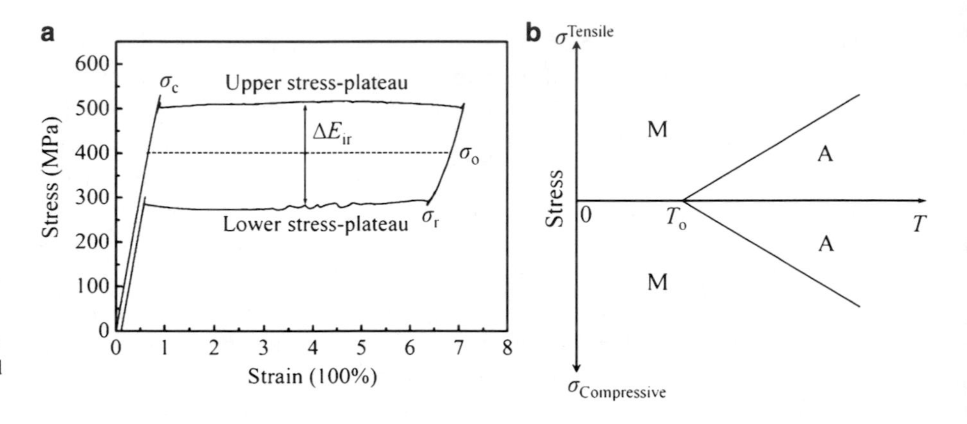

Superelasticity and the Shape Memory Effect, Fig. 3 (**a**) A typical stress–strain curve of NiTi SMA showing superelasticity. (**b**) Schematic illustration of the stress–temperature relationship of an SMA. The stress can be both tensile and compressive stress

martensite transformations as the ambient temperature increases. The stress–temperature relationship in stress-induced martensitic transformations is schematically illustrated in Fig. 3b.

As shown in Figs. 2b and 3a, the forward and reverse deformation paths do not overlap, i.e., temperature and stress hysteresis exist during thermo- and stress-induced martensitic transformation. The austenite start temperature is higher than the martensite start temperature ($A_s > M_s$, $A_s > T_o$ and $M_s < T_o$), and the unloading stress plateau is lower than the loading stress plateau. This is because the generation of irreversible energies (ΔE^{ir}), including the frictional energy of the martensite–austenite phase interface propagation, the acoustic emission, and the production and motion of dislocations, is ineluctable during the martensitic transformation [11]. Therefore, additional chemical or mechanical energy is required to promote the forward and reverse martensitic transformations.

Crystallography of Stress-Induced Martensitic Transformation

Critical Transformation Stress

Two important characteristics, the critical transformation stress and transformation strain, define the stress-induced martensitic transformation. It has been suggested that the selection of a martensitic variant under external stress (uniaxial tension and compression) comes from satisfying Schmid's law, as illustrated in Fig. 4 [12]. The stress component along the shear direction in the habit plane can be expressed as:

$$\tau_s = \frac{F_s}{A_s} = \frac{F\cos\lambda}{A_o/\cos\varphi} = \frac{F}{A_o}\cos\lambda\cdot\cos\varphi = \sigma\cos\lambda\cdot\cos\varphi \quad (3)$$

where $m = \cos\lambda \cdot \cos\varphi$ is defined as the Schmid factor, λ is the angle between the loading axis and the

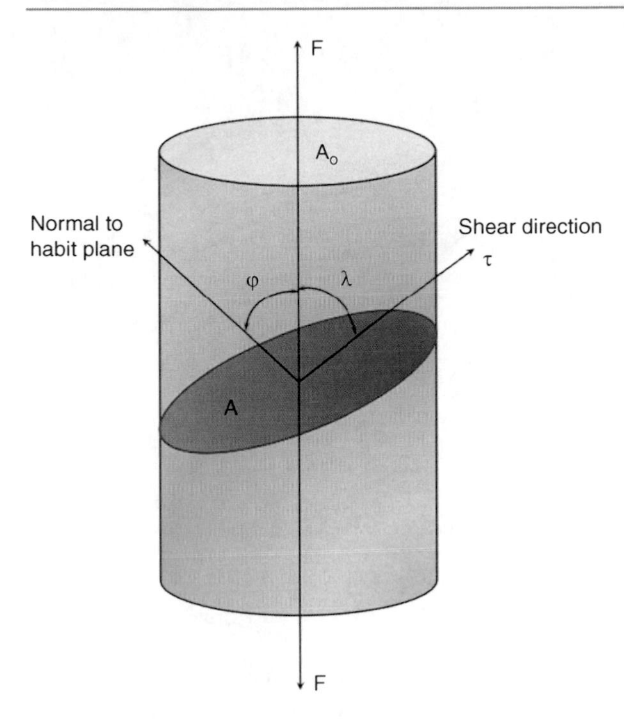

Superelasticity and the Shape Memory Effect, Fig. 4 Schematic illustration of Schmid's law

shear direction, and φ is the angle between the loading axis and the normal of the habit plane.

According to Eq. 3, the variant with the highest Schmid factor and, consequently, the highest resolved shear stress will be triggered in one grain, which shows that the critical transformation stress of martensite is strongly related to the grain/crystal orientation.

Maximum Recoverable Strain

The two phases of NiTi alloys, austenite and martensite, are known to have cubic (B2) and monoclinic (B19') structures. Figure 5 shows a schematic illustration of the lattice distortion of the B2–B19' transformation in NiTi SMA. A martensitic unit cell is identified by the dashed line within the austenite lattice, as shown in Fig. 5a. The embryo may be transformed to the martensitic unit cell by a simple "bain distortion," as indicated in Fig. 5b, which can be expressed as an expansion in c, contractions in the a and b directions, and a change of the beta angle from 90° to 96.8°.

Such a transformation can happen in 12 equivalent lattice correspondence martensitic variants (LCMVs) in NiTi [13]. The lattice distortion shown in Fig. 5 corresponds to LCMV #1 with lattice correspondences of $[001]_m$–$[001]_a$, $[010]_m$–$[011]_a$, and $[001]_m$–$[0\text{–}11]_a$. The lattice deformation matrix M' in the coordinates of the martensite (i', j', k') can be expressed using the lattice constants of the parent and the product phases. For the martensitic transformation in near-equiatomic NiTi, the lattice constant of the austenite is $a_o = 0.3015$ nm and those of the martensite are $a_m = 0.2889$ nm, $b_m = 0.412$ nm, $c_m = 0.4622$ nm, and $\beta = 96.8^o$ [14]; thus:

$$M' = \begin{bmatrix} \frac{a_m}{a_o} & 0 & \frac{c_m \cos\beta}{\sqrt{2}a_o} \\ 0 & \frac{b_m}{\sqrt{2}a_o} & 0 \\ 0 & 0 & \frac{c_m \sin\beta}{\sqrt{2}a_o} \end{bmatrix} = \begin{bmatrix} 0.9582 & 0 & 0.1283 \\ 0 & 0.9663 & 0 \\ 0 & 0 & 1.0763 \end{bmatrix} \tag{4}$$

The lattice deformation matrix M in the coordinates of austenite (i, j, k) can then be transformed from M' with a coordinate transformation matrix R from the martensite to the austenite via:

$$M = RM'R^{\mathrm{T}} \tag{5}$$

where R^{T} is the transpose of R. Using the lattice correspondence of austenite and martensite, the coordinate transformation matrix R for the variant shown in Fig. 5 is expressed as:

$$R = \begin{bmatrix} 1 & 0 & 0 \\ 0 & 1/\sqrt{2} & -1/\sqrt{2} \\ 0 & 1/\sqrt{2} & 1/\sqrt{2} \end{bmatrix} \tag{6}$$

The lattice deformation matrix M is then calculated to be:

$$M = RM'R^T = \begin{bmatrix} \alpha & -\beta & \beta \\ 0 & \omega & -\gamma \\ 0 & -\gamma & \omega \end{bmatrix} \tag{7}$$

where $\alpha = 0.9582$, $\beta = 0.0907$, $\omega = 1.0213$, and $\gamma = 0.0550$.

With this, a vector $\boldsymbol{x}$ in the austenite is transformed to $\mathbf{x}'$ upon the martensitic transformation by the following equation:

$$\mathbf{x}' = Mx \tag{8}$$

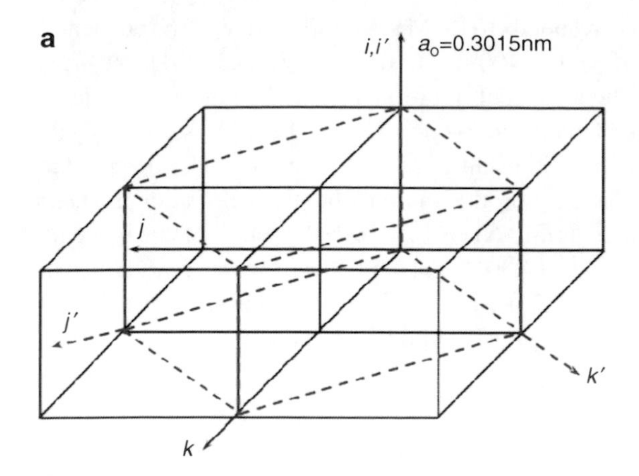

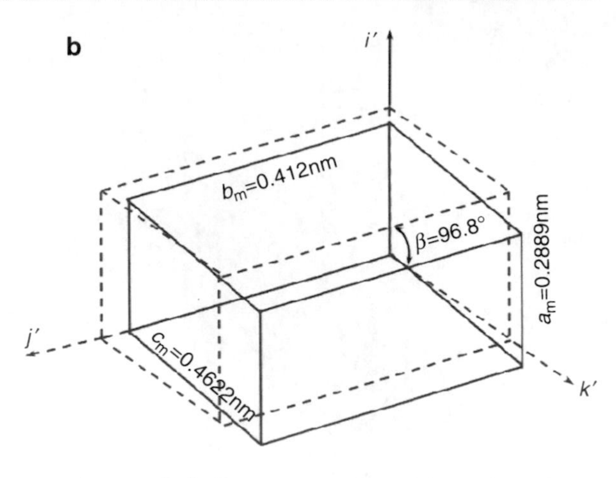

Superelasticity and the Shape Memory Effect, Fig. 5 Schematic illustration of the lattice distortion of the B2–B19' martensitic transformation in NiTi: (**a**) atomic coordinate systems, with (i, j, k) representing the reference frame in the austenite and (i', j', k') representing the reference frame in the martensite, (**b**) lattice distortion from B2 (*dashed lattice*) to B19' (*solid lattice*)

Consequently, the transformation strain can be calculated by:

$$\varepsilon = \frac{|\mathbf{x}'| - |\mathbf{x}|}{|\mathbf{x}|} \tag{9}$$

According to thermodynamic principles, the variant with the maximum strain in the direction of the applied load produces the maximum driving force for the stress-induced martensitic (SIM) transformation [15] and is triggered to form first.

Mechanisms of the Shape Memory Effect and Superelasticity

The unique properties, superelasticity and the shape memory effect, of SMA can be expressed by a simplified 2D geometric depiction, as illustrated in Fig. 6. Upon cooling to a temperature $T < M_f$, two martensitic variants, defined as A and B, with the same crystal structure but different orientations were cooperatively triggered to accommodate and minimize the martensitic transformation strain. The interface between the thermally induced martensite, composed of the matrix and its twin, and the austenite is the habit plane. The martensite variant at the habit plane is then referred to as the habit plane variant. Under external stress ($T < M_f$), martensitic variant A grows at the expense of variant B by the movement of the interface of variants A and B. The deformation process is referred to as detwinning or reorientation of the martensites. The twinned or detwinned martensitic variants can fully transform back to austenite when heating the specimen to a temperature above A_f corresponding to the recovery of the macroscopic SMA to its original shape. There are two types of SME: one-way and two-way. The difference is whether the SMA remembers its low-temperature shape. The critical stress to drive the reorientation of martensite is known to be very small (about 100 MPa); therefore, it is very easy to set the shape of SMA at low temperatures ($T < M_f$). For the one-way SME, the shape of SMA will remain stable until heated above the reverse transformation start temperature (A_s) and reverse to its original when the temperature is higher than A_f. Upon the second cooling to a temperature lower than M_f, no macroscopic shape change happens, indicating that only the high-temperature shape is "remembered."

The two-way SME is an effect where both the low-temperature and high-temperature shapes can be "remembered." Thus, when cooling the SMA to a temperature $T < M_f$, the high-temperature shape will change to the low-temperature shape; likewise, the low-temperature shape will transform back to the high-temperature shape on heating to $T > A_f$. The shape variations between the two shapes are stress free. It is an intrinsic property of SMA to "remember" the high-temperature shape; however, additional training is required for SMA to "remember" the low-temperature shape.

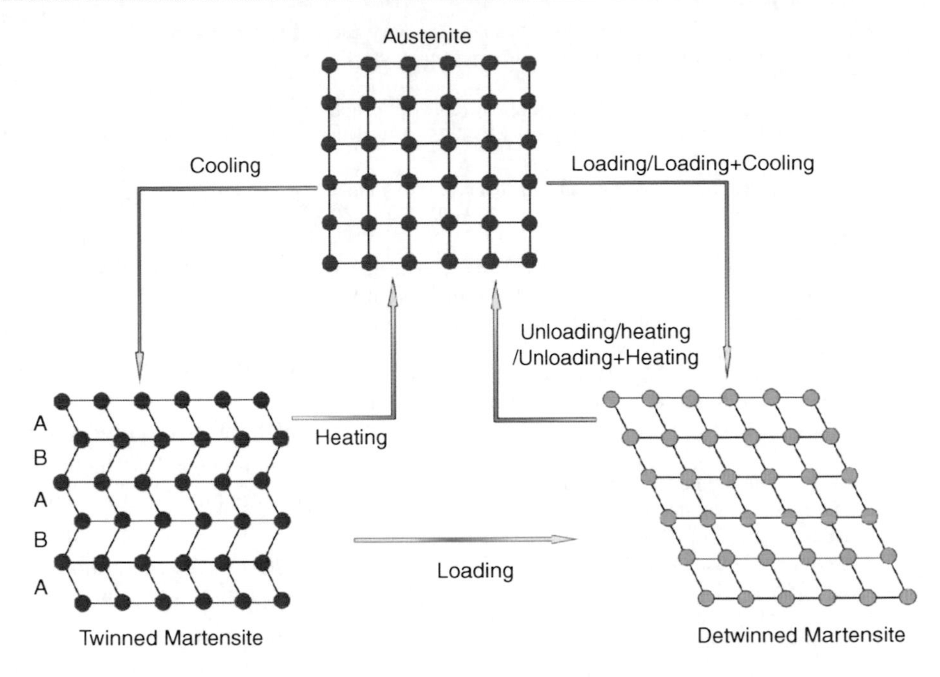

Superelasticity and the Shape Memory Effect, Fig. 6 Schematic illustration of the martensitic structure upon loading, unloading, heating, and cooling

In addition to the SME, SMAs have another unique property known as superelasticity, or pseudoelasticity, which occurs at temperatures above A_f and distinguishes SMA from other metallic alloys. Under external mechanical loading, detwinned martensite nucleates in the austenite matrix after reaching the critical resolved shear stress. Martensite continuously propagates with increasing external strain in a Lüders-type manner under a nearly constant stress (see the upper stress plateau in Fig. 3). The nucleation of martensite is so concentrated that a region nearly fully transformed to martensite is the first to form in the SMA. There is a clear edge between the martensite and austenite, with a shear angle of about 55° to the loading axis [15]. The martensite band propagates by new nucleation of martensite ahead of the martensite–austenite edge [16]. The inhomogeneous martensitic transformation is similar to the stress-induced slip band in low-carbon steels. This transformation behavior was first reported by Guillaume Piobert and W. Lüders in 1864, and the martensitic band is then referred to as a Lüders-like deformation band (LBD).

The stress-induced martensite is unstable at the testing temperature (*higher than* A_f) and will transform back to austenite when the external stress is unloaded. The superelasticity of SMA is associated with the reverse martensitic transformation and recovery of the macroscopic deformation strain. Upon unloading to the lower stress plateau (see Fig. 3), the martensite plates inside the LDB gradually transform back to austenite. At the macroscopic scale, the reverse transformation is characterized by shrinkage of the LDB.

Cross-References

- ▶ Nanomechanical Properties of Nanostructures
- ▶ Plasticity Theory at Small Scales
- ▶ Size-Dependent Plasticity of Single Crystalline Metallic Nanostructures
- ▶ Surface Tension Effects of Nanostructures

References

1. Otsuka, K., Wayman, C.M.: Mechanism of shape memory effect and superelasticity. In: Otsuka, K., Wayman, C.M. (eds) Shape memory materials, pp. 27–48. Cambridge, Cambridge University Press (1998)
2. Saburi, T., Nenno, S.: The shape memory effect and related phenomena. In: Proceedings of the International Conference on Solid-Solid Phase Transitions, pp. 1455–1479. Pittsburg (1981)
3. Otsuka, K., Ren, X.: Physical metallurgy of Ti–Ni-based shape memory alloys. Prog. Mater. Sci. **50**, 511–678 (2005)
4. Tanaka, Y., Himuro, Y., Kainuma, R., Sutou, Y., Omori, T., Ishida, K.: Ferrous polycrystalline shape-memory alloy showing huge superelasticity. Science **327**, 1488–1490 (2010)
5. Yue, Y.H., Liu, P., Zhang, Z., Han, X.D., Ma, E.: Approaching the theoretical elastic limit in Cu nanowires. Nano Lett. **11**, 3151 (2011)
6. Wong, E.W., Sheehan, P.E., Lieber, C.M.: Nanobeam mechanics: elasticity, strength and toughness of nanorods and nanotubes. Science. **277**, 1971 (1997)

7. Park, H., Gall, S.K., Zimmerman, J.A.: Shape memory and pseudoelasticity in metal nanowires. Phys. Rev. Lett. **95**, 255504 (2005)
8. Humbeeck, J.V.: Non-medical applications of shape memory alloys. Mater. Sci. Eng. A **273–275**, 134–148 (1999)
9. Duerig, T., Pelton, A., St□ckel, D.: An overview of nitinol medical applications. Mater. Sci. Eng. A **273–275**, 149–160 (1999)
10. Wollants, P., Roos, J.R., Delaey, L.: Thermally- and stress-induced thermoelastic martensitic trans- formations in the reference frame of equilibrium thermodynamics. Prog. Mater. Sci. **37**, 227–288 (1993)
11. Liu, Y., McCormick, P.G.: Thermodynamic analysis of the martensitic transformation in NiTi—I. Effect of heat treatment on transformation behaviour. Acta Metal. Mater. **42**, 2401–2406 (1994)
12. Gall, K., Sehitoglu, H.: The role of texture in tension–compression asymmetry in polycrystalline NiTi. Int. J. Plasticity **15**, 69–92 (1999)
13. Matsumoto, O., Miyazaki, S., Otsuka, K., Tamura, H.: Crystallography of martensitic transformation in Ti-Ni single crystals. Acta Metall. **35**, 2137–2144 (1987)
14. Wollants, P., Bonte, M.D., Roos, J.R.: A thermodynamic analysis of the stress-induced martensitic transformation in a single crystal. Z. Metallkd. **70**, 113–117 (1979)
15. Shaw, J.A.: Thermomechanical simulations of localized thermo-mechanical behavior in a NiTi shape memory alloy. Int. J. Plast. **16**, 541–562 (2000)
16. Mao, S.C., Luo, J.F., Zhang, Z., Wu, M.H., Liu, Y., Han, X. D.: EBSD studies of the stress-induced B2-B19'martensitic transformation in NiTi tubes under uniaxial tension and compression. Acta Mater. **58**, 3357–3366 (2010)

Superhydrophobicity

► Lotus Effect

Superoleophobicity of Fish Scales

Lei Jiang[1] and Ling Lin[2]
[1]Center of Molecular Sciences, Institute of Chemistry Chinese Academy of Sciences, Beijing, People's Republic of China
[2]Beijing National Laboratory for Molecular Sciences (BNLMS), Key Laboratory of Organic Solids, Institute of Chemistry Chinese Academy of Sciences, Beijing, People's Republic of China

Synonyms

Oil-repellency of fish scales

Definition

Superoleophobicity of fish scales is a wetting phenomenon that fish scales show oil-repellency in water, with a static oil contact angle (CA) higher than 150°. Generally, underwater superoleophobicity is defined as a static oil CA higher than 150° on solid surface in an oil/water/solid three-phase system.

Chemical and Physical Principles

Wettability is a fundamental property of solid surfaces, which not only affects the behavior of the creatures in nature, but also plays an important role in all aspects of our life. A direct expression of wetting behavior is a static CA of a liquid droplet sitting on a solid surface when the surface tensions at multiphase interface reach thermodynamic equilibrium.

For an ideal flat surface in a liquid/gas/solid system, the CA (θ) is given by the Young's equation [1]:

$$\cos\theta = \frac{\gamma_{sg} - \gamma_{sl}}{\gamma_{lg}} \quad (1)$$

where γ_{sg} is the solid/gas interface tension, γ_{sl} is the solid/liquid interface tension, and γ_{lg} is the liquid/gas interface tension.

For a rough surface in the air atmosphere, the situation is more complex. The surface topographic structure has a great influence on the wettability. Two distinct models, Wenzel model and Cassie–Baxter model, are commonly used to explain the effect of roughness on the apparent CAs of liquid drops. As described by Wenzel's model [2], liquid completely penetrates into the rough structures, leading the increase of contact area of solid/liquid interface. Therefore, the apparent CA (θ_w) can be described as:

$$\cos\theta_w = r\cos\theta \quad (2)$$

where the surface roughness factor (r) is defined as the ratio of the actual contact area of solid/liquid interface to the projected contact area ($r \geq 1$), and θ is the intrinsic CA on the flat surface. As described by Cassie–Baxter's model [3], liquid suspends on the rough surface rather than completely penetrates. The rough surface therefore can be considered as a solid/gas composite surface, and the apparent CA (θ_c) can be described as:

$$\cos\theta_c = f\cos\theta - (1 - f) \quad (3)$$

where f is the area fraction of solid/ liquid interface, $(1 - f)$ is that of liquid/gas interface. Based on both theories, scientists can better understand different wetting phenomena in nature, which would help the design of artificial materials with functional surfaces [4–6].

Key Research Findings

Oil-Wetting Behaviors on Fish Scales with Micro/ Nanostructures

Nature abounds with mysterious living organisms which have special surface properties. A famous example is the lotus leaf, a typical superhydrophobic surface in nature. When a water droplet falls on the lotus leaf, it keeps bead shape and rolls off the surface immediately taking away the adherent dirt particles. This so-called "lotus effect" has been understood as a result of the complementary roles of low surface free energy and micro/nanostructures on the leaf surface [8]. Inspired by this phenomenon, tremendous artificial superhydrophobic surfaces have been fabricated and have facilitated practical applications in broad fields [9–11].

In a similar manner to the lotus effect in the air atmosphere, fish can resist oil pollution in water, keeping itself antifouling. This fascinating phenomenon shows great potential for many applications in an oil/water/solid system, such as marine antifouling, prevention of oil spills, microfluidic technology, and bioadhesion [12, 13]. Liu et al. first reported the oil-wetting behavior on fish scales in the water environment [7]. It is common knowledge that fish scales are covered by a thin layer of mucus that leads to their hydrophilic nature. Liu et al. revealed that the hydrophilic surface of fish scales showed superoleophilicity in air (Fig. 1a) [14]. However, the surface of fish scales turned to be superoleophobic once it was immersed in water, with an oil CA larger than 150° (Fig. 1b). To figure out the reason of oil-wetting reversion, they firstly observed surface structures on fish scales in detail. Figure 1c–f display typical images of fish scales (Crucian Carp, *Carassius carassius*) using scanning electronic microscopy (SEM). The fan-shaped fish scales with diameters of 4–5 mm are densely arranged (Fig. 1c). The magnifying images disclose that there are oriented micropapillae on each fish scale. Each micropapillea is in length of 100–300 mm and in width of 30–40 mm (Fig. 1d). In high-magnification SEM images (Fig. 1e–f), nanoscale roughness is clearly observed on the surface of micropapillae. It was suggested that these hierarchical structures could trap water and form a composite interface on fish scales to resist oil, which might play an important role on the oil-wetting reversion.

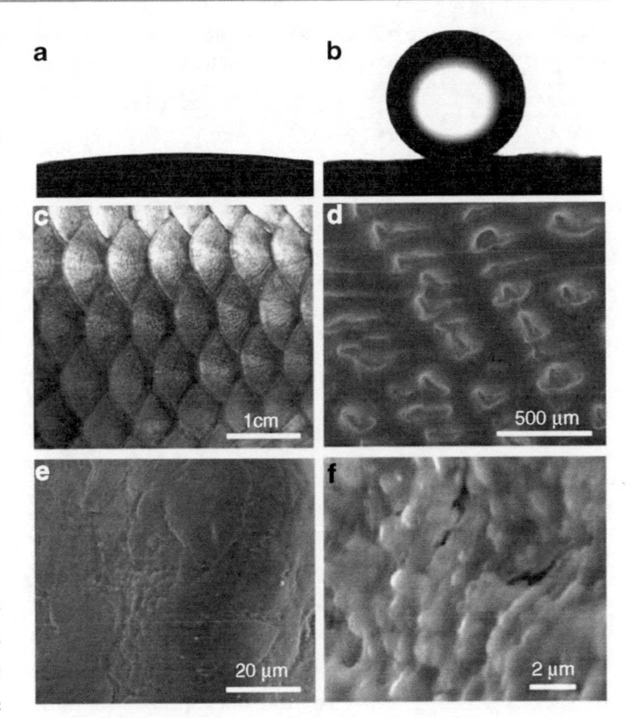

Superoleophobicity of Fish Scales, Fig. 1 (**a**) Fish scales show superoleophilicity in air (1,2-dichloroethane (*DCE*) as a detecting oil, density: 1.245 g cm^{-3}, surface tension at 25°C: 31.86 mN m^{-1}). (**b**) Fish scales become superoleophobic once it is immersed in water (droplet of DCE as in (a)). (**c–f**) SEM images of fish scales disclose the surface micro/nanostructures with the increase of magnification (Reproduced with permission. Copyright Wiley-VCH Verlag GmbH & Co. KGaA (2009))

Mechanism of Wetting Behaviors on Fish Scales in Oil/Water/Solid System

To better understand the reversion of oil-wetting behavior on fish scales, Liu et al. chose three kinds of designed surfaces of silicon wafers as the models: smooth surface, microstructured surface, and micro/nanostructured

Superoleophobicity of Fish Scales, Table 1 Contact angles of oil (1,2-dichloroethane, DCE) or water on three solid surfaces with different surface structures in the air or water environment, respectively

Surfaces systems	Smooth silicon	Microstructured silicon	Micro/nanostructured silicon
$C_2H_4Cl_2$ droplets (in air)	<5°	<5°	<5°
Water droplets (in air)	52.5 ± 1.4°	<5°	<5°
$C_2H_4Cl_2$ droplets (in water)	134.8 ± 1.6°	151.5 ± 1.8°	174.8 ± 2.3°

Source: Reproduced with permission. Copyright Wiley-VCH Verlag GmbH & Co. KGaA (2009)

surface [7]. Table 1 lists the contact angles of oil (DCE) or water on these three solid surfaces in the air or water environment, respectively. In air, the smooth surface is hydrophilic with a CA of 52.5 ± 1.4°, while microstructured and micro/nanostructured surfaces are both superhydrophilic (CAs < 5°). For the oil-wetting behaviors, all surfaces are superoleophic, with the oil CAs smaller than 5°. However, in the water environment, all three kinds of surfaces become oleophobic, or even superoleophobic. For the smooth surface, the oil CA is 134.8 ± 1.6°; for the microstructured surface, the oil CA is 151.5 ± 1.8°; for the micro/nanostructured surface, the oil CA is 174.8 ± 2.3°. The reversions of oil-wetting behaviors on these three surfaces are similar to that on fish scales. Comparing the different environments of two systems, the oil-wetting reversion is likely caused by the surrounding water. With these model silicon surfaces, the mechanism can be further analyzed through classical theories.

Although Young's equation is originally applied in a liquid/gas/solid three-phase system, it can be extended to an oil/water/solid system, in which an oil droplet sits on a solid surface in a water environment (Fig. 2). In this case, Young's equation is expressed as follows:

$$\cos\theta_{oil/water} = \frac{\gamma_{water-s} - \gamma_{oil-s}}{\gamma_{oil-water}} \quad (4)$$

where $\theta_{oil/water}$ is an oil CA on solid surface in oil/water/solid three-phase system, $\gamma_{water-s}$ is the water/solid interface tension, γ_{oil-s} is the oil/solid interface tension, and $\gamma_{oil-water}$ is the oil/water interface tension. Considering the liquid/gas/solid system, γ_{oil-s} and $\gamma_{water-s}$ can be described as:

$$\gamma_{s-g} = \gamma_{oil-s} + \gamma_{oil-g}\cos\theta_{oil} \quad (5)$$

$$\gamma_{s-g} = \gamma_{water-s} + \gamma_{water-g}\cos\theta_{water} \quad (6)$$

where θ_{oil}, θ_{water} is the oil or water CA in the air atmosphere, respectively, γ_{oil-g} is the oil/gas

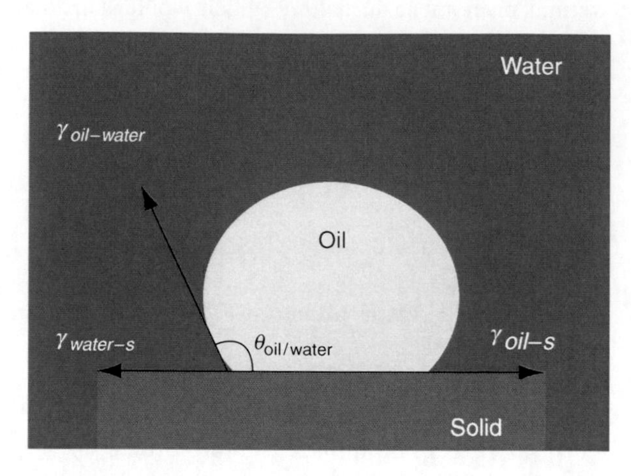

Superoleophobicity of Fish Scales, Fig. 2 Schematic illustration of Young's equation in an oil/water/solid three-phase system, where a liquid droplet (*oil*) sits on smooth surface (*solid*) in another liquid (*water*) phase

interface tension, and $\gamma_{water-g}$ is the water/gas interface tension.

So (4) can also be expressed as follows:

$$\cos\theta_{oil/water} = \frac{\gamma_{oil-g}\cos\theta_{oil} - \gamma_{water-g}\cos\theta_{water}}{\gamma_{oil-water}} \quad (7)$$

Through (7), it can be easily understood why an oleophilic surface in air becomes oleophobic in water. Taking DCE for example, the surface tension (γ_{oil-g}) of DCE is 24.1 mN m^{-1}, the water surface tension ($\gamma_{water-g}$) is 73 mN m^{-1}, and the interfacial tension ($\gamma_{oil-water}$) of DCE/water is 28.1 mN m^{-1} [14]. As mentioned in the above experimental results, in air, the water CA (θ_{water}) on smooth silicon surface is 52.5 ± 1.4° and the oil CA (θ_{oil}) is nearly 0°. Therefore, it can be calculated that $\cos\theta_{oil/water} = -0.72$, and $\theta_{oil/water} \approx 136°$. This result imparts that the reversion of oil-wetting behavior certainly happens when different three-phase systems are involved, which is in great consistence with the experimental oil CA (134.8 ± 1.6°) in water.

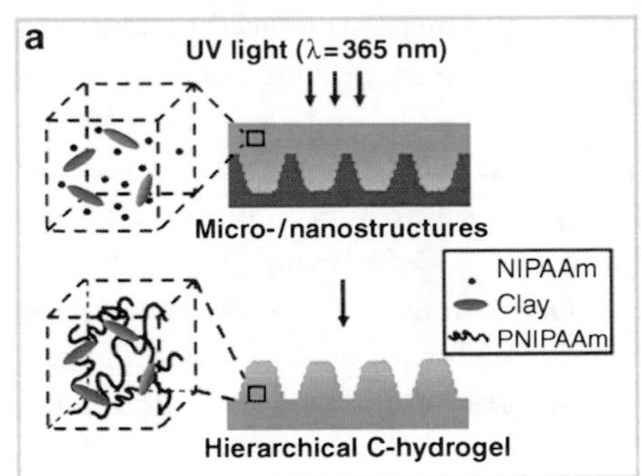

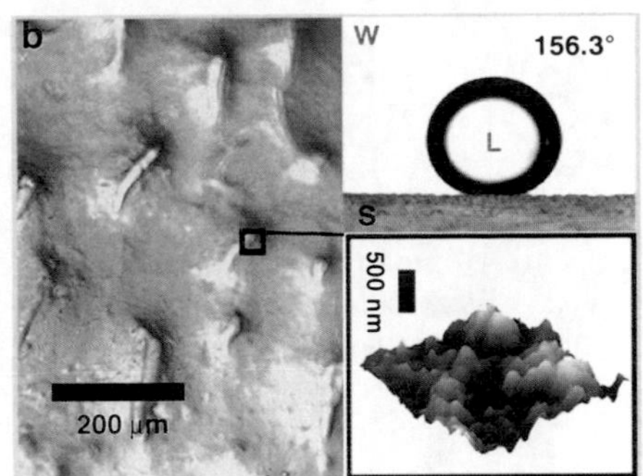

Superoleophobicity of Fish Scales, Fig. 3 Biomimetic design of PNIPAAm–nanoclay hydrogels (C-hydrogels) with hierarchical surface structures. (**a**) Schematic illustration of the fabricating process. (**b**) Optical and AFM images indicate micro/nanostructured surface of fish-scale replica shows underwater superoleophilicity. S = hydrogel, L = 1, 2-dichloroethane (DCE), W = water (Reproduced with permission. Copyright Wiley-VCH Verlag GmbH & Co. KGaA (2010))

Moreover, when it comes to micro/nanostructured surfaces, the oil CA is up to 174.8 ± 2.3° in the oil/water/solid system. This phenomenon is likely caused by the new forming interface on the micro/nanostructured surface. According to the Cassie–Baxter's model, micro/nanostructures could trap air forming a composite solid/gas surface in liquid/gas/solid three-phase system. As to oil/water/solid three-phase system, hierarchical structures can also trap abundant water forming a composite water/solid interface, therefore, the apparent oil CA ($\cos\theta'_{oil/water}$) in the water environment can be extended as:

$$\cos\theta'_{oil/water} = f'\cos\theta_{oil/water} - (1 - f') \quad (8)$$

where f' is the area fraction of oil/solid interface, $(1 - f')$ is that of the oil/water interface, $\theta_{oil/water}$ is the intrinsic CA of oil on a flat surface in oil/water/solid system. Once the composite surface forms, the oil droplet can rarely touch the solid surface, leading to the large increase of apparent oil CA. This effect endows the micro/nanostructured silicon surface with superoleophobicity and as well contributes to the oil resistance of micro/nanostructured fish scales.

Biomimetic Hydrogels for Robust Underwater Superoleophobicity

By understanding the fish oil-repellency nature, Lin et al. began to design a bionic artificial surface with robust underwater superoleophobicity [15]. From a practical perspective, the robustness of superoleophobic surface is crucial for underwater applications. Inspired by fish scales, hydrophilic hydrogel, biophysically similar to mucus, was chosen to construct fish scale-like surface with micro/nanostructures. Meanwhile, hydrophilic clay (synthetic hectorite), a rigid nanolayered structure, was selected as a composite component to enhance mechanical strength. In the experiment, the mixture of *N*-isopropylacrylamide (NIPAAm) and nanoclay were poured onto a polydimethylsiloxane (PDMS) template which was firstly molded from dried fish scales (Grass Carp, *Ctenopharyngodon idella*) (Fig. 3a). Then, a photo-initiated in situ radical polymerization ($\lambda = 365$ nm, 40 min) was taken to fabricate hybrid PNIPAAm–nanoclay hydrogels (C-hydrogels) with hierarchical surface structures. Figure 3b displays that micro/nanostructures are well molded on the surface of fish-scale replica. Furthermore, note that the surface exhibits superoleophobicity with a static oil (CA) of 156.3 ± 1.4° in oil/water/solid three-phase system (DCE as the detecting oil). This result showed that hybrid hydrogels with hierarchical surface structures successfully mimicked the oil-repellent fish scales.

To further investigate the robustness of superoleophobicity in a complex water environment, underwater oil-adhesion experiments were dynamically measured by a high sensitivity microelectromechanical balance system [13]. In each process, an oil droplet (DCE) was controlled to squeeze against the surface with a constant preload and then released, during which the adhesion force between oil and surface was recorded (Fig. 4a). It was found that, in the case of no preload on surfaces, the adhesion forces between oil and hierarchical surfaces were much lower than those between oil and smooth surfaces (Fig. 4b). As the preload increased, the oil-adhesion behaviors on

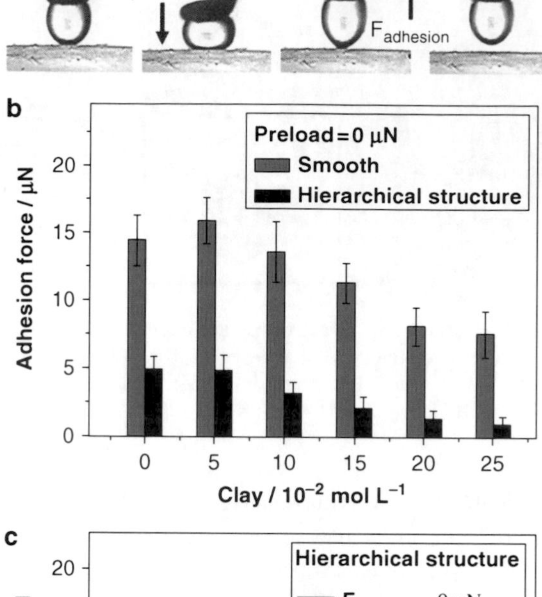

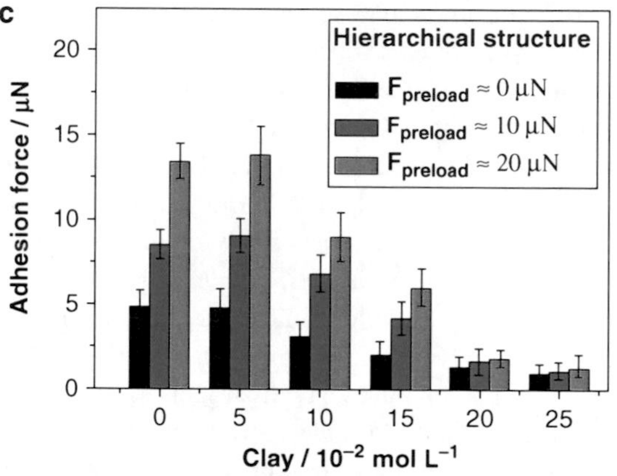

Superoleophobicity of Fish Scales, Fig. 4 Robustness of superoleophobicity on C-hydrogel surfaces measured by dynamic underwater oil-adhesion measurements (oil: DCE). (**a**) A measurement process. (**b**) Oil-adhesion forces on hierarchical surfaces are much lower than those on smooth surface (preload = 0). (**c**) As the preload increases, the oil-adhesion force increases only on the low clay content C-hydrogel, while C-hydrogel with high clay content retains ultralow oil adhesion (Reproduced with permission. Copyright Wiley-VCH Verlag GmbH & Co. KGaA (2010))

hierarchical surfaces became different. The adhesion force increased only on the surface of low clay content C-hydrogel, while the C-hydrogel surface with high clay content retained excellent low oil adhesion (Fig. 4c). These results were attributed to the synergetic effects of rigid nanoclays and flexible macromolecules. It is known that PNIPAAm possesses hydrophilic macromolecule chains which can trap water to prevent oil adhesion. But this effect is weakened by the frangibility of hierarchical structures on PNIPAAm surface. On the other hand, hybrid hydrogels with rigid nanoclays can enhance mechanical strength of surface micro/nanostructures and, thus, keep the stability of trapped water on the surface. As a result, hybrid hydrogels with high clay content achieve robust underwater superoleophobicity. This study would bring a new concept to the design and fabrication of underwater antifouling materials.

Future Directions for Research

The research on superoleophobic fish scales may open a new branch of the wettability field and make strong impact on underwater applications. The studies in oil/water/solid three-phase systems have just begun, and many challenges remain on their way of development. First of all, the wetting theories in oil/water/solid system need to be further established. It is based on the tremendous wettability data of different materials with specific chemical components and surface structures. Secondly, inspired by fish scale's effect, underwater superoleophobic materials with environment-friendly, durable properties need be further explored, which show great potential for underwater antifouling application. Finally, in oil/water/solid system, intelligent responsive materials with switchable oil-wetting behavior will attract great attention due to the promising applications in artificial muscles, actuators, sensors. In the future, learning from nature will be a primary principle to design biomimetic or bioinspired functional materials with special wettability.

Cross-References

▶ Biomimetics of Marine Adhesives
▶ Shark Skin Effect

References

1. Young, T.: An essay on the cohesion of fluids. Philos. Trans. R. Soc. Lond. A. **95**, 65–87 (1805)
2. Wenzel, R.N.: Resistance of solid surfaces to wetting by water. Ind. Eng. Chem. **28**, 988–994 (1936)
3. Cassie, A.B.D., Baxter, S.: Wettability of porous surfaces. Trans. Faraday. Soc. **40**, 0546–0550 (1944)

4. Öner, D., McCarthy, T.J.: Ultrahydrophobic surfaces. Effects of topography length scales on wettability. Langmuir **16**(20), 7777–7782 (2000)
5. Extrand, C.W.: Model for contact angles and hysteresis on rough and ultraphobic surfaces. Langmuir **18**(21), 7991–7999 (2002)
6. Quéré, D.: Wetting and roughness. Annu. Rev. Mater. Res. **38**, 71–99 (2008)
7. Liu, M.J., Wang, S.T., Wei, Z.X., Song, Y.L., Jiang, L.: Bioinspired design of a superoleophobic and low adhesive water/solid interface. Adv. Mater. **21**(6), 665–669 (2009)
8. Barthlott, W., Neinhuis, C.: Purity of the sacred lotus, or escape from contamination in biological surfaces. Planta **202**(1), 1–8 (1997)
9. Blossey, R.: Self-cleaning surfaces – virtual realities. Nat. Mater. **2**(5), 301–306 (2003)
10. Gennes, P-Gd, Brochard-Wyart, F., Quéré, D.: Capillarity and wetting phenomena: drops, bubbles, pearls, waves. Springer, New York (2004)
11. Yao, X., Song, Y.L., Jiang, L.: Recent developments in bio-inspired special wettability. Adv. Mater. **23**, 719–734 (2011). doi:10.1002/adma.201002689
12. Nosonovsky, M., Bhushan, B.: Multiscale effects and capillary interactions in functional biomimetic surfaces for energy conversion and green engineering. Philos. Trans. R. Soc. Lond. A. **367**(1893), 1511–1539 (2009)
13. Liu, M.J., Zheng, Y.M., Zhai, J., Jiang, L.: Bioinspired super-antiwetting interfaces with special liquid-solid adhesion. Acc. Chem. Res. **43**(3), 368–377 (2010)
14. Lide, D.R.: CRC handbook of chemistry and physics, 84th edn. CRC Press, Boca Raton (2003–2004)
15. Lin, L., Liu, M.J., Chen, L., Chen, P.P., Ma, J., Han, D., Jiang, L.: Bio-inspired hierarchical macromolecule-nanoclay hydrogels for robust underwater superoleophobicity. Adv. Mater. **22**(43), 4826–4830 (2010)

Surface Electronic Structure

Regina Ragan
The Henry Samueli School of Engineering,
Chemical Engineering and Materials Science
University of California, Irvine, Irvine, CA, USA

Synonyms

Local density of states

Definition

Surface electronic structure is defined by filled and empty electronic states of the system near the surface of a solid material. In both bulk systems and nanosystems, the type of atoms in the system, the atomic arrangement, and atomic order versus disorder influence electronic structure. Tight Binding, an early method for calculating electronic structure using a single particle Hamiltonian, approximates the electronic wave functions in a crystal as a linear combination of the atomic wave functions of constituent atoms. Tight Binding analysis shows that electrons in solid materials exhibit collective behavior that deviates from discrete electronic states in isolated atoms. Coupling of atomic wave functions in crystals leads to a continuum of energy states, called energy bands. The electronic structure at the surface often differs from that in the bulk due to broken bonds and atomic surface reconstructions. If the surface to volume ratio is high, as in nanosystems, then surface effects can dominate observed electronic properties.

An important physical parameter to characterize electronic structure is the Fermi energy. The Fermi energy is often referred to as the highest occupied energy state since the probability of electron occupation of states having higher energy than the Fermi energy decays exponentially at equilibrium as determined from the Fermi-Dirac distribution function. The Fermi energy level is typically measured indirectly by the work function of the material. The work function is the energy needed to remove an electron from a material and is equal to the difference between the Fermi energy and the vacuum energy level. The position of the Fermi energy with respect to the energy bands determines whether the material is a metal, semiconductor, or insulator. Consider a simplified intuitive view. If atoms in the crystal have an odd number of valence electrons, then the energy bands that arise due to splitting of atomic energy levels are half filled. In this case, as found in metals, there is a continuum of available electronic states near the Fermi energy. A low resistance to electron motion is exhibited and thus metals are good conductors. In comparison, from this simple view, if there is an even number of valence electrons in constituent atoms of the crystal, then the (valence) energy bands are completely filled at zero Kelvin and the Fermi energy is in the energy bandgap. Near zero Kelvin, in both semiconductors and insulators, electrons do not occupy states in the conduction band and thus will exhibit negligible conductivity. Near room temperature, semiconductors have few electrons at energy levels available for conduction.

In the case of nanosystems, the number of atoms in the system also affects electronic structure due to quantum confinement of electrons. In metals the "space"an electron occupies in the crystal is defined by the Fermi wavelength that has an inverse relationship with the electron density. In semiconductors, where charge carriers are electron and holes, the Bohr exciton radius defines the "space" for an electron hole pair. The Bohr exciton radius is a function of the dielectric constant of the material and the effective mass of the charge carriers. It is typically much larger than the Fermi wavelength in magnitude. When dimensions of nanoscale systems decrease below the Fermi wavelength (metals) or Bohr exciton radius (semiconductors) unique physical properties not observed in bulk materials arise due to quantum-size effects.

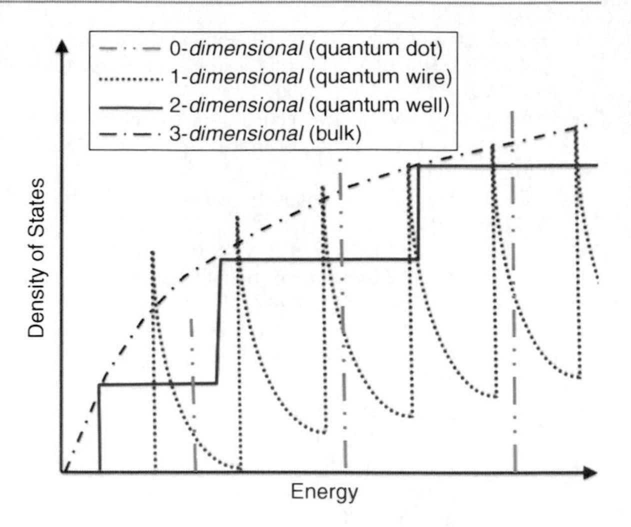

Surface Electronic Structure, Fig. 1 Schematic of density of states for 3-dimensional, bulk system (*black dot-dashed curve*), 2-dimensional, quantum well (*blue solid curve*), 1-dimensional, quantum wire (*red dashed curve*), 0-dimensional, quantum dot (*cyan double dot- dashed curve*)

Overview

Drude theory provides a simple, intuitive analysis of the collective behavior of electrons in metallic systems. After J.J. Thompson discovered the electron in 1897, Drude treated valence electrons in metals as an electron gas and applied the kinetic theory of gases to describe electrical and thermal conductivity in metals. In this model, valence electrons are assumed to move freely in the solid and the potential exerted on electrons in the crystal lattice is assumed to be uniform due to nondirectional bonding of atoms in the crystal (metals have high coordination numbers). Electrons are only assumed to interact with ion cores during finite scattering events. Thus, the model is called the free electron model and the approximation for finite scattering events is called the relaxation time approximation. Although Drude theory only works well for alkali metals, it provides a qualitative understanding of thermal and electrical conductivity in metals. Early on Sommerfeld addressed some of the shortcomings of Drude theory to evaluate thermal conductivity by using Fermi-Dirac statistics to model the velocity distributions of electrons in metals in the context of Drude theory. Quantum mechanical corrections are needed to accurately estimate the number of electrons that contribute to conductivity in most metals. Despite the need for these corrections, the approach of the electron gas to model observed electrical, thermal, and optical properties in materials was pivotal in our understanding of experimental observations in metals and semiconductors. The density of states, that is the number of available states per unit energy at a particular energy level, is derived from the electron gas approach. A schematic of how the density of states varies between bulk systems and quantum-confined systems is shown in Fig. 1. Van Hove singularities (discontinuity in the density of states) can be seen in quantum-confined systems. The concept of an electron gas is still used and describes quantum-confined systems such as metallic, single-walled carbon nanotubes (*one-dimensional* electron gas) that exhibit Van Hove singularities [1] and graphene (*two-dimensional* electron gas) that exhibits the quantum hall effect where the Hall conductivity exhibits quantized values [2].

When analyzing systems of atoms that do not form metallic bonds, ionic, and covalent systems, the approximations that bonding is nondirectional and the potential is uniform is no longer valid. Kronig and Penny provided early intuition of energy levels and bands in crystals using both Bloch's theorem and a simple one-dimensional, periodic square wave potential that roughly approximates the potential of atoms in a periodic crystalline system. Via this simple analysis it is found that there are energy levels that yield nonphysical solutions for the electron wave functions and thus are not allowed for electrons in the system. These forbidden energy levels are defined as an energy

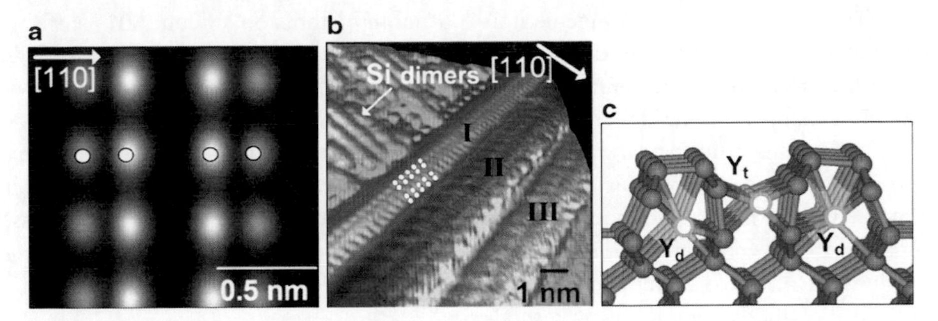

Surface Electronic Structure, Fig. 2 (**a**) DFT simulation of the calculated electronic structure on the surface of an yttrium disilicide nanowire having a width of 1.1 nm. (**b**) STM image of dysprosium disilicide nanowires on Si(001) substrate. Nanowire labeled I has a width of 1.1 nm, the calculated surface electronic structure matches the STM image as indicated by the white round circles. (**c**) Cross-sectional view of the relaxed atomic struture in the disilicide nanowires determined from DFT calculations and STM images (Printed with permission from Ref. [4])

bandgap. If the Fermi energy is in the bandgap, there are few electrons occupying energy states in the conduction band (above the Fermi energy) and thus conductivity is lower than in metals, hence the name semiconductor. As mentioned prior, the main difference between metals and semiconductors/insulators is that the Fermi energy sits within a band in a metal and in the energy bandgap in a semiconductor/insulator. The main difference between semiconductors and insulators is the magnitude of the bandgap. If greater than approximately 4 eV, then the probability of electrons occupying states in the conduction band is negligible at room temperature.

Basic Methodology

Computational Approaches

Significant advances in computational abilities coupled with important fundamental physical simplifications have allowed for first principles, ab initio, calculations of electronic structures of many body systems more closely modeling solid macroscopic and nanosystems and thus increasing accuracy. Since the Schrodinger equation cannot be solved analytically for these many electron systems, it was pivotal that Hohenberg and Kohn proved that the ground state energy of a many body system is a unique function of the charge density distribution [3]. Charge density distributions have a relationship with the atomic structure and periodicity in the crystal and thus can be evaluated by the types of atoms in the crystal and the crystal structure (atomic arrangement). Later, using an exchange correlation potential acquired from the theory of a homogeneous electron gas, Kohn and Sham determined how a many body system could be reduced to a single particle equation for input into the Schrodinger equation. This is referred to as the local density approximation [3].

Due to these critical advances in understanding of many body systems, density functional theory (DFT) is widely used to calculate electronic structure that can be directly compared to experimental systems for understanding of experimental observations or used to predict physical properties to guide experiments. For example, DFT provided fundamental understanding of low-dimensional magnetism such as giant magnetic moments found in two-dimensional systems, such as monolayers of Mn and Cr. Giant magnetic moments arise due to an increase in the density of electronic states near the Fermi energy [3]. The phenomenon of Giant Magnetoresistance allowed for an increase in storage density in magnetic hard disk drives. Furthermore, low-dimensional metallic nanowires and nanoparticles can exhibit ballistic transport or unique chemical activity, respectively, and typically require feature sizes smaller than achievable with lithography. Thus self-organization is needed for fabrication. Atomic arrangements and driving forces for self-organization of low-dimensional metallic systems can be determined using atomic scale imaging in conjunction with DFT [4, 5]. Figure 2 shows how the correlation between DFT simulations and scanning tunneling microscopy (STM) measurements allowed for an understanding of the atomic structure in disilicide nanowires that exhibit *one-dimensional* electron transport [4, 5]. The correct atomic arrangement is also critical for understanding the charge density distribution and in turn the surface electronic structure that is relevant, for example, to the development of nanocatalysts [6]. Improving efficiency and selectivity of catalysts will have significant economic benefit for the chemical industry. DFT has also been used to design electrode materials for lithium ion batteries. For example, Meng et al. have correlated structure with performance in electrodes using ab initio methods [7]. Materials design for batteries is crucial for development of plug-in electric vehicles that meet consumer performance requirements. Overall DFT allows for fundamental understanding of the relationship between atomic structure and electronic structure that is critical for understanding and utilizing physical properties of nanoscale systems.

Measurement Techniques

Photoelectron Spectroscopy

Photoelectron spectroscopy (PES) is a traditional surface analysis technique that uses a photon beam to provide energy for electrons to escape the potential of atoms or molecules near the surface. Kinetic energy distributions of emitted photoelectrons provide information regarding ionization energies or work function of the surface that reflects the composition and electronic states on the surface. Typical methods are x-ray photoelectron spectroscopy (XPS) that uses soft X-rays with energy on the order of 200–2,000 eV and ultraviolet photoelectron spectroscopy (UPS) that uses ultraviolet light with energy in the range of 10–45 eV. The energy of the photon beam affects the type of electrons that are emitted from the sample. During XPS core electrons are emitted from the atoms and during UPS the energy is sufficient to emit only valence electrons. Angle resolved photoemission spectroscopy (ARPES) provides additional information about the momentum of the emitted electrons. Since momentum is conserved, one can obtain the energy-momentum relationship (called the dispersion relationship) of the electrons in the crystal that reflects the energy band structure as a function of crystallographic direction. ARPES has been an important tool to measure the electronic band structure in bulk materials and in quantum-confined systems. For example, Yeom et al. measured a charge density wave in linear chains of indium atoms on silicon (001) surfaces [8]. Charge density waves are a coupling between electrons and lattice vibrations that exhibit a periodic modulation of charge that can be observed on a *one-dimensional* metallic surface. In the same study, a measured temperature dependent metal to semiconductor transition for in atomic chains on silicon (001) was attributed to a Peierls transition that is another signature of *one-dimensional* quantum confinement.

Scanning Probe Microscopy

Scanning probe microscopy (SPM) includes techniques such as scanning tunneling microscopy/spectroscopy (STM/STS), atomic force microscopy (AFM), electrostatic force microscopy (EFM), and Kelvin probe force microscopy (KPFM) that can be used to measure atomic structure and/or surface electronic structure. STM, invented in 1981 by Binnig and Rohrer, probes both atomic and electronic structure on surfaces by measuring tunneling current from the probe tip to sample surface across a narrow vacuum (dielectric) gap. STM measurements typically achieve atomic scale spatial resolution. The AFM was invented shortly after the STM, 1986, to measure topography on nonconducting surfaces and is also capable of atomic and molecular resolution on surfaces. AFM measures van der Waals and electrostatic forces between cantilever tip and surface and thus does not require a conductive sample surface. KPFM is a variant of AFM in which conducting tips are used; KPFM allows for determination of the local surface potential with nanometer spatial resolution.

Tunneling current, as measured in STM, is extremely sensitive to both electron density and surface topography and convolutes the two properties. The tip-sample polarity affects whether electrons tunnel from sample to tip or from tip to sample and thus determines if the measurement probes filled or empty surface electronic states, respectively. A combination or empty and filled states imaging is often used in order to deconvolute the surface electronic structure from the atomic structure. STM can resolve corrugations heights on the sub-angstrom level, i.e., 0.2 Å corrugations between atoms on a clean platinum surface have been measured. The atomic arrangement on clean Pt (111) and how the surface structure evolves after depositing self-assembled monolayers on the surface have been measured using STM [9]. Figure 3a shows an STM image of Pt(111) where the hexagonal close

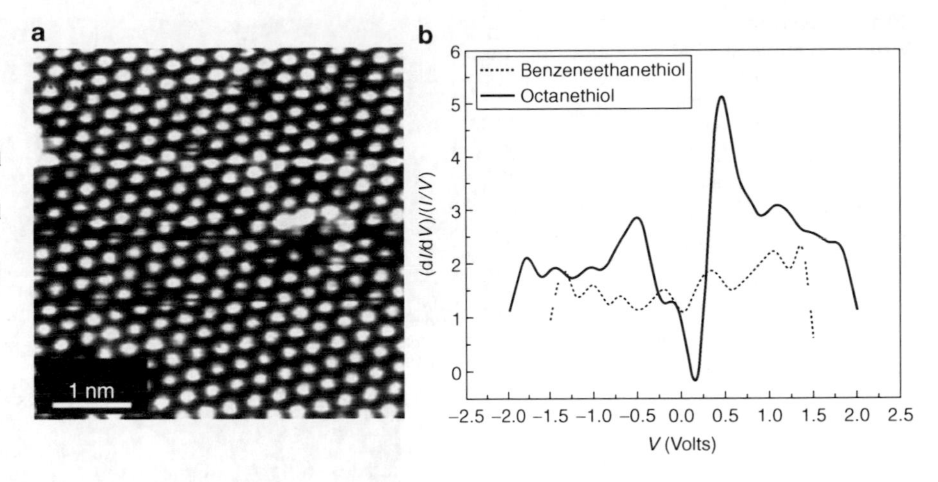

Surface Electronic Structure, Fig. 3 (**a**) STM image of Pt(111) surface. *R. Ragan unpublished results* (**b**) Normalized differential conductance of self-assembled monolayers of benzenethiol (*dashed curve*) and octanethiol (*solid curve*) on Pt(111) (Printed with permission from Ref. [10])

packing of the atoms on the surface is easily observed. Scanning tunneling spectroscopy (STS) is a derivative of STM that measures the density of electronic states in the vicinity of the atomic scale STM tip, again yielding high spatial resolution. During STS, the current–voltage spectrum is measured between tip and surface while the voltage is swept across a specified range usually at constant tip-to-sample distance. Normalization of the differential conductance (dI/dV) with the conductance (I/V) can reflect the local density of states near the Fermi energy. Figure 3b shows normalized differential conductance data for benzene ethanethiol and octanethiol self-assembled monolayers on Pt(111) determined from STS data. Note that the Fermi energy is set at zero volts on the x-axis. The octanethiol/Pt(111) junction has zero conductance at the Fermi energy and this represents an energy bandgap, i.e., no available states near the Fermi energy. For the benzene ethanethiol/Pt(111) junction, there is finite conductance at the Fermi energy and this is an indication of metallic behavior. A higher conductivity near the Fermi energy across benzene ethanethiol/Pt(111) junctions in comparison to octanethiol/Pt(111) junctions can be attributed to the fact that the benzene ethanethiol molecule has conjugated bonds in the molecule [9].

AFM measures the forces between AFM tip and surface by optically measuring the deflection of an AFM cantilever. With this basic mechanism, the type of signal feedback provides a wealth of information about the sample surface. Amplitude and frequency modulated AFM monitors the variation of the amplitude or frequency, respectively, of the AFM cantilever in response to forces between the tip and the surface and measures topography of the surface. In intermittent contact mode, the phase shift of the free cantilever resonance frequency provides nanometer scale information of the viscoelastic properties and adhesion force of the surface since it represents energy dissipation between tip and surface. For example, when imaging under a repulsive tip-sample condition, regions of the surface with the higher elastic modulus appears darker in a phase contrast AFM image. One can observe a contrast reversal when the tip changes from repulsive to attractive mode. Variations in local topography also induce a phase shift in the cantilever frequency and thus topography and phase images need to be analyzed in conjunction to understand the physical properties of the surface.

Derivatives of AFM can be used to measure electrical properties when using a conducting cantilever. EFM measures electrostatic forces between surface and cantilever and is modeled by treating the vacuum, air or any dielectric gap between tip and surface as a capacitor. The force is dependent on the tip-surface distance and the potential difference between tip and surface. In particular, KPFM directly measures the contact potential difference (difference between sample surface work function and tip work function) using a lock-in technique to null the electrostatic forces between tip and sample surface. Recently, Ragan and Wu et al. have demonstrated that measured values of work function obtained from KPFM compare quantitatively with DFT calculations and together provide information on the atomic arrangement and termination of atoms on alloy surfaces [6]. Work function variations on surfaces reflect charge transfer [11, 12], quantum-size effects [13], and localized surface

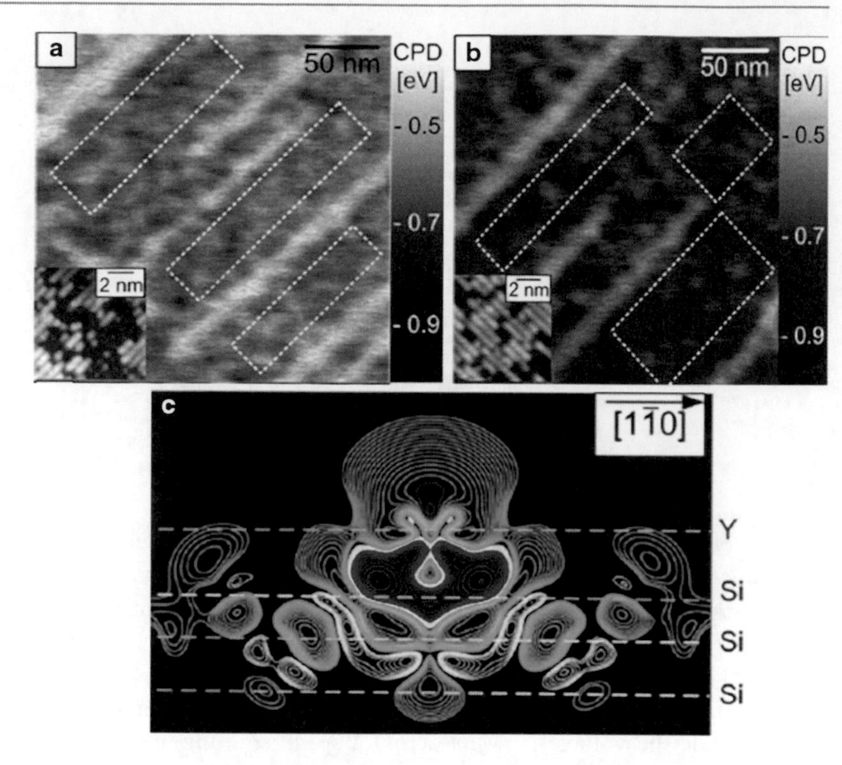

Surface Electronic Structure, Fig. 4 KPFM images of dysprosium nanowires on Si(001) that have been annealed post-growth at (**a**) 600°C and (**b**) 680°C. (**c**) Simulated charge density difference image for a single metal adatom on Si(001) with the cross section perpendicular to the surface. Charge accumulation increases from yellow to pink contour lines, whereas charge depletion increases from green to blue contour lines. The greatest charge depletion is seen at the metal adatom location and the greatest charge accumulation is seen in the region between the adatom and the subsurface Si atoms (Printed with permission from Ref. [11])

charge [14]. Figure 4 demonstrates how the work function on silicon changes when metal adatoms are on the semiconductor surface due to charge transfer between metal adatoms and substrate atoms. Figure 4a is a KPFM image after deposition of dysprosium on Si(001) and annealing the sample at 600°C. Metallic disilicide nanowires form on the surface and appear as bright lines in the KPFM image since the disilicide nanowires have a lower work function than the Si(001) substrate. The Si(001) substrate in between nanowires is highlighted with white dashed boxes. A comparison of the substrate regions in Fig. 4a with those in Fig. 4b shows that the work function is lower on the substrate regions of the former. The STM images shown as insets in the lower left corner show that the Si(001) surface in Fig. 4a has more metal adatoms (dark regions in STM image) on the surface than the Si(001) surface of Fig. 4b due to the lower annealing temperature. The DFT simulation of Fig. 4c demonstrates that metal adatoms transfer charge to the Si atoms on the surface. This creates a dipole on the surface that lowers the work function [11]. Overall, KPFM measurements provide information about material behavior in devices; KPFM across heterojunctions provides information regarding band offsets, device performance of diodes and chemical sensitive field effect transistors and trapped charge at interfaces in high electron mobility transistors.

Ion Scattering

Low energy ion scattering (LEIS) is a method complimentary to scanning probe techniques that also provides information about surface composition, electronic structure, and atomic structure. LEIS has less stringent requirements on surface conditions than SPM as data can be acquired using LEIS from rough and contaminated materials. The energy of scattered ions depends on the ratio of the projectile and target masses, providing a measure of the atomic mass distribution on the surface. The degree of charge exchange that occurs during scattering is dependent on the surface electronic properties when using projectiles with low ionization energies, such as alkalis. When an alkali-metal atomic particle is in the vicinity of a surface, its ionization level shifts up due to the image charge interaction, while it broadens due to overlap of the ion and surface wave functions. The measured neutral fraction depends on the ionization potential, the degree that the level shifts near the surface, and the work function at a point just above the scattering site. The charge

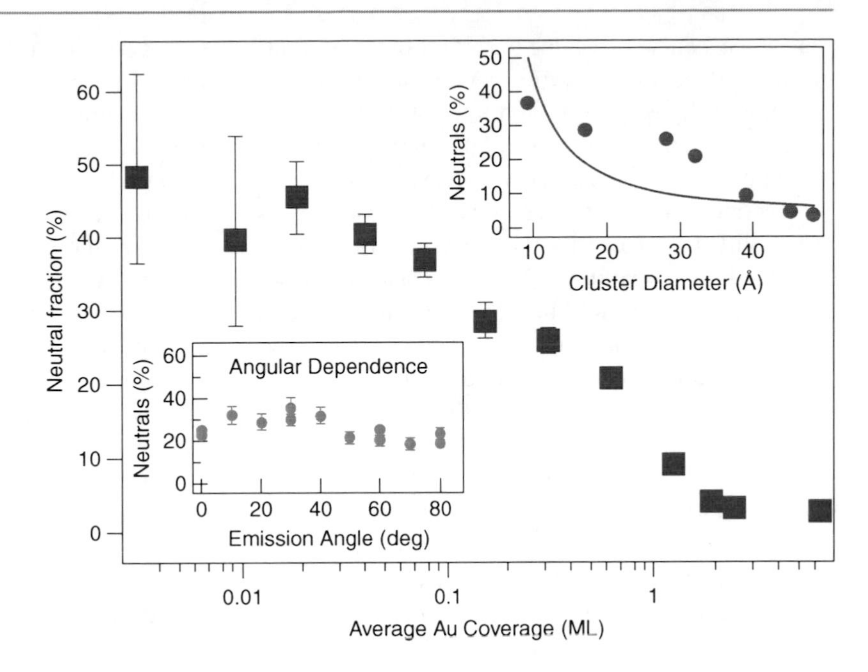

Surface Electronic Structure, Fig. 5 Neutral fractions (NF) of singly scattered 2.0 keV Na ions shown sas a function of the average Au coverage. The right side inset shows NF versus cluster diameter, with the symbols indicating experimental data and the solid line a theoretical fit. The left side inset shows NF for Na+ scattered from a 0.15 ML Au coverage as a function of the emission angle with respect to the surface normal (Printed with permission from Ref. [16])

exchange process is well described by a non-adiabatic resonant charge transfer model [15]. Since the interaction is local to the site where projectile atoms exit the surface, the neutralization of low energy alkali ions provides a unique method for measuring the local work function and quantum-size behavior in nanomaterials.

Yarmoff et al. has used LEIS to measure quantum-size effects in the electronic structure of gold nanoclusters on TiO_2 [16] to provide insight regarding how catalytic activity and surface electronic structure are correlated. Au or other heavy metal nanocrystals are ideal for ion scattering experiments, as the large mass of the cluster atoms enables a complete separation of the ions that impact the nanoclusters from those that impact the substrate. The integrated single scattering peaks in "Neutrals" and "Total Yield" spectra are divided to obtain the neutral fraction for scattering from the clusters. The neutral fraction for Na^+ ions scattered from Au nanocrystals as a function of Au coverage is directly correlated with the size of the Au nanocrystals. The neutral fraction goes from about 50% for the smallest clusters down to about 3% for the film [16]. The enhanced neutralization from small clusters is due to participation of quantum-confined states in the non-adiabatic resonant charge transfer process. Bulk Au has a relatively high work function (5.1 eV), so that the Fermi level is degenerated with the Na ionization level (also ~5.1 eV) and most of the scattered Na remains ionic. The neutral fraction for small Au clusters is considerably larger than for bulk Au because filled states associated with the clusters provide electrons that can tunnel to the outgoing projectile. The existence of such filled states is consistent with reports that small Au clusters are negatively charged. The additional filled states above the Fermi level depend on the size of the clusters, and thus provide a measure of the quantum-size behavior (Fig. 5).

Examples of Application

Nanowire Sensors

Chemical and biological nanowire sensors that use electronic transduction are an example of a class of devices that uses changes in surface electronic structure to measure local molecular binding events or adsorption of molecules on surfaces. A chemical or biological sensor can be reduced to some basic components: a receptor (biological recognition element if selectivity is required), a transducer, and a means for processing whether or not target molecules of interest have bound to the surface and relaying this information to the user. Transducers can sense molecular interactions electronically (e.g., changes in localized charge) or optically (e.g., changes in dielectric constant). Changes in surface electronic structure can affect the conductance and this is easily measured in an electronic platform. A high surface to volume ratio in nanowires leads to high sensitivity for measuring changes in conductance due to surface binding events. If the surface of nanowires is functionalized with receptors that selectively bind to the target molecule of interest then selectivity can be engineered into the sensor platform. Since an early demonstration of semiconducting carbon nanotubes as chemical sensors, many different materials that form nanowires have exhibited the capability to sense molecules. In_2O_3 nanowires, SnO_2 nanoribbons, Si and ZnO nanowires have all been used as chemical and/or biological sensors. Si nanowires have been used for the detection of DNA hybridization, protein–protein interactions, cancer markers, and viruses. Some recent review articles that discuss nanowires-based sensor devices are referenced here [17, 18].

Nanowire sensors based on electronic transduction are commonly fabricated in a field effect transistor (FET) architecture with a back gate configuration. A single or multiple nanowires are connected to a source and drain and nanowires serve as the conducting channel. The conductance in the channel will change as molecules bind to the surface. Molecular binding events on semiconducting nanowire surfaces induce accumulation (depletion) of charge carriers that can be measured as an increase (decrease) lateral conductivity along nanowires in the FET architecture. The onset of accumulation or depletion of carriers depends on the surface charge induced and the carrier-type in the nanowire, *n-type* signifies that the majority carriers are negatively charged electrons and *p-type* signifies that the majority carriers are positively charged holes. Thus in a FET architecture, molecular binding events on surfaces can be monitored in an electronic circuit. For example, in a study by Cui et al. Si nanowires were functionalized with biotin receptors. When streptavidin, target molecule, was introduced into the system at a concentration of 250 nM an increase in conductance was observed. Streptavidin is well known to have a high binding affinity with biotin. The increase in conductance due to the introduction of strepavidin could be measured for concentrations as low as 25 pM [17, 18]. This early study demonstrated both selectivity and high sensitivity of Si nanowire sensors.

Si has some advantages as a sensing platform since microelectronic circuits are typically made from Si. This approach provides a strategy for fabricating high-density, high-quality nanoscale sensors that can be integrated with Si-based circuits. Si nanowire sensors have been fabricated by either "top-down" or "bottom-up" fabrication methods where the former uses lithographic methods and the latter uses self-organization routes. Previously, Li et al. reported an approach to configure FET device architectures using Si nanowires for DNA sensors using a standard "top-down" semiconductor process [19]. Reactive ion etching is used to fabricate Si nanowires on silicon-on-insulator substrates that have been patterned using electron beam lithography. Figure 6a is an optical image of the entire sensor system. Figure 6b, c are high-resolution SEM images showing a Si nanowire in the system. The chemical functionalization process to bind a single strand DNA receptors on nanowire surfaces is illustrated in the schematic of Fig. 6d. The chemical modification process can be monitored after each step by measuring changes in surface potential using the surface photovoltage technique. When solutions containing complementary strands of DNA with concentrations of 25 pmol are introduced to *p-type* Si nanowire surfaces, there is an increase in conductance due to accumulation of carriers. In the case of *n-type* Si nanowires, the DNA binding event leads to a decrease in conductance due to depletion of carriers. Accumulation (*p-type*) or depletion (*n-type*) of carriers is associated with the negative charge on the backbone of DNA molecules. Quitoriano et al. introduced a different fabrication method for Si nanowire FET devices. This method is a combined "bottom-up" and "top-down" fabrication process utilizing the vapor-liquid–solid growth mechanism and optical lithography to fabricate FET devices [20, 21]. In this method, Au nanoparticles are deposited on the bottom surfaces of lithographically defined Si electrodes that overhang over a recessed trench of silicon dioxide. The growth of Si nanowires is guided from the bottom of the electrode at the site of the Au catalyst along the silicon dioxide layer to the opposite electrode. A benefit of combining "bottom-up" methods with "top-down" methods is to integrate nanowires into microelectronic circuits using high-throughput methods. In both cases, Si nanowires sensors can be connected directly to the adjoining circuitry for signal amplification and automated data acquisition.

Nanoscale Catalysts

Metallic and bimetallic surfaces and nanostructures with tunable physical and chemical properties have attracted particular attention in recent years due to

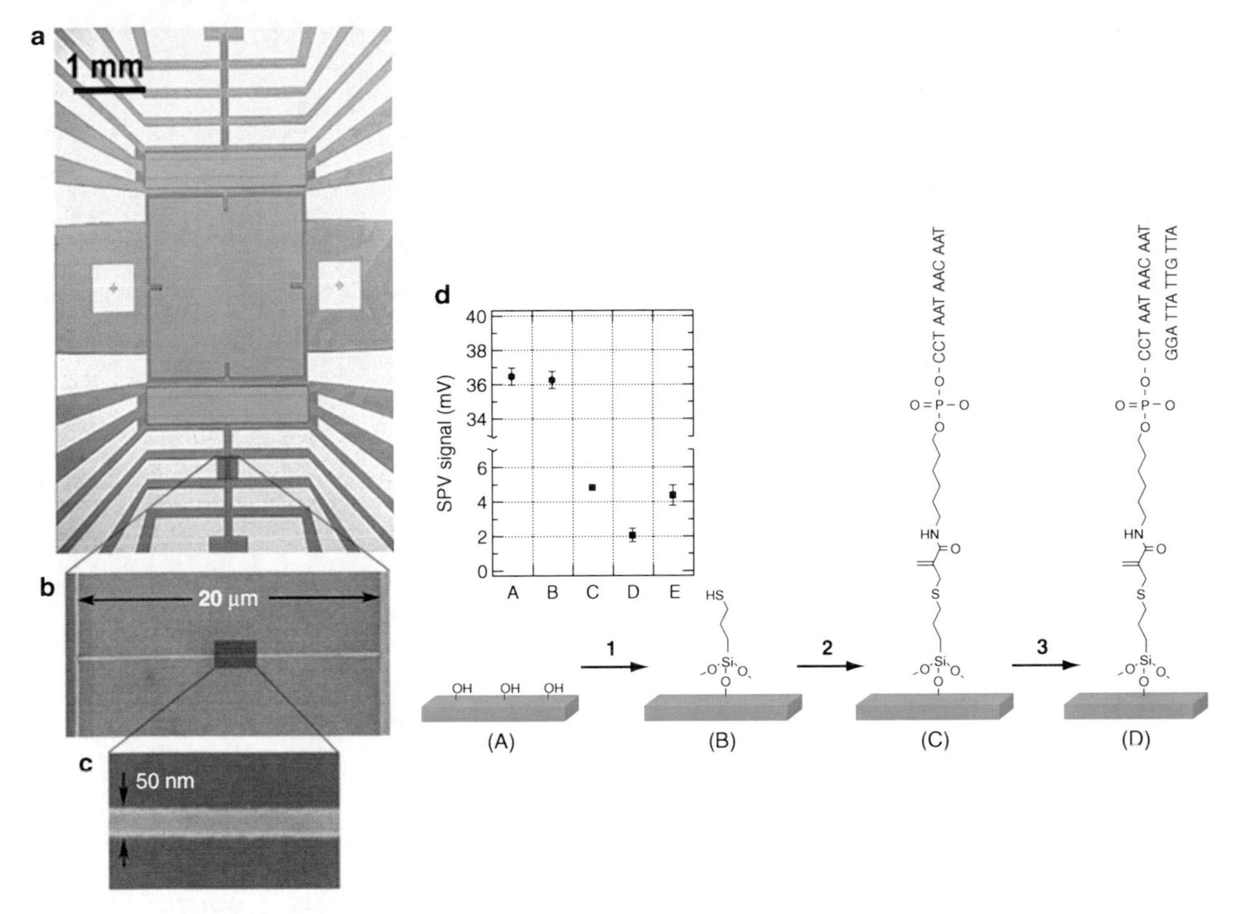

Surface Electronic Structure, Fig. 6 (**a**) Optical image of lab on chip system using Si nanowires for signal transduction. (**b**) SEM image of Si nanowire spanning electrical leads. (**c**) High-resolution SEM image of Si nanowire. (**d**) Chemical functionalization of Si nanowire surfaces, steps labeled A-D, for sequence specific detection of DNA. The inset shows how the change in surface photovoltage (SPV) signal monitors each chemical functionalization step. (Printed with permission from Ref. [19])

their potential for use in a broad range of applications. The discovery by Haruta, et al. that gold nanoparticles on oxide supports exhibit surprisingly high catalytic activity for reactions such as CO oxidation and propylene epoxidation has inspired an enormous wave of research in the quest for innovative nanocatalysts. Chemically active Au nanoparticles have been prepared on reducible (TiO_2, ZrO_2, NiO, or Fe_2O_3) oxides, and irreducible (SiO_2 or Al_2O_3) oxides in order to gain insight on the role of quantum-size effects, facets and steps at the edges of nanoclusters, and charge transfer to the substrate in chemical activity. A review of mechanisms involved in enhanced activity of Au nanoclusters can be found here [22]. Bimetallic surfaces of Pt and Pd, and trimetallic nanoparticles of Au/Pt/Rh were also found to be highly effective in promoting a variety of reactions, typically higher than corresponding monometallic nanoparticles. Thus there appears to be a variety of parameters affecting chemical activity.

Surface electronic structure has been identified as playing an important role in enhanced catalytic activity in nanoscale metallic systems. Electronic structure can be modified in nanosystems in many ways such as by charge transfer between nanocatalyst and substrate support or due to alloy effects in bimetallic systems. Both experimental and theoretical results show the strong coupling between electronic structure and catalytic activity. The Hammer-Nørskov model predicted that chemisorption energy correlates with the d-band center in transition metals and this is experimentally observed for oxygen and sulfur chemisorption energies on different metal surfaces [23]. The effect of electronic structure and its relationship to catalytic activity is also observed in experiments. For example, Au clusters on TiO_2 were measured by STS to undergo a metal to insulator transition at nanometer length scales. The nanocluster size were the metal to insulator transition occurs exhibited the highest turnover frequency for CO oxidation, while larger clusters having no band gap had lower activity [24]. It has also been shown that catalytic reaction rates, in the context of electrochemical catalysis, exhibit an exponential dependence on the catalyst work function and catalytic rate enhancements of up to a factor of 60 having been reported. Changes in work function as small as 200 meV can lead to an increase in rate enhancement by a factor of 10 [25]. The work function of Au nanowires can be varied by surface alloying [6] in order to optimize catalytic properties. Heterogeneous metal nanocatalysts with clusters that are a few nanometers in size hold great promise because of their large surface area to volume ratios, the availability of an enormous number of active sites and their enhanced resistance to poisoning from products of the reactions.

Cross-References

- ▶ Ab initio DFT Simulations of Nanostructures
- ▶ Atomic Force Microscopy
- ▶ Kelvin Probe Force Microscopy
- ▶ Nanomaterials for Electrical Energy Storage Devices
- ▶ Nanostructure Field Effect Transistor Biosensors
- ▶ Scanning Tunneling Microscopy
- ▶ Scanning Tunneling Spectroscopy
- ▶ Self-assembly

References

1. Wildoer, J.W.G., Venema, L.C., Rinzler, A.G., Smalley, R.E., Dekker, C.: Electronic structure of atomically resolved carbon nanotubes. Nature **391**, 59 (1998)
2. Berger, C., Song, Z.M., Li, T.B., Li, X.B., Ogbazghi, A.Y., Feng, R., Dai, Z.T., Marchenkov, A.N., Conrad, E.H., First, P.N., de Heer, W.A.: Ultrathin epitaxial graphite: 2D electron gas properties and a route toward graphene-based nanoelectronics. J. Phys. Chem. B **108**, 19912 (2004)
3. Freeman, A.J., Wu, R.Q.: Electronic-structure theory of surface, interface and thin-film magnetism. J. Magn. Magn. Mater. **100**, 497 (1991)
4. Shinde, A., Wu, R., Ragan, R.: Thermodynamic driving forces governing assembly of disilicide nanowires. Surf. Sci. **604**, 1481 (2010)
5. Zeng, C., Kent, P.R.C., Kim, T., Li, A., Weitering, H.H.: Charge-order fluctuations in one-dimensional silicides. Nat. Mater. **7**, 539 (2008)
6. Ouyang, W., Shinde, A., Zhang, Y., Cao, J., Ragan, R., Wu, R.: Structural and chemical properties of gold rare earth disilicide core – shell nanowires. ACS Nano **5**, 477 (2011)
7. Meng, Y.S., Arroyo-de Dompablo, M.E.: First principles computational materials design for energy storage materials in lithium ion batteries. Energy Environ. Sci. **2**, 589 (2009)
8. Yeom, H.W., Takeda, S., Rotenberg, E., Matsuda, I., Horikoshi, K., Schaefer, J., Lee, C.M., Kevan, S.D., Ohta, T., Nagao, T., Hasegawa, S.: Instability and charge density wave of metallic quantum chains on a silicon surface. Phys. Rev. Lett. **82**, 4898 (1999)
9. Ragan, R., Ohlberg, D., Blackstock, J.J., Kim, S., Williams, R.S.: Atomic surface structure of UHV-prepared template- stripped platinum and single-crystal platinum (111). J. Phys. Chem. B **108**, 20187 (2004)
10. Lee, S., Park, J., Ragan, R., Kim, S., Lee, Z., Lim, D.K., Ohlberg, D.A.A., Williams, R.S.: Self-assembled monolayers on Pt(111): molecular packing structure and strain effects observed by scanning tunneling microscopy. J. Am. Chem. Soc. **128**, 5745 (2006)
11. Shinde, A., Cao, J.X., Lee, S.Y., Wu, R.Q., Ragan, R.: An atomistic view of structural and electronic properties of rare earth ensembles on Si(001) substrates. Chem. Phys. Lett. **466**, 159 (2008)
12. He, T., Ding, H.J., Peor, N., Lu, M., Corley, D.A., Chen, B., Ofir, Y., Gao, Y.L., Yitzchaik, S., Tour, J.M.: Silicon/molecule interfacial electronic modifications. J. Am. Chem. Soc. **130**, 1699 (2008)
13. Lee, S., Shinde, A., Ragan, R.: Morphological work function dependence of rare-earth disilicide metal nanostructures. Nanotechnology **20**, 6 (2009)
14. Rosenwaks, Y., Shikler, R., Glatzel, T., Sadewasser, S.: Kelvin probe force microscopy of semiconductor surface defects. Phys. Rev. B **70**, 085320 (2004)
15. Kimmel, G.A., Goodstein, D.M., Levine, Z.H., Cooper, B.H.: Local adsorbate-induced effects on dynamic charge-transfer in ion-surface interactions. Phys. Rev. B **43**, 9403 (1991)
16. Liu, G.F., Sroubek, Z., Yarmoff, J.A.: Detection of quantum confined states in Au nanoclusters by alkali ion scattering. Phys. Rev. Lett. **92**, 216801 (2004)
17. Patolsky, F., Lieber, C.M.: Nanowire nanosensor. Mater. Today **8**, 20 (2005)
18. Kolmakov, A., Moskovits, M.: Chemical sensing and catalysis by one-dimensional metal-oxide nanostructures. Annu. Rev.Mater. Res. **34**, 152 (2005)
19. Li, Z., Chen, Y., Li, X., Kamins, T.I., Nauka, K., Williams, R.S.: Sequence-specific label-free DNA sensors based on silicon nanowires. Nano Lett. **4**, 245 (2004)
20. Quitoriano, N.J., Kamins, T.I.: Integratable nanowire transistors. Nano Lett. **8**, 4410 (2008)
21. Quitoriano, N.J., Wu, W., Kamins, T.I.: Guiding vapor-liquid-solid nanowire growth using SiO_2. Nanotechnology **20**, 145303 (2009)

22. Min, B.K., Friend, C.M.: Heterogeneous gold-based catalysis for green chemistry: low-temperature CO oxidation and propene oxidation. Chem. Rev. **107**, 2709 (2007)
23. Greeley, J., Norskov, J.K., Mavrikakis, M.: Electronic structure and catalysis on metal surfaces. Annu. Rev. Phys. Chem. **53**, 319 (2002)
24. Valden, M., Lai, X., Goodman, D.W.: Onset of catalytic activity of gold clusters on titania with the appearance of nonmetallic properties. Science **281**, 1647 (1998)
25. Vayenas, C.G., Bebelis, S., Ladas, S.: Dependence of catalytic rates on catalyst work function. Nature **343**, 625 (1990)

Surface Energy and Chemical Potential at Nanoscale

Vanni Lughi
DI3 – Department of Industrial Engineering and Information Technology, University of Trieste, Trieste, Italy

Synonyms

Interfacial energy and chemical potential at nanoscale; Nanothermodynamics; Surface energy density and chemical potential at nanoscale; Surface free energy and chemical potential at nanoscale; Surface tension and chemical potential at nanoscale

Definition

Chemical potential is the energy increment of a system, associated to the addition of one single element of the substance – e.g., one atom or molecule. *Surface energy* is commonly defined as the energy increment associated to the formation of a unit area of a new surface. Although the latter is by far the most common and practical definition of surface energy and serves its purpose in most physical situations, it is subject to a number of ambiguities, which will be sorted out in the following section.

In nanoscale systems, a large portion of the atoms or molecules constituting the system is at or near surfaces. Knowledge of the surface properties, and in particular of surface energy, is therefore a key step in understanding and controlling nanostructures and nanostructured materials. On the other hand, surface geometry affects the chemical potential of the system, as shall be seen below, and this effect becomes most important at the nanoscale. Chemical potential and surface energy are therefore not only key elements when it comes to describing systems at the nanoscale, but they are also somewhat interrelated and are therefore presented here together.

In most of current studies and applications, the subjects of surface energy and chemical potential can be properly discussed within a classical approach – and this is done in most of this entry. The more consistent and modern, but less agile, approach of nanothermodynamics is briefly introduced in the last paragraph.

Fundamental Considerations on Surface Energy

Before turning to the specifics of the subject, it is important to sort out some ambiguities that normally arise when discussing surface energy. In Gibbs' approach, the surface of a solid or liquid is considered as a transition volume between the bulk and its vapor, as depicted in Fig. 1. This model can in principle be generalized to multicomponent systems. All extensive thermodynamic quantities that characterize the system – such as volume, V; number of atomic or molecular constituents, N; internal energy, U; and entropy, S – can therefore be partitioned and a specific portion V_S, N_S, U_S, and S_S is assigned to the "surface volume" (Fig. 1). As these quantities represent extra amounts with respect to the bulk values, they are known as *excess* quantities. By combining these surface excesses, the Gibbs free energy $G_S = U_S - TS_S + PV_S$ *of the surface* and the Helmholtz free energy $F_S = U_S - TS_S$ *of the surface* can be defined – in complete analogy with the bulk. G_S and F_S are also excess quantities. Here T and P are the temperature and the pressure, respectively. Without loss of generality [1, 2], it is possible to assign the excess quantities that originally pertain to the surface volume to a representative, two-dimensional interface of area A (dotted line in Fig. 1). In this case, the surface energies can finally be defined:

$$\text{Surface energy} \qquad u_S = \frac{U_S}{A} \tag{1}$$

$$\text{Gibbs surface free energy} \qquad g_S = \frac{G_S}{A} \tag{2}$$

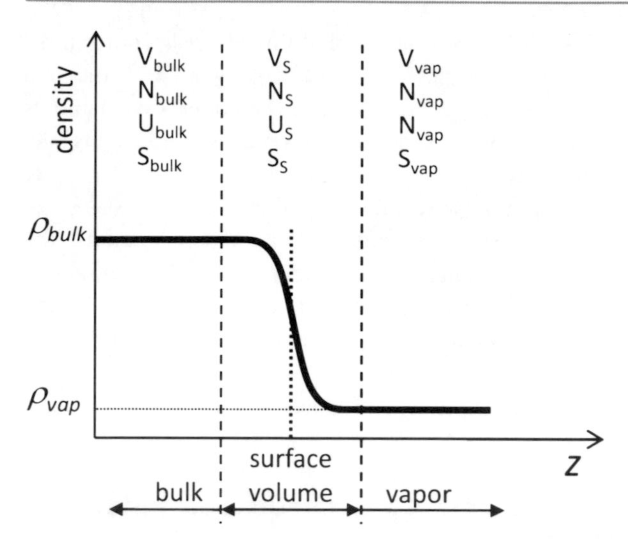

Surface Energy and Chemical Potential at Nanoscale, Fig. 1 Gibbs model of a solid in equilibrium with its vapor (one-component system). The density, ρ, of the component is plotted as a function of the distance normal to the surface, z. The volume partition and the associated extensive quantities are indicated by the *dashed lines*

Helmhotlz surface free energy $$f_S = \frac{F_S}{A} \quad (3)$$

Hence, strictly speaking, surface energy is defined as the internal energy *of the surface* per unit area, and the surface free energy is defined as the free energy (Helmholtz or Gibbs) *of the surface* per unit area, where in this case the "surface" is defined as the transition *volume* between the semi-infinite, bulk phases of the system as illustrated in Fig. 1.

In a more common approach, one can consider the entire free energy, G, of the generic one-component solid system characterized by the presence of a surface [2, 3] (the equation can be generalized to multicomponent systems by adding a $\mu_i N_i$ term for each ith component):

$$G = U - TS + PV = \mu N + \gamma A \quad (4)$$

Here μ and γ are the intensive quantities *chemical potential* and *surface tension*, which are therefore thermodynamically defined as:

$$\gamma = \left.\frac{\partial G}{\partial A}\right|_{T,P,N} \quad (5)$$

$$\mu = \left.\frac{\partial G}{\partial N}\right|_{T,P,A} \quad (6)$$

Adsorption and Relationship Between Surface Tension and Surface Energy

Unfortunately, in most materials science, nanoscience, and nanotechnology literature, the distinction between surface energy, surface free energy, and surface tension is not always maintained. Although strictly speaking γ as defined in Eq. 5 is the surface tension and can be used as such for solids and liquids, the terms "surface energy" and "surface free energy" are also commonly used to identify the same quantity. In general, however, surface tension (Eq. 5) and surface (free) energy (Eqs. 1–3) are different, and the difference depends on the surface excess of species: This can be in general the absorption of foreign species, or for the case of multicomponent systems simply the surface excess per unit area of the ith component of the system $\Gamma_i = N_{S,i}/A$. As shown by Gibbs [1], one finds:

$$\gamma = fs - \sum_i \Gamma_i \mu_{S,i} \quad (7)$$

Equation 7 can be used to relate the concentration of a solute at the surface with a change of the surface energy. In pure one-component systems the summation is nil and $\gamma = f_S$, but this is a rather ideal case. In common systems, it is virtually impossible to obtain perfectly clean surfaces. In nanosystems foreign species are often added on purpose – such as in the case of colloidal nanoparticles, where capping agents are added to avoid the formation of clusters.

In the following, the term "surface energy" will be used as a generic term indicating the relevant one among "surface tension" (Eq. 5) and the different surface free energies defined in Eqs. 1–3. These are all intensive quantities. Throughout this entry, the term "energy *of the surface*" will be used to indicate the relevant one among the extensive quantities γA, $g_S A$, or $f_S A$, where A is the surface area of the system. Note that some authors prefer to use the term "surface (free) energy *density*" when referring to these intensive quantities, in order to avoid confusion with the extensive quantities.

Surface Stress

In the definition of surface energy given above, the implicit mechanism for the formation of a new surface is cleaving of a bulk material. However, the area of an existing surface can be increased by other mechanisms, i.e., by applying a deformation (stretching, distortion). In this case a term needs to be added to Eq. 4:

$$G = U - TS + PV = \mu N + \gamma A + \xi A \tag{8}$$

Here ξ is the surface deformation energy density. For an elastic solid it is $\frac{1}{2}\sum_{i,j}\sigma_{ij}\varepsilon_{ij}$ where σ_{ij} and ε_{ij} are the tensor components of the *surface* stress and strain, respectively, and are reciprocally correlated by the compliance properties of the material. (Note that in this case σ_{ij} and ε_{ij} are defined as surface tensors in complete analogy with the bulk, and have units of force per unit length.) Some mathematical manipulation (see for example [2]) leads to:

$$A\,dy + S_S\,dT + V_S\,dP + N_S\,d\mu + A\sum_{i,j}(\gamma\,\delta_{ij} - \sigma_{ij})d\varepsilon_{ij} = 0 \tag{9}$$

where S_S, V_S, and N_S are the excess quantities as defined previously and δ_{ij} is the Dirac delta function. Equation 9 is one of the key results in surface thermodynamics: On one hand, it can be regarded as a general form of Eq. 7 (some literature reports it as the *Gibbs adsorption equation*); moreover, it shows the relationship between chemical potential, surface energy, and surface stress (note that, thanks to the Gibbs-Duhem equation $SdT - VdP + Nd\mu = 0$, which defines the general relationship between the intensive thermodynamic parameters, of the five variables in Eq. 9 γ, μ, ε, P, T, only three are independent). Finally, from Eq. 9 it can be shown [2] that:

$$\sigma_{i,j} = \gamma\,\delta_{i,j} + \left.\frac{\partial\gamma}{\partial\varepsilon_{ij}}\right|_T \tag{10}$$

Equation 10 shows that surface stress is rigorously equal to surface tension only when there is no dependence of the latter on the surface deformation. This is normally only true in liquids, where the surface atoms can rapidly rearrange in response to a deformation, whereas in solids surface stresses have to be relieved by opportune mechanisms, such as dislocations or buckling of the surface. In solid nanosystems, due to the predominance of surfaces, these effects can have a rather large impact.

Typical Values

When a surface is created, a number of atomic bonds are broken. This consideration enables a practical estimation of the surface energy, since for a solid-vapor interface, the energy per unit area associated to the broken bonds is simply $\frac{1}{2}\varepsilon\,N_{bb}\,\rho_S$ where $\varepsilon/2$ is half of the bond energy, N_{bb} is the number of broken bonds per surface atom, and ρ_S is the planar atomic density at the surface (atoms per unit area). The bond energy can in turn be estimated – quite precisely in some instances such as the case of pure metals – from the latent heat of sublimation: $L_{sub} = \frac{1}{2}\varepsilon\,N_A N_{nn}$ where N_A is Avogadro constant and N_{nn} is the number of atom's nearest neighbors in the bulk [3]. L_{sub} is a known quantity that can be found in thermodynamic tables. The one described here, however, only provides a rough estimate. An entropic term should also be considered, and a number of structural rearrangements of the system, which shall be described in the following section, contribute to the actual, lower value of the surface energy.

Surface energy and surface tension are reported indistinctly with units of $J\ m^{-2}$ or $N\ m^{-1}$, although the latter is often preferred for expressing surface tension in liquids. Common values range from a few tens (in liquids, in polymers, and in ceramics) to a few thousands (in metals) of $mJ\ m^{-2}$.

Surface Energy at the Nanoscale

Nanosystems have very large surface extensions with respect to bulk systems. Based on geometrical considerations, it is easy to show that once the total amount of material is fixed, the total surface area of an ensemble of objects is inversely proportional to the linear size of the object, and so will be the contribution of the energy *of the surface* to the total energy system. For instance, an ensemble of nanoparticles with 10 nm diameter has 10^6 times more surface area (and energy of the surface) than the same amount of material organized in a single 1-cm particle. From a different standpoint, one can observe that the fraction of atoms that lie at the surface increases at the nanoscale. Figure 2 in particular shows a very sharp increase when reducing the size below

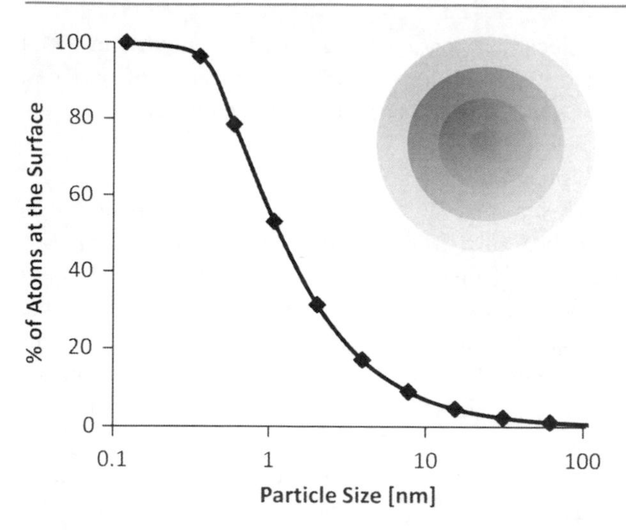

Surface Energy and Chemical Potential at Nanoscale, Fig. 2 Geometrical approximation of the fraction of atoms at the surface of a spherical particle as a function of the sphere diameter. The *curve* is constructed by approximating the number of surface atoms with the volume of concentric spherical shells (*inset*)

~20 nm. Materials properties that are strongly dependent on the energy of the system are expected to undergo dramatic changes in a similar size range. A number of such effects are briefly discussed in the next paragraph dedicated to the chemical potential.

Minimization Mechanisms for the Energy of the Surface: Role in Nanostructures

As part of the total energy minimization process, a system will in general undergo a number of rearrangements at all scales in order to reduce the energy of the surface [4]. In nanosystems, this is particularly important for two reasons: Firstly, because the energy of the surface constitutes a large portion of the total system's energy. Moreover, even small rearrangements that would hardly affect a macroscopic system's morphology and/or properties can strongly impact structures with nanometric size. There are three categories of mechanisms that can contribute to reducing the total energy of the surface in a system, involving: (a) local surface phenomena, which act on reducing the surface energy of the structures; (b) the individual structure; (c) the overall system. These mechanisms have general validity, but will be discussed in the following from the standpoint of nanosystems, for which they are most relevant.

Local Surface Mechanisms. The first and unavoidable mechanism is *surface relaxation*: Atoms at the surface cannot maintain the position they would have in the bulk, as they would be subject to asymmetric forces, and are therefore bound to find a different equilibrium position. This normally results in an inward and/or lateral shifts of the near-surface atomic layers, depending on the crystal symmetry. Whereas in bulk materials the associated reduction of the average lattice constant is essentially negligible, it can be noticeable in some small nanoparticle systems. If there is more than one broken bond per atom, relaxation of the surface can occur by *reconstruction*, where new bonds form at the surface and the geometry of the surface changes. A classic case is the 7 × 7 reconstruction of clean (111) silicon surfaces. If foreign species are available, *physical or chemical adsorption* also reduces surface energy. In atmosphere, essentially all surfaces are covered by adsorbed species, most commonly hydrogen or hydroxyl groups. Finally, changes of the chemical composition at the surface, for instance by *segregation* of impurities, are another effective way to reduce surface energy. This mechanism can have a strong impact on the properties of nanoparticles and nanowires: In these cases segregation at the surface can fully deplete any impurity in the interior of the nanostructure. If on one hand this is one of the reasons for the high purity of nanocrystals, on the other it is a major, fundamental obstacle for achieving effective electronic doping in semiconducting nanostructures.

Individual Structure Mechanisms. An individual structure – e.g., nanoparticles, nanowires, etc. when considering the nanoscale – can reduce the energy of the surface by opportunely changing shape. In isotropic systems, such as liquids or amorphous solids, surface area, A and total energy of the surface, γA, are strictly proportional, therefore minimizing the former also minimizes the latter.

For anisotropic systems, such as crystals, the surface energy varies from one crystal plane to the other. The quantity that needs to be minimized is $\sum_i A_i\gamma_i$ (the index i refers to the different crystallographic planes in the crystal) so that minimization of the area is not in general the optimal solution. This problem is solved by using the *Wulff construction*, which predicts the equilibrium shape of crystals. As illustrated in Fig. 3a, a *vicinal surface* – i.e., a crystal plane forming a small angle θ with the close-packed plane – will have additional broken bonds with respect to a close-packed plane. The number of broken bonds on such a vicinal

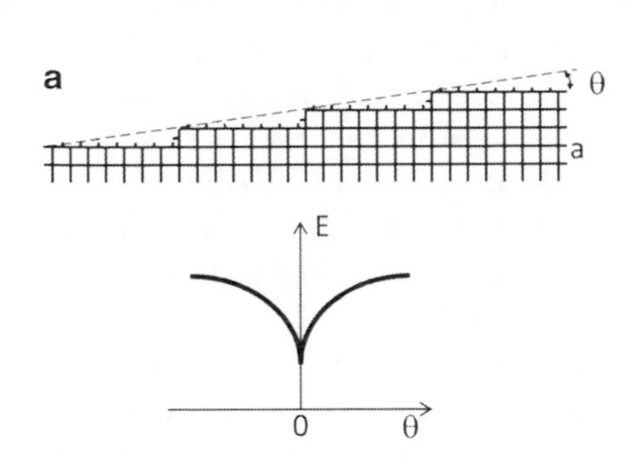

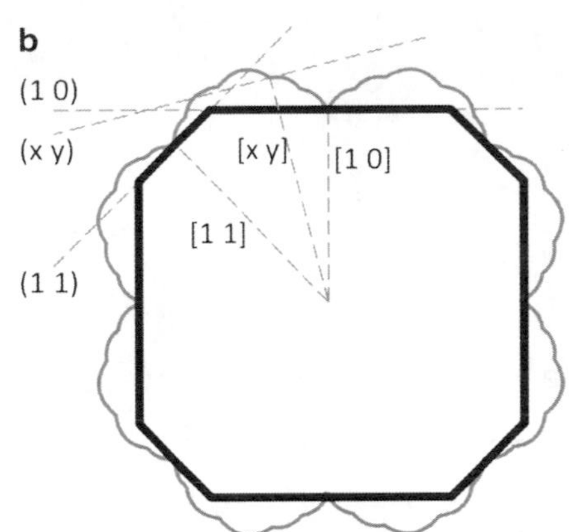

Surface Energy and Chemical Potential at Nanoscale, Fig. 3 (**a**) Schematic cross section of a vicinal surface forming an angle θ with a close-packed plane, and diagram of the surface energy as a function of the angle. (**b**) Example of a two-dimensional polar plot of the surface energy (*gray solid line*). The *dashed lines* indicate the Wulff construction of the planes perpendicular to selected crystallographic directions. The *black solid line* indicates the expected equilibrium shape of the crystal. Note that the generic (xy) does not contribute to the final equilibrium shape because of the high surface energy associated to it

surface can be derived from simple geometrical considerations:

$$N_{bb} = \frac{\cos\theta + \sin|\theta|}{2a^2} \qquad (11)$$

where a is the lattice parameter [2]. In proximity of a close-packed plane the surface energy will then have the shape shown in Fig. 3a, in accordance with Eq. 11: The close-packed plane, characterized in general by small Miller indexes, is associated to a local minimum. The surface energy for all crystal directions can be plotted in a polar diagram, as shown in Fig. 3b for a two-dimensional cross section. One can draw a radius vector for each crystal direction, and then, at the intersection point, the plane that is perpendicular to such vector. Wulff demonstrated that the internal envelope of all such planes defines the equilibrium shape of a crystal, as it minimizes the quantity $\sum_i A_i\gamma_i$. While in macroscopic crystals the shape predicted by the Wulff construction is rarely observed because the kinetics to equilibrium are rather slow at the temperatures of interest, nanosystems often do reach equilibrium due to the short spatial scales involved, and the approach described here can be a useful tool for predicting the shape of nanocrystals.

Mechanisms Involving the Overall System. A system consisting of an ensemble of structures can reduce the overall energy of the surface by simply *aggregating* such structures to one another, thus reducing the surface area. An important case is that of suspensions colloidal particles, where stabilization mechanisms (such as electrostatic charging, or steric stabilization by coating the particles with organic ligands) are needed especially for nanosized particles. Lack of proper stabilization leads to aggregation of the nanoparticles with the formation of large clusters, and the benefits of the colloidal suspension as well as the nanoscale properties of the particles are lost.

An aggregate, e.g., of particles, can further reduce the energy of the surface by *sintering*. During sintering, atoms move toward concave surfaces driven by diffusion mechanisms, which in turn are controlled by the radius of curvature of the target surface: The lower the radius (negative values correspond to concave surfaces), the faster the diffusion toward that area. This results in the progressive merging of adjacent particles, as depicted in Fig. 4. The sources of atoms can be the interior volume of the particles, the surface of the particles, or the boundaries between particles (i.e., as sintering proceeds, the *grain boundaries*). When the source of atoms is the interior volume or a grain boundary, the particles get closer, and densification can occur. If properly conducted, sintering can lead to the formation of a fully dense solid from an aggregate of particles. When the source of atoms is the surface, then no densification can occur, and the final structure will have a high porosity – and therefore still a high surface area and energy of the surface. The dominant source of atoms strongly depends on the sintering conditions and can in

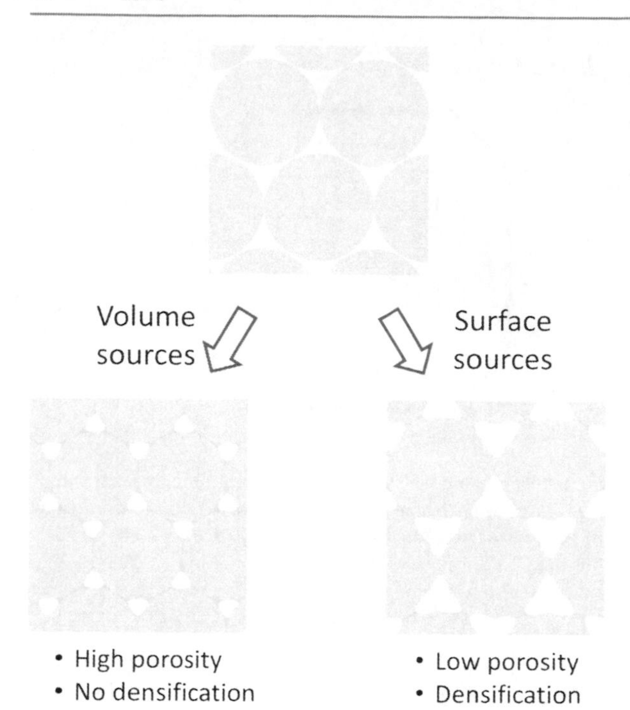

Surface Energy and Chemical Potential at Nanoscale, Fig. 4 Schematic of a sintering process and of the resulting microstructures. (*Left*) If the source of the diffusing species is the volume of the particles or the grain boundaries, densification occurs: The distance between the particle centers decreases, and the porosity is reduced. (*Right*) If the source of the diffusing species is the particle surface, no densification can occur: The material is just redistributed along the surfaces, favoring the necks that are now forming between particles, and the particle centers do not change position

principle be controlled, thus providing a tool for tailoring the material's morphology at the nanoscale and consequently engineering its properties: On one hand, obtaining a fully dense solid with grains of nanoscale size has generated quite a bit of interest because it leads to excellent mechanical properties and, in the case of materials subject to phase transformation, to enhanced phase stability (as illustrated in the following paragraph). On the other hand, obtaining a highly porous material, where the pore size is at the nanoscale, is of interest for a number of situations including: high-performance gas sensors; separation of pollutants in environmental applications; catalysis and photocatalysis, including innovative routes for producing and storing hydrogen for energy applications; scaffolds in biological and biomedical applications; electrodes for fuel cells and batteries; high-performance thermal insulators. However, sintering in nanosystems can also be an unwanted effect. This is especially true when the nanosystem is subject to heat treatment, even at rather low temperatures: The reason is that the diffusion mechanisms that govern sintering become well active at temperatures close to 50–70% of the melting temperature, which in turn drops dramatically at the nanoscale as mentioned in the following paragraph.

Finally, *Ostwald ripening* is another important mechanism where interaction between the elements of the entire system leads to reduction of the overall surface area and of the energy associated to it. In this case, larger particles grow at the expenses of smaller particles. The reason is that atoms at the surface of a small particle are less stable than those at the surface of a larger particle. (More details on this mechanism will be given in the following section.) Ostwald ripening is of major importance when dealing with nanostructures, where small radii of curvature are common. Important examples include sintering processes of nanopowders – in particular densification processes, where the ripening should be avoided as it conflicts with the need of keeping grain size small and uniformly distributed. (See for example the phenomenon of "exaggerated grain growth" typical of ceramic materials processing.) Also, Ostwald ripening is often observed during the synthesis of colloidal nanoparticles – where it can either be deleterious or, if opportunely controlled, lead to a better size distribution.

Chemical Potential at the Nanoscale and Relationship with Surface Energy

The definition of chemical potential was given above as the energy increment of a system associated to the addition of one atom or molecule, and was expressed quantitatively in Eq. 6. One requirement that is explicit in this equation is that the area of the system must remain constant. In general, this is not true, as addition of even one atom or molecule to the material does change, in principle, the surface area. This effect is only important when the size of the system is small. It can be shown [3, 4] that the difference between the chemical potential, μ_r, of a spherical particle of radius r and the chemical potential, μ_∞, of a semi-infinite bulk with a flat surface made of the same material (Fig. 5) is:

$$\mu_r - \mu_\infty = \frac{2\gamma\Omega}{r} \tag{12}$$

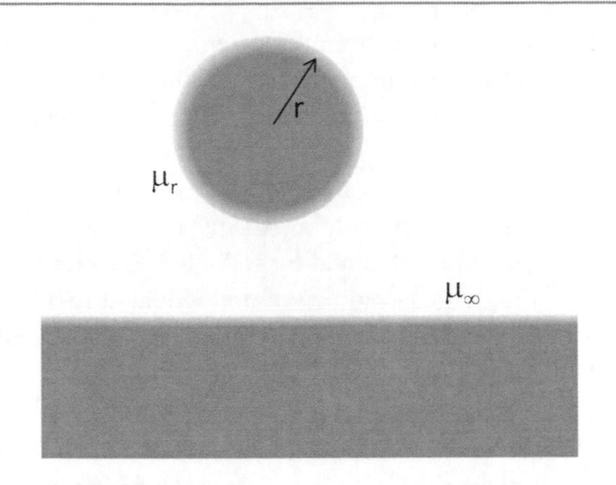

Surface Energy and Chemical Potential at Nanoscale, Fig. 5 Model for calculating the chemical potential difference, $\mu_r - \mu_\infty$, between a particle of radius r and a semi-infinite bulk with a flat surface

where Ω is the atomic volume. This relationship, named after Gibbs and Thomson who first independently derived it, is obtained by realizing that the energy change due to the addition of a single atom to the particle, $\mu_r \partial N$, must be equal to the energy change associated to the change in surface area, $\gamma \partial A$, and mathematically manipulating the balance equation. The term $2\gamma/r$ is known as the Laplace pressure, and it can be shown that the pressure difference, Δp, between the material (solid or fluid) inside and outside of the curved surface is:

$$\Delta p = \frac{2\gamma}{r} \tag{13}$$

Known as the *Young-Laplace equation*, this is also often regarded as a mechanical definition of the surface tension, and is one of the key equations when studying micro- and nanofluidics. Note that, in the case of a solid, a consequence of this relationship is that in a small crystal, there should be a contraction of the lattice parameter, an effect that is actually observed but only for extremely small particles (below about 1 nm). Equations 12 and 13 can be extended to any curved surface, by substituting $2/r$ with the generalized local curvature of the surface, $(1/r_1 + 1/r_2)$, where r_1 and r_2 are the principal radii of curvature.

The Gibbs-Thomson relationship shows the coupling between surface energy and chemical potential, and provides a solid framework for deriving their combined effects on mechanisms and properties of systems, especially at the nanoscale where $\mu_r - \mu_\infty$ can be very large and the effects are most evident. Some important examples are briefly described in the following [4].

First, the Gibbs-Thomson can be extended to the *vapor pressure* of the system. Assuming an ideal gas behavior for the vapor, one can show the chemical potential of an atom in the vapor phase is $\mu_v = \mu_\infty - kT \ln P_\infty$ where P_∞ is the equilibrium vapor pressure of a flat surface, T the temperature, and k Boltzmann's constant; an analogous relationship holds when a curved surface is involved and P_r is the equilibrium vapor pressure of a curved surface. It is therefore possible to calculate $\mu_r - \mu_\infty$ and to use Eq. 12 to find the *Kelvin equation*:

$$\ln \frac{P_r}{P_\infty} = \frac{2}{r} \frac{\gamma \Omega}{kT} \tag{14}$$

The vapor pressure increases as the inverse of the particle (or droplet in the case of a liquid) radius. Once again, the effect becomes sensible at the nanoscale. An identical relationship can be derived for the *solubility* of solids, S, revealing that nanoparticles are much more soluble than the bulk counterpart:

$$\ln \frac{S_r}{S_\infty} = \frac{2}{r} \frac{\gamma \Omega}{kT} \tag{15}$$

The solubility difference between particles of different radius is also strictly correlated with *Ostwald ripening*, a phenomenon that is observed in colloidal suspensions of nanocrystals or during sintering processes of ceramic nanopowders. The driving force for this mechanism is the difference in chemical potential between small and large particles. Considering for example a suspension of particles in a solvent, the equilibrium between precipitate and solute will initially be established when the chemical potentials of the solution, μ_{SOL}, and that of the *smallest* particle, $\mu_{r,small}$, are the same. However, precipitation will still occur locally at the surface of the larger particles, since $\mu_{r,large} < \mu_{r,small}$ and $\mu_{r,small} = \mu_{SOL}$. This precipitation, however, will further reduce the chemical potential in the solution, so now $\mu_{SOL} < \mu_{r,small}$, leading to further dissolution of the smaller particles. The process continues, and smaller particles dissolve while large particles grow.

One of the most important effects of the coupling between size and chemical potential, especially because of its practical consequences in nanotechnology processes, is the dramatic drop of the melting temperature, T_m, with size. For a spherical particle, mathematical manipulation of the Gibbs-Thomson equation leads to the following relationship:

$$\frac{T_m(r)}{T_{m,\infty}} = \left(1 - \frac{2\gamma\Omega}{H_f r}\right) \tag{16}$$

where r is the particle radius, $T_{m,\infty}$ is the melting temperature of the bulk, H_f is the enthalpy of fusion. As mentioned, this has a major effect also on the sintering processes of nanopowders and in general of nanostructured materials.

The size of a system can also influence *thermodynamic phase stability* at a given temperature. This can again be derived by considering that the energy of the surface increases when reducing size. Consider a bulk material that is thermodynamically stable in a phase A at higher temperature and in a phase B at lower temperature. Considering now a spherical particle of radius r, the free energy change associated to the transformation from phase A to phase B must include surface terms as well as bulk terms:

$$\Delta G_{\mathrm{A}\to\mathrm{B}} = G_{\mathrm{B}} - G_{\mathrm{A}} = \left(\frac{4}{3}\pi r^3 \frac{\rho \mathrm{N_A}}{\mathrm{M}} \mu_{\mathrm{B}} + 4\pi r^2 \gamma_{\mathrm{B}}\right) - \left(\frac{4}{3}\pi r^3 \frac{\rho \mathrm{N_A}}{\mathrm{M}} \mu_{\mathrm{A}} + 4\pi r^2 \gamma_{\mathrm{A}}\right) \tag{17}$$

Here the factor $(4\pi r^3/3)(\rho N_{\mathrm{A}}/M)$ multiplying the chemical potential is simply the number N of atomic or molecular constituents in the particle, as defined before; M is the molar mass, ρ the density, and N_{A} the Avogadro number. First, one can observe that $\mu_{\mathrm{B}} < \mu_{\mathrm{A}}$ because in the bulk, below the transformation temperature, A will spontaneously transform to B. Then, by setting $\Delta G_{\mathrm{A}\to\mathrm{B}} = 0$, simple mathematical manipulation leads to the result that, if $\gamma_{\mathrm{B}} > \gamma_{\mathrm{A}}$, there exists a critical radius:

$$\mathrm{r_c} = 3\frac{M}{\rho N_{\mathrm{A}}}\frac{\gamma_{\mathrm{B}} - \gamma_{\mathrm{A}}}{\mu_{\mathrm{A}} - \mu_{\mathrm{B}}} \tag{18}$$

such that if $r < r_c$ then $\Delta G_{\mathrm{A}\to\mathrm{B}} > 0$. In other words, particles smaller than the critical radius are thermodynamically stable, and the transformation from A to B does not occur even if the system is below the bulk transformation temperature. Clearly, this effect is only important when the surface terms in Eq. 17 are of the same magnitude as the volume terms, and this can only happen for small radii. Considering typical values of γ and g, r_c is in the order of a few nanometers. A typical example [5] is the stabilization, obtained by reducing the grain size below the critical radius, of tetragonal zirconia at room temperature – where the thermodynamically stable phase for the bulk would be monoclinic. The critical size for pure zirconia powders is in the range of 5–10 nm, while in a polycrystalline zirconia (where stress effects arise) it is about 30 nm. However, for the practical use of zirconia, chemical stabilizers are commonly added, too, such that stabilization is achieved for larger grain sizes, compatible with common processing.

Nanothermodynamics

Although quite effective in achieving useful and realistic results, the approach outlined above for studying the thermodynamics of small systems – consisting in adding surface-related *ad-hoc* terms to the energy equations – has a certain degree of inconsistency: When dealing with macroscopic systems, energy is an extensive quantity, i.e., proportional to the amount of material that constitutes the system. This does not hold anymore in nanoscopic systems, since a large portion of the total energy of the system is associated to the surface, which is dependent on the size of the objects that constitute the system. Hill has proposed a consistent thermodynamic approach [6, 7], overcoming these inconsistencies.

One can consider a macroscopic system comprising η equivalent and non-interacting, smaller subsystems (a good example would be a colloidal suspension of nanocrystals), characterized by the extensive variables U, S, V, N as defined before. Then, for the entire system $U_{tot} = \eta\ U$; $V_{tot} = \eta\ V$; $S_{tot} = \eta\ S$, $N_{tot} = \eta\ N$. Hill modifies the standard thermodynamic equation for the entire system as:

$$dU_{tot} = TdS_{tot} - PdV_{tot} + \mu dN_{tot} + Ed\eta \tag{19}$$

Essentially, he adds the term $Ed\eta$, where $E = \partial U_{tot}/\partial\eta|_{S_{tot},V_{tot},N_{tot}}$ is the *subdivision potential*. The procedure followed by Hill is analogous to what

Gibbs had done by introducing the chemical potential. The analogy is even stronger, however: While the chemical potential determines how the energy of the system varies when changing the number N_{tot} of atoms or molecules that constitute the system, the subdivision potential determines the energy change upon a variation of the number of subsystems η – which in turn corresponds to a variation of the size of the subsystems if the total amount of material in the system is kept constant. In nanoscopic systems, changes in the number of subsystems imply large variations to the total energy, and the subdivision potential contributes significantly to Eq. 19. However, the subdivision potential will be essentially zero for macroscopic systems, as a change in the number of subsystems will have little effect on the total energy (considering for example the surface energy contribution, this will be negligible because in the surface area variations are small as long as the system size is not in the nanoscale regime). In this case, Eq. 19 reduces to the standard thermodynamic equation, so that in this view nanothermodynamics can be viewed as a generalization of standard thermodynamics. Without recurring to ad-hoc additions to the thermodynamic equations, the subdivision potential enables one to naturally and consistently treat all size-related phenomena, such as surface effects, edge effects, rotation and translation of the systems, etc., thus becoming particularly useful in nanosystems and especially when such phenomena occur together and possibly when coupled. A number of complex nanoscale problems have been successfully treated by using nanothermodynamics, most notably: open linear biological aggregates, local correlations in bulk magnetic materials, glassy systems, and metastable liquid droplets in vapor. More formulations of nanothermodynamics have been formulated over the past couple of decades, with some advantages in terms of the treatment of fluctuations – which might be important in nanosystems – or for the treatment of complex systems, but are essentially equivalent to Hill's.

Cross-References

► Applications of Nanofluidics
► Capillary Flow
► Lotus Effect
► Mechanical Properties of Nanocrystalline Metals
► Nanomechanical Properties of Nanostructures
► Nanoscale Properties of Solid–Liquid Interfaces
► Nanostructures for Surface Functionalization and Surface Properties
► Nanotribology
► Self-Assembled Monolayers for Nanotribology
► Self-Assembly
► Self-Assembly of Nanostructures
► Surface-Modified Microfluidics and Nanofluidics
► Surface Tension Effects of Nanostructures
► Wetting Transitions

References

1. Gibbs, J.W.: Collected Works, vol. 1. Yale University Press, Hew Haven (1948)
2. Zangwill, A.: Physics at Surfaces. Cambridge University Press, New York (1988)
3. Porter, D.A., Easterling, K.E.: Phase Transformations in Metals and Alloys. Van Nostrand Reinhold, Wokingham (1981)
4. Cao, G.: Nanostructures & Nanomaterials: Synthesis, Properties & Applications. Imperial College Press, London (2004)
5. Garvie, R.C.: Stabilization of the tetragonal structure in zirconia microcrystals. J Phys. Chem. **82**, 218 (1978)
6. Hill, T.L.: A different approach to nanothermodynamics. Nano. Lett. **1**, 273–275 (2001)
7. Hill, T.L.: Thermodynamics of Small Systems. Dover, New York (1994)

Surface Energy Density

► Surface Tension Effects of Nanostructures

Surface Energy Density and Chemical Potential at Nanoscale

► Surface Energy and Chemical Potential at Nanoscale

Surface Force Balance

► Surface Forces Apparatus

Surface Forces Apparatus

Carlos Drummond[1] and Marina Ruths[2]
[1]Centre de Recherche Paul Pascal, CNRS–Université Bordeaux 1, Pessac, France
[2]Department of Chemistry, University of Massachusetts Lowell, Lowell, MA, USA

Synonyms

Surface force balance

Definition

The surface forces apparatus (SFA) is an instrument for sensitive measurements of normal and lateral forces between two macroscopic surfaces in contact or separated by a thin film. The surface separation distance can be measured and independently controlled to 0.1 nm. The surfaces typically form a single asperity contact where the substrates deform elastically, and time- and rate-dependent effects in the measured forces can be ascribed to phenomena in the thin films or adsorbed monolayers confined between the surfaces.

Basics of the SFA Technique

The Surfaces and the Apparatus

In SFA experiments, normal and lateral interaction forces are measured between two surfaces across air or a medium (a confined film). The most commonly used surface substrates are back-silvered, molecularly smooth muscovite mica sheets glued to half-cylindrical fused-silica disks. These half-cylindrical surfaces are mounted in the SFA in a crossed-cylinder configuration (cf. Figs. 1 and 2). At surface separations (distances D) much smaller than the radius of curvature (R) of the surfaces, this is equivalent to a sphere-on-flat geometry, which presents several advantages over that of two parallel plates: Alignment issues and edge effects are avoided, and different points of contact between the two surfaces can be easily investigated by displacing the disks laterally (for example, to check the repeatability of measurements at different regions of a pair of surfaces, or to move away from wear debris or contamination). In addition, the normal force F between cylindrical surfaces is related to the interaction energy between flat surfaces W according to the Derjaguin approximation [1], $F(D)/R = 2\pi W$, which provides a convenient mean for comparison of experimental data with model predictions. For this reason, and to allow quantitative comparison of data from different experiments, normal force data are typically presented normalized by R.

Several different SFA setups have been developed for the measurements of normal and lateral forces. Early designs by Tabor, Winterton, and Israelachvili [2] were refined for measurements in liquids and vapors (the Mk II model) [1, 3]. More recent models are easier to assemble and clean (the Mk IV [4]), or have improved distance controls and many different attachments (the Mk III/SFA3 [5] in Figs. 1 and 2, and the SFA2000 [6]). The newer designs are also more user-friendly and have improved stability against thermal and mechanical drifts.

Normal Distance and Force Measurements

The interaction forces normal to the surfaces are determined from the deflection of a spring supporting the lower surface (Fig. 1). Typically, a double cantilever spring is used to minimize tilting and/or sliding of the surfaces. Its spring constant, k, is determined by putting small weights on its end and measuring the resulting deflection with a traveling optical microscope. During an experiment to determine normal force, F, as a function of distance (surface separation), D, the spring deflection is found by monitoring the change in distance with multiple beam interferometry (MBI) [7] as the surfaces are moved toward or away from one another using motorized stages (for moving the base of the spring, the "spring mount" in Fig. 1) or a piezoelectric actuator (for moving the top surface). Any deviation in measured distance from the one expected from a calibration of the movement at large separations (where no forces act between the surfaces) represents a deflection of the spring, ΔD. The change in force when moving from one position to another is $\Delta F = k\Delta D$, according to Hooke's law.

Regions of a force vs. distance curve where the gradient of the force, dF/dD, exceeds k are inaccessible to the technique due to mechanical instability, and the surfaces will spontaneously jump from one stable region to the next. The choice of spring stiffness is

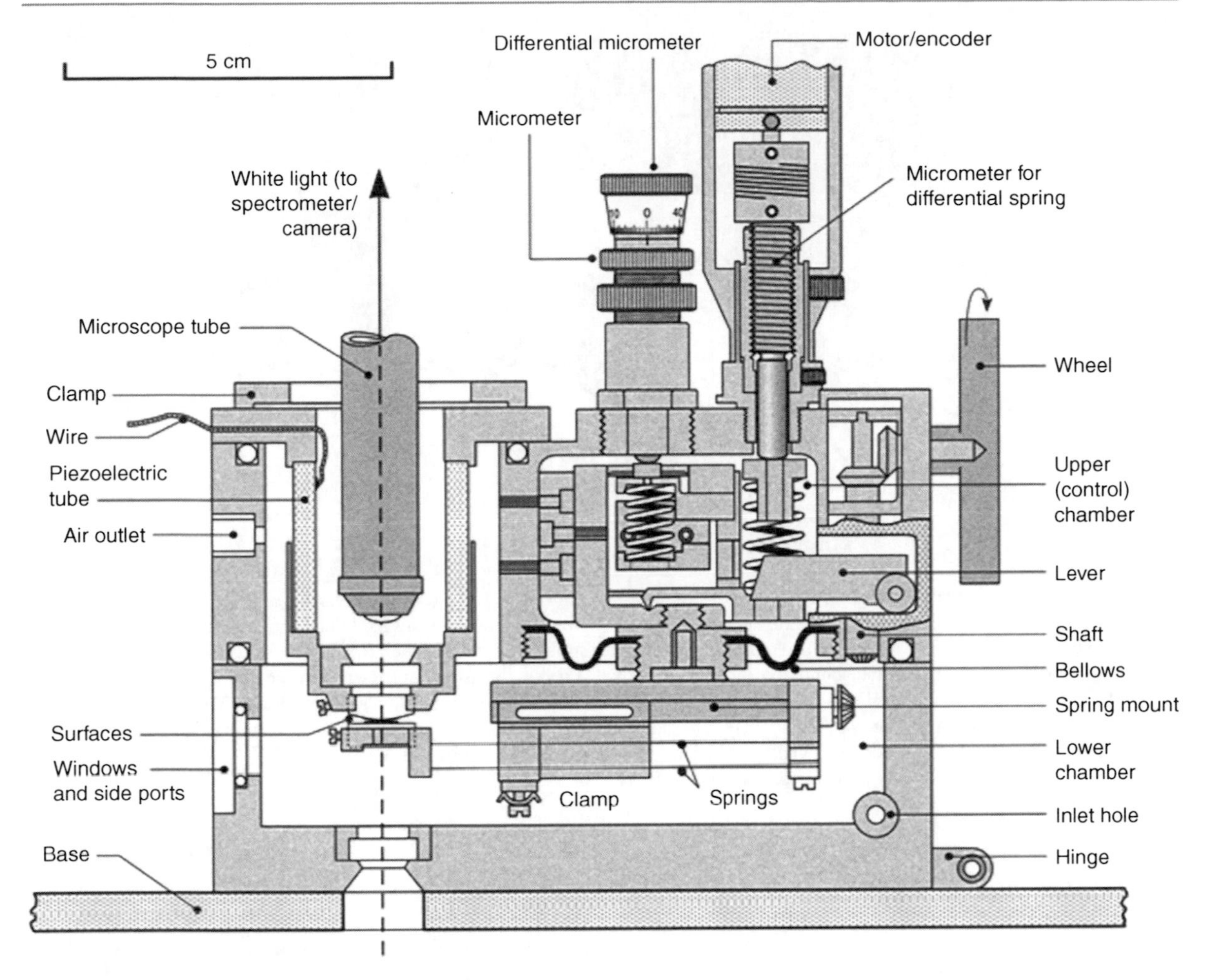

Surface Forces Apparatus, Fig. 1 Schematic drawing of the SFA3 (Mk III) configured for measurements of forces normal to the surfaces [5]. The lower surface is supported on a "force-measuring" double cantilever spring with movable clamping, and the upper surface is mounted on a holder containing a piezoelectric tube for fine control of the separation distance between the surfaces, D. The mechanical distance control system (a system of weak and stiff springs reducing the movement of the manual and motor-driven micrometers) is separated from the main chamber of the instrument by Teflon bellows. The path of the light used for multiple beam interferometry (MBI) is indicated with an arrow (Reprinted with permission from Ref. [2])

therefore of importance for the detection of different regions of the force curve, and several SFA designs incorporate springs whose length can be changed during the experiment (by moving a clamp along the spring, cf. Fig. 1). The use of force feedback to control the force applied to the surfaces independently of the displacement has also been suggested, to maintain a constant deflection of the spring. Implementations of this include the use of a magnetic force transducer and a bimorph deflection sensor [8], or a feedback system utilizing capacitive displacement transducers [9].

Multiple Beam Interferometry (MBI)

The use of multiple beam interferometry (MBI) [7, 10] makes it possible to determine the distance D between the surfaces with subnanometer resolution (0.1 nm), and to measure the refractive index of confined films to 0.01. MBI is a particular useful feature of the standard SFA setup, since it allows real time, in situ monitoring of the geometry of the interacting surfaces and of the region of contact. Parameters such as the film thickness (including the thickness of individual molecular layers) and the diameter or radius of the contact area (to a lateral resolution of about 1 μm) can be readily

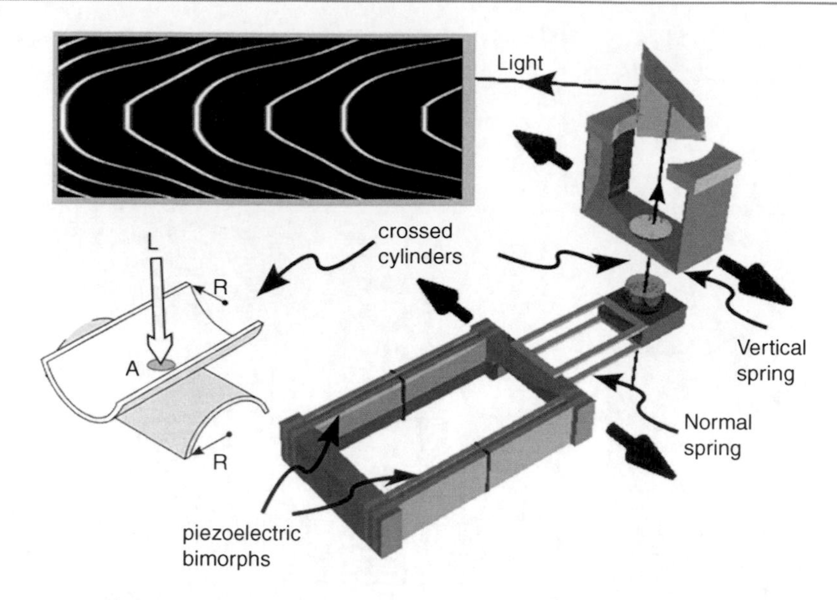

Surface Forces Apparatus, Fig. 2 Schematic illustration of the SFA tribometer design by Israelachvili and coworkers. As in Fig. 1, two back-silvered mica surfaces with a radius of curvature *R* are mounted in a crossed-cylinder configuration. The FECO arising from flattening of these surfaces over a contact area *A* at a given load *L* are shown in the *top* graph. (The alternating shapes of the fringes arise from odd and even orders of interference.) The lower surface is mounted on a double-cantilever "normal spring" (spring constant *k*) attached to a piezoelectric bimorph device allowing lateral movement of the lower surface in the direction of the arrows. The upper surface is mounted on a friction sensing device with vertical springs whose deflection is monitored with resistance strain gauges. The direction of incoming and emerging light is indicated with arrows (Adapted with permission from Ref. [11])

measured. The occurrence of protrusions or regions of different refractive index can also be detected, which may indicate wear or accumulation of material in some region of the contact.

The mica surfaces, any material confined between them, and the semitransparent, highly reflective silver layers on their back sides form a built-in Fabry–Perot interferometer. White light is passed through the lower surface, and the wavelengths that interfere constructively after multiple reflections between the silver mirrors emerge through the top surface. This light is focused onto the slit of a grating spectrometer. The resulting pattern, fringes of equal chromatic order (FECO, cf. top left of Fig. 2), represents the shape and thickness of the space between the silver layers, and depends on the optical path length through the materials in the interferometer (the thicknesses and refractive indexes of the mica sheets and any air gap or material confined between them). A measurement of the interfering wavelengths from surfaces in mica–mica contact in dry N_2 gas is used as the basis for calculation of distances (*D*) between the front sides of the mica surfaces (and thus of film thicknesses of confined materials), since an increase in optical path causes the fringes to shift to larger wavelengths with respect to this value. Relationships between the measured wavelengths, refractive indexes, and *D* for a symmetric interferometer (two mica sheets of equal thickness) [7] have later been extended to more complicated arrangements such as asymmetric, absorbing, anisotropic, or more complicated multilayer systems [10].

SFA-Based Tribometers

Control of Film Thickness and Contact Area

Since the contact geometry in the SFA closely resembles a sphere-on-flat contact and the surfaces can be very smooth (atomically smooth in the case of bare mica surfaces), a single-asperity contact can be formed whose size, position, and deformation is easily monitored using MBI. This contact geometry is also advantageous for comparisons with contact mechanics models [1]. In nanotribological experiments the film thickness and applied load (normal force) are typically regulated with the motor-driven distance controls. If the surfaces adhere or if the load is sufficiently high, the glue layers under the mica sheets deform elastically [1, 2] and a flat, circular contact area of uniform film thickness T and area of contact A is formed, as shown schematically in Fig. 2 (drawing on left). The corresponding FECO (Fig. 2, top left) shows a characteristic flattening indicating a region of uniform film thickness. The diameter of the contact area can be measured directly (to about 1 μm) from this flattened region of the FECO (observations can be made in two orthogonal directions by rotating the light directed to the spectrometer with a dove prism, not shown here). Surfaces with a radius of curvature of $R \approx 2$ cm typically give rise to contacts with a diameter of a few μm to a few hundred micrometers, depending on the load and strength of adhesion. The resulting maximum pressure in the middle of the contact is typically at most a few tens of megapascals.

Any changes in shape of the surfaces or in thickness of the confined film can be observed directly on the FECO and measured with the same resolution as during a measurement of the normal force. Thickening and thinning of the confined film can be observed, as well as shear-induced hydrodynamic deformations. Furthermore, any damage of the surfaces can be easily detected as soon as it occurs, which enables one to identify sliding conditions that lead to wear, and investigate wearless friction versus friction with wear.

Identification and accurate measurements of the contact area are very important for the analysis of nanotribological experiments. In many cases, the measured friction force mainly arises from the region of highest confinement, i.e., from the film in the flattened contact. Certain systems, especially ones where there is adhesion between the surfaces, give rise to a friction force that increases nonlinearly with load and appears to be proportional to this contact area [2]. Unlike many nonadhesive systems, the friction of adhesive contacts is thus not well described by the commonly used friction coefficient, μ (the slope of the friction force as a function of load). They are better characterized by their shear stress, σ, which is the friction force normalized by the contact area. In SFA experiments on a variety of thin films, the shear stress depended on the film structure and thickness and typically remained constant over the investigated range of loads. However, there are situations where the shear stress may vary with pressure, and it is important to recognize that because of the curved surfaces in the SFA, the pressure in the direction normal to the surfaces is not a constant over the flattened contact area, but varies in a manner that can be described by contact mechanics models [1].

One can also envision a more complicated situation where regions of the surfaces outside the flattened area contribute to the friction response. This may occur, for example, if molecules are able to bridge a relatively large gap between the surfaces, and their "bonds" to the surfaces have to be broken and reformed during sliding. One possibility to define the contact area is to adopt a cutoff length, and assume that the contributions to the frictional force are negligible over larger separation distances. Information on the extension of molecules from the surface can be obtained from measurements of the normal force, but it is still difficult to estimate how a region outside the contact will contribute to the total measured friction force. However, this problem is shared by most experimental techniques in nanotribology, and the SFA coupled with the MBI technique provides the most direct way to directly observe the contact geometry during sliding.

Shear and Friction Attachments and Measurements

Several types of shear and friction attachments for the SFA have been developed in the past 30 years, each with its own capabilities and limitations. These setups can be used to investigate different regimes of sliding velocity, magnitude of the friction forces, and sliding distance (amplitude). The most commonly used setups are discussed below. All of these setups have in common that the mechanical properties of the system (i.e., its compliance and inertial mass) will influence the results, and these factors have to be taken into account in order to obtain meaningful information from the measured signals. Because of the mechanical simplicity and easily characterized mechanical properties of the SFA, this can be done in a more straightforward manner than in many other devices used for nanotribological investigations. Detailed descriptions of the devices and of experimental data obtained on different systems can be found in the original publications.

Several devices for lateral shear and friction measurements have been developed by Israelachvili and coworkers [6, 11]. A schematic illustration of a current version is shown in Fig. 2. The lower surface is mounted on a double cantilever spring (of known spring constant k) used to apply and measure the load (normal force) based on expected (calibrated) versus directly measured distance changes. This spring is attached to a "bimorph slider" consisting of sectored piezoelectric elements (electromechanical transducers) [11]. As a constant slope voltage ramp (triangular waveform) is applied over the electrodes of certain sectors of the bimorph elements, the lower surface is translated laterally in a linear manner. The maximum distance (amplitude) of the sliding motion is typically tens of micrometers, depending on the characteristics of the bimorph elements. The driving speed (sliding velocity of the lower surface) can be varied between a few Å/s to about 0.1 mm/s. The bimorph slider can also be used for nanorheological measurements if a constant frequency sinusoidal input is chosen instead of a triangular wave.

The detection of friction forces or viscoelastic responses is done with the device holding the upper surface [6, 11]. This "friction device" consists of a vertical double-cantilever metal spring (with a known spring constant K), whose deflection is measured using strain gauges forming the arms of a Wheatstone bridge. The force experienced by the upper surface due to the movement of the lower can thus be calculated from this spring deflection using Hooke's law. The detection limit depends on the stiffness of the spring and sensitivity of the strain gauges, and is typically a few μN. In cases where larger displacements are desired than the ones available with the bimorph slider, the friction device can be used as both an actuator and detector: The base of the vertical double cantilever can be translated laterally (approx. ±5 mm) using a reversible, variable speed motor-driven micrometer (not shown), and the deflection of the vertical springs recorded simultaneously. A combination of the bimorph slider and the friction device, or the friction device alone, has been used by Israelachvili and coworkers to study the shear or friction response of a wide range of systems such as highly confined simple liquids, polymer melts and solutions, self-assembled surfactant and polymer layers, and liquid crystals (see Refs. [1, 2]).

A different design, used to investigate smaller deformations and mainly the linear response of confined films, has been developed by Granick and coworkers [12]. Small deformations are advantageous for investigations of long-time relaxation processes that might be occurring in the contact region. Although mainly developed for the study of deformations of the order of the film thickness or less, the displacement range goes from less than 1 nm to a few hundred nanometers, with a reported force sensitivity of about 5 μN. This type of device is illustrated schematically in Fig. 3. The load (normal force) is regulated by adjusting the vertical position of the lower surface, and this surface remains stationary in the lateral direction during shear experiments.

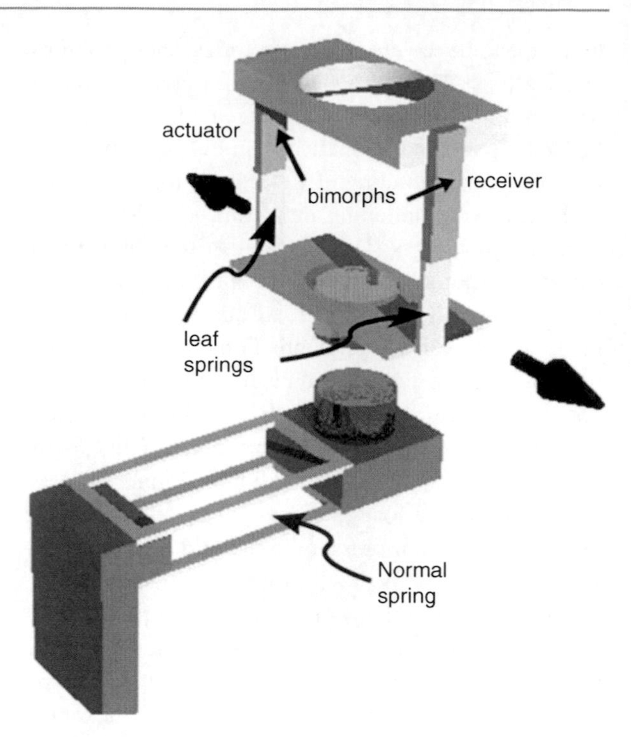

Surface Forces Apparatus, Fig. 3 Schematic illustration of the SFA tribometer design by Granick and coworkers. Bending of one of the bimorph elements (the actuator) causes the upper surface to move laterally with a displacement detected with the other bimorph element (the receiver). The distance (film thickness) and load are controlled by moving the base of the spring holding the lower surface, as in the design in Fig. 1 (Adapted with permission from Ref [12])

The upper surface is mounted on a double-cantilever device. Different from the design in Fig. 2, the double cantilever consists not of metal springs with strain gauges but of metal springs with attached piezoelectric bimorph elements. One element is used as an actuator and the other as a detector (receiver). In a typical experiment, a constant frequency sinusoidal input causes bending of the actuator bimorph. The output voltage from the detector bimorph is used to measure the actual displacement of the upper surface. The viscoelastic properties of the confined film is extracted by comparing these signals to those

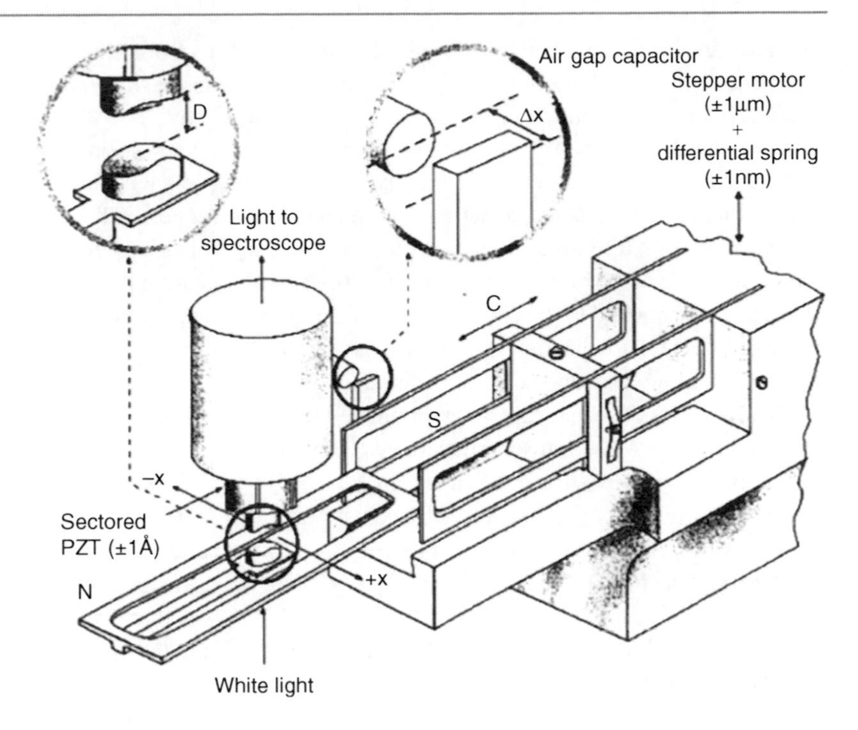

Surface Forces Apparatus, Fig. 4 Schematic drawing of the Surface Force Balance (SFB) designed by Klein and coworkers and configured for nanotribological measurements. The distance between the surfaces and the lateral displacement of the upper surface is controlled with a sectored piezoelectric tube. The deflection of the shear force spring (S) is detected by an air-gap capacitor (Reprinted with permission from Ref. [13])

measured with the surfaces in strongly adhesive (mica–mica) contact and when separated, i.e., when the device is moving freely. The electromechanical characteristics of the system are modeled as a combination of effective masses, springs, and dashpots representing the different components of the apparatus [12]. In practice, the response in mica–mica contact (from the mica and glue layers) is typically purely elastic. This setup has been used by Granick and coworkers to study a variety of systems, including simple liquids, polymer melts, adsorbed layers and solutions.

A third type of shear device, developed by Klein and coworkers [13], is shown schematically in Fig. 4. In this design, the displacement is induced by the upper surface, and the friction force is measured at the lower. The upper surface is attached to a sectored piezoelectric tube that is used to produce a normal or lateral displacement. The lower surface is mounted on a single-cantilever spring (for measurements of normal forces, N in Fig. 4), attached to a double cantilever shear force spring (S in Fig. 4). The lateral displacement of this double-cantilever spring is measured using an air-gap capacitor, and the friction force calculated using Hooke's law. The reported sensitivity of this device is 50 nN and the maximum displacement of the upper surface is a few tens of micrometers. In this device, the sensitivity to the measured friction forces is greatly improved with respect to the designs discussed above. The high sensitivity arises from the detection method, and has proven to be very valuable for investigations by Klein and coworkers of films of water (where the friction can be very low) and simple liquids, as well as polymer melts and solutions.

Recent Technique Developments

Advances in Distance and Force Measurements

Many different improvements to the original techniques and setups have been proposed and implemented in the past decades. Significant steps have been made to automate the normal and lateral force measurement, and noninterferometric techniques have been introduced to measure the separation between the surfaces to enable the use of opaque substrates. Several designs for noninterferometric control and measurements of distance have incorporated piezoelectric bimorph elements [14, 15]. These devices have been used for thin-film viscosity measurements and studies of time- and rate-dependent adhesion in the laboratories of Israelachvili and Granick, respectively. Generally, piezoelectric elements are inadequate for long or quasi-static measurements because of their intrinsic drift and signal decay. The use of an ultrahigh impedance amplifier to lengthen the decay time of the bimorph sensor has to some extent helped overcome this difficulty. An entirely different method of distance detection is based on measuring the capacitance either between the silver layers on the back-side of the mica sheets, or between one plate of a capacitor attached to the chamber of the instrument and one attached to the moving surface [16]. These techniques allow fast and accurate measurements of the displacement of the surfaces and eliminate the constraint of having to use transparent surfaces. However, when opaque surfaces are used, it is not possible to obtain a real-time, in situ image of the contact region during shear, which is one of the major strengths of the standard SFA technique.

Local Structural Information: Combination of SFA with Other Techniques

The data gathered in a conventional SFA experiment is the result of an average response of the confined film to shear and compression. Information on the molecular-level film structure during these processes would add significantly to the interpretation of the measured forces. Obtaining such information is very challenging because of the nature of the confined region: The response arises from a relatively small number of molecules, which implies that any characteristic spectroscopic signal from this region will be of low intensity. The confined film is surrounded by thicker layers of materials that will give rise to spectroscopic signals of their own, typically much stronger than that from the film. Despite these experimental difficulties, interesting observations on film structure and molecular alignment have been made by combining SFA with several different techniques, as briefly described below. Further development of techniques suitable for studying the molecular properties of highly confined films can be expected to significantly improve our understanding of friction phenomena.

Extensions of standard MBI have been used to obtain structural information on the orientation and intermolecular interactions of optically active molecules in confined films [17]. Since the light adsorption of optically active (dye) molecules is enhanced by the multiple reflections in the optical cavity, very thin films can be studied, and local information in the contact area can be obtained.

Combining the SFA with other optical techniques has been limited by the presence of the reflective silver layers needed for the standard MBI approach. These layers strongly reduce the illumination of confined films, limiting the performance of spectrometric techniques. This has been successfully overcome by replacing the silver layers with multilayer dielectric coatings [18], with transparency to different wavelengths, so that information on the structure of ultrathin films under shear can be obtained. For example, the SFA has been combined with fluorescence correlation spectroscopy to measure the molecular diffusion coefficient in thin films within spots of submicron size, obtaining spatially resolved measurements [18]. A drawback is that fluorescent molecules have to be added to the film to be investigated. A reduction in the diffusion coefficient by about two orders of magnitude was found with confinement of films of simple liquids, and the diffusion coefficient was found to decrease when going from the edges toward the center of the contact region. Results have also been reported from combining the SFA with confocal Raman spectroscopy [18], which avoids the problem of the bulk contributing to the scattered signal. Using multilayer reflective coatings, the geometry of the contact area could be monitored simultaneously with the Raman signal, showing the effects of shear on the orientations of molecules within the confined film.

The SFA has also been combined with x-ray diffraction measurements (the XSFA) [19]. Although this technique has thus far been limited to films thicker than 500 nm, the shear-induced alignment of liquid crystal molecules in confined films has been shown. Because of the size of the x-ray beam, the results represent an average over the contact area. Active research in this area is aiming at reducing the investigated region and film thickness, recently indicating that structural information on confined molecular films can be obtained with x-ray reflectivity [20].

The range of shear rates has been extended by incorporating a standard quartz crystal resonator as the lower surface in the SFA [21], giving a high oscillation frequency (MHz). A back-silvered mica piece was attached to the planar quartz crystal, and the top surface was mica glued to a half-spherical disk, so that a sphere-on-flat geometry could be obtained. Standard MBI was used to measure the diameter of the contact area and the radius of curvature of the sphere-on-flat contact geometry.

Mica and Beyond: Modified and Alternative Substrates

Muscovite mica is the most commonly used substrate in SFA experiments, because of its transparency and the relative ease by which large areas of atomically smooth, step-free sheets of uniform thickness (typically 2–5 μm) can be cleaved from larger crystals [3, 7]. The compressibility of mica is quite low, and it is chemically inert, i.e., in its native state it does not expose functional groups to which chemical reactions can occur. On the one hand, this implies that its properties are quite well known from one experiment to the other. On the other hand, a substantial effort is needed to modify mica for different purposes by physisorption or chemical reactions. The substrate plays a large role in the surface phenomena investigated by SFA, particularly in nanotribology, and unmodified (bare) mica is not necessarily representative of many surfaces of engineering or biomedical interest. A large number of procedures have been developed for the purpose of modifying or replacing the mica surfaces with, for example, organic thin films, metals, or metal oxides to get surfaces with different chemical properties and surface energies (cf. ref. [1, 2]).

Important considerations when modifying or replacing the mica substrates are the smoothness of the resulting surface and its transparency to light. An increase in roughness might change the standard single-asperity contact to a more complex, multi-asperity one. However, in many of the approaches described below, the roughness of the deposited layers can be controlled or modified at least to some extent, and in cases where rougher surfaces are formed this allows investigations of the effects of surface roughness on the measured forces, which in itself is an important field of research.

Surface modifications by deposition of thin organic layers (often monolayers) can often be done so that the smoothness of the surface is close to that of the mica substrates themselves. Many different types of surface-active molecules (surfactants, lipids, block copolymers, proteins) have been deposited by self-assembly from solution or by Langmuir–Blodgett deposition [1, 2]. For example, in aqueous solution, the mica surface becomes negatively charged (because of solvation of K^+ ions), and positively charged species spontaneously adsorb onto it. The friction response of the modified surfaces depends strongly on the properties of the adsorbed layers such as their surface energy, morphology, and inherent stiffness [1, 2]. The attachment of organic layers to the mica surface can be enhanced by chemical modification of the mica itself, for example by water vapor plasma treatment. This enables the chemical binding of chlorosilane species to form robust, molecularly smooth, hydrophobic surfaces.

Mica surfaces have also been used as substrates for deposition of various inorganic materials, including metals and dielectrics. In many of these cases, the inertness of the native mica increases the risk of dewetting of the deposited layer and some pretreatment of the mica (for example, plasma treatment, or the deposition of a separate adhesion layer) is necessary.

The analysis of the resulting FECO becomes more complicated because of the larger number of optical layers in the interferometer, but algorithms for this are available [10]. Thin deposited layers of materials such as silver, gold, platinum, silica, alumina, and zirconia with various film thicknesses and smoothness have been investigated [1, 2].

The interface between a supporting mica sheet and a material deposited on it has a roughness similar to that of the mica sheet itself. Because of the inertness of the mica, the bond between the mica and the deposited material is weak and this can be utilized to form so-called template-stripped surfaces for use in the SFA [22]. The material of interest, for example gold, is deposited on a freshly cleaved mica sheet to a desired thickness, its exposed surface is attached to another substrate (for example, by gluing), and the mica sheet is peeled off with tweezers or with adhesive tape to expose a very smooth gold substrate that can be further modified through chemical reactions.

Substrates not based on mica or on a sacrificial mica substrate have also been developed and successfully investigated (cf. Ref. [1] and references therein): Among these, the earliest example is sapphire (aluminum oxide) single crystals grown from the vapor phase. More recently silicon nitride surfaces have been formed by plasma enhanced chemical vapor deposition onto rigid silica disks coated with a reflective layer [23]. A more easily prepared substrate, sheets of silica (quartz glass) or borosilicate glass [24], collected from a bubble with a thickness of a few to 10 μm formed by standard glassblowing techniques, has been used in a couple of studies. These flexible glass substrates, which are quite robust and have a roughness of only a fraction of a nanometer due to the surface tension of the glass when molten, can be silvered on their back-side and used in a similar manner to mica sheets. Surface modifications developed for glass and silica substrates (e.g., chemical bonding of silanes) can be readily applied to these substrates. Thin sheets of polymers have also been successfully prepared by casting and stretching for use in adhesion studies [25].

Cross-References

- ▶ Atomic Force Microscopy
- ▶ Disjoining Pressure and Capillary Adhesion
- ▶ Friction Force Microscopy
- ▶ Nanotribology

References

1. Israelachvili, J.N.: Intermolecular and Surface Forces, 3rd edn. Academic, Amsterdam (2011), and references therein
2. Ruths, M., Israelachvili, J.N.: Surface forces and nanorheology in molecularly thin films. In: Bhushan, B. (ed.) Springer Handbook of Nanotechnology, 3rd edition, pp. 857–922. Springer, Berlin/Heidelberg (2010), and references therein
3. Israelachvili, J.N., Adams, G.E.: Measurements of forces between two mica surfaces in aqueous electrolyte solutions in the range 0–100 nm. J. Chem. Soc., Faraday Trans. I **74**, 975–1001 (1978)
4. Parker, J.L., Christenson, H.K., Ninham, B.W.: Device for measuring the force and separation between two surfaces down to molecular separations. Rev. Sci. Instrum. **60**, 3135–3138 (1989)
5. Israelachvili, J.N., McGuiggan, P.M.: Adhesion and short-range forces between surfaces. Part 1: new apparatus for surface force measurements. J. Mater. Res. **5**, 2223–2231 (1990)
6. Israelachvili, J., Min, Y., Akbulut, M., Alig, A., Carver, C., Greene, W., Kristiansen, K., Meyer, E., Pesika, N., Rosenberg, K., Zeng, H.: Recent advances in the surface forces apparatus (SFA). Rep. Progr. Phys. **73**, 036601 (2010), and references therein
7. Israelachvili, J.N.: Thin film studies using multiple-beam interferometry. J. Colloid Interface Sci. **44**, 259–272 (1973)
8. Stewart, A.M., Parker, J.L.: Force feedback surface force apparatus: principles of operation. Rev. Sci. Instrum. **63**, 5626–5633 (1992)
9. Tonck, A., Georges, J.M., Loubet, J.L.: Measurements of intermolecular forces and the rheology of dodecane between alumina surfaces. J. Colloid Interface Sci. **126**, 150–163 (1988)
10. Heuberger, M.: The extended surface forces apparatus. Part I. Fast spectral correlation interferometry. Rev. Sci. Instrum. **72**, 1700–1707 (2001), and references therein
11. Luengo, G., Schmitt, F.-J., Hill, R., Israelachvili, J.: Thin film rheology and tribology of confined polymer melts: contrasts with bulk properties. Macromolecules **30**, 2482–2494 (1997), and references therein
12. Peachey, J., Van Alsten, J., Granick, S.: Design of an apparatus to measure the shear response of ultrathin liquids films. Rev. Sci. Instrum. **62**, 463–473 (1991), and references therein
13. Raviv, U., Tadmor, R., Klein, J.: Shear and frictional interactions between adsorbed polymer layers in a good solvent. J. Phys. Chem. B **105**, 8125–8134 (2001), and references therein
14. Israelachvili, J.N., Kott, S.J., Fetters, L.J.: Measurements of dynamic interactions in thin films of polymer melts: the transition from simple to complex behavior. J. Polymer Sci. B **27**, 489–502 (1989)
15. Dhinojwala, A., Granick, S.: New approaches to measure interfacial rheology of confined films. J. Chem. Soc., Faraday Trans. **92**, 619–623 (1996)
16. Stewart, A.M.: Capacitance dilatometry attachment for a surface-force apparatus. Meas. Sci. Technol. **11**, 298–304 (2000)
17. Mächtle, P., Müller, C., Helm, C.A.: A thin absorbing layer at the center of a Fabry-Perot interferometer. J. Phys. II **4**, 481–500 (1994)
18. Bae, S.C., Wong, J.S., Kim, M., Jiang, S., Hong, L., Granick, S.: Using light to study boundary lubrication: spectroscopic study of confined films. Phil. Trans. R. Soc. A **366**, 1443–1454 (2008), and references therein
19. Idziak, S.H.J., Koltover, I., Israelachvili, J.N., Safinya, C.R.: Structure in a confined smectic liquid crystal with competing surface and sample elasticities. Phys. Rev. Lett. **76**, 1477–1480 (1996)
20. Seeck, O.H., Kim, H., Lee, D.R., Shu, D., Kaendler, I.D., Basu, J.K., Sinha, S.K.: Observation of thickness quantization in liquid films confined to molecular dimension. Europhys. Lett. **60**, 376–382 (2002)
21. Berg, S., Ruths, M., Johannsmann, D.: High-frequency measurements of interfacial friction using quartz crystal resonators integrated into a surface forces apparatus. Phys. Rev. E **65**, 026119 (2002)
22. Chai, L., Klein, J.: Interaction between molecularly smooth gold and mica surfaces across aqueous solutions. Langmuir **25**, 11533–11540 (2009), and references therein
23. Golan, Y., Alcantar, N.A., Kuhl, T.L., Israelachvili, J.: Generic substrate for the surface forces apparatus: deposition and characterization of silicon nitride surfaces. Langmuir **16**, 6955–6960 (2000)
24. Horn, R.G., Smith, D.T., Haller, W.: Surface forces and viscosity of water measured between silica sheets. Chem. Phys. Lett. **162**, 404–408 (1989)
25. Merrill, W.W., Pocius, A.V., Thakker, B.V., Tirrell, M.: Direct measurement of molecular level adhesion forces between biaxially oriented solid polymer films. Langmuir **7**, 1975–1980 (1991)

Surface Free Energy and Chemical Potential at Nanoscale

▶ Surface Energy and Chemical Potential at Nanoscale

Surface Plasmon Enhanced Optical Bistability and Optical Switching

Weiqiang Mu and John B. Ketterson
Department of Physics and Astronomy, Northwestern University, Evanston, IL, USA

Definition

A surface plasmon is a collective oscillation involving the coupled motion of the electrons in a metal and the associated electromagnetic field in small structures and at interfaces. In small structures (of order a wavelength and smaller), the oscillations are localized, whereas at a planar interface separating a metal and a dielectric a propagating mode exists that decays exponentially on both sides of the interface. Plasmons can be excited by fast electrons or, in the right geometry, an incoming electromagnetic wave. Here, optical excitation is considered. An optically bistable system can have two different but stable outputs for the same input signal over some range. By changing the input beyond some threshold, the output of an optically bistable system can be switched between the two stable states, a feature that can be used to switch the output between two channels.

Optical Bistability and Optical Switching

The development of modern optical communications and optical signal processing requires ultrafast switching of optically encoded information. All-optical switching, which uses light to control light, is a very promising candidate for obtaining high operating speeds. One strategy to achieve all-optical switching is to exploit optical bistability. This phenomenon was first observed in sodium vapor in the 1970s [1]. Soon after that, many materials, gas, liquid, and solid, were shown to produce optical bistability [2–4].

To make a system bistable, both nonlinearity and feedback are required. For an optical system having these two properties, the response function, F, of the system can be defined as

$$F(I_i, I_o) = \frac{I_o}{I_i} \tag{1}$$

where I_i and I_o are respectively the input and output signal. The response function can have various forms, depending on system parameters, which in turn affect the relation between the input and output.

As a simple but important example, consider the effect of the bell-shaped response function shown in Fig. 1a relative to the output signal. Then I_i and I_o are related as

$$I_i = \frac{I_o}{F(I_o)}. \tag{2}$$

Since $F(I_o)$ has a single valued bell-shaped behavior, it is a simple matter to plot I_i using I_o as the independent variable, as shown in Fig 1b. The corresponding behavior of I_o as a function of I_i is then shown in Fig. 1c. The system is clearly nonlinear, but more importantly for a special range of the inputs the output has three possible solutions, as can be seen from Fig. 1c. Those involving a negative slope are unstable and hence the system can be bistable in the region between the two points where the derivative diverges. When continuously varying the input, the output will select one of the stable branches, depending on its history; that is, a kind of hysteresis is achieved as shown in Fig. 1d. As noted, the dashed lines correspond to a region of unstable solutions. The sudden changes and the history dependence of the output can be utilized to make optical switches.

There are many ways to introduce feedback in an optical system. As examples, bistability has been demonstrated in devices based on Fabry–Perot etalons [5], microdisk resonators [6], photonic-crystal cavities [7], etc. The nonlinearities can arise if either refractive index or absorption is intensity dependent. Devices based on the former are said to be dispersive, while the latter are called dissipative. The intensity dependent nonlinearities, which arise from electronic contributions to the third order susceptibility, $\chi^{(3)}$, such as the Kerr effect, can respond in a picosecond or less. While such response times are fast enough for presently envisioned applications, the magnitude of these nonlinearities is very small; very high laser powers are then required to generate the desired optical bistability.

Surface Plasmons

The Dielectric Constant of Metals

When free electrons near a metal surface are electromagnetically excited, the charge fluctuations are typically

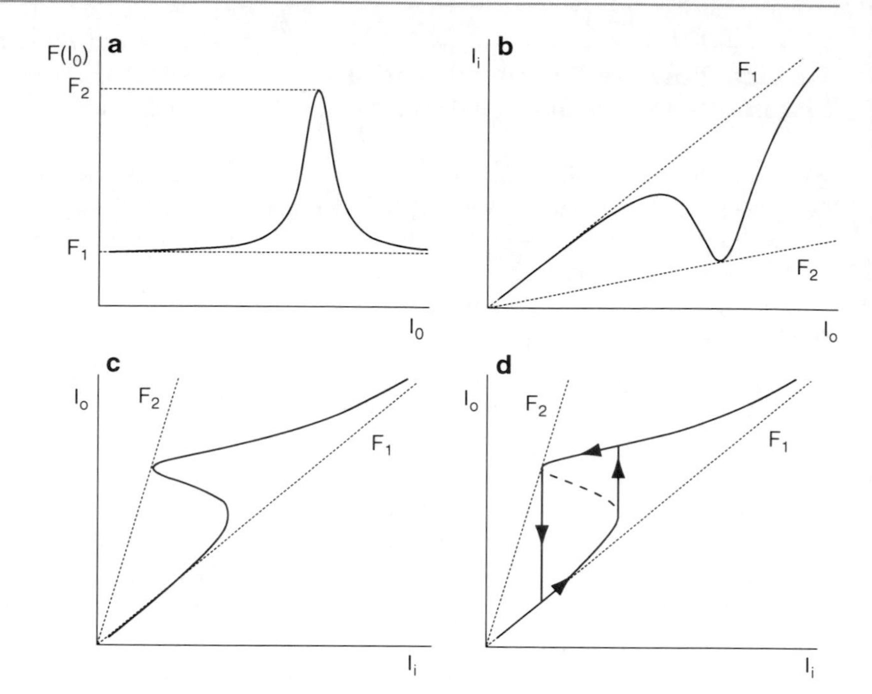

Surface Plasmon Enhanced Optical Bistability and Optical Switching, Fig. 1 (**a**) Typical response curve of an optically bistable system; (**b**) the corresponding I_i relative to I_o behavior; (**c**) the corresponding I_o relative to I_i behavior. (**d**) When I_i continuously increases or decreases, I_o shows a hysteresis; the *dashed line* denotes the unstable region

confined within a Thomas-Fermi screening length of the metal surface (~1 Å). The metals used in plasmonic devices are typically gold, silver, copper, or aluminum. It is common to distinguish two types of surface plasmons: localized surface plasmons (LSP) and propagating surface plasmons (PSP); this primarily depends on the geometry and the distinction is not always sharp. The main difference is that the LSP does not propagate (although, as in the case of a strip resonator, the mode can be considered as counter propagating waves). On the other hand, the PSP transports energy present within the interface. Both originate from the special dielectric properties of metals: a negative real part and a relatively small imaginary part. In the very simplest model the dielectric constant, ε_m, can be described by the Drude-Lorentz form:

$$\varepsilon(\omega) = 1 - \frac{{\omega_p}^2}{\omega^2 + i\omega\gamma} \tag{3}$$

here $\omega_p = 4\pi ne^2/m$ is the electron plasma frequency, with n and m being the electron density and mass respectively, while γ is the collision rate that governs the damping. The dielectric response is clearly strongly frequency dependent. In visible or near infrared, γ is generally much smaller than ω. In this limit, the real and imaginary parts of ε_m can be approximated as

$$\begin{aligned} \varepsilon_r(\omega) &= 1 - \frac{\omega_p^2}{\omega^2} \\ \varepsilon_i(\omega) &= \frac{\gamma\omega_p^2}{\omega^2} \end{aligned} \tag{4}$$

From Eq. (4), it is clear that when $\omega < \omega_p$, the real part, ε_r, is negative.

Equation (3) neglects the host dielectric response, which can be incorporated by replacing the one on the right hand side by a parameter ε_∞; more complicated forms, involving the addition of Lorentzian-like poles, have been used which give a much better overall representation of the measured optical response [8]. Most metals have a plasma frequency in the ultraviolet or visible. The real part of $\varepsilon(\omega)$ for the noble metals gold and silver is negative in the visible and infrared while the imaginary part is small. In addition, they are very stable chemically which makes them the best candidates for plasmonics. Besides metals, some doped semiconductors are good plasmonic materials in the infrared region.

Localized Surface Plasmons

For LSP, the simplest case is a spherical metal particle excited by an optical field $\mathbf{E}_0(\omega)$, where the sphere radius, a, is much smaller than the optical wavelength, λ.

Using a simple electrostatic model the induced dipole moment is given by

$$\vec{p}(\omega) \propto \frac{\varepsilon_m - \varepsilon_d}{\varepsilon_m + 2\varepsilon_d} \vec{E}_0(\omega). \quad (5)$$

Here ε_d is the dielectric constant of the medium. For the model based on Eq. (2), it is easy to see that with $\varepsilon_d = 1$, the real part of the dominator in Eq. (5) will vanish for $\omega = \omega_p/\sqrt{3}$. Similar behavior happens with more complex models. Because of the relative small value of ε_i for Ag and Au, the induced dipole moment can be very large. The scattering from such metal particles is then strong. Clearly, the position of the resonance is determined by the real part of ε_m, but the magnitude and bandwidth are determined by the imaginary part of ε_m.

Propagating Surface Plasmons

As a surface wave, the propagating surface plasmon can only exist as a transverse-magnetic (TM) wave, as shown the inset of Fig. 2. The electric fields in the two media have the form [9]:

$$\begin{aligned} \vec{E}_d &= (E_{dx}, 0, E_{dx}) \cdot \exp\left[i \cdot (k_{sp}x - \omega t) - \gamma_d z\right] \quad (z > 0) \\ \vec{E}_m &= (E_{mx}, 0, E_{mx}) \cdot \exp\left[i \cdot (k_{sp}x - \omega t) + \gamma_m z\right] \quad (z < 0) \end{aligned} \quad (6)$$

where k_{sp} is the surface plasmon wavevector, and γ_d and γ_m satisfy:

$$\gamma_d = \sqrt{k_{sp}^2 - \varepsilon_d k_0^2}; \quad \gamma_m = \sqrt{k_{sp}^2 - \varepsilon_m k_0^2} \quad (7)$$

where $k_0 = \omega/c$ is the free space wavevector. In order that the EM wave decays in both sides of the boundary (a requirement for a surface wave), the real part of γ_m and γ_d must be positive. From the conditions that $E_{//}$ and $D_\perp$ be continuous, yielding $E_{dx} = E_{mx}$, and $\varepsilon_d E_{dz} = \varepsilon_m E_{mz}$, the surface plasmon dispersion relation can be derived:

$$k_{sp}(\omega) = k_0 \sqrt{\frac{\varepsilon(\omega) \cdot \varepsilon_d}{\varepsilon(\omega) + \varepsilon_d}} \quad (8)$$

The necessary condition for a propagating surface plasmon is that the real part of the quotient under the square root sign in Eq. (8) be positive. This imposes a restriction that $\varepsilon_r(\omega) < -\varepsilon_d$. The real part of k_{sp}, is the propagation wavevector of the surface plasmon. The characteristic plasmon propagation distance is the inverse of the imaginary part of k_{sp}. To achieve long propagation distances, it is necessary to have small ε_i or larger $|\varepsilon_r|$. In this entry, silver is selected as the plasmonic material. Figure 2 shows the dispersion relation for the surface plasmons propagating at the interface between silver and fused silica (n = 1.4607 [10]). By fitting the experimental values of the dielectric constant of silver to polynomials, the dielectric constant of silver in visible region can be well represented using the following equation [11]:

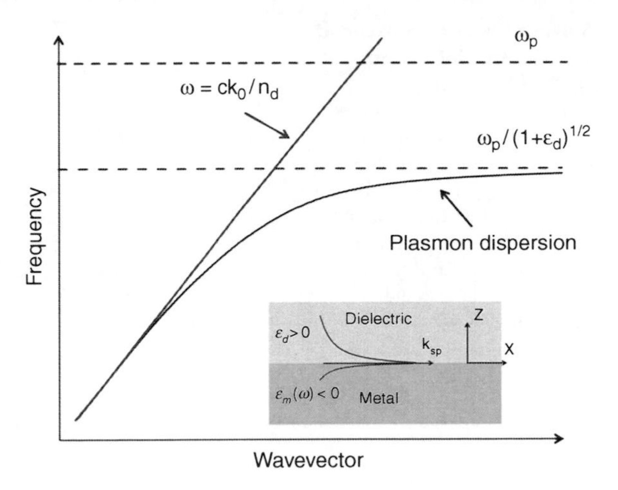

Surface Plasmon Enhanced Optical Bistability and Optical Switching, Fig. 2 Dispersion relation of propagating surface plasmons based on Eqs. (3) and (8) with $\varepsilon_d = 1$. The *red line* is the light dispersion in the dielectric. The plasmon dispersion asymptotically approaches the frequency $\omega_p/(1+\varepsilon_d)^{1/2}$. The inset is the schematic representation of surface plasmon at the metal-dielectric interface. The wave is TM wave, and the fields decay exponentially in both sides of the interface

$$\begin{aligned} \varepsilon_r &= -255.32 + 198.63\omega - 60.79\omega^2 + 8.38\omega^3 - 0.43\omega^4 \\ \varepsilon_i &= 83.26 - 132.79\omega + 90.47\omega^2 - 32.88\omega^3 + 6.66\omega^4 \\ &\quad - 0.71\omega^5 + 0.03\omega^6 \end{aligned} \quad (9)$$

where $\omega = 2\pi \cdot c/\lambda \cdot 10^{-15} S^{-1}$.

Because the PSP wavevector is larger than that in the dielectric, the mode cannot be directly excited from the dielectric. Two popular methods to excite PSP are grating coupling and prism coupling. The grating coupling [12–14] utilizes a diffraction grating to provide

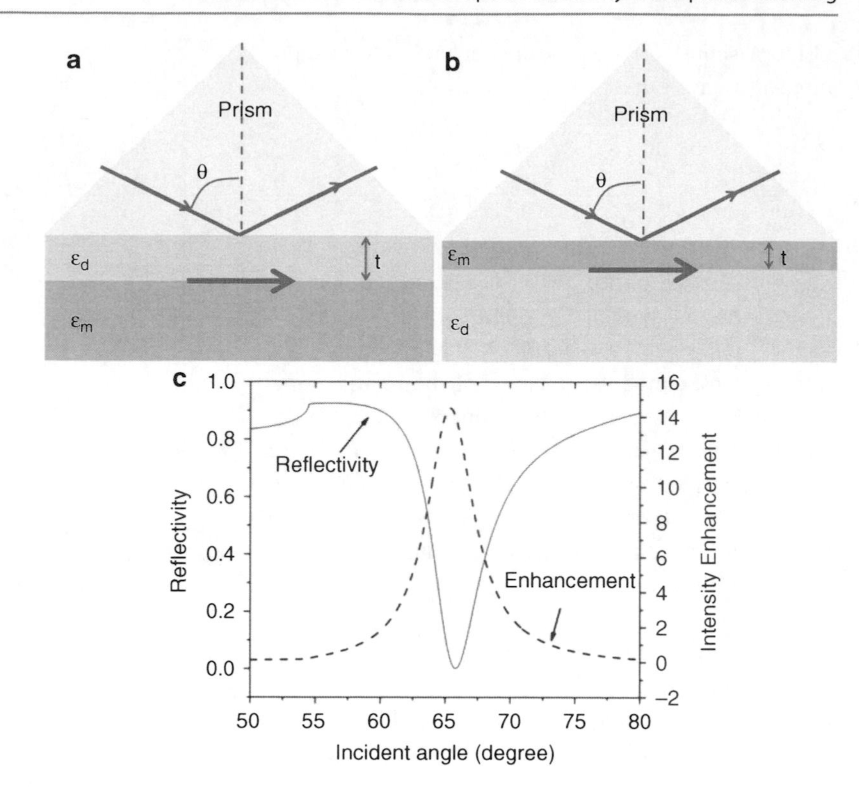

Surface Plasmon Enhanced Optical Bistability and Optical Switching, Fig. 3 (**a**) The Otto geometry and (**b**) the Kretschmann configuration for coupling to a propagating surface plasmon mode with a TM optical wave by total internal reflection. (**c**) The reflectivity and the intensity enhancement of the surface plasmon with Kretschmann configuration

the extra contribution to the wavevector of the photons propagating in the dielectric side to match the real part of k_{sp} (this is called an umklapp process in solid state physics). The grating coupling occurs when

$$k_{//} = k_{sp}^{r} + mK_{g} \qquad m = 0, \pm 1, \pm 2... \qquad (10)$$

where $k_{//}$ is the in-plane wavevector of the incoming wave, k_{sp}^{r} is the real part of the surface plasmon wavevector, and K_g is the grating wavevector. The coupling efficiency depends on the parameters of the grating such as the ditch depth and ditch shape.

The two common prism coupling schemes, the Otto [15] and the Kretschmann [16] geometries (named after their originators), are shown in Fig. 3a, b. For both configurations, the coupling prisms must have a higher refractive index than the dielectric at the metal interface. By correctly choosing the incident beam parameters, such as the incident angle or the wavelength, the in-plane wavevector $k_{//}$ of TM mode incoming beams can be matched to k_{sp}^{r} for the surface plasmon between metal and the dielectric. If the coupling layer between the prism surface and the plasmon interface is not too thick, the evanescent field from the prism surface will excite the metal surface. In the Otto configuration, the coupling from the prism to the metal film is through a layer with dielectric constant smaller than that of the prism, as in Fig. 3a. In the Kretschmann configuration, the coupling occurs through the metal film itself.

When light is coupled into the PSP with prism coupling, the reflectivity, which occurs in a regime that would otherwise correspond to total internal reflection (TIR), will have a dip; this is called *attenuated total internal reflection* (ATR). This results from part of the energy of the incoming beam has been coupled into the PSP wave, and partially from dissipation by the ohmic losses in the metal. For both methods, the coupling efficiency, or the dip depth, depends on the thickness of the coupling layer. If the coupling layer is too thick, only a small portion of the incoming beam is coupled into surface plasmon; this is referred to as undercoupling. On the other hand, if the coupling layer is too thin, a large portion of the excited surface plasmon will be coupled back into the reflection beam and the buildup of the wave is limited; this is called overcoupling. To get the optimum coupling, which occurs where the reflected beam intensity vanishes and all the energy goes into the mode, the coupling layer has to have just the right thickness.

The reflectivity and the intensity enhancement associated with the PSP can be modeled with the multilayer Fresnel reflection method. The intensity enhancement is defined as the ratio of the intensity at the metal-dielectric boundary to the incoming intensity in the prism. Figure 3c shows simulation results with the Kretschmann configuration. In the simulation, the light wavelength was fixed at 532 nm, corresponding to the second harmonic frequency of a Nd:YAG laser, and the incident angle is scanned from 50° to 80°. The refractive indices of prism (SF11) and dielectric material (fused silica) are 1.7948 and 1.4607 respectively. The silver thickness has been optimized as 44.5 nm to obtain the best coupling.

Surface Plasmon–Enhanced Optical Bistability

Surface plasmons can be exploited to significantly enhance optical bistability. First, as shown in the simulation results of Fig. 3c, the local optical field at the interface is greatly enhanced. This reduces the needed laser driving power. Secondly, the high sensitivity of surface plasmons to the change in the refractive index of the adjacent dielectric makes for a fast and sensitive switch. Surface plasmons (in many geometries) can then be utilized to produce optical bistability. In what follows, several approaches are described to achieve plasmon-based bistability.

The first approach that exploits plasmon induced optical bistability utilizes the Kretschmann configuration [17, 18]. If the dielectric material adjacent to the silver surface supporting the plasmons is replaced by a nonlinear Kerr medium, then the refractive index will depend on the local optical intensity. The refractive index n_k of Kerr material is written as (When the surface plasmon is excited, the actual optical intensity in the Kerr medium is exponentially decaying. In the simulations presented here, the value of n_k is taken to be that calculated with the intensity *at the interface* as a simplified model and neglect its change in the Kerr medium, as in Ref. [17]. In Ref. [18], it is shown that if the intensity distribution is considered, the bistability shape will be same, but the required optical intensity will be higher than the value predicted by this simple model)

$$n_k = n_0 + n_2 \cdot I = n_0 + n_2 \cdot A \cdot I_{in}; \qquad (11)$$

here, n_0 and n_2 are the linear and second order nonlinear refractive index of the Kerr material, I the light intensity in the Kerr medium, I_{in} the input optical intensity, and A is the intensity enhancement due to the surface plasmon wave. The Kerr medium examined here is a mixture of nitrobenzene and ethanol. The values of n_0 and n_2 for nitrobenzene are 1.553 and 2.766×10^{-20} $(m/V)^2$ respectively [19]; the refractive index of ethanol is 1.362. By mixing the two liquids in the right ratio, the linear refractive index n_0 can be adjusted to lie between 1.362 and 1.553. The effective n_2 will be assumed to be proportional to the volume percentage of the nitrobenzene.

Using the same geometry as for the simulation of Fig. 3c but with the incident angle fixed at 65.81° (which corresponds to the best coupling in Fig. 3c), and scanning the refractive index of the dielectric medium from 1.3 to 1.65, the corresponding intensity enhancement factor $A(n_k)$ is plotted as the dashed line in Fig. 4a. With the calculated $A(n_k)$, the values of n_k based on Eq. (11) will exhibit bistability relative to the incident intensity, I_{in}. This is graphically depicted in Fig. 4b [20]. Equation (11) is first rewritten as:

$$A(n_k) = \frac{n_k - n_0}{n_2 I_{in}} = \frac{n_k - n_0}{I_d} \qquad (12)$$

here I_d is defined as the dimensionless input intensity. The left side of Eq. (12) $A(n_k)$ is plotted as the black curve in Fig. 4b. With fixed value of n_0, the right side of Eq. (12) is straight line, the slope of which is proportional to the reciprocal of I_{in} (or I_d). The values of n_k are then determined by the intersections of the black curve and the straight line. For the simulation shown in Fig. 4b, the value of n_0 is selected to be 1.38. On the other hand, the resonance, which is the peak position of curve $A(n_k)$, occurs at $n_k = 1.46$. The difference between these two values corresponds to three times the Δn associated with the full width at half maximum (FWHM) of the resonance. Note all of the lines, resulting from different I_{in}, intersect at $n_k = n_0$, as seen in Fig. 4b. As I_{in} increases from zero, the slope of the lines continuously decreases, from $a \rightarrow b \rightarrow c \rightarrow d \rightarrow e$. For the regions with very low and very high input intensity (the region between lines a and b and the region beyond line d), there is only one solution for n_k. Line b and line d correspond to critical points, where there are two solutions to the equation. Between line b and line d, there are three possible values of n_k for the same input, line c being an example. When there are multiple solutions, the

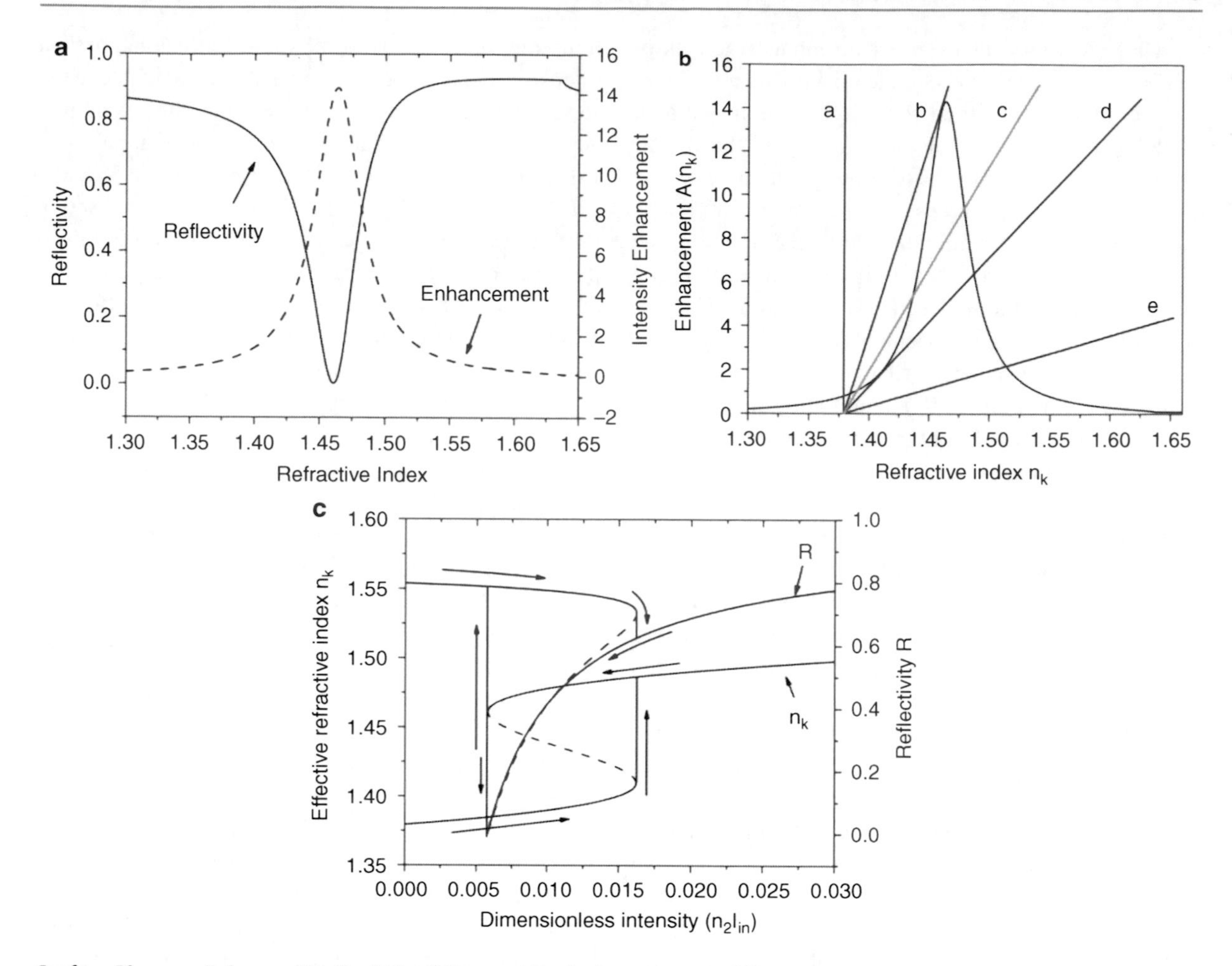

Surface Plasmon Enhanced Optical Bistability and Optical Switching, Fig. 4 (**a**) The reflectivity and intensity enhancement relative to the refractive index of the dielectric in the Kretschmann configuration. (**b**) Graphic solutions for n_k with different optical input intensities. The *black curve* is the intensity enhancement $A(n_k)$ relative to n_k; the different straight lines represent different input intensities; the intersections between the *straight lines* and the *enhancement curve* give the solution for n_k. (**c**) Solutions for n_k and the corresponding values of reflectivity for different input intensities. By continuously changing the input intensity, both n_k and the reflectivity exhibit bistability

physical system selects one, depending on its history. Figure 4c shows the effective index of refraction (n_k) and the reflectivity (R) as a function of the dimensionless input intensity; note the bistability in both n_k and R is clearly presented. The dashed parts of the line are the unstable solutions.

Long-Range Surface Plasmon–Based Optical Bistability

For most of the nonlinear materials, the Kerr coefficient n_2 in Eq. (11) is very small. As in Fig. 4c, to achieve the bistability, the required dimensionless intensity needs to be at least 0.017, corresponding to an input optical intensity on the order of 10^{12} W/cm^2. Therefore, to have a strong optical bistability, it is better to have a sharper resonance, and with it a higher enhancement of the local field. Long-range surface plasmon polaritons (LRSPP), a mode that exists under special conditions, possesses these two characteristics. Of course, a narrower resonance comes at the cost of a longer response time.

If a thin metal film is sandwiched between dielectric materials having the same optical properties, surface plasmon waves will exist on *both* sides of the metal

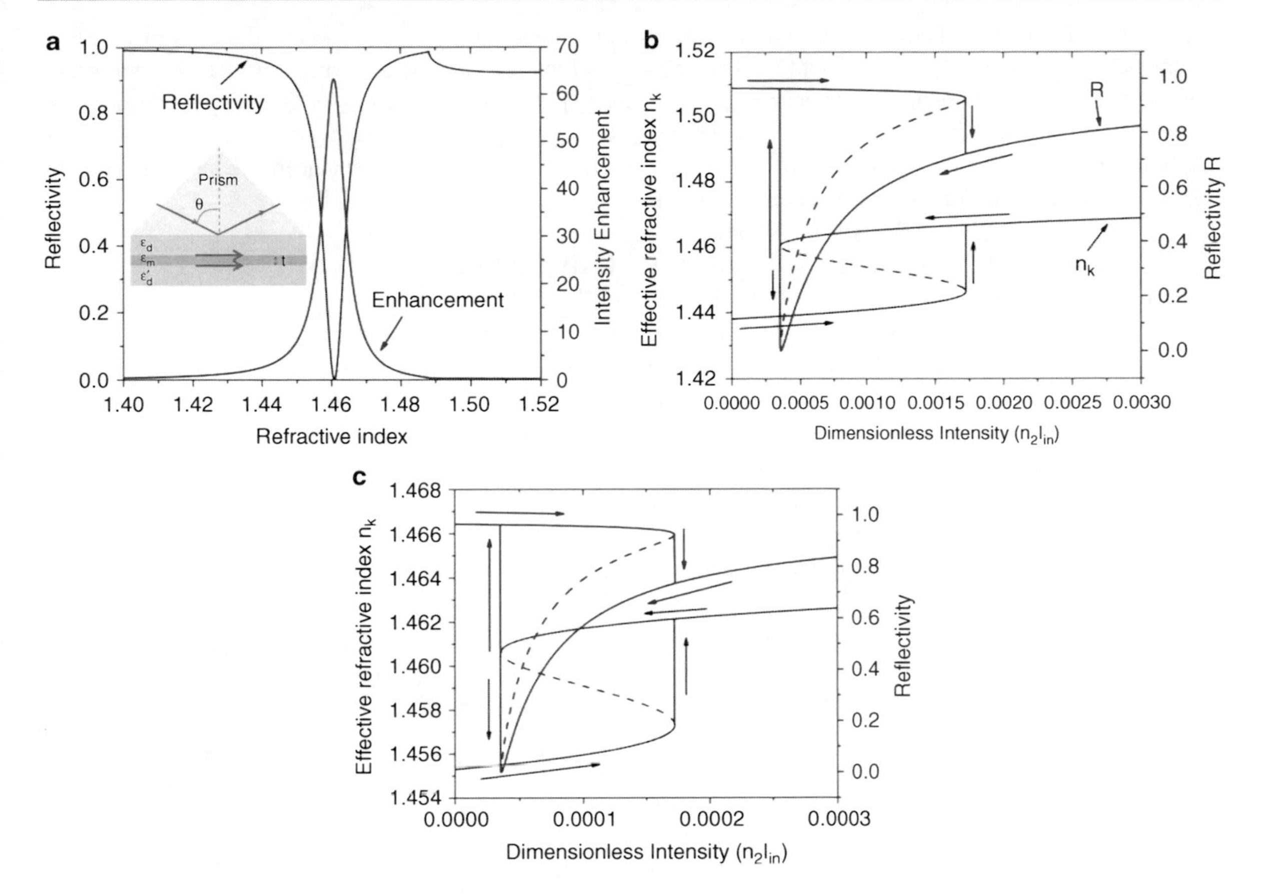

Surface Plasmon Enhanced Optical Bistability and Optical Switching, Fig. 5 (**a**) The reflectivity and intensity enhancement of the LRSPP when it is excited via the ATR method. The *inset* shows the geometry used to excite the LRSPP. (**b**) The bistability of refractive index of the Kerr material and the reflectivity relative to the dimensionless intensity for the case where the silver film thickness is 20 nm. (**c**) The bistability of refractive index of the Kerr material and the reflectivity relative to the dimensionless intensity when silver film thickness is 10 nm

film. When the metal film is thin enough, the two surface plasmons will interact with each other and split into two modes: a symmetric mode and an antisymmetric mode relative to the in-plane electric field at the film center. The electric fields on the two surfaces are in phase for symmetric mode and 180° out of phase for the antisymmetric mode. In a manner similar to that used to obtain the dispersion relation for SPP, and after imposing the boundary conditions, the wavevectors for these two surface plasmon modes are as [21]:

$$\tanh\left(\frac{\gamma_m t}{2}\right) = -\frac{\varepsilon_m \gamma_d}{\varepsilon_d \gamma_m} \quad \text{(anti - symmetric mode)}$$
$$\tanh\left(\frac{\gamma_m t}{2}\right) = -\frac{\varepsilon_d \gamma_m}{\varepsilon_m \gamma_d} \quad \text{(symmetric mode)} \tag{13}$$

where t is the thickness of the metal layer, and $\gamma_i = \sqrt{k_{sp}^2 - \varepsilon_i k_{0i}^2}$. For the symmetric mode, the optical fields constructively interfere inside the metal, which, in turn, results in an increase in the ohmic damping. Therefore, the plasmon lifetime (and with it the decay length) is relatively small; hence it is called a short-range surface plasmon polariton (SRSPP). For the antisymmetric mode, the electric fields from the two surfaces destructively interfere inside the metal, thereby decreasing the ohmic losses. The plasmon lifetime (and the decay length) then become much longer; this mode is called the long-range surface plasmon polariton (LRSPP). As in Eq. (13), the wavevectors of these surface plasmons depend on the metal thickness t. In the thin film limit, the damping of LRSPP is approximately proportional to t^2. Therefore, in principle, by continuously decreasing the metal thickness, we should have very small losses and hence very large local optical intensities.

In order to be able to compare with the optical bistability induced by conventional PSP shown in Fig. 4, the same calculations were repeated for the LRSPPs. The inset of Fig. 5a shows the geometry of the excitation of LRSPP using the ATR coupling method. When the silver thickness is 20 nm, perfect coupling (the absence of reflection) happens when the coupling layer, here a fused silica layer between the silver and the prism, is 555 nm, and the incident angle is 56.02°. Then the last layer of the dielectric is replaced with a material for which the refractive index ranges between 1.40 and 1.52. The reflectivity and the intensity enhancement are the black and blue curves respectively. Clearly, compared with Fig. 4a, the peak intensity enhancement resulting from the LRSPP is much larger, while the width of the bell-shape enhancement is much narrower. To directly compare the enhancement between the LRSPP and conventional PSP, the difference between n_0 and the resonance of intensity enhancement curve A (n_k) in Fig. 5b is still set to be three time the Δn associated with the full width at half maximum (FWHM) of the resonance of A(n_k). Figure 5b shows the resulting bistability of the refractive index n_k for the Kerr medium and the total reflectivity. Clearly, the peak laser intensity needed to produce the optical bistability here is much smaller than the value needed in Fig. 4c, decreasing by one order of magnitude (from 0.017 to 0.0015). By further decreasing the silver film thickness, the needed peak intensity to produce bistability will decreases even more; this is demonstrated in Fig. 5c for the case where the silver film is 10 nm. Here, the optimized coupling layer thickness is 965 nm and coupling angle 54.87°.

Conclusions

The resonant behavior exhibited by conventional propagating surface plasmon polaritons and especially the long-range propagating plasmon polaritons can be exploited to enhance the optical intensity in an adjacent dielectric media. When combined with a nonlinear Kerr medium, one can produce optical bistability and with it the ability to form an optical switch. The effect can be manifested as a change in the reflectivity or a phase delay from plasmons excited in an ATR geometry [22], a change in the scattering intensity from localized surface plasmons [23], a shift in the transmission or reflection coefficient in a grating-coupled surface plasmon geometry [24], and so on. To lower the pumping threshold for bistability, geometries leading to a sharper resonance, and with it a larger intensity enhancement, must be exploited.

Cross-References

- ► Active Plasmonic Devices
- ► Nonlinear Optical Absorption and Induced Thermal Scattering Studies in Organic and Inorganic Nanostructures
- ► Surface Plasmon-Polariton-Based Detectors

References

1. Gibbs, H.M., McCall, S.L., Venkatesan, T.N.C.: Differential gain and bistability using a Sodium-filled Fabry–Perot interferometer. Phys. Rev. Lett. **36**, 1135–1138 (1976)
2. Abraham, E., Smith, S.D.: Optical bistability and related devices. Rep. Prog. Phys. **45**, 815–885 (1982)
3. Gibbs, H.M.: Optical Bistability: Controlling Light with Light. Academic, Orlando (1985)
4. Gibbs, H.M., McCall, S.L., Venkatesan, T.N., Gossard, A.C., Passner, A., Wiegmann, W.: Optical bistability in semiconductors. Appl. Phys. Lett. **36**, 451–453 (1979)
5. Felber, F.S., Marburger, J.H.: Theory of nonresonant multistable optical devices. Appl. Phys. Lett. **28**, 731–733 (1976)
6. Borselli, M., Johnson, T.J., Painter, O.: Beyond the Rayleigh scattering limit in high-Q silicon microdisks, theory and experiment. Opt. Express **13**, 1515–1530 (2005)
7. Notomi, M., Shinya, A., Mitsugi, S., Kira, G., Kuramochi, E., Tanabe, T.: Optical bistable switching action of Si high-Q photonic-crystal nanocavities. Opt. Express **13**, 2678–2687 (2005)
8. Sukharev, M., Sievert, P.R., Seideman, T., Ketterson, J.B.: Perfect coupling of light to surface plasmons with ultra-narrow linewidths. J. Chem. Phys. **131**, 034708 (2009)
9. Rather, H.: Surface Plasmons on Smooth and Rough Surfaces and on Gratings. Springer, Berlin/New York (1988)
10. Palik, E.D.: Handbook of Optical Constants of Solids. Academic, New York (1985)
11. Chen, Z., Hooper, I.R., Sambles, J.R.: Srongly coupled surface plasmons on thin shallow metallic gratings. Phys. Rev. B **77**, 161405 (2008)
12. Mu, W., Buchholz, D.B., Sukharev, M., Jang, J.I., Chang, R.P.H., Ketterson, J.B.: One-dimensional long-range plasmonic-photonic structures. Opt. Lett. **35**, 550–552 (2010)
13. Chen, Y.J., Koteles, E.S., Seymour, R.J., Sonek, G.J., Ballantyne, J.M.: Solid State Commun. **46**, 95 (1983)
14. Gruhlke, R.W., Holland, W.R., Hall, D.G.: Optical emission from coupled surface plasmons. Opt. Lett. **12**, 364–366 (1987)

15. Otto, A.: Excitation of nonradiative surface plasma waves in silver by method of frustrated total reflection. Z. Phys. **216**, 398–410 (1968)
16. Kretschmann, E.: Z. Phys. **241**, 313 (1971)
17. Wysin, G.M., Simon, H.J., Deck, R.T.: Optical bistability with surface plasmon. Opt. Lett. **6**, 30–32 (1981)
18. Hickernell, R.K., Sarid, D.: Optical bistability using prism-coupled, long-range surface plasmons. J. Opt. Soc. Am. B **3**, 1059–1069 (1986)
19. Boyd, R.W.: Nonlinear Optics. Academic, Boston (2003)
20. Saleh, B.E.A., Teich, M.C.: Fundamentals of Photonics. Wiley, New York (1991)
21. Sarid, D.: Long- range surface-plasma waves on very thin metal films. Phys. Rev. Lett. **47**, 1927–1931 (1981)
22. Nazvanov, V.F., Kovalenko, D.I.: Phase optical bistability in structures with surface plasmons. Tech. Phys. Lett. **24**, 650–651 (1998)
23. Leung, K.M.: Optical bistability in the scattering and absorption of light from nonlinear microparticles. Phys. Rev. A **33**, 2461–2464 (1986)
24. Wurtz, G.A., Pollard, R., Zayats, A.V.: Optical bistability in nonlinear surface-plasmon polaritonic crystals. Phys. Rev. Lett. **97**, 057402 (2006)

Surface Plasmon Nanophotonics

► Optical Properties of Metal Nanoparticles

Surface Plasmon-Polariton Photodetectors

► Surface Plasmon-Polariton-Based Detectors

Surface Plasmon-Polariton-Based Detectors

Pierre Berini
School of Information Technology and Engineering (SITE), University of Ottawa, Ottawa, ON, Canada

Synonyms

Surface plasmon-polariton photodetectors

Definition

A surface plasmon polariton is a transverse-magnetic optical surface wave propagating along the interface of a metal and a dielectric; it is a coupled excitation formed from electromagnetic fields coupled to a charge density wave in the metal. A surface plasmon-polariton detector is a device capable of detecting surface plasmons or involving surface plasmons in the detection process.

Notation

An $e^{+j\omega t}$ time-harmonic dependence is assumed. The relative permittivity is denoted ε_r, and is written for a metal in terms of real and imaginary parts as $\varepsilon_{r,m} = -\varepsilon_R - j\varepsilon_I$. $\boldsymbol{k}$ denotes a wavevector (in general). $k_0 = 2\pi/\lambda_0 = \omega/c_0$ is the wavenumber of plane waves in vacuum, λ_0 the wavelength in vacuum, c_0 the speed of light in vacuum, and $\omega = 2\pi\nu$ is the angular frequency.

Single-Interface Surface Plasmon-Polariton

A surface plasmon-polariton (SPP) waveguide is a metallo-dielectric structure along which SPP modes are guided. The simplest structure supporting SPPs is an interface between an optically semi-infinite dielectric and an optically semi-infinite metal [1], as sketched in Fig. 1 (the relative permittivity of the metal is denoted $\varepsilon_{r,m}$, and that of the dielectric $\varepsilon_{r,d}$.) This structure is termed the single-interface.

Highly conductive metals and good dielectrics are used to implement SPP waveguides. Ag is often preferred because it has the lowest optical loss among the metals over a broad wavelength range, but it is reactive, so care must be taken during fabrication and use in order to avoid degradation (which can occur if Ag is exposed to, e.g., air or water). Au is also a good choice given its chemical stability and its good optical performance. A good metal for supporting SPPs normally satisfies $\varepsilon_R >> \varepsilon_I$.

Metals are dispersive at optical wavelengths. Away from interband transitions, the Drude model for the permittivity captures the dispersive character of metals [1]:

$$\begin{aligned}\varepsilon_{r,m} &= -\varepsilon_R - j\varepsilon_I \\ &= 1 - \frac{\omega_p^2}{\omega^2 + 1/\tau^2} - j\frac{\omega_p^2/\tau}{\omega(\omega^2 + 1/\tau^2)}\end{aligned} \quad (1)$$

In the above, ω_p is the plasma frequency and τ the relaxation time. The “Drude region” corresponds to

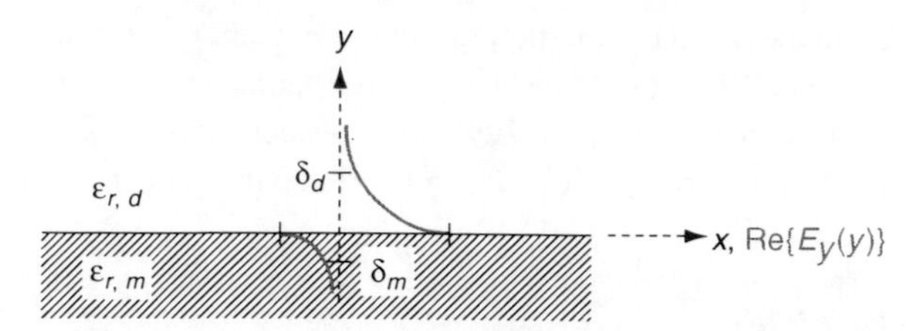

Surface Plasmon-Polariton-Based Detectors, Fig. 1 Single-interface surface plasmon-polariton waveguide; the distribution of the main transverse electric field component (E_y) of the SPP is sketched on the structure as the thick gray curve

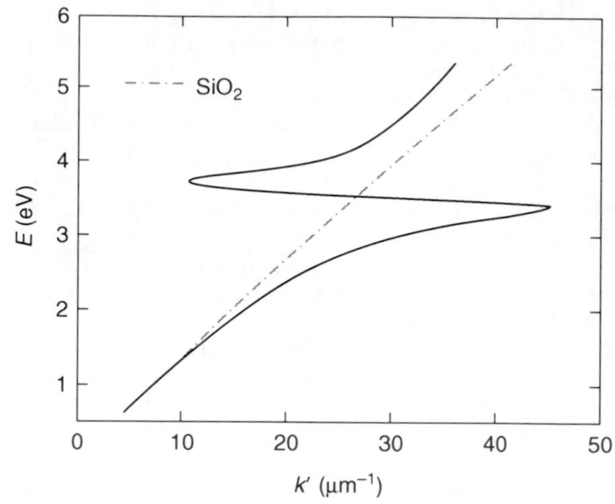

Surface Plasmon-Polariton-Based Detectors, Fig. 2 Dispersion of the SPP along a Ag/SiO_2 interface. The light line in SiO_2 is plotted as the dash-dot curve

that portion of the electromagnetic spectrum where Eq. 1 holds. For many metals, this region spans the range from visible wavelengths to the infrared. The metal approaches a perfect electric conductor as the wavelength increases through the infrared and beyond.

The single-interface (Fig. 1) supports one purely bound (nonradiative) SPP mode. This mode is transverse magnetic (TM) and may propagate at any angle in the x-z plane (e.g., along + z). The SPP fields (E_y, E_z, and H_x) are confined along the y direction, peaking at the interface and decaying exponentially into both media. The distribution of the main transverse electric field component (E_y) of the SPP is sketched on the structure of Fig. 1 as the thick gray curve. The field penetration depth in the metal δ_m is much smaller than the field penetration depth in the dielectric δ_d. Confinement of the SPP arises because the metal and dielectric have Re$\{\varepsilon_r\}$ of opposite sign at the wavelength of operation (indeed over a large wavelength range).

The wavenumber of the single-interface SPP is [1]:

$$k^{\mathrm{SPP}} = k_0 \left(\frac{\varepsilon_{r,m}\varepsilon_{r,d}}{\varepsilon_{r,m} + \varepsilon_{r,d}} \right)^{\frac{1}{2}} \tag{2}$$

For a lossless dielectric cladding (Im$\{\varepsilon_{r,d}\} = 0$), the above simplifies to the following approximate expressions for the real (phase) and imaginary (attenuation) parts of k^{SPP} [1]:

$$k' \cong k_0 \left(\frac{\varepsilon_R \varepsilon_{r,d}}{\varepsilon_R - \varepsilon_{r,d}} \right)^{\frac{1}{2}} \text{ and}$$
$$k'' \cong k_0 \frac{\varepsilon_I}{2\varepsilon_R^2} \left(\frac{\varepsilon_{r,d}\varepsilon_R}{\varepsilon_R - \varepsilon_{r,d}} \right)^{\frac{3}{2}} \tag{3}$$

Figure 2 plots the dispersion curve of the SPP ($E = hv$ in eV versus k', h is Planck's constant) on a Ag/SiO_2 interface [2]; the light line in SiO_2 is also plotted as the dash-dot curve. As the frequency decreases (increasing wavelength), the metal approaches a PEC, the confinement of the SPP decreases, and its dispersion curve merges with the light line (SPPs are not supported at the interface between a smooth PEC and a semi-infinite dielectric). As the frequency increases, the SPP approaches an "energy asymptote," readily observed in Fig. 2 near $E \sim 3.4$ eV ($\lambda_0 \sim 360$ nm) where k' diverges. The group velocity decreases and the optical density-of-states increases as the asymptote is approached. The SPP bend-back is also observed for wavelengths shorter than the asymptote ($\lambda_0 < 360$ nm, $E > 3.4$ eV) and links the nonradiative SPP on the right of the light line to the radiative one on the left.

The nonradiative SPP is located to the right of the light line (Fig. 2) and so cannot be directly excited by an incident beam. An additional structure is required to increase the in-plane momentum of the beam to match that of the SPP. A prism or a corrugated grating is commonly used to accomplish this task; alternatively, the SPP can be excited by end-fire coupling [1].

Detection Mechanisms

Figure 3a gives a schematic of a conventional Schottky photodiode on n-type Si (n-Si) [3], adopted here to

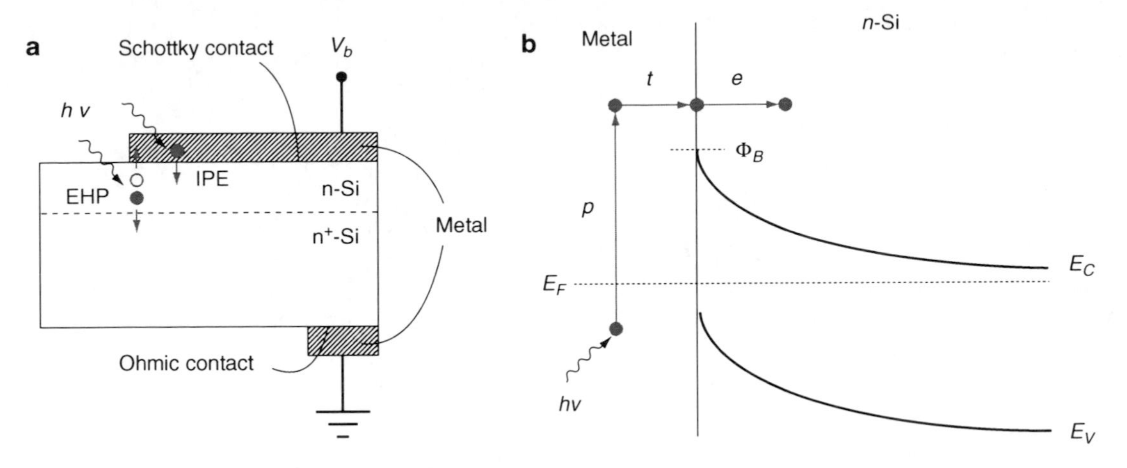

Surface Plasmon-Polariton-Based Detectors, Fig. 3 (**a**) Schottky diode on n-Si illuminated by light of photon energy $h\nu$. Reverse biasing ($V_b < 0$) is assumed. Electrons are depicted by filled circles and holes by unfilled ones. (**b**) Energy band diagram of a Schottky contact on n-Si and the three-step internal photoemission process: p photoexcitation, t transport, e emission. E_C and E_V are the conduction and valence band edges, respectively, E_F is the Fermi level, and Φ_B is the Schottky barrier height

describe generic detection mechanisms that can also be used for the detection of SPPs. A rectifying Schottky contact is formed at the abrupt interface between the top metal and the semiconductor, and an Ohmic (non-rectifying) contact is formed at the interface between the bottom metal and the heavily doped body. The structure is reverse biased (for detection) by applying $V_b < 0$. The complementary structure, obtained by exchanging the dopants ($n \leftrightarrow p$, $n^+ \leftrightarrow p^+$), is also of interest. The Schottky diode is a convenient structure with which to detect SPPs because SPPs may propagate along its metal surfaces.

Figure 3a shows two mechanisms used for photodetection. The first consists of the creation of electron–hole pairs (EHPs) in the semiconductor due to absorption therein of incident radiation of energy $h\nu$ greater than the bandgap energy of the semiconductor E_g. This mechanism, sketched in Fig. 3a and labeled EHP, involves three steps: optical absorption and creation of EHPs, separation of EHPs and transport across the absorption region (with or without gain) under reverse bias, and collection of EHPs into the photocurrent at the device contacts.

The second mechanism consists of the internal photoemission (IPE) of hot carriers created in the metal due to absorption therein of incident radiation of energy $h\nu$. This mechanism is sketched in Fig. 3a, labeled IPE, and is described in greater detail via the energy band diagram of Fig. 3b. IPE is a 3-step process consisting of the photoexcitation of hot (energetic) carriers in the metal by optical absorption (p), the transport and scattering of hot carriers toward the Schottky contact (t), and the emission of hot carriers over the Schottky barrier into the semiconductor (e) where they are collected under a reverse bias as the photocurrent. This mechanism requires that $h\nu$ be greater than the Schottky barrier energy Φ_B, and is useful when $\Phi_B < h\nu < E_g$; i.e., for detection at energies below the bandgap of the semiconductor (because the creation of EHPs in the semiconductor is more efficient and dominates over IPE when $h\nu > E_g$).

The internal quantum efficiency η_i (internal photoyield) is a useful measure to characterize and optimize the detection mechanism. It is defined as the number of carriers that contribute to the photocurrent I_p per absorbed photon per second:

$$\eta_i = \frac{I_p/q}{S_{abs}/h\nu} \tag{4}$$

where S_{abs} is the absorbed optical power and q is the elemental charge. $\eta_i \sim 1$ for detection via the creation of EHPs in high-quality (defect-free) direct bandgap semiconductors (assuming absorption in the semiconductor only). In the case of detection via IPE, assuming absorption in the metal only near the Schottky contact, η_i is given approximately by [4]:

$$\eta_i = \frac{1}{2}\left(1 - \sqrt{\frac{\Phi_B}{h\nu}}\right)^2 \quad (5)$$

Typical Schottky barrier heights are $\Phi_B = 0.34$, 0.8, 0.58, and 0.72 eV for Au/p-Si, Au/n-Si, Al/p-Si, and Al/n-Si, respectively [3]; for detection at, e.g., $\lambda_0 = 1{,}310$ nm η_i ranges from about 0.3% to 9%. Metal silicides can also be used, providing lower Schottky barriers.

The external quantum efficiency η_e (external photoyield) and the responsivity R describe how well the detector performs when inserted into a system. η_e is defined similarly to η_i except that it depends on the incident optical power S_{inc}:

$$\eta_e = \frac{I_p/q}{S_{inc}/h\nu} \quad (6)$$

η_e and η_i are related by:

$$\eta_e = A\eta_i \quad (7)$$

where A is the optical absorptance, defined as:

$$A = \frac{S_{abs}}{S_{inc}} \quad (8)$$

The responsivity R is given by the ratio of the photocurrent to the incident optical power, and can be expressed in terms of η_e and η_i:

$$R = \frac{I_p}{S_{inc}} = \frac{\eta_e q}{h\nu} = \frac{A\eta_i q}{h\nu} \quad (9)$$

SPP detectors, or SPP-enhanced detectors, typically combine a metallic structure that supports SPPs with a detector structure such as that shown in Fig. 3.

Grating-Coupled Detectors

An SPP detector structure of interest combines a Schottky detector with a grating designed to couple incoming optical waves to SPPs, as shown in Fig. 4 [5]. The structure consists of a top Al electrode on a thin

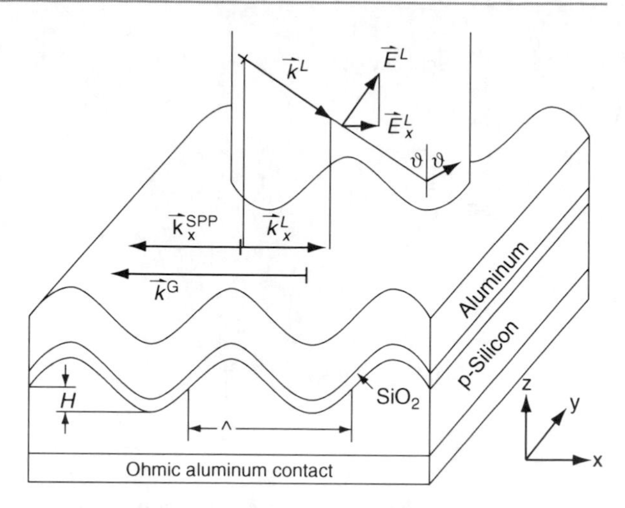

Surface Plasmon-Polariton-Based Detectors, Fig. 4 SPP Schottky detector integrated with a grating coupler to excite x-propagating SPPs along the top surface of the top Al contact (Adapted from [5]. © 1989 Optical Society of America.)

SiO_2 tunneling barrier on p-Si, formed as a sinusoidal corrugated grating of height H, period Λ, and grating vector $\boldsymbol{k^G} = \boldsymbol{x}2\pi/\Lambda$ ($\boldsymbol{x}$ is the x-directed unit vector). p-polarized light having a wavevector $\boldsymbol{k^L}$ and an electric field $\boldsymbol{E^L}$ is incident onto the grating at the angle ϑ. The incident light couples to the SPP propagating along the air/Al interface in the x direction when the following momentum conservation equation is satisfied:

$$k_x^L + mk^G = k_x^{SPP} \quad (10)$$

where $k_x{}^L$ is the x-directed component of $\boldsymbol{k^L}$, m is an integer, and $k_x{}^{SPP}$ is the SPP wavenumber (following the notation of [5]). Equation 10 holds for shallow gratings – H is a small perturbation to the surface. The Al Schottky contact is thin enough (35 nm) for the SPP to tunnel through and leak into the high refractive index p-Si layer for detection via the creation of EHPs at the wavelength of operation ($\lambda_0 = 646$ nm).

Grating-coupled SPP detectors are sensitive to the angle of incidence, polarization, and wavelength of the incident light through Eq. 10. The polarization sensitivity, for instance, can be exploited in an arrangement of two detectors to determine the polarization angle of linearly polarized normally incident ($\vartheta = 0$) light [5].

Grating-coupled SPPs have also been used to increase the absorptance in the metal contact of Schottky detectors based on IPE (for sub-bandgap detection) leading to increased responsivity. For example, a 30× increase in responsivity was observed for a Au/p-InP Schottky detector by exciting SPPs along the air/Au interface at $\lambda_0 = 1150$ nm via an integrated corrugated grating, relative to the same structure without the grating [6].

Prism coupling, in the Otto arrangement [1] for instance, has also been used to excite SPPs on detector structures as an alternative to grating coupling. For example, a 10–20× increase in IPE was observed for Au/n-GaAs Schottky detectors by exciting SPPs along the Au/GaAs interface at $\lambda_0 = 1150$ nm in an Otto arrangement implemented monolithically through the substrate [7].

Hole-Coupled Detectors

Optical transmission through a single sub-wavelength hole in a metal film can be significantly larger than predicted by Bethe's theory for a hole in a perfect conductor [8]. This larger (extraordinary) transmission is due to the excitation of SPPs on the metal surface near the hole that then propagate through the hole and couple to radiation on the other side; i.e., the transfer of energy through the hole is mediated by SPPs. Transmission at visible wavelengths is largest for a hole that is 150–300 nm in diameter in 200–300 nm thick Au or Ag films [8].

Structuring the metal film into a corrugation, e.g., around a sub-wavelength hole or slit leads to increased collection of incident light and to greater transmission at resonant wavelengths [8]. Applying this concept to detectors leads to devices that may offer improved performance [9–11].

Ishi et al. [9] reported a Schottky detector on n-Si where the metal forming the Schottky contact was structured into a circular corrugated grating surrounding a 300 nm diameter hole (and covered by SiO_2). The contact metal was a Ag/Cr stack 200/10 nm thick. Detection was reported at 840 nm based on absorption in a 300 nm diameter n-Si mesa located directly below the hole, and the creation of EHPs therein. A prospective advantage of the structure is high-speed operation due to the small size of the detection volume while maintaining good responsivity due to the light collection ability of the grating/hole combination; thus the detector exhibits a favorable trade-off between these two parameters (compared to conventional detectors). It has also been predicted that the grating/hole (or slit) combination can lead to detectors having an improved signal-to-noise ratio [10, 11].

Arrays of holes in metal films also couple incident light to SPPs, in a manner analogous to a corrugated grating [8]. The momentum conservation equation for a two-dimensional (low perturbation) periodic structure on a metal surface is [8, 12]:

$$\boldsymbol{k}^{SPP} = \boldsymbol{k}_{\|} + l\boldsymbol{G}_x + m\boldsymbol{G}_y \tag{11}$$

where $\boldsymbol{k}_{\|}$ is the in-plane wavevector of the incident light, $\boldsymbol{G_x}$ and $\boldsymbol{G_y}$ are reciprocal lattice vectors of the periodic structure, and l, m are integers. This equation models approximately the coupling of light to SPPs on a two-dimensional hole array, although the holes are actually neglected along with their associated effects such as scattering and transmission. For example, in the case of a hexagonal or triangular lattice of holes of lattice constant a, illuminated at normal incidence ($\boldsymbol{k}_{\|} = 0$) and neglecting all losses, the above yields the wavelengths λ_c for which strong coupling to SPPs on the array occurs:

$$\lambda_c = a\left[\frac{4}{3}\left(l^2 + lm + m^2\right)\right]^{\frac{-1}{2}}\left(\frac{\varepsilon_R \varepsilon_{r,d}}{\varepsilon_R - \varepsilon_{r,d}}\right)^{\frac{1}{2}} \tag{12}$$

In the above, $\varepsilon_{r,d}$ corresponds to the dielectric bounding the array on the illuminating side. If the structure is symmetric (same dielectric on both sides of the array), then these wavelengths also correspond to those at which transmission peaks occur. If the structure is asymmetric (different dielectrics bounding the array) then two sets of transmission peaks are observed.

Hole arrays can be integrated with broadband detectors to introduce wavelength and polarization selectivity, and potentially, to enhance the responsivity [12, 13]. This is achieved by structuring the top or bottom contact of a detector into an array of sub-wavelength holes and then illuminating at normal incidence. The photocurrent is then largest at wavelengths for which extraordinary transmission occurs, which depend on the lattice constants of the array, e.g., via Eq. 12 for the case of a hexagonal array, and so are easily tunable. Polarization selectivity can be introduced by using elliptical or rectangular holes.

The responsivity can be enhanced for a thin detection layer by enhancing the absorptance (A) via

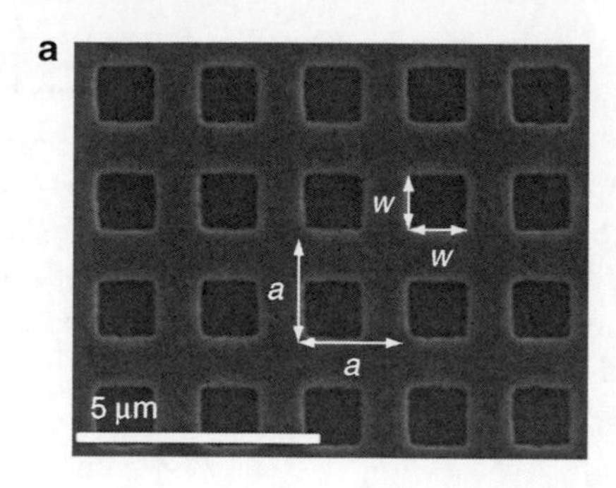

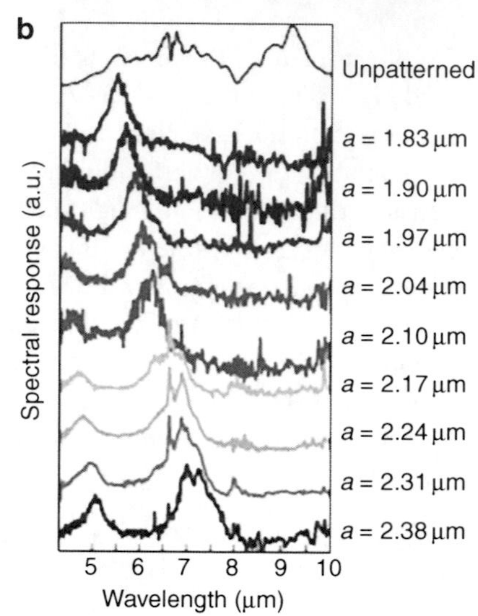

Surface Plasmon-Polariton-Based Detectors, Fig. 5 (**a**) Image of a square lattice of square holes; the lattice constant is a and the hole size is w. (**b**) Spectral response of quantum dot detectors each bearing a hole array of the indicated lattice constant; $w = 0.6a$ for all detectors (Adapted from [13]. © 2009 American Institute of Physics.)

mediation by SPPs. This is achieved by placing the detection layer close to the hole array such that SPPs propagating along the array overlap with the layer and are absorbed therein; thus the enhancement arises because SPPs propagate along the absorption region rather than through it perpendicularly.

Wavelength selectivity and responsivity enhancement are particularly attractive for mid-infrared detector arrays, which are typically broadband and have a low responsivity due to the low absorption of the detector materials used in this wavelength range. Applications such as multi-color (multi-spectral) night vision or medical imaging would benefit from inexpensive spectral filtering integrated at the pixel level. Figure 5a shows a square array (lattice constant a) of square holes (dimensions w) patterned onto the top contact (150-nm thick Ag) of an InAs mid-infrared quantum dot detector [13]. Figure 5b shows the measured spectral response of such detectors as a function of a, compared to an unpatterned control detector (upper trace). The peak wavelength response is observed to range from $\lambda_0 = 5.5$–7.2 µm as the lattice constant ranges from $a = 1.83$–2.38 µm.

Detectors Incorporating Nanoparticles

Small metal particles exhibit resonant responses under optical excitation, characteristic of particle shape, size, and composition, the dielectric environment in which they find themselves, and the wavelength of illumination [14]. Resonances occur when the electron charge density oscillates coherently with the illuminating optical (electrical) field. The fundamental resonant mode of, e.g., a spherical metal nanoparticle (small compared to the illuminating wavelength) is dipolar with densities of opposite charge forming at opposite spherical caps of the particle along the polarization of the illuminating electric field. The wavelength of the dipolar resonance red-shifts as the radius of the particle increases or as the index of the background increases. If the particle is large enough then higher-order resonant modes can exist (e.g., quadrupolar). On resonance, the electric field near the particle (in the dielectric) is strongly enhanced compared to the illuminating field. Resonant features appear in the measured extinction and scattering spectra of metal nanoparticles, in correlation with the plasmon resonances that are excited thereon.

Metal nanoparticles can be integrated with a detector most readily via deposition onto the surface through which light penetrates into the detector [15, 16]. As light propagates through, resonances can be excited, and strong scattering can occur, leading to improved detector performance.

Figure 6a shows a scanning electron microscope image of Ag-island nanoparticles on a 30-nm thick LiF layer on a Si-on-insulator (SOI) detector [15]. These approximately hemispherical particles are 108 nm across, on average. They were formed by first depositing a thin Ag film, then annealing the film to induce islandization. This process has the advantage of being simple but it generates particles that are non-uniform in shape and size, albeit, with a controllable

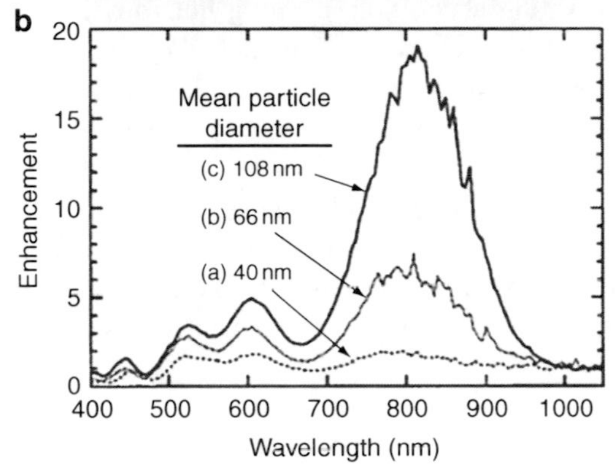

Surface Plasmon-Polariton-Based Detectors, Fig. 6 (a) Scanning electron microscope image of Ag-island nanoparticles on LiF-coated Si-on-insulator (SOI) detector. The particles are 108 nm across, on average. (b) Spectral response of pn junction SOI photodetectors coated with Ag islands similar to those of (a) as a function of average island size. The enhancement is defined as the ratio of the photocurrent from a detector with Ag islands to the photocurrent from a detector without the islands (Adapted from [15]. © 1998 American Institute of Physics.)

average size. Detectors were formed in the thin Si slab (165 nm thick) above the insulator but beneath the LiF (and the nanoparticles) as pn junction detectors. Figure 6b shows measured spectral responses of detectors coated with Ag islands similar to those of Fig. 6a as a function of average island size. The enhancement is defined as the ratio of the photocurrent from a detector with Ag islands to the photocurrent from a detector without islands. A strongly enhanced photocurrent ($\sim 20\times$) is noted at $\lambda_0 = 800$ nm in the case of islands that are 108 nm in average size. The enhancement is attributed to the large scattering cross-section of the islands, particularly to mediation by the islands whereby dipole resonances excited thereon radiate into guided modes of the thin underlying Si slab. These modes propagate longitudinally and thus are absorbed more effectively by the detector compared to perpendicularly incident light.

Chemically synthesized (colloidal) nanoparticles can also be deposited and attached to detector surfaces. Spectral responses of detectors coated with 50, 80, and 100 nm diameter spherical Au nanoparticles showed strong photocurrent enhancement at wavelengths where peaks were measured in the extinction spectra of the particles [16]. In contrast to the devices of [15], the enhanced photocurrent was attributed to enhanced absorption in the Si regions in contact with resonating nanoparticles, around which the resonating electric fields are significantly enhanced compared to the illuminating fields.

The effects induced by metal nanoparticles, i.e., increased scattering into detectors and increased detector absorption on resonance, are useful to increase the absorptance of detectors, particularly those based on a thin absorption layer, or at wavelengths where the absorption is low (e.g., near the bandgap of an indirect semiconductor). These effects can potentially improve the efficiency of, e.g., thin-film Si solar cells [17].

Waveguide Detectors

The single-interface (Fig. 1) is one example of an SPP waveguide; other popular 1-D geometries include the thin metal slab bounded by dielectrics and the thin dielectric slab bounded by metals [2]. The latter are often termed the IMI and MIM, respectively (I – insulator, M – metal). The IMI and MIM support super-modes formed from the coupling of single-interface SPPs through the thin intervening layer. The structures can be designed such that they support super-modes that are low loss but weakly confined (symmetric IMI) or strongly confined but high loss (MIM), and thus that are at opposite ends of the confinement-attenuation trade-off [2].

SPP waveguides can be formed into SPP detectors and conveniently integrated with other plasmonic or photonic waveguide structures. SPP waveguide detectors based on absorption in, e.g., organics [18], semiconductors [19], or metals [20] have been reported.

The detector described in [18] consists of a thick Ag film on which SPPs are excited by illuminating a slit in the film. The SPPs then propagate along the Ag film to the detector section, which consists of a pn junction fabricated from organic polymers directly on the Ag

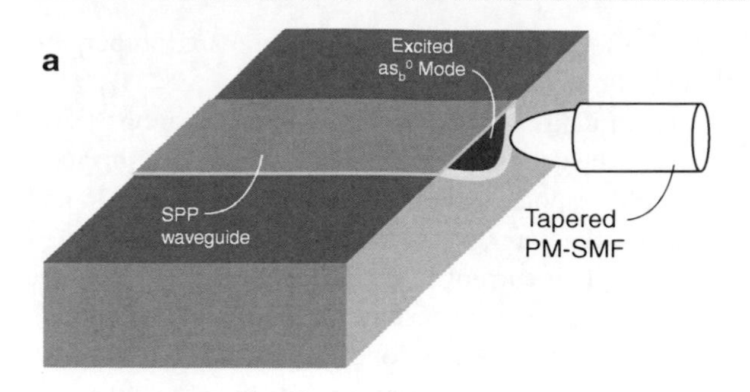

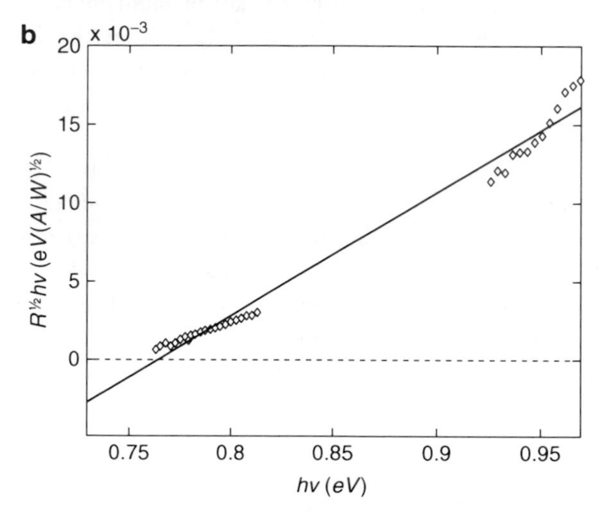

Surface Plasmon-Polariton-Based Detectors, Fig. 7 (**a**) Sketch of an SPP waveguide detector consisting of a metal stripe forming a Schottky contact on Si, excited (in the as_b^0 mode) via butt-coupling to a tapered polarization-maintaining single mode fiber (PM-SMF). (**b**) Spectral response of a Au on n-Si detector (as in (**a**)), plotted in Fowler form (Adapted from [20]. © 2010 Optical Society of America.)

film. The pn junction is covered by another Ag film, thereby creating an MIM-type detector. SPPs incident from the Ag film onto this MIM detector excite SPPs therein, which are absorbed by the pn junction creating EHPs that are collected as the photocurrent. Current maps correlating the photocurrent with the excitation of SPPs at $\lambda_0 = 632.8$ nm along the Ag film are reported.

The structure described in [19] consists of an MIM (Au/HSQ/Au) waveguide with slits, fabricated on a GaAs wafer. A slit in the top Au layer is used to couple incident light to SPPs propagating in the MIM. A slit in the bottom Au layer, placed a short distance away from the input, couples the SPP into radiation directed into the GaAs wafer, which is absorbed therein creating EHPs that are collected as the photocurrent. Spectral responses and photocurrent maps demonstrating the operation of the detector are reported.

Figure 7a gives a sketch of a SPP waveguide detector consisting of a thin narrow metal stripe on semiconductor (with air on top) [20]. The thin metal stripe is similar to the asymmetric IMI, but it supports SPPs with additional confinement along the lateral dimension (stripe width). Detectors have been fabricated as Au or Al stripes on n-Si forming Schottky contacts thereon [20]. The as_b^0 mode, localized to the bottom metal-Si interface, is excited via butt-coupling to a tapered polarization-maintaining single mode fiber (PM-SMF). This mode propagates along the stripe with strong absorption, creating hot carriers therein, some of which may cross the Schottky barrier and be collected as the photocurrent; thus detection occurs via IPE. Figure 7b gives the spectral response of a Au on n-Si detector, plotted in Fowler form. The intercept with the abscissa yields the cutoff photon energy (~0.765 eV), corresponding in this case to a cutoff wavelength of $\lambda_0 \sim 1620$ nm. Responsivities up to 1 mA/W were reported with this detector scheme.

Concluding Remarks

SPP detectors, or SPP-enhanced detectors, typically combine a metallic structure that supports SPPs, such as a planar or grating structure, metallic nanoparticles, or holes in metal films, with a semiconductor detector structure such as a Schottky or a pn junction. Properties inherent to SPPs are exploited to convey additional characteristics to a detector such as polarization, angular or spectral selectivity, or to enhance the absorptance (A) of the detector. The involvement of SPPs has led to detectors with improved performance and greater functionality.

References

1. Maier, M.A.: Plasmonics: Fundamentals and Applications. Springer, New York (2007)
2. Berini, P.: Figures of merit for surface plasmon waveguides. Opt. Express **14**, 13030–13042 (2006)
3. Sze, S.M.: Physics of Semiconductor Devices. Wiley, New York (1981)
4. Scales, C., Berini, P.: Thin-film Schottky barrier photodetector models. IEEE J. Quant. Electr. **46**, 633–643 (2010)
5. Jestl, M., Maran, I., Kock, A., Beinstingl, W., Gornik, E.: Polarization-sensitive surface plasmon Schottky detectors. Opt. Lett. **14**, 719–721 (1989)
6. Brueck, S.R.J., Diadiuk, V., Jones, T., Lenth, W.: Enhanced quantum efficiency internal photoemission detectors by

grating coupling to surface-plasma waves. Appl. Phys. Lett. **46**, 915–917 (1985)
7. Daboo, C., Baird, M.J., Hughes, H.P., Apsley, N., Emeny, M.T.: Improved surface plasmon enhanced photodetection at an Au-GaAs Schottky junction using a novel molecular beam epitaxy grown Otto coupling structure. Thin Solid Films **201**, 9–27 (1991)
8. Genet, G., Ebbesen, T.W.: Light in tiny holes. Nature **445**, 39–46 (2007)
9. Ishi, T., Fujikata, J., Makita, K., Baba, T., Ohashi, K.: Si Nano-photodiode with a surface plasmon antanna. Jpn. J. Appl. Phys. **44**, L364–L366 (2005)
10. Yu, Z., Veronis, G., Fan, S., Brongersma, M.L.: Design of midinfrared photodetectors enhanced by surface plasmons on grating structures. Appl. Phys. Lett. **89**, 151116 (2006)
11. Bhat, R.D.R., Panoiu, N.C., Brueck, S.J.R., Osgood, R.M.: Enhancing the signal-to-noise ratio of an infrared photodetector with a circular metal grating. Opt. Express **16**, 4588–4596 (2008)
12. Chang, C.Y., Chang, H.Y., Chen, C.Y., Tsai, M.W., Chang, Y.T., Lee, S.C., Tang, S.F.: Wavelength selective quantum dot infrared photodetector with periodic metal hole arrays. Appl. Phys. Lett **91**, 163107 (2007)
13. Rosenburg, J., Shenoi, R.V., Vandervelde, T.E., Krishna, S., Painter, O.: A multispectral and polarization-selective surface-plasmon resonant midinfrared detector. Appl. Phys. Lett. **95**, 161101 (2009)
14. Kelly, K.L., Coronado, E., Zhao, L.L., Schatz, G.C.: The optical properties of metal nanoparticles: The influence of size, shape, and dielectric environment. J. Phys. Chem. B **107**, 668–677 (2003)
15. Stuart, H.R., Hall, D.G.: Island size effects in nanoparticle-enhanced photodetectors. Appl. Phys. Lett. **73**, 3815–3817 (1998)
16. Schaadt, D.M., Feng, B., Yu, E.T.: Enhanced semiconductor optical absorption via surface plasmon excitation in metal nanoparticles. Appl. Phys. Lett. **86**, 063106 (2005)
17. Atwater, H.A., Polman, A.: Plasmonics for improved photovoltaic devices. Nat. Mater. **9**, 205–213 (2010)
18. Ditlbacher, H., Aussenegg, F.R., Krenn, J.R., Lamprecht, B., Jakopic, G., Leising, G.: Organic diodes as monolithocally integrated surface plasmon polariton detectors. Appl. Phys. Lett. **89**, 161101 (2006)
19. Neutens, P., Van Dorpe, P., De Vlaminck, I., Lagae, L., Borghs, G.: Electrical detection of confined gap plasmons in metal-insulator-metal waveguides. Nat. Photon. **3**, 283–286 (2009)
20. Akbari, A., Tait, R.N., Berini, P.: Surface plasmon waveguide Schottky detector. Opt. Express **18**, 8505–8514 (2010)

Surface Properties

► Nanostructures for Surface Functionalization and Surface Properties

Surface Tension and Chemical Potential at Nanoscale

► Surface Energy and Chemical Potential at Nanoscale

Surface Tension Effects of Nanostructures

Ya-Pu Zhao and Feng-Chao Wang
State Key Laboratory of Nonlinear Mechanics (LNM), Institute of Mechanics, Chinese Academy of Sciences, Beijing, China

Synonyms

Surface energy density; interface excess free energy

Definition

The surface tension is the reversible work per unit area needed to elastically stretch/compress a preexisting surface. In other words, it is a property of the surface that characterizes the resistance to the external force. Surface tension has the dimension of force per unit length or of energy per unit area. For nanostructures with a large surface to volume ratio, the surface tension effects dominate the size-dependent mechanical properties.

Overview

Nanostructures have a sizable surface to volume ratio as compared to bulk materials, which leads its mechanical properties to be quite different from those of bulk materials [1]. The size-dependent mechanical properties of nanostructures are generally attributed to the surface effects, in which surface tension is one of the most predominant factors.

In the framework of the thermodynamics, the surface (interface) between two phases is first modeled as a bidimensional geometrical boundary of zero thickness, that is, the mathematical surface, as shown in Fig. 1a. The physical quantities between the two phases are discontinuous, and the surface (interfacial)

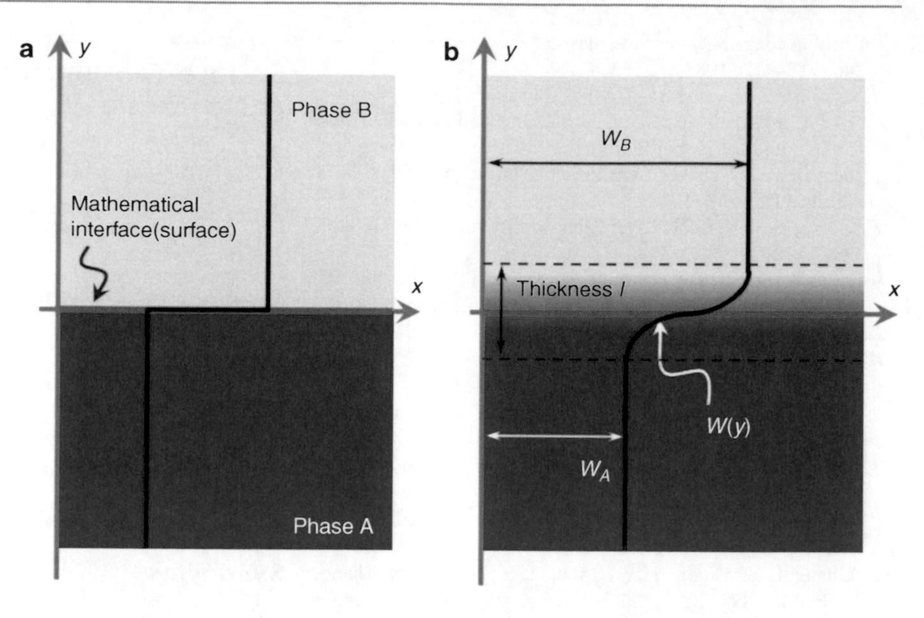

Surface Tension Effects of Nanostructures, Fig. 1 An illustration for the surface (interface) models. (**a**) The mathematical surface of zero thickness, the physical quantities are discontinuous. (**b**) Gibbs's surface with an infinitesimal thickness, the physical quantities are continuous. Cahn–Hilliard model for surface with a finite thickness is also illustrated

tension is a jump in stress. This idealization was then extended by Gibbs [2], who proposed that the physical quantities should undergo a smooth transition at the surface (interface) while the surface is still modeled as an infinitesimal thin boundary layer, as shown in Fig. 1b. In order to preserve the total physical properties of the system, the excess physical properties have to be assigned to the geometrical surface. The surface tension, which is defined based on the interface excess free energy, can be written as

$$\gamma = \int_0^{\infty} [w(y) - w_A]\mathrm{d}y + \int_{-\infty}^{0} [w(y) - w_B]\mathrm{d}y \quad (1)$$

where w is the free energy distribution of the actual surface, w_A and w_B are free energy in the two phases of A and B.

In some other theoretical models, the surface is treated as an extended interfacial region with a nonzero thickness. According to Cahn–Hilliard theory [3], the surface tension can be derived as

$$\gamma = N_V \int_{-\infty}^{\infty} \left[w_0(c) + k(\mathrm{d}c/\mathrm{d}y)^2 - c\mu_B - (1-c)\mu_A\right]\mathrm{d}y, \quad (2)$$

in which N_V is the number of atoms per unit volume, c is one of the intensive scalar properties, such as composition or density, $w_0(c)$ is the free energy per atom of a solution of uniform composition c, k reflects the crystal symmetry, μ_A and μ_B are the chemical potentials per atom in the A or B phase. The surface (interfacial) thickness can be obtained by

$$l = 2\Delta c_e \sqrt{\frac{k}{\Delta w_{\max}}}, \quad (3)$$

where $2\Delta c_e = c_B - c_A$ is the difference of the uniform composition in the two phases, $\Delta w_{\max}$ is the maximum of the free energy referred to a standard state of an equilibrium mixture. Cahn–Hilliard model gives a finite thickness through Eq. 3, which has been used in many applications [4].

From the standpoint of molecular theory, the surface tension effects arise due to the difference of the atomic interactions in the bulk and on the surface. Atoms are energetically favorable to be surrounded by others. At the surface, the atoms are only partially surrounded by others and the number of the adjacent atoms is smaller than in the bulk. Thus the atoms at the surface are energetically unfavorable. If an atom moves from the bulk to the surface, work has to be done. With this view, the surface tension can be interpreted as the energy required to bring atoms from the bulk to the surface [5]. Therefore the term "surface energy density" is often used to when the surface tension is referred to. For the surface of solid, Gibbs pointed out that surface tension and surface energy density are not identical. The surface tension is the reversible work per unit area needed to elastically stretch/compress a preexisting surface. The surface energy density is the reversible work per unit area needed to create a new surface. The surface tension can be positive or negative, while the surface energy density is usually positive. The surface tension for liquid surface is a property that characterizes its resistance to the external force, which is identical to the surface energy density.

Basic Methodology

Since the surface to volume ratio increases as the dimension scale decreases, surface tension effects of nanostructures can be overwhelming. In the absence of external loading, the surface tension effects would induce a residual stress field in bulk materials. The relations between surface stress and surface tension for small deformations can be described by the Shuttleworth-Herring equation [6, 7],

$$\sigma_{ij} = \gamma\delta_{ij} + \frac{\partial\gamma}{\partial\varepsilon_{ij}}, \tag{4}$$

where γ is the surface tension, δ_{ij} is the Kronecher delta, σ_{ij} and ε_{ij} are the surface stress tensor and the surface strain tensor, respectively. The Shuttleworth-Herring equation interprets that the difference between the surface stress and surface tension is equal to the variation of surface tension with respect to the elastic strain of the surface.

With the development of computational materials science, molecular dynamics (MD) simulations are wildly performed to investigate the surface tension effects on the mechanical properties of nanostructures. Especially for the surface elastic constant, atomistic simulation is almost the only way to get them up to now. According to the Gibbs's definition, the surface tension of a solid is given by $\gamma = (E_S - nE_B)/A_0$, where A_0 is the total area of the surface considered, E_S is the total energy of a n-layer slab and E_B is the bulk energy per layer of an infinite solid. In cases where the surfaces of the slab are polar, the electrostatic energy of the slab contains an energy contribution, E_{pol}, proportional to the substrate thickness and the surface energy needs corrections. Thus there is an alternative way to calculate the surface tension $\gamma = \left(E_S - nE_B - E_{\text{pol}}\right)/A_0$, which does not rely on an exact knowledge of the lattice or polar energy. MD simulations have identified that the surface tension induced surface relaxation is proved to be a dominate factor of the size-dependent mechanical properties. When the surface stress is negative, the surface relaxation is inward; otherwise, the relaxation is outward.

Surface stress has been used as an effective molecular recognition mechanism. Surface stresses due to DNA hybridization and receptor-ligand binding induce the deflection of a cantilever sensor [8]. The curvature of bending beam under a surface stress is governed by Stoney's formula [9]. Stoney's formula serves as a cornerstone for curvature-based analysis and a technique for the measurement of surface stress, which is given as follows as a general form for a film/substrate system,

$$\sigma = \frac{Et_s^2 f}{6(1-v)} \tag{5}$$

in which E is the effective Young's modulus, v is Poisson's ratio of the sensor material, t_s is the substrate thickness, $f = 3\Delta z/2\,L^2$ is the sensor curvature, L is the length and Δz is the deflection. The applicability of the above Stoney's formula relies on several assumptions, which are well summarized as the following six: (1) both the film and substrate thicknesses are small compared to the lateral dimensions; (2) the film thickness is much less than the substrate thickness; (3) the substrate material is homogeneous, isotropic, and linearly elastic, and the film material is isotropic; (4) edge effect near the periphery of the substrate are inconsequential and all physical quantities are invariant under change in position parallel to the interface; (5) all stress components in the thickness direction vanish throughout the material; and (6) the strains and rotations are infinitesimally small. However, the one or several of above six assumptions can be easily violated in reality, which is to say that the Stoney's formula needs to be revised to fit in the real applications.

To summarize for the solid cases, the behaviors of nanostructures can be affected significantly by either of the two distinct parameters, surface tension and surface stress. The relation between the two parameters can be obtained by the Shuttleworth-Herring equation. For the surface stress induced deflection of cantilever sensors, the curvature is by Stoney's formula. MD simulation is helpful to understand the surface effects of nanostructures and partial results are comparable to the experiments.

Surface Tension Effects of Nanostructures, Table 1 Characteristic time related to the surface tension effects

Name	Expression	Meaning
Capillary characteristic time	$t_c = \sqrt{m/\gamma_{LV}}$	Characteristic time derived from the droplet mass and the liquid–vapor surface tension
Lord Rayleigh's period	$t_p = \frac{\pi}{4}\sqrt{\rho d_0^3/\gamma_{LV}}$	The period of a free droplet in free oscillation
Lord Rayleigh's characteristic time	$t_R \sim \sqrt{\rho l^3/\gamma_{LV}}$	Characteristic time of droplet dynamics
Viscous characteristic time	$t_{vis} \sim \eta l/\gamma_{LV}$	Characteristic time related to the liquid viscosity

m mass of the droplet, d_O diameter, ρ density, η viscosity, *l* characteristic length.

For liquid droplets in contact with the surface of a nanostructure, surface tension is responsible for the shape of liquid droplets in the equilibrium state, as well as the dynamics response in wetting and dewetting. There are several characteristic time which are related to the surface tension effects, listed in Table 1. Dimensionless number related to the surface tension effects are listed in Table 2. Surface tension is dependent on temperature T, concentration of surfactants c, and the electric field V. The gradient of surface tension caused by these factors can be described by [10]

$$d\gamma = \frac{\partial\gamma}{\partial T}dT + \frac{\partial\gamma}{\partial c}dc + \frac{\partial\gamma}{\partial V}dV. \tag{6}$$

To the first order, the dependency of the surface tension on temperature is given by Guggenheim–Katayama formula with power index $n = 1$, $\gamma = \gamma_0(1 - T/T_c)$. Here γ_O is a constant for each liquid and T_c is the critical temperature. For the dependency of the surface tension on concentration c, the surface tension can be expressed as a linear function of the concentration, $\gamma = \gamma_0[1 + \beta(c - c_0)]$. The solid–liquid surface tension can be changed by applying a voltage V, $\gamma_{SL} = \gamma_{SL}^0 - \frac{\varepsilon_0\varepsilon_D}{2d}V^2$, where γ_{SL}^0 is the solid–liquid surface tension in the absence of the applied voltage, ε_O is the permittivity of vacuum, ε_D is the relative permittivity of the dielectric layer with a thickness d separating the bottom electrode from the liquid.

The wetting properties of a solid surface can be described by the introduction of the contact angle, which is defined as the angle at which the liquid–vapor interface meets the solid surface. The contact angle is affected by various factors, including the surface tension, the line tension, the applied voltage as well as the molecular interactions between the liquid and solid surfaces. It is proposed that the dependence of the contact angle on these factors can be represented by the generalized Young's equation, which has a form of

$$\cos\theta = \cos\theta_0 - \frac{\tau}{\gamma_{LV}R} + \frac{\varepsilon_0\varepsilon_D}{2d\gamma_{LV}}V^2 + \frac{A}{12\pi h^2\gamma_{LV}}, \tag{7}$$

where

$$\cos\theta_0 = \frac{\gamma_{SV} - \gamma_{SL}}{\gamma_{LV}} \tag{8}$$

is the classical Young's equation, θ_0 is the equilibrium contact angle (also the Young contact angle); The subscripts S, L, and V denote solid, liquid, and vapor, respectively. The second term on the right-hand side of Eq. 7 is related to the line tension. τ is the line tension and R is the radius of the contact area. For a more generalized case, R can be replaced by $1/\kappa$, in which κ is the geodesic curvature of the triple contact line. $\eta_e = \frac{\varepsilon_0\varepsilon_D}{2\gamma_{LV}d}V^2$ is defined as the dimensionless electrowetting number. The last term in Eq. 7 measures the strength of the effective interaction energy compared to surface tension. The interaction energy is related to the disjoining pressure $\Pi(h)$, which can be obtained by $W(h) = \int_h^\infty \Pi(h')dh'$, where h is the film thickness and A is the Hamaker constant [11, 12]. As discussed in the following paragraphs, each term would be illustrated in detail.

If only the classical Young's equation is referred to, the wetting properties of the surface can be obtained directly if the three tensions are known. The spreading parameter S determines the type of spreading, which is defined as $S = \gamma_{SV} - (\gamma_{SL} + \gamma_{LV})$. If $S > 0$, the liquid spreads on the solid surface and forms a macroscopic liquid layer covers the whole solid surface; it is the case for complete wetting. If $S < 0$, the contact angle $0° < \theta_0 < 180°$, which means the liquid forms a droplet with a finite contact angle; it is the case for speak of partial wetting. The wettability of the solid surface could be distinguished by the contact angle θ_0, as shown in Fig. 2. If $0° \leq \theta_0 < 90°$, the solid substrate is hydrophilic. If $90° < \theta_0 \leq 180°$, the solid substrate is hydrophobic. Especially, if $150° < \theta_0 \leq 180°$, the solid substrate is superhydrophobic.

Surface Tension Effects of Nanostructures, Table 2 Dimensionless number related to the surface tension effects

Name	Expression	Meaning
Adhesion number	$N_a = \cos\theta_a - \cos\theta_r$	θ_a and θ_r are the advancing and receding angle.
Bond number	$\mathrm{Bo} = \rho g l^2/\gamma_{LV}$	The capillary length $l_c = \sqrt{\gamma_{LV}/(\rho g)}$ can be determined
Capillary number	$\mathrm{Ca} = \eta v/\gamma_{LV}$	$\mathrm{Ca} = \mathrm{On}\cdot\sqrt{\mathrm{We}}$; the capillary velocity $v = \gamma_{LV}/\eta$ can be obtained
Deborah number	$\mathrm{De} = t/t_R = t\Big/\sqrt{\rho l^3/\gamma_{LV}}$	t is the relaxation time, Lord Rayleigh characteristic time can be derived
Elasto-capillary number	$\mathrm{Ec} = t\gamma_{LV}/(\eta l)$	t is the relaxation time, viscous characteristic time can be obtained
Electrowetting number	$\eta_e = \varepsilon_0\varepsilon_D V^2/(2d\gamma_{LV})$	The ratio of the electrostatic energy to the surface tension.
Laplace number	$\mathrm{La} = (1/\mathrm{On})^2$	See Ohnersoge number
Marangoni number	$\mathrm{Ma} = -\dfrac{d\gamma_{LV}}{dT}\dfrac{L\cdot\Delta T}{\eta\alpha}$	Thermal surface tension force divided by viscous force
Ohnersorge number	$\mathrm{On} = \eta\Big/\sqrt{\rho l\gamma_{LV}}$	$\dfrac{\mathrm{Viscousforce}}{\sqrt{\mathrm{Inertiaforce}\times\mathrm{Surfacetension}}}$
Weber number	$\mathrm{We} = \rho v^2 l/\gamma_{LV}$	The inertia force compared to its surface tension

g gravitational acceleration, α thermal diffusivity, ΔT temperature difference.

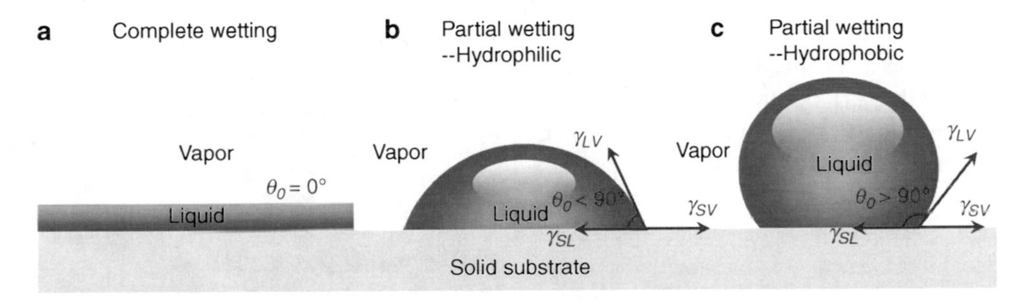

Surface Tension Effects of Nanostructures, Fig. 2 Wetting properties of the surface

The physics behind wettability is that, the solid surfaces have been divided into high energy and low-energy types [11, 12]. The relative energy of a solid has to do with the bulk nature of the solid itself. (1) High-energy surfaces such as metals, glasses, and ceramics are bound by the strong chemical bonds, for example, covalent, ionic, or metallic, for which the chemical binding energy E_{binding} is of the order of 1 eV. The solid–liquid interface tension is given by $\gamma_{SV} \approx E_{\mathrm{binding}}/a^2 \sim 0.5 - 5\mathrm{N/m}$, in which a^2 is the effective area per molecule. Most high-energy surfaces are hydrophilic, some can permit complete wetting. (2) For low-energy surfaces, such as weak molecular crystals (bound by van der Waals forces or in some special cases, by hydrogen bonds), the chemical binding energy is of the order of k_BT. In this category, the surface tension is $\gamma_{SV} \approx k_BT/a^2 \sim 0.01 - 0.05\mathrm{N/m}$. Depending on the type of liquid chosen, low-energy surfaces can be either hydrophobic or hydrophilic.

When the liquid droplet comes to the nanoscale, the classical Young's equation seems to be not applicable, since it has been derived for a triple line without consideration of the interactions near the triple contact line. The molecules close to the triple line experience a different set of interactions than at the interface [10]. To take into account this effect, the "line tension" term has been introduced in the generalized Young's equation. A sketch of line tension is shown in Fig. 3.

Line tension was first introduced by Gibbs [2]. The line tension τ was introduced as an analogue of the surface tension:

$$F = \gamma_{LV}\int_{S'} \mathrm{d}S' + (\gamma_{LV} - W)\int_{S^*} \mathrm{d}S^* + \tau\int_{C^*} \mathrm{d}C^*, \quad (9)$$

where F, S', S^*, and C^* are the free energy, the free surface, the adhering interface, and the contact line of the droplet. $W = \gamma_{LV} + \gamma_{SL} - \gamma_{SV}$ is the adhesion potential. The line tension, depending on the radius R of the contact line, should be connected to the classical Young's equation to include the interactions near the triple contact line. In the three-phase equilibrium

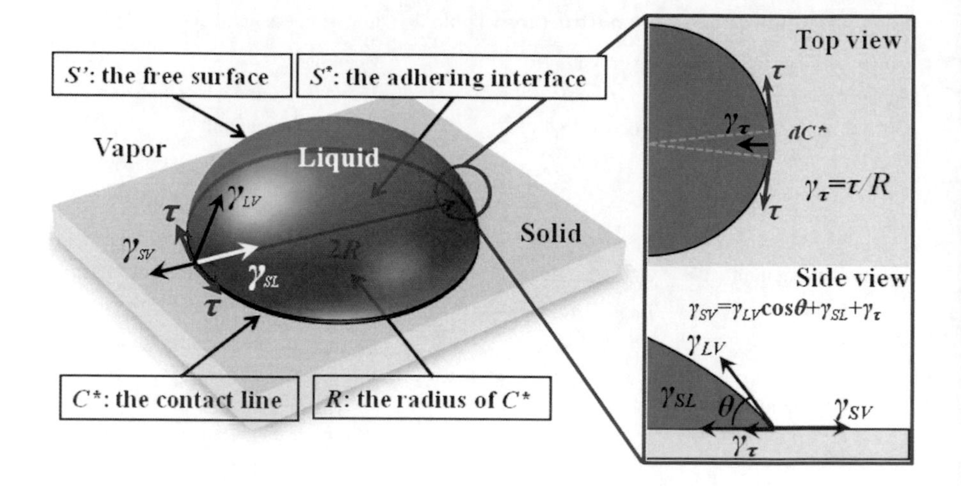

Surface Tension Effects of Nanostructures, Fig. 3 Illustration of the line tension τ

systems when R decreases, the contact angle θ will increase at $\tau > 0$ and will decrease at $\tau < 0$. However, in accordance to the mechanical equilibrium stability conditions, while the surface tension can only be positive, the line tension can have either positive or negative values. The characteristic length of this problem $l = |\tau|/\gamma_{LV}$ ($|\tau| < 10^{-9}$ N, $\gamma_{LV} \sim 0.1$ N/m for water) ranges from 10^{-8} to 10^{-6} m, which means small droplet (with typical dimension of $|l|$) should appreciate the line tension effect. The line tension can be indirectly measured through the contact angle, $\tau \approx 4\delta\sqrt{\gamma_{SV}\gamma_{LV}}\cot\theta$, in which δ denotes the average distance between liquid and solid molecules. Thus the line tension is negative for an obtuse contact angle, while it is positive for an acute contact angle.

In the presence of a charged interface, which can be achieved by applying a direct or alternating-current electric field, the wetting properties of solid surface will be modified. The physics describing the electric forces on interfaces of conducting liquids and on triple contact lines is called "electrowetting" [10]. In the year of 1875, Gabriel Lippmann observed the capillary depression of mercury in contact with an electrolyte solution could be varied by applying a voltage between the mercury and electrolyte. This phenomenon is called electrocapillarity, which is the basis of modern electrowetting. Then, the idea was developed to isolate the liquid droplet from the substrate using a dielectric layer in order to avoid electrolysis. This concept has subsequently become known as electrowetting on dielectric (EWOD) and involves applying a voltage to modify the wetting behavior of a liquid in contact with a hydrophobic, insulated electrode. When an

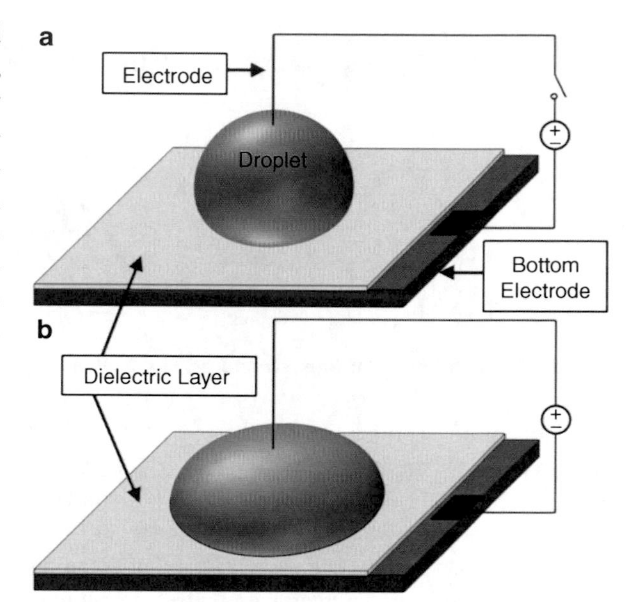

Surface Tension Effects of Nanostructures, Fig. 4 Illustration of EWOD. The external voltage is applied between a thin electrode (the Pt wire) and the bottom electrode (typically the indium-tin-oxide glass). Partially wetting liquid droplet in the absence of an applied voltage (**a**) and after the voltage is applied (**b**)

electric field was applied to the system (as shown in Fig. 4), electric charges gather at the interface between the conductive electrodes and the dielectric material; the surface becomes increasingly hydrophilic (wettable), the contact angle is reduced and the contact line moves. The change in contact angle over the buried electrodes can be evaluated by the Lippmann–Young equation $\cos\theta = \cos\theta_0 + \varepsilon_0\varepsilon_D V^2/(2d\gamma_{LV})$. The parabolic variation of cos θ in the Lippmann–Young equation is only applicable when the applied voltage lies below a threshold value. If the voltage is increased above this threshold, then contact angle saturation starts to occur and cos θ eventually becomes independent of the applied voltage.

The physics of why a liquid film wets or dewets is found in the derivative of the effective interfacial potential $W = \gamma_{LV} + \gamma_{SL} - \gamma_{SV}$ with respect to film thickness h, called the disjoining pressure $\Pi(h)$ [11, 12]. $\Pi(h) = \mathrm{d}W(h)/\mathrm{d}h$, where the surface area remains constant in the derivative. The effective interface potential $W(h)$, which arises from the interaction energies of molecules in a film being different from that in the bulk, is the excess free energy per unit area of the film. If the interactions between the molecules in the film and the solid substrate are more attractive than the interactions between molecules in the bulk liquid, $W(h) > 0$. Consequently, a liquid film with a thickness in a range where $\Pi(h) > 0$ can lower its free energy by becoming thicker in some areas while thinning in others, that is, by dewetting. When $\Pi(h) < 0$, wetting or spreading occurs.

The van der Waals interaction $w(r) \propto 1/r^6$ includes all intermolecular dipole–dipole, dipole–induced dipole, and induced dipole–induced dipole interactions. Performing a volume integral over all molecules present in the two half spaces bounding the film one finds a corresponding decay $\Pi(h) \sim A/6\pi h^3$ [11, 12], where the Hamaker constant A ($\sim 10^{-19}$ J) gives the amplitude of the interaction. In the "attractive" case, in which the layer tends to thin, $A < 0$. In the "repulsive" case, in which the layer tends to thicken, $A > 0$. Because of the disjoining pressure, there is a precursor film ahead of the nominal contact line of the liquid droplet [13]. The thickness of the precursor film can be defined by a molecular length [11, 12] $h_{PF} \sim \sqrt{A/6\pi\gamma_{LV}}$, which is on the order of several Å.

Key Research Findings

Elastic Models for the Nanostructures

Gurtin and Murdoch established the theoretical framework of the surface elasticity under the classical theory of membrane [14]. Recently, studies [15] have shown that, even in the case of infinitesimal deformations, one should distinguish between the reference and the current configurations; otherwise the out-plane terms of surface displacement gradient, associated with the surface tension, may sometimes be overlooked in the Eulerian descriptions, particularly for curved and rotated surfaces. By combining elastic models for surface and bulk, the size-dependent elastic and properties of nanomaterials have been investigated. Usually, in the absence of external mechanical or thermal loadings, the surfaces of a nanostructure will be subjected to residual surface stresses, and an elastic field in the bulk materials will be induced by such residual surface stresses induce from the point of view of equilibrium conditions. This self-equilibrium state without external loadings is usually chosen as the reference configuration, from which nanostructures will deform (see Fig. 5). That is to say, the bulk will deform from the residual stress states. However, in the prediction of elastic and thermoelastic properties of nanostructures, the elastic response of the bulk is usually described by classical Hooke's law, in which the aforementioned residual stress was neglected in the existing literatures.

Considering a bulk material with the surface properties aforementioned, the surface tension would induce a stress field in the bulk. According to Young–Laplace equation, the surface tension will result in a nonclassical boundary condition. The boundary condition together with the equations of classical elasticity forms a coupled system of field equations to determine the stress distribution. To solve a problem considering the surface properties, the surface model and the bulk model are established separately and using the Young–Laplace equation to bridge the two models together.

Here, an example named surface elasticity would be used to show how to establish a model combined the surface and bulk together. Assuming the surface to be isotropic and homogeneous, the constitutive relations of the surface in the Lagrangian description can be written as [15]

$$\mathbf{S}_s = \gamma_0^*\mathbf{I}_0 + \left(\gamma_0^* + \gamma_1^*\right)(\mathrm{tr}\mathbf{E}_s)\mathbf{I}_0 - \gamma_0^*(\bar{\nabla}_{0s}\mathbf{u}_0) + \gamma_1\mathbf{E}_s + \gamma_0^*\mathbf{F}_s^{(o)}, \quad (10)$$

where $\mathbf{S}_s$ is the first kind Piola–Kirchhoff stress of the surface, $\mathbf{I}_0$ is the identity tensor on the tangent planes of the surface in the reference configuration; the constants γ_0^*, γ_1^*, and γ_1 are the surface tension and the surface Lame moduli; $\mathbf{E}_s$, $\bar{\nabla}_{0s}\mathbf{u}_0$, and $\mathbf{F}_s^{(o)}$ denote, respectively, the surface strain tensor, the in-plane

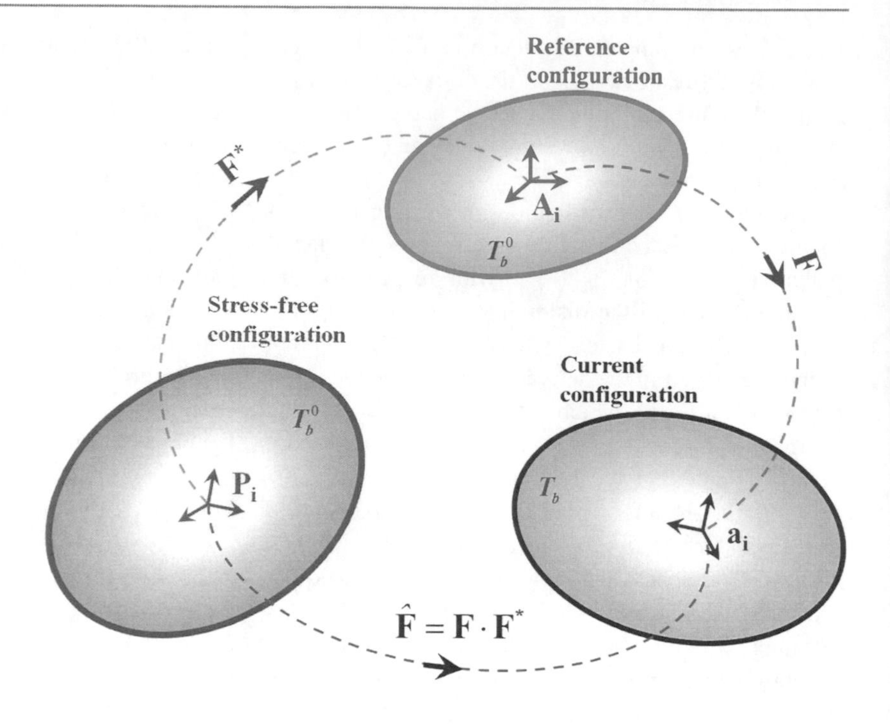

Surface Tension Effects of Nanostructures, Fig. 5 The choice of the reference configuration: the stressed state

part of surface displacement gradient and the out-plane term of surface deformation gradient. In view of the importance of the linearization of the general constitutive equations, the linear elastic constitutive relations of the bulk with residual stresses can be written as follows:

$$\mathbf{S} = \mathbf{T}_R + \mathbf{u}\nabla \cdot \mathbf{T}_R + \lambda \mathrm{tr}(\mathbf{E})\mathbf{1} + 2\mu\mathbf{E}, \qquad (11)$$

where $\mathbf{S}$ is the first Piola–Kirchhoff stress, $\mathbf{T}_R$ is the residual stress in the reference configuration, $\mathbf{u}\nabla$ is the displacement gradient calculated from the reference configuration, $\mathbf{E}$ is the infinitesimal strain, and λ and μ are material elastic constants.

In the absence of external loading, surface tension will induce a compressive residual stress field in the bulk of the nanoplate and there may be self-equilibrium states which correspond to plate self-buckling. The self-instability of nanoplates is investigated and the critical self-instability size of simplify supported rectangular nanoplates is proposed. The critical size for self-buckling is $b = \pi h\sqrt{(\alpha_A^{-2} + 1)Eh/[24\gamma_0^*(1 - v^2)]}$, where b and h are the width and thickness of the nanoplate, respectively; E and $v$denote the Young's modulus and Poisson's ratio, $\alpha_A = l/b$ is the aspect ratio [15].

Surface Tension Effects on the Mechanical Properties of Nanostructures

Many models are developed to relax one or some of the six above assumptions to extend Stoney's formula to a more generalized and realistic application. For example, the effects such as axial force, the damaged/nonideal interface effect and gradient stress, which violates one or several of the above assumptions, are analyzed and the extended/revised Stoney's formulas are given. Surface stress physically is a distributed one and this characteristic is not emphasized in many studies. The analysis by Zhang et al. shows that the Stoney's formula is obtained by assuming the influence of surface stress as a concentrated moment applied at the free end of a cantilever beam [16]. However, if the influence of surface stress is modeled as a distributed axial load and bending moment, the following nonlinear governing equation is obtained:

$$EI\frac{\mathrm{d}^4 z}{\mathrm{d}x^4} - \sigma b(L - x)\frac{\mathrm{d}^2 z}{\mathrm{d}x^2} + \sigma b\frac{\mathrm{d}z}{\mathrm{d}x} = 0, \qquad (12)$$

in which EI is the cantilever effective bending stiffness; b, L, and z are the beam width, length, and deflection, respectively. To solve the above nonlinear equation with the (given) boundary conditions at the two ends of the beam is a two-point-boundary value

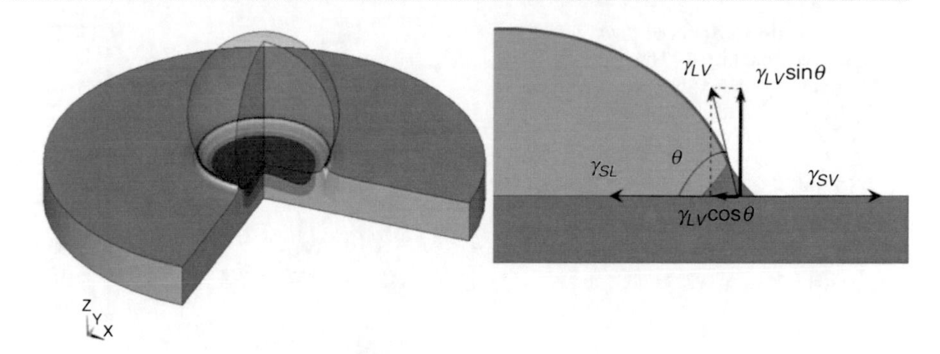

Surface Tension Effects of Nanostructures, Fig. 6 Sketch of deformation of PDMS membrane induced by a water droplet

problem [16], which is rather difficult. The semi-analytical series solutions can more or less ease the difficulty of solving Eq. 12. One implication of the Stoney's formula is that because the beam curvature is constant, the beam deflection under a surface stress is an arc of a circle (or a parabola if the approximate curvature definition is used). There is no mechanism to guarantee such kind of deflection. In general the curvature of a beam under a surface stress is not a constant, which has been verified in the experiments.

The Stoney's formula has been used to explain recent experimental results, in which a hybrid device based on a microcantilever interfaced with bacteriorhodopsin (bR), undergoes controllable and reversible bending when the light-driven proton pump protein, bR, on the microcantilever surface is activated by visible light [17]. It should be pointed that the Young's modulus of a nanostructure is size-dependent, that is, it would be enhanced or softened with decreasing the size of the nanostructure, which is generally attributed to the surface effects. Surface tension is one of the most important factors that cause the size effects of the Young's modulus of a nanostructure. The surface tension can be introduced into mechanical model via energy method. Using the relation of energy equilibrium, the effective elastic modulus of nanobeams are dependent on the surface tension [18].

Surface Tension Effects Induced the Deformation of Nanostructures

The classical Young's equation describes the equilibrium of forces in the direction parallel to the solid surface, while the vertical component of liquid–vapor interfacial tension is ignored. That is, there is a net force $\gamma_{LV} \sin\theta$ acting normal to the smooth solid surface at the solid–liquid–vapor contact line. Due to the unbalance force, there will be a surface deformation, as shown in Fig. 6. Indeed, several decades ago, surface deformation of semi-infinite solid was theoretically analyzed with the physical assumption that the liquid–vapor has a finite thickness (maybe at the order of tens nanometers) and the liquid–vapor interfacial tension acts uniformly in this region. Their research suggests there is a wetting ridge at the three-phase contact line. Later, Shanahan and Carré used dimensional analysis to characterize the maximum height at the order of γ_{LV}/G, where G is the shear modulus of solid [19]. For the material widely used at that time were very rigid (at the order of at least 100 GPa), such a deformation is too small to be considered. However, to meet with the rapid development of microelectromechanical systems (MEMS) and nanoelectromechanical systems (NEMS), polydimethylsiloxane (PDMS) is widely fabricated to channels or membrane, which has at least one dimension on the order of submillimeters or even nanometers. The surface deformation might no longer be neglected. Moreover, it should be noted that whether the theoretical solution for semi-infinite case can be extended to the case of thin flexible membrane. Recently, Yu and Zhao considered the deformation of thin elastic membrane induced by sessile droplet and gave a theoretical solution correspondingly [20]. There are two important conclusions. The first is that there exists a saturated membrane thickness at the order of millimeter, if the solid is thicker than this, it can be taken regard as semi-infinite; otherwise, it is better to consider the effect of membrane thickness. The second is that if the membrane has a very low Young's modulus (for example, on the order of MPa or much less), the effect of membrane thickness will become significant.

Apart from theoretical analysis, experimental investigations on surface deformation induced by droplet have also been reported. Because of the surface resolving ratio, it is difficult to get the detailed

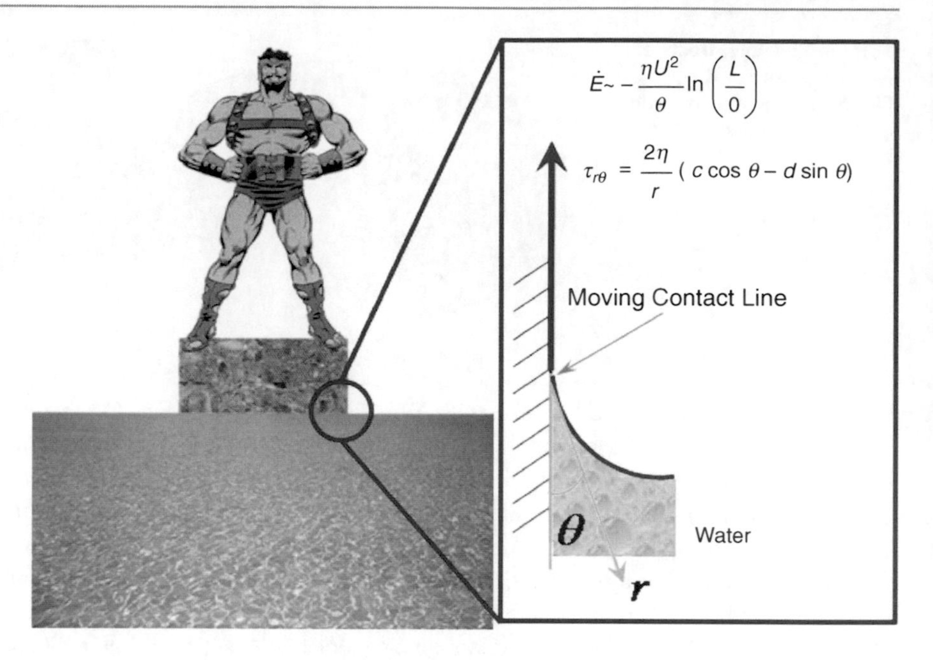

Surface Tension Effects of Nanostructures, Fig. 7 Huh and Scriven's paradox: Not even Herakles could sink a solid

information of surface deformation at the contact line. Moreover, there might be a highly stressed zone near the contact line so that such a question should be studied further when necessary.

Surface Tension Effects of Wetting at the Three-Phase Contact Line

When considering the surface tension effects at the three-phase contact line, there is a famous paradox named Huh–Scriven paradox. It was first pointed out by Huh and Scriven [21] that there is a conflict between the moving contact line and the conventional no-slip boundary condition between a liquid and a solid. The interface meets the solid boundary at some finite contact angle θ. Owing to the no-slip condition, the fluid at the bottom moves with constant velocity U and viscosity η, while the flux through the cross section is zero. The energy dissipation per unit time and unit length of the contact line is obtained, $\dot{E} \sim -\eta U^2 \ln(L/0)$, where L is an outer length scale like the radius of the spreading droplet. Stresses are unbounded at the contact line, and the force exerted by the liquid on the solid becomes infinite. The energy dissipation is logarithmically diverging, "not even Herakles could sink a solid" (Fig. 7)

In reality, dynamic wetting occurs at a finite rate with changes in the wetted area and liquid shape. These processes are thermodynamically irreversible and therefore dissipative. But, the energy dissipation is finite. There are two typical theories in identifying the effective channel of energy dissipation for small Capillary and Reynolds number. One of the two approaches is the hydrodynamic theory emphasizes energy dissipation caused by viscous flow within the wedge of liquid near the moving contact line. The other is the molecular kinetic theory emphasizes energy dissipation caused by of attachment (or detachment) of fluid molecules to (or from) the solid surface [22].

The Huh–Scriven paradox is raised from four ideal assumptions summarized as: incompressible Newtonian fluid, smooth solid surface, impenetrable liquid/solid interface, and no-slip boundary. Hence, the typical methods proposed to relieve the dynamical singularity near the contact line are the precursor film, surface roughness, diffuse interface, and nonlinear slip boundary, aiming the four assumptions respectively, as shown in Fig. 8.

Examples of Application

Nanostructures of Silicon Used as the Anode Material of Lithium Ion Batteries

Rechargeable lithium ion batteries become the most suitable energy carrier for portable electro-equipments, electromobiles, and high performance computing, not only for the high energy density and low cost, but also for the environmental needs for energy storage. Silicon

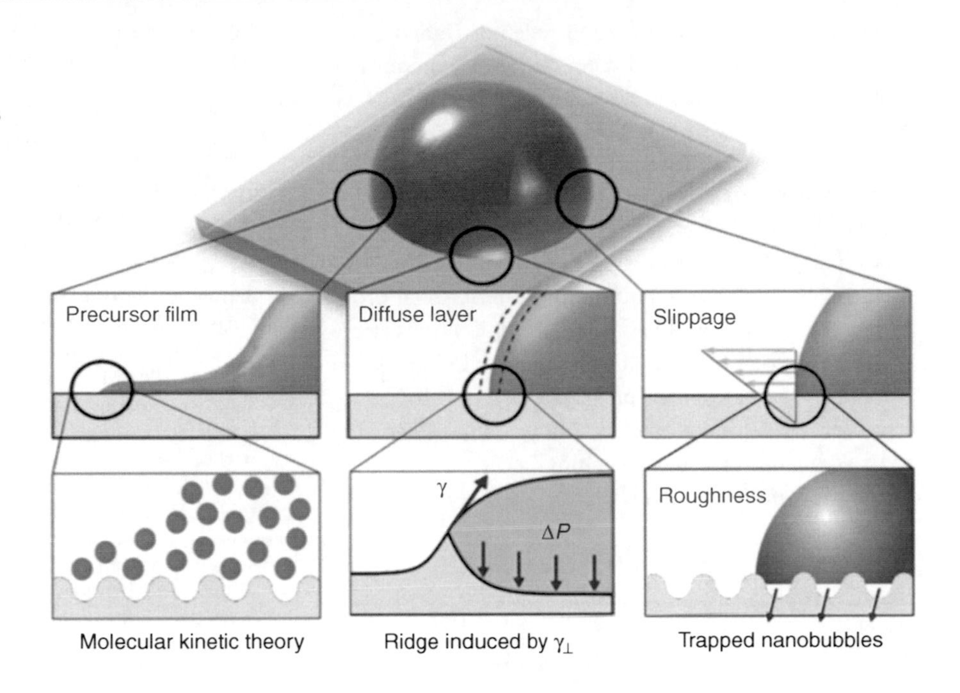

Surface Tension Effects of Nanostructures, Fig. 8 Possible mechanisms to solve the Huh and Scriven's paradox

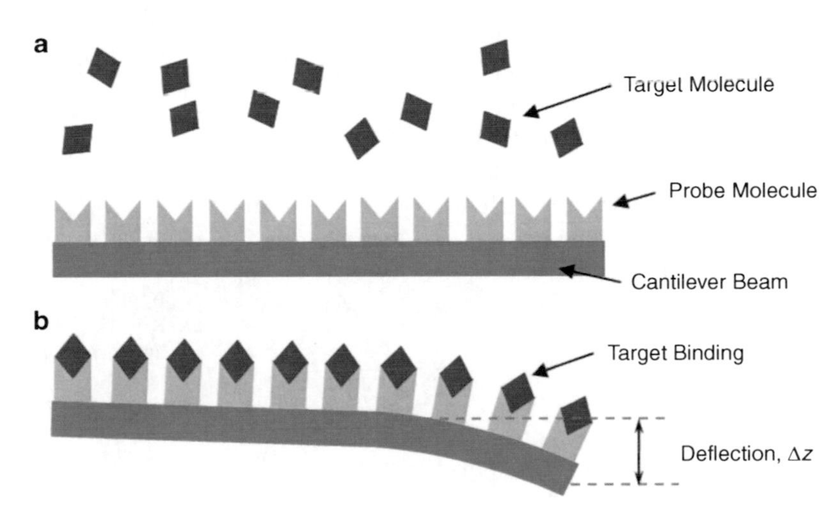

Surface Tension Effects of Nanostructures, Fig. 9 Schematic illustration of the deflection of a cantilever sensor due to the surface stress induced by surface tension effects [8, 17]. (**a**) The cantilever is functionalized on one side with the probe molecules. (**b**) After the target binding, the cantilever bends and the deflection can be measured

is selected as a promising anode material of the lithium ion batteries due to the high-energy density (about 4,200 $mAhg^{-1}$). Nevertheless, a loss of electrical contact due to the fracture and crack of the bulk silicon induced by huge volume change (~400%) in charging and discharging cycles hinders its applications. In recent years, nanostructured silicon is investigated as the material for electrode effectively circumvented the fragmentation, since surface tension effects of the nanostructured silicon on diffusion induced stresses would have much effect on the stress distribution [23].

Surface Tension Effects Induced the Deflection of the Cantilever Sensor

Surface stresses scale linearly with dimension and surface to volume ratio increases as micro/nanostructure scale decreases. Therefore, surface stress can be very important in microstructures in the size domain of MEMS/NEMS. Surface stress has been used as an effective molecular recognition mechanism. For a microcantilever used in a bioactuator, some biomaterials and biomolecules, which are immobilized on the microcantilever surface, are used to convert chemical energy into mechanical energy. When excitations such as DNA hybridization and receptor-ligand binding are applied, the microcantilever generates a nanomechanical deflection response due to the surface tension effects, as shown in Fig. 9.

Surface Tension Effects Used in the Application of Electrowetting

One particularly promising application area for electrowetting is the manipulation of individual droplets in digital microfluidic systems. Applications range from "lab-on-a-chip" devices to adjustable lenses and new kinds of electronic displays. Besides, electrowetting has been used for the application to displays showing video content since the switching speed is very high (only a few milliseconds) [10].

Summary

The surface tension effects become particularly predominant in nanostructures since the surface to volume ratio is sizable. In this entry, the surface tension effects that lead to the size-dependent mechanical properties of nanostructures are considered. Some key research findings related to this issue were listed and several examples of application were given, which are expected to be helpful to readers. Understanding and controlling these surface tension effects is a basic goal in the design and application of nanodevices.

Cross-References

▶ Disjoining Pressure and Capillary Adhesion
▶ Electrowetting
▶ Nanoscale Properties of Solid–Liquid Interfaces
▶ Surface Energy and Chemical Potential at Nanoscale
▶ Wetting Transitions

References

1. Bhushan, B. (ed.): Handbook of Nanotechnology. Springer, New York (2010)
2. Gibbs, J.W.: On the equilibrium of heterogeneous substances. In: Gibbs, J.W. (ed.) The Scientific Papers of J. Willard Gibbs. Volume 1: Thermodynamics, pp. 55–353. Dover, New York (1961)
3. Cahn, J.W., Hilliard, J.E.: Free energy of a nonuniform system. I. Interfacial free energy. J. Chem. Phys. **28**, 258–267 (1957)
4. Lu, W., Suo, Z.: Dynamics of nanoscale pattern formation of an epitaxial monolayer. J. Mech. Phys. Solids. **49**, 1937–1950 (2001)
5. Butt, H.J., Graf, K., Kappl, M.: Physics and Chemistry of Interface. Wiley-VCH, Weinheim (2003)
6. Shuttleworth, R.: The surface tension of solids. Proc. Phys. Soc. A. **63**, 444–457 (1950)
7. Herring, C.: Surface tension as a motivation for sintering. In: Kingston W.E. (ed.) The Physics of Powder Metallurgy. pp. 143–179, McGraw Hill, New York (1951)
8. Fritz, J., Baller, M.K., Lang, H.P., Rothuizen, H., Vettiger, P., Meyer, E., Guntherodt, H.J., Gerber, C., Gimzewski, J.K.: Translating biomolecular recognition into nanomechanics. Science **288**, 316–318 (2000)
9. Stoney, G.: The tension of metallic films deposited by electrolysis. Proc. R. Soc. Lond. A. **82**, 172–175 (1909)
10. Berthier, J.: Microdrops and Digital Microfluidics. William Andrew, Norwich (2008)
11. De Gennes, P.G.: Wetting: statics and dynamics. Rev. Mod. Phys **57**, 827–863 (1985)
12. De Gennes, P.G., Brochard-Wyart, F., Quere, D.: Capillarity and Wetting Phenomena. Springer, New York (2004)
13. Yuan, Q.Z., Zhao, Y.P.: Precursor film in dynamic wetting, electrowetting and electro-elasto-capillarity. Phys. Rev. Lett. **104**, 246101 (2010)
14. Gurtin, M.E., Murdoch, A.I.: A continuum theory of elastic material surfaces. Arch. Ration. Mech. Anal. **57**, 291–323 (1975)
15. Wang, Z.Q., Zhao, Y.P., Huang, Z.P.: The effects of surface tension on the elastic properties of nano structures. Int. J. Eng. Sci. **48**, 140–150 (2010)
16. Zhang, Y., Ren, Q., Zhao, Y.P.: Modelling analysis of surface stress on a rectangular cantilever beam. J. Phys. D: Appl. Phys. **37**, 2140–2145 (2004)
17. Ren, Q., Zhao, Y.P.: A nanomechanical device based on light-driven proton pumps. Nanotechnol. **17**, 1778–1785 (2006)
18. Guo, J.G., Zhao, Y.P.: The size-dependent bending elastic properties of nanobeams with surface effects. Nanotechnol. **18**, 295701 (2007)
19. Shanahan, M.E.R., Carré, A.: Nanometric solid deformation of soft materials in capillary phenomena. In: Rosoff, M. (ed.) Nano-Surface Chemistry. CRC Press, New York (2001)
20. Yu, Y.S., Zhao, Y.P.: Elastic deformation of soft membrane with finite thickness induced by a sessile liquid droplet. J. Colloid Interf. Sci. **339**, 489–494 (2009)
21. Huh, C., Scriven, L.: Hydrodynamic model of steady movement of a solid/liquid/fluid contact line. J. Colloid Interface Sci. **35**, 85–101 (1971)
22. Wang, F.C., Zhao, Y.P.: Slip boundary conditions based on molecular kinetic theory: The critical shear stress and the energy dissipation at the liquid-solid interface. Soft Matter **7**, 8628–8634 (2011)
23. Cheng, Y.T., Verbrugge, M.W.: Evolution of stress within a spherical insertion electrode particle under potentiostatic and galvanostatic operation. J. Power Sources **190**, 453–460 (2009)

Surface Tension–Driven Flow

► Capillary Flow

Surface Tension–Powered Self-Assembly

► Capillary Origami

Surface-Modified Microfluidics and Nanofluidics

Shaurya Prakash
Department of Mechanical and Aerospace Engineering, The Ohio State University, Columbus, OH, USA

Synonyms

Heterogeneous walls; Lab-on-a-chip (LOC); Micro-total analytical systems (μ-TAS)

Definition

Fluid phenomena in micrometer- or nanometer-sized channels are governed by coupled principles from fluid mechanics, surface chemistry, electrochemistry, and electrostatics. The smaller length scales compared to traditional fluid mechanics provide several new and interesting phenomena due to significant enhancement of the surface-area-to-volume ratio, which makes surface-mediated flows important to the overall field of microfluidics and nanofluidics.

Introduction to Surfaces in Microfluidics and Nanofluidics

Microfluidic and nanofluidic devices and systems are characterized by high surface-area-to-volume (*SA*/*V*) ratio. For example, a rectangular cross-section channel with 100 μm width and 100 nm depth and 1 cm length will have a *SA*/*V* ratio on the order of 10^6 m^{-1}. In fact, operational devices incorporating components with *SA*/*V* ratio on the order of 10^9 m^{-1} have already been reported [1]. Consequently, the influence of device and system walls in affecting phenomena within confined spaces can no longer be ignored since the walls (i.e., surfaces) of the devices and systems interact extensively and directly with the species contained within these devices. Therefore, the surface of interest here is defined as an interface or thin region in space (often only a few nm in extent) that influences transport and reaction phenomena in its vicinity. In this surface region, properties such as chemical composition, refractive index, mechanical strength, and conductivity can significantly differ from the bulk material of the underlying substrate.

Scaling of Surface Forces in Microchannels and Nanochannels

One non-dimensional parameter often used to characterize flow regimes, is the ratio of the inertial forces of the fluid to the viscous (or frictional) forces, i.e. the Reynolds number, Re, and is given by

$$\mathrm{Re} = \frac{\rho V L_c}{\mu}, \tag{1}$$

where ρ is the fluid density, μ is the fluid viscosity, and V is the fluid velocity.

As seen from Eq. 1, Re is directly proportional to the characteristic length, L_c, of the flow. With L_c decreasing (and *SA*/*V* increasing), the inertial forces decrease, and as a result the Re becomes smaller. The direct consequence is that for given flow parameters of fixed velocity and fluid type, a decreasing L_c implies increasing influence of viscous forces in contrast to the inertial forces.

In most microfluidics and nanofluidics applications, a particle (ion, colloid, biomolecule, AFM probe tip, etc.) will interact with the channel walls. Considering three important forces, electrostatic or Coulombic, F_{el}, van der Waals, F_{vdw}, and hydrodynamic forces, F_h, will provide a better insight into surface-particle interactions and subsequent phenomena. For an infinite flat surface separated by a distance D from a particle of radius R, F_{el} is given by [2]

$$F_{el} = -\frac{2\pi R L_D}{\varepsilon\varepsilon_0}\left[2\sigma_S\sigma_P \exp\left(-\frac{D}{\lambda_D}\right) + \left(\sigma_S^2 + \sigma_P^2\right)\exp\left(-\frac{2D}{\lambda_D}\right)\right] \tag{2}$$

where λ_D is the Debye length and σ_s and σ_p are the surface charge densities of sample and particle, respectively. F_{vdw} is given by [2]

$$F_{vdw} = \frac{A_H R}{6D^2}, \tag{3}$$

where A_H is the Hamaker constant, which serves as an indicator of the interaction between the particle and the sample surface, and F_h is given by [2]

$$F_h = b_s V = -f^* \frac{6\pi\mu R^2 V}{D}, \tag{4}$$

where V is the particle velocity, b_s is the hydrodynamic damping coefficient, and μ is the viscosity of the fluid. The coefficient f^* is related to the boundary slip properties. If the no-slip assumption holds true at the solid/water interface, then $f^* = 1$ and if slip exists then $f^* < 1$. As an aside, it should be noted that Eq. 4 is often used in literature for determining slip lengths using a colloidal AFM probe tip. As the operational length scales (here, characteristic length scale is separation distance, D) decrease, it can be seen from Eqs. 2–4 that contributions for F_{el} displays an exponential dependence, F_{vdw} an inverse quadratic dependence, and F_h an inverse linear dependence on the separation distance. Therefore, consideration of each force term at the microscale, and even more so at the nanoscale is essential toward completely describing importance of the wall-species interactions.

Most microfluidic and nanofluidic devices with liquid flows are driven by electric fields and fall under the category of electrokinetic flows. The main reason for using applied voltages to generate electric fields for driving flows is the scaling of pressure forces. For example, for a circular nanochannel with laminar, incompressible, Poiseuille flow with water as the working fluid, the necessary pressure drop across a 100 μm long channel which is 1 nm in diameter for only an attoliter (10^{-18} m^3) per second would be greater than 3 GPa, which is impractical for any device. Electrokinetic flows can sustain higher flow rates through nanometer channels without excessive pressures [1].

Methods for Surface Modification

Surface modification methods can be divided in two broad categories: physical and chemical methods [3]. Physical methods, in most cases, do not change the chemical composition of the surface but may change the surface roughness, grain sizes and grain boundaries, and faceting. Physical methods often use lasers, plasmas, temperature, ion beams, ball-milling, and polishing or grinding, to alter the surface state of a material of interest. While, the main intent with physical modification methods is to not alter the chemical composition of the material, in some cases, physical surface modification methods can lead to changes in the chemical composition of the surface due to removal or addition of material or chemical reactions on surfaces, as in the case of selective or ion-beam sputtering, or by selective cross-linking in presence of plasmas. Temperature gradients and thermal treatments have been used to change surface roughness, grain sizes, and grain boundaries, and create nanoscale features, facets, textures, and nanoparticles on ceramics, metals, polymers, and semiconductors. Thermal treatments in the presence of gases such as oxygen or water vapor can cause creation of steps or induce other forms of nanostructures. For example, as was discussed in a recent review [3] that for crystalline α-Al_2O_3 surfaces heat-treated at 1,500°C in Ar/O_2 and $H_2/He/O_2$ led to step formation and roughening as quantified through AFM topology images.

Chemical methods introduce a change in the chemical composition at the surface by introducing chemical properties (e.g., surface charge density or surface energy) different from the bulk material. Among the methods for chemical modification, formation of surface layers, either covalently bonded or physisorbed, has been most common. Other chemical methods include treatment with UV light and reactive plasmas. Modification schemes are governed by a wide range of parameters including sample type (polymers, metals, ceramics, etc.), stability to treatment conditions (e.g., thermal or structural), and eventual applications. For example, polymeric surfaces are often modified by photochemical methods of which UV irradiation in air, other reactive atmospheres such as ozone combined with lasers, and grafting surface layers is fairly common. Use of reactive plasmas has been gaining popularity due the wide compatibility of

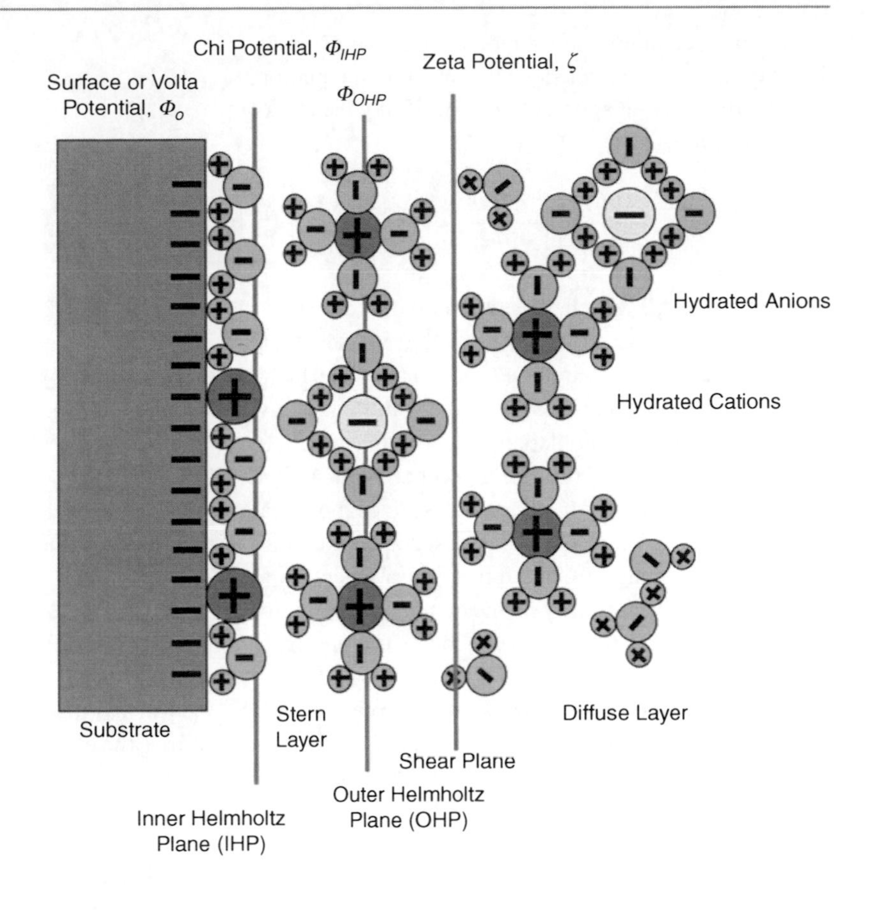

Surface-Modified Microfluidics and Nanofluidics, Fig. 1 Schematic representation of the electric double layer. The surface charge and potentials are depicted based on the theories developed over the past century (Figure from [1])

materials and integration to microfabrication processes for device development. Many gas plasmas have been used such as air, oxygen, water vapor, ammonia, and argon for modification of polymer surfaces. Plasma modification processes generate new chemical species on polymer surfaces. The new chemical species can arise due to surface reactions with reactive gases or due to physical sputtering (such as with Ar plasma) caused by active gas-phase species. These new surface chemical species can provide an anchor for attaching a series of different molecules that display different properties from the underlying bulk polymer. For example, preferential hydroxylation of poly(methyl methacrylate) or PMMA surfaces by use of a water vapor plasma as opposed to an oxygen plasma has been used to activate surfaces toward trichlorosilane modification and subsequent "click" chemistries to form surface scaffolds of desired functionalities. The "click" chemistry method has also been used to modify glass channels for systematic control over electroosmotic flow velocity [4].

Applications

In this section, applications utilizing modified surfaces are presented along with short discussions of the underlying physics. Chemically modified surfaces are influenced by changes to surface charge density, as discussed above. Any charged surface in contact with a liquid forms an electric double layer (EDL). A schematic diagram illustrating the classic EDL structure is shown in Fig. 1, depicting the positions to which different characteristic surface charge–related potentials are referenced within the EDL.

One direct consequence of surface modification is a change in the interaction forces between the surface and the surroundings. Specifically for microfluidics and nanofluidics, changes to surface charge density, surface roughness, and surface energy alter the electrostatic and van der Waals forces that determine flow (ionic and fluid) phenomena. Following the scaling of pressure forces discussed above, electric fields and capillary forces are the two most common tools used to fill microchannels and nanochannels with liquids. For capillary forces, consider the Washburn equation, which provides one approach to quantifying the rate of channel filling,

$$\frac{dl_f}{dt} = \frac{a\gamma \cos\theta_c}{4\mu l_f}, \tag{5}$$

where l_f is the fill length, a is the radius (or characteristic length) of the channel, θ_c is the contact angle between the fluid wall and the fluid, γ is the interfacial (or surface) tension, and μ is the fluid viscosity. Equation 5 shows the dependence of the capillary (or micro/nanochannel) filling as a function of surface properties governed by θ_c. Chemical surface modification can easily affect a change to the surface wettability and therefore change θ_c. Therefore, for a given liquid such as water in a micro- or nanochannel the rate of filling decreases as the surface becomes progressively more hydrophobic.

Given the importance of walls for microfluidics and nanofluidics, a contrast to the boundary conditions with traditional fluid mechanics should be considered. For example, macroscale fluid mechanics assumes a no-slip condition at the wall, which implies that the velocity of the fluid layer in contact with the walls is the same as that of the walls. However, in the early 1800s, Navier postulated that fluid slip may be possible and suggested the slip boundary condition with u_w, the velocity at the wall (tangential component), can be expressed by the relationship in Eq. 6:

$$u_w = b\frac{\partial u_b}{\partial y} \tag{6}$$

where b is the slip length, u_b is the bulk velocity, and y the axis perpendicular to the wall. Figure 2 shows a conceptual schematic for defining the physical effect of slip flow. Conceptually, one can imagine the slip length as the additional distance the wall must be extended to where the fluid velocity would be zero.

Literature has noted that there are three types of slip [5]. First is molecular or intrinsic slip which occurs when molecular forces (e.g., van der Waals) are balanced by viscous forces. Intrinsic slip is usually observed at very high shear rates (e.g., shear rate of $10^{12}\ s^{-1}$ for water). Second is apparent slip in which there is a finite distance between the slip and no-slip planes. Electrokinetically driven microfluidics and

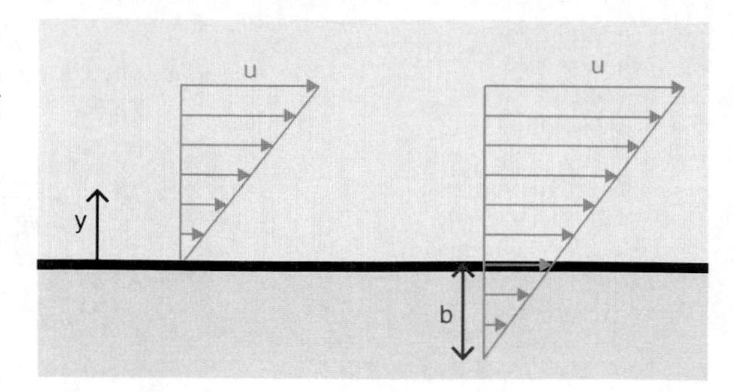

Surface-Modified Microfluidics and Nanofluidics, Fig. 2 Conceptual schematic for the slip boundary condition. The *left panel* shows the no-slip boundary condition for a stationary wall, with the wall velocity and the tangential component of fluid velocity being zero. The *right panel* shows a case for slip flow with a finite slip length (Figure based on [5, 6], and assistance from Mr. M. Hansen in drawing the figure is acknowledged)

nanofluidics follows this type of slip with the slip plane (location where ζ potential is defined in Fig. 1) being at a finite distance from the physical channel wall. Third is the effective slip which refers to the estimate of either intrinsic or apparent slip by averaging a measurement over the length scale of the experiment. The main factors that affect slip are roughness of the surface and wettability, which presents a measure of the surface energy of the surface. Other possible factors are gases trapped in the fluid, nanobubbles, shear rates of fluid at the surface, and absorbates in the fluid [5–7], all of which can be influenced by surface modification. Recently, it was also demonstrated that surface charge can alter the drainage velocity for confined fluids indicating possibly a role for surface charge in mediating slip flow [2]. Additional use of slip phenomena has been in developing drag reduction configurations both for laminar and turbulent flows, self-cleaning superhydrophobic surfaces [8], and enhanced mixing in microfluidic devices [9].

Continuing with a discussion of applications relevant to microfluidics and nanofluidics, valving (or metering of fluids) is one critical operation for sample manipulation for lab-on-chip devices. Valving by surface-mediated flows is a passive method due to a fixed surface state and can be achieved by the tendency of the fluid to favor a channel direction based on the wettability or the surface charge of the channel. Examples include use of hydrophobic-hydrophilic interfaces to generate boundaries for directing water [10] and use of chemically modified surfaces for exploiting differences in surface charge for selectively driving electrokinetically driven flows in microchannels [11].

Summary

Microfluidic and nanofluidic phenomena are governed by large *SA/V* ratios. Therefore, interactions between fluids, species, and the channel walls are critical. Toward the development of devices and systems enabling applications and related fluid control, the ability to systematically modify and control wall properties provides a powerful tool for scientists and engineers.

Cross-References

▶ AC Electroosmosis: Basics and Lab-on-a-Chip Applications
▶ Applications of Nanofluidics
▶ Electrokinetic Fluid Flow in Nanostructures

References

1. Prakash, S., Piruska, A., Gatimu, E.N., Bohn, P.W., Sweedler, J.V., Shannon, M.A.: Nanofluidics: systems and applications. IEEE Sens. J. **8**, 441–450 (2008)
2. Wu, Y., Misra, S., Karacor, M.B., Prakash, S., Shannon, M.A.: Dynamic response of AFM cantilevers to dissimilar functionalized silica surfaces in aqueous electrolyte solutions. Langmuir **26**(22), 16963–16972 (2010)
3. Prakash, S., Karacor, M.B., Banerjee, S.: Surface modification in microsystems and nanosystems. Surf. Sci. Rep. **64**(7), 233–254 (2009)
4. Prakash, S., Long, T.M., Selby, J.C., Moore, J.S., Shannon, M.A., Shannon, M.A.: 'Click' modification of silica surfaces and glass microfluidic channels. Anal. Chem. **79**(4), 1661–1667 (2007)
5. Lauga, E., Stone, H.A., Brenner, M.P.: Microfluidics: the no-slip boundary condition. In: Tropea, C., Yarin, A., Foss, J.F. (eds.) Handbook of Experimental Fluid Dynamics, pp. 1219–1240. Springer, New York (2007)
6. Granick, S., Zhu, Y., Lee, H.: Slippery questions about complex fluids flowing past solids. Nat. Mater. **2**, 221–227 (2003)
7. Neto, C., Evans, D.R., Bonaccurso, E., Butt, H.-J., Craig, V.S. J.: Boundary slip in Newtonian liquids: a review of experimental studies. Rep. Progr. Phys. **68**, 2859–2897 (2005)
8. Bhushan, B., Jung, Y.C.: Natural and artificial surfaces for superhydrophobicity, self-cleaning, low adhesion, and drag reduction. Prog. Mater. Sci. **56**, 1–108 (2011)
9. Rothstein, J.P.: Slip on superhydrophobic surfaces. Annu. Rev. Fluid Mech. **42**, 89–109 (2010)
10. Zhao, B., Moore, J.S., Beebe, D.J.: Surface-directed liquid flow inside microchannels. Science **291**(5506), 1023–1026 (2001)
11. Prakash, S., Karacor, M.B.: Surface mediated flows in glass nanofluidic channels. In: Solid state sensors, actuators, and microsystems workshop. Transducer Research Foundation, Hilton Head Island, SC (2010)

Surface-Plasmon-Enhanced Solar Energy Conversion

▶ Plasmonic Structures for Solar Energy Harvesting

Synthesis of Carbon Nanotubes

Simon J. Henley, José V. Anguita and S. Ravi P. Silva
Nano Electronics Center, Advanced Technology Institute, University of Surrey, Guildford, Surrey, UK

Synonyms

Growth of carbon nanotubes; Growth of CNTs

Definition

Carbon nanotubes (CNTs) are allotropes of graphitic carbon with a cylindrical structure and diameters <100 nm. A CNT can consist of one or many concentric graphene sheets rolled up as cylinders. CNTs are of interest for a wide variety of technological applications. In this entry, the various experimental methods to synthesize carbon nanotubes are introduced.

Introduction

Carbon nanotubes (CNTs) were first brought to worldwide attention by Iijima in 1991 [1] after he analyzed, by electron microscopy, the samples produced during an electrical arc discharge between carbon rods held in a helium atmosphere. He observed nanoscale hollow tubes, similar to those seen by Russian researchers in the 1950s [2]. Carbon nanotubes can be visualized as graphene sheets rolled up to form tubes; if one sheet is rolled up a single-walled carbon nanotube (SWCNT) is formed. The structures formed when two or more concentric tubes are present are termed multiwalled carbon nanotubes (MCWNTs) [3–5]. The high temperatures formed during an electrical arc give an idea as to the extreme conditions that were thought to be required to form these structures. Since then, a wide variety of different growth technologies have been developed covering a broad range of different environmental conditions, from high-temperature laser vaporization, down to catalytically enhanced growth in rarefied gas mixtures at much lower temperatures.

Initially, the arc discharge method [2, 6] was mainly employed to produce carbon nanotubes. Later, many more techniques such as laser ablation [7] or chemical vapor deposition (CVD) [8] were developed, researched, and optimized toward the large-scale synthesis of CNTs with repeatable and controllable morphologies. The synthesis of CNTs has now moved from a small-scale research activity to an industrial scale, with production facilities producing hundreds of tons per year. However, CNTs are not only synthesized in laboratories; they can be formed in flames produced by burning organic chemicals such as methane and benzene. They have been found also in soot in the air, likely from industrial or automotive processes, and have been generated by metallurgical process such as those used to make Damascus Steel. These naturally occurring CNTs are typically of low quality though. In this entry the main methods used to synthesize CNTs are introduced.

Synthesis Methods

Arc Discharge

The arc discharge technique (see Fig. 1) generally involves the use of two high-purity graphite electrodes. The electrodes are brought in close proximity and an electric arc is struck between them by connection to a high-current, low-voltage power supply [3, 6, 9]. Similar electric arcs were used by Roger Bacon in the early 1960s to synthesize large carbon fibers. The synthesis is carried out in a controlled environment at a low pressure (50–700 millibar) of an inert gas, typically helium or argon.

The distance between the electrodes is controlled to keep a large current (50–150 A) flowing between them. The conditions in the region between the electrodes are extreme, and a high-temperature plasma is formed. The temperature is high enough that carbon evaporates (sublimes) from the anode. After the reaction time (typically 30 s to a couple of minutes) has expired and the system has cooled, the products can be collected. The products take different forms depending on the region of the reactor that they are collected from. Carbon soot is found to coat the chamber walls, and a more solid deposit is left on the cathode. The soot contains mainly fullerenes while MWCNTs and graphitic nanoparticles are found in the solid deposit. The yield of the process can be quite high and the CNTs are of high quality, typically, with few defects. The rate of CNT synthesis can be high, producing around 20–100 mg/min [3]. The rate of synthesis has been shown to increase with increasing gas pressure. Although higher currents during the arc would be expected to produce more material, it has been shown that the deposit becomes fused at higher current, actually reducing the effective yield [3, 6]. As no metallic catalyst particles are required for the growth of MWCNTs, arc discharge tubes can be used without acidic treatment to remove the metal.

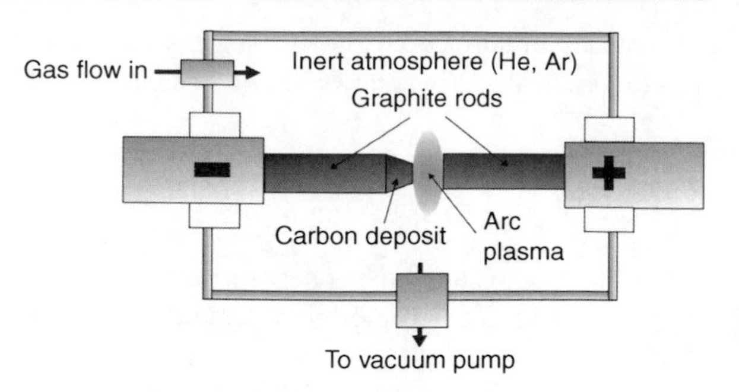

Synthesis of Carbon Nanotubes, Fig. 1 Schematic diagram of an arc-discharge system to synthesize CNTs

When metal catalysts (e.g., iron, nickel, or cobalt) are incorporated into the anode and evaporated along with the graphite, the nature of the deposits changes and SWCNTs can be synthesized. The deposit on the electrode is found to contain SWCNTs, MWCNTs, metal-filled MWCNTs, and other graphitic contaminants, while the soot contains MWNTs and SWNTs. In addition, in some situations a "collar" is observed around the deposit containing mainly SWNTs (80%) with diameters of around 1 nm. The exact nature of the growth products is determined strongly by the physical conditions during the arc such as the gas pressure, the carbon flux, the arc current, and the composition of

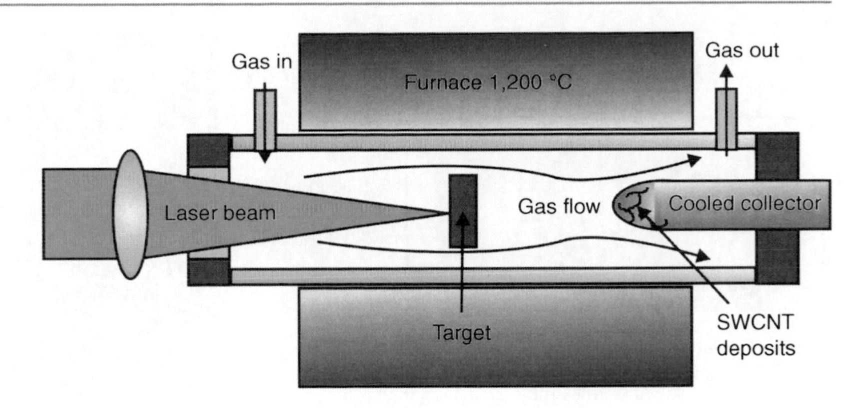

Synthesis of Carbon Nanotubes, Fig. 2 Schematic diagram of an arc-discharge system to synthesize CNTs

the catalyst. Growth models have been proposed by a range of groups in order to try and explain the process [3, 6].

Laser Ablation

The high power densities produced by focusing a short duration laser pulse onto a target leads to very rapid heating of the material at the focal point. Subsequent thermal and other direct sputtering processes ablate material from the target and the interaction between the laser pulse and the ablated material forms a highly energetic, non-equilibrium, plasma plume [10]. The presence of ambient gas can have a dramatic effect on the expansion dynamics of these laser-ablated plasmas. At higher pressures plume confinement, and plume heating, occurs. The high collision rates enhance conversion of the plume stream velocity into thermal energy, producing a confined, high-temperature plasma, with short mean free paths for species. In this regime, the self-assembly of nanoclusters in the gas phase can occur. By ablating a target containing a mixture of the growth material and a suitable catalyst, and by controlling the gas chemistry, nanomaterials such as CNTs can be self-assembled in the gas phase.

In the laser ablation CNT synthesis process (see Fig. 2), a pulsed laser vaporizes a graphite target held at a high temperature (typically 1,000–1,200°C) inside a tube, within a furnace. An inert gas flows through the tube driving the ablation products toward one end. SWCNTs that grow in the gas phase are deposited in the cooler parts of the reactor as the vaporized carbon condenses. A water-cooled collector may be included in the system to collect the CNTs [3, 5, 9].

The laser ablation synthesis method was demonstrated by Smalley et al. [7, 11] where the growth of MWCNTs was demonstrated. Later this method was developed further to produce SWCNTs by ablating graphite targets mixed with metallic elements such as cobalt and nickel [3, 11–13]. The laser ablation method with a catalyst present has a yield around 70% and produces primarily single-walled carbon nanotubes with a diameter determined by the temperature of the furnace. The quality of the SWCNTs is very high, although there is typically a significant quantity of catalyst nanoparticles present in the product, which may need to be removed by acid treatment producing additional defects.

Chemical Vapor Deposition

Chemical vapor deposition (CVD) is a vacuum deposition technique that allows the deposition of materials uniformly over large surface areas (typically flat surfaces) with high levels of purity. For this reason, CVD equipment (and its variants) is commonly found in semiconductor manufacturing foundries, and is used for depositing thin layers of materials on silicon wafers. For CNT growth, a catalyst is often used, in the form of a thin metal layer on top of the substrate, in order to lower the temperature that is required for CNT growth [3, 8, 9]. This method is sometimes is referred to as catalytic CVD (CCVD) in the literature due to the involvement of catalytic activity during the CVD growth process [3, 8]. In this method, the substrate plus catalyst is mounted on a heated stage, inside a vacuum chamber. A schematic of the CVD arrangement is shown in Fig. 3.

The temperature of the substrate is raised, typically to between 400°C and 1,000°C in a low pressure of hydrogen gas. At elevated temperatures, the catalyst thin film breaks up spontaneously and forms

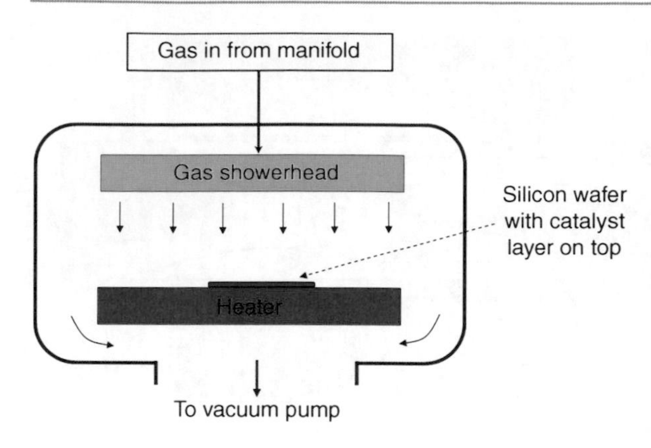

Synthesis of Carbon Nanotubes, Fig. 3 Schematic of a CVD system

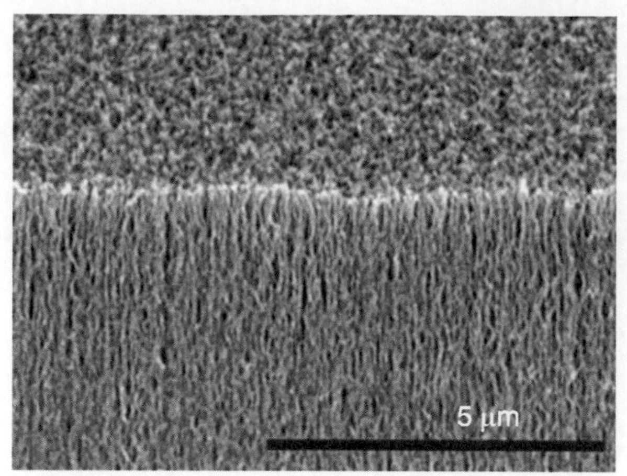

Synthesis of Carbon Nanotubes, Fig. 4 SEM image of long CNTs grown by water-assisted CVD

nanoparticles on the surface of the substrate. A carbon-containing gas, typically a hydrocarbon such as acetylene or methane, is then bled into the chamber, together with a carrier gas, typically hydrogen. Hydrocarbon concentrations of typically between 1% to 10% are used. Upon contact with the catalyst nanoparticles, the hydrocarbon gas decomposes, and releases carbon to the individual nanoparticles. The nanoparticles then extrude this carbon in the form of concentric graphitic sheets (the nanotube). A dynamic equilibrium is set up at the nanoparticles between the hydrocarbon decomposition rate and the growth speed of the CNTs, which ensures the continual growth of the CNTs. The exact growth mechanism and events that take place at the nanoparticles during growth is still the subject of debate [8].

Although commercial CVD equipment has been available for more than half a century for the deposition of traditional semiconductors and dielectrics, the development of systems solely for CNT growth is only recent, and commercial systems (e.g., Surrey NanoSystems NanoGrowth system.) have only been released to the market within the last 5 years. This is due to the more recent discovery of the techniques that allow for growth processes that are compatible with the already existing semiconductor infrastructure. In particular, techniques for reducing the growth temperatures of CNTs have been sought, as silicon processing only allows maximum operating temperatures of up to around 450°C before degradation of the silicon diffusion layers starts to take place [14, 15].

Water-Assisted CVD

Another area that is still subject of debate is the mechanism by which CNT growth termination occurs when using CCVD techniques. It is noted that CNT growth stops abruptly sometime after growth initiation, typically after around 5–10 min. It is believed that this takes place due to the buildup of amorphous carbon at the surface of the catalyst particles, which effectively "poisons" the catalytic properties of the particles, inhibiting the dynamic equilibrium required for CNT growth. This theory was enlightened when samples of CNTs where growth had terminated were cleaned using an oxygen plasma to remove this amorphous carbon. It was noted that CNT growth could resume after performing this step.

Despite this success, it was clear that introducing oxygen gas as a cleaning agent during CNT growth (which involves hydrogen and hydrocarbon gas mix) would be unsafe. For this reason water vapor is currently used for this purpose, which although is less effective than oxygen at removing amorphous carbon, it is compatible with the CNT growth gas mix. It is now well accepted that introducing a small amount of water vapor to the hydrogen and hydrocarbon gas mix reduces the growth termination effect (see Fig. 4). This allows growing CNTs in lengths up to several millimeters, reaching the centimeter scale, in a single run [16].

Plasma-Enhanced CVD

A further variation of the CVD process is plasma-enhanced CVD (PECVD) [17]. In this case, a plasma is struck between the gas showerhead and the substrate electrode (with the heater), by means of the application of a radio frequency to the showerhead electrode. This arrangement is depicted in Fig. 5. In this case, a fraction

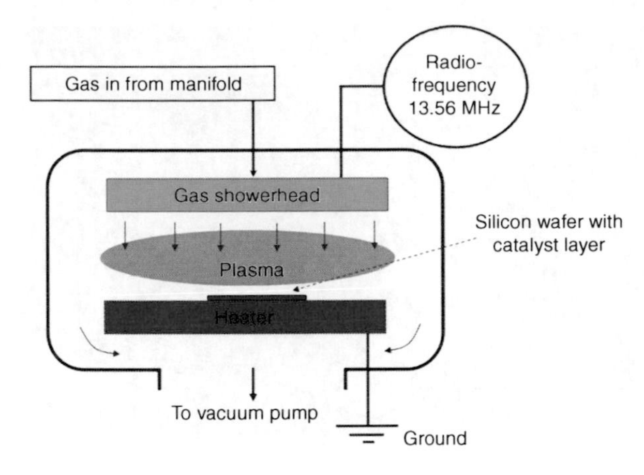

Synthesis of Carbon Nanotubes, Fig. 5 Simplified schematic of a PECVD system

of the hydrocarbon gas decomposes in the plasma phase to produce smaller molecules and energetic species which provide additional energy at the catalytic nanoparticles, allowing a further reduction of the temperature of the substrate. Additionally, this plasma provides a strong electric field between the surface of the substrate and the plasma sheath (a region of the plasma that is located only a few millimeters above the substrate electrode). This electric field provides a vertically upward pull force on the nanoparticles, which results in a high degree of CNT verticality, as depicted in Fig. 6a. This allows for the formation of vertical CNT "forests" even when the CNT density is low. Although (PE)CVD can produce large quantities of CNTs, the quality of the material is lower than that produced by arc discharge and laser ablation, with many structural defects present in the tubes (see Fig. 6b).

As for the case of CVD, PECVD process equipment is also common in silicon manufacturing foundries, and has been commercially available for many decades. However, specific equipment dedicated to CNT growth has only been available in recent years. This is partly because of specialized techniques that are involved in striking a stable plasma at the higher gas pressures required for CNT growth, typically 5 Torr and above, up to several hundred Torr. These pressures are much higher than those typically associated with the common PECVD processes, of a few hundred millitorr.

Fluidized Bed CVD

The CVD process is a promising route for the bulk production of CNTs, however producing truly industrial scale volumes of material is challenging. Fluidized bed CVD (FBCVD) reactors are becoming one of the most widely targeted methods to achieve these scales. However, scale-up of the reactor is still a significant challenge, and this field has become an active area of research [18, 19]. In the FBCVD process, the CVD reaction occurs within a fluidized bed of catalyst particles. The setup consists of a reaction tube inside a cylindrical furnace. The contiguous mixing that occurs in the catalyst beds allows efficient utilization of all of the catalyst particles, which are all exposed to the feed gases. One such catalytic process, CoMoCAT®, is briefly introduced, but a detailed discussion of the mechanisms can be found in a recent review article [19].

The CoMoCAT® technique uses active cobalt stabilized in a nonmetallic state by interaction with molybdenum oxide (MoO_3) in a fluidized bed. When the mixture is exposed to carbon monoxide, the Co–Mo dual oxide is carburized, producing molybdenum carbide and small Co nanoparticles which are of very uniform sizes and well dispersed, allowing the growth of SWCNTs with small diameters and a high selectivity for certain chiral indices.

Floating Catalyst CVD

Similar to the fluidized bed CVD, a high degree of catalyst utilization can be achieved if the nanoparticles are kept suspended in the gas phase. If a narrow size distribution of small catalyst nanoparticles can be generated, growth of SWCNTs can be achieved. This idea leads to a simple method for production whereby the catalyst is introduced into a CVD reactor by either (1) a syringe process using the catalyst dissolved in a carbon source, (2) by sublimation of the catalyst at elevated temperatures, or (3) using a gaseous catalyst source, e.g., $Fe(CO)_5$. The catalyst and the carbon source are directly reacted in the gas phase. One such process that has attracted much attention is the HiPCo technique, as it has been scaled to produce industrial quantities of SWCNTs. We discuss this process in more detail here.

The high-pressure carbon monoxide disproportionation process (HiPco) is a method for the production of SWCNTs using a continuous flow of high-pressure carbon monoxide as the carbon source and an iron carbonate $Fe(CO)_5$ as the source of catalyst nanoparticles [20]. This process was developed at Rice University by the group of the late R.E. Smalley. SWCNTs are produced by flowing CO, mixed with

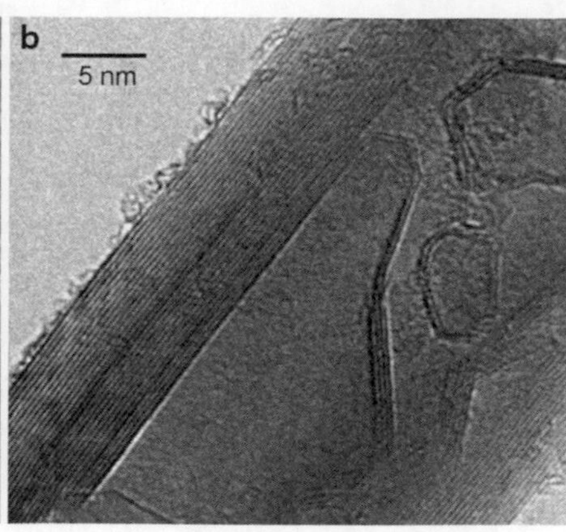

Synthesis of Carbon Nanotubes, Fig. 6 (**a**) SEM image of CNTs grown using the PECVD process, showing a high degree of verticality. (**b**) High-resolution TEM image of a CVD MWCNT showing internal structural defects and surface contaminants

Synthesis of Carbon Nanotubes, Table 1 Summary of the three main growth methods

Method	Arc discharge	Laser ablation	CVD
CNT type:	*SWCNTs*: Short nanotubes with diameters of around a nanometer *MWCNTs*: Short nanotubes with inner diameters of a few nm and outer diameter around 10 nm	*SWCNTs*: Formation of bundles of long (many microns) nanotubes, with diameters of a few nm *MWCNTs*: Not typically grown by this method	*SWCNTs*: Growth of isolated tubes with diameters in the range 1–4 nm *MWCNTs*: Growth of long (many microns) nanotubes with diameters from a few nm up to hundreds of nm (carbon fibers)
Yield	30–90%	Up to 70%	Up to 100%
Pros	Relatively simple equipment required and produces large quantities of SWCNTs or MWCNTs. CNTs produced are of a good quality, with few defects. Yield is reasonable	SWCNTs produced by this method have the highest quality of all the methods and diameter control is good. Yield is good	Can be grown over large areas and in large quantities (good for industrial scale-up) Diameters can be controlled relatively easily by size of catalyst. Low-temperature growth possible. Purity can be high
Cons	Size control is difficult and significant purification is often required	Expensive experimental setup (high peak power lasers). Purification is often required	Nanotubes are highly defective typically

a small amount of $Fe(CO)_5$ through a heated reactor. The HiPco process produces SWCNTs with diameters of approximately 1.1 nm. The yield achieved is approximately 70% SWCNTs with 97% purity at rates of up to 450 mg/h [20].

Summary of Growth Mechanisms

In this entry the main routes to synthesize CNTs have been detailed. A summary of the important points is shown in Table 1.

Purification

The technical challenges in the production of usable CNTs often do not end when the material is first grown. After growth the yield and purity of the product is often not high enough for direct utilization; so purification to remove unwanted contaminants such as metallic catalyst particles and non-CNT carbonaceous materials is required. The techniques employed typically are strong gaseous oxidation or chemical treatments such as acid refluxing, which will both have an effect (often detrimental) on the structure of the tubes [3]. Following the demonstration that carbon nanotubes could be attacked by oxidizing gases, Ebbessen et al. realized that the more defective carbon nanoparticle contaminants, such as amorphous carbon, would be oxidized more readily than the more perfect nanotubes. They found that a significant purification of CNTs could be achieved this way, but with a significant loss in yield. [6] For the chemical purification routes, a typical treatment method would involve: (a) dispersal in organic solvent and filtration to remove large particulates,

(b) treatment with concentrated acids to remove fullerenes and catalyst particles, (c) centrifugal separation, and (d) microfiltration. Such a chemical treatment can cause damage to the surface layer of MWCNTs and possible total destruction of SWCNTs, if too extreme.

Cross-References

- ▶ Carbon Nanotube-Metal Contact
- ▶ Carbon Nanotubes
- ▶ Carbon Nanotubes for Chip Interconnections
- ▶ Functionalization of Carbon Nanotubes

References

1. Iijima, S.: Helical microtubules of graphitic carbon. Nature **354**, 56 (1991)
2. Radushkevich, L.V., Lukyanovich, V.M.: Zurn. Fisic. Chim. **26**, 88 (1952)
3. Harris, P.F.: Carbon Nanotube Science: Synthesis, Properties and Applications, 2nd edn. Cambridge University Press, Cambridge (2009)
4. Ebbesen, T.W. (ed.): Carbon Nanotubes, Preparation and Properties. CRC Press, Boca Raton (1996)
5. Saito, R., Dresselhaus, G., Dresselhaus, M.S. (eds.): Physical Properties of Carbon Nanotubes. World Scientific, Singapore (1998)
6. Ebbesen, T.W.: Carbon nanotubes. Annu. Rev. Mater. Sci. **24**, 235 (1994)
7. Guo, T., Nikolaev, P., Rinzler, A.G., Tomanek, D., Colbert, D.T., Smalley, R.E.: Self-assembly of tubular fullerenes. J. Phys. Chem. **99**, 10694–10697 (1995)
8. Kumar, M., Ando, Y.: Chemical vapor deposition of carbon nanotubes: a review on growth mechanism and mass production. J. Nanosci. Nanotechnol. **10**, 3739–3758 (2010)
9. Dresselhaus, M.S., Dresselhaus, G., Avouris, P. (eds.): Carbon Nanotubes: Synthesis, Structure, Properties, and Applications. Topics in Applied Physics, vol. 80. Springer, Berlin (2001)
10. Ashfold, M.N.R., Claeyssens, F., Fuge, G.M., Henley, S.J.: Pulsed laser ablation and deposition of thin films. Chem. Soc. Review. **33**, 23–31 (2004)
11. Yakobson, B.I., Smalley, R.E.: Fullerene nanotubes: C-1,000,000 and beyond. Am. Sci. **85**, 324 (1997)
12. Guo, T., Nikolaev, P., Thess, A., Colbert, D.T., Smalley, R. E.: Catalytic growth of single-walled nanotubes by laser vaporization. Chem. Phys. Lett. **243**, 49–54 (1995)
13. Thess, A., Lee, R., Nikolaev, P., Dai, H., Petit, P., Robert, J., Xu, C., Lee, Y.H., Kim, S.G., Rinzler, A.G., Colbert, D.T., Scuseria, G.E., Tománek, D., Fischer, J.E., Smalley, R.E.: Crystalline ropes of metallic carbon nanotubes. Science **273**, 483 (1996)
14. Boskovic, B.O., Stolojan, V., Khan, R.U.A., Haq, S., Silva, S.R.P.: Large area synsthesis of carbon nanofibres at room temperature. Nat. Mater. **1**, 165–168 (2002)
15. Chen, G., Jensen, B., Stolojan, V., Silva, S.R.P.: Growth of carbon nanotubes at temperatures compatible with integrated circuit technologies. Carbon **49**, 280–285 (2011)
16. Amama, P.B., Pint, C.L., McJilton, L., Kim, S.M., Stach, E. A., Murray, P.T., Hauge, R.H., Maruyama, B.: Role of water in super growth of single-walled carbon nanotube carpets. Nano Lett. **9**, 44–49 (2009)
17. Meyyappan, M., Delzeit, L., Cassell, A., Hash, D.: Carbon nanotube growth by PECVD: a review. Plasma Sources Sci. Technol. **12**, 205–216 (2003)
18. See, C.H., Harris, A.T.: A review of carbon nanotube synthesis via fluidized-bed chemical vapor deposition. Ind. Eng. Chem. Res. **46**, 997–1012 (2007)
19. MacKenzie, K.J., Dunens, O.M., Harris, A.T.: An updated review of synthesis parameters and growth mechanisms for carbon nanotubes in fluidized beds. Ind. Eng. Chem. Res. **49**, 5323–5338 (2010)
20. Nikolaev, P.: Gas-phase production of single-walled carbon nanotubes from carbon monoxide: a review of the hipco process. J. Nanosci. Nanotechnol. **4**, 307 (2004)

Synthesis of Gold Nanoparticles

Munish Chanana[1], Cintia Mateo[1], Verónica Salgueirino[2] and Miguel A. Correa-Duarte[1]
[1]Departamento de Química Física, Universidade de Vigo, Vigo, Spain
[2]Departamento de Física Aplicada, Universidade de Vigo, Vigo, Spain

Synonyms

Production of gold nanoparticles

Definition

Description of different synthetic approaches for the fabrication of gold nanoparticles.

Overview

Since the beginning of recorded history, gold has always held the majestic throne among all the noble metals [1, 2]. It has been known by artisans since the Chalcolithic (Copper Age) and has become a highly coveted metal for coinage, jewelery, and other arts since that time. The first extraction of gold has been dated back to the fifth millennium B.C. in Bulgaria, but "soluble" gold (colloidal gold) first appeared around the fifth century B.C. in Egypt and China [1, 2]. Colloidal gold was used to make ruby glass and to color ceramics. Perhaps the most famous examples are the Lycurgus Cup (manufactured around fifth/fourth century B.C, exhibited in British Museum) and the pigment "Purple of Cassius" (invented by Andreas Cassius, seventeenth century) [1, 2]. But scientific research on gold sol started with Michael Faraday. In 1857, Faraday reported the formation of deep red solutions of colloidal gold by the reduction of an aqueous solution of chloroaurate ($HAuCl_4$) using phosphorus in CS_2 (a two-phase system) [1, 2].

The most relevant properties of gold colloids are based on the presence of a strong absorption band in the visible-NIR, which is the origin of the observed brilliant red/purple colors of certain gold nanoparticles in solution. This absorption band results from the collective oscillation of the conduction band electrons in resonance with the frequency of the incident electromagnetic field and is known as surface plasmon resonance (SPR) absorption. The SPR frequency, and thus the color of the gold NP mainly depends on the particle size, shape, the nature of the surrounding medium, and the interparticle distance. The influence of shape and interparticle distance is in general even greater than that of size [1, 2]. While a single absorption band is present for spherically symmetric gold particles, multiple absorption bands correlated with their various axes appear for nonspherical ones. Such structures can support both propagating and localized surface plasmons. For instance, gold nanorods possess two different resonance modes due to electron oscillation across and along the long axis of the nanorod and are commonly labeled the transverse and longitudinal modes, respectively, the latter of which is extremely sensitive to the aspect ratio of the rod [3]. Gold nanotriangles, for instance, exhibit three different resonances in the UV-Vis absorption spectra, one SPR absorption out of plane and two in-plane SPR absorptions at longer wavelengths [4, 5]. In the case of branched, platonic, or platelet nanostructures, the plasmon resonance shifts to longer wavelengths with respect to the common resonance of spherical nanoparticles at 520 nm [4–6]. Special attention requires the case of gold nanoshells which, usually due to their synthetic approach, are composed of a dielectric core coated with a thin metallic layer. Although these structures have similar properties to spherical gold nanoparticles in terms that they only present a single SPR absorption, they offer, however, the ability of tuning the SPR over a full range of wavelengths from the visible to the infrared region [4–6].

Since particle shape has a tremendous effect on the physical, chemical, optical, electronic, and catalytic properties of nanoparticles, gold nanoparticles have been synthesized in various shapes, including rods, cubes, plates, polyhedrons, and wires following different and ingenious techniques [1–8]. In the literature, there are multiple excellent reviews [1–6] and books [7, 8] regarding the synthesis of gold nanoparticles of different sizes and shapes, where their different physical properties (e.g., optical, electronic, or mechanical) are also discussed. Therefore, in this chapter, we provide a comprehensive overview on the major wet chemistry synthetic approaches for the preparation of gold nanoparticles attending to the particle morphology. Hence, the main criterion on which this chapter is structured is the shape of the nanoparticles, predominantly according to the isotropy of the particles.

Isotropic Gold Nanoparticles

Spheres

Spherical is the most thermodynamically favored shape for all kinds of colloidal particles synthesized in bulk solutions. Spherical gold nanoparticles can be synthesized in solution in various sizes, ranging from 1 nm up to several hundreds of nanometers, and with different capping agents. The capping agent mainly dictates the chemical and physicochemical behavior of the particles, and this in turn determines the employment of the particles [1, 2].

One of the most popular synthetic methods for the preparation of gold nanospheres is based on the reduction of $HAuCl_4$ by citrate in water, which was first described in 1951 by Turkevich [1, 2]. In this method, citrate serves as both a reducing agent and an anionic stabilizer. It yields 15 nm nanoparticles with a fairly narrow size distribution. The Turkevich method has been modified by a number of groups to produce spherical gold nanoparticles with diameters ranging from 15 to 150 nm by either controlling the ratio of citrate to $HAuCl_4$ or employing a γ-radiation method. However, nanoparticles larger than 20 nm synthesized by this method usually lack isotropy and size uniformity [1, 2].

Other reduction methods have also been developed to achieve a better control over the size and monodispersity, including the "Schmid method" published in 1981 and the "Brust–Schiffrin method" reported in 1994 [1, 2]. The Schmid's cluster $[Au_{55}(PPh_3)_{12}Cl_6]$ remained unique for a long period of time with its narrow dispersity (1.4 ± 0.4 nm) for the study of a quantum-dot nanomaterial, despite its arduous synthesis. The synthesis requires rigorously anaerobic conditions and diborane gas as a reducing agent. As a result, phosphine-stabilized gold nanoparticles have lost favor since the development of a more convenient, scalable synthesis of thiol-stabilized air-stable gold nanoparticles of reduced polydispersity and controlled size by the method of Brust et al. [1, 2].

In the Brust–Schiffrin method, the synthesis is carried out in a two-phase system and thiol ligands that strongly bind to gold $AuCl_4^-$ is transferred to toluene using tetraoctylammonium bromide as the phase-transfer reagent and reduced by $NaBH_4$ in the presence of dodecanethiol. The organic phase changes color from orange to deep brown within a few seconds upon addition of $NaBH_4$. The particle size ranges between 1.5 and 5.2 nm. Additionally, these gold NPs can be repeatedly isolated and dispersed in common organic solvents without any aggregation or decomposition. Using the principle of this synthesis method, modified procedures have been developed to synthesize gold nanoparticles in the size range of 4–10 nm capped with thiol-functionalized molecules (small molecules or polymers) in one phase system using methanol as the solvent [1, 2].

The use of microemulsions, copolymer micelles, and inverse micelles is a significant research field for the synthesis of stabilized gold NPs. Typically, these syntheses involve a two-phase system with a surfactant that causes the formation of the microemulsion or the micelle maintaining a suitable microenvironment of controlled dimensions. The surfactants create small pockets of a water phase in an organic solvent or vice versa where the surfactant faces the aqueous phase with its polar group, and the tail faces the organic phase. Varying the ratio of the two solvents affects the dimensions of the micelles, which allows for tuning the size of the resulting nanoparticles with good monodispersity [1, 2].

All of the methods described above lead to uniform nanoparticles, however, only in the size range of below 20 nm. For the synthesis of larger spherical NPs (Fig. 1), the seeded-growth procedure is the most popular technique that has been intensively studied in recent years [1, 2]. A strong chemical reducing agent (usually sodium borohydride) is used to generate small, generally spherical, nanoparticles (commonly denoted as *seeds*), which are then added to a growth solution composed of surfactants and Au metal ions to induce particle growth. The growth solution employs a weaker reducing agent (often ascorbic acid) to reduce the metal salt to an intermediate state so that only catalyzed reduction on the nanoparticle surface is allowed, avoiding secondary nucleation. During the growth process, usually a small amount of undesired nonspherical particles is also generated, which can be removed through a CTAB-assisted shape-selective separation method [1, 2, 4]. The gold nanospheres obtained after such purification can be further grown by means of the same procedure described here up to a diameter of 200 nm or more. The final size of the nanospheres synthesized by this method depends strongly on the size of the employed seeds and the ratio between the seeds and the gold salt. To achieve predetermined final sizes of the nanospheres, the concentrations can be adjusted using the following equation [9]:

$$R_{final} = R_{seeds}\left(\frac{[Au^{3+}]_{final} + [Au_{seeds}]_{final}}{[Au_{seeds}]_{final}}\right)^{\frac{1}{3}}$$

where $[Au_{seeds}]_{final}$ is the value to be calculated:

$$[Au_{seeds}]_{final} = -\frac{[Au^{3+}]_{final}}{1 - \left(R_{final}/R_{seeds}\right)^3}$$

Shells

Nanoshells have similar properties to spherical gold nanoparticles in the sense that they also exhibit a single surface plasmon resonance (SPR) absorption. However, the SPR absorption band of nanoshells can be tuned across the visible and infrared region over a range of wavelengths spanning hundreds of nanometers, far exceeding the spectral range of spherical particles. This has a huge advantage for SERS-based applications, because the plasmon resonance can be tuned to the excitation of common laser radiation sources optimizing the electromagnetic enhancement

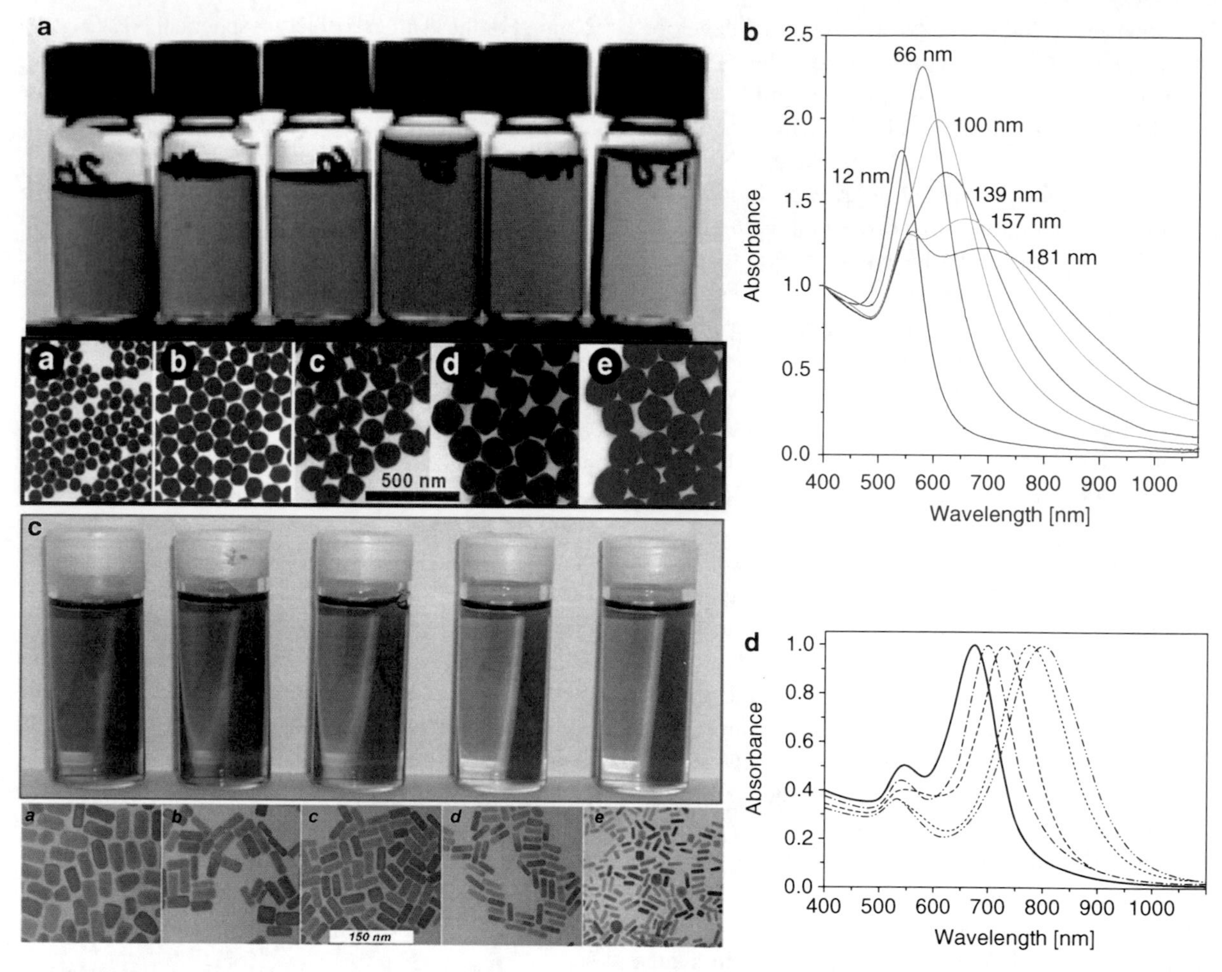

Synthesis of Gold Nanoparticles, Fig. 1 (**a**) Representative TEM micrographs of Au spheres (*bottom*) obtained after subsequent growth steps with their respective color solution (*top*). Average diameters are 66 (a), 100 (b), 139 (c), 157 (d), and 181 (e) nm. (**b**) UV-Vis spectra of Au spheres with various average diameters (*right*). (Reprinted with permission from [24]. Copyright 2006 American Chemical Society.) (**c**) Photographs (*top*) and the respective TEM micrographs (*bottom*) of gold nanorod solutions with average aspect ratios of 1.94 (a), 2.35 (b), 2.48 (c), 3.08 (d), and 3.21 (e). (**d**) UV-Vis absorption spectra of different gold nanorod samples stabilized with CTAB with increasing aspect ratio from *left to right* (Reprinted with permission from [3, 25]. Copyright 2004 Elsevier B.V.)

mechanism [5, 10, 11]. Apart from their optical properties, these nanostructures also present inert biological activities which make them appropriate for biomedical applications, such as targeted drug delivery, photothermal therapeutic applications, and molecular imaging, as contrast agents for optical coherence tomography (OCT).

Nanoshells are usually composed of dielectric cores (e.g., silica particles) coated with a single or multiple thin metal layers [10]. The growth of metal nanoshells on core particles combines techniques of molecular self-assembly with the reduction chemistry of metal colloid synthesis. This approach is general and can potentially be adapted to a variety of core and shell materials. The common synthesis route of gold nanoshells involves the synthesis of the dielectric core material, typically silica or polystyrene and their surface functionalization with terminal amine groups to facilitate attachment of small colloidal gold, which are usually formed in a separate process and act as nucleation sites for the subsequent gold salt reduction and shell growth. The absorbed seed colloids increase in size as reduction ensues, followed by the coalescence of gold particles on the surface, until finally the apparent formation of a polycrystalline continuous metallic nanoshell occurs [10].

Although this common two-step process of seeding, i.e., formation of small colloidal gold followed by attachment to the core is effective, it is also laborious and costly, especially in terms of time. Another alternative and also very effective approach is the deposition-precipitation (DP) method, which employs the in situ formation of gold seeds directly on the support. The functionalized core particles are placed in contact with a basic solution of gold (III) chloride and upon heating, oxidic precursor particles ($Au[OH]_3$) are formed on the support. These precursor seed particles then act as nucleation sites for the formation of gold nanoshells. The size and density of the seeds on the substrate is easily controlled by the concentration of gold (III) chloride, support surface functionalization, reaction temperature and time, and the pH of reaction. This DP method is relatively easy to handle and is currently used for producing commercial gold catalysts in a highly controlled manner [11].

Anisotropic Gold Nanoparticles

Since the last century, colloid chemists have gained excellent control over the synthesis of spherical gold particles in terms of size and uniformity. However, synthetically controlling particle shape, particularly in solutions, has been a bigger challenge and has met somewhat limited success. Nevertheless, especially in the last 10 years, various methods have been developed to synthesize a variety of anisotropic gold nanoparticles with somewhat controllable shape and morphology. Anisotropic gold nanoparticles have been synthesized in symmetric shapes, such as cylindrical, cubic, polyhedral and plate-like shapes, but more interestingly, there are methods that produce predominately asymmetric branched morphologies such as nanostars. In the following, the most representative synthesis techniques for various types of anisotropic gold nanoparticles will be described in ascending order of shape complexity [3–6].

Nanorods

In the field of anisotropic nanoparticles, the synthesis of gold nanorods is perhaps the most established protocol, in terms of the degree of control of the size, shape, and monodispersity [3, 4]. There are three main methods used to produce gold rods through wet chemistry, namely, the template method, electrochemical method, and seeded growth method [3]. The template method for the synthesis of gold nanorods is based on the electrochemical deposition of Au within the pores of nanoporous polycarbonate or alumina template membranes. For this, Ag or Cu is first sputtered onto the alumina template membrane to provide a conductive film for electrodeposition, i.e., the growth of Au nanorods. Subsequently, both the template membrane and the copper or silver film are selectively dissolved in the presence of a polymeric stabilizer such as poy(vinylpyrrolidone) (PVP), which allows for dispersing the rods either in water or in organic solvents by means of sonication or agitation. The dimensions of these synthesized nanorods coincide with the pore diameter of the template membranes which can be tuned by controlling the pore sizes [3].

The electrochemical synthesis of gold nanorods is conducted within a simple two-electrode type electrochemical cell, where a gold metal plate is used as a sacrificial anode while the cathode is a platinum plate with similar dimensions. Both electrodes are immersed in an electrolytic solution containing a mixture of cationic surfactants, hexadecyltrimethylammonium bromide (CTAB), tetradodecylammonium bromide (TDDAB), acetone, and cyclohexane The electrolytic cell containing the mixture is then placed inside an ultrasonic bath at 36°C and controlled-current electrolysis is performed with a typical current of 3 mA for a typical electrolysis time of 30 min. The particle sizes can be tuned either via electrolysis time or by addition of silver ions (silver plate) [3].

The seeded growth method is perhaps the most facile and most applied method to synthesize gold nanorods. During the early attempts of seed-mediated approaches of growing pre-synthesized spherical gold nanoparticles in solution, formation of a distinct population (5–10%) of colloidal gold rods was observed. In the beginning of this century, it was found that by controlling the growth conditions in aqueous surfactant media, it was possible to synthesize gold nanorods with tunable aspect ratio (Fig. 1). Moreover, addition of $AgNO_3$ influences not only the yield and aspect ratio control of the gold nanorods, but also the mechanism for gold nanorod formation and correspondingly its crystal structure morphology and optical properties. A fine-tuning of the aspect ratio of the nanorods can be achieved by simply varying the pH in silver-free

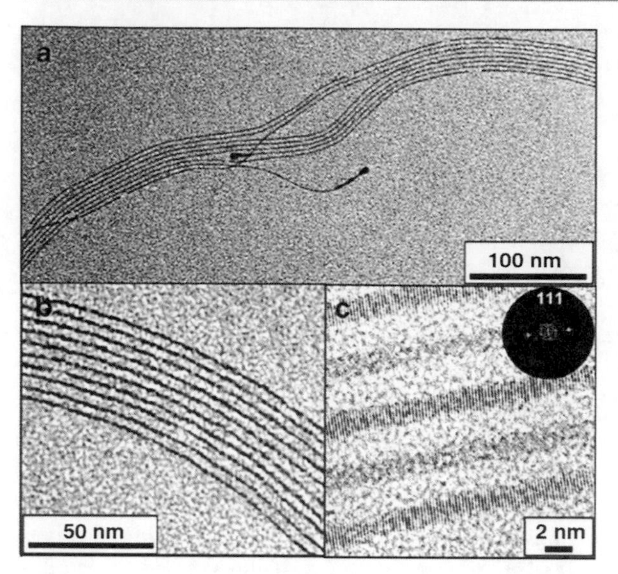

Synthesis of Gold Nanoparticles, Fig. 2 TEM images of Au nanowires at different magnifications (**a**, **b**) and an HRTEM image (**c**) showing the single-crystalline structure (Reprinted with permission from [13]. Copyright 2008 American Chemical Society)

syntheses or by adjusting one of the parameters in the silver-mediated procedures, namely, the concentration of silver ions, gold ions, seeds or that of the reducing agent (ascorbic acid), while keeping the other parameters constant [3, 4].

Nanowires

In comparison to nanorods, nanowires exhibit much larger aspect ratios, and hence, larger anisotropy, which makes them ideal candidates for many applications, such as biosensors, logic circuits, field-effect transistors, and nonvolatile memory elements [4, 5]. Being conductive materials, metal nanowires have driven a multitude of theoretical studies on their conductance behavior in the fabrication of nanoelectronics. Inspired from the synthesis of gold nanorods, various methods have been developed to synthesize gold nanowires. These include the pore-template method, block-polymer or DNA template methods, patterned lithography, and chemical reduction methods in bulk solutions. Seed-mediated growth in aqueous CTAB solutions [12] and seedless growth in oleylamine solutions of organic solvents are prime examples of the chemical synthesis of gold nanowires in solutions (Fig. 2).

Depending on the synthesis method and the growth mechanism, gold nanowires can feature different dimensions (diameter, length, and aspect ratio), crystallinity (monocrystalline or polycrystalline or twinned) and shape (straight or twisted). Inspired by the seeded growth method, Kim et al. reported an acidic solution route to synthesize gold nanowires at room temperature with diameters tunable between 16 and 66 nm and lengths up to 10 μm with an aspect ratio larger than 200 [12]. On the other hand, the method described by Giersig and coworkers, [13] ensures the synthesis of thin gold nanowires with diameters of $\sim$1.6 nm and lengths ranging from 10 nm to $\geq$3.5 μm. Control over Au wire lengths is realized by tuning the oleylamine/$HAuCl_4$ volume ratio, the reaction time, and the addition of a second solvent.

Platonic Nanoparticles (Polyhedrons)

Platonic and quasi-platonic nanocrystals with multiple faces and vertices have been successfully synthesized in relative high yields using both aqueous and nonaqueous approaches [4, 5, 14, 15]. In the method developed by Sau et al., [14] platonic nanoparticles are synthesized in a seeded growth procedure in aqueous solutions at room temperature. A fine control of nanoparticle shape is achieved by systematic variation of experimental parameters. The experimental procedures involve addition of an appropriate quantity of the pre-synthesized gold seeds to aqueous growth solutions containing desired quantities of CTAB, $HAuCl_4$, ascorbic acid, and in some cases, a small quantity of silver ions. The morphology and dimension of the gold nanoparticles depend on the concentrations of the seed particles and CTAB, in addition to the reactants (Au^{3+} and ascorbic acid).

The synthesis of platonic gold nanoparticles in nonaqueous media is mainly based on the so-called polyol process. This process involves the thermal reduction of a gold salt in an organic solvent with a relatively high boiling point such as poly(ethylene glycol) or N,N-dimethylformamide (DMF) in the presence of a polymeric stabilizer, usually poly(vinyl pyrrolidone). The ratio of gold salt to PVP is an important parameter to control the shape of the polyhedral particles. Tetrahedral particles are formed at higher gold/PVP ratios and icosahedrons at lower ratios. However, the presence of silver ions induces the formation of uniform gold hexahedrons, i.e., nanocubes. Octahedral, truncated octahedral, cubo-octahedral, and cubic particles (Fig. 3) were synthesized by Seo et al. by conversion of cubo-octahedral particles into

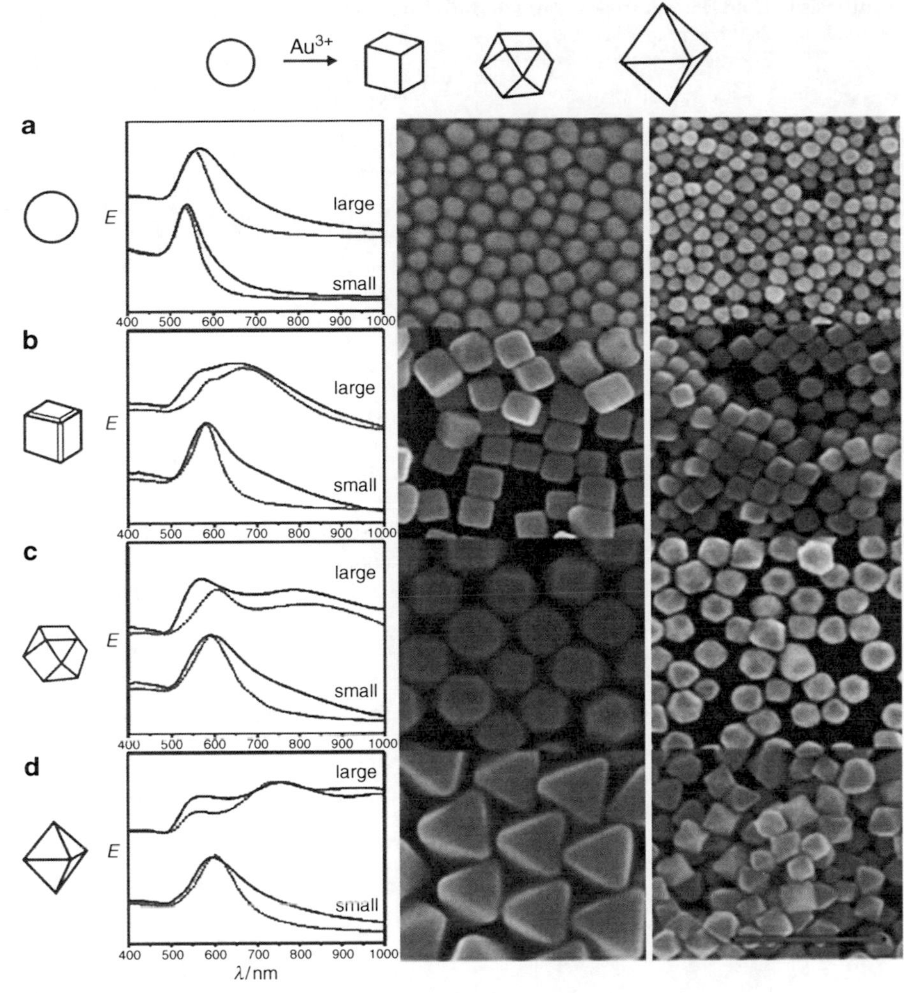

Synthesis of Gold Nanoparticles, Fig. 3 Size and shape control of Au polyhedral nanocrystals from spherical seeds. UV-Vis extinction spectra (*solid lines*) and results of DDA calculations (*dotted lines*) of large (**a**) and small spherical seeds (**a**), large (edge length 116 nm) and small (67 nm) cubes (**b**), large (edge length 122 nm) and small (54 nm) cuboctahedrons (**c**), and large (edge length 236 nm) and small (88 nm) octahedrons (**d**). SEM images (on the *right*, from *top to bottom*) of spherical seeds (818 and 495 nm), cubes (11,611 and 676 nm), cuboctahedrons (12,213 and 548 nm), and octahedrons (23,619 and 8,810 nm). The *bar* represents 500 nm (Reprinted with permission from [15]. Copyright 2004 Wiley-VCH)

different shaped particles with the aid of silver ions, which seems to be related to selective blocking of the growth along certain crystallographic facets [15]. Also, gold nanorods, synthesized in aqueous media, can be completely reshaped into perfect, single-crystalline octahedrons in a DMF–PVP solution [4]. Besides reshaping and interconversion methods, facetted seeds can be also employed to grow polyhedral nanoparticles, as in the case of synthesizing very regular and nearly monodisperse decahedrons from preformed penta-twinned gold nanoparticle seeds. The synthesis methods for different platonic nanoparticles along with their chemical and physical properties are summarized in Table 1.

Gold Nanoplates

Other very interesting types of anisotropic nanoparticles are triangular or hexagonal gold nanoplates (Fig. 4), which are highly attractive for a number of applications such as optical biosensing and surface-enhanced Raman spectroscopy (SERS), due to their sharp edges, which lead to high local electric field gradients under illumination [4, 5]. The generation of gold nanoplates is often observed during modifications employed for the synthesis of gold nanorods via the seeded-growth method. Small changes in the synthetic procedure, such as an increase in the pH or surfactant concentration, or the addition of extra halide ions, can lead to the formation of planar nanostructures. To date, all current procedures result in rather low yields (40–65%) compared to the synthesis of gold nanorods (~99%); hence, purification steps are needed in order to eliminate isotropic and nonplanar structures. In the last few years, other techniques rather than seed-growth or polyol methods have been employed to synthesize nanoplates. Very recently,

Synthesis of Gold Nanoparticles, Table 1 An overview of various synthesis methods of anisotropic gold nanoparticles, with their physical and chemical properties

Particle Shape	Synthesis method	Max. Leng. (nm)	SPR Max. (nm)	Synthesis medium	Reduction agent	Capping agent
Rods [18]	Seeded growth	≤500 nm	λ_{max}tr: 520–530 λ_{max}long: 750–800	H_2O	Ascorbic acid	CTAB
Wires [12]	Seeded growth	≤10,000	λ_{max}tr: ~540 λ_{max}long: >1,200 nm	H_2O	Ascorbic acid	CTAB
Wires [13]	Seedless growth	≥3,500	λ_{max}tr: – λ_{max}long: > 1,200 nm	Oleylamine	Oleylamine	Oleylamine
Tetrahedra [19]	Polyol	~210	~626 and ~950	Ethylene glycol	Ethylene glycol	PVP
Cubes [14]	Seeded growth	~90	~555	H_2O	Ascorbic acid	CTAB
Cubes [19]	Polyol	~150	~621	Ethylene glycol	Ethylene glycol	PVP
Octahedra [20]	Polyol	20–400	530–1,100	Ethylene glycol	Ethylene glycol	PDDA
Decahedra [21]	Seeded growth	36–65	600–700	H_2O/DMF	PVP	PVP
Decahedra [22]	Polyol	48–88	600–700	Diethylene glycol	Diethylene glycol	PVP
Icosahedra [19]	Polyol	230	~613 and ~950	Ethylene glycol	Ethylene glycol	PVP
Triangles [23]	Seeded growth	90–210	750–1,300	H_2O/DMF	Ascorbic acid	CTAB
Hexagons [16]	Oligoamine surfactant	20–15000	550–900 broad	H_2O	Oligoamine surfactant	Oligoamine surfactant
Stars [14]	Seeded growth	~350	Broad	H_2O	Ascorbic acid	CTAB
Stars [17]	Nonaqueous seeded growth	60–80	Broad	DMF	PVP	PVP

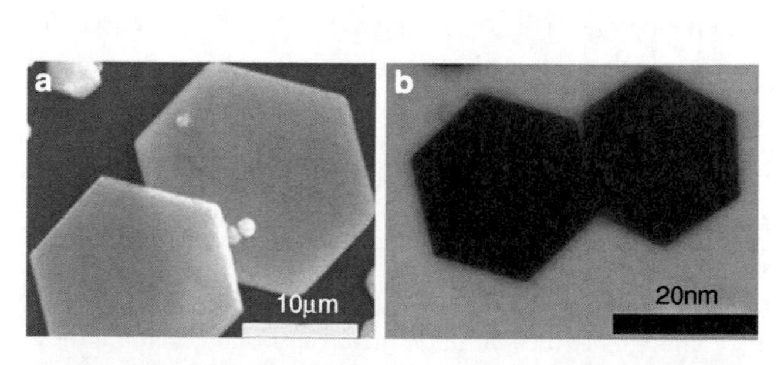

Synthesis of Gold Nanoparticles, Fig. 4 Influence of concentration of C18N3 on the size of gold nanoplates at standard conditions. SEM and TEM images of gold nano- and microplates obtained at different concentrations of C18N3, 2 mM (**a**) and 0.1 mM (**b**). On the right it is shown the UV-Vis absorption spectra of gold nanospheres (*A*), gold decahedrons (*B*), and gold microplates (*C*) (Reprinted with permission from [16]. Copyright 2010 American Chemical Society)

a more facile synthesis procedure for the fabrication of gold nano- and microplates was reported by Lin et al. [16]. They used a single tree-type multiple-head surfactant, bis (amidoethyl-carbamoylethyl) octadecylamine (C18N3), which functions as both the reducing and capping agent in the reaction system. The triangle and hexagonal plate, polyhedron, and sphere morphology of the gold nanoparticles could be easily

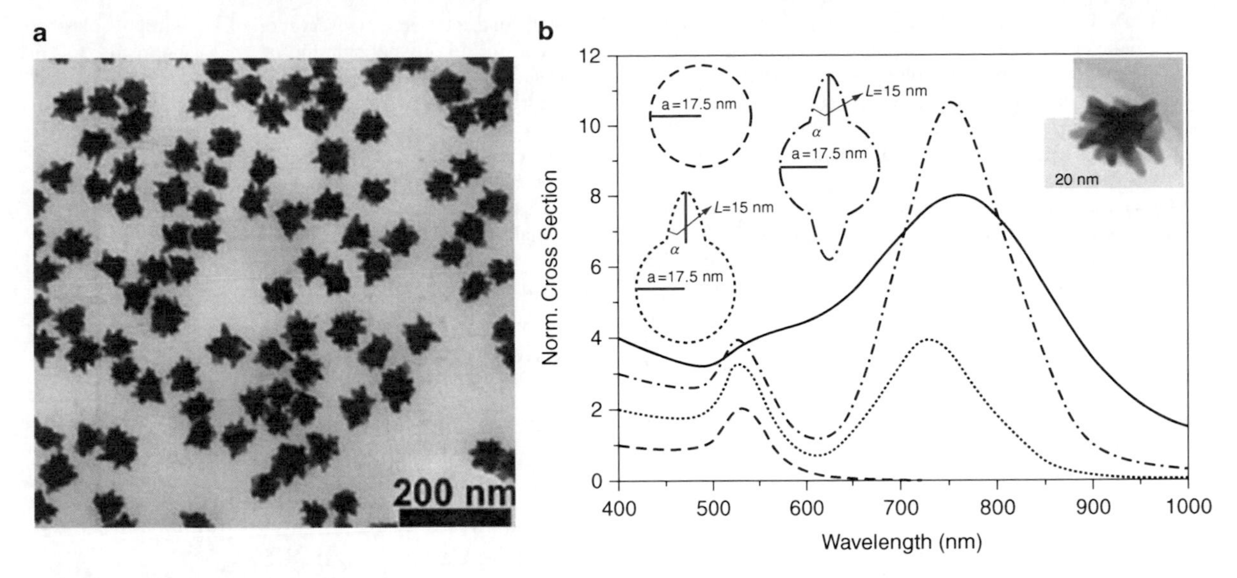

Synthesis of Gold Nanoparticles, Fig. 5 Representative transmission electron microscopy (TEM) image of Au nanostars (**a**), experimental (*solid line*) and calculated (*dashed line*) absorption spectra for gold nanostars; the inset shows a TEM image of a single nanostar (**b**) (Reprinted with permission from [17]. Copyright IOP Publishing 2008)

controlled simply by changing the molar ratio of C18N3/$HAuCl_4$ (Fig. 4).

Branched NPs

Coming from symmetric isotropic and anisotropic gold nanoparticles, another important group of gold particles are asymmetric nanoparticles, the so-called branched nanoparticles or nanostars (Fig. 5) [6]. Although these obviously more complex structures are not yet well understood, they generate interest because of their sharp edges and the correspondingly high localization of surface plasmon modes. This makes them potential candidates for a number of applications, including SERS-based detection and analysis, as well as the case of nanoplates. Sau et al. proposed the synthesis of branched nanoparticles via the seeded-growth method in the presence of silver ions by varying the seed-to-gold salt ratio and the amount of reducing agent in order to increase the rate of gold ion reduction and thereby induce branching (see Table 1). However, an easier way to synthesize gold nanostars comprises the use of a nonaqueous approach [6, 17]. Branched "nanostars" were synthesized in a concentrated solution of PVP in DMF at room temperature, where PVP acts as a stabilizer and reductant. Nanostars were not only obtained in a seed-mediated growth process, but could also be synthesized in a seedless growth process. The fascinating aspect of these particles is that each tip is a single crystal, so the challenge remains to understand the underlying growth mechanism of the selective tip growth on the seed particle's surface.

In summary, the most relevant synthetic approaches mentioned above for the preparation of different anisotropic nanoparticles are presented in Table 1. This table offers the opportunity to take a look at the wide range of synthetic possibilities for the production of anisotropic nanoparticles, but at the same time gives a good idea of the physical and chemical properties as well as limitations of these synthetic approaches for the production of particular morphologies with specific sizes. That is why further synthetic research efforts are still needed.

Acknowledgments Authors acknowledge the financial support of the Spanish Ministerio de Ciencia e Innovación (MAT2008-06126), Xunta de Galicia (INCITE09209101PR, INCITE08PXIB209007PR and 2008/077), and EU (METACHEM, grant number CP-FP 228762–2).

References

1. Eustis, S.E.-S., El-Sayed, M.A.: Why gold nanoparticles are more precious than pretty gold: noble metal surface plasmon resonance and its enhancement of the radiative and nonradiative properties of nanocrystals of different shapes. Chem. Soc. Rev. **35**(3), 209–217 (2006)

2. Astruc, D., Daniel, M.C.: Gold nanoparticles: assembly, supramolecular chemistry, quantum-size-related properties, and applications toward biology, catalysis, and nanotechnology. Chem. Rev. **104**(1), 293–346 (2004)
3. Perez-Juste, J., et al.: Gold nanorods: synthesis, characterization and applications. Coord. Chem. Rev. **249**(17–18), 1870–1901 (2005)
4. Grzelczak, M., et al.: Shape control in gold nanoparticle synthesis. Chem. Soc. Rev. **37**(9), 1783–1791 (2008)
5. Treguer-Delapierre, M., et al.: Synthesis of non-spherical gold nanoparticles. Gold Bull. **41**(2), 195–207 (2008)
6. Guerrero-Martínez, A., et al.: Nanostars shine bright for you: colloidal synthesis, properties and applications of branched metallic nanoparticles. Curr. Opin. Colloid Interf. Sci. **16**(2), 118–127 (2011)
7. Chow, P.E. (ed.):Gold Nanoparticles: Properties, Characterization and Fabrication Nanotechnology Science and Technology. Nova Science Publishers, New York (2010)
8. Feldheim, D.L. (ed.): Metal Nanoparticles: Synthesis Characterization and Applications. CRC Press, Baco Raton (2001)
9. Jana, N.R., Gearheart, L., Murphy, C.J.: Evidence for seed-mediated nucleation in the chemical reduction of gold salts to gold nanoparticles. Chem. Mater. **13**(7), 2313–2322 (2001)
10. Oldenburg, S.J., et al.: Nanoengineering of optical resonances. Chem. Phys. Lett. **288**(2–4), 243–247 (1998)
11. Kah, J.C.Y., et al.: Synthesis of gold nanoshells based on the deposition-precipitation process. Gold Bull. **41**(1), 23–36 (2008)
12. Kim, F., et al.: Chemical synthesis of gold nanowires in acidic solutions. J. Am. Chem. Soc. **130**(44), 14442–14443 (2008)
13. Pazos-Perez, N., et al.: Synthesis of flexible, ultrathin gold nanowires in organic media. Langmuir **24**(17), 9855–9860 (2008)
14. Sau, T.K., Murphy, C.J.: Room temperature, high-yield synthesis of multiple shapes of gold nanoparticles in aqueous solution. J. Am. Chem. Soc. **126**(28), 8648–8649 (2004)
15. Seo, D., et al.: Directed surface overgrowth and morphology control of polyhedral gold nanocrystals. Angew. Chem. Int. Edit. **47**(4), 763–767 (2008)
16. Lin, G.H., et al.: A simple synthesis method for gold nano- and microplate fabrication using a tree-type multiple-amine head surfactant. Cryst. Growth Des. **10**(3), 1118–1123 (2010)
17. Kumar, P.S., et al.: High-yield synthesis and optical response of gold nanostars. Nanotechnology **19**(1), 015606 (2008)
18. Liu, M.Z., Guyot-Sionnest, P.: Mechanism of silver(I)-assisted growth of gold nanorods and bipyramids. J. Phys. Chem. B **109**(47), 22192–22200 (2005)
19. Kim, F., et al.: Platonic gold nanocrystals. Angew. Chem. Int. Edit. **43**(28), 3673–3677 (2004)
20. Li, C.C., et al.: A facile polyol route to uniform gold octahedra with tailorable size and their optical properties. ACS Nano **2**(9), 1760–1769 (2008)
21. Sanchez-Iglesias, A., et al.: Synthesis and optical properties of gold nanodecahedra with size control. Adv. Mater. **18**(19), 2529–2534 (2006)
22. Seo, D., et al.: Shape adjustment between multiply twinned and single-crystalline polyhedral gold nanocrystals: decahedra, icosahedra, and truncated tetrahedra. J. Phys. Chem. C **112**(7), 2469–2475 (2008)
23. Millstone, J.E., et al.: Observation of a quadrupole plasmon mode for a colloidal solution of gold nanoprisms. J. Am. Chem. Soc. **127**(15), 5312–5313 (2005)
24. Rodriguez-Fernandez, J., et al.: Seeded growth of submicron Au colloids with quadrupole plasmon resonance modes. Langmuir **22**(16), 7007–7010 (2006)
25. Perez-Juste, J., Correa-Duarte, M.A., Liz-Marzan, L.M.: Silica gels with tailored, gold nanorod-driven optical functionalities. Appl. Surf. Sci. **226**(1–3), 137–143 (2004)

Synthesis of Graphene

Swastik Kar[1] and Saikat Talapatra[2]
[1]Department of Physics, Northeastern University, Boston, MA, USA
[2]Department of Physics, Southern Illinois University Carbondale, Carbondale, IL, USA

Synonyms

Fabrication of graphene or graphene fabrication; Graphene synthesis; How to synthesize/make/manufacture/fabricate/produce graphene; Making graphene; Manufacturing graphene or graphene manufacturing; Production of graphene or graphene production

Definition

► Graphene (or monolayer graphene or single-layer graphene) is a single-atom thick, quasi-infinite, sp^2-hybridized allotrope of carbon in which the atoms are packed in a planar honeycomb crystal lattice (see Fig. 1). It can be visualized as a single sheet of graphite and its lattice structure is related to those of ► fullerenes for drug delivery and ► carbon nanotubes. Few-layered graphene, multilayered graphene, or multigraphene refers to a few layers of graphene stacked (with weak interlayer attraction) in a manner similar to graphite. Figure 1a, b are schematic representations of a sheet of single-layer graphene.

Overview

Compiling a review on the diverse methods available for the synthesis of graphene is a daunting task, chiefly

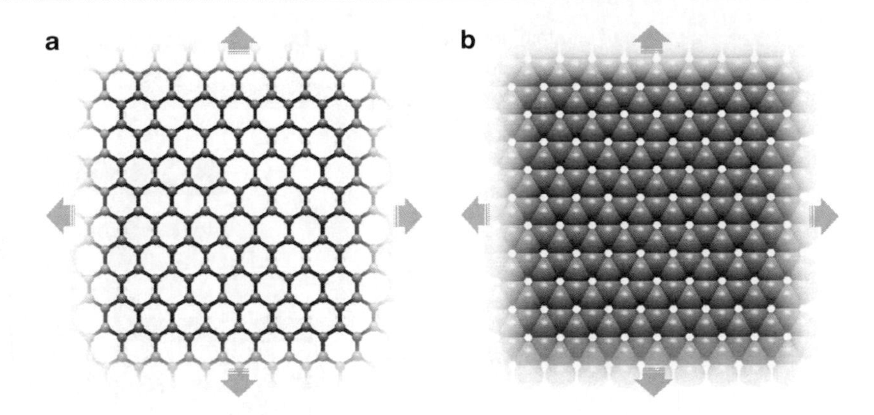

Synthesis of Graphene, Fig. 1 Schematic representation of carbon atoms in a graphene lattice: (**a**) a ball-and-stick model and (**b**) a space-filling model. The arrows indicate that the sheets are quasi-infinitely extended in all directions. The edge carbon atoms are usually hydrogen terminated

due to the explosive intensity at which this field has grown over the past few years. Numerous techniques for graphene synthesis have been developed, based on both conventional and innovative techniques, and the graphene thus produced can have vastly different properties. Quite often, this diversity stems from a need for the production of graphene that is tailored for a broad variety of experiment-specific and application-specific research. It is quite impossible to enlist all these methods into a few categories, and this article is not an attempt to individually address all of them. In general, this work attempts to track and identify some of the most popular, novel, and/or promising methods for graphene synthesis. An attempt has been made to categorize these methods into groups. Since the search for new and/or improved methods for graphene synthesis is in its ascent, the material presented here will not reflect published reports beyond May of 2011. In most cases, generic descriptions of the synthesis conditions have been presented. The aim is to provide the reader with a starting point rather than a complete recipe in each case. It should be noted that experimental conditions and parameters presented here represent typical synthesis conditions as reported in literature, since what constitutes the "best practice" is both subjective and continuously evolving. It should also be noted that this review does not deal with the many electronic, optical, thermal, mechanical or other fascinating properties of graphene, for which excellent review articles already exist in literature.

Historical Perspective

The concept of graphene has been around for quite a long time, and the descriptive presence of this material appear in a variety of reports in a number of languages. For example, "very thin layers of graphite" including "single foils of carbon" were reported as early as 1962, prepared through a partial reduction of graphite oxide [1]. Monolayer graphite/carbon was identified to form at elevated temperatures on surfaces of Pt in 1969 [2], C-doped Ni (111) crystals in 1979 [3], and other transition metal surfaces [4]. The term "graphene" was already in use in the early 1990s [5], to describe not only monolayer graphite, but also the structure of carbon fibers and fullerenes. An earnest endeavor to obtain isolated high-quality graphene from graphite started in 1999, using mechanical exfoliation of patterned graphite pillars [6, 7]. This approach matured in 2004 [8], the same year that synthesis of graphene was also reported by a high-temperature thermal annealing of SiC [9]. These two methods opened up the floodgate of graphene research and catapulted it into becoming the millennium material. In the following paragraphs, a flavor of these and some of the other popular graphene synthesis methods that have been developed since then has been captured.

Recent Advances in Graphene Synthesis

Micromechanical Exfoliation of Graphene from Graphite

Strictly not a synthesis process, the micromechanical exfoliation of graphene (both monolayer and few-layer graphene) from highly oriented pyrolytic graphite (HOPG), as reported by Novoselov et al. enabled, for the first time, a direct measurement of its unique 2D electronic properties. This ended the decades-long debate whether pure monatomic two-dimensional crystals can indeed be stable in an isolated state

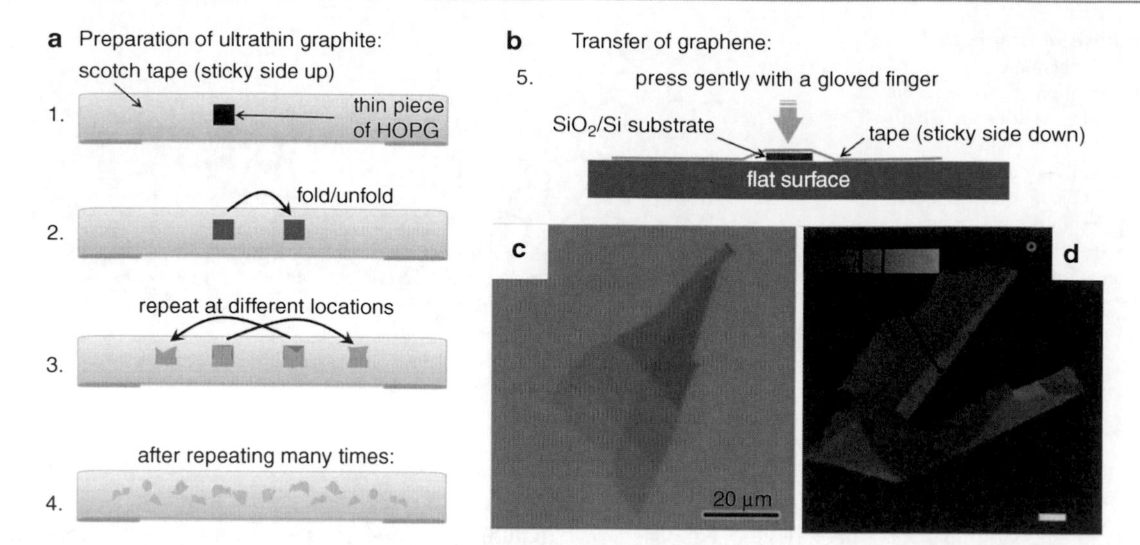

Synthesis of Graphene, Fig. 2 A simple recipe for obtaining single and few-layered graphene crystals using the micromechanical cleavage of graphite. (**a**) 1. Attach a very thin piece of HOPG onto a scotch tape. 2–4. Fold and unfold as shown to obtain ultrathin (almost invisible) flakes of graphite on the tape. (**b**) Select a region with very thin flakes and transfer on a suitable substrate (e.g. SiO_2/Si) as shown. Carefully remove the tape from the substrate. (**c**) Typical image of a few-layered flake with monolayer extensions as seen under an optical microscope (Reprinted with permission from Ref. [13]; copyright © 2010 Elsevier). (**d**) A typical AFM image of graphene flakes (scale bar = 1 μm) (Reprinted with permission from Ref. [12]; copyright © 2005 National Academy of Sciences, U.S.A.)

[10, 11]. The few tens-of-microns sized graphene and few-layered graphene flakes could be easily transferred onto SiO_2/Si and other substrates, and enabled researchers to perform a wide variety of electronic, optical, thermal and mechanical property measurements. In their first report [8], Novoselov et al. described an elaborate process that started from millimeter-thick platelets of HOPG, and involved patterning, etching, photoresist, an adhesive-tape peel off, and a subsequent ultrasonic rinse. A simplified prescription for obtaining graphite was reported by the same group a year later, whereby a freshly cleaved HOPG surface was mechanically rubbed against any solid surface. This invariably left behind debris of crystallites on the surface, and among them, one can usually find single- and few-layered graphene [12]. Owing to the simplicity of this method, a variety of alternatives have become popular, and Fig. 2a, b describes a protocol that simply uses an adhesive tape, which is favored by many. This method has become known as the scotch-tape method. As a receiving substrate, it is helpful to use a Si wafer that has ~300-nm thick oxide layer: graphene itself is nearly transparent, but using this oxide thickness enables one to "see" graphene even using an optical microscope due to the suitable amount of interferometric phase contrasting, as seen in Fig. 2c [13]. Figure 2d is an ► atomic force microscopy (AFM) image of a typical flake of graphene, showing the graphene layer thicknesses in flat and folded regions [12].

Epitaxial Growth of Graphene on Silicon Carbide

In 2004, another technique for graphene synthesis was reported by Berger et al. [9]. In this method, graphene crystals were produced on the Si-terminated (0001) face of single-crystal 6H-SiC by thermal desorption of Si. The SiC surfaces were initially prepared by oxidation or H_2 etching. The oxide was then removed by electron bombardment-assisted heating in ultrahigh vacuum ($\sim 10^{-10}$Torr) to 1,000°C (repeating the oxidation/deoxidation cycle improves the surface quality, and the best initial surface quality was obtained with H_2 etching). The deoxidized samples were heated to temperatures ranging from 1,250°C to 1,450°C for 1–20 min; during which time, thin graphite layers were formed. Figure 3 shows a scanning tunneling microscope (STM) image of graphene produced this way by heating SiC for 8 min at 1,400°C. Over the years, this process has been fine-tuned so as to enable one to controllably produce monolayer graphene on SiC substrates [14]. It turns out that graphene grown

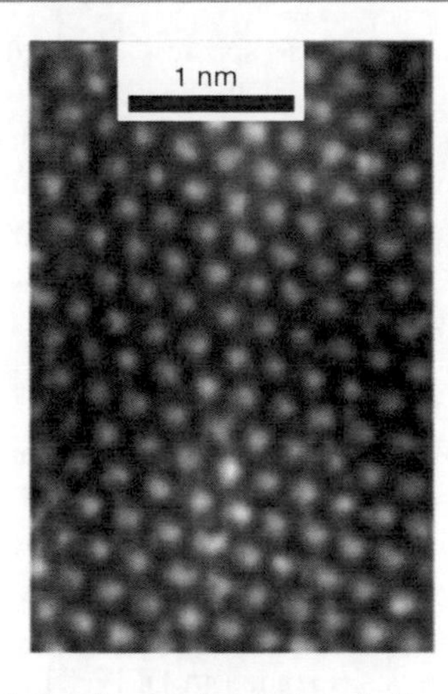

Synthesis of Graphene, Fig. 3 Atomically resolved STM image of a graphene sample grown on SiC(0001) at 1,400°C for 8 min (Reprinted with permission from Ref. [9]; copyright © 2004 American Chemical Society)

from the SiC(000$\bar{1}$) (C-terminated) surface are of higher quality than those grown on SiC(0001) [15]. Growth on the C face renders graphene with domain sizes more than three times larger than those grown on the Si face, and with significantly reduced disorder. In 2009, it was reported that an ex situ graphitization of Si-terminated SiC (0001) in an argon atmosphere of about 1 bar produces monolayer graphene films with much larger domain sizes than those previously attained. This method for the direct synthesis of large-domain graphene (~10s of microns in length, with ~micron sized widths that were limited by the step-sizes of the underlying substrate) on an insulating underlayer formed an important step toward realizing graphene-based nanoelectronics on a semiconductor-processable substrate.

In the initial years, most samples were produced using one of the two above-mentioned methods, which produced high-quality samples of graphene appropriate for fundamental studies such as electric- and magnetic-field-modulated electronic transport, ARPES studies, Raman studies, and a variety of related experiments. At the same time, as it became evident that there are several other important properties of graphene including their gigantic specific surface area, extreme mechanical strength values, and other thermal and optical properties, several other methods have been developed for large-scale production of graphene. The large-scale production can be categorized into two directions. One route is to develop methods for mass production, often using liquid-phase exfoliation of graphene from graphite, while the other is headed toward wafer-scale production of single- or few-layered graphene, targeted toward electrical, electromechanical, and optoelectronic applications.

Liquid-Phase Exfoliation

Reduced Graphene Oxide and Chemically Modified Graphene

A significant amount of effort has gone into obtaining graphene from the exfoliation of graphite oxide into graphene oxide (GO), followed by a reduction step [16]. The resultant material is more appropriately termed reduced graphene oxide (rGO), and it belongs to the broader category of chemically modified graphenes. Typically, graphite is oxidized using techniques based on the Hummer's method [17]. GO, as produced by the Hummers' method, is composed of functionalized graphene sheets decorated by strongly bound oxidative debris, which acts as a surfactant to stabilize aqueous GO suspensions [18]. The oxide can hence be dispersed in water using an ultrasonic bath till it becomes a clear-colored liquid with no visible particulates. The parent graphite oxide and the dispersed graphene oxide are both insulators, and hence are unsuitable for any applications where electrical conductivity is important. One approach to overcome this problem is a liquid-or gas-phase reduction of the oxidized carbon. Among reducing agents, the use of hydrazine hydrate has become quite popular, and this can be added to the aqueous GO dispersion and heated in an oil bath at 100°C for a day, during which time, the reduced graphene oxide gradually precipitates out as a black solid. This precipitate is a high-surface-area material which can be separated out using a simple vacuum filtration technique that leaves behind a thick film of rGO. Such films can be transferred onto other substrates in the form of large-area ultrathin films of reduced graphene oxide, which are useful for the development of transparent and flexible electronic materials [19] (see also ▶ Flexible Electronics) and other applications [20, 21].

Although the reduction process restores its conductivity by several orders of magnitude compared to that of GO [22], the oxidation of graphene seems to be quite robust and can only be partially removed. Solid-state ^{13}C NMR spectroscopy of these materials suggests that the sp^2-bonded carbon network of graphite is strongly disrupted, and a significant fraction of this carbon network is bonded to hydroxyl groups or participates in epoxide groups [23], with some carboxylic or carbonyl groups present at the edges. Further works have shown the presence of five- and six-membered-ring lactols [24]. As a result, this material has a poor electrical conductivity compared to that of pristine graphene and does not display most of the exciting quantum properties of pure graphene. These groups can be significantly removed by using a sodium borohydride and sulfuric acid treatment, followed by thermal annealing [24], which is effective in restoring the π-conjugated structure and to a significant extent the conductive nature of the graphene materials. While the effort to restore the electronic properties of rGO to a level similar to that of pristine graphene has yet to be reported, rGO-related materials can have several important properties of their own, including large specific surface area and other mechanical properties that enable them to be used as various kinds of energy storage devices (▶ Nanomaterials for Electrical Energy Storage Devices), sensors, and composite structures.

Liquid-Phase Exfoliation of Graphene Directly from Graphite

In some applications, where it is important to have large volumes of *chemically unmodified* graphene, a different approach can be adopted, involving liquid-phase exfoliation of graphene directly from graphite. The idea of liquid-phase exfoliation is essentially a combination of the two methods – mechanical exfoliation from graphite, and the liquid-phase environment that is usually applied to GO. In case of GO, the process is pretty straightforward, owing to the weaker interplanar interactions and the strong hydrophilic nature of the individual graphene sheets, and simple ultrasonication of GO causes it to easily disperse stably in water. In contrast, graphene sheets in graphite interact more strongly, and being hydrophobic, do not disperse readily into aqueous media.

Early efforts to exfoliate graphite were mostly through the intercalation of graphite using a range of compounds [25]. With the resurging interest in graphene, this idea has been taken forward to produce large-scale exfoliation of graphene from commercially available expandable graphite [26]. Expandable graphite can be prepared by chemical intercalation of sulfuric acid and nitric acid. Upon heating, the volatile gaseous species released from the intercalant help exfoliate graphite very rapidly into multilayer graphene. This product is then reintercalated with oleum, (fuming sulpfuric acid with 20% free SO_3), and tetrabutylammonium hydroxide (TBA, 40% solution in water) is inserted into the oleum-intercalated graphite. The TBA-inserted oleum-intercalated graphite is sonicated in a N,N-dimethylformamide (DMF) solution of 1,2-distearoyl-sn-glycero-3-phosphoethanolamine-N-[methoxy(polyethyleneglycol)-5000](DSPE-mPEG) for 60 min to form a homogeneous suspension. A process of centrifugation removes large pieces of material from the supernatant and results in large amounts of graphene sheets suspended in DMF. The exfoliation–reintercalation–expansion of graphite can produce high-quality single-layer graphene sheets stably suspended in organic solvents and possess high electrical conductance and can be collected as large transparent conducting films by Langmuir–Blodgett assembly in a layer-by-layer manner.

In a more simple approach, stable dispersions up to ~0.01 mg/ml can be directly exfoliated from graphite by ultrasonication in organic solvents such as N-methyl pyrrolidone [27]. This method is physicochemically possible because the energy required to exfoliate graphene is balanced by the solvent–graphene interaction for solvents whose surface energies match that of graphene. Typically, graphite is dispersed in a relevant solvent by sonicating in a low-power ultrasonic bath for 30 min. The resultant dispersion is then centrifuged for 90 min at 500 rpm and decanted by pipetting off the top half of the dispersion. The decanted product usually contains a mixture of single- and few-layer graphene flakes that can be made into thin films by passing them through nanoporous membranes and drying them off.

A different method of obtaining stable dispersions in a liquid medium was reported by An et al. [28]. In this method, 1-pyrenecarboxylic acid (PCA) was used as a method to "wedge" out graphene sheets from graphite in suitably polarity-controlled combination of media to give rise to noncovalently functionalized graphene sheets that were stably dispersed in water. In a controlled medium, initially, PCA serves as a "molecular wedge" that cleaves the individual graphene flakes from the parent graphite pieces and then forms stable polar functional groups on the

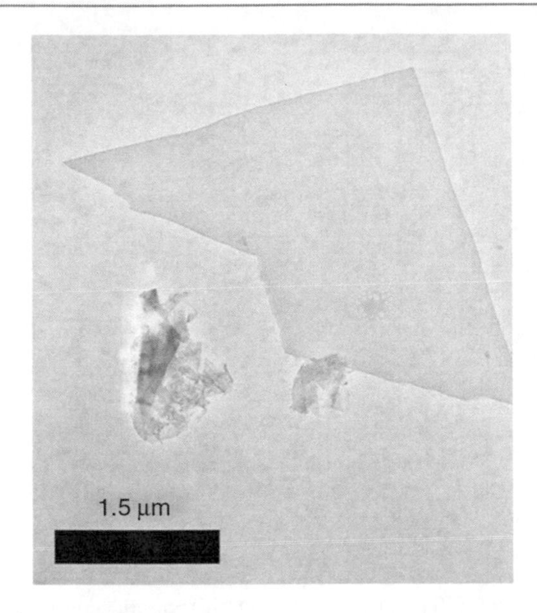

Synthesis of Graphene, Fig. 4 (a) A TEM image of a typical graphene flake exfoliated from graphite using a molecular wedging technique [28]

graphene surface via a noncovalent π–π stacking mechanism that does not disrupt its sp^2 hybridization. Figure 4 shows a typical TEM (▶ Transmission Electron Microscopy) image of graphene obtained this way. The hydrophilic –COOH group of PCA facilitates the formation of stable aqueous dispersions of graphene, in a manner similar to that of graphene oxide, but without degrading the sp^2 structure. The advantage of this method lies in the fact that the final product is stable in water – opening up the possibilities of experiments where an aqueous medium could be important, such as interactions with biological samples. At the same time, a simple methanol wash will remove the excess PCA from the surface and the pure graphene sheet precipitates out with time – to be utilized for experiments where pure, unfunctionalized graphene sheets are required. Similar methods have been demonstrated by other groups to obtain dispersions of graphene from graphite in surfactant – water solutions [29], through which significant levels of exfoliation and monolayer formation can be achieved.

Epitaxial and Large-Area Graphene on Metal Surfaces

High-temperature annealing of Pt single crystals produces an atomically thin entity on their surfaces, and this layer has been identified as monolayer graphite back in 1969 through low-energy electron diffraction (LEED) experiments [2]. Since then, it became clear that graphene in this morphology appears either due to the surface desegregation of carbon or through decomposition of hydrocarbons. Graphene has been reported to grow on a variety of metals such as Co, Ni, Ru, Rh, Pd, Ir, and Pt, with varying degrees of grain size, epitaxial alignment with the underlying metal, defect densities and surface morphologies [30]. For example, it was shown in 2006 that ethylene, absorbed at room temperature on Ir (111) planes, could be desegregated at 1,450K to produce graphene flakes of ~100 nm size that covers about 30% of the surface [31]. More recently, the desegregation method was successfully utilized to graphene on Ru (0001) surfaces, where macroscopic single-crystalline domains with linear dimensions exceeding 200 μm could be achieved [32, 33]. Figure 5a shows a high-resolution STM (▶ Scanning Tunneling Microscopy) image of graphene grown on Ru surface. By controlling the chamber pressure, growth time, and temperature, millimeter-scale graphene was reported on Ru surfaces using a controlled desegregation process in 2009 [34].

In 2008, the desegregation method received a boost when a low-pressure ▶ chemical vapor deposition (CVD) technique was utilized to synthesize graphene on an Ir (111) surface in a controlled manner over a range of temperatures from 1,120K to 1,320K [35]. Scanning tunneling microscopy of the grown structures reveals that graphene prepared using this technique exhibits a continuity of its carbon rows over terraces and step edges, enabling the possibility of large-scale growth that is not limited by surface features of metals. The thermodynamics of carbon segregation and graphene formation involves a complex interplay of carbon diffusion from the bulk to the surface of the host metal, followed by surface diffusion of C atoms that lead to nucleation and growth. In a typical CVD process, a metal substrate is placed in a flow-type furnace which is heated to a temperature between 750°C and 1,000°C in the presence of a mixture of Ar and H_2 at a fixed chamber pressure P. Once the target temperature stabilizes, the flow is maintained for some time (30–60 min) to clean and prepare the surface of the metal. At the end of this preparation time, the carbon source, typically ethylene or methane is introduced for a fixed time, at the end of which the furnace is allowed to cool down. Once

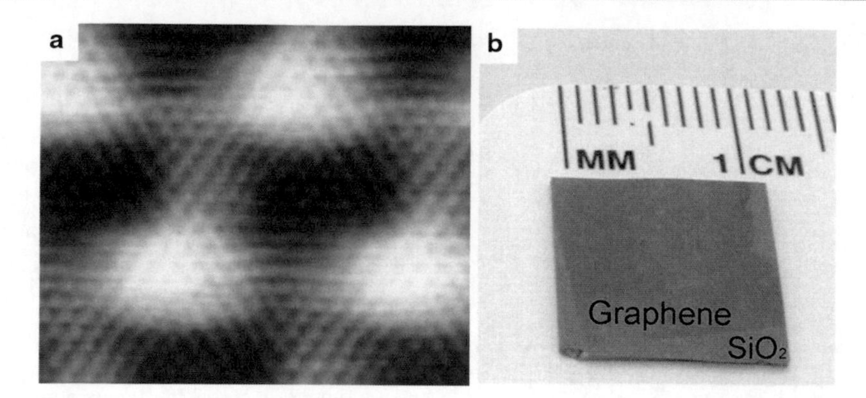

Synthesis of Graphene, Fig. 5 (**a**) Atomically resolved STM image of the graphene layer grown on Ru (0001) (Reprinted with permission from Ref. [32]; copyright © (2007) by the American Physical Society). (**b**) A digital photograph of a graphene film grown on a Cu foil using a CVD method, and transferred onto a SiO_2/Si substrate (Reprinted with permission from Ref. [38]; copyright © (2009) by the American Association for the Advancement of Science)

cooled, the surface is found to be covered with single- and multilayered graphene. Factors that appear to control layer thickness and domain sizes of graphene obtained such as temperature, feed-gas concentration, chamber pressure and time appear to be vastly different for different metals. For example, Yu et al. used CH_4:H_2:Ar = 0.15:1:2 with a total gas flow rate of 315 sccm and found that the type of graphene obtained on Ni foils depended strongly on the cooling rate, with multilayer graphene obtained only when the cooling rate was controlled to 10°C/s [36]. A great benefit of this technique is that thin layers of pre-patterned Ni could be utilized to fabricated large-scale patterned growth of graphene films that could be transferred to arbitrary substrates [37].

At elevated temperatures, the solubility of C atoms in Ni is much higher compared to that in Cu, and the resulting graphene formation can have different morphologies for these two metals. Centimeter-scale graphene films were shown to grow rather easily on Cu substrates in 2009 through a similar CVD process [38], with more than 95% of the surface was found to be covered with single-layer graphene that are continuous across copper surface steps and grain boundaries. Figure 5b shows an example of a macroscopic sheet of graphene grown on a Cu foil and transferred onto a SiO_2/Si substrate. In case of Cu, it seems that the cooling rate is unimportant, most probably due to the low solubility of carbon in copper which help to make the growth process self-limiting. There were no discernible differences in the graphene morphology when the cooling rate was varied from >300°C/min to about 40°C/min. The ability to grow high-quality large-scale graphene on a low-cost material (Cu) at such ease has been rapidly taken advantage of to produce graphene at a roll-to-roll capability [39]. This, along with a wet chemical doping and layer-by-layer stacking process, can be used to devise many-layer macroscopic graphene films with sheet resistance at values as low as $\sim 30\ \Omega\ \square^{-1}$ at ~90% transparency, which is superior to commercial transparent electrodes such as indium tin oxides and opens up several new possibilities in the fields of optics, optoelectronics, and photovoltaics.

Other Novel Advances in Gaphene Synthesis

In addition to the categories described above, there are numerous innovative methods being reported on a regular basis that are either extensions of the above-mentioned categories, or are completely new methods that can become significantly useful and/or popular in future. In addition to graphene, there is a whole array of related materials, including doped and functionalized graphene and graphene nanoribbons, each of which is unique and important by its own rights. It is impossible to either enlist all or describe these methods individually within the scope of this article. Some of the more popular and/or innovative ones are enlisted in Table 1 in order to provide the reader with a starting point. It should be borne in mind that a keyword search of existing databases through available searching tools was employed to obtain these references, and omission of any significant report(s) is purely accidental and inadvertent.

Synthesis of Graphene, Table 1 Some recent, novel and uncategorized graphene synthesis techniques

Method	Publication details
Solvothermal synthesis	Choucair et al., Nat. Nanotechnol. **4**, 30–33 (2009)
Substrate-free synthesis	Dato et al., Nano Lett. **8**, 2012–2016 (2008)
Electrochemical synthesis	1. Liu et al., Adv Func. Mat. **18**, 1518–1525 (2008) 2. Guo et al., ACS Nano **3**, 2653–2659 (2009)
Microwave plasma enhanced CVD	Malesevic et al., Nanotech. **19**, 305604.1–305604.6 (2008)
Arc discharge exfoliation	Wu et al., ACS Nano **3**, 411–417 (2009)
Microwave-solvothermal synthesis	Murugan et al., Chem. Mat. **21**, 5004–5006 (2009)
Doped graphene	1. Panchokarla et al., Adv Mat **21**, 4726–4730 (2009) 2. Ci et al., Nat. Mater. **9**, 430–435 (2010) 3. Li et al., Carbon **48**, 255–259 (2010)
Graphite exfoliation in ionic liquids	Lu et al., ACS Nano **3**, 2367–2375 (2009)
Hydrophilic and organophilic graphene	Wang et al., Carbon **47**, 1359–1364 (2009)
Graphene nanoribbons	1. Han et al., Phys. Rev. Lett. **98**, 206805.1–206805.4 (2007) 2. Li et al., Science **319**, 1229–1232 (2008) 3. Jiao et al., Nat. Nanotech. 5, 321–325 (2010)
Reducing sugar	Zhu et al., ACS Nano **4**, 2429–2437 (2010)

Conclusion

It is evident that within a very short time frame of a few years, the scientific community has witnessed a tremendous advancement in terms of various techniques of graphene synthesis. This advancement has progressed hand-in-hand with a number of extremely important fundamental discoveries related to this novel 2D crystalline material and has also opened up several avenues for groundbreaking technological directions. Despite the rapid success that this field has experienced, the interest in developing better and more cost-effective synthesis methods keeps growing. In the near term, production of high-quality graphene with reproducible electrical as well as mechanical properties will be a challenge. Similarly, techniques and parameters for size-specific, controlled synthesis of graphene such as graphene nanoribbons, graphene quantum dots etc., need to be well understood. Processes for large-scale synthesis of monolayer graphene (e.g., through chemical exfoliation) needs to perfected in order to utilize these materials in variety of bulk applications such as electrodes in electrical energy storage systems, fillers in composite etc. Overall, the current level of progress as well as the ongoing research efforts in graphene synthesis has been quite promising for certain niche area applications, and the scientific community is quite optimistic that in future, graphene will eventually emerge as the next ultimate engineering material.

Cross-References

▶ Atomic Force Microscopy
▶ Carbon Nanotubes
▶ Chemical Vapor Deposition (CVD)
▶ Flexible Electronics
▶ Fullerenes for Drug Delivery
▶ Graphene
▶ Nanomaterials for Electrical Energy Storage Devices
▶ Scanning Tunneling Microscopy
▶ Transmission Electron Microscopy

References

1. Boehm, H.P., Clauss, A., Fischer, G.O., Hofmann, U.: Das Adsorptions verhalten sehr dünner Kohlenstoffolien. Zeitschrift für anorganische und allgemeine Chemie **316**, 119–127 (1962)
2. May, J.W.: Platinum surface LEED rings. Surf. Sci. **17**, 267–270 (1969)
3. Eizenberg, M., Blakely, J.M.: Carbon monolayer phase condensation on Ni(111). Surf. Sci. **82**, 228–236 (1979)
4. Aizawa, T., Souda, R., Otani, S., Ishizawa, Y., Oshima, C.: Anomalous bond of monolayer graphite on transition-metal carbide surfaces. Phys. Rev. Lett. **64**, 768–771 (1990)
5. Dresselhaus, M.S., Dresselhaus, G., Saito, R.: Carbon fibers based on C60 and their symmetry. Phys. Rev. B **45**, 6234–6242 (1992); Aizawa, T., Hwang, Y., Hayami, W., Souda, R., Otani, S., Ishizawa, Y.: Phonon dispersion of monolayer graphite on Pt(111) and NbC surfaces: bond softening and interface structures. Surf. Sci. **260**, 311–318 (1992)
6. Lu, X., Yu, M., Huang, H., Ruoff, R.S.: Tailoring graphite with the goal of achieving single sheets. Nanotechnology **10**, 269–272 (1999)

S

7. Lu, X., Huang, H., Nemchuk, N., Ruoff, R.S.: Patterning of highly oriented pyrolytic graphite by oxygen plasma etching. Appl. Phys. Lett. **75**, 193–195 (1999)
8. Novoselov, K.S., Geim, A.K., Morozov, S.V., Jiang, D., Zhang, Y., Dubonos, S.V., Grigorieva, I.V., Firsov, A.A.: Electric field effect in atomically thin carbon films. Science **306**, 666–669 (2004)
9. Berger, C., Song, Z., Li, T., Li, X., Ogbazghi, A.Y., Feng, R., Dai, Z., Marchenkov, A.N., Conrad, E.H., First, P.N., de Heer, W.A.: Ultrathin epitaxial graphite: 2D electron gas properties and a route toward graphene-based nanoelectronics. J. Phys. Chem. B **108**, 19912–19916 (2004)
10. Peierls, R.E.: Quelques proprieties typiques des corpses solides. Ann. I. H. Poincare **5**, 177–222 (1935)
11. Landau, L.D.: Zur Theorie der phasenumwandlungen II. Phys. Z. Sowjetunion **11**, 26–35 (1937)
12. Novoselov, K.S., Jiang, D., Schedin, F., Booth, T.J., Khotkevich, V.V., Morozov, S.V., Geim, A.K., Rice, T.M.: Two-dimensional atomic crystals. Proc. Natl. Acad. Sci. U.S. A. **102**, 10451–10453 (2005)
13. Soldano, C., Mahmood, A., Dujardin, E.: Production, properties and potential of graphene. Carbon **48**, 2127–2150 (2010)
14. Rollings, E., Gweon, G.-H., Zhou, S.Y., Mun, B.S., McChesney, J.L., Hussain, B.S., Fedorov, A.N., First, P.N., de Heer, W.A., Lanzara, A.: Synthesis and characterization of atomically thin graphite films on a silicon carbide substrate. J. Phys. Chem. Sol **67**, 2172–2177 (2006)
15. Hass, J., Feng, R., Li, T., Li, X., Zong, Z., de Heer, W.A., First, P.N., Conrad, E.H., Jeffrey, C.A., Berger, C.: Highly ordered graphene for two dimensional electronics. Appl. Phys. Lett. **89**, 143106–143108 (2006)
16. Stankovich, S., Dikin, D.A., Piner, R.D., Kohlhaas, K.A., Kleinhammes, A., Jia, Y., Wu, Y., Nguyen, S.T., Ruoff, R. S.: Synthesis of graphene-based nanosheets via chemical reduction of exfoliated graphite oxide. Carbon **45**, 1558–1565 (2007)
17. Hummers., W.S., Jr., Offeman, R.E.: Preparation of graphitic oxide. J. Am. Chem. Soc. **80**, 1339–1339 (1958)
18. Rourke, J.P., Pandey, P.A., Moore, J.J., Bates, M., Kinloch, I. A., Young, R.J., Wilson, N.R.: The real graphene oxide revealed: stripping the oxidative debris from the graphene-like sheets. Angew. Chem. Int. Ed. **50**, 3173–3177 (2011)
19. Eda, G., Fanchini, G., Manish Chhowalla, M.: Large-area ultrathin films of reduced graphene oxide as a transparent and flexible electronic material. Nat. Nanotechnol. **3**, 270–274 (2008)
20. Gilje, S., Han, S., Wang, M., Wang, K.L., Kaner, R.B.: A chemical route to graphene for device applications. Nano Lett. **7**, 3394–3398 (2007)
21. Tung, V.C., Allen, M.J., Yang, Y., Kaner, R.B.: High-throughput solution processing of large-scale graphene. Nat. Nanotechnol. **4**, 25–29 (2009)
22. Si, Y., Samulski, E.T.: Synthesis of water soluble graphene. Nano Lett. **8**, 1679–1682 (2008)
23. Park, S., Ruoff, R.S.: Chemical methods for the production of graphenes. Nat. Nanotechnol. **4**, 217–224 (2009)
24. Gao, W., Alemany, L.B., Ci, L., Ajayan, P.M.: New insights into the structure and reduction of graphite oxide. Nat. Chem. **1**, 403–408 (2009)
25. Dresselhaus, M.S., Dresselhaus, G.: Intercalation compounds of graphite. Adv. Phys. **30**, 139–326 (1981)
26. Li, X., Zhang, G., Bai, X., Sun, X., Wang, X., Wang, E., Dai, H.: Highly conducting graphene sheets and Langmuir–Blodgett films. Nat. Nanotechnol. **3**, 538–542 (2008)
27. Hernandez, Y., Nicolosi, V., Lotya, M., Blighe, F.M., Sun, Z., De, S., McGovern, I.T., Holland, B., Byrne, M., Gun'Ko, Y.K., Boland, J.J., Niraj, P., Duesberg, G., Krishnamurthy, S., Goodhue, R., Hutchison, J., Scardaci, V., Ferrari, A.C., Coleman, J.N.: High-yield production of graphene by liquid-phase exfoliation of graphite. Nat. Nanotechnol. **3**, 563–568 (2008)
28. An, X., Simmons, T., Shah, R., Wolfe, C., Lewis, K.M., Washington, M., Nayak, S.K., Talapatra, S., Kar, S.: Stable aqueous dispersions of noncovalently functionalized graphene from graphite and their multifunctional high-performance applications. Nano Lett. **10**, 4295–4301 (2010)
29. Lotya, M., Hernandez, Y., King, P.J., Smith, R.J., Nicolosi, V., Karlsson, L.S., Blighe, F.M., De, S., Wang, Z., McGovern, I.T., Duesberg, G.S., Coleman, J.N.: Liquid phase production of graphene by exfoliation of graphite in surfactant/water solutions. J. Am. Chem. Soc. **131**, 3611–3620 (2009)
30. Wintterlin, J., Bocquet, M.-L.: Graphene on metal surfaces. Surf. Sci. **603**, 1841–1852 (2009)
31. N'Diaye, A.T., Bleikamp, S., Feibelman, P.J., Michely, T.: Two-dimensional Ir cluster lattice on a graphene moire on Ir(111). Phys. Rev. Lett. **97**, 215501.1–215501.4 (2006)
32. Marchini, S., Gunther, S., Wintterlin, J.: Scanning tunneling microscopy of graphene on Ru(0001). Phys. Rev. B **76**, 075429.1–075429.9 (2007)
33. Sutter, P.W., Flege, J.-I., Sutter, E.A.: Epitaxial graphene on ruthenium. Nat. Mater. **7**, 406–411 (2008)
34. Pan, Y., Zhang, H.G., Shi, D.X., Sun, J., Du, S., Liu, F., Gao, H.-J.: Highly ordered, millimeter-scale, continuous, single-crystalline graphene monolayer formed on Ru (0001). Adv. Mater. **21**, 2777–2780 (2009)
35. Coraux, J., N'Diaye, A.T., Busse, C., Michely, T.: Structural coherency of graphene on Ir(111). Nano Lett. **8**, 565–570 (2008)
36. Yu, Q., Lian, J., Siriponglert, S., Li, H., Chen, Y.P., Pei, S.-S.: Graphene segregated on Ni surfaces and transferred to insulators. Appl. Phys. Lett. **93**, 113103–113105 (2008)
37. Kim, K.S., Zhao, Y., Jang, H., Lee, S.Y., Kim, J.M., Kim, K.S., Ahn, J.-H., Kim, P., Choi, J.-Y., Hong, B.H.: Large-scale pattern growth of graphene films for stretchable transparent electrodes. Nature **457**, 706–710 (2009)
38. Li, X., Cai, W., An, J., Kim, S., Nah, J., Yang, D., Piner, R., Velamakanni, A., Jung, I., Tutuc, E., Banerjee, S.K., Colombo, L., Ruoff, R.S.: Large-area synthesis of high-quality and uniform graphene films on copper foils. Science **324**, 1312–1314 (2009)
39. Bae, S., Kim, H., Lee, Y., Xu, X., Park, J.-S., Zheng, Y., Balakrishnan, J., Lei, T., Kim, H.R., Song, Y.I., Kim, Y.-J., Kim, K.S., Özyilmaz, B., Ahn, J.-H., Hong, B.H., Iijima, S.: Roll-to-roll production of 30-inch graphene films for transparent electrodes. Nat. Nanotechnol. **5**, 574–578 (2010)

Synthesis of Nanoparticles

▶ Nanoparticles

Synthesis of Subnanometric Metal Nanoparticles

Javier Calvo Fuentes[1], José Rivas[2,3] and
M. Arturo López-Quintela[2]
[1]NANOGAP SUB-NM-POWDER S.A.,
Milladoiro – Ames (A Coruña), Spain
[2]Laboratory of Magnetism and Nanotechnology,
Institute for Technological Research, University of
Santiago de Compostela, Santiago de Compostela,
Spain
[3]INL – International Iberian Nanotechnology
Laboratory, Braga, Portugal

Synonyms

Atomic cluster; Cluster; Nanocluster; Quantum cluster; Quantum dot

Definition

Metal atomic clusters consist of groups of atoms with well-defined compositions and one or very few stable geometric structures. They represent the most elemental building blocks in nature – after atoms – and are characterized by their size, comparable to the Fermi wavelength of an electron, which makes them a bridge between atoms and nanoparticles or bulk metals, with properties very different from both of them.

Introduction

Typical metal nanoparticles with dimensions from two to several tens of nanometers show smoothly size-dependent properties. However, when particle size becomes comparable to the Fermi wavelength of an electron (~0.52 nm for gold and silver), properties of metal clusters are dramatically different from what should be expected if they were due only to their high surface-to-volume ratio. In these subnanometric species, quantum effects are responsible for totally new chemical, optical, and electronic properties such as, for example, magnetism, photoluminescence, or catalytic activity.

The term *nanoparticle* usually refers to any particle of bulk metal with dimensions in the nanoscale. Nanoparticles usually present a core-shell structure with a core of bulk metal surrounded by a shell of disordered atoms, and their enhanced properties are due mainly to their high surface/volume ratio. The term *cluster* is used in reference to *subnanometric species* consisting of well-defined structures of metal atoms stabilized by different types of protecting ligands, with sizes below approximately 1–2 nm. In general, clusters can be divided into (1) large clusters, consisting of a core formed by a number of metal atoms in the range ≈10–20 to 100–200 and a protecting shell of strong ligands such as phosphines or thiols; and (2) small clusters formed by a reduced number of atoms (≈2 to 10–20), which do not need any strong stabilizing ligand and have almost all their atoms on the surface. Due to quantum effects, both kinds of clusters – large and small – present discrete energy levels and an increasing band gap with decreasing size (Fig. 1).

Because of this splitting of energies at the Fermi level, clusters show different properties to those of nanoparticles. Cluster properties are highly dependent on their sizes. As an example, the typical surface plasmon band observed in Ag-metal nanoparticles (Fig. 2a) disappears for clusters below approximately 1–2 nm (Figs. 2b and c corresponding to large and small clusters, respectively), indicating that all the conducting electrons are now *frozen* and the *metal* silver loses its typical *metallic character*. At the same time, a clear difference between large and small clusters can be observed: large clusters show a continuous decrease of the absorption band with some small bumps, similar to the absorption displayed by semiconductors (Fig. 2b), and small clusters display well-defined absorption bands indicating a *molecular-like* behavior (Fig. 2c).

The presence of such size-dependent properties can be well represented on a 3D periodic table of elements – schematically depicted in Fig. 3 – where, between the atomic and the bulk state of every element, a whole range of new materials with their different size-dependent properties appear.

Cluster Structure

It is very difficult to do precise calculations of the electronic structure of metal clusters, particularly when they are larger than just a small number of

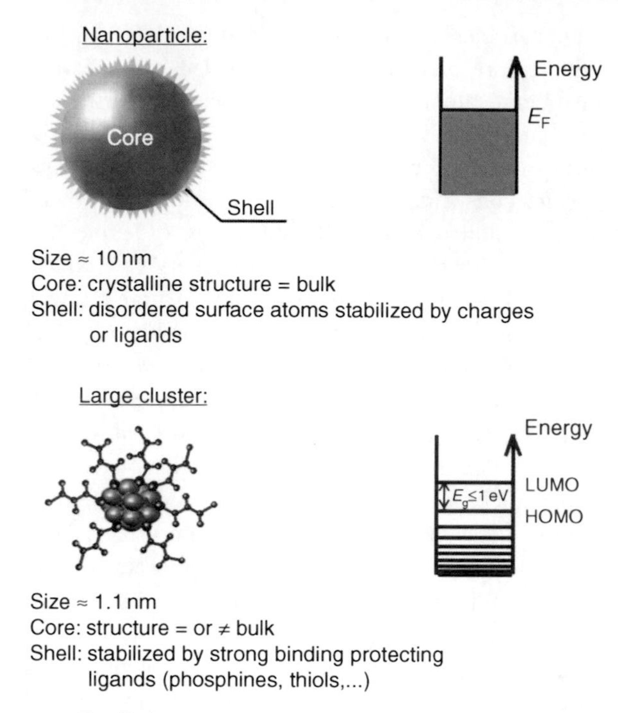

Synthesis of Subnanometric Metal Nanoparticles, Fig. 1 Schematic properties of nanoparticles and subnanometric clusters

atoms, and they are in solution stabilized by some capping molecules. Such difficulty can be avoided using simple models, such as the Jellium model [1] firstly developed for clusters in the gas phase. In this model, the real cluster is replaced by an electronic shell structure consisting of a uniform, positively charged *jellium* sphere surrounded by valence electrons. The electrons are considered to be like moving in a mean-field potential occupying energy levels according to the Aufbau principle ($1S^2$ | $1P^6$ | $1D^{10}$ | $2S^2$ $1F^{14}$ | $2P^6$ $1G^{18}$ | ...) as represented in Fig. 4. This model gives a considerably good approximation, preserving many of the physicochemical characteristics of clusters.

The total energy as a function of cluster size, calculated by this approximation, has discontinuities corresponding to electronic shell closures that are consistent with the positions of the peaks appearing in the mass spectrum of the vaporized metal. These peaks are associated with the abundance of exceptionally stable clusters with certain number of atoms: 8, 18, 20, 34, 40, 58, 92, 138..., usually referred to as *magic numbers*.

As it was mentioned above, one simple consequence of this quantum-size regime with the presence of discrete states in metal clusters is the appearance of a sizable HOMO-LUMO band gap similar to that of semiconductors. Such semiconductor-like behavior is particularly significant for smaller clusters (i.e., those with a low number of atoms) with band gaps widely exceeding 1 eV, as it is shown in Fig. 5 for some selected Ag and Cu clusters.

Some efforts have been made in the last years to find prediction models for ligand-protected gold-cluster species in solution. In such models, clusters are assumed to be electronically stabilized by ligands that either withdraw electrons from the metal core or are attached as weak Lewis-base ligands coordinated to the core surface by dative bonds. In this case, the requirement for an electronically closed shell, [3] formulated as $(L_S{\cdot}A_NX_M)^z$, is:

$$n^* = N\nu_A - M - z \tag{1}$$

where n^* represents the number of electrons for shell closing of the metallic core, which has to match one of the magic numbers corresponding to strong electron shell closures in an anharmonic mean-field potential giving rise to stable clusters; N stands for the number of core metal atoms (A); ν_A is the atomic valence; M is the number of electron-localizing (or electron-withdrawing) ligands X, assuming a withdrawal of one electron per ligand molecule; and z represents the overall charge of the complex. The weak ligands represented by L_S may be needed for completing the steric surface protection of the cluster core.

Trying to combine at the same time (1) the mentioned magic numbers required for electronic stabilization of the cluster, (2) the requirements of the Eq. 1, and (3) the fact that solution-phase clusters require a sterically complete protective ligand shell compatible with a compact atomic shell structure for the metallic core, complicates enormously the calculations of the structures satisfying such conditions. Furthermore, in cases such as some gold and silver thiolate clusters, the identity of the actual X groups is not clear due to

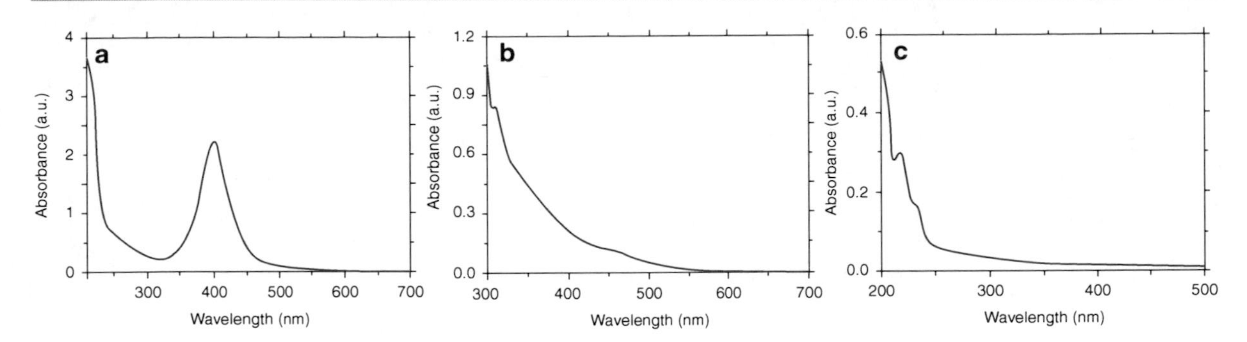

Synthesis of Subnanometric Metal Nanoparticles, Fig. 2 Absorption spectra of (**a**) silver nanoparticles displaying the typical plasmon band at ≈400 nm, (**b**) surfactant-protected large silver clusters showing a typical continuum decrease of the absorption, and (**c**) *strong-ligand free* small silver clusters with well-defined *molecular-like* absorption bands

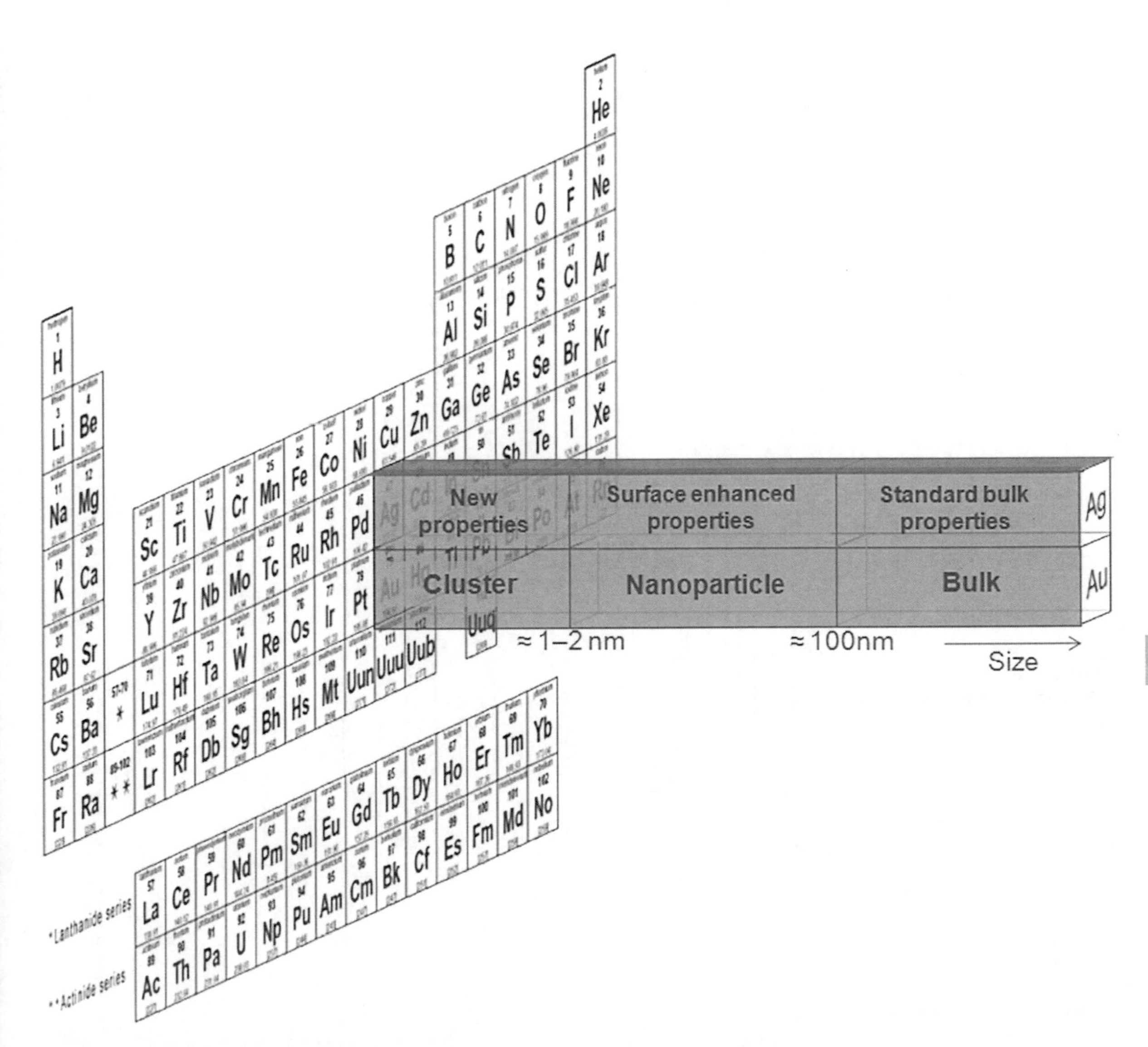

Synthesis of Subnanometric Metal Nanoparticles, Fig. 3 Size-dependent 3D periodic table of elements

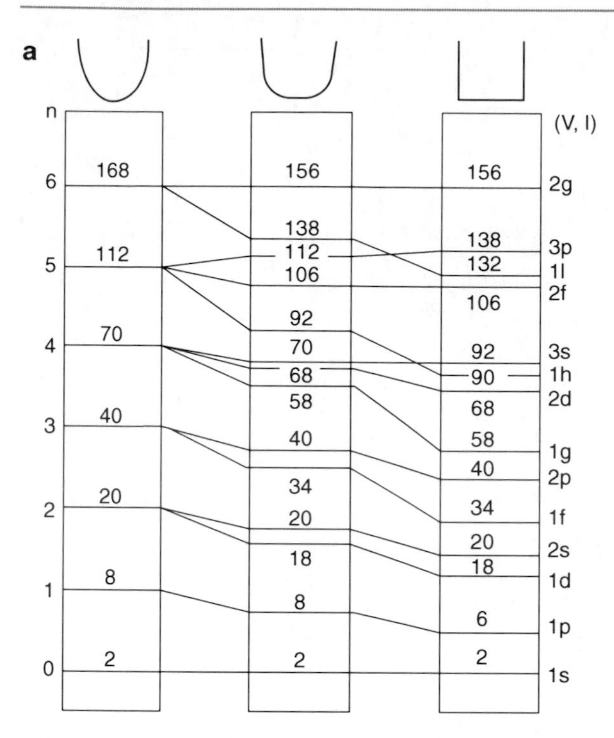

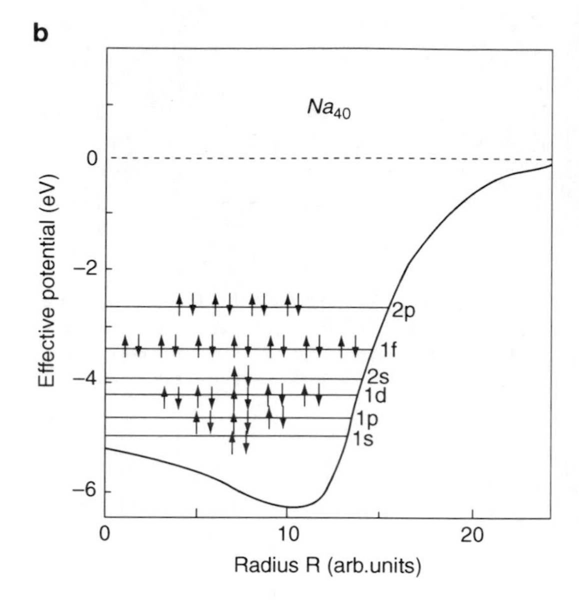

Synthesis of Subnanometric Metal Nanoparticles, Fig. 4 (**a**) Energy-level occupations for spherical three-dimensional, harmonic, intermediate, and square-well potentials. (**b**) Self-consistent effective potential of a Jellium sphere corresponding to Na_{40} with the electron occupation of the energy levels. Reprinted with permission from reference [2]. Copyright 1993 by the American Physical Society

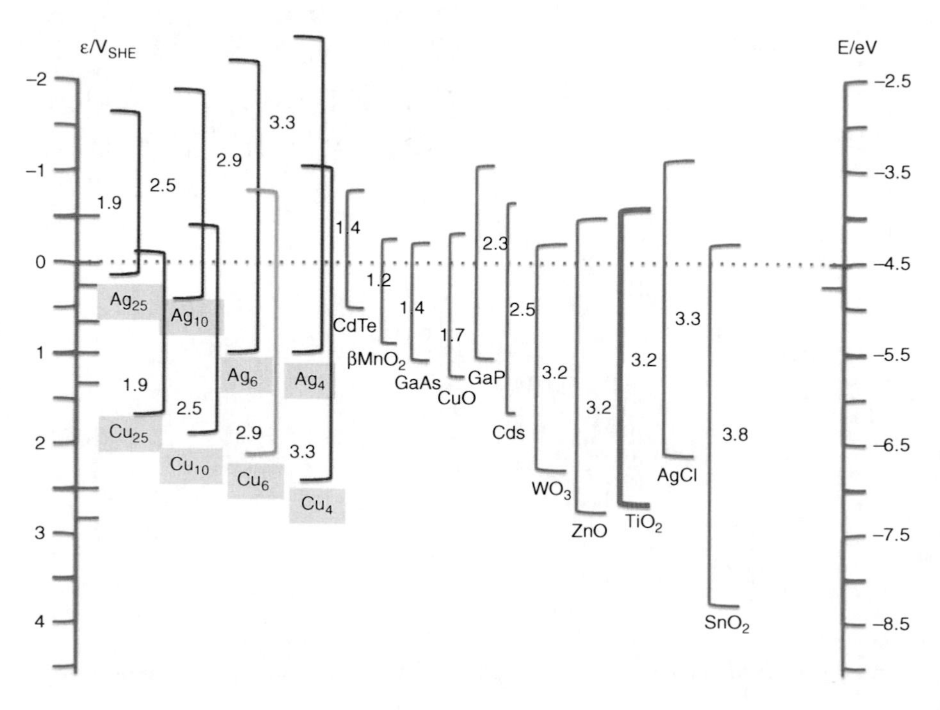

Synthesis of Subnanometric Metal Nanoparticles, Fig. 5 Schematic comparison between bandgaps of some silver and copper clusters (M_N, N = number of atoms) and those of well-known semiconductors. Bandgaps (E_g) were calculated from the spherical Jellium model ($E_g = E_F/N^{1/3}$; E_F = Fermi level), and the position of the conduction band, E_{CB}, was estimated by the formula $E_{CB} = \chi - E_F - ½ E_g$ (χ = electronegativity)

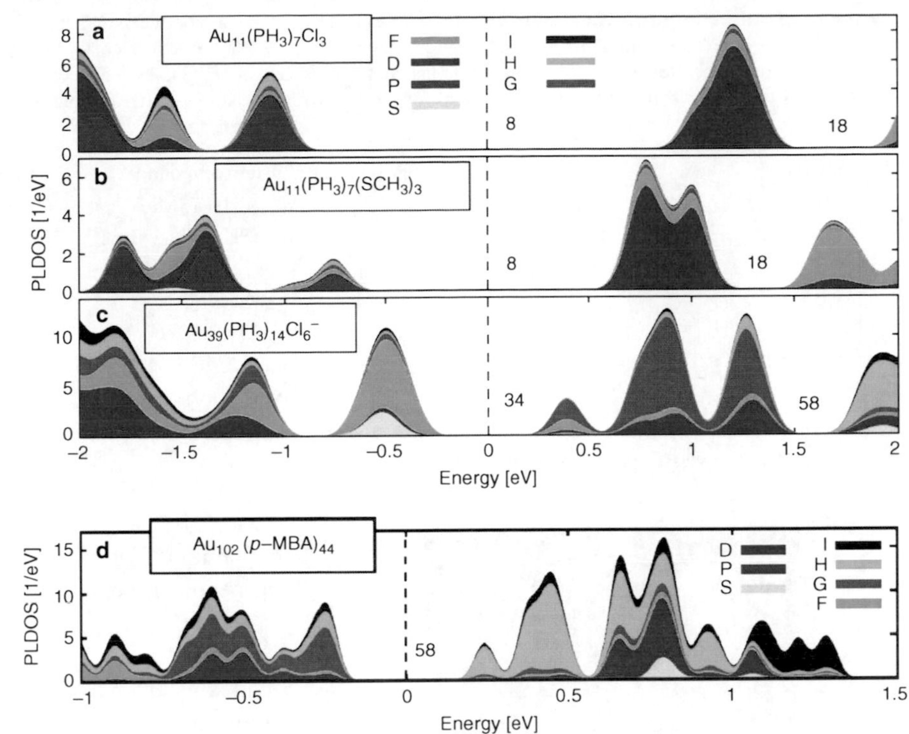

Synthesis of Subnanometric Metal Nanoparticles, Fig. 6 Electronic structure analysis of gold clusters. The angular-momentum-projected local electron density of states (PLDOS) (projection up to the l symmetry, i.e., $l = 6$) for the gold core in $Au_{11}(PH_3)_7Cl_3$ (**a**), $Au_{11}(PH_3)_7(SCH_3)_3$ (**b**), $Au_{39}(PH_3)_{14}Cl_6^-$ (**c**), and $Au_{102}(p\text{-MBA})_{44}$ (**d**). The zero energy corresponds to the middle of the HOMO-LUMO gap (dashed line). For plotting, each individual electron state is displayed by a Gaussian smoothing of 0.07 eV (0.03 eV in **d**). Shell-closing number is indicated for each case. For more details see reference [3]. Copyright 2008 National Academy of Sciences, U.S.A.

the undefined nature of the surface chemical bonds. There have been only relatively few studies so far to resolve the cluster structures by ab initio calculations. Very recently, Häkkinen's group managed to determine, through large-scale Density Functional Theory (DFT) calculations, [3] the electronic structure of $Au_{102}(p\text{-MBA})_{44}$ (*p*-MBA, *para*-mercaptobenzoic acid, $SC_7O_2H_5$), an all-thiolate-protected cluster formed by 102 gold atoms, as well as some smaller phosphine-halide- and phosphine-thiolate-protected clusters, as it is shown in Fig. 6.

In all cases, the characteristic bandgap of the clusters is clearly observed. In Table 1, it is shown how this bandgap changes for different cluster sizes and ligands. For comparison purposes, the bangap calculated from the simple spherical Jellium model is included (for clusters with sizes $\approx\geq 25$ atoms, a correction of -0.4 eV was used because of the anharmonicity observed in large clusters) [4]. It can be observed that, in general, bandgap increases when the cluster size decreases, being a nice agreement between the DFT-calculated bandgap values and those predicted by the simple Jellium model, showing at the same time that ligands play a relatively minor role.

Properties of Metal Clusters

Among the properties that make metal clusters unique and useful in many applications, the following ones can be highlighted here as examples:

Photoluminescence

The molecular-like behavior of few-atom noble metal clusters and the mentioned bandgap similar to that of semiconductors, allows the possibility of size-tunable electronic intraband transitions [5].

The appearance of these optical transitions with energies ranging from the UV-visible to the infrared region makes noble metal quantum dots ideal potential candidates for applications such as fluorescent labeling in biology or light-emitting sources in nanoelectronics. As an example, Fig. 7 shows the fluorescence of aqueous solutions of Cu_N clusters ($N < 14$) excited at 296 nm with quantum yields over 10%, which are stable for more than 2 years.

Even though, in cases where the quantum yield is lower than that cited in previous example, fluorescent noble metal clusters are a very interesting alternative to semiconductor quantum dots as biological labels, due not only to their small hydrodynamic size and inert

Synthesis of Subnanometric Metal Nanoparticles, Table 1 Comparison between: (1) experimentally determined bandgaps for free gas-phase gold cluster anions from photoelectron spectroscopy; (2) theoretical (Density Functional Theory) values for HOMO-LUMO gaps of passivated gold cluster compounds that correspond to 8, 34, and 58 conduction-electron shell closings; and (3) bandgaps calculated through the Jellium model for clusters with the same number of gold atoms

	Experiment		Theory		Jellium model
Shell closing	Cluster	Gap (eV)	Cluster compound	Gap (eV)	$E_g = 5.32/N^{1/3}$ (eV)
8e ($1S^2 1P^6$)			$Au_{11}(PH_3)_7(SMe)_3$	1.5	2.4
8e			$Au_{11}(PH_3)_7Cl_3$	2.1	2.4
8e			$Au_{13}(PH_3)_{10}Cl_2^{3+}$	1.8	2.3
8e			$Au_{25}(SMe)_{18}^-$	1.2	1.3[a]
34e (8e + $1D^{10}2S^2 1F^{14}$)	$Au34^-$	1.0	$Au_{39}Cl_6(PH_3)_{14}^-$	0.8	1.0[a]
58e (34e + $2P^6 1G^{18}$)	$Au58^-$	0.6	Au_{102}(p-MBA)$_{44}$	0.5	0.6[a]
58e			$Au_{102}(SMe)_{44}$	0.5	0.6[a]

Adapted from reference [3]. Copyright 2008 National Academy of Sciences, U.S.A.
[a]For these clusters, a correction of −0.4 eV was used – see text

Synthesis of Subnanometric Metal Nanoparticles, Fig. 7 Blue emission observed for an aqueous solution of copper clusters ($N \leq 14$ atoms) under an excitation wavelength of 296 nm. Reprinted with permission from reference [6]. Copyright 2010 American Chemical Society

nature, but especially because of their biocompatibility and also they present reduced photobleaching compared to organic fluorophores. Water-soluble gold clusters have been, therefore, explored as fluorescent labels in biological experiments. Recently, for example, Parak's group described [7] the use of stable gold clusters capped with dihydrolipoic acid (DHLA) and conjugated through EDC chemistry to biologically relevant molecules such as PEG, BSA, avidin, or streptavidin. Unconjugated clusters can also be nonspecifically uptaken by living cells (see Fig. 8).

Catalysis

Small clusters have been found to own a catalytic activity not observed in their bulk analogues or even nanoparticles, which makes them very attractive as new catalytic materials. Quantum chemical calculations indicate that such high reactivity is due to undercoordination of the metal atoms forming the cluster [8]. Several families of metal clusters have been found to possess surprisingly high and selective catalytic activity when immobilized on a support. For example, platinum clusters with 8–10 atoms can be used as catalysts for the oxidative dehydrogenation of propane [9], while gold clusters with 6–10 atoms have been shown to be highly active for catalyzing propene epoxidation [10]. Recently, Harding et al. [11] studied the control and tunability of the catalytic oxidation of CO by Au_{20} clusters deposited on MgO surfaces, finding that the active site on the cluster is characterized by enhanced electron density, which activates the adsorbed O_2 molecule and promotes the bonding of CO.

Small metal clusters have been also found to display electrocatalytic activities different from the material in bulk or as nanoparticles [12]. Although this is a field not too much explored so far, these electrocatalytic properties make metal clusters to be promising materials in fuel-cell applications. As another interesting example of such properties, a recent report of their use to prevent pathologies, such as fetal alcohol syndrome, [13] is highlighted here: It has been proved that some

Control

50 nM AuNC@DHLA Normal medium

50 nM AuNC@DHLA Condition medium

Synthesis of Subnanometric Metal Nanoparticles, Fig. 8 Nonspecific uptake of unconjugated fluorescent DHLA-capped Au clusters by human endothelial cells. Cell nuclei were stained to yield blue fluorescence. Red fluorescence corresponds to the clusters. See reference [7] for more information. Copyright 2009 American Chemical Society. Reprinted with permission

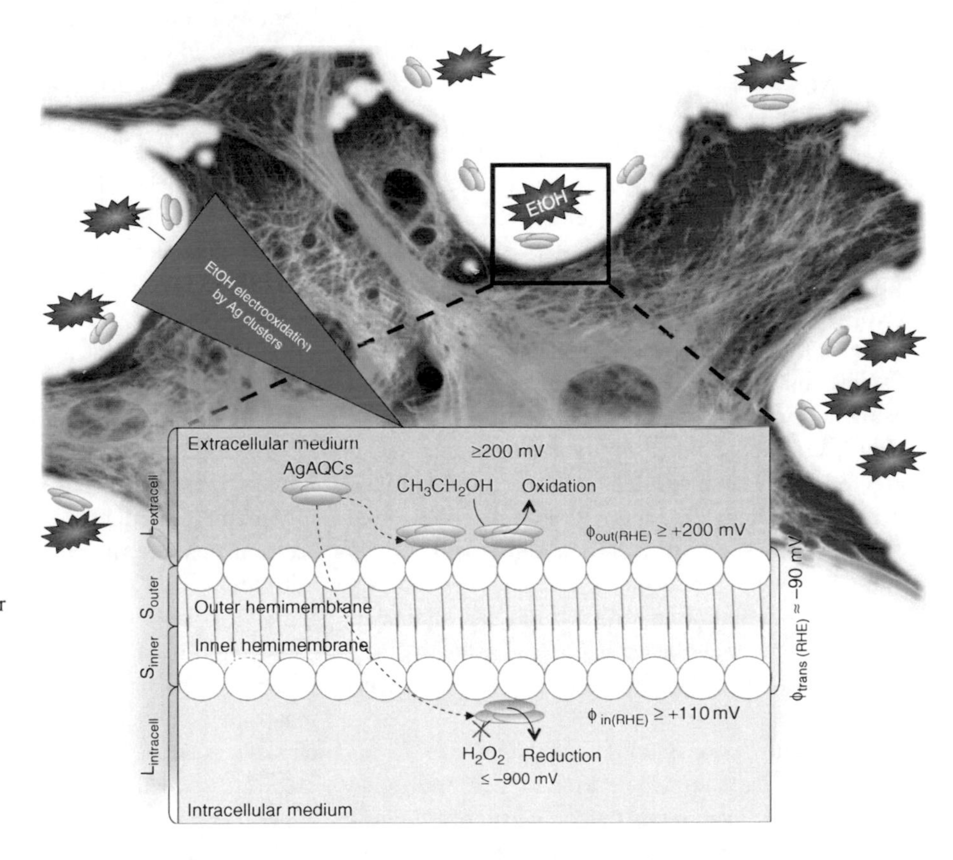

Synthesis of Subnanometric Metal Nanoparticles, Fig. 9 Schematic representation of the electrocatalysis of the oxidation of ethanol by silver clusters on the cellular membrane preventing the alcohol toxicity in living cells. For more details, see reference [13]. Copyright 2010 American Chemical Society. Reprinted with permission

clusters are able to electrocatalyze the oxidation of alcohols and prevent its cytotoxicity under physiological conditions and at very low potentials, like those found on living cells, as it is schematically represented in Fig. 9.

Synthesis of Metal Clusters

As it also occurs in the case of nanoparticles, there are two main approaches to the synthesis of metal clusters: top-down and bottom-up.

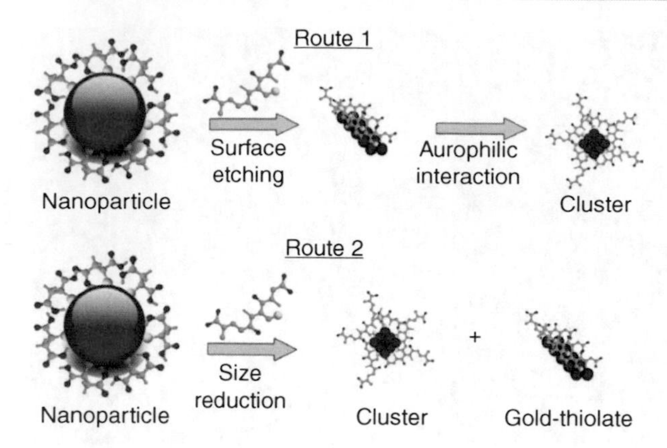

Synthesis of Subnanometric Metal Nanoparticles, Fig. 10 Schematic representation of two possible routes for the formation of gold clusters by etching of mercaptosuccinic acid-capped gold nanoparticles. Reprinted from reference [14] with kind permission from Springer Science + Business Media B.V.

Top-Down

In top-down approaches, clusters are synthesized from larger precursors such as nanoparticles or bulk metal. The most common top-down technique used for the synthesis of clusters is etching of nanoparticles (Fig. 10). In this technique, clusters are synthesized using the etching capacity of some ligands (like thiols), which remove the surface atoms of metallic nanoparticles or break them into smaller pieces leading to stable quantum clusters characterized by shell-closing magic numbers.

For example, etching of mercaptosuccinic acid-protected gold nanoparticles with excess glutathione has been used to yield small photoluminescent gold clusters [14]. In such work, clusters, either with 8 or 25 gold atoms, were produced by etching of previously synthesized nanoparticles with 4–5 nm core diameter. Selection of the final cluster size was possible by only varying the etching pH from ~3 (for 25 atom clusters) to 7–8 (for 8 atom clusters).

Multivalent coordinating polymers such as polyethylenimine can also be used to etch preformed colloidal gold nanocrystals producing highly fluorescent water-soluble nanoclusters formed by 8 gold atoms [15].

Bottom-Up

In bottom-up approaches, clusters are built from smaller structures at atomic level such as single atoms or ions (Fig. 11). In this case, the most

Metal precursor + Reductor (+ Stabilizer) ⇨

Synthesis of Subnanometric Metal Nanoparticles, Fig. 11 Schematic representation of bottom-up synthesis

commonly used techniques differ depending on the size of the clusters to be synthesized. Wet chemical preparation of clusters includes the chemical reduction of metal salts in the presence of strong binding ligands [4, 16, 17] or cages [18], and by kinetic control using, e.g., microemulsions [19–21], electrochemical methods [6, 12, 22], etc.

For the synthesis of large clusters, a large variety of strong, protecting ligands can be used to control the growth of the primarily formed clusters and to stabilize them. For small clusters, the kinetic control, based on slow reaction rates provided either by soft reducing agents or low concentrations (as e.g., those provided by microemulsions, electrochemical methods, etc.), is the key point to their synthesis. It is generally believed that the synthesis of small clusters is extremely difficult to achieve because of the highly precise control of experimental conditions required to stop and then isolate the clusters as soon as they are formed. That belief is based on the theory of nucleation and growth, according to which, nuclei formed during the first steps of the chemical synthesis are only stable when they grow over a specific size called critical nucleus. Below that size, nuclei dissolve because of their large Laplace pressure. Above that size, they continue growing in order to reduce their surface energy by different mechanisms like autocatalysis, Ostwald ripening, etc., until the growth is stopped by capping agents or other templates. However, such arguments, which correctly can be applied to the synthesis of nanoparticles and larger clusters having the same structure of the bulk material, are not correct when they apply to very small clusters, as those produced by kinetic control. As it was previously mentioned, numerous theoretical and experimental reports indicate that clusters can be especially stable because of their particular electronic and geometric structures different from the bulk. In such a case, the use of macroscopic thermodynamic arguments, like those used in the theory of nucleation and growth, cannot be applied. To clarify this important aspect, in Fig. 12, a scheme is shown of the variation of the free energy throughout a reaction of nanoparticle's

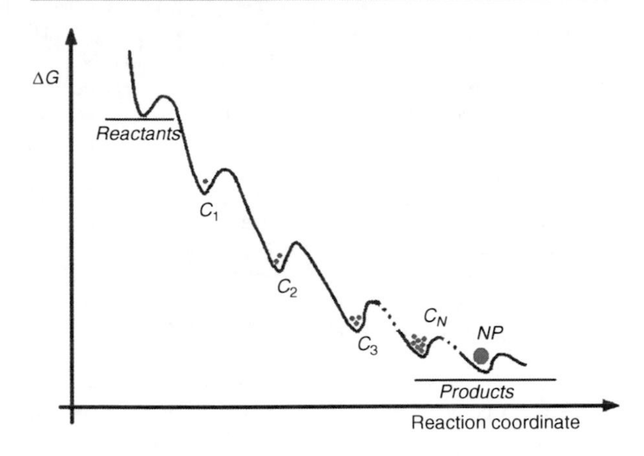

Synthesis of Subnanometric Metal Nanoparticles, Fig. 12 Schematic representation of the variation of free energy during the chemical synthesis of metal nanoparticles

formation in solution. The initial reagents at the top of the curve represent the metal ions in the presence of a reducer (or, e.g., a cathode, were reduction of metal ions takes place), while the products at the bottom represent the final stable nanoparticles (NP). In a typical synthesis of nanoparticles, a fast reduction reaction is used, which means that the driving force (which can be represented by the difference in the electrochemical potentials of the metal to be synthesized and the reducing agent) is enough to drive the system directly from a maximum energy (reactants) to a minimum energy (products), represented by the first stable nucleus (C_N) having the same structure of the bulk. However, when a small reaction rate is used, stable clusters like C_2, C_3, etc. are produced and, with an adequate control of the reaction kinetics, it is possible to stop the reduction and to obtain stable metal clusters with the desired size.

Following this scheme, silver clusters with less than 10 atoms have been synthesized, for example, in water-in-oil microemulsions consisting of a mixture of sodium bis(2-ethylhexyl) sulfosuccinate, isooctane, and water [20]. Synthesis conditions necessary for the appearance and isolation of stable clusters were achieved using small concentrations and a mild reduction agent as sodium hypophosphite monohydrate. Silver clusters produced in this way were found not only to be stable for years, but also to have a large bandgap ($\approx$2.3 eV) and to present molecular-like paramagnetic properties and an intense photoluminescence.

In the same way, using electrochemical techniques, very small clusters can be synthesized, providing a good control of the cluster size. For example, stable poly(*N*-vinylpyrrolidone), PVP-protected gold clusters with only 2 or 3 atoms could be synthesized through a simple electrochemical method based on the anodic dissolution of a gold electrode in the presence of the homopolymer PVP and subsequent electroreduction of the Au-PVP complexes [22]. The resulting clusters (Au_2 and Au_3) are the smallest ones that can be prepared and they display fluorescent (Au_2) and paramagnetic (Au_3) properties. By a similar procedure, highly fluorescent copper clusters stabilized with tetrabutylammonium nitrate, with less than 14 atoms, have also been reported [6]. These clusters can be dispersed in different solvents (both polar and apolar), are stable for several years and display also similar fluorescent properties than those synthesized in microemulsions [21].

Finally, it is worth mentioning the use of hard cages to inhibit the growth and synthesize small clusters, as it was demonstrated very recently by Royon et al. [18], showing that silver-containing glasses can be used for preparing fluorescent materials reducing the Ag ions in the matrix with laser pulses. In this way, they were able to prepare perennial high capacity 3D-optical recording media.

In summary, it can be said that by appropriate use of the typical techniques developed for the synthesis of NPs, clusters of different sizes down to 2 atoms can be produced, isolated, and used as stable materials with novel properties, which differ very strongly from the bulk materials and nanoparticles due to their totally different electronic and geometrical structure.

Acknowledgements We would like to thank the Spanish Ministry of Science and Innovation (MAT2008-06503/NAN; MAT2010-20442; NanoBioMed CONSOLIDER INGENIO) and the Regional Government of Galicia (Competitive Reference Groups 2010/41, financed with FEDER funds and Project INCITE08PXIB209049PR) for providing US financial support of different parts related to this work.

Cross-References

▶ Dry Etching
▶ Nanoparticles
▶ Optical and Electronic Properties
▶ Wet Etching

References

1. Chou, M.Y., Cleland, A., Cohen, M.L.: Total energies, abundances, and electronic shell structure of lithium, sodium, and potassium clusters. Solid State Commun. **52**, 645–648 (1984)
2. de Heer, W.A.: The physics of simple metal clusters: experimental aspects and simple models. Rev. Mod. Phys. **65**, 611–676 (1993)
3. Walter, M., Akola, J., Lopez-Acevedo, O., Jadzinsky, P.D., Calero, G., Ackerson, C.J., Whetten, R.L., Grönbeck, H., Häkkinen, H.: A unified view of ligand-protected gold clusters as superatom complexes. Proc. Natl. Acad. Sci. **105**, 9157–9162 (2008)
4. Zheng, J., Zhang, C., Dickson, R.M.: Highly fluorescent, water-soluble, size-tunable gold quantum dots. Phys. Rev. Lett. **93**, 077402 (2004)
5. Zheng, J., Nikovich, P.R., Dickson, R.M.: Highly fluorescent noble metal quantum dots. Annu. Rev. Phys. Chem. **58**, 409–431 (2007)
6. Vilar-Vidal, N., Blanco, M.C., López-Quintela, M.A., Rivas, J., Serra, C.: Electrochemical synthesis of very stable photoluminescent copper clusters. J. Phys. Chem. C **114**, 15924–15930 (2010)
7. Lin, C.-A.J., Yang, T.-Y., Lee, C.-H., Huang, S.H., Sperling, R.A., Zanella, M., Li, J.K., Shen, J.-L., Wang, H.-H., Yeh, H.-I., Parak, W.J., Chang, W.H.: Synthesis, characterization, and bioconjugation of fluorescent gold nanoclusters toward biological labeling applications. ACS Nano **3**, 395–401 (2009)
8. Heiz, U., Landman, U.: Nanocatalysis (Nanoscience and Technology). Springer, Berlin (2007)
9. Vajda, S., Pellin, M.J., Greeley, J.P., Marshall, C.L., Curtiss, L.A., Ballentine, G.A., Elam, J.W., Catillon-Mucherie, S., Redfern, P.C., Mehmood, F., Zapol, P.: Subnanometre platinum clusters as highly active and selective catalysts for the oxidative dehydrogenation of propane. Nat. Mater. **8**, 213–216 (2009)
10. Lee, S., Molina, L.M., López, M.J., Alonso, J.A., Hammer, B., Lee, B., Seifert, S., Winans, R.E., Elam, J.W., Pellin, M. J., Vajda, S.: Selective propene epoxidation on immobilized Au_{6-10} clusters: the effect of hydrogen and water on activity and selectivity. Angew. Chem. Int. Ed. **48**, 1467–1471 (2009)
11. Harding, C., Habibpour, V., Kunz, S., Farnbacher, A.N.-S., Heiz, U., Yoon, B., Landman, U.: Control and manipulation of gold nanocatalysis: effects of metal oxide support thickness and composition. J. Am. Chem. Soc. **131**, 538–548 (2009)
12. Rodríguez-Vázquez, M.J., Blanco, M.C., Lourido, R., Vázquez-Vázquez, C., Pastor, E., Planes, G.A., Rivas, J., López-Quintela, M.A.: Synthesis of atomic gold clusters with strong electrocatalytic activities. Langmuir **24**, 12690–12694 (2008)
13. Selva, J., Martínez, S.E., Buceta, D., Rodríguez-Vázquez, M.J., Blanco, M.C., López-Quintela, M.A., Egea, G.: Silver sub-nanoclusters electrocatalyze ethanol oxidation and provide protection against ethanol toxicity in cultured mammalian cells. J. Am. Chem. Soc. **132**, 6947–6954 (2010)
14. Habeeb Muhammed, M.A., Ramesh, S., Sinha, S.S., Pal, S. K., Pradeep, T.: Two distinct fluorescent quantum clusters of gold starting from metallic nanoparticles by pH-dependent ligand etching. Nano Res. **1**, 333–340 (2008)
15. Duan, H., Nie, S.: Etching colloidal gold nanocrystals with hyperbranched and multivalent polymers: a new route to fluorescent and water soluble atomic clusters. J. Am. Chem. Soc. **129**, 2412–2413 (2007)
16. Schaeffer, N., Tan, B., Dickinson, C., Rosseinsky, M.J., Laromaine, A., McComb, D.W., Stevens, M.M., Wang, Y., Petit, L., Barentin, C., Spiller, D.G., Cooper, A.I., Lévy, R.: Fluorescent or not? Size-dependent fluorescence switching for polymer-stabilized gold clusters in the 1.1–1.7 nm size range. Chem. Commun. **34**, 3986–3988 (2008)
17. Negishi, Y., Nobusada, K., Tsukuda, T.: Glutathione-protected gold clusters revisited: bridging the gap between gold(I)-thiolate complexes and thiolate-protected gold nanocrystals. J. Am. Chem. Soc. **127**, 5261–5270 (2005)
18. Royon, A., Bourhis, K., Bellec, M., Papon, G., Bousquet, B., Deshayes, Y., Cardinal, T., Canioni, L.: Silver clusters embedded in glass as a perennial high capacity optical recording medium. Adv. Mater. **22**, 5282–5286 (2010)
19. López-Quintela, M.A.: Synthesis of nanomaterials in microemulsions: formation mechanisms and growth control. Curr. Opin. Colloid Interface Sci. **8**, 137–144 (2003)
20. Ledo-Suárez, A., Rivas, J., Rodríguez-Abreu, C.F., Rodríguez, M.J., Pastor, E., Hernández-Creus, A., Oseroff, S.B., López-Quintela, M.A.: Facile synthesis of stable subnanosized silver clusters in microemulsions. Angew. Chem. Int. Ed. **46**, 8823–8827 (2007)
21. Vázquez-Vázquez, C., Bañobre-López, M., Mitra, A., López-Quintela, M.A., Rivas, J.: Synthesis of small atomic copper clusters in microemulsions. Langmuir **25**, 8208–8216 (2009)
22. Santiago González, B., Rodríguez, M.J., Blanco, C., Rivas, J., López-Quintela, M.A., Martinho, J.M.G.: One step synthesis of the smallest photoluminescent and paramagnetic PVP-protected gold atomic clusters. Nano Lett. **10**, 4217–4221 (2010)

Synthesized Conductance Injection

▶ Dynamic Clamp

Synthesized Ionic Conductance

▶ Dynamic Clamp

Synthesized Synaptic Conductance

▶ Dynamic Clamp

Synthetic Biology

Soichiro Tsuda
Exploratory Research for Advanced Technology, Japan Science and Technology Agency, Osaka, Japan

Synonyms

Synthetic genomics

Definition

Synthetic biology is the design and construction of biological components (e.g., enzymes, gene circuits, and whole cells) from scratch or from standardized parts.

Overview

The term "Synthetic Biology" was first used in 1980 by Barbara Hobom to describe genetically engineered bacteria using recombinant DNA technology. In 2000, the term reappeared again by Eric Kool and others at the annual meeting of the American Chemical Society to refer to synthesis of unnatural chemical products (e.g., organic molecules) as an extension of synthetic chemistry [1].

Although the scope of synthetic biology has explosively expanded in the last decade, synthetic biology can be best described as "efforts to redesign life," that is, the design and construction of biological components (e.g., enzymes, gene circuits, and whole cells) using living cells. Similar approaches have already been taken long before the advent of synthetic biology, for example, by genetic engineering, which is one of the roots of synthetic biology. However, what makes synthetic biology distinctive from other preceding biological research fields is its strong engineering perspective. While genetic engineering focused on modifying individual genes and pathways, synthetic biology does on whole systems of genes and gene products. It aims at designing and constructing the behavior of organisms to perform new tasks [2]. Computer engineering analogy is often used to explain the goal of synthetic biology: In order to build a computer, it involves several hierarchical layers. The most fundamental layer of computers is the electrical components, such as resistor, transistor, capacitor, etc. Based on them, a higher level structure, logic gates (AND, OR, NOT, XOR, etc) are built, which are further integrated into even upper layer, integrated circuits, and so on. Biological systems can be described similarly: DNA, RNA, and proteins are the most fundamental components of biological cells. These components form gene circuit networks, which are parts of intracellular signal pathways, and so on. Thus, synthetic biology research involves the developments of biological components in each layer for engineering new biological functions: standardized fundamental biological parts, gene circuit construction, and whole biological cells, and biological and chemical products synthesized by living cells.

Standardized Biological Parts

To engineer complex systems, the standardization of components used in the systems (in this case biological systems) is necessary. Tom Knight and Drew Endy proposed the BioBrick standard biological parts as basic units for synthetic biological systems and set up an online registry of BioBrick parts where anyone can contribute in an open-source manner [3–5]. It provides a collection of well-characterized and standardized building blocks that can be used for the "plug-and-play" design and construction of unnatural biological systems. Using BioBrick standard parts, a biological engineer can program a living cell, as an electrical engineer can program an electrical circuit using electrical components. BioBrick parts are mostly DNA sequences with functionality, such as promoters, plasmids, and ribosome binding sites. Not only natural components, biologically inspired molecules can be BioBrick parts: For example, DNA analog PNA, in which the sugar-phosphate backbone is altered by N(2-aminoethyl)-glycine units linked by peptide bonds [6].

Gene Circuit Design and Construction

Having set the standards for biological components, the next step is to develop minimum functional components from the standard parts. They are typically gene circuits that express particular proteins. These single circuits are connected together to construct a gene regulatory network. Network motifs, such as cascades, feedfoward, and feedback, are commonly used for the design of artificial regulatory network (and they are also commonly found in natural biological systems) [7]. Cascades are series-connected gene circuits that output (protein) of an upstream gene circuit regulates expression of its immediately downstream gene. Feedforward motif involves a master regulatory gene that influences downstream genes through non-circular pathways. When the feedforward loop is introduced in a network, for example, it is possible to construct a network that shows a nonlinear non-monotonic change in the output. For example, a sigmoid function with a steep transition can be implemented with two gene circuits connected by negative (inhibitory) feedforward motif [8]. Basu and co-workers implemented a pulse-generator system using the motif [9]. When the input stimulus is monotonously changed from low to high, the output of the pulse-generator network changes from low to high and then back to low. In addition, the pulse amplitude and delay reflect not only the input concentration but also its rate of change. Feedback is a common biological regulatory motif not only for intracellular gene networks but also for intercellular control (e.g., negative feedback regulation of hormonal control). Feedback loop allows various complex regulation over expression of genes: noise reduction [10], expression level control [11], bistable gene toggle switch [12], autonomous genetic oscillator [13], etc. In principle, any artificial regulatory scheme can be implemented with the combination of above network motifs.

Although standardization and other engineering concepts, such as abstraction modularity, reliability, greatly contribute to the speed and tractability of gene circuit design, one should note the liquid nature of biological systems and components. In contrast to solid electrical components, biological components, DNA, RNA, and proteins, all operate in the liquid solution which provide an appropriate "context" (e.g., pH, temperature, materials, energy, etc) for the systems to function. This cellular context dependence and sensitivity make it difficult for biological components to be modular and interchangeable. A gene circuit from naturally occurring system is likely to be optimized to a specific cellular context through evolution. Thus, it may not function when it is placed in an artificial context. Ron Weiss pointed out the need to understand "how the function of a module or an entire biological system can be derived from the function of its component parts 'and establish' the biological rules of composition" in order to build biological components and modules [2].

Minimal Life

Craig Venter and his co-workers took the synthetic gene design to its extreme. They proposed to synthesize a complete genome from scratch and create "artificial life." In 2003, they have synthesized the whole genome of Phi-X 174, a 5386 base pair bacteriophase, from synthetic oligonucleotides, which took them only 14 days to complete the whole process [14]. The synthesized genome was injected into *E. coli* bacteria and confirmed to be infectious. After the success in synthesized virus, they took a further step to create a synthesized organism that has a minimal genome. A bacterial genome of *Mycoplasma genitalium*, which is one of the smallest known genomes in any living organisms (517 genes, 580,000 DNA base pairs), was chosen as a base for the synthesis. They removed genes in the genome that are considered redundant and injected into a *M. capricolum* cell in which the original genome was removed in order to test if the cell with the modified genome can be alive. In 2010, they reported that the size of the genome was eventually reduced to approximately one mega base pair that are minimal and sufficient to be considered alive. They called the cell *Mycoplasma laboratorium* which is the first artificial life, of which parent is a digital computer [15].

Another important aspect of minimal life in synthetic biology is "cells as chassis." The context dependence of biological components requires the constant maintenance of appropriate cellular context. Artificially synthesized biological components and modules are not exception and therefore require cellular context and its maintenance to perform tasks. Apart from the synthetic genome mentioned above, thus, all the other parts of protein expression process (transcription, RNA processing, translation, protein folding, amino-acid modifications) can be subjects of synthetic biology research [16]. The construction of a minimal cell from scratch, as a basic unit for proving cellular context, is also an active area in this sub-research field. Lipid vesicles (liposomes) are commonly used as minimal bioreactors to simulate biological cells surrounded by lipid bilayer membrane. Functions of biological cells, such as membrane division, fusion, protein expression, and self-replication, have been attempted to replicate inside vesicles [17]. In recent years, microfluidic techniques, called digital microfluidics, have been applied to the production of lipid vesicles [18]. These techniques allow high-throughput production and screening of protein expression in the microfluidic devices, and are expected to facilitate the more precise control of experimental conditions.

Applied Biological and Chemical Synthesis

Practical applications of synthetic biology are expected to be quite enormous because one of ultimate goals in synthetic biology is to create synthetic organisms that perform commercially useful tasks. For example, Craig Venter's group developed the synthetic genome described above so that it could be served as a platform for future synthetic organisms on which new functions (i.e., gene circuits) can be added. Since 2009, Venter's company, Synthetic Genomics Incorporated (SGI), has been working with Exxon Mobil Corporation to develop synthetic algae that produce next generation biofuel, such as ethanol or hydrogen [19]. They also claim it would be possible to synthesize organisms which can fix carbon dioxide using photosynthesis to mitigate climate change [20].

Another example is medical and pharmaceutical applications of synthetic biology. Microbes, typically bacteria and yeast, are engineered and used as "chemical factory" to produce chemical products that are difficult to synthesize artificially. This process inevitably involves metabolic pathway engineering of living cells and therefore requires management of complex and emergent metabolic processes [21]. Keasling and co-workers developed synthetic bacteria for the anti-malarial drug production [1, 22]. The drug, artemisinin, extracted from *Artemisia annua L* (sweet wormwood) is known to be the most effective drug for the treatment of malaria, but expensive to produce because of short supply of the wood. The engineered bacteria incorporate several genes originally from bacteria *E. coli* and yeast *S. cerevisiae* to produce artemisinic acid, a bio-genetic precursor of artemisinin. First, MevT operon encodes enzymes that transform acetyl-CoA into mevalonate. Enzymes encoded by MBIS operon then transform mevalonate into farmesyl pyrophosphate (FPP). Additionally, genes for amor-phadiene synthase (transforming FPP into amorphadiene) and oxidase and redox partners (amorphadiene into artemisinic acid) were introduced. Although *E. coli* has a native pathway, 1-deoxy-D-xylulose 5-phosphate (DXP) pathway, to produce FPP, it has been found that the engineered mevalonate pathway produces more amorphadiene than the native pathway. Artemisinic acid purified from bacterial extracts is transformed to artemisinin using established chemistry.

Cross-References

▶ Computational Systems Bioinformatics for RNAi
▶ Liposomes
▶ Nanotechnology Applications in Polymerase Chain Reaction (PCR)
▶ Structure and Stability of Protein Materials

References

1. Benner, S.A., Sismour, A.M.: Synthetic biology. Nat. Rev. Genet. **6**(7), 533–543 (2005)
2. Andrianantoandro, E., Basu, S., Karig, D.K., and Weiss, R.: Synthetic biology: new engineering rules for an emerging discipline. Mol. Syst. Biol. 2(1), 2006.0028 (2006)
3. The BioBricks Foundation http://bbf.openwetware.org/
4. Knight, T.: Idempotent vector design for standard assembly of biobricks standard biobrick sequence interface. MIT Synthetic Biology Working Group Report, MIT Artificial Intelligence Laboratory, 1–11 (2003)
5. Endy, D.: Foundations for engineering biology. Nature **438**(7067), 449–453 (2005)
6. Nielsen, P.E., Egholm, M.: An introduction to peptide nucleic acid. Curr. Issues Mol. Biol. **1**(1–2), 89–104 (1999)
7. McDaniel, R., Weiss, R.: Advances in synthetic biology: on the path from prototypes to applications. Curr. Opin. Biotechnol. **16**(4), 476–483 (2005)
8. Hooshangi, S., Thiberge, S., Weiss, R.: Ultrasensitivity and noise propagation in a synthetic transcriptional cascade. Proc. Natl Acad. Sci. U.S.A. **102**(10), 3581–3586 (2005)
9. Basu, S., Mehreja, R., Thiberge, S., Chen, M.-T., Weiss, R.: Spatiotemporal control of gene expression with pulse-generating networks. Proc. Natl Acad. Sci. U.S.A. **101**(17), 6355–6360 (2004)
10. Becskei, A., Serrano, L.: Engineering stability in gene networks by autoregulation. Nature **405**, 590–593 (2000)
11. Nevozhay, D., Adams, R.M., Murphy, K.F., Josic, K., Balzsi, G.: Negative autoregulation linearizes the

dose-response and suppresses the heterogeneity of gene expression. Proc. Natl Acad. Sci. U.S.A. **106**(13), 5123–5128 (2009)
12. Gardner, T.S., Cantor, C.R., Collins, J.J.: Construction of a genetic toggle switch in *Escherichia coli*. Nature **403**(6767), 339–342 (2000)
13. Judd, E.M., Laub, M.T., McAdams, H.H.: Toggles and oscillators: new genetic circuit designs. BioEssays **22**(6), 507–509 (2000)
14. Smith, H.O., Hutchison, C.A., Pfannkoch, C., Venter, J.C.: Generating a synthetic genome by whole genome assembly: x174 bacteriophage from synthetic oligonucleotides. Proc. Natl Acad. Sci. U.S.A. **100**(26), 15440–15445 (2003)
15. Gibson, D.G., Glass, J.I., Lartigue, C., Noskov, V.N., Chuang, R.-Y., Algire, M.A., Benders, G.A., Montague, M.G., Ma, L., Moodie, M.M., Merryman, C., Vashee, S., Krishnakumar, R., Assad-Garcia, N., Andrews-Pfannkoch, C., Denisova, E.A., Young, L., Qi, Z.-Q., Segall-Shapiro, T.H., Calvey, C.H., Parmar, P.P., Hutchison, C.A., Smith, H.O., Venter, J.C.: Creation of a bacterial cell controlled by a chemically synthesized genome. Science **329**(May), 52–56 (2010)
16. Forster, A.C., Church, G.M.: Towards synthesis of a minimal cell. Mol. Syst. Biol. **2**(45), 45 (2006)
17. Luisi, P.L., Ferri, F., Stano, P.: Approaches to semi-synthetic minimal cells: a review. Naturwissenschaften **93**(1), 1–13 (2006)
18. Theberge, A.B., Courtois, F., Schaerli, Y., Fischlechner, M., Abell, C., Hollfelder, F., Huck, W.T.S.: Microdroplets in microfluidics: an evolving platform for discoveries in chemistry and biology. Angew. Chem. Int. Ed. **49**(34), 5846–5868 (2010)
19. Synthetic Genomics Inc. Press Release http://www.syntheticgenomics.com/media/press/71409.html, July 14 2009
20. Etc group: Extreme genetic engineering: an introduction to synthetic biology, Etc Group (2007)
21. Khosla, C., Keasling, J.D.: Metabolic engineering for drug discovery and development. Nat. Rev. Drug Discov. **2**(12), 1019–1025 (2003)
22. Keasling, J.: Synthetic biology for synthetic chemistry. ACS Chem. Biol. **3**(1), 64–76 (2008)

Synthetic Genomics

► Synthetic Biology

Synthetic Lubricants

► Boundary Lubrication

Systems Level Data Mining for RNAi

► Computational Systems Bioinformatics for RNAi

T

TEM

► Electron Microscopy of Interactions Between Engineered Nanomaterials and Cells

Terahertz

► Terahertz Technology for Nano Applications

Terahertz Technology for Nano Applications

Nezih Pala and Ahmad Nabil Abbas
Department of Electrical and Computer Engineering, Florida International University, Miami, FL, USA

Synonyms

Terahertz; THz; T-rays

Definition

The terahertz (THz) region of the electromagnetic spectrum is generally defined as the frequency range of 0.1–10 THz (10^{12} cycles per second) corresponding to quantum energy of 0.4 meV–0.4 eV (see Fig. 1). THz electromagnetic waves (also known as T-rays) have several properties that could promote their use as sensing and imaging tool. There is no ionization hazard for biological tissue and Rayleigh scattering of electromagnetic radiation is many orders of magnitude less for THz wavelengths than for the neighboring infrared and optical regions of the spectrum. THz radiation can also penetrate nonmetallic materials such as fabric, leather, and plastic which makes it useful in security screening for concealed weapons. The THz frequencies correspond to energy levels of molecular rotations and vibrations of DNA and proteins, as well as explosives, and these may provide characteristic fingerprints to differentiate biological tissues in a region of the spectrum not previously explored for medical use or detect and identify trace amount of explosives. THz wavelengths are particularly sensitive to water and exhibit absorption peaks which makes the technique very sensitive to hydration state and can indicate tissue condition. THz radiation has also been used in the characterization of semiconductor materials, and in testing and failure analysis of VLSI circuits. THz techniques also allowed art historians to see murals hidden beneath coats of plaster or paint in centuries-old building, without harming the artwork.

The lack of efficient sources and detectors in THz frequency range compared to the relatively well-developed lower-frequency RF/microwave and higher-frequency infrared (IR)/far infrared (FIR) ranges was referred as "THz gap" in the scientific community about a decade ago. However, the unique position of THz frequencies in the electromagnetic spectrum has allowed an intense multidisciplinary approach in the development of THz emission and detection techniques as well as their applications, closing the "THz gap" from both sides. Especially rapidly evolving nanosciences and nanotechnology

B. Bhushan (ed.), *Encyclopedia of Nanotechnology*, DOI 10.1007/978-90-481-9751-4,

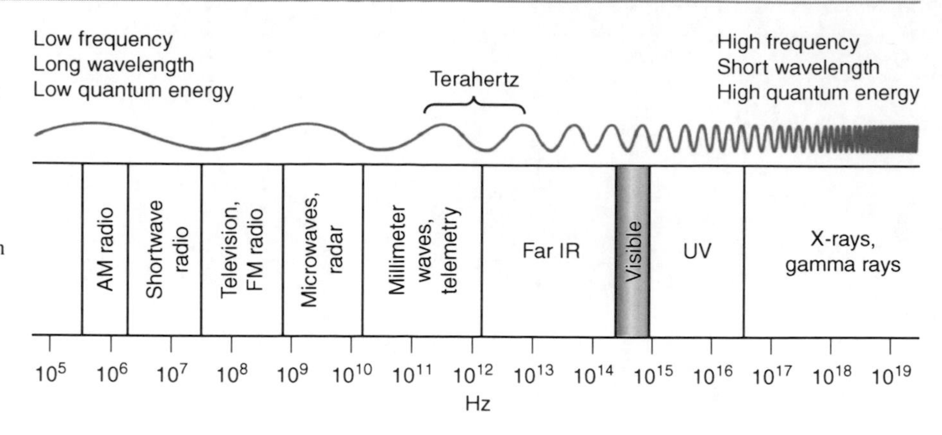

Terahertz Technology for Nano Applications, Fig. 1 A schematic showing the THz region within the electromagnetic spectrum. Although there is no strict definition of what qualify as THz waves, it is commonly considered the region between 100 GHz and 10 THz

immensely contribute to the advancement of THz technology. These advancements lead to creation of many unique applications some of which in turn contributes to nanotechnology. Therefore, the interaction between THz and nanotechnologies is reciprocal.

Principles of THz Technology

Terahertz Sources

Taking advantage the above-mentioned unique spectral position THz frequencies, four major approaches have been adopted to develop THz sources. The first approach is pushing the operation frequencies of the existing microwave and millimeter-wave devices into the THz range. The second one is the free-electron-based sources like free-electron lasers and backward wave oscillators (BWOs). The third is the utilization of optical methods which has been the major technique for the demonstration of many THz applications. And the fourth one is the newly developed Quantum Cascade Lasers (QCLs). THz emission power of selected sources are shown in Fig. 2 as a function of frequency. Despite the tremendous research and development efforts, currently available THz source technologies, particularly the tunable emitters, such as free-electron lasers, backward wave oscillators, gas lasers, heterodyne photomixers, optical parametric converters are usually large, complex, require high power, and hence costly. Semiconductor electronic devices such as Gunn oscillators or Schottky diodes need frequency multipliers which severely impact the output power. Quantum cascade lasers which are the true compact solid-state THz emitters are still in their infancy and operate only at cryogenic temperatures.

Solid-State Electronic THz Sources

Solid-state electronic emitters are limited in frequency due to the transient time of carriers which results in high-frequency roll-off. This intrinsic limitation prevented the development of a single electronic device which could oscillate in the entire THz bandwidth of 0.1–10 THz. Gunn diodes (also known as transferred electron devices), IMPATT (Impact Avalanche Transit Time) diodes, and TUNNETT (Tunnel Injection Transit Time) diodes are well-developed high-frequency sources. They take advantage of negative differential resistance (NDR) in current voltage characteristics. When a DC bias large enough to drive such a device into NDR is applied, unstable state triggers oscillations with frequencies up into the mm-wave range. Specific mechanism resulting the NDR determines the upper limit of the oscillations frequency. Current Gunn diodes and IMPATT diodes can reach oscillating frequencies of 400–500 GHz. InP Gunn devices with graded doping profiles generated RF output powers of 283 μW at 412 GHz, 203 μW at 429 GHz and the highest third-harmonic frequency of 455 GHz with an output power of 23 μW. GaAs TUNNETs, on the other hand, produced 10 mW at 202 GHz. Resonant tunneling diodes (RTDs) are semiconductor devices with double barrier structures which forms a quantum well in between. Electrons in this quantum well are at quantized discrete energy levels. At a particular DC bias, the energy levels at the bottom of the conduction band on one side align with the lowest energy level in the well, resulting in a resonant condition in which electrons can tunnel through the barriers. As the bias increases further, the resonant condition disappears and the current falls. This NDR process combined with the inherently fast tunneling mechanism allows RTDs working at high operation

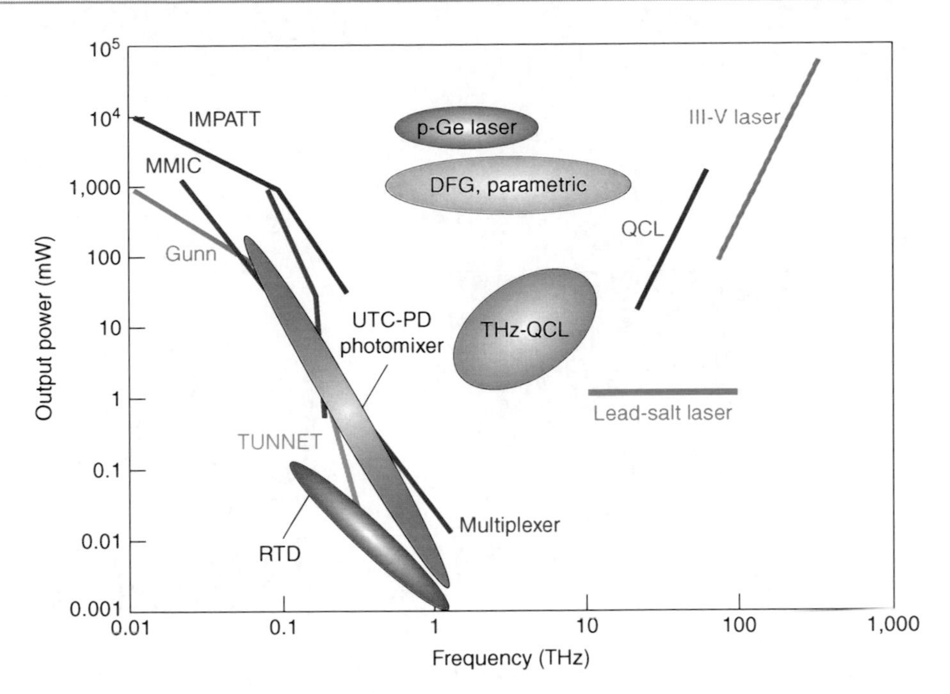

Terahertz Technology for Nano Applications, Fig. 2 THz-emission power as a function of frequency. Solid lines are for the conventional THz sources; IMPATT diode stands for impact ionization avalanche transit-time diode, MMIC stands for microwave monolithic integrated circuit, TUNNET stands for tunnel injection transit time, and the multiplexer is an SBD frequency multiplier. Ovals denote recent THz sources. The values of the last two are indicated by peak power; others are by c.w. power (After [1])

frequencies. RTDs based on InGaAs/AlAs material system have reached the fundamental oscillation frequency of 915 GHz with, however, quite low output level of a few tens of nanowatts.

Recently, in parallel to the advancements in material- and device-processing technologies, transistors with very high cutoff frequencies (f_T) have also been reported. Northrop Grumman announced an InP high electron mobility transistor (HEMT) with extrapolated $f_T > 1$ THz. This device had sub-50 nm gate length, 25% increase in electron mobility, and reduced contact resistance. Similarly, heterojunction bipolar transistors (HBTs) based on InP materials systems achieved a cutoff frequency f_T of 765 GHz at room temperature and 855 GHz at 218 K.

As a newly emerged alternative approach, ballistic deflection transistors based on InGaAs-InAlAs heterostructure on an InP substrate were developed at University of Rochester with a theoretical maximum f_T of approximately 1.02 THz (Fig. 3a). Another novel approach is the plasma wave instabilities in the FET channels with high sheet carrier concentrations. Such instabilities were predicted by Dyakonov and Shur in 1991. Recently, a group of researchers from Japan and Europe reported room temperature generation of radiation at 0.75 and 2.1 THz from AlGaN/GaN HEMTs (Fig. 3b).

Due to the limited frequency performance of solid-state emitters, they are often used in combination with multiplication circuits to reach THz frequencies. A multiplier consists of a nonlinear electronic device, such as a Schottky varactor diode placed between an input and an output-matching network. Unfortunately, the output power is much lower than that of the input, which is a serious drawback for the THz frequency range. An input with a power of 200–300 mW at 100 GHz can be produced by HEMT amplifiers, but a multiplier with a high-order of multiplication from 100 GHz up to 1–3 THz is not feasible due to the very high losses. Much lower losses are achievable only in multipliers with a low-order of multiplication, i.e., doublers (2x) and triplers (3x), so that a THz multiplier could consist of a sequence of doublers and triplers of the frequency up to the desired THz frequency.

Free-Electron-Based Sources

Free-electron lasers have quite different working principles than more traditional optical lasers such as gas lasers. They can generate either CW or pulsed high-power THz radiation, but they are very costly and have very large dimensions, functioning in large rooms containing many additional facilities which limits their availability to only limited number of facilities around the world. However, backward wave oscillators (BWO) (also called carcinotron) are based

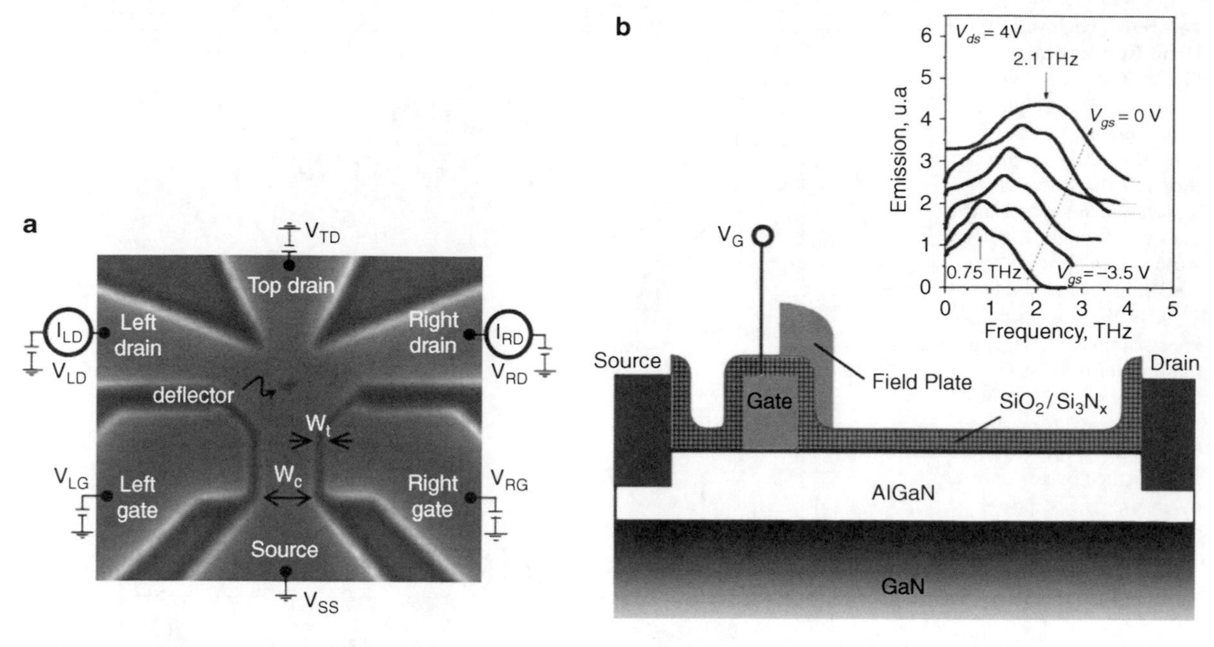

Terahertz Technology for Nano Applications, Fig. 3 (**a**) SEM Image of top view of ballistic deflection transistors. After [2]. (**b**) Schematics of the GaN/AlGaN plasmonic HEMT with gate covered by a field plate reported in [3]. The inset shows the emission spectra of the device

on the same principles as an electron laser and are able to deliver a few milliwatts in a tunable range of 0.3–1.3 THz with a high sweeping rate. Although BWO requires a water-cooling system due to its high bias voltages of 1–6 kV their size and weight is much smaller (~15 kg) which makes them convenient and affordable for many research labs [4].

Recently, there has been an intensifying research effort to develop micromachined Vacuum Electronic Devices, or "micro-VEDs (μVEDs)." Vacuum technology offers a feasible solution for efficient THz transmitters but requires a significant invention to overcome the problem of complex and difficult size scaling necessary to achieve THz operation which calls for advanced fabrication methods and stringent tolerances for interaction structures. Critical developments include the fabrication of the interaction circuit, a high current density integratable cathode structure, and achieving stable high-power electron beam transport. Fortunately, today's microfabrication methods such as Deep Reactive Ion Etching (DRIE) and LIGA are now capable of manufacturing slow-wave interaction structures at these frequencies with the required resolution and surface roughness characteristics. Further, significant progress has been made in the development of new cathode materials and structures.

Folded waveguide (FWG) approach has emerged as promising approach to design micro-VED high-power amplifiers (HPAs) and THz-regime oscillators. FWG offers sufficient gain and bandwidth to achieve 60 mW at 670 GHz and 10 mW at 1 THz and 56 mW at 560 GHz as THz Oscillator. Another approach is the Extended Interaction Klystron (EIK) approach which applies the concept of multiple interaction structures at offset frequencies to simultaneously achieve high gain from the superposition of multiple structures interacting with the signal and high bandwidth from the gradual offset design of the same structures. This concept is expected to allow design of an HPA that would deliver output power >20 dBm and gain >20 dB at 1.03 THz. The EIK approach requires machining optical quality surfaces over macroscopic areas because the skin depth in the metal at 1 THz is comparable to the previously achievable surface roughness and will result in significant ohmic losses. The issue is complicated by the severe thermal constraints placed on the interaction structure which limits material choices.

Optical THz Emission Techniques

Gas lasers are probably the most conventional CW THz sources in the frequency range of 0.9–3 THz and with output powers in the range of 1–30 mW. They are typically pumped by a carbon dioxide laser, and the output frequency is determined by the gas in the cavity (CH_4, N_2, etc.). Although gas lasers are not tunable, they have several discrete emission lines.

Photomixing, also known as optical heterodyne down-conversion, is a technique used to generate CW THz radiation by using ultrafast lasers. In its most commonly used version, photoconductive emitters biased at ~5 kV/cm are excited by Ti-sapphire lasers, which typically deliver pulse lengths of 50–100 fs, and can be as short as 10 fs with the repetition rate of the order of 50–100 MHz and with the average power in the range of 0.2–2 W. Low-temperature grown (LT), GaAs is usually the material of choice due to its photocarrier lifetime (<0.5 ps) bandgap of 1.42 eV, allowing absorption of the 800 nm radiation from Ti-sapphire lasers. When a laser pulse hits on a semiconductor, the absorbed photons generate photocarriers. These photocarriers are accelerated by the bias field, while simultaneously their density changes under the varying laser intensity. As a result, ultrashort high-peak currents are generated in the semiconductor, which radiate into free space at broad range of THz frequencies.

Another use of the ultrafast lasers for THz generation is in optical rectification. Optical rectification is the result of the transient polarization which occurs when a short, high-intensity laser pulse interacts with the electrooptic medium. The THz power generated is proportional to the square of the optical power, is determined by the second-order optical nonlinear coefficient ($\chi^{(2)}$), and varies with the relative orientation of the laser polarization and the crystallographic axes. Typical THz bandwidths obtained from optical rectification in EO crystals are 0.1–3 THz, with a total average THz power of a few milliwatts. However, much broader THz emission has been achieved. In both ZnTe and GaSe bandwidths of up to 40 THz have been demonstrated.

Difference frequency generation (DFG) is another second-order nonlinear optical process which is used to produce THz electromagnetic waves by mixing two optical beams in a nonlinear crystal, such that the output frequency is the difference between the two input frequencies. DFG method allowed the demonstration of THz peak powers of the order of hundreds of milliwatts in the frequency range, including frequencies above 3 THz, with the ability to produce a tunable narrow-line single frequency. Tuning over 0.5–7 THz has been obtained in a GaP crystal, with peak THz powers of around 100 mW. In an organic DAST crystal, continuous tuning between 2 and 20 THz has been achieved, with peak powers of >10 W.

Quantum Cascade Lasers

One of the most exciting approaches to generate tunable THz radiation is the quantum cascade lasers (QCLs). QCL idea was proposed in 1971 as a FIR radiation source and experimentally demonstrated in 1994. In a QCL, the light produced by one carrier transition between two levels is amplified due to photon-assisted tunneling of a single type of carriers in a sequence of coupled quantum wells (superlattice) that has a staircase-like band energy. Therefore, it is a unipolar laser where the carriers can be either electrons or holes. The number of amplification stages determines the output power. The discreteness of energy levels, named sub-bands, inside the same band is a result of the spatial confinement of carriers inside the heterostructure, and the radiation frequency is determined by the energy difference of sub-bands between which radiative/lasing transitions occur. The first QCL working in the THz range was reported in 2002 [5]. This laser delivers about 2 mW power at 4.4 THz and operates at 50 K. The output power decreases dramatically with increasing temperature and becomes nearly zero at room temperature. The realization of a QCL at THz frequencies encounters a series of difficulties and limitations due to the very large values of the wavelength. Among them are very large free-carrier absorption losses and the necessity of growing a very thick heterostructure. In 2009, a resonant-phonon terahertz QCL operating up to a record braking heat-sink temperature of 186 K was demonstrated (see Fig. 4). At the lasing frequency of 3.9 THz, 63 mW of peak optical power was measured at 5 K, and approximately 5 mW could still be detected at 180 K. All these THz cascade lasers are based on n-type carriers (electrons), and the photon emission is parallel to the heterostructure plane (edge-emission).

Terahertz Detection Systems

Present imaging techniques in THz applications are divided between Pulsed Time Domain (PTD) and Continuous Wave (CW) modalities. THz time-domain spectroscopy (THz-TDS) has been the primary PTD

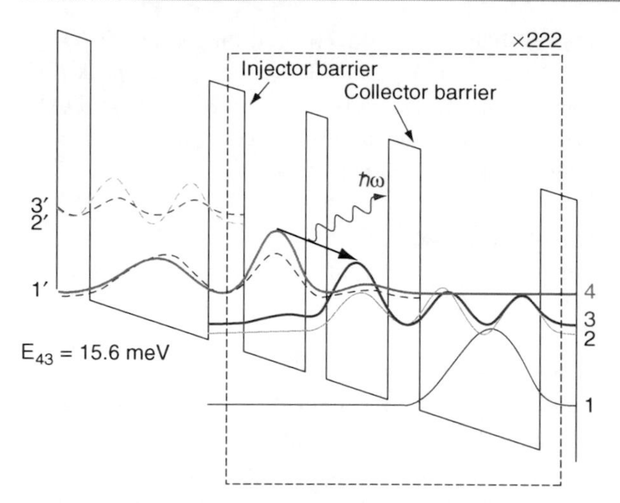

Terahertz Technology for Nano Applications, Fig. 4 Conduction band diagram of a diagonal design quantum cascade laser with the lasing power of approximately 5 mW at 180 K. 1' is the injector level from the preceding module. The figure shows that the upper- and lower-state wave functions are localized in separate wells with little spatial overlap and the radiative transition is from 4 → 3 ($f = E_{43}/h \cong 3.8$ THz). Therefore, this scheme is called diagonal design (Adapted from [6])

method for demonstration of the potential application in THz technology. A typical THz-TDS system, as shown in Fig. 5, includes an ultrafast laser (e.g., Ti-Sapphire laser) which produces optical pulses in fs duration. Each pulse is separated into two optical paths. One path travel through a time delay stage and hits the emitter such as a photoconductive antenna or a nonlinear crystal in which the optical pulses are converted into ultrashort electromagnetic pulses. The generated EM pulses are focused onto the sample under test and collimated again to be focused onto the detector. The other part of the pulse is also delivered onto the detector directly. Amplitude of the electromagnetic waves are measured by the detector to calculate the change in the waveform due to the sample. By changing delay time between the two beams, it is possible to scan the THz pulse and construct its electric field as a function of time. Subsequently, a Fourier transform is used to extract the frequency spectrum from the time-domain data. Comparison of the waveforms with and without the sample allows the estimation of the complex refractive index of the material, which results other parameters, such as the dielectric constant, conductivity and surface impedance. Although they have been widely used in research facilities, their size and complexity prevent widespread deployment for out-of-lab applications.

On the other hand, CW imaging systems can be passive, where the terahertz or sub-terahertz energy analyzed is present in the environment or emitted by the object itself (i.e., the human body), or active, where the energy is supplied by a dedicated source. In active systems, a major concern is the presently only limited available terahertz sources. This limitation places significant demands on detector capabilities. For CW systems, two primary modalities exist for detection of signals in the terahertz range. One involves the reduction of the terahertz signal to sufficiently low frequencies to allow amplification and signal processing. This (heterodyne receiver) approach typically uses nonlinear diodes to mix the terahertz signal with a local oscillator reference. Heterodyne detection imposes many restrictions on the system design and makes it complex and expensive. These restrictions may be difficult or impossible to fulfill. The second approach is direct detection, where the terahertz signal impinging upon a detecting device results in a measurable response. Available direct detection technologies with key parameters are summarized in Table 1.

As it is shown on Table 1, today the most sensitive THz detectors are Golay cells, pyroelectric detectors, bolometers, and Schottky diodes. However, they are not portable, not tunable or, most of the time, are very slow. Schottky diodes, for instance, can reach very low noise equivalent power (NEP) and high responsivity at millimeter waves. However, their responsivity drops orders of magnitude (to $<10^3$ V/W) at frequencies greater than 1 THz and they are not tunable.

Dyakonov and Shur proposed a novel idea of using plasmonic resonances in the two-dimensional electron gas (2DEG) for tunable emission and detection of terahertz radiation, which are being explored and proven experimentally, culminating in the recent demonstration of resonant detection of the THz radiation at room temperature. The theory [7, 8] predicts that such detectors should have very high sensitivities (10^7 V/W at 77 K and 10^4 V/W at 300 K). The plasmonic detectors are capable for operating even at zero bias current, thus minimizing shot noise and allowing unprecedented sensitivity. GaN-based detectors with noise equivalent power (NEP) smaller than 10^{-8} W/Hz$^{1/2}$ have been demonstrated. This value is slightly higher than for such commercial detectors as Golay cell, pyroelectric detectors, and Schottky diodes, with the potential advantage of operation at

Terahertz Technology for Nano Applications, Fig. 5 Simplified schematic of a typical THz-TDS system

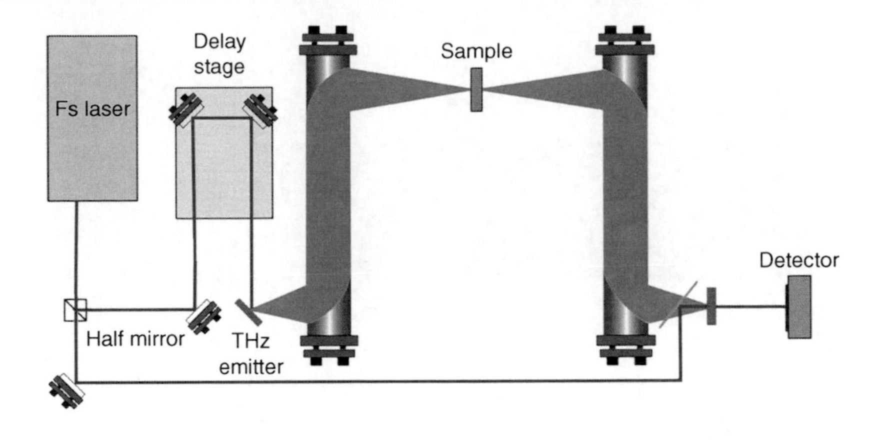

Terahertz Technology for Nano Applications, Table 1 Comparison of THz direct detection technologies

Detector Type	NEP (W/Hz$^{1/2}$)	Responsivity (V/W)	Resp. time (s)	Op. tmp. (K)	Size	Tunable
Bolometers	10^{-16}–10^{-13}	10^5–10^7	10^{-3}–10^{-2}	≤4.2	Large	No
Hot-e microbolometers	10^{-19}–10^{-17}	10^9	10^{-8}	≤0.3	Compact	No
STJ detectors	10^{-16}	10^9	10^{-3}	≤0.8	Compact	No
Golay cells	10^{-10}	10^5	10^{-2}	300	Large	No
Pyroelectric detectors	10^{-10}	10^5	10^{-2}	240–350	Medium	No
Schottky diodes	10^{-12}	10^3	10^{-12}	10–420	Compact	No
Plasmonic HEMTs	10^{-10*}	10^{3*}	10^{-10*}	10–420	Compact	YES

*Expected

very high sampling frequencies. Indeed, theoretical analysis of the detection of modulated sub-terahertz and terahertz radiation by a short channel plasmonic detectors predicts a very high upper limit for modulation frequency up to 100 GHz or even higher. These detectors can detect not only the intensity but also polarization and direction of a terahertz beam. Most intriguing advantage of all, the plasmonic terahertz detectors are tunable by DC bias over a wide frequency range.

Plasmonic detection of THz radiation is based on the excitation of plasma oscillations in the 2-dimensional electron gas (2DEG) channels of FET structures by the incoming radiation. Such FETs support two different types of plasma oscillations. The first is the plasma oscillations in the gated region of the 2DEG channel. If an infinite perfectly conductive plane is located at distance d from the infinite 2D electron sheet and the conducting plane is close enough to 2D electron sheet (i.e., $kd \ll 1$), then the dispersion relation is given by $\omega_p = k\sqrt{(e^2Nd/m^*\varepsilon_0\varepsilon_1)}$, where ω_p and k are the frequency and wave vector of plasma wave, respectively, N is the sheet electron density, e and m^* are the charge and effective mass of electron, ε_0 is the dielectric permittivity, and ε_1 is the dielectric constant of the insulator separating the 2D electron layer from the perfectly conductive plane. The second type of plasma oscillations exist in ungated regions. For negligible electron scattering in 2D electron gas, the dispersion relation for ungated plasma oscillations in an infinite homogenous 2D electron sheet is $\omega_p = \sqrt{k}\sqrt{(e^2Nd/m^*\varepsilon_0\bar{\varepsilon}}$ where $\bar{\varepsilon}$ is the effective dielectric function which depends on the geometry of the structure. In both cases, the resonant frequency ω_p depends on the sheet carrier concentration N which can be easily controlled by an applied gate bias V_g. Encouraging results have been reported for THz detection by plasmonic nanoscale single-gate FETs. However, response of a single FET plasma detector is limited by its small area compared to the beam cross section. Also, gated plasmons in a single-gate FET are weakly coupled to terahertz radiation because of the strong screening of the gate plasmons by the metal gate

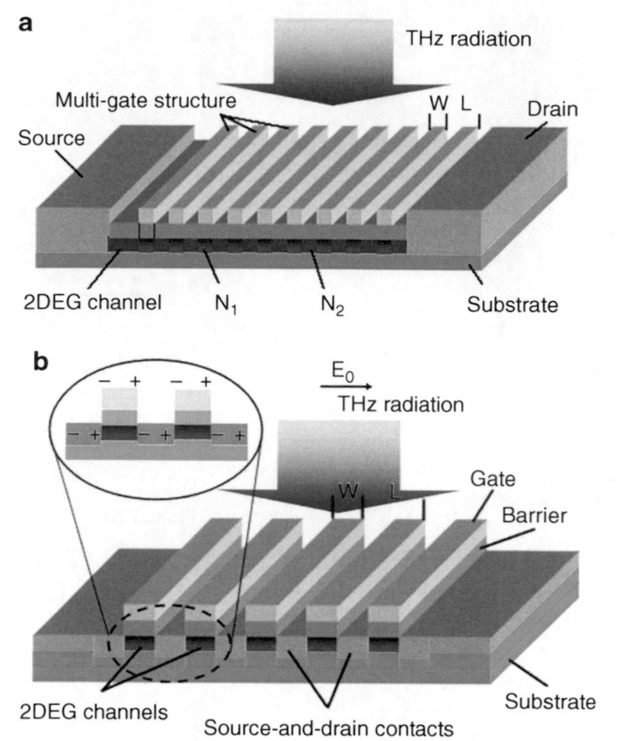

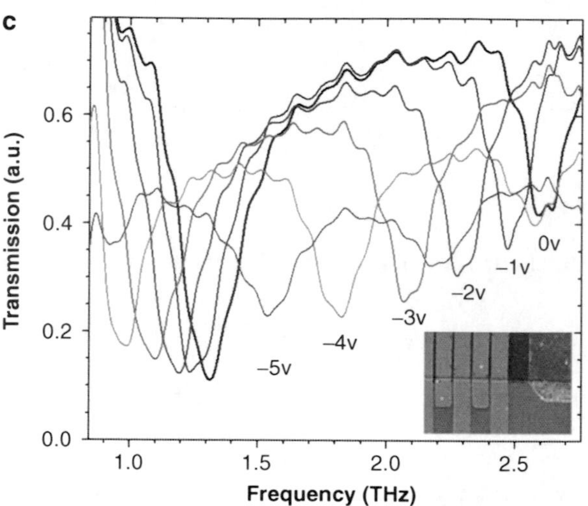

Terahertz Technology for Nano Applications, Fig. 6 Schematics of the plasmonic devices with (**a**) multiple gates (grating gate) on a common channel, (**b**) multichannel FETs having an array of separate channels with combined intrinsic source and drain contacts, (**c**) transmission spectra of a AlGaN/GaN single-channel multi-gate plasmonic device at different gate voltages. The data shows the possibility of tunable THz detection. The inset shows the close-up SEM image of the gate elements of the measured device (Adapted from [9])

electrode and their vanishingly small net dipole moment due to their acoustic nature. Coupling efficiency can be greatly enhanced by utilizing multiple gates (grating gate) on a common channel (Fig. 6a) or multichannel FETs having an array of separate channels with combined intrinsic source and drain contacts (Fig. 6b). Resonant coupling of THz radiation into the plasmons in 2DEG of a single-channel multi-gate plasmonic device have been demonstrated which could be used for precise tuning by the applied gate voltage (Fig. 6c).

THz Applications

Distinctive position of THz frequencies in the electromagnetic spectrum with their lower quantum energy compared to IR and higher frequency compared to MW range calls for many potential applications unique to them. The former characteristic allows detection and identification of substances based on their spectral signatures due to molecular vibration and rotation, while the latter one allows wide bandwidth for wireless communication. Indeed, spectroscopic analysis, sensing, and imaging represent the major application category of today's THz technology. Near-field imaging for subwavelength-scale small features is a notably fast-advancing field in this category. Communication, although still at infancy, is also promising application for THz frequencies.

Spectroscopic Sensing and Imaging

The unique advantage of THz waves for many applications can be attributed to their low frequency (or low quantum energy) compared to the commonly used parts of the optical spectrum. The low energy of terahertz photons can perhaps be better appreciated when compared to average thermal energy, 1 THz corresponds to just 4 meV, or about 1/6th of the energy in one degree of freedom at 300 K. Therefore, terahertz

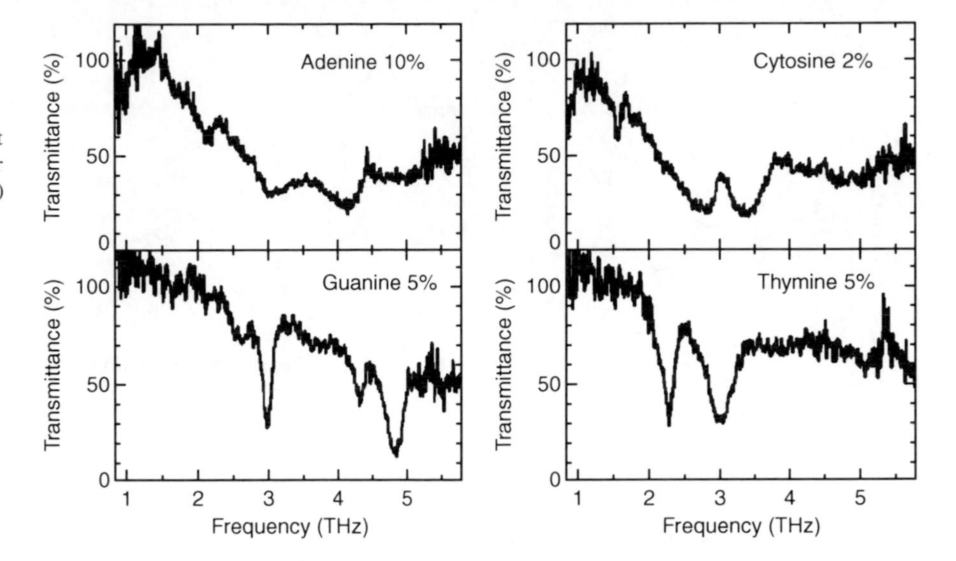

Terahertz Technology for Nano Applications, Fig. 7 Transmission spectra of the nucleobases (adenine, cytosine, guanine, thymine) at THz frequencies identify their spectral signatures (After [12])

radiation is useful for studying very low-frequency phenomena such as vibrations and rotations in molecules and soft lattice vibrations in dielectrics. Since materials, such as polymers, clothes, plastic, wood, and soil are nearly transparent to terahertz rays, while water, window-glass, and most metals are opaque; many novel imaging applications are available with T-rays. To date, THz waves have been used to read letters in closed envelopes, to detect hidden explosives and weaponry, pharmaceutical quality control, and to detect tumors in breast and epidermal tissue.

THz frequencies are particularly important for biosensing applications since numerous characteristic vibrational modes of macromolecules like proteins and DNA are located in this spectral range. This correspondence opens up the possibility of label-free biosensing. Although millimeter wave spectroscopy has been known since the 1950s, the cost was prohibitive and studies did not extend to large biological proteins such as DNA. In 1995, Hu and Nuss demonstrated the first THz imaging system based on THz-TDS which has become the method of choice for demonstration of many sensing and imaging applications [10]. Shortly after, Woolard et al. proposed to use millimeter-THz spectrocopy for detecting DNA mutagenesis. In this first attempt, they successfully demonstrated the lesion-induced vibrational modes in DNA observed from 80 to 1,000 GHz. The observed modes were associated with localized defects of the DNA polymers [11]. Low-frequency collective vibrational modes of lyophilized powder samples of calf thymus DNA, bovine serum albumin (BSA), and collagen in the 0.06–2.00 THz frequency range have been studied, showing that a large number of the low-frequency collective modes for these systems are active in this range. Further studies demonstrated the feasibility of THz-TDS to (a) identify biomolecular species, (b) identify the conformational state, and (c) identify the mutation of biomolecules. The understanding of conformational change and flexibility is particularly critical for bioengineering of biomolecules for drug discovery and therapeutic purposes. Transmission spectra of the nucleobases (adenine, cytosine, guanine, thymine) have also been measured to identify the spectral signatures at THz frequencies (see Fig. 7).

In the security side of the THz sensing applications, absorption spectra of explosives and related compounds (ERCs) by using THz-TDS has also been measured. The obtained absorption spectra in the range of 0.1–3 THz show that most of the ERCs have THz fingerprints which are caused by both the intramolecular and intermolecular vibrational modes of these materials (Fig. 8). Since commonly used nonpolar dielectric materials (e.g., fabric, leather) are almost transparent in the THz range, THz fingerprints makes THz technology a competitive technique for detecting hidden explosives. In most of these studies, the samples under investigation were either simply deposited on a THz-transparent (e.g., polyethylene) substrate, placed in a cell consisting of the same or, in rare cases, formed into durable free standing films.

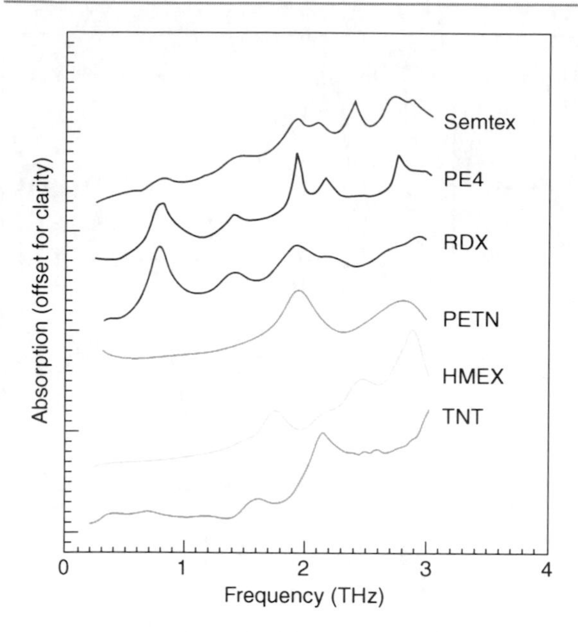

Terahertz Technology for Nano Applications, Fig. 8 Terahertz transmission spectra of the raw explosive materials TNT, HMX, PETN, and RDX together with the spectra of the compound explosives PE4 and Semtex H [13]

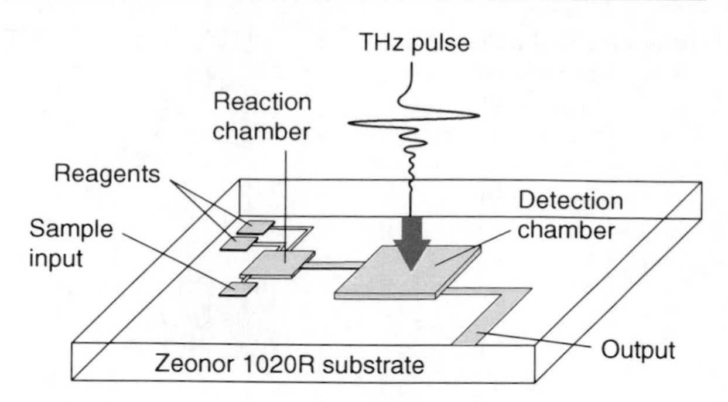

Terahertz Technology for Nano Applications, Fig. 9 A system for on-chip THz sensing. The device contains sample and reagent inputs, a reaction chamber, and a detection chamber for THz, as well as optical and IR spectroscopy (After [14])

By the time, more sophisticated micro-/nanoscale devices have been developed for advanced THz sensing applications. Nagel et al. developed functionalized planar waveguide sensors to detect hybridization of unknown DNA molecules to complementary immobilized probe molecules with a known base sequence. Thin-film microstrip (TFMS) line waveguides building a ring resonator have resonance frequencies depending on the effective permittivity of the TFMS line. In order to use the ring resonator as a sensor for DNA molecule detection, its metallic signal line surface is chemically functionalized with single-stranded probe molecules. Target molecules can hybridize to these capture probes consisting of complementary single-stranded DNA oligonucleotides. At frequencies f_r, where the resonance condition is fulfilled, the number of transmission events through the sample is approximately multiplied by the quality factor $Q = f_r/\Delta f_r$ of the resonator. The ring resonator amplifies the interaction between the THz signal and the probed sample in a controlled and efficient way. The complex effective permittivity of the TFMS lines and with it the characteristic parameters of the resonator, such as resonance frequency or quality factor, is altered in dependence on the dielectric properties and amount of sample material brought to the resonator. Such sensing systems allows DNA detection without complex and costly labeling (tagging) procedure. Researchers in Cornell University developed a microfluidic chip for terahertz spectroscopy of biomolecules in aqueous solutions. The chip is fabricated out of Zeonor 1020R, a plastic material that is both mechanically rigid and optically transparent with near-zero dispersion in the terahertz frequency range. Microfluidic channel devices could make possible the real-time THz spectroscopy of chemical reactions and allow the use of low-power THz-TDS systems by avoiding excess water absorption and to enable the spectroscopy of picomole quantities of biomolecules. Moreover, such devices can easily be integrated with photonic components to realize multifunctional spectroscopy platforms. A schematic of a THz sensing platform based on microfluidic channels is shown in Fig. 9.

One notable application which might be particularly important for nanotechnology is the THz imaging of integrated circuits (ICs) for quality control and testing. Laser terahertz emission microscope (LTEM) for noncontact, nondestructive inspection of electrical failures in circuits has been proposed and demonstrated that it could localize a broken line in an operational amplifier. LTEM is based on the measurement of the THz emission in free space from the ultrafast transient photocurrent in a circuit by scanning it with femtosecond (fs) laser pulses. The THz waves can be emitted from areas in the circuits such as photoconductive switch structures with external bias voltage, various unbiased interfaces with built-in electric fields – such as p-n junctions, Schottky contacts

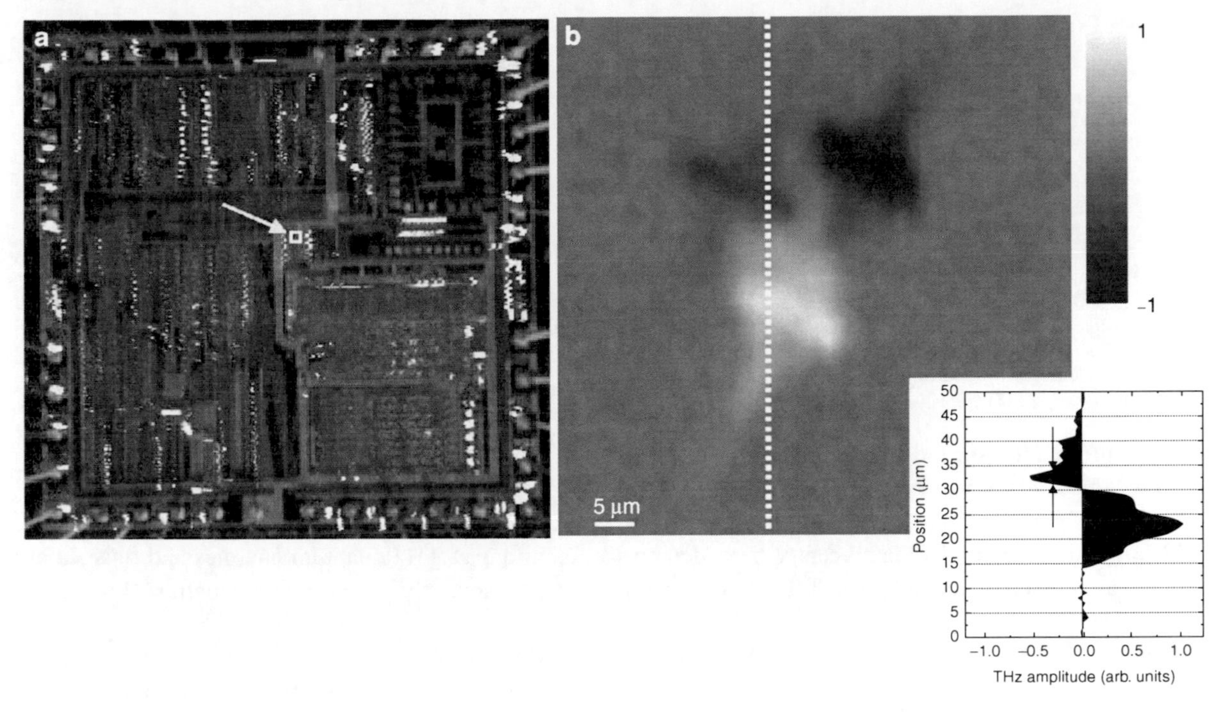

Terahertz Technology for Nano Applications, Fig. 10 (**a**) The THz emission image an LSI 8-bit microprocessor chip. The red and blue regions indicate THz emissions with positive and negative amplitudes, respectively, which correspond to the direction of the electric field. (**b**) The magnified THz emission image of the area indicated by the white square in Fig. 10a. The inset shows the cross-sectional distribution of the THz emission image at the dotted line (After [15])

(metal/semiconductor interfaces) – or just the surface of semiconductor. The THz emission amplitude is proportional to the local electric field in the photo-excited area. By comparing the THz emission images of a normal and damaged chip, the abnormal electrical field resulting from an electrical failure in the circuit can be visualized (Fig. 10). However, the spatial resolution of LTEM was strongly affected by the long focal length of the focusing lens and the aperture size in the system, limiting it to about 20 μm. Therefore, noncontact testing of ICs using THz radiation required near-field imaging techniques which will be discussed in the next section.

THz Near-Field Imaging and THz Microscopy

Low-energy photons of THz radiation make a rich variety of light-matter interactions possible by exciting molecular vibrations and phonons as well as plasmons and electrons of nonmetallic conductors. Therefore, T-rays offer new ways for material and device characterization. However, diffraction limits the spatial resolution to the order of wavelength and prevents the use of THz waves for imaging beyond ~100 μm. Conventional imaging methods cannot be employed for THz mapping of micro- or nanoelectronic devices, low-dimensional semiconductor nanostructures, cellular entities, or single molecules. This problem can be circumvented by invoking the near-field scanning optical microscopy technique for THz imaging applications. One way to overcome the diffraction limit is to limit the detection area with an aperture. It has been shown that using a detector with an integrated aperture, near-field images can be made with a spatial resolution determined by the aperture size, and not by the THz wavelength. This method, however, suffered from microfabrication complexity and waveguide effects which lead to strong attenuation for long wavelength components. A widely explored alternative method is the use of sharp metal tips as local field enhancers. The sharp metal tip is held in close contact to the sample under study, while light is scattered by the tip. Acting as antennas, the wires capture incident THz waves and convert them into strongly confined near fields at the wire tip apex which has submicrometer dimensions. When this confined field becomes modified by a close-by scanned sample, the scattered radiation

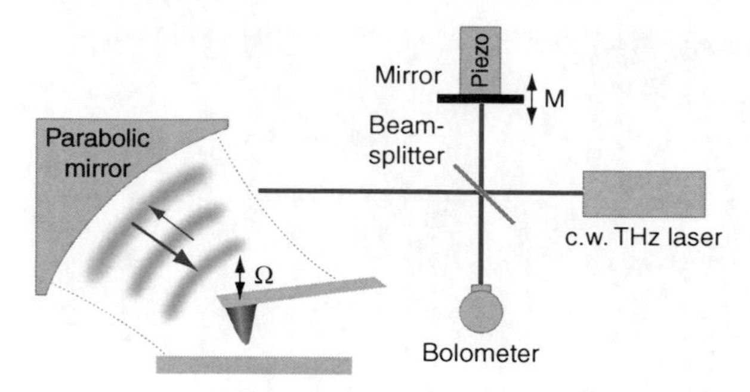

Terahertz Technology for Nano Applications, Fig. 11 Schematic of apertureless THz scanning near-field optical microscope (THz-SNOM) (After [17])

carries information on the local dielectric properties of the sample. This apertureless near-field imaging, which was originally developed at visible and mid-infrared wavelengths, has recently been demonstrated at terahertz wavelengths. THz images with a subwavelength resolution can be obtained by recording the scattered radiation by a distant THz receiver. A big advantage of this imaging technique is that it does not suffer from a waveguide cutoff, which limits the resolution in aperture techniques. Attempts of realizing such THz-scattering near-field optical microscopy (THz-SNOM), however, suffer from weak signals and faint material contrasts owing to strong background scattering.

In the first demonstration of THz near-field imaging at THz frequencies, Hunsche et al. used ultrashort, broadband pulses which were focused into a tapered metal tip with a small exit aperture and scanned a sample in the near field of this aperture. Their method allowed spatial resolution better than $\lambda/4$ for the corresponding average wavelength of 220 mm [16]. Huber et al. demonstrated high-resolution THz near-field microscopy using the experimental setup shown in Fig. 11 [17]. The THz signal was incident through a parabolic mirror with an angle to allow total internal reflection over the sample surface. On top of the sample, a cantilever of 20 μm length holding a metallic tip of 30 nm radius vibrated to modulate the incident wave in an On and Off manner. The cantilever mechanically oscillated at the mechanical resonance frequency of 35 kHz. A Michelson interferometer detector and hot electron bolometer were used to measure backscattered THz radiation. This technique allowed 30–150 nm spatial resolution imaging, corresponding to $\lambda/4{,}000$ for 2.54 THz. Moreover, near-field THz contrast between materials with different dielectric properties allowed imaging of a polished cut through nanoscale transistor structure and simultaneous recognition of materials and concentration of mobile carriers in the range of 10^{16}–10^{19} cm^{-3}. Since the spatial resolution of 40 nm infers that the volume probed by the THz near-field is about (40 nm)^3, it can be concluded that an average of less than 100 electrons in the probed volume suffices to evoke significant THz contrast. This opens the possibility of THz studies of single electrons by improving the used setup and, in conjunction with ultrafast techniques, even their dynamics. The reported results are encouraging to use THz near-field microscopy for studying other charged particles and quasiparticles in condensed matter, for example, in superconductors, low-dimensional electron systems, or conducting biopolymers, which possess intrinsic excitations at THz quantum energies and thus should exhibit resonantly enhanced THz contrast (Fig. 12).

THz Communication

T-rays are very attractive for communication applications for the large bandwidth at these frequencies which could provide very high data transmission rates. Moreover, frequencies above 300 GHz are currently not allocated by the Federal Communications Commission. On the down side, strong atmospheric absorption due to the water vapor together with the low available power of the current THz sources limits the range of communication. A 10 mW source and 1 pW detection sensitivity would be required for 1 km communication range considering the typical attenuation rate of ~100 dB/km rate in the atmospheric absorption windows (Fig. 13).

The strong atmospheric attenuation, however, does not pose a real threat for a number of communication applications. Space communication, for instance, can benefit from the high data transmission rates and smaller antenna size required for THz frequencies which would make possible the development of smaller, lighter therefore satellites with longer lifetime. Large bandwidth of THz frequencies can also be used to for short-range indoor communication exceeding Gigabit/s transmission rates. Rayleigh scattering of electromagnetic radiation is many orders of magnitude less for THz wavelengths than for the alternative infrared frequencies.

One unique and promising potential for THz frequencies is the on-chip and communication, especially systems-on-chip (SoC) applications. With the

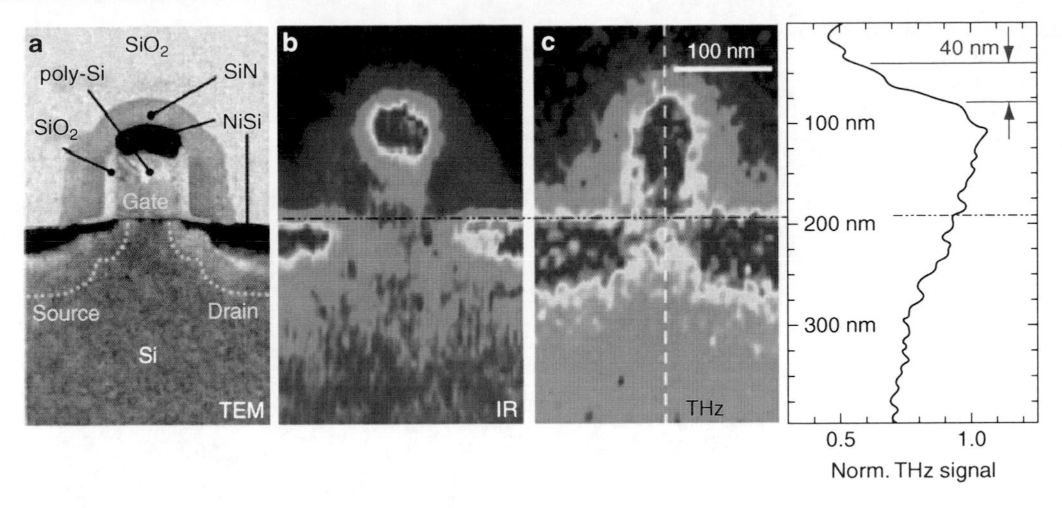

Terahertz Technology for Nano Applications, Fig. 12 (**a**) TEM image of a single transistor. The highly doped regions below the source and drain NiSi contacts are marked by dashed yellow lines. (**b**) Infrared image of the single transistor 12a ($\lambda \approx$ 11 μm). (**c**) High-resolution THz image of the single showing all essential parts of the transistor: source, drain, and gate. The THz profile extracted along the dashed white line (averaged over a width of 12 nm and normalized to the signal obtained on the metallic NiSi gate contact) allows the estimation of a spatial resolution of about 40 nm, from the strong signal change at the SiO_2/SiN/NiSi transition [17]

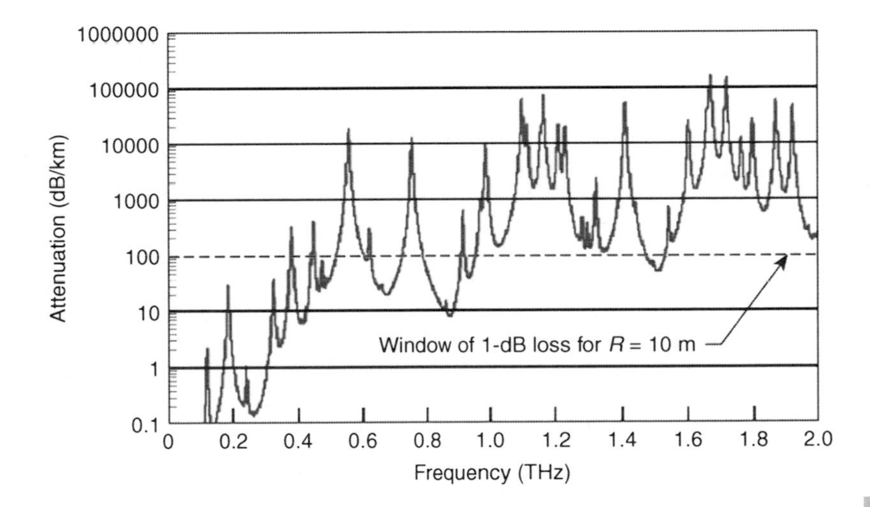

Terahertz Technology for Nano Applications, Fig. 13 Atmospheric attenuation in the THz frequency range [18]

ever-increasing complexity of integrated circuits, larger and larger number of functional and storage cores are squeezed into a single die. Extrapolating from the current CMOS scaling trends, conventional on-chip interconnect technologies have been projected to reach their limits to meet the performance needs of the next-generation multi-core and SoC systems. Interconnect delays do not scale as well as local wires with respect to gate delays, and global interconnects have an increasing impact on the performance of overall SoCs. A radical alternative to the existing metal/dielectric interconnect infrastructures is to use transmission of signals via wireless interconnects. On-chip wireless interconnects were demonstrated first for clock signal distribution at 15 GHz carrier frequency [19]. If the THz frequencies are employed for on-chip communication, corresponding antenna sizes decrease occupying much less chip real estate. Nanowires and carbon nanotubes (CNTs) can serve as antenna structures in such applications. CNT bundles are predicted to enhance performance of antenna modules by up to 40 dB in radiation efficiency and provide

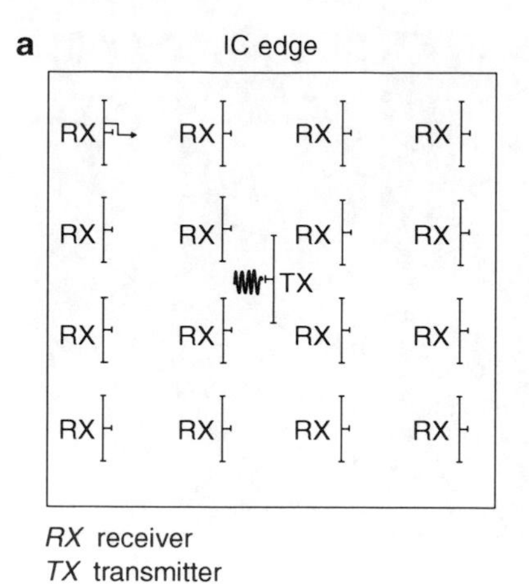

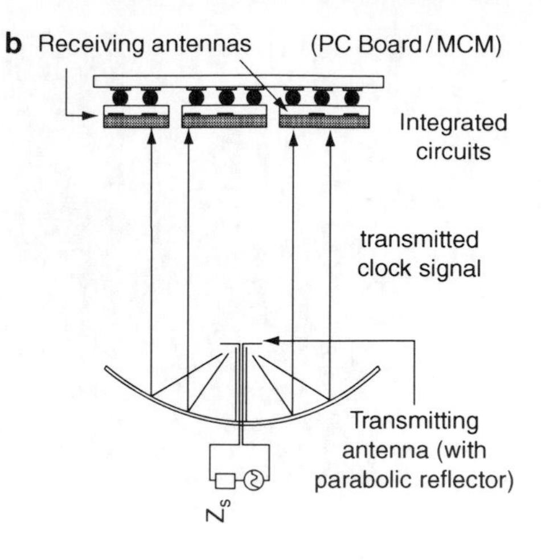

Terahertz Technology for Nano Applications, Fig. 14 Conceptual system illustrations of (**a**) intra-chip and (**b**) inter-chip wireless interconnect systems for clock signal distribution (After [19])

excellent directional properties in far field patterns and can achieve a bandwidth of 500 GHz and hence can achieve much higher data rates (Fig. 14).

Cross-References

▶ Biosensors
▶ Nanostructure Field Effect Transistor Biosensors

Further Reading

1. Tonouchi, M.: Cutting-edge terahertz technology. Nat. Photonics **1**, 97–105 (2007)
2. Diduck, Q., Irie, H., Margala, M.: A room temperature ballistic deflection transistor for high performance applications. Int. J. High Speed Electron. Syst. **19**, 23–31 (2009)
3. El Fatimy, A., Dyakonova, N., Meziani, Y., Otsuji, T., Knap, W., Vandenbrouk, S., Madjour, K., Théron, D., Gaquiere, C., Poisson, M.A., Delage, S., Prystawko, P., Skierbiszewski, C.: AlGaN/GaN high electron mobility transistors as a voltage-tunable room temperature terahertz sources. J. Appl. Phys. **107**, 024504 (2010)
4. Dragoman, D., Dragoman, M.: Terahertz fields and applications. Prog. Quantum Electron. **28**, 1–66 (2004)
5. Kohler, R., Tredicucci, A., Beltram, F., Beere, H.E., Linfield, E.H., Gilles Davies, A., Ritchie, D.A., Iotti, R.C., Rossi, F.: Terahertz semiconductor-heterostructure laser. Nature **417**, 156–159 (2002)
6. Kumar, S., Hu, Q., Reno, J.L.: 186 K operation of terahertz quantum-cascade lasers based on a diagonal design. Appl. Phys. Lett. **94**, 131105 (2009)
7. Dyakonov, M.I., Shur, M.S.: Shallow water analogy for a ballistic field effect transistor: new mechanism of plasma wave generation by dc current. Phys. Rev. Lett. **71**, 2465 (1993)
8. Dyakonov, M.I., Shur, M.S.: Detection, mixing, and frequency multiplication of terahertz radiation by two dimensional electronic fluid. IEEE Trans. Electron Devices **43**, 380 (1996)
9. Muravjov, A.V., Veksler, D.B., Popov, V.V., Polischuk, O. V., Pala, N., Hu, X., Gaska, R., Saxena, H., Peale, R.E., Shur, M.S.: Temperature dependence of plasmonic terahertz absorption in grating-gate GaN HEMT structures. Appl. Phys. Lett. **96**, 042105 (2010)
10. Hu, B.B., Nuss, M.C.: Imaging with terahertz waves. Opt. Lett. **20**, 1716–1718 (1995)
11. Woolard, D.L., Koscica, T., Rhodes, D.L., Cuj, H.L., Pastore, R.A., Jensen, J.O., Jensen, J.L., Loerop, W.R., Jacobsen, R.H., Mittleman, D., Nuss, M.C.: Millimeter wave-induced vibrational modes in DNA as a possible alternative to animal tests to probe for carcinogenic mutations. J. Appl. Toxicol. **17**, 243–246 (1997)
12. Nishizawa, J.: Development of THz wave oscillation and its application to molecular sciences. Proc. Jpn. Acad. B **80**, 74 (2004)
13. Kemp, M.C., Taday, P.F., Cole, B.E., Cluff, J.A., Fitzgerald, A.J., Tribe, W.R.: Security applications of terahertz technology. In: Hwu, R.J., Woolard, D.L. (eds.) Terahertz for Military and Security Applications, vol. 5070, pp. 44–52. SPIE, Bellingham (2003)
14. George, P.A., Hui, W., Rana, F., Hawkins, B.G., Smith, A.E., Kirby, B.J.: Microfluidic devices for terahertz spectroscopy of biomolecules. Opt. Express **16**, 1577 (2008)
15. Yamashita, M., Kawase, K., Otani, C.: Imaging of large-scale integrated circuits using laser terahertz emission microscopy. Opt. Express **13**, 115 (2005)
16. Hunsche, S., Koch, M., Brener, I., Nuss, M.C.: THz near-field imaging. Opt. Commun. **150**, 22 (1998)
17. Huber, A.J., Keilmann, F., Wittborn, J., Aizpurua, J., Hillenbrand, R.: Terahertz near-field nanoscopy of mobile carriers in single semiconductor Nanodevices. Nano Lett. **8**, 3766–3770 (2008)

18. Fitch, M.J., Osiander, R.: Terahertz waves for communications and sensing. Johns Hopkins APL Tech. Dig. **25**, 348 (2004)
19. Floyd, B.A., Hung, C.M., Kenneth, K.O.: Intra-chip wireless interconnect for clock distribution implemented with integrated antennas, receivers and transmitters. IEEE J. Solid-State Circuits **37**, 543 (2002)

Theoretical Elasticity

▶ Superelasticity and the Shape Memory Effect

Theory of Artificial Electromagnetic Materials

▶ Theory of Optical Metamaterials

Theory of Optical Metamaterials

Carsten Rockstuhl, Christoph Menzel, Stefan Mühlig and Falk Lederer
Institute of Condensed Matter Theory and Solid State Optics, Abbe Center of Photonics, Friedrich-Schiller-Universität Jena, Jena, Germany

Synonyms

Composite materials; Effective media; Left-handed materials; Theory of artificial electromagnetic materials

Definition

Optical metamaterials can be understood as artificial media consisting of suitably assembled micro- or nanostructured materials such as dielectrics or metals and acquiring their essential properties from the structure rather than from the underlying material. The purpose of metamaterials is to affect the characteristics of light propagation in a manner inaccessible with natural available media. Their theoretical description evokes various concepts which are outlined in this essay.

Introduction

Optical metamaterials (MMs) can be understood as artificial media consisting of suitably assembled micro- or nanostructured materials such as dielectrics or metals [1–3] and acquiring their essential properties from the structure rather than from the underlying material. The purpose of MMs is to affect the characteristics of light propagation in a manner inaccessible with natural available media; hence the term material was linked with the prefix *meta*, being derived from Greek and literally meaning beyond or after. This rather generic and broad definition comprises many kinds of structured materials. Examples thereof would be photonic crystals or plasmonic structures [4, 5]. In a stricter sense, as the definition is further refined here, MMs are thought to be made from unit cells a few times or much smaller than the wavelengths in the spectral domain of interest. Here restriction is made specifically to the latter case where the propagating light does not probe for the fine details of the unit cells [6]. Its propagation can rather be fully understood by replacing the complex-shaped MM with a homogenous medium to which effective properties are assigned [7]. In the alternative case where the unit cell is only a few times less than the relevant wavelength, that is, the metamaterials represent a mesoscopic system, nonlocal effects come into play and the assignment of both an effective permittivity and permeability ceases to be meaningful. Alternative approaches for describing light propagation have to be used which are beyond the scope of this entry.

Depending on which approach is chosen to discuss MMs, the innovation comes with the deliberation from restrictions imposed by nature. The linear optical properties of natural materials at optical frequencies are entirely described by considering the permittivity function only, that is, a complex frequency-dependent function characterizing the response of matter to an electric field. The permeability function, which characterizes the response of matter to a magnetic field, can be safely disregarded since it marginally deviates from that of vacuum. For any frequency considered both permeability and permittivity represent material parameters and are linked to wave parameters, as for example, the wave vector or the refractive index, by the dispersion relation of the normal- or eigenmodes which is of pivotal importance to describe the propagation of electromagnetic fields in any structure. These eigenmodes are self-consistent solutions to Maxwell's equations in the respective medium. The propagation of an arbitrary field in this medium can then be written as a linear superposition of such eigenmodes. The eigenmodes of a homogenous, dispersive achiral and isotropic medium are elliptically polarized plane waves characterized by wave parameters, such as impedance and refractive index. For natural materials both quantities are linked by inversion. Hence, if one parameter is determined in a given material, the other is equally fixed. Thus, independent of whether material or wave properties are considered, there is only a single degree of freedom [8, 9].

Metamaterials have the potential to circumvent this limitation and to provide independent control over permittivity and permeability, or refractive index and impedance. Both sets of parameters are connected by defining suitable constitutive relations. They have to be introduced into Maxwell's equations to specify the matter responses to an electric or magnetic field. Having materials at hand in which these material or wave parameters can be freely adjusted, many future applications can be envisaged which at first glance contradict the daily experience of how light propagates. Examples are the perfect lens which reconstructs an object in the image plane with unprecedented resolution or a cloaking device that allows concealing objects from external observers. It requires precise guidelines for designing MMs to achieve the necessary effective properties for bringing such applications to reality. This entry documents such approaches.

However, it has to be underlined again that any description of MMs in terms of effective material parameters is necessarily an approximation. It may serve as a model which is valid under certain constraints and where a finite number of free parameters mimic the complex structure. This greatly facilitates the description of current MMs and allows considering them in a subsequent design process for functional devices. The model is valid if the derived effective parameters are such that they do not violate the assumptions made in deriving them (for details, see Ref. [10]). A simple canonical example which illustrates the case is the retrieval of isotropic material properties for a MM whose unit cell fulfills all geometrical requirements to be considered as isotropic in the quasi-static (long wavelength) limit. Obviously, while being isotropic such material parameters must not depend on the angle of incidence. Unfortunately, this may be observed in the spectral domain where the dispersive effects of interest occur. Then, obviously, the material cannot be strictly considered as isotropic, though the deviations might be miniscule for suitably designed MMs and might be acceptable for a given application [10]. What is acceptable or not cannot be rigorously defined and it will depend on the explicit application and purpose of the effective properties, and it is moreover subject to a constantly heated debate. It is therefore not the intention of this entry to discuss the details of effective properties of various MMs but rather to explain and summarize the framework and the means to derive effective properties of optical MMs on theoretical grounds. Nevertheless, to illustrate representative unit cells for MMs and to provide an idea on how such structures look like, Fig. 1 shows examples of canonical unit cells for optical MMs along with an indication of which effective properties are affected.

The entry is structured as follows. In the second section, the idea of assigning constitutive relations to a MM is outlined. In the third section, an overview of theoretical approaches is provided that can be used to discuss the properties of optical MM on theoretical grounds. The limitations of such descriptions are finally discussed and eventually some conclusions are provided.

Constitutive Relations and Effective Properties

Any theoretical description of optical MMs has necessarily to start with Maxwell's equations which read in time domain as

$$\mathbf{rot}\,\mathbf{E}(\mathbf{r},t) = -\frac{\partial \mathbf{B}(\mathbf{r},t)}{\partial t}, \qquad \mathbf{div}\,\mathbf{D}(\mathbf{r},t) = 0,$$
$$\mathbf{rot}\,\mathbf{H}(\mathbf{r},t) = \frac{\partial \mathbf{D}(\mathbf{r},t)}{\partial t} + \mathbf{j}(\mathbf{r},t), \qquad \mathbf{div}\,\mathbf{B}(\mathbf{r},t) = 0$$

where $\mathbf{E}(\mathbf{r},t)$ and $\mathbf{H}(\mathbf{r},t)$ are the electric and magnetic fields; $\mathbf{D}(\mathbf{r},t)$ and $\mathbf{B}(\mathbf{r},t)$ are the electric displacement and the magnetic induction, respectively; and $\mathbf{j}(\mathbf{r},t)$ is the conductive current. The latter depend both, in general, on the electric and magnetic field in a complicated manner, specified by the respective constitutive relations. It has to be stressed that in this special form of Maxwell's equations the external charge carrier density is assumed to be zero. Assuming the realm of linear optics to be valid, it is convenient to write Maxwell's equations in temporal Fourier space by assuming a time harmonic oscillation of all fields. This leads to

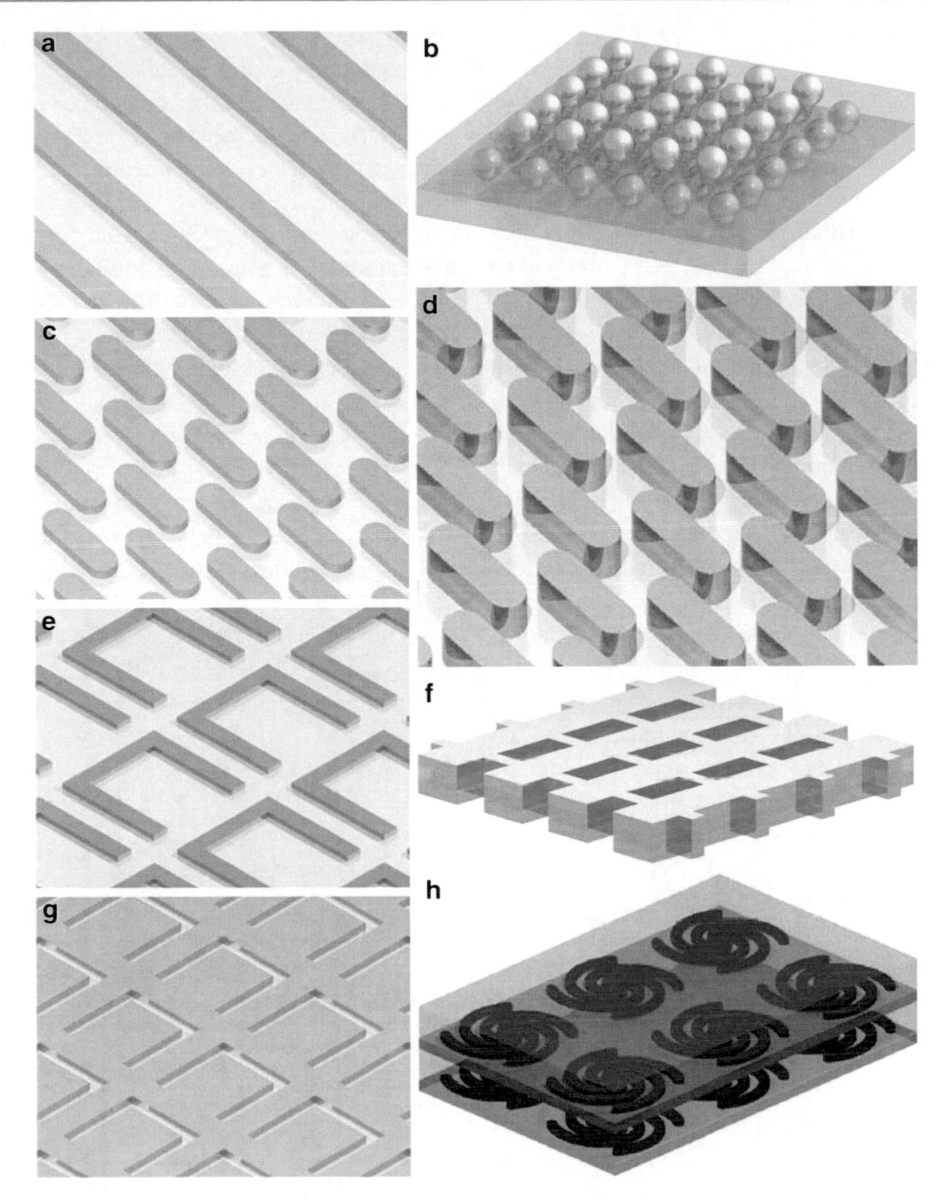

Theory of Optical Metamaterials, Fig. 1 Characteristic optical MMs consisting of different unit cells. (**a**) Metallic wire grid which acts as a diluted metal with a Drude-type effective permittivity; (**b**) and (**c**) MMs made from small metallic nanoparticles which will induce a Lorentzian-type effective permittivity around the resonance frequency of the localized plasmon polariton; (**c**) indicates that the resonance frequency can be tuned by modifying the geometry of the nanoparticles; (**d**) cut-wire pair where two metallic particles separated by a dielectric are strongly coupled, leading to a symmetric and an antisymmetric resonance where either the permittivity or the permeability exhibits a Lorentzian resonance; (**e**) split-ring resonator supports complex resonances depending on the illumination direction and the polarization, but being best known as a unit cell possessing a magnetic resonance; (**f**) fishnet structure where a cut-wire pair is combined with a metallic wire, and the resulting medium may show both a dielectric and a magnetic resonance in the same spectral range leading to a negative refractive index; (**g**) complementary structures, as the complementary split-ring resonator, are equally important; (**h**) gammadion structure which acts as an effective chiral medium

$$\mathbf{rot}\,\overline{\mathbf{E}}(\mathbf{r},\omega) = \mathrm{i}\omega\overline{\mathbf{B}}(\mathbf{r},\omega), \qquad \mathbf{div}\,\overline{\mathbf{D}}(\mathbf{r},\omega) = 0,$$
$$\mathbf{rot}\,\overline{\mathbf{H}}(\mathbf{r},\omega) = -\mathrm{i}\omega\overline{\mathbf{D}}(\mathbf{r},\omega) + \overline{\mathbf{j}}(\mathbf{r},\omega),$$
$$\mathbf{div}\,\overline{\mathbf{H}}(\mathbf{r},\omega) = 0.$$

Arbitrary solutions to Maxwell's equations in time domain are then written as a superposition of those time harmonic solutions weighted with appropriate amplitudes. Quantum mechanical or phenomenological models allow the derivation of frequency-dependent permittivity and permeability functions or tensors, where usually the electrical conductivity is absorbed into the complex permittivity which includes contributions from bound states and free electrons. Whereas the general dependency of **D(E,H)**, **B(E,H)**, and **j(E,H)** can be arbitrary complex, in most cases there is a linear relationship between these quantities mediated by permittivity, permeability, and electromagnetic coupling functions. The most general constitutive relation that is yet amenable to a theoretical treatment is that of a bi-anisotropic material. Assuming from now on an infinite homogenous space and dropping the dependency on the spatial coordinate, the resulting local constitutive relations read as [11] provided that the complex conductivity is absorbed into the complex permittivity:

$$\overline{\mathbf{D}}(\omega) = \varepsilon_0\overline{\overline{\varepsilon}}(\omega)\overline{\mathbf{E}}(\omega) + \frac{1}{c}\overline{\overline{\xi}}(\omega)\overline{\mathbf{H}}(\omega)$$
$$\overline{\mathbf{B}}(\omega) = \frac{1}{c}\overline{\overline{\zeta}}(\omega)\overline{\mathbf{E}}(\omega) + \mu_0\overline{\overline{\mu}}(\omega)\overline{\mathbf{H}}(\omega).$$

The four tensorial material functions are the permittivity $\overline{\overline{\varepsilon}}(\omega)$, the permeability $\overline{\overline{\mu}}(\omega)$, and two further tensors $\overline{\overline{\xi}}(\omega)$ and $\overline{\overline{\zeta}}(\omega)$ which mediate the electromagnetic coupling. Since the magnetic (electric) field depends on the curl of the electric (magnetic) field, the equations can also be understood as a Taylor series of a more complex functional dependency of the electric (magnetic) displacement on the electric (magnetic) fields. In this expansion only the lowest order coefficients are retained. For a medium with sufficiently high symmetry, that is, unit cells composed of three mirror-symmetric orthogonal planes, electromagnetic coupling vanishes. Consequently, in this Taylor expansion a next higher order term also needs to be considered, that is, media possessing a weak nonlocal response (weak spatial dispersion). If this holds, terms would appear where the displacement fields depend on the second derivative of the fields. Under appropriate conditions, which are assumed here to be valid, this term in the electric displacement field can be transformed as to appear in the dispersive permeability $\overline{\overline{\mu}}(\omega)$ in the magnetic constitutive relation and does not need to be considered explicitly. Such transformation can be understood as a gauge transformation since physical observable fields are not affected. Moreover, such transformation is useful to explain how it is possible to observe artificial magnetism at optical frequencies for which all available materials possess usually a nondispersive permeability. An artificial medium which supports a weak nonlocal electric response, that is, a nonlocal current appearing in mesoscopic metallic structures, is hence required to observe artificial magnetism, leading to the above constitutive relations [10, 12].

At this point it has to be stressed that the constitutive relations are very general. If they can be furthermore assumed to be reciprocal, they reduce to

$$\overline{\mathbf{D}}(\omega) = \varepsilon_0\overline{\overline{\varepsilon}}(\omega)\overline{\mathbf{E}}(\omega) + i\frac{1}{c}\overline{\overline{\kappa}}(\omega)\overline{\mathbf{H}}(\omega),$$
$$\overline{\mathbf{B}}(\omega) = -i\frac{1}{c}\overline{\overline{\kappa}}^T(\omega)\overline{\mathbf{E}}(\omega) + \mu_0\overline{\overline{\mu}}(\omega)\overline{\mathbf{H}}(\omega).$$

If, additionally, electromagnetic coupling effects may be excluded, it is possible to arrive at the relation for anisotropic crystal.

$$\overline{\mathbf{D}}(\omega) = \varepsilon_0\overline{\overline{\varepsilon}}(\omega)\overline{\mathbf{E}}(\omega),$$
$$\overline{\mathbf{B}}(\omega) = \mu_0\overline{\overline{\mu}}(\omega)\overline{\mathbf{H}}(\omega).$$

If isotropy holds the simplest form of the constitutive relations is achieved as

$$\overline{\mathbf{D}}(\omega) = \varepsilon_0\varepsilon(\omega)\overline{\mathbf{E}}(\omega),$$
$$\overline{\mathbf{B}}(\omega) = \mu_0\mu(\omega)\overline{\mathbf{H}}(\omega)$$

It has to be noted that the introduction of more complex constitutive relations, for example, explicitly containing strong spatial dispersion, when compared to the departing point is rather pointless, though they can be written down phenomenologically. The boundary conditions that need to be imposed onto the fields at the interface between two such media are quite involved (additional boundary conditions). They have to be applied in addition to ordinary Maxwell's boundary conditions and are the subject of ongoing research in the field of MMs [13].

These material parameters can be related to wave parameters describing the propagation of an eigenmode in a MM that is characterized by any of the constitutive relations above. Assuming the simplest scenario of a homogenous isotropic medium, the wave parameters of the eigenmodes are found by plugging the ansatz for an elliptically polarized plane wave $\mathbf{E}(\mathbf{r},t) = \tilde{\mathbf{E}}(\omega)e^{i[k(\omega)r-\omega t]}$ as the possibly simplest solution into Maxwell's equations where the propagation vector $\mathbf{k}$ is one of the wave parameters. The dispersion relation relates it to the frequency ω which appears in the material parameters

$$\mathbf{k}^2 = k_x^2 + k_y^2 + k_z^2 = \frac{\omega^2}{c^2}\varepsilon(\omega)\mu(\omega)$$

The dispersion relation defines the set of plane waves that are allowed to propagate. The second wave parameter derives from the ratio of the moduli of electric and magnetic fields and is the impedance Z

$$Z = \sqrt{\frac{\mu(\omega)}{\varepsilon(\omega)}}$$

Whereas the propagation vector describes essentially the wave evolution of the normal mode in the respective medium, the impedance determines the reflection and transmission properties at interfaces to other media. For media with more complicated constitutive relations the corresponding dispersion relations can be equally derived on analytical grounds and can be found in literature.

With these governing equations the light propagation through an arbitrary complex material can be described exploiting a set of simple quantities. The challenge to theoretically describe optical MMs is to link a MM made of mesoscopic unit cells to a homogenized medium in which the fine details of the structures may be neglected. It is therefore required that at least the fields propagating either in the actual MM or in an effectively homogeneous material possess the same wave parameters. Both materials would then be indistinguishable since their optical action is fully preserved while the complexity is significantly reduced. This link has to be established first. By assuming furthermore feasible constitutive relations, these wave parameters can be linked to effective material parameters. Established approaches for exactly this purpose are documented in the following chapter.

Retrieval of Effective Properties

The retrieval of effective properties of optical MMs is usually based on an analysis of the interaction of light with the structure under investigation; therefore, it is not possible to define the properties in an absolute sense but they are rather probed for a particular illumination scenario. In theory, the interaction and the propagation of such probing fields with the MM is numerically analyzed by solving rigorously Maxwell's equations and taking into account the fine details of the spatially distributed materials forming the unit cells, usually given by $\varepsilon(\mathbf{r},\omega)$. The spatial distribution of the permeability does not need to be considered since the intrinsic permeability of natural materials is negligible. Various numerical means exist that can be applied for this purpose such as the Finite-Difference Time-Domain (FDTD) method, Finite-Element Method (FEM), or plane wave expansion techniques [14]. Generally, depending on whether the unit cells are periodically or randomly arranged in space, various strategies can be applied to retrieve the effective properties. In the following between these two categories will be distinguished.

Periodic Structures

Most of current MMs are fabricated by top-down technologies such as electron beam lithography or direct laser beam writing. These technologies allow fabricating unit cells with critical dimensions in the order of a few tenths of nanometers, in most cases as planar nanostructured thin films on substrates. Repeating these techniques in a sequential or parallel process allows manufacturing MMs made of periodically arranged unit cells in all spatial dimensions. To eventually observe effective properties beyond those achievable by a mere spatial average of the intrinsic material properties of the involved constituents, resonances in the unit cells are often exploited. The required sub-wavelength size of the unit cells usually requires the coupling of the electromagnetic field to the free electrons in metals; hence plasmonic materials are often used in the design of optical MMs. In most cases gold is preferred since it is chemically stable, but also silver or aluminum can be used. Such MMs are therefore comparable to metallic photonic crystals, however, with much smaller periods, and consequently the language used to discuss their properties is intimately linked to it. Most notably, light propagation in bulk MMs is characterized by a dispersion relation as it will be discussed at first. The dispersion relation only allows determining the wave vector **k** that characterizes the evolution of a Bloch mode (for a definition see below), from which an effective index can be derived. As argued below, from the dispersion relation further quantities can be derived that are important to describe the propagation of finite beams (either spatially or temporally confined) through the bulk MM. From the tangential fields of an eigenmode the impedance linked to it can be derived. With these two quantities the entire set of wave parameters would be at hand. Moreover, for a deviating geometry and by relying on simulated reflection and transmission coefficients of an incident plane wave at a MM slab with finite thickness, both wave parameters can be equally retrieved. Such a procedure is part of a second succeeding section. Eventually, only if the wave parameters as retrieved by both approaches are identical the MM can be considered homogenous.

Retrieval Using the Dispersion Relation

An appropriate description of bulk properties of MMs made of periodically arranged unit cells is based on the dispersion relation. The periodic structure is characterized by the dielectric function

$$\varepsilon(\mathbf{r}, \omega) = \varepsilon(\mathbf{r} + m\Lambda, \omega)$$

with Λ and m being the period of the structure and an integer, respectively. For simplicity it has been assumed that the MM's unit cells are made from isotropic materials. Since in the infinite medium two points displaced by one period in space are indistinguishable, the eigenmodes have to feature identical properties, that is, it is said they are Bloch periodic

$$\mathbf{u}_{\mathbf{k}}(\mathbf{r}, \omega) = \mathbf{u}_{\mathbf{k}}(\mathbf{r} + m\Lambda, \omega)$$

The Bloch modes amplitude is periodic with the lattice and its phase evolves like a plane wave

$$\mathbf{E}_{\mathbf{k}}(\mathbf{r}, \omega) = \mathbf{u}_{\mathbf{k}}(\mathbf{r}, \omega) e^{i[\mathbf{k}(\omega)\mathbf{r} - \omega t]}$$

As for any normal mode the Bloch wave vector **k** and the frequency ω are interrelated, that is, they have to obey a specific dispersion relation. This dispersion relation to be solved requires in general numerical means. There are a discrete number of Bloch modes which can be excited in the bulk MM. Nevertheless, from a conceptual point of view the approach is identical to that of a homogeneous, dispersive medium, that is, the governing dispersion relation is a single equation with four interrelated parameters. It suggests that three of them can be chosen freely. Alternatively, fixing three parameters by external constraints only excites Bloch modes with a given fourth parameter. In dealing with photonic crystals it is common to fix all components of the wave vector and then solve for the frequency at which the solutions are allowed to oscillate. However, a more rational approach is preferred here that is linked to an experiment. Therefore, the frequency ω (imposed by selecting a light source) and the tangential wave vector components at interfaces k_x and k_y are fixed. This is regarded as the referential plane where the evolution of the Bloch modes in the bulk starts (imposed by selecting the angle of incidence for an external plane wave since the tangential wave vector components are conserved). The dispersion relation is solved for the longitudinal wave vector component k_z. This allows to reveal by numerical means the entire functional dependency of k_z on all parameters and allows to disclose the entire dispersion relation $k_z = k_z\,(k_x, k_y, \omega)$ [15].

The number of modes that can be solved for is generally infinite. For an empty sub-wavelength lattice, however, only a single mode has a real valued propagation constant whereas all the others are evanescent. The same holds for a MM which is expected to be homogenizable, that is, there is a single mode with an imaginary part much less than those of all higher order modes. The imaginary part of this dominating mode is not strictly zero as soon as materials are incorporated into the unit cells that are intrinsically absorptive (e.g., metals). It furthermore has to be said that the eigenmodes appear always pairwise with opposite sign corresponding to forward and backward propagating modes, respectively. The proper branch has to be chosen in accordance with considerations imposed by the finiteness of the amplitudes, that is, by considering as the principal propagation the positive z-axis, only those eigenmodes can be excited which decay exponentially upon propagation.

The corresponding eigenmode with the negative propagation constant has to be taken into account for opposite propagation direction. A MM supporting a backward wave is easily identified in such analysis by unraveling the parameter space where the imaginary part of the propagation constant is opposite to the real part. In this spectral domain the MM is said to have a negative index, since the index is defined as the length of the wave vector in the MM normalized to the length of the wave vector in vacuum.

From the dispersion relation of the modes in the center of the Brillouin zone, that is, $k_x = k_y = 0$, and while considering additionally the symmetry of the unit cell, many conclusions can be already drawn on the polarization properties of the MM. For example, a C_4 symmetric unit cell (identical under rotation by $90°$ with respect to the z-axis) leading to a pair of identical propagation constants will show a polarization independent response. The field of the two modes will then be dominated by a component either polarized in the x- or in the y-direction and an arbitrary superposition of both modes is an eigenmode, in particular, also circularly polarized light. If the two propagation constants are different while still being C_4 symmetric, the medium is chiral and the chirality constant κ can be extracted from the difference of both propagation constants. A complete discussion of such properties is beyond the scope of this entry but can be found in literature.

The question on whether the medium is isotropic or anisotropic (biaxial or arbitrary) can only be answered while disclosing the functional dependency of k_z on k_x and k_y at a fixed frequency of interest. As argued before, this iso-frequency curve has a fixed shape for a given medium, that is, it is spherical for an isotropic medium, ellipsoidal for a biaxial anisotropic medium, but in most cases it possesses a complexity beyond those simple shapes and the medium is anisotropic in the literal sense, that it is just said to be non-isotropic. An example of such a dispersion relation depending on the frequency for a fixed tangential wave vector and another one depending on a tangential wave vector for a fixed frequency are shown in Fig. 2 for the fishnet MM. As can be seen in Fig. 2, the fishnet MM is neither isotropic nor biaxial anisotropic, but the iso-frequency curve shows a much more complicated functional dependency on the frequency where the relevant dispersive effects occur, that is, where the effective index takes negative values for the real part. This has dramatic consequences on the ability of the structure to refract or diffract light. While the direction into which light is refracted is linked to the refractive index only for an isotropic material, it is given in such more complicated mesoscopic structures by the normal vector of the iso-frequency curve. This is an extension of a well-known concept for biaxial anisotropic materials, where the direction of the Poynting vector, that is, the direction into which energy flows or a bundle of rays is refracted, is no more parallel to the wave vector but normal of the iso-frequency curve for the respective wave vector.

Similarly, the next higher order derivative is linked to the diffraction coefficient, being a measure how strong a finite bundle spreads upon propagation (normal diffraction – positive curvature of the iso-frequency curve) or whether the opposite holds and a diffractively broadened beam can be re-focused while propagating through a suitably designed MM (anomalous diffraction – negative curvature of the iso-frequency curve). Such compensation of the normal diffraction of free space with a suitable anomalous diffraction explains how it is possible to use a MM slab as an imaging device.

The conclusion can be drawn that in order to describe the propagation of pulsed beams in an arbitrarily structured bulk medium it suffices to know the dispersion relation of the normal modes. Effects like diffraction and dispersive spreading have their physical origin in the different phase evolution in propagation direction for different frequencies and transverse wave vector components. A simple example for this statement is pulse propagation in optical fibers. There diffraction is arrested (only one transverse wave component) and the first term of the Taylor expansion of the dispersion relation provides the group velocity whereas the second one provides group velocity dispersion (dispersive spreading). And similarly, just as dispersion managed fiber (alternating segments of normal and anomalous group velocity dispersion) allows for a complete pulse restoration, a MM slab that compensates the diffraction in a homogeneous medium may act as a perfect lens that enables to restore a spatially confined input field in some output plane.

However, the dispersion holds only in an infinite bulk medium. It contains no information on the transition between different media (conventional medium-MM, MM-MM). The missing link is the field impedance which is a measure for the coupling

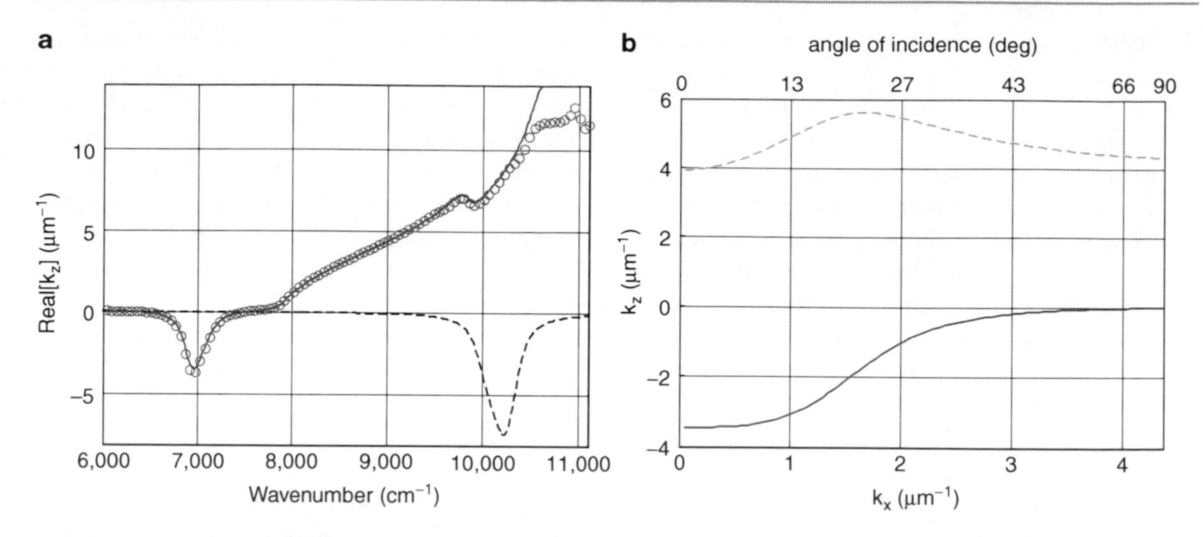

Theory of Optical Metamaterials, Fig. 2 Dispersion relation of a referential optical metamaterial, i.e., the fishnet as shown in Fig. 1f. (**a**) Shows the real part of the propagation constant of the two lowest order modes (*blue solid line* and *black dashed line*) as a function of the frequency at $k_x = k_y = 0$. The *dotted points* show for comparison results as retrieved from the finite structure which requires a retrieval using the complex reflection and transmission coefficient as discussed in the entry further below. (**b**) Shows the real (*blue solid line*) and the imaginary part (*green dashed line*) of the lowest order mode as a function of the tangential wavevector component k_x at $k_y = 0$. The frequency was fixed as to correspond to the spectral domain where the propagation constant is the largest negative, i.e., 7,000 cm^{-1}

strength of normal modes propagating in different media. Various definitions with different complexity are feasible. Here a definition that is based on the coupling efficiency of an input to a homogenizable MM (field propagation may be described by only one Bloch mode) at a certain interface is used. This is the so-called tangential impedance which is defined as the ratio of the spatial average of the tangential electric to the tangential magnetic fields across a given interface. Assuming a mirror-symmetric unit cell the average of the x-component of the electric field vanishes and the MM impedance reads as (the superscript hints to a Bloch mode)

$$Z = \frac{\langle E_y^B \rangle}{\langle H_x^B \rangle}.$$

To sum up to only describe light transition between media and light propagation in an arbitrary, but homogenizable, MM it suffices to know the dispersion relation and the impedance of the normal modes, which represent so-called wave parameters. In general, it is impossible, but also not required, to assign effective material parameters to MMs. However, if this is desired the consistent validity of the respective constitutive relations is required. This aspect will be discussed in the succeeding section when the parameter retrieval is performed using the reflection and transmission coefficients of a plane wave at a MM slab. This method, known as NRW-method, is particularly popular since it allows for exploiting measured reflection/transmission data for the retrieval thus not requiring propagation simulations [16]. But, since both complex coefficients need to be known with high precision, most notably in the measured phase, the retrieval using solely experimental results still constitutes a major challenge.

Retrieval Using Reflection and Transmission Data from a Slab

The appealing aspect of the parameter retrieval using reflection/transmission data relies on its simplicity. For a MM characterized by two parameters (propagation constant and impedance), two independent sets of measurements are required. Since most MMs are fabricated as slabs with a finite thickness, their explicit consideration is attractive. To render such simple approach to be applicable, it has to be assumed that the MM does not generate depolarized light, that is, electromagnetic coupling should not occur, and the light propagating through the slab has to remain in the plane of incident while traversing the structure. If both conditions are valid, the complex reflection and transmission coefficients can be computed under the assumption that the field propagating in the MM is characterized by a single Bloch mode (homogenization), that has a characteristic dispersion relation, and the coupling from a surrounding media occurs equally only to that single eigenmode, that can be expressed using the tangential impedance. Then reflection and transmission coefficients of a MM slab (labeled by F) with thickness d, sandwiched between a substrate (labeled by S) and a cladding (labeled by C) material where the dispersion relation for isotropic media holds, read as

$$T = \frac{2k_S \xi A}{\xi(k_S + k_C)\cos(kd) - i(\xi^2 + k_S k_C)\sin(kd)},$$

$$R = \frac{\xi(k_S - k_C)\cos(kd) + i(\xi^2 - k_S k_C)\sin(kd)}{\xi(k_S + k_C)\cos(kd) - i(\xi^2 + k_S k_C)\sin(kd)}.$$

Here it has been assumed that the z-coordinate is normal to the slab and that only the x-component of the tangential wave vector is nonzero. Moreover, some abbreviations have been introduced to facilitate the notation, that is (for details, see Ref. [10]),

$$k_{S,C} = \alpha^{S,C} k_z^{S,C},\ k = k_z^F,\ \xi = \alpha^F k_z^F,\ \alpha_{TE}^i = 1/\mu^i,\ \alpha_{TM}^i = 1/\varepsilon^i,\ A_{TE} = 1,\ A_{TM} = \sqrt{\frac{\mu^C \varepsilon^S}{\varepsilon^C \mu^S}}$$

Here a TE and TM polarization is introduced where the electric field is normal to or in the plane of incidence, respectively. Here a restriction is imposed to the TE case. The equations can be uniquely inverted as to provide the desired wave vector component k in longitudinal direction inside the MM slab and the tangential impedance ξ. These equations read as

$$k(k_x, \omega) = \pm\cos^{-1}\left\{\frac{k_z^S(1-R^2) + k_z^C T^2}{T\left[k_z^S(1-R) + k_z^C(1+R)\right]}\right\} + 2m\pi,$$

$$\xi(k_x, \omega) = \pm\sqrt{\frac{(k_z^S)^2(1-R)^2 - k_z^C T^2}{\left[(1+R)^2 - T^2\right]}}.$$

The different signs account for forward and backward propagation. They have to be chosen such that the waves decay upon propagation (imaginary part of k) and the tangential impedance takes positive values at normal incidence where it is linked to the impedance via $\xi_{TE} = \frac{\omega}{c}\frac{1}{Z}$ and $\xi_{TM} = Z\frac{\omega}{c}$. The branch number m occurring in the expression for the longitudinal wave number reflects that the field may exhibit a different number of nodes in a finite slab. The correct determination of the proper branch number can be achieved by various means. For example, it is possible to solve for the longitudinal wave number at marginally small frequencies where independent of the absolute thickness of the slab a field node does not appear because of the long wavelength. An alternative approach is to solve for the longitudinal wave for varying thickness. A consistent solution must not depend on the thickness. Another popular approach, the authors prefer too, is to solve the problem at the frequency of interest where all geometrical MM dimensions are scaled sufficiently down and the branch number can then be unambiguously fixed as to be $m = 0$. Up-scaling the geometry hitherto and requiring continuity of the longitudinal wave number while the scaling is performed requires the permanent adjustment of the branch number. The process is repeated until the geometry of interest is finally reached.

The wave parameters as retrieved by considering the reflection and transmission data are only comparable to those retrieved from the dispersion relation if the light propagation in the MM is dominated by a single, dominating Bloch mode and if the external illumination excites only this mode. If both conditions are met, the MM is said to be homogenizable since all the complex details are projected on two quantities, the wave number and the tangential impedance. If MMs are homogenizable their properties will be independent of the MM thickness. Since the tangential impedance is linked to the field distributions in a particular plane of the unit cell, it is furthermore required that the unit cell is symmetric for both propagation directions. Extensions of the parameter retrieval taking other properties into account, for example, electromagnetic coupling, are reported in literature and shall be not discussed here [17].

Crucial to an effective description of MMs is the fact that the tangential impedance and the longitudinal wave vector component will depend explicitly on the tangential wave vectors and the frequency.

The frequency dependence is usually not an issue since any known material possesses a frequency dependent response, that is, it is said to be dispersive. To further reduce the complexity of the wave parameters and to remove the dependency on the tangential wave vectors, the wave parameters as retrieved can be linked in some cases to material parameters. This requires, however, the strict validity of the constitutive relations. For example, if the material is assumed to be isotropic, the scalar and only frequency dependent material properties are linked to the wave number and the impedance for the scenario above for TE polarization as

$$\mu(\omega) = \frac{k(k_x, \omega)}{\xi(k_x, \omega)} \text{ and } \varepsilon(\omega) = \frac{n^2(\omega)}{\mu(\omega)}$$

and for TM polarization as

$$\varepsilon(\omega) = \frac{k(k_x, \omega)}{\xi(k_x, \omega)} \text{ and } \mu(\omega) = \frac{n^2(\omega)}{\varepsilon(\omega)}$$

Here the effective index of the MM defined as $n^2(\omega) = \frac{k_x^2+k^2}{(\omega/c)^2}$ has been introduced.

Please note, the assumption on the validity of the isotropy enters this framework of equation by assuming that the iso-frequency curve derived from the dispersion relation is a circle in the 2D case and the index will not depend explicitly on the tangential wave vector. Moreover, the functional dependency of k and ξ on k_x is also intimately linked and it will cancel out for an isotropic MM, that is, it will provide always the same material parameter independent of the tangential wave vector component. Of course, this does not necessarily need to be valid and deviations can be taken as a clear signature that the MM does not act as an isotropic material.

Similar procedures exist for biaxial materials, though then the response of the matter at two tangential wave vector components needs to be considered to disclose some selected entries of the material tensors. Other illumination scenarios are then furthermore required (i.e., modifying k_y or the considered polarization) to disclose all entries. Also, retrieval procedures that reveal the entire tensors taking into account also electromagnetic coupling effects exist in literature [16]. More complicated constitutive relations cannot be considered in most cases and in the case of contradictions, that is, different material properties are revealed while considering waves with different tangential wave vector components, a description in terms of material properties is not suggested and instead considerations should be restricted to the wave parameters.

Amorphous Structures

Whereas in the approach discussed in the previous section it was assumed that the unit cells are periodically arranged and a discussion in terms of Bloch modes could be performed, MMs can be also fabricated using bottom-up technologies that rely on self-organization mechanisms, for example, based on colloidal nanochemistry. Then, control over the geometry of the individual unit cell might be possible but not over the way they arrange. Likely, the structures will be randomly or amorphously arranged. For MMs of such geometry various mixing rules have been established in the past that allow to assign effective properties while considering only the scattering properties of the individual unit cells and their volume filling fraction. Such approaches are discussed in this section. Since the importance of such bottom-up MMs is still less when compared to MMs fabricated with top-down technologies, this section is kept rather concise to reflect its weight.

Retrieving Properties of Individual Meta-atoms

The properties of individual meta-atoms that eventually form a bulk MM are best accessed by investigating the scattered field generated by these meta-atoms. This scattered field can be expanded into multipoles which can be written either in a Cartesian or a spherical base. The idea of such an expansion is best understood in analogy to a Fourier series, where an arbitrary wave field can be written as a superposition of plane waves. Here the fields are decomposed into terms that are not spatially delocalized as plane waves, but they are linked to a definite origin where the scattered field emanates from. Since the multipoles are orthonormal functions, the expansion is unique and characteristic for a certain meta-atom. An expansion in spherical coordinates is well established in other fields of photonics, that is, in Mie theory. There the field scattered by a sphere is decomposed into spherical harmonics [18]. The expansion of a field scattered from an arbitrary object using such spherical harmonics reads as

$$\mathbf{E}_{\mathrm{Sca}}(r,\theta,\varphi)=\sum_{n=1}^{\infty}\sum_{m=-n}^{n}k^2E_{nm}\left[a_{nm}\mathbf{N}_{nm}(r,\theta,\varphi)+b_{nm}\mathbf{M}_{nm}(r,\theta,\varphi)\right]$$

where k is the free space wave number of the surrounding medium, E_{nm} are some normalizing amplitudes, $\mathbf{N}_{nm}$ and $\mathbf{M}_{nm}$ are the vector spherical harmonics which are defined in literature, and a_{nm} and b_{nm} are the complex amplitudes at which they are excited to generate a particular scattered field. It has to be stressed that any scattered field can be expanded around a chosen origin in this way. Occasionally, the center of the coordinate system might be chosen inappropriate, leading to some ambiguity in the expansion. Therefore, the coordinate system should be appropriately chosen such that the smallest number of scattering coefficients attains nonzero values. In most practical cases this is either the geometrical center of the meta-atom or the center of gravity. Moreover, the vector spherical harmonics have a physical meaning. Roughly speaking, the vector spherical harmonics $\mathbf{N}_{nm}$ describe the scattered electric field in terms of contributions evoked by electric multipole situated in the origin [19]. The first index n designates the respective Cartesian multipole. The electric field generated by their magnetic counterparts is associated to the coefficients $\mathbf{M}_{nm}$, respectively. Assembling the scattering properties of a single meta-atom is best performed by establishing the so-called T-matrix. The entries of this T-matrix suggest that illuminating the meta-atom with an incident field corresponding to a purely spherical harmonic of the order $n'm'$ will generate scattered multipoles of all the orders with the index nm and their respective amplitude strengths are the entries of one column of the T-matrix.

The retrieval of the contributions to the scattered field generated by a given meta-atom is performed by computing its scattered field $\mathbf{E}(\mathbf{r}, \omega)$ with a rigorous method of choice (e.g., FDTD or FEM) upon illumination with a suitably chosen illumination and solving for the mode contents of the scattered field. Computing the overlap integral over a sphere at a certain radius a yields all coefficients of the multipole expansion

$$a_{nm}=\frac{\int_0^{2\pi}\int_0^{\pi}\mathbf{E}(r=a)\mathbf{N}^*_{nm}(r=a)\sin\theta d\theta d\varphi}{\int_0^{2\pi}\int_0^{\pi}\mathbf{N}_{nm}(r=a)\mathbf{N}^*_{nm}(r=a)\sin\theta d\theta d\varphi},$$

$$b_{nm}=\frac{\int_0^{2\pi}\int_0^{\pi}\mathbf{E}(r=a)\mathbf{M}^*_{nm}(r=a)\sin\theta d\theta d\varphi}{\int_0^{2\pi}\int_0^{\pi}\mathbf{M}_{nm}(r=a)\mathbf{M}^*_{nm}(r=a)\sin\theta d\theta d\varphi},$$

where the asterisks designate complex conjugated fields.

From these coefficients it can be seen whether the structure indeed causes a magnetic response, which likely will induce a resonance in the magnetic dipole coefficients linked to b_{1m}, or to an electric dipole coefficient linked to a_{1m}. It also allows revealing the possible contribution of higher order multipoles, the appearance of which prevents the assignment of effective material parameters. Nevertheless, with such an analysis all properties of interest can be evaluated.

An example where the scattered field of a split-ring resonator is expanded into multipoles at optical frequencies can be seen in Fig. 3. Since the symmetry of the split-ring resonator suggests the usage of a Cartesian coordinate system, the spherical multipoles are transferred into Cartesian ones. Only multipoles with significant nonzero amplitude are shown. It can be seen that the lowest order resonance is linked to a magnetic dipole oriented normal to the plane of the split-ring resonator and to an electric dipole oriented in this plane. The second order resonance is linked entirely to an electric dipole where currents oscillate in the two arms of the split-ring resonator in phase.

Mixing Rules

By assuming that the MM is finally composed of complex meta-atoms which generate upon the excitation by an external field secondary electric or magnetic dipole, effective properties can be assigned to such materials on the base of suitably chosen mixing rules [20]. Then, the meta-atoms attain the position of artificial polarizable entities and an averaging procedure can be applied, similar to the transition of microscopic to macroscopic Maxwell's equations, just with the only noticeable distinction that the transition is now made from mesoscopic to macroscopic Maxwell's equations where this mesoscopic system is now man-made. Many of such mixing rules exist and usually they have been developed for a particular MM geometry. For example, for MMs made of interleaved domains the Bruggemann formula was used; a MM made of a sequence of slabs requires for the

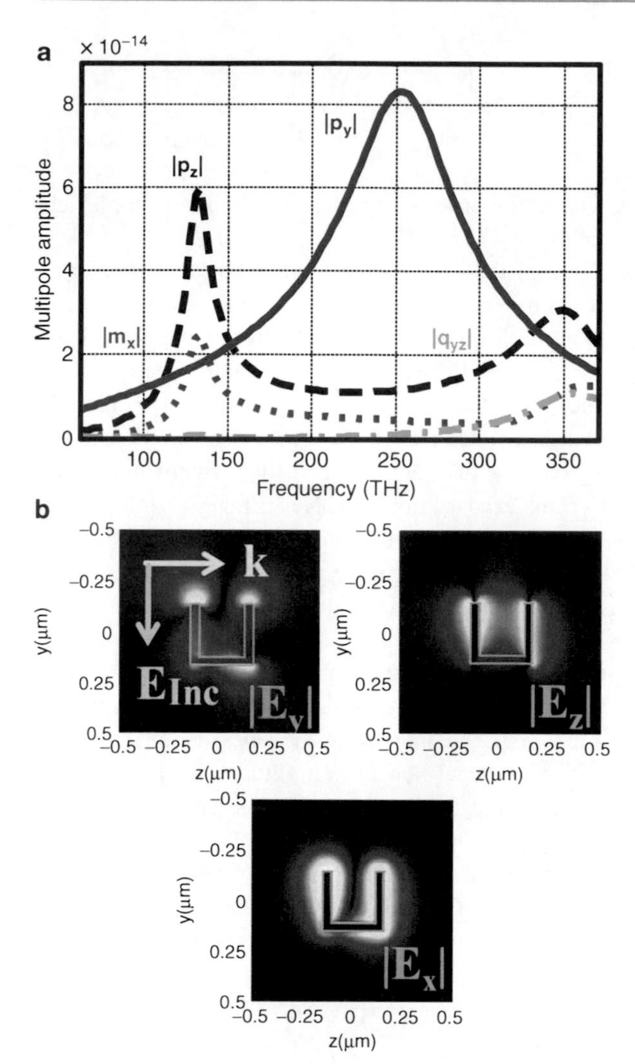

Theory of Optical Metamaterials, Fig. 3 (**a**) Amplitude of the dominant electromagnetic multipole contributions to the scattered field if a split ring is illuminated with a plane wave propagating in the plane of the split-ring resonator and the electric field is also polarized in the plane of the split-ring resonator, as shown in the top of (**b**). (**b**) Amplitude of the scattered electric field from an individual meta-atom in the y–z plane at 130 THz

introduction of an effective property either to take the geometrical mean of the intrinsic material properties (in the direction where the electric field is parallel to the slab, i.e., TE polarization) or the reciprocal of the geometrical mean of the reciprocal material properties (in the direction where the electric field is perpendicular to the slab, i.e., TM polarization). However, all these approaches represent approximations which have to be verified by suitable experiments. Thus, the application of particular mixing rules might be suitable but occasionally it might also be inappropriate. An exhaustive overview over such mixing rules can be found in the respective literature.

In the context of effective properties as discussed here where it is assumed that the MM is made of strongly scattering localized meta-atoms for which the scattered field can be expanded in terms of electromagnetic multipoles, the introduction of effective properties using the Clausius-Mossotti formula is most appropriate. In its simplest formulation it assumes that the MM consists of scattering entities characterized by both an electric and magnetic polarizability $\alpha_{E,M}$ which can be obtained by normalizing the electric and magnetic dipole moments to the dipole coefficients of the incident fields. By assuming that all meta-atoms are arranged in the same direction, the Cartesian polarizabilities can be used to retrieve the particular entries of the respective material tensors which are valid by assuming the constitutive relations of an anisotropic crystal. If the surrounding medium is vacuum the effective parameters are given by

$$\varepsilon_{ii,\mathrm{Eff}}(\omega) = \frac{3 + 2V\alpha_{ii,\mathrm{E}}(\omega)}{3 - V\alpha_{ii,\mathrm{E}}(\omega)} \quad \text{and}$$

$$\mu_{ii,\mathrm{Eff}}(\omega) = \frac{3 + 2V\alpha_{ii,\mathrm{M}}(\omega)}{3 - V\alpha_{ii,\mathrm{M}}(\omega)}.$$

Here V enters the equation as the volume filling fraction which is required to smaller, or even significantly smaller, than the percolation limit. For a completely statistical orientation of meta-atoms, an appropriate averaging procedure over the polarizability for all possible configurations has to be performed at first to extract the isotropic material properties emerging in this case. By this final step, effective material properties can eventually be assigned to a medium made from such strongly scattering entities.

However, the aspects last discussed here are not yet fully explored since the availability of suitable MMs made from strongly scattering elements is currently the subject of ongoing research. The approach discussed above has been extensively verified for MMs made from small metallic nanoparticles where all the assumptions imposed here are valid. But in any case the outlined procedure can be used to assign effective properties to such MM once sufficient information concerning their structure and scattering properties is available.

Conclusion

In conclusion, an introduction has been presented here that will facilitate the understanding of optical metamaterials. The entry is written from a perspective of a theoretician who may basically pursue two different aims. On the one hand he starts from a medium at hand that exhibits particular constitutive relations. Whereas for most optical materials available in nature, the complexity is limited by a dispersive tensorial permittivity, metamaterials may provide more complex relations which include dispersive tensorial permittivity and permeability as well as electromagnetic coupling. Then a genuine theoretical goal consists in the investigation of exciting implications for light propagation. Optics investigated under such assumption with the addition of a spatial dependent material is now frequently termed transformation optics and bears a great potential for innovative applications in optical systems.

On the other hand, at present it is feasible to fabricate almost arbitrarily nanostructured unit cells with an ever increasing complexity. The complex details tend to be important if light interacts resonantly with the unit cells, leading to unusual and extraordinary effects which allow for the control of light propagation in MMs formed from these unit cells. In this respect it is required that the transition from/to the environment and the propagation within the MM may be described by a simple set of wave parameters as the propagation vector and the impedance of the respective normal modes. Under certain conditions it might be possible to relate these parameters to effective material parameters the introduction of which facilitates appreciably the description of optical effect in MMs. In this case a very close connection to the first issue can be established. The complete understanding of the second issue is currently an important aim in theoretical optics.

The entry is actually devoted to shed some light on these issues and to provide a first guide of orientation to discuss them. Many more questions could likely be raised and many aspects necessarily had to be kept quite concise. Many details just mentioned can be found in the literature referred to here. The more detailed references that have been included hopefully will serve the valuable reader as a valuable source for further information beyond the basic ideas discussed in here.

Cross-References

► Active Plasmonic Devices
► Light Localization for Nano-Optical Devices
► Nanostructures for Photonics
► Optical Properties of Metal Nanoparticles
► Optical Techniques for Nanostructure Characterization

References

1. Engheta, N., Ziolkowski, R.W.: Electromagnetic Metamaterials: Physics and Engineering Explorations. Wiley-IEEE Press, Piscataway (2006)
2. Cai, W., Shalaev, S.: Optical Metamaterials: Fundamentals and Applications. Springer, New York (2009)
3. Marqués, R., Martín, F., Sorolla, M.: Metamaterials with Negative Parameters: Theory, Design and Microwave Applications. Wiley, Hoboken (2008)
4. Sakoda, K.: Optical Properties of Photonic Crystals. Springer, Berlin (2004)
5. Maier, S.A.: Plasmonics: Fundamentals and Applications. Springer, New York (2010)
6. Sihvola, A.: Metamaterials in electromagnetics. Metamater. **1**, 2–11 (2007)
7. Simovski, C.A.: On electromagnetic characterization and homogenization of nanostructured metamaterials. J. Opt. **13**, 013001 (2010)
8. Jackson, J.D.: Classical Electrodynamics. Wiley, New York (1998)
9. Born, M., Wolf, E.: Principles of Optics. Cambridge University Press, Cambridge (1999)
10. Menzel, C., Paul, T., Rockstuhl, C., Pertsch, T., Tretyakov, S., Lederer, F.: Validity of effective material parameters for optical fishnet metamaterials. Phys. Rev. B **81**, 035320 (2010)
11. Tretyakov, S.: Analytical Modeling in Applied Electromagnetics. Artech House, London (2003)
12. Serdyukov, A., Semchenko, I., Tretyakov, S., Sihvola, A.: Electromagnetics of Bi-anisotropic Materials–Theory and Applications. Gordon and Breach, Amsterdam (2001)
13. Agranovich, V.M., Yudson, V.I.: Boundary conditions in media with spatial dispersion. Opt. Commun. **7**, 121–124 (1973)
14. Taflove, A., Hagness, S.C.: Computational Electrodynamics: The Finite-Difference Time-Domain Method. Artech House, Norwood (2005)
15. Rockstuhl, C., Menzel, C., Paul, T., Pertsch, T., Lederer, F.: Light propagation in a fishnet metamaterial. Phys. Rev. B **78**, 155102 (2008)
16. Morits, D., Simovski, C.: Electromagnetic characterization of planar and bulk metamaterials: a theoretical study. Phys. Rev. B **82**, 165114 (2010)
17. Kwon, D.-H., Werner, D.H., Kildishev, A.V., Shalaev, V.M.: Material parameter retrieval procedure for general

bi-isotropic metamaterials and its application to optical chiral negative-index metamaterial design. Opt. Express. **16**, 11822–11829 (2008)
18. van de Hulst, H.C.: Light Scattering by Small Particles. Dover, New York (1981)
19. Raab, R.E., De Lange, O.L.: Multipole Theory in Electromagnetism. Oxford University Press, New York (2005)
20. Sihvola, A.E.: Electromagnetic Mixing Formulas and Applications. The Institution of Electrical Engineers, London (1999)

Thermal Actuators

Joseph J. Brown and Victor M. Bright
Department of Mechanical Engineering,
University of Colorado, Boulder, CO, USA

Synonyms

Electrothermomechanical actuators; Heatuators; Thermomechanical actuators

Definition

Thermal actuators are mechanical systems that use the thermally induced expansion and contraction of materials as a mechanism for the creation of motion. These devices are compliant structures, using elastic deformation and mechanical constraints, that frequently are designed to amplify the motion generated by thermal expansion or contraction. Temperature changes that result in thermal actuation are most commonly provided by environmental changes or by Joule heating from electrical current flow. In the context of nanotechnology, thermal actuators refer to microscale and nanoscale devices used to mechanically interact with nanoscale structures, with motion generated by the thermally induced expansion and contraction of materials.

Key Principles, Concepts, and Phenomena

Thermal actuators are useful for applications where low voltage, small footprint, and high force output are desirable device characteristics. In the context of microscale and nanoscale engineering, thermal actuators can be used as microelectromechanical systems (MEMS) capable of translating electrical signals into controllable displacements with resolution smaller than 1 μm. Displacement for thermal actuators can be designed as small as several nanometers [1]. For the discussions below, it should be noted that simple calculations discussed in this entry have been provided in order to enable quick initial evaluation of possible actuator designs. For more accurate understanding and operation of these actuators, multiphysics simulations and calculations, and experimental characterization of constructed devices are recommended.

Surface Contact and Connections in MEMS/NEMS Design

At size scales of hundreds of microns or smaller, contacting surfaces can have significant friction and stiction capable of inhibiting device motion. Stiction is the phenomenon where static surfaces contact each other, preventing motion and separation of the surfaces. Stiction may be caused by electrostatic and van der Waals forces. This can cause devices to pull into contact with underlying surfaces and substrates. In contact, these surfaces experience friction, the resistance of contacted surfaces to sliding motion, and which can be strong enough to impede all motion. In pinned joints, contact surfaces are smaller, but friction can lead to significant wear, resulting in eventual failure of a device. For design of solid microscale and nanoscale actuators, contact surfaces are minimized to only those surfaces where contact is absolutely necessary for device function. For maximum reliability, actuators themselves are typically designed as suspended structures with clamped end conditions. The fabrication of pinned beam connections is typically imprecise using common microfabrication techniques. Additionally, they present regions subject to wear and friction which may decrease the reliability of actuator motion.

Compliant Structures

Compliant structures are arrangements of elastic bodies with constrained joints that generate or transmit motion. Because pinned joints are not optimal for microscale and most nanoscale thermal actuator designs, these designs employ elastic deformation to allow motion. As such, thermal actuators are a subset of compliant mechanisms used for the generation of motion. Repeatable operation of the thermal actuator requires operation within the elastic limits of the actuator material. Excessive heating and mechanical strain can ultimately lead to failure of the thermal actuator mechanism. In particular, it is possible to "burn out" the thermal actuator due to excessive input power, which leads to creep, softening, and eventual melting of the actuator material.

Thermal Expansion

Thermal expansion and contraction of materials is the fundamental mechanism by which motion is generated in thermal actuators. Thermal expansion in solids originates from the nature of atomic or molecular bonding within solids. A rise in temperature increases the bond lengths between atoms or molecules in a solid due to increased energy of the populations of phonons and electrons within the solid.

Thermal expansion can be treated as a continuum behavior in classical solid mechanics. The local three-dimensional strain tensor ε_{ij} is given by Eq. 1 for isotropic materials, where E is the Young's Modulus, v is the Poisson's ratio, α is the linear thermal expansion coefficient, σ_{ij} is the local stress tensor, σ_{kk} is sum of the three normal stresses ($\sigma_{kk} = \sigma_{11} + \sigma_{22} + \sigma_{33}$), the term $(T-T_0)$ defines the temperature rise above a reference temperature T_0, and δ_{ij} is the Kronecker delta [2]. In quasi steady-state behavior, the actuator establishes a thermal equilibrium with its environment, which develops a constant temperature profile throughout the actuator. The corresponding thermal strain distribution in the actuator drives motion to a mechanical equilibrium.

$$\varepsilon_{ij} = \frac{1+v}{E}\sigma_{ij} - \frac{v}{E}\sigma_{kk}\delta_{ij} + \alpha(T-T_0)\delta_{ij} \tag{1}$$

Joule Heating and Power Dissipation

The flow of electricity through an electrically resistive material dissipates power according to Ohm's law. Microscopically, the local density of power dissipation $\dot{q}$ is given by the local resistivity ρ, current density $\vec{J}$, and electric field $\vec{E}$ according to Eq. 2.

$$\dot{q} = \vec{J}\cdot\vec{E} = \frac{1}{\rho}\left|\vec{E}\right|^2 = \rho\left|\vec{J}\right|^2 \tag{2}$$

Equation 2 is valid for DC currents and the instantaneous power dissipation from time-varying currents and electric fields. For average power dissipation from time-varying fields, the instantaneous power dissipation is averaged in time t', according to Eq. 3a, where t'_0 may, for instance, be defined as the length of time for one period of oscillation. In practice, the result is Eq. 3b for harmonic, time-dependent electric fields and currents [3]. Note that although most thermal actuators rely on heating due to controlled electrical currents, other heating mechanisms exist such as absorbance of electromagnetic waves which may create similar thermal responses.

$$\dot{q}_{ave} = \frac{1}{t'_0}\int_0^{t'_0} \dot{q}dt' = \frac{1}{t'_0}\int_0^{t'_0} \frac{1}{\rho}\left|\vec{E}\right|^2 dt' = \frac{1}{t'_0}\int_0^{t'_0} \rho\left|\vec{J}\right|^2 dt' \tag{3a}$$

$$\dot{q}_{\omega} = \frac{1}{2\rho}\left|\vec{E}\right|^2 = \frac{1}{2}\rho\left|\vec{J}\right|^2 \tag{3b}$$

Integrating the local power density $\dot{q}$ through the volume of the actuator gives the lumped form of Ohm's law, Eq. 4, where total power dissipation P is readily calculated from measurement of the current flow I, voltage drop V, and total electrical resistance R in the actuator.

$$P = IV = \frac{V^2}{R} = I^2R \tag{4}$$

Frequently the thermal actuator is constructed from materials whose resistivity changes with temperature and strain (i.e., due to piezoresistive effects), such as silicon. In this case, the local resistivity ρ cannot be assumed to be constant for all output displacements and levels of mechanical loading, nor can the total device resistance R. For this reason, power dissipation in thermal actuators should be derived from measurements of current and voltage without assuming a constant resistivity or resistance. Control based on current or voltage alone, rather than total power dissipation, should be validated experimentally before reliance on these simpler mechanisms to control the thermal actuator mechanical output.

Heat Transfer: Fundamental Concepts

The temperature rise in the actuator is determined by the heat generation and heat transfer throughout the thermal actuator. This behavior is governed by Eq. 5, the heat equation for stationary materials, and Eq. 6, the Fourier heat law [4]. The heat equation is properly written with the full time derivative, $\frac{DT}{Dt'} = \frac{\partial T}{\partial t'} + \vec{v} \cdot \nabla T$, where $\vec{v}$ is the local velocity, and t' is time. From a macroscopic view, there is no convective mass transport within the solid structures, so the velocity components $\vec{v}$ in the full derivative are reduced to zero, and the full derivative reduces to the partial time derivative.

Understanding of the temperature profile of the thermal actuator is developed by solving Eqs. 5 and 6 for the boundary conditions appropriate to specific thermal actuator designs. The local heat generation $\dot{q}$ was defined in Eq. 2 for Joule heating. The average temperature rise ($\Delta T = T - T_a$) in a beam element can be found using Eq. 7.

$$\eta c \frac{\partial T}{\partial t'} = k_\theta \nabla^2 T + \dot{q} \tag{5}$$

$$\vec{q} = -k_\theta \nabla T \tag{6}$$

$$\Delta T = \frac{1}{L_i} \int_0^{L_i} T dx' - T_a \tag{7}$$

T temperature (K)
T_a initial temperature (K)
ΔT average temperature change (K)
k_θ thermal conductivity (W/m-K)
$\vec{q}$ heat flow vector (W/m^2)
η density (kg/m^3)
c heat capacity (J/kg-K)
x' local distance coordinate along the length of the beam (m)
L_i length of beam element over which the average is calculated (m)

Heat Transfer: Environment Effects on Actuation

By modifying the boundary conditions for heat transfer from the thermal actuator, the ambient environment surrounding a thermal actuator affects its output motion. Heat transfer mechanisms modify the temperature profile and corresponding thermal strain of the thermal actuator. In vacuum, heat transfer from the thermal actuator, for MEMS dimensions, occurs primarily through conduction through the actuator clamps. Radiative heat transfer is a secondary mechanism for heat loss from the thermal actuator, with energy transferred from the actuator to its substrate and surrounding environment. For initial design considerations, analysis of conductive heat transfer in the solid actuator material in vacuum allows estimation of the order of magnitude of actuator power requirements and temperatures, and is one step to understanding of the thermal behavior of the actuator.

In non-vacuum environments, heat transfer to a surrounding fluid can significantly alter the temperature profile of the thermal actuator. Because of the small dimensions of most thermal actuators, conduction usually dominates over convection in air, and published models generally only account for conduction through air [5–7]. Operation of thermal actuators in environments such as air may require significantly more power than used in a vacuum environment in order to obtain the same output displacement [5]. As an example, 1.4 mA applied to one design in vacuum may lead to the same temperature rise as 5.0 mA applied to the same device in air [5]. This result implies that more than 12 times the power consumption may be needed in air as in vacuum in order to achieve similar output from some thermal actuator designs.

A lower bound on the time constant for response of the thermal actuating system in vacuum can be found by considering a lumped model of the heat capacity and the thermal resistance of the beam design, Eq. 8. The time constant τ_θ is calculated from the lumped heat capacity C and thermal resistance R_θ of the actuator beam that experiences the maximum current density. From this time constant calculation, it may be expected that the thermal actuator will not move faster than a frequency of f_θ.

$$R_\theta = \frac{L_i}{k_\theta A_i} \tag{8a}$$

$$C = c\eta L_i A_i \tag{8b}$$

$$\tau_\theta = R_\theta C \tag{8c}$$

$$f_\theta = \frac{1}{2\pi\tau_\theta} = \frac{1}{2\pi R_\theta C} \tag{8d}$$

A_i beam cross-section area (m^2)
f_θ thermal bandwidth maximum frequency (Hz)
τ_θ thermal time constant (s)
C lumped heat capacity (J/K)
R_θ lumped thermal resistance of a beam (K/W)

Statically Indeterminate Beams

The output motion of thermal actuators derives from statically indeterminate beams subjected to mechanical constraints. Design of these constraints can determine whether motion will occur in-plane or out-of-plane. A number of texts are available that further discuss the concepts of indeterminate structures [8]. Several detailed models have been developed for the output motion of constrained beams within thermal actuators [9, 10]. Some of these analyses will be examined in the examples below.

Displacement Output and Force Output

Experimental measurement of an output displacement curve recorded versus input power, voltage, or current can provide a fit curve that allows prediction of the freely moving thermal actuator displacement d_0 without requiring detailed analysis of the thermomechanical system. When the thermal actuator is used to provide an output force, this force may be estimated using the spring constant k calculated for the beams of the thermal actuator. The output force F provides a perturbation δd to the free-moving displacement curve, giving actual displacement d. Force is estimated (Eq. 9) according to Hooke's law applied to this perturbation from the free-moving displacement d_0 [11]. Simulations indicate that for some device mechanisms, the output force may differ significantly from these predicted values in cases when displacement is constrained to nearly zero [12]. Experimental measurement of this effect and calibration of the output forces of thermal actuators remains an ongoing active area of research.

$$F = k\,\delta d = k(d_0 - d) \tag{9}$$

The spring constant k_i of many individual thermal actuator beams can be estimated (Eq. 10) according to the solid mechanics of a beam with one end clamped and one end "guided" (prevented from rotation but allowed to displace in one direction), where L_i is the beam length and I_z is its moment of inertia. For beams with width w and thickness t, I_z is given by Eq. 11, and k_i is estimated with Eq. 12. The total spring constant is found (Eq. 13) by multiplying the number of beams N by the spring constant k_i for each beam.

$$k_i = \frac{12EI_z}{L_i^3} \tag{10}$$

$$I_z = \frac{tw^3}{12} \tag{11}$$

$$k_i = \frac{Etw^3}{L_i^3} \tag{12}$$

$$k = Nk_i \tag{13}$$

Using k and the mass m calculated based on the density η and volume of the whole actuator assembly (for example, $m = N\eta L_i wt$), a lower bound to the natural frequency f_0 can be found (Eq. 14).

$$f_0 = \frac{1}{2\pi}\sqrt{\frac{k}{m}} \tag{14}$$

For operation in laboratory environments, it is important that the natural frequency not be too low; otherwise, the actuator may be significantly perturbed by environmentally induced vibrations. A good rule of thumb is $f_0 > 20$ kHz, which implies attenuation of most sounds and other laboratory disturbances.

Applications

Thermal actuators are one of the key mechanisms for effecting motion in microscale and nanoscale solid structures. Other common microelectromechanical system (MEMS) mechanisms typically use electrostatic forces, hydrostatic forces, magnetic forces, or piezoactuation [6]. The advantages of thermal actuators over these other mechanisms lie in low operating voltage, high output force, and compact footprint. Disadvantages include comparatively high power consumption, comparatively slow response time, and dependence of actuation output on ambient environment.

Because thermal actuators dissipate heat, thermal management of the environment surrounding the actuator may be an important design consideration. The substrate supporting a thermal actuator can experience temperature changes due to heating from the actuator. In vacuum, the substrate to which the actuator is anchored becomes the primary mechanism for dissipation of heat from the actuator. Material selection for the substrate should allow for temperature changes due to heating, and adequate thermal engineering to prevent overheating. Furthermore, placement of temperature-sensitive materials near the actuator should be carefully planned.

Thermal actuators also face high temperature and low temperature limits to operation. At high temperatures, material creep and failure may occur. At low temperatures such as cryogenic environments, heating of the environment due to the actuator may disrupt the environment. Furthermore, some materials such as semiconductors may be insufficiently conductive to allow much current flow. Other materials such as many metals may develop superconductivity at cryogenic temperatures, which leads to insufficient power dissipation to create significant temperature changes for thermal expansion and actuation.

In the examples discussed below, the V-type actuator design has an advantage over the hot arm, U-type design in that all beams contribute to actuation motion in the V-type design. In the U-type design, motion is generated by the hot arms, but resisted by the cold arms. As a result, a V-type design requires a smaller footprint to achieve the same output as a given U-type design. Furthermore, the inherently linear motion output of the V-type designs makes them easier to incorporate in device designs.

History

The hot arm, U-type actuators were originally demonstrated in the early 1990s and quickly became a staple for generation of mechanical motion in solid MEMS structures. In the late 1990s, V-type actuators were initially demonstrated and this design has been subject to much further development in the subsequent years. In particular, V-type actuators have been explored for driving compliant mechanisms, microgripper devices, and micropositioning devices. Bimorph thermal actuators (Example 3, below) have been in use for submillimeter scale applications since the late 1980s. Extensive review of the history of microfabricated thermal actuators may be found in Ref. [6].

Uses for thermal actuators have included optical and electrical switches in addition to micropositioning, microgripping, microscale mechanical testers, and microhandling and nanohandling devices [6, 11, 12]. Thermal actuators may be seen in many reports as the drivers of mechanical motion in many geared and pivoting MEMS systems, particularly in the time period of 1994–2002. Such systems are fundamentally limited by frictional wear, and consequently, they have not been the subject of much development in more recent years. Newer applications focus more on structures for manipulating and interacting with objects at the microscale and nanoscale.

Examples

In application, three main mechanisms have been implemented as the most common thermal actuators: V-type, hot arm/U-type, and bimorph thermal actuators. Other types of thermal actuators have been created [1], but most work with thermal actuators has focused on mechanisms discussed in the examples presented here. Calculations are provided with these examples in order to walk through the basic design of common thermal actuators. These calculations are intended to help the designer to understand the range of actuator performance and the effect of device geometry on the output behavior. Final design should use multiphysics simulations and experimental characterization of fabricated devices in order to understand the real performance of any particular design.

In order to facilitate design calculations, Table 1 provides some published values on the properties of polycrystalline silicon (or "polysilicon"), a common material for thermal actuators. Because it is a semiconductor, polycrystalline silicon has electrical resistivity that varies with dopant concentration, temperature, and applied strain. The values reported in this table for resistivity and other properties reflect the range of values that have been used in research literature as inputs to simulations of polysilicon layers generated by the PolyMUMPS foundry service (MEMSCAP Inc., NC, USA). In general, many of the properties of polycrystalline silicon are process-dependent. Therefore, designs should ultimately be based upon measurements of the properties of material created according to a specific, defined process, and validation of device models against measured experimental output.

Example 1: V-Type Actuators

The first common thermal actuator mechanism has been called, variously: V-type, chevron, bent-beam, or symmetric thermal actuators. In this mechanism (Fig. 1), suspended, opposing beams are connected to a suspended shuttle structure. The ends of the actuator

Thermal Actuators, Table 1 Some material properties of PolyMUMPS polycrystalline silicon, approximated as an isotropic material [5, 9, 11, 13]

Property	Symbol	Value	Units
Electrical resistivity			
Constant:	ρ	1.7×10^{-5}–4×10^{-4}	Ω-m
Temperature dependent:	$\rho = \rho_0[1 + \xi(T - T_0)]$		
		$\rho_0 = 3.4\times10^{-5}$	Ω-m
		$\xi = 1.25\times10^{-3}$	K^{-1}
		$T_0 = 293$	K
Thermal conductivity			
Constant value, polycrystalline Si	k_θ	29–34	W/m-K
Constant value, single crystal Si	k_θ	150	W/m-K
Density	η	2.33×10^3	kg/m^3
Heat capacity			
Constant value	c	705	J/kg-K
Temperature dependent:	$c = 1.976\times10^{-6}T^3 - 3.767\times10^{-3}T^2 + 2.623T + 215.0$		J/kg-K
Young's modulus	E	162±14	GPa
Poisson's ratio	ν	0.29	
Linear coefficient of thermal expansion			
Constant value (at 293 K)	α	2.7×10^{-6}	K^{-1}
Temperature dependent:	$\alpha = \left[3.725\left(1 - e^{-5.88\times10^{-3}(T-125)}\right) + 5.548\times10^{-4}T\right]\times10^{-6}$		K^{-1}

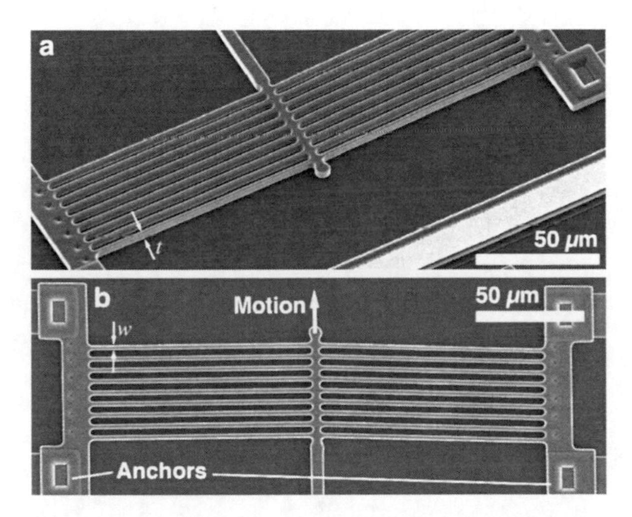

Thermal Actuators, Fig. 1 Examples of V-type, "chevron" thermal actuators. (**a**) Tilted, oblique view in scanning electron microscope (SEM) image. (**b**) SEM image of the top view of a fabricated V-type thermal actuator. Beams are angled 1° off perpendicular to the direction of motion of the central shuttle (Image credit: J.J. Brown and V.M. Bright)

beams are anchored to a substrate that provides electrical current through the beams. Power dissipation leads to expansion of the beams, which experience a buckling motion. When the beams are placed symmetrically, linear motion is achieved in the shuttle. Engineering of the beams guides the buckling motion that they experience. The aspect ratio of the beam is defined asymmetrically in order to confine beam bending motion to one plane, typically chosen as motion in the plane of fabrication. If the beam has a cross section with thickness t, defined by the thickness of the film used for fabrication of the beam, and a width w, defined by patterning perpendicular to the film, in-plane motion is achieved when $t > w$. In this case, the beam has a moment of inertia that is smaller for in-plane motion than for out-of-plane motion, so in-plane motion is more probable as buckling is induced. Another technique used to guide the buckling is to angle the beams at their attachment point to the shuttle. Typically the beams are angled a few degrees off perpendicular to the shuttle, and the device is built with mirror symmetry across the shuttle in order to cancel bending moments that would cause deviations from uniaxial motion. With elongation of the actuator beams, buckling and shuttle motion is achieved in a predictable direction. The choice of angle is made dependent on whether displacement or output force is more important. Smaller angles from perpendicular provide more output displacement for a given actuator power; larger angles provide more output force.

Electrical Calculation

The goals of electrical calculations associated with thermal actuator design are the following:

1. Find the total resistance R of the actuator system in order to allow design of an appropriate power supply.
2. Understand how R relates to the actuator geometry.
3. Allow the use of I^2R to determine the total power generation and by extension the thermal power that must be managed by the anchors and other structures supporting the thermal actuator.

Regarding the third goal, the packaging environment surrounding the actuator must be able to remove the steady-state power generated by the actuator in order to establish stable operation.

The fundamental unit of a V-type thermal actuator is a symmetric beam pair as seen in Fig. 2. For simplicity of this example calculation, power consumption in the central shuttle will be omitted from the calculations. Because the actuator is formed from symmetric beams, the calculation only necessitates examination of the power consumption in one beam. The geometry of the beam is defined as in Fig. 3. The beam has an electrical resistivity ρ, which is a temperature-dependent property. For the purposes of this initial calculation, ρ will be held constant, and the material of the beam will be defined as polycrystalline silicon, using the property values defined in Table 1. The beam resistance is calculated according to Eqs. 15, using example dimensions for 1 beam: $L_i = 100\ \mu\text{m}, t_i = 3.5\ \mu\text{m}, w_i = 2\ \mu\text{m}, A_i = w_i t_i = 7\ \mu\text{m}^2$.

$$R_i = \frac{\rho L_i}{A_i} \tag{15a}$$

$$R_i = \frac{(3.4 \times 10^{-5}\ \Omega\text{m})(100\ \mu\text{m})}{(7\ \mu\text{m}^2)} = 490\ \Omega \tag{15b}$$

In Fig. 4, the electrical model of the pair of beams in the actuator indicates connection in series, so the resistance of the pair, $R_p = 2R_i$. For an actuator built from N_p beam pairs, the beam pair resistance adds in parallel, and the total resistance R of the actuator may be found according to Eq. 16. For $N_p = 10$ and the result from Eq. 15b, R=98 Ω (Fig. 5).

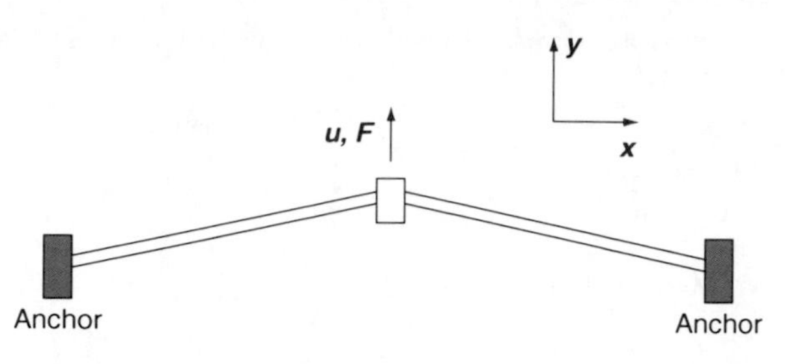

Thermal Actuators, Fig. 2 Fundamental unit of the V-type actuator, a pair of angled beams. Motion u and output force F occur in direction y. A loading force $-F$ may oppose motion in the y direction

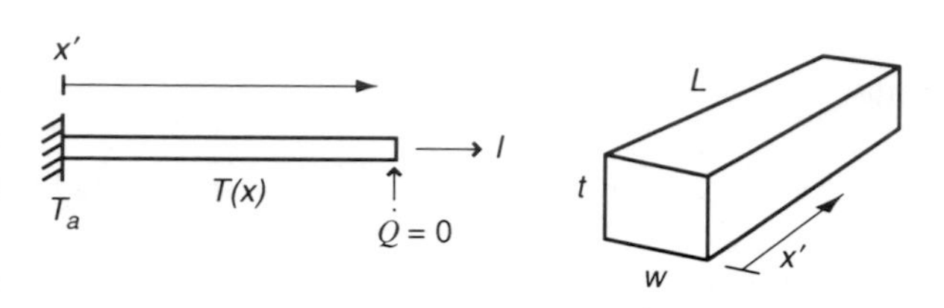

Thermal Actuators, Fig. 3 Basic geometry of one of the V-type actuator beams, for electrical and thermal calculations. The distance coordinate along the length of the angled beam is defined by x'

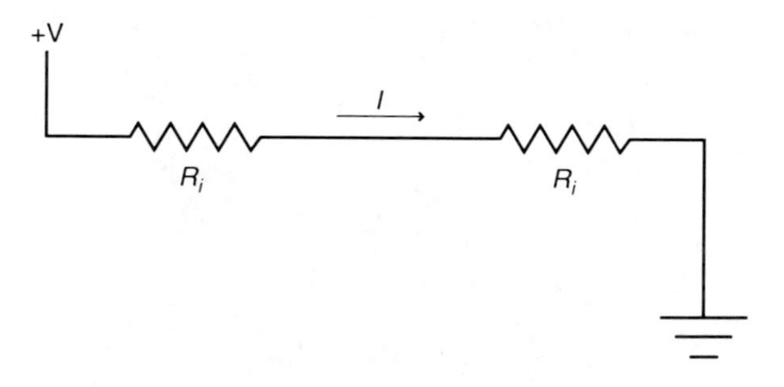

Thermal Actuators, Fig. 4 Electrical circuit for the pair of beams in the thermal actuator. Resistance in the central shuttle is omitted for simplicity

$$R = \frac{R_p}{N_p} = \frac{2R_i}{N_p} \tag{16}$$

Thermal Calculation

The goals of thermal calculations for the V-type actuator are the following:

1. Find an average temperature rise ΔT at a given input current in order to estimate the thermal strain in subsequent mechanical calculations.

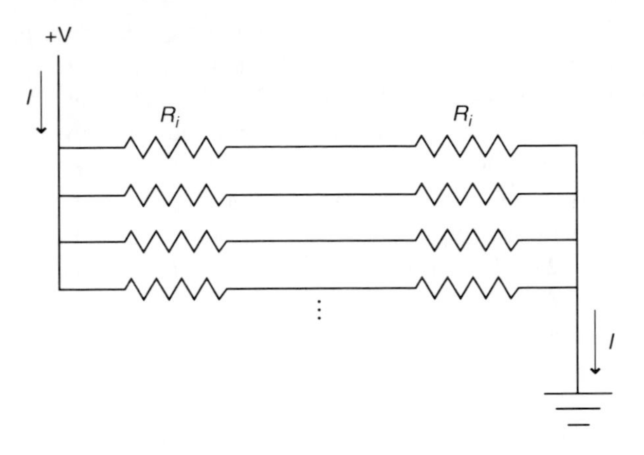

Thermal Actuators, Fig. 5 Electrical circuit diagram for an assembly of beam pairs such as that seen in Fig. 1. The resistances of each beam pair add in parallel. The central shuttle makes electrical connections in the center of this diagram, but it is omitted from the diagram because there is assumed to be an identical voltage drop in each beam. Therefore, there should be no current flow from one beam pair to another through the central shuttle

2. Find maximum temperature as a function of input current, in order to understand the performance limits to the actuator.
3. Find boundaries on the thermal time constant in order to understand the limit to the actuator response time.

The calculation of the temperature profile in the thermal actuator can focus on the profile within a single beam, similar to the electrical calculation. As discussed earlier, for initial design purposes, it is necessary only to examine thermal conduction. Other heat transfer effects should be considered for a complete understanding of the actuator thermal behavior. Simulations based on the coupled, temperature-dependent electrical, thermal, and mechanical behavior show that V-type thermal actuators may develop nonlinear output force behavior when subjected to displacement constraints [12]. However, for basic understanding, conductive heat transfer generally suffices. As with the electrical calculation, thermal conductivity k_θ and heat capacitance c are temperature dependent, but for initial design, they are considered to be constant.

The calculation can approximate the beam as a one-dimensional element with a temperature varying along position x. The boundary conditions are defined in two locations: The anchored end maintains a constant temperature due to significant heat transfer from the anchor to the supporting substrate. The suspended end of the beam has no heat flow, due to the symmetry of the suspended structure. Consider the diagram in Fig. 3, where T_a is the temperature at point **a**, the beam has a temperature profile that varies along the beam with position x', heat leaves the beam only through point **a**, and a total current I flows through the beam. This diagram can be simplified into a one-dimensional model (Fig. 6) with the following boundary conditions, where $\dot{Q}$ is one-dimensional heat flow with units of watts (W):

$$T(0) = T_a \quad \dot{Q}(L_i) = 0$$

In this case, all variables in the thermal model are considered constant across the beam cross section at any given x' position. Using Eqs. 17 to set up the one-dimensional model and assuming steady-state operation, the heat equation (Eq. 5) can be rewritten as Eqs. 18a and 18b, given for cases without and with air conduction, respectively. In Eq. 18a, constant thermal conductivity is also assumed. For the model with conduction through air to a nearby substrate, S is a shape factor, z_g is the distance between the suspended device and the substrate, $k_{\theta,air}$ is the thermal conductivity of air, and Z is a thermal insulance term defined by the layers through which heat flows (with thickness z_i, thermal conductivity $k_{\theta,i}$, and total number p) from the suspended beam to a substrate held at a constant temperature [5–7, 13].

$$k_\theta A_i \frac{\partial T}{\partial x'} = -\dot{Q} \tag{17a}$$

$$\frac{\partial \dot{Q}}{\partial x'} = \dot{q} A_i \tag{17b}$$

$$\dot{q} = \rho \left|\vec{J}\right|^2 = \rho \frac{I^2}{A_i^2} \tag{17c}$$

$$S = \frac{t}{w}\left(\frac{2z_g}{t} + 1\right) + 1 \tag{17d}$$

$$Z = \sum_{j=1}^{p} \frac{z_i}{k_{\theta,i}} = \frac{z_g}{k_{\theta,air}} \tag{17e}$$

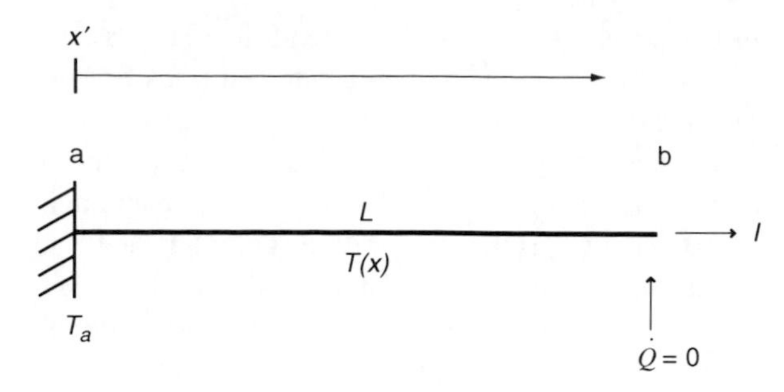

Thermal Actuators, Fig. 6 One-dimensional model of the symmetric thermal system for the V-type actuator

$$0 = k_\theta A_i \frac{\partial^2 T}{\partial x'^2} + \dot{q} A_i = k_\theta A_i \frac{\partial^2 T}{\partial x'^2} + \rho \frac{I^2}{A_i} \tag{18a}$$

$$0 = \frac{\partial}{\partial x'}\left(k_\theta \frac{\partial T}{\partial x'}\right) + \rho \frac{I^2}{A_i^2} - \frac{S(T - T_a)}{tZ} \tag{18b}$$

The temperature profile in the beam in vacuum, Eq. 19, is found by integrating Eq. 18a twice and applying the boundary conditions to determine the constants of integration. A more precise analytical solution to Eq. 18a may be found when k_θ and ρ are allowed to vary with temperature. Ref [7] provides a solution for Eq. 18b for constant k_θ and temperature-dependent ρ. Elimination of air conduction terms in that solution provides a solution for Eq. 18a with temperature-dependent ρ. The maximum temperature $T_{\max}$ is found at the location where the first derivative of temperature is zero, which is where heat flow $\dot{Q}$ is zero. In this case, heat flow is zero at $x' = L_i$ and $T_{\max}$ is given by Eq. 20. The average temperature rise in the beam can be calculated by the application of Eq. 19 to Eq. 7, yielding Eq. 21.

$$T = T_a + \rho \frac{I^2}{A_i^2 k_\theta}\left(L_i x' - \frac{x'^2}{2}\right) \tag{19}$$

$$T_{\max} = T_a + \rho \frac{I^2 L_i^2}{2A_i^2 k_\theta} \tag{20}$$

$$\Delta T = \frac{\rho I^2 L_i^2}{3A_i^2 k_\theta} \tag{21}$$

For an upper bound on the thermal time constant, the lumped thermal resistance R_θ can be calculated using the beam length L_i and assuming that the substrate is an infinitely large thermal reservoir that can adequately remove the generated thermal power P from the actuator without undergoing significant changes in temperature, thereby maintaining T_a at the beam anchor point. Equations 22a and 22b show the resulting time constant and the minimum bandwidth for operation in vacuum, derived from Eqs. 8, polysilicon properties from Table 1, and example dimensions for 1 beam: $L_i = 100\ \mu m$, $t_i = 3.5\ \mu m$, $w_i = 2\ \mu m$, $A_i = w_i t_i = 7\ \mu m^2$. In air, the overall thermal resistance is reduced, thereby decreasing τ_θ and increasing the bandwidth.

$$\tau_\theta = R_\theta C = \left(\frac{L_i}{k_\theta A_i}\right)(c\eta L_i A_i) = \frac{c\eta L_i^2}{k_\theta}$$
$$= \frac{(705)(2330)(10^{-4})^2}{30} = 5.5 \times 10^{-4}\ \text{s} \tag{22a}$$

$$f_\theta = \frac{1}{2\pi\tau_\theta} = 290\ \text{Hz} \tag{22b}$$

Analysis of Mechanical Behavior

For basic design of a thermal actuator, modeling of the mechanical behavior can help the designer to understand the maximum force and range of displacements that are available from a given actuator design. Mechanical analysis can be performed with the following goals:

1. Find an expression for the maximum force output available from the actuator.
2. Find an expression for the maximum displacement from the actuator.
3. Develop an understanding of how device geometry impacts the force and displacement.

Several contrasting mechanical analyses of V-type actuators have been published [9, 10, 14, 15]. The analysis presented here follows that of Ref. [9]. As with the electrical and thermal analyses, the symmetric device geometry allows simplification of the mechanical analysis to an examination of only one beam. Begin by simplifying the actuator design in Fig. 2 to the free body diagram seen in Fig. 7a, and consider the coordinate system presented in Fig. 7b. At point **a**, the beam is anchored and restricted from displacement or rotation. At point **b**, the end of the beam is restricted from rotation or motion in the x direction, but it can move in the y direction. A reaction force $F_{x,R}$ opposes motion in the x direction. Figure 7b depicts two Cartesian coordinate systems.

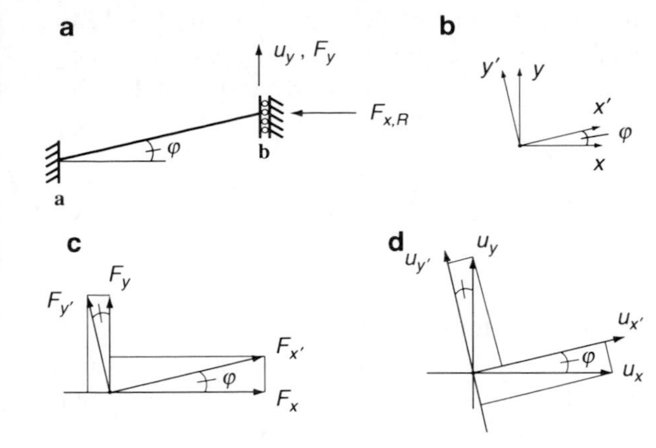

Thermal Actuators, Fig. 7 Diagrams for mechanical analysis of V-type actuators. (**a**) Free body diagram. (**b**) Coordinate systems for output motion in xy and motion in the reference frame of the beam $x'y'$. (**c**) Bending and extensional forces in the beam can be projected into forces in the xy coordinate system. (**d**) Motion in the xy coordinates is projected into motion in the $x'y'$ coordinates

The xy system rests in the frame of the substrate and actuator support. The actuator output motion u_y is defined in the y direction of the xy coordinate system. The rotated $x'y'$ coordinate system is a local system oriented to the reference frame of the angled actuator beam. Consequentially, the $x'y'$ system is rotated at an angle of φ from the xy system.

An external force F_i may be applied to point **b** in the y direction, affecting the u_y displacement. In the reference frame of the beam, motion at point **b** can be decomposed to motion $u_{x'}$ in the x' direction and motion $u_{y'}$ in the y' direction (Fig. 7d). Motion in the x' direction is pure elongation, described by Eq. 23a to include thermal strain. Motion in the y' direction is pure bending, described by Eq. 23b for the case where rotation at the beam ends is not allowed.

$$F_{x'} = EA\left(\frac{u_{x'}}{L} - \alpha\Delta T\right) = \frac{EA}{L}u_{x'} - \alpha\Delta TEA \quad (23a)$$

$$F_{y'} = \frac{12EI_z}{L^3}u_{y'} \quad (23b)$$

The forces $F_{x'}$ and $F_{y'}$ can be projected (Fig. 7c) to forces in x and y according to Eqs. 24 for elongation and Eqs. 25 for bending.

$$F_{x,e} = F_{x'}\cos\varphi \quad (24a)$$

$$F_{y,e} = F_{x'}\sin\varphi \quad (24b)$$

$$F_{x,b} = -F_{y'}\sin\varphi \quad (25a)$$

$$F_{y,b} = F_{y'}\cos\varphi \quad (25b)$$

Summing the F_x forces and F_y forces yields Eqs. 26.

$$\begin{aligned}-F_{x,R} + F_{x,b} + F_{x,e} = &- F_{x,R} - F_{y'}\sin\varphi \\ &+ F_{x'}\cos\varphi = 0\end{aligned} \quad (26a)$$

$$\begin{aligned}&- F_{x,R} - \sin\varphi\frac{12EI_z}{L^3}u_{y'} \\ &+ \left(\frac{EA}{L}u_{x'} - \alpha\Delta TEA\right)\cos\varphi = 0\end{aligned} \quad (26b)$$

$$\begin{aligned}F_i + F_{y,b} + F_{y,e} = &F_i + F_{y'}\cos\varphi \\ &+ F_{x'}\sin\varphi = 0\end{aligned} \quad (26c)$$

$$F_i + \cos\varphi\frac{12EI_z}{L^3}u_{y'} + \left(\frac{EA}{L}u_{x'} - \alpha\Delta TEA\right)\sin\varphi = 0 \quad (26d)$$

The displacements $u_{x'}$ and $u_{y'}$ can be written in terms of u_x and u_y by projecting u_x and u_y onto the rotated coordinate system (Fig. 7d).

$$u_{x'} = u_y\sin\varphi - u_x\cos\varphi \quad (27a)$$

$$u_{y'} = u_y\cos\varphi - u_x\sin\varphi \quad (27b)$$

However, u_x=0 at point **b**, so Eqs. 27 simplify to Eqs. 28.

$$u_{x'} = u_y\sin\varphi \quad (28a)$$

$$u_{y'} = u_y\cos\varphi \quad (28b)$$

With substitution of Eqs. 28 into Eq. 26d and some algebraic manipulation, a very useful final expression Eq. 29 can be derived that relates the average actuator temperature rise ΔT, the actuator output displacement u_y, and the force F_i that may be produced by the actuator and which may constrain the displacement. Substitution of Eqs. 28 into Eq. 26b would yield an expression for the x direction reaction force, if desired. Ref. [9] appeared to show reasonable agreement between this model and measured unconstrained experimental displacement output. However, an exploration of force and displacement behavior performed in Ref. [12] demonstrated that V-type actuators undergo asymmetric buckling which complicates the relationship between force and displacement at different power or temperature levels. Therefore, Eq. 29 may be useful for basic design, but it underrepresents the interaction between F_i and u_y when u_y is constrained near zero or when large force is applied.

$$u_y = \frac{\alpha\Delta TEA\sin\varphi}{\left(\frac{EA}{L}\sin^2\varphi + \frac{12EI_z}{L^3}\cos^2\varphi\right)} + \frac{-F_i}{\left(\frac{EA}{L}\sin^2\varphi + \frac{12EI_z}{L^3}\cos^2\varphi\right)} \tag{29}$$

If the force constraint F_i is left at zero, Eq. 29 reduces to an expression for the free-moving actuator displacement. If instead the displacement u_y is constrained to zero, the magnitude of the maximum force generated by the beam (Eq. 30) is seen to be the result of plane strain in the beams. A detailed thermomechanical derivation of stresses due to plane strain and thermal expansion applied to Eq. 1 may be found in Ref. [2]. The result is that, under the normal strain condition $\varepsilon_{33} = 0$ and no external traction or internal body forces, no stresses are present except for the compressive normal stress $\sigma_{33} = -\alpha E\Delta T$.

$$F_{i,\max} = \alpha\Delta TEA\sin\varphi \tag{30}$$

If the thermal strain $\alpha\Delta T$ is held to zero, then a spring constant k_i for the beam (Eq. 31) can be determined using Hooke's Law. This spring constant reduces to Eq. 10 in the small angle approximation where $\varphi \to 0$. For actuators that contain N angled beams, the total spring constant k can be found according to Eq. 13 by multiplying k_i by N, and the maximum output force from the actuator would be found by multiplying Eq. 30 by N.

$$k_i = \frac{-F_i}{u_y} = \frac{EA}{L}\sin^2\varphi + \frac{12EI_z}{L^3}\cos^2\varphi \tag{31}$$

A final comment about the mechanical analysis of V-type actuators is that it may sometimes be necessary to include a temperature-dependent coefficient of thermal expansion in the average thermal strain calculated as an input to Eq. 29. Because the mechanical analysis presented here only concerns motion at the end of the actuator beam, the thermal strain can be determined as an average over the length of the beam using a modified version of Eq. 7, as shown here in Eq. 32.

$$\overline{\alpha\Delta T} = \frac{1}{L}\int_0^L \alpha(T - T_a)dx' \tag{32}$$

Example 2: Hot Arm Actuators

The second common thermal actuation mechanism may be referred to by any of the following names: pseudo-bimorph, U-type, asymmetric, hot arm, or flexure electrothermal actuators. In these structures (Fig. 8), two parallel beams are provided. These beams are fixed to a substrate at one end, and linked to each other at the opposite end. The beams are designed to experience differential thermal expansion such that one beam expands further than the other beam. These actuators can be designed for motion out of a plane of manufacture, but they are more commonly seen for motion within planes. For in-plane motion with one common design, a "hot arm" and a "cold arm" are defined. The cold arm typically consists of a narrow flexure near the attachment point of the arm, and a wider distal portion. The hot arm is a narrow beam of uniform cross section. Electric current flows through this structure. In the wide portion of the cold arm, the current density and thermal resistance are lower than in the other portions of this actuator due to a larger cross-section area. This region is thus subject to less temperature change and less thermal expansion than the other regions. With the flow of electric current, the cold arm elongates less than the hot arm. Because the two arms are constrained to move together, the two arms sweep an arc of motion due to the difference in thermal expansion between them. By symmetrically placing these U-type actuators around a central shuttle or yoke, and linking them to the shuttle with compliant structures, linear motion can be generated.

Electrical Analysis

Electrical and thermal analysis of these actuators can be performed by dividing the actuator into four segments, as seen in Fig. 9. Ignoring small discrepancies at the boundaries to the segments, the actuator can be converted into a one-dimensional model similar to

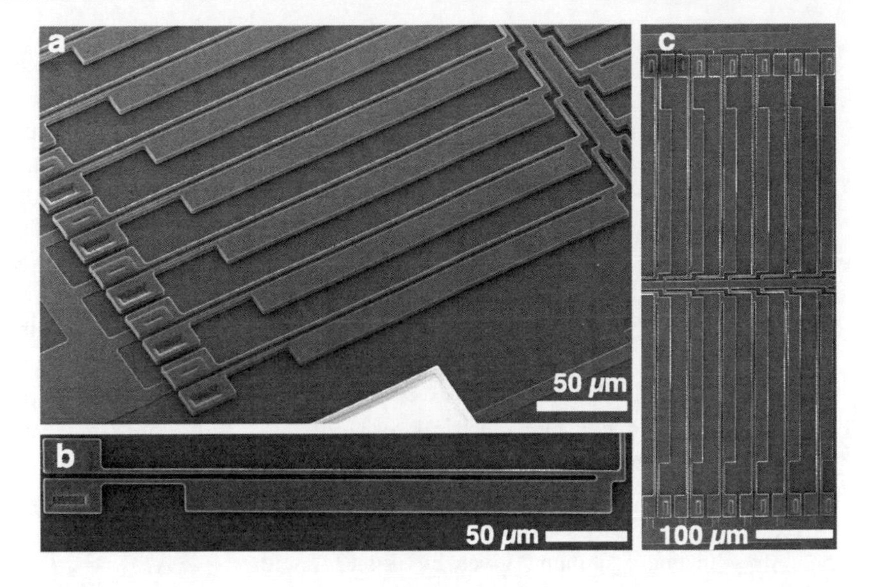

Thermal Actuators, Fig. 8 SEM images of U-type, "hot arm" thermal actuators. (**a**) Tilted, oblique view of an actuator array. (**b**) Magnified view of one of the actuators in this array. The actuator is anchored at left and greatest motion occurs in the parts at right. (**c**) Top view of an array of fabricated U-type thermal actuators (Image credit: J.J. Brown and V.M. Bright)

that of Example 1, but constructed from a series connection of the four distinct segments, as seen in the electrical circuit diagram, Fig. 10.

The total electrical resistance can be calculated, Eq. 33, by summing in series the electrical resistance for each segment and assuming constant resistivity. The resistance for each segment can be found according to Eq. 15a. In most cases, the thickness will be identical for all segments due to fabrication from a surface-micromachined material layer, so the thickness term can be moved outside of the summation along with the resistivity. For a polycrystalline silicon device with typical dimensions of $L_1 = 100\ \mu m$, $L_2 = 5\ \mu m$, $L_3 = 80\ \mu m$, $L_4 = 20\ \mu m$, $w_1 = w_4 = 2\ \mu m$, $w_2 = 5\ \mu m$, $w_3 = 20\ \mu m$ and $t = 3.5\ \mu m$, and with $\rho = 3.4\times10^{-5}\ \Omega m$, the total resistance R is $632\ \Omega$.

$$R = \sum_{i=1}^{n} R_i = \sum_{i=1}^{n} \frac{\rho L_i}{A_i} = \rho \sum_{i=1}^{n} \frac{L_i}{w_i t_i} \qquad (33)$$

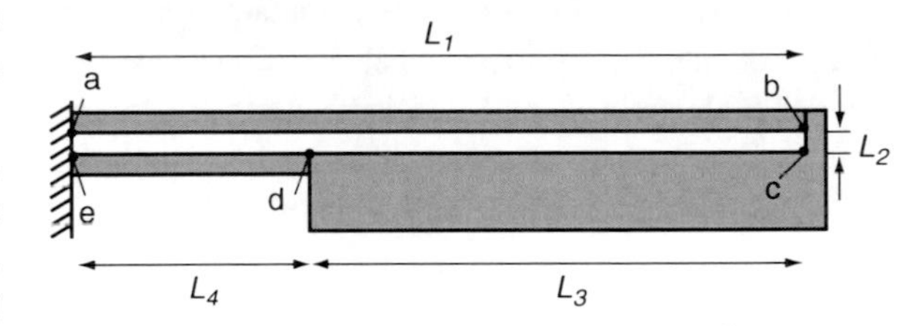

Thermal Actuators, Fig. 9 Schematic of one hot arm actuator

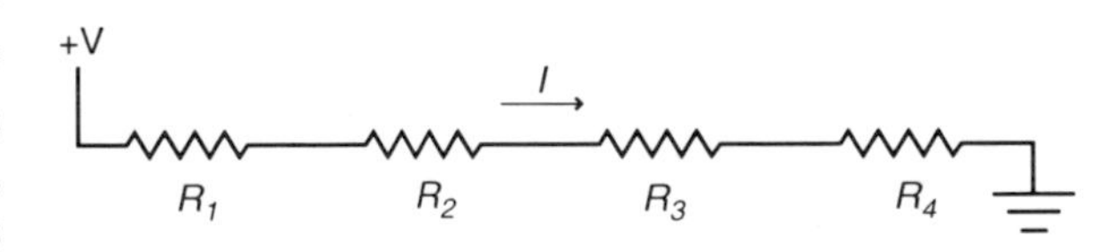

Thermal Actuators, Fig. 10 Simplified electrical schematic for a hot arm actuator

Thermal Analysis

The thermal analysis of the hot arm actuator can proceed similarly to the electrical analysis, dividing the actuator into four segments according to the schematic in Fig. 11. The four segments are coupled together with temperature T_i and heat flow $\dot{Q}_i$ in each segment i, and with temperature and heat flow boundary conditions described in Table 2. The mechanical anchor points are maintained at constant temperature T_a. Uniform

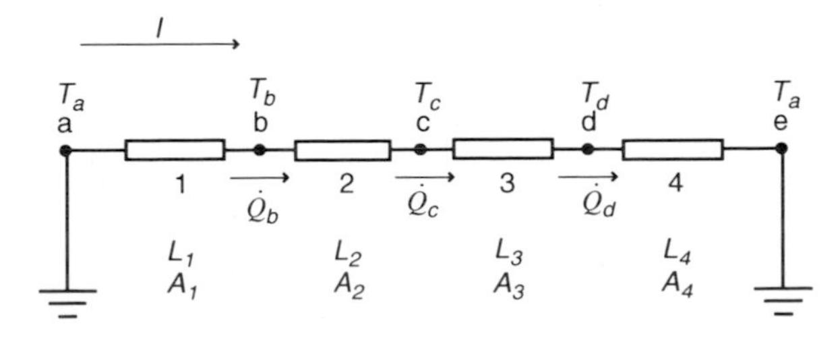

Thermal Actuators, Fig. 11 Schematic for thermal analysis of the hot arm actuator

Thermal Actuators, Table 2 Boundary conditions for the hot arm actuator thermal model

Temperature boundary conditions	Heat flow boundary conditions
$T_1(0) = T_a$	$\dot{Q}_1(L_1) = \dot{Q}_2(0)$
$T_1(L_1) = T_b = T_2(0)$	$\dot{Q}_2(L_2) = \dot{Q}_3(0)$
$T_2(L_2) = T_c = T_3(0)$	$\dot{Q}_3(L_3) = \dot{Q}_4(0)$
$T_3(L_3) = T_d = T_4(0)$	
$T_4(0) = T_a$	

current flow I and constant electrical resistivity are assumed through the actuator.

Solution of Eq. 18a following a similar approach as in Example 1 leads to equations for temperature and heat flow in each segment, given by Eqs. 34, with constants C_{1i} and C_{2i} determined for each ith segment by algebraic manipulation of the boundary conditions. The constants are reported in Table 3. This analysis is similar to the approach of Ref. [7], with the difference that Ref. [7] also examines heat transfer to air in solving the heat equation for the hot arm actuator.

$$T_i = -\frac{\rho I^2}{A_i^2 k_\theta}\frac{x_i^2}{2} + C_{1i}x_i + C_{2i} \tag{34a}$$

$$\dot{Q}_i = -k_\theta A_i \frac{\partial T}{\partial x_i} = -k_\theta A_i C_{1i} + \frac{\rho I^2}{A_i} x_i \tag{34b}$$

The maximum temperature $T_{\max}$ and its location $x_{\max}$ in the hot arm segment 1 can be found by setting the first derivative of temperature equal to zero. The first derivative of the temperature is proportional to the heat flow, given in Eq. 34b using C_{11}. The resulting temperature and position are given in Eq. 35 and Eq. 36, respectively.

$$\begin{aligned} T_{\max} &= T_1(x_{\max}) = T_a + \frac{I^2\rho}{2k_\theta}\frac{x_{\max}^2}{A_1^2} \\ &= T_a + \frac{I^2\rho}{2k_\theta}\left(\frac{1}{4}\right)\left(\frac{L_1}{A_1} + \frac{L_2}{A_2} + \frac{L_3}{A_3} + \frac{L_4}{A_4}\right)^2 \end{aligned} \tag{35}$$

$$x_{\max} = \frac{A_1}{2}\left(\frac{L_1}{A_1} + \frac{L_2}{A_2} + \frac{L_3}{A_3} + \frac{L_4}{A_4}\right) \tag{36}$$

The average temperature rise ΔT and the thermal time constant can be calculated as described earlier, subject to the discretion of the designer in choosing the complexity of the system to be represented. One option is to focus solely on the hot arm because this is the segment that experiences the highest temperatures and the most significant thermal strain. In the hot arm, the thermal behavior is most important for the actuator operation. In this case, the thermal capacitance C is calculated using the volume of the hot arm $V_m = L_1 A_1$. The thermal resistance can be calculated using the location $x_{\max}$ of maximum temperature or the total length L_1 of the hot arm, and the hot arm cross-section area A_1. For the average temperature rise ΔT, the average is taken over the length L_1 of the hot arm only, giving Eq. 37.

$$\begin{aligned} \Delta T &= \frac{1}{L_1}\int_0^{L_1} T_1 dx - T_a \\ &= \frac{I^2\rho}{4k_\theta}\left(\frac{L_1}{A_1}\right)\left(\frac{L_1}{3A_1} + \frac{L_2}{A_2} + \frac{L_3}{A_3} + \frac{L_4}{A_4}\right) \end{aligned} \tag{37}$$

Analysis of Mechanical Behavior

Few mechanical analyses of hot arm actuators have been published. The results of analyses based on energy methods may be found in Refs. [7, 10]. The analysis presented here uses a more straightforward technique based on superposition. Begin by dividing the device depicted in Fig. 9 into three sections: two cantilevers and one rigid frame. As seen in Fig. 12, section 1 becomes a cantilever with clamped attachment at left and applied force F_{1y} and moment M_1 at right. Similarly, section 4 becomes a cantilever with shorter length, and with clamped attachment at left and applied force F_{4y} and moment M_4 at right. Sections 2 and 3 become an L-shaped rigid frame that links the two cantilevers together with moment and force reactions at the joining surfaces. Additionally, an external force F_Y is applied at point **b** of this L-frame, and the frame also experiences a bending moment M due to thermal stresses in the cantilevers. In order to maintain simplicity in calculating the coupled bending behavior of the cantilevers for this example, this thermal bending moment is calculated separately rather than combining the axial deformation behavior of the cantilevers with the rotation of the rigid frame and the bending of the cantilevers.

To calculate M from the thermal strain in cantilever 1, use the sum of forces in the x direction, the sum

Thermal Actuators, Table 3 Algebraic constants for Eqs. 34 in each segment of the hot arm actuator

C_{1i}	C_{2i}
$C_{11} = \frac{I^2\rho}{2k_\theta A_1}\left(\frac{L_1}{A_1}+\frac{L_2}{A_2}+\frac{L_3}{A_3}+\frac{L_4}{A_4}\right)$	$C_{21} = T_a$
$C_{12} = \frac{I^2\rho}{2k_\theta A_2}\left(-\frac{L_1}{A_1}+\frac{L_2}{A_2}+\frac{L_3}{A_3}+\frac{L_4}{A_4}\right)$	$C_{22} = T_a + \frac{I^2\rho}{2k_\theta}\left(\frac{L_1}{A_1}\right)\left(\frac{L_2}{A_2}+\frac{L_3}{A_3}+\frac{L_4}{A_4}\right)$
$C_{13} = \frac{I^2\rho}{2k_\theta A_3}\left(-\frac{L_1}{A_1}-\frac{L_2}{A_2}+\frac{L_3}{A_3}+\frac{L_4}{A_4}\right)$	$C_{22} = T_a + \frac{I^2\rho}{2k_\theta}\left(\frac{L_1}{A_1}+\frac{L_2}{A_2}\right)\left(\frac{L_3}{A_3}+\frac{L_4}{A_4}\right)$
$C_{14} = \frac{I^2\rho}{2k_\theta A_4}\left(-\frac{L_1}{A_1}-\frac{L_2}{A_2}-\frac{L_3}{A_3}+\frac{L_4}{A_4}\right)$	$C_{24} = T_a + \frac{I^2\rho}{2k_\theta}\left(\frac{L_1}{A_1}+\frac{L_2}{A_2}+\frac{L_3}{A_3}\right)\left(\frac{L_4}{A_4}\right)$

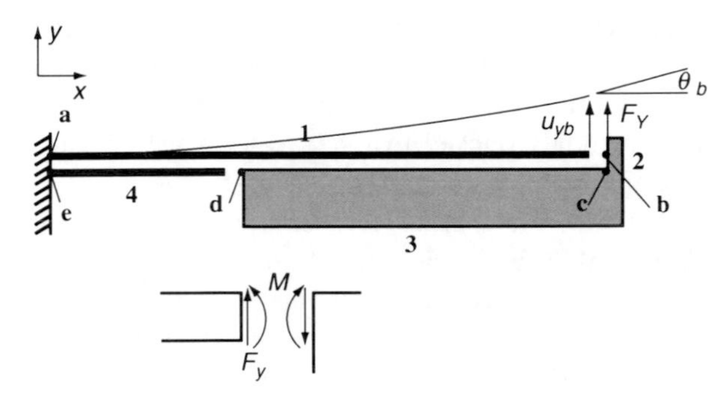

Thermal Actuators, Fig. 12 Diagram for mechanical analysis of the hot arm actuator. The lower part of this diagram illustrates the matching of moments and forces at points **b** and **d** where the rigid frame is cut from the two cantilevers

of moments at point **c**, and Hooke's law applied to elongation of the two cantilevers in order to generate Eqs. 38. For simplicity, thermal strain is assumed only to be present in cantilever 1. Algebraic manipulation leads to an expression for the displacement of the rigid frame, Eq. 39a, and an expression, Eq. 39b, for the thermal bending moment that drives operation of the hot arm actuator.

$$F_{xb} + F_{xd} = 0 \tag{38a}$$

$$M - F_{xb}L_2 = 0 \tag{38b}$$

$$F_{xb} = k_1 e_1 = \frac{A_1 E}{L_1}(u_x - \alpha\Delta T L_1) \tag{38c}$$

$$F_{xd} = k_4 e_4 = \frac{\Lambda_4 E u_x}{L_4} \tag{38d}$$

$$u_x = \frac{\alpha\Delta T A_1}{\left(\frac{A_1}{L_1}+\frac{A_4}{L_4}\right)} \tag{39a}$$

$$M = \frac{-\alpha\Delta T A_1 L_2 E}{\left(1+\frac{A_1 L_4}{A_4 L_1}\right)} \tag{39b}$$

In order to calculate the actuator motion, begin by examining the rigid frame and recording a set of force and moment equilibrium equations, and compatibility equations that describe the relationship between the rotation and displacements of the cantilever ends (Eqs. 40). Note that the small angle approximation is used to simplify Eq. 40d to the relation in Eq. 40e.

$$F_Y - F_{1y} - F_{4y} = 0 \tag{40a}$$

$$-M_4 - M_1 + M + L_3(F_Y - F_1) = 0 \tag{40b}$$

$$\theta_b = \theta_d = \theta \tag{40c}$$

$$u_{yd} + L_3 \sin\theta = u_{yb} \tag{40d}$$

$$u_{yd} = u_{yb} - L_3\theta \tag{40e}$$

The bending of cantilevers is a well-characterized problem, and nodal equations may be written that relate the force and moment applied to the free end of the cantilever to the displacement and rotation that are present at that end [8]. For the cantilevers of sections 1 and 4 of the hot arm actuator, these nodal equations are provided in Eqs. 41.

$$F_{1y} = \frac{EI_1}{L_1^3}\left(12u_{yb} - 6L_1\theta\right) \tag{41a}$$

$$M_1 = \frac{EI_1}{L_1^3}\left(-6L_1 u_{yb} + 4L_1^2\theta\right) \tag{41b}$$

$$F_{4y} = \frac{EI_4}{L_4^3}\left(12u_{yd} - 6L_4\theta\right) \tag{41c}$$

$$M_4 = \frac{EI_4}{L_4^3}\left(-6L_4u_{yd} + 4L_4^2\theta\right) \tag{41d}$$

By combining Eqs. 40 and 41, expressions may be derived that relate the force and moment applied in the rigid frame to the displacement and rotation experienced by the end of the actuator. These relations, Eqs. 42, may be further manipulated to provide an expression for the actuator displacement as a function of applied force and applied moment, Eq. 43. Using the equation for M found in Eq. 39b, the expression in Eq. 43 relates input temperature and output force and displacement for the hot arm actuator. Similar to Eq. 29 in Example 1, Eq. 43 may be further manipulated to find a spring constant for the actuator and a maximum output force.

$$\frac{F_Y}{E} = 12u_{yb}\left(\frac{I_1}{L_1^3} + \frac{I_4}{L_4^3}\right) - 6\theta\left(\frac{I_1}{L_1^2} + \frac{2L_3I_4}{L_4^3} + \frac{I_4}{L_4^2}\right) \tag{42a}$$

$$\frac{M}{E} = -6u_{yb}\left(\frac{I_1}{L_1^2} + \frac{2L_3I_4}{L_4^3} + \frac{I_4}{L_4^2}\right) + 2\theta\left[2\left(\frac{I_1}{L_1} + \frac{I_4}{L_4}\right) + 6I_4\left(\frac{L_3}{L_4^2} + \frac{L_3^2}{L_4^3}\right)\right] \tag{42b}$$

$$u_{yb} = \frac{\frac{M}{2E}\left(\frac{I_1}{L_1^2} + \frac{I_4}{L_4^2} + \frac{2L_3I_4}{L_4^3}\right) + \frac{F_Y}{6E}\left[2\left(\frac{I_1}{L_1} + \frac{I_4}{L_4}\right) + 6I_4\left(\frac{L_3}{L_4^2} + \frac{L_3^2}{L_4^3}\right)\right]}{\left(\frac{I_1}{L_1^2} + \frac{I_4}{L_4^2}\right)^2 + \frac{4I_1I_4}{L_1L_4}\left(\frac{1}{L_1} - \frac{1}{L_4}\right)^2 + \frac{12L_3I_1I_4}{L_1^2L_4^2}\left(\frac{1}{L_1} - \frac{1}{L_4} + \frac{L_3}{L_1L_4}\right)} \tag{43}$$

Experimental verification of this or related hot arm actuator models has not been clearly performed, as most hot arm actuator experiments have recorded data in air environments, which require different thermal analysis from that performed in this example. In Ref. [7], experimental displacement measurements appear to correspond fairly well with a curve predicted from input current, but with ΔT calculated to include the effects of heat loss to air. A final point about this mechanical analysis of hot and cold arm actuators is that variations of these actuators exist, but similar analytic approaches can apply to these variations. For instance, if L_3 is set equal to zero, then Eq. 43 describes the motion of a two arm actuator that has no wide region but does have different length cantilevers and different temperature profiles in each cantilever due to the length difference [16]. The out-of-plane pseudo-bimorph actuator could be analyzed similarly [6].

Example 3: Thermal Bimorph Actuators

The bimorph thermal actuator predates the designs explored in the previous two examples. The principle of operation of this device is the generation of bending due to a differential in thermal expansion between two materials stacked together to form the actuator. The bimorph actuator is typically constructed as shown in Fig. 13, as a cantilevered beam composed of two (bimorph) or more (multimorph) dissimilar materials, and with a resistive heater patterned on top of the beam or formed from one of the beam materials. Advantages of bimorph structures are high force, simple design, and low operating voltage. The major disadvantages of these structures are relatively high electrical power consumption, curling motion, low operating frequency, and laminated construction. Many examples of implementation of bimorph thermal actuators, including more discussion of simulation and analytical models, may be found in Ref. [6].

Power consumption and average temperature rise can be calculated by applying concepts from Example 1. Others have extensively described the mechanical behavior of bimorph thermal actuators [6, 10, 17]. This example will review the relevant mechanical concepts and results. According to Fig. 13, assume the lengths of the two layers in the actuator are identical length L. Using the small angle approximation, deflection due to uniform curvature in the cantilever is given by Eq. 44, where r is a radius of curvature in the cantilever that is generated due to the thermal expansion mismatch of the two materials. Superimposed on this displacement is the deflection due to force applied at the end of a cantilever, Eq. 45, where $\overline{EI_i}$ is the average flexural rigidity of the composite cantilever. Together, these give a total displacement described by Eq. 46, analogous to Eqs. 29 and 43 presented earlier.

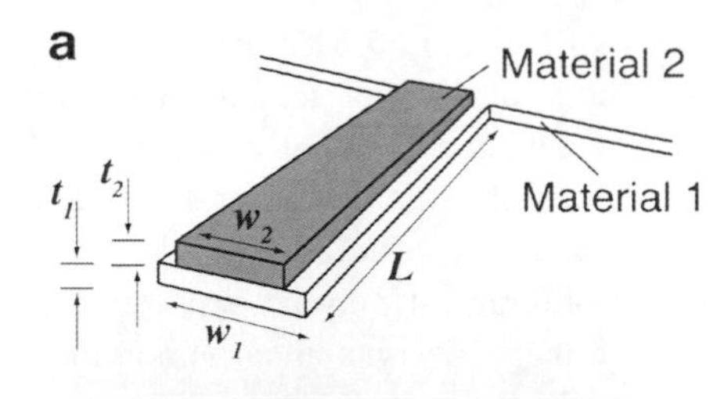

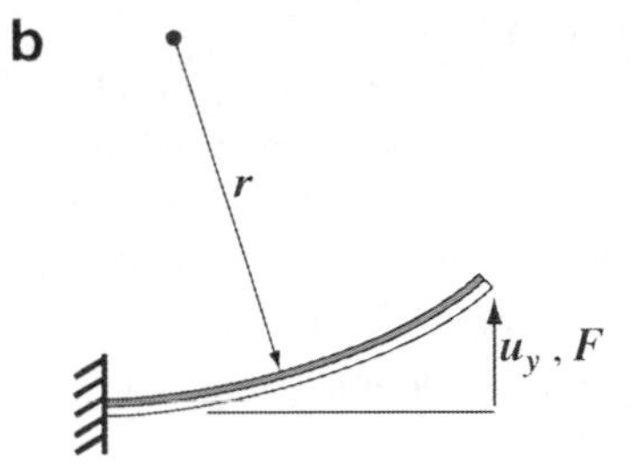

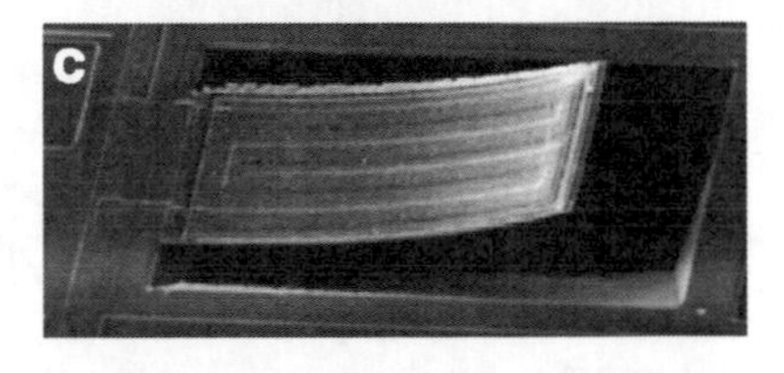

Thermal Actuators, Fig. 13 Illustration of bimorph thermal actuators. (**a**) Bimorph cantilever consisting of two materials with identical length, but different thicknesses, widths, and composition. (**b**) Diagram of radius of curvature r and cantilever tip displacement u_y used in analysis of the output motion of the bimorph actuator. (**c**) SEM image of a bimorph actuator ~300 μm length, with a heater trace patterned on its top surface (Image credit: B. Read and V.M. Bright, Air Force Institute of Technology, 1994)

$$u_{y1} = \frac{L^2}{2r} \tag{44}$$

$$u_{y2} = \frac{FL^3}{3\overline{EI_i}} \tag{45}$$

$$u_y = \frac{L^2}{2r} + \frac{FL^3}{3\overline{EI_i}} \tag{46}$$

The radius of curvature r is determined (Eq. 47) using force and moment equilibrium in the composite beam, and strain compatibility at the bonded surface between the two materials [10, 17]. The combined flexural rigidity $\overline{EI_i}$ is found for the nonhomogeneous beam according to Eq. 48 after the moments I_1 and I_2 are first calculated for the different layers in relation to the neutral surface of the beam [8, 10]. The result is Eq. 49. In Ref. [17], Eqs. 47 and 44 are derived and shown to lead to agreement with previously reported experimental data. Eqs. 45, 46, and 49 are derived through continuum solid mechanics analysis [8].

$$r = \frac{\left(E_1 w_1 t_1^2\right)^2 + \left(E_2 w_2 t_2^2\right)^2 + 2E_1 w_1 t_1 E_2 w_2 t_2 \left(2t_1^2 + 3t_1 t_2 + 2t_2^2\right)}{6E_1 w_1 t_1 E_2 w_2 t_2 (\alpha_1 - \alpha_2)\Delta T} \tag{47}$$

$$\overline{EI_i} = E_1 I_1 + E_2 I_2 \tag{48}$$

$$\overline{EI_i} = \frac{\left(E_1 w_1 t_1^2\right)^2 + \left(E_2 w_2 t_2^2\right)^2 + 2E_1 w_1 t_1 E_2 w_2 t_2 \left(2t_1^2 + 3t_1 t_2 + 2t_2^2\right)}{12(E_1 w_1 t_1 + E_2 w_2 t_2)} \tag{49}$$

Future Directions

Miniaturization

The phenomena of electrical power dissipation, heat flow, and thermal expansion are valid concepts at submicron size scales. Continued miniaturization may require consideration of additional phenomena such as the effects of dimensional constraints on the motion of phonons and electrons, increased importance of surface forces with smaller sizes, and the Thomson effect in systems where electrical current is flowing [18]. It may be expected that thermal actuator designs will be demonstrated in submicron sizes as fabrication technology improves. Ultimately, thermal actuator designs formed from engineered macromolecular structures may be possible. As molecular-scale thermal actuators are developed, such devices may not be subject to the same friction conditions as current microscale technology, and thus pinned connections and sliding surfaces may be within the range of designs allowed at the nanoscale.

Environmental Robustness

As discussed earlier, the temperature profile of the thermal actuator, and the corresponding output displacement or force, can vary due to the environment in which the actuator is operated. In vacuum environments at ambient temperature, this may be a small effect, but operation in ambient air and gaseous environments can require significant modification to the actuator power. For reliable and predictable operation across a range of environments, improved control systems for thermal actuators are needed. Additionally, many possible studies of environment effects comparing experiment with simulations and analytical models remain to be explored for thermal actuators. These may, for instance, include effects of vacuum, liquid, high temperature, high radiative heat transfer, low temperature, packaging design, and environmental transients.

Device Control and Sensor Integration

One approach to improved displacement control and robustness would be implementation of closed-loop control systems that rely on data from a displacement sensor integrated to the actuator. In practice, integration of accurate in-plane displacement sensors for actuator control presents a significant design and fabrication challenge. Capacitive sensors have been integrated to thermal actuators, but these sensors typically require a large footprint to achieve useful sensitivity at small displacements [9]. Operation of capacitive sensors must be driven by sophisticated electronics placed physically close to the sensor in order to avoid signal losses due to parasitic capacitance. Piezoresistive sensors require smaller device footprints than capacitive sensors, but may be subject to distorted readings due to temperature variation resulting from heat dissipation from the thermal actuator. Furthermore, readings from piezoresistive sensors can be distorted by the electrical operation of the thermal actuator if the sensor is monolithically linked and thus electrically connected to the thermal actuator. Strategies for electrical isolation of piezoresistive sensors are needed. Optical sensors provide the most accurate options for displacement measurement, but they are not yet easily integrated with other chip scale devices for in-plane displacement measurements.

Application Development

As one of the key mechanisms for generating motion in microscale and nanoscale electromechanical systems (MEMS and NEMS), it may be expected that a major thrust of thermal actuator development will track the overall development and use of MEMS and NEMS. Thermal actuators are an important part of the toolkit available to microsystem designers. As new designs proliferate, it may be expected that a certain portion of these designs incorporate thermal actuators. As discussed earlier, current usages include optical and electrical switches, microgrippers, nano manipulation structures, and tensile test devices. As these tools move into broader use, thermal actuators are likely to remain key components of these devices. Other key opportunities for new application development using thermal actuators include control of optical gratings, lock and security systems, and controlled motion in microscale positioning stages. Many MEMS and NEMS applications may yet be conceived requiring low actuation voltage, compact footprint, and high force output, and thermal actuators provide a useful design option for all such applications.

Force Output Measurement and Calibration

The use of thermal actuators in applications where force is applied to microscale and nanoscale objects raises the question, "Exactly how much force is applied?" Answering this question requires calibration and measurement of the force applied with the actuator. No standardized, traceable approaches exist for microscale force calibration, and current best practices rely on measurement of the displacement of a well-characterized spring. Such an approach relies on the characterization of the spring, which itself is not a traceable, standardized practice for microscale and nanoscale devices. Furthermore, many analytical and finite element mechanical models have been presented in literature, but more experimental verification of these models is needed, as the observations of Ref. [12] show. As thermal actuators and other actuators are employed for generation of motion in microscale and nanoscale devices, further strategies are needed for understanding the magnitude of forces within these devices.

Cross-References

▸ Basic MEMS Actuators
▸ Compliant Mechanisms
▸ Thermally Actuated Silicon Resonators

References

1. Lu, S., Dikin, D.A., Zhang, S., Fisher, F.T., Lee, J., Ruoff, R.S.: Realization of nanoscale resolution with a micromachined thermally-actuated testing stage. Rev. Sci. Instrum. **75**, 2154–2162 (2004)
2. Fung, Y.C., Tong, P.: Classical and Computational Solid Mechanics. Advanced Series in Engineering, vol. 1. World Scientific, Singapore (2001)
3. Jackson, J.D.: Classical Electrodynamics, 2nd edn. Wiley, New York (1975)
4. Mills, A.F.: Heat Transfer, 2nd edn. Prentice Hall, Upper Saddle River (1999)
5. Lott, C.D., McLain, T.W., Harb, J.N., Howell, L.L.: Modeling the thermal behavior of a surface-micromachined linear-displacement thermomechanical microactuator. Sens. Actuators A Phys. **101**, 239–250 (2002)
6. Geisberger, A.A., Sarkar, N.: Techniques in MEMS microthermal actuators and their applications. In: Leondes, C.T. (ed.) MEMS/NEMS Handbook: Techniques and Applications, Sensors and Actuators, vol. 4, pp. 201–261. Springer, New York (2006)
7. Huang, Q.-A., Lee, N.K.S.: Analysis and design of polysilicon thermal flexure actuator. J. Micromech. Microeng. **9**, 64–70 (1999)
8. Craig R.R. Jr.,: Mechanics of Materials, 2nd edn. Wiley, Hoboken (2000)
9. Zhu, Y., Corigliano, A., Espinosa, H.D.: A thermal actuator for nanoscale in situ microscopy testing: design and characterization. J. Micromech. Microeng. **16**, 242–253 (2006)
10. Lobontiu, N., Garcia, E.: Mechanics of Microelectromechanical Systems. Kluwer, New York (2005)
11. Brown, J.J., Suk, J.W., Singh, G., Baca, A.I., Dikin, D.A., Ruoff, R.S., Bright, V.M.: Microsystem for nanofiber electromechanical measurements. Sens. Actuators A Phys. **155**, 1–7 (2009)
12. Wittwer, J.W., Baker, M.S., Howell, L.L.: Simulation, measurement, and asymmetric buckling of thermal microactuators. Sens. Actuators A Phys. **128**, 395–401 (2006)
13. Howell, L.L., McLain, T.W., Baker, M.S., Lott, C.D.: Techniques in the design of thermomechanical actuators. In: Leondes, C.T. (ed.) MEMS/NEMS Handbook: Techniques and Applications, Sensors and Actuators, vol. 4, pp. 187–200. Springer, New York (2006)
14. Gianchandani, Y.B., Najafi, K.: Bent-beam strain sensors. J. Microelectromech. Syst. **5**, 52–58 (1996)
15. Que, L., Park, J.-S., Gianchandani, Y.: Bent-beam electrothermal actuators – part I. Single beam and cascaded devices. J. Microelectromech. Syst. **10**, 247–254 (2001)
16. Huang, Q.-A., Lee, N.K.S.: Analytical modeling and optimization for a laterally-driven polysilicon thermal actuator. Microsyst. Technol. **5**, 133–137 (1999)
17. Chu, W.-H., Mehregany, M., Mullen, R.L.: Analysis of tip deflection and force of a bimetallic cantilever microactuator. J. Micromech. Microeng. **3**, 4–7 (1993)
18. Jungen, A., Pfenniger, M., Tonteling, M., Stampfer, C., Hierold, C.: Electrothermal effects at the microscale and their consequences on system design. J. Micromech. Microeng. **16**, 1633–1638 (2006)

Thermal Cancer Ablation Therapies Using Nanoparticles

Steven Curley
Department of Surgical Oncology, The University of Texas M. D. Anderson Cancer Center, Houston, TX, USA
Department of Mechanical Engineering and Materials Science, Rice University, Houston, TX, USA

Synonyms

Heat; Magnetic electromagnetic radiation; Near infrared; Radiofrequency

Definition

Nanoparticle-sized materials have interesting physical, chemical, and electronic properties. Among these is the release of heat upon exposure to various types of electromagnetic radiation. For example, iron and other magnetic nanoparticles can be injected directly into malignant tumors and then release heat in response to frictional heating in an alternating magnetic field. The intrinsic plasmon resonance of gold has been utilized to produce near-infrared laser-induced thermal activation of gold nanoparticles to treat malignant lesions on the skin or mucosal surfaces. Finally, several types of conducting and semiconducting nanoparticles have been conjugated successfully to antibodies or peptides to target the nanoparticles to malignant tumor cells or to the tumor neovasculature. The malignant cells are then exposed to a noninvasive shortwave radiofrequency field which results in intracellular heating and thermal cytotoxicity to the malignant cells. Cancer cells generally are more sensitive to thermal injury compared to normal cells; thus, nanoparticle-based thermal tumor ablation has the potential to be an important and hopefully less toxic addition to current anticancer treatments.

Overview

Hyperthermic cancer therapy was first described in Egyptian papyri over 4,000 years ago by applying hot oil or cautery to tumors. Modern hyperthermic treatments are now applied to both premalignant and malignant tumors in various locations such as the peritoneal cavity, liver, kidneys, lungs, and prostate. Hyperthermia also potentiates the effects of cytotoxic chemotherapeutic agents and improves the response of tumors to ionizing radiation. Thus, regional or systemic hyperthermia has been induced as part of a treatment regimen in patients with certain types of malignancies. Hyperthermic isolated limb perfusion with melphalan with or without dactinomycin is performed to treat in-transit melanoma metastases and is being investigated as a treatment for advanced extremity soft tissue sarcoma [1]. In the last several years, there have been a growing number of reports on hyperthermic intraperitoneal chemotherapy combined with cytoreductive surgery for patients with diffuse peritoneal spread of colorectal cancer, gastric cancer, mesothelioma, or pseudomyxoma peritoneii [2].

Although effective at treating several cancer types, recent studies have shown that some mammalian cancer cells are sensitive, whereas others are significantly more resistant to hyperthermia compared to healthy cells [3]. The reasons for the variable sensitivity to hyperthermia in cancer are multifactorial, but are primarily due to the role of heat shock proteins (HSPs) in cancer's resistance to the physical stress of increasing temperature [3]. Mammalian cancer cells that are heat sensitive frequently have mutated or underexpressed HSPs, whereas deregulation in some malignant cells leads to increased expression of HSPs which confers increased thermal protection and reduces apoptosis following thermal stress.

Many hyperthermic devices have been designed to produce low-level total-body hyperthermia (42–46°C) and have been used to treat a wide variety of advanced malignancies combined with systemic cytotoxic agents. Both regional and systemic hyperthermic treatment systems elevate tissue temperatures from 3°C to 7°C above normal over a 1–12-h period. Some populations of malignant cells are sensitive to relatively long durations of exposures to low-level hyperthermia, but results to date suggest that these regional and systemic hyperthermic treatments have produced little benefit in overall patient outcomes at the expense of major regional and systemic toxicities and side effects. Interestingly, exposure of mammalian cells to temperatures of 55–60°C for periods of a few seconds to less than a minute has been shown to produce apoptosis and necrosis, so a treatment system that can achieve such brief periods of heating in cancer cells would be highly desirable.

Invasive Thermal Ablation Therapies

Thermal tumor ablation is routinely practiced, particularly for malignant liver tumors, through direct intratumoral insertion of radiofrequency (RF) needle electrodes, microwave-emitting probes, or end- and side-firing laser fibers [4]. RF ablation therapy is an invasive treatment that is implemented by inserting needle electrodes directly into the tumor(s) to be treated and applying RF current through the wire into the tumor, resulting in local tumor and adjacent tissue necrosis (Fig. 1). The basis for the observed heating phenomena is understood by the rotational response of the charged ions and proteins in the tissue under the RF field. This rapid molecular rotation results in local frictional heating of the aqueous environment of the tumor and death of tumor cells through protein denaturation, melting of lipid bilayers, and destruction of nuclear DNA.

Although invasive RFA has demonstrated efficacy in cancer therapy, there are several limitations to this approach. First, clinical studies have shown incomplete tumor destruction in 5–40% of the treated lesions, allowing significant probability of local cancer recurrence and metastasis. Another limitation of RFA is that the treatment is not targeted (i.e., nonspecific). Placement and insertion of the needle electrode is usually guided by an imaging system with poor spatial resolution (e.g., ultrasonography ~ mm), and therefore, this method is dependent on the experience of the physician that implements RFA. Furthermore, RFA treatments result in significant complications in approximately 10% of patients due to thermally induced necrosis of healthy noncancerous tissues. The final and the most problematic limiting factor with respect to wire-based RFA is that it is a clinical treatment that is indicated for just a few select organ sites and specific tumor types (e.g., liver, kidney, breast, lung). On the other hand, tissue penetration by noninvasive RF energy fields is known to be excellent. Theoretically, noninvasive RF treatment of malignant tumors at any site in the body should be possible if agents that preferentially respond to RF energy with

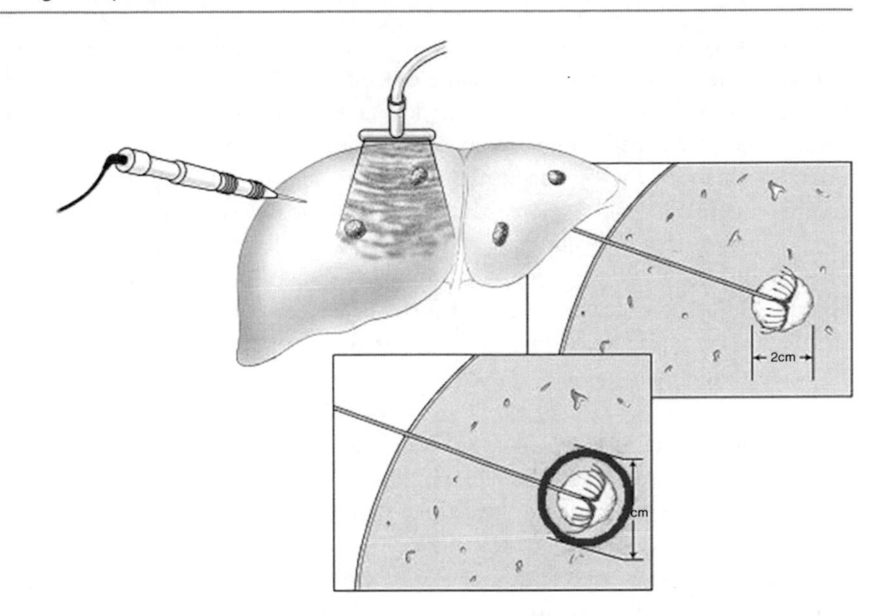

Thermal Cancer Ablation Therapies Using Nanoparticles, Fig. 1 Schematic illustration demonstrating invasive radiofrequency ablation of malignant liver tumors. A multiple array needle electrode is inserted directly into the tumor to be treated using ultrasound guidance to assure proper placement of the needle. An alternating current is then passed across the array resulting in frictional ionic heating in the tumor and tissue surrounding the array

intracellular release of heat can be delivered to the malignant cells.

Roles for Nanotechnology in Cancer Diagnosis and Therapy

Nanotechnology is a term used to describe the engineering of materials that utilize the unique physical properties of nanometer scale matter. Nanomaterials have been used across all the major scientific disciplines including chemistry, physics, and medicine. It is in the medical realm that nanotechnology is widely heralded as one of the most promising and important approaches to diagnose and treat cancer [5]. Some nanoscale materials are already FDA approved and are used clinically. For example, iron oxide nanoparticles (Feridex™) have been shown to enhance the diagnostic capability of magnetic resonance imaging (MRI) to detect cancer [6]. Conjugating iron nanoparticles to antibodies that target proteins expressed on the surface of human cancer cells may further enhance the accuracy of MRI to diagnose early stage cancer [6].

Other nanoparticles such as carbon or polymeric nanoparticles labeled with fluorine-18 deoxyglucose have been studied in preclinical models to enhance tumor diagnosis and detection rates using positron emission tomography. Surface modification of quantum dots, which are semiconducting nanocrystals that fluoresce upon absorption of visible light, is being investigated to improve detection of lymph nodes and other sites of metastases during surgical procedures [7]. Additionally, conjugation of quantum dots with tumor-specific peptides or antibodies may improve targeting of cancer cells and thus improve the diagnostic accuracy of this technique. Optimal imaging techniques using fluorescent nanoparticles targeted to a variety of types of human cancer with immunoconjugation to antibodies are being studied to permit in vivo localization of malignant cells. It is hoped that such imaging techniques will improve the diagnostic accuracy in numerous types of imaging modalities used to detect and follow patients with cancer. It is possible that these techniques may also allow earlier detection of cancer in high-risk populations and guide the duration and type of therapy in patients with more advanced stages of malignant disease. Finally, immunocomplexes consisting of gold nanoparticles and labeled antibodies have been demonstrated to improve the detection of several known serum tumor markers, including carcinoembryonic antigen, carcinoma antigen 125, and carbohydrate antigen 19–9, in a more rapid and accurate fashion than currently available techniques.

Nanoparticles are being used as intravascular carriers to enhance delivery of anticancer agents to malignant cells. Given that most anticancer drugs have a narrow therapeutic index with significant acute and cumulative toxicities in normal tissues, a variety of types of nanoparticles are being studied in preclinical and in vitro experiments to increase the delivery of cytotoxic chemotherapy drugs to cancer cells while reducing toxicity by limiting exposure of the drug to normal tissues. For example, nanoparticles loaded with hydroxycamptothecin, 5-fluorouracil, docetaxel, and gemcitabine have been used in preclinical studies to improve chemotherapy-induced cytotoxicity in lung, colon, squamous, and pancreas cancers. Colloidal gold nanoparticles bearing tumor necrosis factor-α (TNF-α) molecules are being used in early phase human clinical trials to treat several types of cancer. Preclinical studies demonstrated that delivery of TNF-α to malignant tumors was enhanced using the nanoparticle delivery system, while avoiding the systemic toxicities that usually limit the clinical utility of this biologic agent. Nanoparticles conjugated with targeting agents are also being investigated to deliver gene therapy payloads to malignant cells. Carbon nanotubes have been used to deliver genes or proteins through nonspecific endocytosis in cancer cells. Similar to some cytotoxic chemotherapy drugs, gold or other metal nanoparticles have been shown to improve the therapeutic efficacy of external beam ionizing radiation in preclinical models. In addition to being used to deliver pharmaceutical agents, nanoparticles may be used as intrinsic chemotherapeutic agents. Gold nanoparticles 5–10 nm in diameter have been shown to have antiangiogenic properties in solid tumors and appear to induce apoptosis in chronic lymphocytic B cell leukemias through mechanisms not elucidated entirely.

Nanoparticle-Based Thermal Tumor Ablation Strategies

Recent studies have shown that nanoscale materials are capable of enhancing thermal deposition in cancer cells by absorbing optical energy. Gold-silica nanoshells 150 nm in diameter have a particular plasmon resonance in near-infrared (NIR) wavelengths (650–950 nm) of light and release significant heat when exposed to NIR laser sources (Fig. 2) [8]. In a murine colon carcinoma subcutaneous tumor model, it was demonstrated that intravenous injection of polyethylene glycol-coded gold-silica nanoshells followed 6 h later by illumination of the tumor with a diode laser (8.0 nm, 4 W/cm^2, 3 min) led to complete destruction of the malignant tumors. Animals treated with laser treatment alone without prior intravenous injection of the nanoshells continued to grow at a rate similar to control animals who received neither nanoshell injection or laser treatment. The authors point out that this is a nontargeted delivery of the gold-silica nanoshells to tumors, and the size of the nanoparticles (130–150 nm) corresponds to the size of the enlarged pores and fenestrations in the neovasculature of tumors. The gold-silica nanoshells circulate and a significant number arrest in these pores, thus allowing thermal destruction of the vasculature of the tumor under laser irradiation. A subsequent study of prostate cancer cells using 110-nm gold nanoshells injected intravenously demonstrated that there was a nanoshell dose-related response. The tumors were again treated with a near-infrared laser at 4 W/cm^2 for 3 min. Tumors treated with laser therapy alone continued to grow at a rate similar to untreated control tumors, while tumors treated at the highest dose of gold nanoshells were over 93% necrotic following a single laser treatment. A clinical trial using gold-silica nanoshell hyperthermia following NIR light exposure has been initiated for patients with oropharyngeal malignancies. Thermal therapy with NIR energy is limited to the treatment of superficial tumors (less than 1–2-cm deep) due to the significant attenuation of NIR light by biologic tissues.

Iron oxide nanoparticles are also being evaluated to produce alternating current magnetic field-based hyperthermia in tumors, but the limited thermal enhancement of these nanoparticles in a magnetic field requires extremely high concentrations of iron oxide [9]. Human breast carcinomas implanted into mice underwent intratumoral injection of colloidal suspensions of iron oxide particles 10–200 nm in diameter. Twenty minutes after the injection, the animals were exposed to an alternating current magnetic field (6.5kA/m; frequency, 400 kHz) for 4 min. Real-time temperature monitoring revealed that temperatures increased from 12°C to 73°C at various areas in the tumor. Subsequent analysis of the tumor showed some areas of coagulative necrosis, but other areas of viable tumor remained. Subsequent studies have attempted to address the issue of variable distribution of the superparamagnetic iron oxide nanoparticles through different regions of the tumor. Tumor cells near blood vessels are more difficult to treat with the magnetic nanoparticles because of the heat-sink phenomena of heat being carried away by blood flow in adjacent vessels coupled with the diffusion of nanoparticles in the tumor microenvironment into the

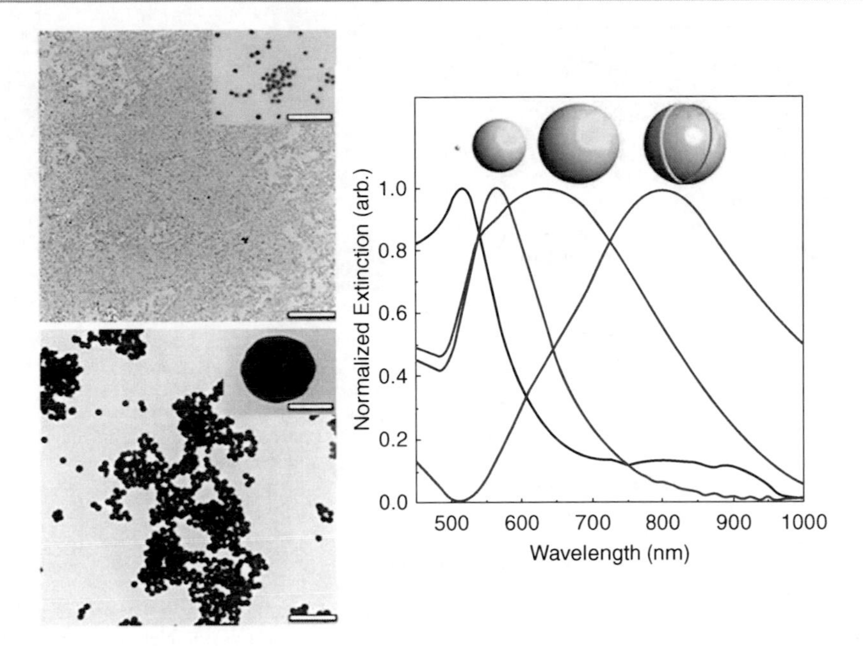

Thermal Cancer Ablation Therapies Using Nanoparticles, Fig. 2 *Left column* – Transmission electron micrographs of 10-nm diameter solid gold nanoparticles (top) and 150-nm diameter gold nanoshells (bottom). TEM images are ×20,000 magnification and ×500,000 (inset) magnification, and scale bars = 1,000 nm (main) and 100 nm (inset). *Right column* – Diameter-dependent visible spectra of solid gold nanoparticles (*black* = 10 nm, *red* = 100 nm, *blue* = 150 nm) and the red-shifted gold nanoshell spectrum (*green* = 150 nm). Nanoparticle illustrations above each corresponding spectra are drawn to scale. Note, spectra have been normalized, and background subtracted to ease viewing of relative peak positions

blood vessels. One of the principle issues and limitations is that the iron oxide particles must be injected directly into the tumor. While this allows high concentrations of the magnetic nanoparticles for subsequent magnetic field thermal treatment, it is an invasive technique and allows for only treatment of tumors that can be identified through imaging techniques or during an operation. The magnetic particles can be used as imaging vehicles for tumors, but there are problems with diffusion of the nanoparticles out of the tumor and into the surrounding normal tissues. These high concentrations and the difficulty targeting the iron oxide nanoparticles to only cancer cells lead to destruction of normal (nonmalignant) cells surrounding the tumor, and diffusion of nanoparticles out of some areas of the tumor may cause insufficient heating to produce thermal cytotoxicity.

It was recently demonstrated that gold or carbon nanoparticles and a confined, noninvasive shortwave RF field produces significant local heat release that can be used to produce thermal cytotoxicity in cancer cells in vitro and in vivo with RF field treatments of 2 min or less [10, 11]. The heat release from gold nanoparticles exposed to shortwave RF fields is particularly impressive and is dependent on both the concentration and diameter of gold nanoparticles (Fig. 3). Solid gold nanoparticles smaller in diameter (5–10 nm) heated at a more rapid rate than gold nanoparticles 50–250 nm in diameter.

In addition, this method does not suffer the limitations of other nanoparticle-based hyperthermia systems given that shortwave radiofrequency (RF) energy has excellent deep tissue penetration, low tissue specific absorption rates (SAR), and documented safety for brief exposures of humans to this form of electromagnetic radiation [12]. Concerns have been expressed about possible mutagenic alterations in cells exposed to nonionizing electromagnetic radiation in the GHz range. However, studies of shortwave RF from 10 to 100 MHz has not been associated with any cytogenetic or cytotoxic damage [13]. Exposure of volunteer test subjects to prolonged periods of shortwave RF irradiation can produce minimal thermoregulatory effects (vasodilation, sweating) but has not

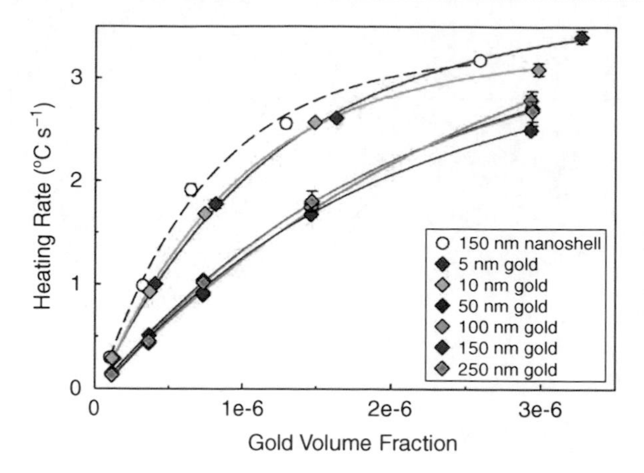

Thermal Cancer Ablation Therapies Using Nanoparticles, Fig. 3 At gold nanoparticle concentrations less than 10 ppm, the heating rate of deionized water in a 13.56 MHz RF field was almost 3°C/s. This remarkable heating rate at very low concentrations of gold nanoparticles represents a heating efficiency of >30,000 W/g of nanoparticles and indicates the generation of elevation temperatures in the nanoscale environment surrounding the gold nanoparticles sufficient to denature proteins, melt lipid bilayers, and produce irreparable damage to intracellular structures and organelles

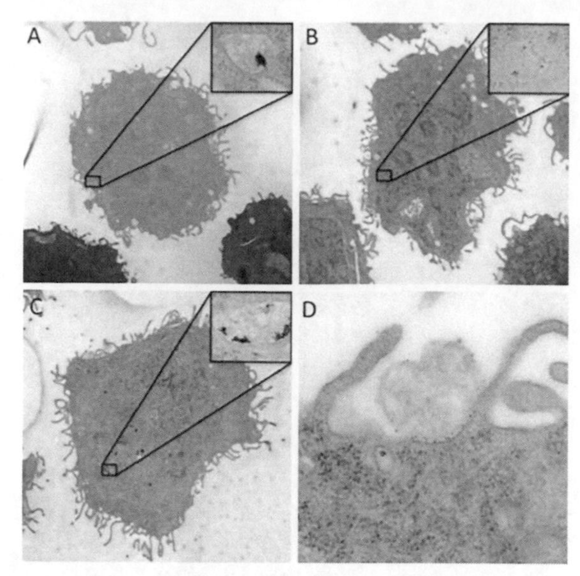

Thermal Cancer Ablation Therapies Using Nanoparticles, Fig. 4 (**a**). Transmission electron microscope (TEM) image of Panc-1 human pancreatic adenocarcinoma cells (×7,500 magnification) that were incubated for 30 min with unconjugated, nontargeted 5-nm gold nanoparticles (control group). Almost no gold nanoparticles are seen within the cytoplasm of the cell, and it is only at a magnification of ×100,000 (inset upper right) that an intracytoplasmic vesicle containing a small number of gold nanoparticles is seen. Incubation of Panc-1 cells for 60 or 120 min with the unconjugated gold nanoparticles also produced only a few scattered vesicles in the cytoplasm containing gold nanoparticles. (**b**). TEM image of Panc-1 cells incubated for 30 min with nonspecific IgG-conjugated gold nanoparticles (control group). Similar to the unconjugated gold nanoparticles, the nonspecific IgG-conjugated gold nanoparticles were not readily taken into the Panc-1 cells. The larger image (×7,500 magnification) demonstrates few or no obvious intracytoplasmic vesicles containing gold nanoparticles. At higher magnification (×100,000, inset upper right), an occasional intracytoplasmic vesicle with a few gold nanoparticles is evident. (**c**). TEM image of Panc-1 pancreatic adenocarcinoma cells incubated for 30 min with C225-conjugated 5-nm gold nanoparticles. At 30 min, a number of intracytoplasmic vesicles (dense black vesicles) containing gold nanoparticles are evident in the larger image (×7,500 magnification). At a higher magnification (×100,000), numerous vesicles containing dense collections of gold nanoparticles are found throughout the cytoplasm. At 60 and 120 min of incubation with the C225-targeted gold nanoparticles, the number of intracytoplasmic vesicles containing gold nanoparticles was seen to increase. (**d**). A higher magnification TEM image (×100,000) showing details of the cell surface of Panc-1 cells incubated for 30 min with C225-conjugated gold nanoparticles. Numerous gold nanoparticles are seen at the surface of the cell consistent with binding of the C225 to epidermal growth factor receptor (EGFR). Pretreatment of the cells with C225 alone completely blocked any binding of targeted gold nanoparticles to the cell surface and led to no gold nanoparticles appearing in intracytoplasmic vesicles 30, 60, or 120 min after incubation

been found to cause any acute or long-term damage to normal cells and tissues. Thus, the finding that charged gold nanoparticles release heat in response to shortwave RF exposure for durations of seconds to a few minutes has important implications. The physical mechanism of this phenomenon is being investigated, and data suggests that conductive polarizations of metal nanoparticles may explain the high thermal efficiency and tremendous heat release at low concentrations of nanoparticles exposed to the RF field. Importantly, since tissue penetration by shortwave RF fields is excellent with low attenuation of the signal, it should be possible to treat tumors at almost any site in the body. The crucial component of this treatment will be successful delivery of adequate numbers of gold nanoparticles to cancer cells to produce intracellular temperatures of at least 55°C following an RF field exposure of no more than a few minutes duration, while ideally limiting uptake of the nanoparticles by normal cells.

It has shown that by directly conjugating 5-nm gold nanoparticles to cetuximab to target epidermal growth factor receptors (EGFRs) expressed on cancer cells, there is enhanced uptake of the gold nanoparticles (Fig. 4). In contrast, cells expressing little or no

EGFR should have no enhanced uptake of cetuximab-conjugated gold nanoparticles and, thus, should not be significantly affected by treatment in the RF field. Targeting gold nanoparticles approximately 5–10 nm in diameter to cancer cells to create RF-induced hyperthermic cytotoxicity has several advantages: gold nanoparticles are simple and inexpensive to synthesize; they are easily characterized due to the presence of a characteristic surface plasmon resonance band; their surface chemistry permits manipulation of charge and shape relatively easily; attaching cancer cell targeting molecules, including antibodies, peptides, or pharmacologic agents, is easily achieved; and they are biocompatible and not associated with any acute or chronic toxicities in preclinical studies [14]. In addition to producing significant heat upon exposure to shortwave RF fields, 5–10-nm gold charged nanoparticles easily penetrate through pores and fenestrations in the neovasculature of solid tumors. This method may be promising for focused local or regional application of a noninvasive RF field to treat primary or established metastatic malignant tumors overexpressing EGFR, but would be problematic for whole-body RF therapy to treat diffuse micrometastatic disease. Constitutive expression of EGFR at normal levels occurs throughout many tissues in the body, and the results in SN12PM6 renal cancer cells (low-level expression of EGFR) show that even with low levels of EGFR expression, there is enough uptake of cetuximab-targeted gold nanoparticles to produce low levels of thermal apoptosis following exposure to the RF field. It may be possible to modulate or minimize effects on normal tissues expressing a target molecule like EGFR by use of brief duration pulsed RF therapy or repeated short duration exposures, but these effects must be studied in greater detail to confirm this theoretical consideration. Nonetheless, these results indicate that it is possible to use overexpression of a specific cell surface moiety, e.g., EGFR, to enhance delivery and uptake of gold nanoparticles to produce thermal cytotoxicity in the malignant cells following brief exposure to a noninvasive RF field.

There are numerous other possibilities for targeted delivery of nanoparticles into malignant cells. Radiolabeled carbon nanotubes have been conjugated with rituximab and lintuzumab to diagnose and potentially treat lymphoma cells [15]. Recently, aptamers, nucleic acid ligands much smaller than monoclonal antibodies, have been shown to target specific molecules in the neovasculature of tumors or on the surface of different types of cancer cells [16]. These aptamers have been conjugated to gold nanoparticles initially for use as an improved diagnostic technique. However, co-conjugation of cytotoxic agents, or use of our noninvasive RF field treatment technique, could be used as an actual anticancer therapy using this targeting technique. So-called cell-penetrating peptides, generally less than 100 amino acids in length, also have been shown to target certain types of cancer cells. The 86-amino acid HIV-1 Tat protein has been conjugated to gold nanoparticles, with subsequent rapid intracellular uptake of the nanoparticles and localization to the nucleus [17, 18]. The monoclonal antibody-conjugated gold nanoparticles used by our group are found in the cytoplasm of cells within 30 min of adding these nanoparticle conjugates to cancer cells in culture. Localization of the nanoparticles to the nucleus with thermal destruction of DNA following RF field exposure must be investigated as a technique to enhance cancer cell destruction. Chlorotoxin is another small peptide (36 amino acids) cancer cell agent that can be easily conjugated with gold nanoparticles and is currently being investigated as a cancer cell targeting agent in several human malignant cell lines [19]. Identification of cancer-specific ligands not expressed on normal cells will permit targeting of gold nanoparticles to only malignant cells for treatment in our noninvasive RF field generator and should allow treatment of measurable tumors and micrometastatic disease with minimal therapy-related toxicities. Targeting ligands, like EGFR, that are overexpressed but not unique to cancer cells are feasible but will require careful evaluation to determine acute and long-term thermal effects in normal cells and organs following RF field therapy.

Nanoscale thermal therapy of targeted cancer cells is a promising new weapon in the battle against cancer. By combining nonionizing radiation and targeted metallic nanoparticles, noninvasive targeted thermal therapy has the promise of benefiting cancer patients that need a nontoxic alternative to the debilitating effects of conventional chemotherapeutics. Using noninvasive RF fields, magnetic fields, or laser irradiation to heat nanoscale materials that are internalized into cancer cells or the tumor microvasculature has the promise of reducing toxicity and maximizing therapeutic benefit for cancer patients.

References

1. Beasley, G.M., Ross, M.I., Tyler, D.S.: Future directions in regional treatment strategies for melanoma and sarcoma. Int. J. Hyperthermia **24**, 301–309 (2008)
2. Stewart, J.H.T., Shen, P., Russell, G., Fenstermaker, J., McWilliams, L., Coldrun, F.M., Levine, K.E., Jones, B.T., Levine, E.A.: A phase I trial of oxaliplatin for intraperitoneal hyperthermic chemoperfusion for the treatment of peritoneal surface dissemination from colorectal and appendiceal cancers. Ann. Surg. Oncol. **15**, 2137–2145 (2008)
3. Kampinga, H.H.: Cell biological effects of hyperthermia alone or combined with radiation or drugs: a short introduction to newcomers in the field. Int. J. Hyperthermia **22**, 191–196 (2006)
4. Arciero, C.A., Sigurdson, E.R.: Diagnosis and treatment of metastatic disease to the liver. Semin. Oncol. **35**, 147–159 (2008)
5. Hartman, K.B., Wilson, L.J., Rosenblum, M.G.: Detecting and treating cancer with nanotechnology. Mol. Diagn. Ther. **12**, 1–14 (2008)
6. Neumaier, C.E., Baio, G., Ferrini, S., Corte, G., Daga, A.: MR and iron magnetic nanoparticles. Imaging opportunities in preclinical and translational research. Tumori **94**, 226–233 (2008)
7. Misra, R.D.: Quantum dots for tumor-targeted drug delivery and cell imaging. Nanomed **3**, 271–274 (2008)
8. Gobin, A.M., Lee, M.H., Halas, N.J., James, W.D., Drezek, R.A., West, J.L.: Near-infrared resonant nanoshells for combined optical imaging and photothermal cancer therapy. Nano Lett. **7**, 1929–1934 (2007)
9. Kalambur, V.S., Longmire, E.K., Bischof, J.C.: Cellular level loading and heating of superparamagnetic iron oxide nanoparticles. Langmuir **23**, 12329–12336 (2007)
10. Gannon, C.J., Cherukuri, P., Yakobson, B.I., Cognet, L., Kanzius, J.S., Kittrell, C., Weisman, R.B., Pasquali, M., Schmidt, H.K., Smalley, R.E., Curley, S.A.: Carbon nanotube-enhanced thermal destruction of cancer cells in a noninvasive radiofrequency field. Cancer **110**, 2654–2665 (2007)
11. Gannon, C.J., Patra, C.R., Bhattacharya, R., Mukherjee, P., Curley, S.A.: Intracellular gold nanoparticles enhance non-invasive radiofrequency thermal destruction of human gastrointestinal cancer cells. J. Nanobiotechnol. **6**, 2 (2008)
12. Adair, E.R., Blick, D.W., Allen, S.J., Mylacraine, K.S., Ziriax, J.M., Scholl, D.M.: Thermophysiological responses of human volunteers to whole body RF exposure at 220 MHz. Bioelectromagnetics **26**, 448–461 (2005)
13. Klima, J., Scehovic, R.: The field strength measurement and SAR experience related to human exposure in 110 MHz to 40 GHz. Meas. Sci. Rev. **6**, 40–44 (2006)
14. Daniel, M.C., Astruc, D.: Gold nanoparticles: assembly, supramolecular chemistry, quantum-size-related properties, and applications toward biology, catalysis, and nanotechnology. Chem. Rev. **104**, 293–346 (2004)
15. McDevitt, M.R., Chattopadhyay, D., Kappel, B.J., Jaggi, J.S., Schiffman, S.R., Antczak, C., Njardarson, J.T., Brentjens, R., Scheinberg, D.A.: Tumor targeting with antibody-functionalized, radiolabeled carbon nanotubes. J. Nucl. Med. **48**, 1180–1189 (2007)
16. Javier, D.J., Nitin, N., Levy, M., Ellington, A., Richards-Kortum, R.: Aptamer-targeted gold nanoparticles as molecular-specific contrast agents for reflectance imaging. Bioconjug. Chem. **19**, 1309–1312 (2008)
17. Berry, C.C.: Intracellular delivery of nanoparticles via the HIV-1 tat peptide. Nanomed **3**, 357–365 (2008)
18. Berry, C.C., de la Fuente, J.M., Mullin, M., Chu, S.W., Curtis, A.S.: Nuclear localization of HIV-1 tat functionalized gold nanoparticles. IEEE Trans. Nanobioscience **6**, 262–269 (2007)
19. Mamelak, A.N., Jacoby, D.B.: Targeted delivery of antitumoral therapy to glioma and other malignancies with synthetic chlorotoxin (TM-601). Expert Opin. Drug Deliv. **4**, 175–186 (2007)

Thermal Chemical Vapor Deposition

▶ Chemical Vapor Deposition (CVD)

Thermal Conductance

▶ Thermal Conductivity and Phonon Transport

Thermal Conductivity and Phonon Transport

Yee Kan Koh
Department of Mechanical Engineering, National University of Singapore room: E2 #02-29, Singapore, Singapore

Synonyms

Heat conduction; Heat conductivity; Nanoscale heat transport; Thermal conductance; Thermal resistance; Thermal resistivity; Thermal transport

Definition

Thermal conductivity (symbols: κ or Λ) is an empirical material property that relates local heat flux (J) across a real or imaginary surface to the temperature gradient ($\partial T/\partial x$) at the same point in direction x normal to the surface. Normally, the concept of thermal conductivity is used for heat conduction not mediated by massive advection of atoms and molecules, or propagation of electromagnetic waves; such heat transport is traditionally referred to as convective and radiative heat transfer, respectively.

Thermal conductivity is defined according to the Fourier's law of heat conduction given below,

$$J = -\kappa \frac{\partial T}{\partial x}$$

The negative sign indicates that heat flow against temperature gradient, i.e., from the hot side to the cold side. The units commonly used for thermal conductivity include watts per meter-Kelvin (W m^{-1} K^{-1}) and watts per centimeter-Kelvin (W cm^{-1} K^{-1}). The reciprocal of thermal conductivity is called the thermal resistivity.

Overview

As summarized by the second law of thermodynamics, Nature seeks to approach equilibrium in an isolated system, be it a mechanical, chemical, or thermal system. In the context of heat transfer, heat flows from the hot side to the cold side whenever the system is not in thermal equilibrium (i.e., temperature is not uniform), until all temperature difference is ultimately eliminated. Although the second law of thermodynamics is very general and applies to any thermal system, it does not however stipulate how equilibrium is reached. In other words, although heat transfer occurs whenever temperature difference exists according to the second law of thermodynamics, the rate of heat transfer differs from systems to systems depending on the properties of materials or mediums in the thermal system.

Historically, heat transfer is classified as conduction, convection, and radiation, depending on the mediums involved and the mechanisms in which heat is transported across the mediums. Conduction usually refers to heat transfer in solids, in which atoms and molecules are practically immobile and diffusion of atoms and molecules is relatively slow (as long as the solids are not at highly elevated temperatures). In such cases, heat is predominantly carried by vibration of atoms, or other quasiparticles in the crystals such as electrons and holes (see more discussions on quasiparticles below). Also, in still fluids, heat transfer is due to vibration and collisions of atoms and molecules, and thus is also categorized as conduction. On the other hand, the term convection is mainly used for heat transfer from a solid surface to fluids, mediated by diffusion or advection (macroscopic motion) of molecules and atoms. In convection, fluids are driven either by external means such as pumps (forced convection) or by buoyancy forces (natural convection). Lastly, when heat is carried by electromagnetic waves, the heat transfer is called radiation. Unlike conduction and convection, no mediums are required for radiation, since electromagnetic waves can travel through vacuum. For example, heat is transported from the Sun to the Earth by visible or infrared light through radiation. Of course, radiation could also occur through transparent mediums (e.g., glass, air etc.).

For heat conduction, the rate of thermal energy transfer per unit time and unit area is proportional to temperature gradient in the respective direction. This empirical relationship is known as the Fourier's law of heat conduction. Mathematically,

$$J = -\kappa \frac{\partial T}{\partial x}$$

where J is the heat flux across a real or imaginary interface (SI unit: W m^{-2}) and $\partial T/\partial x$ is the local temperature gradient (SI unit: K m^{-1}). Thermal conductivity (κ) is defined as the proportional constant in the Fourier's law of heat conduction and has the SI unit of W m^{-1} K^{-1}. Thermal conductivity for fully condensed solids spans for more than five orders of magnitude, from κ >1,000 W m^{-1} K^{-1} for carbon allotropes [1, 2], to κ <0.01 W m^{-1} K^{-1} for disordered crystals [3]. Theoretically, thermal conductivity is an intrinsic material property and thus should not depend on the geometry and size of the materials. This theoretical definition of thermal conductivity is somewhat relaxed recently for the thermal conductivity of nanostructures. Thermal conductivity of nanostructures depends on the characteristic lengths of the nanostructures, see the discussions below.

The effect of thermal conductivity on heat conduction in solids is illustrated in Fig. 1. For a constant heat

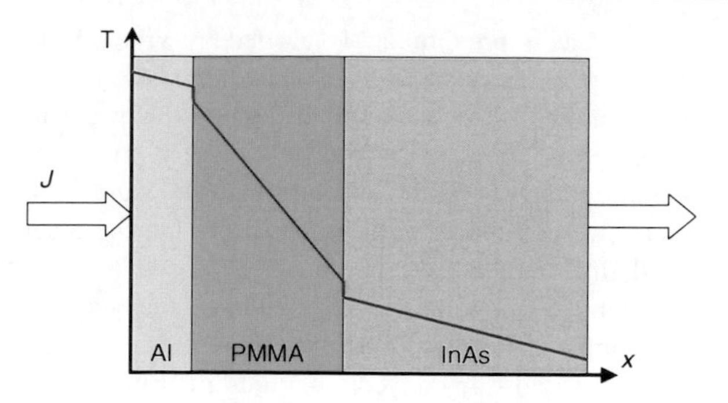

Thermal Conductivity and Phonon Transport, Fig. 1 Schematics of a hypothetical multilayered solid of aluminum (Al) with κ >200 W m^{-1} K^{-1}, amorphous polymer PMMA with $\kappa \approx 0.2$ W m^{-1} K^{-1}, and semiconductor indium arsenide (InAs) with $\kappa \approx 27$ W m^{-1} K^{-1}, being heated uniformly with a constant heat flux of J. The red solid line is the temperature profile in the hypothetical solid. Different slopes of the temperature profile are due to different thermal conductivity of each layer. Temperature drops at the interfaces are due to finite Kapitza thermal conductance of the interfaces

flux applied on a hypothetical multilayered solid as in Fig. 1, temperature drop across each layer is inversely proportional to the thermal conductivity of the layer. For a high thermal conductivity material (e.g., aluminum), the temperature drop is minimal since heat could be easily transported across the material. In other words, thermal energy could easily spread across high thermal conductivity materials, resulting in a near-uniform temperature distribution. On the other hand, low thermal conductivity impedes heat transport, and thus higher temperature difference could be maintained in the materials. This simple understanding of thermal conductivity is crucial for many modern applications (such as thermal barrier coatings and thermoelectrics), in which high temperature differences enhance the performance of the devices. Researchers in these fields seek to modify the existing materials, often through nanostructuring, to reduce the thermal conductivity so that a higher temperature difference could be maintained in the devices.

While the thermal conductivity is a bulk property, interfacial or Kapitza thermal conductance [4] (symbol: G) is a property of interfaces. As heat flows across an interface, a finite temperature drop is observed at the interface, as illustrated in Fig. 1, due to resistance of the interface to thermal transport. Kapitza thermal conductance (G) relates this temperature drop (ΔT) to heat flux across the interface; $J = -G\,\Delta T$. Typically, thermal conductance ranges from $G > 1$ GW m^{-2} K^{-1} for metal/metal interfaces [5] to $G \approx 10$ MW m^{-2} K^{-1} for interfaces with highly dissimilar materials [6]. It is important to realize that the Kapitza thermal conductance is not solely due to imperfections of interfaces (e.g., imperfect contact as a result of surface roughness, diffuse interfaces due to intermixing, etc.). Even for atomically perfect interfaces, heat flow is still impeded by finite thermal conductance because differences in the properties (e.g., density of phonon states) of two mediums reduce the probability for heat carriers to transmit across the interfaces. In other words, since interfaces break the reflection and translational symmetry, even perfect interfaces are imperfections by definition. Thus, heat transport is retarded at interfaces due to this break of symmetry engendered by the interfacial imperfections.

Heat Transport by Phonons in Nonmetallic Solids

In a microscopic view, heat conduction in solids is due to quasiparticles that carry heat across the solids. In simple words, quasiparticles (especially those important for heat conduction in solids) are quantized excitation modes resulting from quantum-mechanical interaction of atoms and electrons in the periodic lattices. Strictly speaking, some of these quasiparticles are only theoretically defined for perfect crystals, but this restriction is usually relaxed for imperfect crystals. Quasiparticles behave like real particles with properties and behavior modified from their real particle counterparts.

Common quasiparticles that are important for heat conduction in solids include phonons, electron quasiparticles and holes. Phonons are quantized lattice vibration modes due to interaction of sound waves in solids and are responsible for heat conduction in most nonmetallic solids (dielectrics and semiconductors). Electron quasiparticles are quantized excitation modes due to interaction between electrons and periodic electrical potential in the lattices. Electron quasiparticles are the main carriers for both heat and charges in metals, and as a result of electron quasiparticles being the common carriers, the thermal conductivity and electrical conductivity of metals are related by a fundamental relationship called the

Wiedemann–Franz law. Electron quasiparticles also contribute significantly to heat conduction in highly n-doped semiconductors and semimetals. Holes, on the other hand, are empty states available in the crystals that are not filled by electrons. Holes behave similar to electron quasiparticles, except that they carry positive charges and have different effective masses. Holes can contribute significantly to heat conduction in highly p-doped semiconductors and semimetals. Note that electron quasiparticles have different behavior (e.g., different effective masses) from stand-alone electrons, even though both carry negative charges and both are fermions. Consistent with the literature and for simplicity, electron quasiparticles are thereafter referred to as "electrons" throughout the text.

To understand heat conduction by phonons in crystalline dielectrics and semiconductors, it is important to realize that phonons do not carry same amounts of heat due to different polarizations (longitudinal or transverse) and a broad distribution of frequencies (ω) and lifetimes (τ) of phonons. Heat conduction by phonons depends on three properties of phonons:

1. The volumetric heat capacity (C) of phonon modes, which affects the amount of heat carried by phonons. The volumetric heat capacity is a function of phonon frequency, polarization, and temperature. At high temperatures, the heat capacity of each phonon mode approaches a classical value of k_B, where $k_B = 1.38 \times 10^{-23}$ J K^{-1} is the Boltzmann constant.
2. The group velocity (υ) of phonons, which determines how fast phonons propagate in the materials and is defined as

$$\upsilon = \frac{\partial \omega}{\partial k}$$

where k is the wavenumber of phonons. The group velocity depends on the slope of phonon dispersion and thus is a function of phonon frequency (ω) and polarization. At low frequencies, the group velocity of acoustic phonons approach the speed of sounds of the materials, while at high frequencies, the group velocity of acoustic phonons approach zero due to interference of reflected waves near the Brillouin zone boundaries.
3. The mean-free-path (ℓ) of phonons, which is the average distance phonons propagate without being scattered. In crystals, phonons are scattered by imperfections including impurities, anharmonicity, and grain boundaries. This scattering of phonons is the source of retardation for heat conduction in crystals, since perfect crystals have infinitely high thermal conductivity. Normally, the scattering process is quantified by the relaxation time (τ) or the mean-free-path (ℓ) of phonons; the relaxation time and mean-free-path are related by $\ell = \upsilon\tau$, where υ is the group velocity of phonons. Both the relaxation time and the mean-free-path strongly depend on the frequency of phonon modes.

At high temperatures, most phonon modes are excited, but heat is carried mainly by a small fraction of these excited modes (long-wavelength acoustic phonons). Acoustic phonons near the edge and the center of the Brillouin zone, and optical phonons are inefficient in carrying heat due to their low group velocity or heat capacity. The mean-free-paths of these heat-carrying acoustic phonons, however, are not constant but span more than an order of magnitude in most crystals depending on phonon frequency. For example, calculations [7] show that $\approx 90\%$ of heat could be carried by phonons with $0.1 \leq \ell(\omega) \leq 1$ μm in Si. The mean-free-paths of phonons are also sensitive to the purity and harmonicity of materials, since imperfections, such as impurities, point defects and anharmonicity, scatter phonons. In fact, the disparity in the thermal conductivity of different materials is mainly due to differences in the mean-free-paths (instead of the heat capacity or the group velocity) of phonons in these materials, since the heat capacity and phonon velocity of different materials do not vary by more than an order of magnitude.

Despite the importance, understanding of the mean-free-paths of phonons is still incomplete, mainly due to lack of a direct method to measure the mean-free-paths of phonons [8]. The mean-free-paths of phonons are usually inferred from systematic measurements of the thermal conductivity by comparing the experimental results to theoretical calculations. Figure 2 shows an example of calculations of the mean-free-paths of acoustic phonons in bulk silicon, a 200 nm silicon thin film and bulk $Si_{0.1}Ge_{0.9}$ alloy, as a function of phonon frequency. Note that different imperfections are efficient in scattering phonons of different frequency. At room temperature, phonons in bulk Si are predominantly scattered by anharmonicity, which is

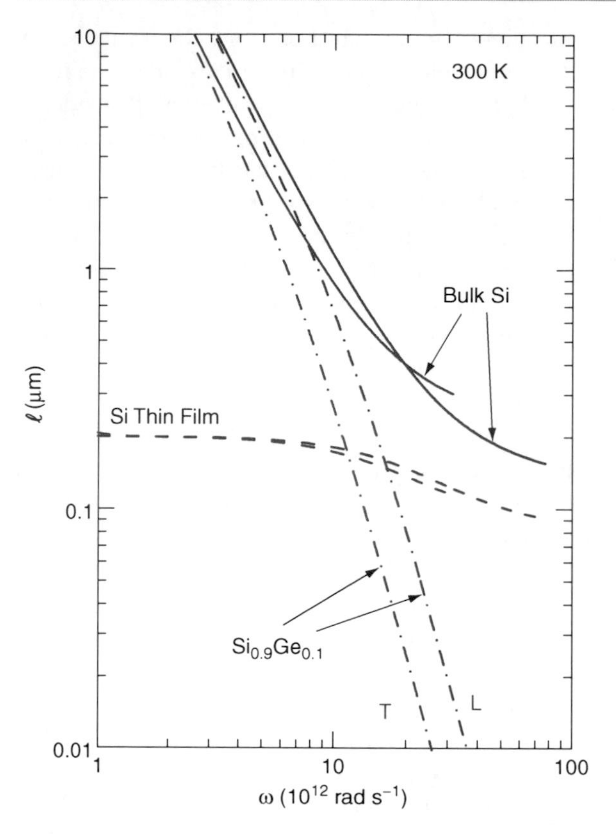

Thermal Conductivity and Phonon Transport, Fig. 2 Mean-free-path (ℓ) of longitudinal (*red*) and transverse (*blue*) acoustic phonons in bulk Si (*solid lines*), a 200 nm Si thin film (*dashed lines*) and bulk $Si_{0.9}Ge_{0.1}$ alloy (*dash-dotted lines*) calculated from a Callaway-type model [8]. ω is the frequency of phonons. The dominant scattering mechanisms in bulk Si, the SiGe alloy, and the Si thin film are three-phonon process (anharmonicity), Rayleigh scattering, and boundary scattering, respectively

effective to scatter high-frequency phonons but not low-frequency phonons, see Fig. 2. Similarly, Rayleigh scattering, which is the dominant scattering mechanism for phonons in alloys, is more effective to scatter high-frequency phonons than low-frequency phonons. On the other hand, boundary scattering is effective to scatter low- and medium-frequency phonons. Over the past decade, nanostructures, such as thin films, superlattices, and nanowires, have been widely used to scatter the low-energy phonons to further reduce the thermal conductivity. It is stressed that the scattering rate and mean-free-paths of phonons in most materials are *not* known – calculations as in Fig. 2 are at best just an approximation. In fact, polarization of the dominant heat carriers in silicon was determined as recent as a decade ago [9].

Thermal Conductivity Reduction in Nanostructures

Over the past decade, researchers advanced the understanding of heat conduction in nanostructures through systematic measurements of the thermal conductivity of nanostructures. This understanding on nanoscale heat transport facilitates the design of more efficient thermoelectric materials with reduced thermal conductivity. In this section, prior experimental studies on heat conduction by phonons in different nanostructures, including crystalline thin films, superlattices, nanowires, and semiconductor embedded with nanoscale precipitates, are briefly and selectively review.

Goodson and coworkers [9, 10] reported the in-plane thermal conductivity of Si thin films over a wide film thickness (74–1.6 μm) and temperature (20–320 K) range. The in-plane thermal conductivity was measured by the 3ω method [11], which is discussed below. Heat was previously perceived to be carried by transverse acoustic phonons in Si, but Ju and Goodson [9] find from the in-plane thermal conductivity measurements of Si thin films that longitudinal acoustic phonons with mean-free-path $\ell \approx 300$ nm are the dominant carriers in Si at 300 K.

Yao [12] first reported the cross-plane thermal conductivity of AlAs/GaAs superlattices, measured using an ac calorimetric method. He finds that the thermal conductivity of a $(AlAs)_{5\ nm}(GaAs)_{5\ nm}$ superlattice approaches the thermal conductivity of AlGaAs alloys at room temperature. The observation that the thermal conductivity of superlattices approaches the thermal conductivity of the alloys is not universal. For example, Lee and coworkers [13] reported the thermal conductivity of fully strained Si/Ge superlattices, measured by the 3ω method. They find that for period <6 nm, the thermal conductivity decreases with decreasing period and is governed by the thermal conductance of interfaces. For longer period of >10 nm, however, the defect density is so high that the thermal conductivity approaches the amorphous limit. Further measurements [14, 15] on other Si/Ge nanostructures are consistent with measurements by Lee et al.

Venkatasubramanian [16] reported the thermal conductivity of Bi_2Te_3/Sb_2Te_3 superlattices as a function of superlattice period and find that the thermal conductivity of short-period superlattice is minimum when the period is ≈ 4 nm. Subsequent measurements on similar samples by Touzelbaev and coworkers [17], however, do not present similar minimum thermal conductivity. The minimum thermal conductivity is also not observed on AlN/GaN superlattices [18], in which the AlN/GaN interfaces are atomically sharp and chemically abrupt.

Li and coworkers [19] first measured the thermal conductivity of Si nanowires using a microfabricated device. Li et al. find that the thermal conductivity of nanowires is reduced by more than an order of magnitude when the diameter of the nanowire is 22 nm. The approach has subsequently been applied to study the thermal conductivity of other nanostructures, including rough Si nanowires [20]. The reduction of thermal conductivity, in most cases, is due to enhanced boundary scattering.

Harman and coworkers [21] estimated the thermal conductivity of PbTe/PbSe nanodot superlattices from their thermoelectric devices. They found that the thermal conductivity is 0.33 W m^{-1} K^{-1}, a factor of 3 lower than the thermal conductivity of the corresponding alloys. More complete and systematic measurements [22] on similar materials, however, indicate that the thermal conductivity of PbTe/PbSe nanodot superlattices is on the order of 1 W m^{-1} K^{-1}, comparable with the thermal conductivity of PbTeSe alloys.

The thermal conductivity of InGaAs with ErAs nanoparticles randomly distributed in the InGaAs matrix was first reported by Kim and coworkers [23]. The thermal conductivity of the nanostructured materials is found to be reduced from the thermal conductivity of the alloys by up to a factor of 2, even when the concentration of the ErAs nanoparticles is relatively small (0.3%). This reduction of thermal conductivity is explained by enhanced Rayleigh scattering when the size of the nanoparticles better match the wavelength of phonons. Subsequent measurements [24] on InGaAs with higher concentration of ErAs nanoparticles indicate that the thermal conductivity could be further reduced with higher concentration of ErAs nanoparticles.

Modern Techniques for Measurements of the Thermal Conductivity of Nanostructures

For measurements of heat conduction on submicron or nanometer length scales, temperature profile has to be monitored within the similar characteristic length scales (~1 μm) or equivalently the temperature decay profile in nanosecond time scales. Conventional approaches to measure the thermal conductivity are incapable to reach these small length and time scales, and thus are not suitable for measurements of the thermal conductivity of nanostructures. In this section, two important techniques that are capable of characterizing the thermal properties of nanostructures, namely, the 3ω method and the time-domain thermoreflectance (TDTR), are introduced.

The 3ω method [11] is widely used to measure the thermal conductivity of bulk solids and thin films. In 3ω measurements, a narrow metal line (often Au or Pt with Cr or Ti as the adhesion layer) of ~400 nm thickness and ~30 μm width is patterned on a sample. Electrical current $\boldsymbol{i}$ of frequency ω is applied to the metal line with electrical resistance $\boldsymbol{r}$, generating joule heating of $\boldsymbol{i}^2\boldsymbol{r}$ within the metal line with a frequency component at 2ω. As a result of this oscillating heat source, a temperature oscillation and a corresponding resistance oscillation at frequency 2ω are induced in the metal line. Hence, a component of the voltage oscillation ($\boldsymbol{v} = \boldsymbol{ir}$) across the metal line contains a third harmonic, 3ω. The thermal conductivity of the sample can be deduced from this 3ω voltage oscillation.

The cross-plane thermal conductivity of thin films can be measured by the differential 3ω method [16]. A reference sample without the thin film of interest is prepared simultaneously with the sample containing the film of interest such that the metal line patterned on both samples has the same thickness and width. Both the thin film sample and the reference are measured using similar heating power and the same range of heater frequencies. The temperature drop across the thin film ΔT_f can then be derived from the difference in the amplitude of the temperature oscillation of the sample and the reference. Assuming one-dimensional (1D) heat conduction, the cross-plane thermal conductivity of the thin film is derived from ΔT_f. If the thin film is semiconducting, a thin dielectrics layer is required to electrically insulate the metal lines from the thin films. For semiconducting crystalline thin films, the differential 3ω method is usually only suitable to measure the thermal conductivity of the films if the films are >1 μm thick [24].

Another important technique for measurements of heat conduction on nanometer length scale is the time-domain thermoreflectance (TDTR) [25]. Figure 3

T

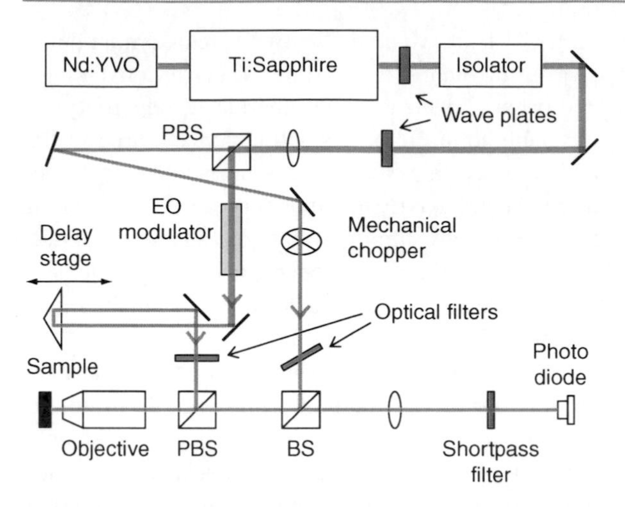

Thermal Conductivity and Phonon Transport, Fig. 3 A schematic [26] of the TDTR setup. The optical filters and the shortpass filter are optional and can eliminate the artifacts due to diffusely scattered pump beam leaked into the Si photodiode during TDTR measurements of rough samples (Reprinted with permission from [26]. Copyright 2008, American Institute of Physics)

shows a schematic diagram [26] of the TDTR setup. Before TDTR measurements, the samples are usually coated with a thin layer (~80 nm) of metal with high thermoreflectance (e.g., Al) by magnetron sputtering or thermal evaporation. The metal film serves as a transducer to absorb the heating pump beam and to convert the temperature excursions at the surface into changes in the intensity of the reflected probe beam. Usually, Al is used as the metal transducer due to the high thermoreflectance. However, for high temperature measurements, Al is not suitable due to its low melting point; Pt, Ta, or AuPd could be used for high temperature measurements, see [27] for further discussion. The thickness of the metal transducer layer is normally simultaneously determined during TDTR measurements by picosecond acoustics.

In TDTR measurements, the output of a mode-locked Ti:sapphire laser oscillator is split into a pump beam and a probe beam, with the relative delay time between the pump and probe pulses being adjusted via a mechanical stage. Samples are heated by the pump beam, which is modulated by an electro-optic modulator at frequency f, $0.1 < f < 10$ MHz. Cooling of the surface after being heated by pump pulses is then monitored through changes in the intensity of the reflected probe beam using a Si photodiode and a radio-frequency (rf) lock-in amplifier. If the samples are rough, TDTR measurements should be performed using a two-tint configuration [26], see Fig. 3. Normally, heat waves only penetrate <1 μm in TDTR measurements, while the $1/e^2$ radius of the laser beam is usually >5 μm. Hence, heat flow in TDTR is predominantly one dimensional.

Data analysis in TDTR is considerably complicated. The changes in reflected intensity at frequency f have both an in-phase V_{in} and out-of-phase V_{out} components. Normally, the ratio V_{in}/V_{out} is used for the analysis (instead of just the V_{in} component) to make use of the additional information in the out-of-phase signal and eliminate artifacts created by unintended variations in the diameter or position of the pump beam created by the optical delay line. Measurements of the ratios V_{in}/V_{out} as a function of delay time are compared to numerical solutions of the diffusion equation in cylindrical coordinates using a thermal model [28]. The thermal model normally has two free parameters: the thermal conductance of the Al/sample interface and Λ of the sample. For most cases, these two parameters can be separated from the fitting of the model calculations to the measurements [29].

Future Directions for Research

Detailed understanding of heat transport in crystalline solids, especially on how phonons are scattered in nanostructures, is crucial for the design of new thermoelectric materials and thermal management of electronic devices. In this regard, future research will focus on experimental and theoretically studies on the details of how phonons carry heat across different kinds of nanostructures. One example is recent efforts by researchers in the field to investigate how heat is transported along [1, 2] and across [30] monolayer and few-layer graphenes. By carefully comparing the experimental results to theoretical calculations, these studies advance our knowledge on how phonons are scattered by impurities and nanostructures and thus facilitate future design of more efficient devices.

Cross-References

▸ Finite Element Methods for Computational Nano-optics
▸ Nanostructured Thermoelectric Materials
▸ Nanostructures for Energy
▸ Thermoelectric Heat Convertors

References

1. Balandin, A.A., et al.: Nano Lett. **8**, 902 (2008)
2. Seol, J.H., et al.: Science **328**, 213 (2010)
3. Chiritescu, C., et al.: Science **315**, 351 (2007)
4. Swartz, E.T., Pohl, R.O.: Rev. Mod. Phys. **61**, 605 (1989)
5. Gundrum, B.C., Cahill, D.G., Averback, R.S.: Phys. Rev. B **72**, 245426 (2005)
6. Lyeo, H.-K., Cahill, D.G.: Phys. Rev. B **73**, 144301 (2006)
7. Koh, Y. K.: Ph.D thesis, University of Illinois at Urbana-Champaign (2010)
8. Koh, Y.K., Cahill, D.G.: Phys. Rev. B **76**, 075207 (2007)
9. Ju, Y.S., Goodson, K.E.: Appl. Phys. Lett. **74**, 3005 (1999)
10. Asheghi, M., Leung, Y.K., Wong, S.S., Goodson, K.E.: Appl. Phys. Lett. **71**, 1798 (1997)
11. Cahill, D.G., Pohl, R.O.: Phys. Rev. B **35**, 4067 (1987)
12. Yao, T.: Appl. Phys. Lett. **51**, 1798 (1987)
13. Lee, S.-M., Cahill, D.G., Venkatasubramanian, R.: Appl. Phys. Lett. **70**, 2957 (1997)
14. Huxtable, S.T., Abramson, A.R., Tien, C.-L., Majumdar, A., LaBounty, C., Fan, X., Zeng, G., Bowers, J.E., Shakouri, A., Croke, E.T.: Appl. Phys. Lett. **80**, 1737 (2002)
15. Yang, B., Liu, J.L., Wang, K.L., Chen, G.: Appl. Phys. Lett. **80**, 1758 (2002)
16. Venkatasubramanian, R.: Phys. Rev. B **61**, 3091 (2000)
17. Touzelbaev, M.N., Zhou, P., Venkatasubramanian, R., Goodson, K.E.: J. Appl. Phys. **90**, 763 (2001)
18. Koh, Y.K., Cao, Y., Cahill, D.G., Jena, D.: Adv. Funct. Mater. **19**, 610 (2009)
19. Li, D., Wu, Y., Kim, P., Shi, L., Yang, P., Majumdar, A.: Appl. Phys. Lett. **83**, 2934 (2003)
20. Hochbaum, A.I., Chen, R., Delgado, R.D., Liang, W., Garnett, E.C., Najarian, M., Majumdar, A., Yang, P.: Nature **451**, 163 (2008)
21. Harman, T.C., Taylor, P.J., Walsh, M.P., LaForge, B.E.: Science **297**, 2229 (2002)
22. Koh, Y.K., Vineis, C.J., Calawa, S.D., Walsh, M.P., Cahill, D.G.: Appl. Phys. Lett. **94**, 153101 (2009)
23. Kim, W., Zide, J., Gossard, A., Klenov, D., Stemmer, S., Shakouri, A., Majumdar, A.: Phys. Rev. Lett. **96**, 045901 (2006)
24. Koh, Y.K., Singer, S.L., Kim, W., Zide, J.M.O., Lu, H., Cahill, D.G., Majumdar, A., Gossard, A.C.: J. Appl. Phys. **105**, 054303 (2009)
25. Paddock, C.A., Eesley, G.L.: J. Appl. Phys. **60**, 285 (1986)
26. Kang, K., Koh, Y.K., Chiritescu, C., Zheng, X., Cahill, D. G.: Rev. Sci. Instrum. **79**, 114901 (2008)
27. Wang, Y., Park, J.-Y., Koh, Y.K., Cahill, D.G.: J. Appl. Phys. **108**, 043507 (2010). Accepted
28. Cahill, D.G.: Rev. Sci. Instrum. **75**, 5119 (2004)
29. Costescu, R.M., Wall, M.A., Cahill, D.G.: Phys. Rev. B **67**, 054302 (2003)
30. Koh, Y.K., Bae, M.-H., Cahill, D.G., Pop, E.: Nano Lett. **10**, 4363 (2010)

Thermal Evaporation

▶ Physical Vapor Deposition

Thermal Infrared Detector

▶ Insect Infrared Sensors

Thermal Management Materials

▶ Modeling Thermal Properties of Carbon Nanostructure Composites

Thermal Resistance

▶ Thermal Conductivity and Phonon Transport

Thermal Resistivity

▶ Thermal Conductivity and Phonon Transport

Thermal Transport

▶ Thermal Conductivity and Phonon Transport

Thermally Actuated MEMS Resonators

▶ Thermally Actuated Silicon Resonators

Thermally Actuated Micromechanical Resonators

▶ Thermally Actuated Silicon Resonators

Thermally Actuated Nanoelectromechanical Resonators

▶ Thermally Actuated Silicon Resonators

Thermally Actuated Nanomechanical Resonators

► Thermally Actuated Silicon Resonators

Thermally Actuated Resonators

► Thermally Actuated Silicon Resonators

Thermally Actuated Silicon Resonators

Siavash Pourkamali
Department of Electrical and Computer Engineering, University of Denver, Denver, CO, USA

Synonyms

Thermally actuated MEMS resonators; Thermally actuated micromechanical resonators; Thermally actuated nanoelectromechanical resonators; Thermally actuated nanomechanical resonators; Thermally actuated resonators; Thermal-piezoresistive resonators

Definition

Thermally actuated electromechanical resonators are micro- to nanoscale structures whose specific mechanical resonant modes can be actuated by periodic thermal expansion forces resulting from joule heating in parts of their structure.

Introduction

Thermal actuation based on joule-heating-induced thermal expansion of solids is a well-known mechanism that can be conveniently implemented at micro- and nanoscale. An electrothermal actuator simply consists of a conductive element that heats up as a result of electrical current flowing through it and therefore can be implemented without any fabrication challenges or the need for material integration. Electrothermal actuators also have great properties such as large actuation force, low operating voltage, and simplicity of design and integration. On the downside, their considerable power consumption makes them an undesirable choice for many applications. Furthermore, thermal actuators are generally considered slow actuators only suitable for DC or very-low-frequency applications. This is mainly based on the general perception emanating from the more familiar marco-world where thermal phenomena have relatively large time constants in the few seconds or even minutes and hours range. However, as will be shown in this section, thermal actuation along with piezoresistive sensing provides an interesting option for transduction of high-frequency resonant modes of micro- and nanostructures.

Micro- and nanoscale electromechanical resonant devices have received a lot of attention over the past decade as a potential emerging technology for next generation integrated frequency references, electronic filters, and resonant sensors. As a result, a wide variety of high-frequency microelectromechanical resonator technologies have been developed over the past few years, most of which use piezoelectric [1, 2] or electrostatic (capacitive) [3–5] transduction. Despite significant improvements in both piezoelectric and capacitive microresonator technologies, each approach still has certain challenges to overcome. Piezoelectric microresonators require integration of piezoelectric and metallic thin films generally resulting in lower quality factors and frequency and quality control issues. In case of air-gap capacitive resonators, the weak electromechanical coupling necessitates deep submicron transduction gaps leading to fabrication and reliability challenges, power handling limitations, and excessive squeezed film damping when operating in air. Solid dielectric resonators [5] use electrostatic forces in dielectric thin films (preferably a high-K dielectric) to take advantage of the larger dielectric coefficients and improve the electromechanical coupling strength. The fully solid structure of such device relaxes several problems associated with aggressive reduction of the air gaps such as stiction and pull-in issues and allows reduction of the dielectric gap sizes to a few nanometers. However, similar to the piezoelectric resonators, such devices require integration and physical contact of dielectric thin films and conductive electrodes with the resonant structures which defeats some of the main purposes of moving away from piezoelectric devices. Moreover, large dielectric constants of the very thin dielectric films result in large parasitic capacitors to be associated with such devices.

Frequency tuning and compensation of the temperature-induced frequency drift are among other formidable challenges for both piezoelectric and capacitive resonators. Finally, in order to achieve a strong enough electromechanical transduction and consequently a large signal to noise ratio, both capacitive and piezoelectric resonators require relatively large electrode areas (thousands of μm^2) consuming costly real estate on the chip and limiting the maximum number of devices that can be integrated on a single chip.

Electrothermal actuation presents a third option for actuation of resonant modes of micro- and nanoscale structures that has been neglected to some extent so far, especially for high-frequency applications. Different versions of lower frequency thermally actuated micromechanical resonant devices have been utilized as mass sensors for chemical [6] or particulate [7] sensing. Utilization of thermal actuation for high-frequency resonant modes has been gaining more attention lately. A 10-MHz MEMS oscillator was demonstrated in [8] via electrothermal actuation of resonant modes in a suspended polysilicon membrane. In [9], thermal actuation and piezoresistive detection of flexural resonant modes of a silicon-carbide nanobeam with frequencies up to 1.1 GHz has been demonstrated. Recent studies and experiments have shown promising results and plenty of unexplored potentials for thermally actuated high-frequency resonators especially as the structural dimensions are scaled down into the nanoscale [10]. Both modeling and experimental measurements suggest that as opposed to the general presumption, thermal actuation becomes a more efficient actuation approach for higher-frequency resonant modes provided that higher frequency is achieved by shrinking the structural dimensions down. Unlike electrostatic and piezoelectric resonators, thermal-piezoresistive resonators perform better as their size is shrunk down and therefore could be a much stronger candidate for realization of highly integrated nanomechanical arrays of resonant signal processors. Self-sustained oscillation capability [11, 12] and robustness and design flexibility for sensory applications [7] are among other unique properties that thermal-piezoresistive transduction offers.

High-Frequency Thermal-Piezoresistive Resonators: Principles of Operation and Modeling

High-Frequency Thermal Actuation: As mentioned earlier, based on the perception from the macro-world, it is hard to imagine a thermal system with small enough thermal time constants to respond to high-frequency excitations. Similar to electrical RC circuits, thermal time constant of a thermal actuator, for example, a suspended conducting beam, is the product of its thermal capacitance and thermal resistance: $\tau_{th} = R_{th}C_{th}$. Thermal resistance can be calculated using a very similar equation to electrical resistance, that is, $R_{th} = \sigma_{th}^{-1}L/A$, where σ_{th} is the thermal conductivity of the structural material and L and A are the length and cross-sectional area of the element. Similar to electrical resistance, thermal resistance increases proportionally as all the dimensions of the element are scaled down. This implies that if a thermal actuator is scaled by a factor of S, its thermal resistance changes by a factor of S^{-1}. On the other hand, thermal capacitance is proportional to the mass and consequently the volume of the element and scales by a factor of S^3. Hence, overall, the thermal time constant of the element scales by a factor of S^2.

On the other hand, if a mechanical structure is scaled by a factor S, its mechanical resonant frequency (ω_m) changes by a factor of S^{-1}, that is, its mechanical time constant changes by a factor of S. The overall conclusion is that if the dimensions of a thermally actuated resonant structure are scaled down, its thermal time constant shrinks faster than its mechanical time constant. To be more specific, both thermal and mechanical responses of the system become faster upon scaling the dimensions down; however, the increase in the speed of the thermal response is sharper (proportional to S^{-2}) than the increase in the mechanical resonant frequency (proportional to S^{-1}). Therefore, as the structural dimensions are scaled down, the performance of thermally actuated mechanical resonators is expected to improve (more actuation force for the same actuation power) as the thermal response of the system can catch up with the mechanical vibrations more effectively.

Thermal-Piezoresistive Transduction: Piezoresistive sensing is the most convenient sensing mechanism to be integrated and utilized along with thermal actuation.

Similar to thermal actuation, piezoresistive sensing requires only a conductive element preferably with a high piezoresistive coefficient. To add to the simplicity, the same elements can be used for both thermal actuation and piezoresistive sensing simultaneously.

Thermal-piezoresistive resonators can come in different shapes and sizes. Figure 1 shows the 3D schematic view of a resonant structure, known as I-shaped bulk acoustic wave resonator or IBAR (also known as dog-bone resonator), which is very suitable for thermal-piezoresistive transduction while being capable of providing relatively high resonance frequencies. In the in-plane extensional resonance mode, the two blocks hanging on the two sides of the structure move back and forth in opposite directions subjecting the two beams connecting them together to periodic tensile and compressive stress. Such resonance mode can simply be actuated by applying the actuation voltage (or current) between the two support pads on the two sides of the structure. The ohmic loss caused by the current flow is maximized in the narrower parts of its path, that is, the extensional pillars in the middle of the structure (the two beams connecting the two blocks to each other). Upon application of a fluctuating actuation voltage, the fluctuating ohmic power generation results in a fluctuating temperature gradient and therefore periodic thermal expansion of the extensional actuator beams. The alternating extensional force resulting from the fluctuating temperature in the beams can actuate the resonator in its in-plane extensional resonance mode. If the ohmic power and consequently the resulting temperature fluctuations have the same frequency as the mechanical resonance frequency of the structure, the mechanical vibration amplitude is amplified by the mechanical quality factor (Q) of the resonator. The amplified alternating stress in the pillars leads to considerable fluctuations in their electrical resistance due to the piezoresistive effect. When biased with a DC voltage, such resistance fluctuations modulate the current passing through the structure resulting in an AC current component known as the motional current.

Thermo-electromechanical modeling: Operation of a thermal-piezoresistive resonator involves phenomena in three different domains of thermal, mechanical, and electrical nature. Ohmic loss turns the electrical input voltage into a temperature, thermal expansion turns the temperature into a mechanical force that causes a mechanical displacement, and finally the piezoresistive effect turns the mechanical displacement back into an electrical signal (motional current).

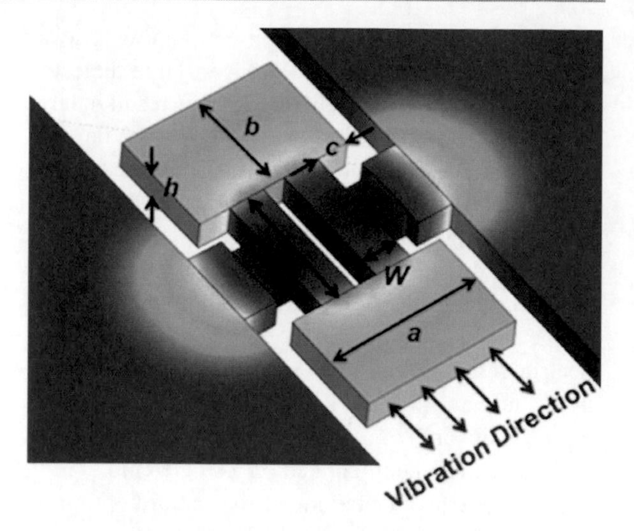

Thermally Actuated Silicon Resonators, Fig. 1 3D schematic view of a thermally actuated IBAR (dog-bone resonator) showing the qualitative distribution of temperature fluctuation amplitude when subject to a fluctuating current flow (*red* shows the maximum and *blue* shows the minimum)

Figure 2 shows the equivalent electrical circuit models for the three domains. In the thermal domain, the actuator beams act like a low-pass RC combination. The equivalent circuit consists of a current source representing the alternating heating power along with a capacitance (C_{th}) and a resistance (R_{th}) representing the effective thermal capacitance and thermal resistance of the thermal actuators, respectively (Fig. 2a). The resulting voltage across the parallel RC combination (T_{ac}) represents the temperature fluctuation amplitude.

Typically, the fluctuating actuation voltage is provided by a combination of a DC and an AC voltage component. Due to the square relationship between ohmic power generation and electrical voltage (or current), application of only an AC actuation voltage with frequency of f_a results in an ohmic power component with frequency of $2f_a$. In order to have the same frequency as the input AC actuation voltage for the thermal actuation force, a combination of AC and DC voltages (V_{dc} and v_{ac}) needs to be applied between the two terminals of the resonators. The resulting power loss component at the same frequency as v_{ac} in this case will be $P_{ac} = \frac{2V_{dc}v_{ac}}{R_A}$, where R_A is the electrical resistance of the actuator elements (extensional beams

Thermally Actuated Silicon Resonators, Fig. 2 Equivalent electrical circuit for a thermal-piezoresistive resonator: (**a**) the thermal domain; (**b**) the mechanical domain; (**c**) the electrical domain

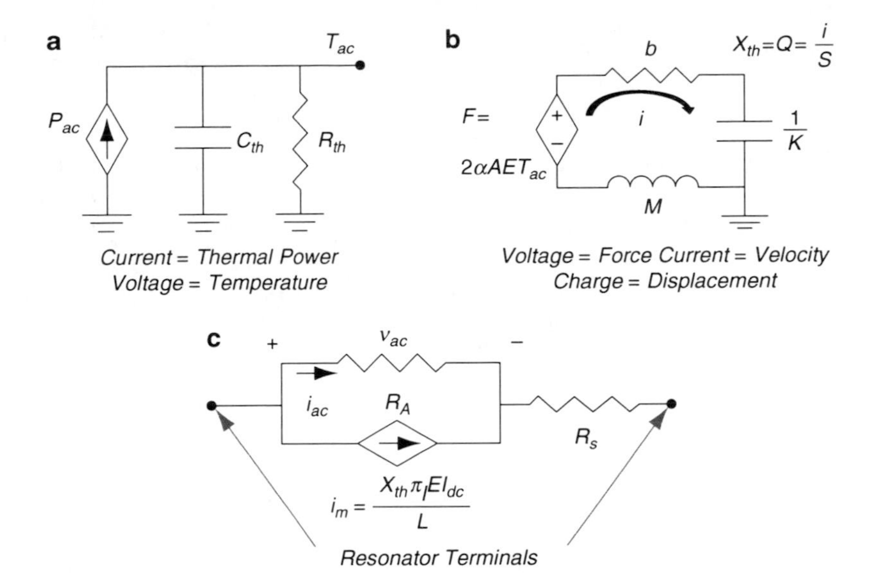

in IBARs) and V_{dc} and v_{ac} are the applied DC and AC actuation voltages, respectively.

The RLC combination of Fig. 2b represents the mechanical resonance behavior of the structure. The voltage source represents the thermally generated mechanical force which is the product of thermal expansion coefficient (α) of the structural material, thermal actuator cross-sectional area (A), and Young's modulus (E) of the thermal actuator along its length. The coefficient "2" has been added since there are two actuator beams in each resonator contributing to the actuation force. The inductor, capacitor, and resistor represent the effective mechanical mass (M), stiffness (K), and damping (b) of the resonator, respectively. Consequently, current represents velocity and electrical charge ($Q = \int i.dt$) represents displacement (X_{th}), which is the elongation of the thermal actuators (resonator vibration amplitude). The vibration amplitude, which is the output of the mechanical subsystem, acts as an input to the electrical subsystem (Fig. 2c) causing a change in the resistance of the thermal actuators due to the piezoresistive effect. This change in resistance modulates the current flow through the resonator which is shown as a current source in the equivalent circuit. R_s in Fig. 2c represents the parasitic electrical resistances in series with the actuator resistance which mainly includes resistance of resonator support beams.

The overall transfer function that relates the AC motional current (output current) to the AC input voltage is the product of the three transfer functions from the thermal, mechanical, and electrical domains. The simplified transfer function at resonance frequency is

$$H_T\big|_{s=j\omega_0} = g_m = 4\alpha E^2 \pi_l Q \frac{AI_{dc}^2}{KLC_{th}\omega_0} \quad (1)$$

where π_l is the longitudinal piezoresistive coefficient of the structural material, L is the thermal actuator length, and Q and ω_0 are the quality factor and angular resonant frequency of the resonator.

For IBAR structures, knowing that $K = 2EA/L$, Eq. 1 can be further simplified to

$$g_m = 2\alpha E \pi_l Q \frac{I_{dc}^2}{C_{th}\omega_0}. \quad (2)$$

A more detailed version of the modeling and derivations above is available in [10]. The transfer function in Eqs. 1 and 2 can be referred to as small signal voltage to current gain, or motional conductance (g_m) of the resonator at its resonance frequency, which is one of the most important parameters of a thermal-piezoresistive resonator when utilized as an electronic circuit component.

Resonator equivalent electrical circuit: For the resonators discussed here, where the extensional beams acting as thermal actuators are also utilized for piezoresistive readout, the physical resistance of the resonator connects the input and output of the devices.

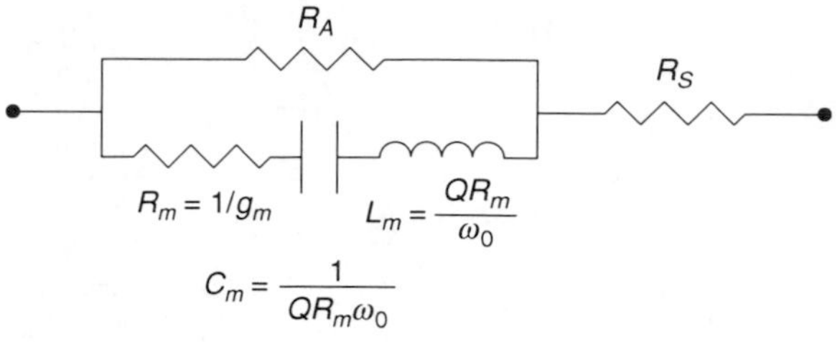

R_A R_S

$R_m = 1/g_m$ $L_m = \frac{QR_m}{\omega_0}$

$C_m = \frac{1}{QR_m\omega_0}$

Thermally Actuated Silicon Resonators, Fig. 3 Compact overall equivalent electrical circuit for one-port thermally actuated resonators with piezoresistive readout

The equivalent electrical circuit in this case includes a resistance $R_A + R_s$ connected between the two terminals of the resonator, which is the overall resistance in the current path between the two pads of the resonator. In parallel to the static resistance of the actuators, there is a series RLC combination that represents the mechanical resonant behavior of the structure. The value of R_m in the RLC has to be set so that at resonance, a motional current of $i_m = g_m.v_{ac} = v_{ac}/R_m$ is added to the feedthrough current passing through R_A. Therefore, $R_m = g_m^{-1}$ and L_m and C_m values can be calculated based on the value of R_m as shown in Fig. 3 according to the resonance frequency and quality factor of the resonator.

To calculate the motional conductance (g_m) of a thermal-piezoresistive IBAR, which is the most important parameter to be extracted from the model using Eq. 2, all the parameters except the effective thermal capacitance of the actuators (C_{th}) are known for every set of resonator dimensions and structural material. Due to the distributed nature of the thermal parameters including thermal generator (electrical resistance of the structure), thermal capacitance, and thermal conductance, analytical derivations or finite element analysis should be used to find an accurate value for C_{th}. In [10, 13], finite element analysis has been used to simulate the transient thermal response of the actuators and calculate the fluctuating temperature amplitude (T_{ac}) at different points along the length of the actuator beams. The mean value of the small signal temperature amplitudes at different points along the actuators is the effective temperature fluctuation amplitude for the actuator from which the effective thermal capacitance value can be extracted. As shown in [13], the effective thermal capacitance for a thermal actuator beam is equal to 0.92 of the static lumped element thermal capacitance of the beam with a good approximation.

Resonator Scaling Behavior and Optimization

As an electronic component, it is desirable to maximize the resonator motional conductance (g_m) (for improved signal to noise ratio) while minimizing power consumption. Therefore, a figure of merit for the resonators can be defined as the ratio of the resonator g_m to the overall DC power consumption:

$$F.M. = \frac{g_m}{P_{DC}} = \frac{2\alpha E\pi_l Q}{C_{th}\omega_0(R_A + R_s)} \tag{3}$$

where $P_{DC} = I_{DC}^2.(R_A + R_S)$ is the static power consumption of the resonator. Different parameters that can be used to maximize $F.M.$ are as follows:

Actuator Beam Dimensions: Generally, smaller actuator dimensions lead to smaller C_{th} improving resonator figure of merit. However, the effect of actuator dimensions on its electrical resistance (R_A) should also be taken into account. Since the extensional stiffness of the actuator beams defines the resonance frequency of the IBAR, in order to maintain the same resonance frequency for an IBAR while reducing its actuator thermal capacitance, both length and width of the actuator should be scaled down simultaneously. Such scaling does not affect the actuator electrical resistance (R_A). Therefore, scaling down both the length and width of the actuator beams by a scale S_a results in an improvement in the resonator $F.M.$ by a factor of S_a^2 while maintaining an almost constant resonance frequency for the device.

Resonance Frequency and Resonator Scaling: At a first glance at Eq. 3, higher resonance frequencies seem to have a deteriorating effect on resonator figure of merit. However, if higher resonant frequencies are achieved by shrinking the resonator size, at the same time C_{th} will be shrinking sharply. If all resonator dimensions are scaled down proportionally by a factor S, $C_{th} \propto S^{-3}$, $\omega_0 \propto S$, and $(R_A + R_S) \propto S$. Therefore, $F.M. \propto S$ that is, $F.M.$ increases proportionally if the resonator dimensions are scaled down, while its resonance frequency increases at the same time. This confirms the validity of the explanation provided previously claiming suitability of thermal actuation for high-frequency applications. It can be concluded from the last two discussions that the capability to fabricate resonator actuator beams with very small (nanoscale) dimensions is the key to achieving very high

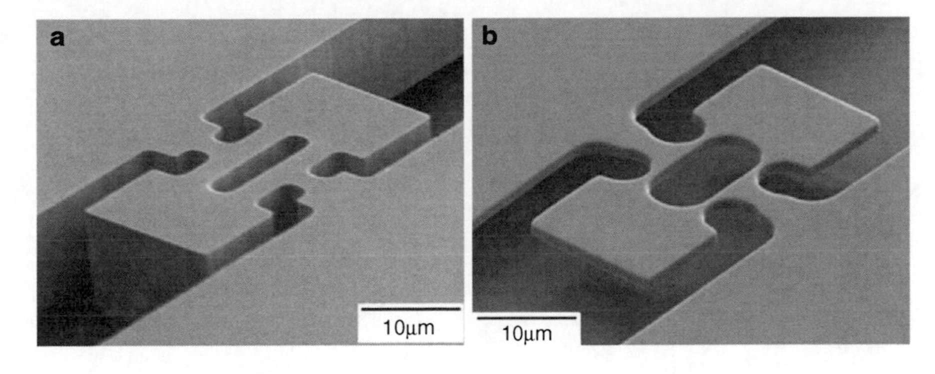

Thermally Actuated Silicon Resonators, Fig. 4 SEM view of single crystalline silicon thermal-piezoresistive IBARs (**a**) a 61-MHz 15-μm-thick IBAR fabricated on a low-resistivity p-type SOI substrate; (**b**) a 41-MHz 3-μm-thick IBAR fabricated on a low-resistivity n-type SOI substrate followed by thermal oxidation to thin down its thermal actuator beams

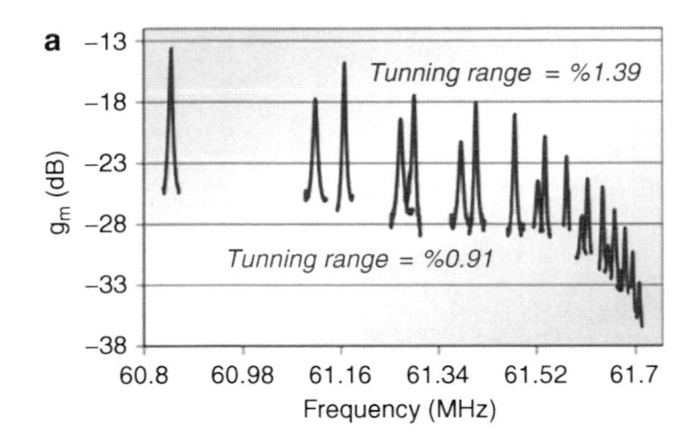

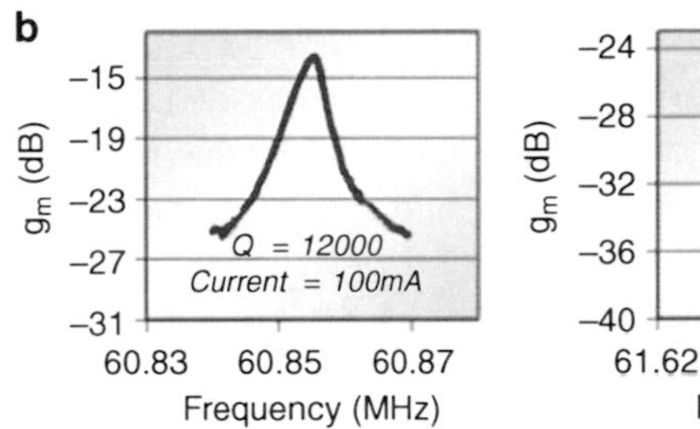

Thermally Actuated Silicon Resonators, Fig. 5 Measured frequency response of the thermally actuated 61-MHz resonator of Fig. 4a with different bias currents. The vertical axis in the frequency response plots shows the motional conductance of the resonators (g_m) in dB with respect to 50-Ω network analyzer terminations. Red and blue plots refer to vacuum and air testing conditions, respectively. Current range is 45–100 mA in vacuum and 55–100 mA in air

performance levels for thermal-piezoresistive resonators, both at low and high frequencies.

Electrical Resistivity and Other Material Properties: One of the most important parameters that can also be controlled for some of the most popular resonator structural materials (e.g., Si or SiC) is the electrical resistivity of the structural material. Lower electrical resistivity of the structural material improves the transduction figure of merit by lowering $R_A + R_s$. This can be explained by the fact that, with lower electrical resistivity, the same amount of DC bias current and therefore the same motional conductance can be maintained while burning less ohmic power in the structure.

A number of other structural material properties also affect the figure of merit for thermal-piezoresistive resonators. Higher thermal expansion coefficient and lower specific heat capacity improve the thermal actuation resulting in proportionally higher *F.M*. Higher piezoresistive coefficient improves the piezoresistive readout having the same effect on *F.M*. One interesting and promising fact is the giant piezoresistive effect found in single crystalline silicon nanowires that has been observed and reported by different researchers [14]. Therefore, as the resonators are scaled down to reach higher resonant frequencies in the hundreds of MHz and GHz range, it is expected that the actuator beams with deep submicron width exhibit such strong piezoresistivity significantly improving resonator performances. Finally, using structural materials with higher Young's modulus (E) can also improve the resonator figure of merit while increasing its resonant frequency.

Figure 4 shows the scanning electron micrograph (SEM) of two different microscale single crystalline silicon IBAR structures. Figure 5 shows the motional conductance frequency plots extracted from the measured frequency response of the 61-MHz resonator in Fig. 4a. The resonator quality factor ranges from 12,000 to 14,000 under vacuum and drops to

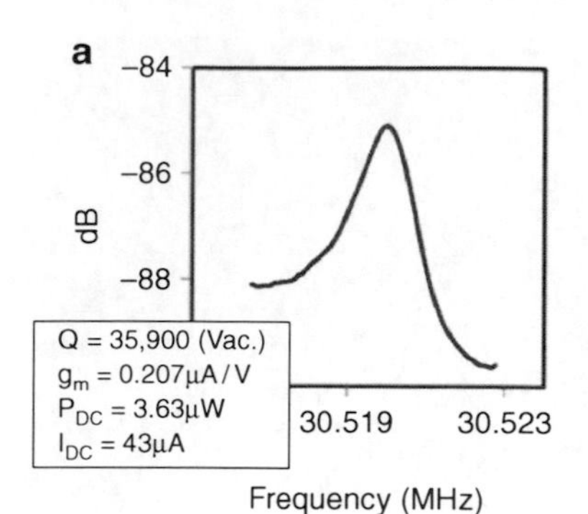

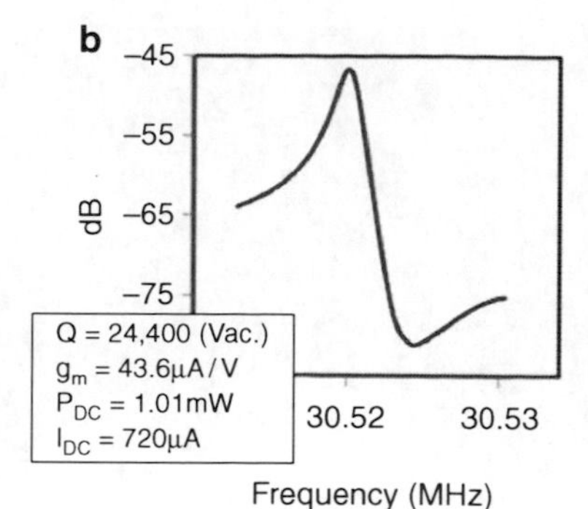

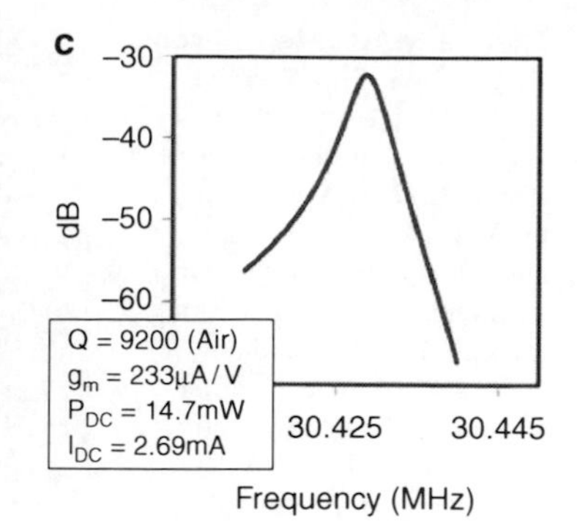

Thermally Actuated Silicon Resonators, Fig. 6 Measured motional conductance frequency plots for a 3-μm-thick 30.5-MHz IBAR. The vertical axis in the plots shows the motional conductance of the resonators (g_m) in dB with respect to 50-Ω network analyzer terminations. Due to its very narrow (750 nm wide) actuator beams, clear resonance peaks with motional conductance in the tens of μA/V have been measured for this resonator with sub-mA DC bias currents (sub-mW power consumption)

6,000–8,000 under atmospheric pressure. A large-frequency tuning range is demonstrated for this resonator by changing its DC bias current. This is mainly due to the raising temperature of the resonating body, especially its extensional beams that provide most of the structural stiffness in the in-plane extensional mode. As expected, as the DC bias current and consequently resonator motional conductance at resonance (g_m) increases, due to higher static temperature and softening of the structural material, the resonance frequency decreases. Despite having high-quality factors and relaively large motional conductance, the large current requirement and power consumption for the resonator in Fig. 5 could be a major drawback against using such as electronic components.

Figure 6 shows a few frequency response plots measured for a 30.5-MHz resonator with narrowed down actuator beams (~750 nm wide) similar to the structure in Fig. 4b. Submicron actuator beams can be achieved by controlled oxidation of the silicon structures followed by oxide removal leaving behind a narrower beam. Because of its much thinner actuators, a resonance peak can be detected for this resonator with currents as low as 43 μA translating into DC power consumptions as low as a 3.6 μW. At the DC bias current of 720 μA (DC power of 1 mW), motional conductance of 44 μA/V was measured for this resonator (Fig. 6b). Maximum currents tolerable for such narrow actuator beams are in the few mA range leading to motional conductances as high as 0.23 mA/V in air with Q of 9,200 (Fig. 6c). According to the derived model, much higher motional conductances can be

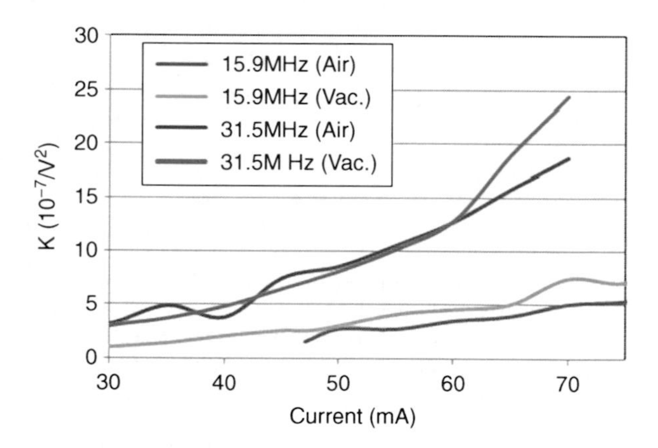

Thermally Actuated Silicon Resonators, Fig. 7 Extracted transduction figure of merit (*K*) for two IBARs versus resonator bias current. The higher-frequency (31.5 MHz) resonator, which is a 2X-scaled-down version of the 15.9-MHz resonator, has a higher transduction strength. The increase in the transduction figure of merit with bias current, which is not directly predicted by the derivations, is expected to be a result of the effect of increased static temperature at higher currents on different material properties, for example, increased thermal expansion coefficient at higher temperatures

achieved by consuming the same amount of power using lower-resistivity structural material allowing higher bias currents.

To verify better suitability of thermal actuation for higher-frequency resonators with smaller dimensions, Fig. 7 shows the measured transduction figure of merit for two thermal-piezoresistive resonators with the effect of quality factor taken out ($K = \frac{F.M.}{Q}$) under different bias currents. The two resonators are fabricated on the same substrate and have similar horizontal dimension aspect ratios with the 32-MHz resonator being a 2X-scaled-down version of the 16-MHz resonator. As expected based on the previously discussed scaling trend and the derived model, the higher-frequency resonator has a higher transduction figure of merit. This further confirms that thermal actuation is not only a potentially suitable actuation mechanism for high-frequency applications, but also it can provide higher performances at higher rather than lower frequencies. Therefore, much higher transduction efficiencies are expected to be achievable for higher-frequency versions of such devices with resonant frequencies in the hundreds of megahertz and even GHz range, provided that the dimensions can be scaled down proportionally into the nanoscale. The derived resonator model backed up by measurement results shows that with sub-100-nm actuator width and sub-micron actuator length, g_m as high as 1mA/V can be achieved for thermal-piezoresistive silicon IBARs with resonant frequencies in the gigahertz range with power consumptions as low as a few microwatts [10]. This is not including the giant piezoresistive effect in crystalline silicon that could further improve this by over an order of magnitude [14].

Self-Sustained Thermal-Piezoresistive Oscillators

Generally, in order to implement an electronic oscillator with a mechanical resonator as its frequency reference, the resonator needs to be engaged in a positive feedback loop consisting of amplifying circuitry with appropriate phase shift. One of the interesting aspects of the thermal-piezoresistive mechanical resonators that use the same elements as both thermal actuators and piezoresistive sensors is the internal feedback mechanism formed between thermal actuation force and piezoresistivity. This feedback results from the fact that the piezoresistive motional current of the structure that passes through the thermal actuators also acts as an excitation for the resonator by affecting the ohmic loss in the actuator beams. Depending on the sign of the structural material piezoresistive coefficient, electrical terminations of the resonator, and the phase shift resulting from thermal delays, this could act as a positive feedback stimulating mechanical vibrations or a negative feedback limiting vibration amplitude of the structure or reducing Brownian motion of the structure at that specific frequency (refrigeration [11]). For example, ohmic power generation in a thermal actuator biased with a constant current (terminated with a large electrical resistance) decreases if the actuator is subjected to tensile stress, due to the decreased electrical resistance. As part of a resonant structure, when such actuator is fully stretched and is about to start to compress, its decreased ohmic power and resulting cooling assist the actuator going into compression. In the same manner, after compression of the actuator, increased ohmic power generation helps its expansion via raising its temperature. Mechanical and thermal time delays play a key role here in determining whether the feedback has an amplifying or attenuating effect.

It was demonstrated by researchers at NXP semiconductors [11] that such positive feedback can be strong enough to initiate and maintain mechanical oscillations upon providing a DC electrical power source to the structure without the need for any external AC stimulation. Figure 8 shows the scanning electron micrograph of the suspended N-type single crystalline silicon structure for which such effect was demonstrated for the first time [11]. The structure is made of N-type silicon because N-type silicon has a negative longitudinal piezoresistive coefficient, which is required for such oscillation under constant current conditions.

Figure 9 schematically shows the feedback loop and the transduction mechanisms connecting the three physical parameters: displacement (x), temperature (T_{ac}), and voltage (v_{ac}) leading to self-sustained oscillation [11]. As shown in Fig. 9b, physically this can be explained as follows: upon application of a DC bias, the thermal actuator heats up and expand, pushing the mass to the right side. Due to the mass (inertia) of the plate, the nanobeam actuator experiences an overexpansion after pushing the mass and undergoes tensile stress. The negative piezoresistive coefficient turns the tensile stress into reduced electrical resistance for the nanobeam. When biased with a constant DC bias current, this means reduced ohmic power in the actuator forcing it to contract. After structural overcontraction (due to the mass of the plate), the actuator will be under compressive stress and will have increased electrical resistance. This translates into a higher thermal power forcing the structure to expand again. If the resulting driving force (due to heating and cooling) in each cycle is large enough

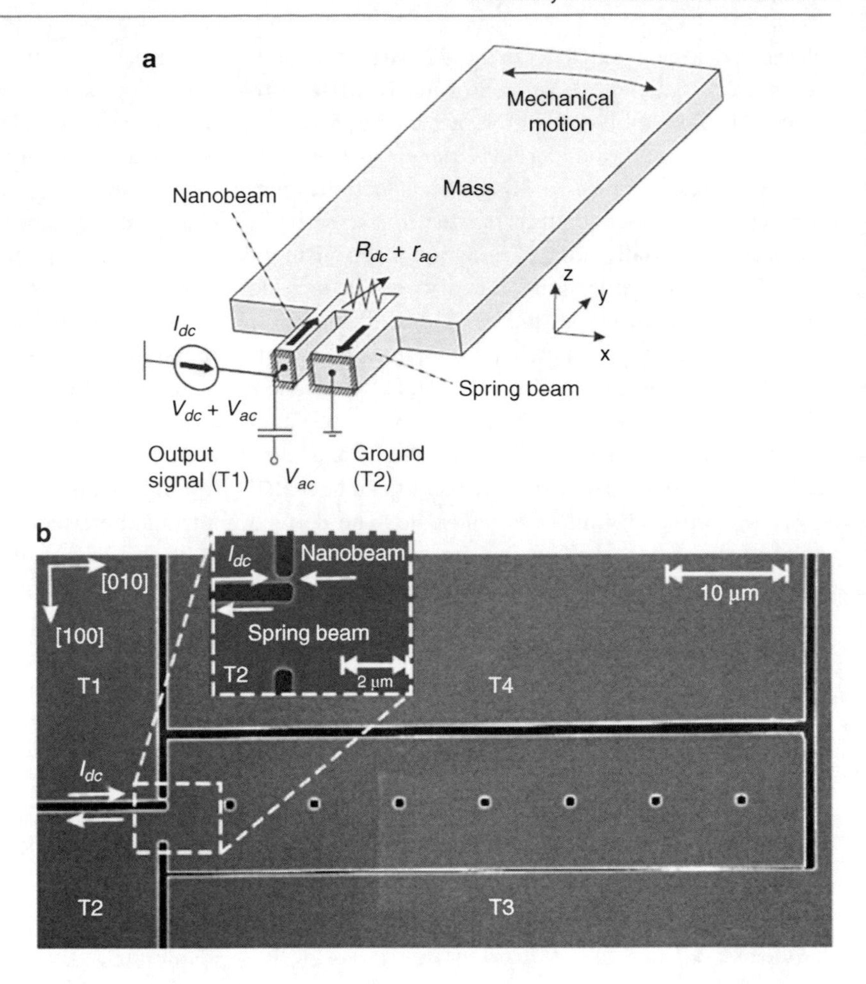

Thermally Actuated Silicon Resonators, Fig. 8 Schematic view and SEM top view of a fully micromechanical crystalline silicon oscillator. To operate the oscillator, a DC current is applied between the terminals of the device, and the output AC voltage resulting from the resonator's spontaneous mechanical vibrations is measured. The inset shows a zoomed-in view of the nanobeam actuator and the wide beam providing the flexural stiffness of the structure. Output electrical signal amplitude of 27 mV with frequency of 1.26 MHz was measured for this resonator when biased with a current of 1.2 mA (1.2-mW power consumption) under vacuum

to compensate for mechanical losses of the structure, the same sequence is repeated over and over in a self-sustained manner and the vibration amplitude keeps increasing until it is limited by nonlinearities. The 90° phase lags (Fig. 9b) resulting from the mechanical and thermal response delays lead to a perfect timing for the abovementioned sequence of physical phenomena to continue.

Figure 10 shows the SEM view and time response of an extensional mode 6.7-MHz self-sustained thermal-piezoresistive oscillator operating based on the same principle. Such structure is a derivative of the IBAR structures discussed in the previous sections with relatively larger plates. It operates in an in-plane resonant mode with its two plates moving back and forth in opposite directions and the extensional actuator beams being stretched and compressed periodically. Extensional resonant modes can provide higher frequencies as well as higher-quality factors both under vacuum and in air. Dual plate oscillators with frequencies in the few megahertz range, power consumption of a few to tens of milliwatts, and operating under both vacuum and atmospheric pressure have been demonstrated [12]. The output voltage amplitudes range from tens to hundreds of millivolts.

It can be shown that the requirement for self-sustained oscillation in such devices is that the absolute value of the motional conductance of the resonator becomes larger than the physical static conductance of its actuators. The minimum DC power required to achieve this condition is

$$P_{dc_{Min}} = \frac{KLC_{th}\omega_m}{4\alpha E^2|\pi_l|QA}. \quad (4)$$

Again, the key to having low power consumption is scaling down the resonator dimensions that reduce the power requirement with a square relationship leading

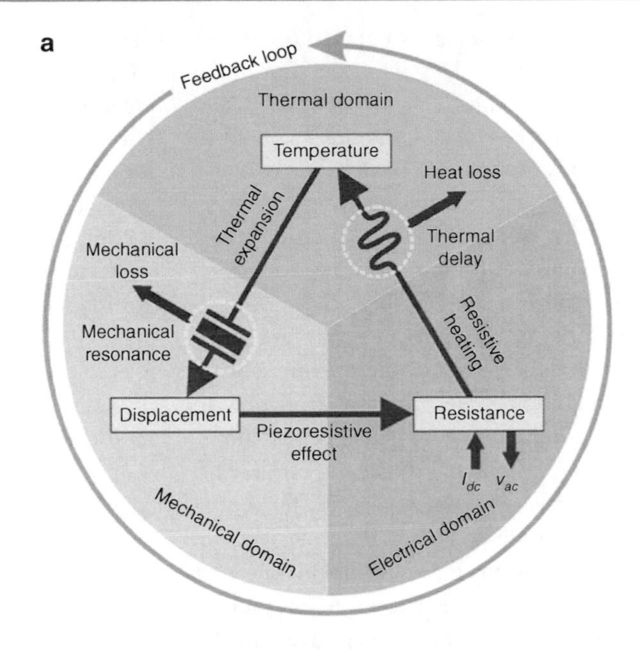

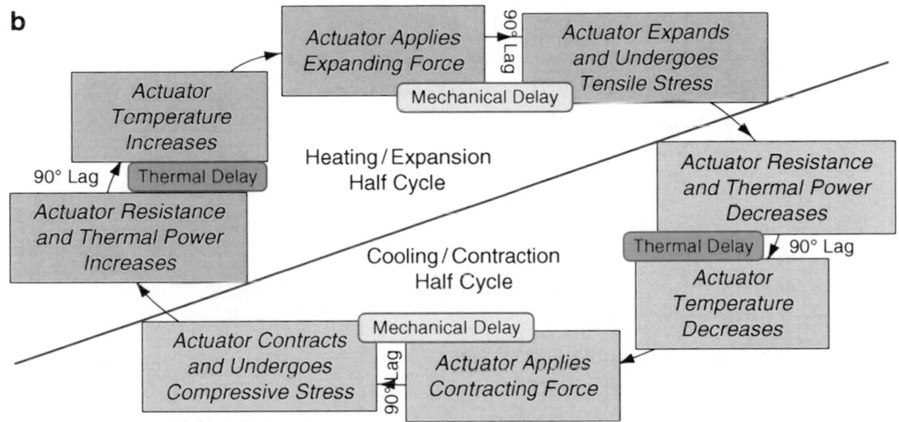

Thermally Actuated Silicon Resonators, Fig. 9 (**a**) Schematic demonstration of the feedback loop and the transduction mechanisms linking the thermal, mechanical, and electrical parameters to each other. Thermal delay and mechanical resonance provide the required phase shift for self-sustained oscillation. (**b**) Sequence of phenomena causing an internal positive feedback loop in thermal-piezoresistive resonators biased with a constant current that lead to self-sustained oscillation

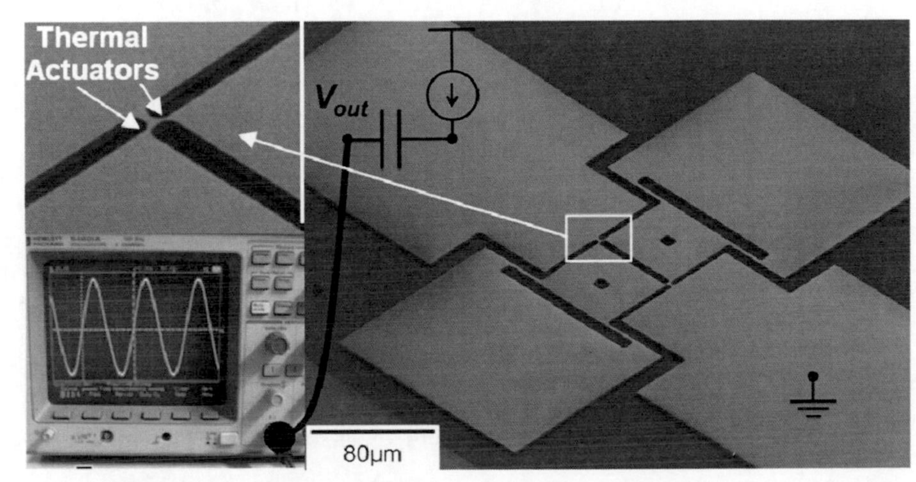

Thermally Actuated Silicon Resonators, Fig. 10 SEM view and output signal of a 6.6-MHz single crystalline silicon thermal-piezoresistive dual-plate resonator capable of self-sustained oscillation. Only a DC bias current of 3.0 mA is being applied to the resonator, and the resonator is operating in air. The power consumption of the oscillator is about 20 mW, and its output voltage amplitude is 70 mV

T

to significantly lower power consumption for scaled-down versions of such devices while having higher oscillation frequencies.

Another area where such devices need significant improvement is the output voltage amplitude, which is a small fraction of the DC voltage across the resonator (typically 0.01–0.1). What limits the oscillation amplitude of the resonators is nonlinearities (most likely mechanical nonlinearities) limiting the feedback loop gain to unity after reaching a certain amplitude. Therefore, a thorough study on minimization of the mechanical nonlinearities can possibly lead to fully micromechanical oscillators with higher output amplitudes. Piezoresistive coefficient of the structural material is another parameter that has a direct effect on the output voltage amplitude of such oscillators. With the same mechanical vibration amplitude and bias current, higher piezoresistive coefficient leads to proportionally higher output voltage amplitude. This further shows the significance of fabricating such devices with sub-100-nm actuator width taking advantage of the huge piezoresistivity observed in silicon nanowires [14] to increase the output amplitude.

Elimination of the sustaining amplifier by the self-sustained fully micromechanical oscillators described here might not be a significant improvement as electronics can be cheap and small and readily necessary to support other functions within the systems employing the electromechanical oscillators. However, for applications where large oscillator arrays are required, for example, sensors arrays, elimination of the amplifiers and the associated interconnections could be highly beneficial.

Thermal-Piezoresistive Resonant Mass Sensors

Electromechanical resonators have great potential as highly sensitive mass sensors for gas, biomolecular, or particulate sensing applications. For such sensory applications where the resonator has to be in contact with the surrounding environment, resonator robustness becomes a critical factor. This makes thermal-piezoresistive resonators made of a single monolithic structural material a great candidate for mass sensing applications. Especially for particulate sensing for air quality monitoring or environmental research, thermal-piezoresistive resonators seem to be the most suitable choice [7, 15]. Air-gap capacitive resonators are extremely vulnerable to contaminants or particulates entering and getting stuck in their nanogaps making it practically impossible to use them as particulate sensors. For piezoelectric resonators, it is very hard (if not impossible) to have a uniform mass sensitivity all over the resonator sensing surface or at least a large portion of it. This is due to the fact that such devices generally operate in their bulk resonance modes resulting in different vibration amplitudes and therefore different effective resonator mass at different locations on the surface. Uniform mass sensitivity is a necessity when targeting mass measurement and counting of individual particles. Furthermore, metallic and piezoelectric thin films in a piezoelectric resonant structure are typically less robust than a monolithic crystalline structure, especially when subjected to impact by high-speed airborne particles.

Thermal-piezoresistive resonators can continue to work flawlessly with tens to thousands of particles deposited on them [7]. For the IBAR and dual-plate structures, the vibrating counter masses (plates) act as rigid bodies having almost the same vibration amplitude on their whole surface. This allows measuring the mass of individual deposited particles as they arrive providing the overall count and mass distribution of particles present in air samples. There is still the possibility of particles landing on the actuator beams and experiencing different mass sensitivities. However, the surface area of the actuator beams is typically much smaller than that of the plates making that probability small enough for the resulting error to be within the acceptable range for most applications. Furthermore, the miniaturization capability of thermal-piezoresistive resonators into the nanoscale allows much higher mass sensitivities to be achievable enabling detection and mass measurement of individual nanoparticles not feasible using the larger capacitive or piezoelectric resonant devices. Figure 11 shows the SEM view of an IBAR exposed to a flow of artificially generated airborne particles with the same size and composition along with its measured resonance frequency during the deposition [15]. The effect of arrival of each particle on the resonance frequency is clearly detectable and the same for all the particles. This confirms uniform mass sensitivity on the resonator sensing plates.

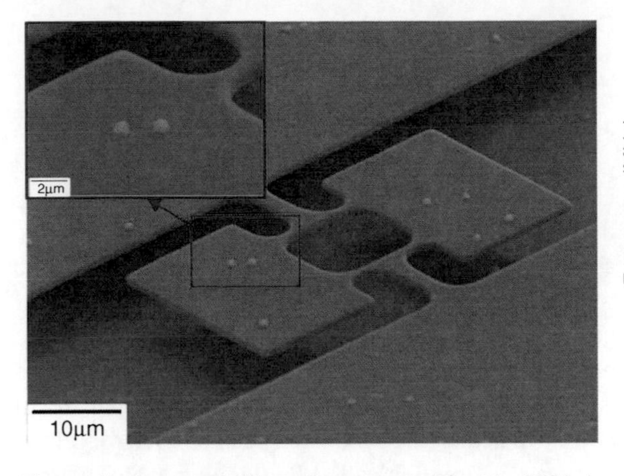

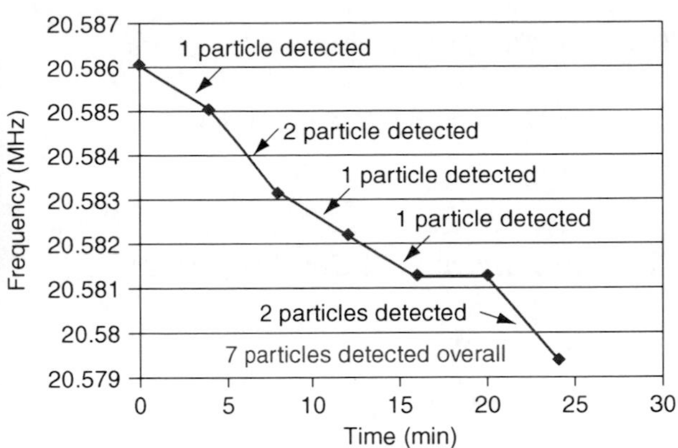

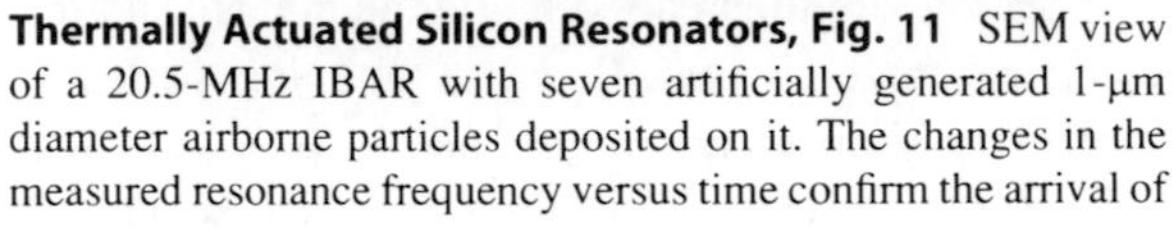
Thermally Actuated Silicon Resonators, Fig. 11 SEM view of a 20.5-MHz IBAR with seven artificially generated 1-μm diameter airborne particles deposited on it. The changes in the measured resonance frequency versus time confirm the arrival of seven particles. There are five frequency reduction steps, of which two have a slope twice that of the rest which show that two particles have been deposited on the device during those periods

Cross-References

- ▶ Laterally Vibrating Piezoelectric Resonators
- ▶ NEMS Resonant Chemical Sensors
- ▶ Optomechanical Resonators
- ▶ Piezoresistivity
- ▶ Thermal Actuators

References

1. Piazza, G.: Integrated aluminum nitride piezoelectric microelectromechanical system for radio front ends. J. Vac. Sci. Tech. A **27**(4), 776–784 (2009)
2. Rinaldi, M., Zuniga, C., Chengjie, Z., Piazza, G.: Super-high-frequency two-port AlN contour-mode resonators for RF applications. IEEE Trans. Ultrason. Ferroelectr. Freq. Control **57**(1), 38–45 (2010)
3. Nguyen, C.T.C.: MEMS technology for timing and frequency control. IEEE Trans. Ultrason. Ferroelectr. Freq. Control **54**(2), 251–270 (2007)
4. Pourkamali, S., Ho, G.K., Ayazi, F.: Low-impedance VHF and UHF capacitive silicon bulk acoustic wave resonators. IEEE Trans. Electron Dev. **54**(8), 2017–2023 (2007)
5. Weinstein, D., Bhave, S.A.: Internal dielectric transduction in bulk mode resonators. J. Microelectromech. Syst. **18**(6), 1401–1408 (2009)
6. Seo, J.H., Brand, O.: High Q-factor in-plane-mode resonant microsensor platform for gaseous/liquid environment. J. Microelectromech. Syst. **17**(2), 483–493 (2008)
7. Hajjam, A., Wilson, J.C., Rahafrooz, A., Pourkamali, S.: Fabrication and characterization of thermally actuated micromechanical resonators for airborne particle mass sensing, part II: device fabrication and characterization. J. Micromech. Microeng. (JMM), **20**, 125019 (2010)
8. Reichenbach, R.B., Zalalutdinov, M., Parpia, J.M., Craighead, H.G.: RF MEMS oscillator with integrated resistive transduction. IEEE Electron Dev. Lett. **27**, 805–807 (2006)
9. Bargatin, I., Kozinsky, I., Roukes, M.L.: Efficient electrothermal actuation of multiple modes of high-frequency nanoelectromechanical resonators. Appl. Phys. Lett. **90**, 093116 (2007)
10. Rahafrooz, A., Pourkamali, S.: High frequency thermally actuated electromechanical resonators with piezoresistive readout. IEEE Transactions on Electron Devices. **58**(4), 1205–1214 (2011)
11. Steeneken, P.G., Phan, K.L., Goossens, M.J., Koops, G.E.J., Verheijden, G.J.A.M., van Beek, J.T.M.: Piezoresistive heat engine and refrigerator. Nature Phys. **7**, 354 (2012). doi:10.1038/nphys1871
12. Rahafrooz, A., Pourkamali, S.: Fully micromechanical piezo-thermal oscillators. In: Proceedings of the IEEE international Electron Device Meeting (IEDM), San Francisco, doi: 10.1109/IEDM.2010.5703314 (2010)
13. Rahafrooz, A., Pourkamali, S.: Fabrication and characterization of thermally actuated micromechanical resonators for airborne particle mass sensing, part I: resonator design and modeling. J. Micromech. Microengin. (JMM) **20**, 125018 (2010)
14. He, R., Yang, P.: Giant piezoresistance effect in silicon nanowires. Nat. Nanotechnol. **1**(1), 42–46 (2006)
15. Hajjam, A., Wilson, J. C., & Pourkamali, S: Individual air-borne particle mass measurement using high frequency micromechanical resonators. IEEE Sensors Journal, **11**(11), 2883–2890 (2011)

Thermal-Piezoresistive Resonators

► Thermally Actuated Silicon Resonators

Thermodynamics of Small Systems

► Nanocalorimetry

Thermoelectric Cooler

► Thermoelectric Heat Convertors

Thermoelectric Device

► Thermoelectric Heat Convertors

Thermoelectric Generator

► Thermoelectric Heat Convertors

Thermoelectric Heat Convertors

Joseph P. Heremans and Audrey M. Chamoire
Department of Mechanical and Aerospace Engineering, The Ohio State University, E443 Scott Laboratory, Columbus, OH, USA

Synonyms

Electric cooler; Peltier coefficient; Peltier cooler; Peltier couple; Peltier effect; Peltier element; Peltier module; Peltier-Seebeck effect; Seebeck coefficient; Seebeck effect; Thermoelectric cooler; Thermoelectric device; Thermoelectric generator; Thermoelectric heat pump; Thermoelectric module; Thermoelectric power; Thermopower; Thomson effect

Definition

Reversible solid-state heat engines that use heat applied between a hot and cold reservoir to generate electrical power and can be inverted to use an electrical current to pump heat out of the cold reservoir and into the hot reservoir. In both cases, they function like conventional heat engines, but use electrons as the working fluid instead of steam or refrigerants or gases. They can be used in power generation mode to recover heat wasted, for instance, in the exhaust of an automobile and convert it directly into useful electrical power. In heat pump mode, they can be used for refrigeration, and are then called Peltier elements. Thermoelectric heat converters have no moving parts, are extremely simple and robust, and have a very high specific power density, meaning that they can convert large amounts of power per gram or cubic centimeter of material. Using the thermoelectric materials commercially available until now, their efficiency is lower than that of conventional mechanical heat engines, and therefore are mostly used in low-power applications, but recent research developments promise to change that. Thermoelectric heat converters are based on two complementary effects. The generators use the Seebeck effect, which shows that a temperature difference applied to a conducting solid results in the development of a voltage; the ratio between voltage and temperature difference is the Seebeck coefficient, also known as thermoelectric power or thermopower. In heat pump mode, they are based on the Peltier effect, which shows that when an electrical current passes through a conductor, it results in the development of a heat flow; the ratio between the heat flow and the current is the Peltier coefficient. The efficiency of thermoelectric heat converters depends on the temperature differences, as it does for all heat engines, but also on a material parameter called the thermoelectric figure of merit. Recent developments in nanotechnologies have resulted in a doubling of the efficiency of modern materials compared to the state of the art a decade ago.

Introduction

Thermoelectric heat convertors are *solid-state heat engines*, i.e., solids that convert heat to electricity or

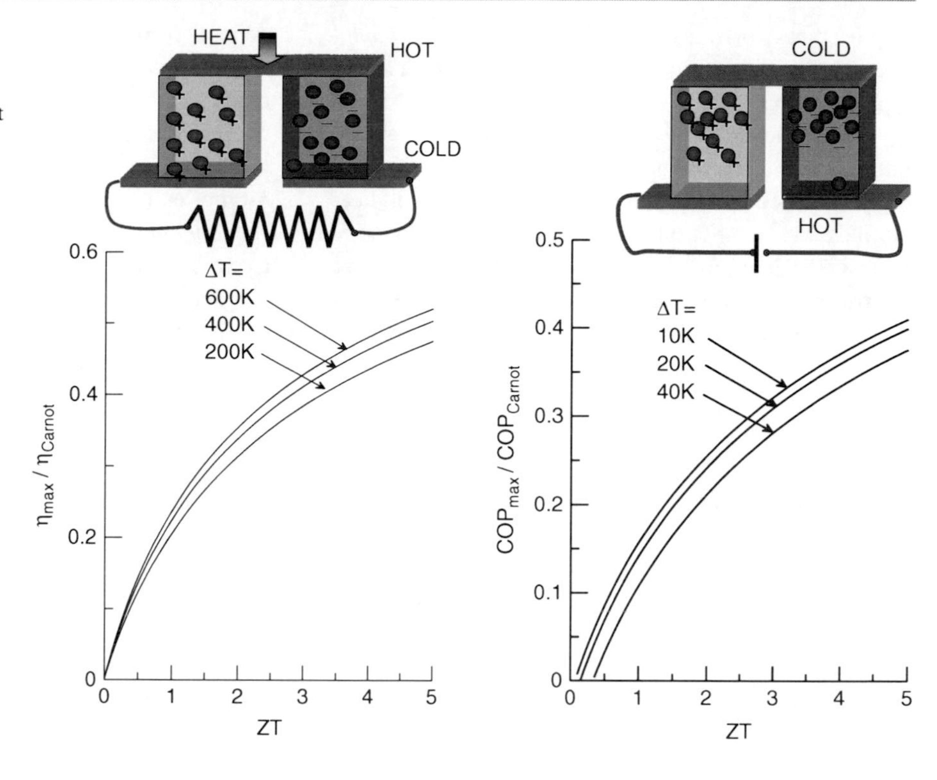

Thermoelectric Heat Convertors, Fig. 1 Configuration of a thermoelectric couple as heat engine (*left*) or heat pump (*right*), and efficiency of coefficient of performance (COP) of both normalized to their Carnot value

inversely use electrical power to pump heat. They behave like conventional steam engines and refrigeration machines [1], except that they have no moving parts. They are extremely reliable, simple, and compact, but their applications have been limited by the fact that their efficiency is lower than that of conventional mechanical systems. That efficiency is a function of a single material parameter, the *thermoelectric figure of merit, ZT*.

Thermoelectric convertors consist of two semiconductor elements, one doped p-type and the other doped n-type, connected as shown in Fig. 1, electrically in series but thermally in parallel. When heat is applied to one side, as shown in the top left panel of Fig. 1, while the other side is kept cold, the temperature difference $\Delta T = T_{\mathrm{HOT}} - T_{\mathrm{COLD}}$ creates a voltage V_p across the p-type element that is positive on the cold side, and a voltage V_n on the n-type element that is negative on the cold side; in Fig. 1, the voltages add across the couple. Many couples can be connected in series to make a pile, which works as a *thermoelectric power generator* and can be connected to a load (the resistor in Fig. 1). The voltages V_p and V_n are proportional to ΔT. For each material, the proportionality constant is defined as *the thermoelectric power*, also known as *thermopower* or *Seebeck coefficient S*:

$$S \equiv \frac{V}{\Delta T} \tag{1}$$

The "engine"in Fig. 1 can be reversed, as shown in its top right panel. If a current I passes through each element, as shown, that element pumps a heat flow Q proportional to the current. The proportionality constant is the *Peltier coefficient* Π:

$$\Pi \equiv \frac{Q}{I} = \frac{q}{j} \tag{2}$$

and can also be written as the ratio of the thermal to electrical fluxes, q/j. The *Kelvin relation* relates Π and S with the absolute temperature T:

$$\Pi = S.T \tag{3}$$

The couple shown on the right therefore has the exact same construction as the couple shown on the left, and several couples can also be connected to form a pile that works as a *thermoelectric heat pump*, cooling down its top surface and pumping the heat into its bottom surface.

The efficiency, η of the couple connected as a generator and the *coefficient of performance, COP* of the couple connected as a heat pump, like those of all heat engines, are a function first of the Carnot efficiency and *COP*. The ratios η/η_{Carnot} and COP/COP_{Carnot} are shown in Fig. 1 as a function of one material parameter, the thermoelectric figure of merit, *ZT*, defined by:

$$ZT \equiv \frac{S^2\sigma}{\kappa}T, \tag{4}$$

where σ and κ are the *electrical* and the *thermal conductivity* of the material. The numerator of Eq. 4 is labeled the *thermoelectric power factor* $P = S^2\sigma$, and has much the same significance as the area under the (I,V) curve of a photovoltaic solar cell. Indeed, for a given temperature gradient ΔT, the voltage $V = S.\Delta T$ and the current $I = S/R$ (R = resistance of the thermoelectric elements, or the source resistance of the generator) so that the power produced is proportional to the intrinsic properties $S^2\sigma$. It is often the case that the *n*-type and *p*-type materials have closely matched values of ZT. Maximizing *ZT* is the goal of research on thermoelectrics. *ZT* is temperature dependent, and is therefore optimized to reach maximum near the desired operating temperature. The temperature ranges, and the semiconductors used in them, are as follows [2]:

1. Materials used in thermoelectric coolers have to operate around room temperature: These materials are typically semiconductors of the *tetradymite class*, typified by the binary semiconductor Bi_2Te_3. Their general formula is $(Bi_{1-x}Sb_x)_2(Te_{1-y}Se_y)_3$. One class of material, the I-V-III$_2$ compounds, such as $AgSbTe_2$ that crystallize as rocksalts, can have $ZT \sim 1.3$ around 200°C, a temperature range intermediate between this and the next.
2. Materials used in waste-heat recovery applications generate electrical power from combustion gases that are typically at 350–500°C. Several classes of materials are suitable for this and have been described in more specialized review articles [2]. The classical semiconductor systems used in these applications are the *rocksalt lead chalcogenides*, typified by the binary compound PbTe and of general formula $(Pb_{1-x}Sn_x)(Se_yTe_{1-y})$. The negative image of lead with the public at large, while not really applicable to chemically stable and insoluble compounds such as the lead chalcogenide salts mentioned here, has led to the development of *n*-type lead-free alternatives based on *skutterudite semiconductors* typified by the binary compound $CoSb_3$, which can reach $ZT \geq 1.5$, based on a reduction of the thermal conductivity by means of the creation of a special phonon mode that absorbs but does not conduct heat. *P*-type skutterudites exist, but to date they have much lower *ZT*s than *p*-type PbTe-based materials. Another class of materials that is being investigated now, but has not yet given $ZT > 1.1$, are *Zintl-phase semiconductor* $Mg_2Si_{1-x}Sn_x$ alloys.
3. At the high end, thermoelectric power generators use radioisotopes at 1,000°C as the source of heat, and the commonly useful thermoelectric semiconductors are $Si_{1-x}Ge_x$ alloys, often with a second intermediate-temperature (500°C) stage made from skutterudites or lead salts. Other materials, not discussed here, are semiconducting rare-earth chalcogenides and pnictides.
4. At the low end, cryogenic heat pumps operating in the range of 10–200 K would be very useful to cool diode lasers and various detectors. Here $Bi_{1-x}Sb_x$ *group V alloys* are the semiconductors of choice, now in a cascaded configuration with tetradymite semiconductors at the higher temperature (>160–300 K) end.

Not all materials benefit from nanostructuring, in particular, not those which come with inherently low thermal conductivities such as skutterudites and $AgSbTe_2$. Therefore, only five classes of materials based on Bi_2Te_3, PbTe, $Mg_2Si_{1-x}Sn_x$, $Si_{1-x}Ge_x$, and $Bi_{1-x}Sb_x$ will be discussed here.

Figure 2 shows a summary [2] of the recent great progress achieved in values for *ZT*, compared to the value $ZT = 1$ that used to be the historical limit. *ZT* consists of mutually "counter-indicated" properties, i.e., such that most mechanisms that improve one property in *ZT* also are deleterious to another. Explicitly, *ZT* can be written as the product of two sets of counter-indicated properties shown here in parentheses:

$$ZT = \left(S^2 n\right).\left(\frac{\mu}{\kappa}\right)qT. \tag{5}$$

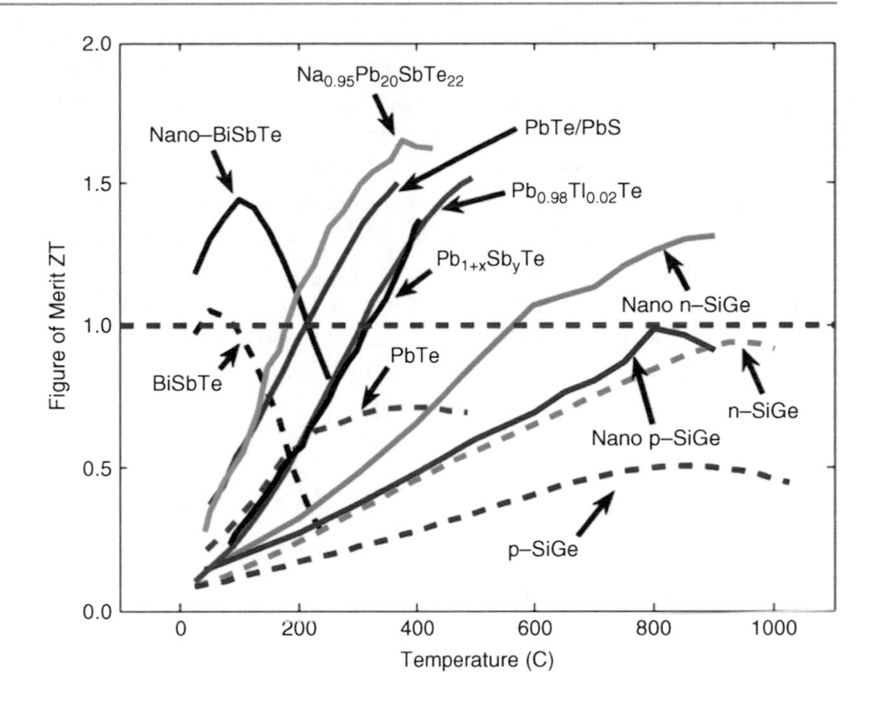

Thermoelectric Heat Convertors, Fig. 2 *ZT* as a function of temperature for a number of classical ($ZT < 1$) and recently developed ($ZT > 1$) thermoelectric materials, after [2] (Reproduced by permission of The Royal Society of Chemistry)

Equation 5 was derived assuming that the electrical conductivity is given by:

$$\sigma = nq\mu \tag{6}$$

where n is the *carrier concentration*, q its charge, and μ its *mobility*. The mobility itself is defined as the ratio between the average *drift velocity* of electrons in a sample under the influence of an applied electric field and that field. The ratio (μ/κ) of the mobility to the thermal conductivity is counter-indicated because defects and impurities that affect one of these properties usually also affect the other. The other counter-indicated property is the product (S^2n): Indeed, it is a general rule in doped semiconductors and metals that the higher the carrier concentration, the lower the thermopower, labeled the *Pisarenko relation* in Ioffe's seminal monograph on thermoelectricity [1].

Nanotechnology can increase both (S^2n) and (μ/κ), because lowering the dimensionality adds a new parameter that can be tuned to optimize each counter-indicated pair of properties. Early work devoted in the last decade focused on improving the ratio (μ/κ) by reducing the lattice thermal conductivity using phonon scattering mechanisms; more recently focus shifted to (S^2n). This entry will therefore be divided into two sections each addressing one of those factors.

Improving the Mobility-to-Thermal Conductivity (μ/κ) Ratio

Theory

The thermal conductivity of semiconductors is the sum of two contributions, the *lattice thermal conductivity* κ_L due to lattice vibrations or phonons and the *electronic thermal conductivity* κ_E due to heat conducted by charge carriers:

$$\kappa = \kappa_L + \kappa_E \tag{7}$$

The kinetic formula developed for heat conduction through ideal gases can be applied to the phonon thermal conductivity, which is then written in terms of the specific heat C of the solid at constant pressure and per unit volume, the phonon or sound velocity, v_P, and the *phonon mean free path* ℓ_P:

$$\kappa_L = \frac{1}{3} C v_P \ell_P. \tag{8}$$

The electronic thermal conductivity κ_E is directly related to the electrical conductivity, σ by the *Wiedemann-Franz law*. The latter is derived from the fact that each electron carries not only a charge q but also an amount of heat $\frac{3}{2}k_BT$, where k_B is the Boltzmann constant, and is:

T

$$\kappa_E = L\sigma T, \quad (9)$$

where L is the *Lorenz ratio*, which collects universal constants and, for free electrons, is 2.45×10^{-8} V^2K^{-2}. In practice, when electrons are scattered inelastically and thus lose a certain amount of energy in each collision, $L < 2.45 \times 10^{-8}$ V^2K^{-2}.

Expressing the electrical conductivity Eq. 6 in terms of the mean free path ℓ_E of the particles (electrons or holes) that carry the charge, ZT for metals and degenerately doped semiconductors, which most practical thermoelectric materials are, can be regrouped as:

$$zT = \left(S^2 n\right) \frac{1}{\left(\frac{\ell_P}{\ell_E}\right) \frac{C_{VP}}{3q^2 T} \left(2m^* E_F\right)^{1/2} + Ln}, \quad (10)$$

where m^* is the *effective mass* and E_F is the *Fermi energy* of the charge carriers, the difference between their electrochemical potential and the extremum of the conduction of valence band they are in. In so doing, electrons or holes are assumed to behave like quasi-classical particles at the Fermi energy in the solid, with an energy (E_F) – velocity (v_F) relation given by $E_F = \frac{m^* v_E^2}{2}$, and $\ell_E = v_F \tau$ where $\tau = m^* \mu / q$ is the *relaxation time*, the average time between collisions.

Equation 10 illustrates how nanostructures can improve the (μ/κ) ratio in ZT. The formula is the following:

1. Select a solid and a temperature regime where $\ell_E << \ell_{P,B}$. Here $\ell_{P,B}$ is the phonon mean free path in the bulk solid, limited by phonon-phonon or phonon-defect interactions.
2. Create a nanostructure of characteristic length scale d chosen so that $\ell_E < d < \ell_{P,B}$.
3. Create interfaces between the nanostructure and the solid such that it scatters phonons, without affecting the electron mean free path too much, i.e., achieve the condition $\ell_E < d \approx \ell_P < \ell_{P,B}$. Such nanostructure will decrease the denominator of Eq. 10 compared to the case of a similar solid without the nanostructure. When $\kappa_E < \kappa_L$, the second term Ln in the denominator of Eq. 10 is smaller than the first, and the effect of reducing ℓ_P is more pronounced.

Several types of nanostructures have been tried experimentally for this purpose over the course of the last decade. Two pioneering articles were published that claimed achieving this goal in MBE or MOCVD grown superlattices (SL), one on Bi_2Te_3/Sb_2Te_3 SL's [3], and one on "quantum-dot superlattices" (QDSLs) made in the $PbTe/PbSe_{1-x}Te_x$ system [4]. In practice, other groups have had difficulties in attempts to reproduce the results of the first paper, and separate measurements of the thermal conductivity [5] of the $PbTe/PbSe_{1-x}Te_x$ QDSLs failed to confirm the high ZT of Ref. [4]. The two pioneering articles [3, 4] above required heroic experimental efforts to measure the thermoelectric properties of the superlattices, and the reported data should therefore be considered only as an indication of the scientific route to follow and not as definitive values of ZT. Both publications certainly have the merit of having started the field of nano-thermoelectrics, which subsequently yielded the easily verifiable and reproducible results of Fig. 2. Work on nanostructures and quantum wires in $Bi_{1-x}Sb_x$ is still ongoing, and the latter will be described in the next section of this entry.

Because thermoelectric technology needs to convert substantial power levels to have a significant impact on the world energy issues, the amounts of materials required exceed what can realistically be prepared using thin-film technology. Therefore, this entry concentrates on materials that can be prepared in bulk. Those are classified according to the stage in the preparation method at which nanostructuring takes place, as prepared either using a "bottom-up approach" or a "top-down" approach. In the bottom-up approach, the materials are made from nanostructured precursors, such as nm-fine powders prepared either by chemical precipitations or ball milling (BM), which are then consolidated by pressing and sintering using either hot pressing (HP) or spark-plasma sintering (SPS). In the top-down approach, the materials are first prepared using a metallurgical process, and nanostructures are subsequently grown into them by an appropriate heat treatment akin to the formation of perlitic steels or Guinnier-Preston zone in aluminum. Both approaches work, and materials with ZT far exceeding unity have been prepared by either.

Bottom-Up Approaches

- *Ball Milling and Sintering Powder-Metallurgical Processes*

  Ball milling (BM) is relatively easy to implement and allows for the quick and large-scale synthesis of materials; it is quite amenable to industrial fabrication. BM can be used either to synthesize a material from pure elements (mechanical alloying) or to obtain nanopowders from bulk pre-alloyed or pre-compounded ingots.

Pure elements are directly introduced in a bowl containing balls, which is tightly enclosed usually inside a glove box to keep the reactor under inert gas (Ar or N_2) to minimize oxidation. Sometimes a liquid is used to promote the reaction of elements in the reactor. Two kinds of ball mills exist, shaker mills, where reactors are fixed and rotate around a horizontal axis, and planetary ball mills, where bowls are fixed in a satellite position on a turning table and rotate in the opposite direction and at a different speed than the table. The balls present in the bowl undergo two different types of movement: either they roll and are tacked on the wall of the reactor, or they roll and are projected onto the opposite wall. Friction or shock caused by the balls on the matter will lead to the formation of either powder or desired compound. Parameter such as the number and the size of the balls, the ball-to-material load ratio as well as the milling time can be adjusted. The results of such a technique are very fine and homogeneous powder with nanometer-sized grains.

Spark-Plasma Sintering (SPS) is then a method of choice to compact the powder. This sintering technique is similar to conventional hot pressing, but without an external source of heat. An alternating current applied via electrodes passes through the conductive wall of pressing (in graphite) and through the sample (if conductive). This technology achieves heating rates of over 600 K/min. The applied pressure is uniaxial and can reach values higher than 200 MPa. Compared to hot pressing, this method can compact samples up to 95–99% of theoretical densities in a very short time and also allows control over grain growth.

- *Bi_2Te_3 Based* [6, 7, 8]
  Nanostructuring in n-type Bi_2Te_3 has been obtained by hot pressing a mixture of 100 μm powder of n-type zone melted Bi_2Te_3 alloys with Bi_2Te_3 nanotubes obtained by hydrothermal synthesis in a weight ratio 85:15. Electrical conductivity increases while lattice thermal conductivity is strongly decreased especially at high temperature and is below 0.5 W/m.K in all the temperature range. A $ZT > 1$ is achieved at higher temperature.

  ZT of p-type $Bi_{0.5}Sb_{1.5}Te_3$ has been increased from 1 for the state-of-the-art (SOA) material up to $ZT = 1.4$ at 373 K by nanostructuring, obtained by ball milling the bulk material followed by hot pressing. Transmission electron microscopy (TEM) observation indicates mostly Sb-rich nanodots of about of 2–10 nm with fuzzy boundaries, and some Te nanoprecipitates. Compared to SOA material σ is slightly higher at all temperatures and S is lower at low temperature and then higher from 433 to 523 K. κ is lowered at high temperature due to the large number of interfaces (nanograins, nanoprecipitates, nanodots) that increase phonon scattering; p-type $Bi_{0.5}Sb_{1.5}Te_3$ shows ZT improvements of 20–40% over SOA material, with a maximum shifted to higher temperature.
- *SiGe* [7, 8]
  Boron-doped p-type doped Si_xGe_{1-x} was prepared from nanoparticles of the constituting element either hot pressed or using SPS. TEM shows particles of about 5–10 nm of Si in a Ge host after sintering process. Property studies as a function of compaction level show that a small change of the density induces a change of σ by orders of magnitude. When κ_L of $Si_{0.8}Ge_{0.2}B_{0.016}$ is compared with a 96 h ball milled sample, the latter shows a strongly lowered value κ_L compared to SOA material. It also exhibits a higher S and only a slight decrease of σ due to strong interface scattering. This leads to an increase in the power factor $P = S^2\sigma$ over the 300–1,000 K temperature range to a maximum of about 18 μW/cm.K^2 at 800 K. $Si_{0.8}Ge_{0.2}B_{0.016}$ finally gives a higher $ZT > 0.7$ at 1,000 K compared to 0.6 at 1,100 K for SOA SiGe.

  Phosphorous-doped n-type nanostructured $Si_{0.8}Ge_{0.2}P_{0.02}$ was prepared by BM followed by HP (direct current–induced hot pressing method) between 1,273 and 1,473 K. $ZT \sim 1.3$ at 1,173 K has been achieved. Average grain size is about 10–20 nm after HP and adjacent grains have similar crystal structure, but different crystalline orientations that could promote the phonons scattering. Both κ and κ_L are reduced compared to SOA material due to the enhancement of phonon scattering related to the increased density of nanograin boundaries. At 300 K, $\kappa = 2.5$ W/m.K and $\kappa_L = 1.8$ W/m.K. Again, σ is much less reduced than κ compared to the SOA alloy up to 1,023 K and shows similar values above this point, even though the electron concentration $n \sim 2.2 \times 10^{20}cm^{-3}$ at 300 K is about the same, indicating a slightly lower electron mobility in the nanostructured samples. S is not much affected by the nanostructuring except for a slight increase in the 673–973 K temperature region. Final ZT shows an improvement of about 40% compared to the SOA sample.

- *Silicon* [9]
  A huge reduction of κ_L by ~90% is observed in BM n-type silicon ($Si_{0.98}P_{0.02}$), compared to single-crystal (SC) material, due to high density of grain boundaries. Nanocrystalline domains from 10 to 100 nm were observed on samples hot pressed at 1,475 K for several hours. Both S and resistivity ($\rho = 1/\sigma$) increase compared to SC, the latter related to a decrease in μ possibly due to the creation of a potential barrier which could originate from grain-boundary scattering or defects at grain boundary or even both. The peak observed on both S and ρ between 800 and 1,000 K has been attributed to the precipitation of phosphorous. κ_L has been reduced from 80 W/m.K for heavily doped Si down to 10 W/m.K for nanostructured Si with moderate BM and even smaller value (5 W/m.K) for extended milling and higher ball-to-load ratios' sample. The μ/κ_L ratio increases from 0.05 for SC up to 0.24 for nanostructured bulk Si. Finally, for best sample, ZT has been increased by a factor of 3.5 compared to Si SC, reaching $ZT = 0.7$ at 1,200 K.
- *PbTe* [7, 8]
  100-nm PbTe nanocrystals Ag-doped with Ag_2Te (p-type) were synthesized via a solution phase technique and densified by SPS and compared to undoped sample. At room temperature they all show an important $S \sim 325$ and 200 μV/K for undoped and Ag-doped compounds respectively which is higher than polycrystalline or SC PbTe at similar carrier concentration. σ is strongly enhanced compared to undoped samples while κ remains comparable. The temperature dependence of the resistivity and mobility shows a unique behavior, suggesting that in addition to the phonon scattering, grain-boundary potential barrier scattering also exists and acts as a dominant scattering mechanism. This scattering could arise from the formation of an oxide layer around nanoparticles that filter the conduction of carrier with energy lower than that of the barrier. Those results suggest that interfacial energy barrier carrier scattering is a good means to enhanced TE performance. Different results obtained by solidification from the melt will be discussed in the top-down approach section.
- $Mg_2Si_{1-x}Sn_x$ [10, 11]
  Antimony-doped n-type $Mg_2Si_{0.4-x}Sn_{0.6}Sb_x$ $0 \leq x \leq 0.015$ alloys show natural nanostructuring after melting, due to compositional fluctuations and structural modulation (~16 nm) and nanodots (~10 nm) naturally formed by phase separation in a fashion that will be described in the following paragraph. Preparation by BM and HP compaction [10] results in a $ZT \sim 1.1$ at 800 K for $x = 0.0075$, which is similar to that of the SOA. However, a new approach is described for this system: Nanostructuring of Mg_2Si [11] was achieved by using microwave heating. Nanopowders of both precursor Mg and Si were first obtained by high energy BM in dry condition, then the mixture was cold pressed and heated by microwaves. While BM influences the size of the particles and the behavior of particles under irradiation, irradiation power and time influence grain growth. Indeed the reaction is initiated by the exclusive absorption of microwaves by Si, making the size of Si particle an important parameter to control. Smaller particles means higher power which in turn induce the sublimation of Mg. Using precursor powders prepared by 2 h BM subjected to microwave power ~175 W for 2 min, Mg_2Si nanoparticles smaller than 100 nm can be prepared. It remains to be seen if such large particles can affect κ much.

Top-Down Approaches

Different solidification processes can be used to achieve nanoinclusions embedded in a matrix and are mainly due to phase separation. The study of phase diagrams is essential to this approach, and examples of the metallurgical techniques used are based on the phase diagrams shown in Figs. 3 and 4. In summary, the techniques used fall into five categories:

(a) The crossing from a single- (solid solution, eutectic composition) to a two-phase region is shown in Fig. 3. In the case of the solid solution, the decreasing of the solubility with temperature is an advantage to obtain fine structure.

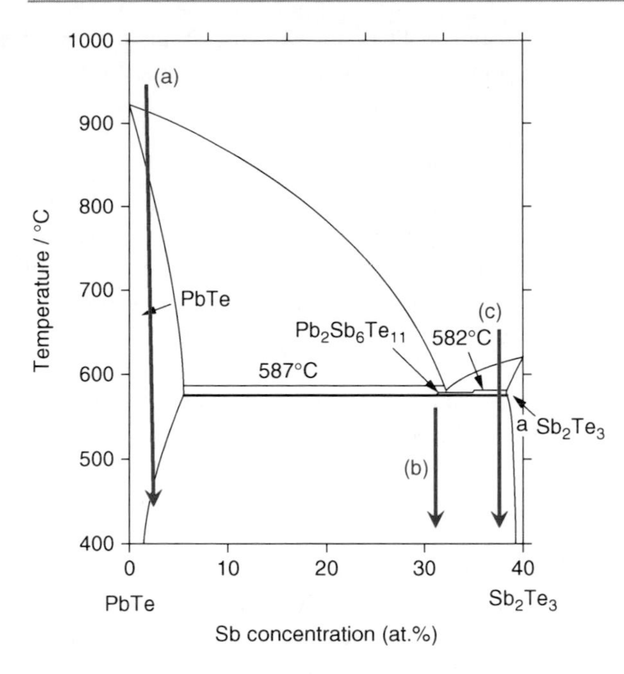

Thermoelectric Heat Convertors, Fig. 3 A pseudo-binary phase diagram showing the different cooling schemes that can result in nanoparticle formation from the melt

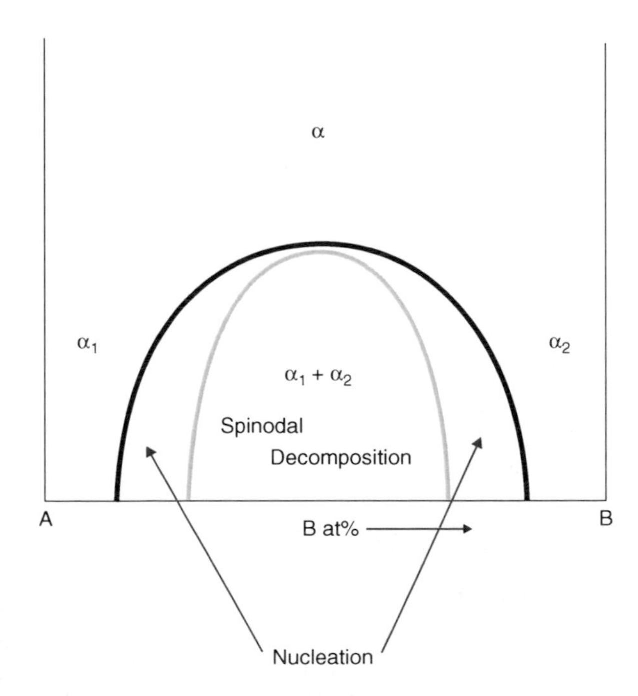

Thermoelectric Heat Convertors, Fig. 4 A pseudo-binary phase diagram showing the solid regions where anneals can result in nanoparticle formation

(b) Decomposition of a metastable phase: In Fig. 3, $Pb_2Sb_6Te_{11}$ decomposes into PbTe and Sb_2Te_3 if annealed at appropriate temperature.
(c) In Fig. 3, solidification from the melt allows control of crystallite size with the cooling rate.
(d) In specific phase diagrams such as shown in Fig. 4, spinodal decomposition or nucleation and growth can be used in specific regions. Both phase separation processes occur in the solid state.
(e) Liquid encapsulation techniques involve solubility of both phases at high temperature (liquid state) and no or very low solubility at room temperature (solid state). During cooling, the matrix solidifies first (higher melting point) and traps the nanoinclusion that solidify later.

- *PbTe Nanostructuring*
- *PbTe/Tetradymites* [8, 12]

Different methods were explored to obtain nanostructures in the system PbTe-Sb_2Te_3 [8] by solidification from liquid based on (a) the presence of an eutectic reaction, (b) eutectoid decomposition of $Pb_2Sb_6Te_{11}$, and (c) precipitation.

The first process was investigated using eutectic, hypereutectic (Sb_2Te_3 rich), and hypoeutectic (PbTe rich) compositions. Microscopic analysis indicates that PbTe-rich material shows a dendritic microstructure of PbTe in a $Pb_2Sb_6Te_{11}$ matrix, while Sb_2Te_3-rich compounds have lamellar microstructure of Sb_2Te_3 in the same matrix. The eutectic composition shows both lamellae of Sb_2Te_3 and PbTe dendrites. The study of the microstructure as a function of the cooling rate indicates that a finer structure is obtained by increasing the solidification rate, and that nanometer scale structures can be reached with ultrarapid solidification techniques.

Crystallographically oriented lamellar structures with alternating (PbTe/Sb_2Te_3) composition similar to those observed in thin-film superlattices [3] were obtained by using the decomposition of metastable $Pb_2Sb_6Te_{11}$ into PbTe and Sb_2Te_3. Interlamellar spacings around 180 nm can be controlled by the temperature and time of the decomposition process. Those alloys are *p*-type with high carrier concentration.

Widmanstätten precipitates were observed in a PbTe matrix by precipitation of Sb_2Te_3 in a PbTe solid solution during unidirectional (Bridgman) solidification of $Pb_{10.5}Sb_{31.6}Te_{57}$. The formation of those plate-like and elongated ribbon-like precipitates is due to the decrease of the solubility with decreasing temperature.

Similar work has been done with the PbTe/Bi_2Te_3 system [12] off and near the eutectic region of the pseudo-binary $(PbTe)_x(Bi_2Te_3)_y$ system with x:y = 2:1,1:1,1:2, and 1:3. After water quenching from the melt, phase separation occurs. The 1:1 sample is mainly $PbBi_2Te_4$ with micrometer-sized precipitates of Bi_2Te_3 containing lamellar PbTe. Bi_2Te_3-rich samples (1:2,1:3) show randomly oriented lamellar microstructures with alternating composition of PbTe-, Bi_2Te_3-, $PbBi_2Te_4$-, and Te-rich phases. PbTe-rich samples (2:1) present a strong PbTe dendritic microstructure in a $PbBi_2Te_4$ matrix. All these compounds have $S < 0$ with the lowest value S = −75 μV/K at 550 K for 1:2 composition. Neither ρ nor κ shows a strong temperature dependence, and ρ is lower than in binaries which was attributed to the lower resistance of $PbBi_2Te_4$. At 300 K, κ_L of the alloys is the same as that of Bi_2Te_3 but becomes smaller with increasing temperature, presumably by scattering of phonons on the increased number of heterointerfaces due to phase separation.

- *PbTe – M (M = Sb, Bi, InSb) by Liquid Encapsulation* [7, 8]
  PbTe-x%M systems where M = Sb, Bi, or InSb and x ranges from 2% to 16% were investigated, and gave similar SEM and XRD results. No solubility of M in the matrix was observed, and nanoinclusions start to form for compositions as low as 2%M. Below x < 4%, nanostructuring is achieved with 2–6-nm inclusions evenly dispersed and probably coherently embedded, with a size that increases with x. At higher x-values, the second phase agglomerates, thus forming larger microstructures at first, and a network at yet higher x.

  Measurements of κ_L indicate that among the PbTe-x%Sb samples, the lowest value is achieved for the 2% sample from 300 to 700 K (κ_L = 0.8 W/m.K at 300 K). κ_L increases with x and at x = 16%, T > 500 K, exhibits a higher value than PbTe itself. The authors attribute this high κ_L to a percolation phenomenon.

  In the PbTe-4%M series, the lowest κ_L is achieved for M = InSb while κ_L is higher than that of binary PbTe for M = Bi: A large contrast in average mass between matrix and guest atoms is needed to obtain strong phonon scattering.
- *$Pb_{1-y}Sb_yTe_{1-x}Se_x$* [13]
  The series $Pb_{9.6}Sb_{0.2}Te_{10-x}Se_x$ (x = 0–10) shows Sb-rich nanophases widely distributed in the matrix. No systematic trend is observed in the variation of either S or ρ, suggesting the presence of inhomogeneities, microscopic cracks, or different doping levels in $Pb_{9.6}Sb_{0.2}Te_{10-x}Se_x$. The highest S (−282 μV/K at 700 K) is found for x = 2; κ_L of samples with x = 1–8 are below 1 W/m.K for 300 K < T < 800 K and the lowest $\kappa_L \sim 0.40$ W/m.K was measured for $Pb_{9.6}Sb_{0.2}Te_3Se_7$. ZT of this n-type compound exceeded that of PbTe around 500 K and reaches $ZT = 1.2$ at 650 K.
- *PbTe-Pb_xSb_y by Liquid Encapsulation* [7]
  An excess of Pb in PbTe prepared from the melt (p- and n-type doped with Tl and Ag respectively) create 40-nm Pb nanoinclusions: S is increased (see Section "Improving the $S(n)$ or Pisarenko Relation"), but counterbalanced by a decrease in μ and σ. Using the matrix encapsulation technique co-nanostructured PbTe containing both x% Pb and y% Sb excess (0.5% ≤ x, y ≤ 3%) was prepared with a majority of 2–10 nm nanoinclusions and some microprecipitates. Above 2–3% excess, Pb and Sb precipitate as a eutectic at the grain boundaries of the matrix. The transport properties and their temperature dependence show important changes compared to SOA PbTe, that are complicated functions of the z = Pb/Sb ratio. Classifying the changes in order of decreasing electrical conductivity at all temperatures: (1) for $z = 0.25$, σ decreases with increasing temperature; (2) for $z = 0.66$, σ increases up to 430 K, then decreases with temperature; (3) for $z = 1$ (two samples), σ is constant up to 450 K and then decreases with temperature; (4) finally for $z = 1$, σ increases with increasing temperature. On the other hand, S shows n-type behavior and the absolute value of $|S|$ roughly decreases with increasing z, and increases linearly with increasing temperature, suggesting a constant carrier concentration over the temperature range; this is confirmed by Hall coefficient measurements. This carrier concentration increases with the (Pb + Sb) total concentration, consistently with the known donor behavior of

Sb substitutions on Pb sites, and of Te vacancies. The best properties were found for the composition PbTe-$Pb_{0.005}Sb_{0.02}$ at $z = 0.25$, where $P \sim 30\ \mu W/cm.K^2$ ($T = 700$ K) is increased by about 71% over SOA-PbTe. ZT is smaller than SOA-PbTe for $T < 475$ K but reaches $ZT \sim 1.5$ at 700 K.

- *PbTe/Ge and PbTe/Si_xGe_{1-x}* [7]
  n-doped PbTe/Ge and PbTe/Si_xGe_{1-x} with eutectic (20% Ge) and hypereutectic (2.5, 5, and 10% Ge) compositions were prepared. For all compositions, Ge appeared to be only slightly soluble in PbTe, inducing the formation of small Pb precipitates and rodlike Ge nanoprecipitates not coherently embedded and randomly oriented. Those provide stronger mechanical properties and reduced brittleness compared to SOA-PbTe. Hypereutectic compositions show lower κ_L, while eutectic compositions show κ_L comparable to that of SOA-PbTe. A minimum $\kappa_L < 1.5$ W/m.K is obtained for $PbTe_{0.95}Ge_{0.05}$. Further reductions of κ were realized by alloying Ge with Si and the use of PbI_2 as an *n*-type dopant to optimize the carrier concentration independently: $ZT \sim 1.3$ at 780 K was obtained for PbTe $(Ge_{0.8}Si_{0.2})_{0.05}$.
- *PbTe/PbS by Spinodal Decompositions* [7]
  Spinodal decompositions and the nucleation and growth phenomena (Fig. 4) were explored as a way to nanostructure $(Pb_{0.95}Sn_{0.05}Te)_{1-x}(PbS)_x$ with x = 0.04, 0.08, 0.16, and 0.30. PbI_2 was used as *n*-type dopant. The Sn substitution helped simplify the synthesis. The nucleation and growth occurs and increases with x up to x = 0.08. The phenomenon totally disappears for x = 0.3. The 4% samples tend to form solid solutions with some sign of nucleation and growth of nanodots; the 8% samples contain 3–10 nm nanocrystals, and spinodal decomposition occurs leading to compositional fluctuations shaped as periodical parallel stripes of PbS- and PbTe-rich regions on a nm scale. Spinodal decomposition becomes more prevalent with increasing x. The 16% sample shows coexisting regions with 3–10 nm PbS nanoparticles embedded in a PbTe matrix outside of the spinodal region. Transport property measurements indicate that both σ and S increase with x up to 8%; κ_L is reduced compared to bulk material. The smallest values are obtained for 8% ($\kappa_L \sim 0.38$ W/m.K) and 16% ($\kappa_L \sim 0.40$ W/m.K) samples where dots from the nucleation and growth and stripes from spinodal decomposition coexist. It is further suggested that larger interface densities and a larger composition variation help to enhance acoustic phonon scattering. The highest ZT (1.50 at 640 K) is achieved by the 8% sample.
- *PbTe-SrTe* [14]
  Endotaxy can be defined as the bulk version of *epitaxy*, and consists of the growth of oriented precipitates with aligned crystallographic planes and directions embedded in a substrate. Na-doped *p*-type PbTe containing 0.5–2% endotaxially arranged SrTe nanoprecipitates has been shown to have a κ_L reduced by selective phonon scattering, while charge carriers are much less affected. The 1–15 nm nanoprecipitates are spherical or ellipsoidal in shape; the smaller ones are coherently strained while the larger ones exhibit interfacial dislocation. μ, σ, κ and S are not really affected by increasing the SrTe content. The Hall coefficient is positive with a maximum at about 430 K ascribed to the appearance of a second heavy-hole valence band in PbTe that seems rather independent of the SrTe content, which in turn suggests that the Sr does not go into a solid solution. The presence of this second band is also confirmed by a change of slope in $\mu(T)$ and the high value of S at high temperature. Only the sample with 0.5% SrTe shows a slight increase of $\mu(T = 300\ K)$ from 350 $cm^2/V.s$ to 500 $cm^2/V.s$. The 1% and 2% SrTe samples show a small decrease of S at high temperature. The highest P (25 $\mu W/cm.K^2$) is reached by the 2% sample at 560 K.

  SrTe strongly affects κ_L, which decreases with increasing nanoprecipitate proportion, almost dividing κ_L by 2 over the whole temperature range. The lowest $\kappa_L \sim 0.45$ W/m.K^2 is reached by the 2% sample at 800 K; it also shows the highest $ZT = 1.7$ at 815 K, and $ZT > 1$ from 550 K upward.
- *$(PbTe)_{1-x}(Ag_2Te)_x$* [15]
  Large (50–200 nm) Ag_2Te plate-like precipitates uniformly dispersed in a PbTe matrix were obtained by stepped solidification from the melt in the $(PbTe)_{1-x}(Ag_2Te)_x$ system with x = 1.3, 2.7, 4.1, and 5.5. Ag_2Te is soluble in the PbTe matrix up to 1%, above which the proportion of Ag_2Te precipitate increases with x while the size remains constant. The series shows consistent semiconducting

behavior with a high $\rho \sim 40m\Omega.cm$ at 300 K that reaches a maximum at about 400 K, indicating a change from an extrinsic to an intrinsic behavior. This change is confirmed by the change of sign of S and Hall coefficient at the same temperature and could be related to the phase transition of Ag_2Te from a monoclinic to a cubic structure that occurs at the same temperature. κ decreases with increases in both concentration of Ag_2Te and temperature; at higher temperature, this is due to the behavior of the κ_E contribution and the dependence of $\rho(T)$. A reduction of κ_L is observed, due to both Ag alloy scattering in PbTe matrix and to phonons scattering on Ag_2Te nanoprecipitates.

Thermoelectric properties have been optimized by doping using La as a donor. The nanostructure remains and La is homogeneously dispersed in the matrix. Compositions near $(Pb_{1-y}La_yTe)_{0.945}(Ag_2Te)_{0.055}$ with y = 0.094, 0.0186, 0.0276, 0.0364 all exhibit heavily n-type doped semiconducting behavior with a strongly reduced ρ <2mΩ.cm (300 K) that peaks above 400 K probably due to the phase transition of Ag_2Te. There is hysteresis between heating-up and cooling-down curves. S has been optimized to 200 μV/K at T > 700 K. κ_L remains low and reaches a minimum of ~0.4 W/m.K at T > 650 K for y = 0.0276. Finally a high ZT = 1.6 at 775 K is obtained for z = 0.0276 and carrier density $n \sim 3 \times 10^{19} cm^{-3}$.

- *$AgPb_mSbTe_{2+m}$ (LAST-m) and Derived Compounds* [7, 16]
  LAST alloys have an average NaCl structure where two Pb^{2+} atoms are isoelectronically substituted by one Ag^+ and one Sb^{3+}, thus giving an average charge of +2 counterbalanced by a −2 charge of chalcogens. These materials are excellent n-type thermoelectrics. Ag deficiency allows to tune the electron concentration and acts as a n-type dopant. By cooling the sample from the melt, spontaneous nanostructuring occurs that consists of coherent and semicoherent nanocrystals of Ag-Sb embedded in a PbTe matrix and compositional fluctuations at nanoscopic level is observed. Those materials are stable up to their melting point (<1,200 K). The absolute value $|S|$ of the thermopower ($S < 0$) of the best composition $Ag_{1-x}Pb_{18}SbTe_{20}$ (~ $(PbTe)_{0.95}(Ag_{1-x}SbTe_2)_{0.05}$) increases from 135 μV/K at room temperature up to 335 μV/K at 700 K giving a high P = 28 μW/cm.K^2. κ is 2.3 W/m.K at 300 K, and decreases to κ = 1 W/m. K at 700 K. A very high electron mobility of 800 cm^2/V.s has been found; the material reaches $ZT \sim 2.1$ at 800 K.

Similar results were obtained on $Ag_{0.8}Pb_{22.5}SbTe_{20}$ prepared by BM + SPS followed by anneals at different times. With annealing time, n is first strongly decreased, but then slightly increases with longer annealing time; ρ, μ, κ and κ_L decrease. Only S shows optimal values after 3 days annealing and reaches $|S|$ = 287 μV/K at 450 K. A metallic to semiconductor transition is observed in the temperature dependence of ρ at 400 K. Transport property changes were attributed to the evaporation of free Pb when annealing that causes a decrease in electron concentration. The highest P and ZT are finally obtained for the longest annealing time and ZT =1.5 at 700 K. The nanostructuring is similar to that of the bulk ingots, with embedded nanocrystals of about 5–20 nm in unannealed samples that increase up to 10–50 nm after annealing. BM + SPS can be considered only as a practical preparation method that leads to more homogenous samples than ingot casting, but not as the origin for the nanostructure. Even here, despite homogeneous X-ray and SEM analysis, scanning Seebeck microprobe measurement shows a huge heterogeneity throughout the samples, with S varying from n- to p-type (from −200 to 200 μV/K) within 100 μm separations in the samples.

A p-type derivative, LASTT-m, is $Ag(Pb_{1-x}Sn_x)_mSbTe_{2+m}$, obtained by substituting Sn for Pb. Similar to LAST material, it contains nanostructures consisting of coherent (Ag, Sb) 5–20 nm nanoprecipitates embedded in a $Pb_{1-x}Sn_xTe$ matrix with compositional fluctuation. Transport properties are controlled by varying the Pb/Sn ratio. The optimum composition is found to be $Ag_{0.5}Pb_6Sn_2Sb_{0.2}Te_{10}$ and reaches a maximum ZT = 1.45 at 630 K due to the significant reduction of κ_L < 0.8 W/m.K over the temperature range.

Substitution of Sb by Bi in $AgPb_m(Sb_{1-x}Bi_x)Te_{2+m}$ alloys does not change the nanostructure, but decreases n, increases μ, and also increases κ_L above that of LAST but it remains below that of

SOA-PbTe. This is attributed to the smaller mass difference between Pb and Bi than between Pb and Sb.

Another variation on LAST is $AgPb_mLaTe_{m+2}$, whose transport properties can be described as intermediate between that of Sb and Bi analogs. La introduces more electrons, which implies higher σ because μ remains high. $AgPb_mLaTe_{m+2}$ transport properties can be described as intermediate between that of Sb and Bi analogs. Compared to LAST itself (the Sb analog), $AgPb_mLaTe_{m+2+}$ has a higher σ and lower S, but comparable power factor and κ_L. Compared to the Bi analog, $AgPb_mLaTe_{m+2+}$ has a higher σ and comparable κ_L, but a comparable S coefficient and therefore a higher P. The higher σ leads to a higher κ_E so that $ZT \sim 0.9$ at ~ 670 K in $AgPb_{25}LaTe_{27}$.

$Na_{1-x}Pb_mSb_yTe_{m+2}$ shows p-type behavior with a ZT above 1 in a wide temperature range from 450 up 750 K and ZT = 1.7 at 700 K for the optimum composition $Na_{0.95}Pb_{20}SbTe_{22}$. κ_L reaches a minimum of 0.55 W/m.K. Distinct Na-Sb-rich nanodomains are observed endotaxially embedded in a Pb-rich matrix. The properties of $NaPb_m(Sb_{1-x}Bi_x)Te_{2+m}$ are less degraded by Bi substitutions for Sb than those of LAST, and $Na_{0.95}Pb_{18}BiTe_{20}$ still shows a low $\kappa_L \sim 0.7$ W/m. K and $ZT = 1.3$ at 670 K.

Substitution of Pb by Sn to form $Na(Pb_{1-y}Sn_y)_mMTe_{2+m}$ (M = Bi,Sb) has been investigated. The Pb/Sn ratio affects both structure and transport properties. Despite a very low nanostructure-limited $\kappa_L \sim 0.4$ W/m.K at 650 K ($Na_{1-x}Pb_2Sn_{16}SbTe_{20}$), n and σ are higher and imply a lower μ and P compared to $Na_{1-x}Pb_mSb_yTe_{m+2}$ system. Nanostructuring is inhomogeneous and consists of a $Pb_{1-y}Sb_yTe$ matrix containing dispersed nanocrystals, and 10 nm lamellar nanostructures related to a composition fluctuation that might be a local ordering between PbTe and SnTe phase are also observed.

$K_{1-x}Pb_mSb_yTe_{m+2}$ (PLAT-m) compounds are n-type and exhibit even lower κ_L than LAST, reaching 0.4 W/m.K at 650 K ($K_{0.95}Pb_{20}Sb_{1.2}Te_{22}$) and a maximum $ZT = 1.6$ at 750 K. The nanostructure consists of K-Sb-rich endotaxial nanocrystallites coherently embedded in a PbTe-rich matrix.

Improving the *S*(*n*) or Pisarenko Relation

Theory

There is a limit to the improvement in ZT that can be achieved by lowering the lattice thermal conductivity, because in bulk solids, there is a limit to how low the phonon conductivity can be brought: The phonon mean free path cannot be smaller than one interatomic distance. Further improvements in ZT must come from increasing the power factor $P = S^2\sigma$ or the product (S^2n). These are purely electronic properties dominated by the electronic band structure and electron scattering, and their optimization is done by *band structure engineering*.

It is useful to first review the origin of the thermopower S, which can also be defined by the differential form of Eq. 1 with respect to the length of the sample as the ratio between the electric field E generated by a temperature gradient ∇T applied across a sample and ∇T: $S = \frac{\mathrm{E}}{\nabla T}$ (refer to one of the elements of the couple in Fig. 1 along the height of the samples). In the absence of an electrical current (a flux of charge carriers), the charge carriers tend to drift toward the cold side of the sample, where they "accumulate." More rigorously, there is an imbalance in the thermal energy between the electronic population at the cold and hot ends of the sample, which must be balanced by a difference in electrostatic energy. Since the particles carry an electrical charge, an electric field E arises from the imbalance in densities. For a single electron, the gradient in thermal energy is $\frac{3}{2}k_B\nabla T$; the gradient in electrical energy is qE. The condition that they be equal dictates that:

$$S = \frac{\mathrm{E}}{\nabla T} = \frac{3}{2}\left(\frac{k_B}{q}\right), \qquad (11)$$

which correctly shows that S reflects the polarity of the majority carrier charge, and that it is on the order of $|K_B/q| = 86$ μV/K for solids with low densities of electrons ($n < 10^{-3}$ per atom). It fails in two notable ways. Firstly, because the thermopower is closely related to the entropy of the electron, Eq. 11 fails the criterion imposed by the third principle of thermodynamics that $\lim_{T\to 0} S(T) = 0$. Secondly, Eq. 11 fails to explain why the thermopower of metals or solids with a concentration of electrons nearly equal to that of atoms is one or two order of magnitude smaller than

86 μV/K. Mott improves Eq. 11 by defining a quantity $\sigma(E)$ as the amount of electrical conductivity contributed by electrons at energies below E, and deriving from it expressions for:

$$\sigma = \int \sigma(E)\left(-\frac{\partial f}{\partial E}\right) dE \qquad (12)$$

and

$$S = \frac{k_B}{q}\frac{1}{\sigma}\int_0^{\infty} \sigma(E)\left(\frac{E - E_F}{k_B T}\right)\left(-\frac{\partial f}{\partial E}\right) dE \qquad (13)$$

where f is the Fermi distribution function. For metals and degenerately doped semiconductors with simple ellipsoidal Fermi surfaces and parabolic energy-momentum relations, such as most thermoelectric materials, these equations simplify to:

$$\begin{aligned} \sigma &= ne\mu(E_F) \\ S &= \frac{\pi^2}{3}\frac{k_B}{q} t(k_B T)\left[\frac{1}{n(E)}\frac{dn(E)}{dE} + \frac{1}{\mu(E)}\frac{d\mu(E)}{dE}\right]_{E=E_F}, \end{aligned} \qquad (14)$$

which avoids both difficulties with Eq. 11 and illustrates the origin of the Pisarenko relation, the decrease of $S(n)$ with n. One additional concept that is useful to introduce is that of *density of states* (DOS)

$$g(E) \equiv \frac{dn(E)}{dE}, \qquad (15)$$

the number of states available for electrons to occupy at energies between E and $E + dE$ in a unit volume of solid. The thermopower becomes:

$$S = \frac{\pi^2}{3}\frac{k_B}{q}(k_B T)\left[\frac{g(E)}{n(E)} + \frac{1}{\mu(E)}\frac{d\mu(E)}{dE}\right], \qquad (16)$$

which consists of two terms, a DOS term $g(E)/n(E)$ and a scattering term $(1/\mu)$ $(d\mu/dE)$. Engineering the shape and energy dependence of $g(E)$ in order to give it sharp features is the main tool used to improve $S(n)$. Often the presence of sharp features in $g(E)$ also results in enhanced energy dependence of the scattering (*resonant scattering*), which can affect the scattering term in Eq. 16. This is the case in so-called Kondo systems, but probably also in the previously mentioned Pb-rich PbTe prepared with Pb 40 nm nanoinclusions, where the nanoinclusions preferentially scatter electrons at a selected energy, thereby increasing $|S|$. The scattering effects will not be considered further here.

Two pioneering theoretical papers are at the basis of recent progress in increasing $S(n)$. Mahan and Sofo [17] concluded in 1996 that the larger g and the stronger its dependence on E, the higher $S(n)$ will be for a given carrier concentration. The limiting case that offers optimal improvement would be that of a delta function like DOS at $E = E_F$, but only if there is no background DOS underneath it. This situation is unattainable in nature. Only electron energy levels on isolated individual atoms give such DOS: For example, and ignoring the excited states, the 1s state on an isolated hydrogen atom in vacuum can accommodate two electrons in two spin states at one single energy value 13.6 eV from vacuum, and so has $g = 2$ at that energy value and zero at all others. Electrons on such levels are *localized*, i.e., they have no conductivity and are not useful in thermoelectrics. That theory did lead to the identification of the usefulness of impurities that, in a thermoelectric solid, could lead to the existence of additional levels at certain energies where they have a DOS somewhat similar to a localized level, but yet conduct electricity somewhat. One known such case is that of *resonant impurity levels*, which have lead to a doubling of the ZT of PbTe [18]. In a real solid, energy levels are naturally broadened by imperfections and lattice vibrations, and thus even a sharply peaked $g(E)$ is expected to be shaped as a Lorentzian function of E.

The seminal paper outlining the potential for nanotechnologies in thermoelectrics was published in 1993 by Hicks and Dresselhaus [19] who predicted that quantum wells and quantum wires could make excellent thermoelectric materials, based on the particular energy dependence of $g(E)$ induced by quantum size effects, and this was confirmed experimentally. The Hicks and Dresselhaus theory requires more explanation, which is reviewed in Ref. [20] and is based on the mathematical definition of the DOS in statistical physics. Properties of ensembles of particles, such as their electrical or thermal conductivity, are functions of integrals over particles of all values of velocity or momentum $\boldsymbol{p}$ a vector. When the momentum is distributed over 3-dimensional (3-D) space, $\boldsymbol{p} = (p_x, p_y, p_z)$, and these are volume integrals, but they are surface integrals in 2-D because $\boldsymbol{p} = (p_x, p_y)$, and line integrals in 1-D systems where $\boldsymbol{p} = (p_x)$. It is much easier to express properties using integrals of the properties of the particles over energy, a scalar. The transformation of variables from 3-D integrals over momentum values of the type $\iiint_{\mathbf{p}} \ldots dp_x dp_y dp_z$ to scalar integrals over energy of the type $\int \ldots g(E)dE$ uses, as transformation variable, the density of states $g(E)$. It is now clear that the dimensionality of the problem will have a very strong influence on the shape of the transformation function $g(E)$, as shown schematically in Fig. 5 for 3-D, 2-D, and 1-D systems. The basic concept behind the extraordinarily large thermopower of quantum wires is that the thermopower should diverge when $g(E)$ diverges as it does in 1-D systems in Fig. 5 at specific values of E_F. In practice, the ratio g/n does not diverge but is strongly enhanced in 1-D systems over 3-D systems, and so great improvements in thermopower are predicted. The only requirement for this to work is the dimensionality, which is defined physically by comparing the critical dimension of the nanostructure (the thickness of the 2-D plane, the diameter d_W of the 1-D quantum wire) to the wavelength of the electron: If the diameter of the quantum wire is smaller than this *de Broglie wavelength* λ_D, $d_W < \lambda_D$, the electrons are subject to *quantum confinement* or subjected to *size quantization* into 1-D, and their S can be dramatically increased. This is verified experimentally.

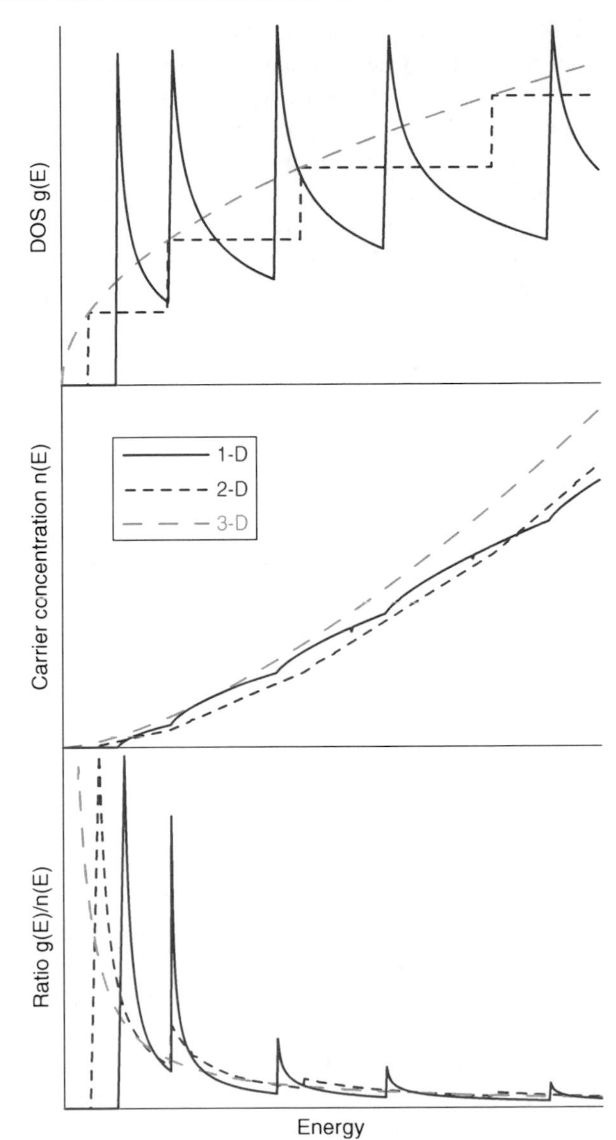

Thermoelectric Heat Convertors, Fig. 5 Density of states $g(E)$ in three-, two-, and one-dimensional solids, $n(E)$ its integrated value over E, and $g(E)/n(E)$ ration, which determines S

Bismuth Nanowires

Bismuth is the element in which the theory above can be most easily proven, because the electrons in Bi have an extremely small effective mass, and consequently the extent of their de Broglie wavelength is on the order of 50 nm. This large threshold diameter below which electrons will be quantum-confined into a 1-D regime avoids nefarious effect such as *localization*. Other advantages in using Bi are: (1) it is an element, and thus its preparation is much easier than that of compound semiconductors; (2) electron mobilities are very high; (3) it is the end-member of the class of $Bi_{1-x}Sb_x$ alloys suitable for cryogenic Peltier cooling. Bi is not a semiconductor, but a semimetal in which electrons in the conduction band coexist with an equal number of holes in the valence band. However, under quantum confinement, Bi nanowires undergo a transition to a semiconductor at $d_W = 50$ nm, and these have an electron structure similar to that of germanium with an energy gap that varies with d_W and T.

$Bi_{1-x}Sb_x$ nanowires (for $x < 0.2$) are predicted to have a better thermoelectric performance than pure Bi nanowires of the same diameter. A phase diagram of 3D bismuth and bismuth nanowires as a function of Sb concentration for both 3D bulk bismuth (a) and bismuth nanowires (b) and as a function of nanowire diameter has been developed theoretically, and identifies one regime where Bi nanowires are predicted to be in a direct gap semiconducting phase, two semimetal regimes, one Bi-like and one Sb-like, and two indirect-gap semiconducting regimes. There is one singular point ($d_W = 60$ nm, $x = 0.13$) where $g(E)$ is particularly high. One expects to obtain a better ZT in $Bi_{1-x}Sb_x$ ($x \sim 0.2$) nanowires than in pure Bi nanowires, but experiments have yet to be carried out.

Experimental Results on Bi Nanowires

The theory is confirmed experimentally on Bi nanowires (Fig. 6) of diameters in the range of 200–4 nm grown in a variety of porous host materials, typically based on alumina or silica. Much work has been done using porous anodic alumina as a host: A layer of anodic Al_2O_3 is grown on metallic Al, and that layer contains a quasi-periodic array of ordered holes that traverse the layer. The Al_2O_3 layer can be chemically detached from the metal substrate, and form a freestanding membrane with 7–200 nm diameter holes across the whole thickness. Other porous host materials, in which the pores form a random network, have been used, typically porous Al_2O_3, silica gel, and porous vycor glass. Three different techniques have been used to grow Bi nanowires in these pores: electrochemical deposition, high pressure liquid injection, and vapor-phase growth which gives nanowires as shown in Fig. 6. All these techniques produce highly crystalline wires, and, in the oriented porous anodic alumina host, even single-crystal wires along a preferred orientation. Figure 7 shows a summary of the data from measurements of electrical conductivity and Fig. 8 shows the Seebeck coefficient [21].

Thermoelectric Heat Convertors, Fig. 6 Bismuth nanowires in porous alumina (*top*) and silica (*bottom*), after [20]

The exact number of wires connected between the measurement electrodes in these experiments is unknown, and thus so is the effective cross section of the sample: Therefore no absolute values of σ can be given, only the value of the resistance $R(T)$ normalized to the room temperature value. This has the unfortunate consequence that neither the power factor nor the one-dimensional figure of merit can be measured; single-wire measurements are needed but are not available to date. The thermopower of the wires of narrower diameters grown in alumina, silica, and vycor glass, three host materials with randomly oriented pores, is also shown in Fig. 8. The resistance data show the semiconductor/semimetal transition as follows. At diameters above 50 nm, the low-temperature slope of $R(T)$ is positive, as the carriers density in a semimetal is fixed by the energy overlap between the conduction and valence band, and the mobility decreases through phonon scattering. At $T > 100$ K, the density of both electrons and holes increases through thermal activation, and this gives a negative temperature coefficient to $R(T)$. In the semiconductor regime, the negative slope prevails over the entire temperature regime. The resistance of the 15 nm wires and the high-temperature resistance of the 9 nm wires can be fit to an activated behavior. The 4 nm wires, and the 9 nm wires below 200 K, have a resistance that follows a $T^{-1/2}$ law; magnetoresistance measurements on narrow wires show that this behavior is due to weak localization effects. The thermopower of bulk Bi (along the binary direction) and of the 200 nm wires is similar. The thermopower of the 15 nm wires has

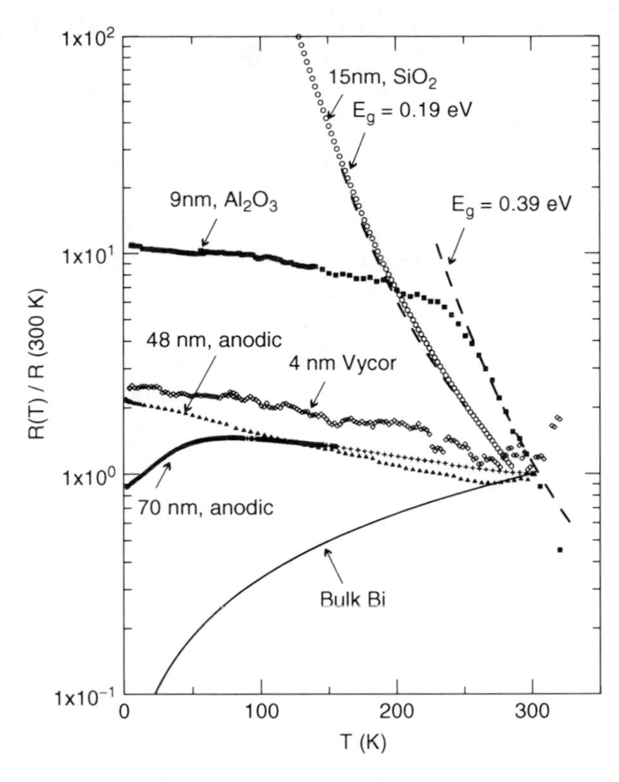

Thermoelectric Heat Convertors, Fig. 7 Normalized resistivity of Bi nanowires of the diameters indicated, after [21]

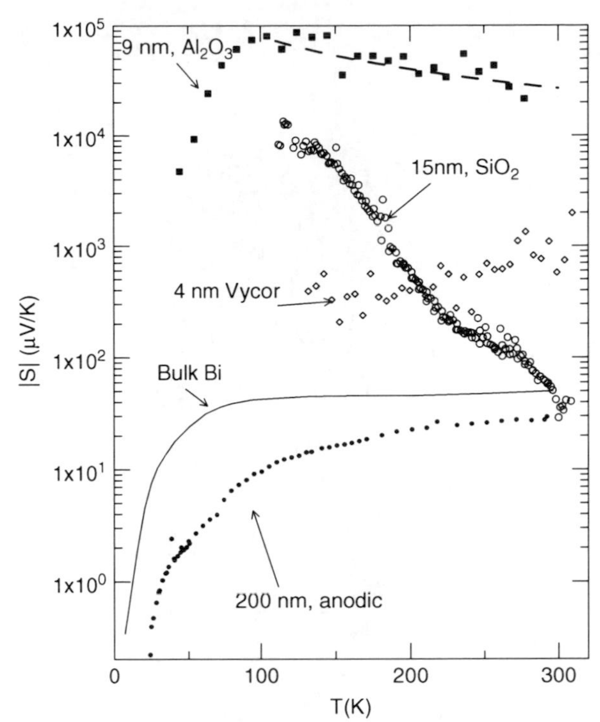

Thermoelectric Heat Convertors, Fig. 8 Thermopower of Bi nanowires of the diameters indicated, showing a 1,000-fold increase of S over bulk that of Bi due to quantum confinement (After [21])

a very pronounced temperature dependence, which was attributed to two-carrier conduction and a compensation between the electron and hole partial thermopowers. The thermopower of the 9 nm wires is much enhanced over that of the bulk, and follows a T^{-1} dependence shown as a dashed line, as expected from an intrinsic semiconductor. Finally, the low-temperature Seebeck coefficient of the 9 nm wires, and that of the 4 nm wires, is strongly decreased, due to localization effects.

While Bi is the material that is most easily subject to size quantization, the nanostructures prepared from Bi nanowires in porous hosts have the inherent disadvantage that the host material constitutes a thermal short: The spectacular thermopowers have not yet resulted in high overall effective ZT of the filled composite.

Metallic Wires

Ideal and practical "bulk-like" nanowires systems would need to be prepared using a top-down approach similar to that in section "Improving the Mobility-to-Thermal Conductivity (μ/κ) Ratio" for nanodots, and should constitute of wires grown naturally inside bulk materials. One recent example has been published by Sugihara et al. [22], where Cu-Ni alloys were deposited by rf magnetron sputtering between Ta and Au layers. Figure 9 shows an atom-probe tomographic analysis of the Cu-Ni layer. Columnar growth of one particular composition is in evidence.

Cu and Ni phase form solid solutions throughout the composition range. Among these alloys is constantan, a thermocouple alloy with very large $S \sim -40\ \mu V\ K^{-1}$ near 300 K. Large values of thermopower are observed over an extended Cu-Ni composition range. The origin for this is that in these alloys, the Fermi level lies close to the edge of the bands that have their origin in the d-levels of Ni. Preliminary thermoelectric measurements were performed on the electrochemically grown material, and at first sight seem to indicate a thermopower value one order of magnitude larger than for bulk alloys. The measurements are very indirect, as the authors measured the heat induced by passing a current through the sample. This heat is the

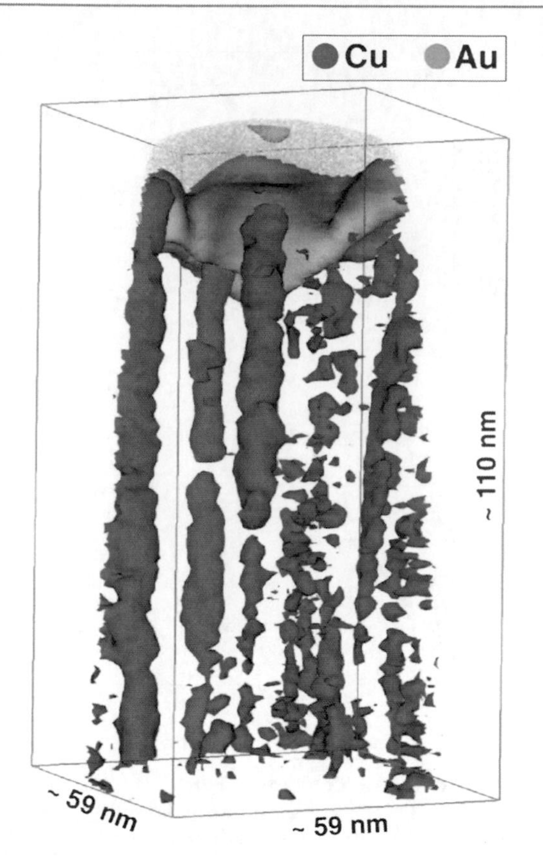

Thermoelectric Heat Convertors, Fig. 9 Ni-rich nanocolumns in Ni-Cu alloys (After [22])

sum of resistively induced Joule heat, and the Peltier heat given by Eq. 2. Joule heating experimentally overwhelms the Peltier effect in all but the very best thermoelectric materials: The data the authors report for the Peltier coefficient are therefore the difference between two large numbers and are in need of further verification. Still, the approach is promising because it is simple.

Conclusions

Nanostructuring has proven to be one of the most versatile and effective tools to increase the thermoelectric figure of merit *ZT* which governs the efficiency of solid-state heat-to-electricity converters. It thus promises to be a major enabler to a major energy-saving technology usable in waste-heat recovery systems. This entry outlines the mechanisms through which nanotechnologies reduce the thermal conductivity of the thermoelectric materials, thus minimizing the heat losses due to conduction. It also shows how in quantum-confined system, i.e., nanowires of diameters smaller than the wavelength of the electron, the thermopower is strongly enhanced, which means that more electrical power can be delivered from a give temperature difference.

In this concluding paragraph, it must be repeated that nanostructuring is not the only approach used: Resonant energy levels can also produce enhancements in thermopower as demonstrated in Tl-doped PbTe, and the use of local phonon modes in $CoSb_3$ or enhanced intrinsic phonon-phonon interactions in thermoelectric solids such as $AgSbTe_2$ can lead to strongly reduced thermal conductivity. Can these effects be additive? Some can, others no. As mentioned, there is a lower limit to the thermal conductivity (in bulk solids), so that in systems where that limit is nearly reached with phonon-phonon interactions or local modes, additional nanostructuring will be of less value. Often thermopower-enhancing techniques such as resonant level doping are sufficiently independent of the role of phonons that nanostructuring can lead to additional gains in *ZT*.

Finally, there are many opportunities that remain unexplored yet. In particular, the phonon mean free paths increase with decreasing temperatures, opening the way for even larger gains to be obtained in *ZT* at cryogenic temperatures. New preparation techniques remain to be explored, for example, the richness and flexibility of organic chemistry can be brought to bear on the problem in organic/inorganic mixed structures. These avenues of research and many others are the objects of active programs. In the future, the field promises to generate many new and important results that will help increase the overall efficiency of heat-to-electricity conversion systems and thus help alleviate the world's energy crisis.

References

1. Ioffe, A.F.: Physics of Semiconductors. Academic, New York (1960) (translated from Russian, "Fizika Poluprovodnikov", Russian Academy of Sciences, Moscow, 1957)
2. Minnich, A.J., Dresselhaus, M.S., Ren, Z.F., Chen, G.: Bulk nanostructured thermoelectric materials: current research and future prospects. Energy Environ. Sci. **2**, 466–479 (2009)

3. Venkatasubramanian, R., Siivola, E., Colpitts, T., O'Quinn, B.: Thin-film thermoelectric devices with high room-temperature figures of merit. Nature **413**, 597–602 (2001)
4. Harman, T.C., Taylor, P.J., Walsh, M.P., LaForge, B.E.: Quantum dot superlattice thermoelectric materials and devices. Science **297**, 2229–2232 (2002)
5. Koh, Y.K., Vineis, C.J., Calawa, S.D., Walsh, M.P., Cahil, D.G.: Lattice thermal conductivity of nanostructured thermoelectric materials based on PbTe. Appl. Phys. Lett. **94**, 153101–153103 (2009)
6. Poudel, B., Hao, Q., Ma, Y., Lan, Y., Minnich, A., Yu, B., Yan, X., Wang, D., Muto, A., Vashaee, D., Chen, X., Liu, J., Dresselhaus, M.S., Chen, G., Ren, Z.: High-thermoelectric performance of nanostructured bismuth antimony telluride bulk alloys. Science. **320**, 634–638 (2008)
7. Kanatzidis, M.G.: Nanostructured thermoelectrics: the new paradigm? Chem. Mater. **22**, 648–659 (2010)
8. Medlin, D.C., Snyder, G.J.: Interfaces in bulk thermoelectric materials A review for current opinion in colloid and interface science. Curr. Opin. Coll. Int. Sci. **14**, 226–235 (2009)
9. Bux, S.K., Blair, R.G., Gogna, P.K., Lee, H., Chen, G., Dresselhaus, M.S., Kaner, R.B., Fleurial, J.-P.: Nanostructured bulk silicon as an effective thermoelectric material. Adv. Funct. Mater. **19**, 2445–2452 (2009)
10. Zaitsev, V.K., Fedorov, M.I., Gurieva, E.A., Eremin, I.S., Konstantinov, P.P., Samunin, A.Y., Vedernikov, M.V.: Highly effective $Mg_2Si_{1-x}Sn_x$ thermoelectric. Phys. Rev. B **74**, 045207–045211 (2006)
11. Savary, E., Gascoin, F., Marinel, S.: Fast synthesis of nanocrystalline Mg_2Si by microwave heating: a new route to nano-structured thermoelectric materials. Dalton Trans. **39**, 1–7 (2010)
12. Yim, J.-H., Jung, K., Kim, H.-J., Park, H.-H., Park, C., Kim, J.-S.: Effect of composition on thermoelectric properties in PbTe-Bi_2Te_3 composites. J. Elec. Mater. **40**(5), 1010–1014 (2011)
13. Poudeu, P.F.P., D'Angelo, J., Kong, H.J., Downey, A., Short, J.L., Pcionek, R., Hogan, T.P., Uher, C., Kanatzidis, M.G.: Nanostructures versus solid solutions: low lattice thermal conductivity and enhanced thermoelectric figure of merit in $Pb_{9.6}Sb_{0.2}Te_{10-x}Se_x$ bulk materials. J. Am. Chem. Soc. **128**, 14347–14355 (2006)
14. Biswas, K., He, J., Zhang, Q., Wang, G., Uher, C., Dravid, V.P., Kanatzidis, M.G.: Strained endotaxial nanostructures with high thermoelectric figure of merit. Nat. Chem. **3**, 160–166 (2011)
15. Pei, Y., Lensch-Falk, J., Toberer, E.S., Medlin, D.L., Snyder, G.J.: High thermoelectric performance in PbTe due to large nanoscale Ag_2Te precipitates and La Doping. Adv. Func. Mater. **21**, 241–249 (2011)
16. Hsu, K.F., Loo, S., Guo, F., Chen, W., Dyck, J.S., Uher, C., Hogan, T., Polychroniadis, E.K., Kanatzidis, M.G.: Cubic AgPbmSbTe2m: bulk thermoelectric materials with high figure of merit. Science. **303**, 818–821 (2004)
17. Mahan, G.D., Sofo, J.O.: The best thermoelectric. Proc. Nat. Acad. Sci. USA **93**, 7436 (1996)
18. Heremans, J.P., Jovovic, V., Toberer, E.S., Saramat, A., Kurosaki, K., Charoenphakdee, A., Yamanaka, S., Snyder, G.J.: Enhancement of thermoelectric efficiency in PbTe by distortion of the electronic density of states. Science **321**, 554–558 (2008)
19. Hicks, L.D., Dresselhaus, M.S.: Thermoelectric figure of merit of a one-dimensional conductor. Phys. Rev. B **47**, 16631 (1993) (1–4)
20. Heremans, J.P.: Low-dimensional thermoelectricity. Acta. Phys. Polon. A **108**, 609–634 (2005)
21. Heremans, J.P., Thursh, C.M., Morelli, D.T., Wu, M.-C.: Thermoelectric power of bismuth nanocomposites. Phys. Rev. Lett. **88**, 216801 (2002) (1–4)
22. Sugihara, A., Kodzuka, M., Yakusiji, K., Kubota, H., Yuasa, S., Yamamoto, A., Ando, K., Takanashi, K., Ohkubo, T., Hono, K., Fukushima, A.: Giant peltier effect in a submicron-sized Cu–Ni/Au junction with nanometer-scale phase separation. Appl. Phys. Express **3**, 065204 (2010) (1–3)

Thermoelectric Heat Pump

▸ Thermoelectric Heat Convertors

Thermoelectric Module

▸ Thermoelectric Heat Convertors

Thermoelectric Nanomaterials

▸ Nanostructured Thermoelectric Materials

Thermoelectric Power

▸ Thermoelectric Heat Convertors

Thermomechanical Actuators

▸ Thermal Actuators

Thermometry

▸ Nanocalorimetry

Thermopower

► Thermoelectric Heat Convertors

Thermostability

► Structure and Stability of Protein Materials

Thickness

► Interfacial Investigation of Protein Films Using Acoustic Waves

Thiols

► Nanostructures for Surface Functionalization and Surface Properties

Third Generation Vectors

► Multistage Vectors

Thomson Effect

► Thermoelectric Heat Convertors

Three-Dimensional Imaging of Human Tissues

► Imaging Human Body Down to Molecular Level

THz

► Terahertz Technology for Nano Applications

TiO_2 Nanotube Arrays: Growth and Application

Karthik Shankar
Department of Electrical and Computer Engineering, W2-083 ECERF, University of Alberta, Edmonton, AB, Canada

Definition

TiO_2 nanotube arrays (TNAs) denote self-organized tubular nanostructures of titanium dioxide grown by the process of electrochemical anodization. When it is titanium metal which is subjected to electrochemical anodization, the TNAs grow directly in a single step without the assistance of a template. When it is aluminum metal which is subjected to anodization, the formation of TNAs is said to be templated and requires multiple steps including (1) the formation of the nanoporous alumina template, (2) a process to coat the walls of the template with TiO_2 or infiltrate a TiO_2 precursor into the template followed by heating, etc. (3) release of the alumina template by selective chemical etching to form TNAs. TNAs produced by the one-step process shall henceforth be referred to as anodic TiO_2 nanotube arrays and TNAs produced by the multistep process as templated TiO_2 nanotube arrays. In either process, the titanium dioxide nanotubes grow in arrays with their long axis aligned orthogonal to the substrate, due to which they are commonly referred to as vertically oriented TiO_2 nanotube arrays.

Introduction

The formation of anodic porous titania by using fluoride-based electrochemistry was first reported by Zwilling in 1999 [1]. However, it was Gong et al. [2] in 2001 that first formed nanoporous TiO_2 with a clearly identifiable nanotubular structure. Pioneering contributions to the field of TNAs were made by the research groups of Craig Grimes [3] at Penn State University and Patrik Schmuki [4] at the University of Erlangen. Initially, anodic TiO_2 nanotube arrays could only be grown on titanium foil substrates. In 2005, Mor et al. reported the successful formation of anodic TNAs on

Ti-coated glass substrates. Since then, anodic TNAs have been formed on silicon wafers and on transparent conducting oxide (TCO) coated glass substrates. Correspondingly, the first report on templated TNAs emerged in 1996 [5], but more detailed studies commenced in late 2001 [6].

TiO_2 is a highly versatile compound with applications as a pigment for paints, nontoxic biocompatible material for orthopedic implants, gas sensor, germicide, photocatalyst, n-type semiconductor scaffold for sensitized solar cells, optical coating, structural ceramic, electrical circuit varistor, spacer material for magnetic spin-valve systems, and more. TiO_2 is also relatively inexpensive owing to its stability and the plentiful abundance of Ti and O in the earth's crust. TiO_2 occurs in three allotropes: anatase, rutile, and the much less common brookite. The metastable anatase phase has superior electronic properties. The key motivation behind the development of TNAs is to enhance and exploit the versatility inherent in TiO_2 using the specific nanostructured morphology of TNAs. The oriented and aligned one-dimensional TNA architecture offers access to the anatase phase, a high surface area and directed electron percolation pathways for electronic and optoelectronic applications. For other applications of TNAs such as their use in stem-cell differentiators, antibiotic elution and energy transfer-based light harvesting, the specific properties of TiO_2 are less important; these applications derive instead from size selectivity, confinement in nanochannels, and other geometrical attributes of the nanoscale TNA architecture.

The research area of TNAs has witnessed an explosion of interest in the last decade with more than 1,000 research papers on this topic. Apart from the versatility of TiO_2 referenced above, another reason for the intense interest is the location of self-organized vertically oriented TiO_2 nanotube arrays at the intersection of three major trends in nanotechnology research over the last decade as shown in the schematic of Fig. 1. One such trend has been the drive to form one-dimensional structures such as nanorods and nanotubes in semiconductors, inspired by the unique properties of carbon nanotubes. Another trend, which is also one of the driving motivations underlying nanotechnology, is the exploitation of self-assembly and self-organization to form complex structures from the "bottom-up" as an alternative to patterning and fabricating structures from the "top down." The third trend has been the use of ordered anodic nanoporous alumina as a template to form a variety of oriented and aligned nanomaterials.

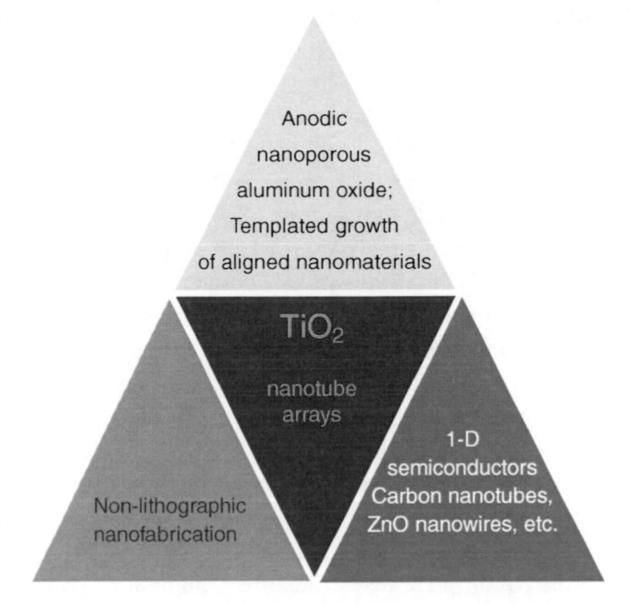

TiO_2 Nanotube Arrays: Growth and Application, Fig. 1 TNAs sit at the intersection of three major nanotechnology trends

The use of electrochemical anodization to form nanoporous alumina has been known since the 1950s. However nanoporous alumina is an insulator which limits its device applications. Anodic aluminum oxide (AAO) is widely used as a template to form functional nanomaterials. Since crystalline TiO_2 is a semiconductor, anodically formed TiO_2 nanotube arrays can be directly used as a functional material after the induction of crystallinity. A major difference between AAO and TNAs is that a wider range of morphologies are achievable in TNAs. Ordered AAO has been formed in a close-packed hexagonal pore architecture and in square and triangular lattices. In TNAs, the anodization electrolyte has a decisive role in determining the morphology of the resulting structure. By anodization in ethylene glycol-based electrolytes, TNAs may be formed in a close-packed hexagonal architecture similar to AAO. At the same time, nanotube arrays with a distinctive cylindrical cross section may be achieved in other electrolytes. The wall thickness can be varied to change the porosity of the architecture. In comparison with nanochannels in AAO or in track-etched polycarbonate membranes, the nanotubular structure has a significantly higher surface area due to the availability of both the inner and outer surfaces of the nanotube walls.

In the course of the last 10 years 2001–2010, a high degree of control has been achieved over the geometrical dimensions of the anodic TNA architecture by optimizing the anodization process. Potentiostatic anodization is most commonly used either by direct application of a constant anodization voltage or though a linear sweep of the potential to a final constant value. The key parameters involved in the fabrication process are the composition of the anodization electrolyte, the anodization voltage applied, the temperature of the anodization bath, and the anodization duration. For applications involving increased pattern order of the TNAs and for TNA growth on nonnative substrates such as glass, Si, etc., the nature of the Ti film or Ti foil specifically its density, roughness, grain size, and crystallographic texture (if any) are also critical. By tuning the various anodization parameters listed above, control over the geometrical dimensions of TNAs has been demonstrated over remarkably broad length scales [4, 7–9]; the diameter of nanotubes has been varied from 16 to 900 nm, the tube length from 250 to 1 mm, the wall thickness from 5 to 45 nm and the inter-tube spacing from 0 (close packed nanotubes) to several micrometers (widely separated nanotubes). Of these geometrical features, only the diameter and the tube length are independently controllable to date. The wall thickness and inter-tube spacing depend on the diameter and the tube length.

Growth of Anodic TiO$_2$ Nanotube Arrays

When a Ti foil or a Ti film coated electrode is subjected to anodization in highly conductive electrolytes, the anodization current experiences a steep drop in the first several seconds of the process. The drop in current occurs due to the rapid formation of a passivating oxide layer (called the barrier layer) through which electronic conduction can only occur through tunneling. In the simplest case, the anodization current density J for Fowler-Nordheim tunneling through the oxide barrier layer (BL) may be expressed as:

$$J = C\left(\frac{\Delta U}{t}\right)^2 e^{-\frac{\beta_{FN} t}{\Delta U}} \tag{1}$$

where ΔU is the portion of the anodization voltage which drops across the barrier layer, t is the oxide thickness and C, β_{FN} are constants. Equation 1 shows that the tunneling current reduces exponentially with increasing BL thickness. As the passivating layer increases in thickness, tunneling weakens and the formation of nascent gases at the anode due to electron transfer reactions is suppressed. Visually, this is indicated by the cessation of bubbling activity at the anode. At this point, current flowing through the oxide consists almost entirely of ions. The solid state migration of ions through the oxide is described by the high-field transport model in which the anodization current density J is given by:

$$J = Ae^{\frac{\beta_{HF} \Delta U}{t}} \tag{2}$$

where A and β_{HF} are constants.

Later in the anodization process, the growth process is under mass-transport control as the anodization current is controlled by diffusion of fluoride ions through the tubes to the pore bottoms and results in an inverse dependence of the anodization current on the tube length [4]. The current density is given as:

$$J = -pD\frac{\partial c}{\partial x}(0 \leq x \leq L) \tag{3}$$

where D, c, x, and L are the diffusion coefficient and concentration of the ionic species, thickness of the diffusion layer, and length of the nanotubes, respectively, and p is the ratio of the weight of dissolved oxide to the weight of produced oxide.

The anodic formation of TNAs is a highly nonequilibrium process which occurs at high overpotentials as a result of the interplay between three competing and essential processes (see Fig. 2): Field-assisted oxide dissolution and cation migration, field-assisted oxidation of Ti and chemical etching. It is chemical etching which is responsible for the development of nanotubes as opposed to mere nanopores. In contrast, the formation of nanoporous alumina is controlled mainly by the field-assisted processes. The field-assisted reactions occur on either side of the barrier layer at the bottom of the nanotubes and are responsible for driving the Ti/TiO$_2$ interface deeper into the Ti foil or Ti film, a process which increases the length of the nanotubes. Chemical

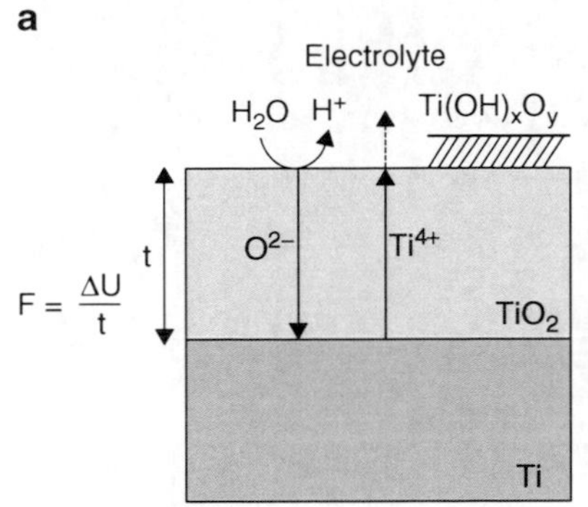

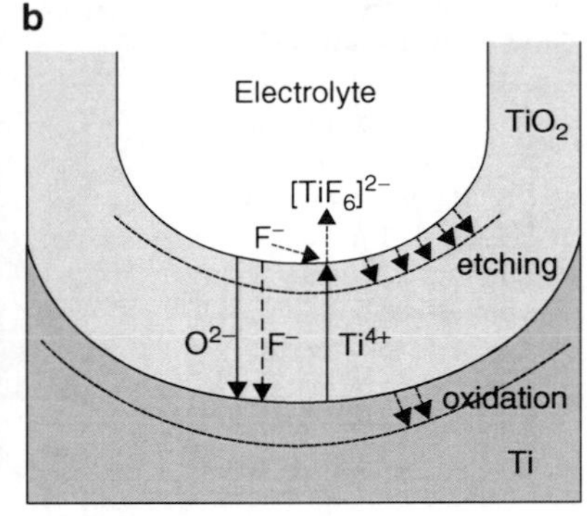

TiO_2 Nanotube Arrays: Growth and Application, Fig. 2 Schematic representation of the Ti anodization (**a**) in absence of fluorides (results in flat layers), and (**b**) in presence of fluorides (results in the tube growth) [4] (Reprinted with permission from Elsevier)

etching shortens the length of the nanotubes. The reactions involved in the form ation of TNAs are:

Field assisted oxidation:

$$Ti + 2H_2O \rightarrow TiO_2 + 4H^+ + 4e^- \quad (4)$$

Field assisted migration: $$Ti^{4+} + 6F^- \rightarrow [TiF_6]^{2-} \quad (5)$$

Field assisted dissolution:

$$TiO_2 + 6F^- + 4H^+ \rightarrow [TiF_6]^{2-} + 2H_2O \quad (6)$$

Chemical dissolution:

$$TiO_2 + 6HF \rightarrow [TiF_6]^{2-} + 2H_2O + 2H^+ \quad (7)$$

The as-anodized nanotube arrays are amorphous and non-insulating. Manifestation of the semiconductor properties of TiO_2 requires induction of crystallinity, which is typically accomplished by annealing the nanotube arrays in air or in an oxygen-rih ambient at elevated temperatures of 350–580°C. For annealing temperatures of up to 500°C, the nanotubes are transformed into polycrystalline anatase alone; at higher temperature, increasing amounts of rutile are present as well. Annealing at temperatures higher than 580°C results in the destruction of the nanotubular architecture.

The evolution of different fabrication recipes for anodic TNAs has occurred in four generations to date [3, 7, 8]. The first three generations were defined by the maximum length of nanotubes that was achieved and they all used fluoride ions to effect the necessary chemical etching. The fourth generation was defined by its nonuse of fluoride ions. First-generation nanotubes were up to 500 nm long and were grown by anodization in strongly acidic aqueous electrolytes containing 0.1–0.5 wt% HF (pH $\leq$ 1). Other additives such as acetic acid were sometimes added to improve the mechanical strength and quality of the nanotubes. The tube length in first-generation TNAs was limited by the strong chemical etching by HF. Typical morphology obtained in first-generation nanotubes is shown in Fig. 3a. The steady-state TNA film thickness (corresponding to the tube length) occurred when the rate of movement of the Ti/TiO_2 interface deeper into the metal was exactly balanced by the chemical etching at the mouths of the nanotubes. In the second generation, fluoride ion-bearing aqueous electrolytes were still used but the rate of chemical etching was tightly controlled by adjusting the pH by adding buffers to the anodization electrolyte. Typically, trisodium citrate and sodium dihydrogen sulfate were used as buffering agents in the electrolyte and salts such as ammonium fluoride, sodium fluoride, and potassium fluoride were used as sources for fluoride ions. As the pH was varied from 1 upward toward 5, the nanotubes grew correspondingly longer due to the reduction in chemical etching. The maximum nanotube length in aqueous electrolytes of 6.6 μm was achieved at a pH of 5. The maximum growth rate was ~0.25 μm/h. When the pH was increased beyond 5, nanotube formation did not occur. Nanotube formation in both the first and second generations occurred in aqueous electrolytes in a narrow window of anodization voltages ranging from 10 to 27 V, which restricted the inner diameters of the nanotubes to the range 22–100 nm. In the third generation (Fig. 3b, c, d), fluoride-bearing organic electrolytes were used to vastly expand the morphological parameter space achievable. Nanotubes as long as 1 mm were grown by complete conversion of the underling Ti foil to

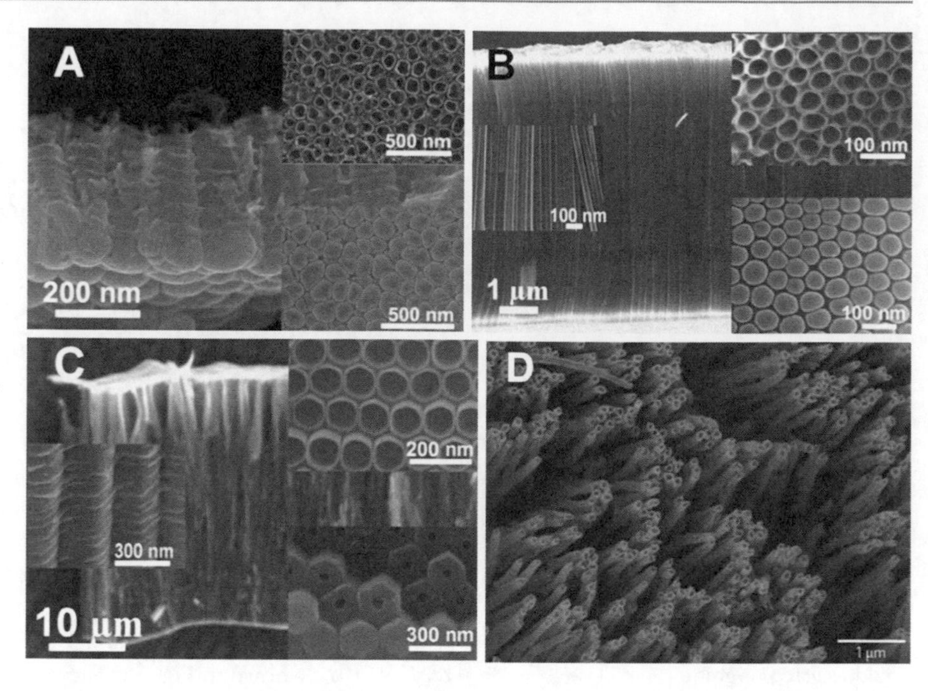

TiO_2 Nanotube Arrays: Growth and Application, Fig. 3 Scanning electron micrographs showing the effect of anodization electrolyte on the morphology of TNAs: (**a**) Typical morphology obtained in aqueous acidic fluoride or HF electrolytes, (**b**) glycerol/fluoride electrolytes, (**c**) ethylene glycol/fluoride electrolytes, and (**d**) DMSO/fluoride electrolytes (Images **a** through **c** were reprinted from [4] with permission from Elsevier. Image **d** was reprinted from [9] with permission from Nature)

TNAs. Thus, the nanotube growth is limited in length only by the thickness of the underlying substrate and in principle can be extended even longer by choosing thicker substrates. The inner diameter of the nanotubes could be varied from 17 to 900 nm. When chemical etching is minimized, the slow solid state ionic transport of reactants through the barrier layer is the rate-limiting step preventing further increase in tube length. From Eq. 7, it is clear that the high-field ionic current is increased by a thinner barrier layer. This key insight motivated the use of organic electrolytes in the third generation which provided more reducing conditions than aqueous electrolytes and ensured the formation of a thinner barrier layer. The thinner barrier layer allows growth rates as high as 15 µm/h to be achieved. Third generation electrolytes consist of a fluoride-ion bearing salt dissolved in an amphiprotic organic solvent of moderate to high viscosity such as formamide, N-methyl formamide, ethylene glycol, glycerol, etc. About 1–5% of water is typically added to form organic electrolytes in order to provide oxygen-bearing hydroxyl ions during anodization for oxide formation. Nanotubes do not grow well when low-viscosity protic solvents such as methanol or ethanol are used as the base solvent to form the organic electrolyte (although mixtures with water and HF have been found to result in anodic TNAs). The same is true of aprotic solvents such as propylene carbonate and acetonitrile. An exception is dimethylsulfoxide, an aprotic solvent which has been used to form high-quality TNAs. By detaching the nanotube arrays from the foil substrate, TNA membranes similar to nanoporous alumina membranes may be produced. The nanotube arrays need to be at least 10 µm long for the detached film to be mechanically robust enough to a form free-standing membrane. Some of the methods used to detach the TNAs from the native Ti foil substrate include the use of natural stresses in the thick TNA films through ultrasonication, selective etching of the Ti substrate by immersion in a solution of CH_3OH/Br_2, and the use of induced stresses by double-step anodization. The membranes have a strong tendency to curl up due to capillary forces. The membranes can be kept planar by processing with low surface tension solvents or by supercritical CO_2 drying. Fourth-generation nanotubes were formed in fluoride-free electrolytes (Table 1).

Growth of Templated TiO_2 Nanotube Arrays

Both hard templates such as anodic nanoporous aluminum oxide (AAO) and soft templates such as polycarbonate membrane filters have been used to form

TiO_2 Nanotube Arrays: Growth and Application, Table 1 Common recipes for TNAs and the resulting morphological parameters

Electrolyte	Anodic potential (V)	Inner diameter (nm)	Length (μm)
Aqueous 0.5% HF	10	22	0.2
Aqueous 0.5% HF	10–23 (nanocones)		0.4
Aqueous buffered 0.1 M KF (pH ~ 5)	25	110	6.6
0.1 M NH_4F in 1:1 DMF:H_2O	15 porous		
0.15 M NH_4F + 5% H_2O in formamide	10	12	3.6
0.15 M NH_4F + 5% H_2O in formamide	15	29	5.4
0.15 M NH_4F + 5% H_2O in formamide	20	70	14.4
0.22 M NH_4F + 5% H_2O in formamide	35	180	30
0.27 M NH_4F + 5% H_2O in formamide	20	65	19.6
0.3% NH_4F in EG + 1% H_2O	30	45	9
0.3% NH_4F in EG + 1% H_2O	40	70	12
0.3% NH_4F in EG + 1% H_2O	65	135	30

Reprinted with permission from [7]

TNAs. Soft templates have two major disadvantages: (1) an elevated temperature (~450–500°C) thermal annealing step is frequently to induce or improve the crystallinity of the templated TNAs. However, most polymers are not able to withstand temperatures above 200°C. Consequently, the polymer template pyrolyzes before the phase transition in TiO_2 thus eliminating the mechanical support provided by the template and/or result in the undesirable incorporation of carbonaceous materials in the TNAs; (2) the pore densities are smaller and the nanochannels in polymeric membrane filters are frequently tilted from the surface normal and sometimes intersect within the membrane [10]. Therefore, discussion shall be restricted to AAO-templated TNAs.

AAO templates are formed by electrochemical anodization of aluminum metal in three common electrolytes based on sulfuric acid, oxalic acid, and phosphoric acid. The choice of electrolyte is determined by the desired pore size. The Al metal maybe a native substrate such as aluminum foil or could be a film deposited by vacuum evaporation or sputtering on nonnative substrates such as glass and silicon. High-ordered templates approaching ideal hexagonal, square, or triangular lattices may be prepared by techniques such as two-step anodization and various imprinting methods to guide pore formation [11].

Methods to synthesize TNAs in AAO templates can be classified into liquid phase and gas-phase methods. Sol-gel hydrolysis, electrochemical hydrolysis and electrophoretic deposition of nanoparticles are examples of liquid phase methods. Atomic layer deposition (ALD) and chemical vapor deposition (CVD) are gas-phase methods. The growth of nanotubes as opposed to nanorods in nanoporous alumina requires preferential formation of TiO_2 on the walls of the template. This is accomplished by preferential wetting of the template walls by TiO_2 precursors in sol-gel methods, by preferential reduction or oxidation at the walls in electrochemical methods, by preferential concentration of the electric field lines at the pore walls in electrophoretic deposition, and by the non-line-of-sight conformal deposition on the walls in gas-phase methods. ALD is a self-limiting film growth process in which alternate saturative surface reactions are made to occur by pulsing the vapors of two or more precursors in the reaction chamber alternately [12]. ALD is capable of producing highly conformal coatings even on substrates with complex features and offers exceptional control in tuning the wall thickness of the nanotubes. Consequently, ALD is the most widely used method to form the templated growth of TNAs. Typically, a flow reactor configuration is employed and the precursors used to grow TiO_2 films are a titanium alkoxide, whose generic formula is $Ti(OR)_4$ and water (H_2O). The deposition temperatures used in ALD of TiO_2 films for templated TNA growth vary widely, from 100°C to 325°C. The exposure times of the template to each precursor pulse are typically limited to a few seconds and the chamber is evacuated prior to the introduction of the next pulse. In addition, in some recipes, an inert gas is used to purge the chamber before the next pulse. The main disadvantages of ALD are the high cost and complexity of ALD equipment and

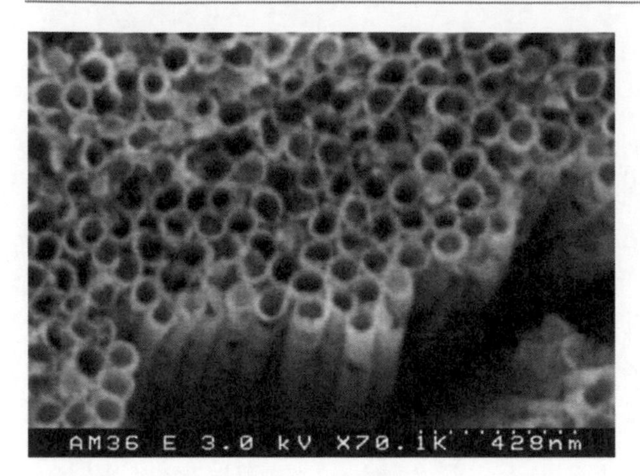

TiO_2 Nanotube Arrays: Growth and Application, Fig. 4 Scanning electron micrograph of templated TNAs formed by sequential infiltration and oxidation of titanium isopropoxide, followed by removal of the alumina matrix in which the tubes were formed (Reprinted with permission from [6])

the extremely slow film growth rate. A brief overview of other processes used to form templated TNAs is now provided. The operating principle of the sol-gel method is the use of capillary forces to drive a TiO_2 sol into the nanochannels of the template which transforms into TNAs upon thermal annealing. A sol-gel process based on TiF_4 hydrolysis with hydroxypropylcellulose added as a gelling agent, has been used to form 200 nm diameter TNAs in an AAO template. However, the process was highly nonuniform over large areas with irregular deposition of the TiO_2 and poor penetration of the sol particles into the template nanochannels. Much better results were obtained using the pressure-assisted infiltration of liquid titanium isopropoxide (TiO_2 precursor) into the AAO nanochannels followed by thermal decomposition method of the precursor at 500°C in an oxygen stream to yield TiO_2 nanotube arrays [6]. In the electrophoretic sol-gel method, an electric field is used to provide significantly larger forces than mere capillary action to drive the charged particles created by sol-gel processing into the nanochannels [13]. A triple step templating process has been used to form TiO_2 by the anodic oxidative hydrolysis of $TiCl_3$ [5]. In this process, the AAO template morphology was transferred to a polymer mold which contained a negative replica of the template pattern. Gold nanorods were formed by electroless deposition on to the polymer mold. Then the gold nanorods served as a conducting template to form TNAs (Fig. 4).

Biological Applications of TiO_2 Nanotube Arrays

Biosensing Applications

Biosensors are devices that combine a biological component (a recognition layer) and a physicochemical detector component (a transducer). The majority of biosensors are targeted to proteins but biosensors for cells and for metabolically important small molecules are also being investigated. Molecular labels have enabled the construction of biosensors of high sensitivity and specificity. However, labels often disrupt the accurate measurement of kinetic constants, particularly binding equilibria, and problems, such as antibody cross reaction or impure solutions, can occur. Due to these drawbacks, label-free biosensors are highly desired for a variety of biotechnology applications such as disease diagnostics, drug discovery, environmental monitoring, and food safety. For label-free optical biosensors, the transduction methods can be divided into two categories: optical interferometric and surface plasmon methods [14]. In interferometric sensors, sensing is achieved by extracting the refractive index of the porous matrix from the Fabry-Perot interference fringes in the reflectance spectrum. Some of the prerequisites for an effective optical interferometric sensing material are a large internal surface area, ease of fabrication, and a chemically modifiable surface. TiO_2 nanotube arrays satisfy the above requirements but in addition possess a high refractive index ($n = 2.5$ for TNAs) and superior chemical stability over a wide pH range in comparison with other candidate materials such as porous silicon and nanoporous alumina. The larger index contrast between the porous host and the aqueous matrix in which the biomolecular binding measurement is carried out is expected to provide greater contrast in the interferometric spectrum, leading to lower noise and higher sensitivity [14]. Mun et al. [14] demonstrated the highly specific label-free interferometric sensing of rabbit immunoglobulin G (IgG) using protein A (derived from *Staphylococcus aureus* bacteria) immobilized on the walls on the nanotubes as the capture probe (Fig. 5).

Orthopedic Implant Applications

Because of their mechanical strength, chemical inertness, and biocompatibility, TiO_2 and its alloys (particularly Ti-6Al-4V) have been used extensively

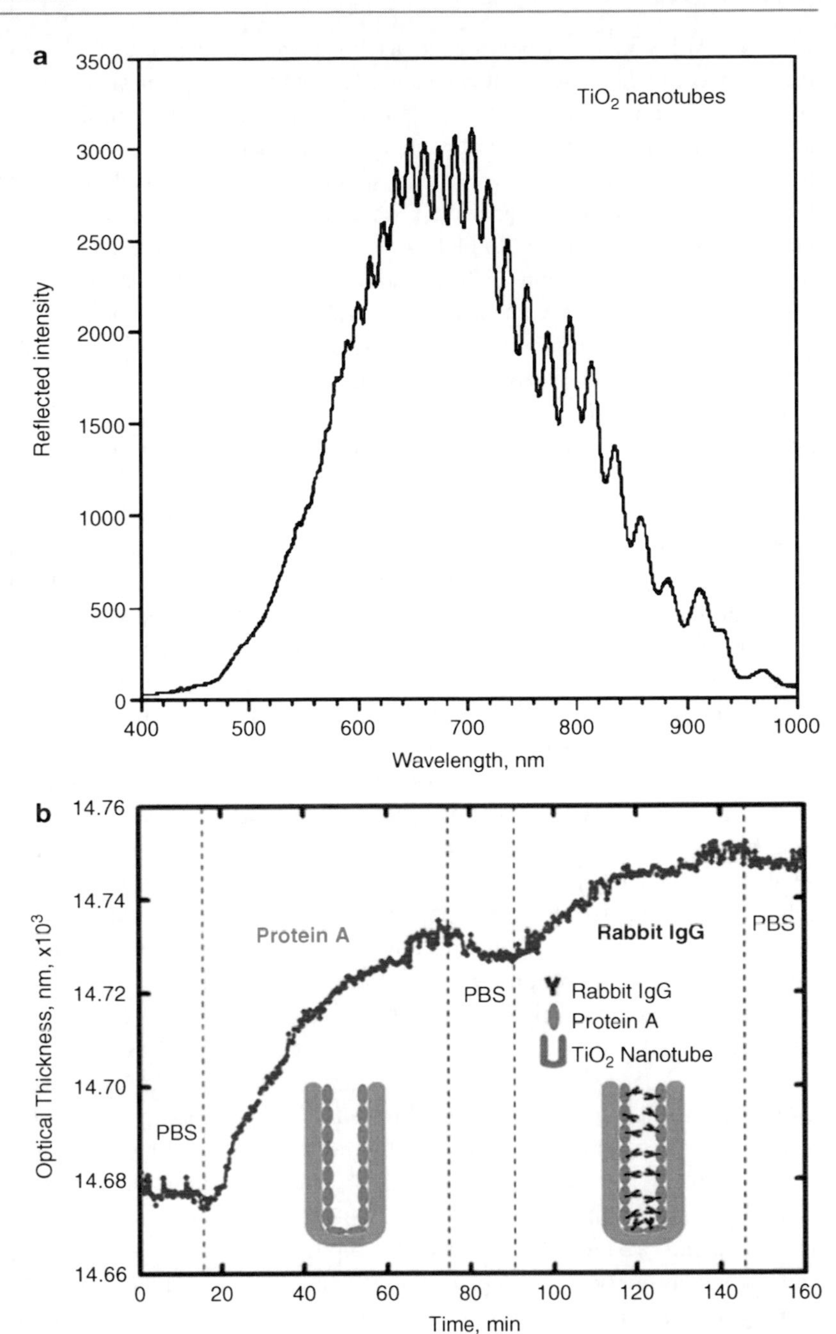

TiO_2 Nanotube Arrays: Growth and Application, Fig. 5 TNA-based interferometric biosensors: (**a**) representative reflectivity spectrum of anodic TNAs showing the characteristic interference fringes and (**b**) time-dependent optical thickness measurements showing sequential binding of Protein A and rabbit IgG within TiO_2 nanotubes (Reprinted with permission from [14])

in orthopedic and dental implants since 1970 [15]. Osteogenesis is the process of formation of new bone tissue by cells called osteoblasts. Ideal implants are those whose surfaces will promote in vitro and in vivo osteogenesis. However, the influence of surface micro- and nanotopography on osteogenesis is still not well understood. High surface area porous materials are used in implants to mimic bone, which is a naturally occurring porous ceramic material composed of nano-sized organic and mineral phases that form a large macrostructure. Proteins in extracellular bone matrix and calcium phosphate, important constituents of the bone matrix, are nanostructured but the porosities in human bone are predominantly in the range 1–100 μm [15]. Spark anodization is a process which leads to the formation of a disordered oxide structure with porosities in the same regime as bone and is therefore commonly used to increase the biocompatibility of titanium and its alloys. However, recent studies using nanoporous alumina and nanotubular titania have revealed that their much smaller pores (20–200 nm) allow bone ingrowth. TNAs, in particular, have been demonstrated to be favorable templates for bone cell growth and differentiation, and also shown to significantly enhance osteoblast activity [8, 15]. The nanoporous architecture of TNAs provides a framework in which osteoblasts produce new bone. The anatase phase of TiO_2 is much more efficient for nucleation and growth of bone than rutile. Another advantage of TNAs is their strong adherence to the underlying Ti substrate. In cell culture experiments, increased adhesion to TNAs of cells which produce and maintain cartilage (chondrocytes) was observed and in animal trials, chronic inflammation was found to be absent in TNAs implanted below the skin in rats. Furthermore, the calcium and phosphorous concentrations were 50% higher on these surfaces, suggesting that deposition of the bone matrix was accelerated on TNAs [8, 15]. SEM images have been used to show that the microspike extensions (filopodia) of migrating osteoblast cells go into the nanotube pores and in general progagate much faster on TNA surfaces. It has been speculated that the nanotube pore and the gap between the nanotubes serve as a useful pathway for continuous supply of ions, nutrients, and proteins, which is likely to positively contribute to the health of the growing cells [8].

Drug Eluting Coatings for Medical Implants

Bacterial infection and acute inflammation at the site of the implant are the most common postsurgical complications of bone implants, dental implants, and vascular stents. To reduce the chance of complications and to prevent early stage implant failure, antibiotic therapy is indicated 6–8 weeks after surgery [15]. Systemic administration of drugs such as antibiotics to avoid or treat the infection is often ineffective at reaching infected tissues near the implant and in addition, can cause toxicity and other adverse side effects. On the other hand, localized drug delivery from the site of an implant has more efficient therapeutic effects and fewer side effects. In general, an ideal drug eluting coating must have the ability to incorporate a drug, preserve it, and deliver it gradually over the time to a specific target site [16]. TiO_2 nanotube arrays are being actively researched for localized drug delivery for implants because they combine the appropriate biointegration and biocompatibility with the ability to controllably release drugs (a process known as elution) confined in their channels to their surroundings. Drug loading is performed through capillary action by either immersing the templates in the concentrated drug solution or dropping the solution slowly on the template surfaces. In addition, simulated body fluid (SBF)-assisted physical adsorption of antibiotics and anti-inflammatory agents was found to be a superior method of drug loading which also resulted in improved prolonged release of both types of drugs [16]. The elution of gentamicin (a powerful aminoglycoside antibiotic used against gram-negative bacteria) from titania nanotubes has been studied [8]. In TNAs of 80 nm pore diameter and 400 nm length loaded with gentamicin (70–85% loading efficiency), drug release kinetics were found to be dependent on initial loading; and there was a sustained release, for 45, 90, and 150 min for loadings of 200, 400, and 600 μg, respectively. In addition, it was found that bacterial adhesion on the surface of gentamicin-loaded titania nanotubes was reduced significantly (in comparison with titanium surface and unloaded nanotubes), whereas normal osteoblast adhesion and proliferation is retained [8, 15]. It has also been shown that TiO_2 nanotubes can control small molecule delivery in the order of weeks, and larger molecules in the order of months. The total drug loading as well as elution was mostly affected by nanotube length, and maximum small molecule elution was reached at $\sim$2 weeks. TNA-based drug eluting coatings are superior to the conventional strategy of drug elution through a drug-loaded polymer coating, because even though the polymer coating is effective in delivering a drug for long periods, it has been known to induce an inflammatory response due to eventual degradation of the polymer. Such inflammation is implicated in late-stage implant failure.

TiO_2 Nanotube Arrays for Stem-Cell Differentiation

Recent studies on cell interactions with TNAs have demonstrated that stem cell fate is dictated by the diameter of the nanotubes. Using amorphous (unannealed) TNAs, Schmuki et al. found the adhesion, proliferation, migration, and differentiation of rat bone marrow mesenchymal stem cells to be highest on nanotubes 15 nm in diameter and dramatically decreased on 70- and 100-nm nanotubes. In contrast, Jin et al. found using crystalline (annealed) TNAs and human mesenchymal stem cells that the optimum length scale for cell vitality and differentiation was 100 nm. More recent studies seem to indicate that nanotube size plays a more important role than crystallinity and nanotube chemical composition although more research is necessary to clarify the nature of topographical cues for cell behavior in this emerging field.

Applications of TiO_2 Nanotube Arrays in Light Harvesting Devices

One of the factors limiting maximum achievable efficiencies in conventional light harvesting devices including photovoltaic cells, photoelectrochemical cells, and photocatalytic films is the ever-present trade-off between light absorption and charge generation on the one hand, and charge separation and carrier collection on the other. This trade-off is best illustrated by considering the case of a planar p-n junction solar cell where only electron-hole pairs generated close to the depletion region are swept apart by the inbuilt electric field and efficiently collected. On the one hand, the light absorbing semiconductor layer needs to be thick enough to absorb a majority of the incident photons and on the other hand, the large thickness means that a portion of the minority charge carriers generated away from the junction are lost to recombination as they diffuse toward the junction. The vertically oriented TiO_2 nanotube array architecture overcomes this trade-off by orthogonalizing the processes of light absorption/charge generation and charge separation/carrier collection so that the trade-off is no longer as severe. The process of charge separation now occurs in the radial direction while light absorption occurs in the longitudinal direction along the axis of the nanotubes. To facilitate charge separation, the diameter of the nanotubes can be made smaller or the nanotube walls may be made thinner while simultaneously making the nanotubes longer to maximize light absorption. The porosity of the ordered structure allows the incident photons to be more effectively absorbed, due to scattering effects, than on a flat electrode. For these reasons, the vertically oriented nanotube array architecture has been posited as nearly ideal for light harvesting applications.

TNAs for the Photocatalytic Reduction of CO_2 to Methane

The free energy for overall methane formation through the chemical reaction is about 801 kJ/mol, a process requiring eight photons:

$$CO_2 + 2H_2O(g) \rightarrow CH_4 + 2O_2 \quad (8)$$

Hydrogen formation from water involves a free energy change (ΔG^0) of 237 kJ/mol and an enthalpy change (ΔH^0) of 285 kJ/mol; the corresponding values for CO formation from CO_2 are 257 and 283 kJ/mol at 25°C (1 atm.). Hence, the minimum energy required for water and CO_2 splitting processes are, respectively, 1.229 and 1.33 eV (per photon). In theory, the bandgap of a photocatalyst used for co-splitting of CO_2 and water should be at least 1.33 eV, which corresponds to absorption of solar photons of wavelengths below about 930 nm. Considering the energy loss associated with entropy change (about 87 J/mol·K) and other losses involved in the CO_2 splitting process (forming CO and O_2), a bandgap between 2 and 2.4 eV is optimal [17]. Using nitrogen-doped TiO_2 nanotube arrays, with a wall thickness low enough to facilitate effective carrier transfer to the adsorbing species, surface-loaded with nano-sized islands of platinum and copper cocatalysts, efficient sunlight-driven conversion of CO_2 into CH_4 and other hydrocarbons was achieved, with hydrocarbon production rates as high as 160 μL/(g h).

The outstanding nature of TNAs as a platform to perform the photocatalytic reduction of CO_2 is shown by the fact that the above rate of CO_2 to hydrocarbon production obtained under outdoor sunlight is at least 20 times higher than previous published reports [8, 17].

TNAs for Oxidative Photochemistry

Due to the large oxidizing power of photogenerated valence band holes in titania, TiO_2 nanotube arrays are excellent catalysts for oxidative photochemistry. OH radicals and superoxide (O_2^-) anions produced by photogenerated charge carriers in TiO_2 can oxidize essentially all organic molecules present in the solution into carbon dioxide and water [4]. It has been shown that crystalline TiO_2 nanotubes show considerably higher efficiency for the decomposition of organic small molecules than a compacted Degussa P25 layer (20–30 nm diameter nanopowder composed of anatase and rutile) under comparable conditions [4]. The TNA architecture is also exceptionally stable with no degradation observed in the samples with exposure to UV/visible light over a course of a several months. For sunlight-driven water photolysis, the titania nanotube array architecture results in a large effective surface area in close proximity with the electrolyte thus enabling diffusive transport of photogenerated holes to oxidizable species in the electrolyte. In the TNA architecture, most minority carriers generated within a "retrieval" length from the material surface, i.e., a distance from the surface equal to the sum of the depletion layer width and the diffusion length, and such carriers escape recombination and reach the electrolyte redox species. Consequently, bulk recombination is greatly reduced and the quantum yield (incident photon-to-electron current efficiency) is close to unity. Under visible light illumination, the efficiency of TNAs is limited by the large bandgap of TiO_2 (3.0–3.2 eV), which renders the TNAs sensitive only to ultraviolet photons. A photoconversion efficiency for photoelectrolysis of 12.25% was obtained under UV (320–400 nm) illumination using second-generation nanotubes 6 μm long annealed at 580°C. High quantum yields >80% were obtained indicating that the incident UV photons were effectively utilized by the nanotube arrays for charge carrier generation. The corresponding hydrogen evolution rate was 76 mL/h W. The water-splitting reaction was confirmed by the 2:1 ratio of evolved hydrogen to oxygen. Using third-generation nanotubes ~200 nm in diameter and 30 μm long formed by anodization at 35 V in a formamide-based electrolyte, the water-splitting efficiency improved to 16.25% [7].

TiO_2 Nanotube Arrays as Scaffolds in Excitonic Solar Cells

In excitonic solar cells, photons of energy larger than the absorption threshold of the excitonic absorber do not create free electron-hole pairs as in a conventional inorganic semiconductor. Instead excitons (electron-hole pairs bound by Coulombic attraction) are generated. Unless these excitons are dissociated at a suitable interface, they are lost to recombination. Therefore, the trade-off in excitonic solar cells is between light absorption and exciton diffusion. As alluded to before, vertically oriented TNAs orthogonalize the processes of exciton generation and exciton diffusion overcoming the trade-off. Also, TNAs provide electron percolation pathways for the directed (nonrandom) movement of electrons in the structure. TNAs provide flexibility in device design since the excitonic absorbers may be coated on the nanotube walls thus exploiting the large internal surface area of the TNA architecture or the absorbers may be volumetrically filled inside the hollow cylindrical spaces of the nanotubes. Due to these reasons, TNAs are very attractive materials for use as n-type semiconducting scaffolds for excitonic solar cells. TNAs are used in two different kinds of excitonic photovoltaic devices: dye sensitized solar cells, where the absorber is a broadly absorbing dye molecule anchored to the surface of the TNAs and ordered bulk heterojunction solar cells, where the absorber is filled inside the nanotubes. Templated TNAs formed by a sol-gel method were found to have very large roughness factors which enabled them to adsorb >500 nmol cm^{-2} of N-719. Using 17.6 μm long transparent anodic TNAs grown on conductive glass substrates, Varghese et al. obtained N-719 sensitized solar cells with an efficiency of 6.9% under AM 1.5 1 sun illumination and quantum yields higher than 65% in the spectral range 400–670 nm [9]. Using a freestanding membrane consisting of 63 μm long TNAs transferred to a FTO: glass substrate, dye-sensitized solar cells subjected to 1 sun AM.15 illumination demonstrated an open circuit photovoltage of 0.77 V, a short-circuit photocurrent of 18.5 mA cm^{-2} and a fill factor of 0.64, resulting in an overall efficiency of 9.1% [18]. There is also increasing research interest in sensitized solar cells where the sensitizers are quantum dots of CdS or CdSe or CdTe instead of molecular dyes. Quantum yields as high as 55% have been demonstrated in quantum dot-sensitized TNA solar cells [19]. An ordered double heterojunction solar cell was constructed by infiltrating a blend of regioregular poly (3-hexylthiophene) (RR-P3HT) and a methanofullerene (PCBM) into transparent anodic TNAs on FTO coated glass. The configuration used is termed a double heterojunction because of the availability of two interfaces for charge separation, namely, the TiO_2/P3HT and PCBM/P3HT interfaces. The resulting solar cell had an overall conversion efficiency under 1 sun AM 1.5 illumination of 4.1% [8] (Fig. 6).

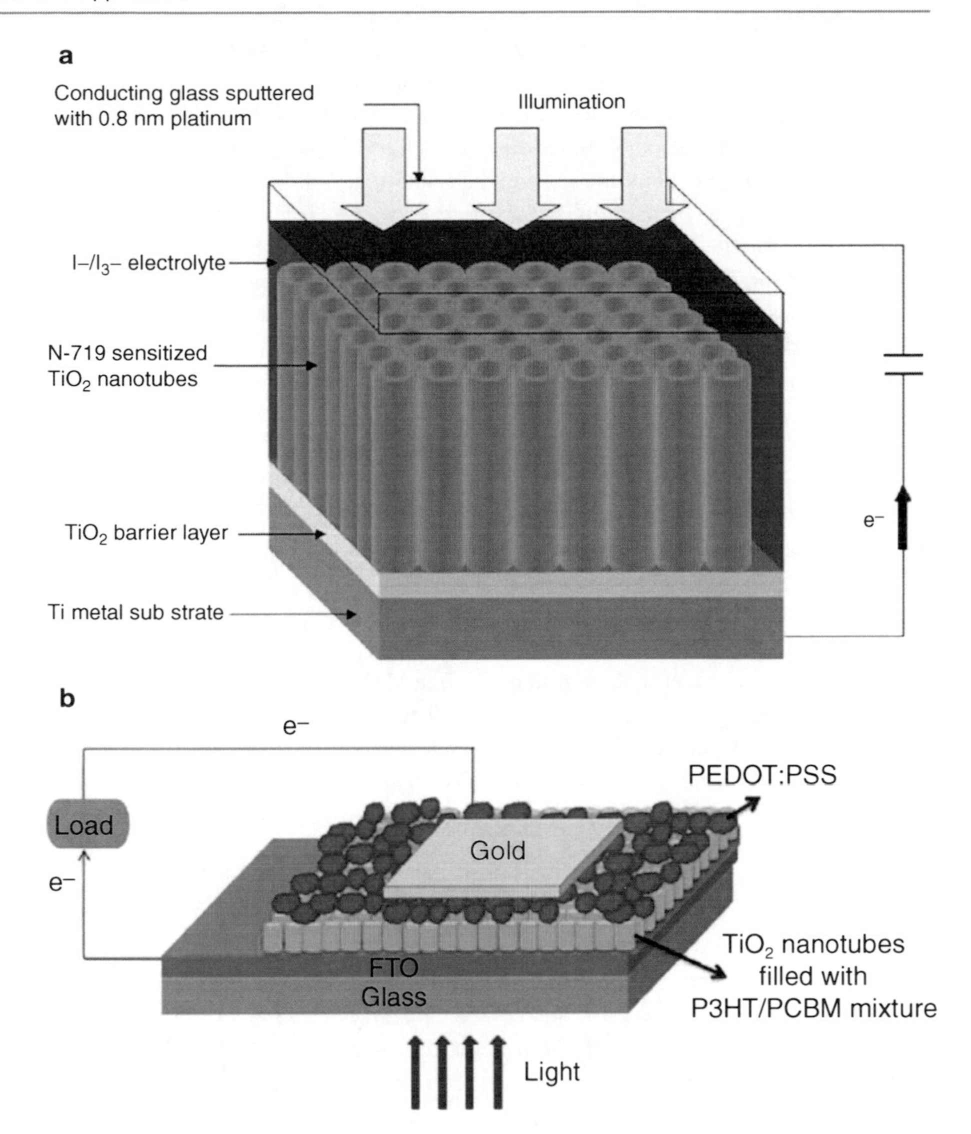

TiO_2 Nanotube Arrays: Growth and Application, Fig. 6 Schematics of TNA-based solar cells (**a**) Liquid junction dye-sensitized solar cells and (**b**) ordered bulk heterojunction solar cells (Reprinted with permission from [8])

Chemical Sensing Applications of TiO_2 Nanotube Arrays

Chemical sensors often suffer from unwanted contamination, or poisoning which limits the sensor's operational lifetime. Volatile organic compounds are a frequent source of sensor contamination. In this respect, an important advantage of TNA-based chemical sensors is their self-cleaning ability in the presence of ultraviolet radiation [8]. The strong oxidizing power of photogenerated holes in crystalline TiO_2 oxidizes organic contaminants and results in self-cleaning.

TNA-Based Ultrasensitive H_2 gas Sensors

One of the first applications of TNAs was in hydrogen gas sensing by measuring a change in the electrical resistance of the nanotubes upon exposure to hydrogen. A metal-semiconductor-metal (MSM) structure was employed for resistive H_2 gas sensing wherein Ti metal underlying the nanotube arrays was used as the ohmic contact and a high work function metal (mainly platinum) was used as the Schottky contact. Initial work used first-generation nanotubes at high operating temperatures (e.g., 290°C) to obtain a three-order change in resistivity when exposed to a few ppm of hydrogen. By sputtering a thin overlayer of Pd on the nanotubes to enable facile dissociation of molecular hydrogen into atomic form, the operating temperature of the resistive gas sensors were brought down to room temperature. By using second-generation nanotubes anodized in aqueous electrolytes at pH 4.5 in combination with Pd sputtering, an unprecedented high sensitivity of 8.7 orders of magnitude was achieved. Room temperature operation coupled with the high sensitivities ushered in applications such as transcutaneous hydrogen sensors. The quantification of ppm-ppb hydrogen gas concentrations escaping from blood through the skin has medical relevance as an indicator of lactose intolerance, fructose malabsorption, microbial activity, bacterial growth, fibromyalgia, diabetic gastroparesis, and neonatal necrotizing enterocolitis. The prevailing explanation for the remarkable change in resistivity is that chemisorption of hydrogen atoms on the nanotube surface results in partial charge transfer to the TiO_2 resulting in an electron accumulation layer on the nanotube surface which drastically improves the electrical conductance of the MSM structure. In addition, hydrogen adsorption reduces the work functions of Pd and Pt. Consequently, the height of the potential barriers at the TiO_2/Pt and TiO_2/Pd interfaces are lowered, thus again increasing the conductance of the MSM structure.

TNA-Based H_2O_2 Sensors

TNAs on which horseradish peroxidise (HRP) and thionine were co-immobilized have been used as electrochemical sensors to detect and quantify hydrogen peroxide [8]. Due to the high electrocatalytic activity of the immobilized HRP for thionine oxidation in the presence of H_2O_2, a large increase of reduction current occurs for the Th/HRP/TiO_2 electrode due to peroxide. The detection limit for H_2O_2 was estimated to be as low as 1.2 μM.

TNA-Based Amine Sensors

Amines are one of several chemical pollutants released in oil and gas spills which contaminate fresh and seawater. In the light of the fact that many amines are proven or suspected to be carcinogenic and have been implicated in inducing bladder cancer, environmental monitoring of amine levels assumes significance. In this regard, TNAs coated with an electrostatically adsorbed layer of ruthenium tris(bipyridinium) ions [$Ru(bpy)_3^{2+}$] have been used in an electrochemiluminescence sensor to detect and quantify the tripropylamine concentrations in water as low as 1 nM [20].

Cross-References

- ► Active Carbon Nanotube-Polymer Composites
- ► Atomic Layer Deposition
- ► Biosensors
- ► Hybrid Solar Cells
- ► Microfabricated Probe Technology
- ► Nanomaterials for Excitonic Solar Cells
- ► Nanostructured Materials for Sensing

References

1. Zwilling, V., Aucouturier, M., Darque-Ceretti, E.: Anodic oxidation of titanium and TA6V alloy in chromic media. An electrochemical approach. Electrochim. Acta **45**(6), 921–929 (1999)
2. Gong, D., Grimes, C.A., Varghese, O.K., Hu, W.C., Singh, R.S., Chen, Z., Dickey, E.C.: Titanium oxide nanotube arrays prepared by anodic oxidation. J. Mater. Res. **16**(12), 3331–3334 (2001)
3. Mor, G.K., Varghese, O.K., Paulose, M., Shankar, K., Grimes, C.A.: A review on highly ordered, vertically oriented TiO_2 nanotube arrays: fabrication, material properties, and solar energy applications. Sol. Energy Mater. Sol. Cells **90**(14), 2011–2075 (2006). doi:10.1016/j.solmat.2006.04.007
4. Macak, J.M., Tsuchiya, H., Ghicov, A., Yasuda, K., Hahn, R., Bauer, S., Schmuki, P.: TiO_2 Nanotubes: self-organized electrochemical formation, properties and applications. Curr. Opin. Solid State Mat. Sci. **11**(1–2), 3–18 (2007). doi:10.1016/j.cossms.2007.08.004
5. Hoyer, P., Formation of a titanium dioxide nanotube array. Langmuir **12**(6), 1411–1413 (1996)
6. Michailowski, A., AlMawlawi, D., Cheng, G.S., Moskovits, M.: Highly regular anatase nanotubule arrays fabricated in porous anodic templates. Chem. Phys. Lett. **349**(1–2), 1–5 (2001)
7. Shankar, K., Basham, J.I., Allam, N.K., Varghese, O.K., Mor, G.K., Feng, X.J., Paulose, M., Seabold, J.A., Choi, K.S., Grimes, C.A.: Recent advances in the Use of TiO_2 nanotube and nanowire arrays for oxidative photoelectrochemistry. J. Phys. Chem. C **113**(16), 6327–6359 (2009). doi:10.1021/jp809385x
8. Grimes, C.A., Mor, G.K.: TiO_2 Nanotube Arrays: Synthesis, Properties, and Applications (Springer, New York, 2009)
9. Varghese, O.K., Paulose, M., Grimes, C.A.: Long vertically aligned titania nanotubes on transparent conducting oxide for highly efficient solar cells. Nat. Nanotechnol. **4**(9), 592–597 (2009). doi:http://www.nature.com/nnano/journal/v4/n9/suppinfo/nnano.2009.226_S1.html
10. Wu, X.J., Zhu, F., Mu, C., Liang, Y.Q., Xu, L.F., Chen, Q.W., Chen, R.Z., Xu, D.S.: Electrochemical synthesis and applications of oriented and hierarchically quasi-1D semiconducting nanostructures. Coord. Chem. Rev. **254**(9–10), 1135–1150 (2010). doi:10.1016/j.ccr.2010.02.014

11. Masuda, H.: Highly ordered nanohole arrays in anodic porous alumina. In: Wehrspohn, R.B. (ed.) Ordered Porous Nanostructures and Applications, pp. 37-56. Springer, New York (2005)
12. Kemell, M., Pore, V., Tupala, J., Ritala, M., Leskela, M., Atomic layer deposition of nanostructured TiO_2 photocatalysts via template approach. Chem. Mat. **19**(7), 1816–1820 (2007). doi:10.1021/cm062576e
13. Ren, X., Gershon, T., Iza, D.C., Munoz-Rojas, D., Musselman, K., MacManus-Driscoll, J.L.: The selective fabrication of large-area highly ordered TiO_2 nanorod and nanotube arrays on conductive transparent substrates via sol–gel electrophoresis. Nanotechnology **20**(36) (2009). doi:365604 10.1088/0957-4484/20/36/365604
14. Mun, K.S., Alvarez, S.D., Choi, W.Y., Sailor, M.J.: A stable, label-free optical interferometric biosensor based on TiO_2 nanotube arrays. ACS Nano **4**(4), 2070–2076 (2010). doi:10.1021/nn901312f
15. Losic, D., Simovic, S.: Self-ordered nanopore and nanotube platforms for drug delivery applications. Expert Opin. Drug Deliv. **6**(12), 1363–1381 (2009). doi:10.1517/17425240903300857
16. Gultepe, E., Nagesha, D., Sridhar, S., Amiji, M.: Nanoporous inorganic membranes or coatings for sustained drug delivery in implantable devices. Adv. Drug Deliv. Rev. **62**(3), 305–315 (2010). doi:10.1016/j.addr.2009.11.003
17. Roy, S.C., Varghese, O.K., Paulose, M., Grimes, C.A.: Toward solar fuels: photocatalytic conversion of carbon dioxide to hydrocarbons. ACS Nano **4**(3), 1259–1278 (2010). doi:10.1021/nn9015423
18. Lin, C.J., Yu, W.Y., Chien, S.H.: Transparent electrodes of ordered opened-end TiO_2-nanotube arrays for highly efficient dye-sensitized solar cells. J. Mater. Chem. **20**(6), 1073–1077 (2010). doi:10.1039/b917886d
19. Baker, D.R., Kamat, P.V.: Photosensitization of TiO_2 nanostructures with CdS quantum dots: particulate versus tubular support architectures. Adv. Funct. Mater. **19**(5), 805–811 (2009). doi:10.1002/adfm.200801173
20. Xu, Z.H., Yu, J.G.: A novel solid-state electrochemiluminescence sensor based on Ru(bpy)(3)(2+) immobilization on TiO2 nanotube arrays and its application for detection of amines in water. Nanotechnology **21**(24) (2010). doi:24550110.1088/0957-4484/21/24/245501

Titanium Dioxide

► Fate of Manufactured Nanoparticles in Aqueous Environment

Total Internal Reflection (Fluorescence) Velocimetry

► Nano(Evanescent-Wave)-Particle Velocimetry

Toward Bioreplicated Texturing of Solar-Cell Anodes

► Toward Bioreplicated Texturing of Solar-Cell Surfaces

Toward Bioreplicated Texturing of Solar-Cell Surfaces

Aditi Risbud[1], Akhlesh Lakhtakia[2] and Michael H. Bartl[3]
[1]Molecular Foundry, Lawrence Berkeley National Laboratory, Berkeley, CA, USA
[2]Department of Engineering Science and Mechanics, Pennsylvania State University, University Park, PA, USA
[3]Department of Chemistry, University of Utah, Salt Lake City, UT, USA

Synonyms

Toward bioreplicated texturing of solar-cell anodes

Definition

Biological species have developed myriad photonic structures to efficiently interact with light. The results of these interactions include large angular fields of view, reduced surface reflection, Bragg diffraction, and multiple scattering. Bioreplicated texturing of solar cell surfaces and photoanodes takes advantage of these excellent optical properties in biology. The aim is to convert biological structures into exact replicas made out of semiconductors, metals, and polymers and incorporate these replicas into solar cells to enhance light harvesting and light–matter interactions.

Introduction

The development of efficient, inexpensive, and environmentally benign energy sources is one of the biggest technological challenges of the twenty-first century. Even the most conservative predictions suggest worldwide demand for energy will triple within the next 40 years [1]. This rapidly growing energy demand, coupled with economic and ecologic impacts of fossil fuel–based energy sources, has sparked research into low-cost and environment-friendly alternatives. Given the immense amount of energy the sun delivers to earth daily, solar power constitutes by far the most promising and attractive alternative to fossil fuels. However, to become the primary energy source for future consumption, highly efficient yet inexpensive strategies for direct conversion of solar power into thermal energy, electricity and chemical fuels are needed. A central challenge in solar energy conversion and photocatalysis is to optimize "light management" at the nanometer scale – from capture of sunlight to transfer and conversion of solar light into energy, fuels, or drivers of chemical reactions.

A prime example of highly evolved light management is photosynthesis – one of the most fundamental processes in biology. Calculations show photosynthesis produces approximately 90 TW of energy annually, approximately six times the current worldwide total energy consumption per year [1]. In addition, the light-independent phase of photosynthesis fixes carbon dioxide from the atmosphere, providing a potential mechanism for carbon dioxide sequestration. Photosynthetic organisms capture sunlight and channel solar energy from light-harvesting outer units to a reaction center by transferring energy along multiple functional units and converting it to fuel. It is this elaborate, integrated energy conversion system consisting of controlled and cooperative interaction between functional nanoscale components that results in efficient capture, transfer and use of solar energy. In contrast, artificial solar conversion units are still far from achieving the sophistication of their counterparts in nature. The goal is thus to examine simple models of biological systems, to learn from nature's complex energy conversion systems and to unlock and mimic its mechanisms [2–4].

For example, in biological materials, such as the wings of a butterfly or the exoskeleton of an insect, chitin and air are organized in ordered structures to produce optical interference leading to structural colors, camouflage and the control of solar radiative heat intake for thermal regulation. An early study calculated the effects of total solar radiation absorption and emissivity in the forms of beam radiation, which travels directly from the sun to a surface; diffuse radiation, which is a uniform distribution of light on a surface from all possible angles; and reflected radiation, which is light reflected from the ground on an insect thin film [5]. The outcome of this study suggests minute changes in the thickness of these biological thin films, typically 100 nm thick, strongly influence intake of energy from solar radiation. For example, changes in butterfly wing scale film thickness could significantly shift absorption levels (by up to 25%) over a span of 80 nm, while diffuse solar absorption and emissivity remain relatively constant. Furthermore, iridescence has been found to be strongly correlated with higher absorption of the solar radiation [6].

In this entry, several novel concepts inspired by biological systems are introduced to enhance the first steps in solar energy conversion: the capture of sunlight through bioreplicated surfaces and light–matter interactions driving the conversion process. In the first section, strategies to increase the light-harvesting capability of conventional solar cells are discussed. These strategies are based on the excellent optical properties of insect eyes. In these structures, called compound eyes, intricate microscale and nanoscale features result in significantly enhanced angular fields of view, while minimizing surface reflectivity. Since these parameters are of utmost importance for optimizing light harvesting in conventional solid-state junction solar cells, researchers have recently started to develop methods for mass-replicating these biological structures as polymer films for deposition onto the surface of solar cells [7, 8]. In the second section, focus will be directed on methods to increase light–matter interactions in a new type of solar cell, the Grätzel-type electrochemical cell [9]. The heart of this low-cost alternative to solid-state junction devices is a nanocrystalline titania photoanode sensitized with visible-light absorbing dyes. Interestingly, coupling this photoanode to photonic crystal structures can increase its optical path, and therefore results in higher solar conversion efficiency [10, 11]. Since some of the most effective photonic crystal structures are found in biological systems, such as colored beetles and butterflies [12], researchers are devising ways to replicate these remarkable architectures into nanocrystalline titania [13]. These efforts are reviewed and several examples of attempts to generate bioreplicated photoanodes are given.

Surface Structuring of Solar Cells

A dominant factor in the performance of any solar conversion device is the amount of incident sunlight coupled into the active module of a given solar cell. The key considerations for efficient coupling of light into solar cells are to maximize the angular field of view of the solar absorption module and to minimize the overall reflection of light at the cell's surface. Increasing the angular field of view aims to harvest as much light as possible, in terms of direct illumination from the sun, as well as diffuse light. Harvesting the direct illumination of sunlight is highest when it hits the solar cell surface at a 90° angle. While, in principle, this can be achieved simply by installing a mechanical device that rotates the solar cell surface in response to the sun's position, such motion devices increase the cost of solar cells, require maintenance and use up part of the cell's energy. In addition, these motion devices would only minimally increase harvesting of diffuse light – scattered and reflected light that hits the solar cell from any possible angle.

During the course of millennia, biological structures have evolved to capture light from a wide range of directions. Indeed, houseflies have optimized their angular field of view through eye placement and underlying structure such that these insects can see 270° around them from a horizontal plane [2]. This remarkable ability stems from a compound eye structure, consisting of an array of microscale cylindrical lenses arranged on a curved surface. These tiny lenses not only capture light along the axis of the cylinder, but also can absorb light propagating in other directions through the cylinder's sidewalls. In addition, compound eyes can have nanometer-scale surface structures, which serve both as an antireflective [7] and a superhydrophobic surface [14]. These properties are the result of a gradual change in refractive index between the air and the eye. While superhydrophobicity keeps the eye clean, reduced reflection from insect eyes protects them from being visible to predators, and it also significantly enhances their photosensitivity in dim environments. As a result, these surfaces are an important enhancement strategy that could be used to boost efficiency in a solar cell.

Although these biological structures have complex nano-to-microscale features, recent advances in engineered biomimicry now allow scientists to replicate these structures using materials optimal for harvesting light for solar cells. Both physical and chemical nanofabrication techniques such as atomic layer deposition, conformal-evaporated-film-by-rotation technique, chemical vapor deposition, soft lithography and nanocasting are now employed by researchers to generate replicas of various biological structures in technologically relevant materials. For example, a research group developed a soft lithographic patterning technique to replicate a nanometer-scale antireflective coating found on the lens of a moth's compound eye [7]. A three-dimensional mold of this coating was designed with perfluoropolyether, an elastomeric polymer material capable of capturing nanometer-scale surface features. This mold was then used to create polyurethane replicas with antireflective behavior. As shown in Fig. 1, this approach results in high-fidelity replicas preserving topographical features as small as 100 nm. In addition, the group successfully used this technique to emboss antireflective nanoscale patterns in a widely used polymer blend for the photoactive region of an organic photovoltaic cell.

In an attempt to scale up compound eye bioreplication methods to mass-produce antireflective/large-field-of-view solar cell components, researchers developed a method to create reusable master negatives [8]. Specifically, corneas of the common blowfly were used as templates and replicated into metallic nickel. To achieve both a high-fidelity imprint of the hierarchically organized cornea surface and a stable and reusable master negative, the replication process was conducted in two steps. First, a modified version of the conformal-evaporated-film-by-rotation method was used to deposit a thin layer of nickel onto the surface relief of the cornea sample. In a second step, additional nickel was deposited by electroforming to strengthen the initial thin film replica. Finally, after removal of the biotemplate, a free-standing negative copy of the original cornea was obtained (Fig. 2). This group also showed this negative master could be successfully used (and reused) to create polymer replicas by a casting technique to faithfully reproduce micrometer-sized surface features. In addition, it was suggested by using stamping instead of casting, nanometer-sized features could be transferred into mass-produced polymer replicas.

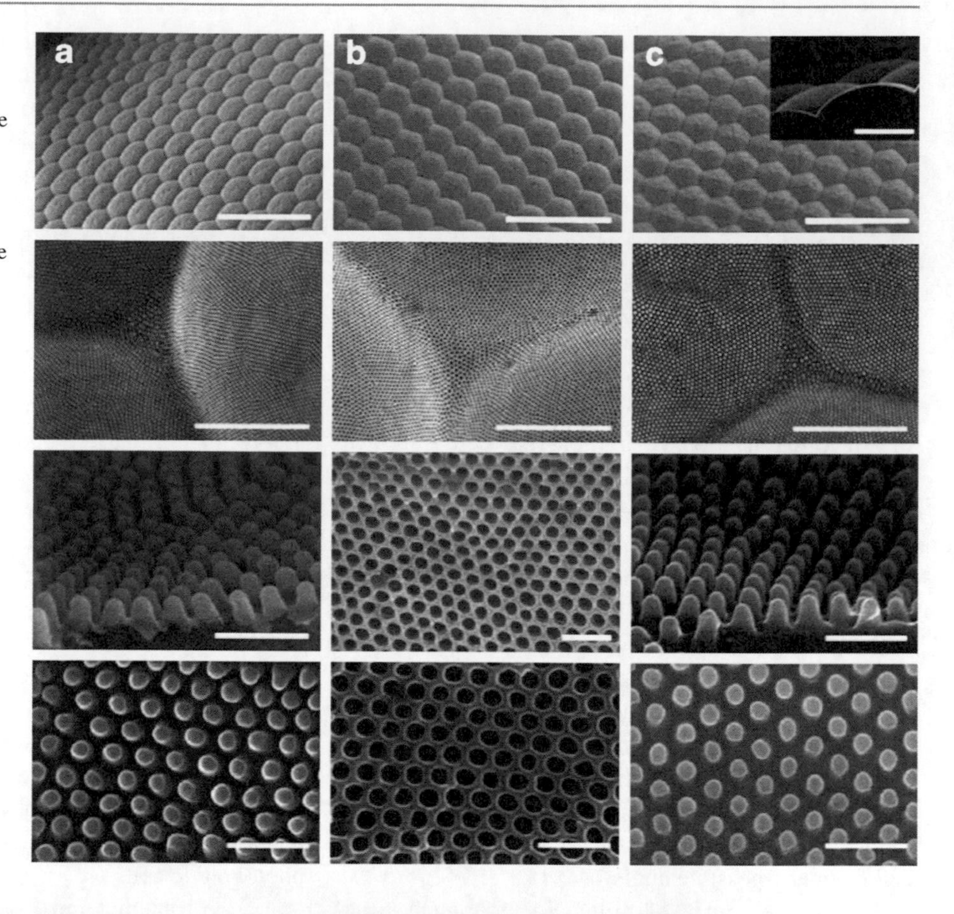

Toward Bioreplicated Texturing of Solar-Cell Surfaces, Fig. 1 Scanning electron micrographs of (**a**) the *Attacus atlas* moth eye structure, (**b**) the negative polymer replica, (**c**) the positive replica made by polymerization in the negative mold. Inset of (**c**) shows curved surface of the replica. Scale bars are 100 μm for the top images (inset: 20 μm), 5 μm for the second row images, and 500 nm for the third and fourth row images. Reproduced from [7] with permission of The Royal Society of Chemistry http://dx.doi.org/10.1039/C1SM05302G

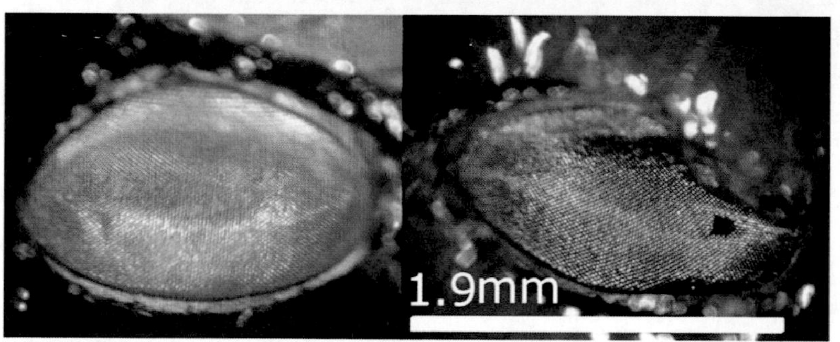

Toward Bioreplicated Texturing of Solar-Cell Surfaces, Fig. 2 Photographs of (*left*) a blowfly's cornea and (*right*) a nickel replica after removal of the biotemplate. The nickel structure is a positive replica when viewed from above (as shown), but a negative replica when viewed from below. Reproduced with permission from [8]. Copyright 2010 IOP Publishing Ltd.

Biotemplating of Photoanodes

Along with light-harvesting techniques, engineered biomimicry is a promising approach to enhance light–matter interactions in solar conversion systems. This is a particularly important area of research in next-generation, low-cost devices based on nanostructured semiconductors, conducting polymers and various hybrid device architectures. For example, theoretical studies recently showed coupling the photoanode of a Grätzel-type solar cell to a photonic crystal could significantly enhance its efficiency [10, 11]. Photonic crystals are periodically ordered structures with lattice parameters comparable to the optical

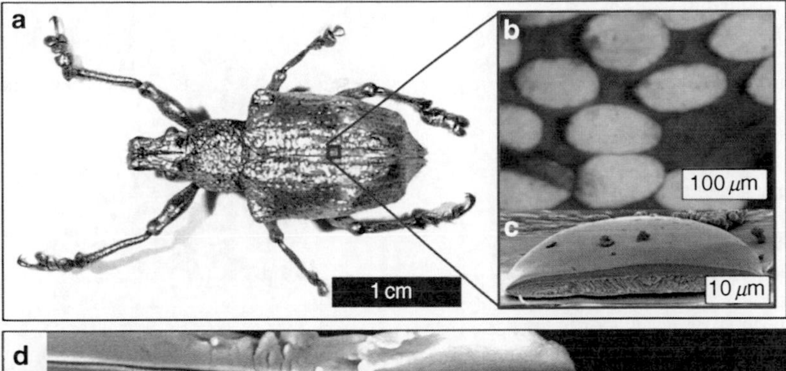

Toward Bioreplicated Texturing of Solar-Cell Surfaces,
Fig. 3 Photographs of (**a**) the weevil *L. augustus* and (**b**) individual scales attached to its exoskeleton. Cross-sectional scanning electron microscopy images of (**c**) a single scale and (**d**) a region of a scale at higher magnification. Reproduced with permission from [16]. Copyright 2008, American Physical Society

wavelength of interest [15]. Due to this periodic variation in the refractive index, light within a photonic crystal experiences a combination of Bragg diffraction and multiple scattering events. As a consequence, light of certain frequencies undergoes multiple reflections as it slowly percolates through this structure. This results in a greatly increased nominal optical path length, an increase in light–photoanode interactions and significantly enhances the probability of light-to-charge conversion.

Interestingly, millions of years before we "discovered" photonic crystals, biological systems created such structures to produce vibrant colors, control solar absorption and enhance bio-fluorescence for a variety of reasons, ranging from camouflaging to frightening predators and attracting mates [12, 13]. Biological photonic structures thus were optimized to produce a wide range of optical effects and function under various illumination conditions. Not surprisingly, the range of photonic structures found in biological species is virtually limitless, ranging from simple one-dimensional multilayers, to two-dimensional lattices found in skeletons and hairs of sea animals and bird feathers and to various three-dimensional architectures within colored cuticle scales of many species of beetles and butterflies. For example, scientists discovered certain beetles obtain their color from diamond-based photonic crystals – one of the most effective lattice structures [16]. What's more, such photonic structures (as an example, Fig. 3 shows the green Brazilian beetle *Lamprocyphus augustus*) are still beyond our engineering capabilities. Similarly, photonic structures found in many butterfly wing scales have an astonishing hierarchy, consisting of ridges, lamellae and ribs, as well as highly periodic honeycomb and gyroid structures [12]. These elaborate hierarchical-scale structures act as optical components interacting with sunlight, such as multilayer reflectors, diffraction gratings and photonic crystals.

Only recently have scientists begun to use this invaluable portfolio of biological photonic crystal lattices for engineered biomimicry approaches. The method of choice is bioreplication: here, a biological photonic structure is used as a template to create positive or negative copies out of photocatalytically active compounds. An example of this approach is shown in Fig. 4 for conversion of the diamond-based photonic structure of the beetle *L. augustus*. Replication is achieved by a sol–gel double-imprint method using a sacrificial silica glass intermediary [17]. Sol–gel templating is quite attractive due to its simplicity and low cost. The original biological templates were backfilled with liquid precursors of the desired inorganic compound (e.g., silica, titania, alumina). After solidification of the inorganic compound, the original template was selectively removed

Toward Bioreplicated Texturing of Solar-Cell Surfaces, Fig. 4 Sol–gel bioreplication route of photonic weevil scales into titania replicas (*top left*). Scanning electron microscopy images of the original biopolymeric photonic structure (*top right*), the negative replica made of silica (*bottom right*), and the positive replica made of titania (*bottom left*). Scale bars are 200 μm (*top left*) and 1 μm (*all others*). Reproduced with permission from [17]. Copyright Wiley-VCH Verlag GmbH & Co. KGaA

by etching or heat treatment. This approach enabled a positive copy ("inverse-of-the-inverse") of the original structure composed of photocatalytically active nanocrystalline titania.

Along with sol–gel chemistry, other physical and chemical deposition methods can be used for replication of biological structures – as long as they operate under mild deposition conditions and temperatures. Successful methods include atomic layer deposition [18] and conformal-evaporated-film-by-rotation techniques [19]. In general, these techniques can be applied under noncorrosive reaction environments, mild pH conditions, and relatively low deposition temperatures (below 100–200 °C). Since these replicas are formed by deposition from a gas phase, these methods are highly efficient in infiltrating even complex three-dimensional frameworks as long as the internal structure is fully accessible. Furthermore, the deposition parameters can readily be controlled, enabling the amount of deposited material to be carefully controlled. As shown in Fig. 5, control of atomic layer deposition can be used to create shell-like replicas of the original biological structures with finely tuned replica shell thickness and, as a result, well-controlled optical properties.

Recently, a group of scientists took such bioreplication approaches one step closer to solar conversion applications by devising strategies to construct a Grätzel-type titania photoanode coupled to arrays of replicated butterfly scales [20]. In this approach, $1 \times 1\ cm^2$ samples of butterfly wings were soaked in a titania sol–gel precursor and arranged on top of a Grätzel-type photoanode (Fig. 6). This composite structure was then heat-treated to allow the top butterfly-wing replica layer to bond to the bottom titania layer. This heat treatment (500 °C in air) also induced titania crystallization into polycrystalline anatase and removed the biopolymeric template. Despite structural shrinkage during the heat-based framework densification-crystallization treatment, these replicated materials displayed improved light absorption behavior as compared to anodes without the replicated butterfly wings. While this outcome is an important step toward bioreplicated solar cells, these researchers also note major challenges remaining in generating such biotemplated anodes, optimizing their electrical conductivity and integrating these structures into functional solar conversion devices.

Concluding Remarks

Engineered biomimicry has arrived as a serious contender in materials research and engineering. While a few years ago the concepts of super-vision from bug eyes, ultrastrong fabrics from spiders, vertically

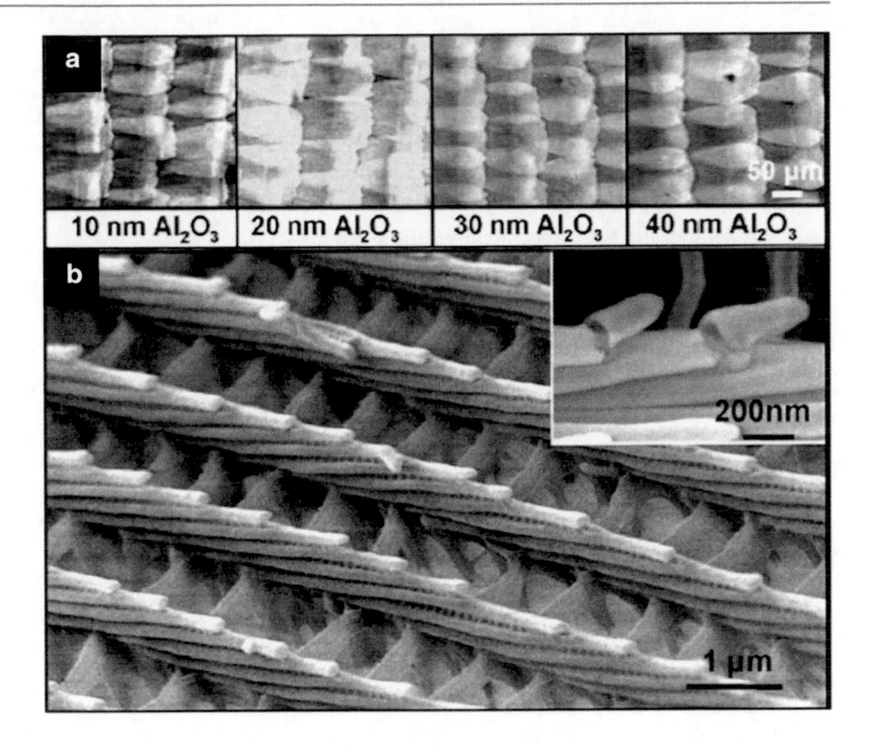

Toward Bioreplicated Texturing of Solar-Cell Surfaces, Fig. 5 (**a**) Optical microscope image of the alumina-coated *Morpho peleides* butterfly wing scales; color changes from original blue to pink. (**b**) Higher magnification scanning electron microscopy image of alumina replicas of butterfly wing scales and two broken rib tips (*inset*). Reproduced with permission from [18]. Copyright 2006 American Chemical Society

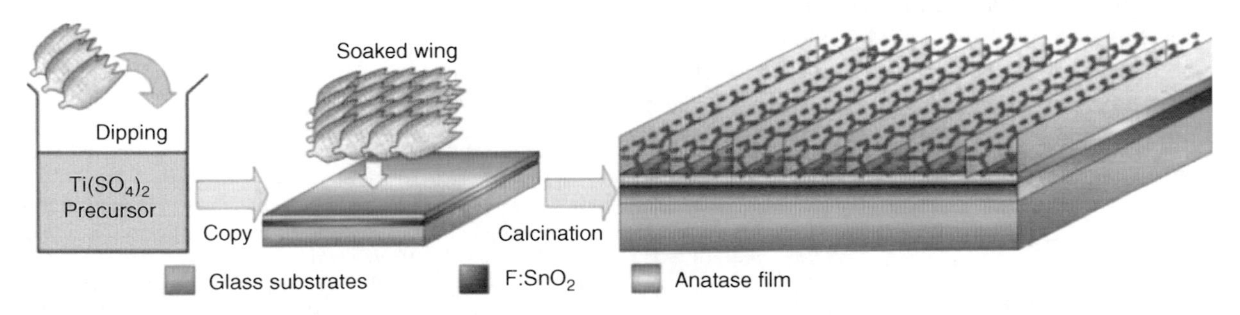

Toward Bioreplicated Texturing of Solar-Cell Surfaces, Fig. 6 Schematic illustration of creating bioreplicated titania photoelectrodes. Step 1, soaking the butterfly wings in the titania precursor; step 2, calcination of soaked butterfly wings on top of an anatase titania film. Reproduced with permission from [20]. Copyright 2009 American Chemical Society

climbing gecko-like robots, or optical devices from beetles was more the stuff of Hollywood science fiction, these ideas have found their way into research laboratories and inspire the development of new devices in electronics, optics and mechanics. Indeed, although the pure beauty of natural iridescence in the form of opal gem stones, jewel beetles and feather ornaments has fascinated and inspired mankind for thousands of years, today, these materials and optical concepts offer revolutionary new ways to manipulate light. In addition, these materials hold great potential for more efficient solar cells and photocatalysts, new optical reflectors and sensors and ultrafast switches.

An exciting area in engineered biomimicry is bioreplication, in which complex biological structures are converted into replicas made of semiconductors, metals or polymers. In this entry, a few examples of solar-related bioreplication strategies to increase light-harvesting ability and light–matter interaction efficiency in various types of solar cells were discussed. While many of these approaches are promising and first successful steps have been achieved, it should be emphasized there are also several limitations and challenges in bioreplication. For example, biological samples have an inherent size limit given by the dimensions of the eyes, scales, or feathers used as templates. Direct use and integration of biological samples, therefore, have limited applications, suggesting more efficient mass-duplication routes have to be developed. In addition, most natural structures have a much higher defect density than what is tolerated by current engineering standards, and no two natural structures are exactly the same. For example, the structural parameters of colored scales taken from different parts of a beetle, or even within a single scale, can vary by as much as 20%.

Consequently, bioreplication must be viewed in its correct and just perspective: Although materials prepared by bioreplication will perhaps never achieve the precision, uniformity and reproducibility of their counterparts engineered by micro- or nano-fabrication methods, bioreplication opens the door to the full portfolio of highly complex hierarchical structures found in nature – structures still far from our engineering capabilities. It will therefore be of vital importance to devise strategies to use and integrate materials created by engineered biomimicry with micro- or nano-fabrication methods and combine the appealing aspects of two worlds – biology and materials science – in a concerted approach. Engineered biomimicry has found its way from science fiction into the laboratories; whether it makes the next step from fundamental research to wide-scale applications will depend on scientific creativity and imagination.

Cross-References

- ▶ Biomimetics of Optical Nanostructures
- ▶ BioPatterning
- ▶ Hybrid Solar Cells
- ▶ Moth-Eye Antireflective Structures
- ▶ Nanomaterials for Excitonic Solar Cells
- ▶ Nanostructures for Energy
- ▶ Nanostructures for Photonics
- ▶ Structural Color in Animals

References

1. Hoffert, M.I., Caldeira, K., Jain, A.K., Haites, E.F., Harvey, L.D.D., Potter, S.D., Schlesinger, M.E., Schneider, S.H., Watts, R.G., Wigley, T.M.L., Wuebbles, D.J.: Energy implications of future stabilization of atmospheric CO_2 content. Nature **395**, 881 (1998)
2. Chiadini, F., Fiumara, V., Scaglione, A., Pulsifer, D.P., Martín-Palma, R.J., Pantano, C.G., Lakhtakia, A.: Insect eyes inspire improved solar cells. Optics Photonics News **22**(4), 38 (2011)
3. Paris, O., Burgert, I., Fratzl, P.: Biomimetics and biotemplating of natural materials. MRS Bull. **35**, 219 (2010)
4. Jin, L., Zhai, J., Heng, L., Wei, T., Wen, L., Jiang, L., Zhao, X., Zhang, X.: Bio-inspired multi-scale structures in dye-sensitized solar cell. J. Photochem. Photobiol. C **10**, 149 (2009)
5. Heilman, B.D., Miaoulis, I.N.: Insect thin film as solar collectors. Appl. Optics **33**, 6642 (1994)
6. Bosi, S.G., Hayes, J., Large, M.C.J., Poladian, L.: Color, iridescence, and thermoregulation in Lepidoptera. Appl. Opt. **47**, 5235 (2008)
7. Ko, D.-H., Tumbleston, J.R., Henderson, K.J., Euliss, L.E., DeSimone, J.M., Lopez, R., Samulski, E.T.: Biomimetic microlens array with antireflective "moth-eye" surface. Soft Matter **7**, 6404 (2011). doi:10.1039/C1SM05302G
8. Pulsifer, D.P., Lakhtakia, A., Martín-Palma, R.J., Pantano, C.G.: Mass fabrication technique for polymeric replicas of arrays of insect corneas. Bioinsp. Biomim. **5**, 036001 (2010)
9. Grätzel, M.: Photoelectrochemical cells. Nature **414**, 338 (2001)
10. Mihi, A., Miguez, H.: Origin of light-harvesting enhancement in colloidal-photonic-crystal-based dye-sensitized solar cells. J. Phys. Chem. B **109**, 15968 (2005)
11. Halaoui, L.I., Abrams, N.M., Mallouk, T.E.: Increasing the conversion efficiency of dye-sensitized TiO_2 photoelectrochemical cells by coupling to photonic crystals. J. Phys. Chem. B **109**, 6334 (2005)
12. Srinivasarao, M.: Nano-optics in the biological world: Beetles, butterflies, birds, and moths. Chem. Rev. **99**, 1935 (1999)
13. Jorgensen, M.R., Bartl, M.H.: Biotemplating routes to three-dimensional photonic crystals. J. Mater. Chem. **21**, 10583 (2011). doi:10.1039/C1JM11037C
14. Koch, K., Bhushan, B., Barthlott, W.: Diversity of structure, morphology and wetting of plant surfaces. Soft Matter **4**, 1943 (2008)
15. Joannopoulos, J.D., Villeneuve, P.R., Fan, S.H.: Photonic crystals: Putting a new twist on light. Nature **386**, 143 (1997)
16. Galusha, J.W., Richey, L.R., Gardner, J.S., Cha, J.N., Bartl, M.H.: Discovery of a diamond-based photonic crystal structure in beetle scales. Phys. Rev. E **77**, 050904 (2008)
17. Galusha, J.W., Jorgensen, M.R., Bartl, M.H.: Diamond-structured titania photonic bandgap crystals from biological templates. Adv. Mater. **22**, 107 (2010)
18. Huang, J., Wang, X., Wang, Z.L.: Controlled replication of butterfly wings for achieving tunable photonic properties. Nano Lett. **6**, 2325 (2006)
19. Martín-Palma, R.J., Pantano, C.G., Lakhtakia, A.: Biomimetization of butterfly wings by the conformal-evaporated-film-by-rotation technique for photonics. Appl. Phys. Lett. **93**, 083901 (2008)
20. Zhang, W., Zhang, D., Fan, T., Gu, J., Ding, J., Wang, H., Guo, Q., Ogawa, H.: Novel photoanode structure templated from butterfly wing scales. Chem. Mater. **21**, 33 (2009)

Toxicity

► In Vitro and In Vivo Toxicity of Silver Nanoparticles

Toxicity of Metal and Metal Oxide Nanoparticles on Prokaryotes (Bacteria), Unicellular and Invertebrate Eukaryotes

► Ecotoxicity of Inorganic Nanoparticles: From Unicellular Organisms to Invertebrates

Toxicity/Ecotoxicity

► Exposure and Toxicity of Metal and Oxide Nanoparticles to Earthworms

Toxicology: Plants and Nanoparticles

Marie Carrière and Camille Larue
Laboratoire Lésions des Acides Nucléiques, Commissariat à l'Energie Atomique, SCIB, UMR-E 3 CEA/UJF-Grenoble 1, INAC, Grenoble, Cedex, France

Synonyms

Ecotoxicity; Nanoparticle; Phytotoxicity

Definition

Impact of engineered nanoparticles (ENP) on plant growth, development, physiology, and morphology.

Key Research Findings

Three research areas are currently being investigated to define engineered nanoparticle (ENP) impacts on plant development (seed germination, root and shoot elongation, etc.), on plant physiology, and ENP uptake and translocation through plants.

Impact on Plant Development

In this research area, regulatory tests are used, such as the ones described in EPA guidelines [1]. These guidelines have been developed for toxicity testing of chemical pollutants, and their relevance for toxicity testing of ENPs is under debate, with several groups working on their implementation to render them accurate. Among these tests, seed germination, root and shoots elongation assays have been intensively used to assess ENP impact on plant development, and the obtained results are either positive, or negative, or inconsequential. For instance multi-walled carbon nanotubes (CNT) have been shown to cause plant biomass decrease, particularly on *Cucurbita pepo* [2], while they showed no impact on rye grass, lettuce, rape, radish, corn, and cucumber [3]. The same is true for TiO_2 nanoparticles, which were shown in some studies to alter leaf growth and transpiration on maize [4], while they did not cause any alteration in the growth of willow tree plantlets [5].

Described results are difficult to compare, since model plant species are often different, and the considered nanoparticles are often different as well: Multi-walled CNTs may differ in their production process, post-synthesis surface treatment, and in the preparation of dispersed suspensions. TiO_2 nanoparticles may differ in their primary diameter, specific surface area, shape (round, elongated, see Fig. 1), crystalline structure, etc. These conflicting effects are easily explained by the broad panel of tested ENPs, and their very different physicochemical properties.

Moreover indirect effects such as ENP dissolution or impact of the solvent (or other environmental pollutants) which may be adsorbed on ENP surface have to be considered (Fig. 2).

The best examples of the impact of ENP dissolution on plant development was demonstrated with Cu and ZnO nanoparticles. Even if the impact of dissolved ions did not explain all the toxic effects observed in the following studies, they greatly contributed to them. Indeed, Cu NPs have been demonstrated to cause inhibition of plant growth and to accumulate in mung beans and wheat [6], but also have some influence on the germination of lettuce [7]. Among all the ENPs tested for their impact on rye grass, lettuce, rape, radish, corn, and cucumber, only Zn and ZnO NPs

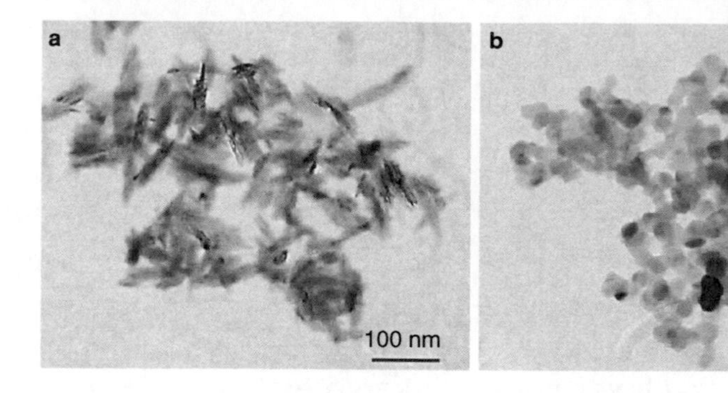

Toxicology: Plants and Nanoparticles, Fig. 1 Titanium dioxide nanoparticles, rutile and elongated-shaped (**a**) or anatase and round-shaped (**b**)

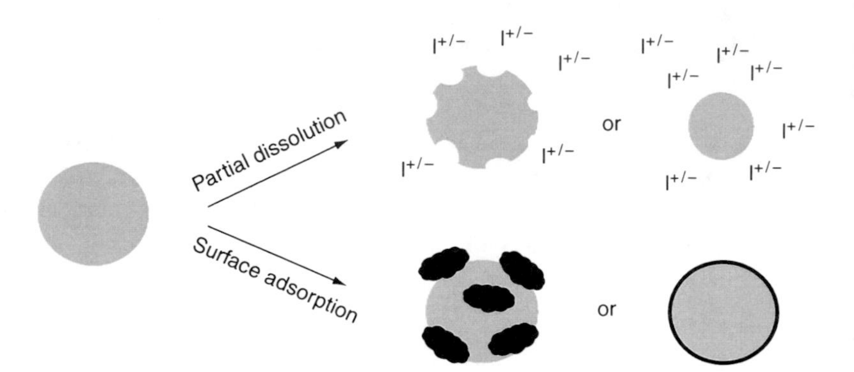

Toxicology: Plants and Nanoparticles, Fig. 2 Schematic representation of potential ENP evolution in the environment: ENPs may partially dissolve, releasing free ions ($I^{+/-}$). They may also adsorb environmental pollutants or solvent on their surface

reduced germination of ryegrass and corn, respectively. These two ENPs also reduced root elongation [3]. ZnO-NP dissolution was then considered by the same team the year after [8].

Uptake and Translocation

Despite the relative impermeability of plant roots to particulate matter, due to their small size, ENPs have been demonstrated to penetrate plant root wall, even if most of the ENPs remain adsorbed on root surface. It has for instance been demonstrated for Fe_3O_4 NPs on pumpkins [9], for TiO_2-alizarin red nanoconjugates on *Arabidopsis thaliana* [10], and CeO_2 nanoparticles on alfalfa, cucumber, corn, and tomato [11]. The consequence of such result is that ENPs may then be transferred to the food chain. Note that the ENP content in these plants is very low: ENP uptake is not an efficient process; consequently the amount of ENPs reaching the food chain would also be low.

Several techniques are available to demonstrate ENP uptake by plants. First, electron microscopy, due to its very high spatial resolution (typically 1 nm or below), is the method of choice to identify the presence of ENPs in biological samples, and particularly in plants. Scanning electron microscopy (SEM) permits to observe ENPs on the surface of plant tissues (roots or leaves), but does not give access to the internalized ENPs. Conversely transmission electron microscopy (TEM), achieved on dehydrated plant samples included in a resin and cut in ultrathin sections, permits the detection of nanosized materials in plant cells or extracellular compartments. TEM has been successfully used to prove that carbon-based ENPs were accumulated in rice plantlets [12]. Coupling these two techniques with energy-dispersive spectroscopy (EDS) allows a local chemical identification of the electron-dense spots observed by TEM or SEM in plant samples, proving that they contain the chemical element constituting the ENP (when EDS analysis is performed, then spatial resolution ranges from 30 to 50 nm).

Another valuable technique is micro-X-ray-fluorescence (μ-XRF), achieved either on a laboratory source or using synchrotron radiation. This technique allows local identification of chemical composition, and basically the identification of hot spots of one particular element constituting the ENP. With this technique, spatial resolution can be as high as 100 nm on some synchrotron beam lines such as ID22

(ESRF, Grenoble, France). On some synchrotron beam lines, this technique is coupled to X-ray absorption spectroscopy (XAS), which allows describing the local environment of the analyzed element, giving the absolute proof that the element is in the nanoparticulate form, and not present as dissolved ions. Moreover it informs on ENP crystal structure, and sometimes on their size (it is the case for TiO_2-NPs). For instance this technique has been used to evidence that alfalfa, cucumber, corn, and tomato plants accumulated CeO_2-NPs in their roots [11].

Regarding the detection of CNTs, TEM enables the observation of single or clusters of NTCs. However, images obtained by TEM can be misinterpreted, since the shape of CNT is close to the structure of natural vegetal compounds. In order to avoid any misinterpretation, very specific techniques such as photothermal and photoacoustic detection have been developed and used for the identification of their accumulation and translocation in tomato plantlets [13]. These techniques are highly sensitive and specific, allowing detecting single MWCNT in plant tissues. Raman spectroscopy is also a very specific technique, which has been used to identify the presence of CNTs inside plant seeds thanks to their G band (1568 cm^{-1}) [14].

Another point is that, when taken up in plant roots, ENPs are in some cases transferred from the roots to the shoots (Fig. 3). This phenomenon, also called translocation, was demonstrated for instance with Au-NPs in tobacco [15] and Fe_3O_4 NPs on pumpkins [9]. Carbon-based NPs have also been demonstrated to translocate from the roots to the shoots and from the shoots to the seeds. The newly developed plantlets, resulting from the germination of these contaminated seeds, also contained carbon-based NPs, meaning that ENPs were transmitted to the next generation [12].

Very recently, thanks to these techniques, gold NPs were shown to be transferred to higher trophic levels: hornworms fed with Au-NP-exposed tobacco were shown to be contaminated with Au-NPs [15]. In this particular case, gold maximal concentration in tobacco reached less than 100 mg Au kg^{-1} dry tobacco.

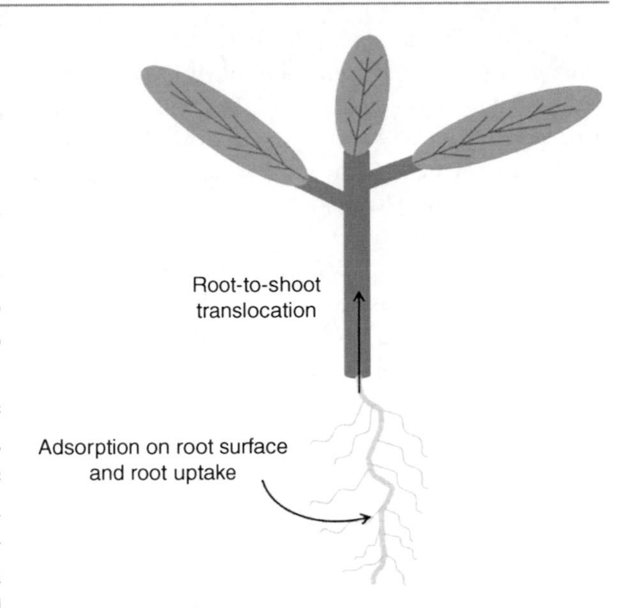

Toxicology: Plants and Nanoparticles, Fig. 3 Schematic representation of ENP adsorption on root surface, root uptake, and transfer to the shoot

Impact on Plant Physiology

The impact of ENP exposure on plant physiology and metabolism is poorly described, underlining that the area of research of ENP phytotoxicity is at its infancy. A very recent study describes the impact of NTC on plant gene expression [13]. Many stress-related genes are activated upon exposure to NTC, and among them the gene encoding the protein constituting water channels (aquaporins) [13]. This channel regulates water permeability, and therefore is significantly implicated in seed germination and plant growth. In one study TiO_2-NPs have also been demonstrated to inhibit root growth and transpiration by interference with water transport [4]. Actually, ENPs blocked apoplastic water flow in plant roots (within a plant, the apoplast is the free diffusional space outside the plasma membrane), certainly by obstructing nanosized cell wall pores, but also by reducing the exclusion diameter of these pores. However TiO_2-NPs are likely to agglomerate when dispersed in a saline, neutral aqueous solution. Therefore, exposing plant roots to TiO_2-NPs dispersed in a $CaCl_2$ solution at near-neutral pH, even if it is more physiologically relevant, means that plants were exposed to highly agglomerated ENPs. It is thus not surprising that water flow through the plant was blocked, and it rather derives from root exposure to a high concentration (0.3 or 1 gl^{-1}) of agglomerated material, than by a specific effect of ENPs.

Several studies reported that ENPs impacted plant photosynthetic activity. In a series of articles, TiO_2-NPs were shown to induce an increase in spinach germination, growth, nitrogen metabolism, and photosynthesis, and protected plants from oxidative damage induced by UV radiation [16]. Indeed TiO_2-NPs promote energy transfer from chlorophyll b and carotenoids to chlorophyll a, but also to the P680 chlorophyll a of D1/D2/Cyt 559 complex, in vitro. The authors conclude that nanosized TiO_2 would enter spinach cells and chloroplasts more easily due to their small size, then bind to the D1/D2/Cyt 559 complex and improve absorption, transfer, and conversion of light energy in chloroplasts [16].

Lastly, genotoxic effects of ENPs are currently investigated on several plant species. Root exposure to Ag-NPs reduce the mitotic index of exposed cells in *Allium cepa*, and more generally impair the cell division process, resulting in chromatin bridges, stickiness, disturbed metaphase, chromosomal breaks, and cell disintegration [17]. In the same test (*A. cepa* test is a standardized assay) TiO_2-NPs show no genotoxic impact, neither chromosomal aberration nor micronuclei [18]. Conversely DNA damage is detected in *A. cepa* and *Nicotiana tabaccum* exposed to TiO_2-NPs, by using comet assay and DNA laddering experiment. This genotoxic effect is correlated with the induction of malondialdehyde production, which is a marker of lipid peroxidation, as a result of oxidative stress [19].

Concluding Remarks

As a conclusion, the impact of ENPs on plants is still poorly documented; however some effects have been reported. Some ENPs, particularly those which partly dissolve, impact plant developments, i.e., seed germination, root and shoot elongation, and plant biomass. Some ENPs are taken up by plant roots, sometimes translocated to the shoots, and even transmitted to the next generation via their transfer to seeds. When accumulated in plant leaves, some ENPs can be transferred to primary consumer, for instance hornworms, meaning that they reach the food chain. The genotoxic potential of some ENPs has been demonstrated, although other articles relate opposite results. Some ENPs would modulate photosynthesis by impacting photosystem activity. Note that only some ENPs have these properties and exert these effects, the results described herein definitely cannot be generalized to all ENPs. Note also that almost all the results were obtained on plants exposed to ENPs in vitro, but only one publication describes the impact of ENPs-amended soil on plant development. Data thus need to be collected on environmentally relevant systems, i.e., plants cultivated on natural soils and contaminated with low doses of ENPs since massive contamination of the environment with ENPs would improbably occur.

Cross-References

► Cellular Mechanisms of Nanoparticle's Toxicity
► Effect of Surface Modification on Toxicity of Nanoparticles

References

1. U.S. Environmental Protection Agency: Ecological effects test guidelines (OPPTS 850.1200): seed germination/root elongation toxicity test. http://www.epa.gov/opptsfrs/publications/OPPTS_Harmonized/850_Ecological_Effects_Test_Guidelines/Drafts/850-4200.pdf (1996)
2. Stampoulis, D., Sinha, S.K., White, J.C.: Assay-dependent phytotoxicity of nanoparticles to plants. Environ. Sci. Technol. **43**(24), 9473–9479 (2009)
3. Lin, D., Xing, B.: Phytotoxicity of nanoparticles: inhibition of seed germination and root growth. Environ. Pollut. **150**(2), 243–250 (2007)
4. Asli, S., Neumann, P.M.: Colloidal suspensions of clay or titanium dioxide nanoparticles can inhibit leaf growth and transpiration via physical effects on root water transport. Plant Cell Environ. **32**(5), 577–584 (2009)
5. Seeger, E., et al.: Insignificant acute toxicity of TiO_2 nanoparticles to willow trees. J. Soils Sediments **9**(1), 46–53 (2009)
6. Lee, W.M., et al.: Toxicity and bioavailability of copper nanoparticles to the terrestrial plants mung bean (*Phaseolus radiatus*) and wheat (*Triticum aestivum*): plant agar test for water-insoluble nanoparticles. Environ. Toxicol. Chem. **27**(9), 1915–1921 (2008)
7. Shah, V., Belozerova, I.: Influence of metal nanoparticles on the soil microbial community and germination of lettuce seeds. Water Air Soil Pollut. **197**(1–4), 143–148 (2009)
8. Lin, D., Xing, B.: Root uptake and phytotoxicity of ZnO nanoparticles. Environ. Sci. Technol. **42**(15), 5580–5585 (2008)
9. Zhu, H., et al.: Uptake, translocation, and accumulation of manufactured iron oxide nanoparticles by pumpkin plants. J. Environ. Monit. **10**(6), 713–717 (2008)
10. Kurepa, J., et al.: Uptake and distribution of ultrasmall anatase TiO_2 Alizarin red S nanoconjugates in *Arabidopsis thaliana*. Nano Lett. **10**(7), 2296–2302 (2010)
11. Lopez-Moreno, M.L., et al.: X-ray absorption spectroscopy (XAS) corroboration of the uptake and storage of CeO_2 nanoparticles and assessment of their differential toxicity

in four edible plant species. J. Agric. Food Chem. **58**(6), 3689–3693 (2010)
12. Lin, S., et al.: Uptake, translocation, and transmission of carbon nanomaterials in rice plants. Small **5**(10), 1128–1132 (2009)
13. Khodakovskaya, M.V., et al.: Complex genetic, photothermal, and photoacoustic analysis of nanoparticle-plant interactions. Proc. Natl Acad. Sci. U.S.A. **108**(3), 1028–1033 (2011)
14. Khodakovskaya, M., et al.: Carbon nanotubes are able to penetrate plant seed coat and dramatically affect seed germination and plant growth. ACS Nano **3**(10), 3221–3227 (2009)
15. Judy, J.D., Unrine, J.M., Bertsch, P.M.: Evidence for biomagnification of gold nanoparticles within a terrestrial food chain. Environ. Sci. Technol. **45**(2), 776–781 (2011)
16. Mingyu, S., et al.: Promotion of nano-anatase TiO_2 on the spectral responses and photochemical activities of D1/D2/Cyt b559 complex of spinach. Spectrochim. Acta A **72**, 1112–1116 (2009)
17. Kumari, M., Mukherjee, A., Chandrasekaran, N.: Genotoxicity of silver nanoparticles in *Allium cepa*. Sci. Total Environ. **407**(19), 5243–5246 (2009)
18. Klancnik, K., et al.: Use of a modified Allium test with nanoTiO_2. Ecotoxicol. Environ. Saf. **74**(1), 85–92 (2011)
19. Ghosh, M., Bandyopadhyay, M., Mukherjee, A.: Genotoxicity of titanium dioxide (TiO_2) nanoparticles at two trophic levels plant and human lymphocytes. Chemosphere **81**(10), 1253–1262 (2010)

Transcutaneous Delivery

▶ Dermal and Transdermal Delivery

Transdermal Permeation

▶ Dermal and Transdermal Delivery

Transduction

▶ Micromachined 3-D Force Sensors

Transfer

▶ Fate of Manufactured Nanoparticles in Aqueous Environment

Transmission Electron Microscopy

Helmut Kohl
Physikalisches Institut, Westfälische Wilhelms-Universität Münster, Wilhelm-Klemm-Straße 10, Münster, Germany

Definition

Transmission electron microscopy is a method to image small thin objects by use of the transmitted electrons.

The Transmission Electron Microscope (TEM)

Similar to the general scheme of a light microscope, a transmission electron microscope [1, 2] consists of an electron source, a condenser system, an objective lens and a projector system as shown in Fig. 1. Many transmission electron microscopes have additional instruments attached to it, such as an X-ray detector and/or an energy loss spectrometer in order to be able to perform elemental analyses. These techniques will be explained later on.

The electron source, commonly called "gun" is composed of a cathode and an accelerating system. There are different types of cathodes being used:

- Thermal cathodes
- Schottky emitters, often misleadingly called (hot) field emission cathodes
- (Cold) field emission cathodes.

Thermal cathodes make use of thermal emission of electrons at high temperatures. Most often they use tungsten or LaB_6 as emitting material. The emission barrier can be lowered by applying a strong electrostatic field to a tip (Schottky effect). This is the basic principle of a Schottky emitter. Using an even stronger electrostatic field at a sharp tip one can extract electrons by means of the quantum mechanical tunnel effect. The resulting field emission cathodes require ultra high vacuum to operate, because otherwise positively charged ions will be accelerated towards the tip eventually leading to its destruction. All of these cathodes are followed by an accelerating system ending with the anode. Most transmission electron microscopes usually operate at voltages of

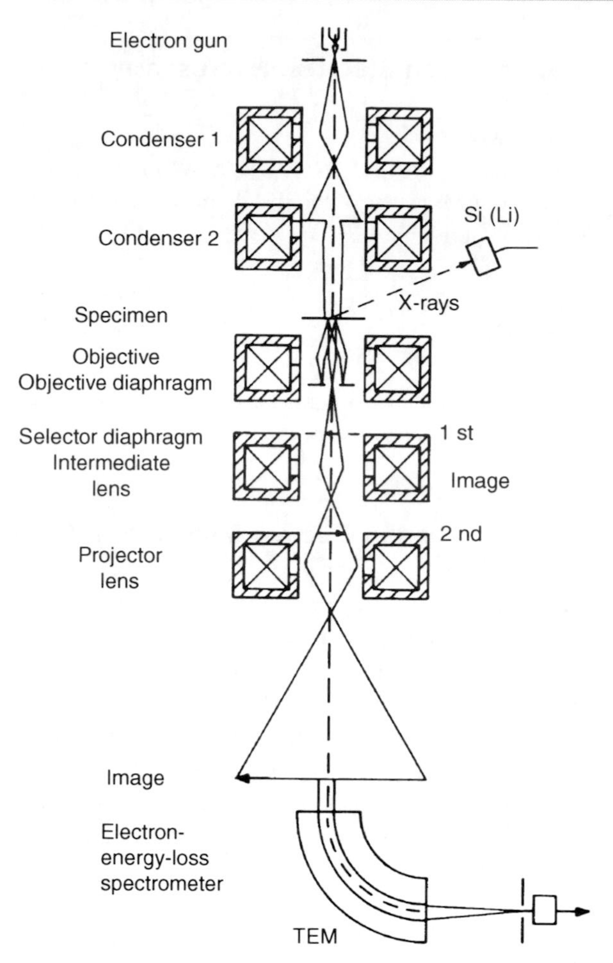

Transmission Electron Microscopy, Fig. 1 Schematic ray paths for a transmission electron microscope (TEM) equipped with an additional x-ray spectrometer and an energy loss spectrometer [1]

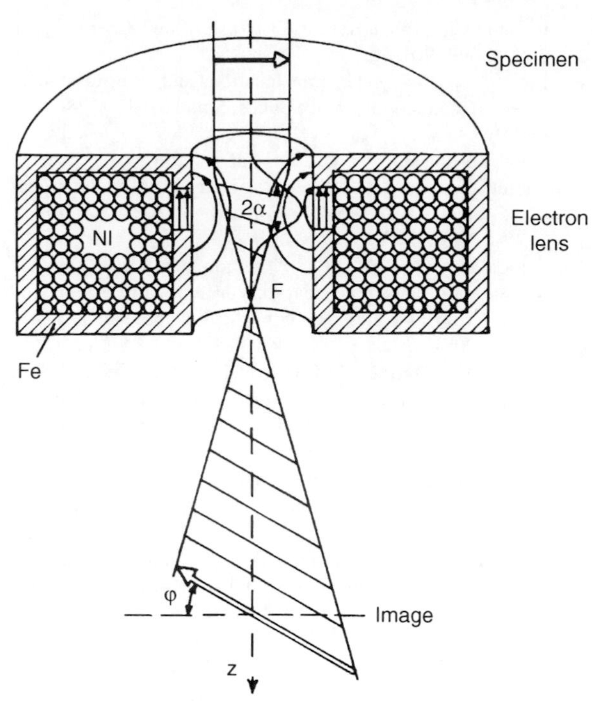

Transmission Electron Microscopy, Fig. 2 Schematic diagram of an electromagnetic lens [1]

100–300 kV, even though there are special low voltage TEMs operating at voltages from 20 kV upwards and high voltage TEMs operating at voltages up to 1.2 MeV.

The role of the condenser system is to transfer the electrons from the gun onto the specimen. It is composed of two to five electron lenses to allow to independently vary the illumination angle and the size of the illuminated area. The transmitted electrons are then focused to a magnified intermediate image which in turn is magnified and transferred into the final image plane.

Magnetic Electron Lenses

To focus an electron beam, we need lenses operating similar to glass lenses in light optics. These can be constructed by coils encapsulated by a ferromagnetic material as shown in Fig. 2. To understand their working principle, let us consider the movement of an electron in the homogeneous magnetic field of a long coil. The Lorentz force on the electron is given by

$$\vec{F} = -e[\vec{v} \times \vec{B}] \tag{1}$$

The velocity $\boldsymbol{v}$ can be separated into a component parallel v_{II} to the magnetic induction $\boldsymbol{B}$ and one perpendicular component $v_{\perp}$. As there is no force operating on the parallel component, v_{II} remains constant. Perpendicular to the magnetic induction $\boldsymbol{B}$ the electron performs a circular motion with a radius r that is determined by the equivalence of the Lorentz force to the centrifugal force

$$\frac{mv_{\perp}^2}{r} = ev_{\perp}B \tag{2}$$

or

$$r = \frac{mv_{\perp}}{eB} \tag{3}$$

An electron starting from the optic axis will cross it again after it has completed one circle – in other words, the electrons starting from the initial plane are all imaged in the latter plane. For a homogeneous magnetic field it can be shown that the focal length f is determined by

$$\frac{1}{f} = \frac{e}{8m\phi^*} B^2 L \tag{4}$$

where $\phi^* = \phi(1 + e\phi/(2E_0))$ is the (relativistically corrected) potential, ϕ the electrostatic potential, $E_0 = 511$ keV the rest energy of the electron, and L the length of the coil. As can be seen from Eq. 4, the focal length depends strongly on the magnetic induction $\boldsymbol{B}$. This general fact is also true for the commonly used "short magnetic lenses" that use an inhomogeneous magnetic field. The focal length of a magnetic electron lens in an electron microscope can therefore be easily modified by adjusting the current through the coil thus changing the magnetic field.

Unfortunately, electron lenses suffer from strong spherical and chromatic aberrations [3, 4] that limit the spatial resolution limit for conventional electron microscopes to about 0.15 nm. Just in the very last years it has become possible to correct for aberrations so that with the most advanced instruments on can achieve a resolution limit of about 0.05 nm.

As the focal length of every lens can be easily varied by changing the lens current, the ray paths in a transmission electron microscope can be chosen within a wide range. In particular, it is easily possible to change the illumination from a large area to a small spot mode. Whereas the illumination of a large area is used for conventional imaging, the small spot mode allows the investigation of small areas whose size may well approach the 0.1 nm level. By changing the excitations of the projector lenses it is easily possible to vary the magnification over a large range.

Image Formation in the Transmission Electron Microscope

Figure 3 shows schematically the image formation process for a periodic specimen. The incident electron beam has to be treated quantum mechanically as a wave that is diffracted by the periodic object. Constructing the ray paths in the usual way, one finds that all the rays diffracted in a particular direction are recombined in one point in the back focal plane of the objective lens. Rays emitted parallel to the optic axis are focused in the focal point. In other words, in the back focal plane the electrons are selected according to the scattering angles or, to put it differently again, one obtains a diffraction pattern in the back focal plane. Therefore the back focal plane is often called the diffraction plane.

For a general, non-periodic specimen the diffraction pattern consists of a continuous distribution rather than of individual spikes. Its value is proportional to the squared modulus of the Fourier transform of the electron wave just below the specimen.

In the case of ideal imaging all rays originating in one point of the specimen will be refocused into one point in the image plane. As in light optics, the image formation in an electron microscope can be viewed as a sequence of two Fourier transforms – the first one between the specimen and the back focal plane and the second one between the back focal and the image plane. Consequently any manipulation in the back focal plane will influence the final image.

One example is the insertion of a circular aperture which prevents electrons scattered by an angle larger than α_0 from entering into the image. If the first diffraction angle of a periodic specimen is larger than α_0, only the undiffracted beam will enter into the image. In that case the diffraction amplitude beyond the aperture is identical to the amplitude without any specimen and thus one obtains an image without any structure. This example demonstrates that the resolution is then limited by the maximum angle passing through the objective aperture. Using Bragg's law

$$2a \sin\theta = n\lambda \tag{5}$$

where a is the lattice spacing and θ the scattering angle for *n-th* order diffraction, one obtains for the lattice resolution limit d

$$d \approx \frac{\lambda}{2 \sin\alpha_0} \tag{6}$$

Contrast Mechanisms in Transmission Electron Microscopy

As the fast incident electrons pass through the thin specimens used for transmission electron microscopy

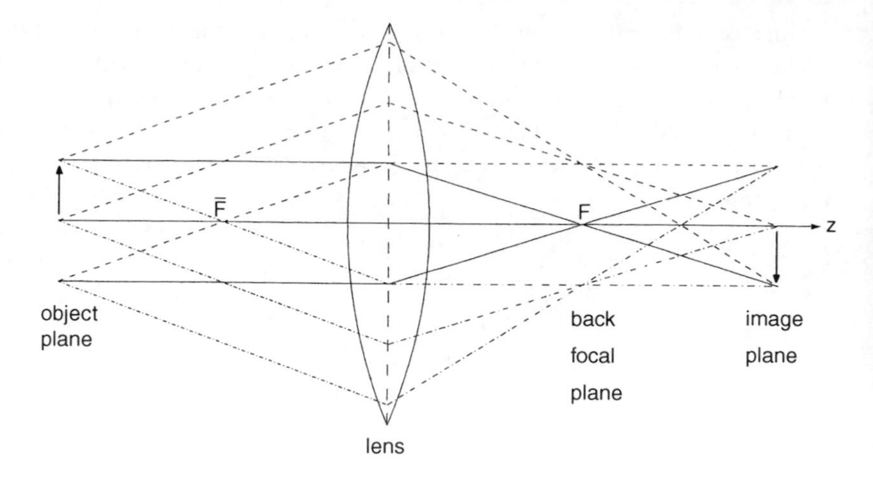

Transmission Electron Microscopy, Fig. 3 Ray paths showing the imaging by a lens. All rays emitted by the specimen in a given direction are focused to an off-axis point in the back focal plane

there is practically no absorption. Therefore an "ideal" image (without any aperture or any other optical element in the back focal plane) does not exhibit any contrast. In order to see anything in the image, the electron wave in the back focal plane has to be influenced in order to discern the structure of the specimen. This can be done in several ways leading to different types of contrast.

Scattering Absorption Contrast

A circular diaphragm, inserted into the back focal plane, intercepts all the electrons scattered at angles larger than $\alpha_0 = r/f$, where r is the radius of the hole in the aperture and f the focal length of the objective lens. Strong scatterers, who scatter a substantial fraction of the incident electrons into angles larger than α_0, appear darker in an otherwise bright image. It has to be stressed that no electrons are absorbed in the specimen. The scattering takes place in the specimen, the absorption in the diaphragm in the back focal plane. The contrast is strongest for small aperture angles, because then a large fraction of the scattered electrons will be removed from the beam. Equation 6 shows that for small scattering angles the resolution deteriorates. Even though this method is very simple, it can only be used for low and medium resolution images (coarse structures). For practical applications a suitable compromise has to be found between contrast and resolution.

Dark Field Imaging

To avoid the conflict between resolution and contrast one can use a completely different way to obtain contrast. One major problem of scattering absorption contrast is the fact that in the image one sees only a minor decrease of intensity, which is hard to discern, in an otherwise bright image. Therefore it is of great interest to visualize strong scatterers as bright spots. To form such an image the unscattered electrons have to be removed so that only the scattered electrons remain in the image. This can be done by use of a beam stop in the back focal plane intercepting all unscattered electrons that have left the specimen in a direction parallel to the optic axis. One way of doing this is to mount the beam stop on a thin wire and place it in the diaphragm in the back focal plane. It is difficult, however, to properly adjust the beam stop.

The essential requirement for the dark field imaging mode is that the unscattered electrons are removed in the back focal plane. As in light microscopy, this can also be achieved by inserting a circular diaphragm in the back focal plane and using a hollow-cone illumination whose angle is larger than the objective aperture angle α_o. In electron microscopy such an illumination can be most easily obtained by using deflection coils in the condenser system to tilt the incident beam and rotate the direction of incidence by properly addressing the deflection coils.

A dark field image exhibits bright spots at the positions where scattering is strong on an otherwise dark background. The resolution is determined by the aperture angle α_0. Usually only a small fraction of the incident electrons is scattered. Therefore the intensity in dark field images is rather low.

Phase Contrast Imaging

In quantum mechanics, the influence of a thin specimen on the incident electron wave can be described by a spatially varying phase shift φ, which is proportional to the projected potential $V(\vec{\rho}, z)$

$$\varphi(\vec{\rho}) = \frac{-i}{\hbar v} \int_{-\infty}^{\infty} V(\vec{\rho}, z)dz, \tag{7}$$

where $\vec{\rho}$ is a two-dimensional vector describing the lateral position in the object plane and $\hbar$ is Planck's constant.

Is it possible to obtain an image whose intensity is proportional to the phase shift? In light optics this can be achieved using Zernikes phase contrast method using a phase plate in the back focal plane to introduce a phase shift of $\pi/2$ between the unscattered and the scattered beam. To achieve this, one has to fabricate a glass plate of the proper thickness. How can this effect be utilized in electron microscopy?

As mentioned earlier, electron lenses suffer from severe spherical aberrations that correspond to a phase shift increasing with scattering angle. Including the effect of a small defocus Δf, this phase shift γ is given by

$$\gamma(\theta) = k\left\{\frac{C_s}{4}\theta^4 - \Delta f\frac{\theta^2}{2}\right\} \tag{8}$$

where C_S is the spherical aberration constant, $k = 2\pi/\lambda$ the wave number of the electrons and θ the scattering angle. By a proper choice of the defocus, one can obtain a phase shift of approximately $\pi/2$ over a large range of scattering angles. This is shown in Fig. 4. To optimize phase contrast, the defocus is chosen so that γ stays close to $\pi/2$ for a large range of scattering angles.

The performance of this method can best be described by use of a transfer function. Such transfer functions are well known in electronics, where they determine to what extend an electric signal of a given frequency f is transmitted through a circuit. To describe the transmission of a general electric signal, the latter is decomposed into its frequencies – in other words, it is Fourier transformed. The concept of a transfer function can readily be generalized to two-dimensional spatial signals. Due to the circular symmetry of an aligned microscope, the transfer function depends only on the modulus of the spatial frequency θ/λ. For phase contrast imaging, the "phase contrast transfer function" is given by the sine of γ

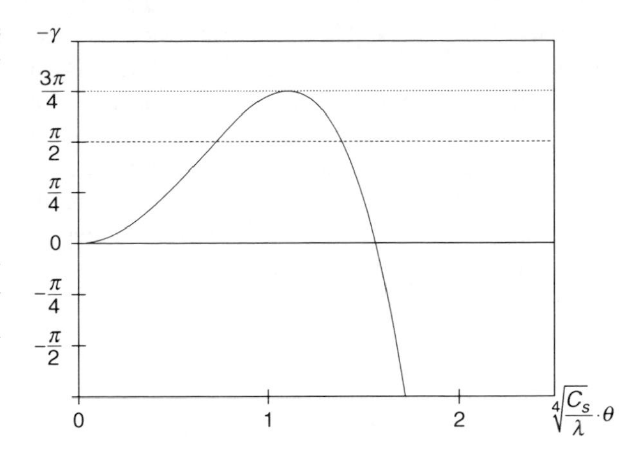

Transmission Electron Microscopy, Fig. 4 Phase shift γ introduced by the spherical aberration of the objective lens

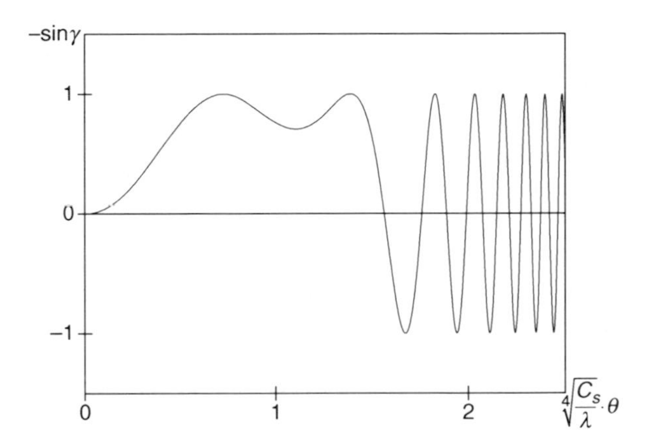

Transmission Electron Microscopy, Fig. 5 Transfer function $L(\theta) = -\sin\gamma(\theta)$ for phase contrast imaging

$$L(\theta) = -\sin\gamma(\theta) \tag{9}$$

Figure 5 shows its behavior for the optimum defocus, the so-called Scherzer focus

$$\Delta f \approx 1.2\sqrt{C_s\lambda} \tag{10}$$

The first zero of the phase contrast transfer function determines the resolution limit d, which is given by

T

$d = 0.67\sqrt[4]{C_s\lambda^3}$. It can be shown, that spherical aberration is unavoidable for electric-magnetic round lenses, provided they are time independent and that there is no electrostatic charge within the lens. Typical values of the spherical aberration constant C_S are about 1 mm. Inserting $C_S = 1$ mm and assuming an accelerating voltage of 300 kV, corresponding to a wave length $\lambda = 20$ pm, one obtains a resolution limit of $d = 0.2$ nm. One can thus obtain high-resolution images with good contrast.

High-Resolution Electron Microscopy

So far the discussion has been confined to thin objects, where multiple scattering can be neglected. Frequently, high resolution electron microscopy is applied to crystals oriented so that the incident electron beam is parallel to a low index axis [5]. Often the resolution is then sufficient to discern atomic columns. In such crystals, however, multiple elastic scattering plays an important role. These processes strongly influence the intensity distribution in the image. It is therefore no longer possible to interpret high-resolution images of crystals directly. Instead, one has to simulate an image of a model crystal, compare the result with the experimental data, modify the model, recalculate, compare, ... until a good match between the simulated and the experimental image has been achieved.

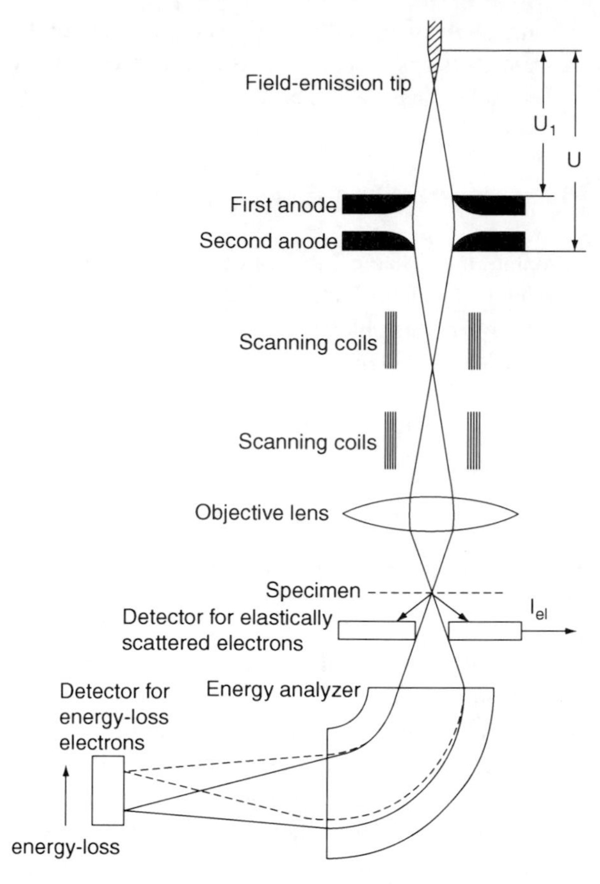

Transmission Electron Microscopy, Fig. 6 Field emission scanning transmission electron microscope equipped with an energy loss spectrometer [1]

Correctors

Due to the spherical aberration of the objective lenses, the resolution limit of electron microscopes has stayed at values well above 0.1 nm for decades. The correction for this aberration has been a challenging task, which has been achieved only very recently by using lenses without rotational symmetry [4]. Nowadays several correctors using sextupole or quadrupole and octupole lenses are commercially available, some of which do not only correct for the spherical aberration, but also for chromatic aberration. In this way a resolution limit of 0.05 nm can be achieved.

Scanning Transmission Electron Microscopy

It is also possible to obtain images by focusing the beam to a small spot that is scanned in a raster like manner over the specimen. The number of counts, e.g., of the scattered electrons, is then stored as intensity value for that pixel. The setup of a scanning transmission electron microscope (STEM) is schematically shown in Fig. 6. To obtain an image in a STEM, one commonly uses an annular detector subtending angles outside the cone of the incident rays. This produces a dark-field image, hence the detector is called an annular dark-field (ADF) or, for a large inner angle, a high-angle ADF (HAADF) detector. A major advantage of scanning transmission electron microscopy is the possibility to use other secondary signals to determine the intensity in the corresponding pixel of the image. In particular, the use of an element specific signal (e.g., X-rays of a particular transition) permits to acquire images showing the distribution of that element within the specimen. Whereas dedicated STEMs are not very widely spread, many modern TEMs have a scanning attachment permitting to

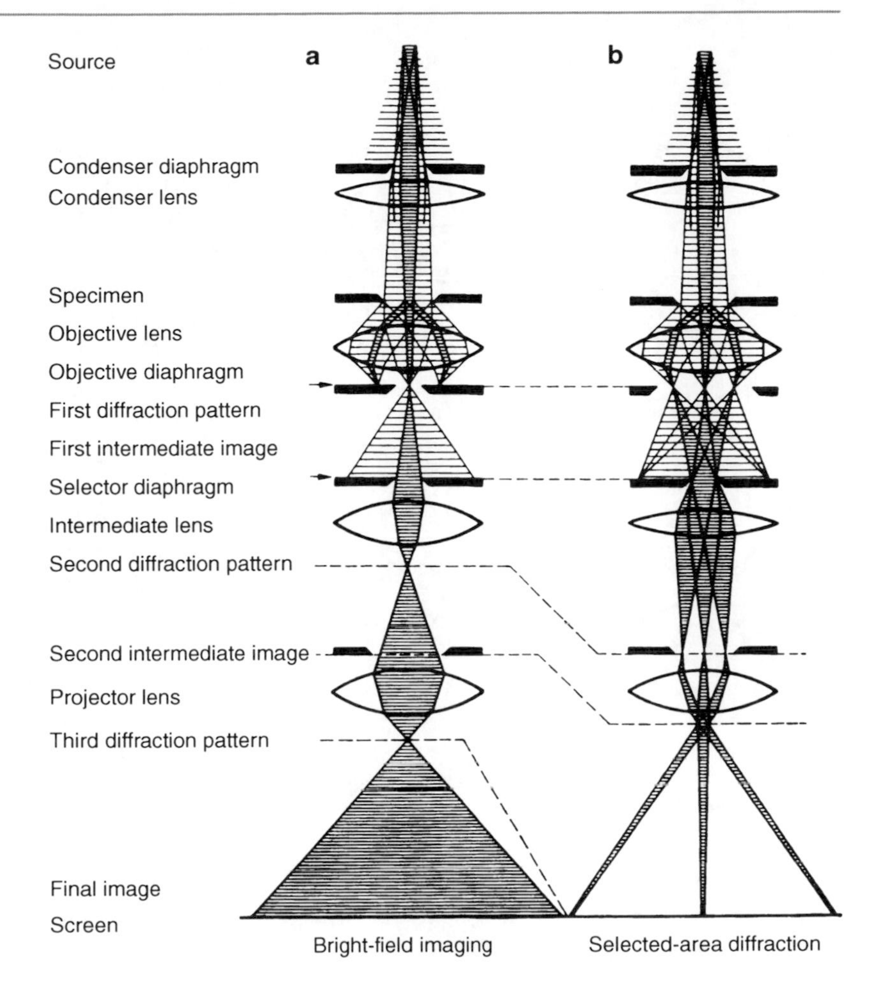

Transmission Electron Microscopy, Fig. 7 Ray diagram for a TEM for (**a**) bright field imaging and (**b**) diffraction [1]

perform scanning transmission electron microscopy in addition to the conventional operating modes.

Electron Diffraction

In the previous chapters the various possibilities to obtain an image in an electron microscope have been discussed. By adjusting the intermediate lens, which is the first lens in the projector system, it is easily possible to switch to the diffraction mode, where the back focal plane, i.e., the diffraction pattern, is imaged on the final screen [6, 7]. Essentially in this mode the second and further intermediate image planes and diffraction planes are interchanged as can be seen from Fig. 7.

Using diffraction mode one can obtain a wealth of information on the structure of a crystallite with any standard TEM. To obtain the position of the diffraction spots, one can use Laues equation which states that the difference between the wave vector $\mathbf{k}_0$ of the incident and $\mathbf{k}_f$ of the diffracted beam must be equal to a reciprocal lattice vector $\mathbf{g}$

$$\mathbf{k}_f - \mathbf{k}_0 = \mathbf{g} \tag{11}$$

Every spot in a diffraction pattern corresponds to a reciprocal lattice vector. Electron diffraction thus permits to measure the reciprocal lattice, which in turn allows determining the Bravais lattice of the illuminated crystallite. The position of the spots can most easily be visualized using Ewalds construction (Fig. 8). There are two conditions for the appearance of a diffraction spot:

1. The energy of the electrons, and therefore the magnitude of their wave vector, remains unchanged during the scattering process (elastic scattering). This can be visualized by drawing a sphere around the origin of the wave vector of the incident electron

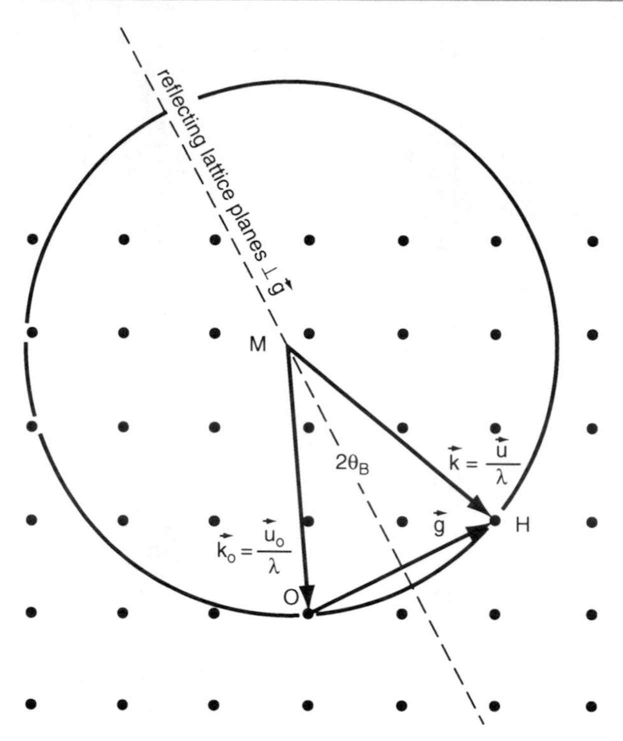

Transmission Electron Microscopy, Fig. 8 Ewald sphere construction to determine the directions of diffracted radiation [1]

with a radius corresponding to its length. The wave vectors of both the incident and the scattered electron must end on the surface of this so-called Ewald sphere.

2. Drawing the wave vector of the incident electron as to end in the origin of the reciprocal lattice, the wave vector of the scattered electron has to end in another reciprocal lattice point.

Putting these two conditions together, one finds that electrons can only be diffracted if the Ewald sphere crosses a reciprocal lattice point. This condition is equally valid for the diffraction of X-rays, neutrons and electrons. Fast electrons with energies of the order of 100 keV or more have wavelengths of a few pm – much smaller than the lattice constant of a crystal (about 1 nm). Consequently, the length of the wave vector of the electron beam is two or three orders of magnitudes larger than the length of a reciprocal lattice vector. In the area near the origin of the reciprocal lattice, the Ewald sphere can then be approximated by its tangential plane. Thus, Ewalds construction states that in electron diffraction near the central spot one observes a planar cut through the reciprocal lattice. The plane through the origin is called the zero order Laue zone. As an example, Fig. 9 shows the arrangements of spots for a face-centered cubic lattice illuminated along different zone axes. For larger scattering angles, this planar approximation is no longer valid and the Ewald sphere cuts through reciprocal lattice vectors of planes parallel to the zero order Laue zone, which are denoted as first, second, etc. order Laue zones.

So far electron diffraction has been discussed for using a parallel beam for illumination. This method results in a spot pattern giving information of the illuminated crystallite. If there are many randomly oriented crystallites in the illuminated area, one obtains Debye-Scherrer rings just like in powder diffraction patterns. The determination of the crystal potential, however, is more complicated due to the fact that in electron diffraction multiple scattering effects cannot be neglected. These depend strongly on the thickness of the crystal and on the exact direction of the incident beam. Instead of varying the incident beam direction with time, one can illuminate the crystal with a range of incident beam directions all at once. This technique is called convergent beam electron diffraction (CBED) (Fig. 10). Instead of the diffraction spots for parallel illumination, one then observes circles that exhibit an internal structure. Comparing the intensity distribution within these circles with the results of simulations, one can determine the crystal potential with high accuracy. As long as the illumination angle is smaller than the Bragg angle, the circles do not overlap and each one corresponds to a specific reciprocal lattice vector.

Analytical Electron Microscopy

Taking a broader view of the scattering of electrons in a thin specimen one can see, that in addition to the elastic scattering processes discussed up to now, many electrons experience inelastic scattering [1, 2, 6]. In doing so, they loose part of their energy. The various processes are shown schematically in Fig. 11. In order to excite an inner shell electron into an unoccupied state, the incident electron has to transfer an energy corresponding to at least the binding energy of the original inner shell electron (Fig. 11a). As these binding energies are characteristic for the element, they can be used as a signature of that particular element. Measuring the energy loss of the scattered

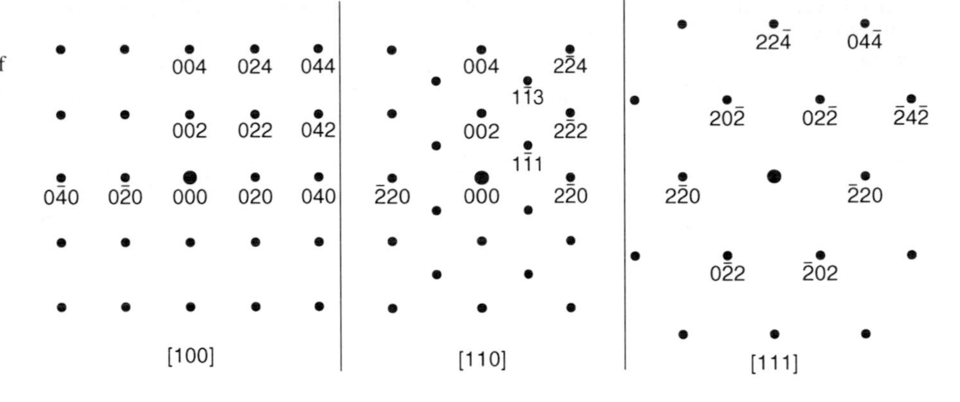

Transmission Electron Microscopy, Fig. 9 Diffraction pattern of a face centered cubic crystal illuminated along the [100], [110], and [111] zone axes near the central spot [1]

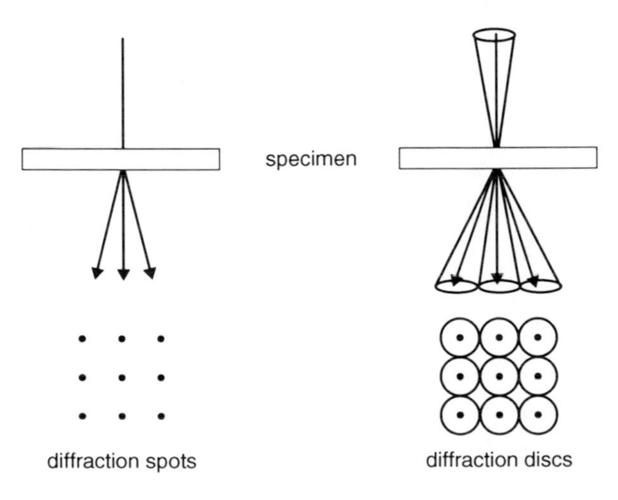

Transmission Electron Microscopy, Fig. 10 Schematic diagram demonstrating the effect of a converging beam on the diffraction pattern (CBED)

electrons thus permits to determine the chemical composition of the illuminated area of the specimen. This is the basis of a technique called EELS for electron energy loss spectroscopy. Shortly after its excitation, the atomic electron decays into its ground state releasing the energy via an X-ray photon (Fig. 11b) or an Auger electron (Fig. 11c). Both of these secondary processes can in principle be used to determine the chemical composition of the specimen. Due to the small mean free path length, the emission of Auger electrons is very surface sensitive. As their detection requires an UHV environment, they are frequently used in surface science, but not in transmission electron microscopy. X-ray spectroscopy, however, is a very common technique in electron microscopy.

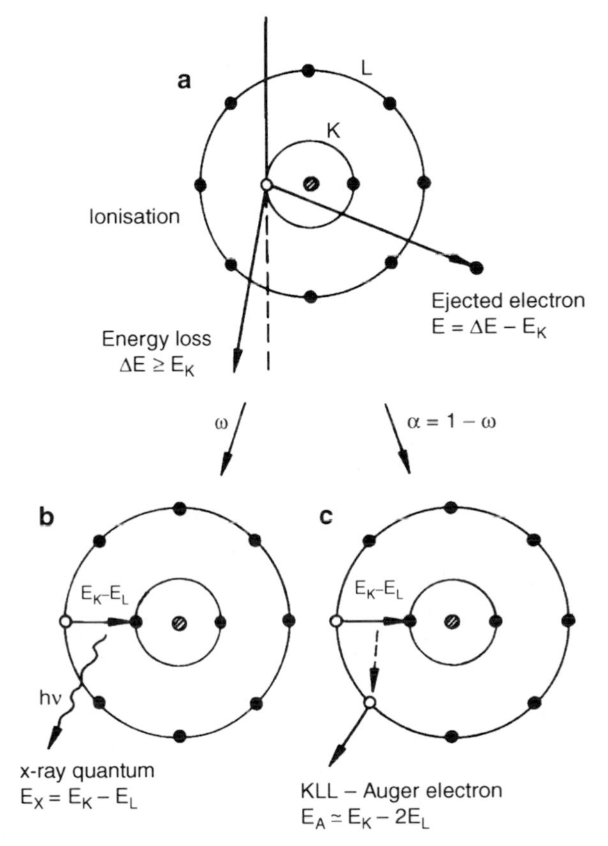

Transmission Electron Microscopy, Fig. 11 Schematic representation of (**a**) the ionization process, (**b**) the x-ray emission and (**c**) the Auger emission [1]

Electron Energy Loss Spectroscopy

Using a monoenergetic incident beam and measuring its energy distribution after it has passed through a thin specimen, one obtains an energy loss spectrum [8].

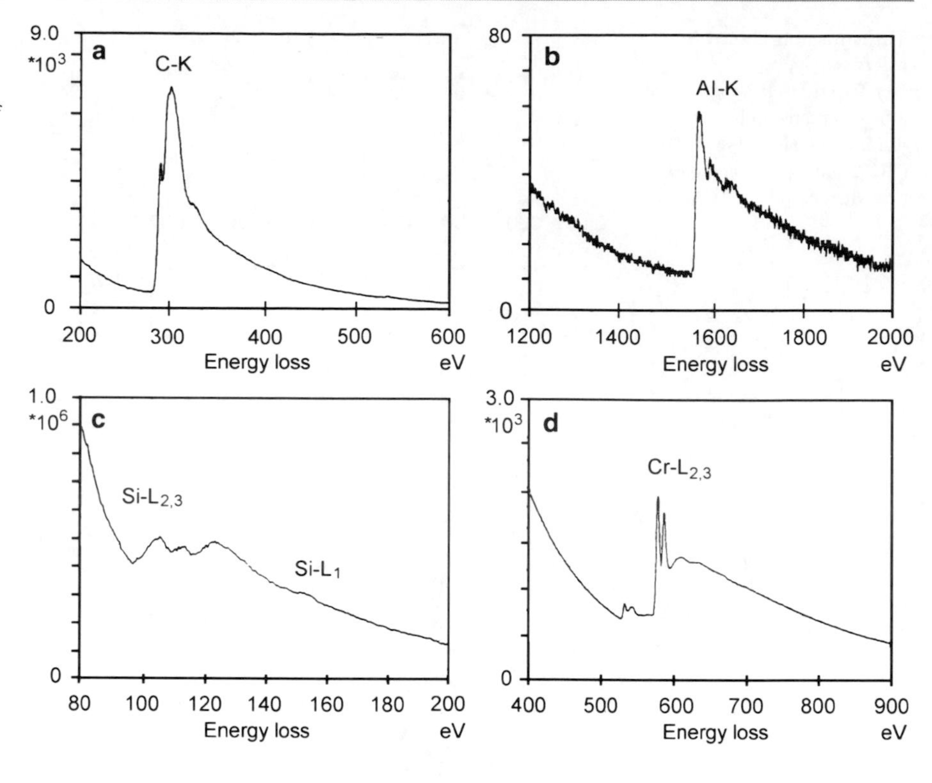

Transmission Electron Microscopy, Fig. 12 Examples of energy loss spectra for the K-edges of (**a**) carbon, (**b**) aluminum and the L23-edges of (**c**) silicon and (**d**) chromium [1]

One example is shown in Fig. 12. At characteristic energies, corresponding to the binding energy of an inner shell, the signal shows a marked increase followed by a slow decay. The latter is due to the fact that it is also possible to excite an electron into any state above the Fermi level. The excitation probability, however, decreases for final states with relatively large energies. In order to obtain a quantitative measure of the concentration of an element in the specimen, the background stemming from the tails of edges with lower energies has to be extrapolated. This is most often done by using a power law model for the background intensity I_b as a function of the energy loss E

$$I_b = \mathrm{AE}^{-r} \tag{12}$$

where A and r are free parameters that need to be fitted to experimental data. The signal intensity remaining after background subtraction is proportional to the area density of the element N_A, the intensity of the incident beam I_0, and the excitation cross section $\sigma_A(\alpha_0, \Delta E)$, which depends on the acceptance angle α_0 and the energy loss window ΔE. For two elements A and B one obtains the two equations

$$I_A = N_A I_0 \sigma_A(\alpha_0, \Delta E) \tag{13}$$

and

$$I_B = N_B I_0 \sigma_B(\alpha_0, \Delta E) \tag{14}$$

Dividing Eq. 13 by Eq. 14 one obtains

$$\frac{I_A}{I_B} = \frac{N_A}{N_B} \frac{\sigma_A(\alpha_0, \Delta E)}{\sigma_B(\alpha_0, \Delta E)} \tag{15}$$

and for the concentration ratio

$$\frac{N_A}{N_B} = \frac{I_A}{I_B} \frac{\sigma_B(\alpha_0, \Delta E)}{\sigma_A(\alpha_0, \Delta E)} \tag{16}$$

The concentration can thus easily be determined from the background corrected intensities above the energy loss edges multiplied by the inverse of the ratio of the cross-sections. To measure an energy loss spectrum it is most convenient to attach a magnetic sector field spectrometer underneath the image plane of a transmission electron microscope. Modern commercial instruments allow measuring a large part of the spectrum in parallel with a diode array.

Using such a spectrometer in a scanning transmission electron microscope it is possible to obtain the spatial distribution of the elements in the specimen for each pixel. This technique results in images called elemental maps with a resolution at the atomic level in favorable cases.

Alternatively one can obtain elemental maps by using an imaging energy filter, which turns the TEM into an energy filtering transmission electron microscope (EFTEM). Such a filter is inserted in the ray path after the first part of the projective system and intercepts all electrons except for those lying within a well defined energy window. Ideally such an energy filter should provide an image without any chromatic aberration, but at the same time it needs an energy dispersion to be able to select the desired energy range. These two conditions are met by a setup which images an intermediate image plane – the filter entrance plane – into an achromatic image plane and images the corresponding diffraction plane – the crossover plane – into the energy dispersive plane (Fig. 13). In the achromatic image plane the energy loss is "coded" by a tilt proportional to the energy loss resulting in a spatial displacement in the corresponding diffraction plane. The energy selection takes place in this energy dispersive plane.

To obtain an elemental map in an EFTEM, one takes a series of at least three images at different energy losses – two below and one above the edge. With the images below the edge one determines the parameters A and r on a pixel by pixel basis, from these the background intensity is computed and subtracted from the image above the edge (Fig. 14).

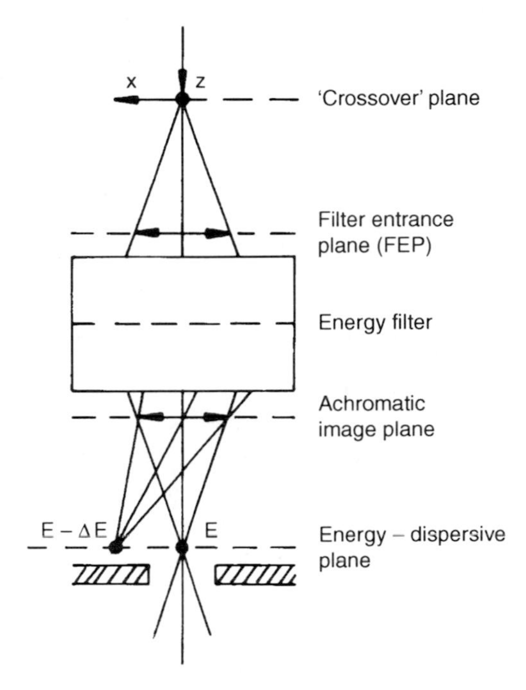

Transmission Electron Microscopy, Fig. 13 Schematic diagram showing the ray paths and the relevant planes in an imaging energy filter [1]

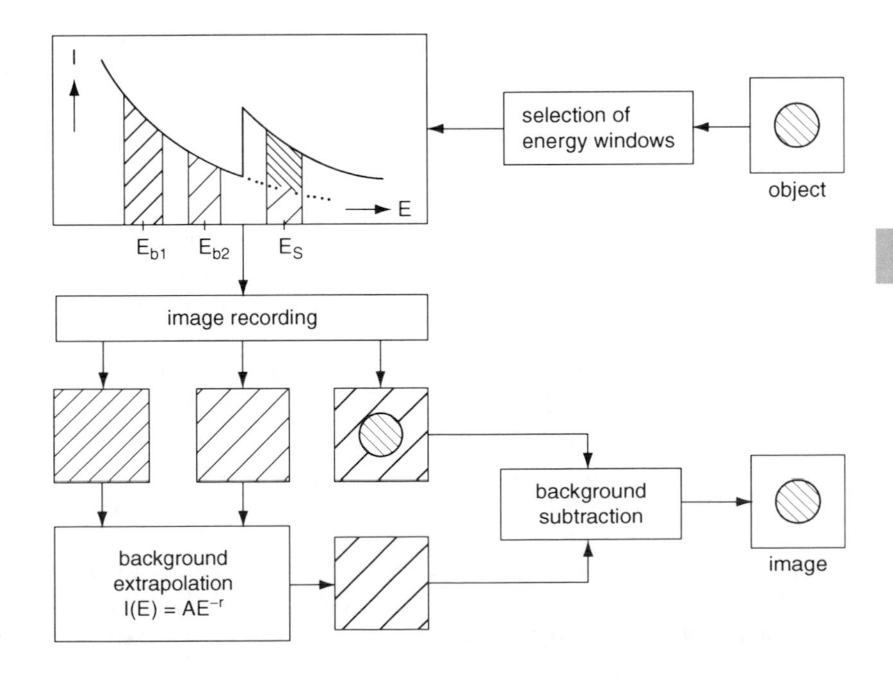

Transmission Electron Microscopy, Fig. 14 The steps necessary to acquire an elemental map

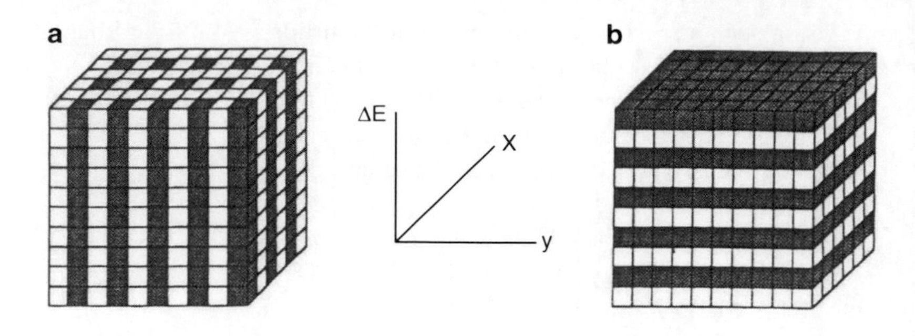

Transmission Electron Microscopy, Fig. 15 The data cube depicting the complete information that can be explored in different ways. (**a**) A STEM acquires spectra point by point (**b**) In EFTEM, the information is obtained energy slice by energy slice

The total information that can be obtained via EELS in a STEM or in an EFTEM is conveniently visualized as a data cube with two spatial coordinates and one energy loss coordinate (Fig. 15). In a STEM many energy channels corresponding to one pixel are measured at the same time (Fig. 15a), whereas in an EFTEM one measures the entire picture for one energy channel (Fig. 15b). The efficiency of both methods depends on the number of pixels and energy channels to be acquired.

Taking a closer look at a spectrum just above an edge, one finds fine details in the structure that depend on the chemical environment. These details can be used to determine the bonding of the element. The energy loss near edge structure (ELNES) is similar to the X-ray absorption near edge structure (XANES).

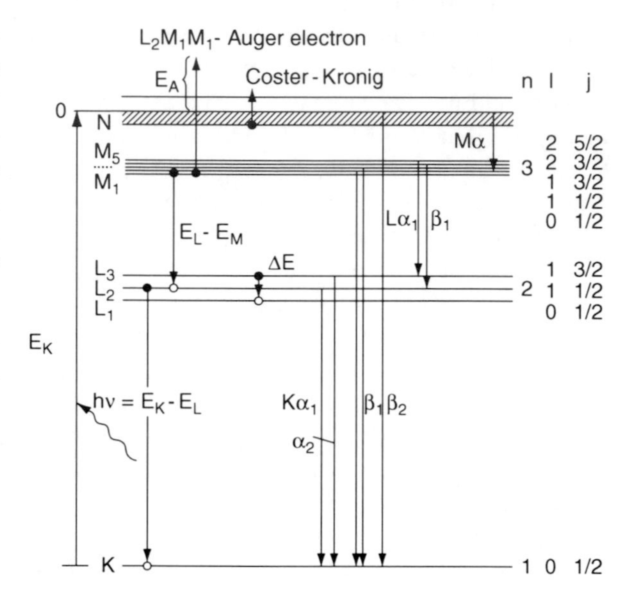

Transmission Electron Microscopy, Fig. 16 Energy levels of the atomic subshells with their quantum numbers n, l, j and possible transitions to fill a hole in the K-shell including the nomenclature of the emitted x-ray lines [1]

Energy Dispersive X-Ray Spectroscopy

Rather than detecting the inelastically scattered electrons, it is also possible to use secondary processes for the determination of the elemental distribution [6]. This is the basis of x-ray spectroscopy for elemental analysis. The inner shell levels of an atom and the corresponding transitions are shown schematically in Fig. 16. The energy difference between the excited and the ground state can be released either by emission of an x-ray or an Auger electron. The probability for the emission of an Auger electron, called Auger yield, and the corresponding fluorescence yield are shown in Fig. 17 for K, L, and M-shell transitions. From these graphs one can clearly see that for large atomic numbers the fluorescence yield dominates.

To determine the chemical composition of the specimen one needs to measure the energy spectrum of the emitted x-rays. This can be done either by using Bragg reflection from a crystal spectrometer to obtain the wavelength or by measuring the number of electron–hole pairs created after absorption of the x-ray photon in a semiconductor detector. Whereas the use of the first method, called wavelength dispersive x-ray spectroscopy (WDX or WDS) is confined to electron microprobes and scanning electron microscopes, the latter method, named energy dispersive x-ray spectroscopy (EDX or EDS) is used in transmission electron microscopes as well as in scanning electron microscopes. An example of an EDX-spectrum, taken with a high-purity Ge-detector, is shown in Fig. 18. The position of the lines reveals the elements present in the illuminated area of the specimen. To determine the chemical composition, one has to establish a relation between the intensity of the x-ray line and the number of corresponding atoms. Even though there are many factors entering into this relation (angular range subtended by the detector, fluorescence yield, detection efficiency, . . .), it is obvious, that the number of counts is proportional to the concentration of the element. For two elements A and B one can lump all these unknown factors into

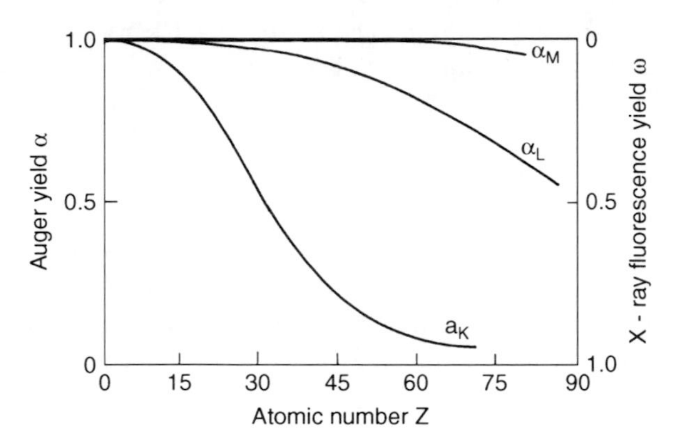

Transmission Electron Microscopy, Fig. 17 Auger electron yield *a* and fluorescence yield $\omega = 1-a$ as a function of atomic number [1]

the two parameters k_A and k_B thus obtaining for the counts I_A and I_B

$$I_A = k_A c_A \text{ and} \tag{17}$$

$$I_B = k_B c_B. \tag{18}$$

Dividing Eq. 17 by Eq. 18 one obtains

$$\frac{C_A}{C_B} = k_{AB} \frac{I_A}{I_B} \tag{19}$$

where the "k-factor" $k_{AB} = k_B/k_A$ is a parameter that needs to be determined. This can be done experimentally by use of well defined standard specimens. In practice one can measure the k-factors with respect to one element, e.g., Fe or Si, and use the relation

$$k_{AB} = k_{AC}/k_{BC}.$$

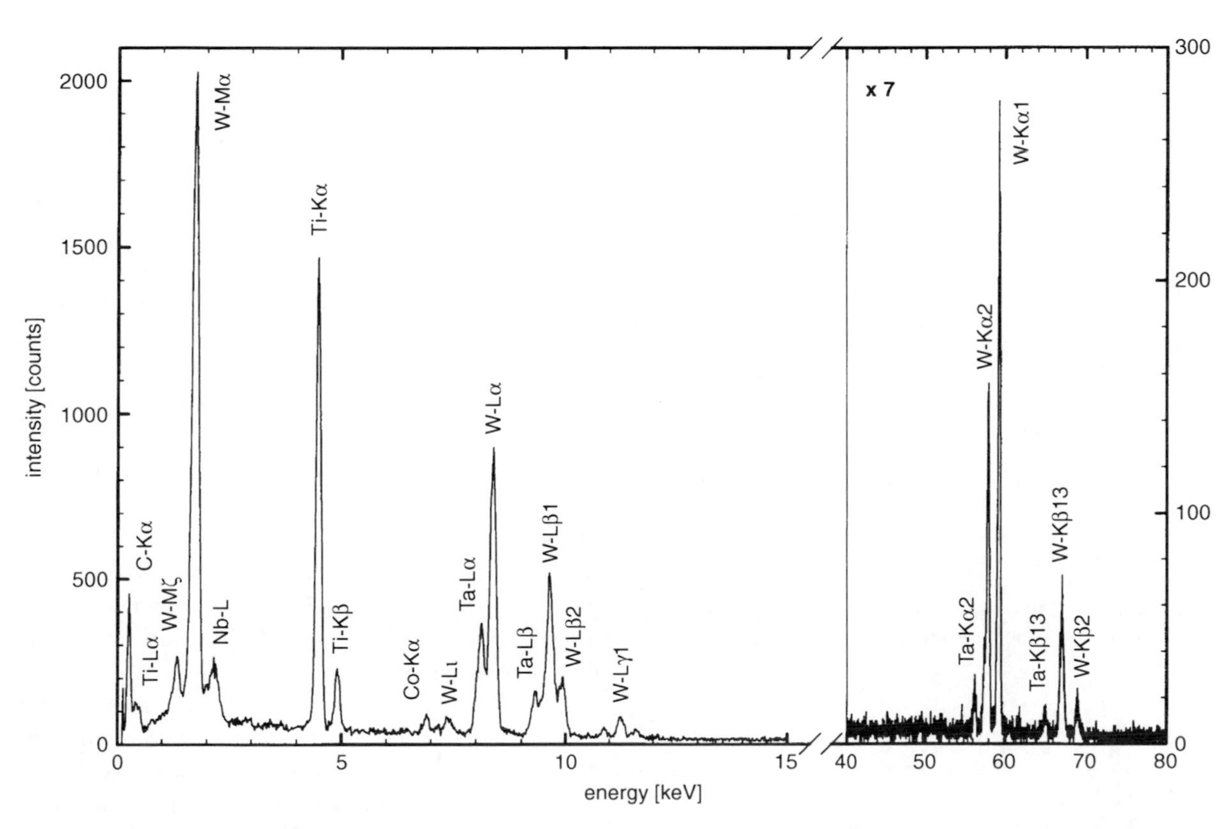

Transmission Electron Microscopy, Fig. 18 Energy dispersive x-ray spectrum of a hard metal (WC-TaC-TiC-Co) recorded with a high-purity germanium detector [1]

This should be done once for every instrument and can then be used as long as the conditions remain the same.

Comparing the limits of EELS and EDX for the detection of an element, one finds that the dependence of the fluorescence yield on the atomic number favors EELS for light elements. The relatively large background in EELS measurements as compared to an X-ray line indicates that the detection of low concentrations is easier with EDX than with EELS. On the other hand, the fine structure in an EELS spectrum yields additional information about the chemical bonding not accessible in an EDX spectrum. Thus both methods have their merits and their drawbacks and the choice strongly depends on the specimen and the information to be gained from it.

Cross-References

- ► Electron Microscopy of Interactions Between Engineered Nanomaterials and Cells
- ► Field Electron Emission from Nanomaterials
- ► Scanning Electron Microscopy

References

1. Reimer, L., Kohl, H.: Transmission Electron Microscopy. Springer, New York (2008)
2. Williams, D., Carter, B.: Transmission Electron Microscopy. Plenum, New York (1996)
3. Hawkes, P., Kasper, E.: Principles of Electron Optics, vol. I & II. Academic, London (1989)
4. Rose, H.: Geometrical Charged-Particle Optics. Springer, Berlin (2009)
5. Spence, J.: High-Resolution Electron Microscopy. Oxford, Oxford (2003)
6. Loretto, M.H.: Electron Beam Analysis of Materials. Chapman & Hall, London (1994)
7. Morniroli, J.P.: Large-Angle Convergent Beam Electron Diffraction (LACBED). Société Française des Microscopies, Paris (2002)
8. Egerton, R.: Electron Energy-Loss Spectroscopy in the Transmission Electron Microscope. Plenum, New York (1996)

T-Rays

► Terahertz Technology for Nano Applications

Tuning Electrical Properties of Carbon Nanotubes via Spark Plasma Sintering

Keqin Yang, Dale Hitchcock, Jian He and Apparao M. Rao
Department of Physics and Astronomy, Clemson University, Clemson, SC, USA

Definition

The spark plasma sintering technique utilizes highly localized intensive Joule heating and discharging generated by a self-adjusted high-density current flow to alter the micromorphology of carbon nanotubes.

Introduction

Carbon comes in many forms, ranging from *sp*-bonded metastable carbene chains, sp^2-bonded graphite, graphene, single wall and multiwall nanotubes, fullerenes, and amorphous soot to sp^3-bonded diamond. In particular, carbon nanotubes (CNTs) exhibit unique electrical, magnetic, thermal, chemical, mechanical, and optical properties that stem from their tubular topology and reduced dimensionality. Topologically, a single-walled nanotube (SWNT) can be viewed as one sp^2-bonded graphene sheet rolled into a seamless tube. A sibling of the SWNT is the multiwalled nanotube (MWNT), made of a few to tens of concentric shells with random chiral registries, spaced by ~0.34 nm and coupled through van der Waals interactions as in graphite. In the past few decades, CNTs have been extensively studied from fundamental research and technological points of view. With a nm scale diameter and a large aspect ratio, CNTs are one of the closest material realizations of a 1-D system, offering unprecedented opportunities to study 1-D physics. Furthermore, CNTs are prime candidates for molecular electronics, one of the most promising directions of nanotechnology.

The electrical properties of CNTs are sensitive not only to intrinsic parameters such as the number of shells, tube diameter, and chirality but also to extrinsic parameters such as defects, dopants, and the environment in which they reside [1]. For example, devices based on macroscopic forms of CNTs such as arrays, yarns, and films are gaining intensive interest. In these devices, the primary factor governing the macroscopic electrical transport is the micromorphology, comprised of three basic aspects, (1) the crystal chemistry of an individual CNT such as chirality, defects, and dopants; (2) the topology of the interconnected CNTs; and (3) the density and nature of inter-tube junctions. It is often the case that aspects (2) and (3) matter more than (1) in determining the electrical properties of a macroscopic CNT network. Control of these intrinsic and extrinsic parameters remains challenging and relies on techniques capable of tailoring CNT structures on the atomic, nm, and μm scales.

Tailoring CNT Morphology to Realize New Functions

Many methods have been used to alter the structure of CNTs, from simply hot pressing to more advanced techniques such as electron or ion irradiation, which gives rise to a seemingly endless number of new phenomena and potential functions. In an early attempt to densify CNTs, Ma et al. employed hot pressing at 2,000 °C and 25 MPa in an argon atmosphere to form a "soft sintered" CNT network without altering the tubular morphology of CNTs [2]. In addition, Rubio et al. showed that it is possible to cut and introduce defects into CNTs with very high precision using an STM tip, which has the salient advantage of real-time monitoring of the change of structure [3].

Using Irradiation to Tailor CNT Morphology

More recent research has focused on altering the structure of CNTs using electron or ion irradiation. How would CNTs respond if atoms were removed from the hexagonal lattice continuously [4]? At first glance, irradiating CNTs should lead to a detrimental effect on their crystal structure and thereby properties; however, it has been demonstrated that proper irradiation conditions can have beneficial effects on the properties of CNTs. This is due mainly to the CNT's innate ability to self-heal. The lattice of CNTs can be rapidly rearranged to accommodate the formation of defects and restore a new local coherent structure from a state that is driven far away from equilibrium during the irradiation process. For example, vacancies in SWNTs lead to the formation of pentagons and heptagons while in MWNTs they result in inter-shell bonds. This extraordinary innate ability renders a way to tailor the nanostructure of CNTs and facilitates a controlled method of self-assembly [5]. The main advantage of using irradiation is to tailor the CNT structure on a nm scale, while the focused electron beam of an aberration-corrected scanning transmission electron microscope can even reach Å scale resolution. The major drawback of this approach is the low throughput of beam-modified CNTs. Lastly, it should be mentioned that when dangling bonds are formed in adjacent tubes it is often energetically favorable to form a junction between the tubes rather than the reconstruction of individual tubes. The goal of this chapter is to introduce the spark plasma sintering (SPS) process as an alternate approach for controlling inter- and intratubular bonding in CNT networks, at nm and μm scale lengths. In electron beam irradiation, it is known that only a small portion of the impinging electron energy can be transferred to and displace nuclei due to conservation of momentum. As such, a very high electron energy (on the order of 10–100 keV) is needed to alter the structure of CNTs. The SPS process utilizes a low-voltage high-density current and the accompanied localized intensive joule heating or discharging to modify CNTs. The SPS process can control the structure of CNTs in much the same way as irradiation but using macroscopic control rather than the microscopic control of irradiation.

Mechanisms of Spark Plasma Sintering

The SPS technique has been used widely in powder metallurgy, and densification of nanomaterials and nanocomposites [6]. The basic idea dates back to 1930s but was not put into wide use until recently, primarily due to the lack of applications and the high cost of equipment. SPS is a pulsed plasma discharge process that can generate intensive heating (up to 2,000 °C and above) in a short time (few minutes). The voltage is merely on the order of 1–10 V, while the current density is typically on the order 10^2-10^3 A/cm^2 and is highly concentrated at the inter-granular contact or interfacial boundaries. The SPS process features a high thermal efficiency because of the direct heating of the sintering mold and sintered materials, and it results in a homogeneously sintered and densified sample because of the "self-adjusting" uniform heating mechanism [7], surface decontamination, and surface activation. The densification mechanism is related to

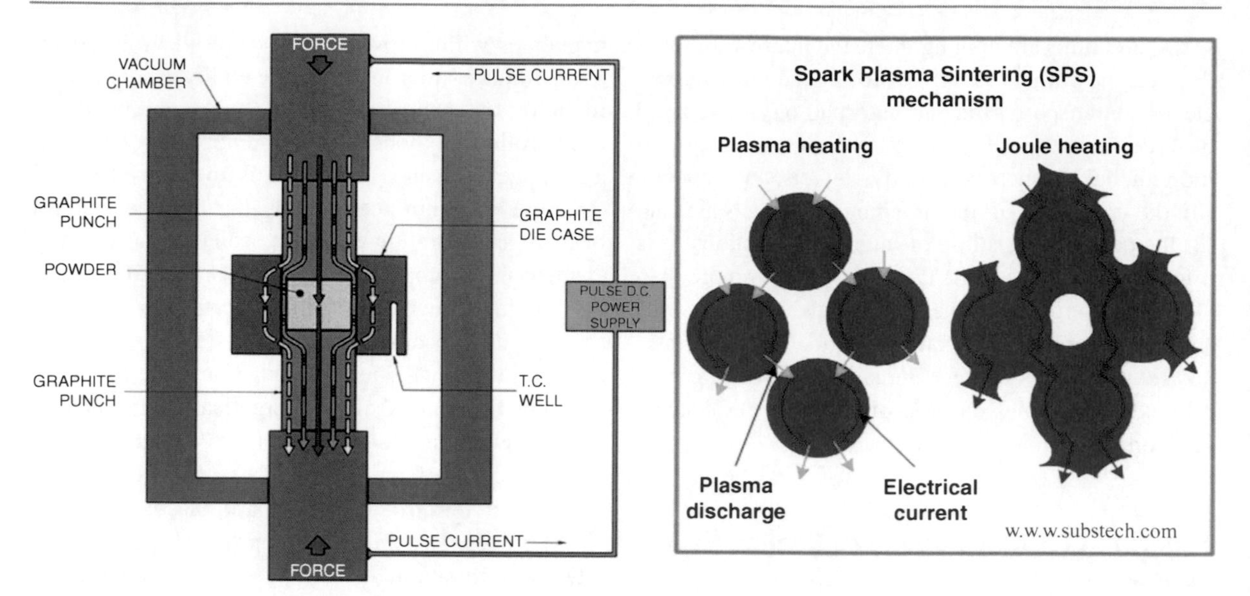

Tuning Electrical Properties of Carbon Nanotubes via Spark Plasma Sintering, Fig. 1 (*left*) The schematic of a SPS system, and (*right*) two major heating mechanisms in the SPS process

plastic deformation and grain rotation coalescence and sliding, aided by the pre-melting of grain boundaries. A schematic diagram of the SPS system and the primary heating mechanisms are shown in Fig. 1.

The control parameters of SPS process are macroscopic: they are the sintering pressure, current density, ON/OFF ratio, and sintering time. It should be noted that heating and current flow are interdependent but are regarded as separate mechanisms in the SPS process. The sintered materials are under a combination of one or more of the following effects [6]: spark plasma, spark impact pressure, Joule heating, and electric field–induced diffusion. In particular, the surface of sintered materials are more easily degassed and chemically activated, promoting strong mass transport at atomic, nano, and macroscopic levels at a relatively lower temperature and in a shorter time than conventional processes such as hot pressing. Mass transport in SPS is commonly believed to be driven by electric field diffusion. In view of the self-healing ability of CNTs, the current density in the SPS process can be varied to control the deviation from equilibrium, yielding a controllable number of permanent defects. At present, a comprehensive detailed model of the impact of SPS process on CNTs is lacking so that most ongoing work is experimental and exploratory in nature.

Using SPS to Tailor CNT Morphology

It has been shown that the SPS process may cause the bonding in CNTs to transform from van der Waals in a rope or yarn, to robust sp^2, then to sp^3 in microdiamonds [8], or lead to graphitization and crystalline–amorphous heterojunctions in amorphous carbon fibers [9]. Depending on whether CNTs survive the SPS process, the SPS process can be *loosely* categorized into two regimes. The first regime is the mid-power range, where most research is done. In this regime, the tubular morphology is preserved, and there is enough energy to introduce intratubular restructuring (including defects and doping) and create coherent intertubular bonding (i.e., junctions). At present, all experimental evidence suggests that the intra-tube restructuring and inter-tube bonding occur simultaneously during the SPS process. The process of densification also occurs in the mid-power regime. Despite the complexity of inter-tube interactions, the inter-tube spacing and relative orientation of CNTs are the two most important factors which determine the nature of the resulting inter-tube junctions and the packing density. Densification of CNTs is of practical importance in a number of applications such as, the electrodes for supercapacitors, biosensors, and artificial bone scaffolds in which one takes advantage of the high surface-to-volume ratio and the outstanding mechanical and electronic properties of CNTs.

The other regime is the high-power regime where the temperature and current density are high enough to destroy tubes, forming other carbon allotropes. The boundary between the mid-power range and high-power range is not clear-cut. Preliminary studies suggest a boundary temperature of 1,000–1,200°C for MWNTs while the boundary temperature is lower for SWNTs and depends more on the perfectness of CNT structure and environment.

This section mainly addresses the effects of SPS in the mid-power range, focusing on the SPS-mediated (1) intra-tube restructuring (defects and doping) and (2) formation of inter-tube and inter-shell junctions. These modifications are crucial for controlling the electrical and thermal properties of CNT networks. When a high-density current is pulsed though a CNT assembly, the contact resistance between tubes is much higher than the intrinsic resistance of the tubes, which leads to a highly localized intensive Joule heating or discharging, and the temperature can be sufficiently high to cause rearrangement of atoms/bonds.

Inter-tube and Inter-shell Junctions

It is known that inter-tube junctions dominate the electrical properties of a CNT network. An important objective of ongoing research is to study to what extent and in what aspects the transport properties of a SPSed network are inherited from individual CNTs. Therefore, the manipulation and formation of inter-tube junctions, especially coherent bonding, is of utmost importance in nanoelectronics (molecular electronics) and transparent conductors [10]. A direct impact of the formation of coherent inter-tube bonding is the change of the effective dimensionality of a CNT assembly. In particular, phonons and electrons may have different effective dimensionality in a SPSed CNT network. With coherent inter-tube junctions, the CNT assembly offers a flexible network architecture, opening up opportunities for several applications.

The modification of inter-tube bonding from van der Waals type to robust sp^2 inter-tube bonding is a nontrivial task, since the topologically complicated arrangement of sp^2 bonds in the inter-tube bonding is not energetically favored. For this reason, high-energy nonequilibrium processes are needed. One example of inter-tube junctions is the coalescence of single- or double-wall CNTs to form a single SWNT of larger diameter [4]. Charlier et al. have proposed a mechanism for the coalescence of tubes using molecular dynamics simulations. The model begins with unzipping of a nanotube to form a carbene chain, and if the metastable carbene chain comes in contact with a neighboring nanotube, it will initiate a reaction to minimize its energy and form a coalesced nanotube. Three steps in Charlier's model are shown in Figs. 2a–c. In spark-plasma-sintered aligned MWNTs, Yang et al. have observed that coherent inter-tube bonding can be formed at a sintering temperature of 1,500 °C for 7 min under 7 MPa axial pressure. The current is applied in the direction perpendicular to the tube direction. As shown in Fig. 2d, the SPS process welds part of two parallel MWNTs together [11].

Several interesting properties of SPSed MWNTs have been uncovered through electrical and thermal transport measurements [11]. First, the transverse resistivity (ρ_T) of a low-dimensional electron system such as CNTs is expected to be much higher than the longitudinal resistivity (ρ_L), as is the case in thin films made of aligned CNTs [12]. The MWNTs SPSed at 1,500 °C, however, showed lower resistivity in the transverse direction than the longitudinal direction (Fig. 3a). This surprising observation may be due to the following reasons: the SPS process creates coherent inter-tube bonds that account for the lower ρ_T; the SPS-induced defects within the tubes may lower ρ_L; and finally lower ρ_T may be due to the larger cross section of the conduction surface in the transverse direction than in the longitudinal direction due to the hollow core and inter-tube spacing of CNTs. Second, although MWNTs are typically metallic, the resistivity in both directions shows nonmetallic behavior ($d\rho/dT < 0$) as in aligned CNT think films, suggesting that inter-tube junctions dominate the overall electrical conductivity of the CNT network. The temperature dependence of the resistivity is also unusual, and cannot be satisfactorily interpreted in the frameworks of the Luttinger liquid model, variable-range hopping, or weak localization theory. Nonetheless, the temperature dependence of resistivity is reminiscent of the results of individual highly disordered MWNTs in which the $T^{1/2}$ dependence has been attributed to electron–electron scattering in the presence of random impurities [11]. Third, the thermopower is not only of different magnitudes in the transverse and longitudinal directions but also shows different temperature dependences, with the longitudinal thermopower showing

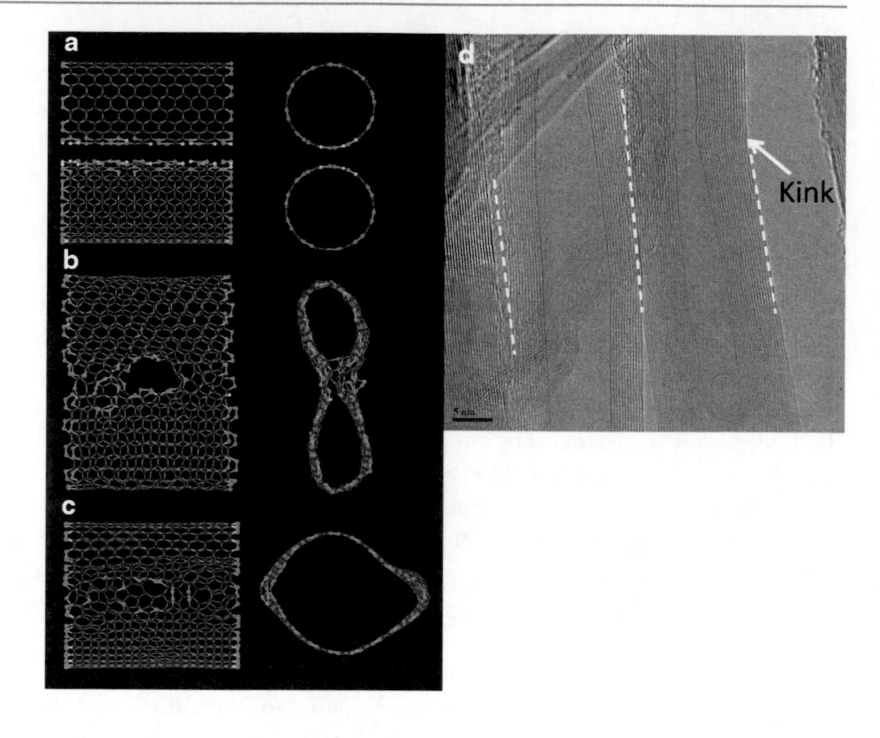

Tuning Electrical Properties of Carbon Nanotubes via Spark Plasma Sintering, Fig. 2 (**a–c**) Three stages from a theoretical model of the coalescence of two SWNTs. (**d**) shows that several MWNTs are welded with some degree of corrugations in the sidewalls

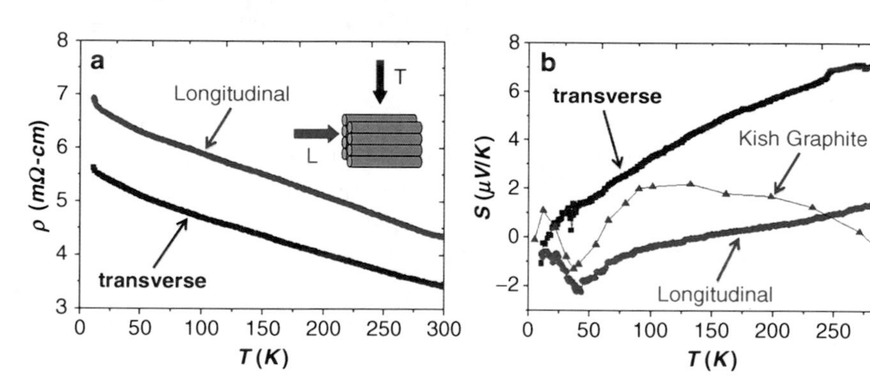

Tuning Electrical Properties of Carbon Nanotubes via Spark Plasma Sintering, Fig. 3 MWNTs SPSed at 1,500°C. (**a**) The resistivity is lower in the transverse direction than the longitudinal direction; (**b**) Thermopower in the longitudinal direction shows a strong phonon drag peak below 50 K, which is similar to Kish graphite

a pronounced phonon drag peak below 50 K very similar to that of Kish graphite (Fig. 3b). In view of electron–phonon coupling, the phonon drag features in the thermopower are due to carriers being preferentially scattered by the phonons in the direction of heat flow, when the mean free path of both electrons and phonons are long enough. For individual CNTs, the quantized electronic band structure and phonon dispersion relations lead to a reduced phase space for electron–phonon coupling. Hence the observed phonon drag feature supports the argument that the SPS process leads to an increased effective dimensionality of the CNT assembly.

Subsequently, Yang et al. [13] showed that the electrical connectivity of a network made of randomly oriented MWNTs can be effectively tuned by varying the SPS temperature. An increase in the SPS temperature from 500°C to 1,600°C leads to an increase in the packing density from ~0.6 g/cm^3 to ~1.25 g/cm^3, reflecting a change in the inter-tube spacing. The concomitant changes in the magnitude (Fig. 4a) and temperature dependence (Fig. 4b) of resistivity strongly suggest a change in the nature of the inter-tube junctions and thereby the electrical connectivity of the network. As shown in Fig. 4a, the room temperature resistivity as a function of SPS temperature follows a power-law T-dependence, suggesting a percolation behavior. Percolation phenomena are important for the operation of electronic devices based on CNT conducting networks. In Fig. 4b, it is noted that the normalized resistivity ρ_N (T) below 100 K can be categorized into three characteristic T-dependencies. The as-prepared (AP) sample and the sample SPSed at 500 °C (S500) exhibit a sub-linear T-dependence, which is labeled as type I. In traces grouped into type II (S800, S1000 and S1200) and type III (S1500) dependencies, there is an up-turn below 60 K. This up-turn may be associated with the formation of coherent inter-tube junctions. In addition, a plateau is evident below 50 K in sample S1500. The 'S' shape anomaly in the thermopower for the S1500 sample (Fig. 4c) resembles the thermopower behavior of Kish graphite (Fig. 3b), as noted above. It is interesting to note that this feature in the thermopower data occurs at the same temperature as a plateau in the resistivity data for the SPS1500 sample, which suggests that there is a single unified mechanism governing both phenomena. The possibility of a Kondo effect has been ruled out [13].

SPS can also induce coherent bonding between CNTs and neighboring graphitic particles. In a CNT bundle where graphitic particles are not bonded to CNTs, Zhang et al. have shown that though the graphitic particles have lower intrinsic resistivity than CNTs, their contribution to the overall conduction will be small [14]. Without bonding to CNTs, the conduction of graphitic particles is restricted to variable range hopping (VRH) conduction, which will be very small as long as the concentration of graphitic particles is small.

Intra-tube Restructuring

It is also possible to introduce intra-tube defects by the SPS process, including creation of vacancies, collapsing/zipping tubes (Fig. 5a), and kinks on the tube wall (Fig. 5b) [11]. The local curvature change of CNTs often indicates the removal of carbon atoms from the lattice and the subsequent reconstruction of a new coherent structure containing non-hexagonal rings and other types of defects. Defects may profoundly affect the electronic properties of CNTs by creating new localized states near the Fermi level. Electrical transport is also affected in view of the quasi-1D molecular nature of CNTs. Finally, these SPS-induced crystal lattice imperfections, especially vacancies, serve as a good platform for doping CNTs since the dopant atoms tend to lodge themselves at these energetically favorable positions.

Doping CNTs

Doping is a powerful tool for controlling the electrical properties of CNT networks because it offers a way to vary the carrier concentration and thereby tune the electron–electron and electron–phonon coupling that basically govern the electrical properties of CNTs. It has been shown that through doping the majority carrier conduction in a CNT network can be varied from p-type to n-type in a controllable way [15]. However, doping CNTs is technically challenging for several reasons. First, it is difficult to attain a uniform doping level, due to the finite length effect, the subtle differences in capping topology, the amount and distribution of defects, and the proximity environment. This situation is worsened by the low solubility limit of dopants in carbon allotropes (e.g., the solubility of B in graphite is limited to ~4 at.% at 1,200°C). Second, in small-diameter CNTs, substitutional doping inevitably compromises the integrity of the hexagonal lattice. Third, electric field doping does not work well in MWNTs and thick films. Apparently, controlling crystal chemistry in the synthesis stage is *not* adequate for controlling the doping of CNTs, some post-synthesis treatment is needed. Ion implantation or low-energy ion bombardment has been often used to dope carbon nanostructures [5, 16], but the application is limited by expensive equipment. It is also possible that by varying the doping conditions, the nature of the doping could be varied from low-energy endohedral (filling) or exohedreal (intercalation) doping to higher energy substitutional doping. The SPS process is a promising candidate for post-growth doping of CNTs.

Recently, Hitchcock et al. have shown that a post-growth treatment of SWNT bundles with boric acid solution followed by SPS process is a viable method for doping CNTs and controlling their carrier concentration. Using boric acid concentration as the control parameter, they were able to alter the carrier concentration of the SPSed CNT networks. As a result, the resistivity, thermopower, and Hall coefficient measurements show a doping-induced shift of the Fermi level and the accompanied changes in electron–phonon coupling [17].

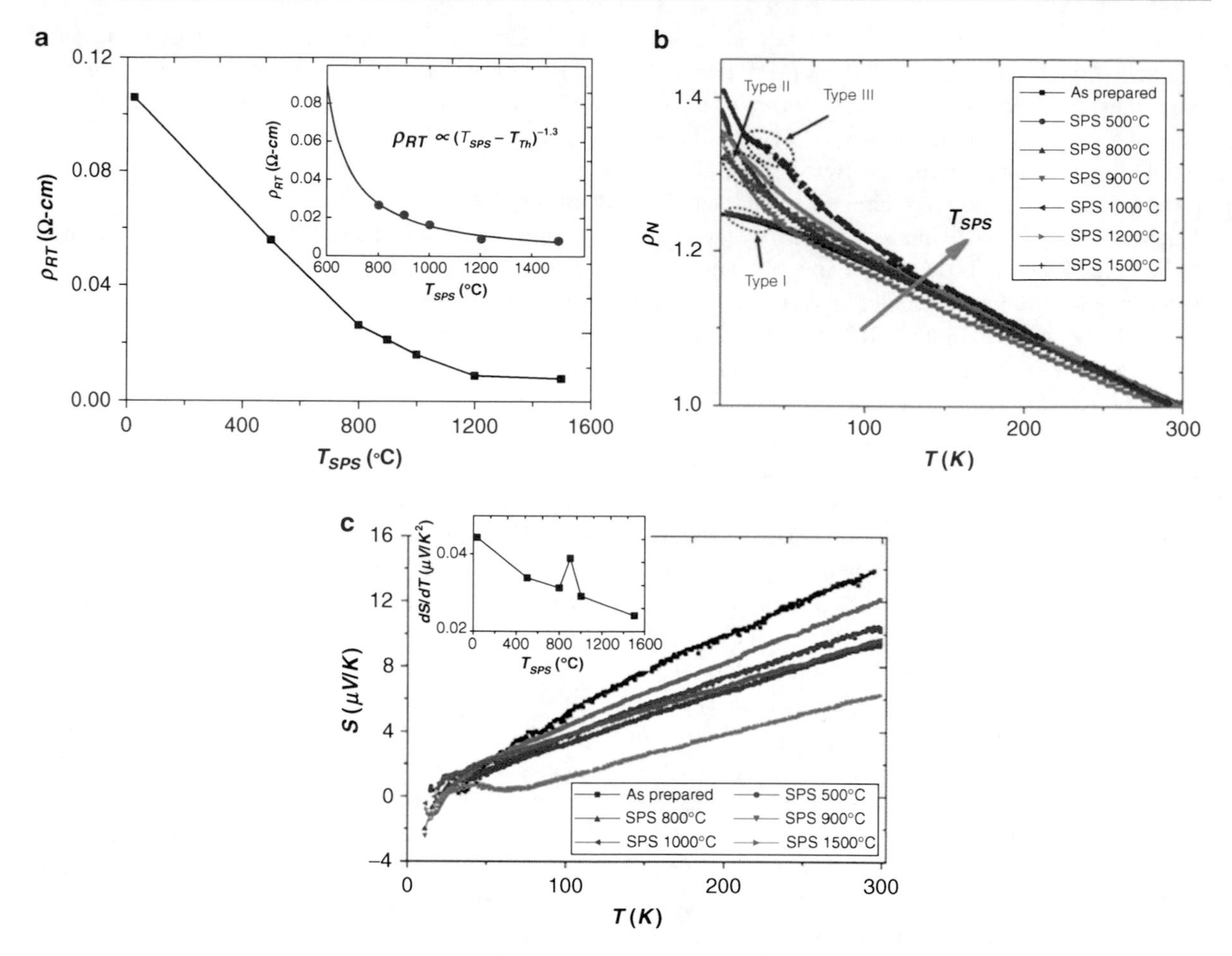

Tuning Electrical Properties of Carbon Nanotubes via Spark Plasma Sintering, Fig. 4 (**a**) Room temperature resistivity value as a function of SPS temperature, the inset shows the power law fitting to a percolative model; and (**b**) The crossover in the temperature dependence of normalized resistivity, type I, II and III, reflects the change of the nature of inter-tube junctions, the normalization is performed by dividing the $\rho(T)$ by its corresponding room temperature ρ_{RT}; (**c**) The temperature dependence of thermopower of the sample SPSed at 1,500°C shows a strong phonon drag peak below 50 K [13]

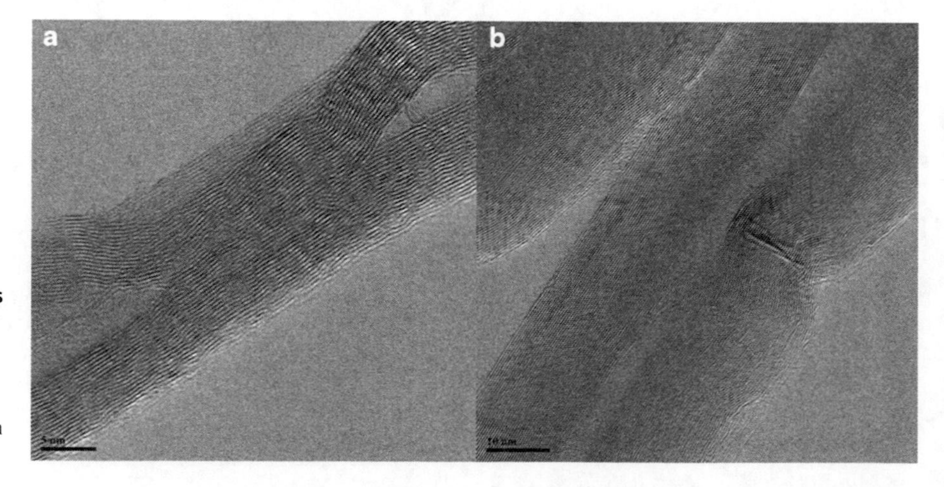

Tuning Electrical Properties of Carbon Nanotubes via Spark Plasma Sintering, Fig. 5 SPS-induced defects in MWNTs (**a**) collapse of a MWNTs (**b**) kink formed on the wall a MWNT

Altering the Morphology of Tubes (Beyond the Soft Sintering)

At high SPS temperatures (>1,500 °C), it is possible to alter the morphology of CNTs. In an extreme case, it is possible to "unzip" CNTs to form graphitic nanostructures (Fig. 6) [18, 19]. This has been proposed as a novel method for generating large amounts of graphene in a short amount of time. Under the proper conditions, SPS can also change the sp^2 bonds of CNTs to sp^3 bonds, forming diamond [8]. Shen et al. [20] have shown that using the SPS, it is possible to grow diamond particles up 100 μm in diameter at very low pressure (80 MPa). It is noteworthy that SPS under a moderate pressure of 10–100 MPa leads to the conversion of CNTs to diamonds, whereas a relatively higher pressure in the GPa range is usually required for diamond formation by other methods. It is proposed that the nucleation of diamond is from the core of carbon nano-onions, which is an intermediate product from the breakage of carbon–carbon bonds and unzipping CNTs [8]. The nano-onions behave as compression cells ("self-compression") to help nucleate and grow diamonds. Hence the SPS process provides for a new path for diamond synthesis, which could offer many applications in industry due to the exceptional hardness and thermal conductivity of diamond.

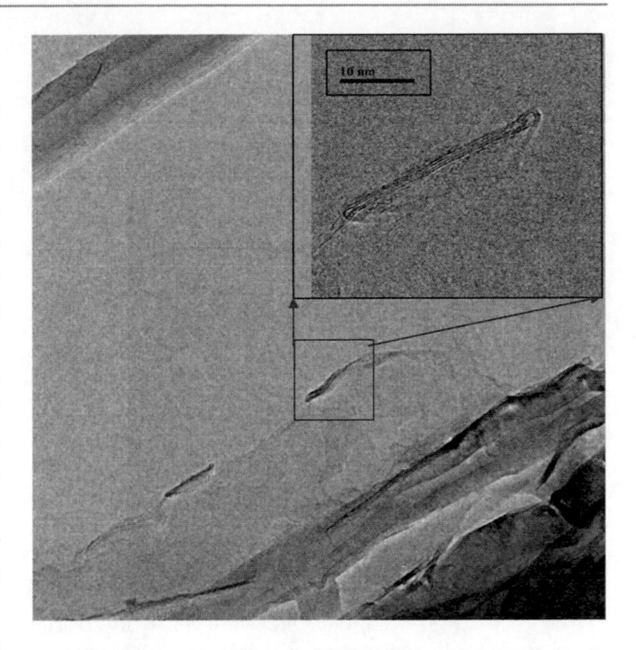

Tuning Electrical Properties of Carbon Nanotubes via Spark Plasma Sintering, Fig. 6 Graphitic nanostructures formed in SPS by unzipping MWNTs [11]

Conclusion

Since their discovery, carbon nanotubes have been a topic of much research due to their amazing physical properties. However, in order to take CNTs from the lab to applications, scalable techniques must be developed to tune their properties. While it is possible to alter the structure and morphology of CNTs with very high precision using an irradiation technique, this method involves microscopic control parameters and the throughput is very low. In this regard, the SPS technique utilizes high-density current without the need to focus or align the electron beam, the control parameters of SPS are macroscopic, and the throughput is very high. One can envisage a combination of these two techniques to give some control of the structure and properties of CNTs at multiple length scales, which is a very promising approach to move CNTs from materials used solely in research to viable materials with real-world applications.

Acknowledgments D. H., J. H., and A. M. R. acknowledge the support from a DOE/EPSCoR Implementation Grant (No. DE-FG02-04ER-46139) and the SC EPSCoR Office/Clemson University cost sharing. K. Y. and A. M. R. also acknowledge the support through U.S. AFOSR under Grant No. FA9550-09-1-0384. All authors thank Prof. Malcolm Skove for scientific discussions and proofreading of the manuscript.

Cross-References

▶ Carbon Nanotubes

References

1. Saito, R., Dresselhaus, G., Dresselhaus, M.S.: Physical Properties of Carbon Nanotubes. Imperial College Press, London (1998)
2. Ma, R.Z., Xu, C.L., Wei, B.Q., Liang, J., Wu, D.H., Li, D.J.: Electrical conductivity and field emission characteristics of hot-pressed sintered carbon nanotubes. Mater. Res. Bull. **34**, 741–747 (1999)
3. Rubio, A., Apell, S.P., Venema, L.C., Dekker, C.: A mechanism for cutting carbon nanotubes with a scanning tunneling microscope. Eur. J. Phys. B. **17**, 301–308 (2000)
4. Charlier, J.C.: Defects in carbon nanotubes. Acc. Chem. Res. **35**, 1063 (2002)

5. Krasheninnikov, A.V., Banhart, F.: Engineering of nanostructured carbon materials with electron or ion beams. Nat. Mater. **6**, 723 (2007)
6. Munir, Z.A., Anselmi-Tamburini, U., Ohyanagi, M.: The effect of electric field and pressure on the synthesis and consolidation of materials: a review of the spark plasma sintering method. J. Mater. Sci. **41**, 763 (2006)
7. Song, X., Liu, X., Zhang, J.: Neck formation and self-adjusting mechanism of neck growth of conducting powders in spark plasma sintering. J. Am. Ceram. Soc. **89**, 494–500 (2006)
8. Zhang, F., Shen, J., Sun, J., Zhu, Y.-Q., Wang, G., McCartney, G.: Conversion of carbon nanotubes to diamond by spark plasma sintering. Carbon **43**, 1254 (2005)
9. Qi, X., Bao, Q., Li, C.M., Gan, Y., Song, Q., Pan, C., Tang, D.Y.: Spark plasma sintering-fabricated one-dimensional nanoscale "crystalline-amorphous" carbon heterojunction. Appl. Phys. Lett. **92**, 113113 (2008)
10. Gruner, G.: Carbon nanotube films for transparent and plastic electronics. J. Mater. Chem. **16**, 3533 (2006)
11. Yang, K., He, J., Su, Z., Reppert, J.B., Skove, M., Tritt, T. M., Rao, A.M.: Inter-tube bonding, graphene formation and anisotropic transport properties in spark plasma sintered multi-wall carbon nanotube arrays. Carbon **48**, 756 (2010)
12. Hone, J., Llaguno, M.C., Nemes, N.M., Johnson, A.T., Fischer, J.E., Walters, D.A., Casavant, M.J., Schmidt, M., Smalley, R.E.: Electrical and thermal transport properties of magnetically aligned single wall carbon nanotube films. Appl. Phys. Lett. **77**, 666–668 (2000)
13. Yang, K., He, J., Puneet, P., Su, Z., Skove, M.J., Gaillard, J., Tritt, T.M., Rao, A.M.: Tuning electrical and thermal connectivity in multiwall carbon nanotube buckypaper. J Phys Condens Matter **22**, 334215 (2010)
14. Zhang, H.-L., Li, J.-F., Zhang, B.-P., Yao, K.-F., Liu, W.-S., Wang, H.: Electrical and thermal properties of carbon nanotube bulk materials: experimental studies for the 328–958 K temperature range. Phys. Rev. B **75**, 205407 (2007)
15. Duclaux, L.: Review of doping of carbon nanotubes (multiwalled and single-walled). Carbon **40**, 1751–1764 (2002)
16. Bangert, U., Bleloch, A., Gass, M.H., Seepujal, A., van der Berg, J.: Doping of few layered graphene and carbon nanotubes using ion implantation. Phys. Rev. B **81**, 245423 (2010)
17. Hitchcock, D., Yang, K., He, J., Rao, A.M.: Electrical transport properties of single-walled carbon nanotube bundles treated with boric acid. Nano: Brief Rep. Rev. **6**(4), 337–341 (2011)
18. Gutirrez, H.R., Kim, U.J., Kim, J.P., Eklund, P.C.: Thermal conversion of bundled carbon nanotubes into graphitic ribbons. Nano Lett. **5**(11), 2195–2201 (2005)
19. Kim, W.S., Moon, S.Y., Bang, S.Y., Choi, B.G., Ham, H., Sekino, T., Shim, K.B.: Fabrication of graphene layers from multiwalled carbon nanotubes using high dc pulse. Appl. Phys. Lett. **95**, 083103 (2009)
20. Shen, J., Zhang, F.M., Sun, J.F., Zhu, Y.Q., McCartney, D.G.: Spark plasma sintering assisted diamond formation from carbon nanotubes at very low pressure. Nanotechnology **17**, 2187–2191 (2006)

U

Ultrahigh Vacuum Chemical Vapor Deposition (UHVCVD)

► Chemical Vapor Deposition (CVD)

Ultralarge Strain Elasticity

► Superelasticity and the Shape Memory Effect

Ultraprecision Machining (UPM)

Suhas S. Joshi
Department of Mechanical Engineering, Indian Institute of Technology Bombay, Mumbai, Maharastra, India

Synonyms

Nanomachining; Nano-mechanical machining

Definition

Ultraprecision machining refers to the ultimate ability of a manufacturing process wherein processing of a material at its lowest scale that is, at the atomic scale, is achieved. It is known that the lattice distances between two atoms are of the order of 0.2–0.4 nm; therefore, the ultraprecision machining refers to processing or removal actions of a manufacturing process in the vicinity of 1 nm. The process is also referred as "atomic bit" processing. To remove or process atomic bits, extremely large energy density is required, which is equivalent to the atomic bonding energy. The conventional cutting tools neither have high strength to sustain high specific cutting energy nor have hardness to sustain the tool wear. Therefore, ultraprecision machining refers to use of single crystal diamond (SCD) tools for ultrafine cutting or very fine abrasives for lapping or polishing. It may also refer to use of high-energy elementary particles like photons, electrons, ions, and reactive atoms to undertake removal [1].

Historical Perspective

In early 1970s, Japanese researcher N. Taniguchi illustrated historical evolution of precision in manufacturing through "Taniguchi curves" [1]. It was realized that it takes about 20 years to improve precision by one decimal point; see Table 1 depicting the progression of precision.

He extrapolated the curves further to postulate that the resolution of machining processes would reach to 1 nm in the year 2000 [1]. True to the expectations, today, a number of ultraprecision machining processes are available.

Applications

Ultraprecision processing is required in the manufacture of high-precision block gages, diamond indenters and tools, 3D metallic mirrors, etc., in the mechanical

B. Bhushan (ed.), *Encyclopedia of Nanotechnology*, DOI 10.1007/978-90-481-9751-4,

Ultraprecision Machining (UPM), Table 1 Evolution of ultraprecision

Year	1900	1920	1940	1960	1980	2000
Resolution (μm)	>10	5	0.5	0.05	0.005	0.001

domain; Si wafers, ICs memory, thin film, ULSI devices in electronics field; and optical flats, diffraction gratings, mirrors, and aspherical lenses in the optical field.

Types of UPM Processes

The ultraprecision machining processes can be broadly classified into three categories [1]:

1. Nanomechanical processing: These processes use either SCD tools to perform diamond turning or very fine abrasives in the bonded or loose form to perform nanogrinding, nanolapping, or nanopolishing. The other processes in this category include progressive mechanochemical polishing for Si wafers, pitch polishing for aspherical lenses.
2. Nanophysical processing: It involves use of high-energy elementary particles like photons, electrons, ions, and reactive atoms to perform direct ablation of substrate or carry out lithography.
3. Nanochemical or electrochemical processing: These processes involve use of chemical or electrochemical principles to effect material removal. They are chemical milling, photochemical machining, and other processes.

In the following sections, principles of nanometric removal in ultraprecision machining processes are elucidated along with their applications.

Nanomechanical Processing

In these processes, the processing unit, defined as the size of chip generated in a single stroke of tool, should be of the order of single or subatomic magnitude.

Principle of Ultraprecision Removal

The removal of atomic bits in nanomechanical processes requires specific cutting energy of varying degree as the scale of the process decreases and cutting tools with certain sharpness and hardness as discussed below:

Ultraprecision Machining (UPM), Table 2 Size-effect in processes as a function of chip thickness

Process	Process scale	Chip thickness (μm)	Resisting shear stress (N/mm^2)
Tension test	Multicrystal grain	300–500	300
Turning	Subcrystal grain	40–50	500
Precision milling	Subcrystal grain	5–10	1,000
Grinding	Atomic cluster region	0.5–1	10,000

Size-Effect Implications

As the processing scale reduces from simple tension testing to grinding, where submicrometric chips are generated, the specific shear or breaking energy becomes extremely large. This effect is called as *size effect*. See Table 2 showing processing scales and shear energies [2].

Extending this trend further, when the machining scale reaches an atomic-bit, the specific shear energy reaches atomic bonding energy [1] or the theoretical shear strength of a defect-free material given by

$$\tau_{th} = \frac{G}{2\pi} \quad (1)$$

For carbon steel with the modulus of rigidity, G = 82 GPa, the theoretical shear strength is given by –

$$\pi_{thCarbonSteel} = 13GPa \quad (2)$$

The reason for increasing the specific shear energy is attributed to the kind of defects available in the crystalline structure at that processing scale. At the atomic bit scale (0.001 μm), the energy required is comparable to the theoretical shear strength (10^5 J/cm^3) [1] (Fig. 1). At the atomic cluster scale, deformation of a material can take place only with the help of point defects (Fig. 1) and requires energy of the order of 10^3–10^4 J/cm^3 [1].

Nevertheless, at some higher scale (0.1–10 μm), called as subcrystal grain size, 2D or 3D defects in the crystalline structure cause a significant reduction in the specific shear energy to 10^2–10^3 J/cm^3 (Fig. 1) [1]. In the crystal grain size range, (>10 μm), all kinds of dislocations and defects at grain boundaries lower the processing energy further to 10^1–10^2 J/cm^3 [1]. Unlike the case of the ductile (metallic) materials discussed above, the brittle materials have a network of

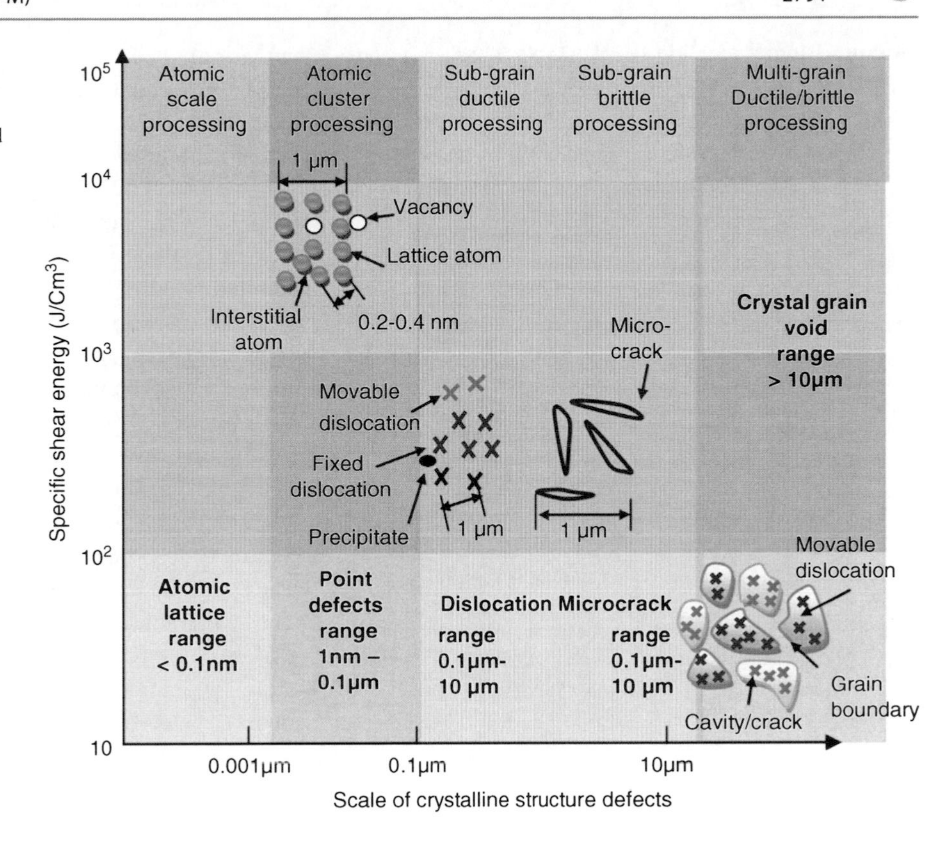

Ultraprecision Machining (UPM), Fig. 1 Defect distribution and scale of specific shear energy required

microcracks in place of the network of dislocations (Fig. 1). In the ductile materials, below the point defect scale (1 nm to 0.1 μm) (Fig. 1), and in the brittle materials below the microcracks scale (0.1–10 μm) (Fig. 1), the specific shearing energy is equal to the theoretical shear strength of the work material.

Cutting Tool Material and Geometry Implications

The cutting tool edge in ultraprecision machining being subjected to extremely high cutting pressure should withstand not only the pressure but also the wear. In addition, the edge should have high degree of sharpness to perform atomic bit removal. The SCD tools, possessing the desired properties, are the most suitable tool materials for ultraprecision machining. However, their limitation stems from the achievable cutting edge sharpness, which normally is limited to hundreds of nanometers. Therefore, in UPM, several atomic layers or atomic clusters are processed than atomic bits, primarily on softer metallic materials like Al and Cu. On the other hand, on hardened steels, and other brittle materials, SCD tools undergo rapid wear, thereby necessitating the use of diamond abrasive process like nanogrinding.

Ultraprecision Machining Systems

It uses time-tested principles of precision engineering along with (1) the recent advancements in the control, feedback, and drive systems, (2) CAD/FEM for designing, followed by (3) fine-tuned assembly as well as system integration. This combination results in a machine that is thermally stable, highly reliable, and flexible and at the same time faster in response [3]. Several features of an ultraprecision machine are summarized in Table 3.

Ultraprecision Machines and Processes

A number of ultraprecision machines and processes have evolved using nanomechanical removal principles. They are discussed below.

Single Point Diamond Turning

It is a nanomechanical processing machine that was developed primarily for 3D shaping and surface finishing of soft metals such as copper and aluminum and polymers. The machine with the features listed in Table 3 uses very sharp SCD tools to perform ultraprecision cutting. The chips formed during the process are extremely thin, which indicates ductile

Ultraprecision Machining (UPM), Table 3 Salient features of an ultraprecision machining system

Machine elements/purpose/types	Main features
Machine base	
• Provides thermal and mechanical stability, damping characteristics	• Made of cast iron, natural or epoxy granite, polymer concrete
Work spindle	
• Spindle motion errors significantly affect surface quality and accuracy of machined features. • Use aerostatic or hydrostatic, recent grooved air bearings	• Both the spindle types have high rotational accuracy and rotational speeds • Aerostatic spindles are for low/medium loads, hydrostatic bearings take heavy loads
Drives	
• Slide drives provide stiffness, acceleration, speed, smoothness of motion, accuracy, and repeatability • Spindle drives are usually AC/DC motors • Slides are usually provided with linear motor or friction drives	• Servo drives are used in contouring operations • Small and precise motions of tools for tool positioning and fine motion are achieved by piezoelectric actuators
Controls	
• Controls are required for linear and rotary drives, limiting, position, and time switches, sensors. • They also control thermal, geometrical, and tool setting errors.	• A multiaxes CNC controllers are used • PC-based controls are used more recently • Feedback controls have a resolution of nm or sub-nm
Measurement and inspection systems	
• Provides rapid and accurate positioning of cutting tool towards work surface • It also monitors the tool wear condition	• Online measurement and error compensation • Laser interferometer for tool position control

deformation of the material during machining. At the same time, the process leaves extremely thin degenerated layer of the work surface as the process initiates shear slip using dislocations (Fig. 1).

Applications of the process include fabrication of metallic mirrors, precision component fabrication in laser, and space and optics fields. A typical diamond turning lathe has T-axis configuration (Fig. 2a). The main machine slide moves along x-axis, and the cutting tool is moves along Z-direction.

Nanogrinding

Single-point diamond turning machine is usually provided with add-on grinding attachment (Fig. 2b) for nanogrinding. This way, harder work materials can also be processed on the machine using the grinding attachment. However, it increases the stiffness requirement of the machine. In nanogrinding process, shear slip is generated due to very low (tens of nm) depth of cut (Fig. 1). This causes crackless ductile failure of hard/brittle materials under the abrasive grains in the diamond grinding wheels. Since the working stress on the abrasives is extremely high, only fine-grained diamond abrasives can be used for this process. The process can achieve mirror like finish on hard and brittle materials like glasses and ceramics. Its other applications involve grinding of mould inserts, glass lenses, and aspherical lenses.

Electrolytic In-process Dressing (ELID) Grinding

It is a type of nanogrinding process in which metal-bonded diamond grinding wheels are used. The process uses an electrolytic solution as a grinding liquid. Upon application of voltage, anodic dissolution of bonding metal from the grinding wheel occurs. This ensures that worn abrasive particles on the grinding wheel are removed quickly (a wheel dressing action), thereby ensuring continuous availability of sharp particles for the grinding process. The sharp particles initiate shear slip at low depth of cut (Fig. 3) causing crackless material removal. The process therefore is capable of achieving mirror like surface finish on glasses and ceramics.

Nanolapping and Nanopolishing

Unlike nanogrinding, which uses fixed abrasives, these processes use free abrasives to achieve mirror-like finish. The processes involve atomic cluster processing.

Nanopolishing involves sliding of soft abrasives between a soft polishing pad and work surface (Fig. 3a). The abrasives are usually soft and dull; they include Fe_2O_3, Cr_2O_3, CeO_2, or MgO [1]. The abrasives get embedded in the soft pad and originate shear slip at the point defects to effect removal [1] (Fig. 3a). This ensures smooth polishing of hard metallic surfaces.

The nanolapping on the other hand uses a medium to hard pad besides extremely sharp and hard abrasives such as diamond, CBN, SiC, SiO_2, or B_4C [1]. The fine and hard abrasives initiate cracking at the point defects to effect removal [1] (Fig. 3b). This helps generate mirror-like finishing on hard and brittle materials.

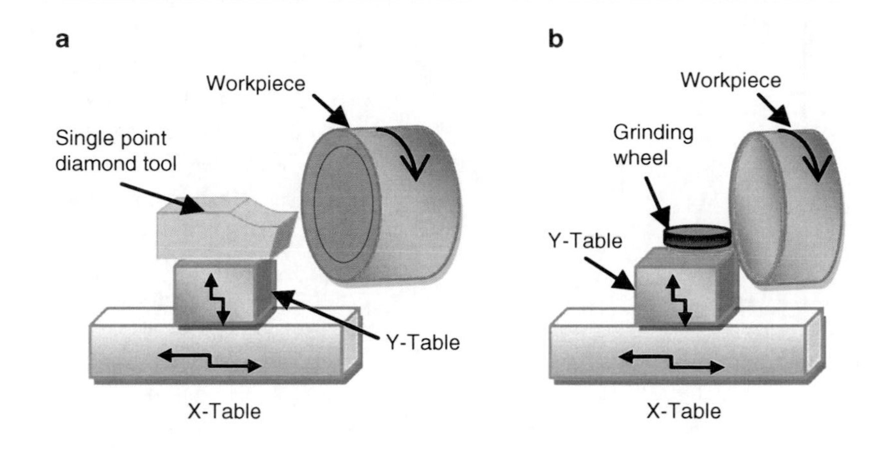

Ultraprecision Machining (UPM), Fig. 2 Typical schematics of (**a**) Single-point diamond turning (**b**) Diamond grinding machine

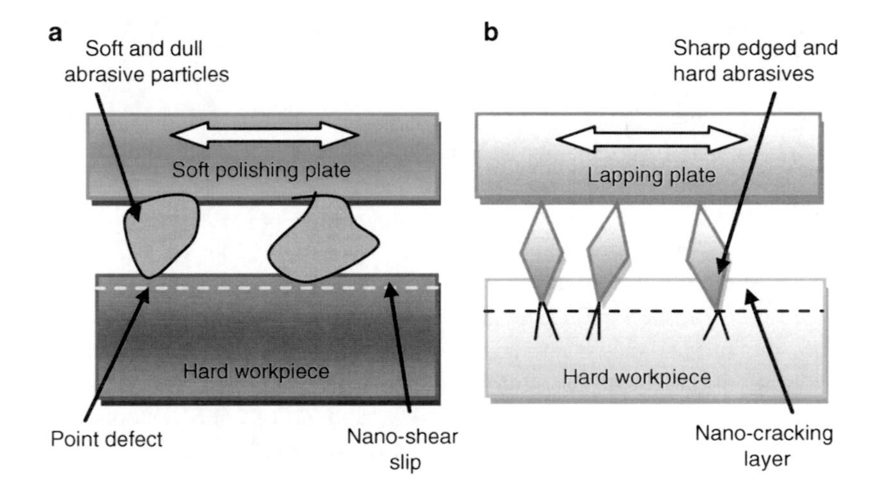

Ultraprecision Machining (UPM), Fig. 3 (a–b) Nanopolishing and nanolapping

P-MAC

The progressive mechanical-chemical (P-MAC) polishing is a hybrid nanopolishing process that uses mechanical as well as chemical action to effect removal to achieve mirror-like finish and damage-free polishing of Si wafers. The process uses ultrafine SiO_2 abrasives (size $\sim$10 nm) suspended in an alkaline (pH $\sim$10) solution along with an artificial leather foam to perform polishing in four stages. The abrasive particle size and the polishing pressure and the stock removal are gradually reduced after each stage [1].

In P-MAC, the greater the chemical action, the higher is the polishing efficiency and the lower is the damaged layer [1] (Fig. 4). However, the increased chemical action reduces control over accuracy (flatness), Fig. 4 [1].

Aspherical Lens Polishing

Aspherical profiles are required at the edges of the lenses to ensure that light beams from the lens edges converge at the focal point. Therefore, the aspherical lens polishing involves polishing over free-form surfaces. The process is often required for polishing of highly accurate X-ray optics and lenses in commercial cameras.

It involves gradual removal using a pitch polishing tool that has very small contact area as compared to the dimensions of the workpiece. The polishing rate is governed by Preston's rule of thumb [1]

$$h = \alpha \cdot v \cdot \Delta t \tag{3}$$

The pitch polishing tool acts as a pad and supports the abrasives. It remains solid at room temperature but

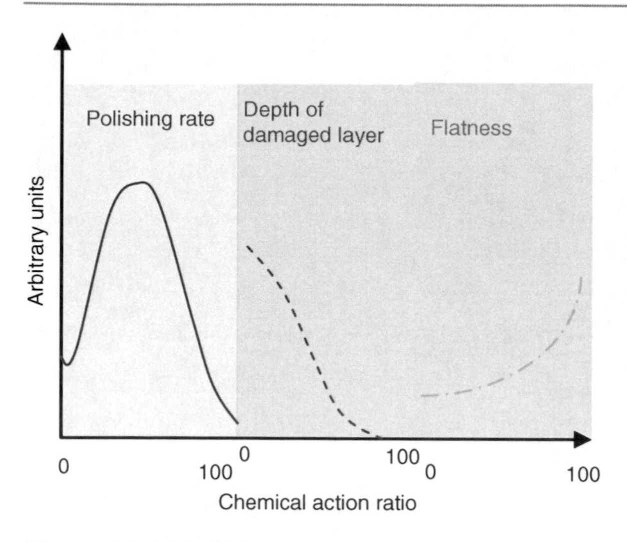

Ultraprecision Machining (UPM), Fig. 4 General processing characteristics of P-MAC

flows with time when heated or under pressure. The composition of the pitch material is usually proprietary of various manufacturers but contains materials like tar, oil, wood, paraffin wax, shellac, and so on. The properties of pitch material include [1] viscosity in the range of 10^7–10^9 Pa·s), softening point, (55–7°C), penetration hardness (60–80 by Shore D). In this process, the surface roughness tends to improve as the penetration number of the pitch increases, i.e., as the pitch becomes softer. It is seen that CeO_2 and ZrO_2 give better surface finish than super hard diamond abrasives.

Nanophysical Processes

The principle of removal in these processes involves photothermal or photochemical interactions with the work surface. The photothermal interaction causes melting and vaporization of work material due to energy of photon or laser beams. On the other hand, the photochemical interaction involves use of photon energy to break bond between work elements [1]. It is understood that usually laser beams with metallic materials perform removal action by the photothermal principle. However, when the polymeric materials interact with laser beams, photochemical reaction takes place. With the ceramics, the interaction is partly by the photothermal and partly by photochemical reactions.

The photon beams owing to their wavelength of the order of 10–0.1 μm and the photon energy of the order of 0.1–10 eV cause thermal energy transfer that is smaller than the atomic bonding energy of materials. Hence, the photon beams are not suitable for nanometric material removal. Similarly, an electron beam due to small size (2.8×10^{-6} nm), mass (9×10^{-31} kg), and low energy of several hundred kilovolts penetrate deep (several micrometers) into the work surface. Therefore, the electron beams are also not suitable for nanometric material removal.

Unlike the electrons, most of the ions interfere with the surface atoms of a workpiece because their mean diameter is ~0.1 nm and the mean atomic distance is ~0.3 nm. Consequently, the projected energized ions frequently collide with the nuclei of atoms of the workpiece and knock out or sputter the surface atoms. Hence, ion beams are most suitable for atomic bit removal. The process is also called as ion sputter machining. In the process, the electrically accelerated inert-gas ions such as Ar ions with average energy of 10 keV (which corresponds to a speed of ~200 km/s) are unidirectionally oriented and projected on to the workpiece surface in a high vacuum (1.3×10^{-4} Pa) [1]. Hence, they adhere firmly on the target surface. A beam of focused ions can also be used for cutting of extremely hard materials like diamond or sharpening of SCD tools.

Nanochemical or Electrochemical Processing

The chemical reactions inherently involve atomic bit processing. A chemical reaction is a change in the atomic combination of reacting molecules, in which the atomic bonding of a reacting molecule is broken and a new molecule is generated.

In the process, a chemically reactive gas or liquid is applied to a specified position on the solid surface of the workpiece. The reacted molecules are removed or diffused into the surrounding reacting gas or liquid. If the reacted molecules are insoluble or not in vapor form, chemically reactive deposition occurs on the workpiece surface. But if the reacted or reagent molecules diffuse into the surface layers of the workpiece and react with the atoms or molecules there, they perform a chemically reactive surface treatment [1]. The dimensional accuracy obtainable in chemical reactions is in the nanometer range when the processing conditions are stable. In addition, the following aspects are necessary for nanometric processing [1]:

1. In-process measurement and feedback control of position of the processing point and control of the processed volume (area) are necessary. However, it is difficult to realize these in practice. In the photoresist method, control of the processing point position or area is achieved by the patterned mask.
2. The control of the processing volume or depth can be done only by adjusting the processing time and flow rate of etchants.

In the electrochemical process, the basic reaction is the same, but activation energy for the reaction is given by electric field potential and differs from the ordinary activation energy of the chemical reactions based on the thermal potential energy.

Summary

Ultraprecision machining uses mechanical, physical, chemical, or electrochemical sources of energy for effecting material removal to the nanometric scale. Grain size, load applied on the grains, and the type of work material govern removal in nanomechanical processes. Size of the energy beam particle governs the processing resolution in nanophysical processes. In-process measurement and feedback control are essential for the nanochemical or electrochemical processes.

Cross-References

- ► Electrochemical Machining (ECM)
- ► Nanochannels for Nanofluidics: Fabrication Aspects
- ► Nanotechnology

References

1. Taniguchi, N. (ed.): Nanotechnology – Integrated Processing Systems for Ultra-precision and Ultra-fine Products. Oxford University Press, New Delhi (2008)
2. Baker, W.R., Marshall, E.R., Shaw, M.C.: The size effect in metal cutting. Trans. ASME **74**, 61 (1952)
3. Lou, X., Cheng, K., Webb, D., Wardle, F.: Design of ultra-precision machine tools with applications to manufacture of miniature and micro components. J. Mater. Process. Technol. **167**, 515–528 (2005)

Ultrashort Carbon Nanotubes

Lesa A. Tran and Lon J. Wilson
Department of Chemistry and the Richard E. Smalley Institute for Nanoscale Science and Technology, Rice University, Houston, TX, USA

Synonyms

US-tubes

Definition

Ultrashort carbon nanotubes (US-tubes) are 20–100 nm segments of single-walled carbon nanotubes (SWNTs).

Overview

In recent years, single-walled carbon nanotubes (SWNTs) have been extensively studied for their use in biomedical applications. Owing to their peculiar physical and chemical properties, SWNTs have become a widely used platform for the design of medical therapeutic and diagnostic agents [1, 2]. Nevertheless, the biocompatibility and biodistribution of SWNTs still remain uncertain and are under current investigation. The ideal length of SWNTs for biomedical applications has yet to be determined, but it has been suggested that discrete, individualized SWNTs of shorter (<300 nm) lengths are more ideal for in vivo use [3]. Therefore, ultrashort carbon nanotubes (US-tubes), which are 20–100 nm segments of SWNTs, may be best suited for biomedical applications (Fig. 1).

Methodology

Synthesis

Chemical Cutting

Generally, the chemical cutting process of SWNTs into US-tubes can be viewed as a two-step procedure: (1) the modification of the SWNT sidewalls with

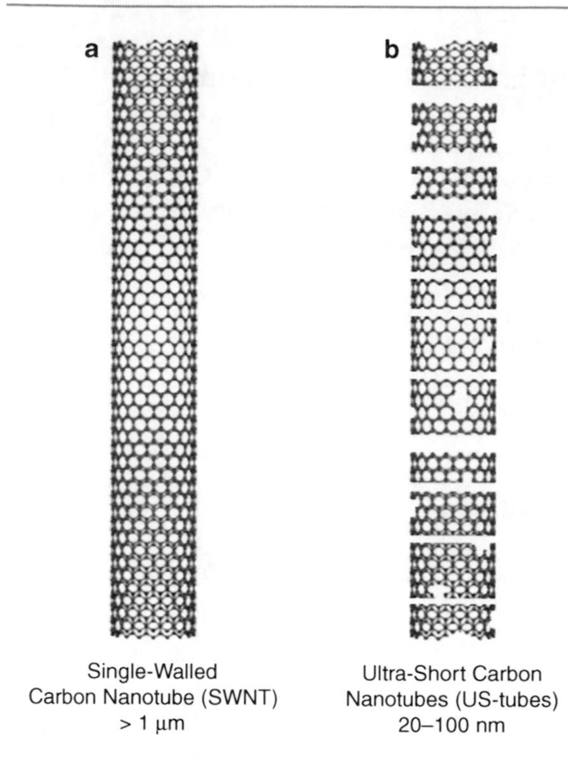

Ultrashort Carbon Nanotubes, Fig. 1 Illustration of (**a**) a single-walled carbon nanotube and (**b**) a collection of ultrashort carbon nanotubes (with sidewall defects) derived from a single-walled carbon nanotube

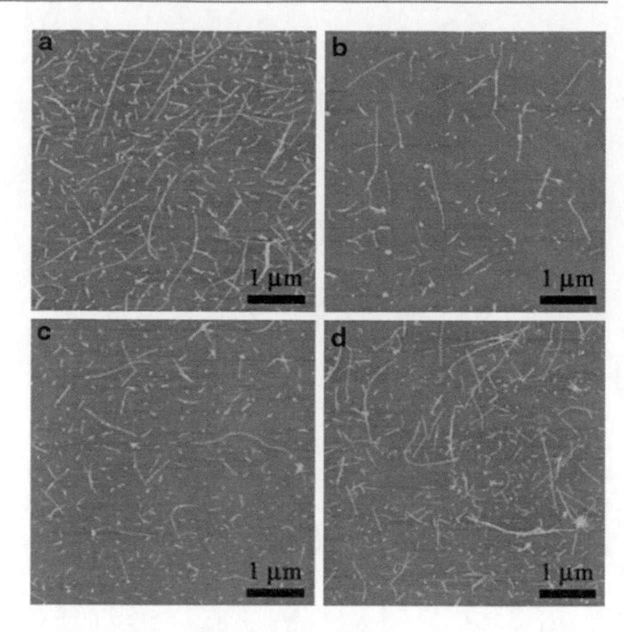

Ultrashort Carbon Nanotubes, Fig. 2 Atomic force microscopy (AFM) images of (**a**) purified SWNTs and SWNTs after piranha treatment at 22°C for (**b**) 1 h, (**c**) 3 h, and (**d**) 9 h (Reproduced with permission from [4])

functional groups or defect sites, and (2) the removal of or the cutting at these sidewall modifications. While the predominant mechanism of chemical SWNT cutting is debatable, there are two proposed methods under consideration: (1) the "fuse burning" mechanism, with carbon etching occurring at the bond-strained tube ends, and (2) defect site propagation, with cutting occurring at the random defect sites on the SWNT walls. Both oxidizing and nonoxidizing methods to chemically cut SWNTs have been explored.

Various oxidative processes using different strong oxidizing agents (HNO_3, H_2SO_4, and H_2O_2) have been proposed for the cutting of SWNTs. However, the efficiency and length distributions of these synthetic methods vary with the type and concentration of the oxidizing agents used, the reaction temperature and time, the method of dispersion during the reaction (refluxing or sonication), and the length and diameter of the SWNTs being cut. Although more severe reaction conditions result in more reaction exposure of SWNTs and, thus, shorter lengths of resulting US-tubes, less severe reaction conditions are preferred in order to minimize amorphous carbon production.

For instance, it has been shown that 4:1 (vol/vol 96% H_2SO_4/30% H_2O_2) piranha solutions are able to attack existing damage sites on the nanotube sidewalls to chemically shorten SWNTs [4]. While hot (70°C) piranha solutions can induce more damage sites, room-temperature (22°C) solutions can react with existing sidewall defects but are incapable of initiating further damage sites, resulting in minimal carbon loss and slow etch rates (Fig. 2). On the other hand, high-temperature conditions produce increasingly shorter US-tubes with increasing reaction time, with approximately half of the starting material destroyed due to chemical etching. Therefore, room-temperature piranha solutions offer the ability to exploit active damage sites in the sidewalls of the nanotube in a controlled manner without the counter-productive destruction of the nanotubes.

The aggressive oxidation of SWNTs with oleum (100% H_2SO_4 with 20% SO_3) and nitric acid can simultaneously shorten and carboxylate the SWNTs, rendering water-soluble US-tubes less than 60 nm in length [5]. Purified SWNTs are first dispersed in oleum to disentangle individual SWNTs from their rope-like bundles. Next, the SWNT-oleum dispersion is extensively oxidized with strong HNO_3 at elevated temperatures to cut the SWNTs into US-tubes. The oxidation of the SWNTs also forms carboxylic acid groups at the sidewall defects of the SWNTs, making US-tubes that are up to 2 wt% soluble in polar organic solvents, acids, and water.

One nonoxidative process to synthesize US-tubes involves the chemical shortening of SWNTs to lengths of less than 50 nm by a fluorination/pyrolysis treatment [6]. Briefly, purified SWNTs are exposed to fluorine gas, during which fluorine atoms tend to arrange around the circumference of the nanotube to form band-like regions along the nanotube sidewalls. Upon pyrolysis of the fluorinated SWNTs at 1,000°C in an inert atmosphere, the chemically bound fluorine atoms are driven off in the form of CF_4 and COF_2. This leaves behind chemically cut US-tubes with partial sidewall defects along the nanotube axis.

Mechanical Cutting

There have been various documented methods to mechanically cut SWNTs into shorter segments, including ultrasonication, grinding, and ball milling. However, these techniques lack length control; such mechanical methods create extensive sidewall damage and are not able to achieve sub-100 nm lengths. Therefore, more precise mechanical cutting techniques have been developed.

Microlithography can be used to cut SWNTs into US-tubes [7]. This technique involves a protective photoresist polymer layered on top of a SWNT-covered solid substrate in a patterned manner to protect portions of nanotubes from lithographic damage and modification. Reactive ion etching using an oxygen plasma is then used to remove the unprotected sections of the SWNTs. After etching, the newly formed nanotube ends remain, with dangling carboxylic acid and ether groups. The photoresist polymer can then be removed, leaving cut segments of SWNTs. Regardless of whether the SWNTs are previously aligned or properly dispersed prior to cutting, microlithography offers a straightforward method to shorten SWNTs with a narrow and selective length distribution.

Electron beam etching has also been used to cut SWNTs with nanometer precision [8]. Using electron energy beams exceeding 100 keV, which is made

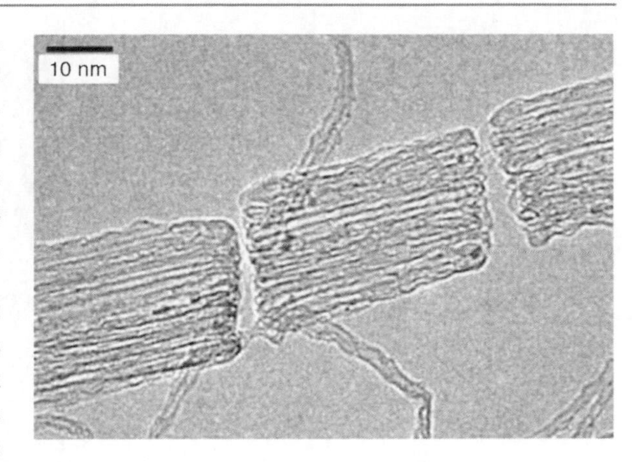

Ultrashort Carbon Nanotubes, Fig. 3 Transmission electron microscopy (TEM) image of the complete cutting of a SWNT bundle into US-tubes using electron beam etching (Reproduced with permission from [8])

possible by currently available electron microscopes, partial and complete cuts into SWNTs can be made with great control without the need of covering portions of SWNTs with a protective layer. Similar methods have been developed using high voltage-bearing scanning tunneling microscope (STM) tips. Interestingly, upon cutting into a SWNT bundle, the newly formed tube ends immediately form hemispherical caps, enhancing the stability of the cut US-tubes (Fig. 3). Another peculiar observation is that forming a second cut within 10–50 nm from a preexisting cut proves more difficult to perform. This increase in stability is thought to be a result of the generation and annealing of vacancy-interstitial atom pairs.

More recently, an ultramicrotome has been used to cut frozen layers of magnetically aligned SWNT membranes, or Buckypapers [9]. Enough mechanical control was obtained to create open-ended US-tubes with minimal sidewall damage, typically seen after chemical oxidative methods. When the Buckypaper was cut into 50 nm segments, the mean length was about 87 nm, with about 60% of the tubes ranging from 50 to 150 nm. However, the large pressure exerted by the diamond knife tip in its proximity can cause deformation and cracking of the nanotubes.

Individualization of US-tubes

Like their SWNT precursors, US-tubes have a large tendency to bundle together, which renders it difficult to form single-tube suspensions. Although it is

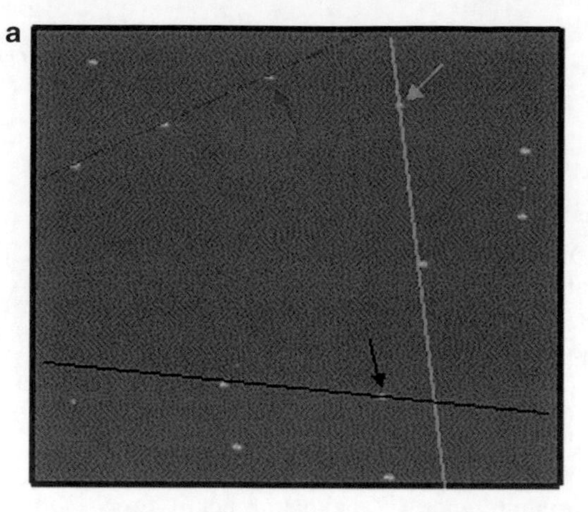

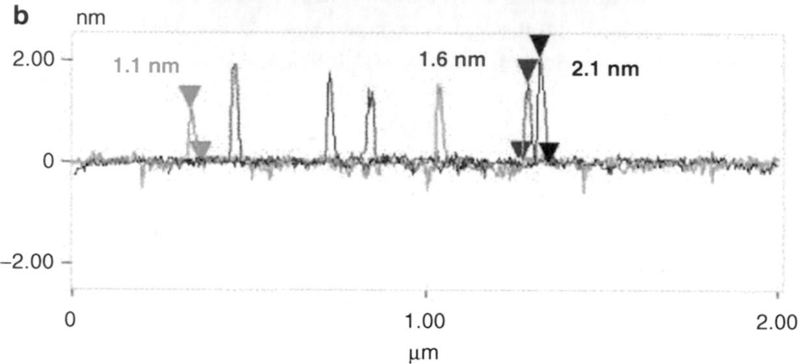

Ultrashort Carbon Nanotubes, Fig. 4 AFM image of individualized US-tubes after chemical reduction (Reproduced with permission from [10])

possible for full-length SWNTs to disperse in solution with the aid of polymers and surfactants, US-tubes cannot disperse as readily in these same conditions [10]. Because of their high aspect ratios, SWNTs are relatively flexible and can "peel" more readily from each other upon introduction to surfactant wrapping. On the other hand, US-tubes have a much lower aspect ratio, therefore making them more rigid and more difficult for surfactant molecules to insert between individual US-tubes. Therefore, chemical means to debundle and individualize US-tubes have been explored to better disperse US-tubes.

Reduction of US-tubes

Chemical reduction can be used to individualize, or debundle, US-tubes [10]. Upon sonication with K^0 metal in THF, the negative charge imparted on the US-tubes electrostatically overcomes the bundling forces to yield individualized US-tubes (Fig. 4). It is thought that the individualization of US-tubes via chemical reduction increases US-tube dispersion in solution, thus allowing for more even chemical functionalization.

Functionalization of US-tubes

The chemical functionalization of US-tubes is thought to occur at carboxylic acid end groups of the sidewall defects. This allows for the attachment of various chemical groups that would render the US-tubes more soluble in different solvent systems. One functionalization method utilizes an in situ Bingel-Hirsch (CBr_4/DBU) cyclopropanation to functionalize US-tubes with the water-soluble malonic acid bis-(3-tert-butoxycarbonylaminopropyl)ester to yield four to five adducts per nm [10]. Another documented functionalization method involves cyclopropanation using diazoacetic ester in the presence of $Rh_6(CO)_{16}$ [11]. This allows for the US-tube to be readily reactive with primary amines, thus making possible the functionalization of US-tubes with a number of amino acid derivatives, such as water-soluble serine, with 3–19 groups per nm. Other biologically relevant moieties, such as antioxidants [12] and oligonucleotides [13], have been successfully attached onto US-tubes and are described below. Alkylation-based reduction can also be performed on US-tubes to improve nanotube dispersion in nonpolar environments [14].

Loading of US-tubes

In addition to being able to chemically functionalize the sidewalls of US-tubes, the hollow interior of the US-tubes can be loaded or filled with various ions and small molecules, such as Gd^{3+} ions [15], molecular I_2 [16], and $^{211}AtCl$ radionuclides [17]. This property is especially relevant for drug design, as the carbon structure of the US-tube serves as an encapsulating sheath to protect potentially toxic moieties from their surrounding environment. Although the loading mechanism is not completely clear, it is likely that ions and small molecules load into the US-tubes through the sidewall defects.

Phenomena

Toxicology

Before US-tubes are considered for biomedical applications, their toxicological profiles must be known. A recent study documented the toxicology as well as the acute and subchronic effects of high doses (50–1,000 g/kg bodyweight) of Tween-suspended US-tubes in Swiss mice [3]. Oral administration of US-tubes did not result in death or any behavioral abnormalities for up to 14 days. Intraperitoneal administration allowed for the US-tubes to reach systemic circulation and various tissues through the lymphatic system. As opposed to unfunctionalized SWNTs, well-individualized US-tubes had the ability to pass through the reticuloendothelial system, excreting via the kidneys and bile ducts. High doses of US-tubes can induce strongly adherent granuloma formation on and inside organs due to large (>10 μm) nanotube aggregation into fiber-like structures. However, small (<2 μm) aggregates of US-tubes can readily enter various cell types without granuloma formation or any sign of toxicity. Neither death nor growth or behavioral abnormalities were observed for the animals used in the experiment.

Applications

Diagnostic Agent Design

Magnetic Resonance Imaging (MRI) Contrast Agents

Gadonanotubes, or US-tubes loaded with Gd^{3+} ions, have been shown to outperform currently used clinical T_1-weighted MRI contrast agents by approximately 40-fold [15]. Although the exact loading mechanism is unclear, it is likely that the Gd^{3+} ions enter the US-tubes through the sidewall defect sites, where they form nanoscale clusters. Because of the lipophilicity of the carbon sidewalls and their short lengths, Gadonanotubes can also readily translocate across cell membrane, allowing for Gadonanotubes to become an effective T_1-weighted cellular MRI contrast agent. It has already been shown that Gadonanotubes can be readily internalized and imaged inside a variety of cells, including mesenchymal stem cells (Fig. 5) [18].

In the absence of Gd^{3+} ions, the US-tubes alone can perform as a T_2-weighted MRI contrast agent [19]. The US-tubes are approximately three times more efficacious than Ferumoxtran®, a clinically used Fe_3O_4-based contrast agent. Although the US-tubes contain much less (<1%) iron catalyst than unpurified SWNTs, the US-tubes exhibit a much higher r_2 performance, suggesting that the carbon nanotube material itself is contributing to the superparamagnetic nature and that the iron particles are not the only contributing factor to the superparamagnetic signature of the US-tubes. The shorter lengths and the sidewall defects of the Gadonanotubes may also contribute to the high performance. This also implies that Gadonanotubes can concurrently perform as both a T_1- and T_2-weighted MRI contrast agents. Gadonanotubes have previously been used as a T_2-weighted contrast agent in vivo, with r_2 relaxivities as high as 578 mM^{-1} s^{-1} at 7 T [20]. This suggests that the Gadonanotubes are useful T_1 positive agents at lower magnetic fields and T_2 negative agents at higher fields.

Computed Tomography (CT) Agents

Molecular I_2 has also been loaded into US-tubes for CT X-ray imaging contrast agent design [16]. After being functionalized with serinol amide groups via the Bingel-Hirsch reaction, the US-tubes were filled with gaseous I_2 at 100°C, then washed with ethanol and reduced with NaH to remove excess I_2 from the nanotube exterior. A 10 wt% loading of I_2 was observed for the US-tubes. Using XPS and X-ray-induced Auger emission spectroscopy, it was confirmed that the molecular I_2 was sequestered within the US-tubes. Micro-CT images were obtained to show that the iodine-loaded US-tubes were more X-ray opaque. These serinol amide-functionalized, I_2-loaded US-tubes are the first water-soluble CT contrast agent derived from SWNT materials.

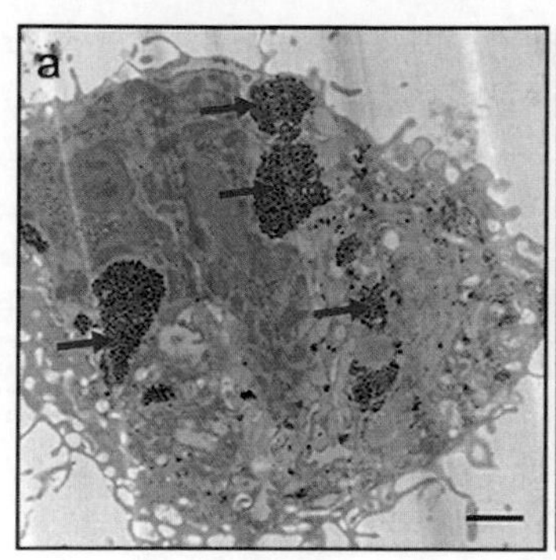

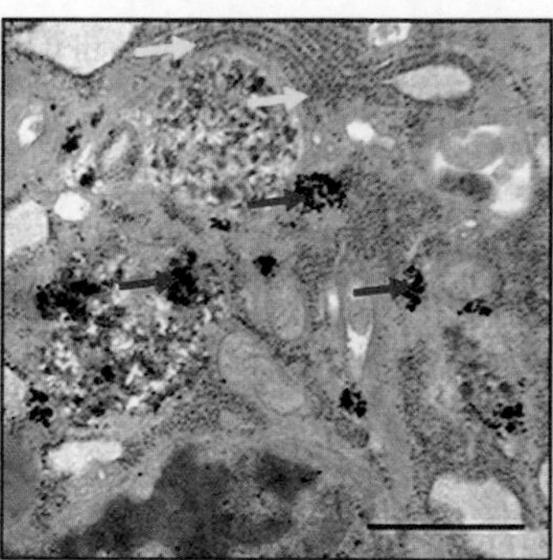

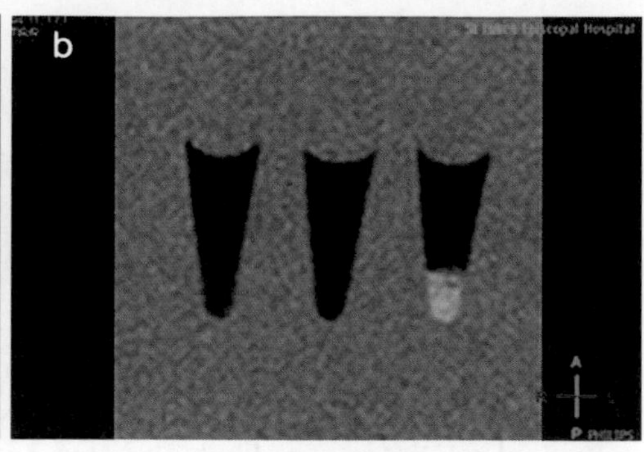

Ultrashort Carbon Nanotubes, Fig. 5 (**a**) TEM images of a Gadonanotube-labeled mesenchymal stem cell (MSC). *Red arrows* point to the Gadonanotube aggregates. *Yellow arrow* points to the ribosomes of the endoplasmic reticulum. (**b**) T_1-weighted MR images at 1.5 T and 25 C of (*left* to *right*) unlabeled MSCs, Gd-DTPA-labeled MSCs, and Gadonanotube-labeled MSCs at $T_i = 500$ ms. Scale bar = 1 μm (Adapted with permission from [18])

Therapeutic Agent Design

Radionuclide Containment and Delivery

Similar to the loading of Gd^{3+} ions and I_2 molecules into US-tubes, $^{211}At^-$ ions can be loaded and oxidized into $^{211}AtCl$ within US-tubes to contain and deliver the alpha-emitting radionuclides in the nanotube capsule [17]. Upon oxidizing the $^{211}At^-$ ions with chloramine-T or N-chlorosuccinimide, the AtCl molecules labeled the US-tubes to a great extent, with 91.3% by aqueous loading. Subsequent PBS washing and serum exposure removed excess AtCl residing on the exterior or at the sidewall defects of the nanotubes, but the US-tubes exhibit reasonable stability toward physiological challenge.

Bone Tissue Scaffold Composites

US-tubes have also been used to dope the highly porous (poly(propylene fumarate)) polymer for bone tissue scaffold engineering [14]. These composites are injectable, thermally cross-linkable, and cytocompatible, making them ideal for trabecular bone tissue scaffolding biomaterials. While high (75–90 vol %) porosity is preferred to allow for cell adhesion, proliferation, and vascularized tissue ingrowth, the mechanical properties of highly porous polymer scaffolds become compromised. Therefore, US-tubes have been added to PPF polymer scaffolds to improve the mechanical properties. It was found that dodecylated US-tubes functionalized with dodecyl groups better dispersed into PPF, improving their reinforcing effects. While there was no significant difference in porosity, pore size, and pore interconnectivity among US-tube-doped PPF and PPF alone, the compressive modulus, offset yield strength, and compressive strength of the dodecylated US-tube nanocomposites were higher than or similar to that of PPF alone. The US-tube/PPF nanocomposite has also been shown to be osteoconductive, allowing marrow stromal cells to attach and proliferate readily on the scaffold.

Gadonanotubes have also been incorporated into poly(lactic-co-glycolic acid) (PLGA) polymer scaffolds in order to explore the degradation process of the scaffold and the biodistribution of the Gadonanotubes upon their release from the polymer matrix in vivo [20]. T_2-weighted MR imaging of the Gadonanotube-PLGA nanocomposites and their degradation process were obtained in vitro, showing the strong influence of the Gadonanotubes on surrounding water proton relaxation (Fig. 6). In addition to providing mechanical reinforcement from the US-tube structure, the Gadonanotube-reinforced PLGA scaffolds may elucidate the mechanism of the biodegradation of the polymer and the release and biodistribution of the Gadonanotubes in vivo.

Free Radical Scavengers

US-tubes have also been functionalized as a scaffold for the phenolic antioxidant, butylated hydroxytoluene (BHT) using DCC/DMAP coupling [12]. As expected, the higher amount of BHT functionalization (either via covalent or noncovalent interactions) to the carboxylic acid groups on the oxidized US-tubes corresponded to higher overall antioxidant activity, with performances more than 200 times better than that of the radioprotective dendritic fullerene, DF-1. Strikingly, PEGylated US-tubes alone without any BHT functional groups

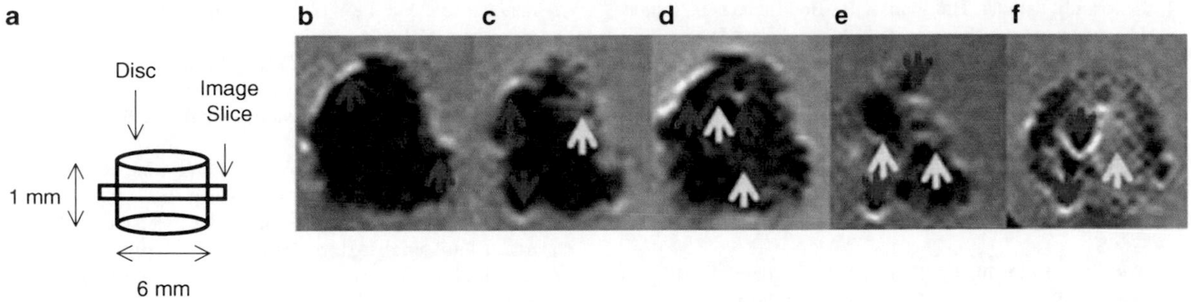

a b c d e f

Disc
Image Slice
1 mm
6 mm

Ultrashort Carbon Nanotubes, Fig. 6 (**a**) Schematic of the nanocomposites sample arrangement within the MRI. Representation two-dimensional images through a nanocomposite disc after (**b**) 2 h, (**c**) 24 h, (**d**) 3 days, (**e**) 5 days, and (**f**) 7 days. The higher (*white*) pixel intensities within the dark disc represent regions of higher water concentration. The high intensity surrounding the disc is the agarose gel (TE = 15 ms) (Reproduced with permission from [20])

are extremely effective antioxidants, performing nearly 40 times greater than DF-1 in oxygen scavenging ability. Even though the US-tubes have heavily compromised sidewalls from the oxidative cutting process, US-tubes are still able to sequester oxygen radicals due to the sp^2-hybridized carbon nanotube framework.

Oligonucleotide Delivery

US-tubes have also served as delivery vehicles for oligonucleotide (ODN) decoys, which are synthetic ODNs containing the DNA-binding sequence of a transcription factor [13]. Because US-tubes can inherently translocate into cells, US-tubes can efficiently facilitate the intracellular delivery of ODN decoys into target cells so that ODN decoys can better target transcription factors of interest for gene therapy. Specifically, US-tubes have successfully delivered a decoy ODN against nuclear factor-κB (NF-κB), which is a transcription factor that regulates genes important in immune and inflammatory responses. Using carbodiimide chemistry, the amino groups at the 5′ end of the double-stranded ODN was covalently coupled to the carboxylic acid groups of the oxidized US-tubes. To test the performance of the US-tubes as a delivery vehicle, NF-κB-dependent gene expression was significantly down-regulated in monocyte-derived human macrophages incubated with ODN-functionalized US-tubes, as compared to macrophages incubated with US-tubes functionalized with nonspecific ODN decoys.

Future Directions

While there have been significant strides made in the development and application of US-tubes, many challenges still must be addressed. In order to produce more uniform US-tubes from SWNTs, better size control during the cutting procedure and size sorting after synthesis must be achieved. Additionally, improving the solubility and dispersion of US-tubes in biological media must be achieved to fully realize their potential in biomedical applications.

Taking advantage of the exterior functionalization and the interior loading capabilities of US-tubes, a nearly limitless number of possibilities abound for advanced US-tube-based therapeutic and diagnostic agent design. US-tubes can either be used as a nanocapsule, sequestering biologically relevant ions or molecules that may be inherently toxic if not sequestered, or as a nanovector, delivering various drug payloads or peptide sequences to target cells. Nevertheless, much work still needs to be completed to better characterize the acute effects, long-term toxicity, and biodistribution of US-tubes, as well as other types of carbon nanostructures, in living systems.

Cross-References

► Carbon Nanotubes
► Fullerenes for Drug Delivery

References

1. Kostarelos, K., Bianco, A., Prato, M.: Promises, facts and challenges for carbon nanotubes in imaging and therapeutics. Nat. Nanotechnol. **4**, 627–633 (2009)
2. Liu, Z., Tabakman, S., Welsher, K., Dai, H.: Carbon nanotubes in biology and medicine: *in vitro* and *in vivo* detection, imaging and drug delivery. Nano Res. **2**, 85–120 (2009)

3. Kolosnjaj-Tabi, J., Hartman, K.B., Boudjemaa, S., Ananta, J.S., Morgant, G., Szwarc, H., Wilson, L.J., Moussa, F.: In vivo behavior of large doses of ultrashort and full-length single-walled carbon nanotubes after oral and intraperitoneal administration to Swiss mice. ACS Nano **4**(3), 1481–1492 (2010)
4. Ziegler, K.J., Gu, Z., Peng, H., Flor, E.L., Hauge, R.H., Smalley, R.E.: Controlled oxidative cutting of single-walled carbon nanotubes. J. Am. Chem. Soc. **127**(5), 1541–1547 (2005)
5. Chen, Z., Kobashi, K., Rauwald, U., Booker, R., Fan, H., Hwang, W.-F., Tour, J.M.: Soluble ultra-short single-walled carbon nanotubes. J. Am. Chem. Soc. **128**(32), 10568–10571 (2006)
6. Gu, Z., Peng, H., Hauge, R.H., Smalley, R.E., Margrave, J.L.: Cutting single-wall carbon nanotubes through fluorination. Nano Lett. **2**(9), 1009–1013 (2002)
7. Lustig, S.R., Boyes, E.D., French, R.H., Gierke, T.D., Harmer, M.A., Hietpas, P.B., Jagota, A., McLean, R.S., Mitchell, G.P., Onoa, G.B., Sams, K.D.: Lithographically cut single-walled carbon nanotubes: controlling length distribution and introducing end-group functionality. Nano Lett. **3**(8), 1007–1012 (2003)
8. Banhart, F., Li, J., Terrones, M.: Cutting single-walled carbon nanotubes with an electron beam: evidence for atom migration inside nanotubes. Small **1**(10), 953–956 (2005)
9. Wang, S., Liang, Z., Wang, B., Zhang, C., Rahman, Z.: Precise cutting of single-walled carbon nanotubes. Nanotechnology **18**(5), 055301 (2007)
10. Ashcroft, J.M., Hartman, K.B., Mackeyev, Y., Hofmann, C., Pheasant, S., Alemany, L.B., Wilson, L.J.: Functionalization of individual ultra-short single-walled carbon nanotubes. Nanotechnology **17**, 5033–5037 (2006)
11. Mackeyev, Y., Hartman, K.B., Ananta, J.S., Lee, A.V., Wilson, L.J.: Catalytic synthesis of amino acid and peptide derivatized gadonanotubes. J. Am. Chem. Soc. **131**, 8342–8343 (2009)
12. Lucente-Schultz, R.M., Moore, V.C., Leonard, A.D., Price, B.K., Kosynkin, D.V., Lu, M., Partha, R., Conyers, J.L., Wilson, L.J.: Antioxidant single-walled carbon nanotubes. J. Am. Chem. Soc. **131**, 3934–3941 (2009)
13. Crinelli, R., Carloni, E., Menotta, M., Giacomini, E., Bianchi, M., Ambrosi, G., Giorgi, L., Magnani, M.: Oxidized ultrashort nanotubes as carbon scaffolds for the construction of cell-penetrating NF-κB decoy molecules. ACS Nano **4**(5), 2791–2803 (2010)
14. Shi, X., Sitharaman, B., Pham, Q.P., Liang, F., Wu, K., Billups, W.E., Wilson, L.J., Mikos, A.G.: Fabrication of porous ultra-short single-walled carbon nanotube nanocomposite scaffold for bone tissue engineering. Biomaterials **28**, 4078–4090 (2007)
15. Sitharaman, B., Kissell, K.R., Hartman, K.B., Tran, L.A., Baikalov, A., Rusakova, I., Sun, Y., Khant, H.A., Ludtke, S.J., Chiu, W., Laus, S., Tóth, É., Helm, L., Merbach, A.E., Wilson, L.J.: Superparamagnetic gadonanotubes are high-performance MRI contrast agents. Chem. Commun. **31**, 3915–3917 (2005)
16. Ashcroft, J.M., Hartman, K.B., Kissell, K.R., Mackeyev, Y., Pheasant, S., Young, S., Van der Heide, P.A.W., Mikos, A.G., Wilson, L.J.: Single-molecule I_2@US-Tube nanocapsules: a new x-ray contrast-agent design. Adv. Mater. **19**, 573–576 (2007)
17. Hartman, K.B., Hamlin, D.K., Wilbur, S., Wilson, L.J.: ^{211}AtCl@US-Tube nanocapsules: a new concept in radiotherapeutic-agent design. Small **3**(9), 1496–1499 (2007)
18. Tran, L.A., Krishnamurthy, R., Muthupillai, R., Cabreira-Hansen, M.G., Willerson, J.T., Perin, E.C., Wilson, L.J.: Gadonanotubes as magnetic nanolabels for stem cell detection. Biomaterials **31**(36), 9482–9491 (2010)
19. Ananta, J.S., Matson, M.L., Tang, A.M., Mandal, T., Lin, S., Wong, K., Wong, S.T., Wilson, L.J.: Single-walled carbon nanotube materials as T_2-weighted MRI contrast agents. J. Phys. Chem. C. **113**, 19369–19372 (2009)
20. Sitharaman, B., Van Der Zande, M., Ananta, J.S., Shi, X., Veltien, A., Walboomers, X.F., Wilson, L.J., Mikos, A.G., Heerschap, A., Jansen, J.A.: Magnetic resonance imaging studies on gadonantube-reinforced biodegradable polymer nanocomposites. J. Biomed. Mater. Res. A **93A**(4), 1454–1462 (2010)

Ultrasonic Atomization

► Acoustic Nanoparticle Synthesis for Applications in Nanomedicine

Ultrasonic Machining

Suhas S. Joshi
Department of Mechanical Engineering, Indian Institute of Technology Bombay, Mumbai, Maharastra, India

Synonyms

Non-conventional machining; USM; Vibration assisted machining

Definition

Ultrasonic machining involves imparting ultrasonic vibrations (of frequency ~20 kHz) to a tool to effect material removal. The process is more effective on materials that have hardness more than RC 40, but it is used on almost all including metallic and nonmetallic materials such as glasses, ceramics, and composites.

Process Variants

The process is configured into two main types (Fig. 1a–b) depending upon how the force is transferred from the vibrating tool to the work surface. If the transfer of vibration energy is achieved by introducing a liquid containing abrasive grits in the gap between the tool and work surface, the process is called "ultrasonic impact grinding" or simply "ultrasonic machining." On the other hand, if the abrasive grits are not present in the carrier liquid, then the tool is rotated besides being vibrated, and the process is called as "rotary ultrasonic machining" [1, 2].

Basic differences in these two processes are illustrated in Table 1.

Basic Equipment

Equipment for ultrasonic machining is available in various forms, like dedicated machines or as an attachment to other machines as well as table-top machines. Typical machine elements for performing both the ultrasonic machining (USM) and rotary ultrasonic machining (RUM) are identical. They include [1, 2]:

1. *Power Supply*: a power supply that provides electrical output at ultrasonic frequency
2. *Transducer*: a transducer that converts high-frequency electrical supply to tool displacement at ultrasonic frequency. The transducers work on magnetostrictive or piezoelectric principles. The magnetostrictive effect involves application of alternating current to a ferromagnetic coil, which induces vibrations at the applied frequency in the coil. The phenomenon allows transfer of vibrations over wider frequency band of (17–23 kHz). However, it also causes high electrical losses (e.g., eddy current loss) and has low energy conversion efficiency (~ 50%). Most of the energy losses appear as heat; hence, these transducers require air or water cooling. Consequently, such systems are bulky. On the other hand, the piezoelectric transducers comprise of a stack of ceramic discs placed between a high-density material base and low-density material radiating face. The energy conversion efficiency can be in the range of (~90–96%). Therefore, they do not require cooling.
3. *Horn*: the horn is referred with various names such as tool holder, sonotrode, stub, concentrator, or acoustic coupler. With appropriate design of horn, the vibration amplitude can be amplified by as high as 600% over its initial value at the transducer surface of ~0.001–0.1 μm. The increase in the vibration amplitude is inversely proportional to the reduction in the area ratio between top and bottom faces of the horn. Usually, the horn is made from Monel, titanium 6-4, stainless steel, aluminum, or aluminum bronze.
4. *Tool*: the tool is shaped such that it is the inverse of the cavity to be made. The tool is attached to the horn either by brazing or soldering. This helps avoid possible fatigue failure or self-loosening of the threaded fasteners. The tool is designed such that the amplitude of vibrations is maximum at its free end. Mass and length of the tool are also important. While the bulky tools absorb ultrasonic vibrations, very long tools will flutter during the process. Therefore, tools with slenderness ratio of less than 20:1 are recommended. Often when the tools are of large cross-sectional area, the tool center faces slurry starvation. Therefore, the tools with larger perimeter for same cross-sectional area are preferred. The tools should be designed to have flutes to aid slurry flow away from the cut. Sometimes the tools are relieved behind the face. In the case of longer tools, availability of center holes for feeding abrasive slurry is of great advantage as it avoids sidewall friction. The tool materials should be tough and ductile and should have high wear and fatigue resistance. The tool materials usually include mild steel, stainless steel, brass, Monel, bearing steel, and molybdenum. Softer materials like aluminum or brass may face significant tool wear.
5. *Abrasives*: the abrasives transfer the force of vibrations to the work surface and impinge on the surface to perform removal operation. The abrasives material should be harder than the work material, and the size of the abrasive grits should be the same as that of the amplitude of vibrations being used. Abrasives are normally carried by slurry containing 50% volume of abrasives and 50% volume of water. The rate of penetration of tool in the work surface increases with increasing abrasive concentration but reaches a maximum, beyond which the jamming of the abrasives takes place, thereby the tool penetration reduces. The abrasives used for the USM process include diamond, cubic boron nitride, boron carbide, silicon carbide, and aluminum

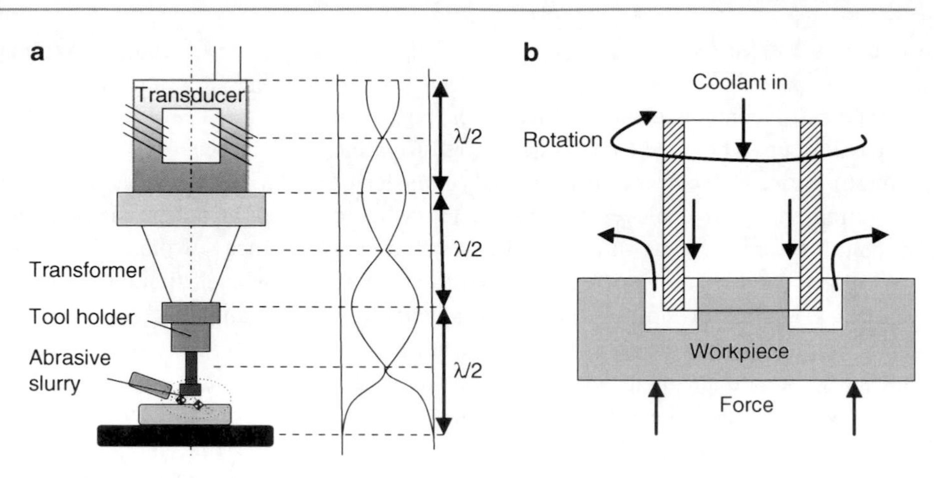

Ultrasonic Machining, Fig. 1 (**a**) Ultrasonic machining (**b**) Rotary ultrasonic machining

Ultrasonic Machining, Table 1 Comparison between the two variants of USM

Parameters	USM	RUM
Tool motions	Reciprocation	Reciprocation and rotation
Tool-work interaction	No contact	Contact
Tool materials	Tough and ductile metals, mild steel, stainless steel	Diamond tools
Tool shape	Mirror image of the desired cavity	Core drill type tool
Abrasives	Diamond, CBN, boron carbide, silicon carbide, aluminum oxide	Diamond particles impregnated in the tool
Carrier liquid functions	Carrying abrasive grits and cooling	No abrasive grits hence only cooling
Force transfer	Tool vibrations force abrasive grits to impinge on work surface	Vibrating diamond tool contacts work surface to cause removal
Removal mechanism	Indentation and chipping	Indentation and abrasion
Shapes generated	Mirrors tool shape	Holes and other axisymmetric shapes
Size of features generated	About 90 mm diameter and 64 mm deep	Maximum diameter 38–50 mm

oxide. Boron carbide is the most widely used abrasive material. It is often used in the processing of tungsten carbide, ceramics, minerals, metals, and precious and semiprecious stones. Silicon carbide is used for low-density ceramics, glasses, Si, Ge, and mineral stones. Aluminum oxide is used for machining of glasses and sintered or hard powder components.

Mechanism of Removal

Typical sequence of events that occurs in ultrasonic machining includes (Fig. 2) [3]:

1. Indentation by abrasive particles leading to generation of an inelastic deformation zone around the particle.
2. At some threshold, deformation induced flow develops into a microcrack, termed as a "median" or "lateral" crack.
3. An increase in load further causes steady growth of the median crack.
4. During unloading action, the median crack begins to close inducing formation of lateral vents.
5. Upon complete unloading, the lateral vents continue their extension towards specimen surface and lead to chipping or forming a fragmented section on the work surface.

The critical load for the initiation of the median crack is given by [4]

$$P_c = \alpha \frac{K_{IC}^4}{H_v^4} \tag{1}$$

where, α is a dimensionless factor. The size of the mean crack (C_s) is given by [5]

$$(C_s)^m = K \cdot P \tag{2}$$

where $m = 1 - 1.5$ and k is a coefficient. This indicates that the size of the median crack grows with an increase in the load and a decrease in the fracture toughness of the work material.

A simple model to evaluate material removal rate (MRR) in ultrasonic machining, which gives

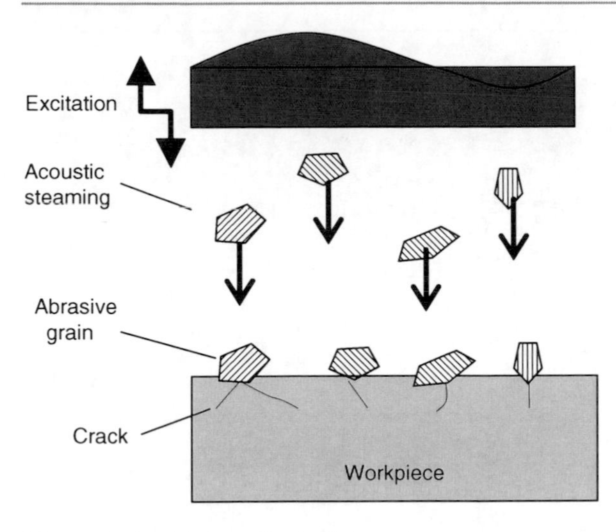

Ultrasonic Machining, Fig. 2 Schematic of crack formation due to impact in USM

contribution of direct impact of the abrasive particles on the work surface in total MRR, was developed by Shaw [6]. The model is primarily based on the assumption that the work volume removed is proportional to the number of particles making impact per cycle and the volume of work material removed per particle impact. The abrasive grains are assumed to be identical and spherical in shape. The MRR therefore is given by [6]

$$Z_w = k \cdot n \cdot f(d_g \cdot \delta)^{\frac{3}{2}} \tag{3}$$

where δ is the depth of indentation, n is the number of impacting particles per cycle, f frequency of ultrasonic vibrations, and k coefficient.

In the above equation, the depth of indentation is given by

$$\delta = \left[\frac{8F_s A d_g}{\pi k_1 H_v C(1+q)}\right]^{\frac{1}{2}} \tag{4}$$

where F_s is static force, A is amplitude of ultrasonic vibration, k_1 is the constant of proportionality, H_v is the hardness of workpiece, C concentration of slurry, and q ratio of the hardness of the workpiece to that of the tool.

Therefore, substituting Eq. 3 in 4, we get, MRR for ultrasonic machining as

$$Z_w = k\,n f (d_g)^{\frac{3}{2}} \left(\frac{8F_s A d_g}{\pi k_1 H_v C(1+q)}\right)^{\frac{3}{4}} \tag{5}$$

The Shaw's model is the most widely used because of its simplicity, despite its limitations. It predicts monotonous increase in the MRR with an increase in static force, F_s. However, in practice, this is not so. After a certain increase in the static force, the abrasive particles tend to crack, thereby reducing the MRR.

The above mechanism prevails mainly during machining of brittle work materials that are more susceptible to fracture. However, when the work materials are ductile, such as metals, the abrasives have tendency to get embed in the surface. The embedded particles work harden the surface, and upon successive impacts, chipping removal from the work surface occurs.

Thus, though USM can be used on both brittle and ductile work materials, it is more effective in the case of former than the latter.

In the RUM, the removal mechanism involves linear as well as rotational motion to the abrasives that are bonded to the tools, (Fig. 3). Therefore, the actions involved include hammering followed by abrasion and extraction [7].

Extensive work on the mechanism of removal shows that besides indentation, multiple mechanisms operate in USM process; these include [8]:

1. Mechanical abrasion by hammering of abrasive particles
2. Microchipping by impact of free abrasives
3. Cavitation effects from abrasive slurries
4. Chemical action associated with the fluid employed

Process Parameters

The theoretical as well as experimental investigations on the process have shown that the MRR in the ultrasonic machining is a function of a large number of parameters such as [9] (1) frequency, (2) amplitude, (3) static force, (4) hardness ratio (q) of the work to tool, (5) grain size, and (6) concentration of abrasive slurry. A pictorial summary of these effects is presented in Fig. 4a–f.

It is understood that the MRR increases with an increase in the frequency and the amplitude of ultrasonic vibrations (Fig. 4a). The MRR increases with an increase in the static force, but after a certain increase in the static force (Fig. 4b), the abrasive particles tend to break thereby reducing the MRR. A decrease in the hardness of workpiece in the ratio (q) causes a rapid reduction in MRR (Fig. 4e). This indicates that the

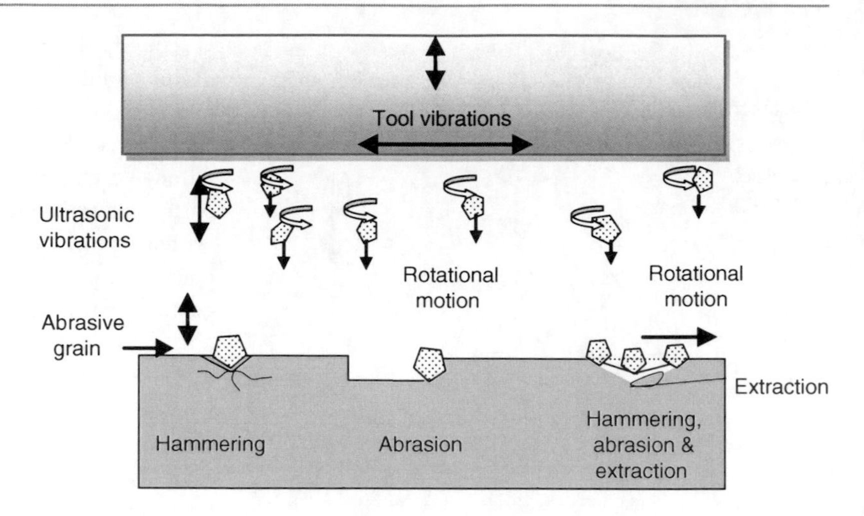

Ultrasonic Machining, Fig. 3 Removal mechanism in RUM

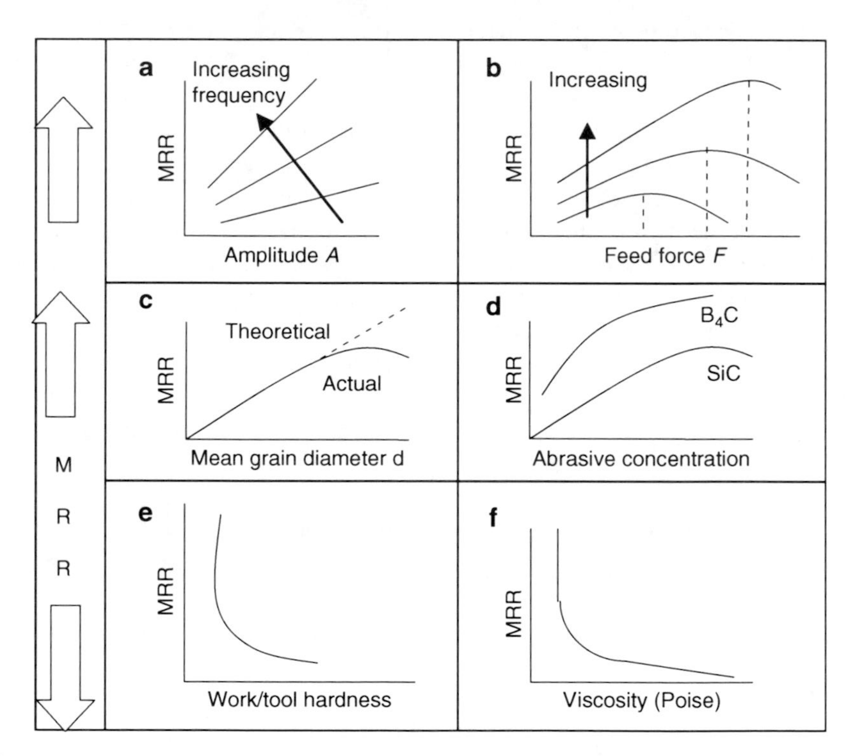

Ultrasonic Machining, Fig. 4 (**a–f**) Variation of MRR with process parameters in USM

brittle work materials are machined rapidly by USM. A table showing relative removal rate of USM in machining of glass by considering it as 100 is presented in Table 2.

The MRR increases in proportion to the increase in the grain size (Fig. 4c) until the grain size becomes more than the amplitude of vibrations, that is, where the particles tend to fracture.

The MRR increases proportional to the one-fourth power of slurry concentration, i.e., $C^{1/4}$ (Fig. 4d). However, the increasing trend continues till concentration reaches 30%, beyond which an increase in the concentration does not help. The MRR decreases significantly with a decrease in the viscosity of the carrier liquid (Fig. 4f).

The machined surface roughness using USM depends upon grain size of abrasives [8]. The smaller the grain size, lower the roughness. In addition, surface roughness is proportional to $\{\text{hardness (H)/elastic modulus (E)}\}^n$ as depicted in Fig. 5a. The work materials that have low H/E ratio are found to give lower surface roughness.

Ultrasonic Machining, Table 2 Relative removal rates in USM (For f = 16.3 kHz, A = 12.3 μm, mesh size of abrasives = 100) [9]

Work Material	Relative Removal rate
Glass	100.0
Brass	6.6
Tungsten	4.8
Titanium	4.0
Steel	3.9
Chrome steel	1.4

Again, in most of the cases, the RUM process is found to give better surface finish than the USM (Fig. 5a). Nevertheless, the values as low as 0.4 μmRa for surface roughness are obtainable by using the process.

The holes made by USM are oversized and have form errors [8]. The oversize is maximum at the entry the of hole, and it increases as the length of the hole increases. Usually, the upper limit on the oversize is equal to the maximum size of the abrasives used. The oversize and out of roundness are found to reduce with increasing the static load, as it suppresses the lateral vibrations of the tool. Again, the work materials with lower H/E ratio again show lower out of roundness (Fig. 5b).

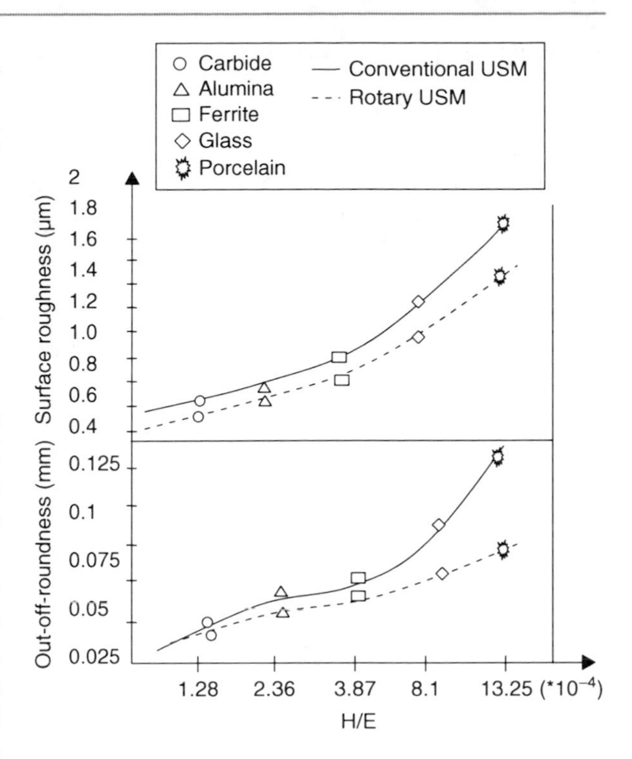

Ultrasonic Machining, Fig. 5 Effect of H/E on (**a**) surface roughness for various materials under USM and RUM (**b**) out of roundness [8]

Tool Wear

Tool wear is an important variable in USM that affects both MRR and hole accuracy [8]. In general, the tool wear occurs along the length (W_L) as well as along the diameter (W_D) of the tool. It is understood that the tool wear is maximum when the static load is maximum. Also, at the maximum static load, the MRR is also maximum. Therefore, the maximum static load may be considered as the optimum point for the tool wear. The tool wear tends to increase with an increase in the hardness and size of the abrasives. W_L is more influenced by the product of hardness (H) and impact strength (K_i) of the tool material. On the other hand, the W_D is more influenced by the hardness (H) of the tool materials. Therefore, Nimonic 80A, thoriated tungsten, or silver steel is highly recommended as tool materials. The tool wear also increases with an increase in the depth of hole drilled and machining time.

A number of times, fatigue failure of the tools is observed in the USM process. Therefore, the tool materials should have high fatigue strength too.

Recent Trends

The recent developments in ultrasonic machining include use of various tools and techniques to improve performance of the USM process and the application of ultrasonic vibrations in various other manufacturing processes to improve their productivity.

These developments include (a) use of automatic and intelligent machining control systems, (b) use of elliptical vibrations during ultrasonic machining to improve MRR, (c) use of numerical tools to design the horn for efficient conversion of longitudinal vibrations to mixed lateral and longitudinal vibrations thereby maximizing the MRR, and (d) combining the ultrasonic vibrations with other machining processes, for example, in electric discharge machining, where the vibrations help better debris disposal.

Summary

- USM is a nonthermal process and is the most preferred method for machining of hard and brittle materials with hardness above 40 HRC.

- Of the two process variants, USM and RUM, the RUM gives better machining performance than USM; however, the process application is limited to axisymmetric shapes.
- The removal mechanism in the process involves mainly indentation, impact, and chipping leading to generation of lateral crack and followed by the materials removal on brittle work surfaces.
- On the other hand, on ductile work surfaces, removal involves impact induced work hardening followed by chipping.
- In USM, the MRR increases mainly with an increase in the amplitude, frequency, and the static force.

Cross-References

► Ultraprecision Machining (UPM)

References

1. Tyrrell, W.R.: Ultrasonic Machining, Machining, ASM Metals Handbook, vol. 16. ASM International, Metals Park (1989)
2. Drozda, T.: Tool and manufacturing engineers handbook. In: Randy, G. (ed.) Ultrasonic Machining, vol. 1, 4th edn. SME, Michigan (1983). Chapter 14
3. Evans, A.G.: Fracture mechanics determinations. In: Bradt, R.C., et al. (eds.) Fracture Mechanics of Ceramics, vol. 1. Plenum, New York (1974)
4. Lawn, B.R., Evans, A.G.: A model for crack initiation in elastic/plastic indentation fields. J. Mater. Sci. **12**, 2195–2199 (1977)
5. Lawn, B.R., Evans, A.G., Marshall, D.B.: Elastic/Plastic indentation damage in ceramics: the median/radial crack system. J. Am. Ceram. Soc. **63**, 574 (1980)
6. Shaw, M.C.: Ultrasonic grinding. Microtechnic **10**(6), 257–265 (1956)
7. Pei, Z.J., Ferreira, P.M., Kapoor, S.G., Haselkorn, M.: Rotary ultrasonic machining for face milling of ceramics. Int. J. Mach. Tool Manufact. **35**(7), 1033–1046 (1995)
8. Thoe, T.B., Aspinwall, D.K., Wise, M.L.H.: Review on ultrasonic machining. Int. J. Mach. Tool Manufact. **38**(4), 239–255 (1998)
9. Ghosh, A., Mallik, A.K.: Ultrasonic machining (USM), Manufacturing Science. Affiliated East-West Press Private Limited, New Delhi (1985)

Unconventional Computing

► Molecular Computing

Uptake/Internalization/Sequestration/Biodistribution

► Exposure and Toxicity of Metal and Oxide Nanoparticles to Earthworms

USM

► Ultrasonic Machining

US-Tubes

► Ultrashort Carbon Nanotubes

V

Vertically Aligned Carbon Nanotubes

► Chemical Vapor Deposition (CVD)

Vertically Aligned Carbon Nanotubes, Collective Mechanical Behavior

Shelby B. Hutchens and Siddhartha Pathak
California Institute of Technology MC 309-81,
Pasadena, CA, USA

Synonyms

CNT arrays; CNT brushes; CNT bundles; CNT foams; CNT forests; CNT mats; CNT turfs

Definition

Vertically aligned carbon nanotubes are complex, hierarchical structures of intertwined tubes arrayed in a nominally vertical alignment due to their perpendicular growth from a stiff substrate. They are a unique class of materials having many of the desirable thermal, electrical, and mechanical properties of individual carbon nanotubes, while exhibiting these properties through the collective interaction of thousands of tubes on a macroscopic scale.

Introduction

While individual CNTs have been announced as the strongest material known and have shown extremely high strength and Young's modulus in tensile tests (► Carbon Nanotubes), VACNTs are more likely to find use in applications requiring large compliance and deformability. Examples of these include microelectromechanical systems (MEMS) and impact mitigation/energy absorption, where they are promising candidates for their multifunctional nature, wide ranging thermal stability, and relative ease of manufacture. A proper understanding of the collective mechanical behavior of these structures, especially instabilities leading to buckling and inhomogeneities which weaken mechanical performance, is thus of great importance for their design and success in these and other future applications.

Several features set VACNTs apart from other CNT structures. First, the tubes that make up the material grow perpendicularly to the support substrate, making them *nominally* vertically aligned. An important characteristic all VACNTs share is that the tubes themselves are long enough to become intertwined with each other during the growth process, making the structure into a kind of fibrous mat. This is in contrast to arrays of vertically aligned CNTs that are short and/or sparse enough that each CNT stands alone. Because of the nominal alignment, VACNTs exhibit a marked anisotropy in structure which is most apparent at magnifications of 1,000 × (see Fig. 1, left). Magnifying one hundred times more reveals their highly interconnected, foam-like structure and the network

B. Bhushan (ed.), *Encyclopedia of Nanotechnology*, DOI 10.1007/978-90-481-9751-4,

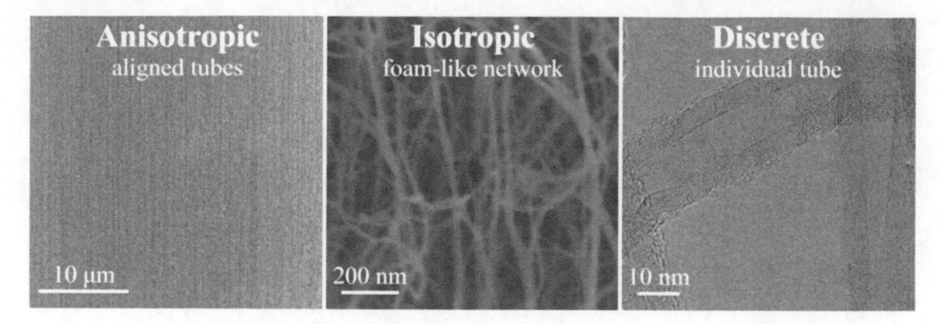

Vertically Aligned Carbon Nanotubes, Collective Mechanical Behavior, Fig. 1 A series of micrographs illustrating the hierarchical nature of the VACNTs. The *left* and *center* images are taken with a scanning electron microscope (SEM). The *right* image is a transmission electron micrograph (TEM)

of CNTs begins to appear nearly isotropic. Magnifying another hundred times, one obtains a view of the individual CNTs themselves. As evidenced by the micrographs in Fig. 1, in the as-grown state, certain segments of the CNT forest may be pre-buckled/pre-bent, with the resulting bending strain energy being balanced by the favorable contact energy between the tubes [1].

The method by which these structures are synthesized is the primary factor affecting their complex, hierarchical morphology. This microstructure, in turn, affects their mechanical behavior, in particular the modulus, buckling strength, and recoverability. Synthesis techniques for VACNTs can be divided into two main categories: the Chemical Vapor Deposition (CVD) synthesis method, and the Carbide-Derived Carbon (CDC) synthesis method. In the CVD process, the VACNT film is coated onto an existing substrate. This is accomplished by depositing a thin layer of catalyst (e.g., Fe) on the substrate (typically Si or Quartz) and flowing a carbon source (e.g., ethylene) over the substrate at atmospheric pressure and temperatures typically around 750°C. In the CDC method [2], on the other hand, carbon is formed by selective extraction of the metal or metalloid atoms in the carbide (e.g., silicon carbide) at high temperatures (>1,600°C), transforming the carbide structure into pure carbon. Since the CNT layer is formed by inward growth, this usually retains the original shape and volume of the precursor.

VACNTs grown by these different techniques demonstrate very different structure and mechanical properties. Even within materials grown via CVD, control of the growth conditions, such as the atmosphere, catalyst activity, and pressure, are known to significantly affect the repeatability of the VACNT's morphology and hence the consistency of mechanical properties [3]. Differences in the growth processes used are revealed in the widely varying VACNT information reported in literature, as seen in the large range of properties in Table 1, not to mention variations in tube diameter (from 2–3 [4] to 20–50 nm [5], to greater than 100 nm [6]), number of walls in the CNT structure, and degree of tube alignment; properties which are sometimes neither measured nor reported. Further, the different stages of the CNT growth process can result in a height-dependent inhomogeneity [7]. This manifests as a gradient in both the density and the alignment of the tubes within the same VACNT structure. As discussed later in this entry, such a structure gradient may lead to a corresponding strength and stiffness gradient along the VACNT height. As an extreme example of the microstructure–property relationship, CDC-VACNTs are known to have a considerably higher average density (roughly 10 times higher than typical VACNTs), due to the conformal transformation of the carbide into carbon. This in turn leads to significantly larger values for the elastic modulus and yield stress in CDC-VACNTs [4] (see Table 1). These promising characteristics, however, are unavailable for applications requiring macroscopic films as currently only VACNTs grown via CVD can reach macroscopic heights (~mm). Growth of CDC-VACNTs remains limited to only a few micrometers in height.

Another marked difference in material behavior is the ability of some microstructures to recover from large deformations, while others deform plastically. Within those materials grown via CVD, several groups have observed excellent recoverability [8–10] in their VACNT samples, seeing less than 15% deformation after thousands of cycles of strain to 85%. Other

Vertically Aligned Carbon Nanotubes, Collective Mechanical Behavior, Table 1 Summary of reported VACNT elastic modulus values

Synthesis technique	Density	Measurement method	E (MPa)	Yield/ buckling strength (MPa)	Reference
CVD	87% porosity	Uniaxial compression	50	12	Cao et al. [8]
	97% porosity	Uniaxial compression	818	14.1	Deck et al. [14]
	NG[d]	DMA[e]	~50	NA[f]	Mesarovic et al. [1]
	NG	Uniaxial compression	<2	NA	Suhr et al. [9]
	10^{10} tubes/cm^{2}[a]	Uniaxial compression	0.22–0.25	NA	Tong et al. [15]
	NG	Nanoindentation – Berkovich uniaxial compression	15	0.2–4.3	Zbib et al. [11]
	NG	Nanoindentation – Berkovich	58	NA	Zhang et al. [12]
	NG	Nanoindentation – Berkovich	50 ± 25[b]	NA	Qiu et al. [16]
	0.018 g/cm^3 0.114[c] g/cm^3	Uniaxial compression	NG	0.12 5.5 [c]	Bradford et al. [13]
CDC	0.95 g/cm^3	Nanoindentation – spherical	18,000	90–590	Pathak et al. [4]

NG not given, *DMA* dynamic mechanical analysis, *NA* not applicable
[a]Density in tubes per unit area. Tubes were 20–30 nm in diameter
[b]Reported as reduced modulus. Indentation was performed with a diamond tip, so difference from sample modulus is small
[c]After post-growth CVD treatment

groups, however, observe that the materials remain plastically deformed [5, 11, 12]. Energy can be dissipated in the former as they behave like viscoelastic rubbers (discussed in a later section), whereas energy can be absorbed in the latter. Both appear to deform via the same structural mechanism (localized buckles) intrinsic to the complex microstructure of these systems. It is still largely unclear what is responsible for a VACNT material displaying plastic versus recoverable behavior. Since individual CNTs are known to recover from large bending angles, the plasticity must be due to their collective interactions. Possible interactions include entanglement, tube-to-tube adhesion through attractive interactions (e.g., van der Waals), or the fracture of the tubes themselves. However, fracture is unlikely as experimental observations of quasistatic mechanical loading have shown no evidence of individual tube failure, most notably in Ref. [5]. Two promising clues arise from recent work by Bradford et al. [13]. They find that VACNTs can be converted from irrecoverable to viscoelastic (recoverable) by a second CVD treatment that increases tube diameter and surface roughness, suggesting either may play a role in determining recoverability.

Instrumented indentation, using a variety of tip geometries such as flat punch, spherical, Berkovich, and cube corner, has been the most common method for studying mechanical properties of VACNTs [6]. Each geometry has its own strengths and weaknesses. While maintaining parallel contact between the indenter and the sample is a major concern for flat punch indentation, it does allow for loading in a direction largely normal to tube growth. Because of the marked anisotropy of VACNTs, the sharper Berkovich or cube corner geometries can cause the CNTs to bend away from the indenter in shorter samples, which results in testing a slightly different mode of behavior by applying more load in the direction perpendicular to growth. In both cases, modulus and hardness are measured from the unloading portion of the test. On the other hand, spherical indentation is advantageous in that it allows indentation stress–strain curves to be extracted from the raw load–displacement data, which enables resolution of the evolution of the mechanical response in the VACNT array: from initial elasticity, to the initiation of buckling, to post-buckling behavior at finite strains [4]. Though indentation is a relatively simple test to perform, analyses of the results, especially nonlinear elastic behavior is difficult due to their highly localized stress fields. It is also limited in total strain.

Another testing geometry, that of uniaxial compression, eliminates these localized, applied stress fields and reveals the existence of a localized deformation mechanism in VACNTs. In these tests, the samples are either large (~1 mm tall) bulk films and compressed between two platens or they are microscale cylinders and compressed using a flat punch indenter. These tests, especially when performed in situ (load–displacement data gathered simultaneously with micrographs in a SEM) can offer valuable insights on the morphological evolution in the VACNTs during deformation [5] and are discussed in detail in the next section.

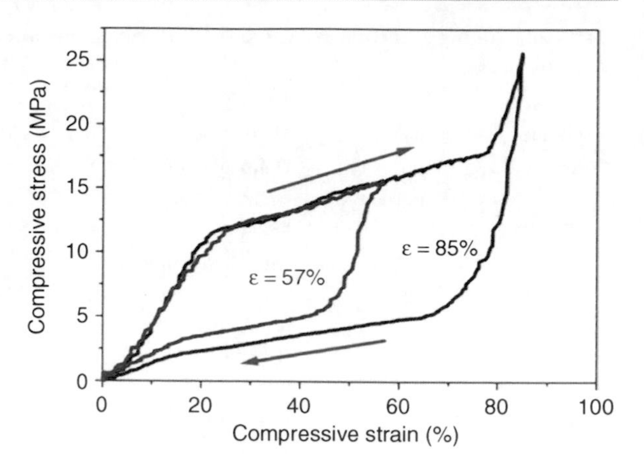

Vertically Aligned Carbon Nanotubes, Collective Mechanical Behavior, Fig. 2 Nominal stress–strain response from the uniaxial compression of a macroscale VACNT sample illustrating the overall embodiment of foam-like elastic-plateau-densification regimes [8] (Reprinted from [8] with permission from AAAS (American Association for the Advancement of Science))

Deformation Under Large Strain

The deformation of VACNTs is governed by their hierarchical microstructure, collective intertube interactions, and inherent property gradient. Taken as a whole, their highly porous nature gives them an overall foam-like response. The idealized stress–strain response of traditional foams is characterized by three distinct loading regimes: an initial elastic loading at low strain, followed by a plateau in the stress during which the supporting struts bend and buckle, and finally a densification regime in which the space between struts has been nearly eliminated and the material begins to approach behavior intrinsic to the struts themselves. For such a response, it is the intermediate plateau regime that is responsible for the bulk of energy absorption in the material, since the area under this region of the stress–strain curve, corresponding to the work done on the material, is largest. This foam-like response of a bulk VACNT film is apparent in the three distinctly differently sloped regions of the stress–strain response shown in Fig. 2. Locally, however, the response of VACNTs is quite unlike that of traditional foams. In VACNTs, the accommodation of strain during uniaxial compression is accomplished entirely through the formation of folds or buckles of small regions of the structure while the remaining portion remains nearly undeformed. This superposition of an overall foam-like response with localized strain accommodation is the key characteristic of VACNT deformation.

Experimental characterization of this buckle formation yields several interesting qualitative results. In their early study, Cao et al. [8] compressed relatively large structures (area: 0.5–2 cm^2, height: 860 μm–1.2 mm) and observed that the buckles formed near the bottom of the structure (the end from which the CNTs grow perpendicularly to the substrate) are more deformed that those that formed near the top (Fig. 3a). Motivated by this observation, they hypothesized that the bottom buckles form first. A reversal of the loading direction, by flipping the sample upside down, resulted in the same deformed morphology, with the tightest buckles forming at what was the end of the sample attached to the growth substrate. These observations point to the idea of an inherent, axial property gradient being responsible for the sequential nature of the buckling. Note that each individual buckle is on the order of 12–25 μm in size (depending on sample height) so that several tens of buckles form during deformation. The sequential, localized buckling phenomenon was later observed in much smaller samples, by Zbib et al. [11] and Hutchens et al. [5] (Fig. 3b, c) illustrating the universality of this response in VACNTs. Buckles in these microscopic studies were 12 and 7 μm in wavelength (measured from the unbuckled conformation) for cylindrical samples with diameters of 30–300 and 50 μm, respectively. The bottom-first buckling mechanism was visually verified by Hutchens et al. [5] through in situ experiments that further revealed the mechanism by which a single, localized buckle evolves. As shown in their study, each individual buckle does not form all at once, but rather nucleates at one point and then propagates laterally

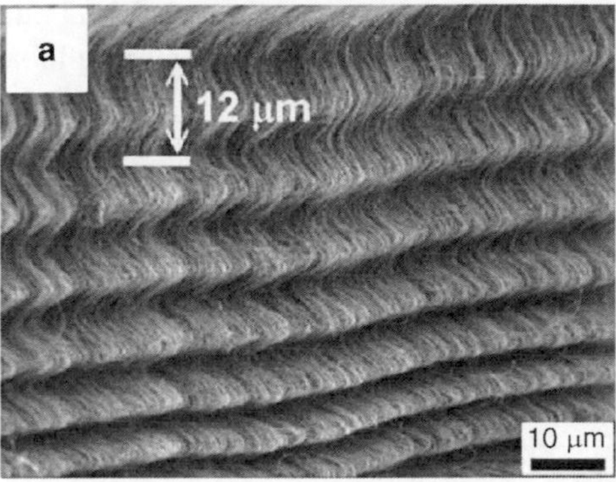

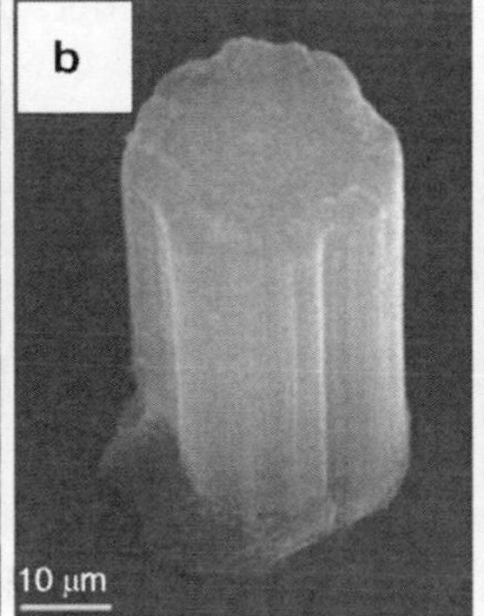

Vertically Aligned Carbon Nanotubes, Collective Mechanical Behavior, Fig. 3 Examples of local buckling in VACNTs. (**a**) Close-up of the wavelike deformation observed after nominal strains of 0.85 over 1,000 cycles from Cao et al. [8] (Reprinted from [8] with permission from AAAS). On the *right*, *bottom*-first localized buckles observed after the uniaxial microcompression of VACNT pillars by (**b**) Zbib et al. [11] (Copyright IOP Publishing Ltd. Reproduced with permission), and (**c**) Hutchens et al. [5] (Copyright Wiley-VCH Verlag GmbH & Co. KGaA. Reproduced with permission)

across the cylindrical sample until completion [5]. In addition, each subsequent buckle begins only after the previous buckle has completely formed, that is, the crease has spread across the entire cylinder. Figure 4 illustrates the evolution and localization of deformation for a uniaxially loaded VACNT cylinder. Cross correlation of the in situ images with the accompanying stress–strain data reveals that the localized buckle formation and propagation correspond to undulations in the nominal stress–strain curve. Specifically, softening corresponds to the first appearance of the buckle and the subsequent hardening coincides with the lateral propagation of the buckle. This localized response overlays the aforementioned overall foam-like behavior seen in these materials.

Notable differences between the classic foam-like stress–strain behavior and the overall response of VACNTs in uniaxial compression studies of VACNT structures have also been observed in both macroscopic [8, 9, 15] and microscopic [5, 11] samples. First, the plateau regime is highly sloped. In typical foams, a sloped plateau indicates some homogeneous variation in strut buckling stress due either to random strut alignment, a distribution of strut sizes (aspect ratio, diameter, etc.), or both. While both kinds of inhomogeneities exist in VACNTs, the plateau regime is too highly sloped to be explained by these small fluctuations and the nonlocal nature of the buckling points to another cause. In fact, the sloped plateau regime actually points toward evidence of an axial property gradient in VACNTs. As discussed previously, a gradient in tube density arises in VACNTs as a result of the CVD growth process. This gradient can be such that there is a lower tube density at the bottom of the structure (i.e., the point at which the substrate attaches) than the top. It follows that such a tube density gradient would result in a corresponding strength and stiffness gradient. This property gradient is evident in the stress–strain response from Fig. 4 in the progressive increase in peak stress values for undulations in the plateau region. Throughout this plateau, buckles are known to form sequentially, bottom-to-top, and therefore each subsequent buckle forms at a higher (and more dense) location within the cylinder than the previous buckle, requiring a larger stress in order to form a new fold. Such property dependence has been modeled by Hutchens et al. [5] in a viscoplastic finite element framework which showed a direct correspondence between the presence and extent of an applied property gradient and the amount of hardening within the plateau. Details of this and other efforts to capture VACNT deformation mechanisms through modeling are summarized in the next section.

Modeling of VACNT Deformation

Only a few preliminary models exist to describe the mechanical deformation of VACNTs. Motivated by both the morphology of the VACNTs (a series of nominally vertical struts) as well as the observed buckling behavior, many researchers utilize an Euler buckling framework to mechanistically describe their findings. In Euler buckling, an ideal column

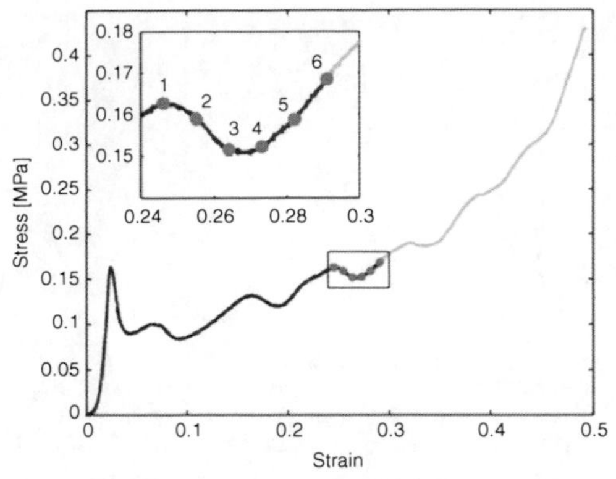

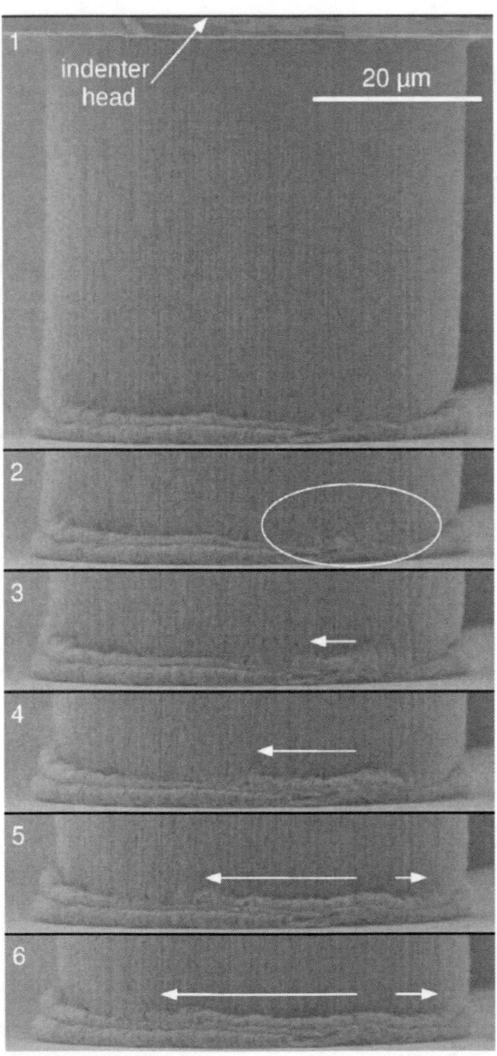

Vertically Aligned Carbon Nanotubes, Collective Mechanical Behavior, Fig. 4 In situ video micrographs and accompanying nominal stress–strain data illustrating bottom-first buckling and buckle initiation and propagation events [5] (Copyright Wiley-VCH Verlag GmbH & Co. KGaA. Reproduced with permission)

(perfectly straight, homogeneous, and free from initial stress) is determined to buckle when the applied load reaches some critical value that causes the column to be in a state of unstable equilibrium. Up to the point of buckling, the column is assumed to be perfectly elastic. Scaling calculations by Cao et al. [8] utilize this critical stress to estimate the transition stress value for departure from linear elastic behavior, that is, the buckling stress, given a reasonable estimate of the tube number density. Similarly, Mesarovic et al. [1] include an additional intertube contact energy to account for favorable van der Waals interactions between tubes in their energetic treatment of VACNTs. Another Euler-based model by Zbib et al. [11], motivated by the formation of buckles they observed, proposes piecewise buckling that assumes the top of the pillar deforms via simple shear while the bottom undergoes collapse. Using this framework, they predict the buckling stress increases asymptotically with decreasing pillar height (for similar aspect ratio pillars). Notably, however, none of these models make predictions concerning the length-scale of the buckles formed or their dependence on material parameters such as density, stiffness, tube alignment, or diameter.

Another subset of mechanical analyses utilizes an alternating hardening-softening-hardening local constitutive relation and are the only theories that attempt to capture material response beyond the initial buckling event. The hardening-softening-hardening behavior is motivated by the anticipated behavior of the individual CNT struts. Locally, the material is expected to transition from elastic loading (hardening) to buckling during which the load carrying capacity of the tubes decreases (softening). Eventually, the material densifies, corresponding to a return to hardening behavior. The first of the analyses to use this general behavior, a hierarchical bistable spring model, captures the *quantitative* stress–strain response of VACNTs in compression [17]. The second, a finite element, viscoplastic solid, captures the *qualitative*, sequential, periodic buckle morphology [18]. The bistable spring model consists of mesoscale elements characterized by elastic-plateau-densification and, most importantly, hysteresis in the unloading curve. These mesoscale elements are the limiting case of infinitely many bistable spring elements in series. Briefly, a bistable spring consists of two thermodynamically stable elastic loading sections (hardening) separated by an unstable, negative stiffness region (softening) of the stress–strain curve across which the material snaps, similar to a phase transition. When placed in series, these mesoscale elements closely capture the hysteretic unloading response seen by Cao et al. [8]. Thus, this model can be utilized to characterize the energy dissipation in VACNTs. In the finite element model, Hutchens et al. [18] postulate a positive–negative–positive sloped stress–strain relation reminiscent of a bistable spring, but rather than being elastic, it is used as a plastic hardening function governing the local deformation of an element post-elastic loading. These latter analyses find that this constitutive relation is capable of producing sequential, periodic buckles in an axisymmetric, circular cylindrical mesh with fixed boundary conditions at the base (similar to VACNTs). In addition, an axial gradient in the strength is not necessary to initiate the bottom-first buckling seen in experiments; rather, the fixed boundary conditions are sufficient for a uniform property distribution. However, a reversal of buckle initiation, top-to-bottom, can be achieved for a sufficient inverse axial gradient (having lower strength at the top of the pillar than the bottom). Both of these analyses capture essential elements of VACNT deformation, but do so for two very different sets of experimental observations: recoverable deformation and plastic deformation, respectively.

Viscoelasticity

In addition to their distinctive buckling behavior and their ability to recover from large deformations, VACNTs have also been reported to demonstrate another case of extreme mechanical performance – a unique viscoelastic response that spans a truly wide temperature range from $-196°C$ to $1{,}000°C$ – something no other material has shown so far. Viscoelastic materials exhibit both viscous and elastic characteristics when subjected to load. Thus, a viscoelastic material is able to both dissipate energy through viscous behavior (as in honey), while storing energy though elasticity (as in rubber band). The stress–strain response of viscoelastic materials is typified by hysteresis in the loading-unloading cycle.

Viscoelastic behavior in a material is generally characterized in terms of its loss (E'') or storage modulus (E') (or by the ratio, known as $\tan \delta = E''/E'$, of these two moduli). E'' relates to the amount of energy dissipated while E' represents the stored energy in the material. The angle, δ, is the phase lag between the oscillatory load and displacement responses under sinusoidal loading. In a typical experiment, the material is loaded to a predetermined strain and the mechanical probe is oscillated across a range of frequencies. By measuring the resultant load amplitude, displacement amplitude, and the phase lag during the test, the values of loss modulus, storage modulus, and $\tan \delta$ can be determined. Similarly, viscoelasticity can also be quantified by the memory or hysteresis effects during load-unload cycles under deformation, where the energy dissipated is given by the area of the hysteresis loop. These viscoelastic effects have recently been documented in the highly intertwined random networks of VACNTs. Viscoelasticity in CVD-VACNTs has been demonstrated by Hutchens et al. [5], where both storage and loss stiffnesses were studied in a frequency range from 1 to 45 Hz at different strain levels (Fig. 5a). Note that these stiffness values are proportional to moduli given a known Poisson's ratio, which is lacking for VACNTs. The elastic response is clearly frequency independent, indicating the VACNT's elastic deformation is likely due to the same mechanism (likely tube bending) over the range of timescales tested. As shown in Fig. 5a, more energy is dissipated (higher values of loss stiffnesses) at larger strain levels when a higher fraction of the VACNT pillar has buckled. Interestingly, for the more dense CDC-VACNTs the opposite seems to be true (Fig. 5b). For the highly dense CDC-VACNTs, Pathak et al. [4] have shown a significant drop in the $\tan \delta$ values of the CNTs after buckling, that is, when the material was highly compacted. These observations indicate that while an increase in density can significantly increase the viscoelastic behavior of CNTs, there appears to be a cutoff beyond which the contacting CNTs became increasingly bundled resulting in a decrease in their ability to dissipate energy.

A recent report by Xu et al. [10] also suggests a rubber-like viscoelastic behavior in a random network of long, interconnected, and tangled CNTs, similar in many ways to VACNTs but significantly less aligned, over a very wide temperature range – from $-196°C$ to $1{,}000\ °C$, with a possibility of extending this behavior beyond $1{,}500\ °C$. While the oxidizing nature of carbon may limit the application of these materials to only vacuum or protective

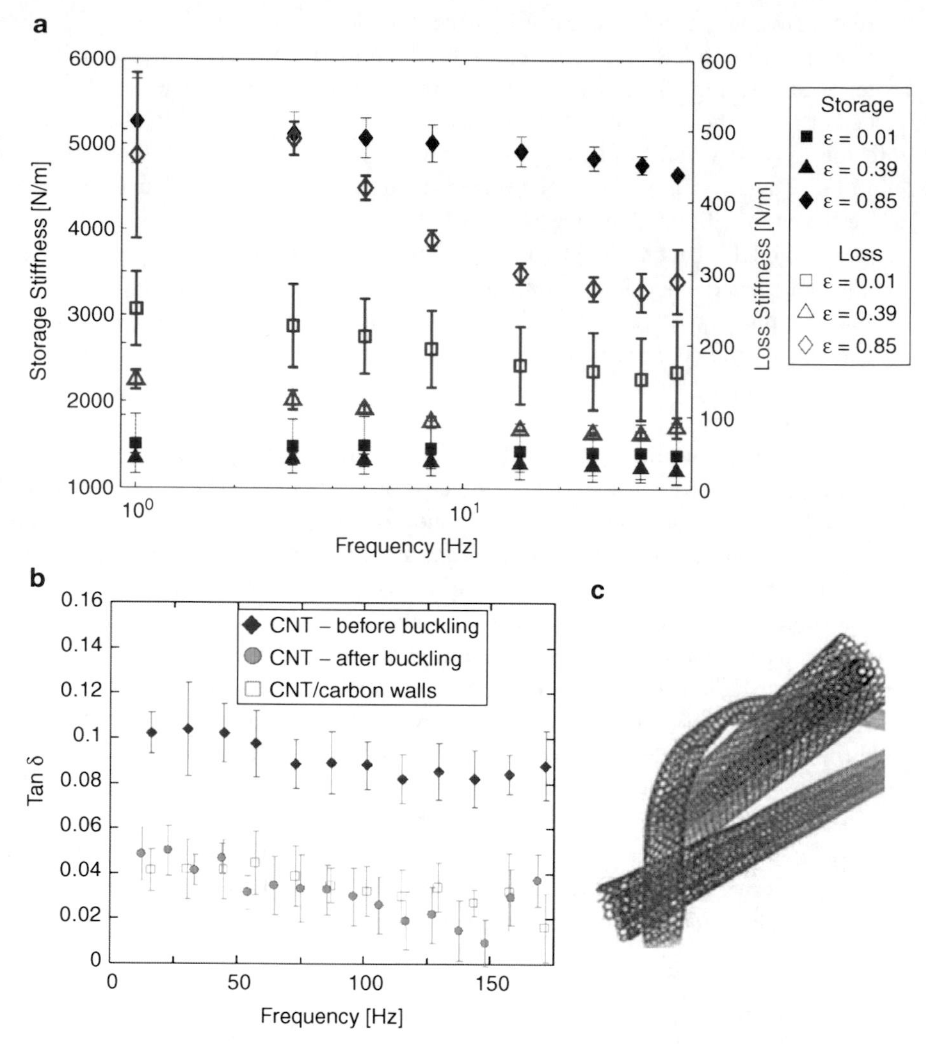

Vertically Aligned Carbon Nanotubes, Collective Mechanical Behavior, Fig. 5 Viscoelastic behavior as seen in: (**a**) CVD-VACNTs showing an increase in the loss stiffness with increasing strains [5] (Copyright Wiley-VCH Verlag GmbH & Co. KGaA. Reproduced with permission) (**b**) CDC-VACNTs showing a significantly decrease in values of tan δ after buckling (Reproduced from [4] with permission from Elsevier) (**c**) Atomistic modeling image of a possible entangled arrangement of single, double, and triple-walled CNTs leading to their rubber-like viscoelastic behavior (Reprinted from [19] [8] with permission from AAAS)

(reducing) environments, CNTs are nevertheless the only known solids to demonstrate such behavior at extremely low or very high temperatures. The authors attribute the reversible dissipation of energy in CNTs over this remarkable range to the zipping and unzipping of the CNTs upon contact (see Fig. 5c) caused by the van der Waals interactions. In this instance, "reversible" means that beyond a critical strain the zipping/unzipping process was no longer possible everywhere as more and more tubes become permanently entangled at higher strain. This would lead to a loss in their viscoelasticity, similar to the observations in the highly dense CDC-VACNTs (Fig. 5b).

The unique combination of superior mechanical properties and the ability to dissipate energy during deformation is expected to have significant impact in damping applications utilizing VACNTs. These, combined with the wide temperature range of its viscoelastic behavior, make VACNT-based materials a promising choice for use in mechanical applications under extreme temperatures or temperature gradients. Possible environments range from the cold of interstellar space to down-to-earth viscoelastic applications, such as MEMS devices to high temperature vacuum furnaces.

Applications

As described in detail in (▶ Carbon-Nanotubes) multiwall carbon nanotubes are well known for a variety of exceptional properties. These include a high tensile modulus, on the order of 1 TPa, high

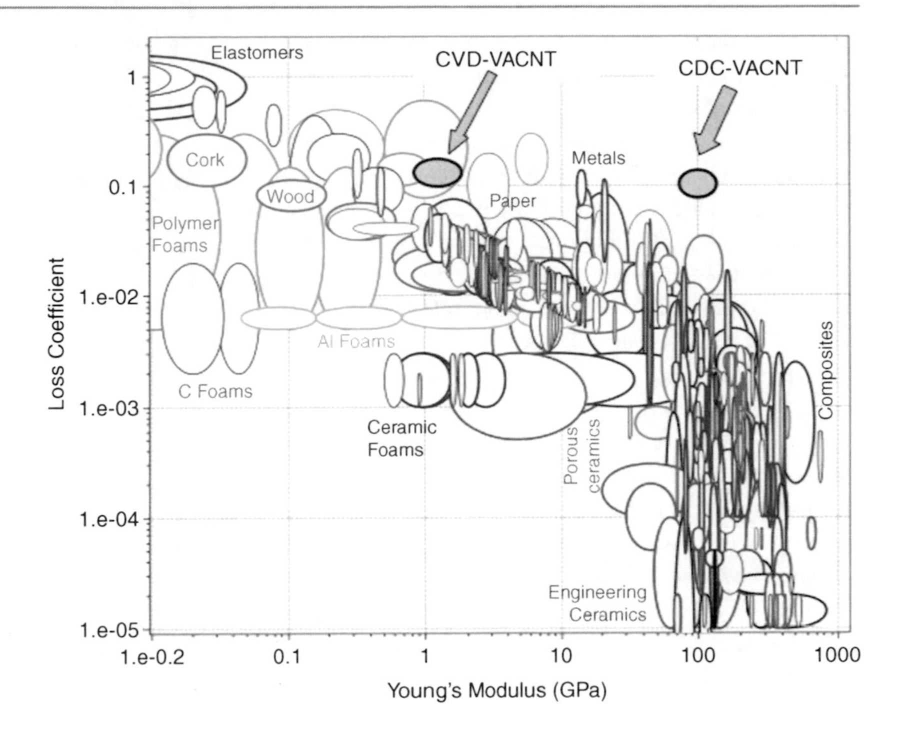

Vertically Aligned Carbon Nanotubes, Collective Mechanical Behavior, Fig. 6 An Ashby plot of loss coefficient (energy dissipated in a stress–strain cycle) versus modulus illustrating the relation of VACNTs to other damping materials

strength, on the order of tens of GPa, high thermal conductivity, recoverability after large bending angle deformation, and a range of electrical properties depending on the chirality of the graphene walls. For these reasons they are currently the subject of a wide range of research (see Function follows shape: Examples from Carbon Nanotubes Research). In the form of VACNTs, these individual tube properties may be significantly altered by the collective interaction of thousands of tubes. Several applications of VACNT structures currently under study rely on the mechanical behavior described earlier in this entry. A remarkable, but incomplete list includes: components of highly compliant thermal contacts for microelectromechanical systems (MEMS) and microelectronics [3, 11] (see Carbon Nanotubes Integrated with MEMS), dry adhesives (see ► Gecko Adhesion), thermally robust energy dissipating rubber [10, 19], and energy absorption or impact mitigation [20]. Other applications, such as optical coatings and cold cathode arrays, may rely less on the mechanical properties for optimization and design, but understanding is still necessary for evaluation of robustness and in-use lifetime analysis.

We go through each application briefly to highlight the VACNT-specific mechanical behavior of interest. First, thermal contacts for delicate electronics devices have two requirements. They must be highly thermally conductive and make conformal contact (to increase heat transfer) while avoiding damage to the components they are transferring heat to and from. CNT foams are ideal for this application due to the high compliance that comes from the reduced load capacity of the CNT struts as they buckle. Dry adhesive applications of VACNTs take advantage of the hierarchical structure which can, as desired in MEMS switches, make conformal contact to surfaces at a variety of roughness length scales, thereby increasing the attractive interactions between the tubes and the surface. Third, energy dissipating and absorbing applications require more in-depth knowledge of the deformation mechanism. As mentioned in the previous section, energy dissipation is currently thought to be due to tube zipping/unzipping or tube-to-tube sliding and rearrangement [1, 4, 9, 10, 19], while energy absorption is certainly maximized during buckle formation. Future applications may be able to take advantage of yet undiscovered traits of the incompletely characterized VACNTs.

Summary and Outlook

Although the mechanisms governing VACNTs collective mechanical behavior are still largely uncharacterized, the special properties of multiwall carbon nanotubes combined with the complex interactions that arise between them in the hierarchical VACNT microstructure have generated significant research interest. Their wide range of properties, mechanical, electrical, and thermal, make them ideal candidates as multifunctional materials, particularly for applications in which soft materials (such as polymers) have traditionally dominated. Notably, VACNTs occupy a unique niche among engineering materials as shown in the Ashby property chart in Fig. 6 with the ability to span a wide range of properties given a tunable microstructure. The authors believe that there are two major hurdles to rationally designing VACNTs for any application: control of the CNT growth process and understanding of the relationship between the microstructure and mechanical response. These hurdles are interrelated through the fact that the design space cannot be systematically probed until the microstructure can be systematically controlled. Mastery of these unknowns will not only further VACNTs place as novel materials in the current applications under study, but may reveal previously undiscovered behaviors in microstructures yet to be created or characterized.

Cross-References

► Carbon-Nanotubes
► Gecko Adhesion

References

1. Mesarovic, S.D., McCarter, C.M., Bahr, D.F., Radhakrishnan, H., Richards, R.F., Richards, C.D., McClain, D., Jiao, J.: Mechanical behavior of a carbon nanotube turf. Scr. Mater. **56**, 157–160 (2007)
2. Presser, V., Heon, M., Gogotsi, Y.: Carbide-derived carbons – from porous networks to nanotubes and graphene. Adv. Funct. Mater. **21**, 810–833 (2011)
3. McCarter, C.M., Richards, R.F., Mesarovic, S.D., Richards, C.D., Bahr, D.F., McClain, D., Jiao, J.: Mechanical compliance of photolithographically defined vertically aligned carbon nanotube turf. J. Mater. Sci. **41**, 7872–7878 (2006)
4. Pathak, S., Cambaz, Z.G., Kalidindi, S.R., Swadener, J.G., Gogotsi, Y.: Viscoelasticity and high buckling stress of dense carbon nanotube brushes. Carbon **47**, 1969–1976 (2009)
5. Hutchens, S.B., Hall, L.J., Greer, J.R.: In situ mechanical testing reveals periodic buckle nucleation and propagation in carbon nanotube bundles. Adv. Funct. Mater. **20**, 2338–2346 (2010)
6. Qi, H.J., Teo, K.B.K., Lau, K.K.S., Boyce, M.C., Milne, W.I., Robertson, J., Gleason, K.K.: Determination of mechanical properties of carbon nanotubes and vertically aligned carbon nanotube forests using nanoindentation. J. Mech. Phys. Solids **51**, 2213–2237 (2003)
7. Bedewy, M., Meshot, E.R., Guo, H.C., Verploegen, E.A., Lu, W., Hart, A.J.: Collective mechanism for the evolution and self-termination of vertically aligned carbon nanotube growth. J. Phys. Chem. C **113**, 20576–20582 (2009)
8. Cao, A.Y., Dickrell, P.L., Sawyer, W.G., Ghasemi-Nejhad, M.N., Ajayan, P.M.: Super-compressible foamlike carbon nanotube films. Science **310**, 1307–1310 (2005)
9. Suhr, J., Victor, P., Sreekala, L.C.S., Zhang, X., Nalamasu, O., Ajayan, P.M.: Fatigue resistance of aligned carbon nanotube arrays under cyclic compression. Nat. Nanotechnol. **2**, 417–421 (2007)
10. Xu, M., Futaba, D.N., Yamada, T., Yumura, M., Hata, K.: Carbon nanotubes with temperature-invariant viscoelasticity from-196° to 1000°C. Science **330**, 1364–1368 (2010)
11. Zbib, A.A., Mesarovic, S.D., Lilleodden, E.T., McClain, D., Jiao, J., Bahr, D.F.: The coordinated buckling of carbon nanotube turfs under uniform compression. Nanotechnology **19**, 175704 (2008)
12. Zhang, Q., Lu, Y.C., Du, F., Dai, L., Baur, J., Foster, D.C.: Viscoelastic creep of vertically aligned carbon nanotubes. J. Phys. D: Appl. Phys. **43**, 315401 (2010)
13. Bradford, P.D., Wang, X., Zhao, H., Zhu, Y.T.: Tuning the compressive mechanical properties of carbon nanotube foam. Carbon **49**, 2834–2841 (2011)
14. Deck, C.P., Flowers, J., McKee, G.S.B., Vecchio, K.: Mechanical behavior of ultralong multiwalled carbon nanotube mats. J. Appl. Phys. 101:23512-23511-23519 (2007)
15. Tong, T., Zhao, Y., Delzeit, L., Kashani, A., Meyyappan, M., Majumdar, A.: Height independent compressive modulus of vertically aligned carbon nanotube arrays. Nano Lett. **8**, 511–515 (2008)
16. Qiu, A., Bahr, D.F., Zbib, A.A., Bellou, A., Mesarovic, S.D., McClain, D., Hudson, W., Jiao, J., Kiener, D., Cordill, M.J.: Local and non-local behavior and coordinated buckling of CNT turfs. Carbon **49**, 1430–1438 (2011)
17. Fraternali, F., Blesgen, T., Amendola, A., Daraio, C.: Multiscale mass-spring models of carbon nanotube foams. J. Mech. Phys. Solids **59**, 89–102 (2011)
18. Hutchens, S.B., Needleman, A., Greer, J.R.: Analysis of uniaxial compression of vertically aligned carbon nanotubes. J. Mech. Phys. Solids **59**, 2227–2237 (2011)
19. Gogotsi, Y.: High-temperature rubber made from carbon nanotubes. Science **330**, 1332–1333 (2010)
20. Misra, A.A.G., Greer, J.R., Daraio, C.: Strain rate effects in the mechanical response of polymer-anchored carbon nanotube foams. Adv. Mater. **20**, 1–5 (2008)

Vibration Assisted Machining

► Ultrasonic Machining

Viscosity

► Interfacial Investigation of Protein Films Using Acoustic Waves

Visual Prosthesis

► Artificial Retina: Focus on Clinical and Fabrication Considerations

Visual Servoing for SEM

► Robot-Based Automation on the Nanoscale

Wavefront Deformation Particle Tracking

► Astigmatic Micro Particle Imaging

Waypoint Detection

Richard P. Mann
Centre for Interdisciplinary Mathematics, Uppsala University, Uppsala, Sweden

Synonyms

Landmark detection; Key point detection

Definition

Animals are frequently required to navigate accurately from one location to another. The means an animal uses to perform a navigational task depends on the available environmental cues, their specific sensory systems, and particularly whether they are in a familiar or unfamiliar area. A *familiar area* can be defined as one of which the animal has extensive experience, leading to detailed knowledge of the environment and available navigational cues.

Waypoints are geostationary locations used by an animal to navigate in a familiar area. Specifically they are small-scale regions of space that the animal returns to as part of the navigational task. A waypoint may be associated with a memorized *landmark* which allows the animal to identify the waypoint from a distance and direct itself toward it. Navigation toward an eventual goal may involve visiting an ordered succession of waypoints finishing at the final location – this mechanism is generally referred to as *pilotage*. Alternatively waypoints may be a by-product of navigational constraints imposed by the environment; if, for example, the goal can be reached only by passing through a narrow passage, that passage will act as a waypoint.

While waypoints may be associated with landmarks – recognizable sensory features such as prominent visual cues or memorable odors – the task of waypoint detection is not in general synonymous with landmark detection. Landmarks may be identified and used by the animal at a distance, without necessarily visiting the landmark's true location. Conversely, as noted above, a waypoint may exist independently of any memorable feature. Instead, waypoints are more accurately defined as importance regions of space and segments of movement paths, whose salience may be the result of a variety of cues and constraints, which cause discernable effects upon the animal's movement.

Waypoint detection is the task of identifying waypoints from observations of the animal's movements. Since the definition of a waypoint requires the animal to return to the specific area, this therefore constitutes a classification task, identifying the segments of the animal's movement trajectory, or *path*, that correspond to waypoint locations. This entry gives an overview introduction to the purpose and methods of waypoint detection, examining three methods for isolating particularly salient elements of movement paths.

B. Bhushan (ed.), *Encyclopedia of Nanotechnology*, DOI 10.1007/978-90-481-9751-4,

Purpose

The principle purpose of detecting waypoints is to understand what drives and facilitates navigation. It is a way of identifying important areas which may contain salient information the animal uses to recall a memorized path or important resources the animal must utilize. Examples of this might include visually striking features, high food concentrations, or unseen constraints on the animal's movement (strong wind patterns acting on flying creatures, for instance). Waypoint detection can be useful for biomimicry, allowing both an understanding of how animals use key features to navigate and replication of this process in artificial navigating agents such as robots.

In the literature, waypoint detection has been used as the basis for understanding which features animals attend to in the landscape [6, 7, 10], determining the extent of an animal's knowledge of the area [5] and determining where animals forage [1, 4]. The technique of isolating important regions of the path for further analysis is a basis for linking animal movement behaviors and the surrounding environment, such as the landscape, sensory cues, and other animals, by determining what the environmental conditions are in the spatial and temporal vicinity of the detected waypoints.

Three Methods for Detecting Waypoints

Specific identification of waypoints from recorded animal movements is a novel task, and there are consequently few formalized examples in the scientific literature that deal explicitly with this problem. However, the likely existence of waypoints in specified regions of space is implicit in several analyses of movement paths. The three principle methods for identifying these spatial regions can be broadly classified as:

- Localized variability *between* multiple movement paths – *Path similarity*
- Changes in speed and direction *within* a single path – *Path complexity*
- Information content of a path segment – *Path predictability*

This entry describes these methods largely in the context of experiments on homing pigeons, where the use of global positioning satellite (GPS) devices has created a huge amount of data calling for novel analysis. This has led to the development of methods for path analysis, which are generalizable and applicable to movement paths from other species.

Path Similarity

Path similarity presents an intuitive method for waypoint detection. The principle property of a waypoint is that it is a *restricted* region of space which the animal will pass through on its way to the objective. A waypoint therefore is likely to present itself as a relatively small region of space to which the animal repeatedly returns. In this case it is likely, though not guaranteed, that the movement paths of the animal will be significantly more similar in the vicinity of the waypoint, since their freedom to vary is constrained, than outside these regions where the animal's movements may vary more widely. Increasing path similarity over time has been linked to the emergence of detailed visual memories of the landscape and the use of pilotage as a navigational strategy [2, 3, 12].

To use path similarity objectively requires an algorithmic measure of the variability between different paths. To identify specific regions of space demands a *localized* measure – one that varies through space – rather than a *global* measure which relates the overall similarity of the paths. Here several options present themselves. None can be considered the sole correct way to measure variability, nor are these options exhaustive. Different measures of variability are necessarily highly correlated but the specific choice depends on the exact nature of the analysis. The basic principle is to measure the spatial separation between paths within some localized region. For this purpose it is helpful to define the concept of the *nearest-neighbor distance* (NND). This is the distance from a point on one path to the closest point on another. Using this distance allows the path separation to be determined in a purely spatial manner. The alternative approach of measuring the distance between paths *at the same point in time* measures a spatiotemporal separation. Ordinarily it is purely spatial separation that best describes the expected effect of waypoints.

A simple means of calculating path separation is to measure the NND from every point on a path to every other path. The value of the separation can then be taken as the mean, median, minimum or maximum of

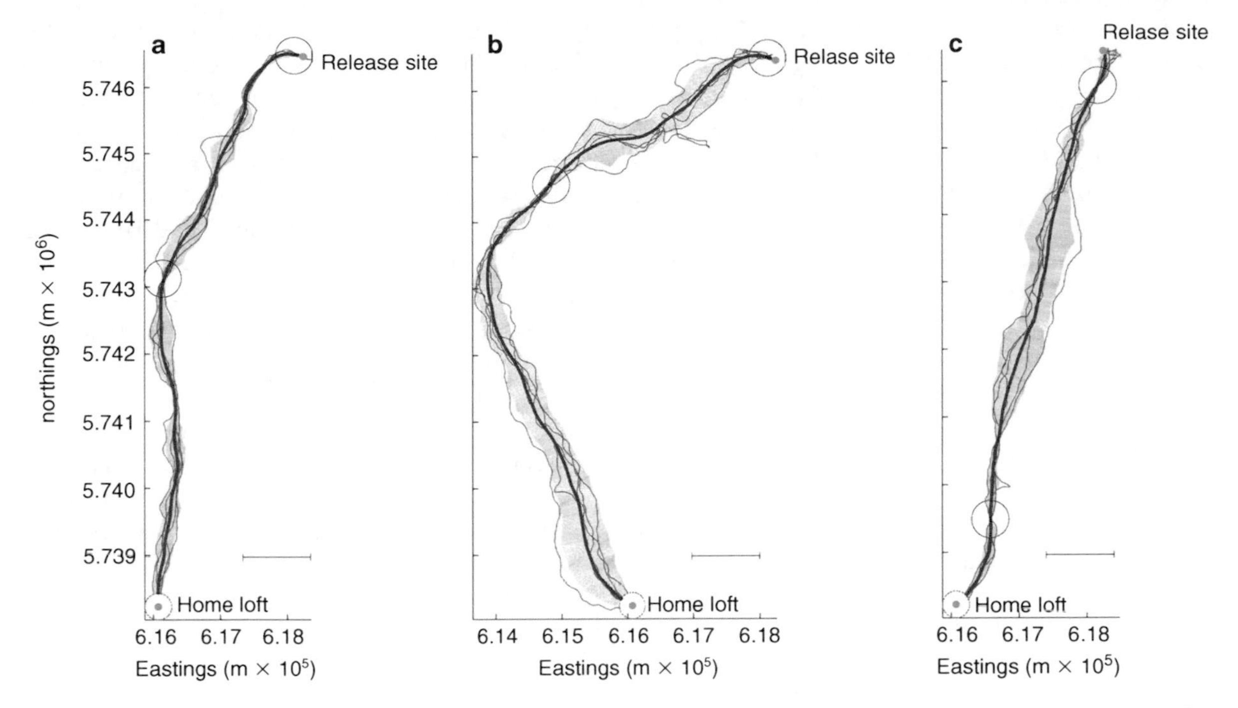

Waypoint Detection, Fig. 1 Five flights each from three pigeons (*light blue*) showing the calculated mean path (*bold red*), variance around the mean path (*grey band*), and location of areas of peak route similarity (*black circles*). The peak similarity regions correspond to variance below a threshold value. Each peak similarity region can contain multiple successive points, but for clarity only one circle is shown per region (Reprinted from [5], Copyright [2010], Creative Commons open access license)

these values as desired, giving a value for separation at every position along the path. This provides a measure of separation along each path, but not a single description of the overall separation of all the paths as a single variable, which may make it difficult to identify single points where variation is low. This may be ameliorated by the fact that areas where the separation is low will also be areas where the paths are very close, so regions of low variability will be spatially restricted.

It is also possible to measure a single value of separation, using variation around a constructed *mean path*. A mean path is a single path that is representative of the full set of paths. As with path separation there is no absolute way to construct such a path. In analogy with the mean of a set of scalar variables, a mean path can be constructed to minimize a particular measure of separation relative to the original paths, much as the scalar mean minimizes the square distance between itself and the original variables. This approach is used, for example, by Freeman et al. [5], where a mean path is generated which minimizes the total nearest-neighbor separation from itself to the true paths. Constructing such a path may require an iterated approximation method rather than a direct calculation. See Fig. 1 for examples of constructed mean paths and mean path similarity in relation to the original paths. Similarly to the case where path similarity is calculated along each route, the mean path will be most representative of the true paths in regions of peak similarity, which correspond to waypoint locations. Therefore, having calculated similarity along the mean path, one can be sure that the waypoints identified from this similarity will be in approximately the correct spatial location.

A related approach is that used by Lipp et al. [8], for investigating the propensity of pigeons to follow roads and other linear features. Here the spatial area containing the paths is segmented into a two-dimensional grid. The number of paths passing through each grid element is recorded and compared with a null hypothesis for the distribution of paths (see [8] for an example of generating an appropriate null hypothesis). In the original paper, the analysis demonstrates that grid elements containing roads also contain significantly more paths than expected. The converse argument can also be applied – grid elements containing more paths than expected (judged by some significance test) are likely to contain waypoints. The disadvantage of this method is the difficulty in striking a balance between the size of the grid and the density of paths. If the grid elements are too small there will be few paths in any single element and it will be impossible to distinguish from the null hypothesis. If the grid elements are too large, the resolution will be limited to the size of the element, making it difficult to pinpoint the location of the waypoint accurately.

Path Complexity

Less intuitive than path variability, but applicable even when only a single path is observed, path complexity segments the movement path into regions of high and low localized complexity. Increased complexity can be an indicator of waypoint location since waypoints may be associated with decision points, where the animal must first adjust its heading to reach the waypoint and subsequently readjust again to direct itself toward the next objective. Conversely when the animal is far from any known locations, it would be expected to follow a smooth path since there is no external cue available to correct the heading and any other variation would lead to unnecessary energy loss.

The complexity of a path segment can be viewed as a measure of how much information is required to describe it. A straight line or a very smooth curve can be described simply by stating the end points and the degree of curvature. A segment whose curvature is constantly changing, which may change direction discontinuously or which returns back to a previously covered point (looping) will require far more information to adequately describe. To understand the relationship between complexity and information, imagine having to describe a path segment to someone who is unable to see it, well enough that they could draw it themselves. The more details that are required to convey that path segment, the more complex it is.

As with path similarity, path complexity can be measured in a variety of ways, which are similarly highly correlated with each other. These range in abstractness from *tortuosity*, the local degree of inefficiency in the path, to the highly abstract concept of spatiotemporal positional entropy introduced by Roberts et al. [13]. Figure 2 shows a plot of a number of recorded movement paths, color-coded by the local degree of complexity, to demonstrate how high complexity and low complexity path segments present themselves visually. Each path is segmented into low-, medium-, and high-complexity regions, using positional entropy as a measure of complexity. In the figure, it can be clearly seen that low complexity (green) regions are smooth, with low curvature and few if any sharp changes of direction. High-complexity (red) regions often include loops, sharp turns, and many changes of direction. The medium-complexity (blue) regions lie between these extremes. The figure also shows that high-complexity regions are often short, indicating some local effect on the path, and occur particularly around the release point, when the pigeons are deciding on a heading to take.

Another similar measure that has been used to detect foraging within movement paths is *passage time*. This is amount of time taken by an animal from the moment when it enters a (typically circular) region of space to the moment when it leaves. This can be measured as the first passage time [4, 11] – the time from first entering to first leaving the region or total passage time [1] – the time from first entering to the final time leaving. This is strongly related to path complexity, since highly complex, inefficient path segments with lots of changes of direction will have high passage times. The use of passage time to detect foraging within directed movement is broadly analogous to waypoint detection, since foraging stops within a directed path will form waypoints by virtue of the concentration of the food source.

Complexity measures have associated spatiotemporal scales which must be appropriate for the detection task. Tortuosity is measured as the inefficiency of a path segment, the length must be selected. Likewise positional entropy measures the complexity of a moving segment of the path. Passage time is measured with respect to a spatial region whose size must be chosen. These scales must bear in mind the classic trade-off between resolution and signal-to-noise ratio; the smaller the scale, the more fine structure will be detected and the more tightly determined will be the position of identified waypoints. Longer scales improve the signal-to-noise ratio by filtering out high-frequency noise from the behavior but also reduce the resolution of the measurement.

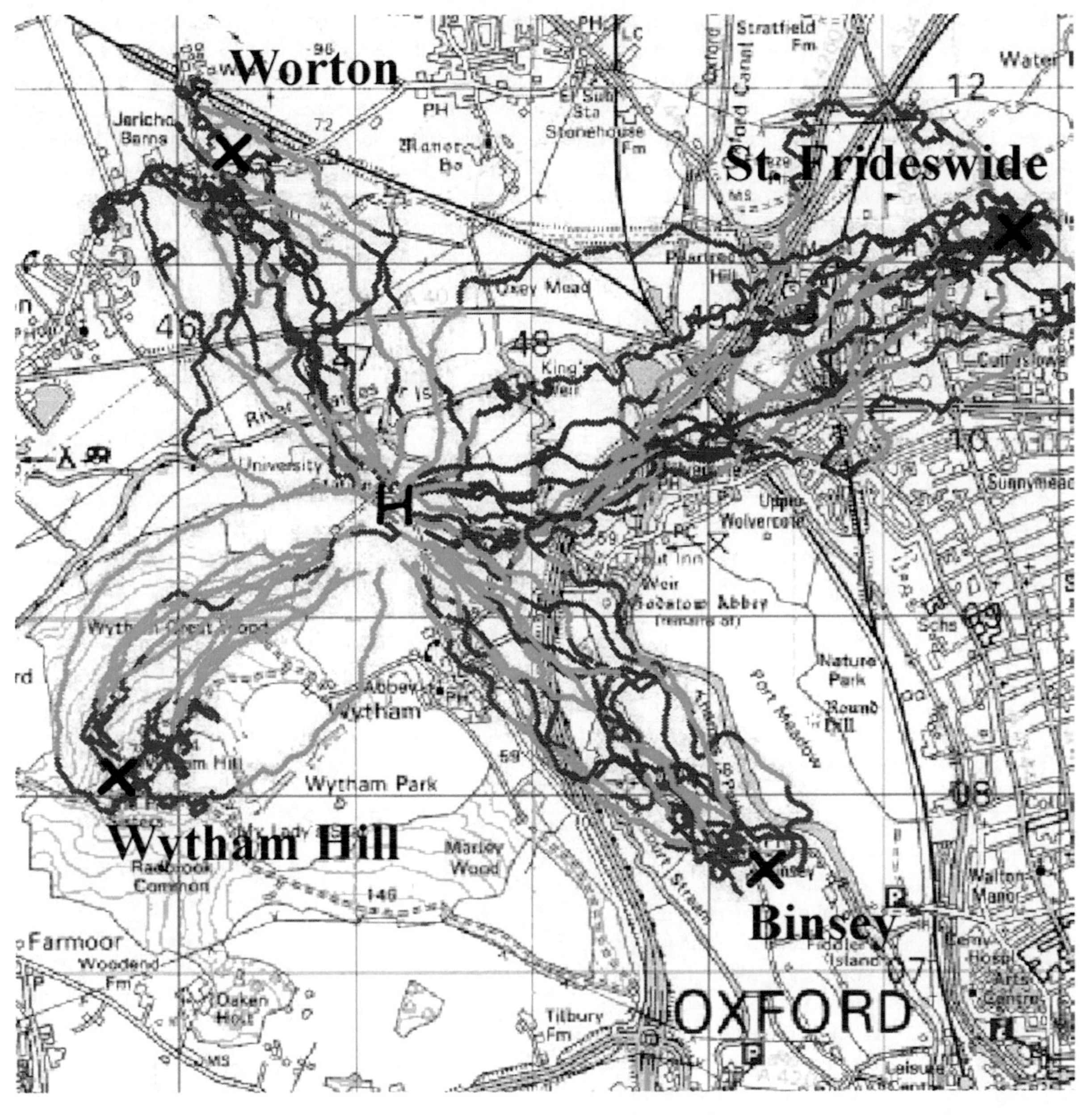

Waypoint Detection, Fig. 2 Identified regions of low (*green*), medium (*blue*), and high (*red*) complexity in a variety of homing pigeon flight paths, as determined by positional entropy (Reprinted from [6], Copyright [2004], with permission from Elsevier)

Variation in complexity measures has been suggested to correlate with changes in the animal's behavioral state [1, 2, 4, 6, 11] and to segment the path into distinct behavioral phases [1, 6, 13] supporting the use of complexity to identify key path segments that may correspond to waypoint locations. Path complexity has also been linked to the surrounding visual environment [7], indicating that the waypoints identified may correspond to visual landmarks, as expected under the pilotage hypothesis.

Path Predictability

Path predictability differs distinctly from the previous two approaches since it uses complete movement paths in a global analysis, rather than examining only localized properties of the paths. Waypoints are selected as those segments of the path that can be used to provide the best prediction of the complete path and other paths (where these exist). The logic of this approach depends on the hypothesis of navigation by *pilotage*. This hypothesis claims that the animal makes it way from its original position to its final objective by successively visiting an ordered sequence of fixed waypoints, as per the definition above. The path predictability method argues that if this hypothesis is accurate, the path(s) of the animal must be defined by the locations of those waypoints, the segments between waypoints being subject to constraints such as energy minimization and random variation from environmental and internal factors. These waypoints are thus analogous to the fixed positions in a curve-fitting exercise, with the constraints given above acting similarly to the constraint of curvature minimization in spline fitting. The upshot of this is that those segments of the path(s) close to the waypoints should provide the maximum amount of information about the rest of path(s).

Similarly to path complexity, path predictability is intimately linked to information. Consider trying to convey to someone not just a path segment but instead a set of complete paths. Now imagine that the only information you are allowed to send is the position of the paths at a small set of times. If your friend subsequently "joins the dots" to link these points, how well will they approximate the true paths? The most informative, or predictive, locations are those that allow the best estimate of the real paths using this technique.

Technically, the information contained in a path can be defined how it affects the probability of another path. If path B is informative about path A then the probability of path A will increase as a result of knowing path B. Mathematically, the information contained in path B about path A is related to the ratio of probability of path A, conditional on knowing path B through some model M, compared to the probability of path A when B is unknown. Information is defined as the log of this probability ratio:

$$\texttt{Information(A | B)/bits} = \texttt{log\_2}\ \texttt{P(path A | path B, M)} - \texttt{log\_2 P(path A | M)} \quad (1)$$

The probability P(path A | path B, M) depends on the choice of model. A simple model might be that each point on path A should be the same as on path B at the equivalent point in time, give or take some amount of Gaussian noise to account for natural variation. More realistic models should take account of the correlations over time that lead to the spatial structure of the path. Mann et al. [10] provide a more sophisticated model choice implemented via the use of Gaussian processes, allowing for autocorrelation within the path over time as well as the correlations between paths. In this implementation, all paths are assumed to be samples from a distribution centered on a mean path – similar to the mean path defined as part of the path similarity measure. Constructing a model to describe the probability distribution of movement paths is a complicated task which is not readily reducible to a simpler scheme. This task is necessary, however, to specify the mutual information contained between paths and between elements of the same path and thus to be able to use path predictability as a measure. See Mann et al. [10] for guidance on model construction.

Having defined a model and therefore a probability distribution from one path to another, the task of identifying the waypoints can be clarified as follows. Path B will be a recorded path, with positions recorded at a set of time points, t. Let t′ be a subset of those timesteps such that path B′ is the recorded positions at times t′. The information in those positions is calculated as

$$\texttt{Information(t}') = \texttt{Information(A | B}') = \texttt{log\_2 P(path A | path B}'\texttt{,M)} \quad (2)$$

Having hypothesized that waypoints correspond to the most informative segments of the path, the task is now rephrased as finding the subset of t, t*, that maximizes Eq. 2,

$$t^* = \texttt{argmax\_t}'\ \texttt{Information(t}') \quad (3)$$

Finding a limited number, m, of waypoints can be accomplished simply by searching for the t* restricted to subsets of size m. Optimization of Eq. 2 can be accomplished by "greedy" forward selection, gradient ascent, Monte Carlo sampling, or other established optimization techniques. Standard texts on learning algorithms, for example, [9], can provide comprehensive details of optimization algorithms.

Predictability between different paths tends to be maximal under two circumstances. The first is in regions

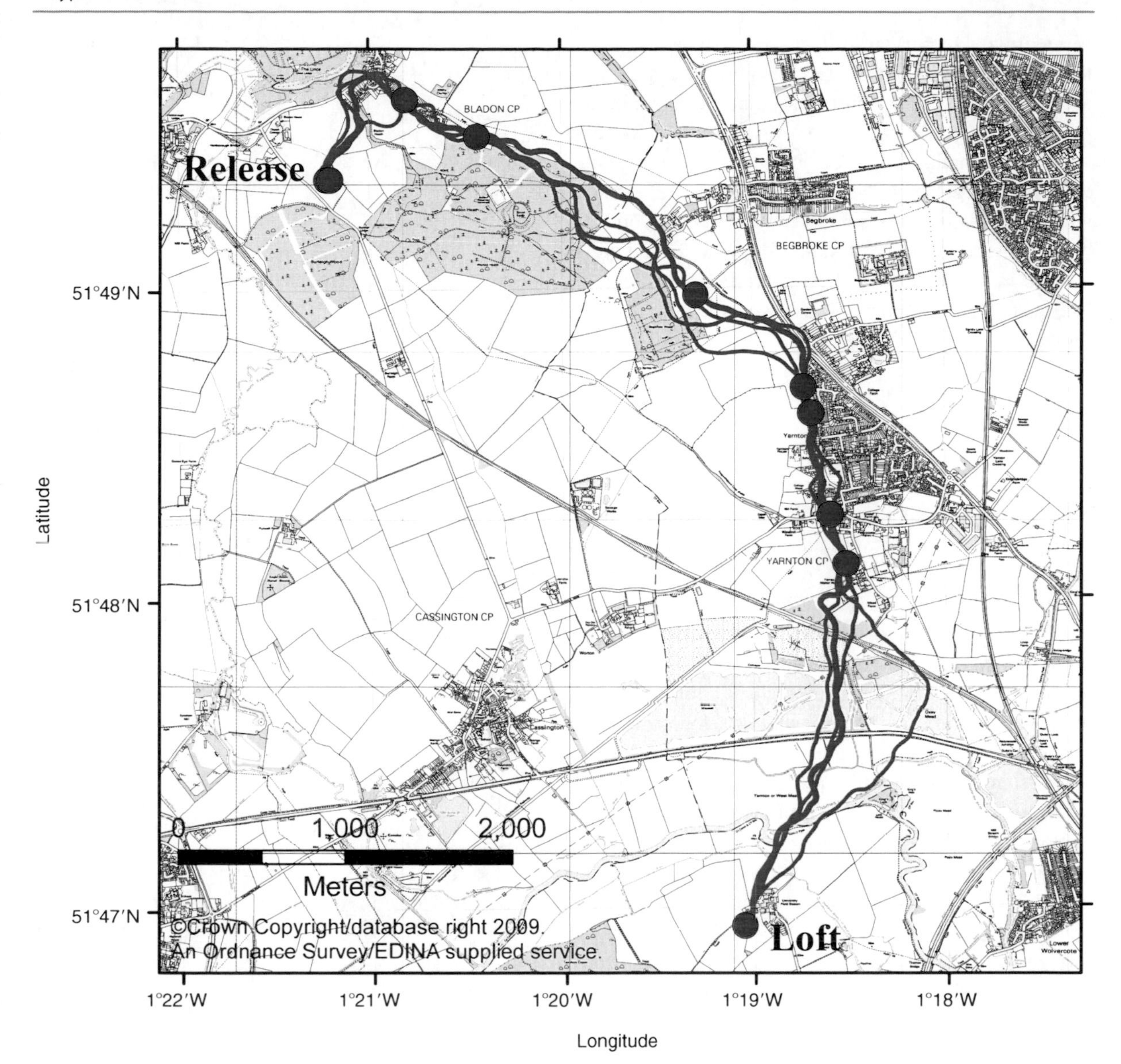

Waypoint Detection, Fig. 3 Ten waypoints identified from five flight paths from a single homing pigeon, using path predictability as a criterion. The waypoints are identified preferentially in flight segments that a similar between paths, where the path exhibits high complexity and where the path is far from the most efficient straight "beeline" between start and finish (Reprinted from [10], Copyright [2010], Creative Commons open access license)

where the two paths are very similar. Here knowing one path gives good information about the likely location of the other. This then is similar to the path similarity criterion. The other case is where the location of the path is highly unpredictable either a priori, or conditional on the other chosen waypoints. Locally this bears a resemblance to the path complexity measure, but also takes account of the global unpredictability – for example finding path segments that are locally smooth but globally far from the straight-line path between start and finish. This corresponds to an intuitive analysis of the path. If the bird consistently flies to a location far from the most efficient route this is unlikely to be due to chance; there must be a waypoint at this location.

Figure 3 shows an example of waypoints selected by maximizing the path predictability, taken from Mann et al. [10]. In this demonstration, the number of selected waypoints is ten and the information in Eq. 2

is maximized by forward selection. It can be seen from this example that the selected waypoints each exhibit at least one of the three properties that maximize information content. Many of the waypoints are located where the paths coalesce into narrow corridors, exhibiting high path similarity. Others are located where the paths turn sharply, which correspond to high path complexity. Finally there are waypoints located in the regions furthest from the most efficient beeline path, at the apex of the curve, which is the most unlikely area for the pigeons to visit a priori.

Summary

Detection of waypoints in recorded movement paths is a novel problem that has potential future applicability in studies of animal movement and biomimicry. The methods described in this entry constitute a basis for identifying important regions of the movement path based on hypotheses about the likely effects of waypoints upon the animal's movement. These methods span a range of mathematical complexity and intuitive applicability. More recent developments such as the path predictability tend toward higher levels of abstraction. The methods explored in this entry are not mutually exclusive and combinations should be explored to achieve the best detection results. With animal movement tracking still developing as an experimental field, methods for the analysis of recorded data are being created in parallel and the development of further methods for identifying salient regions from movement paths can be expected in the near future.

Cross-References

▸ Biomimetics
▸ Nanoparticle Tracking Analysis

References

1. Barraquand, F., Benhamou, S.: Animal movements in heterogeneous landscapes: identifying profitable places and homogeneous movement bouts. Ecology **89**, 3336–3348 (2008)
2. Biro, D., Guilford, T., Dell'Omo, G., Lipp, H.-P.: How the viewing of familiar landscapes prior to release allows pigeons to home faster. J. Exp. Biol. **205**, 3833–3844 (2002)
3. Biro, D., Meade, J., Guilford, T.: Familiar route loyalty implies visual pilotage in the homing pigeon. Proc. Nat. Acad. Sci. **101**(50), 17440–17443 (2004)
4. Fauchald, P., Tveraa, T.: Using first-passage time in the analysis of area-restricted search and habitat selection. Ecology **84**, 282–288 (2003)
5. Freeman, R., Mann, R., Guilford, T., Biro, D.: Group decisions and individual differences: route fidelity predicts flight leadership in homing pigeons (Columba livia). Biol. Lett. **7**(1), 63–66 (2010). Advance online publication
6. Guilford, T., Roberts, S., Biro, D., Rezek, I.: Positional entropy during pigeon homing II: navigational interpretation of Bayesian latent state model. J. Theor. Biol. **227**(1), 25–38 (2004)
7. Lau, K.-K., Roberts, S., Biro, D., Freeman, R., Meade, J., Guilford, T.: An edge-detection approach to investigating pigeon navigation. J. Theor. Biol. **239**(1), 71–78 (2006)
8. Lipp, H.-P., Vyssotski, A.L., Wolfer, D.P., Renaudineau, S., Savini, M., Tröster, M., Dell'Omo, G.: Pigeon homing along highways and exits. Curr. Biol. **14**, 1239–1249 (2004)
9. Mackay, D.: Information Theory, Inference and Learning Algorithms. Cambridge University Press, Cambridge (2003)
10. Mann, R., Freeman, R., Osborne, M., Garnett, R., Armstrong, C., Meade, J., Biro, D., Guilford, T., Roberts, S.: Objectively identifying landmark use and predicting flight trajectories of the homing pigeon using Gaussian processes. J. Roy. Soc. Interface. **8**, 210–219 (2011)
11. McKensie, H.W., Lewis, M.A., Merrill, E.H.: First Passage time analysis of animal movement and insights into the functional response. Bull. Math. Biol. **71**(1), 107–129 (2009)
12. Meade, J., Biro, D., Guilford, T.: Homing pigeons develop local route stereotypy. Proc. Roy. Soc. B. **272**(1558), 17–23 (2005)
13. Roberts, S., Rezek, I., Guilford, T., Biro, D.: Positional entropy during pigeon homing I: application of Bayesian latent state modelling. J. Theor. Biol. **227**(1), 39–50 (2004)

Wear

▸ Nanotribology

Wet Adhesion in Tree Frogs

▸ Adhesion in Wet Environments: Frogs

Wet Chemical Processing

▸ Sol-Gel Method

Wet Etching

Avinash P. Nayak, M. Saif Islam and
V. J. Logeeswaran
Electrical and Computer Engineering, University of California- Davis Integrated Nanodevices & Nanosystems Lab, Davis, CA, USA

Synonyms

Chemical etching; Liquid etching

Definition

Wet etching is a material removal process that uses liquid chemicals or etchants to remove materials from a wafer. The specific patters are defined by photoresist masks on the wafer. Materials that are not protected by this mask are etched away by liquid chemicals. These masks are deposited on the wafer in an earlier fabrication step called lithography.

A wet etching process involves multiple chemical reactions that consume the original reactants and produce new reactants. The wet etch process can be described by three basic steps (1) Diffusion of the liquid etchant to the structure that is to be removed. (2) The reaction between the liquid etchant and the material being etched away. A reduction-oxidation (redox) reaction usually occurs. This reaction entails the oxidation of the material then dissolving the oxidized material. (3) Diffusion of the byproducts in the reaction from the reacted surface.

Isotropy and Anisotropy

When a material is attacked by a liquid or vapor etchant, the material will be removed isotropically (uniformly in all directions) or by anisotropic etching (uniformity in vertical direction). The difference between isotropic etching and anisotropic etching is shown in Fig. 1. Material removal rate for wet-etching is usually faster than the rates for many dry etching processes and can easily be changed by varying temperature or the concentration of active species.

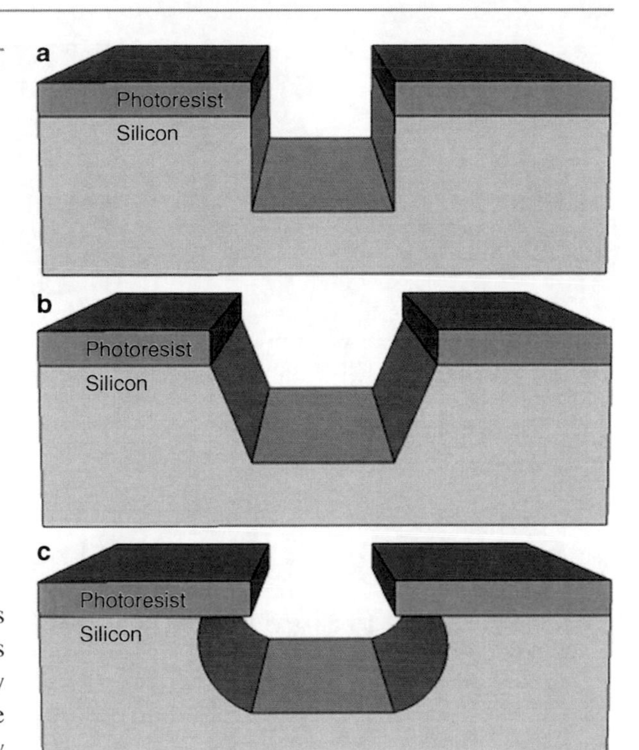

Wet Etching, Fig. 1 (**a**) Completely anisotropic, (**b**) partially anisotropic, and (**c**) isotropic etching

Anisotropic Wet Etching

Liquid etchants etch crystalline materials at different rates depending upon which crystal face is exposed to the etchant. There is a large difference in the etch rate depending on the silicon crystalline plane. In materials such as silicon, this effect can allow for very high anisotropy. Some of the anisotropic wet etching agents for silicon are potassium hydroxide (KOH), ethylenediamine pyrocatechol (EDP), or tetramethylammonium hydroxide (TMAH). Etching a (100) silicon wafer would result in a pyramid-shaped etch pit as shown in Fig. 2. The etched wall will be flat and angled. The angle to the surface of the wafer is 54.7°.

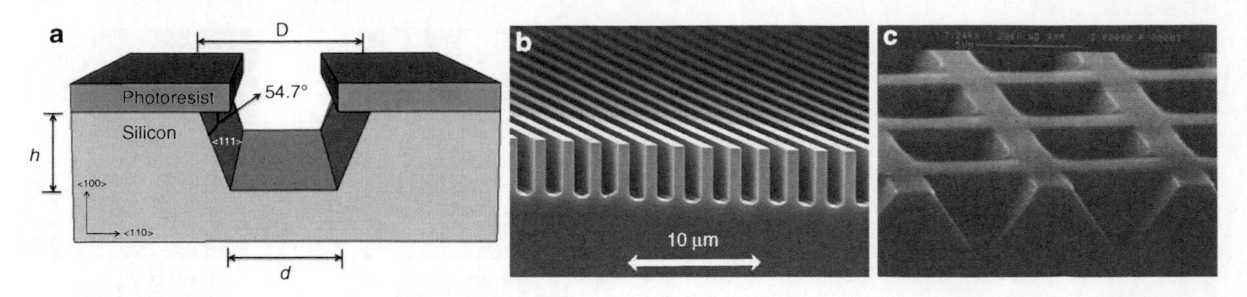

Wet Etching, Fig. 2 Example of an etch profile in anisotropic of (100) silicon wafer. (**a**) Etching profile of silicon where h is the height after etch, D and d are the *top* and *bottom* edge-to-edge distance respectively. (**b**) Complete anisotropic etching of (110)-oriented Si in potassium hydroxide (KOH) solution. (**c**) Partially anisotropic etching of (100)-oriented Si (inverted pyramids) in KOH solution [1] (Note that the directionality of the etching depends heavily on the orientation of the silicon crystal orientation)

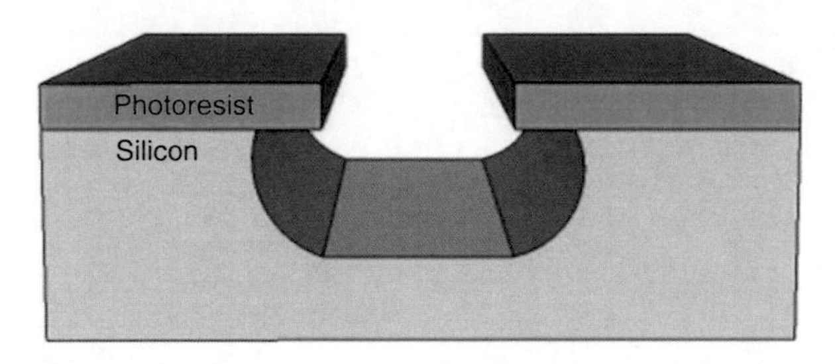

Wet Etching, Fig. 3 Etch profile in isotropic etch of a silicon wafer

The relationship between mask dimensions, etch depth, and the floor width is given in Eq. 1.

$$d = D - \left(\frac{2h}{\tan(54.7^\circ)}\right) \tag{1}$$

Isotropic Wet Etching

For isotropic wet etching, a mixture of hydrofluoric acid, nitric acid, and acetic acid (HNA) is the most common etchant solvent for silicon. The concentrations of each etchant determines the etch rate. Silicon dioxide or silicon nitride is usually used as a masking material against HNA. As the reaction takes place, the material is removed laterally at a rate similar to the speed of etching downward. This lateral and downward etching process takes places even with isotropic dry etching, which is described in the dry etch section. The isotropic wet etch is shown in Fig. 3.

Wet chemical etching is generally isotropic even though a mask is present since the liquid etchant can penetrate underneath the mask. Since directionality is usually important for high-resolution pattern transfer, wet chemical etching is normally not used.

Cross-References

▶ EUV Lithography
▶ Nanofluidics

References

1. Parker, G.J., et al.: Highly engineered mesoporous structures for optical processing. Philos. Trans. Roy. Soc. A Math. Phys. Eng. Sci. **364**, 189–199 (2006)

Wetting Transitions

Edward Bormashenko
Laboratory of Polymers, Applied Physics Department, Ariel University Center of Samaria, Ariel, Israel

Definition

The wetting transition is an abrupt change, spontaneous or induced by external stimulus, in the wetting properties of a flat or rough solid surface.

Wetting of Rough Surfaces and Wetting States

Wetting is a spreading of liquid on a solid or liquid substrate. It is ubiquitous and of primary importance in chemical, automobile and food industries, medicine, soil and climate sciences. Various technological applications call for surfaces with a controlled wettability, in particular water- and oil-repellent surfaces. Rapid progress of *nanotechnologies* has allowed the fabrication of biomimetic surfaces demonstrating pronounced water repellence (the so-called "*lotus effect*" [1]). A variety of natural objects, such as birds' and butterflies' wings and legs of water striders, demonstrate extreme water repellency (superhydrophobicity, which is one of the *surface tension effects of nanostructures*) [2–4]. Various sophisticated experimental techniques, including UV-lithography and plasma etching, were successfully applied for manufacturing lotus-like biomimetic surfaces [4, 5]. Superhydrophobic surfaces demonstrate a potential for a variety of green technologies, including energy conversion and conservation, and environment-friendly self-cleaning underwater surfaces [5].

The wetting of flat and rough surfaces is characterized by a contact angle, which is defined as the angle between the tangent to the liquid-fluid interface and the tangent to the solid surface at the contact line between the three phases (solid, liquid, and vapor) [6, 7]. The Young (or equilibrium) contact angle θ is the angle that a liquid makes with an ideal (smooth, rigid, chemically homogenous, insoluble, and nonreactive) surface. The Young angle θ is given by the well-known Young equation:

$$\cos\theta = \frac{\gamma_{SA} - \gamma_{SL}}{\gamma} \quad (1)$$

where γ, γ_{SL}, γ_{SA} are the surface tensions at the liquid-air (vapor), solid-liquid, and solid-air (vapor) interfaces, respectively. The contact angle of a droplet deposited on a solid depends on external parameters, such as temperature. Change in temperature may stimulate the transition from the partial wetting to the complete wetting of the solid substrate. In this case, one observes a wetting transition (WT) on a smooth solid surface [8].

The Young equation supplies the sole value of a contact angle for a certain combination of solid, liquid and gaseous phases. However, the experimental situation is much more complicated; even on atomically flat solid surfaces a diversity of contact angles is observed. This is due to the long-range interaction between molecules forming the triple (three-phase) line of the droplet and molecules forming the solid substrate. The maximal possible contact angle observed on the surface is called the advancing angle θ_{adv}; the minimal one is called the receding angle θ_{rec} [6, 7]. The difference between advancing and receding contact angles $\theta_{adv} - \theta_{rec}$ is called the contact angle hysteresis [6, 7]. Experimental establishment of advancing and receding angles is a challenging task, and it should be mentioned that the reported contact angles are sensitive to the experimental technique used for their measurement. Both advancing and receding are equilibrium (though metastable) contact angles [7].

The wetting of rough and chemically heterogeneous surfaces is described by the apparent contact angle θ^* (APCA), which is different from the local contact angle given by Young's relation [6, 7]. The APCA is an equilibrium contact angle measured macroscopically. The detailed topography of a rough surface cannot be viewed with regular optical means; therefore, this contact angle is defined as the angle between the tangent to the liquid-vapor interface and the apparent solid surface, as macroscopically observed [7]. Superhydrophobic surfaces are characterized by high values of APCA (usually $>150°$). However, high APCA does not necessarily guarantee the true superhydrophobicity and self-cleaning properties observed in nature. Moreover, high-stick surfaces demonstrating a high APCA were reported recently. Low-contact-angle hysteresis, resulting in low sliding angles and high stability of "lotus-like" wetting state, as well as high APCA supply water-repellent and self-cleaning properties to the surface. The design of such surfaces, allowing emerging green applications, remains a challenging scientific and technological task.

Chemical heterogeneities and roughness strengthen the hysteresis of a contact angle. The wetting of flat, chemically heterogeneous surfaces is characterized by the APCAθ^*, predicted by the Cassie–Baxter wetting model [9]. Consider wetting of a composite flat surface

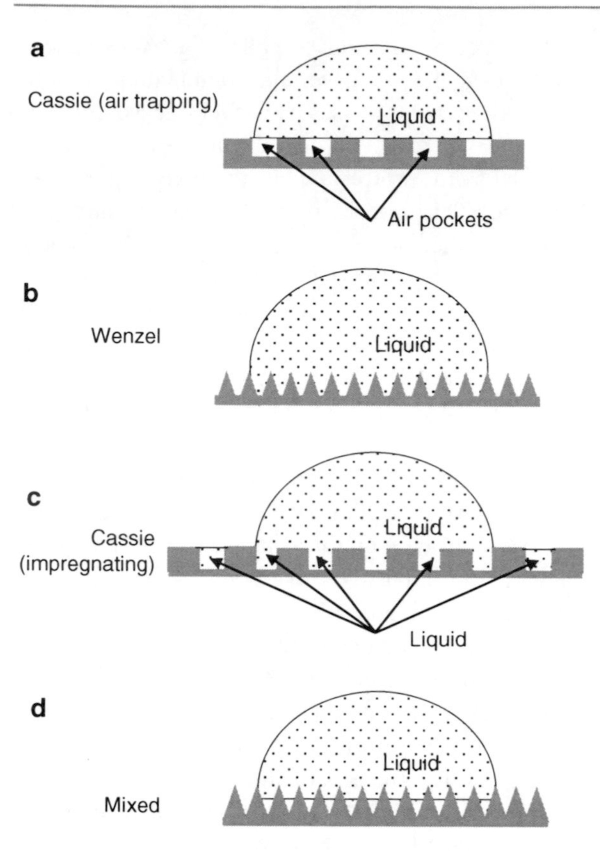

Wetting Transitions, Fig. 1 Various wetting states occurring on rough surfaces: (**a**) Cassie air-trapping state, (**b**) Wenzel state, (**c**) Cassie impregnating wetting state, (**d**) mixed wetting state

comprised of several materials. Each material is characterized by its own surface tension coefficients $\gamma_{i,\mathrm{SL}}$ and $\gamma_{i,\mathrm{SA}}$, and by the fraction f_i in the substrate surface, $f_1 + f_2 + ... + f_n = 1$. The APCA in this case is supplied by the Cassie–Baxter equation:

$$\cos\theta^* = \frac{1}{\gamma}\sum_1^n f_i(\gamma_{i,\mathrm{SA}} - \gamma_{i,\mathrm{SL}}) = \sum_1^n f_i \cos\theta_i \quad (2)$$

where $\theta_i$are equilibrium contact angles for the ith material. The Cassie–Baxter equation can be applied to the solid surface comprised of pores (the contact angle for pores is π and $\cos\theta = -1$; see Fig. 1a). In this case, the Cassie–Baxter equation yields:

$$\cos\theta^* = -1 + f_S(\cos\theta + 1) \quad (3)$$

where f_S and 1-f_S are the relative fractions of the solid and air fractions underneath the drop, respectively. The Cassie air-trapping wetting state is also called the "fakir state." The Cassie-like air-trapping wetting provides easy sliding of water droplets. When the relief is hierarchical, the Cassie–Baxter equation obtains more complicated forms, taking into account the interrelation between the scales constituting the topography of the relief.

The wetting of rough, chemically homogenous surfaces is governed by the Wenzel model [10]. According to the Wenzel model, surface roughness r, defined as the ratio of the real surface in contact with liquid to its projection onto the horizontal plane, always magnifies the underlying wetting properties (see Fig. 1b). Both hydrophilic and hydrophobic properties are strengthened by surfaces textures. The apparent Wenzel contact angle is given by Eq. 4:

$$\cos\theta^* = r\cos\theta \quad (4)$$

The Cassie wetting is featured by low contact angle hysteresis, whereas Wenzel wetted surfaces demonstrate high hysteresis of the contact angle. Actually, pure Cassie and Wenzel wetting situations are rare in occurrence, and a mixed wetting state is often observed [11]. In the mixed wetting state, a droplet partially wets the surface and partially sits on air pockets, as described in Fig. 1d. The APCA $\theta^*$is supplied in this case by Eq. 5:

$$\cos\theta^* = r_f f\cos\theta + f - 1 \quad (5)$$

In this equation, f is the fraction of the projected area of the solid surface that is wetted by the liquid, and r_f is the roughness ratio of the wet area. When $f = 1$, $r_f = r$ and Eq. 5 turns into the Wenzel equation (4); when $r_f = 1, f = f_S$, and Eq. 5 turns into the Cassie–Baxter equation (3).

One more wetting state is possible. This is the Cassie impregnating state, depicted in Fig. 1c. In this case, liquid penetrates into grooves of the solid, and the drop finds itself on a substrate viewed as a patchwork of solid and liquid (solid "islands" ahead of the drop are dry, as shown in Fig. 1c) [6]. The APCA is expressed as

$$\cos\theta^* = 1 - f_S + f_S\cos\theta \quad (6)$$

It should be stressed that Eqs. 3–6 could be applied when the radius of the droplet is much larger than

Wetting Transitions, Fig. 2 Wetting transitions observed by vibration of 15 μl water drop deposited on the micrometrically rough PDMS surface: (**a**) The initial Cassie state. (**b**) The Cassie impregnating state induced by vibrations

the characteristic scale of the surface heterogeneities. The rigorous thermodynamic derivation of Eqs. 3–6 may be obtained within the general variational approach [12].

The physical mechanisms of WT on flat and rough surfaces are quite different. WT on rough surfaces are related to the penetration of liquid into the grooves constituting the relief and result in the change of the APCA (see Fig. 2) [13–20]. Wetting transitions were observed under various experimental techniques utilizing a diversity of factors: gravity, pressure, bouncing and evaporation of droplets, electric field in the electrowetting experiments, and vibration of droplets [13–18]. WT were observed with the use of various experimental techniques, including reflection interference contrast microscopy and environmental scanning electron microscopy [20].

Energetic Approach to Wetting Transitions

Various wetting states featured by very different APCA can coexist on the same heterogeneous surface. The diversity of APCA could be easily understood if one takes into account that the Gibbs energy curve for a droplet on a real surface is characterized by multiple minima points. It could be shown that the Wenzel state (Fig. 1b) is energetically favorable compared to the pure Cassie air trapping one, shown in Fig. 1a, when the inequality (7) holds:

$$\cos\theta > \frac{f_S - 1}{r - f_S} \quad (7)$$

The Wenzel state will be energetically favorable when compared to the mixed trapping state depicted in Fig. 1d, when the inequality (7a) holds:

$$\cos\theta > \frac{f - 1}{r - r_f f} \quad (7a)$$

Obviously for hydrophilic materials, the Wenzel state is always favorable in relation to the Cassie one, whereas for the hydrophobic materials the situation is more complicated. When the hydrophobic material is very rough ($r >> f_s$), the Cassie state will be favorable for any obtuse θ. For single-scale hydrophobic materials, the Cassie air-trapping wetting state usually corresponds to the highest of the multiple minima of the Gibbs energy of the droplet deposited on a rough surface. Thus, for the WT, the energy barrier should be surmounted [18]. It was supposed that this energy barrier is equal to the surface energy variation between the Cassie state and the hypothetical *composite* state, with the almost complete filling of surface asperities by liquid, keeping the liquid-air interface under the droplet and the contact angle constant, as shown in Fig. 3.

Wetting of hydrophobic side surfaces of asperities constituting the relief is energetically unfavorable; this gives rise to the energetic barrier. The collapse of the Cassie state will occur when the liquid filling the grooves will eventually touch the bottoms of the substrate pores, and the liquid-air interface will disappear. Contrasting to the equilibrium mixed wetting state, the composite state is metastable. For the simple topography depicted in Fig. 4, the energy barrier could be calculated as follows:

$$E_{\text{trans}} = 2\pi R^2 h(\gamma_{SL} - \gamma_{SA})/p = -2\pi R^2 h\gamma\cos\theta/p \quad (8)$$

where h and p are the geometrical parameters of the relief, shown in Fig. 4, and R is the radius of the contact area. The numerical estimation of the energetic barrier according to the Formula (8) with the parameters $p = h = 20$ μm, $R = 1$ mm, $\theta = 105°$ (corresponding to low density polyethylene (LDPE)), and $\gamma = 72$ mJ·m^{-2}, gives a value of $E_{\text{trans}} = 120$ nJ. It should be stressed that according to (8), the energy barrier scales as $E_{\text{trans}} \sim R^2$. The validity of this assumption will be discussed below. The energetic barrier is extremely large compared to thermal fluctuations: $\frac{E_{\text{trans}}}{kT} \approx \left(\frac{R}{a}\right)^2 >> 1$, where a is an atomic scale. On the other hand, it is much less than the energy of evaporation of the droplet Q: $kT << E_{\text{trans}} << Q$. Actually, this interrelation between characteristic energies makes wetting transitions possible. If it were

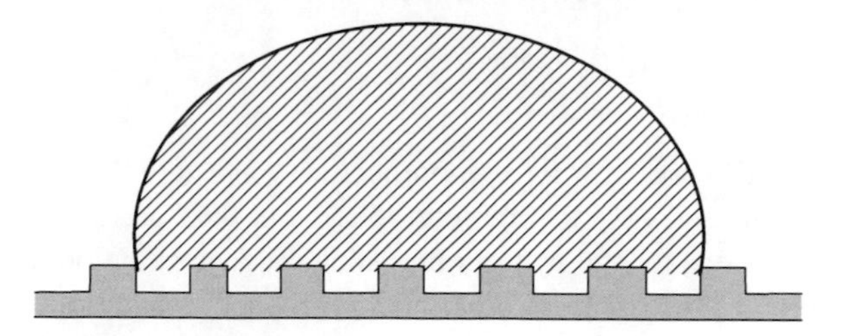

Wetting Transitions, Fig. 3 The composite wetting state

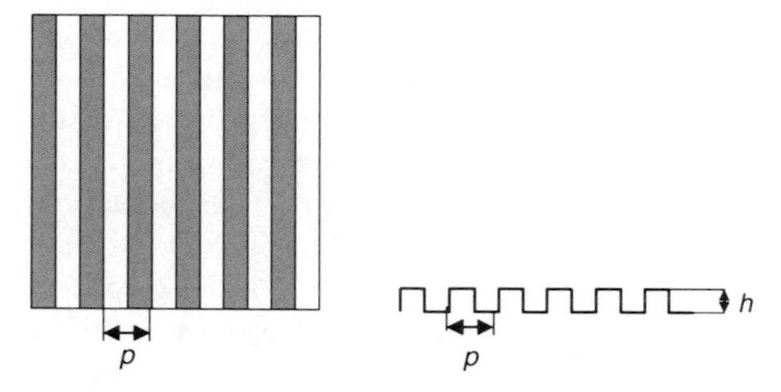

Wetting Transitions, Fig. 4 Geometric parameters of the model relief illustrating the calculation of the Cassie–Wenzel transition energetic barrier

not the case, a droplet exposed to external stimuli might evaporate before the wetting transition.

The energy-barrier height from the side of the metastable (higher-energy) wetting state is always much lower than that from the side of the stable one. The mentioned energy dependence is weak for large apparent angles, and strong for low ones; this leads to a high asymmetry of the energy barrier, i.e., to a large difference between barrier heights. As a result, wetting transitions on rough surfaces are irreversible for both hydrophobic and hydrophilic materials.

Single and two-stage pathways of WT were observed, including Cassie (air-trapping)-Wenzel–Cassie (impregnating), Wenzel–Cassie (impregnating), and Cassie (air-trapping)-Cassie (impregnating) transitions.

The Concept of the Critical Pressure

Various experimental methods used for the study of WT supplied the close values of the pressure necessary for the Cassie–Wenzel transition, which is of the order of magnitude of 100–300 Pa for 10 μl droplets deposited on micrometrically scaled rough surfaces. It is noteworthy that the Cassie air-trapping wetting regime

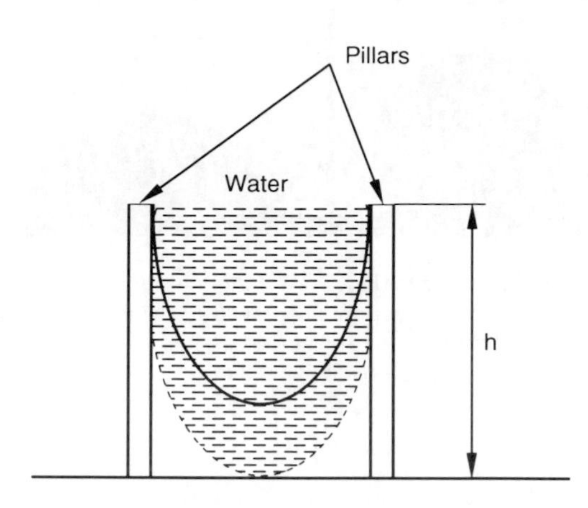

Wetting Transitions, Fig. 5 Pressure-induced displacement of the water front, leading to the collapse of the Cassie wetting state

observed on natural objects (birds' wings) is much more stable when compared to that on artificial surfaces.

The important parameter characterizing WT is the critical pressure under which the transition takes place [19]. For a single-scaled pillar-based biomimetic surface, with pillar width a, and groove width b the critical pressure equals:

$$p_c = -\frac{\gamma f_s \cos\theta}{(1-f_s)\lambda} \tag{9}$$

where $\lambda = \frac{A}{L}$, and A and L are the pillar cross-sectional area and perimeter, respectively. For the Teflon surface$\theta = 114°, a = 50\,\mu m, b = 100\,\mu m$ yield $p_c = 296\,\text{Pa}$, in agreement with experimental results, reported by various groups [15, 17]. Recalling that the dynamic pressure of rain droplets may be as high as 10^4–10^5 Pa, which is much larger than $p_c \approx 300\,\text{Pa}$, it could be inferred that creating biomimetic reliefs with very high critical pressure is of practical importance. The concept of critical pressure leads to the conclusion that reducing the microstructural scales (e.g., the pillars diameters and spacing) is the most efficient means to enlarge the critical pressure [19].

The mechanism of WT based on the concept of critical pressure was proposed. It was supposed that as the pressure applied to the droplet increased, the meniscus will move toward the flat substrate, as shown in Fig. 5. The meniscus will eventually touch the

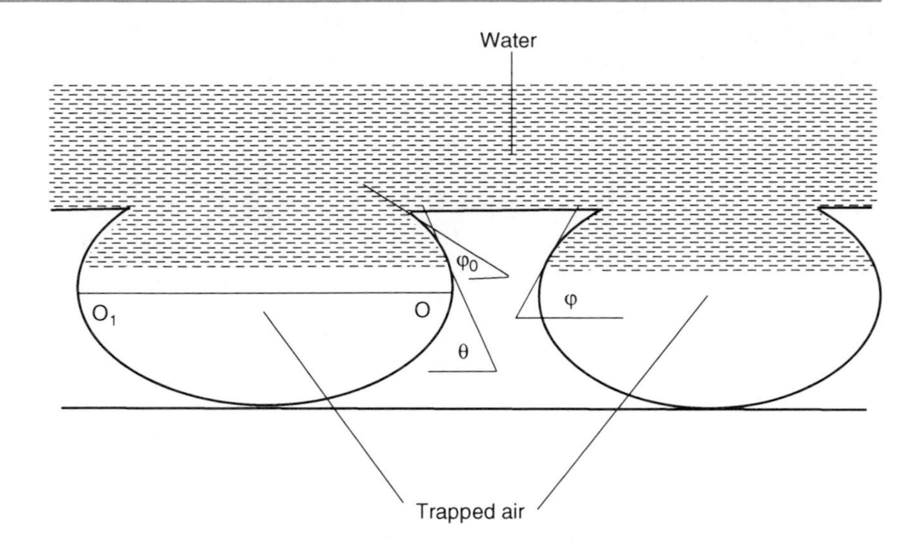

Wetting Transitions, Fig. 6 Geometrical air trapping on the hydrophilic reliefs

substrate; this will cause the collapse of the Cassie wetting and, consequently, lead to a Cassie–Wenzel transition.

Wetting Transitions on Hydrophilic Surfaces

The Cassie wetting was observed on inherently hydrophilic surfaces. For the hydrophilic materials, the difference $\gamma_{SL} - \gamma_{SA}$ is negative, and liquid penetration into pores might have to proceed spontaneously, without the energy barrier. For the explanation of the roughness-induced superhydrophobicity of inherently hydrophilic materials, it was supposed that air is entrapped by cavities constituting the topography of the surface. The simple mechanism of "geometrical" trapping could be explained as follows: consider a hydrophilic surface ($\theta < 90°$) comprised of pores as depicted in Fig. 6. It is seen that air trapping is possible only if $\theta > \varphi_0$, where φ_0 is the angle between the tangent in the highest point of the pattern and the horizontal symmetry axis O_1O. Indeed, when the liquid level is descending, the angle φ is growing (see Fig. 6), and if the condition $\theta > \varphi_0$ is violated; the equilibrium $\theta = \varphi$ will be impossible.

In the equilibrium position, small fluctuations of the contact angle lead to the appearance of curvature on the plane air–water interface that is energetically unfavorable. Below the central plane $O - O_1$ where $\varphi > 90°$, the equilibrium is impossible in the case of $\theta < \frac{\pi}{2}$. For a large pore, small fluctuations of θ can lead to touching the pore bottom near its centre by the curved air-liquid interface, followed by filling the pore and the consequent collapse of the Cassie air trapping-wetting regime.

The Dynamics of Wetting Transitions

The dynamics of WT is not clearly understood. The characteristic time of Cassie–Wenzel transition was established with reflection interference contrast microscopy as less than 20 milliseconds [20]. Two regimes of droplet front displacement were observed: zipping and non-zipping. In the zipping regime, the velocity of the front in one direction (to advance to the next row of pillars) is much smaller than the velocity in the other direction (liquid filling up one row of micropillars). It was established that the velocity of the wetting front increases with increasing gap size, decreasing pillar height, or decreasing contact angle. A velocity of the wetting front as high as 1.5 m/s was registered.

The "Dimension" of Wetting Transitions

One of the unsolved problems in the field of WT is the problem of the "dimension" of the transitions, or, in other words, whether all pores underneath the droplet should be filled by liquid (the "2D scenario"), or, perhaps, only the pores adjacent to the three-phase (triple) line are filled under external stimuli such as pressure, vibrations, or impact (the "1D scenario"). Indeed, the APCA is dictated by the area adjacent to

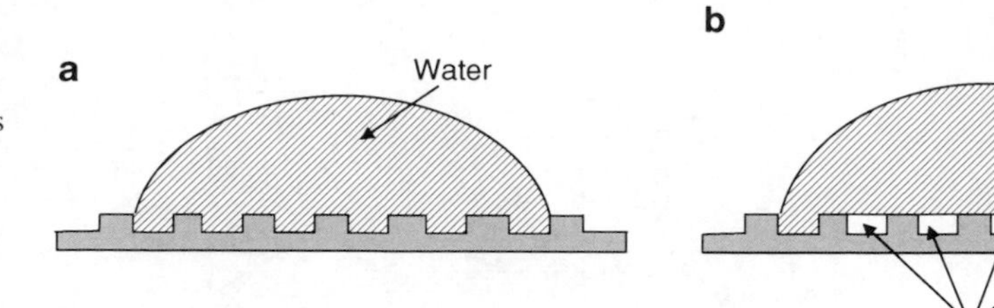

Wetting Transitions, Fig. 7 (**a**) 2D mechanism of WT (all pores are filled). (**b**) 1D scenario of WT (only pores adjacent to the triple line are filled)

the triple line and not by the total area underneath the droplet. 1D and 2D scenario of WT are illustrated by Fig. 7. The experiments carried out with vibrated drops supported the 1D scenario of wetting transitions. It has been established that the transition occurs when the condition $F_{cr} = \text{const}$ is fulfilled, where F_{cr} is the critical force acting on the unit length of the triple line, and the transition is caused by de-pinning of the triple line [17]. The critical value of the de-pinning force F_{cr} has been established experimentally for various microscopically structured surfaces as $F_{\mathrm{cr}} \approx 200 - 350\,\mathrm{mN}\cdot\mathrm{m}^{-1}$ [17]. The energy barrier E_{trans} to be surmounted for the elementary displacement of the triple line δr could be estimated as

$$E_{\mathrm{trans}} \approx 2\pi R F_{\mathrm{cr}} \delta r \tag{10}$$

which scales as R. This scaling law corresponds to results obtained with vibrated drops but contradicts the scaling law given by expression (8). The electrowetting-stimulated WT also supported the 1D mechanism of transitions. De-pinning of the triple line was observed directly with the use of reflection interference contrast microscopy [20]. The potential barrier E_{trans}, according to expression (10) for a drop with a radius of $R \approx 1$ mm deposited on the LDPE relief, presented in Fig. 4 ($F_{\mathrm{cr}} \approx 350\,\mathrm{mN}\cdot\mathrm{m}^{-1}$, the elementary displacement $\delta r \approx p/2 \sim 10^{-5}\,\mathrm{m}$) equals $E_{\mathrm{trans}} \approx 20\,nJ$, smaller than the value predicted by formula (10), but still much larger than thermal fluctuations.

It has been also suggested that the Cassie–Wenzel transition occurs via a nucleation mechanism starting from the drop center. In spite of intensive theoretical and experimental effort expended for the study of wetting transitions, the numerous aspects of these phenomena, including the dynamics of transitions and their "dimension," remain unclear and call for future investigations.

Cross-References

▶ Lotus Effect
▶ Nanotechnology
▶ Surface Tension Effects of Nanostructures

References

1. Koch, K., Bohn, H.F., Barthlott, W.: Hierarchically sculptured plant surfaces and superhydrophobicity. Langmuir **25**, 14116–14120 (2009)
2. Bormashenko, E., Bormashenko, Ye, Stein, T., Whyman, G., Bormashenko, E.: Why do pigeon feathers repel water? Hydrophobicity of pennae, Cassie–Baxter wetting hypothesis and Cassie–Wenzel capillarity-induced wetting transition. J. Colloid Interface Sci. **311**, 212–216 (2007)
3. Zheng, Yo, Gao, Xu, Jiang, L.: Directional adhesion of superhydrophobic butterfly wings. Soft Matter **3**, 178–182 (2007)
4. Quéré, D., Reyssat, M.: Non-adhesive lotus and other hydrophobic materials. Philos. Trans. R. Soc. A **366**, 1539–1556 (2008)
5. Nosonovsky, M., Bhushan, B.: Superhydrophobic surfaces and emerging applications: non-adhesion, energy, green engineering. Curr. Opin. Colloid Interface Sci **14**, 270–280 (2009)
6. de Gennes, P.G., Brochard-Wyart, F., Quéré, D.: Capillarity and Wetting Phenomena. Springer, New York (2004)
7. Marmur, A.: A guide to the equilibrium contact angle maze. In: Mittal, K.L. (ed.) Contact Angle Wettability and Adhesion, vol. 6, pp. 3–18. VSP, Leiden (2009)
8. Bonn, D., Ross, D.: Wetting transitions. Rep. Prog. Phys. **64**, 1085–1163 (2001)
9. Cassie, A.B.D., Baxter, S.: Wettablity of porous surfaces. Trans. Faraday Soc. **40**, 546–551 (1944)
10. Wenzel, R.N.: Resistance of solid surfaces to wetting by water. Ind. Eng. Chem. **28**, 988–994 (1936)
11. Marmur, A.: Wetting on hydrophobic rough surfaces: to be heterogeneous or not to be? Langmuir **19**, 8343–8348 (2003)
12. Bormashenko, E.: Young, Boruvka–Neumann, Wenzel and Cassie–Baxter equations as the transversality conditions for the variational problem of wetting. Colloids Surf. A **345**, 163–165 (2009)
13. Bartolo, D., Bouamrirene, F., Verneuil, E., Buguin, A., Silberzan, B., Moulinet, S.: Bouncing of sticky droplets:

impalement transitions on superhydrophobic micropatterned surfaces. Europhys. Lett. **74**, 299–305 (2006)
14. Lafuma, A., Quéré, D.: Superhydrophobic states. Nat. Mater. **2**, 457–460 (2003)
15. Jung, Y.C., Bhushan, B.: Dynamic effects induced transition of droplets on biomimetic superhydrophobic surfaces. Langmuir **25**, 9208–9218 (2009)
16. McHale, G., Aqil, S., Shirtcliffe, N.J., Newton, G.M.I., Erbil, H.Y.: Analysis of droplet evaporation on a superhydrophobic surface. Langmuir **21**, 11053–11060 (2005)
17. Bormashenko, E., Pogreb, R., Whyman, G., Erlich, M.: Cassie-Wenzel wetting transition in vibrating drops deposited on rough surfaces: is dynamic Cassie-Wenzel wetting transition a 2D or 1D affair? Langmuir **23**, 6501–6503 (2007)
18. Barbieri, L., Wagner, E., Hoffmann, P.: Water wetting transition parameters of perfluorinated substrates with periodically distributed flat-top microscale obstacles. Langmuir **23**, 1723–1734 (2007)
19. Zheng, Q.-S., Yu, Y., Zhao, Z.-H.: Effects of hydraulic pressure on the stability and transition of wetting modes of superhydrophobic surfaces. Langmuir **21**, 12207–12212 (2005)
20. Moulinet, S., Bartolo, D.: Life and death of a fakir droplet: impalement transitions on superhydrophobic surfaces. Eur Phys J **E24**, 251–260 (2007)

Whispering Gallery Mode Biosensor

▶ Whispering Gallery Mode Resonator Biosensors

Whispering Gallery Mode Resonator Biosensors

Kerry Allan Wilson[1] and Frank Vollmer[2]
[1]London Centre For Nanotechnology, University College London, London, UK
[2]Laboratory of Biophotonics and Biosensing, Max Planck Institute for the Science of Light, Erlangen, Germany

Synonyms

Microcavity biosensor; Monolithic total internal reflection biosensor; Optical cavity biosensor; Optical resonance biosensor; Optical ring resonator biosensor; Resonant evanescent wave biosensor; Whispering gallery mode biosensor

Definition

A whispering gallery mode resonator biosensor monitors optical resonances in microcavities for label-free detection of molecules and their interactions.

Overview

Whispering Gallery mode (WGM) resonators are named for the "whispering gallery" of St. Paul's Cathedral in London where sound waves propagate by reflection around the curved conduit of the gallery such that a word whispered at the front end of the gallery can be easily heard by a listener at the far end of the acoustic gallery. Similarly, optical Whispering gallery modes occur when light waves confined by total internal reflection in a dielectric microstructure are reflected back on the same optical path where they interfere constructively. Interference generates the optical resonance signal, which occurs whenever exactly an integer number of wavelengths fit on the confined (circular) light path (Fig. 1). Molecules binding to the microresonator and interacting with the evanescent field generated at the sensor surface are detected from a change in optical resonance frequency. The frequency response is rendered specific for molecular binding events by functionalization of the sensor surface with biorecognition elements such as antibodies or oligonucleotides. WGM biosensors achieve single particle, single virus, and potentially single molecule detection capability by monitoring changes in the optical resonance frequencies in microspheres, microrings, microcapillaries, and microtoroids. Glass microspheres and microtoroids are particularly sensitive WGM biosensors because of their high Quality (Q) factor and their small modal volume (V). However, other types of optical resonators (microcavities) of different shapes and made of different materials are similarly suitable for biosensing applications, examples of these are ring resonators (made of silica, silicon, or transparent polymers) and photonic crystals (fabricated in silicon-on-insulator or in silicon-nitrate wafers). WGM biosensors and optical resonator derivatives have emerged as the one of the most sensitive microsystem biodetection technology that does not require chemical amplification or labelling of the analyte.

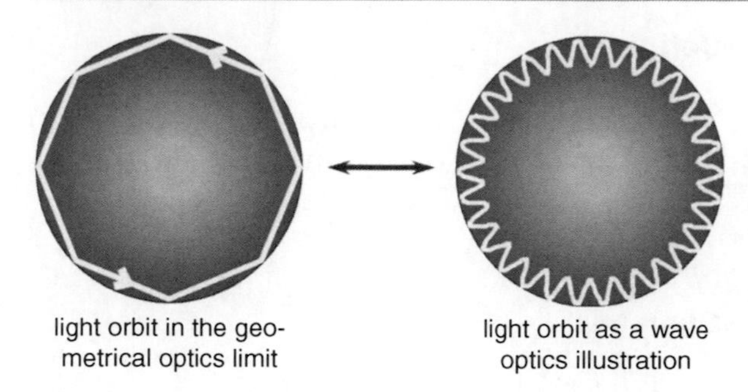

Whispering Gallery Mode Resonator Biosensors, Fig. 1 Schematic representation of light confined inside a resonant microcavity by total internal reflectance

Optical Resonator Physics: Confining Light at High Q-Factor

A high-quality (Q) factor is necessary for achieving sensitive detection using optical resonance in microcavities. Optical resonance is created by confining coherent light inside a miniature dielectric structure such that the returning light wave interferes constructively [1–3]. Figure 1 illustrates the example of an optical resonance confined by total internal reflection in a glass microsphere. Ideally, optical resonators (also referred to as cavities or microcavities) would confine light indefinitely. Real-world divergence from this condition due to damping losses results in a finite resonance lifetime τ. This finite resonance lifetime can be measured, i.e., by charging the microcavity with light, terminating the light source, and then measuring the time it takes for the intensity of the confined light wave to drop to 1/e of its initial value. This exponentially decaying intensity-amplitude of the optical resonator is analogous to the decreasing amplitude of a (nano) mechanical resonator, here illustrated as mass-on-a-spring (Fig. 2b). To compare optical, mechanical, and other resonant systems which can exhibit very different resonance lifetimes, one defines a time-independent and dimensionless figure of merit, the quality (Q)-factor. The Q-factor is proportional to the total number of oscillations stored in the resonator, and this definition is independent from the type of resonator (optical, mechanical, etc.) and the nature of its oscillation. For example, in an optical resonator, the value of the Q-factor is 2π times the number of electromagnetic oscillations before amplitude of the confined light wave drops below 1/e of its initial value. Similarly, the value of the mechanical Q-factor is just 2π times the number of (Pendulum-)swings before that same limit is reached. For any resonant system, the Q-factor is thus defined as $Q = \omega\,\tau$, where τ is the resonator lifetime and where ω is 2π times the resonance frequency (Fig. 2). The resonance frequency for optical microresonators ranges from a few TeraHertz (10^{12} Hz) to one PetaHertz (10^{15} Hz), whereas nanomechanical resonators operate at a few MHz at most. Although resonance lifetimes of very large mechanical resonator (i.e., Feaucault Pendulum) exceed those of optical resonators (many hours compared to microseconds), the optical microresonators exhibit much higher Q values as compared to their large mechanical and nanomechanical counterparts: in fact, microcavities exhibit the highest measured Q-factor of almost $\sim 10^{10}$, whereas a good nanomechanical resonator reaches a Q value of a few million at best and a Pendulum operated in vacuum reaches a Q of about 100,000. In fact, the "ultimate" Q-factor has been measured in a microsphere cavity where an optical resonance (a whispering gallery mode, WGM) is efficiently confined by total internal reflection – with only residual damping due to absorption in the glass material.

Apart from the time-domain response, a resonator also has a frequency-domain response (Fig. 2c) and it is important to realize that a large Q-factor is associated with a very narrow spectral linewidth where $\Delta\omega_{\text{line}} = \omega/Q$ (Fig. 2d). A narrow linewidth allows for precise measurements of resonance frequency and changes thereof, which is of paramount importance for achieving high sensitivity in biosensing applications. Ultimate-Q microsphere resonators are therefore the ideal choice to build a sensitive WGM biosensor [4]. The direct relationship between a high-Q factor and a narrow linewidth can be intuitively understood from the fact that the frequency-domain response of the resonant system (Fig. 2d) is just the Fourier-transform of the time-domain response (Fig. 2c). The Q-factor is not only related to the time-domain and the frequency-domain signal of a resonator, but also to the energy stored in a resonant system. The Q-factor can be calculated from the total energy W_{total} stored in the resonator divided by the energy ΔW lost per oscillation (multiplied by 2 Pi), $Q = 2\pi\ W_{\text{total}}/\Delta W_{\text{per oscillation}}$. For example, the lifetime and Q-factor of an ultimate WGM microsphere resonator is limited only by residual absorption and calculated as $Q = 2\pi\ (W_{\text{total}}/\Delta W)_{\text{per oscillation}} = 2\pi\ n/\lambda\alpha$, where α is the extinction coefficient of glass, about 7 dB/km, λ is the wavelength

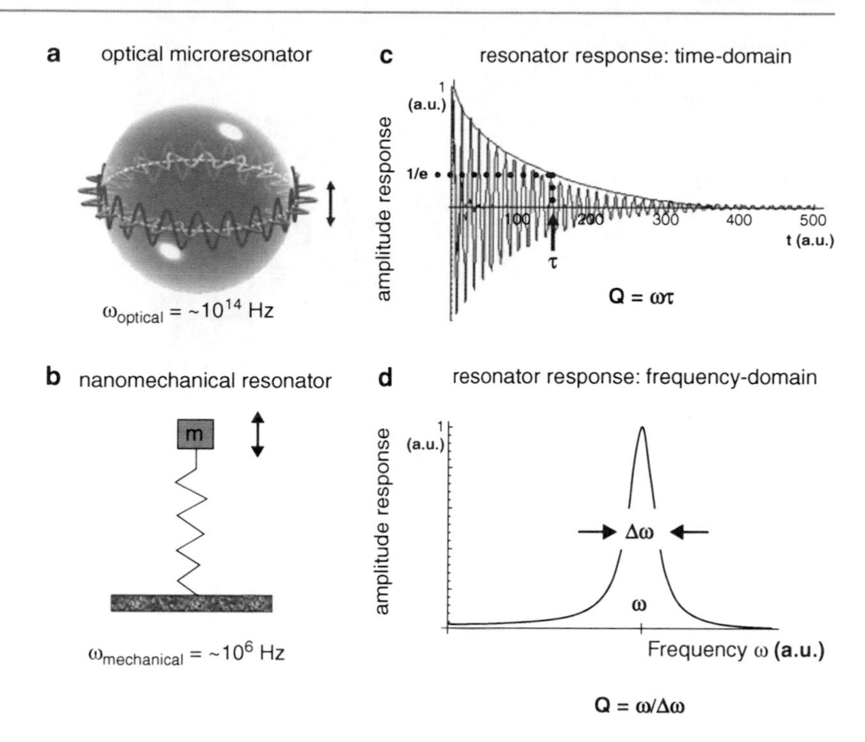

Whispering Gallery Mode Resonator Biosensors, Fig. 2 (**a**) and (**b**) Comparison of optical and nanomechanical resonators. (**c**) Definition of Q-value in the time domain. (**d**) Definition of Q-value in the frequency domain

Whispering Gallery Mode Resonator Biosensors, Table 1 Q-factors of various resonant systems

Integrated circuit filter	20
Mechanical wristwatch	100–300
Surface acoustic waves	2,000
Foucault pendulum (in vacuum)	10^4–10^5
Quartz crystal	10^6
Microsphere optical resonator	10^{10}

(here 633 nm), and n the refractive index of the glass (here ~1.45). Table 1 gives an overview of Q-factors of resonant optical, electrical, and mechanical systems.

Principles of Resonant WGM Biosensing

An optical resonator transduces the binding of molecules into a resonance frequency shift signal which can be detected with great precision due to the high Q-factor of the optical detection system [5]. Optical resonators have no moving parts and maintain a high Q-factor even in a liquid biosensing environment [4] where any mechanical system would be heavily damped (low Q). The change in resonance frequency of an optical resonator occurs in response to the binding of molecules which increase the round-trip optical path length. This is different from a mechanical resonator where the frequency shift would occur due to mass loading by the added molecules (Fig. 3). Such a frequency shift, i.e., for a nanomechanical resonator can be easily calculated: Additional molecular mass loading, Δm, on a system with a nominal mass, m, will cause a resonance frequency shift of $\Delta\omega_{\text{mechanical}} = -\,\omega_{\text{mechanical}}\ \Delta m/2m$ (Fig. 3), where only nanomechanical systems are useful for such measurements due to their small mass. For an optical resonator instead, the binding of molecules add additional optical path length ΔL causing the resonance frequency ω_{optical} to shift by $\Delta\omega_{\text{optical}} = -\,\omega_{\text{optical}}\ \Delta L/2L$, where $2L$ is the round-trip optical path length in the microcavity (Fig. 3). For the special case of a spherical WGM optical resonator where bound molecules form an optically contiguous layer of thickness l (Fig. 3), the resonance frequency shift is calculated as $\Delta\omega_{\text{microsphere}} = -\omega_{\text{microsphere}}\ l/R$, where l is the thickness of the adsorbed molecular layer and R is the radius of the microsphere. This estimate assumes an identical refractive index for microsphere and adsorbed molecular layer, $n_{\text{glass}} = n_{\text{molecules}}$, a good approximation for the case of proteins adsorbing to a glass microsphere ($n_{\text{glass}} = n_{\text{protein}} \sim 1.45$). Further assuming the ability to detect the resonance frequency shift (Fig. 3) with a resolution of S = 1/50 of the linewidth, and assuming a modest Q of 10^7, which can be readily obtained

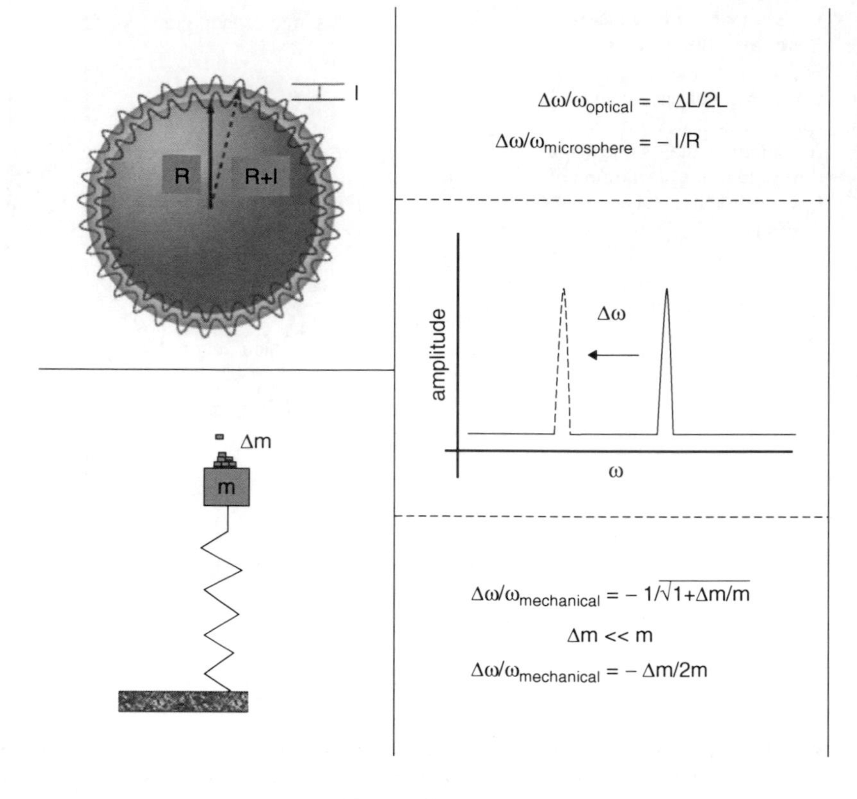

Whispering Gallery Mode Resonator Biosensors, Fig. 3 Reactive sensing principle for an optical resonator and for a nanomechanical resonator

with a 100 μm diameter glass microsphere in a liquid biosensing environment, the smallest detectable thickness l of a molecular layer bound to a WGM microsphere cavity is calculated as $l = S/Q * R = (0.02/10^7)*$ 50 μm $= 10^{-7} = 0.1$ pm! Biomolecules are several nm in diameter (~3.5 nm for a typical protein such as BSA), so detection of a small fraction of a monolayer coverage is an easy task for a WGM biosensor. Achieving such a feat with a (nano)-mechanical system is impossible due to the heavy damping (low Q) of any mechanical resonator after immersion in a liquid biosensing environment. In contrast, high-Q optical resonators have no moving parts and the Q-factor (and corresponding resonance lifetime) is mainly limited by the absorption of the light in the resonator material and in the surrounding medium (ultimate Q). By choosing a wavelength with minimal absorption losses, and by optimizing the material and size of a microsphere resonator, it has been estimated that unprecedented, label-free single molecule detection capability is achievable [6]. This possibility combined with the great variety of materials and geometries available for the design of high-Q optical systems has spurred the tremendous interest in the development and application of microcavity biosensors.

WGM Biosensing: Reactive Sensing Principle

The frequency shift $\Delta\omega$ of an optical resonance due to binding of any kind of molecule to any kind of resonator can be precisely calculated from first-order perturbation theory. Assuming that bound molecules do not change the light field, then the fractional change in resonance frequency $\Delta\omega/\omega$ (and corresponding fractional wavelength $\Delta\lambda/\lambda$) is calculated [6] by comparing the energy it takes to polarize the molecule bound to the microcavity surface at position $\vec{r}_0$, $\Delta W_{\text{particle}} = h\varpi = \frac{1}{2}\alpha\vec{E}(\vec{r}_o)^2$, to the total electromagnetic energy confined in the microcavity $W_{\text{total}} = \int \varepsilon\vec{E}(\vec{r})^2 dV$:

$$\frac{\Delta\lambda}{\lambda} = \frac{\Delta W_{\text{particle}}}{W_{\text{total}}} = \frac{\alpha\vec{E}(\vec{r}_0)^2}{2\int \varepsilon\vec{E}(\vec{r})^2 dV} \quad (1)$$

This so-called reactive sensing principle can be applied to any microcavity structure and to any biomolecule. Furthermore, a large frequency shift and therefore a high sensitivity is obtained for a small denominator in (1), which is achieved by confining the resonant light field within a small modal volume defined as $V = \int \varepsilon \vec{E}(\vec{r})^2 / \vec{E}(\vec{r})^2_{\max} dV$, where $\vec{E}(\vec{r})_{\max}$ is the highest field intensity measured in the cavity. For high-sensitivity biosensing applications, it is therefore important not only to operate at a high Q value, enabling precise measurement of resonance frequency, but equally important to use a miniature optical resonator with small modal volume V, producing a large frequency shift signal due to the binding of molecules. Biosensing applications, therefore, require optimizing the Q/V value. WGM microsphere biosensors typically exhibit modal volumes $\sim$500 um^3 at Q $\sim 10^7$ and therefore achieve Q/V $\sim 2 \times 10^4$ (1/μm^3). Photonic crystal microcavity structures may further boost current Q/V values since these resonators can confine light in record-low modal volumes of $V < (\lambda)^3/n \sim 0.2\ \mu m^3$ (where n is the refractive index of the resonator), a volume which is less than a wavelength cubed! This strong spatial localization of light in a photonic crystal cavity is possible due to multiple Bragg scattering from many lattice elements, different from a WGM resonator where the light field is confined by internal reflection. Although photonic crystals can confine light in record low modal volumes, Q-factors of photonic crystal cavities are typically lower as compared to WGM resonators and reach only ~one to a few million. Here, the Q-factor is mainly limited by the finite precision of the nanoscale photonic crystal fabrication process where residual surface roughness causes additional damping of the resonant light field. With further improvements of the nanoscale fabrication processes, however, photonic crystal microcavities may surpass current record Q/V's of WGM biosensors and are therefore an interesting resonator structure that can be potentially mass produced by deep UV lithography.

WGM biosensors to date are more sensitive as compared to any other commercialized label-free biosensor technology such as surface plasmon resonance sensors (SPR) and quartz crystal microbalances (QCM). WGM biosensors routinely achieve sensitivity far below 1 pg/mm^2 molecular mass loading, approaching the specific detection of the binding of molecules from fM (and less) solution concentration levels. Impressively, WGM technology was used to demonstrate the detection of a single Influenza A virus particle from fM solution concentration [7], a noteworthy feat considering that a single Influenza A virus particle measures only ~50 nm in radius and weighs less than ~500 attograms.

WGM Biosensing: Label-free Analysis of Homogenous Molecular Binding Events

For WGM microsphere resonators, (1) can be used to derive an analytic equation for the observed fractional frequency shift $\Delta\omega/\omega$ due to homogenous binding of nanoscale molecules or particles at random location on the microsphere sensor surface [4, 6]. Then, the frequency shift $\Delta\omega/\omega$ and the corresponding wavelength shift $\Delta\lambda/\lambda$ is calculated from (Fig. 4):

$$\frac{\Delta\lambda}{\lambda} = -\frac{\Delta\omega}{\omega} = \frac{\sigma\alpha}{\varepsilon_0(n_s^2 - n_m^2)R} \tag{2}$$

where σ is the surface density of bound molecules, α is their excess polarizability (in excess to the displaced water), ε_0 is the vacuum permittivity, n_s and n_m are the refractive indices of the glass microsphere and the surrounding water, respectively, and R is the radius of the microsphere. The negative sign for $\Delta\omega/\omega$ results from the fact that a shift to a longer resonance wavelength corresponds to a shift to lower resonance frequency. The excess polarizability α can be estimated from tabulated "*dn*/*dc*" values for each biomolecule [8] measured in aqueous solution. The refractive index increment "*dn*/*dc*" is defined as the increase Δn of the refractive index n measured for the increase Δc of the biomolecule solution concentration c in the limit for c - > 0: $dn/dc = \lim(\Delta n/\Delta c)_{c\to 0}$. The excess polarizability α is then calculated from *dn*/*dc* values according to,

$$\alpha = \varepsilon_O 2n \frac{dn}{dc} m \tag{3}$$

where n is the solvent refractive index (water), and m is the mass of the single biomolecule. Proteins such as bovine serum albumin (BSA) have a typical *dn*/*dc* value of about 0.183 cm^3/g. The excess polarizability of BSA is then calculated as $\alpha_{BSA} = 4\ \pi\ \varepsilon_0\ 3.85 * 10^{-21}\ cm^3$, where the mass of a single BSA molecule is ~66,432 (g/mol) / 6.02 10^{23} (1/mol) = 1.1 10^{-19} g. The excess polarizability α is proportional to the molecular mass, m, of a protein and may vary only slightly between different proteins of similar molecular weight but of different amino acid sequence, such as for BSA and HSA (human serum albumin). Similar arguments can be made for many other

W

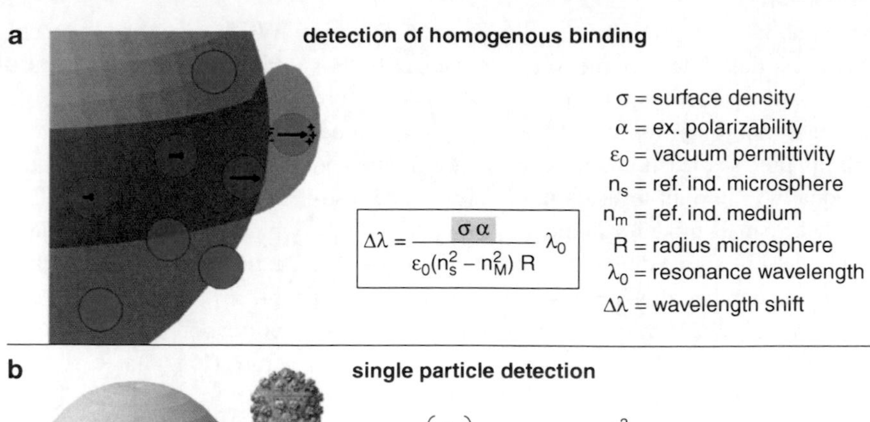

Whispering Gallery Mode Resonator Biosensors, Fig. 4 (**a**) Detection of homogenous binding of biomolecules in the evanescent field of a WGM. (**b**) Detection of single particle interacting with WGM

biopolymers [8]. Table 2 lists typical *dn/dc* values of various biopolymers and biomolecules, in different solvent conditions.

Note that *dn/dc* values depend on many additional parameters such as temperature, solvent, ionic strength, and probing wavelength (Table 2). Interestingly, one can also calculate the wavelength dependence of polarizability values from Kramers-Kronig transformations by integration over the respective absorption spectra [8].

In biosensing experiments, one often measures the resonance wavelength shift $\Delta\lambda$ instead of the frequency shift: $\Delta\lambda = \lambda_{\text{res, after}} - \lambda_{\text{res,before}}$. The wavelength shift $\Delta\lambda$ can be converted in the associated frequency shift Δf by: $\Delta f = c_{light}\ (1/\lambda_{\text{res, after}} - 1/\lambda_{\text{res, before}}) = -\ c_{\text{light}}\ \Delta\lambda/\lambda^2$, where λ is the initial resonance wavelength and c_{light} is the speed of light. In all WGM biosensing experiments, the wavelength shift, $\Delta\lambda$, or frequency shift, Δf, reports on the surface density σ of molecules bound to the microsphere resonator, at a specific time point. Depending on specific cavity material or surface functionalization, the saturation surface density of bound molecules may vary between 0 and $10^{15}/\text{cm}^2$. To achieve a surface coverage of more than $10^{15}/\text{cm}^2$ requires molecules to form multiple layers or to bind within less than a nm distance, certainly a physical limit as the size of an atom is, ~0.1 nm = 1 Angstrom. High-Q microcavities are invaluable tools for studying the binding of molecules to the sensor surface since precise measurements of high-surface densities can reveal interactions between closely spaced molecules, and precise measurements at low surface densities $<10^9/\text{cm}^2$ can reveal intricate mechanisms of adsorption on non-fouling surfaces.

Whispering Gallery Mode Resonator Biosensors, Table 2 Typical dn/dc values of biopolymers and molecules

Biomolecule or biopolymer	~dn/dc (cm^3/g) [at ~633 nm wavelength]	Measurement parameter notes
Bovine serum albumin	0.183	Water at 23°C
Deoxyribonucleic acid from calf thymus	0.166	0.1N tris buffer pH = 8.4 at 23°C
a-chymotrypsinogen	0.171	0.15M NaCl; 0.1 pH buffer, pH = 7.2 at 23°C
Dextran	0.147	Water at 23°C
Gelatin	0.163	0.1N NaCl at 23°C
Lignin sulfonate – Na	0.188	0.1M $NaNO_3$; pH = 6.6 at 26°C
Mucopolysaccharides	0.110	0.1% NaCl at 23°C
Polydimethylsiloxane	−0.091	Toluene at 23°C
Polyvinyl alcohol 98% hydrolyzed	0.150	0.05M $NaNO_3$
RNA	0.160	Water at 23°C

The time scale of resonance wavelength shift depends on the diffusion of molecules and on the time scales associated with their interactions at the

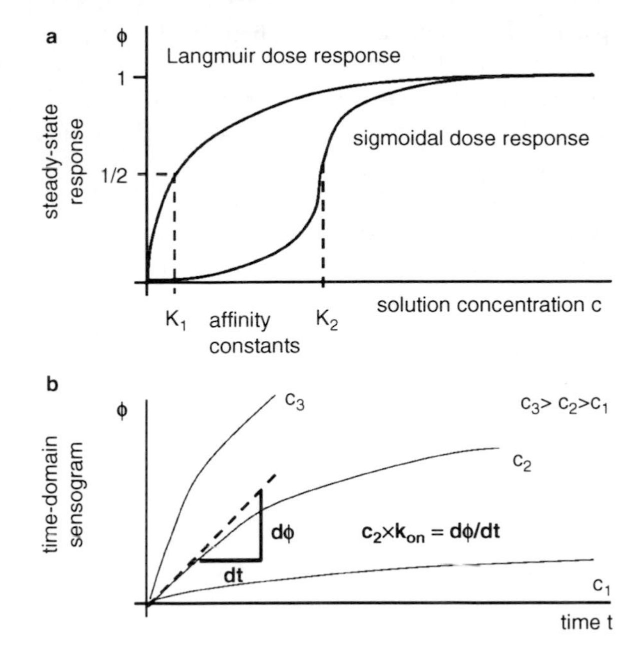

Whispering Gallery Mode Resonator Biosensors, Fig. 5 (**a**) Steady-state response of WGM biosensor. (**b**) Time-domain response (sensogram) of a WGM biosensor

resonator surface. The simplest steady-state kinetic model which can be applied to the binding of molecules to a surface is given by the Langmuir-like model [9] which assumes equilibrium binding between molecules and a fixed number of surface binding sites, further assuming equal affinity and no interaction between the sites. To determine a Langmuir binding isotherm, one first calculates the fractional surface coverage Φ by dividing the equilibrium frequency shift by the shift at saturation (measured at high solution concentration). Then, one plots the measurement of fractional surface coverage Φ versus the different solution concentration levels (Fig. 5a). The rate equation for the Langmuir model is given by $d\Phi/dt = k_{on}*c*(1-\Phi) - k_{off}*\Phi$, so that in steady state $d\Phi/dt = 0$ and $\Phi/(1-\Phi) = k_{on} / k_{off}*c$. For half-maximal surface coverage at $\Phi=1/2$, the solution concentration is just equal to the equilibrium (dissociation) constant $c = K = k_{off}/ k_{on}$, which can be conveniently determined from the Langmuir isotherm (Fig. 5a). The affinity (dissociation) constant is, e.g., measured at ~1 nM for BSA proteins binding to an aminosilanized microsphere [4]. For more complex binding and adsorption processes, different kinetic models can be applied such as the random sequential adsorption (RSA) model, which is derived from the Langmuir model by taking into account the additional exclusion volume of bound molecules. Another kinetic model considers conformational changes by modeling bound proteins as discs with radii that may change in a post-binding step. The reader is referred to standard pharmacokinetic textbooks for further analysis of dose-response curves and their importance in analytical and clinical chemistry. Obtaining such dose-response curves is essential for relating the solution concentration of analyte molecules to the measured surface concentrations as measured by the resonator shift signal.

Instead of analyzing molecular recognition and binding in the steady-state regime, the interaction of biomolecules at a resonator surface can also be analyzed in the time domain by plotting the resonance frequency or wavelength shift signal versus time (sensogram, Fig. 5b). The sensogram and its derivative report on the kinetics and the rate of adsorption or desorption, respectively. It is common to extract the derivative (rate of adsorption or desorption) at the onset of the experiment where $d\Phi/dt = k_{on} * c$ in a binding experiment started at $c > 0$ and $\Phi=0$, and $d\Phi/dt = k_{off}$ for an unbinding experiment started at $c = 0$ and $\Phi>1$. Since time scales for binding can be fast, it is often necessary to start the experiment rapidly using microfluidic techniques that allow a rapid change of solution concentration. In principle, optical resonators follow such a fast response without problem since the time resolution of a microcavity is principally limited only by the resonator photon lifetime τ, which is on the order of micro to nanoseconds, and dependent on the Q-value.

WGM microsphere cavities have a further advantage as compared to SPR and QCM techniques in that they exhibit multiple optical resonances signals, which are useful for detection. In fact, optical resonances appear throughout an optical resonator spectrum since the resonance condition states that a multiple number of resonance wavelengths have to fit the optical path. The precise calculation of the resonances requires Maxwell's equations and has many solutions, and a more versatile sensor uses more than one resonance shift signal to analyze the binding of molecules to the resonator surface. For example, differently polarized (TE and TM) optical resonances can be used to determine the orientation of molecules at the sensor surface [5]. Each one of the two resonances (TE and TM) probe the surface of the microsphere in a different direction, and the specific arrangement of molecules in a monolayer and changes thereof can be measured by comparing the TE and TM wavelength shifts. Different resonance wavelength shifts have already been used to determine conformational changes in photosensitive bacteriorhodopsin membrane and to measure the orientation of the photochrome retinal within a lipid bilayer [5]. It is also possible to compare the resonance wavelength shifts for resonances excited at different wavelength (e.g., at 763 nm and 1,310 nm) in parallel. Using this approach, it is possible to directly determine the thickness and the permittivity (refractive index) of an adsorbed nanolayer, and changes thereof [10].

$$\frac{\left(\frac{\delta\lambda}{\lambda}\right)_1}{\left(\frac{\delta\lambda}{\lambda}\right)_2} \approx \frac{L_1[1-\exp(-l/L_1)]}{L_2[1-\exp(-l/L_2)]} \quad (4)$$

Where l is the thickness of the adsorbed layer and L_1 and L_2 are the evanescent field length at probing wavelength λ_1 and λ_2, with $L \approx (\lambda/4\pi)(n_s^2 - n_m^2)^{-1/2}$.

WGM Biosensing: Label-free Single Particle Analysis

On the single particle level, the frequency shift (or rather "step") associated with a single binding event depends on the location of the particle on the resonator surface. For a microsphere WGM biosensor, the light field is confined along the equator of the sphere (Fig. 4) and the largest frequency shift is recorded for an equatorial binding event. The frequency shift associated with the equatorial binding event can be calculated from the reactive sensing principle [7] (1), where the fractional wavelength shift now depends on particle size, a, and the evanescent field length, L (Fig. 4b). Single bioparticle detection was demonstrated for a single Influenza A virion, and analysis of the discrete frequency shift signals (Fig. 4b) was used to determine the size of the particle (~50 nm radius) as well as its weight (~500 attograms) [7]. The size of a bound nanoparticle can also be determined independently of binding location by measuring resonance frequency shift as well as changes in Q-factor in parallel [11]. In general, single particle detection opens up the possibility to determine size and shape of nanoparticles by monitoring frequency shifts of different optical resonances (TE, TM, etc.) and associated Q-factors in parallel. Even without binding, but already within the evanescent field, the Brownian motion of a single particle causes resonance wavelength fluctuations [5], a signal that can be used to extract single particle diffusion constants. Furthermore, the frequency shift signal associated with a single particle trapped by the optical field of the resonator can be used to determine a chemical surface potential: If a particle of radius a is trapped in the optical field of the resonator, the resonance wavelength shift reports on the radial position of the particle measured as the distance h from the microsphere surface [12]:

$$\frac{\Delta\lambda_r(h+a)}{\Delta\lambda_r(a)} = \frac{\exp[-(h+a)/L]}{\exp[-a/L]} = \exp[-h/L] \quad (5)$$

From this measurement, a histogram of equilibrium particle positions can be replotted as a Boltzmann-distribution in kT units (k = Boltzmann constant, T = Temperature), revealing the trapping potential consisting of a chemical potential on the resonator surface and the optical potential of the evanescent field.

Implementation and Operation of a WGM Resonator Biosensor

In its simplest implementation (Fig. 6), the WGM biosensor device is build by coupling light from a tapered optical fiber to a microsphere optical resonator [4]. The microsphere itself is fabricated from an optical fiber piece where one end is melted in a hot flame, forming a sphere on-a-stem structure. The sphere is then positioned adjacent to the tapered fiber region, where the light can couple from the optical fiber to the resonator. Light from a tunable laser is coupled into one end of the optical fiber from where it is delivered to this tapered coupling region. The output end of the optical fiber is connected to a photodetector, which records the transmitted intensity. As the laser wavelength is tuned to match the resonance wavelength of the microsphere, light starts to couple from the tapered fiber region to the resonator where it remains confined as WGM optical resonance. Due to interference in the coupling region, the light then no longer reaches the photodetector, which now records a minimum in the transmitted intensity. A computer identifies and tracks the minimum in the recorded transmission spectrum and plots the

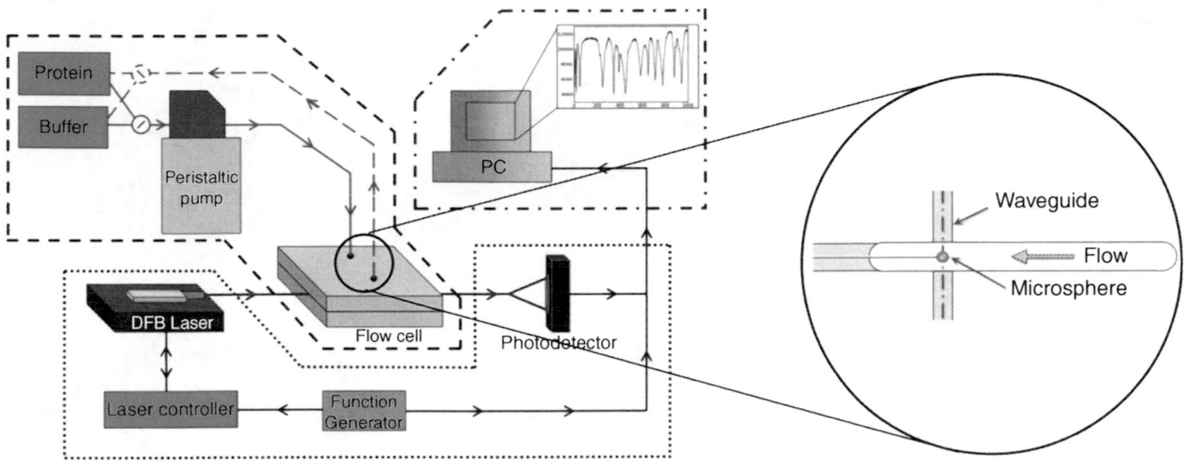

Whispering Gallery Mode Resonator Biosensors, Fig. 6 Schematic representation of WGM biosensor experimental setup comprised of fluidics component (*dashed lines*), excitation source and detector (*dotted line*), and data acquisition software (*dash-dot line*), and sphere-on-stem resonator housed in a flow cell (*inset*)

resonance wavelength versus time, at ms to μs time resolution, typically limited by the speed of the acquisition system, as well as by the speed of the laser control system. The sensogram is analyzed either *post hoc* or in real time and contains information on the binding and interaction of molecules at the WGM biosensor surface.

Several other experimental configurations exist for coupling light into WGM biosensors and for detecting resonances. Earlier studies of WGM resonances even employed "free beam" coupling methods in which the incident laser is guided through external optics and focused onto the microcavity to excite the resonance. Only a small fraction of the incident light, however, can couple radiatively to the microcavity. It is also possible to use light sources other than lasers, e.g., light sources with a broader emission spectrum, such as an ASE (amplified spontaneous emission) light source operating in the telecom band. In this case, however, a high-resolution monochrometer is necessary to resolve the narrow linewidth resonances, certainly a limitation when using a high Q cavity. In other implementations, it is possible to use spheres labeled with a fluorophore and resolve resonances in the emission spectrum, which is collected on a bead-by-bead basis. This implementation limits the operation of the biosensor to lower Q factors, but allows the use of freestanding fluorescent spheres, such as 10 um-diameter polystyrene spheres soaked with rhodamine 6G fluorophore [13].

Most common experimental setups utilize tapered fiber optic waveguides or planar waveguides to deliver the light to various resonator geometries such as discs, rings, toroids, or spheres (Fig. 7). Light sources typically consist of a tunable distributed feedback laser (DFB) with 300 kHz linewidth available at various nominal wavelengths (760–1,550 nm) and which can be tuned by a fraction of a nm simply by ramping the operating (laser diode) current [4]. A simple (silicon or InGaAs) photodetector is sufficient to detect the intensity, and a standard computer can record the spectrum and locate the resonance wavelength. Biosensing experiments often require the integration of the resonator with a microfluidic flow system. Figure 6 shows the integration of a complete microsphere WGM biosensor with a simple flow cell.

Resonator Geometry

One of the primary benefits of WGM biosensors is the unprecedented sensitivity provided by high Q-factors in optical microcavities. High Q-factors are found for a variety of resonator shapes and fabricated by different procedures (Fig. 7). The decision for particular sensor geometry depends on the required sensitivity, cost, multiplexing capability, and need for integration with microfluidics and other photonic or electronic components. Often, the most important factor remains sensitivity and to date the highest Q-factor was

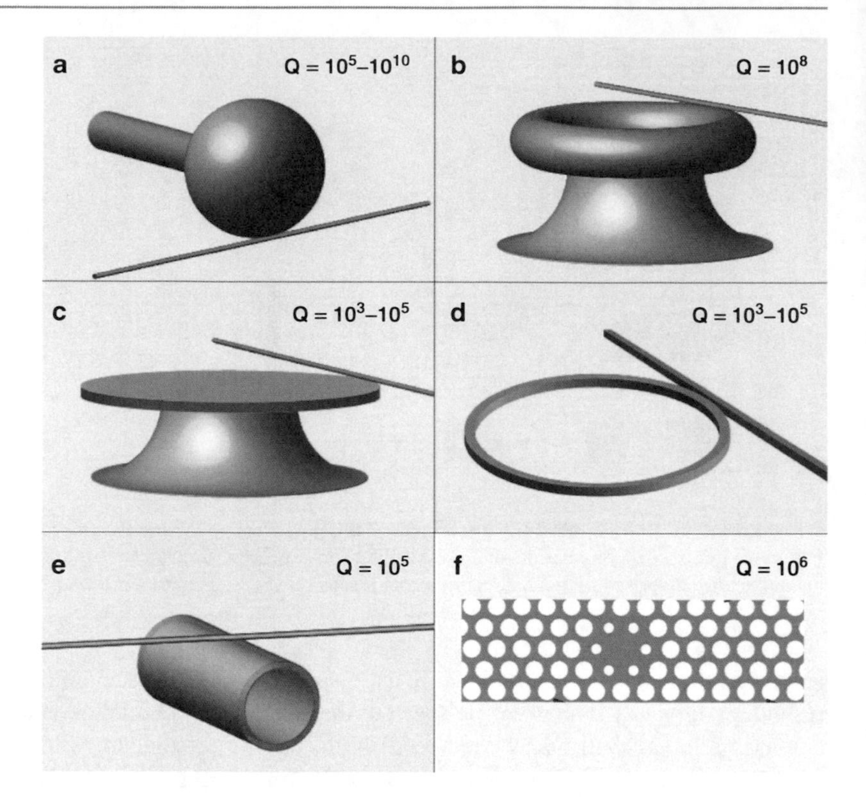

Whispering Gallery Mode Resonator Biosensors, Fig. 7 Schematic illustration of different resonator geometries. (**a**) Sphere, (**b**) Toroid, (**c**) Disk, (**d**) Ring, (**e**) Cylinder, (**f**) Photonic crystal

reported for a microsphere resonator at Q $\sim 10^{10}$ which is thus the ideal choice for building highly sensitive WGM biosensors. The ultimate sensitivity obtained in ultrahigh-Q resonators enables sensing of single virus particle and molecules. However, more modest Q factors ($Q = 1\times10^4$ to 1×10^7) observed in other resonator geometries and materials (Fig. 7) are sufficient in most biosensing applications and can already achieve detection limits superior to SPR sensors (<1 pg/mm^2). As pointed out in previous sections, reducing the modal volume, *V*, of the resonator, thus optimizing the Q/V value, further enhances the sensitivity. High Q/V values are observed in photonic crystal cavities, which are an interesting alternative to micospherical and microtoroidal optical resonators (Fig. 7f).

High-Q Resonators

Disk and Ring Resonators

Ring and disk resonators (Fig. 7c, d) are perhaps the most commonly implemented resonator geometry due to their ease of fabrication [14]. The 2-D and quasi 1-D geometries can be manufactured with photolithographic techniques and in different material systems. For example, rings and disks can be fabricated in SiO2, Si, nanocrystalline diamond, and organic polymers. Photolithography allows for fabrication of large sensor arrays with multiplexed detection capability [15]. Within such an array, individual sensors are used as internal references to eliminate noise from thermal fluctuations and nonspecific molecular binding. This is a significant advantage over systems utilizing only one resonator and greatly simplifies experimentation and data analysis. Due to their ease of fabrication using standard "CMOS techniques," ring and disk resonators are ideal candidates for integration into standardized platforms compatible with portable electronic devices and commercial bench-top systems. Microfabricated ring/disk resonator arrays are also quite easily incorporated with microfluidics due the planar nature of the substrates on which they are fabricated. This feature of chip-based resonators is critical for achieving sensitivities and signal-to-noise ratios that surpass those of currently available systems, as well as allowing handling of the increasingly small samples found in lab-on-chip applications. Commercial systems based on silicon ring-resonator arrays are currently in development and should be available in the near future.

Ring and disk resonators typically possess a low modal volume, which as discussed previously is key to ensure a large resonant shift for detection of molecules. However, rings and disks tend to exhibit relatively low Q-factors, especially after immersion in water ($Q \approx 10^{3-4}$), which makes determination of resonant frequency less precise and limits the ultimate sensitivity of resonators with these geometries. While such Q-factors are sufficient for many biosensing applications, these geometries may not be suitable for single virus or single molecule detection. For the time being, however, ring and disk resonators provide the most attractive option as sensing elements for commercially available bench-top and portable systems.

Liquid Core Optical Ring Resonators (LCORRs)

A relatively new class of microcavity resonators is the liquid core optical ring resonator (LCORR) [16]. This class of resonator utilizes the cylindrical geometry of microcapillaries to excite WGM resonances about the inner and outer circumference of the thin microcavity wall (Fig. 7e). Microcapillaries are fabricated from quartz or glass preforms by drawing under high heat to a final outer capillary diameter of $\sim$100 μm, and a wall thickness of a few microns. The LCORR can then be mounted on a waveguide or tapered fiber, which is used to couple light to the LCORR and excite the WGM resonances. The microcapillary sensors are sensitive to molecules binding to the inside of the capillary and one of the main advantages of this approach is the ease of fluid handling. In the capillary configuration of WGM biosensors, it is not necessary to create a completely separate flow cell, or mount fluidic channels on top of a wafer that holds the resonator structures. Instead, LCORRs can be arranged in parallel on waveguides to create sensor array already integrated with microfluidics. By employing multiple waveguides along a single capillary, it is possible to interrogate the region about each waveguide as an independent sensor. This linear configuration is well suited to such applications as capillary electrophoresis, where sensing regions at defined intervals can be used to monitor the speed of a profile of analyte passing a region. Additionally, LCORRs support high-Q resonances ($\sim 10^5$), which provide them with the necessary sensitivity to compete with established biosensing methods such as SPR and QCM. Investigation into applications of LCORRs is on going and remains an intriguing alternative to other geometries.

Ultrahigh-Q Resonators

Toroids

Silica toroidal resonators (Fig. 7b) exhibit the highest Q-values measured in water [17] in excess of $Q = 10^8$. These types of resonators are fabricated from layers of silica (glass) grown on crystalline silicon substrate, etched into discs with hydrofluoric acid, and then undercut with, i.e., xenon difluoride etch to form a free standing silica discs held on a silicon post. Reflow of the silica disk is achieved in a self-terminating process by heating the silica with a CO_2 laser results in a toroidal (donut) shape. Toroidal resonators exhibit modal volumes lower than those of microspheres of comparable diameter, and exhibit ultrahigh Q-factors due to an atomically smooth surface produced by reflow of glass. Thus, the toroid geometry boasts a high sensitivity and is a good candidate for single molecule detection. To date, only high-Q toroidal resonators made from silica material in the sensing region have been reported, as successful fabrication relies on the reflow of glass under high heat. Therefore, these geometries are difficult to fabricate using higher refractive index materials, such as sapphire, diamond, and alkali halides. Silica-based toroidal resonators are amenable to chemical functionalization by standard silane chemistry as well as other means of covalent and noncovalent modification and are thus useful as ligand-specific biosensors. This resonator geometry can be potentially integrated into on-chip arrays, similarly as described for disks and rings. However, integration with waveguides or optical fiber requires secondary fabrication steps and still remains a challenge. Whether or not practical implementation of microtoroid arrays with integrated waveguides will be achieved still remains a topic of research. As such, they are excellent for bench-top applications that require ultrahigh sensitivity.

Spheres and Spheroids

Next to rings, spherical resonators are perhaps the most common resonator geometry employed. Spherical silica resonators are very easy to fabricate by melting the end of an optical fiber and exhibit ultimate Q-factors of up to 10^{10}. Light can be coupled to a sphere, or sphere-on-stem by tapered fiber, waveguide, prism, or free beam coupling allowing a wide range of possibilities in implementation. Additionally, free standing spheres can be fabricated using a variety of alternative procedures and materials. Silica microspheres are most common, but reports have demonstrated the utility of spheres made from or coated with high refractive index polymers such as polystyrene. Free standing spheres can be used in top-down fabrication schemes that allow a great deal of flexibility in experimental configuration. In addition, the symmetry of a sphere allows for easy analysis and quantification of the WGM resonance signals. The inherent flexibility and simplicity of the spherical geometry makes it perhaps the most promising for future implementations of WGM biosensors. Integration with microfluidics and multiplexing remains challenging, although there are proposals for generating large arrays of microspheres, which can be read out in parallel. Microsphere resonators are currently mainly used in bench-top proof-of-principle experimental applications. However, as the technology develops, the spherical resonator will find applications to commercial instruments as well.

Applications

Protein Detection

Many applications of WGM biosensors are focused on the label-free detection of proteins from solution [18]. In 2002, Vollmer and coworkers [4] demonstrated the quantification of the thickness and surface density of a layer of bovine serum albumin (BSA) adsorbed on a microsphere functionalized with the silane aminopropyltrimethoxysilane (APTES). This result provided the proof-of-principle for using resonance frequency shifts in WGM for real-time biosensing applications. Furthermore, this experiment demonstrated that the glass surface resonators could be functionalized with various recognition elements such as proteins and antibodies, and showed the utility of the reactive sensing principle ((1) and (2)).

Due to the remarkable sensitivity of WGM biosensors, they hold great promise for the detection of low copy numbers of biomarkers that are indicative of early stage disease. Andrea and coworkers published work demonstrating the detection of molecule binding events for Interleukin-2 (IL-2), a molecule indicative of immune response [17]. This combination of sensitivity and the ability to create highly specific sensor surfaces makes WGM sensors ideal candidates for sensor elements in point of care devices and fieldable instrumentation. A similar approach was also demonstrated by Washburn and coworkers in 2009 who reported the detection of the cancer marker carcinoembryonic antigen (CEA) from clinical samples containing serum [19]. This was the first demonstration of detection of a disease marker from a complex biological sample by WGM frequency shift.

DNA Detection

In the post-genomic era, the ability to detect and sequence DNA from low copy number is critical for clinically relevant applications. Despite advances in polymerase chain reaction (PCR) methods, it is still desirable to be able to detect and discriminate DNA sequences directly from unlabelled samples. In this respect, WGM biosensors offer the potential to detect down to single-molecule binding events for DNA using resonators functionalized with complementary strands of DNA. Vollmer and coworkers (2003) demonstrated the sequence-based discrimination of DNA binding to complementary sequences immobilized on a glass microsphere [20]. In these experiments, two microspheres functionalized with different complementary sequences were coupled to a single waveguide. Solutions with target DNA sequences were added and the response from the microspheres was simultaneously measured. It was shown that a single base-pair mismatch could be discriminated. Thus, WGM is a highly suitable technique for the sequence-based detection and discrimination of DNA molecules.

Virus Particle Detection

In addition to the detection of protein and DNA molecules, it is very desirable to be able to detect single particles such as viruses. As a single virus particle has the potential to infect a cell and replicate itself billions of times in a relatively short period of time, it is obviously worthwhile to have sensors capable of detecting virions on the ultimate single particle level. In 2008, Vollmer and coworkers [7] demonstrated the ability to detect the binding and unbinding of single particle of the Influenza A (InfA) virus to an unmodified microsphere.

Sensor Arrays

Due to the ease of fabrication of these types of sensing elements, it is possible to incorporate resonators of tailored geometries into large arrays that can be individually addressed. This is perhaps one of the most promising aspects of WGM biosensors, in that they not only provide sensitivity and specificity better than currently available optical methods, but can also apply these benefits to potentially massive arrays of sensing elements. Washburn and coworkers have taken a first step in this direction by applying arrays of ring resonators for the detection of CEA from complex biological samples [19]. Here it was shown that an array of 16 individually addressable resonators could be interrogated in real time for the detection of their target antigen. Furthermore, due to the redundancy provided by an array of sensors, it was possible to utilize a subset of the resonators as internal controls for compensating for fluctuations in temperature.

Future Directions

The future is bright for the application of WGM biosensors. With the myriad benefits associated with the method and the proof-of-concept work performed to date, much rests on the ability to control surface functionalization and engineering of the associated sensor components (laser source, detectors, electronics, sample handling, etc) to miniaturize the systems to a truly portable handheld format. Similar as electronic devices that have become smaller and more portable, it is inevitable that the components of WGM biosensors will one day be packaged in a form small enough for a primary health care giver or technician to use devices of this type on a regular basis at either the bed side of a patient, or in the field testing for biological agents or environmental contaminants. Further refinement of quantification methods and sensor design (geometry, materials, etc) will provide unmatched sensitivity and accuracy in point-of-care devices.

Cross-References

► Biosensors
► Nanostructured Materials for Sensing
► Nanostructures for Photonics
► Organic Sensors

References

1. Heebner, J., Grover, R., Ibrahim, T.: Optical microresonators: theory, fabrication, and applications, 1st edn. Springer, London (2007)
2. Bloch,S.C.: Introduction to classical and quantum harmonic oscillators, Har/Dsk ed. Wiley-Interscience, Hoboken, NJ, USA (1997)
3. Vahala, K.: Optical microcavities. World Scientific, Singapore (2005)
4. Vollmer, F., Braun, D., Libchaber, A., Khoshsima, M.: Appl. Phys. Lett. **80**, 4057 (2002)
5. Vollmer, F., Arnold, S.: Nat. Methods **5**, 591–596 (2008)
6. Arnold, S., Khoshsima, M., Teraoka, I., Holler, S., Vollmer, F.: Opt. Lett. **28**, 272–274 (2003)
7. Vollmer, F., Arnold, S., Keng, D.: Proc. Natl. Acad. Sci. USA **105**(52), 20701–20704 (2008)
8. Hinchcliffe, A., Munn, R.W.: Molecular electromagnetism. Wiley, Chichester (1985)
9. Hill, T.: An introduction to statistical thermodynamics. Dover Publications, New York (1987)
10. Noto, M., Vollmer, F., Keng, D., Teraoka, I., Arnold, S.: Opt. Lett. **30**(5), 510–512 (2005)
11. Zhu, J., Ozdemir, S.K., Yun-Feng, Xiao, Li, L., He, L., Chen, D.-R., Yang, L.: Nature Photon. **4**, 4 (2010)
12. Arnold, S., Keng, D., Shopova, S., Holler, S., Zurawsky, W., Vollmer, F.: Opt. Express **17**, 6230–6238 (2009)
13. Francois, A., Himmelhaus, M.: Sensors **9**(9), 6836–6852 (2009)
14. De Vos, K., Bartolozzi, I., Schacht, E., Bienstman, P.: Opt. Express **15**, 7610–7615 (2007)
15. Bailey, R.: Future Sci. Bioanalysis, **1**(6), 1043–1047 (2009)
16. Fan, X., White, I., Shopova, S., Zhu, H., Suter, J.: Anal. Chim. Acta **620**, 8–26 (2008)
17. Armani, A., Kulkarni, R., Fraser, S.: Science **317**, 783 (2007)
18. Matsko,A., Savchenkov, A., Strekalov, D.: IPN progress report (2005)
19. Washburn, A.L., Gunn, L.C., Bailey, R.C.: Anal. Chem. **81**(22), 9499–9506 (2009)
20. Vollmer, F., Arnold, S., Braun, D., Teraoka, I., Libchaber, A.: Biophys. J. **85**(3), 1974–1979 (2003)

Wireless Propulsion of Helical Nanobelt Swimmers

► Remotely Powered Propulsion of Helical Nanobelts

Worm Chip

► Lab-on-a-Chip for Studies in *C. elegans*

X-ray Diffraction with Micron-Sized X-Ray Beams

► Selected Synchrotron Radiation Techniques

B. Bhushan (ed.), *Encyclopedia of Nanotechnology*, DOI 10.1007/978-90-481-9751-4,

Young's Modulus

► Nanomechanical Properties of Nanostructures

B. Bhushan (ed.), *Encyclopedia of Nanotechnology*, DOI 10.1007/978-90-481-9751-4,

List of Entries

μMist®
μVED (Micro Vacuum Electronic Devices)
193-nm Lithography
248-nm Lithography
3D Micro/Nanomanipulation with Force Spectroscopy
Ab Initio DFT Simulations of Nanostructures
AC Electrokinetics
AC Electrokinetics of Colloidal Particles
AC Electrokinetics of Nanoparticles
AC Electrokinetics of Sub-micrometer Particles
AC Electroosmosis: Basics and Lab-on-a-Chip Applications
ac-Calorimetry
Accumulator
Acoustic Contrast Factor
Acoustic Nanoparticle Synthesis for Applications in Nanomedicine
Acoustic Particle Agglomeration
Acoustic Separation
Acoustic Trapping
Acoustic Tweezers
Acoustophoresis
Active Carbon Nanotube-Polymer Composites
Active Nanoantenna System
Active Plasmonic Devices
Adhesion
Adhesion in Wet Environments: Frogs
Adsorbate Adhesion
Aerosol-Assisted Chemical Vapor Deposition (AACVD)
AFM
AFM Force Sensors
AFM in Liquids
AFM Probes
AFM Tips
AFM, Non-contact Mode
AFM, Tapping Mode
Aging
ALD
Alkanethiols
AlN
Alteration
Aluminum Nitride
Amphibian Larvae
Anamorphic Particle Tracking
Angiogenesis
Animal Coloration
Animal Reflectors and Antireflectors
Anodic Arc Deposition
Antifogging Properties in Mosquito Eyes
Applications of Nanofluidics
Arc Discharge
Arthropod Strain Sensors
Artificial Muscles
Artificial Retina: Focus on Clinical and Fabrication Considerations
Artificial Synapse
Artificial Vision
Astigmatic Micro Particle Imaging
Atmospheric Pressure Chemical Vapor Deposition (APCVD)
Atomic Cluster
Atomic Force Microscopy
Atomic Force Microscopy in Liquids
Atomic Layer Chemical Vapor Deposition
Atomic Layer Chemical Vapor Deposition (ALCVD)
Atomic Layer CVD (AL-CVD)
Atomic Layer Deposition
Atomic Layer Deposition (ALD)
Atomic Layer Epitaxial (ALE)
Atomic Layer Epitaxy (ALE)
Attocalorimetry
Automatic Data Analysis Workflow for RNAi

Bacterial Electrical Conduction
Band Alignment
Basic MEMS Actuators
Batteries
Bendable Electronics
Bending Strength
Bilayer Graphene
Bioaccessibility/Bioavailability
Bioadhesion
Bioadhesives
Biocalorimetry
Bioderived Smart Materials
Bio-FET
Biofilms in Microfluidic Devices
Biognosis
Bio-inspired CMOS Cochlea
Bioinspired Microneedles
Bioinspired Synthesis of Nanomaterials
Biological Breadboard Platform for Studies of Cellular Dynamics
Biological Nano-crystallization
Biological Photonic Structures
Biological Platform
Biological Sensors
Biological Structural Color
BioMEMS/NEMS
Biomimetic Antireflective Surfaces Arrays
Biomimetic Energy Conversion Devices
Biomimetic Flow Sensors
Biomimetic Infrared Detector
Biomimetic Mosquito-Like Microneedles
Biomimetic Muscles and Actuators
Biomimetic Muscles and Actuators Using Electroactive Polymers (EAP)
Biomimetic Synthesis
Biomimetic Synthesis of Nanomaterials
Biomimetics
Biomimetics of Marine Adhesives
Biomimetics of Optical Nanostructures
Biomimicked Microneedles
Biomimicry
Biomolecular Mechanics
Bionic Ear
Bionic Eye
Bionic Microneedles
Bionics
Bio-optics
BioPatterning
Bio-photonics
Biopolymer
Bioprobes
Biornametics - Architecture Defined by Natural Patterns
Biosensing
Biosensors
Bond-Order Potential
Bone Remodeling
Bone Turnover
Boron- and/or Nitrogen-Doped Carbon Nanotubes
Boron Nitride Nanotubes (BNNTs)
Bottom-Up Nanofabrication
Boundary Lubrication
Brain Implants
Buckminsterfullerene
Bulk Acoustic Wave MEMS Resonators
C_{60}
Cancer Modeling
Capacitive MEMS Switches
Capillarity Induced Folding
Capillary Flow
Capillary Origami
Carbon Nanotube Materials
Carbon Nanotube-Metal Contact
Carbon Nanotube-Metal Interface
Carbon Nanotubes
Carbon Nanotubes (CNTs)
Carbon Nanotubes for Chip Interconnections
Carbon Nanotubes for Interconnects in Integrated Circuits
Carbon Nanotubes for Interconnects in Microprocessors
Carbon Nanowalls
Carbon-Nanotubes
Catalyst
Catalytic Bimetallic Nanorods
Catalytic Chemical Vapor Deposition (CCVD)
Catalytic Janus Particle
Cathodic Arc Deposition
Cavity Optomechanics
Cell Adhesion
Cell Manipulation Platform
Cell Morphology
Cell Patterning
Cellular and Molecular Toxicity of Nanoparticles
Cellular Electronic Energy Transfer
Cellular Imaging
Cellular Mechanisms of Nanoparticle's Toxicity
Cellular Toxicity

Characterization of DNA Molecules by Mechanical Tweezers
Characterizations of Zinc Oxide Nanowires for Nanoelectronic Applications
Charge Transfer on Self-Assembled Monolayer Molecules
Charge Transport in Carbon-Based Nanoscaled Materials
Charge Transport in Self-Assembled Monolayers
Chem-FET
Chemical Beam Epitaxial (CBE)
Chemical Blankening
Chemical Dry Etching
Chemical Etching
Chemical Milling and Photochemical Milling
Chemical Modification
Chemical Solution Deposition
Chemical Vapor Deposition (CVD)
Chemistry of Carbon Nanotubes
Chitosan
Chitosan Nanoparticles
Clastogenicity or/and Aneugenicity
Clinical Adhesives
Cluster
CMOS (Complementary Metal-Oxide-Semiconductor)
CMOS MEMS Biosensors
CMOS MEMS Fabrication Technologies
CMOS-CNT Integration
CMOS-MEMS
CNT Arrays
CNT Biosensor
CNT Brushes
CNT Bundles
CNT Foams
CNT Forests
CNT Mats
CNT Turfs
CNT-FET
Cob Web
Cochlea Implant
Cold-Wall Thermal Chemical Vapor Deposition
Compliant Mechanisms
Compliant Systems
Composite Materials
Computational Micro/Nanofluidics: Unifier of Physical and Natural Sciences and Engineering
Computational Study of Nanomaterials: From Large-Scale Atomistic Simulations to Mesoscopic Modeling
Computational Systems Bioinformatics for RNAi
Computer Modeling and Simulation of Materials
Concentration Polarization
Concentration Polarization at Micro/Nanofluidic Interfaces
Conductance Injection
Conduction Mechanisms in Organic Semiconductors
Confocal Laser Scanning Microscopy
Confocal Scanning Optical Microscopy (CSOM)
Conformal Electronics
Contour-Mode Resonators
Coupling Clamp
Creep
Cutaneous Delivery
Cuticle
Cylindrical Gold Nanoparticles
Decoration of Carbon Nanotubes
Density
Dermal Absorption
Dermal and Transdermal Delivery
Detection of Nanoparticle Biomarkers in Blood
Detection of Nanoparticulate Biomarkers in Blood
Dielectric Force
Dielectrophoresis
Dielectrophoresis of Nucleic Acids
Dielectrophoretic Assembly
Dielectrophoretic Nanoassembly of Nanotubes onto Nanoelectrodes
Differential Scanning Calorimetry
Digital Microfluidics
Dip-Pen Nanolithography
Directed Assembly
Directed Self-Assembly
Disjoining Pressure and Capillary Adhesion
Dislocation Dynamics
Dispersion
Dissociated Adsorption
DNA Computing
DNA FET
DNA Manipulation
DNA Manipulation Based on Nanotweezers
DOLLOP
Doping in Organic Semiconductors
Doping of Organic Semiconductors
Double-Walled Carbon Nanotubes (DWCNTs)

Droplet Microfluidics
Drug Delivery and Encapsulation
Drug Delivery System
Dry Adhesion
Dry Etching
Dual-Beam Piezoelectric Switches
DUV Lithography
DUV Photolithography and Materials
Dye Sensitized Solar Cells
Dye-Doped Nanoparticles in Biomedical Diagnostics
Dye-Doped Nanospheres in Biomedical Diagnostics
Dynamic Clamp
e-Beam Lithography, EBL
ECM
Ecotoxicity
Ecotoxicity of Inorganic Nanoparticles: From Unicellular Organisms to Invertebrates
Ecotoxicology of Carbon Nanotubes Toward Amphibian Larvae
Ecotribology
EC-STM
Effect of Surface Modification on Toxicity of Nanoparticles
Effective Media
Elastic Modulus Tester
Elasto-capillary Folding
Electric Cooler
Electric Double Layer Capacitor
Electric Field–Directed Assembly of Bioderivatized Nanoparticles
Electrical Impedance Cytometry
Electrical Impedance Spectroscopy
Electrical Impedance Tomography for Single Cell Imaging
Electric-Field-Assisted Deterministic Nanowire Assembly
Electrocapillarity
Electrochemical Machining (ECM)
Electrochemical Scanning Tunneling Microscopy
Electrode–Organic Interface Physics
Electrohydrodynamic Forming
Electrokinetic Fluid Flow in Nanostructures
Electrokinetic Manipulation
Electron Beam Evaporation
Electron Beam Lithography (EBL)
Electron Beam Physical Vapor Deposition (EBPVD)
Electron Microscopy of Interactions Between Engineered Nanomaterials and Cells
Electron Transport in Carbon Nanotubes and Graphene Nanoribbons
Electron-Beam-Induced Chemical Vapor Deposition
Electron-Beam-Induced Decomposition and Growth
Electron-Beam-Induced Deposition
Electronic Contact Testing Cards
Electronic Expression
Electronic Pharmacology
Electronic Structure Calculations
Electronic Transport in Carbon Nanomaterials
Electronic Visual Prosthesis
Electroplating
Electroresponsive Polymers
Electrospinning
Electrostatic Actuation of Droplets
Electrostatic MEMS Microphones
Electrostatic RF MEMS Switches
Electrothermomechanical Actuators
Electrowetting
Electrowetting-on-Dielectric (EWOD)
Energy-Level Alignment
Engineered Nanoparticles
Environmental Impact
Environmental Toxicology
Epiretinal Implant
Epiretinal Prosthesis
EUV Lithography
Excitonic Solar Cell
Exoskeleton
Exposure and Toxicity of Metal and Oxide Nanoparticles to Earthworms
Extreme Ultraviolet Lithography (EUVL)
Extrusion Spinning
Fabrication of Graphene or Graphene Fabrication
Fate of Manufactured Nanoparticles in Aqueous Environment
Fatigue Strength
FDTD
Femtocalorimetry
Fiber Reinforced Composite
Fibrillar Adhesion
FIB-SEM
Field Electron Emission from Nanomaterials
Filling of Carbon Nanotubes
Finite Element Methods for Computational Nano-optics
Finite-Difference Frequency-Domain Technique
Finite-Difference Time-Domain Technique

Finite-Element Analysis (FEA)
First Principles Calculations
Flapping-Wing Flight
Flash Evaporation Liquid Atomization
Flexible Electronics
Flexible Sensors
Flexure Mechanisms
Flexures
Floccules
Focused Ion Beam-Induced Deposition (FIB-ID)
Focused-Ion-Beam Chemical-Vapor-Deposition (FIB-CVD)
Force Measurement
Force Modulation in Atomic Force Microscopy
Force Spectroscopy
Force-Detected Magnetic Resonance
Fracture Toughness
Free Flow Aoustophoresis (FFA)
Friction
Friction Force Microscopy
Friction-Reducing Sandfish Skin
Fuel Cell
Fullerenes for Drug Delivery
Functionalization
Functionalization of Carbon Nanotubes
Functionalized Nanomaterials
Fundamental Properties of Zinc Oxide Nanowires
Gallium Arsenide (GaAs)
Gamma-butyrolactone (GBL)
Gammatone Filters
Gas Etching
Gas Phase Nanoparticle Formation
Gas-Phase Molecular Beam Epitaxy (Gas-Phase MBE)
Gecko Adhesion
Gecko Effect
Gecko Feet
Gel Chemical Synthesis
Genotoxicity of Nanoparticles
Gold Nanorods
Graphene
Graphene Synthesis
Graphite
Greases
Green Tribology and Nanoscience
Green's Function Integral Equation Method
Growth of Carbon Nanotubes
Growth of CNTs
Growth of Oxide Nanowires
Growth of Silica Nanowires
Growth of Silicon Dioxide Nanowires
Gyroscopes
Hardbake (HB)
Hardness
Hard-Tip Soft-Spring Lithography
Heat
Heat Capacity
Heat Conduction
Heat Conductivity
Heatuators
Helical Nanobelt
Heterogeneous Integration
Heterogeneous Walls
HGN
Hierarchical Mechanics
High Index Fluid Immersion
High-Density Testing Cards
Higher-Order Plasticity Theory
High-Pressure Carbon Monoxide (HiPCO)
HIV Quantification
Hollow Gold Nanoshells
Hollow Gold Nanospheres
Hot Filament Chemical Vapor Deposition (HFCVD)
Hot-Wall Thermal Chemical Vapor Deposition
How to Synthesize/Make/Manufacture/Fabricate/Produce Graphene
Hybrid Network Method
Hybrid Optoelectric Manipulation
Hybrid Opto-electric Technique (in Viewpoint of Driving Sources Used)
Hybrid Photovoltaics
Hybrid Solar Cells
Hyperelasticity
Imaging Human Body Down to Molecular Level
Immersion Lithography
Immersion Lithography Materials
Impedance Cytometry
Impedance Measurement of Single Cells
Impedance Spectroscopy for Single-Cell Analysis
Imprint Lithography
In Vitro and In Vivo Toxicity of Silver Nanoparticles
In Vitro Toxicity
In Vivo Microscopy Analysis
In Vivo Toxicity of Carbon Nanotubes
In Vivo Toxicity of Titanium Dioxide and Gold Nanoparticles
Indentation
Induced-Charge Electroosmosis

Inductively Coupled-Plasma Chemical Vapor Deposition (ICP-CVD)
Inertial Sensors
Influence of Surface Engineering on Toxicity of Nanoparticles
Influence of Surface Functionalization on Toxicity of Nanoparticles
Injectable Hydrogel
Injectable In Situ-Forming Gel
Insect Flight and Micro Air Vehicles (MAVs)
Insect Infrared Sensors
Insect-Inspired Vision and Visually Guided Behavior
Insects
In-situ STM
Integrated Approach for the Rational Design of Nanoparticles
Integrated Biosensors
Integrated Micro-acoustic Devices
Integration
Integration of Nanostructures within Microfluidic Devices
Intelligent Drug Delivery Microchips
Interface Excess Free Energy
Interfacial Charging Effect
Interfacial Energy and Chemical Potential at Nanoscale
Interfacial Investigation of Protein Films Using Acoustic Waves
Interference Modulator (IMOD) Technology
Intravital Microscopy Analysis
Ion Beam
Iridescent Colors (Organisms other than Animals)
Isothermal Calorimetry
Kelvin Probe Force Microscopy
Key Point Detection
Kinetic Monte Carlo Method
Lab-on-a-Chip
Lab-on-a-Chip (LOC)
Lab-on-a-Chip for Studies in *C. elegans*
Lab-on-a-Chip Whole-Cell Biosensor
Lamb-Wave Resonators
Landmark Detection
Large Unilamellar Vesicle (LUV)
Laser Scanning Confocal Microscopy
Laser Tweezers
Lateral Force Microscopy (LFM)
Laterally Vibrating Piezoelectric Resonators
Lead Zirconate Titanate
Left-Handed Materials
Lesson from Nature
Lifecycle
Ligand-Directed Gold-Phage Nanosystems
Light Localization for Nano-optical Devices
Light-Element Nanotubes and Related Structures
Lipid Vesicles
Liposomes
Lipospheres
Lippmann–Schwinger Integral Equation Method
Liquid Atomization Through Vapor Explosion
Liquid Condensation
Liquid Etching
Liquid Film
Liquid Immersion Lithography
Liquid Meniscus
Lithographie, Galvanoformung, Abformung (LIGA)
LM10
LM20
Local Density of States
Local Surface Plasmon Resonance (LSPR)
Log-Domain
Logic-Embedded Vectors
Lotus Effect
Low Fluid Drag Surface
Low-Power
Low-Pressure Chemical Vapor Deposition (LPCVD)
Lubrication
Lumbricidae/Oligochaeta/Earthworms
Macromolecular Crystallization Using Nano-volumes
Macroscale Platforms Capable of Biological Studies and Manipulations at Nanoscale to Bio-inspired and Bacterial Nanorobotics
Magnetic Assembly
Magnetic Dipole Emitters
Magnetic Dipole Transitions
Magnetic Electromagnetic Radiation
Magnetic Nanostructures and Spintronics
Magnetic Resonance Force Microscopy
Magnetic Self-Assembly
Magnetic-Field-Based Self-Assembly
Magnetron Sputtering
Making Graphene
Manipulating
Manufacturing Graphene or Graphene Manufacturing
Manufacturing Silica Nanowires
Marine Bio-inspired Adhesives
Materiomics
Maxwell–Bloch Simulations or Equations or Studies
Maxwell–Schrödinger Simulations (or Equations)

Maxwell–Wagner Effect
Mean Field Approaches to Many-Body Systems
Mechanical Behavior of Nanostructured Metals
Mechanical Characterization of Nanostructures
Mechanical Properties of Biological Materials
Mechanical Properties of Hierarchical Protein Materials
Mechanical Properties of Nanocrystalline Metals
Mechanical Properties of Nanomaterials
Mechanoreceptors
Medical Adhesives
MEMS
MEMS (Micro-electro-Mechanical Systems)
MEMS = Microelectromechanical Systems
MEMS Adaptive Optics
MEMS Capacitive Microphone
MEMS Condenser Microphone
MEMS Display based on Interferometric Modulation Technology
MEMS Encapsulation
MEMS Fabricated Microwave Devices
MEMS High-Density Probe Cards
MEMS Microphone
MEMS Neural Probes
MEMS on Flexible Substrates
MEMS Packaging
MEMs Pumps
MEMS Vacuum Electronics
MEMS-Based Drug Delivery Devices
Meniscus Adhesion
Mesoscopic Modeling
Metal Plating
Metallic Waveguides
Metallofullerene
Metalorganic Chemical Vapor Deposition (MOCVD)
Metal–Organic Interfaces
Metal-Organic Molecular Beam Epitaxy (MOMBE)
Metropolis Monte Carlo Method
Micro Air Vehicles (MAVs)
Micro- and Nanofluidic Devices for Medical Diagnostics
Micro Coulter
Micro FACS
Micro Flow Cytometry
Micro/Nano Flow Characterization Techniques
Micro/Nano Transport in Microbial Energy Harvesting
Micro/Nanodevices
Micro/Nanofluidic Devices
Micro/Nanomachines
Micro/Nanosystems
Microactuators
Microarrays/Microfluidics for Protein Crystallization
Microbial Aggregations
Microbial Electrolysis Cell
Microbial Fuel Cell
Microcalorimetry
Microcantilever Chemical and Biological Sensors
Microcavity Biosensor
Microcontact Printing
Microelectrodes
MicroElectroMechanical Systems
Micro-fabricated MVED (Microwave Vacuum Electron Devices)
Microfabricated Probe Technology
Microfluidic Device
Microfluidic Pumps
Microfludic Whole-Cell Biosensor
Micromachined 3-D Force Sensors
Micromachined Acoustic Devices
Micromachines
Micromachining
Micromechanical Switches
Micro-optomechanical Systems (MOMS)
Micro-patterning
Micropumps
Microrobotics
Microscale Fluid Mechanics
Microstereolithography
Micro-system
Micro-thermal Analysis
Micro-total Analytical Systems (μ-TAS)
Microtransfer Molding
Micro-trap
Mini Pumps
mirasol® Displays
Modeling Thermal Properties of Carbon Nanostructure Composites
Models for Tumor Growth
Modification of Carbon Nanotubes
Modulus
MOEMS/NOEMS
Molecular Beam Epitaxy (MBE)
Molecular Computation
Molecular Computing
Molecular Dynamics Method
Molecular Dynamics Simulations of Interactions Between Biological Molecules and Nanomaterials

Molecular Dynamics Simulations of Nano-Bio Materials
Molecular Encapsulation
Molecular Layer Deposition (MLD)
Molecular Modeling on Artificial Molecular Motors
Molecular Patterning
Molecular Plasmonics
Molecular Printing
Molecularly Thick Layers
Monolayer Lubrication
Monolithic Integration of Carbon Nanotubes
Monolithic Total Internal Reflection Biosensor
Mosquito Fascicle Inspired Microneedles
Moth-Eye Antireflective Structures
Motion Sensors
Movement of Polynucleotides in Nonuniform Electric Field
Mucoadhesion
Multi-functional Particles
Multilamellar Vesicle (MLV)
Multiscale Fluidics
Multiscale Modeling
Multistage Delivery System
Multistage Vectors
Multiwalled Carbon Nanotubes (MWCNTs)
Nano (Evenescent-Wave)-Particle Image Velocimetry
Nano Manipulation
Nano(Evanescent-Wave)-Particle Velocimetry
Nanoassembly
Nanobelts
Nanobiosensors
Nanobonding
Nanocalorimeter
Nanocalorimetric Sensors
Nanocalorimetry
Nanocarriers
Nanochannels for Nanofluidics: Fabrication Aspects
Nanocluster
Nanocoil
Nanocombs
Nano-Concrete
Nanocrystalline Materials
Nanodentistry
Nanodevices
Nano-ecotoxicology
Nano-effects
Nanoelectrodes
Nanoencapsulation
Nano-engineered Concrete
Nanoengineered Hydrogels for Cell Engineering
Nanofabrication
Nanofatigue Tester
Nano-FET
Nanofibrous Materials and Composites
Nanofluidic Channels
Nanofluidics
Nanofocusing of Light
Nanogap Biosensors
Nanogrippers
Nanohardness
Nanohardness Tester
Nanohelix
Nanoimprint Lithography
Nanoimprinting
Nanoindentation
Nanointerconnection
Nanojoining
Nanomachining
Nanomagnetism
Nanomaterials
Nanomaterials and Nanoproducts
Nanomaterials as Field Emitters
Nanomaterials for Electrical Energy Storage Devices
Nanomaterials for Excitonic Solar Cells
Nanomaterials for Sensors
Nano-mechanical Machining
Nanomechanical Microcantilever Chemical and Biological Sensors
Nanomechanical Properties of Nanostructures
Nanomechanical Resonant Sensors and Fluid Interactions
Nanomechanics
Nanomedicine
Nanomotors
Nano-optical Biosensors
Nano-optics Biosensing
Nano-optics, Nanoplasmonics
Nano-optomechanical Systems (NOMS)
Nanoparticle
Nanoparticle Cytotoxicity
Nanoparticle Self-Assembly
Nanoparticle Tracking Analysis
Nanoparticles
Nanoparticles Toxicity Examined In Vivo in Animals
Nanopatterned Substrata
Nanopatterned Surfaces for Exploring and Regulating Cell Behavior
Nano-patterning

Nanophotonic Structures for Biosensing
Nano-piezoelectricity
Nano-pillars
Nanopolaritonics
Nanopores
Nanopowders
Nanoribbons
Nanorobotic Assembly
Nanorobotic Manipulation of Biological Cells
Nanorobotic Spot Welding
Nanorobotics
Nanorobotics for Bioengineering
Nanorobotics for MEMS and NEMS
Nanorobotics for NEMS Using Helical Nanostructures
Nanorods
Nanoscaffold
Nanoscale Drug Vector
Nanoscale Fluid Mechanics
Nanoscale Heat Transport
Nanoscale Particle
Nanoscale Printing
Nanoscale Properties of Solid–Liquid Interfaces
Nano-scale Structuring
Nanoscale Thermal Analysis
Nanoscale Water Phase Diagram
Nanoscratch Tester
Nano-sized Nanocrystalline and Nano-twinned Metals
Nano-sized Particle
Nanoslits
Nanosoldering
Nanospring
Nanostructure
Nanostructure Field Effect Transistor Biosensors
Nanostructured Antireflective Surfaces Arrays
Nanostructured Functionalized Surfaces
Nanostructured Hydrogels
Nanostructured Materials
Nanostructured Materials for Sensing
Nanostructured Solar Cells
Nanostructured Thermoelectric Materials
Nanostructured Thermoelectrics
Nanostructures Based on Porous Silicon
Nanostructures for Coloration (Organisms other than Animals)
Nanostructures for Energy
Nanostructures for Photonics
Nanostructures for Surface Functionalization and Surface Properties
Nanotechnology
Nanotechnology Applications in Polymerase Chain Reaction (PCR)
Nanotechnology in Cardiovascular Diseases
Nanotechnology in Dental Medicine
Nanothermodynamics
Nanotoxicology
Nanotransfer Printing
Nanotribology
Nano-Tribology Applications in Microprojector Technology
Nanotubes
Nanotweezers
Nano-twinned Materials
Nanowire FET Biosensor
Nanowires
Nature Inspired Microneedle
Near Infrared
Near-Field Molecular Optics
Near-Field Optics
Near-Field Scanning Optical Microscopy (NSOM)
NEMS
NEMS Gravimetric Chemical Sensors
NEMS Gravimetric Sensors
NEMS Mass Sensors
NEMS Piezoelectric Switches
NEMS Resonant Chemical Sensors
NEMS Resonant Mass Sensors
Neural Activity-Dependent Closed-Loop
Neural Interfaces
N-Methyl-2-pyrrolidone (NMP)
Non-conventional Machining
Nonlinear Electrokinetic Transport
Nonlinear Optical Absorption and Induced Thermal Scattering Studies in Organic and Inorganic Nanostructures
NS500
NTA
Nucleation of Silica Nanowires
Nucleic Acid Amplification
Officinal Skink
Oil-Repellency of Fish Scales
Oligonucleotide Amplification
On-Chip Flow Cytometry
Optical and Electronic Properties
Optical and Mechanical Nanostructured Coatings for Future Large-Scale Manufacturing
Optical Cavity Biosensor
Optical Frequency Magnetic Dipole Transitions
Optical Lithography

Optical Lithography by Liquid Immersion
Optical Properties of Metal Nanoparticles
Optical Resonance Biosensor
Optical Response Of Metal Particles
Optical Ring Resonator Biosensor
Optical Techniques for Nanostructure Characterization
Optical Trap
Optical Tweezers
Optimization of Nanoparticles
Optoelectrically Enabled Multi-scale Manipulation
Optoelectrofluidics
Opto-electrokinetic Technique (in Viewpoint of Mechanisms Involved)
Optoelectronic Properties
Optofluidics
Optomechanical Resonators
Optomechanics
Organic Actuators
Organic Bioelectronics
Organic Bioelectronics (OBOE)
Organic Films
Organic Photovoltaics: Basic Concepts and Device Physics
Organic Sensors
Organic Solar Cells
Organic–Inorganic Heterojunctions
Organic–Inorganic Photovoltaics
Organosilanes
Orientation Sensors
Passive Pumping
Patterned Hydrogels
Patterning
Peltier Coefficient
Peltier Cooler
Peltier Couple
Peltier Effect
Peltier Element
Peltier Module
Peltier-Seebeck Effect
Percutaneous Absorption
Perfluorocarbon Nanoparticles
Perfluoropolyethers
Petal Effect
Petroleum Lubricants
PFC Nanoparticles
PFOB Nanoparticles
Photoetching
Photofabrication
Photolithography
Photomilling
Photonic Crystal Nanobeam Cavities
Photonics in Nature
Photovoltaic Devices
Physical Colors
Physical Dry Etching
Physical Modification
Physical Vapor Deposition
Physical-Chemical Etching
Physicochemical Properties of Nanoparticles in Relation with Toxicity
Phytotoxicity
Piezoelectric Effect at Nanoscale
Piezoelectric MEMS Switches
Piezoelectric Switches
Piezoresistance
Piezoresistive Effect
Piezoresistivity
Piezoresponse Force Microscopy and Spectroscopy
Plasma
Plasma Etching
Plasma Grafting
Plasma Polymerization
Plasma-Enhanced Chemical Vapor Deposition (PECVD)
Plasmon Resonance Energy Transfer from Metallic Nanoparticles to Biomolecules
Plasmon Resonance Energy Transfer Nanospectroscopy
Plasmonic Structures for Solar Energy Harvesting
Plasmonic Waveguides
Plasmonics
Plasmonics for Photovoltaics
Plasmonics for Solar Cells
Plasticity of Nanostructured Solids
Plasticity Theory at Small Scales
Plating
Polarization-Induced Effects in Heterostructures
Polyethylene Carbonate (PEC)
Polymer Coatings
Polymer Nanocapsules
Polymer Pen Lithography
Polymer Sensors
Polymeric Surface Modifications
Polymer–Metal Oxide Solar Cells
Polymethyl Methacrylate (PMMA)

Polypropylene Carbonate (PPC)
Post-exposure Bake (PEB)
Precritical Clusters
Prenucleation Clusters
PRINT®
Probe Technology of Scanning Probe Microscopy (SPM) and Its Fabrication Technology
Production of Gold Nanoparticles
Production of Graphene or Graphene Production
Propylene Glycol Methyl Ether Acetate (PGMEA)
Prostheses
Protein Nano-crystallization
Pseudoelasticity
Pulsed Focused Ultrasound
Pulsed HIFU
Pulsed High-Intensity Focused Ultrasound
Pulsed-Laser Deposition (PLD)
Pyroelectricity
PZT
Quantum Cluster
Quantum Dot
Quantum Dot Nanophotonic Integrated Circuits
Quantum Dot Solar Cells
Quantum-Dot Toxicity
Radiofrequency
Rapid Electrokinetic Patterning
Rate Sensors
Reactive Current Clamp
Reactive Empirical Bond-Order Potentials
REBO
Relaxation
Relaxation Calorimetry
Relay
Relay Logic
Reliability of Nanostructures
Remotely Powered Propulsion of Helical Nanobelts
Replica Molding
Resonant Evanescent Wave Biosensor
Retina Implant
Retina Silicon Chip
Retinal Prosthesis
Reversible Adhesion
RF MEMS Switches
RF-MEMS/NEMS
Riblets
Rigorous Maxwell Solver
RNA Interference (RNAi)
RNAi in Biomedicine and Drug Delivery
Robot-Based Automation on the Nanoscale
Robotic Insects
Rolled-Up Nanostructure
Rolled-Up Nanotechnology
Rose Petal Effect
Rotation Sensors
Sand Skink
SANS
SAXS
Scanning Electron Microscopy
Scanning Electron Microscopy and Secondary-Electron Imaging Microscopy
Scanning Force Microscopy in Liquids
Scanning Kelvin Probe Force Microscopy
Scanning Near-Field Optical Microscopy
Scanning Probe Microscopy
Scanning Surface Potential Microscopy
Scanning Thermal Microscopy
Scanning Thermal Profiler
Scanning Tunneling Microscopy
Scanning Tunneling Spectroscopy
Scanning X-Ray Diffraction Microscopy (SXDM)
Scanning-Probe Lithography
Scincus officinalis
Scincus scincus
Scrolled Nanostructure
Seebeck Coefficient
Seebeck Effect
Selected Synchrotron Radiation Techniques
Self-Assembled Monolayers
Self-Assembled Monolayers for Nanotribology
Self-Assembled Nanostructures
Self-Assembled Protein Layers
Self-Assembly
Self-Assembly for Heterogeneous Integration of Microsystems
Self-assembly of Nanostructures
Self-Cleaning
Self-Healing Materials
Self-Organized Layers
Self-Organized Nanostructures
Self-Regeneration
Self-Repairing Materials
Self-repairing Photoelectrochemical Complexes Based on Nanoscale Synthetic and Biological Components
SEM
Semiconductor Nanocrystals

Semiconductor Piezoresistance
Semicrystalline Morphology
Sense Organ
Sensors
Shark Denticles
Shark Skin Drag Reduction
Shark Skin Effect
Shark Skin Separation Control
Short (Low Aspect Ratio) Gold Nanowires
Short-Interfering RNA (siRNA)
Si Nanotubes
Silent Flight of Owls
Silent Owl Wings
Silica Gel Processing
Silicon Microphone
Siloxanes
Silver (Ag)
Simulations of Bulk Nanostructured Solids
Single Cell Analysis
Single Cell Impedance Spectroscopy
Single-Cell Electrical Impedance Spectroscopy
Single-Cell Impedance Spectroscopy
Single-Walled Carbon Nanotubes (SWCNTs)
siRNA Delivery
Size-Dependent Plasticity of Single Crystalline Metallic Nanostructures
Skin Delivery
Skin Penetration
Small angle X-Ray Scattering in Grazing Incidence Geometry
Small Unilamellar Vesicle (SUV)
Small-Angle Neutron Scattering
Small-Angle Scattering
Small-Angle Scattering of Nanostructures and Nanomaterials
Small-Angle X-Ray Scattering
Smart Carbon Nanotube-Polymer Composites
Smart Drug Delivery Microchips
Smart Hydrogels
Soft Actuators
Soft Lithography
Soft X-Ray Lithography
Soft X-Ray Microscopy
Soil/Terrestrial Ecosystem/Terrestrial Compartment
Solar Cells
Sol-Gel Method
Solid Lipid Nanocarriers
Solid Lipid Nanoparticles - SLN
Solid Lipid Nanospheres
Solid Lipid-Based Nanoparticles
Solid–Liquid Interfaces
Sound Propagation in Fluids
Spectromicroscopy
Spectroscopic Techniques
Spider Silk
Spiders
Spintronic Devices
Spontaneous Polarization
Spray Technologies Inspired by Bombardier Beetle
Sputtering
Stable Ion Clusters
Stereolithography
Stimuli-Responsive Drug Delivery Microchips
Stimulus-Responsive Polymeric Hydrogels
Stochastic Assembly
Strain Gradient Plasticity Theory
Structural Color in Animals
Structural Color in Nature
Structural Colors
Structure and Stability of Protein Materials
SU-8 Photoresist
Submicron Structures in Nature
Sub-retinal Implant
Subwavelength Antireflective Surfaces Arrays
Subwavelength Structures
Sub-wavelength Waveguiding
Superelasticity and the Shape Memory Effect
Superhydrophobicity
Superoleophobicity of Fish Scales
Surface Electronic Structure
Surface Energy and Chemical Potential at Nanoscale
Surface Energy Density
Surface Energy Density and Chemical Potential at Nanoscale
Surface Force Balance
Surface Forces Apparatus
Surface Free Energy and Chemical Potential at Nanoscale
Surface Plasmon Enhanced Optical Bistability and Optical Switching
Surface Plasmon Nanophotonics
Surface Plasmon-Polariton Photodetectors
Surface Plasmon-Polariton-Based Detectors
Surface Properties
Surface Tension and Chemical Potential at Nanoscale
Surface Tension Effects of Nanostructures

Surface Tension–Driven Flow
Surface Tension–Powered Self-Assembly
Surface-Modified Microfluidics and Nanofluidics
Surface-Plasmon-Enhanced Solar Energy Conversion
Synthesis of Carbon Nanotubes
Synthesis of Gold Nanoparticles
Synthesis of Graphene
Synthesis of Nanoparticles
Synthesis of Subnanometric Metal Nanoparticles
Synthesized Conductance Injection
Synthesized Ionic Conductance
Synthesized Synaptic Conductance
Synthetic Biology
Synthetic Genomics
Synthetic Lubricants
Systems Level Data Mining for RNAi
TEM
Terahertz
Terahertz Technology for Nano Applications
Theoretical Elasticity
Theory of Artificial Electromagnetic Materials
Theory of Optical Metamaterials
Thermal Actuators
Thermal Cancer Ablation Therapies Using Nanoparticles
Thermal Chemical Vapor Deposition
Thermal Conductance
Thermal Conductivity and Phonon Transport
Thermal Evaporation
Thermal Infrared Detector
Thermal Management Materials
Thermal Resistance
Thermal Resistivity
Thermal Transport
Thermally Actuated MEMS Resonators
Thermally Actuated Micromechanical Resonators
Thermally Actuated Nanoelectromechanical Resonators
Thermally Actuated Nanomechanical Resonators
Thermally Actuated Resonators
Thermally Actuated Silicon Resonators
Thermal-Piezoresistive Resonators
Thermodynamics of Small Systems
Thermoelectric Cooler
Thermoelectric Device
Thermoelectric Generator
Thermoelectric Heat Convertors
Thermoelectric Heat Pump
Thermoelectric Module
Thermoelectric Nanomaterials
Thermoelectric Power
Thermomechanical Actuators
Thermometry
Thermopower
Thermostability
Thickness
Thiols
Third Generation Vectors
Thomson Effect
Three-Dimensional Imaging of Human Tissues
THz
TiO_2 Nanotube Arrays: Growth and Application
Titanium Dioxide
Total Internal Reflection (Fluorescence) Velocimetry
Toward Bioreplicated Texturing of Solar-Cell Anodes
Toward Bioreplicated Texturing of Solar-Cell Surfaces
Toxicity
Toxicity of Metal and Metal Oxide Nanoparticles on Prokaryotes (Bacteria), Unicellular and Invertebrate Eukaryotes
Toxicity/Ecotoxicity
Toxicology: Plants and Nanoparticles
Transcutaneous Delivery
Transdermal Permeation
Transduction
Transfer
Transmission Electron Microscopy
T-Rays
Tuning Electrical Properties of Carbon Nanotubes via Spark Plasma Sintering
Ultrahigh Vacuum Chemical Vapor Deposition (UHVCVD)
Ultralarge Strain Elasticity
Ultraprecision Machining (UPM)
Ultrashort Carbon Nanotubes
Ultrasonic Atomization
Ultrasonic Machining
Unconventional Computing
Uptake/Internalization/Sequestration/Biodistribution
USM
US-Tubes
Vertically Aligned Carbon Nanotubes
Vertically Aligned Carbon Nanotubes, Collective Mechanical Behavior
Vibration Assisted Machining

Viscosity
Visual Prosthesis
Visual Servoing for SEM
Wavefront Deformation Particle Tracking
Waypoint Detection
Wear
Wet Adhesion in Tree Frogs
Wet Chemical Processing
Wet Etching
Wetting Transitions
Whispering Gallery Mode Biosensor
Whispering Gallery Mode Resonator Biosensors
Wireless Propulsion of Helical Nanobelt Swimmers
Worm Chip
X-ray Diffraction with Micron-Sized X-Ray Beams
Young's Modulus